Contents

Discrete
Mathematics
and Its
Applications

Seventh Edition

Kenneth H. Rosen

Monmouth University
(and formerly AT&T Laboratories)

Global Edition Adapted by
Kamala Krithivasan

ALL ITEMS AF *Indian Institute of Technology Madras*
Chennai, India

Connect
Learn
Succeed™

GLOBAL EDITION DISCRETE MATHEMATICS AND ITS APPLICATIONS, SEVENTH EDITION

Published by McGraw-Hill, a business unit of The McGraw-Hill Companies, Inc., 1221 Avenue of the Americas, New York, NY 10020. Copyright © 2013 by Kenneth H. Rosen. All rights reserved. Previous editions © 2007, 2003, and 1999. No part of this publication may be reproduced or distributed in any form or by any means, or stored in a database or retrieval system, without the prior written consent of The McGraw-Hill Companies, Inc., including, but not limited to, in any network or other electronic storage or transmission, or broadcast for distance learning.

Some ancillaries, including electronic and print components, may not be available to customers outside the United States.

This book is printed on acid-free paper.

Printed in China

1 2 3 4 5 6 7 8 9 0 CTP/CTP 1 0 9 8 7 6 5 4 3 2

ISBN   978-0-07-131501-2
MHID   0-07-131501-2

All credits appearing on this page or at the end of the book are considered to be an extension of the copyright page.

Cover image: ©*Dr. Parvinder Sethi*

About the Authors

Kenneth H. Rosen has had a long career as a Distinguished Member of the Technical Staff at AT&T Laboratories in Monmouth County, New Jersey. He currently holds the position of Visiting Research Professor at Monmouth University, where he teaches graduate courses in computer science.

Dr. Rosen received his B.S. in Mathematics from the University of Michigan, Ann Arbor (1972), and his Ph.D. in Mathematics from M.I.T. (1976), where he wrote his thesis in the area of number theory under the direction of Harold Stark. Before joining Bell Laboratories in 1982, he held positions at the University of Colorado, Boulder; The Ohio State University, Columbus; and the University of Maine, Orono, where he was an associate professor of mathematics. While working at AT&T Labs, he taught at Monmouth University, teaching courses in discrete mathematics, coding theory, and data security. He currently teaches courses in algorithm design and in computer security and cryptography.

Dr. Rosen has published numerous articles in professional journals in number theory and in mathematical modeling. He is the author of the widely used *Elementary Number Theory and Its Applications*, published by Pearson, currently in its sixth edition, which has been translated into Chinese. He is also the author of *Discrete Mathematics and Its Applications*, published by McGraw-Hill, currently in its seventh edition. *Discrete Mathematics and Its Applications* has sold more than 350,000 copies in North America during its lifetime, and hundreds of thousands of copies throughout the rest of the world. This book has also been translated into Spanish, French, Greek, Chinese, Vietnamese, and Korean. He is also co-author of *UNIX: The Complete Reference*; *UNIX System V Release 4: An Introduction*; and *Best UNIX Tips Ever*, all published by Osborne McGraw-Hill. These books have sold more than 150,000 copies, with translations into Chinese, German, Spanish, and Italian. Dr. Rosen is also the editor of the *Handbook of Discrete and Combinatorial Mathematics*, published by CRC Press, and he is the advisory editor of the CRC series of books in discrete mathematics, consisting of more than 55 volumes on different aspects of discrete mathematics, most of which are introduced in this book. Dr. Rosen serves as an Associate Editor for the journal *Discrete Mathematics*, where he works with submitted papers in several areas of discrete mathematics, including graph theory, enumeration, and number theory. He is also interested in integrating mathematical software into the educational and professional environments, and worked on several projects with Waterloo Maple Inc.'s Maple^TM software in both these areas. Dr. Rosen has also worked with several publishing companies on their homework delivery platforms.

At Bell Laboratories and AT&T Laboratories, Dr. Rosen worked on a wide range of projects, including operations research studies, product line planning for computers and data communications equipment, and technology assessment. He helped plan AT&T's products and services in the area of multimedia, including video communications, speech recognition, speech synthesis, and image networking. He evaluated new technology for use by AT&T and did standards work in the area of image networking. He also invented many new services, and holds more than 55 patents. One of his more interesting projects involved helping evaluate technology for the AT&T attraction that was part of EPCOT Center.

Kamala Krithivasan has served as a faculty member of the Computer Science and Engineering Department, IIT Madras, Chennai, India since the last two decades. She is currently a professor working in the area of Theoretical Computer Science, especially Formal Language Theory and its applications and Unconventional Models of Computing.

She obtained her bachelor's and master's degrees in Mathematics from Madras Christian College, Tambaram, affiliated to Madras University, where she was the first-rank holder. She also received her PhD from the same place in 1975 for her work on Array Grammars. Thereafter, she joined IIT Madras in 1975 in the Computer Science and Engineering Department (called Computer Center at that time) as a research associate, became a full-time professor in 1990 and held the post of Chairperson of the department during 1992–1995.

She has been associated with researches in the field of Formal Languages and Automata Theory for more than 35 years; her contributions being in the area of array grammars, graph grammars, L-systems, DNA computing and membrane computing. She has also been involved with computational geometry for several years having hands-on experience on optimization problems and path planning problems. She has taught courses on and related to Discrete Mathematical Structures and Theory of Computation numerous times for undergraduate and graduate students at IIT Madras. She has also delivered a series of 40 lectures on Discrete Mathematical Structures under the NPTEL program (National Program on Technology Enhanced learning). These video lectures can be seen via YouTube and the NPTEL website. Dr Krithivasan has guided many PhD students and also, has carried out many sponsored projects and partnered with teams from Germany and Israel on collaborative projects. She has more than 70 publications in international journals and more than 80 publications in conferences. She visited the University of Maryland during the fellowship period, Concordia University, for six months in 1987 and MPI, Saarbrucken, Germany, for short periods in 1992 and 1994.

She was elected as a Fellow of the Indian National Academy of Engineers in 2006. Dr Krithivasan was awarded the Fulbright Fellowship in 1986.

Preface

In writing this book, I was guided by my long-standing experience and interest in teaching discrete mathematics. For the student, my purpose was to present material in a precise, readable manner, with the concepts and techniques of discrete mathematics clearly presented and demonstrated. My goal was to show the relevance and practicality of discrete mathematics to students, who are often skeptical. I wanted to give students studying computer science all of the mathematical foundations they need for their future studies. I wanted to give mathematics students an understanding of important mathematical concepts together with a sense of why these concepts are important for applications. And most importantly, I wanted to accomplish these goals without watering down the material.

For the instructor, my purpose was to design a flexible, comprehensive teaching tool using proven pedagogical techniques in mathematics. I wanted to provide instructors with a package of materials that they could use to teach discrete mathematics effectively and efficiently in the most appropriate manner for their particular set of students. I hope that I have achieved these goals.

I have been extremely gratified by the tremendous success of this text. The many improvements in the seventh edition have been made possible by the feedback and suggestions of a large number of instructors and students at many of the more than 600 North American schools, and at any many universities in parts of the world, where this book has been successfully used.

This text is designed for a one- or two-term introductory discrete mathematics course taken by students in a wide variety of majors, including mathematics, computer science, and engineering. College algebra is the only explicit prerequisite, although a certain degree of mathematical maturity is needed to study discrete mathematics in a meaningful way. This book has been designed to meet the needs of almost all types of introductory discrete mathematics courses. It is highly flexible and extremely comprehensive. The book is designed not only to be a successful textbook, but also to serve as valuable resource students can consult throughout their studies and professional life.

Goals of a Discrete Mathematics Course

A discrete mathematics course has more than one purpose. Students should learn a particular set of mathematical facts and how to apply them; more importantly, such a course should teach students how to think logically and mathematically. To achieve these goals, this text stresses mathematical reasoning and the different ways problems are solved. Five important themes are interwoven in this text: mathematical reasoning, combinatorial analysis, discrete structures, algorithmic thinking, and applications and modeling. A successful discrete mathematics course should carefully blend and balance all five themes.

1. *Mathematical Reasoning:* Students must understand mathematical reasoning in order to read, comprehend, and construct mathematical arguments. This text starts with a discussion of mathematical logic, which serves as the foundation for the subsequent discussions of methods of proof. Both the science and the art of constructing proofs are addressed. The technique of mathematical induction is stressed through many different types of examples of such proofs and a careful explanation of why mathematical induction is a valid proof technique.

2. *Combinatorial Analysis:* An important problem-solving skill is the ability to count or enumerate objects. The discussion of enumeration in this book begins with the basic techniques of counting. The stress is on performing combinatorial analysis to solve counting problems and analyze algorithms, not on applying formulae.

3. *Discrete Structures:* A course in discrete mathematics should teach students how to work with discrete structures, which are the abstract mathematical structures used to represent discrete objects and relationships between these objects. These discrete structures include sets, permutations, relations, graphs, trees, and finite-state machines.

4. *Algorithmic Thinking:* Certain classes of problems are solved by the specification of an algorithm. After an algorithm has been described, a computer program can be constructed implementing it. The mathematical portions of this activity, which include the specification of the algorithm, the verification that it works properly, and the analysis of the computer memory and time required to perform it, are all covered in this text. Algorithms are described using both English and an easily understood form of pseudocode.

5. *Applications and Modeling:* Discrete mathematics has applications to almost every conceivable area of study. There are many applications to computer science and data networking in this text, as well as applications to such diverse areas as chemistry, biology, linguistics, geography, business, and the Internet. These applications are natural and important uses of discrete mathematics and are not contrived. Modeling with discrete mathematics is an extremely important problem-solving skill, which students have the opportunity to develop by constructing their own models in some of the exercises.

Changes in the Seventh Edition

Although the sixth edition has been an extremely effective text, many instructors, including longtime users, have requested changes designed to make this book more effective. I have devoted a significant amount of time and energy to satisfy their requests and I have worked hard to find my own ways to make the book more effective and more compelling to students.

The seventh edition is a major revision, with changes based on input from more than 40 formal reviewers, feedback from students and instructors, and author insights. The result is a new edition that offers an improved organization of topics making the book a more effective teaching tool. Substantial enhancements to the material devoted to logic, algorithms, number theory, and graph theory make this book more flexible and comprehensive. Numerous changes in the seventh edition have been designed to help students more easily learn the material. Additional explanations and examples have been added to clarify material where students often have difficulty. New exercises, both routine and challenging, have been added. Highly relevant applications, including many related to the Internet, to computer science, and to mathematical biology, have been added. The companion website has benefited from extensive development activity and now provides tools students can use to master key concepts and explore the world of discrete mathematics, and many new tools under development will be released in the year following publication of this book.

I hope that instructors will closely examine this new edition to discover how it might meet their needs. Although it is impractical to list all the changes in this edition, a brief list that highlights some key changes, listed by the benefits they provide, may be useful.

More Flexible Organization

- Applications of propositional logic are found in a new dedicated section, which briefly introduces logic circuits.

- Recurrence relations are now covered in Chapter 2.
- Expanded coverage of countability is now found in a dedicated section in Chapter 2.
- Separate chapters now provide expanded coverage of algorithms (Chapter 3) and number theory and cryptography (Chapter 4).
- More second and third level heads have been used to break sections into smaller coherent parts.

Tools for Easier Learning

- Difficult discussions and proofs have been marked with the famous Bourbaki dangerous bend symbol in the margin.
- New marginal notes make connections, add interesting notes, and provide advice to students.
- More details and added explanations, in both proofs and exposition, make it easier for students to read the book.
- Many new exercises, both routine and challenging, have been added, while many existing exercises have been improved.

Enhanced Coverage of Logic, Sets, and Proof

- The satisfiability problem is addressed in greater depth, with Sudoku modeled in terms of satisfiability.
- Hilbert's Grand Hotel is used to help explain uncountability.
- Proofs throughout the book have been made more accessible by adding steps and reasons behind these steps.
- A template for proofs by mathematical induction has been added.
- The step that applies the inductive hypothesis in mathematical induction proof is now explicitly noted.

Algorithms

- The pseudocode used in the book has been updated.
- Explicit coverage of algorithmic paradigms, including brute force, greedy algorithms, and dynamic programing, is now provided.
- Useful rules for big-O estimates of logarithms, powers, and exponential functions have been added.

Number Theory and Cryptography

- Expanded coverage allows instructors to include just a little or a lot of number theory in their courses.
- The relationship between the **mod** function and congruences has been explained more fully.
- The sieve of Eratosthenes is now introduced earlier in the book.
- Linear congruences and modular inverses are now covered in more detail.
- Applications of number theory, including check digits and hash functions, are covered in great depth.

- A new section on cryptography integrates previous coverage, and the notion of a cryptosystem has been introduced.
- Cryptographic protocols, including digital signatures and key sharing, are now covered.

Graph Theory

- A structured introduction to graph theory applications has been added.
- More coverage has been devoted to the notion of social networks.
- Applications to the biological sciences and motivating applications for graph isomorphism and planarity have been added.
- Matchings in bipartite graphs are now covered, including Hall's theorem and its proof.
- Coverage of vertex connectivity, edge connectivity, and n-connectedness has been added, providing more insight into the connectedness of graphs.

Enrichment Material

- Many biographies have been expanded and updated, and new biographies of Bellman, Bézout Bienyamé, Cardano, Catalan, Cocks, Cook, Dirac, Hall, Hilbert, Ore, and Tao have been added.
- Historical information has been added throughout the text.
- Numerous updates for latest discoveries have been made.

Media Resources

- Extensive effort has been devoted to producing valuable web resources for this book.
- Extra examples in key parts of the text have been provided on companion website.
- Interactive algorithms have been developed, with tools for using them to explore topics and for classroom use.
- Student assessment modules are available for key concepts.
- Powerpoint transparencies for instructor use have been developed.
- An extensive collection of external web links is provided.
- Instructor's Resource Guide available on the website for instructors contains full solutions to even-numbered exercises in the text. It also offers sample tests for each chapter and a test bank containing over 1500 exam questions to choose from. Answers to all sample tests and test bank questions are included.

Features of the Book

ACCESSIBILITY   This text has proved to be easily read and understood by beginning students. There are no mathematical prerequisites beyond college algebra for almost all the content of the text. Students needing extra help will find tools on the companion website for bringing their mathematical maturity up to the level of the text. The few places in the book where calculus is referred to are explicitly noted. Most students should easily understand the pseudocode used in the text to express algorithms, regardless of whether they have formally studied programming languages. There is no formal computer science prerequisite.

Each chapter begins at an easily understood and accessible level. Once basic mathematical concepts have been carefully developed, more difficult material and applications to other areas of study are presented.

FLEXIBILITY This text has been carefully designed for flexible use. The dependence of chapters on previous material has been minimized. Each chapter is divided into sections of approximately the same length, and each section is divided into subsections that form natural blocks of material for teaching. Instructors can easily pace their lectures using these blocks.

WRITING STYLE The writing style in this book is direct and pragmatic. Precise mathematical language is used without excessive formalism and abstraction. Care has been taken to balance the mix of notation and words in mathematical statements.

MATHEMATICAL RIGOR AND PRECISION All definitions and theorems in this text are stated extremely carefully so that students will appreciate the precision of language and rigor needed in mathematics. Proofs are motivated and developed slowly; their steps are all carefully justified. The axioms used in proofs and the basic properties that follow from them are explicitly described in an appendix, giving students a clear idea of what they can assume in a proof. Recursive definitions are explained and used extensively.

WORKED EXAMPLES Extensive examples are used to illustrate concepts, relate different topics, and introduce applications. In most examples, a question is first posed, then its solution is presented with the appropriate amount of detail.

APPLICATIONS The applications included in this text demonstrate the utility of discrete mathematics in the solution of real-world problems. This text includes applications to a wide variety of areas, including computer science, data networking, psychology, chemistry, engineering, linguistics, biology, business, and the Internet.

ALGORITHMS Results in discrete mathematics are often expressed in terms of algorithms; hence, key algorithms are introduced in each chapter of the book. These algorithms are expressed in words and in an easily understood form of structured pseudocode, which is described and specified in Appendix 3. The computational complexity of the algorithms in the text is also analyzed at an elementary level.

HISTORICAL INFORMATION The background of many topics is succinctly described in the text. Brief biographies of 83 mathematicians and computer scientists are included as footnotes. These biographies include information about the lives, careers, and accomplishments of these important contributors to discrete mathematics and images, when available, are displayed.

In addition, numerous historical footnotes are included that supplement the historical information in the main body of the text. Efforts have been made to keep the book up-to-date by reflecting the latest discoveries.

KEY TERMS AND RESULTS A list of key terms and results follows each chapter. The key terms include only the most important that students should learn, and not every term defined in the chapter.

EXERCISES There are thousands of exercises in the text, with many different types of questions posed. There is an ample supply of straightforward exercises that develop basic skills, a large number of intermediate exercises, and many challenging exercises. Exercises are stated clearly and unambiguously, and all are carefully graded for level of difficulty. Exercise sets contain special discussions that develop new concepts not covered in the text, enabling students to discover new ideas through their own work.

Exercises that are somewhat more difficult than average are marked with a single star *; those that are much more challenging are marked with two stars **. Exercises whose solutions require calculus are explicitly noted. Exercises that develop results used in the text are clearly identified with the right pointing hand symbol ☞. Answers or outlined solutions to all odd-numbered exercises are provided at the back of the text. The solutions include proofs in which most of the steps are clearly spelled out.

REVIEW QUESTIONS   A set of review questions is provided at the end of each chapter. These questions are designed to help students focus their study on the most important concepts and techniques of that chapter. To answer these questions students need to write long answers, rather than just perform calculations or give short replies.

SUPPLEMENTARY EXERCISE SETS   Each chapter is followed by a rich and varied set of supplementary exercises. These exercises are generally more difficult than those in the exercise sets following the sections. The supplementary exercises reinforce the concepts of the chapter and integrate different topics more effectively.

COMPUTER PROJECTS   Each chapter is followed by a set of computer projects. The computer projects tie together what students may have learned in computing and in discrete mathematics. Computer projects that are more difficult than average, from both a mathematical and a programming point of view, are marked with a star, and those that are extremely challenging are marked with two stars.

COMPUTATIONS AND EXPLORATIONS   A set of computations and explorations is included at the conclusion of each chapter. These exercises are designed to be completed using existing software tools, such as programs that students or instructors have written or mathematical computation packages such as Maple™ or Mathematica™. Many of these exercises give students the opportunity to uncover new facts and ideas through computation.

WRITING PROJECTS   Each chapter is followed by a set of writing projects. To do these projects students need to consult the mathematical literature. Some of these projects are historical in nature and may involve looking up original sources. Others are designed to serve as gateways to new topics and ideas. All are designed to expose students to ideas not covered in depth in the text. These projects tie mathematical concepts together with the writing process and help expose students to possible areas for future study.

APPENDIXES   There are three appendixes to the text. The first introduces axioms for real numbers and the positive integers, and illustrates how facts are proved directly from these axioms. The second covers exponential and logarithmic functions, reviewing some basic material used heavily in the course. The third specifies the pseudocode used to describe algorithms in this text.

SUGGESTED READINGS   A list of suggested readings for the overall book and for each chapter is provided after the appendices. These suggested readings include books at or below the level of this text, more difficult books, expository articles, and articles in which discoveries in discrete mathematics were originally published. Some of these publications are classics, published many years ago, while others have been published in the last few years.

Acknowledgments

I would like to thank the many instructors and students at a variety of schools who have used this book and provided me with their valuable feedback and helpful suggestions. Their input has made this a much better book than it would have been otherwise. I especially want to thank Jerrold Grossman, Jean-Claude Evard, and Georgia Mederer for their technical reviews of the seventh edition and their "eagle eyes," which have helped ensure the accuracy of this book. I also appreciate the help provided by all those who have submitted comments via the website.

I thank the reviewers of this seventh and the six previous editions. These reviewers have provided much helpful criticism and encouragement to me. I hope this edition lives up to their high expectations.

Reviewers for the Seventh Edition

Philip Barry
University of Minnesota, Minneapolis

Miklos Bona
University of Florida

Kirby Brown
Queens College

John Carter
University of Toronto

Narendra Chaudhari
Nanyang Technological University

Allan Cochran
University of Arkansas

Daniel Cunningham
Buffalo State College

George Davis
Georgia State University

Andrzej Derdzinski
The Ohio State University

Ronald Dotzel
University of Missouri-St. Louis

T.J. Duda
Columbus State Community College

Bruce Elenbogen
University of Michigan, Dearborn

Norma Elias
*Purdue University,
Calumet-Hammond*

Herbert Enderton
University of California, Los Angeles

Anthony Evans
Wright State University

Kim Factor
Marquette University

Margaret Fleck
University of Illinois, Champaign

Peter Gillespie
Fayetteville State University

Johannes Hattingh
Georgia State University

Ken Holladay
University of New Orleans

Jerry Ianni
LaGuardia Community College

Ravi Janardan
University of Minnesota, Minneapolis

Norliza Katuk
University of Utara Malaysia

William Klostermeyer
University of North Florida

Przemo Kranz
University of Mississippi

Jaromy Kuhl
University of West Florida

Loredana Lanzani
University of Arkansas, Fayetteville

Steven Leonhardi
Winona State University

Xu Liutong
*Beijing University of Posts and
Telecommunications*

Vladimir Logvinenko
De Anza Community College

Darrell Minor
Columbus State Community College

Keith Olson
Utah Valley University

Yongyuth Permpoontanalarp
*King Mongkut's University of
Technology, Thonburi*

Galin Piatniskaia
University of Missouri, St. Louis

Stefan Robila
Montclair State University

Chris Rodger
Auburn University

Sukhit Singh
Texas State University, San Marcos

David Snyder
Texas State University, San Marcos

Wasin So
San Jose State University

Bogdan Suceava
California State University, Fullerton

Christopher Swanson
Ashland University

Bon Sy
Queens College

Matthew Walsh
*Indiana-Purdue University, Fort
Wayne*

Gideon Weinstein
Western Governors University

David Wilczynski
University of Southern California

I would also like to thank the original editor, Wayne Yuhasz, whose insights and skills helped ensure the book's success, as well as all the many other previous editors of this book.

I want to express my appreciation to the staff of RPK Editorial Services for their valuable work on this edition, including Rose Kernan, who served as both the developmental editor and the production editor, and the other members of the RPK team, Fred Dahl, Martha McMaster, Erin Wagner, Harlan James, and Shelly Gerger-Knecthl. I thank Paul Mailhot of PreTeX, Inc., the compositor, for the tremendous amount to work he devoted to producing this edition, and for his intimate knowledge of LaTeX. Thanks also to Danny Meldung of Photo Affairs, Inc., who was resourceful obtaining images for the new biographical footnotes.

The accuracy and quality of this new edition owe much to Jerry Grossman and Jean-Claude Evard, who checked the entire manuscript for technical accuracy and Georgia Mederer, who checked the accuracy of the answers at the end of the book and the *Instructor's Resource Guide*. As usual, I cannot thank Jerry Grossman enough for all his work authoring this essential ancillary.

I would also express my appreciation the Science, Engineering, and Mathematics (SEM) Division of McGraw-Hill Higher Education for their valuable support for this new edition and the associated media content. In particular, thanks go to Kurt Strand: President, SEM, McGraw-Hill Higher Education, Marty Lange: Editor-in-Chief, SEM, Michael Lange: Editorial Director, Raghothaman Srinivasan: Global Publisher, Bill Stenquist: Executive Editor, Curt Reynolds: Executive Marketing Manager, Robin A. Reed: Project Manager, Sandy Ludovissey: Buyer, Lorraine Buczek: In-house Developmental Editor, Brenda Rowles: Design Coordinator, Carrie K. Burger: Lead Photo Research Coordinator, and Tammy Juran: Media Project Manager.

Kenneth H. Rosen

Changes in the Global Edition

This adapted edition caters to the curriculum requisites of most of the international institutes and universities. It includes a new chapter on Algebraic Structures and Coding Theory which deals with basic properties of **semigroups**, **monoids**, **groups** and **rings**. Since coding theory is also taught in the course on Discrete Structures in some places, especially in mathematics departments, a section on this is added including **group codes**, **Hamming codes**, and **polynomial rings and codes**.

The section on **cardinality** is elaborated and theory related to **lattices** is introduced. A note about **exponential generating functions**; and some **additional methods of solving recurrence relations** have been added. Additional material is given in the website and at appropriate places, the link information is shown. In keeping with the style of the previous edition, key terms and results, exercises, review questions, supplementary exercises, computer projects, computations and explorations, writing projects are given in the new chapter. Exercises are added at the end of added sections. For odd-numbered exercises, solutions have been given.

In a nutshell, the new additions incorporated in this Global Edition are the following:

- Chapter 1 on The Foundations: Logic and Proofs includes a new section on normal forms and expanded coverage of Resolution Principle with its application to Prolog.

- Chapter 2 on Basic Structures: Sets, Functions, Sequences, Sums, and Matrices includes added material on Cardinality of Sets.

- Chapter 8 on Advanced Counting Techniques includes a subsection on Exponential Generating Functions and additional methods of solving recurrence relations by substitution.

- A new chapter on Algebraic Structures and Coding Theory is added as Chapter 12.

- Two chapters are provided at the Global Edition website www.mhhe.com/rosenGE. They are: Boolean Algebra and Modeling Computation.

The Companion Website

The extensive companion website accompanying this text has been substantially enhanced for the seventh edition This website is accessible at *www.mhhe.com/rosenGE*. The homepage shows the *Information Center*, and contains login links for the site's *Student Site* and *Instructor Site*. Key features of each area are described below:

THE INFORMATION CENTER

The Information Center contains basic information about the book including the expanded table of contents (including subsection heads), the preface, descriptions of the ancillaries, and a sample chapter. It also provides a link that can be used to submit errata reports and other feedback about the book.

STUDENT SITE

The Student site contains a wealth of resources available for student use, including the following, tied into the text wherever the special icons displayed below are found in the text:

- *Extra Examples*   You can find a large number of additional examples on the site, covering all chapters of the book. These examples are concentrated in areas where students often ask for additional material. Although most of these examples amplify the basic concepts, more-challenging examples can also be found here.

- *Interactive Demonstration Applets*   These applets enable you to interactively explore how important algorithms work, and are tied directly to material in the text with linkages to examples and exercises. Additional resources are provided on how to use and apply these applets.

- *Self Assessments*   These interactive guides help you assess your understanding of 14 key concepts, providing a question bank where each question includes a brief tutorial followed by a multiple-choice question. If you select an incorrect answer, advice is provided to help you understand your error. Using these Self Assessments, you should be able to diagnose your problems and find appropriate help.

- *Web Resources Guide*   This guide provides annotated links to hundreds of external websites containing relevant material such as historical and biographical information, puzzles and problems, discussions, applets, programs, and more. These links are keyed to the text by page number.

Additional resources in the Student site include:

- *Applications of Discrete Mathematics*   This ancillary contains 24 chapters—each with its own set of exercises—presenting a wide variety of interesting and important applications covering three general areas in discrete mathematics: discrete structures, combinatorics, and graph theory. These applications are ideal for supplementing the text or for independent study.

- *A Guide to Proof-Writing*   This guide provides additional help for writing proofs, a skill that many students find difficult to master. By reading this guide at the beginning of the course and periodically thereafter when proof writing is required, you will be rewarded as your proof-writing ability grows.

- *Common Mistakes in Discrete Mathematics*   This guide includes a detailed list of common misconceptions that students of discrete mathematics often have and the kinds of errors they tend to make. You are encouraged to review this list from time to time to help avoid these common traps.

- *Advice on Writing Projects*   This guide offers helpful hints and suggestions for the Writing Projects in the text, including an extensive bibliography of helpful books and articles for research; discussion of various resources available in print and online; tips on doing library research; and suggestions on how to write well.

INSTRUCTOR SITE

This part of the website provides access to all of the resources on the Student Site, as well as these resources for instructors:

- *Printable Tests*   Printable tests are offered in TeX and Word format for every chapter, and can be customized by instructors.

- *PowerPoints Lecture Slides and PowerPoint Figures and Tables*   An extensive collection of PowerPoint slides for all chapters of the text are provided for instructor use. In addition, images of all figures and tables from the text are provided as PowerPoint slides.

To the Student

What is discrete mathematics? Discrete mathematics is the part of mathematics devoted to the study of discrete objects. (Here *discrete* means consisting of distinct or unconnected elements.) The kinds of problems solved using discrete mathematics include:

- How many ways are there to choose a valid password on a computer system?
- What is the probability of winning a lottery?
- Is there a link between two computers in a network?
- How can I identify spam e-mail messages?
- How can I encrypt a message so that no unintended recipient can read it?
- What is the shortest path between two cities using a transportation system?
- How can a list of integers be sorted so that the integers are in increasing order?
- How many steps are required to do such a sorting?
- How can it be proved that a sorting algorithm correctly sorts a list?
- How can a circuit that adds two integers be designed?
- How many valid Internet addresses are there?

You will learn the discrete structures and techniques needed to solve problems such as these.

More generally, discrete mathematics is used whenever objects are counted, when relationships between finite (or countable) sets are studied, and when processes involving a finite number of steps are analyzed. A key reason for the growth in the importance of discrete mathematics is that information is stored and manipulated by computing machines in a discrete fashion.

WHY STUDY DISCRETE MATHEMATICS?   There are several important reasons for studying discrete mathematics. First, through this course you can develop your mathematical maturity: that is, your ability to understand and create mathematical arguments. You will not get very far in your studies in the mathematical sciences without these skills.

Second, discrete mathematics is the gateway to more advanced courses in all parts of the mathematical sciences. Discrete mathematics provides the mathematical foundations for many computer science courses including data structures, algorithms, database theory, automata theory, formal languages, compiler theory, computer security, and operating systems. Students find these courses much more difficult when they have not had the appropriate mathematical foundations from discrete math. One student has sent me an e-mail message saying that she used the contents of this book in every computer science course she took!

Math courses based on the material studied in discrete mathematics include logic, set theory, number theory, linear algebra, abstract algebra, combinatorics, graph theory, and probability theory (the discrete part of the subject).

Also, discrete mathematics contains the necessary mathematical background for solving problems in operations research (including many discrete optimization techniques), chemistry, engineering, biology, and so on. In the text, we will study applications to some of these areas.

Many students find their introductory discrete mathematics course to be significantly more challenging than courses they have previously taken. One reason for this is that one of the primary goals of this course is to teach mathematical reasoning and problem solving, rather than a discrete set of skills. The exercises in this book are designed to reflect this goal. Although there are plenty of exercises in this text similar to those addressed in the examples, a large percentage of the exercises require original thought. This is intentional. The material discussed in the text provides the tools needed to solve these exercises, but your job is to successfully apply these tools using your own creativity. One of the primary goals of this course is to learn how to attack problems that may be somewhat different from any you may have previously seen. Unfortunately, learning how to solve only particular types of exercises is not sufficient for success in developing the problem-solving skills needed in subsequent courses and professional work. This text addresses many different topics, but discrete mathematics is an extremely diverse and large area of study. One of my goals as an author is to help you develop the skills needed to master the additional material you will need in your own future pursuits.

THE EXERCISES   I would like to offer some advice about how you can best learn discrete mathematics (and other subjects in the mathematical and computing sciences). You will learn the most by actively working exercises. I suggest that you solve as many as you possibly can. After working the exercises your instructor has assigned, I encourage you to solve additional exercises such as those in the exercise sets following each section of the text and in the supplementary exercises at the end of each chapter. (Note the key explaining the markings preceding exercises.)

Key to the Exercises

no marking	A routine exercise
*	A difficult exercise
**	An extremely challenging exercise
☞	An exercise containing a result used in the book (Table 1 on the following page shows where these exercises are used.)
(*Requires calculus*)	An exercise whose solution requires the use of limits or concepts from differential or integral calculus

The best approach is to try exercises yourself before you consult the answer section at the end of this book. Note that the odd-numbered exercise answers provided in the text are answers only and not full solutions; in particular, the reasoning required to obtain answers is omitted in these answers.

WEB RESOURCES   You are ***strongly*** encouraged to take advantage of additional resources available on the Web, especially those on the companion website for this book found at *www.mhhe.com/rosenGE*. You will find many Extra Examples designed to clarify key concepts; Self Assessments for gauging how well you understand core topics; Interactive Demonstration Applets exploring key algorithms and other concepts; a Web Resources Guide containing an extensive selection of links to external sites relevant to the world of discrete mathematics; extra explanations and practice to help you master core concepts; added instruction on writing proofs and on avoiding common mistakes in discrete mathematics; and in-depth discussions of important applications. Places in the text where these additional online resources are available are identified in the margins by special icons. For more details on these and other online resources, see the description of the companion website immediately preceding this "To the Student" message.

THE VALUE OF THIS BOOK   My intention is to make your substantial investment in this text an excellent value. The book, the associated ancillaries, and companion website have taken many years of effort to develop and refine. I am confident that most of you will find that the text and associated materials will help you master discrete mathematics, just as so many previous students have. Even though it is likely that you will not cover some chapters in your

TABLE 1 Hand-Icon Exercises and Where They Are Used			
Section	*Exercise*	*Section Where Used*	*Pages Where Used*
1.1	26	1.3	29
1.1	27	1.3	29
1.3	9	1.6	64
1.3	10	1.6	64
1.3	15	1.6	64
1.3	30	1.6	64, 67
1.7	16	1.7	88
2.3	72	2.3	145
2.3	79	2.5	168
2.5	13	2.5	173
2.5	14	2.5	173
3.1	37	3.1	199
3.2	48	11.2	730
4.1	26	4.2	268
4.3	27	4.1	239
4.4	2	4.6	298
4.4	30	7.2	450
6.4	9	7.2	452
6.4	13	7.4	466
7.2	12	7.2	452
9.1	16	9.4	577
10.4	45	11.1	717
11.1	9	11.1	720
11.1	22	11.1	725
11.1	36	11.2	731
A.2	4	8.3	515

current course, you should find it helpful—as many other students have—to read the relevant sections of the book as you take additional courses. Most of you will return to this book as a useful tool throughout your future studies, especially for those of you who continue in computer science, mathematics, and engineering. I have designed this book to be a gateway for future studies and explorations, and to be comprehensive reference, and I wish you luck as you begin your journey.

Kenneth H. Rosen

List of Symbols

TOPIC	SYMBOL	MEANING	PAGE
LOGIC	$\neg p$	negation of p	3
	$p \wedge q$	conjunction of p and q	4
	$p \vee q$	disjunction of p and q	4
	$p \oplus q$	exclusive or of p and q	6
	$p \rightarrow q$	the implication p implies q	6
	$p \leftrightarrow q$	biconditional of p and q	9
	$p \equiv q$	equivalence of p and q	23
	$\mathbf{T}$	tautology	23
	$\mathbf{F}$	contradiction	23
	$P(x_1, \ldots, x_n)$	propositional function	36
	$\forall x\, P(x)$	universal quantification of $P(x)$	38
	$\exists x\, P(x)$	existential quantification of $P(x)$	40
	$\exists! x\, P(x)$	uniqueness quantification of $P(x)$	41
	$\therefore$	therefore	64
	$p\{S\}q$	partial correctness of S	364
SETS	$x \in S$	x is a member of S	118
	$x \notin S$	x is not a member of S	118
	$\{a_1, \ldots, a_n\}$	list of elements of a set	118
	$\{x \mid P(x)\}$	set builder notation	118
	$\mathbf{N}$	set of natural numbers	118
	$\mathbf{Z}$	set of integers	118
	$\mathbf{Z}^+$	set of positive integers	118
	$\mathbf{Q}$	set of rational numbers	118
	$\mathbf{R}$	set of real numbers	118
	$[a, b], (a, b)$	closed, open intervals	119
	$S = T$	set equality	119
	$\varnothing$	the empty (or null) set	120
	$S \subseteq T$	S is a subset of T	121
	$S \subset T$	S is a proper subset of T	122
	$|S|$	cardinality of S	123
	$\mathcal{P}(S)$	the power set of S	123
	$(a_1, \ldots, a_n)$	n-tuple	124
	(a, b)	ordered pair	124
	$A \times B$	Cartesian product of A and B	125
	$A \cup B$	union of A and B	128
	$A \cap B$	intersection of A and B	129
	$A - B$	the difference of A and B	130
	$\overline{A}$	complement of A	131
	$\displaystyle\bigcup_{i=1}^{n} A_i$	union of A_i, $i = 1, 2, \ldots, n$	135
	$\displaystyle\bigcap_{i=1}^{n} A_i$	intersection of A_i, $i = 1, 2, \ldots, n$	135
	$A \oplus B$	symmetric difference of A and B	138
	$\aleph_0$	cardinality of a countable set	170
	$\mathfrak{c}$	cardinality of $\mathbf{R}$	189

TOPIC	SYMBOL	MEANING	PAGE
FUNCTIONS	$f(a)$	value of the function f at a	140
	$f : A \to B$	function from A to B	140
	$f_1 + f_2$	sum of the functions f_1 and f_2	142
	$f_1 f_2$	product of the functions f_1 and f_2	142
	$f(S)$	image of the set S under f	142
	$\iota_A(s)$	identity function on A	145
	$f^{-1}(x)$	inverse of f	146
	$f \circ g$	composition of f and g	147
	$\lfloor x \rfloor$	floor function of x	150
	$\lceil x \rceil$	ceiling function of x	150
	a_n	term of $\{a_i\}$ with subscript n	156
	$\sum_{i=1}^{n} a_i$	sum of $a_1, a_2, \ldots, a_n$	163
	$\sum_{\alpha \in S} a_\alpha$	sum of a_α over $\alpha \in S$	165
	$\prod_{i=1}^{n} a_n$	product of $a_1, a_2, \ldots, a_n$	168
	$f(x)$ is $O(g(x))$	$f(x)$ is big-O of $g(x)$	207
	$n!$	n factorial	211
	$f(x)$ is $\Omega(g(x))$	$f(x)$ is big-Omega of $g(x)$	216
	$f(x)$ is $\Theta(g(x))$	$f(x)$ is big-Theta of $g(x)$	216
	$\sim$	asymptotic to	219
	$\min(x, y)$	minimum of x and y	265
	$\max(x, y)$	maximum of x and y	265
	$\approx$	approximately equal to	433
INTEGERS	$a \mid b$	a divides b	238
	$a \nmid b$	a does not divide b	238
	a **div** b	quotient when a is divided by b	239
	a **mod** b	remainder when a is divided by b	239
	$a \equiv b \pmod{m}$	a is congruent to b modulo m	240
	$a \not\equiv b \pmod{m}$	a is not congruent to b modulo m	240
	$\mathbf{Z}_m$	integers modulo m	243
	$(a_k a_{k-1} \ldots a_1 a_0)_b$	base b representation	246
	$\gcd(a, b)$	greatest common divisor of a and b	263
	$\operatorname{lcm}(a, b)$	least common multiple of a and b	265
MATRICES	$[a_{ij}]$	matrix with entries a_{ij}	182
	$\mathbf{A} + \mathbf{B}$	matrix sum of $\mathbf{A}$ and $\mathbf{B}$	182
	$\mathbf{AB}$	matrix product of $\mathbf{A}$ and $\mathbf{B}$	182
	$\mathbf{I}_n$	identity matrix of order n	183
	$\mathbf{A}^t$	transpose of $\mathbf{A}$	184
	$\mathbf{A} \vee \mathbf{B}$	join of $\mathbf{A}$ and $\mathbf{B}$	185
	$\mathbf{A} \wedge \mathbf{B}$	the meet of $\mathbf{A}$ and $\mathbf{B}$	185
	$\mathbf{A} \odot \mathbf{B}$	Boolean product of $\mathbf{A}$ and $\mathbf{B}$	185
	$\mathbf{A}^{[n]}$	nth Boolean power of $\mathbf{A}$	186

TOPIC	SYMBOL	MEANING	PAGE
COUNTING AND PROBABILITY	$P(n, r)$	number of r-permutations of a set with n elements	396
	$C(n, r)$	number of r-combinations of a set with n elements	398
	$\binom{n}{r}$	binomial coefficient n choose r	398
	$C(n; n_1, n_2, \ldots, n_m)$	multinomial coefficient	421
	$p(E)$	probability of E	432
	$p(E \mid F)$	conditional probability of E given F	442
	$E(X)$	expected value of the random variable X	463
	$V(X)$	variance of the random variable X	473
	C_n	Catalan number	491
	$N(P_{i_1} \ldots P_{i_n})$	number of elements having properties $P_{i_j}, j = 1, \ldots, n$	541
	$N(P'_{i_1} \ldots P'_{i_n})$	number of elements not having properties $P_{i_j}, j = 1, \ldots, n$	541
	D_n	number of derangements of n objects	545
RELATIONS	$S \circ R$	composite of the relations R and S	560
	R^n	nth power of the relation R	560
	R^{-1}	inverse relation	561
	s_C	select operator for condition C	566
	$P_{i_1, i_2, \ldots, i_m}$	projection	566
	$J_p(R, S)$	join	567
	Δ	diagonal relation	577
	R^*	connectivity relation of R	379
	$a \sim b$	a is equivalent to b	587
	$[a]_R$	equivalence class of a with respect to R	589
	$[a]_m$	congruence class modulo m	590
	(S, R)	poset consisting of the set S and partial ordering R	597
	$a \prec b$	a is less than b	598
	$a \succ b$	a is greater than b	598
	$a \preccurlyeq b$	a is less than or equal to b	598
	$a \succcurlyeq b$	a is greater than or equal to b	598
GRAPHS AND TREES	$G = (V, E)$	graph with vertex set V and edge set E	617
	(u, v)	directed edge	618
	$\{u, v\}$	undirected edge	619
	$\deg v$	degree of the vertex v	627
	$\deg^-(v)$	in-degree of the vertex v	629
	$\deg^+(v)$	out-degree of the vertex v	630
	K_n	complete graph on n vertices	631
	C_n	cycle of size n	631
	W_n	wheel of size n	631
	Q_n	n-cube	631
	$K_{m,n}$	complete bipartite graph of size m, n	633
	$G - e$	subgraph of G with edge e removed	639
	$G + e$	graph produced by adding edge e	639

TOPIC	SYMBOL	MEANING	PAGE
GRAPHS AND	$G_1 \cup G_2$	union of G_1 and G_2	640
TREES (cont.)	$a, x_1, \ldots, x_{n-1}, b$	path from a to b	653
	$a, x_1, \ldots, x_{n-1}, a$	circuit	653
	$\kappa(G)$	vertex connectivity of G	658
	$\lambda(G)$	edge connectivity of G	658
	r	number of regions of the plane	692
	$\deg(R)$	degree of the region R	693
	$\chi(G)$	chromatic number of G	699
	m	greatest number of children of an internal vertex in a rooted tree	718
	n	number of vertices of a rooted tree	722
	i	number of internal vertices of a rooted tree	722
	l	number of leaves of a rooted tree	722
	h	height of a rooted tree	723
LANGUAGES	λ	the empty string	157
AND	xy	concatenation of x and y	344
FINITE-STATE	$l(x)$	length of the string x	344
MACHINES	w^R	reversal of w	352

1

The Foundations: Logic and Proofs

The rules of logic specify the meaning of mathematical statements. For instance, these rules help us understand and reason with statements such as "There exists an integer that is not the sum of two squares" and "For every positive integer n, the sum of the positive integers not exceeding n is $n(n + 1)/2$." Logic is the basis of all mathematical reasoning, and of all automated reasoning. It has practical applications to the design of computing machines, to the specification of systems, to artificial intelligence, to computer programming, to programming languages, and to other areas of computer science, as well as to many other fields of study.

To understand mathematics, we must understand what makes up a correct mathematical argument, that is, a proof. Once we prove a mathematical statement is true, we call it a theorem. A collection of theorems on a topic organize what we know about this topic. To learn a mathematical topic, a person needs to actively construct mathematical arguments on this topic, and not just read exposition. Moreover, knowing the proof of a theorem often makes it possible to modify the result to fit new situations.

Everyone knows that proofs are important throughout mathematics, but many people find it surprising how important proofs are in computer science. In fact, proofs are used to verify that computer programs produce the correct output for all possible input values, to show that algorithms always produce the correct result, to establish the security of a system, and to create artificial intelligence. Furthermore, automated reasoning systems have been created to allow computers to construct their own proofs.

In this chapter, we will explain what makes up a correct mathematical argument and introduce tools to construct these arguments. We will develop an arsenal of different proof methods that will enable us to prove many different types of results. After introducing many different methods of proof, we will introduce several strategies for constructing proofs. We will introduce the notion of a conjecture and explain the process of developing mathematics by studying conjectures.

1.1 Propositional Logic

Introduction

The rules of logic give precise meaning to mathematical statements. These rules are used to distinguish between valid and invalid mathematical arguments. Because a major goal of this book is to teach the reader how to understand and how to construct correct mathematical arguments, we begin our study of discrete mathematics with an introduction to logic.

Besides the importance of logic in understanding mathematical reasoning, logic has numerous applications to computer science. These rules are used in the design of computer circuits, the construction of computer programs, the verification of the correctness of programs, and in many other ways. Furthermore, software systems have been developed for constructing some, but not all, types of proofs automatically. We will discuss these applications of logic in this and later chapters.

Propositions

Our discussion begins with an introduction to the basic building blocks of logic—propositions. A **proposition** is a declarative sentence (that is, a sentence that declares a fact) that is either true or false, but not both.

EXAMPLE 1   All the following declarative sentences are propositions.

Extra
Examples

1. Washington, D.C., is the capital of the United States of America.
2. Toronto is the capital of Canada.
3. $1 + 1 = 2$.
4. $2 + 2 = 3$.

Propositions 1 and 3 are true, whereas 2 and 4 are false.   ◄

Some sentences that are not propositions are given in Example 2.

EXAMPLE 2   Consider the following sentences.

1. What time is it?
2. Read this carefully.
3. $x + 1 = 2$.
4. $x + y = z$.

Sentences 1 and 2 are not propositions because they are not declarative sentences. Sentences 3 and 4 are not propositions because they are neither true nor false. Note that each of sentences 3 and 4 can be turned into a proposition if we assign values to the variables. We will also discuss other ways to turn sentences such as these into propositions in Section 1.4.   ◄

We use letters to denote **propositional variables** (or **statement variables**), that is, variables that represent propositions, just as letters are used to denote numerical variables. The

Links

ARISTOTLE (384 B.C.E.–322 B.C.E.)   Aristotle was born in Stagirus (Stagira) in northern Greece. His father was the personal physician of the King of Macedonia. Because his father died when Aristotle was young, Aristotle could not follow the custom of following his father's profession. Aristotle became an orphan at a young age when his mother also died. His guardian who raised him taught him poetry, rhetoric, and Greek. At the age of 17, his guardian sent him to Athens to further his education. Aristotle joined Plato's Academy, where for 20 years he attended Plato's lectures, later presenting his own lectures on rhetoric. When Plato died in 347 B.C.E., Aristotle was not chosen to succeed him because his views differed too much from those of Plato. Instead, Aristotle joined the court of King Hermeas where he remained for three years, and married the niece of the King. When the Persians defeated Hermeas, Aristotle moved to Mytilene and, at the invitation of King Philip of Macedonia, he tutored Alexander, Philip's son, who later became Alexander the Great. Aristotle tutored Alexander for five years and after the death of King Philip, he returned to Athens and set up his own school, called the Lyceum.

Aristotle's followers were called the peripatetics, which means "to walk about," because Aristotle often walked around as he discussed philosophical questions. Aristotle taught at the Lyceum for 13 years where he lectured to his advanced students in the morning and gave popular lectures to a broad audience in the evening. When Alexander the Great died in 323 B.C.E., a backlash against anything related to Alexander led to trumped-up charges of impiety against Aristotle. Aristotle fled to Chalcis to avoid prosecution. He only lived one year in Chalcis, dying of a stomach ailment in 322 B.C.E.

Aristotle wrote three types of works: those written for a popular audience, compilations of scientific facts, and systematic treatises. The systematic treatises included works on logic, philosophy, psychology, physics, and natural history. Aristotle's writings were preserved by a student and were hidden in a vault where a wealthy book collector discovered them about 200 years later. They were taken to Rome, where they were studied by scholars and issued in new editions, preserving them for posterity.

conventional letters used for propositional variables are $p, q, r, s, \ldots$. The **truth value** of a proposition is true, denoted by T, if it is a true proposition, and the truth value of a proposition is false, denoted by F, if it is a false proposition.

The area of logic that deals with propositions is called the **propositional calculus** or **propositional logic**. It was first developed systematically by the Greek philosopher Aristotle more than 2300 years ago.

Links

We now turn our attention to methods for producing new propositions from those that we already have. These methods were discussed by the English mathematician George Boole in 1854 in his book *The Laws of Thought*. Many mathematical statements are constructed by combining one or more propositions. New propositions, called **compound propositions**, are formed from existing propositions using logical operators.

DEFINITION 1

Let p be a proposition. The *negation of* p, denoted by $\neg p$ (also denoted by $\overline{p}$), is the statement

"It is not the case that p."

The proposition $\neg p$ is read "not p." The truth value of the negation of p, $\neg p$, is the opposite of the truth value of p.

EXAMPLE 3

Find the negation of the proposition

"Michael's PC runs Linux"

Extra Examples

and express this in simple English.

Solution: The negation is

"It is not the case that Michael's PC runs Linux."

This negation can be more simply expressed as

"Michael's PC does not run Linux."

◀

EXAMPLE 4

Find the negation of the proposition

"Vandana's smartphone has at least 32GB of memory"

and express this in simple English.

Solution: The negation is

"It is not the case that Vandana's smartphone has at least 32GB of memory."

This negation can also be expressed as

"Vandana's smartphone does not have at least 32GB of memory"

or even more simply as

"Vandana's smartphone has less than 32GB of memory."

◀

TABLE 1 The Truth Table for the Negation of a Proposition.	
p	$\neg p$
T	F
F	T

Table 1 displays the **truth table** for the negation of a proposition p. This table has a row for each of the two possible truth values of a proposition p. Each row shows the truth value of $\neg p$ corresponding to the truth value of p for this row.

The negation of a proposition can also be considered the result of the operation of the **negation operator** on a proposition. The negation operator constructs a new proposition from a single existing proposition. We will now introduce the logical operators that are used to form new propositions from two or more existing propositions. These logical operators are also called **connectives**.

DEFINITION 2   Let p and q be propositions. The *conjunction* of p and q, denoted by $p \wedge q$, is the proposition "p and q." The conjunction $p \wedge q$ is true when both p and q are true and is false otherwise.

Table 2 displays the truth table of $p \wedge q$. This table has a row for each of the four possible combinations of truth values of p and q. The four rows correspond to the pairs of truth values TT, TF, FT, and FF, where the first truth value in the pair is the truth value of p and the second truth value is the truth value of q.

Note that in logic the word "but" sometimes is used instead of "and" in a conjunction. For example, the statement "The sun is shining, but it is raining" is another way of saying "The sun is shining and it is raining." (In natural language, there is a subtle difference in meaning between "and" and "but"; we will not be concerned with this nuance here.)

EXAMPLE 5   Find the conjunction of the propositions p and q where p is the proposition "Rebecca's PC has more than 16 GB free hard disk space" and q is the proposition "The processor in Rebecca's PC runs faster than 1 GHz."

Solution: The conjunction of these propositions, $p \wedge q$, is the proposition "Rebecca's PC has more than 16 GB free hard disk space, and the processor in Rebecca's PC runs faster than 1 GHz." This conjunction can be expressed more simply as "Rebecca's PC has more than 16 GB free hard disk space, and its processor runs faster than 1 GHz." For this conjunction to be true, both conditions given must be true. It is false, when one or both of these conditions are false. ◀

DEFINITION 3   Let p and q be propositions. The *disjunction* of p and q, denoted by $p \vee q$, is the proposition "p or q." The disjunction $p \vee q$ is false when both p and q are false and is true otherwise.

Table 3 displays the truth table for $p \vee q$.

TABLE 2 The Truth Table for the Conjunction of Two Propositions.		
p	q	$p \wedge q$
T	T	T
T	F	F
F	T	F
F	F	F

TABLE 3 The Truth Table for the Disjunction of Two Propositions.		
p	q	$p \vee q$
T	T	T
T	F	T
F	T	T
F	F	F

The use of the connective *or* in a disjunction corresponds to one of the two ways the word *or* is used in English, namely, as an **inclusive or**. A disjunction is true when at least one of the two propositions is true. For instance, the inclusive or is being used in the statement

"Students who have taken calculus or computer science can take this class."

Here, we mean that students who have taken both calculus and computer science can take the class, as well as the students who have taken only one of the two subjects. On the other hand, we are using the **exclusive or** when we say

"Students who have taken calculus or computer science, but not both, can enroll in this class."

Here, we mean that students who have taken both calculus and a computer science course cannot take the class. Only those who have taken exactly one of the two courses can take the class.

Similarly, when a menu at a restaurant states, "Soup or salad comes with an entrée," the restaurant almost always means that customers can have either soup or salad, but not both. Hence, this is an exclusive, rather than an inclusive, or.

EXAMPLE 6 What is the disjunction of the propositions p and q where p and q are the same propositions as in Example 5?

 Extra Examples

Solution: The disjunction of p and q, $p \vee q$, is the proposition

"Rebecca's PC has at least 16 GB free hard disk space, or the processor in Rebecca's PC runs faster than 1 GHz."

This proposition is true when Rebecca's PC has at least 16 GB free hard disk space, when the PC's processor runs faster than 1 GHz, and when both conditions are true. It is false when both of these conditions are false, that is, when Rebecca's PC has less than 16 GB free hard disk space and the processor in her PC runs at 1 GHz or slower. ◀

As was previously remarked, the use of the connective *or* in a disjunction corresponds to one of the two ways the word *or* is used in English, namely, in an inclusive way. Thus, a disjunction is true when at least one of the two propositions in it is true. Sometimes, we use *or* in an exclusive sense. When the exclusive or is used to connect the propositions p and q, the proposition "p or q (but not both)" is obtained. This proposition is true when p is true and q is false, and when p is false and q is true. It is false when both p and q are false and when both are true.

 Links

GEORGE BOOLE (1815–1864) George Boole, the son of a cobbler, was born in Lincoln, England, in November 1815. Because of his family's difficult financial situation, Boole struggled to educate himself while supporting his family. Nevertheless, he became one of the most important mathematicians of the 1800s. Although he considered a career as a clergyman, he decided instead to go into teaching, and soon afterward opened a school of his own. In his preparation for teaching mathematics, Boole—unsatisfied with textbooks of his day— decided to read the works of the great mathematicians. While reading papers of the great French mathematician Lagrange, Boole made discoveries in the calculus of variations, the branch of analysis dealing with finding curves and surfaces by optimizing certain parameters.

In 1848 Boole published *The Mathematical Analysis of Logic*, the first of his contributions to symbolic logic. In 1849 he was appointed professor of mathematics at Queen's College in Cork, Ireland. In 1854 he published *The Laws of Thought*, his most famous work. In this book, Boole introduced what is now called *Boolean algebra* in his honor. Boole wrote textbooks on differential equations and on difference equations that were used in Great Britain until the end of the nineteenth century. Boole married in 1855; his wife was the niece of the professor of Greek at Queen's College. In 1864 Boole died from pneumonia, which he contracted as a result of keeping a lecture engagement even though he was soaking wet from a rainstorm.

TABLE 4 The Truth Table for the Exclusive Or of Two Propositions.		
p	q	$p \oplus q$
T	T	F
T	F	T
F	T	T
F	F	F

TABLE 5 The Truth Table for the Conditional Statement $p \rightarrow q$.		
p	q	$p \rightarrow q$
T	T	T
T	F	F
F	T	T
F	F	T

DEFINITION 4   Let p and q be propositions. The *exclusive or* of p and q, denoted by $p \oplus q$, is the proposition that is true when exactly one of p and q is true and is false otherwise.

The truth table for the exclusive or of two propositions is displayed in Table 4.

Conditional Statements

We will discuss several other important ways in which propositions can be combined.

DEFINITION 5   Let p and q be propositions. The *conditional statement* $p \rightarrow q$ is the proposition "if p, then q." The conditional statement $p \rightarrow q$ is false when p is true and q is false, and true otherwise. In the conditional statement $p \rightarrow q$, p is called the *hypothesis* (or *antecedent* or *premise*) and q is called the *conclusion* (or *consequence*).

Assessment

The statement $p \rightarrow q$ is called a conditional statement because $p \rightarrow q$ asserts that q is true on the condition that p holds. A conditional statement is also called an **implication**.

The truth table for the conditional statement $p \rightarrow q$ is shown in Table 5. Note that the statement $p \rightarrow q$ is true when both p and q are true and when p is false (no matter what truth value q has).

Because conditional statements play such an essential role in mathematical reasoning, a variety of terminology is used to express $p \rightarrow q$. You will encounter most if not all of the following ways to express this conditional statement:

"if p, then q"   "p implies q"
"if p, q"   "p only if q"
"p is sufficient for q"   "a sufficient condition for q is p"
"q if p"   "q whenever p"
"q when p"   "q is necessary for p"
"a necessary condition for p is q" "q follows from p"
"q unless $\neg p$"

A useful way to understand the truth value of a conditional statement is to think of an obligation or a contract. For example, the pledge many politicians make when running for office is

"If I am elected, then I will lower taxes."

If the politician is elected, voters would expect this politician to lower taxes. Furthermore, if the politician is not elected, then voters will not have any expectation that this person will lower taxes, although the person may have sufficient influence to cause those in power to lower taxes. It is only when the politician is elected but does not lower taxes that voters can say that the politician has broken the campaign pledge. This last scenario corresponds to the case when p is true but q is false in $p \rightarrow q$.

Similarly, consider a statement that a professor might make:

"If you get 100% on the final, then you will get an A."

If you manage to get a 100% on the final, then you would expect to receive an A. If you do not get 100% you may or may not receive an A depending on other factors. However, if you do get 100%, but the professor does not give you an A, you will feel cheated.

Of the various ways to express the conditional statement $p \rightarrow q$, the two that seem to cause the most confusion are "p only if q" and "q unless $\neg p$." Consequently, we will provide some guidance for clearing up this confusion.

To remember that "p only if q" expresses the same thing as "if p, then q," note that "p only if q" says that p cannot be true when q is not true. That is, the statement is false if p is true, but q is false. When p is false, q may be either true or false, because the statement says nothing about the truth value of q. Be careful not to use "q only if p" to express $p \rightarrow q$ because this is incorrect. To see this, note that the true values of "q only if p" and $p \rightarrow q$ are different when p and q have different truth values.

To remember that "q unless $\neg p$" expresses the same conditional statement as "if p, then q," note that "q unless $\neg p$" means that if $\neg p$ is false, then q must be true. That is, the statement "q unless $\neg p$" is false when p is true but q is false, but it is true otherwise. Consequently, "q unless $\neg p$" and $p \rightarrow q$ always have the same truth value.

We illustrate the translation between conditional statements and English statements in Example 7.

> You might have trouble understanding how "unless" is used in conditional statements unless you read this paragraph carefully.

EXAMPLE 7   Let p be the statement "Maria learns discrete mathematics" and q the statement "Maria will find a good job." Express the statement $p \rightarrow q$ as a statement in English.

Solution: From the definition of conditional statements, we see that when p is the statement "Maria learns discrete mathematics" and q is the statement "Maria will find a good job," $p \rightarrow q$ represents the statement

"If Maria learns discrete mathematics, then she will find a good job."

There are many other ways to express this conditional statement in English. Among the most natural of these are:

"Maria will find a good job when she learns discrete mathematics."

"For Maria to get a good job, it is sufficient for her to learn discrete mathematics."

and

"Maria will find a good job unless she does not learn discrete mathematics."   ◄

Note that the way we have defined conditional statements is more general than the meaning attached to such statements in the English language. For instance, the conditional statement in Example 7 and the statement

"If it is sunny, then we will go to the beach."

are statements used in normal language where there is a relationship between the hypothesis and the conclusion. Further, the first of these statements is true unless Maria learns discrete mathematics, but she does not get a good job, and the second is true unless it is indeed sunny, but we do not go to the beach. On the other hand, the statement

"If Juan has a smartphone, then $2 + 3 = 5$"

is true from the definition of a conditional statement, because its conclusion is true. (The truth value of the hypothesis does not matter then.) The conditional statement

"If Juan has a smartphone, then $2 + 3 = 6$"

is true if Juan does not have a smartphone, even though $2 + 3 = 6$ is false. We would not use these last two conditional statements in natural language (except perhaps in sarcasm), because there is no relationship between the hypothesis and the conclusion in either statement. In mathematical reasoning, we consider conditional statements of a more general sort than we use in English. The mathematical concept of a conditional statement is independent of a cause-and-effect relationship between hypothesis and conclusion. Our definition of a conditional statement specifies its truth values; it is not based on English usage. Propositional language is an artificial language; we only parallel English usage to make it easy to use and remember.

The if-then construction used in many programming languages is different from that used in logic. Most programming languages contain statements such as **if** p **then** S, where p is a proposition and S is a program segment (one or more statements to be executed). When execution of a program encounters such a statement, S is executed if p is true, but S is not executed if p is false, as illustrated in Example 8.

EXAMPLE 8   What is the value of the variable x after the statement

if $2 + 2 = 4$ **then** $x := x + 1$

if $x = 0$ before this statement is encountered? (The symbol $:=$ stands for assignment. The statement $x := x + 1$ means the assignment of the value of $x + 1$ to x.)

Solution: Because $2 + 2 = 4$ is true, the assignment statement $x := x + 1$ is executed. Hence, x has the value $0 + 1 = 1$ after this statement is encountered.   ◀

CONVERSE, CONTRAPOSITIVE, AND INVERSE   We can form some new conditional statements starting with a conditional statement $p \rightarrow q$. In particular, there are three related conditional statements that occur so often that they have special names. The proposition $q \rightarrow p$ is called the **converse** of $p \rightarrow q$. The **contrapositive** of $p \rightarrow q$ is the proposition $\neg q \rightarrow \neg p$. The proposition $\neg p \rightarrow \neg q$ is called the **inverse** of $p \rightarrow q$. We will see that of these three conditional statements formed from $p \rightarrow q$, only the contrapositive always has the same truth value as $p \rightarrow q$.

We first show that the contrapositive, $\neg q \rightarrow \neg p$, of a conditional statement $p \rightarrow q$ always has the same truth value as $p \rightarrow q$. To see this, note that the contrapositive is false only when $\neg p$ is false and $\neg q$ is true, that is, only when p is true and q is false. We now show that neither the converse, $q \rightarrow p$, nor the inverse, $\neg p \rightarrow \neg q$, has the same truth value as $p \rightarrow q$ for all possible truth values of p and q. Note that when p is true and q is false, the original conditional statement is false, but the converse and the inverse are both true.

When two compound propositions always have the same truth value we call them **equivalent**, so that a conditional statement and its contrapositive are equivalent. The converse and the inverse of a conditional statement are also equivalent, as the reader can verify, but neither is equivalent to the original conditional statement. (We will study equivalent propositions in Section 1.3.) Take note that one of the most common logical errors is to assume that the converse or the inverse of a conditional statement is equivalent to this conditional statement.

We illustrate the use of conditional statements in Example 9.

> Remember that the contrapositive, but neither the converse or inverse, of a conditional statement is equivalent to it.

EXAMPLE 9   What are the contrapositive, the converse, and the inverse of the conditional statement

"The home team wins whenever it is raining?"

Solution: Because "*q* whenever *p*" is one of the ways to express the conditional statement $p \rightarrow q$, the original statement can be rewritten as

"If it is raining, then the home team wins."

Consequently, the contrapositive of this conditional statement is

"If the home team does not win, then it is not raining."

The converse is

"If the home team wins, then it is raining."

The inverse is

"If it is not raining, then the home team does not win."

Only the contrapositive is equivalent to the original statement.   ◄

BICONDITIONALS   We now introduce another way to combine propositions that expresses that two propositions have the same truth value.

DEFINITION 6

Let *p* and *q* be propositions. The *biconditional statement* $p \leftrightarrow q$ is the proposition "*p* if and only if *q*." The biconditional statement $p \leftrightarrow q$ is true when *p* and *q* have the same truth values, and is false otherwise. Biconditional statements are also called *bi-implications*.

The truth table for $p \leftrightarrow q$ is shown in Table 6. Note that the statement $p \leftrightarrow q$ is true when both the conditional statements $p \rightarrow q$ and $q \rightarrow p$ are true and is false otherwise. That is why we use the words "if and only if" to express this logical connective and why it is symbolically written by combining the symbols $\rightarrow$ and $\leftarrow$. There are some other common ways to express $p \leftrightarrow q$:

"*p* is necessary and sufficient for *q*"
"if *p* then *q*, and conversely"
"*p* iff *q*."

The last way of expressing the biconditional statement $p \leftrightarrow q$ uses the abbreviation "iff" for "if and only if." Note that $p \leftrightarrow q$ has exactly the same truth value as $(p \rightarrow q) \wedge (q \rightarrow p)$.

TABLE 6 The Truth Table for the Biconditional $p \leftrightarrow q$.		
p	*q*	$p \leftrightarrow q$
T	T	T
T	F	F
F	T	F
F	F	T

EXAMPLE 10   Let p be the statement "You can take the flight," and let q be the statement "You buy a ticket." Then $p \leftrightarrow q$ is the statement

"You can take the flight if and only if you buy a ticket."

This statement is true if p and q are either both true or both false, that is, if you buy a ticket and can take the flight or if you do not buy a ticket and you cannot take the flight. It is false when p and q have opposite truth values, that is, when you do not buy a ticket, but you can take the flight (such as when you get a free trip) and when you buy a ticket but you cannot take the flight (such as when the airline bumps you).   ◀

IMPLICIT USE OF BICONDITIONALS   You should be aware that biconditionals are not always explicit in natural language. In particular, the "if and only if" construction used in biconditionals is rarely used in common language. Instead, biconditionals are often expressed using an "if, then" or an "only if" construction. The other part of the "if and only if" is implicit. That is, the converse is implied, but not stated. For example, consider the statement in English "If you finish your meal, then you can have dessert." What is really meant is "You can have dessert if and only if you finish your meal." This last statement is logically equivalent to the two statements "If you finish your meal, then you can have dessert" and "You can have dessert only if you finish your meal." Because of this imprecision in natural language, we need to make an assumption whether a conditional statement in natural language implicitly includes its converse. Because precision is essential in mathematics and in logic, we will always distinguish between the conditional statement $p \rightarrow q$ and the biconditional statement $p \leftrightarrow q$.

Truth Tables of Compound Propositions

We have now introduced four important logical connectives—conjunctions, disjunctions, conditional statements, and biconditional statements—as well as negations. We can use these connectives to build up complicated compound propositions involving any number of propositional variables. We can use truth tables to determine the truth values of these compound propositions, as Example 11 illustrates. We use a separate column to find the truth value of each compound expression that occurs in the compound proposition as it is built up. The truth values of the compound proposition for each combination of truth values of the propositional variables in it is found in the final column of the table.

EXAMPLE 11   Construct the truth table of the compound proposition

$$(p \vee \neg q) \rightarrow (p \wedge q).$$

Solution: Because this truth table involves two propositional variables p and q, there are four rows in this truth table, one for each of the pairs of truth values TT, TF, FT, and FF. The first two columns are used for the truth values of p and q, respectively. In the third column we find the truth value of $\neg q$, needed to find the truth value of $p \vee \neg q$, found in the fourth column. The fifth column gives the truth value of $p \wedge q$. Finally, the truth value of $(p \vee \neg q) \rightarrow (p \wedge q)$ is found in the last column. The resulting truth table is shown in Table 7.   ◀

TABLE 7 **The Truth Table of $(p \vee \neg q) \rightarrow (p \wedge q)$.**					
p	q	$\neg q$	$p \vee \neg q$	$p \wedge q$	$(p \vee \neg q) \rightarrow (p \wedge q)$
T	T	F	T	T	T
T	F	T	T	F	F
F	T	F	F	F	T
F	F	T	T	F	F

Precedence of Logical Operators

TABLE 8
Precedence of Logical Operators.

Operator	Precedence
¬	1
∧	2
∨	3
→	4
↔	5

We can construct compound propositions using the negation operator and the logical operators defined so far. We will generally use parentheses to specify the order in which logical operators in a compound proposition are to be applied. For instance, $(p \lor q) \land (\neg r)$ is the conjunction of $p \lor q$ and $\neg r$. However, to reduce the number of parentheses, we specify that the negation operator is applied before all other logical operators. This means that $\neg p \land q$ is the conjunction of $\neg p$ and q, namely, $(\neg p) \land q$, not the negation of the conjunction of p and q, namely $\neg(p \land q)$.

Another general rule of precedence is that the conjunction operator takes precedence over the disjunction operator, so that $p \land q \lor r$ means $(p \land q) \lor r$ rather than $p \land (q \lor r)$. Because this rule may be difficult to remember, we will continue to use parentheses so that the order of the disjunction and conjunction operators is clear.

Finally, it is an accepted rule that the conditional and biconditional operators $\to$ and $\leftrightarrow$ have lower precedence than the conjunction and disjunction operators, $\land$ and $\lor$. Consequently, $p \lor q \to r$ is the same as $(p \lor q) \to r$. We will use parentheses when the order of the conditional operator and biconditional operator is at issue, although the conditional operator has precedence over the biconditional operator. Table 8 displays the precedence levels of the logical operators, $\neg, \land, \lor, \to$, and $\leftrightarrow$.

Logic and Bit Operations

Truth Value	Bit
T	1
F	0

Links

Computers represent information using bits. A **bit** is a symbol with two possible values, namely, 0 (zero) and 1 (one). This meaning of the word bit comes from *b*inary dig*it*, because zeros and ones are the digits used in binary representations of numbers. The well-known statistician John Tukey introduced this terminology in 1946. A bit can be used to represent a truth value, because there are two truth values, namely, *true* and *false*. As is customarily done, we will use a 1 bit to represent true and a 0 bit to represent false. That is, 1 represents T (true), 0 represents F (false). A variable is called a **Boolean variable** if its value is either true or false. Consequently, a Boolean variable can be represented using a bit.

Computer **bit operations** correspond to the logical connectives. By replacing true by a one and false by a zero in the truth tables for the operators $\land$, $\lor$, and $\oplus$, the tables shown in Table 9 for the corresponding bit operations are obtained. We will also use the notation *OR*, *AND*, and *XOR* for the operators $\lor$, $\land$, and $\oplus$, as is done in various programming languages.

Links

JOHN WILDER TUKEY (1915–2000) Tukey, born in New Bedford, Massachusetts, was an only child. His parents, both teachers, decided home schooling would best develop his potential. His formal education began at Brown University, where he studied mathematics and chemistry. He received a master's degree in chemistry from Brown and continued his studies at Princeton University, changing his field of study from chemistry to mathematics. He received his Ph.D. from Princeton in 1939 for work in topology, when he was appointed an instructor in mathematics at Princeton. With the start of World War II, he joined the Fire Control Research Office, where he began working in statistics. Tukey found statistical research to his liking and impressed several leading statisticians with his skills. In 1945, at the conclusion of the war, Tukey returned to the mathematics department at Princeton as a professor of statistics, and he also took a position at AT&T Bell Laboratories. Tukey founded the Statistics Department at Princeton in 1966 and was its first chairman. Tukey made significant contributions to many areas of statistics, including the analysis of variance, the estimation of spectra of time series, inferences about the values of a set of parameters from a single experiment, and the philosophy of statistics. However, he is best known for his invention, with J. W. Cooley, of the fast Fourier transform. In addition to his contributions to statistics, Tukey was noted as a skilled wordsmith; he is credited with coining the terms *bit* and *software*.

Tukey contributed his insight and expertise by serving on the President's Science Advisory Committee. He chaired several important committees dealing with the environment, education, and chemicals and health. He also served on committees working on nuclear disarmament. Tukey received many awards, including the National Medal of Science.

HISTORICAL NOTE There were several other suggested words for a binary digit, including *binit* and *bigit*, that never were widely accepted. The adoption of the word *bit* may be due to its meaning as a common English word. For an account of Tukey's coining of the word *bit*, see the April 1984 issue of *Annals of the History of Computing*.

TABLE 9 Table for the Bit Operators *OR*, *AND*, and *XOR*.				
x	y	$x \vee y$	$x \wedge y$	$x \oplus y$
0	0	0	0	0
0	1	1	0	1
1	0	1	0	1
1	1	1	1	0

Information is often represented using bit strings, which are lists of zeros and ones. When this is done, operations on the bit strings can be used to manipulate this information.

DEFINITION 7   A *bit string* is a sequence of zero or more bits. The *length* of this string is the number of bits in the string.

EXAMPLE 12   101010011 is a bit string of length nine.   ◄

We can extend bit operations to bit strings. We define the **bitwise *OR***, **bitwise *AND***, and **bitwise *XOR*** of two strings of the same length to be the strings that have as their bits the *OR*, *AND*, and *XOR* of the corresponding bits in the two strings, respectively. We use the symbols $\vee$, $\wedge$, and $\oplus$ to represent the bitwise *OR*, bitwise *AND*, and bitwise *XOR* operations, respectively. We illustrate bitwise operations on bit strings with Example 13.

EXAMPLE 13   Find the bitwise *OR*, bitwise *AND*, and bitwise *XOR* of the bit strings 01 1011 0110 and 11 0001 1101. (Here, and throughout this book, bit strings will be split into blocks of four bits to make them easier to read.)

Solution: The bitwise *OR*, bitwise *AND*, and bitwise *XOR* of these strings are obtained by taking the *OR*, *AND*, and *XOR* of the corresponding bits, respectively. This gives us

```
  01 1011 0110
  11 0001 1101
  ------------
  11 1011 1111    bitwise OR
  01 0001 0100    bitwise AND
  10 1010 1011    bitwise XOR
```
◄

Exercises

1. Which of these sentences are propositions? What are the truth values of those that are propositions?

  a) Boston is the capital of Massachusetts.

  b) Miami is the capital of Florida.

  c) $2 + 3 = 5$.

  d) $5 + 7 = 10$.

  e) $x + 2 = 11$.

  f) Answer this question.

2. Which of these are propositions? What are the truth values of those that are propositions?

  a) Do not pass go.

  b) What time is it?

  c) There are no black flies in Maine.

  d) $4 + x = 5$.

  e) The moon is made of green cheese.

  f) $2^n \geq 100$.

3. What is the negation of each of these propositions?

  a) Mei has an MP3 player.

  b) There is no pollution in New Jersey.

  c) $2 + 1 = 3$.

  d) The summer in Maine is hot and sunny.

4. Let p and q be the propositions

      p : I bought a lottery ticket this week.

      q : I won the million dollar jackpot.

Express each of these propositions as an English sentence.

a) $\neg p$ **b)** $p \vee q$ **c)** $p \to q$
d) $p \wedge q$ **e)** $p \leftrightarrow q$ **f)** $\neg p \to \neg q$
g) $\neg p \wedge \neg q$ **h)** $\neg p \vee (p \wedge q)$

5. Suppose that during the most recent fiscal year, the annual revenue of Acme Computer was 138 billion dollars and its net profit was 8 billion dollars, the annual revenue of Nadir Software was 87 billion dollars and its net profit was 5 billion dollars, and the annual revenue of Quixote Media was 111 billion dollars and its net profit was 13 billion dollars. Determine the truth value of each of these propositions for the most recent fiscal year.

a) Quixote Media had the largest annual revenue.
b) Nadir Software had the lowest net profit and Acme Computer had the largest annual revenue.
c) Acme Computer had the largest net profit or Quixote Media had the largest net profit.
d) If Quixote Media had the smallest net profit, then Acme Computer had the largest annual revenue.
e) Nadir Software had the smallest net profit if and only if Acme Computer had the largest annual revenue.

6. Let p, q, and r be the propositions

p : You have the flu.
q : You miss the final examination.
r : You pass the course.

Express each of these propositions as an English sentence.

a) $p \to q$ **b)** $\neg q \leftrightarrow r$
c) $q \to \neg r$ **d)** $p \vee q \vee r$
e) $(p \to \neg r) \vee (q \to \neg r)$
f) $(p \wedge q) \vee (\neg q \wedge r)$

7. Let p and q be the propositions

p : It is below freezing.
q : It is snowing.

Write these propositions using p and q and logical connectives (including negations).

a) It is below freezing and snowing.
b) It is below freezing but not snowing.
c) It is not below freezing and it is not snowing.
d) It is either snowing or below freezing (or both).
e) If it is below freezing, it is also snowing.
f) Either it is below freezing or it is snowing, but it is not snowing if it is below freezing.
g) That it is below freezing is necessary and sufficient for it to be snowing.

8. Let p, q, and r be the propositions

p : You get an A on the final exam.
q : You do every exercise in this book.
r : You get an A in this class.

Write these propositions using p, q, and r and logical connectives (including negations).

a) You get an A in this class, but you do not do every exercise in this book.
b) You get an A on the final, you do every exercise in this book, and you get an A in this class.
c) To get an A in this class, it is necessary for you to get an A on the final.

d) You get an A on the final, but you don't do every exercise in this book; nevertheless, you get an A in this class.
e) Getting an A on the final and doing every exercise in this book is sufficient for getting an A in this class.
f) You will get an A in this class if and only if you either do every exercise in this book or you get an A on the final.

9. Let p and q be the propositions

p : You drive over 65 miles per hour.
q : You get a speeding ticket.

Write these propositions using p and q and logical connectives (including negations).

a) You do not drive over 65 miles per hour.
b) You drive over 65 miles per hour, but you do not get a speeding ticket.
c) You will get a speeding ticket if you drive over 65 miles per hour.
d) If you do not drive over 65 miles per hour, then you will not get a speeding ticket.
e) Driving over 65 miles per hour is sufficient for getting a speeding ticket.
f) You get a speeding ticket, but you do not drive over 65 miles per hour.
g) Whenever you get a speeding ticket, you are driving over 65 miles per hour.

10. Let p, q, and r be the propositions

p : Grizzly bears have been seen in the area.
q : Hiking is safe on the trail.
r : Berries are ripe along the trail.

Write these propositions using p, q, and r and logical connectives (including negations).

a) Berries are ripe along the trail, but grizzly bears have not been seen in the area.
b) Grizzly bears have not been seen in the area and hiking on the trail is safe, but berries are ripe along the trail.
c) If berries are ripe along the trail, hiking is safe if and only if grizzly bears have not been seen in the area.
d) It is not safe to hike on the trail, but grizzly bears have not been seen in the area and the berries along the trail are ripe.
e) For hiking on the trail to be safe, it is necessary but not sufficient that berries not be ripe along the trail and for grizzly bears not to have been seen in the area.
f) Hiking is not safe on the trail whenever grizzly bears have been seen in the area and berries are ripe along the trail.

11. Determine whether each of these conditional statements is true or false.

a) If $1 + 1 = 2$, then $2 + 2 = 5$.
b) If $1 + 1 = 3$, then $2 + 2 = 4$.
c) If $1 + 1 = 3$, then $2 + 2 = 5$.
d) If monkeys can fly, then $1 + 1 = 3$.

12. Determine whether these biconditionals are true or false.

a) $2 + 2 = 4$ if and only if $1 + 1 = 2$.
b) $1 + 1 = 2$ if and only if $2 + 3 = 4$.

c) $1 + 1 = 3$ if and only if monkeys can fly.

d) $0 > 1$ if and only if $2 > 1$.

13. For each of these sentences, state what the sentence means if the logical connective or is an inclusive or (that is, a disjunction) versus an exclusive or. Which of these meanings of or do you think is intended?

a) To take discrete mathematics, you must have taken calculus or a course in computer science.

b) When you buy a new car from Acme Motor Company, you get $2000 back in cash or a 2% car loan.

c) Dinner for two includes two items from column A or three items from column B.

d) School is closed if more than 2 feet of snow falls or if the wind chill is below -100.

14. For each of these sentences, determine whether an inclusive or, or an exclusive or, is intended. Explain your answer.

a) Experience with C++ or Java is required.

b) Lunch includes soup or salad.

c) To enter the country you need a passport or a voter registration card.

d) Publish or perish.

15. Write each of these statements in the form "if p, then q" in English. [*Hint:* Refer to the list of common ways to express conditional statements.]

a) It snows whenever the wind blows from the northeast.

b) The apple trees will bloom if it stays warm for a week.

c) That the Pistons win the championship implies that they beat the Lakers.

d) It is necessary to walk 8 miles to get to the top of Long's Peak.

e) To get tenure as a professor, it is sufficient to be world-famous.

f) If you drive more than 400 miles, you will need to buy gasoline.

g) Your guarantee is good only if you bought your CD player less than 90 days ago.

h) Jan will go swimming unless the water is too cold.

16. Write each of these propositions in the form "p if and only if q" in English.

a) For you to get an A in this course, it is necessary and sufficient that you learn how to solve discrete mathematics problems.

b) If you read the newspaper every day, you will be informed, and conversely.

c) It rains if it is a weekend day, and it is a weekend day if it rains.

d) You can see the wizard only if the wizard is not in, and the wizard is not in only if you can see him.

17. Write each of these propositions in the form "p if and only if q" in English.

a) If it is hot outside you buy an ice cream cone, and if you buy an ice cream cone it is hot outside.

b) For you to win the contest it is necessary and sufficient that you have the only winning ticket.

c) You get promoted only if you have connections, and you have connections only if you get promoted.

d) If you watch television your mind will decay, and conversely.

e) The trains run late on exactly those days when I take it.

18. State the converse, contrapositive, and inverse of each of these conditional statements.

a) If it snows today, I will ski tomorrow.

b) I come to class whenever there is going to be a quiz.

c) A positive integer is a prime only if it has no divisors other than 1 and itself.

19. How many rows appear in a truth table for each of these compound propositions?

a) $p \rightarrow \neg p$

b) $(p \vee \neg r) \wedge (q \vee \neg s)$

c) $q \vee p \vee \neg s \vee \neg r \vee \neg t \vee u$

d) $(p \wedge r \wedge t) \leftrightarrow (q \wedge t)$

20. Construct a truth table for each of these compound propositions.

a) $p \rightarrow \neg p$ **b)** $p \leftrightarrow \neg p$

c) $p \oplus (p \vee q)$ **d)** $(p \wedge q) \rightarrow (p \vee q)$

e) $(q \rightarrow \neg p) \leftrightarrow (p \leftrightarrow q)$

f) $(p \leftrightarrow q) \oplus (p \leftrightarrow \neg q)$

21. Construct a truth table for each of these compound propositions.

a) $p \wedge \neg p$ **b)** $p \vee \neg p$

c) $(p \vee \neg q) \rightarrow q$ **d)** $(p \vee q) \rightarrow (p \wedge q)$

e) $(p \rightarrow q) \leftrightarrow (\neg q \rightarrow \neg p)$

f) $(p \rightarrow q) \rightarrow (q \rightarrow p)$

22. Construct a truth table for each of these compound propositions.

a) $p \oplus p$ **b)** $p \oplus \neg p$

c) $p \oplus \neg q$ **d)** $\neg p \oplus \neg q$

e) $(p \oplus q) \vee (p \oplus \neg q)$ **f)** $(p \oplus q) \wedge (p \oplus \neg q)$

23. Construct a truth table for each of these compound propositions.

a) $p \rightarrow (\neg q \vee r)$

b) $\neg p \rightarrow (q \rightarrow r)$

c) $(p \rightarrow q) \vee (\neg p \rightarrow r)$

d) $(p \rightarrow q) \wedge (\neg p \rightarrow r)$

e) $(p \leftrightarrow q) \vee (\neg q \leftrightarrow r)$

f) $(\neg p \leftrightarrow \neg q) \leftrightarrow (q \leftrightarrow r)$

24. Construct a truth table for $((p \rightarrow q) \rightarrow r) \rightarrow s$.

25. Construct a truth table for $(p \leftrightarrow q) \leftrightarrow (r \leftrightarrow s)$.

26. Explain, without using a truth table, why $(p \vee \neg q) \wedge (q \vee \neg r) \wedge (r \vee \neg p)$ is true when p, q, and r have the same truth value and it is false otherwise.

27. Explain, without using a truth table, why $(p \vee q \vee r) \wedge (\neg p \vee \neg q \vee \neg r)$ is true when at least one of p, q, and r is true and at least one is false, but is false when all three variables have the same truth value.

28. What is the value of x after each of these statements is encountered in a computer program, if $x = 1$ before the statement is reached?

   a) if $x + 2 = 3$ then $x := x + 1$

   b) if $(x + 1 = 3)$ OR $(2x + 2 = 3)$ then $x := x + 1$

   c) if $(2x + 3 = 5)$ AND $(3x + 4 = 7)$ then $x := x + 1$

   d) if $(x + 1 = 2)$ XOR $(x + 2 = 3)$ then $x := x + 1$

   e) if $x < 2$ then $x := x + 1$

29. Find the bitwise OR, bitwise AND, and bitwise XOR of each of these pairs of bit strings.

   a) 101 1110, 010 0001

   b) 1111 0000, 1010 1010

   c) 00 0111 0001, 10 0100 1000

   d) 11 1111 1111, 00 0000 0000

*30. Is the assertion "This statement is false" a proposition?

Fuzzy logic is used in artificial intelligence. In fuzzy logic, a proposition has a truth value that is a number between 0 and 1, inclusive. A proposition with a truth value of 0 is false and one with a truth value of 1 is true. Truth values that are between 0 and 1 indicate varying degrees of truth. For instance, the truth value 0.8 can be assigned to the statement "Fred is happy," because Fred is happy most of the time, and the truth value 0.4 can be assigned to the statement "John is happy," because John is happy slightly less than half the time. Use these truth values to solve Exercises 31–33.

31. The truth value of the conjunction of two propositions in fuzzy logic is the minimum of the truth values of the two propositions. What are the truth values of the statements "Fred and John are happy" and "Neither Fred nor John is happy?"

32. The truth value of the negation of a proposition in fuzzy logic is 1 minus the truth value of the proposition. What are the truth values of the statements "Fred is not happy" and "John is not happy?"

33. The truth value of the disjunction of two propositions in fuzzy logic is the maximum of the truth values of the two propositions. What are the truth values of the statements "Fred is happy, or John is happy" and "Fred is not happy, or John is not happy?"

34. An ancient Sicilian legend says that the barber in a remote town who can be reached only by traveling a dangerous mountain road shaves those people, and only those people, who do not shave themselves. Can there be such a barber?

*35. The nth statement in a list of 100 statements is "Exactly n of the statements in this list are false."

   a) What conclusions can you draw from these statements?

   b) Answer part (a) if the nth statement is "At least n of the statements in this list are false."

   c) Answer part (b) assuming that the list contains 99 statements.

1.2 Applications of Propositional Logic

Introduction

Logic has many important applications to mathematics, computer science, and numerous other disciplines. Statements in mathematics and the sciences and in natural language often are imprecise or ambiguous. To make such statements precise, they can be translated into the language of logic. For example, logic is used in the specification of software and hardware, because these specifications need to be precise before development begins. Furthermore, propositional logic and its rules can be used to design computer circuits, to construct computer programs, to verify the correctness of programs, and to build expert systems. Logic can be used to analyze and solve many familiar puzzles. Software systems based on the rules of logic have been developed for constructing some, but not all, types of proofs automatically. We will discuss some of these applications of propositional logic in this section and in later chapters.

Translating English Sentences

There are many reasons to translate English sentences into expressions involving propositional variables and logical connectives. In particular, English (and every other human language) is often ambiguous. Translating sentences into compound statements (and other types of logical expressions, which we will introduce later in this chapter) removes the ambiguity. Note that this may involve making a set of reasonable assumptions based on the intended meaning of the sentence. Moreover, once we have translated sentences from English into logical expressions we can analyze these logical expressions to determine their truth values, we can manipulate them, and we can use rules of inference (which are discussed in Section 1.6) to reason about them.

To illustrate the process of translating an English sentence into a logical expression, consider Examples 1 and 2.

EXAMPLE 1   How can this English sentence be translated into a logical expression?

"You can access the Internet from campus only if you are a computer science major or you are not a freshman."

Solution: There are many ways to translate this sentence into a logical expression. Although it is possible to represent the sentence by a single propositional variable, such as p, this would not be useful when analyzing its meaning or reasoning with it. Instead, we will use propositional variables to represent each sentence part and determine the appropriate logical connectives between them. In particular, we let a, c, and f represent "You can access the Internet from campus," "You are a computer science major," and "You are a freshman," respectively. Noting that "only if" is one way a conditional statement can be expressed, this sentence can be represented as

$$a \rightarrow (c \vee \neg f).$$ ◀

EXAMPLE 2   How can this English sentence be translated into a logical expression?

"You cannot ride the roller coaster if you are under 4 feet tall unless you are older than 16 years old."

Solution: Let q, r, and s represent "You can ride the roller coaster," "You are under 4 feet tall," and "You are older than 16 years old," respectively. Then the sentence can be translated to

$$(r \wedge \neg s) \rightarrow \neg q.$$

Of course, there are other ways to represent the original sentence as a logical expression, but the one we have used should meet our needs.   ◀

System Specifications

Translating sentences in natural language (such as English) into logical expressions is an essential part of specifying both hardware and software systems. System and software engineers take requirements in natural language and produce precise and unambiguous specifications that can be used as the basis for system development. Example 3 shows how compound propositions can be used in this process.

EXAMPLE 3   Express the specification "The automated reply cannot be sent when the file system is full" using logical connectives.

Solution: One way to translate this is to let p denote "The automated reply can be sent" and q denote "The file system is full." Then $\neg p$ represents "It is not the case that the automated reply can be sent," which can also be expressed as "The automated reply cannot be sent." Consequently, our specification can be represented by the conditional statement $q \rightarrow \neg p$. ◄

System specifications should be **consistent**, that is, they should not contain conflicting requirements that could be used to derive a contradiction. When specifications are not consistent, there would be no way to develop a system that satisfies all specifications.

EXAMPLE 4   Determine whether these system specifications are consistent:

"The diagnostic message is stored in the buffer or it is retransmitted."
"The diagnostic message is not stored in the buffer."
"If the diagnostic message is stored in the buffer, then it is retransmitted."

Solution: To determine whether these specifications are consistent, we first express them using logical expressions. Let p denote "The diagnostic message is stored in the buffer" and let q denote "The diagnostic message is retransmitted." The specifications can then be written as $p \vee q$, $\neg p$, and $p \rightarrow q$. An assignment of truth values that makes all three specifications true must have p false to make $\neg p$ true. Because we want $p \vee q$ to be true but p must be false, q must be true. Because $p \rightarrow q$ is true when p is false and q is true, we conclude that these specifications are consistent, because they are all true when p is false and q is true. We could come to the same conclusion by use of a truth table to examine the four possible assignments of truth values to p and q. ◄

EXAMPLE 5   Do the system specifications in Example 4 remain consistent if the specification "The diagnostic message is not retransmitted" is added?

Solution: By the reasoning in Example 4, the three specifications from that example are true only in the case when p is false and q is true. However, this new specification is $\neg q$, which is false when q is true. Consequently, these four specifications are inconsistent. ◄

Boolean Searches

Links

Logical connectives are used extensively in searches of large collections of information, such as indexes of Web pages. Because these searches employ techniques from propositional logic, they are called **Boolean searches**.

In Boolean searches, the connective *AND* is used to match records that contain both of two search terms, the connective *OR* is used to match one or both of two search terms, and the connective *NOT* (sometimes written as *AND NOT*) is used to exclude a particular search term. Careful planning of how logical connectives are used is often required when Boolean searches are used to locate information of potential interest. Example 6 illustrates how Boolean searches are carried out.

EXAMPLE 6   **Web Page Searching**   Most Web search engines support Boolean searching techniques, which usually can help find Web pages about particular subjects. For instance, using Boolean searching to find Web pages about universities in New Mexico, we can look for pages matching NEW *AND* MEXICO *AND* UNIVERSITIES. The results of this search will include those pages that contain the three words NEW, MEXICO, and UNIVERSITIES. This will include all of the pages of interest, together with others such as a page about new universities in Mexico. (Note

Extra
Examples

that in Google, and many other search engines, the word "AND" is not needed, although it is understood, because all search terms are included by default. These search engines also support the use of quotation marks to search for specific phrases. So, it may be more effective to search for pages matching "New Mexico" *AND* UNIVERSITIES.)

Next, to find pages that deal with universities in New Mexico or Arizona, we can search for pages matching (NEW *AND* MEXICO *OR* ARIZONA) *AND* UNIVERSITIES. (*Note:* Here the *AND* operator takes precedence over the *OR* operator. Also, in Google, the terms used for this search would be NEW MEXICO *OR* ARIZONA.) The results of this search will include all pages that contain the word UNIVERSITIES and either both the words NEW and MEXICO or the word ARIZONA. Again, pages besides those of interest will be listed. Finally, to find Web pages that deal with universities in Mexico (and not New Mexico), we might first look for pages matching MEXICO *AND* UNIVERSITIES, but because the results of this search will include pages about universities in New Mexico, as well as universities in Mexico, it might be better to search for pages matching (MEXICO *AND* UNIVERSITIES) *NOT* NEW. The results of this search include pages that contain both the words MEXICO and UNIVERSITIES but do not contain the word NEW. (In Google, and many other search engines, the word "NOT" is replaced by the symbol "-". In Google, the terms used for this last search would be MEXICO UNIVERSITIES -NEW.) ◄

Logic Puzzles

Puzzles that can be solved using logical reasoning are known as **logic puzzles**. Solving logic puzzles is an excellent way to practice working with the rules of logic. Also, computer programs designed to carry out logical reasoning often use well-known logic puzzles to illustrate their capabilities. Many people enjoy solving logic puzzles, published in periodicals, books, and on the Web, as a recreational activity.

We will discuss two logic puzzles here. We begin with a puzzle originally posed by Raymond Smullyan, a master of logic puzzles, who has published more than a dozen books containing challenging puzzles that involve logical reasoning. In Section 1.3 we will also discuss the extremely popular logic puzzle Sudoku.

EXAMPLE 7

In [Sm78] Smullyan posed many puzzles about an island that has two kinds of inhabitants, knights, who always tell the truth, and their opposites, knaves, who always lie. You encounter two people *A* and *B*. What are *A* and *B* if *A* says "*B* is a knight" and *B* says "The two of us are opposite types?"

Solution: Let p and q be the statements that *A* is a knight and *B* is a knight, respectively, so that $\neg p$ and $\neg q$ are the statements that *A* is a knave and *B* is a knave, respectively.

We first consider the possibility that *A* is a knight; this is the statement that p is true. If *A* is a knight, then he is telling the truth when he says that *B* is a knight, so that q is true, and *A* and *B* are the same type. However, if *B* is a knight, then *B*'s statement that *A* and *B* are of opposite types, the statement $(p \wedge \neg q) \vee (\neg p \wedge q)$, would have to be true, which it is not, because *A* and *B* are both knights. Consequently, we can conclude that *A* is not a knight, that is, that p is false.

If *A* is a knave, then because everything a knave says is false, *A*'s statement that *B* is a knight, that is, that q is true, is a lie. This means that q is false and *B* is also a knave. Furthermore, if *B* is a knave, then *B*'s statement that *A* and *B* are opposite types is a lie, which is consistent with both *A* and *B* being knaves. We can conclude that both *A* and *B* are knaves. ◄

We pose more of Smullyan's puzzles about knights and knaves in Exercises 12–16.
Next, we pose a puzzle known as the **muddy children puzzle** for the case of two children.

EXAMPLE 8   A father tells his two children, a boy and a girl, to play in their backyard without getting dirty. However, while playing, both children get mud on their foreheads. When the children stop playing, the father says "At least one of you has a muddy forehead," and then asks the children to answer "Yes" or "No" to the question: "Do you know whether you have a muddy forehead?" The father asks this question twice. What will the children answer each time this question is asked, assuming that a child can see whether his or her sibling has a muddy forehead, but cannot see his or her own forehead? Assume that both children are honest and that the children answer each question simultaneously.

Solution: Let s be the statement that the son has a muddy forehead and let d be the statement that the daughter has a muddy forehead. When the father says that at least one of the two children has a muddy forehead, he is stating that the disjunction $s \lor d$ is true. Both children will answer "No" the first time the question is asked because each sees mud on the other child's forehead. That is, the son knows that d is true, but does not know whether s is true, and the daughter knows that s is true, but does not know whether d is true.

After the son has answered "No" to the first question, the daughter can determine that d must be true. This follows because when the first question is asked, the son knows that $s \lor d$ is true, but cannot determine whether s is true. Using this information, the daughter can conclude that d must be true, for if d were false, the son could have reasoned that because $s \lor d$ is true, then s must be true, and he would have answered "Yes" to the first question. The son can reason in a similar way to determine that s must be true. It follows that both children answer "Yes" the second time the question is asked.   ◀

Logic Circuits

Propositional logic can be applied to the design of computer hardware. This was first observed in 1938 by Claude Shannon in his MIT master's thesis. We give a brief introduction to this application here.

A **logic circuit** (or **digital circuit**) receives input signals $p_1, p_2, \ldots, p_n$, each a bit [either 0 (off) or 1 (on)], and produces output signals $s_1, s_2, \ldots, s_n$, each a bit. In this section we will restrict our attention to logic circuits with a single output signal; in general, digital circuits may have multiple outputs.

In Chapter 12 we design some useful circuits.

 Links

RAYMOND SMULLYAN (BORN 1919)   Raymond Smullyan dropped out of high school. He wanted to study what he was really interested in and not standard high school material. After jumping from one university to the next, he earned an undergraduate degree in mathematics at the University of Chicago in 1955. He paid his college expenses by performing magic tricks at parties and clubs. He obtained a Ph.D. in logic in 1959 at Princeton, studying under Alonzo Church. After graduating from Princeton, he taught mathematics and logic at Dartmouth College, Princeton University, Yeshiva University, and the City University of New York. He joined the philosophy department at Indiana University in 1981 where he is now an emeritus professor.

Smullyan has written many books on recreational logic and mathematics, including *Satan, Cantor, and Infinity; What Is the Name of This Book?; The Lady or the Tiger?; Alice in Puzzleland; To Mock a Mockingbird; Forever Undecided;* and *The Riddle of Scheherazade: Amazing Logic Puzzles, Ancient and Modern.* Because his logic puzzles are challenging, entertaining, and thought-provoking, he is considered to be a modern-day Lewis Carroll. Smullyan has also written several books about the application of deductive logic to chess, three collections of philosophical essays and aphorisms, and several advanced books on mathematical logic and set theory. He is particularly interested in self-reference and has worked on extending some of Gödel's results that show that it is impossible to write a computer program that can solve all mathematical problems. He is also particularly interested in explaining ideas from mathematical logic to the public.

Smullyan is a talented musician and often plays piano with his wife, who is a concert-level pianist. Making telescopes is one of his hobbies. He is also interested in optics and stereo photography. He states "I've never had a conflict between teaching and research as some people do because when I'm teaching, I'm doing research." Smullyan is the subject of a documentary short film entitled *This Film Needs No Title.*

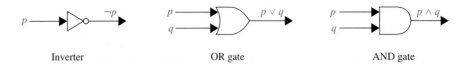

FIGURE 1   **Basic logic gates.**

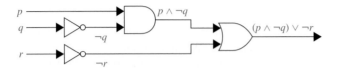

FIGURE 2   **A combinatorial circuit.**

Complicated digital circuits can be constructed from three basic circuits, called **gates**, shown in Figure 1. The **inverter**, or **NOT gate**, takes an input bit p, and produces as output $\neg p$. The **OR gate** takes two input signals p and q, each a bit, and produces as output the signal $p \vee q$. Finally, the **AND gate** takes two input signals p and q, each a bit, and produces as output the signal $p \wedge q$. We use combinations of these three basic gates to build more complicated circuits, such as that shown in Figure 2.

Given a circuit built from the basic logic gates and the inputs to the circuit, we determine the output by tracing through the circuit, as Example 9 shows.

EXAMPLE 9   Determine the output for the combinatorial circuit in Figure 2.

Solution: In Figure 2 we display the output of each logic gate in the circuit. We see that the AND gate takes input of p and $\neg q$, the output of the inverter with input q, and produces $p \wedge \neg q$. Next, we note that the OR gate takes input $p \wedge \neg q$ and $\neg r$, the output of the inverter with input r, and produces the final output $(p \wedge \neg q) \vee \neg r$.   ◄

Suppose that we have a formula for the output of a digital circuit in terms of negations, disjunctions, and conjunctions. Then, we can systematically build a digital circuit with the desired output, as illustrated in Example 10.

EXAMPLE 10   Build a digital circuit that produces the output $(p \vee \neg r) \wedge (\neg p \vee (q \vee \neg r))$ when given input bits p, q, and r.

Solution: To construct the desired circuit, we build separate circuits for $p \vee \neg r$ and for $\neg p \vee (q \vee \neg r)$ and combine them using an AND gate. To construct a circuit for $p \vee \neg r$, we use an inverter to produce $\neg r$ from the input r. Then, we use an OR gate to combine p and $\neg r$. To build a circuit for $\neg p \vee (q \vee \neg r)$, we first use an inverter to obtain $\neg r$. Then we use an OR gate with inputs q and $\neg r$ to obtain $q \vee \neg r$. Finally, we use another inverter and an OR gate to get $\neg p \vee (q \vee \neg r)$ from the inputs p and $q \vee \neg r$.

To complete the construction, we employ a final AND gate, with inputs $p \vee \neg r$ and $\neg p \vee (q \vee \neg r)$. The resulting circuit is displayed in Figure 3.   ◄

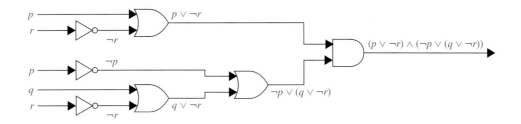

FIGURE 3   **The circuit for** $(p \lor \neg r) \land (\neg p \lor (q \lor \neg r))$.

Exercises

In Exercises 1–3, translate the given statement into propositional logic using the propositions provided.

1. You cannot edit a protected Wikipedia entry unless you are an administrator. Express your answer in terms of e: "You can edit a protected Wikipedia entry" and a: "You are an administrator."

2. You can graduate only if you have completed the requirements of your major and you do not owe money to the university and you do not have an overdue library book. Express your answer in terms of g: "You can graduate," m: "You owe money to the university," r: "You have completed the requirements of your major," and b: "You have an overdue library book."

3. You are eligible to be President of the U.S.A. only if you are at least 35 years old, were born in the U.S.A, or at the time of your birth both of your parents were citizens, and you have lived at least 14 years in the country. Express your answer in terms of e: "You are eligible to be President of the U.S.A.," a: "You are at least 35 years old," b: "You were born in the U.S.A," p: "At the time of your birth, both of your parents where citizens," and r: "You have lived at least 14 years in the U.S.A."

4. Express these system specifications using the propositions p "The message is scanned for viruses" and q "The message was sent from an unknown system" together with logical connectives (including negations).
   a) "The message is scanned for viruses whenever the message was sent from an unknown system."
   b) "The message was sent from an unknown system but it was not scanned for viruses."
   c) "It is necessary to scan the message for viruses whenever it was sent from an unknown system."
   d) "When a message is not sent from an unknown system it is not scanned for viruses."

5. Are these system specifications consistent? "The router can send packets to the edge system only if it supports the new address space. For the router to support the new address space it is necessary that the latest software release

be installed. The router can send packets to the edge system if the latest software release is installed, The router does not support the new address space."

6. Are these system specifications consistent? "Whenever the system software is being upgraded, users cannot access the file system. If users can access the file system, then they can save new files. If users cannot save new files, then the system software is not being upgraded."

7. What Boolean search would you use to look for Web pages about beaches in New Jersey? What if you wanted to find Web pages about beaches on the isle of Jersey (in the English Channel)?

8. An explorer is captured by a group of cannibals. There are two types of cannibals—those who always tell the truth and those who always lie. The cannibals will barbecue the explorer unless he can determine whether a particular cannibal always lies or always tells the truth. He is allowed to ask the cannibal exactly one question.
   a) Explain why the question "Are you a liar?" does not work.
   b) Find a question that the explorer can use to determine whether the cannibal always lies or always tells the truth.

∗9. Each inhabitant of a remote village always tells the truth or always lies. A villager will give only a "Yes" or a "No" response to a question a tourist asks. Suppose you are a tourist visiting this area and come to a fork in the road. One branch leads to the ruins you want to visit; the other branch leads deep into the jungle. A villager is standing at the fork in the road. What one question can you ask the villager to determine which branch to take?

10. When planning a party you want to know whom to invite. Among the people you would like to invite are three touchy friends. You know that if Jasmine attends, she will become unhappy if Samir is there, Samir will attend only if Kanti will be there, and Kanti will not attend unless Jasmine also does. Which combinations of these three friends can you invite so as not to make someone unhappy?

11. When three professors are seated in a restaurant, the hostess asks them: "Does everyone want coffee?" The first professor says: "I do not know." The second professor then says: "I do not know." Finally, the third professor says: "No, not everyone wants coffee." The hostess comes back and gives coffee to the professors who want it. How did she figure out who wanted coffee?

Exercises 12–16 relate to inhabitants of the island of knights and knaves created by Smullyan, where knights always tell the truth and knaves always lie. You encounter two people, A and B. Determine, if possible, what A and B are if they address you in the ways described. If you cannot determine what these two people are, can you draw any conclusions?

12. A says "The two of us are both knights" and B says "A is a knave."

13. A says "At least one of us is a knave" and B says nothing.

14. Both A and B say "I am a knight."

15. A says "I am a knave or B is a knight" and B says nothing.

16. A says "We are both knaves" and B says nothing.

Exercises 17–21 are puzzles that can be solved by translating statements into logical expressions and reasoning from these expressions using truth tables.

17. Steve would like to determine the relative salaries of three coworkers using two facts. First, he knows that if Fred is not the highest paid of the three, then Janice is. Second, he knows that if Janice is not the lowest paid, then Maggie is paid the most. Is it possible to determine the relative salaries of Fred, Maggie, and Janice from what Steve knows? If so, who is paid the most and who the least? Explain your reasoning.

18. The police have three suspects for the murder of Mr. Cooper: Mr. Smith, Mr. Jones, and Mr. Williams. Smith, Jones, and Williams each declare that they did not kill Cooper. Smith also states that Cooper was a friend of Jones and that Williams disliked him. Jones also states that he did not know Cooper and that he was out of town the day Cooper was killed. Williams also states that he saw both Smith and Jones with Cooper the day of the killing and that either Smith or Jones must have killed him. Can you determine who the murderer was if

   a) one of the three men is guilty, the two innocent men are telling the truth, but the statements of the guilty man may or may not be true?

   b) innocent men do not lie?

19. A detective has interviewed four witnesses to a crime. From the stories of the witnesses the detective has concluded that if the butler is telling the truth then so is the cook; the cook and the gardener cannot both be telling the truth; the gardener and the handyman are not both lying; and if the handyman is telling the truth then the cook is lying. For each of the four witnesses, can the detective determine whether that person is telling the truth or lying? Explain your reasoning.

20. Five friends have access to a chat room. Is it possible to determine who is chatting if the following information is known? Either Kevin or Heather, or both, are chatting. Either Randy or Vijay, but not both, are chatting. If Abby is chatting, so is Randy. Vijay and Kevin are either both chatting or neither is. If Heather is chatting, then so are Abby and Kevin. Explain your reasoning.

21. Suppose there are signs on the doors to two rooms. The sign on the first door reads "In this room there is a lady, and in the other one there is a tiger"; and the sign on the second door reads "In one of these rooms, there is a lady, and in one of them there is a tiger." Suppose that you know that one of these signs is true and the other is false. Behind which door is the lady?

22. Freedonia has fifty senators. Each senator is either honest or corrupt. Suppose you know that at least one of the Freedonian senators is honest and that, given any two Freedonian senators, at least one is corrupt. Based on these facts, can you determine how many Freedonian senators are honest and how many are corrupt? If so, what is the answer?

23. Find the output of each of these combinatorial circuits.

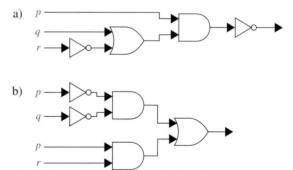

24. Construct a combinatorial circuit using inverters, OR gates, and AND gates that produces the output $((\neg p \vee \neg r) \wedge \neg q) \vee (\neg p \wedge (q \vee r))$ from input bits p, q, and r.

1.3 Propositional Equivalences

Introduction

An important type of step used in a mathematical argument is the replacement of a statement with another statement with the same truth value. Because of this, methods that produce propositions with the same truth value as a given compound proposition are used extensively in the construction of mathematical arguments. Note that we will use the term "compound proposition" to refer to an expression formed from propositional variables using logical operators, such as $p \wedge q$.

We begin our discussion with a classification of compound propositions according to their possible truth values.

DEFINITION 1   A compound proposition that is always true, no matter what the truth values of the propositional variables that occur in it, is called a *tautology*. A compound proposition that is always false is called a *contradiction*. A compound proposition that is neither a tautology nor a contradiction is called a *contingency*.

Tautologies and contradictions are often important in mathematical reasoning. Example 1 illustrates these types of compound propositions.

EXAMPLE 1   We can construct examples of tautologies and contradictions using just one propositional variable. Consider the truth tables of $p \vee \neg p$ and $p \wedge \neg p$, shown in Table 1. Because $p \vee \neg p$ is always true, it is a tautology. Because $p \wedge \neg p$ is always false, it is a contradiction.   ◄

Logical Equivalences

Compound propositions that have the same truth values in all possible cases are called **logically equivalent**. We can also define this notion as follows.

DEFINITION 2   The compound propositions p and q are called *logically equivalent* if $p \leftrightarrow q$ is a tautology. The notation $p \equiv q$ denotes that p and q are logically equivalent.

Remark: The symbol $\equiv$ is not a logical connective, and $p \equiv q$ is not a compound proposition but rather is the statement that $p \leftrightarrow q$ is a tautology. The symbol $\Leftrightarrow$ is sometimes used instead of $\equiv$ to denote logical equivalence.

One way to determine whether two compound propositions are equivalent is to use a truth table. In particular, the compound propositions p and q are equivalent if and only if the columns giving their truth values agree. Example 2 illustrates this method to establish an extremely important and useful logical equivalence, namely, that of $\neg(p \vee q)$ with $\neg p \wedge \neg q$. This logical equivalence is one of the two **De Morgan laws**, shown in Table 2, named after the English mathematician Augustus De Morgan, of the mid-nineteenth century.

EXAMPLE 2   Show that $\neg(p \vee q)$ and $\neg p \wedge \neg q$ are logically equivalent.

TABLE 1 Examples of a Tautology and a Contradiction.			
p	$\neg p$	$p \vee \neg p$	$p \wedge \neg p$
T	F	T	F
F	T	T	F

TABLE 2 De Morgan's Laws.
$\neg(p \wedge q) \equiv \neg p \vee \neg q$
$\neg(p \vee q) \equiv \neg p \wedge \neg q$

Solution: The truth tables for these compound propositions are displayed in Table 3. Because the truth values of the compound propositions $\neg(p \vee q)$ and $\neg p \wedge \neg q$ agree for all possible combinations of the truth values of p and q, it follows that $\neg(p \vee q) \leftrightarrow (\neg p \wedge \neg q)$ is a tautology and that these compound propositions are logically equivalent. ◄

TABLE 3	Truth Tables for $\neg(p \vee q)$ and $\neg p \wedge \neg q$.					
p	q	$p \vee q$	$\neg(p \vee q)$	$\neg p$	$\neg q$	$\neg p \wedge \neg q$
T	T	T	F	F	F	F
T	F	T	F	F	T	F
F	T	T	F	T	F	F
F	F	F	T	T	T	T

EXAMPLE 3 Show that $p \rightarrow q$ and $\neg p \vee q$ are logically equivalent.

Solution: We construct the truth table for these compound propositions in Table 4. Because the truth values of $\neg p \vee q$ and $p \rightarrow q$ agree, they are logically equivalent. ◄

TABLE 4	Truth Tables for $\neg p \vee q$ and $p \rightarrow q$.			
p	q	$\neg p$	$\neg p \vee q$	$p \rightarrow q$
T	T	F	T	T
T	F	F	F	F
F	T	T	T	T
F	F	T	T	T

We will now establish a logical equivalence of two compound propositions involving three different propositional variables p, q, and r. To use a truth table to establish such a logical equivalence, we need eight rows, one for each possible combination of truth values of these three variables. We symbolically represent these combinations by listing the truth values of p, q, and r, respectively. These eight combinations of truth values are TTT, TTF, TFT, TFF, FTT, FTF, FFT, and FFF; we use this order when we display the rows of the truth table. Note that we need to double the number of rows in the truth tables we use to show that compound propositions are equivalent for each additional propositional variable, so that 16 rows are needed to establish the logical equivalence of two compound propositions involving four propositional variables, and so on. In general, 2^n rows are required if a compound proposition involves n propositional variables.

TABLE 5 A Demonstration That $p \vee (q \wedge r)$ and $(p \vee q) \wedge (p \vee r)$ Are Logically Equivalent.							
p	q	r	$q \wedge r$	$p \vee (q \wedge r)$	$p \vee q$	$p \vee r$	$(p \vee q) \wedge (p \vee r)$
T	T	T	T	T	T	T	T
T	T	F	F	T	T	T	T
T	F	T	F	T	T	T	T
T	F	F	F	T	T	T	T
F	T	T	T	T	T	T	T
F	T	F	F	F	T	F	F
F	F	T	F	F	F	T	F
F	F	F	F	F	F	F	F

EXAMPLE 4   Show that $p \vee (q \wedge r)$ and $(p \vee q) \wedge (p \vee r)$ are logically equivalent. This is the *distributive law* of disjunction over conjunction.

Solution: We construct the truth table for these compound propositions in Table 5. Because the truth values of $p \vee (q \wedge r)$ and $(p \vee q) \wedge (p \vee r)$ agree, these compound propositions are logically equivalent.   ◄

The identities in Table 6 are a special case of Boolean algebra identities found in Table 5 of Section 12.1. See Table 1 in Section 2.2 for analogous set identities.

Table 6 contains some important equivalences. In these equivalences, **T** denotes the compound proposition that is always true and **F** denotes the compound proposition that is always

TABLE 6 Logical Equivalences.	
Equivalence	*Name*
$p \wedge \mathbf{T} \equiv p$ $p \vee \mathbf{F} \equiv p$	Identity laws
$p \vee \mathbf{T} \equiv \mathbf{T}$ $p \wedge \mathbf{F} \equiv \mathbf{F}$	Domination laws
$p \vee p \equiv p$ $p \wedge p \equiv p$	Idempotent laws
$\neg(\neg p) \equiv p$	Double negation law
$p \vee q \equiv q \vee p$ $p \wedge q \equiv q \wedge p$	Commutative laws
$(p \vee q) \vee r \equiv p \vee (q \vee r)$ $(p \wedge q) \wedge r \equiv p \wedge (q \wedge r)$	Associative laws
$p \vee (q \wedge r) \equiv (p \vee q) \wedge (p \vee r)$ $p \wedge (q \vee r) \equiv (p \wedge q) \vee (p \wedge r)$	Distributive laws
$\neg(p \wedge q) \equiv \neg p \vee \neg q$ $\neg(p \vee q) \equiv \neg p \wedge \neg q$	De Morgan's laws
$p \vee (p \wedge q) \equiv p$ $p \wedge (p \vee q) \equiv p$	Absorption laws
$p \vee \neg p \equiv \mathbf{T}$ $p \wedge \neg p \equiv \mathbf{F}$	Negation laws

TABLE 7 Logical Equivalences Involving Conditional Statements.
$p \to q \equiv \neg p \vee q$
$p \to q \equiv \neg q \to \neg p$
$p \vee q \equiv \neg p \to q$
$p \wedge q \equiv \neg (p \to \neg q)$
$\neg (p \to q) \equiv p \wedge \neg q$
$(p \to q) \wedge (p \to r) \equiv p \to (q \wedge r)$
$(p \to r) \wedge (q \to r) \equiv (p \vee q) \to r$
$(p \to q) \vee (p \to r) \equiv p \to (q \vee r)$
$(p \to r) \vee (q \to r) \equiv (p \wedge q) \to r$

TABLE 8 Logical Equivalences Involving Biconditional Statements.
$p \leftrightarrow q \equiv (p \to q) \wedge (q \to p)$
$p \leftrightarrow q \equiv \neg p \leftrightarrow \neg q$
$p \leftrightarrow q \equiv (p \wedge q) \vee (\neg p \wedge \neg q)$
$\neg (p \leftrightarrow q) \equiv p \leftrightarrow \neg q$

false. We also display some useful equivalences for compound propositions involving conditional statements and biconditional statements in Tables 7 and 8, respectively. The reader is asked to verify the equivalences in Tables 6–8 in the exercises.

The associative law for disjunction shows that the expression $p \vee q \vee r$ is well defined, in the sense that it does not matter whether we first take the disjunction of p with q and then the disjunction of $p \vee q$ with r, or if we first take the disjunction of q and r and then take the disjunction of p with $q \vee r$. Similarly, the expression $p \wedge q \wedge r$ is well defined. By extending this reasoning, it follows that $p_1 \vee p_2 \vee \cdots \vee p_n$ and $p_1 \wedge p_2 \wedge \cdots \wedge p_n$ are well defined whenever $p_1, p_2, \ldots, p_n$ are propositions.

Furthermore, note that De Morgan's laws extend to

$$\neg (p_1 \vee p_2 \vee \cdots \vee p_n) \equiv (\neg p_1 \wedge \neg p_2 \wedge \cdots \wedge \neg p_n)$$

and

$$\neg (p_1 \wedge p_2 \wedge \cdots \wedge p_n) \equiv (\neg p_1 \vee \neg p_2 \vee \cdots \vee \neg p_n).$$

We will sometimes use the notation $\bigvee_{j=1}^{n} p_j$ for $p_1 \vee p_2 \vee \cdots \vee p_n$ and $\bigwedge_{j=1}^{n} p_j$ for $p_1 \wedge p_2 \wedge \cdots \wedge p_n$. Using this notation, the extended version of De Morgan's laws can be written concisely as $\neg \left(\bigvee_{j=1}^{n} p_j \right) \equiv \bigwedge_{j=1}^{n} \neg p_j$ and $\neg \left(\bigwedge_{j=1}^{n} p_j \right) \equiv \bigvee_{j=1}^{n} \neg p_j$. (Methods for proving these identities will be given in Section 5.1.)

Using De Morgan's Laws

When using De Morgan's laws, remember to change the logical connective after you negate.

The two logical equivalences known as De Morgan's laws are particularly important. They tell us how to negate conjunctions and how to negate disjunctions. In particular, the equivalence $\neg (p \vee q) \equiv \neg p \wedge \neg q$ tells us that the negation of a disjunction is formed by taking the conjunction of the negations of the component propositions. Similarly, the equivalence $\neg (p \wedge q) \equiv \neg p \vee \neg q$ tells us that the negation of a conjunction is formed by taking the disjunction of the negations of the component propositions. Example 5 illustrates the use of De Morgan's laws.

EXAMPLE 5   Use De Morgan's laws to express the negations of "Miguel has a cellphone and he has a laptop computer" and "Heather will go to the concert or Steve will go to the concert."

Solution: Let p be "Miguel has a cellphone" and q be "Miguel has a laptop computer." Then "Miguel has a cellphone and he has a laptop computer" can be represented by $p \wedge q$. By the first of De Morgan's laws, $\neg(p \wedge q)$ is equivalent to $\neg p \vee \neg q$. Consequently, we can express the negation of our original statement as "Miguel does not have a cellphone or he does not have a laptop computer."

Let r be "Heather will go to the concert" and s be "Steve will go to the concert." Then "Heather will go to the concert or Steve will go to the concert" can be represented by $r \vee s$. By the second of De Morgan's laws, $\neg(r \vee s)$ is equivalent to $\neg r \wedge \neg s$. Consequently, we can express the negation of our original statement as "Heather will not go to the concert and Steve will not go to the concert." ◀

Constructing New Logical Equivalences

The logical equivalences in Table 6, as well as any others that have been established (such as those shown in Tables 7 and 8), can be used to construct additional logical equivalences. The reason for this is that a proposition in a compound proposition can be replaced by a compound proposition that is logically equivalent to it without changing the truth value of the original compound proposition. This technique is illustrated in Examples 6–8, where we also use the fact that if p and q are logically equivalent and q and r are logically equivalent, then p and r are logically equivalent (see Exercise 34).

EXAMPLE 6   Show that $\neg(p \rightarrow q)$ and $p \wedge \neg q$ are logically equivalent.

Solution: We could use a truth table to show that these compound propositions are equivalent (similar to what we did in Example 4). Indeed, it would not be hard to do so. However, we want to illustrate how to use logical identities that we already know to establish new logical identities, something that is of practical importance for establishing equivalences of compound propositions with a large number of variables. So, we will establish this equivalence by developing a series of

AUGUSTUS DE MORGAN (1806–1871)   Augustus De Morgan was born in India, where his father was a colonel in the Indian army. De Morgan's family moved to England when he was 7 months old. He attended private schools, where in his early teens he developed a strong interest in mathematics. De Morgan studied at Trinity College, Cambridge, graduating in 1827. Although he considered medicine or law, he decided on mathematics for his career. He won a position at University College, London, in 1828, but resigned after the college dismissed a fellow professor without giving reasons. However, he resumed this position in 1836 when his successor died, remaining until 1866.

De Morgan was a noted teacher who stressed principles over techniques. His students included many famous mathematicians, including Augusta Ada, Countess of Lovelace, who was Charles Babbage's collaborator in his work on computing machines (see page 31 for biographical notes on Augusta Ada). (De Morgan cautioned the countess against studying too much mathematics, because it might interfere with her childbearing abilities!)

De Morgan was an extremely prolific writer, publishing more than 1000 articles in more than 15 periodicals. De Morgan also wrote textbooks on many subjects, including logic, probability, calculus, and algebra. In 1838 he presented what was perhaps the first clear explanation of an important proof technique known as *mathematical induction* (discussed in Section 5.1 of this text), a term he coined. In the 1840s De Morgan made fundamental contributions to the development of symbolic logic. He invented notations that helped him prove propositional equivalences, such as the laws that are named after him. In 1842 De Morgan presented what is considered to be the first precise definition of a limit and developed new tests for convergence of infinite series. De Morgan was also interested in the history of mathematics and wrote biographies of Newton and Halley.

In 1837 De Morgan married Sophia Frend, who wrote his biography in 1882. De Morgan's research, writing, and teaching left little time for his family or social life. Nevertheless, he was noted for his kindness, humor, and wide range of knowledge.

logical equivalences, using one of the equivalences in Table 6 at a time, starting with $\neg(p \rightarrow q)$ and ending with $p \wedge \neg q$. We have the following equivalences.

$$
\begin{aligned}
\neg(p \rightarrow q) &\equiv \neg(\neg p \vee q) && \text{by Example 3} \\
&\equiv \neg(\neg p) \wedge \neg q && \text{by the second De Morgan law} \\
&\equiv p \wedge \neg q && \text{by the double negation law}
\end{aligned}
$$

◀

EXAMPLE 7   Show that $\neg(p \vee (\neg p \wedge q))$ and $\neg p \wedge \neg q$ are logically equivalent by developing a series of logical equivalences.

Solution: We will use one of the equivalences in Table 6 at a time, starting with $\neg(p \vee (\neg p \wedge q))$ and ending with $\neg p \wedge \neg q$. (*Note:* we could also easily establish this equivalence using a truth table.) We have the following equivalences.

$$
\begin{aligned}
\neg(p \vee (\neg p \wedge q)) &\equiv \neg p \wedge \neg(\neg p \wedge q) && \text{by the second De Morgan law} \\
&\equiv \neg p \wedge [\neg(\neg p) \vee \neg q] && \text{by the first De Morgan law} \\
&\equiv \neg p \wedge (p \vee \neg q) && \text{by the double negation law} \\
&\equiv (\neg p \wedge p) \vee (\neg p \wedge \neg q) && \text{by the second distributive law} \\
&\equiv \mathbf{F} \vee (\neg p \wedge \neg q) && \text{because } \neg p \wedge p \equiv \mathbf{F} \\
&\equiv (\neg p \wedge \neg q) \vee \mathbf{F} && \text{by the commutative law for disjunction} \\
&\equiv \neg p \wedge \neg q && \text{by the identity law for } \mathbf{F}
\end{aligned}
$$

Consequently $\neg(p \vee (\neg p \wedge q))$ and $\neg p \wedge \neg q$ are logically equivalent.   ◀

EXAMPLE 8   Show that $(p \wedge q) \rightarrow (p \vee q)$ is a tautology.

Solution: To show that this statement is a tautology, we will use logical equivalences to demonstrate that it is logically equivalent to $\mathbf{T}$. (*Note:* This could also be done using a truth table.)

$$
\begin{aligned}
(p \wedge q) \rightarrow (p \vee q) &\equiv \neg(p \wedge q) \vee (p \vee q) && \text{by Example 3} \\
&\equiv (\neg p \vee \neg q) \vee (p \vee q) && \text{by the first De Morgan law} \\
&\equiv (\neg p \vee p) \vee (\neg q \vee q) && \text{by the associative and commutative} \\
& && \text{laws for disjunction} \\
&\equiv \mathbf{T} \vee \mathbf{T} && \text{by Example 1 and the commutative} \\
& && \text{law for disjunction} \\
&\equiv \mathbf{T} && \text{by the domination law}
\end{aligned}
$$

◀

Propositional Satisfiability

A compound proposition is **satisfiable** if there is an assignment of truth values to its variables that makes it true. When no such assignments exists, that is, when the compound proposition is false for all assignments of truth values to its variables, the compound proposition is **unsatisfiable**. Note that a compound proposition is unsatisfiable if and only if its negation is true for all assignments of truth values to the variables, that is, if and only if its negation is a tautology.

When we find a particular assignment of truth values that makes a compound proposition true, we have shown that it is satisfiable; such an assignment is called a **solution** of this particular satisfiability problem. However, to show that a compound proposition is unsatisfiable, we need to show that *every* assignment of truth values to its variables makes it false. Although we can always use a truth table to determine whether a compound proposition is satisfiable, it is often more efficient not to, as Example 9 demonstrates.

EXAMPLE 9 Determine whether each of the compound propositions $(p \vee \neg q) \wedge (q \vee \neg r) \wedge (r \vee \neg p)$, $(p \vee q \vee r) \wedge (\neg p \vee \neg q \vee \neg r)$, and $(p \vee \neg q) \wedge (q \vee \neg r) \wedge (r \vee \neg p) \wedge (p \vee q \vee r) \wedge (\neg p \vee \neg q \vee \neg r)$ is satisfiable.

Solution: Instead of using truth table to solve this problem, we will reason about truth values. Note that $(p \vee \neg q) \wedge (q \vee \neg r) \wedge (r \vee \neg p)$ is true when the three variable p, q, and r have the same truth value (see Exercise 26 of Section 1.1). Hence, it is satisfiable as there is at least one assignment of truth values for p, q, and r that makes it true. Similarly, note that $(p \vee q \vee r) \wedge (\neg p \vee \neg q \vee \neg r)$ is true when at least one of p, q, and r is true and at least one is false (see Exercise 27 of Section 1.1). Hence, $(p \vee q \vee r) \wedge (\neg p \vee \neg q \vee \neg r)$ is satisfiable, as there is at least one assignment of truth values for p, q, and r that makes it true.

Finally, note that for $(p \vee \neg q) \wedge (q \vee \neg r) \wedge (r \vee \neg p) \wedge (p \vee q \vee r) \wedge (\neg p \vee \neg q \vee \neg r)$ to be true, $(p \vee \neg q) \wedge (q \vee \neg r) \wedge (r \vee \neg p)$ and $(p \vee q \vee r) \wedge (\neg p \vee \neg q \vee \neg r)$ must both be true. For the first to be true, the three variables must have the same truth values, and for the second to be true, at least one of three variables must be true and at least one must be false. However, these conditions are contradictory. From these observations we conclude that no assignment of truth values to p, q, and r makes $(p \vee \neg q) \wedge (q \vee \neg r) \wedge (r \vee \neg p) \wedge (p \vee q \vee r) \wedge (\neg p \vee \neg q \vee \neg r)$ true. Hence, it is unsatisfiable. ◀

Links

AUGUSTA ADA, COUNTESS OF LOVELACE (1815–1852) Augusta Ada was the only child from the marriage of the famous poet Lord Byron and Lady Byron, Annabella Millbanke, who separated when Ada was 1 month old, because of Lord Byron's scandalous affair with his half sister. The Lord Byron had quite a reputation, being described by one of his lovers as "mad, bad, and dangerous to know." Lady Byron was noted for her intellect and had a passion for mathematics; she was called by Lord Byron "The Princess of Parallelograms." Augusta was raised by her mother, who encouraged her intellectual talents especially in music and mathematics, to counter what Lady Byron considered dangerous poetic tendencies. At this time, women were not allowed to attend universities and could not join learned societies. Nevertheless, Augusta pursued her mathematical studies independently and with mathematicians, including William Frend. She was also encouraged by another female mathematician, Mary Somerville, and in 1834 at a dinner party hosted by Mary Somerville, she learned about Charles Babbage's ideas for a calculating machine, called the Analytic Engine. In 1838 Augusta Ada married Lord King, later elevated to Earl of Lovelace. Together they had three children.

Augusta Ada continued her mathematical studies after her marriage. Charles Babbage had continued work on his Analytic Engine and lectured on this in Europe. In 1842 Babbage asked Augusta Ada to translate an article in French describing Babbage's invention. When Babbage saw her translation, he suggested she add her own notes, and the resulting work was three times the length of the original. The most complete accounts of the Analytic Engine are found in Augusta Ada's notes. In her notes, she compared the working of the Analytic Engine to that of the Jacquard loom, with Babbage's punch cards analogous to the cards used to create patterns on the loom. Furthermore, she recognized the promise of the machine as a general purpose computer much better than Babbage did. She stated that the "engine is the material expression of any indefinite function of any degree of generality and complexity." Her notes on the Analytic Engine anticipate many future developments, including computer-generated music. Augusta Ada published her writings under her initials A.A.L. concealing her identity as a woman as did many women at a time when women were not considered to be the intellectual equals of men. After 1845 she and Babbage worked toward the development of a system to predict horse races. Unfortunately, their system did not work well, leaving Augusta Ada heavily in debt at the time of her death at an unfortunately young age from uterine cancer.

In 1953 Augusta Ada's notes on the Analytic Engine were republished more than 100 years after they were written, and after they had been long forgotten. In his work in the 1950s on the capacity of computers to think (and his famous Turing Test), Alan Turing responded to Augusta Ada's statement that "The Analytic Engine has no pretensions whatever to originate anything. It can do whatever we know how to order it to perform." This "dialogue" between Turing and Augusta Ada is still the subject of controversy. Because of her fundamental contributions to computing, the programming language Ada is named in honor of the Countess of Lovelace.

	2	9				4		
			5			1		
	4							
				4	2			
6							7	
5								
7			3					5
	1			9				
							6	

FIGURE 1 **A 9 × 9 Sudoku puzzle.**

Applications of Satisfiability

Many problems, in diverse areas such as robotics, software testing, computer-aided design, machine vision, integrated circuit design, computer networking, and genetics, can be modeled in terms of propositional satisfiability. Although most of these applications are beyond the scope of this book, we will study one application here. In particular, we will show how to use propositional satisfiability to model Sudoku puzzles.

SUDOKU A **Sudoku puzzle** is represented by a 9 × 9 grid made up of nine 3 × 3 subgrids, known as **blocks**, as shown in Figure 1. For each puzzle, some of the 81 cells, called **givens**, are assigned one of the numbers $1, 2, \ldots, 9$, and the other cells are blank. The puzzle is solved by assigning a number to each blank cell so that every row, every column, and every one of the nine 3 × 3 blocks contains each of the nine possible numbers. Note that instead of using a 9 × 9 grid, Sudoku puzzles can be based on $n^2 \times n^2$ grids, for any positive integer n, with the $n^2 \times n^2$ grid made up of n^2 $n \times n$ subgrids.

The popularity of Sudoku dates back to the 1980s when it was introduced in Japan. It took 20 years for Sudoku to spread to rest of the world, but by 2005, Sudoku puzzles were a worldwide craze. The name Sudoku is short for the Japanese *suuji wa dokushin ni kagiru*, which means "the digits must remain single." The modern game of Sudoku was apparently designed in the late 1970s by an American puzzle designer. The basic ideas of Sudoku date back even further; puzzles printed in French newspapers in the 1890s were quite similar, but not identical, to modern Sudoku.

Sudoku puzzles designed for entertainment have two additional important properties. First, they have exactly one solution. Second, they can be solved using reasoning alone, that is, without resorting to searching all possible assignments of numbers to the cells. As a Sudoku puzzle is solved, entries in blank cells are successively determined by already known values. For instance, in the grid in Figure 1, the number 4 must appear in exactly one cell in the second row. How can we determine which of the seven blank cells it must appear? First, we observe that 4 cannot appear in one of the first three cells or in one of the last three cells of this row, because it already appears in another cell in the block each of these cells is in. We can also see that 4 cannot appear in the fifth cell in this row, as it already appears in the fifth column in the fourth row. This means that 4 must appear in the sixth cell of the second row.

Many strategies based on logic and mathematics have been devised for solving Sudoku puzzles (see [Da10], for example). Here, we discuss one of the ways that have been developed for solving Sudoku puzzles with the aid of a computer, which depends on modeling the puzzle as a propositional satisfiability problem. Using the model we describe, particular Sudoku puzzles can be solved using software developed to solve satisfiability problems. Currently, Sudoku puzzles can be solved in less than 10 milliseconds this way. It should be noted that there are many other approaches for solving Sudoku puzzles via computers using other techniques.

To encode a Sudoku puzzle, let $p(i, j, n)$ denote the proposition that is true when the number n is in the cell in the ith row and jth column. There are $9 \times 9 \times 9 = 729$ such propositions, as i, j, and n all range from 1 to 9. For example, for the puzzle in Figure 1, the number 6 is given as the value in the fifth row and first column. Hence, we see that $p(5, 1, 6)$ is true, but $p(5, j, 6)$ is false for $j = 2, 3, \ldots, 9$.

Given a particular Sudoku puzzle, we begin by encoding each of the given values. Then, we construct compound propositions that assert that every row contains every number, every column contains every number, every 3×3 block contains every number, and each cell contains no more than one number. It follows, as the reader should verify, that the Sudoku puzzle is solved by finding an assignment of truth values to the 729 propositions $p(i, j, n)$ with i, j, and n each ranging from 1 to 9 that makes the conjunction of all these compound propositions true. After listing these assertions, we will explain how to construct the assertion that every row contains every integer from 1 to 9. We will leave the construction of the other assertions that every column contains every number and each of the nine 3×3 blocks contains every number to the exercises.

- For each cell with a given value, we assert $p(i, j, n)$ when the cell in row i and column j has the given value n.
- We assert that every row contains every number:

$$\bigwedge_{i=1}^{9} \bigwedge_{n=1}^{9} \bigvee_{j=1}^{9} p(i, j, n)$$

- We assert that every column contains every number:

$$\bigwedge_{j=1}^{9} \bigwedge_{n=1}^{9} \bigvee_{i=1}^{9} p(i, j, n)$$

It is tricky setting up the two inner indices so that all nine cells in each square block are examined.

- We assert that each of the nine 3×3 blocks contains every number:

$$\bigwedge_{r=0}^{2} \bigwedge_{s=0}^{2} \bigwedge_{n=1}^{9} \bigvee_{i=1}^{3} \bigvee_{j=1}^{3} p(3r + i, 3s + j, n)$$

- To assert that no cell contains more than one number, we take the conjunction over all values of n, n', i, and j where each variable ranges from 1 to 9 and $n \neq n'$ of $p(i, j, n) \rightarrow \neg p(i, j, n')$.

We now explain how to construct the assertion that every row contains every number. First, to assert that row i contains the number n, we form $\bigvee_{j=1}^{9} p(i, j, n)$. To assert that row i contains all n numbers, we form the conjunction of these disjunctions over all nine possible values of n, giving us $\bigwedge_{n=1}^{9} \bigvee_{j=1}^{9} p(i, j, n)$. Finally, to assert that every row contains every number, we take the conjunction of $\bigwedge_{n=1}^{9} \bigvee_{j=1}^{9} p(i, j, n)$ over all nine rows. This gives us $\bigwedge_{i=1}^{9} \bigwedge_{n=1}^{9} \bigvee_{j=1}^{9} p(i, j, n)$.

Given a particular Sudoku puzzle, to solve this puzzle we can find a solution to the satisfiability problems that asks for a set of truth values for the 729 variables $p(i, j, n)$ that makes the conjunction of all the listed assertions true.

Solving Satisfiability Problems

A truth table can be used to determine whether a compound proposition is satisfiable, or equivalently, whether its negation is a tautology (see Exercise 38). This can be done by hand for a compound proposition with a small number of variables, but when the number of variables grows, this becomes impractical. For instance, there are $2^{20} = 1,048,576$ rows in the truth table for a compound proposition with 20 variables. Clearly, you need a computer to help you determine, in this way, whether a compound proposition in 20 variables is satisfiable.

When many applications are modeled, questions concerning the satisfiability of compound propositions with hundreds, thousands, or millions of variables arise. Note, for example, that when there are 1000 variables, checking every one of the 2^{1000} (a number with more than 300 decimal digits) possible combinations of truth values of the variables in a compound proposition cannot be done by a computer in even trillions of years. No procedure is known that a computer can follow to determine in a reasonable amount of time whether an arbitrary compound proposition in such a large number of variables is satisfiable. However, progress has been made developing methods for solving the satisfiability problem for the particular types of compound propositions that arise in practical applications, such as for the solution of Sudoku puzzles. Many computer programs have been developed for solving satisfiability problems which have practical use. In our discussion of the subject of algorithms in Chapter 3, we will discuss this question further. In particular, we will explain the important role the propositional satisfiability problem plays in the study of the complexity of algorithms.

Links

Exercises

1. Use truth tables to verify these equivalences.

a) $p \wedge \mathbf{T} \equiv p$ **b)** $p \vee \mathbf{F} \equiv p$
c) $p \wedge \mathbf{F} \equiv \mathbf{F}$ **d)** $p \vee \mathbf{T} \equiv \mathbf{T}$
e) $p \vee p \equiv p$ **f)** $p \wedge p \equiv p$

2. Use a truth table to verify the first De Morgan law
$\neg(p \wedge q) \equiv \neg p \vee \neg q$.

3. Use a truth table to verify the distributive law
$p \wedge (q \vee r) \equiv (p \wedge q) \vee (p \wedge r)$.

4. Use De Morgan's laws to find the negation of each of the following statements.

a) Jan is rich and happy.

b) Carlos will bicycle or run tomorrow.

c) Mei walks or takes the bus to class.

d) Ibrahim is smart and hard working.

5. Show that each of these conditional statements is a tautology by using truth tables.

a) $(p \wedge q) \to p$ **b)** $p \to (p \vee q)$
c) $\neg p \to (p \to q)$ **d)** $(p \wedge q) \to (p \to q)$
e) $\neg(p \to q) \to p$ **f)** $\neg(p \to q) \to \neg q$

Links

HENRY MAURICE SHEFFER (1883–1964)   Henry Maurice Sheffer, born to Jewish parents in the western Ukraine, emigrated to the United States in 1892 with his parents and six siblings. He studied at the Boston Latin School before entering Harvard, where he completed his undergraduate degree in 1905, his master's in 1907, and his Ph.D. in philosophy in 1908. After holding a postdoctoral position at Harvard, Henry traveled to Europe on a fellowship. Upon returning to the United States, he became an academic nomad, spending one year each at the University of Washington, Cornell, the University of Minnesota, the University of Missouri, and City College in New York. In 1916 he returned to Harvard as a faculty member in the philosophy department. He remained at Harvard until his retirement in 1952.

Sheffer introduced what is now known as the Sheffer stroke in 1913; it became well known only after its use in the 1925 edition of Whitehead and Russell's *Principia Mathematica*. In this same edition Russell wrote that Sheffer had invented a powerful method that could be used to simplify the *Principia*. Because of this comment, Sheffer was something of a mystery man to logicians, especially because Sheffer, who published little in his career, never published the details of this method, only describing it in mimeographed notes and in a brief published abstract.

Sheffer was a dedicated teacher of mathematical logic. He liked his classes to be small and did not like auditors. When strangers appeared in his classroom, Sheffer would order them to leave, even his colleagues or distinguished guests visiting Harvard. Sheffer was barely five feet tall; he was noted for his wit and vigor, as well as for his nervousness and irritability. Although widely liked, he was quite lonely. He is noted for a quip he spoke at his retirement: "Old professors never die, they just become emeriti." Sheffer is also credited with coining the term "Boolean algebra" (the subject of Chapter 12 of this text). Sheffer was briefly married and lived most of his later life in small rooms at a hotel packed with his logic books and vast files of slips of paper he used to jot down his ideas. Unfortunately, Sheffer suffered from severe depression during the last two decades of his life.

☞ **6.** Show that each of these conditional statements is a tautology by using truth tables.

   a) $[\neg p \wedge (p \vee q)] \rightarrow q$

   b) $[(p \rightarrow q) \wedge (q \rightarrow r)] \rightarrow (p \rightarrow r)$

   c) $[p \wedge (p \rightarrow q)] \rightarrow q$

   d) $[(p \vee q) \wedge (p \rightarrow r) \wedge (q \rightarrow r)] \rightarrow r$

7. Show that each conditional statement in Exercise 5 is a tautology without using truth tables.

☞ **8.** Determine whether $(\neg q \wedge (p \rightarrow q)) \rightarrow \neg p$ is a tautology.

Each of Exercises 9–12 asks you to show that two compound propositions are logically equivalent. To do this, either show that both sides are true, or that both sides are false, for exactly the same combinations of truth values of the propositional variables in these expressions (whichever is easier).

9. Show that $\neg p \leftrightarrow q$ and $p \leftrightarrow \neg q$ are logically equivalent.

10. Show that $(p \rightarrow q) \wedge (p \rightarrow r)$ and $p \rightarrow (q \wedge r)$ are logically equivalent.

11. Show that $(p \rightarrow r) \wedge (q \rightarrow r)$ and $(p \vee q) \rightarrow r$ are logically equivalent.

12. Show that $(p \rightarrow r) \vee (q \rightarrow r)$ and $(p \wedge q) \rightarrow r$ are logically equivalent.

13. Show that $(p \rightarrow q) \rightarrow (r \rightarrow s)$ and $(p \rightarrow r) \rightarrow (q \rightarrow s)$ are not logically equivalent.

☞ **14.** Show that $(p \vee q) \wedge (\neg p \vee r) \rightarrow (q \vee r)$ is a tautology.

The **dual** of a compound proposition that contains only the logical operators $\vee$, $\wedge$, and $\neg$ is the compound proposition obtained by replacing each $\vee$ by $\wedge$, each $\wedge$ by $\vee$, each **T** by **F**, and each **F** by **T**. The dual of s is denoted by s^*.

15. Find the dual of each of these compound propositions.

   a) $p \wedge \neg q \wedge \neg r$ **b)** $(p \wedge q \wedge r) \vee s$

   c) $(p \vee \mathbf{F}) \wedge (q \vee \mathbf{T})$

16. Find the dual of each of these compound propositions.

   a) $p \vee \neg q$ **b)** $p \wedge (q \vee (r \wedge \mathbf{T}))$

   c) $(p \wedge \neg q) \vee (q \wedge \mathbf{F})$

17. Show that $(s^*)^* = s$ when s is a compound proposition.

18. Find a compound proposition involving the propositional variables p, q, and r that is true when p and q are true and r is false, but is false otherwise. [*Hint:* Use a conjunction of each propositional variable or its negation.]

19. Find a compound proposition involving the propositional variables p, q, and r that is true when exactly two of p, q, and r are true and is false otherwise. [*Hint:* Form a disjunction of conjunctions. Include a conjunction for each combination of values for which the compound proposition is true. Each conjunction should include each of the three propositional variables or its negations.]

☞ **20.** Suppose that a truth table in n propositional variables is specified. Show that a compound proposition with this truth table can be formed by taking the disjunction of conjunctions of the variables or their negations, with one conjunction included for each combination of values for which the compound proposition is true. The resulting compound proposition is said to be in **disjunctive normal form** (See Section 1.7).

A collection of logical operators is called **functionally complete** if every compound proposition is logically equivalent to a compound proposition involving only these logical operators.

21. Show that $\neg$, $\wedge$, and $\vee$ form a functionally complete collection of logical operators. [*Hint:* Use the fact that every compound proposition is logically equivalent to one in disjunctive normal form, as shown in Exercise 20.]

✶ **22.** Show that $\neg$ and $\wedge$ form a functionally complete collection of logical operators. [*Hint:* First use a De Morgan law to show that $p \vee q$ is logically equivalent to $\neg(\neg p \wedge \neg q)$.]

✶ **23.** Show that $\neg$ and $\vee$ form a functionally complete collection of logical operators.

The following exercises involve the logical operators *NAND* and *NOR*. The proposition p *NAND* q is true when either p or q, or both, are false; and it is false when both p and q are true. The proposition p *NOR* q is true when both p and q are false, and it is false otherwise. The propositions p *NAND* q and p *NOR* q are denoted by $p \mid q$ and $p \downarrow q$, respectively. (The operators $\mid$ and $\downarrow$ are called the **Sheffer stroke** and the **Peirce arrow** after H. M. Sheffer and C. S. Peirce, respectively.)

24. Construct a truth table for the logical operator *NAND*.

25. Show that $p \mid q$ is logically equivalent to $\neg(p \wedge q)$.

26. Construct a truth table for the logical operator *NOR*.

27. Show that $p \downarrow q$ is logically equivalent to $\neg(p \vee q)$.

28. In this exercise we will show that $\{\downarrow\}$ is a functionally complete collection of logical operators.

   a) Show that $p \downarrow p$ is logically equivalent to $\neg p$.

   b) Show that $(p \downarrow q) \downarrow (p \downarrow q)$ is logically equivalent to $p \vee q$.

   c) Conclude from parts (a) and (b), and Exercise 49, that $\{\downarrow\}$ is a functionally complete collection of logical operators.

✶ **29.** Find a compound proposition logically equivalent to $p \rightarrow q$ using only the logical operator $\downarrow$.

30. Show that $\{\mid\}$ is a functionally complete collection of logical operators.

31. Show that $p \mid q$ and $q \mid p$ are equivalent.

32. Show that $p \mid (q \mid r)$ and $(p \mid q) \mid r$ are not equivalent, so that the logical operator $\mid$ is not associative.

✶ **33.** How many different truth tables of compound propositions are there that involve the propositional variables p and q?

34. Show that if p, q, and r are compound propositions such that p and q are logically equivalent and q and r are logically equivalent, then p and r are logically equivalent.

35. The following sentence is taken from the specification of a telephone system: "If the directory database is opened, then the monitor is put in a closed state, if the system is not in its initial state." This specification is hard to understand because it involves two conditional statements. Find an equivalent, easier-to-understand specification that involves disjunctions and negations but not conditional statements.

36. How many of the disjunctions $p \vee \neg q$, $\neg p \vee q$, $q \vee r$, $q \vee \neg r$, and $\neg q \vee \neg r$ can be made simultaneously true by an assignment of truth values to p, q, and r?

37. Determine whether each of these compound propositions is satisfiable.

a) $(p \vee \neg q) \wedge (\neg p \vee q) \wedge (\neg p \vee \neg q)$

b) $(p \rightarrow q) \wedge (p \rightarrow \neg q) \wedge (\neg p \rightarrow q) \wedge (\neg p \rightarrow \neg q)$

c) $(p \leftrightarrow q) \wedge (\neg p \leftrightarrow q)$

38. Show that the negation of an unsatisfiable compound proposition is a tautology and the negation of a compound proposition that is a tautology is unsatisfiable.

39. Show how the solution of a given 4×4 Sudoku puzzle can be found by solving a satisfiability problem.

 ## Predicates and Quantifiers

Introduction

Propositional logic, studied in Sections 1.1–1.3, cannot adequately express the meaning of all statements in mathematics and in natural language. For example, suppose that we know that

"Every computer connected to the university network is functioning properly."

No rules of propositional logic allow us to conclude the truth of the statement

"MATH3 is functioning properly,"

where MATH3 is one of the computers connected to the university network. Likewise, we cannot use the rules of propositional logic to conclude from the statement

"CS2 is under attack by an intruder,"

where CS2 is a computer on the university network, to conclude the truth of

"There is a computer on the university network that is under attack by an intruder."

In this section we will introduce a more powerful type of logic called **predicate logic**. We will see how predicate logic can be used to express the meaning of a wide range of statements in mathematics and computer science in ways that permit us to reason and explore relationships between objects. To understand predicate logic, we first need to introduce the concept of a predicate. Afterward, we will introduce the notion of quantifiers, which enable us to reason with statements that assert that a certain property holds for all objects of a certain type and with statements that assert the existence of an object with a particular property.

Predicates

Statements involving variables, such as

"$x > 3$,"   "$x = y + 3$,"   "$x + y = z$,"

and

"computer x is under attack by an intruder,"

and

"computer x is functioning properly,"

are often found in mathematical assertions, in computer programs, and in system specifications. These statements are neither true nor false when the values of the variables are not specified. In this section, we will discuss the ways that propositions can be produced from such statements.

The statement "x is greater than 3" has two parts. The first part, the variable x, is the subject of the statement. The second part—the **predicate**, "is greater than 3"—refers to a property that the subject of the statement can have. We can denote the statement "x is greater than 3" by $P(x)$, where P denotes the predicate "is greater than 3" and x is the variable. The statement $P(x)$ is also said to be the value of the **propositional function** P at x. Once a value has been assigned to the variable x, the statement $P(x)$ becomes a proposition and has a truth value. Consider Examples 1 and 2.

EXAMPLE 1   Let $P(x)$ denote the statement "$x > 3$." What are the truth values of $P(4)$ and $P(2)$?

Solution: We obtain the statement $P(4)$ by setting $x = 4$ in the statement "$x > 3$." Hence, $P(4)$, which is the statement "$4 > 3$," is true. However, $P(2)$, which is the statement "$2 > 3$," is false.   ◀

EXAMPLE 2   Let $A(x)$ denote the statement "Computer x is under attack by an intruder." Suppose that of the computers on campus, only CS2 and MATH1 are currently under attack by intruders. What are truth values of $A(\text{CS1})$, $A(\text{CS2})$, and $A(\text{MATH1})$?

Solution: We obtain the statement $A(\text{CS1})$ by setting $x = \text{CS1}$ in the statement "Computer x is under attack by an intruder." Because CS1 is not on the list of computers currently under attack, we conclude that $A(\text{CS1})$ is false. Similarly, because CS2 and MATH1 are on the list of computers under attack, we know that $A(\text{CS2})$ and $A(\text{MATH1})$ are true.   ◀

We can also have statements that involve more than one variable. For instance, consider the statement "$x = y + 3$." We can denote this statement by $Q(x, y)$, where x and y are variables and Q is the predicate. When values are assigned to the variables x and y, the statement $Q(x, y)$ has a truth value.

EXAMPLE 3   Let $Q(x, y)$ denote the statement "$x = y + 3$." What are the truth values of the propositions $Q(1, 2)$ and $Q(3, 0)$?

Extra
Examples

Solution: To obtain $Q(1, 2)$, set $x = 1$ and $y = 2$ in the statement $Q(x, y)$. Hence, $Q(1, 2)$ is the statement "$1 = 2 + 3$," which is false. The statement $Q(3, 0)$ is the proposition "$3 = 0 + 3$," which is true.   ◀

EXAMPLE 4   Let $A(c, n)$ denote the statement "Computer c is connected to network n," where c is a variable representing a computer and n is a variable representing a network. Suppose that the computer MATH1 is connected to network CAMPUS2, but not to network CAMPUS1. What are the values of $A(\text{MATH1, CAMPUS1})$ and $A(\text{MATH1, CAMPUS2})$?

Solution: Because MATH1 is not connected to the CAMPUS1 network, we see that $A(\text{MATH1, CAMPUS1})$ is false. However, because MATH1 is connected to the CAMPUS2 network, we see that $A(\text{MATH1, CAMPUS2})$ is true.   ◄

Similarly, we can let $R(x, y, z)$ denote the statement "$x + y = z$." When values are assigned to the variables x, y, and z, this statement has a truth value.

EXAMPLE 5   What are the truth values of the propositions $R(1, 2, 3)$ and $R(0, 0, 1)$?

Solution: The proposition $R(1, 2, 3)$ is obtained by setting $x = 1$, $y = 2$, and $z = 3$ in the statement $R(x, y, z)$. We see that $R(1, 2, 3)$ is the statement "$1 + 2 = 3$," which is true. Also note that $R(0, 0, 1)$, which is the statement "$0 + 0 = 1$," is false.   ◄

In general, a statement involving the n variables $x_1, x_2, \ldots, x_n$ can be denoted by

$$P(x_1, x_2, \ldots, x_n).$$

A statement of the form $P(x_1, x_2, \ldots, x_n)$ is the value of the **propositional function** P at the n-tuple $(x_1, x_2, \ldots, x_n)$, and P is also called an n-**place predicate** or a n-**ary predicate**.

Propositional functions occur in computer programs, as Example 6 demonstrates.

CHARLES SANDERS PEIRCE (1839–1914)   Many consider Charles Peirce, born in Cambridge, Massachusetts, to be the most original and versatile American intellect. He made important contributions to an amazing number of disciplines, including mathematics, astronomy, chemistry, geodesy, metrology, engineering, psychology, philology, the history of science, and economics. Peirce was also an inventor, a lifelong student of medicine, a book reviewer, a dramatist and an actor, a short story writer, a phenomenologist, a logician, and a metaphysician. He is noted as the preeminent system-building philosopher competent and productive in logic, mathematics, and a wide range of sciences. He was encouraged by his father, Benjamin Peirce, a professor of mathematics and natural philosophy at Harvard, to pursue a career in science. Instead, he decided to study logic and scientific methodology. Peirce attended Harvard (1855–1859) and received a Harvard master of arts degree (1862) and an advanced degree in chemistry from the Lawrence Scientific School (1863).

In 1861, Peirce became an aide in the U.S. Coast Survey, with the goal of better understanding scientific methodology. His service for the Survey exempted him from military service during the Civil War. While working for the Survey, Peirce did astronomical and geodesic work. He made fundamental contributions to the design of pendulums and to map projections, applying new mathematical developments in the theory of elliptic functions. He was the first person to use the wavelength of light as a unit of measurement. Peirce rose to the position of Assistant for the Survey, a position he held until forced to resign in 1891 when he disagreed with the direction taken by the Survey's new administration.

While making his living from work in the physical sciences, Peirce developed a hierarchy of sciences, with mathematics at the top rung, in which the methods of one science could be adapted for use by those sciences under it in the hierarchy. During this time, he also founded the American philosophical theory of pragmatism.

The only academic position Peirce ever held was lecturer in logic at Johns Hopkins University in Baltimore (1879–1884). His mathematical work during this time included contributions to logic, set theory, abstract algebra, and the philosophy of mathematics. His work is still relevant today, with recent applications of this work on logic to artificial intelligence. Peirce believed that the study of mathematics could develop the mind's powers of imagination, abstraction, and generalization. His diverse activities after retiring from the Survey included writing for periodicals, contributing to scholarly dictionaries, translating scientific papers, guest lecturing, and textbook writing. Unfortunately, his income from these pursuits was insufficient to protect him and his second wife from abject poverty. He was supported in his later years by a fund created by his many admirers and administered by the philosopher William James, his lifelong friend. Although Peirce wrote and published voluminously in a vast range of subjects, he left more than 100,000 pages of unpublished manuscripts. Because of the difficulty of studying his unpublished writings, scholars have only recently started to understand some of his varied contributions. A group of people is devoted to making his work available over the Internet to bring a better appreciation of Peirce's accomplishments to the world.

EXAMPLE 6   Consider the statement

if $x > 0$ **then** $x := x + 1$.

When this statement is encountered in a program, the value of the variable x at that point in the execution of the program is inserted into $P(x)$, which is "$x > 0$." If $P(x)$ is true for this value of x, the assignment statement $x := x + 1$ is executed, so the value of x is increased by 1. If $P(x)$ is false for this value of x, the assignment statement is not executed, so the value of x is not changed.   ◀

PRECONDITIONS AND POSTCONDITIONS   Predicates are also used to establish the correctness of computer programs, that is, to show that computer programs always produce the desired output when given valid input. (Note that unless the correctness of a computer program is established, no amount of testing can show that it produces the desired output for all input values, unless every input value is tested.) The statements that describe valid input are known as **preconditions** and the conditions that the output should satisfy when the program has run are known as **postconditions**. As Example 7 illustrates, we use predicates to describe both preconditions and postconditions. We will study this process in greater detail in Section 5.5.

EXAMPLE 7   Consider the following program, designed to interchange the values of two variables x and y.

```
temp := x
x := y
y := temp
```

Find predicates that we can use as the precondition and the postcondition to verify the correctness of this program. Then explain how to use them to verify that for all valid input the program does what is intended.

Solution: For the precondition, we need to express that x and y have particular values before we run the program. So, for this precondition we can use the predicate $P(x, y)$, where $P(x, y)$ is the statement "$x = a$ and $y = b$," where a and b are the values of x and y before we run the program. Because we want to verify that the program swaps the values of x and y for all input values, for the postcondition we can use $Q(x, y)$, where $Q(x, y)$ is the statement "$x = b$ and $y = a$."

To verify that the program always does what it is supposed to do, suppose that the precondition $P(x, y)$ holds. That is, we suppose that the statement "$x = a$ and $y = b$" is true. This means that $x = a$ and $y = b$. The first step of the program, *temp* $:= x$, assigns the value of x to the variable *temp*, so after this step we know that $x = a$, *temp* $= a$, and $y = b$. After the second step of the program, $x := y$, we know that $x = b$, *temp* $= a$, and $y = b$. Finally, after the third step, we know that $x = b$, *temp* $= a$, and $y = a$. Consequently, after this program is run, the postcondition $Q(x, y)$ holds, that is, the statement "$x = b$ and $y = a$" is true.   ◀

Quantifiers

When the variables in a propositional function are assigned values, the resulting statement becomes a proposition with a certain truth value. However, there is another important way, called **quantification**, to create a proposition from a propositional function. Quantification expresses the extent to which a predicate is true over a range of elements. In English, the words *all*, *some*, *many*, *none*, and *few* are used in quantifications. We will focus on two types of quantification here: universal quantification, which tells us that a predicate is true for every element under consideration, and existential quantification, which tells us that there is one or more element under consideration for which the predicate is true. The area of logic that deals with predicates and quantifiers is called the **predicate calculus**.

Assessment

TABLE 1 Quantifiers.		
Statement	**When True?**	**When False?**
$\forall x\, P(x)$	$P(x)$ is true for every x.	There is an x for which $P(x)$ is false.
$\exists x\, P(x)$	There is an x for which $P(x)$ is true.	$P(x)$ is false for every x.

Assessment

THE UNIVERSAL QUANTIFIER   Many mathematical statements assert that a property is true for all values of a variable in a particular domain, called the **domain of discourse** (or the **universe of discourse**), often just referred to as the **domain**. Such a statement is expressed using universal quantification. The universal quantification of $P(x)$ for a particular domain is the proposition that asserts that $P(x)$ is true for all values of x in this domain. Note that the domain specifies the possible values of the variable x. The meaning of the universal quantification of $P(x)$ changes when we change the domain. The domain must always be specified when a universal quantifier is used; without it, the universal quantification of a statement is not defined.

DEFINITION 1

The *universal quantification* of $P(x)$ is the statement

   "$P(x)$ for all values of x in the domain."

   The notation $\forall x\, P(x)$ denotes the universal quantification of $P(x)$. Here $\forall$ is called the **universal quantifier.** We read $\forall x\, P(x)$ as "for all $x\, P(x)$" or "for every $x\, P(x)$." An element for which $P(x)$ is false is called a **counterexample** of $\forall x\, P(x)$.

The meaning of the universal quantifier is summarized in the first row of Table 1. We illustrate the use of the universal quantifier in Examples 8–13.

EXAMPLE 8   Let $P(x)$ be the statement "$x + 1 > x$." What is the truth value of the quantification $\forall x\, P(x)$, where the domain consists of all real numbers?

Extra Examples

Solution: Because $P(x)$ is true for all real numbers x, the quantification

   $\forall x\, P(x)$

is true.   ◀

Remark: Generally, an implicit assumption is made that all domains of discourse for quantifiers are nonempty. Note that if the domain is empty, then $\forall x\, P(x)$ is true for any propositional function $P(x)$ because there are no elements x in the domain for which $P(x)$ is false.

Remember that the truth value of $\forall x\, P(x)$ depends on the domain!

Besides "for all" and "for every," universal quantification can be expressed in many other ways, including "all of," "for each," "given any," "for arbitrary," "for each," and "for any."

Remark: It is best to avoid using "for any x" because it is often ambiguous as to whether "any" means "every" or "some." In some cases, "any" is unambiguous, such as when it is used in negatives, for example, "there is not any reason to avoid studying."

A statement $\forall x\, P(x)$ is false, where $P(x)$ is a propositional function, if and only if $P(x)$ is not always true when x is in the domain. One way to show that $P(x)$ is not always true when x is in the domain is to find a counterexample to the statement $\forall x\, P(x)$. Note that a single counterexample is all we need to establish that $\forall x\, P(x)$ is false. Example 9 illustrates how counterexamples are used.

EXAMPLE 9   Let $Q(x)$ be the statement "$x < 2$." What is the truth value of the quantification $\forall x\, Q(x)$, where the domain consists of all real numbers?

Solution: $Q(x)$ is not true for every real number x, because, for instance, $Q(3)$ is false. That is, $x = 3$ is a counterexample for the statement $\forall x\, Q(x)$. Thus

$$\forall x\, Q(x)$$

is false.   ◀

EXAMPLE 10   Suppose that $P(x)$ is "$x^2 > 0$." To show that the statement $\forall x\, P(x)$ is false where the universe of discourse consists of all integers, we give a counterexample. We see that $x = 0$ is a counterexample because $x^2 = 0$ when $x = 0$, so that x^2 is not greater than 0 when $x = 0$.   ◀

Looking for counterexamples to universally quantified statements is an important activity in the study of mathematics, as we will see in subsequent sections of this book.

When all the elements in the domain can be listed—say, $x_1, x_2, \ldots, x_n$—it follows that the universal quantification $\forall x\, P(x)$ is the same as the conjunction

$$P(x_1) \wedge P(x_2) \wedge \cdots \wedge P(x_n),$$

because this conjunction is true if and only if $P(x_1), P(x_2), \ldots, P(x_n)$ are all true.

EXAMPLE 11   What is the truth value of $\forall x\, P(x)$, where $P(x)$ is the statement "$x^2 < 10$" and the domain consists of the positive integers not exceeding 4?

Solution: The statement $\forall x\, P(x)$ is the same as the conjunction

$$P(1) \wedge P(2) \wedge P(3) \wedge P(4),$$

because the domain consists of the integers 1, 2, 3, and 4. Because $P(4)$, which is the statement "$4^2 < 10$," is false, it follows that $\forall x\, P(x)$ is false.   ◀

EXAMPLE 12   What does the statement $\forall x\, N(x)$ mean if $N(x)$ is "Computer x is connected to the network" and the domain consists of all computers on campus?

Solution: The statement $\forall x\, N(x)$ means that for every computer x on campus, that computer x is connected to the network. This statement can be expressed in English as "Every computer on campus is connected to the network."   ◀

As we have pointed out, specifying the domain is mandatory when quantifiers are used. The truth value of a quantified statement often depends on which elements are in this domain, as Example 13 shows.

EXAMPLE 13   What is the truth value of $\forall x(x^2 \geq x)$ if the domain consists of all real numbers? What is the truth value of this statement if the domain consists of all integers?

Solution: The universal quantification $\forall x(x^2 \geq x)$, where the domain consists of all real numbers, is false. For example, $(\frac{1}{2})^2 \not\geq \frac{1}{2}$. Note that $x^2 \geq x$ if and only if $x^2 - x = x(x - 1) \geq 0$. Consequently, $x^2 \geq x$ if and only if $x \leq 0$ or $x \geq 1$. It follows that $\forall x(x^2 \geq x)$ is false if the domain consists of all real numbers (because the inequality is false for all real numbers x with $0 < x < 1$). However, if the domain consists of the integers, $\forall x(x^2 \geq x)$ is true, because there are no integers x with $0 < x < 1$.   ◀

THE EXISTENTIAL QUANTIFIER   Many mathematical statements assert that there is an element with a certain property. Such statements are expressed using existential quantification. With existential quantification, we form a proposition that is true if and only if $P(x)$ is true for at least one value of x in the domain.

DEFINITION 2   The *existential quantification* of $P(x)$ is the proposition

"There exists an element x in the domain such that $P(x)$."

We use the notation $\exists x\, P(x)$ for the existential quantification of $P(x)$. Here $\exists$ is called the *existential quantifier*.

A domain must always be specified when a statement $\exists x\, P(x)$ is used. Furthermore, the meaning of $\exists x\, P(x)$ changes when the domain changes. Without specifying the domain, the statement $\exists x\, P(x)$ has no meaning.

Besides the phrase "there exists," we can also express existential quantification in many other ways, such as by using the words "for some," "for at least one," or "there is." The existential quantification $\exists x\, P(x)$ is read as

"There is an x such that $P(x)$,"
"There is at least one x such that $P(x)$,"

or

"For some $x\, P(x)$."

The meaning of the existential quantifier is summarized in the second row of Table 1. We illustrate the use of the existential quantifier in Examples 14–16.

EXAMPLE 14   Let $P(x)$ denote the statement "$x > 3$." What is the truth value of the quantification $\exists x\, P(x)$, where the domain consists of all real numbers?

Solution: Because "$x > 3$" is sometimes true—for instance, when $x = 4$—the existential quantification of $P(x)$, which is $\exists x\, P(x)$, is true.   ◄

Observe that the statement $\exists x\, P(x)$ is false if and only if there is no element x in the domain for which $P(x)$ is true. That is, $\exists x\, P(x)$ is false if and only if $P(x)$ is false for every element of the domain. We illustrate this observation in Example 15.

EXAMPLE 15   Let $Q(x)$ denote the statement "$x = x + 1$." What is the truth value of the quantification $\exists x\, Q(x)$, where the domain consists of all real numbers?

Solution: Because $Q(x)$ is false for every real number x, the existential quantification of $Q(x)$, which is $\exists x\, Q(x)$, is false.   ◄

Remember that the truth value of $\exists x\, P(x)$ depends on the domain!

Remark: Generally, an implicit assumption is made that all domains of discourse for quantifiers are nonempty. If the domain is empty, then $\exists x\, Q(x)$ is false whenever $Q(x)$ is a propositional function because when the domain is empty, there can be no element x in the domain for which $Q(x)$ is true.

When all elements in the domain can be listed—say, $x_1, x_2, \ldots, x_n$— the existential quantification $\exists x \, P(x)$ is the same as the disjunction

$$P(x_1) \vee P(x_2) \vee \cdots \vee P(x_n),$$

because this disjunction is true if and only if at least one of $P(x_1), P(x_2), \ldots, P(x_n)$ is true.

EXAMPLE 16   What is the truth value of $\exists x \, P(x)$, where $P(x)$ is the statement "$x^2 > 10$" and the universe of discourse consists of the positive integers not exceeding 4?

Solution: Because the domain is $\{1, 2, 3, 4\}$, the proposition $\exists x \, P(x)$ is the same as the disjunction

$$P(1) \vee P(2) \vee P(3) \vee P(4).$$

Because $P(4)$, which is the statement "$4^2 > 10$," is true, it follows that $\exists x \, P(x)$ is true.   ◄

It is sometimes helpful to think in terms of looping and searching when determining the truth value of a quantification. Suppose that there are n objects in the domain for the variable x. To determine whether $\forall x \, P(x)$ is true, we can loop through all n values of x to see whether $P(x)$ is always true. If we encounter a value x for which $P(x)$ is false, then we have shown that $\forall x \, P(x)$ is false. Otherwise, $\forall x \, P(x)$ is true. To see whether $\exists x \, P(x)$ is true, we loop through the n values of x searching for a value for which $P(x)$ is true. If we find one, then $\exists x \, P(x)$ is true. If we never find such an x, then we have determined that $\exists x \, P(x)$ is false. (Note that this searching procedure does not apply if there are infinitely many values in the domain. However, it is still a useful way of thinking about the truth values of quantifications.)

THE UNIQUENESS QUANTIFIER   We have now introduced universal and existential quantifiers. These are the most important quantifiers in mathematics and computer science. However, there is no limitation on the number of different quantifiers we can define, such as "there are exactly two," "there are no more than three," "there are at least 100," and so on. Of these other quantifiers, the one that is most often seen is the **uniqueness quantifier**, denoted by $\exists!$ or $\exists_1$. The notation $\exists! x \, P(x)$ [or $\exists_1 x \, P(x)$] states "There exists a unique x such that $P(x)$ is true." (Other phrases for uniqueness quantification include "there is exactly one" and "there is one and only one.") For instance, $\exists! x (x - 1 = 0)$, where the domain is the set of real numbers, states that there is a unique real number x such that $x - 1 = 0$. This is a true statement, as $x = 1$ is the unique real number such that $x - 1 = 0$. Observe that we can use quantifiers and propositional logic to express uniqueness (see Exercise 27 in Section 1.5), so the uniqueness quantifier can be avoided. Generally, it is best to stick with existential and universal quantifiers so that rules of inference for these quantifiers can be used.

Quantifiers with Restricted Domains

An abbreviated notation is often used to restrict the domain of a quantifier. In this notation, a condition a variable must satisfy is included after the quantifier. This is illustrated in Example 17. We will also describe other forms of this notation involving set membership in Section 2.1.

EXAMPLE 17   What do the statements $\forall x < 0 \, (x^2 > 0)$, $\forall y \neq 0 \, (y^3 \neq 0)$, and $\exists z > 0 \, (z^2 = 2)$ mean, where the domain in each case consists of the real numbers?

Solution: The statement $\forall x < 0 \, (x^2 > 0)$ states that for every real number x with $x < 0$, $x^2 > 0$. That is, it states "The square of a negative real number is positive." This statement is the same as $\forall x (x < 0 \rightarrow x^2 > 0)$.

The statement $\forall y \neq 0\,(y^3 \neq 0)$ states that for every real number y with $y \neq 0$, we have $y^3 \neq 0$. That is, it states "The cube of every nonzero real number is nonzero." Note that this statement is equivalent to $\forall y(y \neq 0 \rightarrow y^3 \neq 0)$.

Finally, the statement $\exists z > 0\,(z^2 = 2)$ states that there exists a real number z with $z > 0$ such that $z^2 = 2$. That is, it states "There is a positive square root of 2." This statement is equivalent to $\exists z(z > 0 \wedge z^2 = 2)$. ◀

Note that the restriction of a universal quantification is the same as the universal quantification of a conditional statement. For instance, $\forall x < 0\,(x^2 > 0)$ is another way of expressing $\forall x(x < 0 \rightarrow x^2 > 0)$. On the other hand, the restriction of an existential quantification is the same as the existential quantification of a conjunction. For instance, $\exists z > 0\,(z^2 = 2)$ is another way of expressing $\exists z(z > 0 \wedge z^2 = 2)$.

Precedence of Quantifiers

The quantifiers $\forall$ and $\exists$ have higher precedence than all logical operators from propositional calculus. For example, $\forall x\, P(x) \vee Q(x)$ is the disjunction of $\forall x\, P(x)$ and $Q(x)$. In other words, it means $(\forall x\, P(x)) \vee Q(x)$ rather than $\forall x(P(x) \vee Q(x))$.

Binding Variables

When a quantifier is used on the variable x, we say that this occurrence of the variable is **bound**. An occurrence of a variable that is not bound by a quantifier or set equal to a particular value is said to be **free**. All the variables that occur in a propositional function must be bound or set equal to a particular value to turn it into a proposition. This can be done using a combination of universal quantifiers, existential quantifiers, and value assignments.

The part of a logical expression to which a quantifier is applied is called the **scope** of this quantifier. Consequently, a variable is free if it is outside the scope of all quantifiers in the formula that specify this variable.

EXAMPLE 18   In the statement $\exists x(x + y = 1)$, the variable x is bound by the existential quantification $\exists x$, but the variable y is free because it is not bound by a quantifier and no value is assigned to this variable. This illustrates that in the statement $\exists x(x + y = 1)$, x is bound, but y is free.

In the statement $\exists x(P(x) \wedge Q(x)) \vee \forall x\, R(x)$, all variables are bound. The scope of the first quantifier, $\exists x$, is the expression $P(x) \wedge Q(x)$ because $\exists x$ is applied only to $P(x) \wedge Q(x)$, and not to the rest of the statement. Similarly, the scope of the second quantifier, $\forall x$, is the expression $R(x)$. That is, the existential quantifier binds the variable x in $P(x) \wedge Q(x)$ and the universal quantifier $\forall x$ binds the variable x in $R(x)$. Observe that we could have written our statement using two different variables x and y, as $\exists x(P(x) \wedge Q(x)) \vee \forall y\, R(y)$, because the scopes of the two quantifiers do not overlap. The reader should be aware that in common usage, the same letter is often used to represent variables bound by different quantifiers with scopes that do not overlap. ◀

Logical Equivalences Involving Quantifiers

In Section 1.3 we introduced the notion of logical equivalences of compound propositions. We can extend this notion to expressions involving predicates and quantifiers.

DEFINITION 3

> Statements involving predicates and quantifiers are *logically equivalent* if and only if they have the same truth value no matter which predicates are substituted into these statements and which domain of discourse is used for the variables in these propositional functions. We use the notation $S \equiv T$ to indicate that two statements S and T involving predicates and quantifiers are logically equivalent.

Example 19 illustrates how to show that two statements involving predicates and quantifiers are logically equivalent.

EXAMPLE 19   Show that $\forall x(P(x) \wedge Q(x))$ and $\forall x\, P(x) \wedge \forall x\, Q(x)$ are logically equivalent (where the same domain is used throughout). This logical equivalence shows that we can distribute a universal quantifier over a conjunction. Furthermore, we can also distribute an existential quantifier over a disjunction. However, we cannot distribute a universal quantifier over a disjunction, nor can we distribute an existential quantifier over a conjunction. (See Exercise 30.)

Solution: To show that these statements are logically equivalent, we must show that they always take the same truth value, no matter what the predicates P and Q are, and no matter which domain of discourse is used. Suppose we have particular predicates P and Q, with a common domain. We can show that $\forall x(P(x) \wedge Q(x))$ and $\forall x\, P(x) \wedge \forall x\, Q(x)$ are logically equivalent by doing two things. First, we show that if $\forall x(P(x) \wedge Q(x))$ is true, then $\forall x\, P(x) \wedge \forall x\, Q(x)$ is true. Second, we show that if $\forall x\, P(x) \wedge \forall x\, Q(x)$ is true, then $\forall x(P(x) \wedge Q(x))$ is true.

So, suppose that $\forall x(P(x) \wedge Q(x))$ is true. This means that if a is in the domain, then $P(a) \wedge Q(a)$ is true. Hence, $P(a)$ is true and $Q(a)$ is true. Because $P(a)$ is true and $Q(a)$ is true for every element in the domain, we can conclude that $\forall x\, P(x)$ and $\forall x\, Q(x)$ are both true. This means that $\forall x\, P(x) \wedge \forall x\, Q(x)$ is true.

Next, suppose that $\forall x\, P(x) \wedge \forall x\, Q(x)$ is true. It follows that $\forall x\, P(x)$ is true and $\forall x\, Q(x)$ is true. Hence, if a is in the domain, then $P(a)$ is true and $Q(a)$ is true [because $P(x)$ and $Q(x)$ are both true for all elements in the domain, there is no conflict using the same value of a here]. It follows that for all a, $P(a) \wedge Q(a)$ is true. It follows that $\forall x(P(x) \wedge Q(x))$ is true. We can now conclude that

$$\forall x(P(x) \wedge Q(x)) \equiv \forall x\, P(x) \wedge \forall x\, Q(x).$$ ◀

Negating Quantified Expressions

We will often want to consider the negation of a quantified expression. For instance, consider the negation of the statement

"Every student in your class has taken a course in calculus."

This statement is a universal quantification, namely,

$$\forall x\, P(x),$$

Assessment

where $P(x)$ is the statement "x has taken a course in calculus" and the domain consists of the students in your class. The negation of this statement is "It is not the case that every student in your class has taken a course in calculus." This is equivalent to "There is a student in your class who has not taken a course in calculus." And this is simply the existential quantification of the negation of the original propositional function, namely,

$$\exists x\, \neg P(x).$$

TABLE 2 De Morgan's Laws for Quantifiers.			
Negation	*Equivalent Statement*	*When Is Negation True?*	*When False?*
$\neg \exists x\, P(x)$	$\forall x\, \neg P(x)$	For every x, $P(x)$ is false.	There is an x for which $P(x)$ is true.
$\neg \forall x\, P(x)$	$\exists x\, \neg P(x)$	There is an x for which $P(x)$ is false.	$P(x)$ is true for every x.

This example illustrates the following logical equivalence:

$$\neg \forall x\, P(x) \equiv \exists x\, \neg P(x).$$

To show that $\neg \forall x\, P(x)$ and $\exists x\, P(x)$ are logically equivalent no matter what the propositional function $P(x)$ is and what the domain is, first note that $\neg \forall x\, P(x)$ is true if and only if $\forall x\, P(x)$ is false. Next, note that $\forall x\, P(x)$ is false if and only if there is an element x in the domain for which $P(x)$ is false. This holds if and only if there is an element x in the domain for which $\neg P(x)$ is true. Finally, note that there is an element x in the domain for which $\neg P(x)$ is true if and only if $\exists x\, \neg P(x)$ is true. Putting these steps together, we can conclude that $\neg \forall x\, P(x)$ is true if and only if $\exists x\, \neg P(x)$ is true. It follows that $\neg \forall x\, P(x)$ and $\exists x\, \neg P(x)$ are logically equivalent.

Suppose we wish to negate an existential quantification. For instance, consider the proposition "There is a student in this class who has taken a course in calculus." This is the existential quantification

$$\exists x\, Q(x),$$

where $Q(x)$ is the statement "x has taken a course in calculus." The negation of this statement is the proposition "It is not the case that there is a student in this class who has taken a course in calculus." This is equivalent to "Every student in this class has not taken calculus," which is just the universal quantification of the negation of the original propositional function, or, phrased in the language of quantifiers,

$$\forall x\, \neg Q(x).$$

This example illustrates the equivalence

$$\neg \exists x\, Q(x) \equiv \forall x\, \neg Q(x).$$

To show that $\neg \exists x\, Q(x)$ and $\forall x\, \neg Q(x)$ are logically equivalent no matter what $Q(x)$ is and what the domain is, first note that $\neg \exists x\, Q(x)$ is true if and only if $\exists x\, Q(x)$ is false. This is true if and only if no x exists in the domain for which $Q(x)$ is true. Next, note that no x exists in the domain for which $Q(x)$ is true if and only if $Q(x)$ is false for every x in the domain. Finally, note that $Q(x)$ is false for every x in the domain if and only if $\neg Q(x)$ is true for all x in the domain, which holds if and only if $\forall x\, \neg Q(x)$ is true. Putting these steps together, we see that $\neg \exists x\, Q(x)$ is true if and only if $\forall x\, \neg Q(x)$ is true. We conclude that $\neg \exists x\, Q(x)$ and $\forall x\, \neg Q(x)$ are logically equivalent.

The rules for negations for quantifiers are called **De Morgan's laws for quantifiers**. These rules are summarized in Table 2.

Remark: When the domain of a predicate $P(x)$ consists of n elements, where n is a positive integer greater than one, the rules for negating quantified statements are exactly the same as De Morgan's laws discussed in Section 1.3. This is why these rules are called De Morgan's laws for quantifiers. When the domain has n elements $x_1, x_2, \ldots, x_n$, it follows that $\neg \forall x\, P(x)$

is the same as $\neg(P(x_1) \wedge P(x_2) \wedge \cdots \wedge P(x_n))$, which is equivalent to $\neg P(x_1) \vee \neg P(x_2) \vee \cdots \vee \neg P(x_n)$ by De Morgan's laws, and this is the same as $\exists x \neg P(x)$. Similarly, $\neg \exists x P(x)$ is the same as $\neg(P(x_1) \vee P(x_2) \vee \cdots \vee P(x_n))$, which by De Morgan's laws is equivalent to $\neg P(x_1) \wedge \neg P(x_2) \wedge \cdots \wedge \neg P(x_n)$, and this is the same as $\forall x \neg P(x)$.

We illustrate the negation of quantified statements in Examples 20 and 21.

EXAMPLE 20 What are the negations of the statements "There is an honest politician" and "All Americans eat cheeseburgers"?

Solution: Let $H(x)$ denote "x is honest." Then the statement "There is an honest politician" is represented by $\exists x H(x)$, where the domain consists of all politicians. The negation of this statement is $\neg \exists x H(x)$, which is equivalent to $\forall x \neg H(x)$. This negation can be expressed as "Every politician is dishonest." (*Note:* In English, the statement "All politicians are not honest" is ambiguous. In common usage, this statement often means "Not all politicians are honest." Consequently, we do not use this statement to express this negation.)

Let $C(x)$ denote "x eats cheeseburgers." Then the statement "All Americans eat cheeseburgers" is represented by $\forall x C(x)$, where the domain consists of all Americans. The negation of this statement is $\neg \forall x C(x)$, which is equivalent to $\exists x \neg C(x)$. This negation can be expressed in several different ways, including "Some American does not eat cheeseburgers" and "There is an American who does not eat cheeseburgers." ◀

EXAMPLE 21 What are the negations of the statements $\forall x (x^2 > x)$ and $\exists x (x^2 = 2)$?

Solution: The negation of $\forall x (x^2 > x)$ is the statement $\neg \forall x (x^2 > x)$, which is equivalent to $\exists x \neg(x^2 > x)$. This can be rewritten as $\exists x (x^2 \leq x)$. The negation of $\exists x (x^2 = 2)$ is the statement $\neg \exists x (x^2 = 2)$, which is equivalent to $\forall x \neg(x^2 = 2)$. This can be rewritten as $\forall x (x^2 \neq 2)$. The truth values of these statements depend on the domain. ◀

We use De Morgan's laws for quantifiers in Example 22.

EXAMPLE 22 Show that $\neg \forall x (P(x) \rightarrow Q(x))$ and $\exists x (P(x) \wedge \neg Q(x))$ are logically equivalent.

Solution: By De Morgan's law for universal quantifiers, we know that $\neg \forall x (P(x) \rightarrow Q(x))$ and $\exists x (\neg(P(x) \rightarrow Q(x)))$ are logically equivalent. By the fifth logical equivalence in Table 7 in Section 1.3, we know that $\neg(P(x) \rightarrow Q(x))$ and $P(x) \wedge \neg Q(x)$ are logically equivalent for every x. Because we can substitute one logically equivalent expression for another in a logical equivalence, it follows that $\neg \forall x (P(x) \rightarrow Q(x))$ and $\exists x (P(x) \wedge \neg Q(x))$ are logically equivalent. ◀

Translating from English into Logical Expressions

Translating sentences in English (or other natural languages) into logical expressions is a crucial task in mathematics, logic programming, artificial intelligence, software engineering, and many other disciplines. We began studying this topic in Section 1.1, where we used propositions to express sentences in logical expressions. In that discussion, we purposely avoided sentences whose translations required predicates and quantifiers. Translating from English to logical expressions becomes even more complex when quantifiers are needed. Furthermore, there can be many ways to translate a particular sentence. (As a consequence, there is no "cookbook" approach that can be followed step by step.) We will use some examples to illustrate how to translate sentences from English into logical expressions. The goal in this translation is to produce simple and useful logical expressions. In this section, we restrict ourselves to sentences that can be translated into logical expressions using a single quantifier; in the next section, we will look at more complicated sentences that require multiple quantifiers.

EXAMPLE 23   Express the statement "Every student in this class has studied calculus" using predicates and quantifiers.

Solution: First, we rewrite the statement so that we can clearly identify the appropriate quantifiers to use. Doing so, we obtain:

"For every student in this class, that student has studied calculus."

Next, we introduce a variable x so that our statement becomes

"For every student x in this class, x has studied calculus."

Continuing, we introduce $C(x)$, which is the statement "x has studied calculus." Consequently, if the domain for x consists of the students in the class, we can translate our statement as $\forall x\, C(x)$.

However, there are other correct approaches; different domains of discourse and other predicates can be used. The approach we select depends on the subsequent reasoning we want to carry out. For example, we may be interested in a wider group of people than only those in this class. If we change the domain to consist of all people, we will need to express our statement as

"For every person x, if person x is a student in this class then x has studied calculus."

If $S(x)$ represents the statement that person x is in this class, we see that our statement can be expressed as $\forall x(S(x) \rightarrow C(x))$. [*Caution!* Our statement *cannot* be expressed as $\forall x(S(x) \wedge C(x))$ because this statement says that all people are students in this class and have studied calculus!]

Finally, when we are interested in the background of people in subjects besides calculus, we may prefer to use the two-variable quantifier $Q(x, y)$ for the statement "student x has studied subject y." Then we would replace $C(x)$ by $Q(x, \text{calculus})$ in both approaches to obtain $\forall x\, Q(x, \text{calculus})$ or $\forall x(S(x) \rightarrow Q(x, \text{calculus}))$. ◀

In Example 23 we displayed different approaches for expressing the same statement using predicates and quantifiers. However, we should always adopt the simplest approach that is adequate for use in subsequent reasoning.

EXAMPLE 24   Express the statements "Some student in this class has visited Mexico" and "Every student in this class has visited either Canada or Mexico" using predicates and quantifiers.

Solution: The statement "Some student in this class has visited Mexico" means that

"There is a student in this class with the property that the student has visited Mexico."

We can introduce a variable x, so that our statement becomes

"There is a student x in this class having the property that x has visited Mexico."

We introduce $M(x)$, which is the statement "x has visited Mexico." If the domain for x consists of the students in this class, we can translate this first statement as $\exists x\, M(x)$.

However, if we are interested in people other than those in this class, we look at the statement a little differently. Our statement can be expressed as

"There is a person x having the properties that x is a student in this class and x has visited Mexico."

In this case, the domain for the variable x consists of all people. We introduce $S(x)$ to represent "x is a student in this class." Our solution becomes $\exists x(S(x) \wedge M(x))$ because the statement is that there is a person x who is a student in this class and who has visited Mexico. [*Caution!* Our statement cannot be expressed as $\exists x(S(x) \rightarrow M(x))$, which is true when there is someone not in the class because, in that case, for such a person x, $S(x) \rightarrow M(x)$ becomes either $\mathbf{F} \rightarrow \mathbf{T}$ or $\mathbf{F} \rightarrow \mathbf{F}$, both of which are true.]

Similarly, the second statement can be expressed as

"For every x in this class, x has the property that x has visited Mexico or x has visited Canada."

(Note that we are assuming the inclusive, rather than the exclusive, or here.) We let $C(x)$ be "x has visited Canada." Following our earlier reasoning, we see that if the domain for x consists of the students in this class, this second statement can be expressed as $\forall x(C(x) \vee M(x))$. However, if the domain for x consists of all people, our statement can be expressed as

"For every person x, if x is a student in this class, then x has visited Mexico or x has visited Canada."

In this case, the statement can be expressed as $\forall x(S(x) \rightarrow (C(x) \vee M(x)))$.

Instead of using $M(x)$ and $C(x)$ to represent that x has visited Mexico and x has visited Canada, respectively, we could use a two-place predicate $V(x, y)$ to represent "x has visited country y." In this case, $V(x, \text{Mexico})$ and $V(x, \text{Canada})$ would have the same meaning as $M(x)$ and $C(x)$ and could replace them in our answers. If we are working with many statements that involve people visiting different countries, we might prefer to use this two-variable approach. Otherwise, for simplicity, we would stick with the one-variable predicates $M(x)$ and $C(x)$. ◀

Using Quantifiers in System Specifications

In Section 1.2 we used propositions to represent system specifications. However, many system specifications involve predicates and quantifications. This is illustrated in Example 25.

EXAMPLE 25   Use predicates and quantifiers to express the system specifications "Every mail message larger than one megabyte will be compressed" and "If a user is active, at least one network link will be available."

Remember the rules of precedence for quantifiers and logical connectives!

Solution: Let $S(m, y)$ be "Mail message m is larger than y megabytes," where the variable x has the domain of all mail messages and the variable y is a positive real number, and let $C(m)$ denote "Mail message m will be compressed." Then the specification "Every mail message larger than one megabyte will be compressed" can be represented as $\forall m(S(m, 1) \rightarrow C(m))$.

Let $A(u)$ represent "User u is active," where the variable u has the domain of all users, let $S(n, x)$ denote "Network link n is in state x," where n has the domain of all network links and x has the domain of all possible states for a network link. Then the specification "If a user is active, at least one network link will be available" can be represented by $\exists u A(u) \rightarrow \exists n S(n, \text{available})$. ◀

Examples from Lewis Carroll

Lewis Carroll (really C. L. Dodgson writing under a pseudonym), the author of *Alice in Wonderland*, is also the author of several works on symbolic logic. His books contain many examples of reasoning using quantifiers. Examples 26 and 27 come from his book *Symbolic Logic;* other examples from that book are given in the exercises at the end of this section. These examples illustrate how quantifiers are used to express various types of statements.

EXAMPLE 26 Consider these statements. The first two are called *premises* and the third is called the *conclusion*. The entire set is called an *argument*.

"All lions are fierce."
"Some lions do not drink coffee."
"Some fierce creatures do not drink coffee."

(In Section 1.6 we will discuss the issue of determining whether the conclusion is a valid consequence of the premises. In this example, it is.) Let $P(x)$, $Q(x)$, and $R(x)$ be the statements "x is a lion," "x is fierce," and "x drinks coffee," respectively. Assuming that the domain consists of all creatures, express the statements in the argument using quantifiers and $P(x)$, $Q(x)$, and $R(x)$.

Solution: We can express these statements as:

$$\forall x (P(x) \rightarrow Q(x)).$$
$$\exists x (P(x) \land \neg R(x)).$$
$$\exists x (Q(x) \land \neg R(x)).$$

Notice that the second statement cannot be written as $\exists x (P(x) \rightarrow \neg R(x))$. The reason is that $P(x) \rightarrow \neg R(x)$ is true whenever x is not a lion, so that $\exists x (P(x) \rightarrow \neg R(x))$ is true as long as there is at least one creature that is not a lion, even if every lion drinks coffee. Similarly, the third statement cannot be written as

$$\exists x (Q(x) \rightarrow \neg R(x)). \qquad \blacktriangleleft$$

EXAMPLE 27 Consider these statements, of which the first three are premises and the fourth is a valid conclusion.

"All hummingbirds are richly colored."
"No large birds live on honey."
"Birds that do not live on honey are dull in color."
"Hummingbirds are small."

Let $P(x)$, $Q(x)$, $R(x)$, and $S(x)$ be the statements "x is a hummingbird," "x is large," "x lives on honey," and "x is richly colored," respectively. Assuming that the domain consists of all birds, express the statements in the argument using quantifiers and $P(x)$, $Q(x)$, $R(x)$, and $S(x)$.

Links

CHARLES LUTWIDGE DODGSON (1832–1898)   We know Charles Dodgson as Lewis Carroll—the pseudonym he used in his literary works. Dodgson, the son of a clergyman, was the third of 11 children, all of whom stuttered. He was uncomfortable in the company of adults and is said to have spoken without stuttering only to young girls, many of whom he entertained, corresponded with, and photographed (sometimes in poses that today would be considered inappropriate). Although attracted to young girls, he was extremely puritanical and religious. His friendship with the three young daughters of Dean Liddell led to his writing *Alice in Wonderland*, which brought him money and fame.

Dodgson graduated from Oxford in 1854 and obtained his master of arts degree in 1857. He was appointed lecturer in mathematics at Christ Church College, Oxford, in 1855. He was ordained in the Church of England in 1861 but never practiced his ministry. His writings published under this real name include articles and books on geometry, determinants, and the mathematics of tournaments and elections. (He also used the pseudonym Lewis Carroll for his many works on recreational logic.)

Solution: We can express the statements in the argument as

$$\forall x (P(x) \rightarrow S(x)).$$
$$\neg \exists x (Q(x) \wedge R(x)).$$
$$\forall x (\neg R(x) \rightarrow \neg S(x)).$$
$$\forall x (P(x) \rightarrow \neg Q(x)).$$

(Note we have assumed that "small" is the same as "not large" and that "dull in color" is the same as "not richly colored." To show that the fourth statement is a valid conclusion of the first three, we need to use rules of inference that will be discussed in Section 1.6.) ◀

Logic Programming

An important type of programming language is designed to reason using the rules of predicate logic. Prolog (from *Pro*gramming in *Log*ic), developed in the 1970s by computer scientists working in the area of artificial intelligence, is an example of such a language. Prolog programs include a set of declarations consisting of two types of statements, **Prolog facts** and **Prolog rules**. Prolog facts define predicates by specifying the elements that satisfy these predicates. Prolog rules are used to define new predicates using those already defined by Prolog facts. Example 28 illustrates these notions.

EXAMPLE 28 Consider a Prolog program given facts telling it the instructor of each class and in which classes students are enrolled. The program uses these facts to answer queries concerning the professors who teach particular students. Such a program could use the predicates *instructor*(p, c) and *enrolled*(s, c) to represent that professor p is the instructor of course c and that student s is enrolled in course c, respectively. For example, the Prolog facts in such a program might include:

```
instructor(chan,math273)
instructor(patel,ee222)
instructor(grossman,cs301)
enrolled(kevin,math273)
enrolled(juana,ee222)
enrolled(juana,cs301)
enrolled(kiko,math273)
enrolled(kiko,cs301)
```

(Lowercase letters have been used for entries because Prolog considers names beginning with an uppercase letter to be variables.)

A new predicate *teaches*(p, s), representing that professor p teaches student s, can be defined using the Prolog rule

```
teaches(P,S) :- instructor(P,C), enrolled(S,C)
```

which means that *teaches*(p, s) is true if there exists a class c such that professor p is the instructor of class c and student s is enrolled in class c. (Note that a comma is used to represent a conjunction of predicates in Prolog. Similarly, a semicolon is used to represent a disjunction of predicates.)

Prolog answers queries using the facts and rules it is given. For example, using the facts and rules listed, the query

```
?enrolled(kevin,math273)
```

produces the response

```
yes
```

because the fact *enrolled*(kevin, math273) was provided as input. The query

```
?enrolled(X,math273)
```

produces the response

```
kevin
kiko
```

To produce this response, Prolog determines all possible values of X for which *enrolled*(X, math273) has been included as a Prolog fact. Similarly, to find all the professors who are instructors in classes being taken by Juana, we use the query

```
?teaches(X,juana)
```

This query returns

```
patel
grossman
```

◀

Exercises

1. Let $P(x)$ denote the statement "$x \le 4$." What are these truth values?

  a) $P(0)$ **b)** $P(4)$ **c)** $P(6)$

2. Let $P(x)$ be the statement "the word x contains the letter a." What are these truth values?

  a) $P(\text{orange})$ **b)** $P(\text{lemon})$
  c) $P(\text{true})$ **d)** $P(\text{false})$

3. State the value of x after the statement **if** $P(x)$ **then** $x := 1$ is executed, where $P(x)$ is the statement "$x > 1$," if the value of x when this statement is reached is

  a) $x = 0$.   **b)** $x = 1$.
  c) $x = 2$.

4. Translate these statements into English, where $R(x)$ is "x is a rabbit" and $H(x)$ is "x hops" and the domain consists of all animals.

  a) $\forall x(R(x) \rightarrow H(x))$ **b)** $\forall x(R(x) \wedge H(x))$
  c) $\exists x(R(x) \rightarrow H(x))$ **d)** $\exists x(R(x) \wedge H(x))$

5. Let $P(x)$ be the statement "x spends more than five hours every weekday in class," where the domain for x consists of all students. Express each of these quantifications in English.

  a) $\exists x \, P(x)$ **b)** $\forall x \, P(x)$
  c) $\exists x \, \neg P(x)$ **d)** $\forall x \, \neg P(x)$

6. Let $C(x)$ be the statement "x has a cat," let $D(x)$ be the statement "x has a dog," and let $F(x)$ be the statement "x has a ferret." Express each of these statements in terms of $C(x)$, $D(x)$, $F(x)$, quantifiers, and logical connectives. Let the domain consist of all students in your class.

  a) A student in your class has a cat, a dog, and a ferret.
  b) All students in your class have a cat, a dog, or a ferret.
  c) Some student in your class has a cat and a ferret, but not a dog.
  d) No student in your class has a cat, a dog, and a ferret.
  e) For each of the three animals, cats, dogs, and ferrets, there is a student in your class who has this animal as a pet.

7. Let $P(x)$ be the statement "x can speak Russian" and let $Q(x)$ be the statement "x knows the computer language C++." Express each of these sentences in terms of $P(x)$, $Q(x)$, quantifiers, and logical connectives. The domain for quantifiers consists of all students at your school.

  a) There is a student at your school who can speak Russian and who knows C++.
  b) There is a student at your school who can speak Russian but who doesn't know C++.
  c) Every student at your school either can speak Russian or knows C++.
  d) No student at your school can speak Russian or knows C++.

8. Let $P(x)$ be the statement "$x = x^2$." If the domain consists of the integers, what are these truth values?

  a) $P(0)$ **b)** $P(1)$ **c)** $P(2)$
  d) $P(-1)$ **e)** $\exists x \, P(x)$ **f)** $\forall x \, P(x)$

9. Determine the truth value of each of these statements if the domain consists of all integers.

  a) $\forall n(n + 1 > n)$ **b)** $\exists n(2n = 3n)$
  c) $\exists n(n = -n)$ **d)** $\forall n(3n \le 4n)$

10. Determine the truth value of each of these statements if the domain for all variables consists of all integers.

a) $\forall n (n^2 \geq 0)$ **b)** $\exists n (n^2 = 2)$
c) $\forall n (n^2 \geq n)$ **d)** $\exists n (n^2 < 0)$

11. Suppose that the domain of the propositional function $P(x)$ consists of the integers 0, 1, 2, 3, and 4. Write out each of these propositions using disjunctions, conjunctions, and negations.

a) $\exists x P(x)$ **b)** $\forall x P(x)$ **c)** $\exists x \neg P(x)$
d) $\forall x \neg P(x)$ **e)** $\neg \exists x P(x)$ **f)** $\neg \forall x P(x)$

12. Suppose that the domain of the propositional function $P(x)$ consists of the integers 1, 2, 3, 4, and 5. Express these statements without using quantifiers, instead using only negations, disjunctions, and conjunctions.

a) $\exists x P(x)$ **b)** $\forall x P(x)$
c) $\neg \exists x P(x)$ **d)** $\neg \forall x P(x)$
e) $\forall x ((x \neq 3) \rightarrow P(x)) \vee \exists x \neg P(x)$

13. Translate in two ways each of these statements into logical expressions using predicates, quantifiers, and logical connectives. First, let the domain consist of the students in your class and second, let it consist of all people.

a) Someone in your class can speak Hindi.
b) Everyone in your class is friendly.
c) There is a person in your class who was not born in California.
d) A student in your class has been in a movie.
e) No student in your class has taken a course in logic programming.

14. Translate in two ways each of these statements into logical expressions using predicates, quantifiers, and logical connectives. First, let the domain consist of the students in your class and second, let it consist of all people.

a) Everyone in your class has a cellular phone.
b) Somebody in your class has seen a foreign movie.
c) There is a person in your class who cannot swim.
d) All students in your class can solve quadratic equations.
e) Some student in your class does not want to be rich.

15. Translate each of these statements into logical expressions using predicates, quantifiers, and logical connectives.

a) No one is perfect.
b) Not everyone is perfect.
c) All your friends are perfect.
d) At least one of your friends is perfect.
e) Everyone is your friend and is perfect.
f) Not everybody is your friend or someone is not perfect.

16. Translate each of these statements into logical expressions in three different ways by varying the domain and by using predicates with one and with two variables.

a) Someone in your school has visited Uzbekistan.
b) Everyone in your class has studied calculus and C++.
c) No one in your school owns both a bicycle and a motorcycle.
d) There is a person in your school who is not happy.

e) Everyone in your school was born in the twentieth century.

17. Express each of these statements using logical operators, predicates, and quantifiers.

a) Some propositions are tautologies.
b) The negation of a contradiction is a tautology.
c) The disjunction of two contingencies can be a tautology.
d) The conjunction of two tautologies is a tautology.

18. Translate each of these statements into logical expressions using predicates, quantifiers, and logical connectives.

a) Something is not in the correct place.
b) All tools are in the correct place and are in excellent condition.
c) Everything is in the correct place and in excellent condition.
d) Nothing is in the correct place and is in excellent condition.
e) One of your tools is not in the correct place, but it is in excellent condition.

19. Suppose that the domain of $Q(x, y, z)$ consists of triples x, y, z, where $x = 0, 1,$ or 2, $y = 0$ or 1, and $z = 0$ or 1. Write out these propositions using disjunctions and conjunctions.

a) $\forall y Q(0, y, 0)$ **b)** $\exists x Q(x, 1, 1)$
c) $\exists z \neg Q(0, 0, z)$ **d)** $\exists x \neg Q(x, 0, 1)$

20. Express each of these statements using quantifiers. Then form the negation of the statement so that no negation is to the left of a quantifier. Next, express the negation in simple English. (Do not simply use the phrase "It is not the case that.")

a) All dogs have fleas.
b) There is a horse that can add.
c) Every koala can climb.
d) No monkey can speak French.
e) There exists a pig that can swim and catch fish.

21. Express each of these statements using quantifiers. Then form the negation of the statement, so that no negation is to the left of a quantifier. Next, express the negation in simple English. (Do not simply use the phrase "It is not the case that.")

a) Some old dogs can learn new tricks.
b) No rabbit knows calculus.
c) Every bird can fly.
d) There is no dog that can talk.
e) There is no one in this class who knows French and Russian.

22. Express the negation of these propositions using quantifiers, and then express the negation in English.

a) Some drivers do not obey the speed limit.
b) All Swedish movies are serious.
c) No one can keep a secret.
d) There is someone in this class who does not have a good attitude.

23. Find a counterexample, if possible, to these universally quantified statements, where the domain for all variables consists of all integers.

a) $\forall x(x^2 \geq x)$

b) $\forall x(x > 0 \vee x < 0)$

c) $\forall x(x = 1)$

Exercises 24–25 deal with the translation between system specification and logical expressions involving quantifiers.

24. Express each of these system specifications using predicates, quantifiers, and logical connectives.

    a) When there is less than 30 megabytes free on the hard disk, a warning message is sent to all users.

    b) No directories in the file system can be opened and no files can be closed when system errors have been detected.

    c) The file system cannot be backed up if there is a user currently logged on.

    d) Video on demand can be delivered when there are at least 8 megabytes of memory available and the connection speed is at least 56 kilobits per second.

25. Express each of these statements using predicates and quantifiers.

    a) A passenger on an airline qualifies as an elite flyer if the passenger flies more than 25,000 miles in a year or takes more than 25 flights during that year.

    b) A man qualifies for the marathon if his best previous time is less than 3 hours and a woman qualifies for the marathon if her best previous time is less than 3.5 hours.

    c) A student must take at least 60 course hours, or at least 45 course hours and write a master's thesis, and receive a grade no lower than a B in all required courses, to receive a master's degree.

    d) There is a student who has taken more than 21 credit hours in a semester and received all A's.

26. Determine whether $\forall x(P(x) \rightarrow Q(x))$ and $\forall x P(x) \rightarrow \forall x Q(x)$ are logically equivalent. Justify your answer.

27. Show that $\exists x(P(x) \vee Q(x))$ and $\exists x P(x) \vee \exists x Q(x)$ are logically equivalent.

Exercises 28–29 establish rules for **null quantification** that we can use when a quantified variable does not appear in part of a statement.

28. Establish these logical equivalences, where x does not occur as a free variable in A. Assume that the domain is nonempty.

    a) $\forall x(A \rightarrow P(x)) \equiv A \rightarrow \forall x P(x)$

    b) $\exists x(A \rightarrow P(x)) \equiv A \rightarrow \exists x P(x)$

29. Establish these logical equivalences, where x does not occur as a free variable in A. Assume that the domain is nonempty.

    a) $(\forall x P(x)) \wedge A \equiv \forall x(P(x) \wedge A)$

    b) $(\exists x P(x)) \wedge A \equiv \exists x(P(x) \wedge A)$

30. Show that $\forall x P(x) \vee \forall x Q(x)$ and $\forall x(P(x) \vee Q(x))$ are not logically equivalent.

31. As mentioned in the text, the notation $\exists! x P(x)$ denotes

    "There exists a unique x such that $P(x)$ is true."

    If the domain consists of all integers, what are the truth values of these statements?

    a) $\exists! x(x > 1)$ **b)** $\exists! x(x^2 = 1)$

    c) $\exists! x(x + 3 = 2x)$ **d)** $\exists! x(x = x + 1)$

32. Write out $\exists! x P(x)$, where the domain consists of the integers 1, 2, and 3, in terms of negations, conjunctions, and disjunctions.

33. What are the truth values of these statements?

    a) $\exists! x P(x) \rightarrow \exists x P(x)$

    b) $\forall x P(x) \rightarrow \exists! x P(x)$

    c) $\exists! x \neg P(x) \rightarrow \neg \forall x P(x)$

34. Given the Prolog facts in Example 28, what would Prolog return given these queries?

    a) `?instructor(chan,math273)`

    b) `?instructor(patel,cs301)`

    c) `?enrolled(X,cs301)`

    d) `?enrolled(kiko,Y)`

    e) `?teaches(grossman,Y)`

35. Suppose that Prolog facts are used to define the predicates $mother(M, Y)$ and $father(F, X)$, which represent that M is the mother of Y and F is the father of X, respectively. Give a Prolog rule to define the predicate $sibling(X, Y)$, which represents that X and Y are siblings (that is, have the same mother and the same father).

Exercises 36–38 are based on questions found in the book *Symbolic Logic* by Lewis Carroll.

36. Let $P(x)$, $Q(x)$, and $R(x)$ be the statements "x is a clear explanation," "x is satisfactory," and "x is an excuse," respectively. Suppose that the domain for x consists of all English text. Express each of these statements using quantifiers, logical connectives, and $P(x)$, $Q(x)$, and $R(x)$.

    a) All clear explanations are satisfactory.

    b) Some excuses are unsatisfactory.

    c) Some excuses are not clear explanations.

    *d) Does (c) follow from (a) and (b)?

37. Let $P(x)$, $Q(x)$, and $R(x)$ be the statements "x is a professor," "x is ignorant," and "x is vain," respectively. Express each of these statements using quantifiers; logical connectives; and $P(x)$, $Q(x)$, and $R(x)$, where the domain consists of all people.

    a) No professors are ignorant.

    b) All ignorant people are vain.

    c) No professors are vain.

    d) Does (c) follow from (a) and (b)?

38. Let $P(x)$, $Q(x)$, $R(x)$, and $S(x)$ be the statements "x is a baby," "x is logical," "x is able to manage a crocodile," and "x is despised," respectively. Suppose that the domain consists of all people. Express each of these statements using quantifiers; logical connectives; and $P(x)$, $Q(x)$, $R(x)$, and $S(x)$.

    a) Babies are illogical.

    b) Nobody is despised who can manage a crocodile.

    c) Illogical persons are despised.

    d) Babies cannot manage crocodiles.

    *e) Does (d) follow from (a), (b), and (c)? If not, is there a correct conclusion?

1.5   Nested Quantifiers

Introduction

In Section 1.4 we defined the existential and universal quantifiers and showed how they can be used to represent mathematical statements. We also explained how they can be used to translate English sentences into logical expressions. However, in Section 1.4 we avoided **nested quantifiers**, where one quantifier is within the scope of another, such as

$$\forall x \exists y (x + y = 0).$$

Note that everything within the scope of a quantifier can be thought of as a propositional function. For example,

$$\forall x \exists y (x + y = 0)$$

is the same thing as $\forall x \, Q(x)$, where $Q(x)$ is $\exists y \, P(x, y)$, where $P(x, y)$ is $x + y = 0$.

Nested quantifiers commonly occur in mathematics and computer science. Although nested quantifiers can sometimes be difficult to understand, the rules we have already studied in Section 1.4 can help us use them. In this section we will gain experience working with nested quantifiers. We will see how to use nested quantifiers to express mathematical statements such as "The sum of two positive integers is always positive." We will show how nested quantifiers can be used to translate English sentences such as "Everyone has exactly one best friend" into logical statements. Moreover, we will gain experience working with the negations of statements involving nested quantifiers.

Understanding Statements Involving Nested Quantifiers

To understand statements involving nested quantifiers, we need to unravel what the quantifiers and predicates that appear mean. This is illustrated in Examples 1 and 2.

EXAMPLE 1   Assume that the domain for the variables x and y consists of all real numbers. The statement

$$\forall x \forall y (x + y = y + x)$$

Extra Examples

says that $x + y = y + x$ for all real numbers x and y. This is the commutative law for addition of real numbers. Likewise, the statement

$$\forall x \exists y (x + y = 0)$$

says that for every real number x there is a real number y such that $x + y = 0$. This states that every real number has an additive inverse. Similarly, the statement

$$\forall x \forall y \forall z (x + (y + z) = (x + y) + z)$$

is the associative law for addition of real numbers.   ◀

EXAMPLE 2   Translate into English the statement

$$\forall x \forall y ((x > 0) \wedge (y < 0) \rightarrow (xy < 0)),$$

where the domain for both variables consists of all real numbers.

Solution: This statement says that for every real number x and for every real number y, if $x > 0$ and $y < 0$, then $xy < 0$. That is, this statement says that for real numbers x and y, if x is positive and y is negative, then xy is negative. This can be stated more succinctly as "The product of a positive real number and a negative real number is always a negative real number."   ◀

THINKING OF QUANTIFICATION AS LOOPS   In working with quantifications of more than one variable, it is sometimes helpful to think in terms of nested loops. (Of course, if there are infinitely many elements in the domain of some variable, we cannot actually loop through all values. Nevertheless, this way of thinking is helpful in understanding nested quantifiers.) For example, to see whether $\forall x \forall y P(x, y)$ is true, we loop through the values for x, and for each x we loop through the values for y. If we find that $P(x, y)$ is true for all values for x and y, we have determined that $\forall x \forall y P(x, y)$ is true. If we ever hit a value x for which we hit a value y for which $P(x, y)$ is false, we have shown that $\forall x \forall y P(x, y)$ is false.

Similarly, to determine whether $\forall x \exists y P(x, y)$ is true, we loop through the values for x. For each x we loop through the values for y until we find a y for which $P(x, y)$ is true. If for every x we hit such a y, then $\forall x \exists y P(x, y)$ is true; if for some x we never hit such a y, then $\forall x \exists y P(x, y)$ is false.

To see whether $\exists x \forall y P(x, y)$ is true, we loop through the values for x until we find an x for which $P(x, y)$ is always true when we loop through all values for y. Once we find such an x, we know that $\exists x \forall y P(x, y)$ is true. If we never hit such an x, then we know that $\exists x \forall y P(x, y)$ is false.

Finally, to see whether $\exists x \exists y P(x, y)$ is true, we loop through the values for x, where for each x we loop through the values for y until we hit an x for which we hit a y for which $P(x, y)$ is true. The statement $\exists x \exists y P(x, y)$ is false only if we never hit an x for which we hit a y such that $P(x, y)$ is true.

The Order of Quantifiers

Many mathematical statements involve multiple quantifications of propositional functions involving more than one variable. It is important to note that the order of the quantifiers is important, unless all the quantifiers are universal quantifiers or all are existential quantifiers.

These remarks are illustrated by Examples 3–5.

EXAMPLE 3   Let $P(x, y)$ be the statement "$x + y = y + x$." What are the truth values of the quantifications $\forall x \forall y P(x, y)$ and $\forall y \forall x P(x, y)$ where the domain for all variables consists of all real numbers?

Solution: The quantification

$$\forall x \forall y P(x, y)$$

denotes the proposition

"For all real numbers x, for all real numbers y, $x + y = y + x$."

Because $P(x, y)$ is true for all real numbers x and y (it is the commutative law for addition, which is an axiom for the real numbers—see Appendix 1), the proposition $\forall x \forall y P(x, y)$ is true. Note that the statement $\forall y \forall x P(x, y)$ says "For all real numbers y, for all real numbers x, $x + y = y + x$." This has the same meaning as the statement "For all real numbers x, for all real numbers y, $x + y = y + x$." That is, $\forall x \forall y P(x, y)$ and $\forall y \forall x P(x, y)$ have the same meaning, and both are true. This illustrates the principle that the order of nested universal quantifiers in a statement without other quantifiers can be changed without changing the meaning of the quantified statement. ◀

EXAMPLE 4 Let $Q(x, y)$ denote "$x + y = 0$." What are the truth values of the quantifications $\exists y \forall x \, Q(x, y)$ and $\forall x \exists y \, Q(x, y)$, where the domain for all variables consists of all real numbers?

Solution: The quantification

$$\exists y \forall x \, Q(x, y)$$

denotes the proposition

"There is a real number y such that for every real number x, $Q(x, y)$."

No matter what value of y is chosen, there is only one value of x for which $x + y = 0$. Because there is no real number y such that $x + y = 0$ for all real numbers x, the statement $\exists y \forall x \, Q(x, y)$ is false.

The quantification

$$\forall x \exists y \, Q(x, y)$$

denotes the proposition

"For every real number x there is a real number y such that $Q(x, y)$."

Given a real number x, there is a real number y such that $x + y = 0$; namely, $y = -x$. Hence, the statement $\forall x \exists y \, Q(x, y)$ is true. ◀

Be careful with the order of existential and universal quantifiers!

Example 4 illustrates that the order in which quantifiers appear makes a difference. The statements $\exists y \forall x \, P(x, y)$ and $\forall x \exists y \, P(x, y)$ are not logically equivalent. The statement $\exists y \forall x \, P(x, y)$ is true if and only if there is a y that makes $P(x, y)$ true for every x. So, for this statement to be true, there must be a particular value of y for which $P(x, y)$ is true regardless of the choice of x. On the other hand, $\forall x \exists y \, P(x, y)$ is true if and only if for every value of x there is a value of y for which $P(x, y)$ is true. So, for this statement to be true, no matter which x you choose, there must be a value of y (possibly depending on the x you choose) for which $P(x, y)$ is true. In other words, in the second case, y can depend on x, whereas in the first case, y is a constant independent of x.

From these observations, it follows that if $\exists y \forall x \, P(x, y)$ is true, then $\forall x \exists y \, P(x, y)$ must also be true. However, if $\forall x \exists y \, P(x, y)$ is true, it is not necessary for $\exists y \forall x \, P(x, y)$ to be true. (See Supplementary Exercises 30 and 31.)

Table 1 summarizes the meanings of the different possible quantifications involving two variables.

Quantifications of more than two variables are also common, as Example 5 illustrates.

EXAMPLE 5 Let $Q(x, y, z)$ be the statement "$x + y = z$." What are the truth values of the statements $\forall x \forall y \exists z \, Q(x, y, z)$ and $\exists z \forall x \forall y \, Q(x, y, z)$, where the domain of all variables consists of all real numbers?

Solution: Suppose that x and y are assigned values. Then, there exists a real number z such that $x + y = z$. Consequently, the quantification

$$\forall x \forall y \exists z \, Q(x, y, z),$$

which is the statement

"For all real numbers x and for all real numbers y there is a real number z such that $x + y = z$,"

TABLE 1 **Quantifications of Two Variables.**		
Statement	*When True?*	*When False?*
$\forall x \forall y\, P(x, y)$ $\forall y \forall x\, P(x, y)$	$P(x, y)$ is true for every pair x, y.	There is a pair x, y for which $P(x, y)$ is false.
$\forall x \exists y\, P(x, y)$	For every x there is a y for which $P(x, y)$ is true.	There is an x such that $P(x, y)$ is false for every y.
$\exists x \forall y\, P(x, y)$	There is an x for which $P(x, y)$ is true for every y.	For every x there is a y for which $P(x, y)$ is false.
$\exists x \exists y\, P(x, y)$ $\exists y \exists x\, P(x, y)$	There is a pair x, y for which $P(x, y)$ is true.	$P(x, y)$ is false for every pair x, y.

is true. The order of the quantification here is important, because the quantification

$$\exists z \forall x \forall y\, Q(x, y, z),$$

which is the statement

> "There is a real number z such that for all real numbers x and for all real numbers y it is true that $x + y = z$,"

is false, because there is no value of z that satisfies the equation $x + y = z$ for all values of x and y. ◀

Translating Mathematical Statements into Statements Involving Nested Quantifiers

Mathematical statements expressed in English can be translated into logical expressions, as Examples 6–8 show.

EXAMPLE 6   Translate the statement "The sum of two positive integers is always positive" into a logical expression.

Solution: To translate this statement into a logical expression, we first rewrite it so that the implied quantifiers and a domain are shown: "For every two integers, if these integers are both positive, then the sum of these integers is positive." Next, we introduce the variables x and y to obtain "For all positive integers x and y, $x + y$ is positive." Consequently, we can express this statement as

$$\forall x \forall y((x > 0) \wedge (y > 0) \rightarrow (x + y > 0)),$$

where the domain for both variables consists of all integers. Note that we could also translate this using the positive integers as the domain. Then the statement "The sum of two positive integers is always positive" becomes "For every two positive integers, the sum of these integers is positive. We can express this as

$$\forall x \forall y(x + y > 0),$$

where the domain for both variables consists of all positive integers. ◀

EXAMPLE 7   Translate the statement "Every real number except zero has a multiplicative inverse." (A **multiplicative inverse** of a real number x is a real number y such that $xy = 1$.)

Solution: We first rewrite this as "For every real number x except zero, x has a multiplicative inverse." We can rewrite this as "For every real number x, if $x \neq 0$, then there exists a real number y such that $xy = 1$." This can be rewritten as

$$\forall x((x \neq 0) \rightarrow \exists y(xy = 1)).$$ ◀

One example that you may be familiar with is the concept of limit, which is important in calculus.

EXAMPLE 8   (***Requires calculus***)   Use quantifiers to express the definition of the limit of a real-valued function $f(x)$ of a real variable x at a point a in its domain.

Solution: Recall that the definition of the statement

$$\lim_{x \to a} f(x) = L$$

is: For every real number $\epsilon > 0$ there exists a real number $\delta > 0$ such that $|f(x) - L| < \epsilon$ whenever $0 < |x - a| < \delta$. This definition of a limit can be phrased in terms of quantifiers by

$$\forall \epsilon \exists \delta \forall x (0 < |x - a| < \delta \rightarrow |f(x) - L| < \epsilon),$$

where the domain for the variables δ and ϵ consists of all positive real numbers and for x consists of all real numbers.

This definition can also be expressed as

$$\forall \epsilon > 0 \ \exists \delta > 0 \ \forall x (0 < |x - a| < \delta \rightarrow |f(x) - L| < \epsilon)$$

when the domain for the variables ϵ and δ consists of all real numbers, rather than just the positive real numbers. [Here, restricted quantifiers have been used. Recall that $\forall x > 0 \ P(x)$ means that for all x with $x > 0$, $P(x)$ is true.]   ◀

Translating from Nested Quantifiers into English

Expressions with nested quantifiers expressing statements in English can be quite complicated. The first step in translating such an expression is to write out what the quantifiers and predicates in the expression mean. The next step is to express this meaning in a simpler sentence. This process is illustrated in Examples 9 and 10.

EXAMPLE 9   Translate the statement

$$\forall x (C(x) \vee \exists y (C(y) \wedge F(x, y)))$$

into English, where $C(x)$ is "x has a computer," $F(x, y)$ is "x and y are friends," and the domain for both x and y consists of all students in your school.

Solution: The statement says that for every student x in your school, x has a computer or there is a student y such that y has a computer and x and y are friends. In other words, every student in your school has a computer or has a friend who has a computer.   ◀

EXAMPLE 10   Translate the statement

$$\exists x \forall y \forall z ((F(x, y) \wedge F(x, z) \wedge (y \neq z)) \rightarrow \neg F(y,z))$$

into English, where $F(a,b)$ means a and b are friends and the domain for x, y, and z consists of all students in your school.

Solution: We first examine the expression $(F(x, y) \wedge F(x, z) \wedge (y \neq z)) \rightarrow \neg F(y, z)$. This expression says that if students x and y are friends, and students x and z are friends, and furthermore, if y and z are not the same student, then y and z are not friends. It follows that the original statement, which is triply quantified, says that there is a student x such that for all students y and all students z other than y, if x and y are friends and x and z are friends, then y and z are not friends. In other words, there is a student none of whose friends are also friends with each other. ◀

Translating English Sentences into Logical Expressions

In Section 1.4 we showed how quantifiers can be used to translate sentences into logical expressions. However, we avoided sentences whose translation into logical expressions required the use of nested quantifiers. We now address the translation of such sentences.

EXAMPLE 11   Express the statement "If a person is female and is a parent, then this person is someone's mother" as a logical expression involving predicates, quantifiers with a domain consisting of all people, and logical connectives.

Solution: The statement "If a person is female and is a parent, then this person is someone's mother" can be expressed as "For every person x, if person x is female and person x is a parent, then there exists a person y such that person x is the mother of person y." We introduce the propositional functions $F(x)$ to represent "x is female," $P(x)$ to represent "x is a parent," and $M(x, y)$ to represent "x is the mother of y." The original statement can be represented as

$$\forall x ((F(x) \wedge P(x)) \rightarrow \exists y M(x, y)).$$

Using the null quantification rule in part (b) of Exercise 29 in Section 1.4, we can move $\exists y$ to the left so that it appears just after $\forall x$, because y does not appear in $F(x) \wedge P(x)$. We obtain the logically equivalent expression

$$\forall x \exists y ((F(x) \wedge P(x)) \rightarrow M(x, y)).$$ ◀

EXAMPLE 12   Express the statement "Everyone has exactly one best friend" as a logical expression involving predicates, quantifiers with a domain consisting of all people, and logical connectives.

Solution: The statement "Everyone has exactly one best friend" can be expressed as "For every person x, person x has exactly one best friend." Introducing the universal quantifier, we see that this statement is the same as "$\forall x$(person x has exactly one best friend)," where the domain consists of all people.

To say that x has exactly one best friend means that there is a person y who is the best friend of x, and furthermore, that for every person z, if person z is not person y, then z is not the best friend of x. When we introduce the predicate $B(x, y)$ to be the statement "y is the best friend of x," the statement that x has exactly one best friend can be represented as

$$\exists y (B(x, y) \wedge \forall z ((z \neq y) \rightarrow \neg B(x, z))).$$

Consequently, our original statement can be expressed as

$$\forall x \exists y (B(x, y) \land \forall z ((z \neq y) \rightarrow \neg B(x, z))).$$

[Note that we can write this statement as $\forall x \exists! y\, B(x, y)$, where $\exists!$ is the "uniqueness quantifier" defined in Section 1.4.] ◀

EXAMPLE 13   Use quantifiers to express the statement "There is a woman who has taken a flight on every airline in the world."

Solution: Let $P(w, f)$ be "w has taken f" and $Q(f, a)$ be "f is a flight on a." We can express the statement as

$$\exists w \forall a \exists f (P(w, f) \land Q(f, a)),$$

where the domains of discourse for w, f, and a consist of all the women in the world, all airplane flights, and all airlines, respectively.
   The statement could also be expressed as

$$\exists w \forall a \exists f\, R(w, f, a),$$

where $R(w, f, a)$ is "w has taken f on a." Although this is more compact, it somewhat obscures the relationships among the variables. Consequently, the first solution is usually preferable. ◀

Negating Nested Quantifiers

Statements involving nested quantifiers can be negated by successively applying the rules for negating statements involving a single quantifier. This is illustrated in Examples 14–16.

EXAMPLE 14   Express the negation of the statement $\forall x \exists y (xy = 1)$ so that no negation precedes a quantifier.

Solution: By successively applying De Morgan's laws for quantifiers in Table 2 of Section 1.4, we can move the negation in $\neg \forall x \exists y (xy = 1)$ inside all the quantifiers. We find that $\neg \forall x \exists y (xy = 1)$ is equivalent to $\exists x \neg \exists y (xy = 1)$, which is equivalent to $\exists x \forall y \neg (xy = 1)$. Because $\neg (xy = 1)$ can be expressed more simply as $xy \neq 1$, we conclude that our negated statement can be expressed as $\exists x \forall y (xy \neq 1)$. ◀

EXAMPLE 15   Use quantifiers to express the statement that "There does not exist a woman who has taken a flight on every airline in the world."

Solution: This statement is the negation of the statement "There is a woman who has taken a flight on every airline in the world" from Example 13. By Example 13, our statement can be expressed as $\neg \exists w \forall a \exists f (P(w, f) \land Q(f, a))$, where $P(w, f)$ is "w has taken f" and $Q(f, a)$ is "f is a flight on a." By successively applying De Morgan's laws for quantifiers in Table 2 of Section 1.4 to move the negation inside successive quantifiers and by applying De Morgan's law for negating a conjunction in the last step, we find that our statement is equivalent to each of this sequence of statements:

$$\forall w \neg \forall a \exists f (P(w, f) \land Q(f, a)) \equiv \forall w \exists a \neg \exists f (P(w, f) \land Q(f, a))$$
$$\equiv \forall w \exists a \forall f \neg (P(w, f) \land Q(f, a))$$
$$\equiv \forall w \exists a \forall f (\neg P(w, f) \lor \neg Q(f, a)).$$

This last statement states "For every woman there is an airline such that for all flights, this woman has not taken that flight or that flight is not on this airline." ◀

EXAMPLE 16   (*Requires calculus*) Use quantifiers and predicates to express the fact that $\lim_{x \to a} f(x)$ does not exist where $f(x)$ is a real-valued function of a real variable x and a belongs to the domain of f.

Solution: To say that $\lim_{x \to a} f(x)$ does not exist means that for all real numbers L, $\lim_{x \to a} f(x) \neq L$. By using Example 8, the statement $\lim_{x \to a} f(x) \neq L$ can be expressed as

$$\neg \forall \epsilon > 0 \, \exists \delta > 0 \, \forall x (0 < |x - a| < \delta \to |f(x) - L| < \epsilon).$$

Successively applying the rules for negating quantified expressions, we construct this sequence of equivalent statements

$$\neg \forall \epsilon > 0 \, \exists \delta > 0 \, \forall x (0 < |x - a| < \delta \to |f(x) - L| < \epsilon)$$

$$\equiv \exists \epsilon > 0 \, \neg \exists \delta > 0 \, \forall x (0 < |x - a| < \delta \to |f(x) - L| < \epsilon)$$

$$\equiv \exists \epsilon > 0 \, \forall \delta > 0 \, \neg \forall x (0 < |x - a| < \delta \to |f(x) - L| < \epsilon)$$

$$\equiv \exists \epsilon > 0 \, \forall \delta > 0 \, \exists x \, \neg (0 < |x - a| < \delta \to |f(x) - L| < \epsilon)$$

$$\equiv \exists \epsilon > 0 \, \forall \delta > 0 \, \exists x (0 < |x - a| < \delta \wedge |f(x) - L| \geq \epsilon).$$

In the last step we used the equivalence $\neg(p \to q) \equiv p \wedge \neg q$, which follows from the fifth equivalence in Table 7 of Section 1.3.

Because the statement "$\lim_{x \to a} f(x)$ does not exist" means for all real numbers L, $\lim_{x \to a} f(x) \neq L$, this can be expressed as

$$\forall L \exists \epsilon > 0 \, \forall \delta > 0 \, \exists x (0 < |x - a| < \delta \wedge |f(x) - L| \geq \epsilon).$$

This last statement says that for every real number L there is a real number $\epsilon > 0$ such that for every real number $\delta > 0$, there exists a real number x such that $0 < |x - a| < \delta$ and $|f(x) - L| \geq \epsilon$.   ◀

Exercises

1. Translate these statements into English, where the domain for each variable consists of all real numbers.

   a) $\forall x \exists y (x < y)$
   b) $\forall x \forall y (((x \geq 0) \wedge (y \geq 0)) \to (xy \geq 0))$
   c) $\forall x \forall y \exists z (xy = z)$

2. Let $Q(x, y)$ be the statement "x has sent an e-mail message to y," where the domain for both x and y consists of all students in your class. Express each of these quantifications in English.

   a) $\exists x \exists y \, Q(x, y)$ **b)** $\exists x \forall y \, Q(x, y)$
   c) $\forall x \exists y \, Q(x, y)$ **d)** $\exists y \forall x \, Q(x, y)$
   e) $\forall y \exists x \, Q(x, y)$ **f)** $\forall x \forall y \, Q(x, y)$

3. Let $W(x, y)$ mean that student x has visited website y, where the domain for x consists of all students in your school and the domain for y consists of all websites. Express each of these statements by a simple English sentence.

   a) $W(\text{Sarah Smith}, \text{www.att.com})$
   b) $\exists x \, W(x, \text{www.imdb.org})$
   c) $\exists y \, W(\text{José Orez}, y)$

   d) $\exists y (W(\text{Ashok Puri}, y) \wedge W(\text{Cindy Yoon}, y))$
   e) $\exists y \forall z (y \neq (\text{David Belcher}) \wedge (W(\text{David Belcher}, z) \to W(y, z)))$
   f) $\exists x \exists y \forall z ((x \neq y) \wedge (W(x, z) \leftrightarrow W(y, z)))$

4. Let $T(x, y)$ mean that student x likes cuisine y, where the domain for x consists of all students at your school and the domain for y consists of all cuisines. Express each of these statements by a simple English sentence.

   a) $\neg T(\text{Abdallah Hussein}, \text{Japanese})$
   b) $\exists x \, T(x, \text{Korean}) \wedge \forall x \, T(x, \text{Mexican})$
   c) $\exists y (T(\text{Monique Arsenault}, y) \vee T(\text{Jay Johnson}, y))$
   d) $\forall x \forall z \exists y ((x \neq z) \to \neg(T(x, y) \wedge T(z, y)))$
   e) $\exists x \exists z \forall y (T(x, y) \leftrightarrow T(z, y))$
   f) $\forall x \forall z \exists y (T(x, y) \leftrightarrow T(z, y))$

5. Let $L(x, y)$ be the statement "x loves y," where the domain for both x and y consists of all people in the world. Use quantifiers to express each of these statements.

   a) Everybody loves Jerry.
   b) Everybody loves somebody.
   c) There is somebody whom everybody loves.
   d) Nobody loves everybody.

e) There is somebody whom Lydia does not love.

f) There is somebody whom no one loves.

g) There is exactly one person whom everybody loves.

h) There are exactly two people whom Lynn loves.

i) Everyone loves himself or herself.

j) There is someone who loves no one besides himself or herself.

6. Let $Q(x, y)$ be the statement "student x has been a contestant on quiz show y." Express each of these sentences in terms of $Q(x, y)$, quantifiers, and logical connectives, where the domain for x consists of all students at your school and for y consists of all quiz shows on television.

a) There is a student at your school who has been a contestant on a television quiz show.

b) No student at your school has ever been a contestant on a television quiz show.

c) There is a student at your school who has been a contestant on *Jeopardy* and on *Wheel of Fortune*.

d) Every television quiz show has had a student from your school as a contestant.

e) At least two students from your school have been contestants on *Jeopardy*.

7. Use quantifiers and predicates with more than one variable to express these statements.

a) Every computer science student needs a course in discrete mathematics.

b) There is a student in this class who owns a personal computer.

c) Every student in this class has taken at least one computer science course.

d) There is a student in this class who has taken at least one course in computer science.

e) Every student in this class has been in every building on campus.

f) There is a student in this class who has been in every room of at least one building on campus.

g) Every student in this class has been in at least one room of every building on campus.

8. Let $I(x)$ be the statement "x has an Internet connection" and $C(x, y)$ be the statement "x and y have chatted over the Internet," where the domain for the variables x and y consists of all students in your class. Use quantifiers to express each of these statements.

a) Jerry does not have an Internet connection.

b) Rachel has not chatted over the Internet with Chelsea.

c) Jan and Sharon have never chatted over the Internet.

d) No one in the class has chatted with Bob.

e) Sanjay has chatted with everyone except Joseph.

f) Someone in your class does not have an Internet connection.

g) Not everyone in your class has an Internet connection.

h) Exactly one student in your class has an Internet connection.

i) Everyone except one student in your class has an Internet connection.

j) Everyone in your class with an Internet connection has chatted over the Internet with at least one other student in your class.

k) Someone in your class has an Internet connection but has not chatted with anyone else in your class.

l) There are two students in your class who have not chatted with each other over the Internet.

m) There is a student in your class who has chatted with everyone in your class over the Internet.

n) There are at least two students in your class who have not chatted with the same person in your class.

o) There are two students in the class who between them have chatted with everyone else in the class.

9. Express each of these system specifications using predicates, quantifiers, and logical connectives, if necessary.

a) Every user has access to exactly one mailbox.

b) There is a process that continues to run during all error conditions only if the kernel is working correctly.

c) All users on the campus network can access all websites whose url has a .edu extension.

*d) There are exactly two systems that monitor every remote server.

10. Express each of these statements using mathematical and logical operators, predicates, and quantifiers, where the domain consists of all integers.

a) The sum of two negative integers is negative.

b) The difference of two positive integers is not necessarily positive.

c) The sum of the squares of two integers is greater than or equal to the square of their sum.

d) The absolute value of the product of two integers is the product of their absolute values.

11. Express each of these mathematical statements using predicates, quantifiers, logical connectives, and mathematical operators.

a) The product of two negative real numbers is positive.

b) The difference of a real number and itself is zero.

c) Every positive real number has exactly two square roots.

d) A negative real number does not have a square root that is a real number.

12. Translate each of these nested quantifications into an English statement that expresses a mathematical fact. The domain in each case consists of all real numbers.

a) $\exists x \forall y (xy = y)$

b) $\forall x \forall y (((x < 0) \land (y < 0)) \rightarrow (xy > 0))$

c) $\exists x \exists y ((x^2 > y) \land (x < y))$

d) $\forall x \forall y \exists z (x + y = z)$

13. Determine the truth value of each of these statements if the domain for all variables consists of all integers.

a) $\forall n \exists m (n^2 < m)$ **b)** $\exists n \forall m (n < m^2)$

c) $\forall n \exists m (n + m = 0)$ **d)** $\exists n \forall m (nm = m)$

e) $\exists n \exists m (n^2 + m^2 = 5)$ **f)** $\exists n \exists m (n^2 + m^2 = 6)$

g) $\exists n \exists m (n + m = 4 \land n - m = 1)$

h) $\exists n \exists m (n + m = 4 \land n - m = 2)$

i) $\forall n \forall m \exists p (p = (m + n)/2)$

14. Suppose the domain of the propositional function $P(x, y)$ consists of pairs x and y, where x is 1, 2, or 3 and y is 1, 2, or 3. Write out these propositions using disjunctions and conjunctions.

a) $\forall x \forall y P(x, y)$ **b)** $\exists x \exists y P(x, y)$

c) $\exists x \forall y P(x, y)$ **d)** $\forall y \exists x P(x, y)$

15. Express the negations of each of these statements so that all negation symbols immediately precede predicates.

a) $\forall x \exists y \forall z T(x, y, z)$

b) $\forall x \exists y P(x, y) \vee \forall x \exists y Q(x, y)$

c) $\forall x \exists y (P(x, y) \wedge \exists z R(x, y, z))$

d) $\forall x \exists y (P(x, y) \rightarrow Q(x, y))$

16. Rewrite each of these statements so that negations appear only within predicates (that is, so that no negation is outside a quantifier or an expression involving logical connectives).

a) $\neg \exists y \exists x P(x, y)$ **b)** $\neg \forall x \exists y P(x, y)$

c) $\neg \exists y (Q(y) \wedge \forall x \neg R(x, y))$

d) $\neg \exists y (\exists x R(x, y) \vee \forall x S(x, y))$

e) $\neg \exists y (\forall x \exists z T(x, y, z) \vee \exists x \forall z U(x, y, z))$

17. Rewrite each of these statements so that negations appear only within predicates (that is, so that no negation is outside a quantifier or an expression involving logical connectives).

a) $\neg \forall x \forall y P(x, y)$ **b)** $\neg \forall y \exists x P(x, y)$

c) $\neg \forall y \forall x (P(x, y) \vee Q(x, y))$

d) $\neg (\exists x \exists y \neg P(x, y) \wedge \forall x \forall y Q(x, y))$

e) $\neg \forall x (\exists y \forall z P(x, y, z) \wedge \exists z \forall y P(x, y, z))$

18. Find a common domain for the variables x, y, and z for which the statement $\forall x \forall y ((x \neq y) \rightarrow \forall z ((z = x) \vee (z = y)))$ is true and another domain for which it is false.

19. Express each of these statements using quantifiers. Then form the negation of the statement so that no negation is to the left of a quantifier. Next, express the negation in simple English. (Do not simply use the phrase "It is not the case that.")

a) Every student in this class has taken exactly two mathematics classes at this school.

b) Someone has visited every country in the world except Libya.

c) No one has climbed every mountain in the Himalayas.

d) Every movie actor has either been in a movie with Kevin Bacon or has been in a movie with someone who has been in a movie with Kevin Bacon.

20. Express the negations of these propositions using quantifiers, and in English.

a) Every student in this class likes mathematics.

b) There is a student in this class who has never seen a computer.

c) There is a student in this class who has taken every mathematics course offered at this school.

d) There is a student in this class who has been in at least one room of every building on campus.

21. Find a counterexample, if possible, to these universally quantified statements, where the domain for all variables consists of all integers.

a) $\forall x \forall y (x^2 = y^2 \rightarrow x = y)$

b) $\forall x \exists y (y^2 = x)$

c) $\forall x \forall y (xy \geq x)$

22. Use quantifiers to express the associative law for multiplication of real numbers.

23. Use quantifiers and logical connectives to express the fact that every linear polynomial (that is, polynomial of degree 1) with real coefficients and where the coefficient of x is nonzero, has exactly one real root.

24. Determine the truth value of the statement $\exists x \forall y (x \leq y^2)$ if the domain for the variables consists of

a) the positive real numbers.

b) the integers.

c) the nonzero real numbers.

25. Determine the truth value of the statement $\forall x \exists y (xy = 1)$ if the domain for the variables consists of

a) the nonzero real numbers.

b) the nonzero integers.

c) the positive real numbers.

26. Show that the two statements $\neg \exists x \forall y P(x, y)$ and $\forall x \exists y \neg P(x, y)$, where both quantifiers over the first variable in $P(x, y)$ have the same domain, and both quantifiers over the second variable in $P(x, y)$ have the same domain, are logically equivalent.

*27. Express the quantification $\exists! x P(x)$, introduced in Section 1.4, using universal quantifications, existential quantifications, and logical operators.

1.6   Rules of Inference

Introduction

Later in this chapter we will study proofs. Proofs in mathematics are valid arguments that establish the truth of mathematical statements. By an **argument**, we mean a sequence of statements that end with a conclusion. By **valid**, we mean that the conclusion, or final statement of the argument, must follow from the truth of the preceding statements, or **premises**, of the argument. That is, an argument is valid if and only if it is impossible for all the premises to be true and the conclusion to be false. To deduce new statements from statements we already have, we use rules of inference which are templates for constructing valid arguments. Rules of inference are our basic tools for establishing the truth of statements.

Before we study mathematical proofs, we will look at arguments that involve only compound propositions. We will define what it means for an argument involving compound propositions to be valid. Then we will introduce a collection of rules of inference in propositional logic. These rules of inference are among the most important ingredients in producing valid arguments. After we illustrate how rules of inference are used to produce valid arguments, we will describe some common forms of incorrect reasoning, called **fallacies**, which lead to invalid arguments.

After studying rules of inference in propositional logic, we will introduce rules of inference for quantified statements. We will describe how these rules of inference can be used to produce valid arguments. These rules of inference for statements involving existential and universal quantifiers play an important role in proofs in computer science and mathematics, although they are often used without being explicitly mentioned.

Finally, we will show how rules of inference for propositions and for quantified statements can be combined. These combinations of rule of inference are often used together in complicated arguments.

Valid Arguments in Propositional Logic

Consider the following argument involving propositions (which, by definition, is a sequence of propositions):

"If you have a current password, then you can log onto the network."

"You have a current password."

Therefore,

"You can log onto the network."

We would like to determine whether this is a valid argument. That is, we would like to determine whether the conclusion "You can log onto the network" must be true when the premises "If you have a current password, then you can log onto the network" and "You have a current password" are both true.

Before we discuss the validity of this particular argument, we will look at its form. Use p to represent "You have a current password" and q to represent "You can log onto the network." Then, the argument has the form

$$
\begin{array}{l}
p \rightarrow q \\
\underline{p} \\
\therefore q
\end{array}
$$

where $\therefore$ is the symbol that denotes "therefore."

We know that when p and q are propositional variables, the statement $((p \rightarrow q) \wedge p) \rightarrow q$ is a tautology (see Exercise 6(c) in Section 1.3). In particular, when both $p \rightarrow q$ and p are true, we know that q must also be true. We say this form of argument is **valid** because whenever all its premises (all statements in the argument other than the final one, the conclusion) are true, the conclusion must also be true. Now suppose that both "If you have a current password, then you can log onto the network" and "You have a current password" are true statements. When we replace p by "You have a current password" and q by "You can log onto the network," it necessarily follows that the conclusion "You can log onto the network" is true. This argument is **valid** because its form is valid. Note that whenever we replace p and q by propositions where $p \rightarrow q$ and p are both true, then q must also be true.

What happens when we replace p and q in this argument form by propositions where not both p and $p \rightarrow q$ are true? For example, suppose that p represents "You have access to the network" and q represents "You can change your grade" and that p is true, but $p \rightarrow q$ is false. The argument we obtain by substituting these values of p and q into the argument form is

> "If you have access to the network, then you can change your grade."
> "You have access to the network."
> _____
> $\therefore$ "You can change your grade."

The argument we obtained is a valid argument, but because one of the premises, namely the first premise, is false, we cannot conclude that the conclusion is true. (Most likely, this conclusion is false.)

In our discussion, to analyze an argument, we replaced propositions by propositional variables. This changed an argument to an **argument form**. We saw that the validity of an argument follows from the validity of the form of the argument. We summarize the terminology used to discuss the validity of arguments with our definition of the key notions.

DEFINITION 1   An *argument* in propositional logic is a sequence of propositions. All but the final proposition in the argument are called *premises* and the final proposition is called the *conclusion*. An argument is *valid* if the truth of all its premises implies that the conclusion is true.

An *argument form* in propositional logic is a sequence of compound propositions involving propositional variables. An argument form is *valid* no matter which particular propositions are substituted for the propositional variables in its premises, the conclusion is true if the premises are all true.

From the definition of a valid argument form we see that the argument form with premises $p_1, p_2, \ldots, p_n$ and conclusion q is valid, when $(p_1 \wedge p_2 \wedge \cdots \wedge p_n) \rightarrow q$ is a tautology.

The key to showing that an argument in propositional logic is valid is to show that its argument form is valid. Consequently, we would like techniques to show that argument forms are valid. We will now develop methods for accomplishing this task.

Rules of Inference for Propositional Logic

We can always use a truth table to show that an argument form is valid. We do this by showing that whenever the premises are true, the conclusion must also be true. However, this can be a tedious approach. For example, when an argument form involves 10 different propositional variables, to use a truth table to show this argument form is valid requires $2^{10} = 1024$ different rows. Fortunately, we do not have to resort to truth tables. Instead, we can first establish the validity of some relatively simple argument forms, called **rules of inference**. These rules of inference can be used as building blocks to construct more complicated valid argument forms. We will now introduce the most important rules of inference in propositional logic.

The tautology $(p \wedge (p \rightarrow q)) \rightarrow q$ is the basis of the rule of inference called **modus ponens**, or the **law of detachment**. (Modus ponens is Latin for *mode that affirms*.) This tautology leads to the following valid argument form, which we have already seen in our initial discussion about arguments (where, as before, the symbol $\therefore$ denotes "therefore"):

> p
> $p \rightarrow q$
> _____
> $\therefore q$

Using this notation, the hypotheses are written in a column, followed by a horizontal bar, followed by a line that begins with the therefore symbol and ends with the conclusion. In particular, modus ponens tells us that if a conditional statement and the hypothesis of this conditional statement are both true, then the conclusion must also be true. Example 1 illustrates the use of modus ponens.

EXAMPLE 1 Suppose that the conditional statement "If it snows today, then we will go skiing" and its hypothesis, "It is snowing today," are true. Then, by modus ponens, it follows that the conclusion of the conditional statement, "We will go skiing," is true. ◀

As we mentioned earlier, a valid argument can lead to an incorrect conclusion if one or more of its premises is false. We illustrate this again in Example 2.

EXAMPLE 2 Determine whether the argument given here is valid and determine whether its conclusion must be true because of the validity of the argument.

"If $\sqrt{2} > \frac{3}{2}$, then $\left(\sqrt{2}\right)^2 > \left(\frac{3}{2}\right)^2$. We know that $\sqrt{2} > \frac{3}{2}$. Consequently, $\left(\sqrt{2}\right)^2 = 2 > \left(\frac{3}{2}\right)^2 = \frac{9}{4}$."

Solution: Let p be the proposition "$\sqrt{2} > \frac{3}{2}$" and q the proposition "$2 > (\frac{3}{2})^2$." The premises of the argument are $p \rightarrow q$ and p, and q is its conclusion. This argument is valid because it is constructed by using modus ponens, a valid argument form. However, one of its premises, $\sqrt{2} > \frac{3}{2}$, is false. Consequently, we cannot conclude that the conclusion is true. Furthermore, note that the conclusion of this argument is false, because $2 < \frac{9}{4}$. ◀

There are many useful rules of inference for propositional logic. Perhaps the most widely used of these are listed in Table 1. Exercises 5, 6, 8, and 14 in Section 1.3 ask for the verifications that these rules of inference are valid argument forms. We now give examples of arguments that use these rules of inference. In each argument, we first use propositional variables to express the propositions in the argument. We then show that the resulting argument form is a rule of inference from Table 1.

EXAMPLE 3 State which rule of inference is the basis of the following argument: "It is below freezing now. Therefore, it is either below freezing or raining now."

Solution: Let p be the proposition "It is below freezing now" and q the proposition "It is raining now." Then this argument is of the form

$$\frac{p}{\therefore p \vee q}$$

This is an argument that uses the addition rule. ◀

EXAMPLE 4 State which rule of inference is the basis of the following argument: "It is below freezing and raining now. Therefore, it is below freezing now."

Solution: Let p be the proposition "It is below freezing now," and let q be the proposition "It is raining now." This argument is of the form

$$\frac{p \wedge q}{\therefore p}$$

This argument uses the simplification rule. ◀

EXAMPLE 5 State which rule of inference is used in the argument:

TABLE 1 Rules of Inference.		
Rule of Inference	***Tautology***	***Name***
p $\underline{p \rightarrow q}$ $\therefore q$	$(p \wedge (p \rightarrow q)) \rightarrow q$	Modus ponens
$\neg q$ $\underline{p \rightarrow q}$ $\therefore \neg p$	$(\neg q \wedge (p \rightarrow q)) \rightarrow \neg p$	Modus tollens
$p \rightarrow q$ $q \rightarrow r$ $\therefore \overline{p \rightarrow r}$	$((p \rightarrow q) \wedge (q \rightarrow r)) \rightarrow (p \rightarrow r)$	Hypothetical syllogism
$p \vee q$ $\underline{\neg p}$ $\therefore q$	$((p \vee q) \wedge \neg p) \rightarrow q$	Disjunctive syllogism
$\underline{p}$ $\therefore p \vee q$	$p \rightarrow (p \vee q)$	Addition
$\underline{p \wedge q}$ $\therefore p$	$(p \wedge q) \rightarrow p$	Simplification
p $\underline{q}$ $\therefore p \wedge q$	$((p) \wedge (q)) \rightarrow (p \wedge q)$	Conjunction
$p \vee q$ $\neg p \vee r$ $\therefore \overline{q \vee r}$	$((p \vee q) \wedge (\neg p \vee r)) \rightarrow (q \vee r)$	Resolution

If it rains today, then we will not have a barbecue today. If we do not have a barbecue today, then we will have a barbecue tomorrow. Therefore, if it rains today, then we will have a barbecue tomorrow.

Solution: Let p be the proposition "It is raining today," let q be the proposition "We will not have a barbecue today," and let r be the proposition "We will have a barbecue tomorrow." Then this argument is of the form

$$
\begin{array}{c}
p \rightarrow q \\
\underline{q \rightarrow r} \\
\therefore \ p \rightarrow r
\end{array}
$$

Hence, this argument is a hypothetical syllogism.   ◄

Using Rules of Inference to Build Arguments

When there are many premises, several rules of inference are often needed to show that an argument is valid. This is illustrated by Examples 6 and 7, where the steps of arguments are displayed on separate lines, with the reason for each step explicitly stated. These examples also show how arguments in English can be analyzed using rules of inference.

EXAMPLE 6 Show that the premises "It is not sunny this afternoon and it is colder than yesterday," "We will go swimming only if it is sunny," "If we do not go swimming, then we will take a canoe trip," and "If we take a canoe trip, then we will be home by sunset" lead to the conclusion "We will be home by sunset."

Solution: Let p be the proposition "It is sunny this afternoon," q the proposition "It is colder than yesterday," r the proposition "We will go swimming," s the proposition "We will take a canoe trip," and t the proposition "We will be home by sunset." Then the premises become $\neg p \wedge q, r \rightarrow p, \neg r \rightarrow s$, and $s \rightarrow t$. The conclusion is simply t. We need to give a valid argument with premises $\neg p \wedge q, r \rightarrow p, \neg r \rightarrow s$, and $s \rightarrow t$ and conclusion t.

We construct an argument to show that our premises lead to the desired conclusion as follows.

Step	Reason
1. $\neg p \wedge q$	Premise
2. $\neg p$	Simplification using (1)
3. $r \rightarrow p$	Premise
4. $\neg r$	Modus tollens using (2) and (3)
5. $\neg r \rightarrow s$	Premise
6. s	Modus ponens using (4) and (5)
7. $s \rightarrow t$	Premise
8. t	Modus ponens using (6) and (7)

Note that we could have used a truth table to show that whenever each of the four hypotheses is true, the conclusion is also true. However, because we are working with five propositional variables, p, q, r, s, and t, such a truth table would have 32 rows. ◀

EXAMPLE 7 Show that the premises "If you send me an e-mail message, then I will finish writing the program," "If you do not send me an e-mail message, then I will go to sleep early," and "If I go to sleep early, then I will wake up feeling refreshed" lead to the conclusion "If I do not finish writing the program, then I will wake up feeling refreshed."

Solution: Let p be the proposition "You send me an e-mail message," q the proposition "I will finish writing the program," r the proposition "I will go to sleep early," and s the proposition "I will wake up feeling refreshed." Then the premises are $p \rightarrow q, \neg p \rightarrow r$, and $r \rightarrow s$. The desired conclusion is $\neg q \rightarrow s$. We need to give a valid argument with premises $p \rightarrow q, \neg p \rightarrow r$, and $r \rightarrow s$ and conclusion $\neg q \rightarrow s$.

This argument form shows that the premises lead to the desired conclusion.

Step	Reason
1. $p \rightarrow q$	Premise
2. $\neg q \rightarrow \neg p$	Contrapositive of (1)
3. $\neg p \rightarrow r$	Premise
4. $\neg q \rightarrow r$	Hypothetical syllogism using (2) and (3)
5. $r \rightarrow s$	Premise
6. $\neg q \rightarrow s$	Hypothetical syllogism using (4) and (5)

◀

Resolution Principle

As another way of proving that an argument is correct, we use the resolution principle. This idea is used in the logic programming language Prolog.

A variable or negation of a variable is called a literal.

A disjunction of literals is called a sum and a conjunction of literals is called a product. A clause is a disjunction of literals i.e., it is a sum.

For any two clauses C_1 and C_2, if there is a literal L_1 in C_1 that is complementary to a literal L_2 in C_2, then delete L_1 and L_2 from C_1 and C_2 respectively and construct the disjunction of the remaining clauses. The constructed clause is a resolvent of C_1 and C_2.

EXAMPLE 8   Suppose you have

$$C_1 = P \vee Q \vee R$$

$$C_2 = \neg P \vee \neg S \vee T$$

Then P and $\neg P$ are complementary to each other. The resolvent of C_1 and C_2 is obtained by deleting P and $\neg P$ from C_1 and C_2 respectively and finding the disjunction of the remaining literals. In this case it is $Q \vee R \vee \neg S \vee T$.   ◀

Theorem 1 Given two clauses C_1 and C_2, a resolvent C of C_1 and C_2 is a logical consequence of C_1 and C_2.

For example, consider the rules modus ponens and modus tollens.

Modus ponens rule is $P \wedge (P \rightarrow Q) \rightarrow Q$

Putting in clause form this amounts to

$$C_1 : P$$

$$C_2 : \neg P \vee Q$$

The resolvent of C_1 and C_2 is Q which is the logical consequence of C_1 and C_2.

Similarly modus tollens rule is $(P \rightarrow Q) \wedge \neg Q \rightarrow \neg P$

Putting in clause form

$$C_1 : \neg P \vee Q$$

$$C_2 : \neg Q$$

The resolvent of C_1 and C_2 is $\neg P$ which is the logical consequence C_1 and C_2.

Proof Consider two clauses C_1 and C_2.

C_1 contains L and C_2 contain $\neg L$

Writing C_1 as $L \vee C_1'$ and C_2 as $\neg L \vee C_2'$ resolvent of C_1 and C_2 is $C = C_1' \vee C_2'$

Now we want to show C is the logical consequence of C_1 and C_2. In essence, we want to show that if C_1 and C_2 are true, C is true.

Assume C_1 and C_2 are true.

Now there are two possibilities.

Either L is true or L is false.

If L is true, $\neg L$ is false and in order that C_2 is true, C_2' is true.

Hence $C = C_1' \vee C_2'$ is true.

If L is false, in order that C_1 is true, C_1' is true.

Hence $C = C_1' \vee C_2'$ is true.

So whenever C_1 and C_2 are true C is true.

Hence C is the logical consequence of C_1 and C_2.

The resolution principle: Given a set S of clauses, a (resolution) deduction of C from S is a finite sequence $C_1, C_2, , C_k$ of clauses such that each C_i either is a clause in S or a resolvent of clauses preceding C and $C_k = C$. A deduction of $\square$ (empty clause) is called a refutation or a proof of S.

If you have an argument where $P_1, P_2, \ldots, P_r$ are the premises and C is the conclusion, to get a proof using resolution principle, put $P_1, \ldots, P_r$ in clause form and add to it $\neg C$ in clause form. From this sequence, if $\square$ can be derived, the argument is valid.

For example consider modus ponens P

$$P \to Q$$
$$\therefore \quad \frac{Q}{Q}$$

In the clause form $C_1 = P$

$$C_2 = \neg P \vee Q$$

Take as C_3, the negation of the conclusion, i.e., $\neg Q$, $C_3 = \neg Q$

From C_1 and C_2 by resolution you can get Q and from this and C_3 by resolution $\square$ is arrived at.

$$C_1 \quad P$$
$$C_2 \quad \neg P \vee Q$$
$$C_3 \quad \neg Q$$

from C_1 and C_2 by resolution C_4 Q

from C_3 and C_4 by resolution C_5 $\square$

EXAMPLE 9   We show that the following argument is correct.

If today is Tuesday, I have a test in Mathematics or Economics. If my Economics Professor is sick, I will not have a test in Economics. Today is Tuesday and my Economics Professor is sick. Therefore I have a test in Mathematics.

Converting to logical notation.

Let T denote 'Today is Tuesday'

M denote 'I have a test in Mathematics'

E denote 'I have a test in Economics'

S denote 'My Economics Professor is sick'

So the premises are

$$T \to (M \vee E)$$
$$S \to \neg E$$
$$\therefore \quad \frac{T \wedge S}{M}$$

Putting is clause form

$C_1 : \neg T \lor M \lor E$

$C_2 : \neg S \lor \neg E$

$C_3 : T$

$C_4 : S$

$C_5 : \neg M$ (negation of conclusion)

Now we have to check whether it is possible to derive $\square$ from C_1, C_2, C_3, C_4, C_5.

From C_1 and C_2 $C_6 : \neg T \lor M \lor \neg S$

from C_6 and C_3 $C_7 : M \lor \neg S$

from C_7 and C_4 $C_8 : M$

from C_8 and C_5 $C_9 : \square$

Hence the argument is correct.   ◀

Now we shall consider the connection between resolution principle and Prolog.

A *Horn clause* is a clause with atmost one nonnegated literal. If all literals are negated, it is called a headless Horn clause. If one literal is nonnegated, it is called a headed Horn clause.

Consider the following rule

Rule (1)   Aunt$(x, y) : -$Female(x), Sister(x, z), Parent(z, y)

This means if x is a Female and x is the Sister of z and z is a Parent of y, then x is the Aunt of y.

Suppose we want to conclude Sita is the Aunt of Mohan.

We have the following facts

Female (Sita) $-F_1$
Sister (Sita, Geetha) $-F_2$
Parent (Geetha, Mohan) $-F_3$

Instantiating the rule (1) we get

Aunt (Sita, Mohan) : $-$Female (Sita), Sister (Sita, Geetha), Parent (Geetha, Mohan).

Each F_1, F_2, F_3 is a headed Horn clause.

Rule (1) can be written in logical notation as

$\forall x \forall y \forall z[(\text{Female}(x) \land \text{Sister}(x, z) \land \text{Parent}(z, y)) \rightarrow \text{Aunt}(x, y)]$

Using instantiation we get

(Female (Sita) $\land$ Sister (Sita, Geetha) $\land$ Parent (Geetha, Mohan)) $\rightarrow$ Aunt (Sita, Mohan)

Expressing as a clause this becomes

$\neg$(Female (Sita) $\land$ Sister (Sita, Geetha) $\land$ Parent (Geetha, Mohan)) $\lor$ Aunt (Sita, Mohan)

Using DeMorgan's laws we get

¬Female (Sita) ∨ ¬Sister (Sita, Geetha) ∨ ¬Parent (Geetha, Mohan)) ∨ Aunt (Sita, Mohan)

$-IR_1$ (Instantiation of rule 1)

The conclusion we want to derive is Aunt(Sita, Mohan)
Negating the conclusion we get

¬Aunt(Sita, Mohan) − NC (negation of conclusion)

From F_1, F_2, F_3, IR_1 and NC, it can be seen that using resolution we can derive the empty clause. Hence the conclusion Aunt(Sita, Mohan) is correct. Note that in a rule in Prolog what occurs on the left hand side of : − is the headed portion and what occurs in the right hand portion is the headless portion. Hence a rule corresponds to a Horn clause. A fact consists of a single nonnegated literal and hence a Horn clause. The conclusion derived is : −Aunt(Sita, Mohan) and for resolution we take the negation of the right hand side of : −. This is a headless Horn clause.

The question asked is ? Aunt(Sita, Mohan) and Prolog replies yes.

The goal is Aunt(Sita, Mohan) and Prolog tries to find rules and facts and using proper instantiation tries to see whether the goal is satisfied. This is called backward chaining.

Thus we see that the resolution principle is used in the logic programming language Prolog. It is also used in automatic theorem proving where a program or software package is used to prove theorems, given the axioms.

Fallacies

Several common fallacies arise in incorrect arguments. These fallacies resemble rules of inference, but are based on contingencies rather than tautologies. These are discussed here to show the distinction between correct and incorrect reasoning.

The proposition $((p \rightarrow q) \land q) \rightarrow p$ is not a tautology, because it is false when p is false and q is true. However, there are many incorrect arguments that treat this as a tautology. In other words, they treat the argument with premises $p \rightarrow q$ and q and conclusion p as a valid argument form, which it is not. This type of incorrect reasoning is called the **fallacy of affirming the conclusion**.

EXAMPLE 10   Is the following argument valid?

If you do every problem in this book, then you will learn discrete mathematics. You learned discrete mathematics.

Therefore, you did every problem in this book.

Solution: Let p be the proposition "You did every problem in this book." Let q be the proposition "You learned discrete mathematics." Then this argument is of the form: if $p \rightarrow q$ and q, then p. This is an example of an incorrect argument using the fallacy of affirming the conclusion. Indeed, it is possible for you to learn discrete mathematics in some way other than by doing every problem in this book. (You may learn discrete mathematics by reading, listening to lectures, doing some, but not all, the problems in this book, and so on.)   ◄

The proposition $((p \rightarrow q) \land \neg p) \rightarrow \neg q$ is not a tautology, because it is false when p is false and q is true. Many incorrect arguments use this incorrectly as a rule of inference. This type of incorrect reasoning is called the **fallacy of denying the hypothesis**.

TABLE 2 Rules of Inference for Quantified Statements.	
Rule of Inference	**Name**
$\forall x\, P(x)$ $\therefore\ \overline{P(c)}$	Universal instantiation
$\underline{P(c)\text{ for an arbitrary }c}$ $\therefore\ \forall x\, P(x)$	Universal generalization
$\exists x\, P(x)$ $\therefore\ \overline{P(c)\text{ for some element }c}$	Existential instantiation
$\underline{P(c)\text{ for some element }c}$ $\therefore\ \exists x\, P(x)$	Existential generalization

EXAMPLE 11   Let p and q be as in Example 10. If the conditional statement $p \rightarrow q$ is true, and $\neg p$ is true, is it correct to conclude that $\neg q$ is true? In other words, is it correct to assume that you did not learn discrete mathematics if you did not do every problem in the book, assuming that if you do every problem in this book, then you will learn discrete mathematics?

Solution: It is possible that you learned discrete mathematics even if you did not do every problem in this book. This incorrect argument is of the form $p \rightarrow q$ and $\neg p$ imply $\neg q$, which is an example of the fallacy of denying the hypothesis.   ◄

Rules of Inference for Quantified Statements

We have discussed rules of inference for propositions. We will now describe some important rules of inference for statements involving quantifiers. These rules of inference are used extensively in mathematical arguments, often without being explicitly mentioned.

Universal instantiation is the rule of inference used to conclude that $P(c)$ is true, where c is a particular member of the domain, given the premise $\forall x\, P(x)$. Universal instantiation is used when we conclude from the statement "All women are wise" that "Lisa is wise," where Lisa is a member of the domain of all women.

Universal generalization is the rule of inference that states that $\forall x\, P(x)$ is true, given the premise that $P(c)$ is true for all elements c in the domain. Universal generalization is used when we show that $\forall x\, P(x)$ is true by taking an arbitrary element c from the domain and showing that $P(c)$ is true. The element c that we select must be an arbitrary, and not a specific, element of the domain. That is, when we assert from $\forall x\, P(x)$ the existence of an element c in the domain, we have no control over c and cannot make any other assumptions about c other than it comes from the domain. Universal generalization is used implicitly in many proofs in mathematics and is seldom mentioned explicitly. However, the error of adding unwarranted assumptions about the arbitrary element c when universal generalization is used is all too common in incorrect reasoning.

Existential instantiation is the rule that allows us to conclude that there is an element c in the domain for which $P(c)$ is true if we know that $\exists x\, P(x)$ is true. We cannot select an arbitrary value of c here, but rather it must be a c for which $P(c)$ is true. Usually we have no knowledge of what c is, only that it exists. Because it exists, we may give it a name (c) and continue our argument.

Existential generalization is the rule of inference that is used to conclude that $\exists x\, P(x)$ is true when a particular element c with $P(c)$ true is known. That is, if we know one element c in the domain for which $P(c)$ is true, then we know that $\exists x\, P(x)$ is true.

We summarize these rules of inference in Table 2. We will illustrate how some of these rules of inference for quantified statements are used in Examples 12 and 13.

EXAMPLE 12   Show that the premises "Everyone in this discrete mathematics class has taken a course in computer science" and "Marla is a student in this class" imply the conclusion "Marla has taken a course in computer science."

Solution: Let $D(x)$ denote "x is in this discrete mathematics class," and let $C(x)$ denote "x has taken a course in computer science." Then the premises are $\forall x(D(x) \rightarrow C(x))$ and $D(\text{Marla})$. The conclusion is $C(\text{Marla})$.

The following steps can be used to establish the conclusion from the premises.

Step	Reason
1. $\forall x(D(x) \rightarrow C(x))$	Premise
2. $D(\text{Marla}) \rightarrow C(\text{Marla})$	Universal instantiation from (1)
3. $D(\text{Marla})$	Premise
4. $C(\text{Marla})$	Modus ponens from (2) and (3)

EXAMPLE 13   Show that the premises "A student in this class has not read the book," and "Everyone in this class passed the first exam" imply the conclusion "Someone who passed the first exam has not read the book."

Solution: Let $C(x)$ be "x is in this class," $B(x)$ be "x has read the book," and $P(x)$ be "x passed the first exam." The premises are $\exists x(C(x) \wedge \neg B(x))$ and $\forall x(C(x) \rightarrow P(x))$. The conclusion is $\exists x(P(x) \wedge \neg B(x))$. These steps can be used to establish the conclusion from the premises.

Step	Reason
1. $\exists x(C(x) \wedge \neg B(x))$	Premise
2. $C(a) \wedge \neg B(a)$	Existential instantiation from (1)
3. $C(a)$	Simplification from (2)
4. $\forall x(C(x) \rightarrow P(x))$	Premise
5. $C(a) \rightarrow P(a)$	Universal instantiation from (4)
6. $P(a)$	Modus ponens from (3) and (5)
7. $\neg B(a)$	Simplification from (2)
8. $P(a) \wedge \neg B(a)$	Conjunction from (6) and (7)
9. $\exists x(P(x) \wedge \neg B(x))$	Existential generalization from (8)

Combining Rules of Inference for Propositions and Quantified Statements

We have developed rules of inference both for propositions and for quantified statements. Note that in our arguments in Examples 12 and 13 we used both universal instantiation, a rule of inference for quantified statements, and modus ponens, a rule of inference for propositional logic. We will often need to use this combination of rules of inference. Because universal instantiation and modus ponens are used so often together, this combination of rules is sometimes called **universal modus ponens**. This rule tells us that if $\forall x(P(x) \rightarrow Q(x))$ is true, and if $P(a)$ is true for a particular element a in the domain of the universal quantifier, then $Q(a)$ must also be true. To see this, note that by universal instantiation, $P(a) \rightarrow Q(a)$ is true. Then, by modus ponens, $Q(a)$ must also be true. We can describe universal modus ponens as follows:

$$\forall x(P(x) \rightarrow Q(x))$$
$$\underline{P(a), \text{ where } a \text{ is a particular element in the domain}}$$
$$\therefore Q(a)$$

Universal modus ponens is commonly used in mathematical arguments. This is illustrated in Example 14.

EXAMPLE 14   Assume that "For all positive integers n, if n is greater than 4, then n^2 is less than 2^n" is true. Use universal modus ponens to show that $100^2 < 2^{100}$.

Solution: Let $P(n)$ denote "$n > 4$" and $Q(n)$ denote "$n^2 < 2^n$." The statement "For all positive integers n, if n is greater than 4, then n^2 is less than 2^n" can be represented by $\forall n(P(n) \rightarrow Q(n))$, where the domain consists of all positive integers. We are assuming that $\forall n(P(n) \rightarrow Q(n))$ is true. Note that $P(100)$ is true because $100 > 4$. It follows by universal modus ponens that $Q(100)$ is true, namely that $100^2 < 2^{100}$. ◀

Another useful combination of a rule of inference from propositional logic and a rule of inference for quantified statements is **universal modus tollens**. Universal modus tollens combines universal instantiation and modus tollens and can be expressed in the following way:

$$\forall x(P(x) \rightarrow Q(x))$$
$$\underline{\neg Q(a), \text{ where } a \text{ is a particular element in the domain}}$$
$$\therefore \neg P(a)$$

The verification of universal modus tollens is left as an exercise. Exercises 18–19 develop additional combinations of rules of inference in propositional logic and quantified statements.

Exercises

1. Find the argument form for the following argument and determine whether it is valid. Can we conclude that the conclusion is true if the premises are true?

> If Socrates is human, then Socrates is mortal.
> Socrates is human.
> ─────────────────────────────
> ∴ Socrates is mortal.

2. What rule of inference is used in each of these arguments?

a) Alice is a mathematics major. Therefore, Alice is either a mathematics major or a computer science major.

b) Jerry is a mathematics major and a computer science major. Therefore, Jerry is a mathematics major.

c) If it is rainy, then the pool will be closed. It is rainy. Therefore, the pool is closed.

d) If it snows today, the university will close. The university is not closed today. Therefore, it did not snow today.

e) If I go swimming, then I will stay in the sun too long. If I stay in the sun too long, then I will sunburn. Therefore, if I go swimming, then I will sunburn.

3. Use rules of inference to show that the hypotheses "Randy works hard," "If Randy works hard, then he is a dull boy," and "If Randy is a dull boy, then he will not get the job" imply the conclusion "Randy will not get the job."

4. Use rules of inference to show that the hypotheses "If it does not rain or if it is not foggy, then the sailing race will be held and the lifesaving demonstration will go on," "If the sailing race is held, then the trophy will be awarded," and "The trophy was not awarded" imply the conclusion "It rained."

5. What rules of inference are used in this famous argument? "All men are mortal. Socrates is a man. Therefore, Socrates is mortal."

6. What rules of inference are used in this argument? "No man is an island. Manhattan is an island. Therefore, Manhattan is not a man."

7. For each of these collections of premises, what relevant conclusion or conclusions can be drawn? Explain the rules of inference used to obtain each conclusion from the premises.

a) "If I take the day off, it either rains or snows." "I took Tuesday off or I took Thursday off." "It was sunny on Tuesday." "It did not snow on Thursday."

b) "If I eat spicy foods, then I have strange dreams." "I have strange dreams if there is thunder while I sleep." "I did not have strange dreams."

c) "I am either clever or lucky." "I am not lucky." "If I am lucky, then I will win the lottery."

d) "Every computer science major has a personal computer." "Ralph does not have a personal computer." "Ann has a personal computer."

e) "What is good for corporations is good for the United States." "What is good for the United States is good for you." "What is good for corporations is for you to buy lots of stuff."

f) "All rodents gnaw their food." "Mice are rodents." "Rabbits do not gnaw their food." "Bats are not rodents."

8. For each of these arguments, explain which rules of inference are used for each step.

a) "Linda, a student in this class, owns a red convertible. Everyone who owns a red convertible has gotten at least one speeding ticket. Therefore, someone in this class has gotten a speeding ticket."

b) "Each of five roommates, Melissa, Aaron, Ralph, Veneesha, and Keeshawn, has taken a course in discrete mathematics. Every student who has taken a course in discrete mathematics can take a course in algorithms. Therefore, all five roommates can take a course in algorithms next year."

c) "All movies produced by John Sayles are wonderful. John Sayles produced a movie about coal miners. Therefore, there is a wonderful movie about coal miners."

d) "There is someone in this class who has been to France. Everyone who goes to France visits the Louvre. Therefore, someone in this class has visited the Louvre."

9. For each of these arguments, explain which rules of inference are used for each step.

a) "Doug, a student in this class, knows how to write programs in JAVA. Everyone who knows how to write programs in JAVA can get a high-paying job. Therefore, someone in this class can get a high-paying job."

b) "Somebody in this class enjoys whale watching. Every person who enjoys whale watching cares about ocean pollution. Therefore, there is a person in this class who cares about ocean pollution."

c) "Each of the 93 students in this class owns a personal computer. Everyone who owns a personal computer can use a word processing program. Therefore, Zeke, a student in this class, can use a word processing program."

d) "Everyone in New Jersey lives within 50 miles of the ocean. Someone in New Jersey has never seen the ocean. Therefore, someone who lives within 50 miles of the ocean has never seen the ocean."

10. For each of these arguments determine whether the argument is correct or incorrect and explain why.

a) Everyone enrolled in the university has lived in a dormitory. Mia has never lived in a dormitory. Therefore, Mia is not enrolled in the university.

b) A convertible car is fun to drive. Isaac's car is not a convertible. Therefore, Isaac's car is not fun to drive.

c) Quincy likes all action movies. Quincy likes the movie *Eight Men Out*. Therefore, *Eight Men Out* is an action movie.

d) All lobstermen set at least a dozen traps. Hamilton is a lobsterman. Therefore, Hamilton sets at least a dozen traps.

11. For each of these arguments determine whether the argument is correct or incorrect and explain why.

a) All students in this class understand logic. Xavier is a student in this class. Therefore, Xavier understands logic.

b) Every computer science major takes discrete mathematics. Natasha is taking discrete mathematics. Therefore, Natasha is a computer science major.

c) All parrots like fruit. My pet bird is not a parrot. Therefore, my pet bird does not like fruit.

d) Everyone who eats granola every day is healthy. Linda is not healthy. Therefore, Linda does not eat granola every day.

12. What is wrong with this argument? Let $H(x)$ be "x is happy." Given the premise $\exists x H(x)$, we conclude that $H(\text{Lola})$. Therefore, Lola is happy.

13. Determine whether each of these arguments is valid. If an argument is correct, what rule of inference is being used? If it is not, what logical error occurs?

a) If n is a real number such that $n > 1$, then $n^2 > 1$. Suppose that $n^2 > 1$. Then $n > 1$.

b) If n is a real number with $n > 3$, then $n^2 > 9$. Suppose that $n^2 \leq 9$. Then $n \leq 3$.

c) If n is a real number with $n > 2$, then $n^2 > 4$. Suppose that $n \leq 2$. Then $n^2 \leq 4$.

14. What is wrong with this argument? Let $S(x, y)$ be "x is shorter than y." Given the premise $\exists s S(s, \text{Max})$, it follows that $S(\text{Max}, \text{Max})$. Then by existential generalization it follows that $\exists x S(x, x)$, so that someone is shorter than himself.

15. Determine whether these are valid arguments.

a) If x is a positive real number, then x^2 is a positive real number. Therefore, if a^2 is positive, where a is a real number, then a is a positive real number.

b) If $x^2 \neq 0$, where x is a real number, then $x \neq 0$. Let a be a real number with $a^2 \neq 0$; then $a \neq 0$.

16. Identify the error or errors in this argument that supposedly shows that if $\forall x(P(x) \lor Q(x))$ is true then $\forall x P(x) \lor \forall x Q(x)$ is true.

1. $\forall x(P(x) \lor Q(x))$ Premise
2. $P(c) \lor Q(c)$ Universal instantiation from (1)
3. $P(c)$ Simplification from (2)
4. $\forall x\, P(x)$ Universal generalization from (3)
5. $Q(c)$ Simplification from (2)
6. $\forall x\, Q(x)$ Universal generalization from (5)
7. $\forall x(P(x) \lor \forall x\, Q(x))$ Conjunction from (4) and (6)

17. Identify the error or errors in this argument that supposedly shows that if $\exists x\, P(x) \land \exists x\, Q(x)$ is true then $\exists x(P(x) \land Q(x))$ is true.

1. $\exists x\, P(x) \lor \exists x\, Q(x)$ Premise
2. $\exists x\, P(x)$ Simplification from (1)
3. $P(c)$ Existential instantiation from (2)
4. $\exists x\, Q(x)$ Simplification from (1)
5. $Q(c)$ Existential instantiation from (4)
6. $P(c) \land Q(c)$ Conjunction from (3) and (5)
7. $\exists x(P(x) \land Q(x))$ Existential generalization

18. Justify the rule of **universal transitivity**, which states that if $\forall x(P(x) \to Q(x))$ and $\forall x(Q(x) \to R(x))$ are true, then $\forall x(P(x) \to R(x))$ is true, where the domains of all quantifiers are the same.

19. Use rules of inference to show that if $\forall x(P(x) \to (Q(x) \land S(x)))$ and $\forall x(P(x) \land R(x))$ are true, then $\forall x(R(x) \land S(x))$ is true.

20. Use resolution to show the hypotheses "Allen is a bad boy or Hillary is a good girl" and "Allen is a good boy or David is happy" imply the conclusion "Hillary is a good girl or David is happy."

21. Use resolution to show that the hypotheses "It is not raining or Yvette has her umbrella," "Yvette does not have her umbrella or she does not get wet," and "It is raining or Yvette does not get wet" imply that "Yvette does not get wet."

1.7   Normal Forms

In this section we consider some normal forms for the formulas in propositional and predicate logic.

Disjunctive Normal Form and Conjunctive Normal Form

We have considered propositions, propositional variables and the connectives $\land$, $\lor$, $\neg$, $\to$, $\leftrightarrow$. A well formed formula (wff) of propositional logic, also called prepositional form,' or simply formula, is a string consisting of propositional variables, connectives, and parenthesis used in the proper manner. For example $((p \lor q) \land (\neg p \lor q \lor r)) \land (\neg s \lor q)$ is a wff with propositional variables p, q, r, s. An expression of the form $p \lor q \neg r$ is a disjunction and an expression of the form $p \land \neg q \land r$ is a conjuction. Sometimes it is convenient to use the word 'product' for conjunction and the word 'sum' for disjunction.

A product of the variables, and their negations in a formula is called an elementary product. $\neg p \lor q, q \land r \neg s, q$ are examples of elementary products. A sum of variables and their negations is called an elementary sum. $\neg p \lor q, q \lor p \lor s$, p are examples of elementary sums.

A variable or the negation of a variable is called a literal. Hence an elementary product is a disjunction of literals and an elementary product is a conjunction of literals.

It is easy to observe that

- A necessary and sufficient condition for an elementary product to be identically false (a contradiction) is that it contains at least one pair of literals in which one is the negation of the other.
  If $p \land \neg p \land \ldots$, is the formula, this reduces to $F \land \ldots$, and is equal to F.
- A necessary and sufficient condition for an elementary sum to be identically true (a tautology) is that it contains at least one pair of literals in which one is the negation of the other.
  If $p \lor \neg p \lor \ldots$ is a formula, this reduces $T \lor \ldots$ and is equal to T.

DEFINITION 1   A formula which is equivalent to a given formula and consists of a sum of elementary products is called a disjunctive normal form (DNF).

To bring any formula to disjunctive normal form, first replace $\rightarrow$ and $\leftrightarrow$ using $\wedge$ and $\vee$ and $\neg$. Then De Morgan's law whenever necessary.

EXAMPLE 1 Obtain disjunctive normal forms of

1. $(p \rightarrow q) \wedge \neg q$
2. $\neg(p \wedge q) \leftrightarrow (p \vee q)$

Solution:

1. $(p \rightarrow q) \wedge \neg q$
   $(\neg p \vee q) \wedge \neg q$
   $(\neg q \wedge \neg q) \vee (q \wedge \neg q)$
2. $\neg(p \wedge q) \leftrightarrow (p \vee q)$
   This is equivalent to $[\neg(p \wedge q) \wedge (p \vee q)] \vee [\neg\neg(p \wedge q) \wedge \neg(p \vee q))]$
   i.e., $[(\neg p \vee \neg q) \wedge (p \vee q)] \vee [(p \wedge q) \wedge (\neg p \wedge \neg q)]$
   $[(\neg p \wedge p) \vee (\neg q \wedge p) \vee (\neg p \wedge q) \vee (\neg q \wedge q)] \vee (p \wedge q \wedge \neg p \wedge \neg q)]$
   $(\neg p \wedge p) \vee (\neg q \wedge p) \vee (\neg p \wedge q) \vee (\neg q \wedge q)] \vee (p \wedge q \wedge \neg p \wedge \neg q)$ ◀

The disjunctive normal form is not unique. $(p \vee q) \vee r$ is in disjunctive normal form. This can be as $(p \vee r) \wedge (q \vee r)$. i.e., $(q \wedge q) \vee (r \wedge q) \vee (p \wedge r) \vee (r \wedge r)$ This is an equivalent disjunctive normal form. We can see that a given formula is identically false (a contradiction), if every elementary product appearing in its disjunctive normal form is identically false.

DEFINITION 2   A formula which is equivalent to a given formula and consists of a product of elementary sums is called a conjunctive normal form of the given formula (CNF).

We can bring any formula to conjunctive normal form. It should be noted that conjunctive normal form is not unique.

EXAMPLE 2 Bring to conjunctive normal form):

1. $(p \rightarrow q) \wedge \neg q$
2. $\neg(p \wedge q) \leftrightarrow (p \vee q)$

Solution:

1. $(p \rightarrow q) \wedge \neg q$ is equivalent to $(\neg p \vee q) \wedge \neg q$ This is in CNF
2. $\neg(p \wedge q) \leftrightarrow (p \vee q)$
   This is equal to
   $[\neg(p \wedge q) \rightarrow (p \vee q)] \wedge [(p \vee q) \rightarrow \neg(p \wedge q)]$
   i.e., $[\neg\neg(p \wedge q) \vee (p \vee q)] \wedge [\neg(p \vee q) \vee \neg(p \wedge q)]$
   $[(p \wedge q) \vee (p \vee q)] \wedge [\neg(p \vee q) \vee \neg(p \wedge q)]$
   $[(p \vee (p \vee q)) \wedge (q \vee (p \vee q))] \wedge [(\neg p \wedge \neg q) \vee (\neg p \vee \neg q)]$
   $[(p \vee p \vee q) \wedge (q \vee p \vee q)] \vee [(\neg p \vee (\neg p \vee \neg q) \vee (\neg q \vee (\neg p \vee \neg q)]$
   $(p \vee p \vee q) \wedge (q \vee p \vee q) \wedge (\neg p \vee \neg p \vee \neg q) \vee (\neg q \vee \neg p \vee \neg q)$ ◀

It should be noted that a given formula is identically true (a tautology) if every elementary sum appearing in the formula has at least two literals, of which one is the negation of the other.

Principal Disjunctive Normal Form

Let p and q be propositional variables. Consider the four conjunctions given below.

$$p \wedge q, \quad p \wedge \neg q, \quad \neg p \wedge q, \quad \neg p \wedge \neg q$$

represent conjunctions in which p or $\neg p$ appears as also q or $\neg q$. Each variable occurs either negated or nonnegated but both negated and non negated forms of a variable do not occur together in the conjunction. Also $p \wedge q$ and $q \wedge p$ are treated as the same. These four conjunctions are called minterms of p and q. In general if there are n variables, there will be 2^n minterms. Each minterm is a conjunction in which each variable occurs once either in the negated form or in the non negated form.

DEFINITION 3   For a given formula, an equivalent formula consisting of disjunctions of minterms only is known as its principal disjunctive normal form. Such a normal form is also called sum-of-products canoical form.

Consider the following truth table

p	q	$p \vee q$	$p \vee \neg q$	$\neg p \vee q$	$\neg p \vee \neg q$
T	T	T	F	F	F
T	F	F	T	F	F
F	T	F	F	T	F
F	F	F	F	F	T

We find that in the columns corresponding to minterms only one entry is T and the other entries are false.

Consider the truth table for $p \rightarrow q$ and $p \leftrightarrow q$

p	q	$p \rightarrow q$	$p \leftrightarrow q$
T	T	T	T
T	F	F	F
F	T	T	F
F	F	T	T

The principal disjunctive normal form for $p \rightarrow q$ will be $(p \wedge q) \vee (\neg p \wedge q) \vee (\neg p \wedge \neg q)$. This disjunction corresponds to the disjunction of these minterms having T in the respective rows.

Similarly $p \leftrightarrow q$ has the following principal disjunctive normal form $(p \wedge q) \vee (\neg p \wedge \neg q)$. It should be noted that the number of minterms appearing in the normal form of a formula is the same as the number of T entries in the column corresponding to the formula. A formula which is a tautology will have all minterms. A contradiction will have no minterms. Hence for any formula which is not a contradiction we can have a principal disjunctive normal form. A formula in which the minterms are rearranged is considered as an equivalent formula. We have also seen that if the literals in a minterm are rearranged we get an equivalent minterm, Recall that a literal is a variable or negation of a variable. If two given formulas are equivalent, then both of them must have identical principal disjunctive normal forms.

In order to obtain the principal disjunctive normal form) of a given formula without constructing its truth table, one may first replace implication ($\rightarrow$) and equivalence ($\leftrightarrow$) by using $\wedge$, $\vee$, $\neg$. Then De Morgan's laws are used wherever necessary and distributive laws are also used in bringing to disjunctive normal form. An elementary product which is a contradiction is dropped. Minterms are obtained in the disjunctions by introducing the missing factors. Duplications are avoided. Let us consider a few examples.

EXAMPLE 3 Obtain principal disjunctive normal form for $p \vee \neg q$

Solution:

$$p \vee \neg q = [p \wedge (q \vee \neg q)] \vee [\neg q \wedge (p \vee \neg p)]$$
$$= (p \wedge q) \vee (p \wedge \neg q) \vee (\neg q \wedge p) \vee (\neg q \wedge \neg p)$$
$$= (p \wedge q) \vee (p \wedge \neg q) \vee (\neg p \wedge \neg q)$$

◄

EXAMPLE 4 Obtain principal disjunctive normal form for $(p \wedge q) \vee (\neg p \wedge r) \vee (q \wedge r)$

Solution:

$$(p \wedge q) = (p \wedge q) \wedge (r \vee \neg r)$$
$$= (p \wedge q \wedge r) \vee (p \wedge q \wedge \neg r)$$
$$(\neg p \wedge r) = (\neg p \wedge r) \wedge (q \vee \neg q)$$
$$= (\neg p \wedge r \wedge q) \vee (\neg p \wedge r \wedge \neg q)$$
$$= (\neg p \wedge q \wedge r) \vee (\neg p \wedge \neg q \wedge r)$$
$$(q \wedge r) = (q \wedge r) \vee (p \vee \neg p)$$
$$= (q \wedge r \wedge p) \vee (q \wedge r \wedge \neg p)$$
$$= (p \wedge q \wedge r) \vee (\neg p \wedge q \wedge r)$$

Avoiding duplication, the given expression is equivalent to

$$(p \wedge q \wedge r) \vee (p \wedge q \wedge \neg r) \vee (\neg p \wedge q \wedge r) \vee (\neg p \wedge \neg q \wedge r)$$

◄

If we introduce ordering among variables, as $p_1, p_2, \ldots, p_n$ then there are 2^n minterms and they can be assigned values from 0 to $2^n - 1$. The number i corresponds to the minterm m_i as follows. Let $b_1, b_2, \ldots b_n$ be the binary representation of i, Then in the minterm m_i, the variable p_i occurs as p_i if $b_i = 1$ and occurs as $\neg p + i$ if $b_i = 0$

$00 \ldots 0$ corresponds to $\neg p_1 \wedge \neg p_2 \wedge \ldots \neg p_n$
$11 \ldots 1$ corresponds to $p_1 \wedge p_2 \ldots \wedge p_n$

Let us look at the examples considered earlier. In the first one, there are two variables and the minterms occurring in the expression correspond to 0, 2, and 3, and this is written as $\sum 0, 2.3$. In the second example, the minterms correspond to 7, 6, 3, 1 respectively and the expression is written as $\sum 1, 3, 6, 7$ as a shortened form.

Principal Conjunctive Normal Form

We define maxterm as a dual to minterm, For a given number of variables the maxterm consists of disjunctions in which each variable or its negation, but not both, appears only once. It can be seen that each of the maxterms has the truth value F for exactly one combination of the truth values of the variables. This is illustrated for two variables below:

p	q	$p \vee q$	$\neg p \vee q$	$p \vee \neg q$	$\neg p \vee \neg q$
T	T	T	T	T	F
T	F	T	F	T	T
F	T	T	T	F	T
F	F	F	T	T	T

DEFINITION 4   For a given formula, an equivalent formula consisting of conjunction of maxterms only is known as its principal conjunctive normal form. This normal form is also called the product of sums canoical form.

Every formula which is not a tautology has an equivalent principal conjunctive normal form which is unique except for the rearrangement of the factors in the maxterms as well as rearrangement of the maxterms. The method for obtaining principal conjunctive normal form for a given formula is similar to the one described earlier for the principal disjunctive normal form. In each disjunction, missing variable is provided in the negated and non negated forms,

EXAMPLE 5   Find principal conjunctive normal form for $(p \leftrightarrow q)$

Solution:

$$p \leftrightarrow q = (p \rightarrow q) \wedge (q \rightarrow p)$$
$$= (\neg p \vee q) \wedge (\neg q \vee p)$$ ◀

EXAMPLE 6   Find principal conjunctive normal form for $[(p \vee q) \wedge \neg p \rightarrow \neg q]$

Solution:

$$[(p \vee q) \wedge \neg p \rightarrow \neg q] = [(p \wedge \neg p) \vee (q \wedge \neg p)] \rightarrow \neg q$$
$$= (q \wedge \neg p) \rightarrow \neg q$$
$$= \neg(q \wedge \neg p) \vee \neg q$$
$$= \neg q \vee \neg p \vee \neg q$$
$$= \neg q \vee p$$
$$= p \vee \neg q$$ ◀

If we order the variables as $p_1, p_2, \ldots, P_n$ there are 2^n maxterms, which can be assigned values from 0 to $2^n - 1$ as follows. A number i represents the maxterm M_i, if the binary representation of i is $b_1, b_2, \ldots, b_n$ (with leading zeros permitted) and the variable p_i is present in the non negated form if $b_i = 0$ and is present in the negated form if $b_i = 1$.

The maxterms corresponding to the variables p and q are M_0, M_1, M_2, M_3 given by $p \vee q, p \vee \neg q, \neg p \vee q, v \neg q$ respectively. A formula $(\neg p \vee q) \wedge (p \vee \neg q)$ represents the conjunction if the two maxterms M_1 and M_2 and is denoted by $\prod 1, 2$. In general. if there are maxterms $M_0, \ldots M_{2^n-1}$ a formula which is the conjunction of $M_{i_1}, M_{i_2}, \ldots M_{i_r}$, is denoted by $\prod i_1, i_2, \ldots, i_r$.

To show two formulas are equivalent, we can bring them to principal conjunctive normal form (principal disjunctive normal form). If both the formulas give rise to the same principal conjunctive normal form (principal disjunctive normal form) they are equivalent.

In considering the ordering of variables and the representation of minterms and maxterms by integers in the range 0 to $2^n - 1$. we find that we follow different convention for minterms and maxterms.

In the minterm corresponding to i with binary representation $b_1, \ldots b_n$, the variable p_i is present in the non negated form if $b_i = 1$ and in the negated form if $b_i = 0$. In a maxterm, it is the other way round.

Consider the following formula

$$(p \to q) \to (q \to p)$$

$$(\neg p \vee q) \to (\neg q \vee p)$$

$$\neg(\neg p \vee q) \vee (\neg q \vee p)$$

$$(p \wedge \neg q) \vee (\neg q \vee p)$$

$$(p \vee (\neg q \vee p)) \wedge (\neg q \vee (\neg q \vee p))$$

$$(p \vee \neg q \vee p) \wedge (\neg q \vee \neg q \vee p)$$

$$(p \vee \neg q) \wedge (p \vee \neg q)$$

$$(p \vee \neg q)$$

$$\prod 1 \text{ This is p.c.n.f}$$

$$(p \to q) \to (q \to p)$$

$$(\neg p \vee q) \to (\neg q \vee p)$$

$$\neg(\neg p \vee q) \vee (\neg q \vee p)$$

$$(p \wedge \neg q) \vee (q \vee p)$$

$$(p \wedge \neg q) \vee (\neg q) \vee p$$

$$(p \wedge \neg q) \vee (\neg q \wedge (p \vee \neg p)) \vee (p \wedge (q \vee \neg q))$$

$$(p \wedge \neg q) \vee (\neg q \wedge p) \vee (\neg q \wedge \neg p) \vee (p \wedge q) \vee (p \wedge \neg q)$$

$$(p \wedge q) \vee (p \wedge \neg q) \vee (\neg p \wedge \neg q)$$

$$\sum 0, 2, 3$$

It should be noted that when the same formula with n variables is represented in the $\sum$ and $\prod$ notations, the numbers appearing in the $\sum$ notation will not appear in the $\prod$ notation and $\prod$ will consist of numbers between 0 and $2^n - 1$ which do not appear in the $\sum$ notation.

Normal Forms for First Order Logic

In this section. we consider normal forms for formulas in the first order logic which is called the "prenex normal form."

DEFINITION 5 A formula F in the first order logic is said to be in a prenex normal form if and only if the formula F is in the form of $(Q_1 x_1) \ldots (Q_n x_n)(M)$ where every $(Q_i x_i), i = 1, \ldots, n$ is either $(\forall x_i)$ or $(\exists x_i)$, and M is a formula containing no quantifiers. $(Q_1 x_1) \ldots (Q_n, x_n)$ is called the prefix and M is called the matrix of the formula F.

EXAMPLE 7 $(\forall x)(\forall y)(P(x, y) \land Q(y))$

$\forall x \exists y \forall z (Q(x, y) \rightarrow R(z))$ are in the prenex normal form.   ◀

Let us now see how to convert a given formula in first order logic to prenex normal form.

We denote two formulas F_1, F_2 as equivalent by $F_1 \leftrightarrow F_2$, if and only if the truth values of F_1 and F_2 are the same under every interpretation. We know that

$$\neg \forall x\, P(x) \leftrightarrow \exists x \neg P(x) \qquad\qquad (1)$$
$$\neg \exists x\, P(x) \leftrightarrow \forall x \neg P(x) \qquad\qquad (2)$$

Also $\forall$ distributes over $\land$ and $\exists$ over $\lor$.

$\forall$ does not distribute over $\lor$ and $\exists$ over $\land$. Also if F has a variable x and G does not contain x, then

$$(Qx)F(x) \lor G \leftrightarrow Q(x)(F(x) \lor G) \qquad\qquad (3)$$
$$(Qx)F(x) \land G \leftrightarrow Q(x)(F(x) \land G) \qquad\qquad (4)$$

We see that if F_1 and F_2 have variable x,

$$\forall x\, F_1(x) \lor \forall x\, F_2(x) \neq \forall x (F_1(x) \lor F_2(x))$$

But in $F_2(x)$ we can rename the variable x as z and get $\forall x\, F_1(x) \lor \forall z\, F_2(z)$ which can be brought to the form $\forall x \forall z (F_1(x) \lor F_2(z))$. Note that F_2 does not contain x and F_1 does not contain z.

Similarly, $\exists x\, F_1(x) \land \land x\, F_2(x)$ can be brought to the following form by renaming of variable x as z in F_2.

$$\exists x\, F_1(x) \land \exists x\, F_2(x)$$
$$= \exists x\, F_1(x) \land \exists z\, F_2(z)$$
$$= \exists x \exists z (F_1(x) \land F_2(z))$$

Hence it is possible to bring the quantifiers to the left of the formula. The following steps are carried out to bring a formula of first order logic to prenex normal form:

Step 1: Replace $\leftrightarrow$ and $\rightarrow$ using $\land$, $\lor$, $\neg$
Step 2: Use double negation and De Morgan's laws repeatedly and the laws (1) and (2).
Step 3: Rename variables if necessary
Step 4: Use rules (3) and (4) to bring the quantifiers to the left.

The following examples will illustrate the procedure.

EXAMPLE 8   Transform the formula into prenex normal form:

$$\forall x\, P(x) \rightarrow \exists x\, Q(x)$$

Solution:

$$\neg \forall x\, P(x) \lor \exists x\, Q(x)$$
$$\exists x(\neg P(x)) \lor \exists x\, Q(x)$$
$$\exists x(\neg P(x) \lor Q(x)) \qquad\qquad ◀$$

EXAMPLE 9   Obtain prenex normal form for the formula

$$(\forall x)(\forall y)((\exists z)(P(x, z) \wedge P(y, z)) \rightarrow (\exists u)Q(x, y, u))$$

Solution:

$$(\forall x)(\forall y)(\neg(\exists z)(P(x, z) \wedge P(y, z)) \vee (\exists u)Q(x, y, u))$$

$$(\forall x)(\forall y)((\forall z)\neg(P(x, z) \wedge P(y, z) \vee (\exists u)Q(x, y, u))$$

$$(\forall x)(\forall y)((\forall z)(\neg P(x, z) \vee \neg P(y, z)) \vee (\exists u)Q(x, y, u))$$

$$(\forall x)(\forall y)(\forall z)(\exists u)((\neg P(x, z) \vee \neg P(y, z)) \vee Q(x, y, u))$$

$$(\forall x)(\forall y)(\forall z)(\exists u)(\neg P(x, z) \vee \neg P(y, z) \vee Q(x, y, u)) \qquad \blacktriangleleft$$

Exercises

Obtain disjunctive normal form, conjunctive normal form, principal disjunctive normal form, and principal conjunctive normal form for the following expressions given in Exercises 1–4.

1. $Q \wedge (P \vee \neg q)$

2. $P \rightarrow (p \wedge (Q \rightarrow P))$

3. $(Q \rightarrow P) \wedge (\neg P \wedge Q)$

4. $(\neg P \vee \neg Q) \rightarrow (P \leftrightarrow \neg Q)$

**5. Show how to transform an arbitrary statement to a statement in prenex normal form that is equivalent to the given statement. (Note: A formal solution of this exercise requires use of structural induction, covered in Section 5.3.)

*6. Put these statements in prenex normal form. [*Hint:* Use logical equivalence from Tables 6 and 7 in Section 1.3, Table 2 in Section 1.4, Example 19 in Section 1.4, Exercises 45 and 46 in Section 1.4, and Exercises 48 and 49.]

   a) $\exists x P(x) \vee \exists x Q(x) \vee A$, where A is a proposition not involving any quantifiers.

   b) $\neg(\forall x P(x) \vee \forall x Q(x))$

   c) $\exists x P(x) \rightarrow \exists x Q(x)$

1.8   Introduction to Proofs

Introduction

In this section we introduce the notion of a proof and describe methods for constructing proofs. A proof is a valid argument that establishes the truth of a mathematical statement. A proof can use the hypotheses of the theorem, if any, axioms assumed to be true, and previously proven theorems. Using these ingredients and rules of inference, the final step of the proof establishes the truth of the statement being proved.

In our discussion we move from formal proofs of theorems toward more informal proofs. The arguments we introduced in Section 1.6 to show that statements involving propositions and quantified statements are true were formal proofs, where all steps were supplied, and the rules for each step in the argument were given. However, formal proofs of useful theorems can be extremely long and hard to follow. In practice, the proofs of theorems designed for human consumption are almost always **informal proofs**, where more than one rule of inference may be used in each step, where steps may be skipped, where the axioms being assumed and the rules of inference used are not explicitly stated. Informal proofs can often explain to humans why theorems are true, while computers are perfectly happy producing formal proofs using automated reasoning systems.

The methods of proof discussed in this chapter are important not only because they are used to prove mathematical theorems, but also for their many applications to computer science. These applications include verifying that computer programs are correct, establishing that operating systems are secure, making inferences in artificial intelligence, showing that system specifications are consistent, and so on. Consequently, understanding the techniques used in proofs is essential both in mathematics and in computer science.

Some Terminology

Links

Formally, a **theorem** is a statement that can be shown to be true. In mathematical writing, the term theorem is usually reserved for a statement that is considered at least somewhat important. Less important theorems sometimes are called **propositions**. (Theorems can also be referred to as **facts** or **results**.) A theorem may be the universal quantification of a conditional statement with one or more premises and a conclusion. However, it may be some other type of logical statement, as the examples later in this chapter will show. We demonstrate that a theorem is true with a **proof**. A proof is a valid argument that establishes the truth of a theorem. The statements used in a proof can include **axioms** (or **postulates**), which are statements we assume to be true (for example, the axioms for the real numbers, given in Appendix 1, and the axioms of plane geometry), the premises, if any, of the theorem, and previously proven theorems. Axioms may be stated using primitive terms that do not require definition, but all other terms used in theorems and their proofs must be defined. Rules of inference, together with definitions of terms, are used to draw conclusions from other assertions, tying together the steps of a proof. In practice, the final step of a proof is usually just the conclusion of the theorem. However, for clarity, we will often recap the statement of the theorem as the final step of a proof.

A less important theorem that is helpful in the proof of other results is called a **lemma** (plural *lemmas* or *lemmata*). Complicated proofs are usually easier to understand when they are proved using a series of lemmas, where each lemma is proved individually. A **corollary** is a theorem that can be established directly from a theorem that has been proved. A **conjecture** is a statement that is being proposed to be a true statement, usually on the basis of some partial evidence, a heuristic argument, or the intuition of an expert. When a proof of a conjecture is found, the conjecture becomes a theorem. Many times conjectures are shown to be false, so they are not theorems.

Understanding How Theorems Are Stated

Extra
Examples

Before we introduce methods for proving theorems, we need to understand how many mathematical theorems are stated. Many theorems assert that a property holds for all elements in a domain, such as the integers or the real numbers. Although the precise statement of such theorems needs to include a universal quantifier, the standard convention in mathematics is to omit it. For example, the statement

"If $x > y$, where x and y are positive real numbers, then $x^2 > y^2$."

really means

"For all positive real numbers x and y, if $x > y$, then $x^2 > y^2$."

Furthermore, when theorems of this type are proved, the first step of the proof usually involves selecting a general element of the domain. Subsequent steps show that this element has the property in question. Finally, universal generalization implies that the theorem holds for all members of the domain.

Methods of Proving Theorems

Proving mathematical theorems can be difficult. To construct proofs we need all available ammunition, including a powerful battery of different proof methods. These methods provide the overall approach and strategy of proofs. Understanding these methods is a key component of learning how to read and construct mathematical proofs. One we have chosen a proof method, we use axioms, definitions of terms, previously proved results, and rules of inference to complete the proof. Note that in this book we will always assume the axioms for real numbers found in Appendix 1. We will also assume the usual axioms whenever we prove a result about geometry. When you construct your own proofs, be careful not to use anything but these axioms, definitions, and previously proved results as facts!

To prove a theorem of the form $\forall x(P(x) \to Q(x))$, our goal is to show that $P(c) \to Q(c)$ is true, where c is an arbitrary element of the domain, and then apply universal generalization. In this proof, we need to show that a conditional statement is true. Because of this, we now focus on methods that show that conditional statements are true. Recall that $p \to q$ is true unless p is true but q is false. Note that to prove the statement $p \to q$, we need only show that q is true if p is true. The following discussion will give the most common techniques for proving conditional statements. Later we will discuss methods for proving other types of statements. In this section, and in Section 1.9, we will develop a large arsenal of proof techniques that can be used to prove a wide variety of theorems.

When you read proofs, you will often find the words "obviously" or "clearly." These words indicate that steps have been omitted that the author expects the reader to be able to fill in. Unfortunately, this assumption is often not warranted and readers are not at all sure how to fill in the gaps. We will assiduously try to avoid using these words and try not to omit too many steps. However, if we included all steps in proofs, our proofs would often be excruciatingly long.

Direct Proofs

A **direct proof** of a conditional statement $p \to q$ is constructed when the first step is the assumption that p is true; subsequent steps are constructed using rules of inference, with the final step showing that q must also be true. A direct proof shows that a conditional statement $p \to q$ is true by showing that if p is true, then q must also be true, so that the combination p true and q false never occurs. In a direct proof, we assume that p is true and use axioms, definitions, and previously proven theorems, together with rules of inference, to show that q must also be true. You will find that direct proofs of many results are quite straightforward, with a fairly obvious sequence of steps leading from the hypothesis to the conclusion. However, direct proofs sometimes require particular insights and can be quite tricky. The first direct proofs we present here are quite straightforward; later in the text you will see some that are less obvious.

We will provide examples of several different direct proofs. Before we give the first example, we need to define some terminology.

DEFINITION 1   The integer n is *even* if there exists an integer k such that $n = 2k$, and n is *odd* if there exists an integer k such that $n = 2k + 1$. (Note that every integer is either even or odd, and no integer is both even and odd.) Two integers have the *same parity* when both are even or both are odd; they have *opposite parity* when one is even and the other is odd.

EXAMPLE 1   Give a direct proof of the theorem "If n is an odd integer, then n^2 is odd."

Solution: Note that this theorem states $\forall n\, P((n) \to Q(n))$, where $P(n)$ is "n is an odd integer" and $Q(n)$ is "n^2 is odd." As we have said, we will follow the usual convention in mathematical proofs by showing that $P(n)$ implies $Q(n)$, and not explicitly using universal instantiation. To begin a direct proof of this theorem, we assume that the hypothesis of this conditional statement is true, namely, we assume that n is odd. By the definition of an odd integer, it follows that $n = 2k + 1$, where k is some integer. We want to show that n^2 is also odd. We can square both sides of the equation $n = 2k + 1$ to obtain a new equation that expresses n^2. When we do this, we find that $n^2 = (2k + 1)^2 = 4k^2 + 4k + 1 = 2(2k^2 + 2k) + 1$. By the definition of an odd integer, we can conclude that n^2 is an odd integer (it is one more than twice an integer). Consequently, we have proved that if n is an odd integer, then n^2 is an odd integer.   ◄

EXAMPLE 2   Give a direct proof that if m and n are both perfect squares, then nm is also a perfect square. (An integer a is a **perfect square** if there is an integer b such that $a = b^2$.)

Solution: To produce a direct proof of this theorem, we assume that the hypothesis of this conditional statement is true, namely, we assume that m and n are both perfect squares. By the definition of a perfect square, it follows that there are integers s and t such that $m = s^2$ and $n = t^2$. The goal of the proof is to show that mn must also be a perfect square when m and n are; looking ahead we see how we can show this by substituting s^2 for m and t^2 for n into mn. This tells us that $mn = s^2 t^2$. Hence, $mn = s^2 t^2 = (ss)(tt) = (st)(st) = (st)^2$, using commutativity and associativity of multiplication. By the definition of perfect square, it follows that mn is also a perfect square, because it is the square of st, which is an integer. We have proved that if m and n are both perfect squares, then mn is also a perfect square.   ◄

Proof by Contraposition

Direct proofs lead from the premises of a theorem to the conclusion. They begin with the premises, continue with a sequence of deductions, and end with the conclusion. However, we will see that attempts at direct proofs often reach dead ends. We need other methods of proving theorems of the form $\forall x (P(x) \rightarrow Q(x))$. Proofs of theorems of this type that are not direct proofs, that is, that do not start with the premises and end with the conclusion, are called **indirect proofs**.

An extremely useful type of indirect proof is known as **proof by contraposition**. Proofs by contraposition make use of the fact that the conditional statement $p \rightarrow q$ is equivalent to its contrapositive, $\neg q \rightarrow \neg p$. This means that the conditional statement $p \rightarrow q$ can be proved by showing that its contrapositive, $\neg q \rightarrow \neg p$, is true. In a proof by contraposition of $p \rightarrow q$, we take $\neg q$ as a premise, and using axioms, definitions, and previously proven theorems, together with rules of inference, we show that $\neg p$ must follow. We will illustrate proof by contraposition with two examples. These examples show that proof by contraposition can succeed when we cannot easily find a direct proof.

EXAMPLE 3   Prove that if n is an integer and $3n + 2$ is odd, then n is odd.

Solution: We first attempt a direct proof. To construct a direct proof, we first assume that $3n + 2$ is an odd integer. This means that $3n + 2 = 2k + 1$ for some integer k. Can we use this fact to show that n is odd? We see that $3n + 1 = 2k$, but there does not seem to be any direct way to conclude that n is odd. Because our attempt at a direct proof failed, we next try a proof by contraposition.

Extra
Examples

The first step in a proof by contraposition is to assume that the conclusion of the conditional statement "If $3n + 2$ is odd, then n is odd" is false; namely, assume that n is even. Then, by the definition of an even integer, $n = 2k$ for some integer k. Substituting $2k$ for n, we find that $3n + 2 = 3(2k) + 2 = 6k + 2 = 2(3k + 1)$. This tells us that $3n + 2$ is even (because it is a multiple of 2), and therefore not odd. This is the negation of the premise of the theorem. Because the negation of the conclusion of the conditional statement implies that the hypothesis is false, the original conditional statement is true. Our proof by contraposition succeeded; we have proved the theorem "If $3n + 2$ is odd, then n is odd."   ◄

EXAMPLE 4   Prove that if $n = ab$, where a and b are positive integers, then $a \leq \sqrt{n}$ or $b \leq \sqrt{n}$.

Solution: Because there is no obvious way of showing that $a \le \sqrt{n}$ or $b \le \sqrt{n}$ directly from the equation $n = ab$, where a and b are positive integers, we attempt a proof by contraposition.

The first step in a proof by contraposition is to assume that the conclusion of the conditional statement "If $n = ab$, where a and b are positive integers, then $a \le \sqrt{n}$ or $b \le \sqrt{n}$" is false. That is, we assume that the statement $(a \le \sqrt{n}) \vee (b \le \sqrt{n})$ is false. Using the meaning of disjunction together with De Morgan's law, we see that this implies that both $a \le \sqrt{n}$ and $b \le \sqrt{n}$ are false. This implies that $a > \sqrt{n}$ and $b > \sqrt{n}$. We can multiply these inequalities together (using the fact that if $0 < s < t$ and $0 < u < v$, then $su < tv$) to obtain $ab > \sqrt{n} \cdot \sqrt{n} = n$. This shows that $ab \ne n$, which contradicts the statement $n = ab$.

Because the negation of the conclusion of the conditional statement implies that the hypothesis is false, the original conditional statement is true. Our proof by contraposition succeeded; we have proved that if $n = ab$, where a and b are positive integers, then $a \le \sqrt{n}$ or $b \le \sqrt{n}$. ◄

VACUOUS AND TRIVIAL PROOFS We can quickly prove that a conditional statement $p \to q$ is true when we know that p is false, because $p \to q$ must be true when p is false. Consequently, if we can show that p is false, then we have a proof, called a **vacuous proof**, of the conditional statement $p \to q$. Vacuous proofs are often used to establish special cases of theorems that state that a conditional statement is true for all positive integers [i.e., a theorem of the kind $\forall n P(n)$, where $P(n)$ is a propositional function]. Proof techniques for theorems of this kind will be discussed in Section 5.1.

EXAMPLE 5 Show that the proposition $P(0)$ is true, where $P(n)$ is "If $n > 1$, then $n^2 > n$" and the domain consists of all integers.

Solution: Note that $P(0)$ is "If $0 > 1$, then $0^2 > 0$." We can show $P(0)$ using a vacuous proof. Indeed, the hypothesis $0 > 1$ is false. This tells us that $P(0)$ is automatically true. ◄

Remark: The fact that the conclusion of this conditional statement, $0^2 > 0$, is false is irrelevant to the truth value of the conditional statement, because a conditional statement with a false hypothesis is guaranteed to be true.

We can also quickly prove a conditional statement $p \to q$ if we know that the conclusion q is true. By showing that q is true, it follows that $p \to q$ must also be true. A proof of $p \to q$ that uses the fact that q is true is called a **trivial proof**. Trivial proofs are often important when special cases of theorems are proved (see the discussion of proof by cases in Section 1.9) and in mathematical induction, which is a proof technique discussed in Section 5.1.

EXAMPLE 6 Let $P(n)$ be "If a and b are positive integers with $a \ge b$, then $a^n \ge b^n$," where the domain consists of all nonnegative integers. Show that $P(0)$ is true.

Solution: The proposition $P(0)$ is "If $a \ge b$, then $a^0 \ge b^0$." Because $a^0 = b^0 = 1$, the conclusion of the conditional statement "If $a \ge b$, then $a^0 \ge b^0$" is true. Hence, this conditional statement, which is $P(0)$, is true. This is an example of a trivial proof. Note that the hypothesis, which is the statement "$a \ge b$," was not needed in this proof. ◄

A LITTLE PROOF STRATEGY We have described two important approaches for proving theorems of the form $\forall x(P(x) \to Q(x))$: direct proof and proof by contraposition. We have also given examples that show how each is used. However, when you are presented with a theorem of the form $\forall x(P(x) \to Q(x))$, which method should you use to attempt to prove it? We will provide a few rules of thumb here; in Section 1.9 we will discuss proof strategy at greater length. When you want to prove a statement of the form $\forall x(P(x) \to Q(x))$, first evaluate whether a direct proof looks promising. Begin by expanding the definitions in the hypotheses. Start to reason using these hypotheses, together with axioms and available theorems. If a direct proof does not seem to go anywhere, try the same thing with a proof by contraposition. Recall that in a proof by contraposition you assume that the conclusion of the conditional statement is false and use a direct proof to show this implies that the hypothesis must be false. We illustrate this strategy in Examples 7 and 8. Before we present our next example, we need a definition.

DEFINITION 2   The real number r is *rational* if there exist integers p and q with $q \neq 0$ such that $r = p/q$. A real number that is not rational is called *irrational*.

EXAMPLE 7   Prove that the sum of two rational numbers is rational. (Note that if we include the implicit quantifiers here, the theorem we want to prove is "For every real number r and every real number s, if r and s are rational numbers, then $r + s$ is rational.)

Extra
Examples

Solution: We first attempt a direct proof. To begin, suppose that r and s are rational numbers. From the definition of a rational number, it follows that there are integers p and q, with $q \neq 0$, such that $r = p/q$, and integers t and u, with $u \neq 0$, such that $s = t/u$. Can we use this information to show that $r + s$ is rational? The obvious next step is to add $r = p/q$ and $s = t/u$, to obtain

$$r + s = \frac{p}{q} + \frac{t}{u} = \frac{pu + qt}{qu}.$$

Because $q \neq 0$ and $u \neq 0$, it follows that $qu \neq 0$. Consequently, we have expressed $r + s$ as the ratio of two integers, $pu + qt$ and qu, where $qu \neq 0$. This means that $r + s$ is rational. We have proved that the sum of two rational numbers is rational; our attempt to find a direct proof succeeded. ◄

EXAMPLE 8   Prove that if n is an integer and n^2 is odd, then n is odd.

Solution: We first attempt a direct proof. Suppose that n is an integer and n^2 is odd. Then, there exists an integer k such that $n^2 = 2k + 1$. Can we use this information to show that n is odd? There seems to be no obvious approach to show that n is odd because solving for n produces the equation $n = \pm\sqrt{2k + 1}$, which is not terribly useful.

Because this attempt to use a direct proof did not bear fruit, we next attempt a proof by contraposition. We take as our hypothesis the statement that n is not odd. Because every integer is odd or even, this means that n is even. This implies that there exists an integer k such that $n = 2k$. To prove the theorem, we need to show that this hypothesis implies the conclusion that n^2 is not odd, that is, that n^2 is even. Can we use the equation $n = 2k$ to achieve this? By squaring both sides of this equation, we obtain $n^2 = 4k^2 = 2(2k^2)$, which implies that n^2 is also even because $n^2 = 2t$, where $t = 2k^2$. We have proved that if n is an integer and n^2 is odd, then n is odd. Our attempt to find a proof by contraposition succeeded. ◄

Proofs by Contradiction

Suppose we want to prove that a statement p is true. Furthermore, suppose that we can find a contradiction q such that $\neg p \rightarrow q$ is true. Because q is false, but $\neg p \rightarrow q$ is true, we can conclude that $\neg p$ is false, which means that p is true. How can we find a contradiction q that might help us prove that p is true in this way?

Because the statement $r \wedge \neg r$ is a contradiction whenever r is a proposition, we can prove that p is true if we can show that $\neg p \rightarrow (r \wedge \neg r)$ is true for some proposition r. Proofs of this type are called **proofs by contradiction**. Because a proof by contradiction does not prove a result directly, it is another type of indirect proof. We provide three examples of proof by contradiction. The first is an example of an application of the pigeonhole principle, a combinatorial technique that we will cover in depth in Section 6.2.

EXAMPLE 9   Show that at least four of any 22 days must fall on the same day of the week.

Solution: Let p be the proposition "At least four of 22 chosen days fall on the same day of the week." Suppose that $\neg p$ is true. This means that at most three of the 22 days fall on the same day of the week. Because there are seven days of the week, this implies that at most 21 days could have been chosen, as for each of the days of the week, at most three of the chosen days could fall on that day. This contradicts the premise that we have 22 days under consideration. That is, if r is the statement that 22 days are chosen, then we have shown that $\neg p \rightarrow (r \wedge \neg r)$. Consequently, we know that p is true. We have proved that at least four of 22 chosen days fall on the same day of the week. ◄

Extra
Examples

EXAMPLE 10   Prove that $\sqrt{2}$ is irrational by giving a proof by contradiction.

Solution: Let p be the proposition "$\sqrt{2}$ is irrational." To start a proof by contradiction, we suppose that $\neg p$ is true. Note that $\neg p$ is the statement "It is not the case that $\sqrt{2}$ is irrational," which says that $\sqrt{2}$ is rational. We will show that assuming that $\neg p$ is true leads to a contradiction.

If $\sqrt{2}$ is rational, there exist integers a and b with $\sqrt{2} = a/b$, where $b \neq 0$ and a and b have no common factors (so that the fraction a/b is in lowest terms.) (Here, we are using the fact that every rational number can be written in lowest terms.) Because $\sqrt{2} = a/b$, when both sides of this equation are squared, it follows that

$$2 = \frac{a^2}{b^2}.$$

Hence,

$$2b^2 = a^2.$$

By the definition of an even integer it follows that a^2 is even. We next use the fact that if a^2 is even, a must also be even, which follows by Exercise 16. Furthermore, because a is even, by the definition of an even integer, $a = 2c$ for some integer c. Thus,

$$2b^2 = 4c^2.$$

Dividing both sides of this equation by 2 gives

$$b^2 = 2c^2.$$

By the definition of even, this means that b^2 is even. Again using the fact that if the square of an integer is even, then the integer itself must be even, we conclude that b must be even as well.

We have now shown that the assumption of $\neg p$ leads to the equation $\sqrt{2} = a/b$, where a and b have no common factors, but both a and b are even, that is, 2 divides both a and b. Note that the statement that $\sqrt{2} = a/b$, where a and b have no common factors, means, in particular, that 2 does not divide both a and b. Because our assumption of $\neg p$ leads to the contradiction that 2 divides both a and b and 2 does not divide both a and b, $\neg p$ must be false. That is, the statement p, "$\sqrt{2}$ is irrational," is true. We have proved that $\sqrt{2}$ is irrational. ◄

Proof by contradiction can be used to prove conditional statements. In such proofs, we first assume the negation of the conclusion. We then use the premises of the theorem and the negation of the conclusion to arrive at a contradiction. (The reason that such proofs are valid rests on the logical equivalence of $p \rightarrow q$ and $(p \wedge \neg q) \rightarrow \mathbf{F}$. To see that these statements are equivalent, simply note that each is false in exactly one case, namely when p is true and q is false.)

Note that we can rewrite a proof by contraposition of a conditional statement as a proof by contradiction. In a proof of $p \rightarrow q$ by contraposition, we assume that $\neg q$ is true. We then show that $\neg p$ must also be true. To rewrite a proof by contraposition of $p \rightarrow q$ as a proof by contradiction, we suppose that both p and $\neg q$ are true. Then, we use the steps from the proof of $\neg q \rightarrow \neg p$ to show that $\neg p$ is true. This leads to the contradiction $p \wedge \neg p$, completing the proof. Example 11 illustrates how a proof by contraposition of a conditional statement can be rewritten as a proof by contradiction.

EXAMPLE 11   Give a proof by contradiction of the theorem "If $3n + 2$ is odd, then n is odd."

Solution: Let p be "$3n + 2$ is odd" and q be "n is odd." To construct a proof by contradiction, assume that both p and $\neg q$ are true. That is, assume that $3n + 2$ is odd and that n is not odd. Because n is not odd, we know that it is even. Because n is even, there is an integer k such that $n = 2k$. This implies that $3n + 2 = 3(2k) + 2 = 6k + 2 = 2(3k + 1)$. Because $3n + 2$ is $2t$, where $t = 3k + 1$, $3n + 2$ is even. Note that the statement "$3n + 2$ is even" is equivalent to the statement $\neg p$, because an integer is even if and only if it is not odd. Because both p and $\neg p$ are true, we have a contradiction. This completes the proof by contradiction, proving that if $3n + 2$ is odd, then n is odd.   ◄

Note that we can also prove by contradiction that $p \rightarrow q$ is true by assuming that p and $\neg q$ are true, and showing that q must be also be true. This implies that $\neg q$ and q are both true, a contradiction. This observation tells us that we can turn a direct proof into a proof by contradiction.

PROOFS OF EQUIVALENCE   To prove a theorem that is a biconditional statement, that is, a statement of the form $p \leftrightarrow q$, we show that $p \rightarrow q$ and $q \rightarrow p$ are both true. The validity of this approach is based on the tautology

$$(p \leftrightarrow q) \leftrightarrow (p \rightarrow q) \wedge (q \rightarrow p).$$

EXAMPLE 12   Prove the theorem "If n is an integer, then n is odd if and only if n^2 is odd."

Solution: This theorem has the form "p if and only if q," where p is "n is odd" and q is "n^2 is odd." (As usual, we do not explicitly deal with the universal quantification.) To prove this theorem, we need to show that $p \rightarrow q$ and $q \rightarrow p$ are true.

We have already shown (in Example 1) that $p \rightarrow q$ is true and (in Example 8) that $q \rightarrow p$ is true.

Because we have shown that both $p \rightarrow q$ and $q \rightarrow p$ are true, we have shown that the theorem is true.   ◄

Sometimes a theorem states that several propositions are equivalent. Such a theorem states that propositions $p_1, p_2, p_3, \ldots, p_n$ are equivalent. This can be written as

$$p_1 \leftrightarrow p_2 \leftrightarrow \cdots \leftrightarrow p_n,$$

which states that all n propositions have the same truth values, and consequently, that for all i and j with $1 \le i \le n$ and $1 \le j \le n$, p_i and p_j are equivalent. One way to prove these mutually equivalent is to use the tautology

$$p_1 \leftrightarrow p_2 \leftrightarrow \cdots \leftrightarrow p_n \leftrightarrow (p_1 \rightarrow p_2) \wedge (p_2 \rightarrow p_3) \wedge \cdots \wedge (p_n \rightarrow p_1).$$

This shows that if the n conditional statements $p_1 \rightarrow p_2, p_2 \rightarrow p_3, \ldots, p_n \rightarrow p_1$ can be shown to be true, then the propositions $p_1, p_2, \ldots, p_n$ are all equivalent.

This is much more efficient than proving that $p_i \rightarrow p_j$ for all $i \ne j$ with $1 \le i \le n$ and $1 \le j \le n$. (Note that there are $n^2 - n$ such conditional statements.)

When we prove that a group of statements are equivalent, we can establish any chain of conditional statements we choose as long as it is possible to work through the chain to go from any one of these statements to any other statement. For example, we can show that p_1, p_2, and p_3 are equivalent by showing that $p_1 \rightarrow p_3$, $p_3 \rightarrow p_2$, and $p_2 \rightarrow p_1$.

EXAMPLE 13 Show that these statements about the integer n are equivalent:

$\quad p_1$: n is even.
$\quad p_2$: $n - 1$ is odd.
$\quad p_3$: n^2 is even.

Solution: We will show that these three statements are equivalent by showing that the conditional statements $p_1 \rightarrow p_2$, $p_2 \rightarrow p_3$, and $p_3 \rightarrow p_1$ are true.

We use a direct proof to show that $p_1 \rightarrow p_2$. Suppose that n is even. Then $n = 2k$ for some integer k. Consequently, $n - 1 = 2k - 1 = 2(k - 1) + 1$. This means that $n - 1$ is odd because it is of the form $2m + 1$, where m is the integer $k - 1$.

We also use a direct proof to show that $p_2 \rightarrow p_3$. Now suppose $n - 1$ is odd. Then $n - 1 = 2k + 1$ for some integer k. Hence, $n = 2k + 2$ so that $n^2 = (2k + 2)^2 = 4k^2 + 8k + 4 = 2(2k^2 + 4k + 2)$. This means that n^2 is twice the integer $2k^2 + 4k + 2$, and hence is even.

To prove $p_3 \rightarrow p_1$, we use a proof by contraposition. That is, we prove that if n is not even, then n^2 is not even. This is the same as proving that if n is odd, then n^2 is odd, which we have already done in Example 1. This completes the proof. ◀

COUNTEREXAMPLES   In Section 1.4 we stated that to show that a statement of the form $\forall x P(x)$ is false, we need only find a **counterexample**, that is, an example x for which $P(x)$ is false. When presented with a statement of the form $\forall x P(x)$, which we believe to be false or which has resisted all proof attempts, we look for a counterexample. We illustrate the use of counterexamples in Example 14.

EXAMPLE 14 Show that the statement "Every positive integer is the sum of the squares of two integers" is false.

Solution: To show that this statement is false, we look for a counterexample, which is a particular integer that is not the sum of the squares of two integers. It does not take long to find a counterexample, because 3 cannot be written as the sum of the squares of two integers. To show this is the case, note that the only perfect squares not exceeding 3 are $0^2 = 0$ and $1^2 = 1$. Furthermore, there is no way to get 3 as the sum of two terms each of which is 0 or 1. Consequently, we have shown that "Every positive integer is the sum of the squares of two integers" is false. ◀

Mistakes in Proofs

There are many common errors made in constructing mathematical proofs. We will briefly describe some of these here. Among the most common errors are mistakes in arithmetic and basic algebra. Even professional mathematicians make such errors, especially when working with complicated formulae. Whenever you use such computations you should check them as carefully as possible. (You should also review any troublesome aspects of basic algebra, especially before you study Section 5.1.)

Each step of a mathematical proof needs to be correct and the conclusion needs to follow logically from the steps that precede it. Many mistakes result from the introduction of steps that do not logically follow from those that precede it. This is illustrated in Examples 15–17.

EXAMPLE 15 What is wrong with this famous supposed "proof" that $1 = 2$?

"Proof:" We use these steps, where a and b are two equal positive integers.

Step	**Reason**
1. $a = b$	Given
2. $a^2 = ab$	Multiply both sides of (1) by a
3. $a^2 - b^2 = ab - b^2$	Subtract b^2 from both sides of (2)
4. $(a - b)(a + b) = b(a - b)$	Factor both sides of (3)
5. $a + b = b$	Divide both sides of (4) by $a - b$
6. $2b = b$	Replace a by b in (5) because $a = b$ and simplify
7. $2 = 1$	Divide both sides of (6) by b

Solution: Every step is valid except for one, step 5 where we divided both sides by $a - b$. The error is that $a - b$ equals zero; division of both sides of an equation by the same quantity is valid as long as this quantity is not zero.   ◄

EXAMPLE 16   What is wrong with this "proof?"

"Theorem:" If n^2 is positive, then n is positive.

"Proof:" Suppose that n^2 is positive. Because the conditional statement "If n is positive, then n^2 is positive" is true, we can conclude that n is positive.

Solution: Let $P(n)$ be "n is positive" and $Q(n)$ be "n^2 is positive." Then our hypothesis is $Q(n)$. The statement "If n is positive, then n^2 is positive" is the statement $\forall n(P(n) \to Q(n))$. From the hypothesis $Q(n)$ and the statement $\forall n(P(n) \to Q(n))$ we cannot conclude $P(n)$, because we are not using a valid rule of inference. Instead, this is an example of the fallacy of affirming the conclusion. A counterexample is supplied by $n = -1$ for which $n^2 = 1$ is positive, but n is negative.   ◄

EXAMPLE 17   What is wrong with this "proof?"

"Theorem:" If n is not positive, then n^2 is not positive. (This is the contrapositive of the "theorem" in Example 16.)

"Proof:" Suppose that n is not positive. Because the conditional statement "If n is positive, then n^2 is positive" is true, we can conclude that n^2 is not positive.

Solution: Let $P(n)$ and $Q(n)$ be as in the solution of Example 16. Then our hypothesis is $\neg P(n)$ and the statement "If n is positive, then n^2 is positive" is the statement $\forall n(P(n) \to Q(n))$. From the hypothesis $\neg P(n)$ and the statement $\forall n(P(n) \to Q(n))$ we cannot conclude $\neg Q(n)$, because we are not using a valid rule of inference. Instead, this is an example of the fallacy of denying the hypothesis. A counterexample is supplied by $n = -1$, as in Example 16.   ◄

Finally, we briefly discuss a particularly nasty type of error. Many incorrect arguments are based on a fallacy called **begging the question**. This fallacy occurs when one or more steps of a proof are based on the truth of the statement being proved. In other words, this fallacy arises when a statement is proved using itself, or a statement equivalent to it. That is why this fallacy is also called **circular reasoning**.

EXAMPLE 18   Is the following argument correct? It supposedly shows that n is an even integer whenever n^2 is an even integer.

Suppose that n^2 is even. Then $n^2 = 2k$ for some integer k. Let $n = 2l$ for some integer l. This shows that n is even.

Solution: This argument is incorrect. The statement "let $n = 2l$ for some integer l" occurs in the proof. No argument has been given to show that n can be written as $2l$ for some integer l. This is circular reasoning because this statement is equivalent to the statement being proved, namely, "n is even." Of course, the result itself is correct; only the method of proof is wrong. ◀

Making mistakes in proofs is part of the learning process. When you make a mistake that someone else finds, you should carefully analyze where you went wrong and make sure that you do not make the same mistake again. Even professional mathematicians make mistakes in proofs. More than a few incorrect proofs of important results have fooled people for many years before subtle errors in them were found.

Just a Beginning

We have now developed a basic arsenal of proof methods. In the next section we will introduce other important proof methods. We will also introduce several important proof techniques in Chapter 5, including mathematical induction, which can be used to prove results that hold for all positive integers. In Chapter 6 we will introduce the notion of combinatorial proofs.

In this section we introduced several methods for proving theorems of the form $\forall x(P(x) \to Q(x))$, including direct proofs and proofs by contraposition. There are many theorems of this type whose proofs are easy to construct by directly working through the hypotheses and definitions of the terms of the theorem. However, it is often difficult to prove a theorem without resorting to a clever use of a proof by contraposition or a proof by contradiction, or some other proof technique. In Section 1.9 we will address proof strategy. We will describe various approaches that can be used to find proofs when straightforward approaches do not work. Constructing proofs is an art that can be learned only through experience, including writing proofs, having your proofs critiqued, and reading and analyzing other proofs.

Exercises

1. Use a direct proof to show that the sum of two odd integers is even.

2. Show that the square of an even number is an even number using a direct proof.

3. Prove that if $m + n$ and $n + p$ are even integers, where m, n, and p are integers, then $m + p$ is even. What kind of proof did you use?

4. Use a direct proof to show that every odd integer is the difference of two squares.

5. Use a proof by contradiction to prove that the sum of an irrational number and a rational number is irrational.

6. Prove that if n is a perfect square, then $n + 2$ is not a perfect square.

7. Prove or disprove that the product of two irrational numbers is irrational.

8. Prove that if x is irrational, then $1/x$ is irrational.

9. Show that if n is an integer and $n^3 + 5$ is odd, then n is even using

   a) a proof by contraposition.

   b) a proof by contradiction.

10. Prove that if m and n are integers and mn is even, then m is even or n is even.

11. Prove the proposition $P(0)$, where $P(n)$ is the proposition "If n is a positive integer greater than 1, then $n^2 > n$." What kind of proof did you use?

12. Show that if you pick three socks from a drawer containing just blue socks and black socks, you must get either a pair of blue socks or a pair of black socks.

13. Let $P(n)$ be the proposition "If a and b are positive real numbers, then $(a + b)^n \geq a^n + b^n$." Prove that $P(1)$ is true. What kind of proof did you use?

14. Show that at least ten of any 64 days chosen must fall on the same day of the week.

15. Use a proof by contradiction to show that there is no rational number r for which $r^3 + r + 1 = 0$. [*Hint:* Assume that $r = a/b$ is a root, where a and b are integers and a/b is in lowest terms. Obtain an equation involving integers by multiplying by b^3. Then look at whether a and b are each odd or even.]

16. Prove that if n is a positive integer, then n is odd if and only if $5n + 6$ is odd.

17. Prove or disprove that if m and n are integers such that $mn = 1$, then either $m = 1$ and $n = 1$, or else $m = -1$ and $n = -1$.

18. Prove that $m^2 = n^2$ if and only if $m = n$ or $m = -n$.

19. Are these steps for finding the solutions of $\sqrt{x + 3} = 3 - x$ correct? (*1*) $\sqrt{x + 3} = 3 - x$ is given; (*2*) $x + 3 = x^2 - 6x + 9$, obtained by squaring both sides of (1); (*3*) $0 = x^2 - 7x + 6$, obtained by subtracting $x + 3$ from both sides of (2); (*4*) $0 = (x - 1)(x - 6)$, obtained by factoring the right-hand side of (3); (*5*) $x = 1$ or $x = 6$,

which follows from (4) because $ab = 0$ implies that $a = 0$ or $b = 0$.

20. Is this reasoning for finding the solutions of the equation $\sqrt{2x^2 - 1} = x$ correct? (*1*) $\sqrt{2x^2 - 1} = x$ is given; (*2*) $2x^2 - 1 = x^2$, obtained by squaring both sides of (1); (*3*) $x^2 - 1 = 0$, obtained by subtracting x^2 from both sides of (2); (*4*) $(x - 1)(x + 1) = 0$, obtained by factoring the left-hand side of $x^2 - 1$; (*5*) $x = 1$ or $x = -1$, which follows because $ab = 0$ implies that $a = 0$ or $b = 0$.

21. Prove that if n is an integer, these four statements are equivalent: (*i*) n is even, (*ii*) $n + 1$ is odd, (*iii*) $3n + 1$ is odd, (*iv*) $3n$ is even.

22. Show that if the first 10 positive integers are placed around a circle, in any order, there exist three integers in consecutive locations around the circle that have a sum greater than or equal to 17.

1.9   Proof Methods and Strategy

Introduction

In Section 1.8 we introduced many methods of proof and illustrated how each method can be used. In this section we continue this effort. We will introduce several other commonly used proof methods, including the method of proving a theorem by considering different cases separately. We will also discuss proofs where we prove the existence of objects with desired properties.

In Section 1.8 we briefly discussed the strategy behind constructing proofs. This strategy includes selecting a proof method and then successfully constructing an argument step by step, based on this method. In this section, after we have developed a versatile arsenal of proof methods, we will study some aspects of the art and science of proofs. We will provide advice on how to find a proof of a theorem. We will describe some tricks of the trade, including how proofs can be found by working backward and by adapting existing proofs.

When mathematicians work, they formulate conjectures and attempt to prove or disprove them. We will briefly describe this process here by proving results about tiling checkerboards with dominoes and other types of pieces. Looking at tilings of this kind, we will be able to quickly formulate conjectures and prove theorems without first developing a theory.

We will conclude the section by discussing the role of open questions. In particular, we will discuss some interesting problems either that have been solved after remaining open for hundreds of years or that still remain open.

Exhaustive Proof and Proof by Cases

Sometimes we cannot prove a theorem using a single argument that holds for all possible cases. We now introduce a method that can be used to prove a theorem, by considering different cases separately. This method is based on a rule of inference that we will now introduce. To prove a conditional statement of the form

$$(p_1 \lor p_2 \lor \cdots \lor p_n) \to q$$

the tautology

$$[(p_1 \lor p_2 \lor \cdots \lor p_n) \to q] \leftrightarrow [(p_1 \to q) \land (p_2 \to q) \land \cdots \land (p_n \to q)]$$

can be used as a rule of inference. This shows that the original conditional statement with a hypothesis made up of a disjunction of the propositions $p_1, p_2, \ldots, p_n$ can be proved by proving each of the n conditional statements $p_i \rightarrow q$, $i = 1, 2, \ldots, n$, individually. Such an argument is called a **proof by cases**. Sometimes to prove that a conditional statement $p \rightarrow q$ is true, it is convenient to use a disjunction $p_1 \vee p_2 \vee \cdots \vee p_n$ instead of p as the hypothesis of the conditional statement, where p and $p_1 \vee p_2 \vee \cdots \vee p_n$ are equivalent.

EXHAUSTIVE PROOF   Some theorems can be proved by examining a relatively small number of examples. Such proofs are called **exhaustive proofs**, or **proofs by exhaustion** because these proofs proceed by exhausting all possibilities. An exhaustive proof is a special type of proof by cases where each case involves checking a single example. We now provide some illustrations of exhaustive proofs.

EXAMPLE 1   Prove that $(n + 1)^3 \geq 3^n$ if n is a positive integer with $n \leq 4$.

Solution: We use a proof by exhaustion. We only need verify the inequality $(n + 1)^3 \geq 3^n$ when $n = 1, 2, 3,$ and 4. For $n = 1$, we have $(n + 1)^3 = 2^3 = 8$ and $3^n = 3^1 = 3$; for $n = 2$, we have $(n + 1)^3 = 3^3 = 27$ and $3^n = 3^2 = 9$; for $n = 3$, we have $(n + 1)^3 = 4^3 = 64$ and $3^n = 3^3 = 27$; and for $n = 4$, we have $(n + 1)^3 = 5^3 = 125$ and $3^n = 3^4 = 81$. In each of these four cases, we see that $(n + 1)^3 \geq 3^n$. We have used the method of exhaustion to prove that $(n + 1)^3 \geq 3^n$ if n is a positive integer with $n \leq 4$.   ◀

Extra
Examples

EXAMPLE 2   Prove that the only consecutive positive integers not exceeding 100 that are perfect powers are 8 and 9. (An integer is a **perfect power** if it equals n^a, where a is an integer greater than 1.)

Solution: We use a proof by exhaustion. In particular, we can prove this fact by examining positive integers n not exceeding 100, first checking whether n is a perfect power, and if it is, checking whether $n + 1$ is also a perfect power. A quicker way to do this is simply to look at all perfect powers not exceeding 100 and checking whether the next largest integer is also a perfect power. The squares of positive integers not exceeding 100 are 1, 4, 9, 16, 25, 36, 49, 64, 81, and 100. The cubes of positive integers not exceeding 100 are 1, 8, 27, and 64. The fourth powers of positive integers not exceeding 100 are 1, 16, and 81. The fifth powers of positive integers not exceeding 100 are 1 and 32. The sixth powers of positive integers not exceeding 100 are 1 and 64. There are no powers of positive integers higher than the sixth power not exceeding 100, other than 1. Looking at this list of perfect powers not exceeding 100, we see that $n = 8$ is the only perfect power n for which $n + 1$ is also a perfect power. That is, $2^3 = 8$ and $3^2 = 9$ are the only two consecutive perfect powers not exceeding 100.   ◀

Proofs by exhaustion can tire out people and computers when the number of cases challenges the available processing power!

    People can carry out exhaustive proofs when it is necessary to check only a relatively small number of instances of a statement. Computers do not complain when they are asked to check a much larger number of instances of a statement, but they still have limitations. Note that not even a computer can check all instances when it is impossible to list all instances to check.

PROOF BY CASES   A proof by cases must cover all possible cases that arise in a theorem. We illustrate proof by cases with a couple of examples. In each example, you should check that all possible cases are covered.

EXAMPLE 3   Prove that if n is an integer, then $n^2 \geq n$.

Solution: We can prove that $n^2 \geq n$ for every integer by considering three cases, when $n = 0$, when $n \geq 1$, and when $n \leq -1$. We split the proof into three cases because it is straightforward to prove the result by considering zero, positive integers, and negative integers separately.

Case (i): When $n = 0$, because $0^2 = 0$, we see that $0^2 \geq 0$. It follows that $n^2 \geq n$ is true in this case.

Case (ii): When $n \geq 1$, when we multiply both sides of the inequality $n \geq 1$ by the positive integer n, we obtain $n \cdot n \geq n \cdot 1$. This implies that $n^2 \geq n$ for $n \geq 1$.

Case (iii): In this case $n \leq -1$. However, $n^2 \geq 0$. It follows that $n^2 \geq n$.

Because the inequality $n^2 \geq n$ holds in all three cases, we can conclude that if n is an integer, then $n^2 \geq n$. ◄

EXAMPLE 4   Use a proof by cases to show that $|xy| = |x||y|$, where x and y are real numbers. (Recall that $|a|$, the absolute value of a, equals a when $a \geq 0$ and equals $-a$ when $a \leq 0$.)

Solution: In our proof of this theorem, we remove absolute values using the fact that $|a| = a$ when $a \geq 0$ and $|a| = -a$ when $a < 0$. Because both $|x|$ and $|y|$ occur in our formula, we will need four cases: *(i)* x and y both nonnegative, *(ii)* x nonnegative and y is negative, *(iii)* x negative and y nonnegative, and *(iv)* x negative and y negative. We denote by p_1, p_2, p_3, and p_4, the proposition stating the assumption for each of these four cases, respectively.

(Note that we can remove the absolute value signs by making the appropriate choice of signs within each case.)

Case (i): We see that $p_1 \rightarrow q$ because $xy \geq 0$ when $x \geq 0$ and $y \geq 0$, so that $|xy| = xy = |x||y|$.

Case (ii): To see that $p_2 \rightarrow q$, note that if $x \geq 0$ and $y < 0$, then $xy \leq 0$, so that $|xy| = -xy = x(-y) = |x||y|$. (Here, because $y < 0$, we have $|y| = -y$.)

Case (iii): To see that $p_3 \rightarrow q$, we follow the same reasoning as the previous case with the roles of x and y reversed.

Case (iv): To see that $p_4 \rightarrow q$, note that when $x < 0$ and $y < 0$, it follows that $xy > 0$. Hence, $|xy| = xy = (-x)(-y) = |x||y|$.

Because $|xy| = |x||y|$ holds in each of the four cases and these cases exhaust all possibilities, we can conclude that $|xy| = |x||y|$, whenever x and y are real numbers. ◄

LEVERAGING PROOF BY CASES   The examples we have presented illustrating proof by cases provide some insight into when to use this method of proof. In particular, when it is not possible to consider all cases of a proof at the same time, a proof by cases should be considered. When should you use such a proof? Generally, look for a proof by cases when there is no obvious way to begin a proof, but when extra information in each case helps move the proof forward. Example 5 illustrates how the method of proof by cases can be used effectively.

EXAMPLE 5   Formulate a conjecture about the final decimal digit of the square of an integer and prove your result.

Solution: The smallest perfect squares are 1, 4, 9, 16, 25, 36, 49, 64, 81, 100, 121, 144, 169, 196, 225, and so on. We notice that the digits that occur as the final digit of a square are 0, 1, 4, 5, 6, and 9, with 2, 3, 7, and 8 never appearing as the final digit of a square. We conjecture this theorem: The final decimal digit of a perfect square is 0, 1, 4, 5, 6 or 9. How can we prove this theorem?

We first note that we can express an integer n as $10a + b$, where a and b are positive integers and b is 0, 1, 2, 3, 4, 5, 6, 7, 8, or 9. Here a is the integer obtained by subtracting the final decimal digit of n from n and dividing by 10. Next, note that $(10a + b)^2 = 100a^2 + 20ab + b^2 = 10(10a^2 + 2b) + b^2$, so that the final decimal digit of n^2 is the same as the final decimal digit of b^2. Furthermore, note that the final decimal digit of b^2 is the same as the final decimal digit of $(10 - b)^2 = 100 - 20b + b^2$. Consequently, we can reduce our proof to the consideration of six cases.

Case (i): The final digit of n is 1 or 9. Then the final decimal digit of n^2 is the final decimal digit of $1^2 = 1$ or $9^2 = 81$, namely 1.

Case (ii): The final digit of n is 2 or 8. Then the final decimal digit of n^2 is the final decimal digit of $2^2 = 4$ or $8^2 = 64$, namely 4.

Case (iii): The final digit of n is 3 or 7. Then the final decimal digit of n^2 is the final decimal digit of $3^2 = 9$ or $7^2 = 49$, namely 9.

Case (iv): The final digit of n is 4 or 6. Then the final decimal digit of n^2 is the final decimal digit of $4^2 = 16$ or $6^2 = 36$, namely 6.

Case (v): The final decimal digit of n is 5. Then the final decimal digit of n^2 is the final decimal digit of $5^2 = 25$, namely 5.

Case (vi): The final decimal digit of n is 0. Then the final decimal digit of n^2 is the final decimal digit of $0^2 = 0$, namely 0.

Because we have considered all six cases, we can conclude that the final decimal digit of n^2, where n is an integer is either 0, 1, 2, 4, 5, 6, or 9. ◀

Sometimes we can eliminate all but a few examples in a proof by cases, as Example 6 illustrates.

EXAMPLE 6   Show that there are no solutions in integers x and y of $x^2 + 3y^2 = 8$.

Solution: We can quickly reduce a proof to checking just a few simple cases because $x^2 > 8$ when $|x| \geq 3$ and $3y^2 > 8$ when $|y| \geq 2$. This leaves the cases when x equals $-2, -1, 0, 1$, or 2 and y equals $-1, 0$, or 1. We can finish using an exhaustive proof. To dispense with the remaining cases, we note that possible values for x^2 are 0, 1, and 4, and possible values for $3y^2$ are 0 and 3, and the largest sum of possible values for x^2 and $3y^2$ is 7. Consequently, it is impossible for $x^2 + 3y^2 = 8$ to hold when x and y are integers. ◀

WITHOUT LOSS OF GENERALITY   In the proof in Example 4, we dismissed case *(iii)*, where $x < 0$ and $y \geq 0$, because it is the same as case *(ii)*, where $x \geq 0$ and $y < 0$, with the roles of x and y reversed. To shorten the proof, we could have proved cases *(ii)* and *(iii)* together by assuming, **without loss of generality**, that $x \geq 0$ and $y < 0$. Implicit in this statement is that we can complete the case with $x < 0$ and $y \geq 0$ using the same argument as we used for the case with $x \geq 0$ and $y < 0$, but with the obvious changes.

In a proof by cases be sure not to omit any cases and check that you have proved all cases correctly!

In general, when the phrase "without loss of generality" is used in a proof (often abbreviated as WLOG), we assert that by proving one case of a theorem, no additional argument is required to prove other specified cases. That is, other cases follow by making straightforward changes to the argument, or by filling in some straightforward initial step. Proofs by cases can often be made much more efficient when the notion of without loss of generality is employed. Of course, incorrect use of this principle can lead to unfortunate errors. Sometimes assumptions are made that lead to a loss in generality. Such assumptions can be made that do not take into account that one case may be substantially different from others. This can lead to an incomplete, and possibly unsalvageable, proof. In fact, many incorrect proofs of famous theorems turned out to rely on arguments that used the idea of "without loss of generality" to establish cases that could not be quickly proved from simpler cases.

We now illustrate a proof where without loss of generality is used effectively together with other proof techniques.

EXAMPLE 7   Show that if x and y are integers and both xy and $x + y$ are even, then both x and y are even.

Solution: We will use proof by contraposition, the notion of without loss of generality, and proof by cases. First, suppose that x and y are not both even. That is, assume that x is odd or that y is odd (or both). Without loss of generality, we assume that x is odd, so that $x = 2m + 1$ for some integer k.

To complete the proof, we need to show that xy is odd or $x + y$ is odd. Consider two cases: (i) y even, and (ii) y odd. In (i), $y = 2n$ for some integer n, so that $x + y = (2m + 1) + 2n = 2(m + n) + 1$ is odd. In (ii), $y = 2n + 1$ for some integer n, so that $xy = (2m + 1)(2n + 1) = 4mn + 2m + 2n + 1 = 2(2mn + m + n) + 1$ is odd. This completes the proof by contraposition. (Note that our use of without loss of generality within the proof is justified because the proof when y is odd can be obtained by simply interchanging the roles of x and y in the proof we have given.)   ◀

COMMON ERRORS WITH EXHAUSTIVE PROOF AND PROOF BY CASES   A common error of reasoning is to draw incorrect conclusions from examples. No matter how many separate examples are considered, a theorem is not proved by considering examples unless every possible case is covered. The problem of proving a theorem is analogous to showing that a computer program always produces the output desired. No matter how many input values are tested, unless all input values are tested, we cannot conclude that the program always produces the correct output.

EXAMPLE 8   Is it true that every positive integer is the sum of 18 fourth powers of integers?

Solution: To determine whether a positive integer n can be written as the sum of 18 fourth powers of integers, we might begin by examining whether n is the sum of 18 fourth powers of integers for the smallest positive integers. Because the fourth powers of integers are $0, 1, 16, 81, \ldots$, if we can select 18 terms from these numbers that add up to n, then n is the sum of 18 fourth powers. We can show that all positive integers up to 78 can be written as the sum of 18 fourth powers. (The details are left to the reader.) However, if we decided this was enough checking, we would come to the wrong conclusion. It is not true that every positive integer is the sum of 18 fourth powers because 79 is not the sum of 18 fourth powers (as the reader can verify).   ◀

Another common error involves making unwarranted assumptions that lead to incorrect proofs by cases where not all cases are considered. This is illustrated in Example 9.

EXAMPLE 9   What is wrong with this "proof?"

"Theorem:" If x is a real number, then x^2 is a positive real number.

"Proof:" Let p_1 be "x is positive," let p_2 be "x is negative," and let q be "x^2 is positive." To show that $p_1 \rightarrow q$ is true, note that when x is positive, x^2 is positive because it is the product of two positive numbers, x and x. To show that $p_2 \rightarrow q$, note that when x is negative, x^2 is positive because it is the product of two negative numbers, x and x. This completes the proof.

Solution: The problem with this "proof" is that we missed the case of $x = 0$. When $x = 0$, $x^2 = 0$ is not positive, so the supposed theorem is false. If p is "x is a real number," then we can prove results where p is the hypothesis with three cases, p_1, p_2, and p_3, where p_1 is "x is positive," p_2 is "x is negative," and p_3 is "$x = 0$" because of the equivalence $p \leftrightarrow p_1 \vee p_2 \vee p_3$.   ◀

Existence Proofs

Many theorems are assertions that objects of a particular type exist. A theorem of this type is a proposition of the form $\exists x P(x)$, where P is a predicate. A proof of a proposition of the form $\exists x P(x)$ is called an **existence proof**. There are several ways to prove a theorem of this type. Sometimes an existence proof of $\exists x P(x)$ can be given by finding an element a, called a **witness**, such that $P(a)$ is true. This type of existence proof is called **constructive**. It is also possible to give an existence proof that is **nonconstructive**; that is, we do not find an element a such that $P(a)$ is true, but rather prove that $\exists x\, P(x)$ is true in some other way. One common method of giving a nonconstructive existence proof is to use proof by contradiction and show that the negation of the existential quantification implies a contradiction. The concept of a constructive existence proof is illustrated by Example 10 and the concept of a nonconstructive existence proof is illustrated by Example 11.

EXAMPLE 10 **A Constructive Existence Proof**   Show that there is a positive integer that can be written as the sum of cubes of positive integers in two different ways.

Extra Examples

Solution: After considerable computation (such as a computer search) we find that

$$1729 = 10^3 + 9^3 = 12^3 + 1^3.$$

Because we have displayed a positive integer that can be written as the sum of cubes in two different ways, we are done.

There is an interesting story pertaining to this example. The English mathematician G. H. Hardy, when visiting the ailing Indian prodigy Ramanujan in the hospital, remarked that 1729, the number of the cab he took, was rather dull. Ramanujan replied "No, it is a very interesting number; it is the smallest number expressible as the sum of cubes in two different ways." ◀

EXAMPLE 11 **A Nonconstructive Existence Proof**   Show that there exist irrational numbers x and y such that x^y is rational.

Links

GODFREY HAROLD HARDY (1877–1947)   Hardy, born in Cranleigh, Surrey, England, was the older of two children of Isaac Hardy and Sophia Hall Hardy. His father was the geography and drawing master at the Cranleigh School and also gave singing lessons and played soccer. His mother gave piano lessons and helped run a boardinghouse for young students. Hardy's parents were devoted to their children's education. Hardy demonstrated his numerical ability at the early age of two when he began writing down numbers into the millions. He had a private mathematics tutor rather than attending regular classes at the Cranleigh School. He moved to Winchester College, a private high school, when he was 13 and was awarded a scholarship. He excelled in his studies and demonstrated a strong interest in mathematics. He entered Trinity College, Cambridge, in 1896 on a scholarship and won several prizes during his time there, graduating in 1899.

Hardy held the position of lecturer in mathematics at Trinity College at Cambridge University from 1906 to 1919, when he was appointed to the Sullivan chair of geometry at Oxford. He had become unhappy with Cambridge over the dismissal of the famous philosopher and mathematician Bertrand Russell from Trinity for antiwar activities and did not like a heavy load of administrative duties. In 1931 he returned to Cambridge as the Sadleirian professor of pure mathematics, where he remained until his retirement in 1942. He was a pure mathematician and held an elitist view of mathematics, hoping that his research could never be applied. Ironically, he is perhaps best known as one of the developers of the Hardy–Weinberg law, which predicts patterns of inheritance. His work in this area appeared as a letter to the journal *Science* in which he used simple algebraic ideas to demonstrate errors in an article on genetics. Hardy worked primarily in number theory and function theory, exploring such topics as the Riemann zeta function, Fourier series, and the distribution of primes. He made many important contributions to many important problems, such as Waring's problem about representing positive integers as sums of kth powers and the problem of representing odd integers as sums of three primes. Hardy is also remembered for his collaborations with John E. Littlewood, a colleague at Cambridge, with whom he wrote more than 100 papers, and the famous Indian mathematical prodigy Srinivasa Ramanujan. His collaboration with Littlewood led to the joke that there were only three important English mathematicians at that time, Hardy, Littlewood, and Hardy–Littlewood, although some people thought that Hardy had invented a fictitious person, Littlewood, because Littlewood was seldom seen outside Cambridge. Hardy had the wisdom of recognizing Ramanujan's genius from unconventional but extremely creative writings Ramanujan sent him, while other mathematicians failed to see the genius. Hardy brought Ramanujan to Cambridge and collaborated on important joint papers, establishing new results on the number of partitions of an integer. Hardy was interested in mathematics education, and his book *A Course of Pure Mathematics* had a profound effect on undergraduate instruction in mathematics in the first half of the twentieth century. Hardy also wrote *A Mathematician's Apology*, in which he gives his answer to the question of whether it is worthwhile to devote one's life to the study of mathematics. It presents Hardy's view of what mathematics is and what a mathematician does.

Hardy had a strong interest in sports. He was an avid cricket fan and followed scores closely. One peculiar trait he had was that he did not like his picture taken (only five snapshots are known) and disliked mirrors, covering them with towels immediately upon entering a hotel room.

Solution: By Example 10 in Section 1.8 we know that $\sqrt{2}$ is irrational. Consider the number $\sqrt{2}^{\sqrt{2}}$. If it is rational, we have two irrational numbers x and y with x^y rational, namely, $x = \sqrt{2}$ and $y = \sqrt{2}$. On the other hand if $\sqrt{2}^{\sqrt{2}}$ is irrational, then we can let $x = \sqrt{2}^{\sqrt{2}}$ and $y = \sqrt{2}$ so that $x^y = (\sqrt{2}^{\sqrt{2}})^{\sqrt{2}} = \sqrt{2}^{(\sqrt{2} \cdot \sqrt{2})} = \sqrt{2}^2 = 2$.

This proof is an example of a nonconstructive existence proof because we have not found irrational numbers x and y such that x^y is rational. Rather, we have shown that either the pair $x = \sqrt{2}$, $y = \sqrt{2}$ or the pair $x = \sqrt{2}^{\sqrt{2}}$, $y = \sqrt{2}$ have the desired property, but we do not know which of these two pairs works! ◀

Nonconstructive existence proofs often are quite subtle, as Example 12 illustrates.

EXAMPLE 12

Chomp is a game played by two players. In this game, cookies are laid out on a rectangular grid. The cookie in the top left position is poisoned, as shown in Figure 1(a). The two players take turns making moves; at each move, a player is required to eat a remaining cookie, together with all cookies to the right and/or below it (see Figure 1(b), for example). The loser is the player who has no choice but to eat the poisoned cookie. We ask whether one of the two players has a winning strategy. That is, can one of the players always make moves that are guaranteed to lead to a win?

SRINIVASA RAMANUJAN (1887–1920)   The famous mathematical prodigy Ramanujan was born and raised in southern India near the city of Madras (now called Chennai). His father was a clerk in a cloth shop. His mother contributed to the family income by singing at a local temple. Ramanujan studied at the local English language school, displaying his talent and interest for mathematics. At the age of 13 he mastered a textbook used by college students. When he was 15, a university student lent him a copy of *Synopsis of Pure Mathematics*. Ramanujan decided to work out the over 6000 results in this book, stated without proof or explanation, writing on sheets later collected to form notebooks. He graduated from high school in 1904, winning a scholarship to the University of Madras. Enrolling in a fine arts curriculum, he neglected his subjects other than mathematics and lost his scholarship. He failed to pass examinations at the university four times from 1904 to 1907, doing well only in mathematics. During this time he filled his notebooks with original writings, sometimes rediscovering already published work and at other times making new discoveries.

Without a university degree, it was difficult for Ramanujan to find a decent job. To survive, he had to depend on the goodwill of his friends. He tutored students in mathematics, but his unconventional ways of thinking and failure to stick to the syllabus caused problems. He was married in 1909 in an arranged marriage to a young woman nine years his junior. Needing to support himself and his wife, he moved to Madras and sought a job. He showed his notebooks of mathematical writings to his potential employers, but the books bewildered them. However, a professor at the Presidency College recognized his genius and supported him, and in 1912 he found work as an accounts clerk, earning a small salary.

Ramanujan continued his mathematical work during this time and published his first paper in 1910 in an Indian journal. He realized that his work was beyond that of Indian mathematicians and decided to write to leading English mathematicians. The first mathematicians he wrote to turned down his request for help. But in January 1913 he wrote to G. H. Hardy, who was inclined to turn Ramanujan down, but the mathematical statements in the letter, although stated without proof, puzzled Hardy. He decided to examine them closely with the help of his colleague and collaborator J. E. Littlewood. They decided, after careful study, that Ramanujan was probably a genius, because his statements "could only be written down by a mathematician of the highest class; they must be true, because if they were not true, no one would have the imagination to invent them."

Hardy arranged a scholarship for Ramanujan, bringing him to England in 1914. Hardy personally tutored him in mathematical analysis, and they collaborated for five years, proving significant theorems about the number of partitions of integers. During this time, Ramanujan made important contributions to number theory and also worked on continued fractions, infinite series, and elliptic functions. Ramanujan had amazing insight involving certain types of functions and series, but his purported theorems on prime numbers were often wrong, illustrating his vague idea of what constitutes a correct proof. He was one of the youngest members ever appointed a Fellow of the Royal Society. Unfortunately, in 1917 Ramanujan became extremely ill. At the time, it was thought that he had trouble with the English climate and had contracted tuberculosis. It is now thought that he suffered from a vitamin deficiency, brought on by Ramanujan's strict vegetarianism and shortages in wartime England. He returned to India in 1919, continuing to do mathematics even when confined to his bed. He was religious and thought his mathematical talent came from his family deity, Namagiri. He considered mathematics and religion to be linked. He said that "an equation for me has no meaning unless it expresses a thought of God." His short life came to an end in April 1920, when he was 32 years old. Ramanujan left several notebooks of unpublished results. The writings in these notebooks illustrate Ramanujan's insights but are quite sketchy. Several mathematicians have devoted many years of study to explaining and justifying the results in these notebooks.

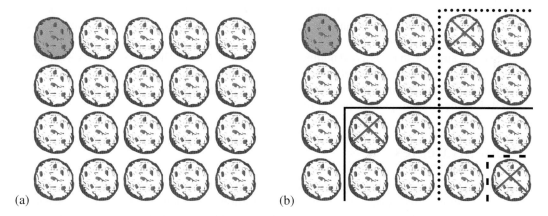

FIGURE 1   **(a) Chomp (Top Left Cookie Poisoned).**   **(b) Three Possible Moves.**

Solution: We will give a nonconstructive existence proof of a winning strategy for the first player. That is, we will show that the first player always has a winning strategy without explicitly describing the moves this player must follow.

First, note that the game ends and cannot finish in a draw because with each move at least one cookie is eaten, so after no more than $m \times n$ moves the game ends, where the initial grid is $m \times n$. Now, suppose that the first player begins the game by eating just the cookie in the bottom right corner. There are two possibilities, this is the first move of a winning strategy for the first player, or the second player can make a move that is the first move of a winning strategy for the second player. In this second case, instead of eating just the cookie in the bottom right corner, the first player could have made the same move that the second player made as the first move of a winning strategy (and then continued to follow that winning strategy). This would guarantee a win for the first player.

Note that we showed that a winning strategy exists, but we did not specify an actual winning strategy. Consequently, the proof is a nonconstructive existence proof. In fact, no one has been able to describe a winning strategy for that Chomp that applies for all rectangular grids by describing the moves that the first player should follow. However, winning strategies can be described for certain special cases, such as when the grid is square and when the grid only has two rows of cookies (see Exercises 11 and 12 in Section 5.2).   ◀

Uniqueness Proofs

Some theorems assert the existence of a unique element with a particular property. In other words, these theorems assert that there is exactly one element with this property. To prove a statement of this type we need to show that an element with this property exists and that no other element has this property. The two parts of a **uniqueness proof** are:

Existence: We show that an element x with the desired property exists.

Uniqueness: We show that if $y \neq x$, then y does not have the desired property.

Equivalently, we can show that if x and y both have the desired property, then $x = y$.

Remark: Showing that there is a unique element x such that $P(x)$ is the same as proving the statement $\exists x (P(x) \wedge \forall y (y \neq x \to \neg P(y)))$.

We illustrate the elements of a uniqueness proof in Example 13.

EXAMPLE 13   Show that if a and b are real numbers and $a \neq 0$, then there is a unique real number r such that $ar + b = 0$.

Solution: First, note that the real number $r = -b/a$ is a solution of $ar + b = 0$ because $a(-b/a) + b = -b + b = 0$. Consequently, a real number r exists for which $ar + b = 0$. This is the existence part of the proof.

Second, suppose that s is a real number such that $as + b = 0$. Then $ar + b = as + b$, where $r = -b/a$. Subtracting b from both sides, we find that $ar = as$. Dividing both sides of this last equation by a, which is nonzero, we see that $r = s$. This means that if $s \neq r$, then $as + b \neq 0$. This establishes the uniqueness part of the proof.   ◀

Extra Examples

Proof Strategies

Finding proofs can be a challenging business. When you are confronted with a statement to prove, you should first replace terms by their definitions and then carefully analyze what the hypotheses and the conclusion mean. After doing so, you can attempt to prove the result using one of the available methods of proof. Generally, if the statement is a conditional statement, you should first try a direct proof; if this fails, you can try an indirect proof. If neither of these approaches works, you might try a proof by contradiction.

FORWARD AND BACKWARD REASONING   Whichever method you choose, you need a starting point for your proof. To begin a direct proof of a conditional statement, you start with the premises. Using these premises, together with axioms and known theorems, you can construct a proof using a sequence of steps that leads to the conclusion. This type of reasoning, called *forward reasoning*, is the most common type of reasoning used to prove relatively simple results. Similarly, with indirect reasoning you can start with the negation of the conclusion and, using a sequence of steps, obtain the negation of the premises.

Unfortunately, forward reasoning is often difficult to use to prove more complicated results, because the reasoning needed to reach the desired conclusion may be far from obvious. In such cases it may be helpful to use *backward reasoning*. To reason backward to prove a statement q, we find a statement p that we can prove with the property that $p \rightarrow q$. (Note that it is not helpful to find a statement r that you can prove such that $q \rightarrow r$, because it is the fallacy of begging the question to conclude from $q \rightarrow r$ and r that q is true.) Backward reasoning is illustrated in Examples 14 and 15.

EXAMPLE 14   Given two positive real numbers x and y, their **arithmetic mean** is $(x + y)/2$ and their **geometric mean** is $\sqrt{xy}$. When we compare the arithmetic and geometric means of pairs of distinct positive real numbers, we find that the arithmetic mean is always greater than the geometric mean. [For example, when $x = 4$ and $y = 6$, we have $5 = (4 + 6)/2 > \sqrt{4 \cdot 6} = \sqrt{24}$.] Can we prove that this inequality is always true?

Solution: To prove that $(x + y)/2 > \sqrt{xy}$ when x and y are distinct positive real numbers, we can work backward. We construct a sequence of equivalent inequalities. The equivalent inequalities are

$$(x + y)/2 > \sqrt{xy},$$
$$(x + y)^2/4 > xy,$$
$$(x + y)^2 > 4xy,$$
$$x^2 + 2xy + y^2 > 4xy,$$
$$x^2 - 2xy + y^2 > 0,$$
$$(x - y)^2 > 0.$$

Extra Examples

Because $(x - y)^2 > 0$ when $x \neq y$, it follows that the final inequality is true. Because all these inequalities are equivalent, it follows that $(x + y)/2 > \sqrt{xy}$ when $x \neq y$. Once we have carried out this backward reasoning, we can easily reverse the steps to construct a proof using forward reasoning. We now give this proof.

Suppose that x and y are distinct positive real numbers. Then $(x - y)^2 > 0$ because the square of a nonzero real number is positive (see Appendix 1). Because $(x - y)^2 = x^2 - 2xy + y^2$, this implies that $x^2 - 2xy + y^2 > 0$. Adding $4xy$ to both sides, we obtain $x^2 + 2xy + y^2 > 4xy$. Because $x^2 + 2xy + y^2 = (x + y)^2$, this means that $(x + y)^2 \geq 4xy$. Dividing both sides of this equation by 4, we see that $(x + y)^2/4 > xy$. Finally, taking square roots of both sides (which preserves the inequality because both sides are positive) yields $(x + y)/2 > \sqrt{xy}$. We conclude that if x and y are distinct positive real numbers, then their arithmetic mean $(x + y)/2$ is greater than their geometric mean $\sqrt{xy}$.   ◀

EXAMPLE 15   Suppose that two people play a game taking turns removing one, two, or three stones at a time from a pile that begins with 15 stones. The person who removes the last stone wins the game. Show that the first player can win the game no matter what the second player does.

Solution: To prove that the first player can always win the game, we work backward. At the last step, the first player can win if this player is left with a pile containing one, two, or three stones. The second player will be forced to leave one, two, or three stones if this player has to remove stones from a pile containing four stones. Consequently, one way for the first person to win is to leave four stones for the second player on the next-to-last move. The first person can leave four stones when there are five, six, or seven stones left at the beginning of this player's move, which happens when the second player has to remove stones from a pile with eight stones. Consequently, to force the second player to leave five, six, or seven stones, the first player should leave eight stones for the second player at the second-to-last move for the first player. This means that there are nine, ten, or eleven stones when the first player makes this move. Similarly, the first player should leave twelve stones when this player makes the first move. We can reverse this argument to show that the first player can always make moves so that this player wins the game no matter what the second player does. These moves successively leave twelve, eight, and four stones for the second player.   ◀

ADAPTING EXISTING PROOFS   An excellent way to look for possible approaches that can be used to prove a statement is to take advantage of existing proofs of similar results. Often an existing proof can be adapted to prove other facts. Even when this is not the case, some of the ideas used in existing proofs may be helpful. Because existing proofs provide clues for new proofs, you should read and understand the proofs you encounter in your studies. This process is illustrated in Example 16.

EXAMPLE 16   In Example 10 of Section 1.8 we proved that $\sqrt{2}$ is irrational. We now conjecture that $\sqrt{3}$ is irrational. Can we adapt the proof in Example 10 in Section 1.8 to show that $\sqrt{3}$ is irrational?

Solution: To adapt the proof in Example 10 in Section 1.8, we begin by mimicking the steps in that proof, but with $\sqrt{2}$ replaced with $\sqrt{3}$. First, we suppose that $\sqrt{3} = d/c$ where the fraction c/d is in lowest terms. Squaring both sides tells us that $3 = c^2/d^2$, so that $3d^2 = c^2$. Can we use this equation to show that 3 must be a factor of both c and d, similar to how we used the equation $2b^2 = a^2$ in Example 10 in Section 1.8 to show that 2 must be a factor of both a and b? (Recall that an integer s is a factor of the integer t if t/s is an integer. An integer n is even if and only if 2 is a factor of n.) In turns out that we can, but we need some ammunition from number theory, which we will develop in Chapter 4. We sketch out the remainder of the proof, but leave the justification of these steps until Chapter 4. Because 3 is a factor of c^2, it must also be a factor of c. Furthermore, because 3 is a factor of c, 9 is a factor of c^2, which means that 9 is a factor of $3d^2$. This implies that 3 is a factor of d^2, which means that 3 is a factor of that d. This makes 3 a factor of both c and d, which contradicts the assumption that c/d is in lowest terms. After we have filled in the justification for these steps, we will have shown that $\sqrt{3}$ is irrational by adapting the proof that $\sqrt{2}$ is irrational. Note that this proof can be extended to show that $\sqrt{n}$ is irrational whenever n is a positive integer that is not a perfect square. We leave the details of this to Chapter 4. ◀

A good tip is to look for existing proofs that you might adapt when you are confronted with proving a new theorem, particularly when the new theorem seems similar to one you have already proved.

Looking for Counterexamples

In Section 1.8 we introduced the use of counterexamples to show that certain statements are false. When confronted with a conjecture, you might first try to prove this conjecture, and if your attempts are unsuccessful, you might try to find a counterexample, first by looking at the simplest, smallest examples. If you cannot find a counterexample, you might again try to prove the statement. In any case, looking for counterexamples is an extremely important pursuit, which often provides insights into problems. We will illustrate the role of counterexamples in Example 17.

EXAMPLE 17   In Example 14 in Section 1.8 we showed that the statement "Every positive integer is the sum of two squares of integers" is false by finding a counterexample. That is, there are positive integers that cannot be written as the sum of the squares of two integers. Although we cannot write every positive integer as the sum of the squares of two integers, maybe we can write every positive integer as the sum of the squares of three integers. That is, is the statement "Every positive integer is the sum of the squares of three integers" true or false?

Extra
Examples

Solution: Because we know that not every positive integer can be written as the sum of two squares of integers, we might initially be skeptical that every positive integer can be written as the sum of three squares of integers. So, we first look for a counterexample. That is, we can show that the statement "Every positive integer is the sum of three squares of integers" is false if we can find a particular integer that is not the sum of the squares of three integers. To look for a counterexample, we try to write successive positive integers as a sum of three squares. We find that $1 = 0^2 + 0^2 + 1^2$, $2 = 0^2 + 1^2 + 1^2$, $3 = 1^2 + 1^2 + 1^2$, $4 = 0^2 + 0^2 + 2^2$, $5 = 0^2 + 1^2 + 2^2$, $6 = 1^2 + 1^2 + 2^2$, but we cannot find a way to write 7 as the sum of three squares. To show that there are not three squares that add up to 7, we note that the only possible squares we can use are those not exceeding 7, namely, 0, 1, and 4. Because no three terms where each term is 0, 1, or 4 add up to 7, it follows that 7 is a counterexample. We conclude that the statement "Every positive integer is the sum of the squares of three integers" is false.

We have shown that not every positive integer is the sum of the squares of three integers. The next question to ask is whether every positive integer is the sum of the squares of four positive integers. Some experimentation provides evidence that the answer is yes. For example, $7 = 1^2 + 1^2 + 1^2 + 2^2$, $25 = 4^2 + 2^2 + 2^2 + 1^2$, and $87 = 9^2 + 2^2 + 1^2 + 1^2$. It turns out the conjecture "Every positive integer is the sum of the squares of four integers" is true. For a proof, see [Ro10]. ◀

Proof Strategy in Action

Mathematics is generally taught as if mathematical facts were carved in stone. Mathematics texts (including the bulk of this book) formally present theorems and their proofs. Such presentations do not convey the discovery process in mathematics. This process begins with exploring concepts and examples, asking questions, formulating conjectures, and attempting to settle these

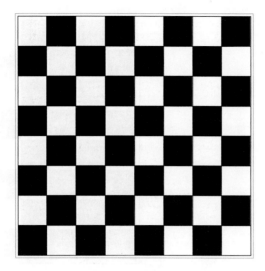

FIGURE 2   The Standard Checkerboard.

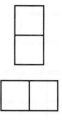

FIGURE 3
Two Dominoes.

conjectures either by proof or by counterexample. These are the day-to-day activities of mathematicians. Believe it or not, the material taught in textbooks was originally developed in this way.

People formulate conjectures on the basis of many types of possible evidence. The examination of special cases can lead to a conjecture, as can the identification of possible patterns. Altering the hypotheses and conclusions of known theorems also can lead to plausible conjectures. At other times, conjectures are made based on intuition or a belief that a result holds. No matter how a conjecture was made, once it has been formulated, the goal is to prove or disprove it. When mathematicians believe that a conjecture may be true, they try to find a proof. If they cannot find a proof, they may look for a counterexample. When they cannot find a counterexample, they may switch gears and once again try to prove the conjecture. Although many conjectures are quickly settled, a few conjectures resist attack for hundreds of years and lead to the development of new parts of mathematics. We will mention a few famous conjectures later in this section.

Tilings

We can illustrate aspects of proof strategy through a brief study of tilings of checkerboards. Looking at tilings of checkerboards is a fruitful way to quickly discover many different results and construct their proofs using a variety of proof methods. There are almost an endless number of conjectures that can be made and studied in this area too. To begin, we need to define some terms. A **checkerboard** is a rectangle divided into squares of the same size by horizontal and vertical lines. The game of checkers is played on a board with 8 rows and 8 columns; this board is called the **standard checkerboard** and is shown in Figure 2. In this section we use the term **board** to refer to a checkerboard of any rectangular size as well as parts of checkerboards obtained by removing one or more squares. A **domino** is a rectangular piece that is one square by two squares, as shown in Figure 3. We say that a board is **tiled** by dominoes when all its squares are covered with no overlapping dominoes and no dominoes overhanging the board. We now develop some results about tiling boards using dominoes.

EXAMPLE 18   Can we tile the standard checkerboard using dominoes?

Solution: We can find many ways to tile the standard checkerboard using dominoes. For example, we can tile it by placing 32 dominoes horizontally, as shown in Figure 4. The existence of one

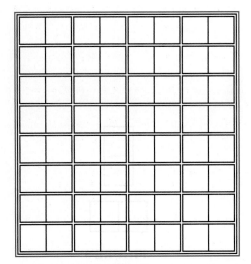

FIGURE 4   Tiling the Standard Checkerboard.

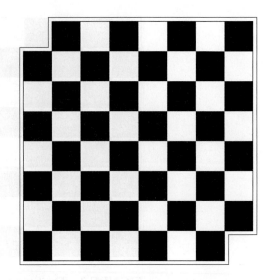

FIGURE 5   The Standard Checkerboard with the Upper Left and Lower Right Squares Removed.

such tiling completes a constructive existence proof. Of course, there are a large number of other ways to do this tiling. We can place 32 dominoes vertically on the board or we can place some tiles vertically and some horizontally. But for a constructive existence proof we needed to find just one such tiling.   ◀

EXAMPLE 19   Can we tile a board obtained by removing one of the four corner squares of a standard checkerboard?

Solution: To answer this question, note that a standard checkerboard has 64 squares, so removing a square produces a board with 63 squares. Now suppose that we could tile a board obtained from the standard checkerboard by removing a corner square. The board has an even number of squares because each domino covers two squares and no two dominoes overlap and no dominoes overhang the board. Consequently, we can prove by contradiction that a standard checkerboard with one square removed cannot be tiled using dominoes because such a board has an odd number of squares.   ◀

We now consider a trickier situation.

EXAMPLE 20   Can we tile the board obtained by deleting the upper left and lower right corner squares of a standard checkerboard, shown in Figure 5?

Solution: A board obtained by deleting two squares of a standard checkerboard contains $64 - 2 = 62$ squares. Because 62 is even, we cannot quickly rule out the existence of a tiling of the standard checkerboard with its upper left and lower right squares removed, unlike Example 19, where we ruled out the existence of a tiling of the standard checkerboard with one corner square removed. Trying to construct a tiling of this board by successively placing dominoes might be a first approach, as the reader should attempt. However, no matter how much we try, we cannot find such a tiling. Because our efforts do not produce a tiling, we are led to conjecture that no tiling exists.

We might try to prove that no tiling exists by showing that we reach a dead end however we successively place dominoes on the board. To construct such a proof, we would have to consider all possible cases that arise as we run through all possible choices of successively placing dominoes. For example, we have two choices for covering the square in the second column of the first row, next to the removed top left corner. We could cover it with a horizontally placed tile or a vertically placed tile. Each of these two choices leads to further choices, and so on. It does not take long to see that this is not a fruitful plan of attack for a person, although a computer could be used to complete such a proof by exhaustion. (Exercise 22 asks you to supply such a proof to show that a 4×4 checkerboard with opposite corners removed cannot be tiled.)

We need another approach. Perhaps there is an easier way to prove there is no tiling of a standard checkerboard with two opposite corners removed. As with many proofs, a key observation can help. We color the squares of this checkerboard using alternating white and black squares, as in Figure 2. Observe that a domino in a tiling of such a board covers one white square and one black square. Next, note that this board has unequal numbers of white square and black squares. We can use these observations to prove by contradiction that a standard checkerboard with opposite corners removed cannot be tiled using dominoes. We now present such a proof.

Proof: Suppose we can use dominoes to tile a standard checkerboard with opposite corners removed. Note that the standard checkerboard with opposite corners removed contains $64 - 2 = 62$ squares. The tiling would use $62/2 = 31$ dominoes. Note that each domino in this tiling covers one white and one black square. Consequently, the tiling covers 31 white squares and 31 black squares. However, when we remove two opposite corner squares, either 32 of the remaining squares are white and 30 are black or else 30 are white and 32 are black. This contradicts the assumption that we can use dominoes to cover a standard checkerboard with opposite corners removed, completing the proof. ◀

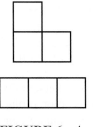

FIGURE 6 A Right Triomino and a Straight Triomino.

We can use other types of pieces besides dominoes in tilings. Instead of dominoes we can study tilings that use identically shaped pieces constructed from congruent squares that are connected along their edges. Such pieces are called **polyominoes**, a term coined in 1953 by the mathematician Solomon Golomb, the author of an entertaining book about them [Go94]. We will consider two polyominoes with the same number of squares the same if we can rotate and/or flip one of the polyominoes to get the other one. For example, there are two types of triominoes (see Figure 6), which are polyominoes made up of three squares connected by their sides. One type of triomino, the **straight triomino**, has three horizontally connected squares; the other type, **right triominoes**, resembles the letter L in shape, flipped and/or rotated, if necessary. We will study the tilings of a checkerboard by straight triominoes here; we will study tilings by right triominoes in Section 5.1.

EXAMPLE 21   Can you use straight triominoes to tile a standard checkerboard?

Solution: The standard checkerboard contains 64 squares and each triomino covers three squares. Consequently, if triominoes tile a board, the number of squares of the board must be a multiple of 3. Because 64 is not a multiple of 3, triominoes cannot be used to cover an 8×8 checkerboard. ◀

In Example 22, we consider the problem of using straight triominoes to tile a standard checkerboard with one corner missing.

EXAMPLE 22   Can we use straight triominoes to tile a standard checkerboard with one of its four corners removed? An 8×8 checkerboard with one corner removed contains $64 - 1 = 63$ squares. Any tiling by straight triominoes of one of these four boards uses $63/3 = 21$ triominoes. However, when we experiment, we cannot find a tiling of one of these boards using straight triominoes. A proof by exhaustion does not appear promising. Can we adapt our proof from Example 20 to prove that no such tiling exists?

Solution: We will color the squares of the checkerboard in an attempt to adapt the proof by contradiction we gave in Example 20 of the impossibility of using dominoes to tile a standard

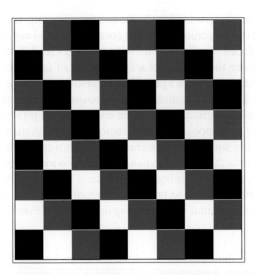

FIGURE 7 **Coloring the Squares of the Standard Checkerboard with Three Colors.**

checkerboard with opposite corners removed. Because we are using straight triominoes rather than dominoes, we color the squares using three colors rather than two colors, as shown in Figure 7. Note that there are 21 blue squares, 21 black squares, and 22 white squares in this coloring. Next, we make the crucial observation that when a straight triomino covers three squares of the checkerboard, it covers one blue square, one black square, and one white square. Next, note that each of the three colors appears in a corner square. Thus without loss of generality, we may assume that we have rotated the coloring so that the missing square is colored blue. Therefore, we assume that the remaining board contains 20 blue squares, 21 black squares, and 22 white squares.

If we could tile this board using straight triominoes, then we would use $63/3 = 21$ straight triominoes. These triominoes would cover 21 blue squares, 21 black squares, and 21 white squares. This contradicts the fact that this board contains 20 blue squares, 21 black squares, and 22 white squares. Therefore we cannot tile this board using straight triominoes. ◀

The Role of Open Problems

Many advances in mathematics have been made by people trying to solve famous unsolved problems. In the past 20 years, many unsolved problems have finally been resolved, such as the proof of a conjecture in number theory made more than 300 years ago. This conjecture asserts the truth of the statement known as **Fermat's last theorem**.

THEOREM 1 **FERMAT'S LAST THEOREM** The equation

$$x^n + y^n = z^n$$

has no solutions in integers x, y, and z with $xyz \neq 0$ whenever n is an integer with $n > 2$.

Links

Remark: The equation $x^2 + y^2 = z^2$ has infinitely many solutions in integers x, y, and z; these solutions are called Pythagorean triples and correspond to the lengths of the sides of right triangles with integer lengths. See Exercise 18.

This problem has a fascinating history. In the seventeenth century, Fermat jotted in the margin of his copy of the works of Diophantus that he had a "wondrous proof" that there are no integer solutions of $x^n + y^n = z^n$ when n is an integer greater than 2 with $xyz \neq 0$. However, he never published a proof (Fermat published almost nothing), and no proof could be found in the papers he left when he died. Mathematicians looked for a proof for three centuries without success, although many people were convinced that a relatively simple proof could be found. (Proofs of special cases were found, such as the proof of the case when $n = 3$ by Euler and the proof of the $n = 4$ case by Fermat himself.) Over the years, several established mathematicians thought that they had proved this theorem. In the nineteenth century, one of these failed attempts led to the development of the part of number theory called algebraic number theory. A correct proof, requiring hundreds of pages of advanced mathematics, was not found until the 1990s, when Andrew Wiles used recently developed ideas from a sophisticated area of number theory called the theory of elliptic curves to prove Fermat's last theorem. Wiles's quest to find a proof of Fermat's last theorem using this powerful theory, described in a program in the *Nova* series on public television, took close to ten years! Moreover, his proof was based on major contributions of many mathematicians. (The interested reader should consult [Ro10] for more information about Fermat's last theorem and for additional references concerning this problem and its resolution.)

We now state an open problem that is simple to describe, but that seems quite difficult to resolve.

EXAMPLE 23

Links

The $3x + 1$ Conjecture   Let T be the transformation that sends an even integer x to $x/2$ and an odd integer x to $3x + 1$. A famous conjecture, sometimes known as the **$3x + 1$ conjecture**, states that for all positive integers x, when we repeatedly apply the transformation T, we will eventually reach the integer 1. For example, starting with $x = 13$, we find $T(13) = 3 \cdot 13 + 1 = 40$, $T(40) = 40/2 = 20$, $T(20) = 20/2 = 10$, $T(10) = 10/2 = 5$, $T(5) = 3 \cdot 5 + 1 = 16$, $T(16) = 8$, $T(8) = 4$, $T(4) = 2$, and $T(2) = 1$. The $3x + 1$ conjecture has been verified using computers for all integers x up to $5.6 \cdot 10^{13}$.

The $3x + 1$ conjecture has an interesting history and has attracted the attention of mathematicians since the 1950s. The conjecture has been raised many times and goes by many other names, including the Collatz problem, Hasse's algorithm, Ulam's problem, the Syracuse problem, and Kakutani's problem. Many mathematicians have been diverted from their work to spend time attacking this conjecture. This led to the joke that this problem was part of a conspiracy to slow down American mathematical research. See the article by Jeffrey Lagarias [La10] for a fascinating discussion of this problem and the results that have been found by mathematicians attacking it.   ◀

Watch out! Working on the $3x + 1$ problem can be addictive.

In Chapter 4 we will describe additional open questions about prime numbers. Students already familiar with the basic notions about primes might want to explore Section 4.3, where these open questions are discussed. We will mention other important open questions throughout the book.

Additional Proof Methods

Build up your arsenal of proof methods as you work through this book.

In this chapter we introduced the basic methods used in proofs. We also described how to leverage these methods to prove a variety of results. We will use these proof methods in all subsequent chapters. In particular, we will use them in Chapters 2, 3, and 4 to prove results about sets, functions, algorithms, and number theory and in Chapters 9, 10, and 11 to prove results in graph theory. Among the theorems we will prove is the famous halting theorem which states that there is a problem that cannot be solved using any procedure. However, there are many important proof methods besides those we have covered. We will introduce some of these methods later in this book. In particular, in Section 5.1 we will discuss mathematical induction, which is an extremely useful method for proving statements of the form $\forall n\, P(n)$, where the domain consists of all positive integers. In Section 5.3 we will introduce structural induction, which can be used to prove results about recursively defined sets. We will use the Cantor diagonalization method, which can be used to prove results about the size of infinite sets, in Section 2.5. In Chapter 6 we will introduce the notion of combinatorial proofs, which can be used to prove results by counting arguments. The reader should note that entire books have been devoted to the activities discussed in this section, including many excellent works by George Pólya ([Po61], [Po71], [Po90]).

Finally, note that we have not given a procedure that can be used for proving theorems in mathematics. It is a deep theorem of mathematical logic that there is no such procedure.

Exercises

1. Prove that $n^2 + 1 \geq 2^n$ when n is a positive integer with $1 \leq n \leq 4$.

2. Prove that there are no positive perfect cubes less than 1000 that are the sum of the cubes of two positive integers.

3. Prove that if x and y are real numbers, then $\max(x, y) + \min(x, y) = x + y$. [*Hint:* Use a proof by cases, with the two cases corresponding to $x \geq y$ and $x < y$, respectively.]

4. Prove the **triangle inequality**, which states that if x and y are real numbers, then $|x| + |y| \geq |x + y|$ (where $|x|$ represents the absolute value of x, which equals x if $x \geq 0$ and equals $-x$ if $x < 0$).

5. Prove that there are 100 consecutive positive integers that are not perfect squares. Is your proof constructive or nonconstructive?

6. Prove that there is a positive integer that equals the sum of the positive integers not exceeding it. Is your proof constructive or nonconstructive?

7. Prove that there exists a pair of consecutive integers such that one of these integers is a perfect square and the other is a perfect cube.

8. Prove that either $2 \cdot 10^{500} + 15$ or $2 \cdot 10^{500} + 16$ is not a perfect square. Is your proof constructive or nonconstructive?

9. Prove or disprove that there is a rational number x and an irrational number y such that x^y is irrational.

10. Prove or disprove that if a and b are rational numbers, then a^b is also rational.

11. Show that each of these statements can be used to express the fact that there is a unique element x such that $P(x)$ is true. [Note that we can also write this statement as $\exists!x\, P(x)$.]
  a) $\exists x \forall y (P(y) \leftrightarrow x = y)$
  b) $\exists x P(x) \wedge \forall x \forall y (P(x) \wedge P(y) \rightarrow x = y)$
  c) $\exists x (P(x) \wedge \forall y (P(y) \rightarrow x = y))$

12. Show that if a, b, and c are real numbers and $a \neq 0$, then there is a unique solution of the equation $ax + b = c$.

13. Suppose that a and b are odd integers with $a \neq b$. Show there is a unique integer c such that $|a - c| = |b - c|$.

14. Use forward reasoning to show that if x is a nonzero real number, then $x^2 + 1/x^2 \geq 2$. [*Hint:* Start with the inequality $(x - 1/x)^2 \geq 0$ which holds for all nonzero real numbers x.]

15. Prove that given a real number x there exist unique numbers n and ϵ such that $x = n - \epsilon$, n is an integer, and $0 \leq \epsilon < 1$.

16. The **quadratic mean** of two real numbers x and y equals $\sqrt{(x^2 + y^2)/2}$. By computing the arithmetic and quadratic means of different pairs of positive real numbers, formulate a conjecture about their relative sizes and prove your conjecture.

17. Prove that there are no solutions in integers x and y to the equation $2x^2 + 5y^2 = 14$.

18. Prove that there are infinitely many solutions in positive integers x, y, and z to the equation $x^2 + y^2 = z^2$. [*Hint:* Let $x = m^2 - n^2$, $y = 2mn$, and $z = m^2 + n^2$, where m and n are integers.]

19. Prove that between every two rational numbers there is an irrational number.

20. Prove or disprove that if you have an 8-gallon jug of water and two empty jugs with capacities of 5 gallons and 3 gallons, respectively, then you can measure 4 gallons by successively pouring some of or all of the water in a jug into another jug.

21. Prove that you can use dominoes to tile a rectangular checkerboard with an even number of squares.

22. Use a proof by exhaustion to show that a tiling using dominoes of a 4×4 checkerboard with opposite corners removed does not exist. [*Hint:* First show that you can assume that the squares in the upper left and lower right corners are removed. Number the squares of the original checkerboard from 1 to 16, starting in the first row, moving right in this row, then starting in the leftmost square in the second row and moving right, and so on. Remove squares 1 and 16. To begin the proof, note that square 2 is covered either by a domino laid horizontally, which covers squares 2 and 3, or vertically, which covers squares 2 and 6. Consider each of these cases separately, and work through all the subcases that arise.]

23. Show that by removing two white squares and two black squares from an 8×8 checkerboard (colored as in the text) you can make it impossible to tile the remaining squares using dominoes.

*24. Prove that when a white square and a black square are removed from an 8×8 checkerboard (colored as in the text) you can tile the remaining squares of the checkerboard using dominoes. [*Hint:* Show that when one black and one white square are removed, each part of the partition of the remaining cells formed by inserting the barriers shown in the figure can be covered by dominoes.]

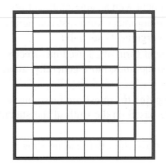

Key Terms and Results

TERMS

proposition: a statement that is true or false

propositional variable: a variable that represents a proposition

truth value: true or false

$\neg p$ (negation of p): the proposition with truth value opposite to the truth value of p

logical operators: operators used to combine propositions

compound proposition: a proposition constructed by combining propositions using logical operators

truth table: a table displaying all possible truth values of propositions

$p \lor q$ (disjunction of p and q): the proposition "p or q," which is true if and only if at least one of p and q is true

$p \land q$ (conjunction of p and q): the proposition "p and q," which is true if and only if both p and q are true

$p \oplus q$ (exclusive or of p and q): the proposition "p XOR q," which is true when exactly one of p and q is true

$p \to q$ (p implies q): the proposition "if p, then q," which is false if and only if p is true and q is false

converse of $p \to q$: the conditional statement $q \to p$

contrapositive of $p \to q$: the conditional statement $\neg q \to \neg p$

inverse of $p \to q$: the conditional statement $\neg p \to \neg q$

$p \leftrightarrow q$ (biconditional): the proposition "p if and only if q," which is true if and only if p and q have the same truth value

bit: either a 0 or a 1

Boolean variable: a variable that has a value of 0 or 1

bit operation: an operation on a bit or bits

bit string: a list of bits

bitwise operations: operations on bit strings that operate on each bit in one string and the corresponding bit in the other string

logic gate: a logic element that performs a logical operation on one or more bits to produce an output bit

logic circuit: a switching circuit made up of logic gates that produces one or more output bits

tautology: a compound proposition that is always true

contradiction: a compound proposition that is always false

contingency: a compound proposition that is sometimes true and sometimes false

consistent compound propositions: compound propositions for which there is an assignment of truth values to the variables that makes all these propositions true

satisfiable compound proposition: a compound proposition for which there is an assignment of truth values to its variables that makes it true

logically equivalent compound propositions: compound propositions that always have the same truth values

predicate: part of a sentence that attributes a property to the subject

propositional function: a statement containing one or more variables that becomes a proposition when each of its variables is assigned a value or is bound by a quantifier

domain (or universe) of discourse: the values a variable in a propositional function may take

$\exists x\, P(x)$ (existential quantification of $P(x)$): the proposition that is true if and only if there exists an x in the domain such that $P(x)$ is true

$\forall x P(x)$ (universal quantification of $P(x)$): the proposition that is true if and only if $P(x)$ is true for every x in the domain

logically equivalent expressions: expressions that have the same truth value no matter which propositional functions and domains are used

free variable: a variable not bound in a propositional function

bound variable: a variable that is quantified

scope of a quantifier: portion of a statement where the quantifier binds its variable

argument: a sequence of statements

argument form: a sequence of compound propositions involving propositional variables

premise: a statement, in an argument, or argument form, other than the final one

conclusion: the final statement in an argument or argument form

valid argument form: a sequence of compound propositions involving propositional variables where the truth of all the premises implies the truth of the conclusion

valid argument: an argument with a valid argument form

rule of inference: a valid argument form that can be used in the demonstration that arguments are valid

fallacy: an invalid argument form often used incorrectly as a rule of inference (or sometimes, more generally, an incorrect argument)

circular reasoning or begging the question: reasoning where one or more steps are based on the truth of the statement being proved

theorem: a mathematical assertion that can be shown to be true

conjecture: a mathematical assertion proposed to be true, but that has not been proved

proof: a demonstration that a theorem is true

axiom: a statement that is assumed to be true and that can be used as a basis for proving theorems

lemma: a theorem used to prove other theorems

corollary: a proposition that can be proved as a consequence of a theorem that has just been proved

vacuous proof: a proof that $p \rightarrow q$ is true based on the fact that p is false

trivial proof: a proof that $p \rightarrow q$ is true based on the fact that q is true

direct proof: a proof that $p \rightarrow q$ is true that proceeds by showing that q must be true when p is true

proof by contraposition: a proof that $p \rightarrow q$ is true that proceeds by showing that p must be false when q is false

proof by contradiction: a proof that p is true based on the truth of the conditional statement $\neg p \rightarrow q$, where q is a contradiction

exhaustive proof: a proof that establishes a result by checking a list of all possible cases

proof by cases: a proof broken into separate cases, where these cases cover all possibilities

without loss of generality: an assumption in a proof that makes it possible to prove a theorem by reducing the number of cases to consider in the proof

counterexample: an element x such that $P(x)$ is false

constructive existence proof: a proof that an element with a specified property exists that explicitly finds such an element

nonconstructive existence proof: a proof that an element with a specified property exists that does not explicitly find such an element

rational number: a number that can be expressed as the ratio of two integers p and q such that $q \neq 0$

uniqueness proof: a proof that there is exactly one element satisfying a specified property

RESULTS

The logical equivalences given in Tables 6, 7, and 8 in Section 1.3.

De Morgan's laws for quantifiers.

Rules of inference for propositional calculus.

Rules of inference for quantified statements.

Review Questions

1. **a)** Define the negation of a proposition.

   b) What is the negation of "This is a boring course"?

2. **a)** Define (using truth tables) the disjunction, conjunction, exclusive or, conditional, and biconditional of the propositions p and q.

   b) What are the disjunction, conjunction, exclusive or, conditional, and biconditional of the propositions "I'll go to the movies tonight" and "I'll finish my discrete mathematics homework"?

3. **a)** Describe at least five different ways to write the conditional statement $p \rightarrow q$ in English.

   b) Define the converse and contrapositive of a conditional statement.

   c) State the converse and the contrapositive of the conditional statement "If it is sunny tomorrow, then I will go for a walk in the woods."

4. **a)** What does it mean for two propositions to be logically equivalent?

   b) Describe the different ways to show that two compound propositions are logically equivalent.

   c) Show in at least two different ways that the compound propositions $\neg p \vee (r \rightarrow \neg q)$ and $\neg p \vee \neg q \vee \neg r$ are equivalent.

5. *(Depends on the Exercise Set in Section 1.3)*

   a) Given a truth table, explain how to use disjunctive normal form to construct a compound proposition with this truth table.

   b) Explain why part (a) shows that the operators $\wedge$, $\vee$, and $\neg$ are functionally complete.

   c) Is there an operator such that the set containing just this operator is functionally complete?

6. What are the universal and existential quantifications of a predicate $P(x)$? What are their negations?

7. **a)** What is the difference between the quantification $\exists x \forall y P(x, y)$ and $\forall y \exists x P(x, y)$, where $P(x, y)$ is a predicate?

   b) Give an example of a predicate $P(x, y)$ such that $\exists x \forall y P(x, y)$ and $\forall y \exists x P(x, y)$ have different truth values.

8. Describe what is meant by a valid argument in propositional logic and show that the argument "If the earth is flat, then you can sail off the edge of the earth," "You cannot sail off the edge of the earth," therefore, "The earth is not flat" is a valid argument.

9. Use rules of inference to show that if the premises "All zebras have stripes" and "Mark is a zebra" are true, then the conclusion "Mark has stripes" is true.

10. a) Describe what is meant by a direct proof, a proof by contraposition, and a proof by contradiction of a conditional statement $p \rightarrow q$.

b) Give a direct proof, a proof by contraposition and a proof by contradiction of the statement: "If n is even, then $n + 4$ is even."

11. a) Describe a way to prove the biconditional $p \leftrightarrow q$.

b) Prove the statement: "The integer $3n + 2$ is odd if and only if the integer $9n + 5$ is even, where n is an integer."

12. To prove that the statements p_1, p_2, p_3, and p_4 are equivalent, is it sufficient to show that the conditional statements $p_4 \rightarrow p_2$, $p_3 \rightarrow p_1$, and $p_1 \rightarrow p_2$ are valid? If not, provide another collection of conditional statements that can be used to show that the four statements are equivalent.

13. a) Suppose that a statement of the form $\forall x\, P(x)$ is false. How can this be proved?

b) Show that the statement "For every positive integer n, $n^2 \geq 2n$" is false.

14. What is the difference between a constructive and nonconstructive existence proof? Give an example of each.

15. What are the elements of a proof that there is a unique element x such that $P(x)$, where $P(x)$ is a propositional function?

16. Explain how a proof by cases can be used to prove a result about absolute values, such as the fact that $|xy| = |x||y|$ for all real numbers x and y.

Supplementary Exercises

1. Let p be the proposition "I will do every exercise in this book" and q be the proposition "I will get an "A" in this course." Express each of these as a combination of p and q.

a) I will get an "A" in this course only if I do every exercise in this book.

b) I will get an "A" in this course and I will do every exercise in this book.

c) Either I will not get an "A" in this course or I will not do every exercise in this book.

d) For me to get an "A" in this course it is necessary and sufficient that I do every exercise in this book.

2. Find the truth table of the compound proposition $(p \lor q) \rightarrow (p \land \neg r)$.

3. Show that these compound propositions are tautologies.

a) $(\neg q \land (p \rightarrow q)) \rightarrow \neg p$

b) $((p \lor q) \land \neg p) \rightarrow q$

4. Give the converse, the contrapositive, and the inverse of these conditional statements.

a) If it rains today, then I will drive to work.

b) If $|x| = x$, then $x \geq 0$.

c) If n is greater than 3, then n^2 is greater than 9.

5. Find a compound proposition involving the propositional variables p, q, r, and s that is true when exactly three of these propositional variables are true and is false otherwise.

6. Show that these statements are inconsistent: "If Miranda does not take a course in discrete mathematics, then she will not graduate." "If Miranda does not graduate, then she is not qualified for the job." "If Miranda reads this book, then she is qualified for the job." "Miranda does not take a course in discrete mathematics but she reads this book."

7. Suppose that you meet three people Aaron, Bohan, and Crystal. Can you determine what Aaron, Bohan, and Crystal are if Aaron says "All of us are knaves" and Bohan says "Exactly one of us is a knave."?

8. Show that if S is a proposition, where S is the conditional statement "If S is true, then unicorns live," then "Unicorns live" is true. Show that it follows that S cannot be a proposition. (This paradox is known as *Löb's paradox*.)

9. (Adapted from [Sm78]) Suppose that on an island there are three types of people, knights, knaves, and normals (also known as spies). Knights always tell the truth, knaves always lie, and normals sometimes lie and sometimes tell the truth. Detectives questioned three inhabitants of the island—Amy, Brenda, and Claire—as part of the investigation of a crime. The detectives knew that one of the three committed the crime, but not which one. They also knew that the criminal was a knight, and that the other two were not. Additionally, the detectives recorded these statements: Amy: "I am innocent." Brenda: "What Amy says is true." Claire: "Brenda is not a normal." After analyzing their information, the detectives positively identified the guilty party. Who was it?

10. Let $P(x)$ be the statement "Student x knows calculus" and let $Q(y)$ be the statement "Class y contains a student who knows calculus." Express each of these as quantifications of $P(x)$ and $Q(y)$.

a) Some students know calculus.

b) Not every student knows calculus.

c) Every class has a student in it who knows calculus.

d) Every student in every class knows calculus.

e) There is at least one class with no students who know calculus.

11. Let $P(m, n)$ be the statement "m divides n," where the domain for both variables consists of all positive integers. (By "m divides n" we mean that $n = km$ for some integer k.) Determine the truth values of each of these statements.

a) $P(4, 5)$ **b)** $P(2, 4)$

c) $\forall m\, \forall n\, P(m, n)$ **d)** $\exists m\, \forall n\, P(m, n)$

e) $\exists n\, \forall m\, P(m, n)$ **f)** $\forall n\, P(1, n)$

12. Use existential and universal quantifiers to express the statement "No one has more than three grandmothers" using the propositional function $G(x, y)$, which represents "x is the grandmother of y."

13. Use existential and universal quantifiers to express the statement "Everyone has exactly two biological parents" using the propositional function $P(x, y)$, which represents "x is the biological parent of y."

14. The quantifier $\exists_n$ denotes "there exists exactly n," so that $\exists_n x P(x)$ means there exist exactly n values in the domain such that $P(x)$ is true. Determine the true value of these statements where the domain consists of all real numbers.

   a) $\exists_0 x (x^2 = -1)$ **b)** $\exists_1 x (|x| = 0)$

   c) $\exists_2 x (x^2 = 2)$ **d)** $\exists_3 x (x = |x|)$

15. Let $P(x, y)$ be a propositional function. Show that $\exists x \forall y P(x, y) \rightarrow \forall y \exists x P(x, y)$ is a tautology.

16. If $\forall y \exists x P(x, y)$ is true, does it necessarily follow that $\exists x \forall y P(x, y)$ is true?

17. If $\forall x \exists y P(x, y)$ is true, does it necessarily follow that $\exists x \forall y P(x, y)$ is true?

18. Find the negations of these statements.

   a) If it snows today, then I will go skiing tomorrow.

b) Every person in this class understands mathematical induction.

c) Some students in this class do not like discrete mathematics.

d) In every mathematics class there is some student who falls asleep during lectures.

19. Express this statement using quantifiers: "Every student in this class has taken some course in every department in the school of mathematical sciences."

20. Express this statement using quantifiers: "There is a building on the campus of some college in the United States in which every room is painted white."

21. Prove that given a nonnegative integer n, there is a unique nonnegative integer m such that $m^2 \leq n < (m + 1)^2$.

22. Prove that if x^3 is irrational, then x is irrational.

23. Prove that there exists an integer m such that $m^2 > 10^{1000}$. Is your proof constructive or nonconstructive?

24. Disprove the statement that every positive integer is the sum of the cubes of eight nonnegative integers.

25. Disprove the statement that every positive integer is the sum of 36 fifth powers of nonnegative integers.

26. Assuming the truth of the theorem that states that $\sqrt{n}$ is irrational whenever n is a positive integer that is not a perfect square, prove that $\sqrt{2} + \sqrt{3}$ is irrational.

Computer Projects

Write programs with the specified input and output.

1. Given the truth values of the propositions p and q, find the truth values of the conjunction, disjunction, exclusive or, conditional statement, and biconditional of these propositions.

2. Given two bit strings of length n, find the bitwise *AND*, bitwise *OR*, and bitwise *XOR* of these strings.

∗3. Give a compound proposition, determine whether it is satisfiable by checking its truth value for all positive assignments of truth values to its propositional variables.

4. Given the truth values of the propositions p and q in fuzzy logic, find the truth value of the disjunction and the conjunction of p and q (see Exercises 46 and 47 of Section 1.1).

∗5. Given positive integers m and n, interactively play the game of Chomp.

∗6. Given a portion of a checkerboard, look for tilings of this checkerboard with various types of polyominoes, including dominoes, the two types of triominoes, and larger polyominoes.

Computations and Explorations

Use a computational program or programs you have written to do these exercises.

1. Look for positive integers that are not the sum of the cubes of nine different positive integers.

2. Look for positive integers greater than 79 that are not the sum of the fourth powers of 18 positive integers.

3. Find as many positive integers as you can that can be written as the sum of cubes of positive integers, in two different ways, sharing this property with 1729.

∗4. Try to find winning strategies for the game of Chomp for different initial configurations of cookies.

5. Construct the 12 different pentominoes, where a pentomino is a polyomino consisting of five squares.

6. Find all the rectangles of 60 squares that can be tiled using every one of the 12 different pentominoes.

Writing Projects

Respond to these with essays using outside sources.

1. Discuss logical paradoxes, including the paradox of Epimenides the Cretan, Jourdain's card paradox, and the barber paradox, and how they are resolved.

2. Describe how fuzzy logic is being applied to practical applications. Consult one or more of the recent books on fuzzy logic written for general audiences.

3. Describe some of the practical problems that can be modeled as satisfiability problems.

4. Describe some of the techniques that have been devised to help people solve Sudoku puzzles without the use of a computer.

5. Describe the basic rules of *WFF'N PROOF*, *The Game of Modern Logic*, developed by Layman Allen. Give examples of some of the games included in *WFF'N PROOF*.

6. Read some of the writings of Lewis Carroll on symbolic logic. Describe in detail some of the models he used to represent logical arguments and the rules of inference he used in these arguments.

7. Extend the discussion of Prolog given in Section 1.4, explaining in more depth how Prolog employs resolution.

8. Discuss some of the techniques used in computational logic, including Skolem's rule.

9. "Automated theorem proving" is the task of using computers to mechanically prove theorems. Discuss the goals and applications of automated theorem proving and the progress made in developing automated theorem provers.

10. Describe how DNA computing has been used to solve instances of the satisfiability problem.

11. Look up some of the incorrect proofs of famous open questions and open questions that were solved since 1970 and describe the type of error made in each proof.

12. Discuss what is known about winning strategies in the game of Chomp.

13. Describe various aspects of proof strategy discussed by George Pólya in his writings on reasoning, including [Po62], [Po71], and [Po90].

14. Describe a few problems and results about tilings with polyominoes, as described in [Go94] and [Ma91], for example.

2

Basic Structures: Sets, Functions, Sequences, Sums, and Matrices

Much of discrete mathematics is devoted to the study of discrete structures, used to represent discrete objects. Many important discrete structures are built using sets, which are collections of objects. Among the discrete structures built from sets are combinations, unordered collections of objects used extensively in counting; relations, sets of ordered pairs that represent relationships between objects; graphs, sets of vertices and edges that connect vertices; and finite state machines, used to model computing machines. These are some of the topics we will study in later chapters.

The concept of a function is extremely important in discrete mathematics. A function assigns to each element of a first set exactly one element of a second set, where the two sets are not necessarily distinct. Functions play important roles throughout discrete mathematics. They are used to represent the computational complexity of algorithms, to study the size of sets, to count objects, and in a myriad of other ways. Useful structures such as sequences and strings are special types of functions. In this chapter, we will introduce the notion of a sequence, which represents ordered lists of elements. Furthermore, we will introduce some important types of sequences and we will show how to define the terms of a sequence using earlier terms. We will also address the problem of identifying a sequence from its first few terms.

In our study of discrete mathematics, we will often add consecutive terms of a sequence of numbers. Because adding terms from a sequence, as well as other indexed sets of numbers, is such a common occurrence, a special notation has been developed for adding such terms. In this chapter, we will introduce the notation used to express summations. We will develop formulae for certain types of summations that appear throughout the study of discrete mathematics. For instance, we will encounter such summations in the analysis of the number of steps used by an algorithm to sort a list of numbers so that its terms are in increasing order.

The relative sizes of infinite sets can be studied by introducing the notion of the size, or cardinality, of a set. We say that a set is countable when it is finite or has the same size as the set of positive integers. In this chapter we will establish the surprising result that the set of rational numbers is countable, while the set of real numbers is not. We will also show how the concepts we discuss can be used to show that there are functions that cannot be computed using a computer program in any programming language.

Matrices are used in discrete mathematics to represent a variety of discrete structures. We will review the basic material about matrices and matrix arithmetic needed to represent relations and graphs. The matrix arithmetic we study will be used to solve a variety of problems involving these structures.

2.1 Sets

Introduction

In this section, we study the fundamental discrete structure on which all other discrete structures are built, namely, the set. Sets are used to group objects together. Often, but not always, the objects in a set have similar properties. For instance, all the students who are currently enrolled in your school make up a set. Likewise, all the students currently taking a course in discrete mathematics at any school make up a set. In addition, those students enrolled in your school who are taking a course in discrete mathematics form a set that can be obtained by taking the elements common to the first two collections. The language of sets is a means to study such collections in an organized fashion. We now provide a definition of a set. This definition is an intuitive definition, which is not part of a formal theory of sets.

DEFINITION 1   A *set* is an unordered collection of objects, called *elements* or *members* of the set. A set is said to *contain* its elements. We write $a \in A$ to denote that a is an element of the set A. The notation $a \notin A$ denotes that a is not an element of the set A.

It is common for sets to be denoted using uppercase letters. Lowercase letters are usually used to denote elements of sets.

There are several ways to describe a set. One way is to list all the members of a set, when this is possible. We use a notation where all members of the set are listed between braces. For example, the notation $\{a, b, c, d\}$ represents the set with the four elements $a, b, c,$ and d. This way of describing a set is known as the **roster method**.

EXAMPLE 1   The set V of all vowels in the English alphabet can be written as $V = \{a, e, i, o, u\}$.   ◄

EXAMPLE 2   The set O of odd positive integers less than 10 can be expressed by $O = \{1, 3, 5, 7, 9\}$.   ◄

EXAMPLE 3   Although sets are usually used to group together elements with common properties, there is nothing that prevents a set from having seemingly unrelated elements. For instance, $\{a, 2, \text{Fred}, \text{New Jersey}\}$ is the set containing the four elements a, 2, Fred, and New Jersey.   ◄

Sometimes the roster method is used to describe a set without listing all its members. Some members of the set are listed, and then *ellipses* ($\ldots$) are used when the general pattern of the elements is obvious.

EXAMPLE 4   The set of positive integers less than 100 can be denoted by $\{1, 2, 3, \ldots, 99\}$.   ◄

Another way to describe a set is to use **set builder** notation. We characterize all those elements in the set by stating the property or properties they must have to be members. For instance, the set O of all odd positive integers less than 10 can be written as

$$O = \{x \mid x \text{ is an odd positive integer less than } 10\},$$

or, specifying the universe as the set of positive integers, as

$$O = \{x \in \mathbf{Z}^+ \mid x \text{ is odd and } x < 10\}.$$

We often use this type of notation to describe sets when it is impossible to list all the elements of the set. For instance, the set $\mathbf{Q}^+$ of all positive rational numbers can be written as

$$\mathbf{Q}^+ = \{x \in \mathbf{R} \mid x = \tfrac{p}{q}, \text{ for some positive integers } p \text{ and } q\}.$$

Beware that mathematicians disagree whether 0 is a natural number. We consider it quite natural.

These sets, each denoted using a boldface letter, play an important role in discrete mathematics:

$\mathbf{N} = \{0, 1, 2, 3, \ldots\}$, the set of **natural numbers**
$\mathbf{Z} = \{\ldots, -2, -1, 0, 1, 2, \ldots\}$, the set of **integers**
$\mathbf{Z}^+ = \{1, 2, 3, \ldots\}$, the set of **positive integers**
$\mathbf{Q} = \{p/q \mid p \in \mathbf{Z}, q \in \mathbf{Z}, \text{and } q \neq 0\}$, the set of **rational numbers**
$\mathbf{R}$, the set of **real numbers**
$\mathbf{R}^+$, the set of **positive real numbers**
$\mathbf{C}$, the set of **complex numbers**.

(Note that some people do not consider 0 a natural number, so be careful to check how the term *natural numbers* is used when you read other books.)

Recall the notation for **intervals** of real numbers. When a and b are real numbers with $a < b$, we write

$$[a, b] = \{x \mid a \leq x \leq b\}$$
$$[a, b) = \{x \mid a \leq x < b\}$$
$$(a, b] = \{x \mid a < x \leq b\}$$
$$(a, b) = \{x \mid a < x < b\}$$

Note that $[a, b]$ is called the **closed interval** from a to b and (a, b) is called the **open interval** from a to b.

Sets can have other sets as members, as Example 5 illustrates.

EXAMPLE 5 The set $\{\mathbf{N}, \mathbf{Z}, \mathbf{Q}, \mathbf{R}\}$ is a set containing four elements, each of which is a set. The four elements of this set are $\mathbf{N}$, the set of natural numbers; $\mathbf{Z}$, the set of integers; $\mathbf{Q}$, the set of rational numbers; and $\mathbf{R}$, the set of real numbers. ◀

Remark: Note that the concept of a datatype, or type, in computer science is built upon the concept of a set. In particular, a **datatype** or **type** is the name of a set, together with a set of operations that can be performed on objects from that set. For example, *boolean* is the name of the set $\{0, 1\}$ together with operators on one or more elements of this set, such as AND, OR, and NOT.

Because many mathematical statements assert that two differently specified collections of objects are really the same set, we need to understand what it means for two sets to be equal.

DEFINITION 2 Two sets are *equal* if and only if they have the same elements. Therefore, if A and B are sets, then A and B are equal if and only if $\forall x (x \in A \leftrightarrow x \in B)$. We write $A = B$ if A and B are equal sets.

EXAMPLE 6 The sets $\{1, 3, 5\}$ and $\{3, 5, 1\}$ are equal, because they have the same elements. Note that the order in which the elements of a set are listed does not matter. Note also that it does not matter if an element of a set is listed more than once, so $\{1, 3, 3, 3, 5, 5, 5, 5\}$ is the same as the set $\{1, 3, 5\}$ because they have the same elements. ◀

Links

GEORG CANTOR (1845–1918)   Georg Cantor was born in St. Petersburg, Russia, where his father was a successful merchant. Cantor developed his interest in mathematics in his teens. He began his university studies in Zurich in 1862, but when his father died he left Zurich. He continued his university studies at the University of Berlin in 1863, where he studied under the eminent mathematicians Weierstrass, Kummer, and Kronecker. He received his doctor's degree in 1867, after having written a dissertation on number theory. Cantor assumed a position at the University of Halle in 1869, where he continued working until his death.

Cantor is considered the founder of set theory. His contributions in this area include the discovery that the set of real numbers is uncountable. He is also noted for his many important contributions to analysis. Cantor also was interested in philosophy and wrote papers relating his theory of sets with metaphysics.

Cantor married in 1874 and had five children. His melancholy temperament was balanced by his wife's happy disposition. Although he received a large inheritance from his father, he was poorly paid as a professor. To mitigate this, he tried to obtain a better-paying position at the University of Berlin. His appointment there was blocked by Kronecker, who did not agree with Cantor's views on set theory. Cantor suffered from mental illness throughout the later years of his life. He died in 1918 from a heart attack.

THE EMPTY SET   There is a special set that has no elements. This set is called the **empty set**, or **null set**, and is denoted by ∅. The empty set can also be denoted by { } (that is, we represent the empty set with a pair of braces that encloses all the elements in this set). Often, a set of elements with certain properties turns out to be the null set. For instance, the set of all positive integers that are greater than their squares is the null set.

{∅} has one more element than ∅.

A set with one element is called a **singleton set**. A common error is to confuse the empty set ∅ with the set {∅}, which is a singleton set. The single element of the set {∅} is the empty set itself! A useful analogy for remembering this difference is to think of folders in a computer file system. The empty set can be thought of as an empty folder and the set consisting of just the empty set can be thought of as a folder with exactly one folder inside, namely, the empty folder.

Links

NAIVE SET THEORY   Note that the term *object* has been used in the definition of a set, Definition 1, without specifying what an object is. This description of a set as a collection of objects, based on the intuitive notion of an object, was first stated in 1895 by the German mathematician Georg Cantor. The theory that results from this intuitive definition of a set, and the use of the intuitive notion that for any property whatever, there is a set consisting of exactly the objects with this property, leads to **paradoxes**, or logical inconsistencies. This was shown by the English philosopher Bertrand Russell in 1902 (see Exercise 29 for a description of one of these paradoxes). These logical inconsistencies can be avoided by building set theory beginning with axioms. However, we will use Cantor's original version of set theory, known as **naive set theory**, in this book because all sets considered in this book can be treated consistently using Cantor's original theory. Students will find familiarity with naive set theory helpful if they go on to learn about axiomatic set theory. They will also find the development of axiomatic set theory much more abstract than the material in this text. We refer the interested reader to [Su72] to learn more about axiomatic set theory.

Venn Diagrams

Sets can be represented graphically using Venn diagrams, named after the English mathematician John Venn, who introduced their use in 1881. In Venn diagrams the **universal set** U, which contains all the objects under consideration, is represented by a rectangle. (Note that the universal set varies depending on which objects are of interest.) Inside this rectangle, circles or other geometrical figures are used to represent sets. Sometimes points are used to represent the particular elements of the set. Venn diagrams are often used to indicate the relationships between sets. We show how a Venn diagram can be used in Example 7.

Assessment

EXAMPLE 7   Draw a Venn diagram that represents V, the set of vowels in the English alphabet.

Solution: We draw a rectangle to indicate the universal set U, which is the set of the 26 letters of the English alphabet. Inside this rectangle we draw a circle to represent V. Inside this circle we indicate the elements of V with points (see Figure 1).   ◄

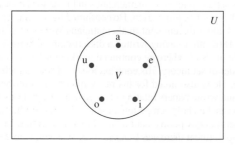

FIGURE 1   Venn Diagram for the Set of Vowels.

Subsets

It is common to encounter situations where the elements of one set are also the elements of a second set. We now introduce some terminology and notation to express such relationships between sets.

DEFINITION 3   The set A is a *subset* of B if and only if every element of A is also an element of B. We use the notation $A \subseteq B$ to indicate that A is a subset of the set B.

We see that $A \subseteq B$ if and only if the quantification

$$\forall x (x \in A \rightarrow x \in B)$$

is true. Note that to show that A is not a subset of B we need only find one element $x \in A$ with $x \notin B$. Such an x is a counterexample to the claim that $x \in A$ implies $x \in B$.

We have these useful rules for determining whether one set is a subset of another:

Showing that A is a Subset of B To show that $A \subseteq B$, show that if x belongs to A then x also belongs to B.

Showing that A is Not a Subset of B To show that $A \not\subseteq B$, find a single $x \in A$ such that $x \notin B$.

EXAMPLE 8   The set of all odd positive integers less than 10 is a subset of the set of all positive integers less than 10, the set of rational numbers is a subset of the set of real numbers, the set of all computer science majors at your school is a subset of the set of all students at your school, and the set of all people in China is a subset of the set of all people in China (that is, it is a subset of itself). Each of these facts follows immediately by noting that an element that belongs to the first set in each pair of sets also belongs to the second set in that pair.   ◀

EXAMPLE 9   The set of integers with squares less than 100 is not a subset of the set of nonnegative integers because -1 is in the former set [as $(-1)^2 < 100$], but not the later set. The set of people who have taken discrete mathematics at your school is not a subset of the set of all computer science majors at your school if there is at least one student who has taken discrete mathematics who is not a computer science major.   ◀

BERTRAND RUSSELL (1872–1970)   Bertrand Russell was born into a prominent English family active in the progressive movement and having a strong commitment to liberty. He became an orphan at an early age and was placed in the care of his father's parents, who had him educated at home. He entered Trinity College, Cambridge, in 1890, where he excelled in mathematics and in moral science. He won a fellowship on the basis of his work on the foundations of geometry. In 1910 Trinity College appointed him to a lectureship in logic and the philosophy of mathematics.

Russell fought for progressive causes throughout his life. He held strong pacifist views, and his protests against World War I led to dismissal from his position at Trinity College. He was imprisoned for 6 months in 1918 because of an article he wrote that was branded as seditious. Russell fought for women's suffrage in Great Britain. In 1961, at the age of 89, he was imprisoned for the second time for his protests advocating nuclear disarmament.

Russell's greatest work was in his development of principles that could be used as a foundation for all of mathematics. His most famous work is *Principia Mathematica,* written with Alfred North Whitehead, which attempts to deduce all of mathematics using a set of primitive axioms. He wrote many books on philosophy, physics, and his political ideas. Russell won the Nobel Prize for literature in 1950.

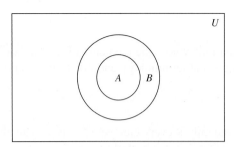

FIGURE 2   Venn Diagram Showing that A Is a Subset of B.

Theorem 1 shows that every nonempty set S is guaranteed to have at least two subsets, the empty set and the set S itself, that is, $\emptyset \subseteq S$ and $S \subseteq S$.

THEOREM 1   For every set S, $(i)\, \emptyset \subseteq S$ and $(ii)\, S \subseteq S$.

Proof: We will prove (i) and leave the proof of (ii) as an exercise.

Let S be a set. To show that $\emptyset \subseteq S$, we must show that $\forall x(x \in \emptyset \rightarrow x \in S)$ is true. Because the empty set contains no elements, it follows that $x \in \emptyset$ is always false. It follows that the conditional statement $x \in \emptyset \rightarrow x \in S$ is always true, because its hypothesis is always false and a conditional statement with a false hypothesis is true. Therefore, $\forall x(x \in \emptyset \rightarrow x \in S)$ is true. This completes the proof of (i). Note that this is an example of a vacuous proof.   ◁

When we wish to emphasize that a set A is a subset of a set B but that $A \neq B$, we write $A \subset B$ and say that A is a **proper subset** of B. For $A \subset B$ to be true, it must be the case that $A \subseteq B$ and there must exist an element x of B that is not an element of A. That is, A is a proper subset of B if and only if

$$\forall x(x \in A \rightarrow x \in B) \wedge \exists x(x \in B \wedge x \notin A)$$

is true. Venn diagrams can be used to illustrate that a set A is a subset of a set B. We draw the universal set U as a rectangle. Within this rectangle we draw a circle for B. Because A is a subset of B, we draw the circle for A within the circle for B. This relationship is shown in Figure 2.

A useful way to show that two sets have the same elements is to show that each set is a subset of the other. In other words, we can show that if A and B are sets with $A \subseteq B$ and $B \subseteq A$, then $A = B$. That is, $A = B$ if and only if $\forall x(x \in A \rightarrow x \in B)$ and $\forall x(x \in B \rightarrow x \in A)$ or equivalently if and only if $\forall x(x \in A \leftrightarrow x \in B)$, which is what it means for the A and B to be equal. Because this method of showing two sets are equal is so useful, we highlight it here.

JOHN VENN (1834–1923)   John Venn was born into a London suburban family noted for its philanthropy. He attended London schools and got his mathematics degree from Caius College, Cambridge, in 1857. He was elected a fellow of this college and held his fellowship there until his death. He took holy orders in 1859 and, after a brief stint of religious work, returned to Cambridge, where he developed programs in the moral sciences. Besides his mathematical work, Venn had an interest in history and wrote extensively about his college and family.

Venn's book *Symbolic Logic* clarifies ideas originally presented by Boole. In this book, Venn presents a systematic development of a method that uses geometric figures, known now as *Venn diagrams*. Today these diagrams are primarily used to analyze logical arguments and to illustrate relationships between sets. In addition to his work on symbolic logic, Venn made contributions to probability theory described in his widely used textbook on that subject.

Showing Two Sets are Equal To show that two sets A and B are equal, show that $A \subseteq B$ and $B \subseteq A$.

Sets may have other sets as members. For instance, we have the sets

$$A = \{\emptyset, \{a\}, \{b\}, \{a, b\}\} \qquad \text{and} \qquad B = \{x \mid x \text{ is a subset of the set } \{a, b\}\}.$$

Note that these two sets are equal, that is, $A = B$. Also note that $\{a\} \in A$, but $a \notin A$.

The Size of a Set

Sets are used extensively in counting problems, and for such applications we need to discuss the sizes of sets.

DEFINITION 4   Let S be a set. If there are exactly n distinct elements in S where n is a nonnegative integer, we say that S is a *finite set* and that n is the *cardinality* of S. The cardinality of S is denoted by $|S|$.

Remark: The term *cardinality* comes from the common usage of the term *cardinal number* as the size of a finite set.

EXAMPLE 10   Let A be the set of odd positive integers less than 10. Then $|A| = 5$.   ◀

EXAMPLE 11   Let S be the set of letters in the English alphabet. Then $|S| = 26$.   ◀

EXAMPLE 12   Because the null set has no elements, it follows that $|\emptyset| = 0$.   ◀

We will also be interested in sets that are not finite.

DEFINITION 5   A set is said to be *infinite* if it is not finite.

EXAMPLE 13   The set of positive integers is infinite.   ◀

We will extend the notion of cardinality to infinite sets in Section 2.5, a challenging topic full of surprising results.

Power Sets

Many problems involve testing all combinations of elements of a set to see if they satisfy some property. To consider all such combinations of elements of a set S, we build a new set that has as its members all the subsets of S.

DEFINITION 6   Given a set S, the *power set* of S is the set of all subsets of the set S. The power set of S is denoted by $\mathcal{P}(S)$.

EXAMPLE 14   What is the power set of the set $\{0, 1, 2\}$?

Solution: The power set $\mathcal{P}(\{0, 1, 2\})$ is the set of all subsets of $\{0, 1, 2\}$. Hence,

$$\mathcal{P}(\{0, 1, 2\}) = \{\emptyset, \{0\}, \{1\}, \{2\}, \{0, 1\}, \{0, 2\}, \{1, 2\}, \{0, 1, 2\}\}.$$

Note that the empty set and the set itself are members of this set of subsets.   ◀

EXAMPLE 15   What is the power set of the empty set? What is the power set of the set $\{\emptyset\}$?

Solution: The empty set has exactly one subset, namely, itself. Consequently,

$$\mathcal{P}(\emptyset) = \{\emptyset\}.$$

The set $\{\emptyset\}$ has exactly two subsets, namely, $\emptyset$ and the set $\{\emptyset\}$ itself. Therefore,

$$\mathcal{P}(\{\emptyset\}) = \{\emptyset, \{\emptyset\}\}.$$ ◀

If a set has n elements, then its power set has 2^n elements. We will demonstrate this fact in several ways in subsequent sections of the text.

Cartesian Products

The order of elements in a collection is often important. Because sets are unordered, a different structure is needed to represent ordered collections. This is provided by **ordered n-tuples**.

DEFINITION 7   The *ordered n-tuple* $(a_1, a_2, \ldots, a_n)$ is the ordered collection that has a_1 as its first element, a_2 as its second element, $\ldots$, and a_n as its nth element.

We say that two ordered n-tuples are equal if and only if each corresponding pair of their elements is equal. In other words, $(a_1, a_2, \ldots, a_n) = (b_1, b_2, \ldots, b_n)$ if and only if $a_i = b_i$, for $i = 1, 2, \ldots, n$. In particular, ordered 2-tuples are called **ordered pairs**. The ordered pairs (a, b) and (c, d) are equal if and only if $a = c$ and $b = d$. Note that (a, b) and (b, a) are not equal unless $a = b$.

RENÉ DESCARTES (1596–1650)   René Descartes was born into a noble family near Tours, France, about 200 miles southwest of Paris. He was the third child of his father's first wife; she died several days after his birth. Because of René's poor health, his father, a provincial judge, let his son's formal lessons slide until, at the age of 8, René entered the Jesuit college at La Flèche. The rector of the school took a liking to him and permitted him to stay in bed until late in the morning because of his frail health. From then on, Descartes spent his mornings in bed; he considered these times his most productive hours for thinking.

Descartes left school in 1612, moving to Paris, where he spent 2 years studying mathematics. He earned a law degree in 1616 from the University of Poitiers. At 18 Descartes became disgusted with studying and decided to see the world. He moved to Paris and became a successful gambler. However, he grew tired of bawdy living and moved to the suburb of Saint-Germain, where he devoted himself to mathematical study. When his gambling friends found him, he decided to leave France and undertake a military career. However, he never did any fighting. One day, while escaping the cold in an overheated room at a military encampment, he had several feverish dreams, which revealed his future career as a mathematician and philosopher.

After ending his military career, he traveled throughout Europe. He then spent several years in Paris, where he studied mathematics and philosophy and constructed optical instruments. Descartes decided to move to Holland, where he spent 20 years wandering around the country, accomplishing his most important work. During this time he wrote several books, including the *Discours,* which contains his contributions to analytic geometry, for which he is best known. He also made fundamental contributions to philosophy.

In 1649 Descartes was invited by Queen Christina to visit her court in Sweden to tutor her in philosophy. Although he was reluctant to live in what he called "the land of bears amongst rocks and ice," he finally accepted the invitation and moved to Sweden. Unfortunately, the winter of 1649–1650 was extremely bitter. Descartes caught pneumonia and died in mid-February.

Many of the discrete structures we will study in later chapters are based on the notion of the *Cartesian product* of sets (named after René Descartes). We first define the Cartesian product of two sets.

DEFINITION 8 Let A and B be sets. The *Cartesian product* of A and B, denoted by $A \times B$, is the set of all ordered pairs (a, b), where $a \in A$ and $b \in B$. Hence,

$$A \times B = \{(a, b) \mid a \in A \land b \in B\}.$$

EXAMPLE 16 Let A represent the set of all students at a university, and let B represent the set of all courses offered at the university. What is the Cartesian product $A \times B$ and how can it be used?

Solution: The Cartesian product $A \times B$ consists of all the ordered pairs of the form (a, b), where a is a student at the university and b is a course offered at the university. One way to use the set $A \times B$ is to represent all possible enrollments of students in courses at the university. ◀

EXAMPLE 17 What is the Cartesian product of $A = \{1, 2\}$ and $B = \{a, b, c\}$?

Solution: The Cartesian product $A \times B$ is

$$A \times B = \{(1, a), (1, b), (1, c), (2, a), (2, b), (2, c)\}.$$ ◀

Note that the Cartesian products $A \times B$ and $B \times A$ are not equal, unless $A = \emptyset$ or $B = \emptyset$ (so that $A \times B = \emptyset$) or $A = B$ (see Exercise 24). This is illustrated in Example 18.

EXAMPLE 18 Show that the Cartesian product $B \times A$ is not equal to the Cartesian product $A \times B$, where A and B are as in Example 17.

Solution: The Cartesian product $B \times A$ is

$$B \times A = \{(a, 1), (a, 2), (b, 1), (b, 2), (c, 1), (c, 2)\}.$$

This is not equal to $A \times B$, which was found in Example 17. ◀

The Cartesian product of more than two sets can also be defined.

DEFINITION 9 The *Cartesian product* of the sets $A_1, A_2, \ldots, A_n$, denoted by $A_1 \times A_2 \times \cdots \times A_n$, is the set of ordered n-tuples $(a_1, a_2, \ldots, a_n)$, where a_i belongs to A_i for $i = 1, 2, \ldots, n$. In other words,

$$A_1 \times A_2 \times \cdots \times A_n = \{(a_1, a_2, \ldots, a_n) \mid a_i \in A_i \text{ for } i = 1, 2, \ldots, n\}.$$

EXAMPLE 19   What is the Cartesian product $A \times B \times C$, where $A = \{0, 1\}$, $B = \{1, 2\}$, and $C = \{0, 1, 2\}$?

Solution: The Cartesian product $A \times B \times C$ consists of all ordered triples (a, b, c), where $a \in A$, $b \in B$, and $c \in C$. Hence,

$$A \times B \times C = \{(0, 1, 0), (0, 1, 1), (0, 1, 2), (0, 2, 0), (0, 2, 1), (0, 2, 2),$$
$$(1, 1, 0), (1, 1, 1), (1, 1, 2), (1, 2, 0), (1, 2, 1), (1, 2, 2)\}. \quad \blacktriangleleft$$

Remark: Note that when A, B, and C are sets, $(A \times B) \times C$ is not the same as $A \times B \times C$ (see Exercise 25).

We use the notation A^2 to denote $A \times A$, the Cartesian product of the set A with itself. Similarly, $A^3 = A \times A \times A$, $A^4 = A \times A \times A \times A$, and so on. More generally,

$$A^n = \{(a_1, a_2, \ldots, a_n) \mid a_i \in A \text{ for } i = 1, 2, \ldots, n\}.$$

EXAMPLE 20   Suppose that $A = \{1, 2\}$. It follows that $A^2 = \{(1, 1), (1, 2), (2, 1), (2, 2)\}$ and $A^3 = \{(1, 1, 1), (1, 1, 2), (1, 2, 1), (1, 2, 2), (2, 1, 1), (2, 1, 2), (2, 2, 1), (2, 2, 2)\}$. $\quad \blacktriangleleft$

A subset R of the Cartesian product $A \times B$ is called a **relation** from the set A to the set B. The elements of R are ordered pairs, where the first element belongs to A and the second to B. For example, $R = \{(a, 0), (a, 1), (a, 3), (b, 1), (b, 2), (c, 0), (c, 3)\}$ is a relation from the set $\{a, b, c\}$ to the set $\{0, 1, 2, 3\}$. A relation from a set A to itself is called a relation on A.

EXAMPLE 21   What are the ordered pairs in the less than or equal to relation, which contains (a, b) if $a \leq b$, on the set $\{0, 1, 2, 3\}$?

Solution: The ordered pair (a, b) belongs to R if and only if both a and b belong to $\{0, 1, 2, 3\}$ and $a \leq b$. Consequently, the ordered pairs in R are (0,0), (0,1), (0,2), (0,3), (1,1), (1,2), (1,3), (2,2), (2, 3), and (3, 3). $\quad \blacktriangleleft$

We will study relations and their properties at length in Chapter 9.

Using Set Notation with Quantifiers

Sometimes we restrict the domain of a quantified statement explicitly by making use of a particular notation. For example, $\forall x \in S(P(x))$ denotes the universal quantification of $P(x)$ over all elements in the set S. In other words, $\forall x \in S(P(x))$ is shorthand for $\forall x(x \in S \rightarrow P(x))$. Similarly, $\exists x \in S(P(x))$ denotes the existential quantification of $P(x)$ over all elements in S. That is, $\exists x \in S(P(x))$ is shorthand for $\exists x(x \in S \land P(x))$.

EXAMPLE 22   What do the statements $\forall x \in \mathbf{R}\ (x^2 \geq 0)$ and $\exists x \in \mathbf{Z}\ (x^2 = 1)$ mean?

Solution: The statement $\forall x \in \mathbf{R}(x^2 \geq 0)$ states that for every real number x, $x^2 \geq 0$. This statement can be expressed as "The square of every real number is nonnegative." This is a true statement.

The statement $\exists x \in \mathbf{Z}(x^2 = 1)$ states that there exists an integer x such that $x^2 = 1$. This statement can be expressed as "There is an integer whose square is 1." This is also a true statement because $x = 1$ is such an integer (as is -1). $\quad \blacktriangleleft$

Truth Sets and Quantifiers

We will now tie together concepts from set theory and from predicate logic. Given a predicate P, and a domain D, we define the **truth set** of P to be the set of elements x in D for which $P(x)$ is true. The truth set of $P(x)$ is denoted by $\{x \in D \mid P(x)\}$.

EXAMPLE 23   What are the truth sets of the predicates $P(x)$, $Q(x)$, and $R(x)$, where the domain is the set of integers and $P(x)$ is "$|x| = 1$," $Q(x)$ is "$x^2 = 2$," and $R(x)$ is "$|x| = x$."

Solution: The truth set of P, $\{x \in \mathbf{Z} \mid |x| = 1\}$, is the set of integers for which $|x| = 1$. Because $|x| = 1$ when $x = 1$ or $x = -1$, and for no other integers x, we see that the truth set of P is the set $\{-1, 1\}$.

The truth set of Q, $\{x \in \mathbf{Z} \mid x^2 = 2\}$, is the set of integers for which $x^2 = 2$. This is the empty set because there are no integers x for which $x^2 = 2$.

The truth set of R, $\{x \in \mathbf{Z} \mid |x| = x\}$, is the set of integers for which $|x| = x$. Because $|x| = x$ if and only if $x \geq 0$, it follows that the truth set of R is $\mathbf{N}$, the set of nonnegative integers. ◀

Note that $\forall x\, P(x)$ is true over the domain U if and only if the truth set of P is the set U. Likewise, $\exists x\, P(x)$ is true over the domain U if and only if the truth set of P is nonempty.

Exercises

1. List the members of these sets.
   a) $\{x \mid x$ is a real number such that $x^2 = 1\}$
   b) $\{x \mid x$ is a positive integer less than 12$\}$
   c) $\{x \mid x$ is the square of an integer and $x < 100\}$
   d) $\{x \mid x$ is an integer such that $x^2 = 2\}$

2. Use set builder notation to give a description of each of these sets.
   a) $\{0, 3, 6, 9, 12\}$
   b) $\{-3, -2, -1, 0, 1, 2, 3\}$
   c) $\{m, n, o, p\}$

3. For each of these pairs of sets, determine whether the first is a subset of the second, the second is a subset of the first, or neither is a subset of the other.
   a) the set of airline flights from New York to New Delhi, the set of nonstop airline flights from New York to New Delhi
   b) the set of people who speak English, the set of people who speak Chinese
   c) the set of flying squirrels, the set of living creatures that can fly

4. Determine whether each of these pairs of sets are equal.
   a) $\{1, 3, 3, 3, 5, 5, 5, 5, 5\}$, $\{5, 3, 1\}$
   b) $\{\{1\}\}$, $\{1, \{1\}\}$ **c)** $\emptyset$, $\{\emptyset\}$

5. Suppose that $A = \{2, 4, 6\}$, $B = \{2, 6\}$, $C = \{4, 6\}$, and $D = \{4, 6, 8\}$. Determine which of these sets are subsets of which other of these sets.

6. For each of the following sets, determine whether 2 is an element of that set.
   a) $\{x \in \mathbf{R} \mid x$ is an integer greater than 1$\}$
   b) $\{x \in \mathbf{R} \mid x$ is the square of an integer$\}$
   c) $\{2, \{2\}\}$
   d) $\{\{2\}, \{\{2\}\}\}$
   e) $\{\{2\}, \{2, \{2\}\}\}$
   f) $\{\{\{2\}\}\}$

7. For each of the sets in Exercise 7, determine whether $\{2\}$ is an element of that set.

8. Determine whether each of these statements is true or false.
   a) $0 \in \emptyset$ **b)** $\emptyset \in \{0\}$
   c) $\{0\} \subset \emptyset$ **d)** $\emptyset \subset \{0\}$
   e) $x \in \{x\}$ **f)** $\{x\} \subseteq \{x\}$
   g) $\{x\} \in \{x\}$ **h)** $\{x\} \in \{\{x\}\}$
   i) $\emptyset \subseteq \{x\}$

9. Use a Venn diagram to illustrate the subset of odd integers in the set of all positive integers not exceeding 10.

10. Use a Venn diagram to illustrate the set of all months of the year whose names do not contain the letter R in the set of all months of the year.

11. Use a Venn diagram to illustrate the relationships $A \subset B$ and $B \subset C$.

12. Suppose that A, B, and C are sets such that $A \subseteq B$ and $B \subseteq C$. Show that $A \subseteq C$.

13. Find two sets A and B such that $A \in B$ and $A \subseteq B$.

14. What is the cardinality of each of these sets?
   a) $\{a\}$ **b)** $\{\{a\}\}$
   c) $\{a, \{a\}\}$ **d)** $\{a, \{a\}, \{a, \{a\}\}\}$

15. Find the power set of each of these sets, where a and b are distinct elements.
   a) $\{a\}$ **b)** $\{a, b\}$ **c)** $\{\emptyset, \{\emptyset\}\}$

16. Can you conclude that $A = B$ if A and B are two sets with the same power set?

17. How many elements does each of these sets have where a and b are distinct elements?

    a) $\mathcal{P}(\{a, b, \{a, b\}\})$

    b) $\mathcal{P}(\{\emptyset, a, \{a\}, \{\{a\}\}\})$

    c) $\mathcal{P}(\mathcal{P}(\emptyset))$

18. Determine whether each of these sets is the power set of a set, where a and b are distinct elements.

    a) $\emptyset$ **b)** $\{\emptyset, \{a\}\}$

    c) $\{\emptyset, \{a\}, \{\emptyset, a\}\}$ **d)** $\{\emptyset, \{a\}, \{b\}, \{a, b\}\}$

19. Let $A = \{a, b, c, d\}$ and $B = \{y, z\}$. Find

    a) $A \times B$.   **b)** $B \times A$.

20. What is the Cartesian product $A \times B$, where A is the set of courses offered by the mathematics department at a university and B is the set of mathematics professors at this university? Give an example of how this Cartesian product can be used.

21. Suppose that $A \times B = \emptyset$, where A and B are sets. What can you conclude?

22. Let $A = \{a, b, c\}$, $B = \{x, y\}$, and $C = \{0, 1\}$. Find

    a) $A \times B \times C$.   **b)** $C \times B \times A$.

    c) $C \times A \times B$.   **d)** $B \times B \times B$.

23. How many different elements does $A \times B$ have if A has m elements and B has n elements?

24. Show that $A \times B \neq B \times A$, when A and B are nonempty, unless $A = B$.

25. Explain why $A \times B \times C$ and $(A \times B) \times C$ are not the same.

26. Translate each of these quantifications into English and determine its truth value.

    a) $\forall x \in \mathbf{R} \ (x^2 \neq -1)$ **b)** $\exists x \in \mathbf{Z} \ (x^2 = 2)$

    c) $\forall x \in \mathbf{Z} \ (x^2 > 0)$ **d)** $\exists x \in \mathbf{R} \ (x^2 = x)$

27. Find the truth set of each of these predicates where the domain is the set of integers.

    a) $P(x): x^2 < 3$ **b)** $Q(x): x^2 > x$

    c) $R(x): 2x + 1 = 0$

*28. The defining property of an ordered pair is that two ordered pairs are equal if and only if their first elements are equal and their second elements are equal. Surprisingly, instead of taking the ordered pair as a primitive concept, we can construct ordered pairs using basic notions from set theory. Show that if we define the ordered pair (a, b) to be $\{\{a\}, \{a, b\}\}$, then $(a, b) = (c, d)$ if and only if $a = c$ and $b = d$. [*Hint:* First show that $\{\{a\}, \{a, b\}\} = \{\{c\}, \{c, d\}\}$ if and only if $a = c$ and $b = d$.]

*29. This exercise presents **Russell's paradox**. Let S be the set that contains a set x if the set x does not belong to itself, so that $S = \{x \mid x \notin x\}$.

    a) Show the assumption that S is a member of S leads to a contradiction.

    b) Show the assumption that S is not a member of S leads to a contradiction.

By parts (a) and (b) it follows that the set S cannot be defined as it was. This paradox can be avoided by restricting the types of elements that sets can have.

2.2   Set Operations

Introduction

Two, or more, sets can be combined in many different ways. For instance, starting with the set of mathematics majors at your school and the set of computer science majors at your school, we can form the set of students who are mathematics majors or computer science majors, the set of students who are joint majors in mathematics and computer science, the set of all students not majoring in mathematics, and so on.

Links

DEFINITION 1   Let A and B be sets. The *union* of the sets A and B, denoted by $A \cup B$, is the set that contains those elements that are either in A or in B, or in both.

An element x belongs to the union of the sets A and B if and only if x belongs to A or x belongs to B. This tells us that

$$A \cup B = \{x \mid x \in A \vee x \in B\}.$$

The Venn diagram shown in Figure 1 represents the union of two sets A and B. The area that represents $A \cup B$ is the shaded area within either the circle representing A or the circle representing B.

We will give some examples of the union of sets.

EXAMPLE 1 The union of the sets $\{1, 3, 5\}$ and $\{1, 2, 3\}$ is the set $\{1, 2, 3, 5\}$; that is, $\{1, 3, 5\} \cup \{1, 2, 3\} = \{1, 2, 3, 5\}$. ◀

EXAMPLE 2 The union of the set of all computer science majors at your school and the set of all mathematics majors at your school is the set of students at your school who are majoring either in mathematics or in computer science (or in both). ◀

DEFINITION 2 Let A and B be sets. The *intersection* of the sets A and B, denoted by $A \cap B$, is the set containing those elements in both A and B.

An element x belongs to the intersection of the sets A and B if and only if x belongs to A and x belongs to B. This tells us that

$$A \cap B = \{x \mid x \in A \land x \in B\}.$$

The Venn diagram shown in Figure 2 represents the intersection of two sets A and B. The shaded area that is within both the circles representing the sets A and B is the area that represents the intersection of A and B.

We give some examples of the intersection of sets.

EXAMPLE 3 The intersection of the sets $\{1, 3, 5\}$ and $\{1, 2, 3\}$ is the set $\{1, 3\}$; that is, $\{1, 3, 5\} \cap \{1, 2, 3\} = \{1, 3\}$. ◀

EXAMPLE 4 The intersection of the set of all computer science majors at your school and the set of all mathematics majors is the set of all students who are joint majors in mathematics and computer science. ◀

DEFINITION 3 Two sets are called *disjoint* if their intersection is the empty set.

EXAMPLE 5 Let $A = \{1, 3, 5, 7, 9\}$ and $B = \{2, 4, 6, 8, 10\}$. Because $A \cap B = \emptyset$, A and B are disjoint. ◀

Be careful not to overcount!

We are often interested in finding the cardinality of a union of two finite sets A and B. Note that $|A| + |B|$ counts each element that is in A but not in B or in B but not in A exactly once, and each element that is in both A and B exactly twice. Thus, if the number of elements that

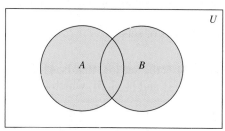

$A \cup B$ is shaded.

FIGURE 1 **Venn Diagram of the Union of A and B.**

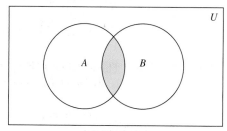

$A \cap B$ is shaded.

FIGURE 2 **Venn Diagram of the Intersection of A and B.**

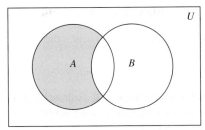

$A - B$ is shaded.

FIGURE 3   Venn Diagram for the Difference of A and B.

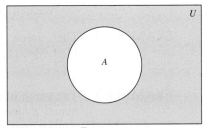

$\overline{A}$ is shaded.

FIGURE 4   Venn Diagram for the Complement of the Set A.

are in both A and B is subtracted from $|A| + |B|$, elements in $A \cap B$ will be counted only once. Hence,

$$|A \cup B| = |A| + |B| - |A \cap B|.$$

The generalization of this result to unions of an arbitrary number of sets is called the **principle of inclusion–exclusion**. The principle of inclusion–exclusion is an important technique used in enumeration. We will discuss this principle and other counting techniques in detail in Chapters 6 and 8.

There are other important ways to combine sets.

DEFINITION 4   Let A and B be sets. The *difference* of A and B, denoted by $A - B$, is the set containing those elements that are in A but not in B. The difference of A and B is also called the *complement of B with respect to A.*

Remark: The difference of sets A and B is sometimes denoted by $A \backslash B$.

An element x belongs to the difference of A and B if and only if $x \in A$ and $x \notin B$. This tells us that

$$A - B = \{x \mid x \in A \wedge x \notin B\}.$$

The Venn diagram shown in Figure 3 represents the difference of the sets A and B. The shaded area inside the circle that represents A and outside the circle that represents B is the area that represents $A - B$.

We give some examples of differences of sets.

EXAMPLE 6   The difference of $\{1, 3, 5\}$ and $\{1, 2, 3\}$ is the set $\{5\}$; that is, $\{1, 3, 5\} - \{1, 2, 3\} = \{5\}$. This is different from the difference of $\{1, 2, 3\}$ and $\{1, 3, 5\}$, which is the set $\{2\}$. ◀

EXAMPLE 7   The difference of the set of computer science majors at your school and the set of mathematics majors at your school is the set of all computer science majors at your school who are not also mathematics majors. ◀

Once the universal set U has been specified, the **complement** of a set can be defined.

DEFINITION 5   Let U be the universal set. The *complement* of the set A, denoted by $\overline{A}$, is the complement of A with respect to U. Therefore, the complement of the set A is $U - A$.

An element belongs to $\overline{A}$ if and only if $x \notin A$. This tells us that

$$\overline{A} = \{x \in U \mid x \notin A\}.$$

In Figure 4 the shaded area outside the circle representing A is the area representing $\overline{A}$.
We give some examples of the complement of a set.

EXAMPLE 8   Let $A = \{a, e, i, o, u\}$ (where the universal set is the set of letters of the English alphabet). Then $\overline{A} = \{b, c, d, f, g, h, j, k, l, m, n, p, q, r, s, t, v, w, x, y, z\}$.   ◄

EXAMPLE 9   Let A be the set of positive integers greater than 10 (with universal set the set of all positive integers). Then $\overline{A} = \{1, 2, 3, 4, 5, 6, 7, 8, 9, 10\}$.   ◄

It is left to the reader (Exercise 11) to show that we can express the difference of A and B as the intersection of A and the complement of B. That is,

$$A - B = A \cap \overline{B}.$$

Set Identities

Table 1 lists the most important set identities. We will prove several of these identities here, using three different methods. These methods are presented to illustrate that there are often many different approaches to the solution of a problem. The proofs of the remaining identities will be left as exercises. The reader should note the similarity between these set identities and the logical equivalences discussed in Section 1.3. (Compare Table 6 of Section 1.6 and Table 1.) In fact, the set identities given can be proved directly from the corresponding logical equivalences. Furthermore, both are special cases of identities that hold for Boolean algebra (discussed in Chapter 12).

Set identities and propositional equivalences are just special cases of identities for Boolean algebra.

One way to show that two sets are equal is to show that each is a subset of the other. Recall that to show that one set is a subset of a second set, we can show that if an element belongs to the first set, then it must also belong to the second set. We generally use a direct proof to do this. We illustrate this type of proof by establishing the first of De Morgan's laws.

TABLE 1 Set Identities.

Identity	Name
$A \cap U = A$ $A \cup \emptyset = A$	Identity laws
$A \cup U = U$ $A \cap \emptyset = \emptyset$	Domination laws
$A \cup A = A$ $A \cap A = A$	Idempotent laws
$\overline{(\overline{A})} = A$	Complementation law
$A \cup B = B \cup A$ $A \cap B = B \cap A\grave{}$	Commutative laws
$A \cup (B \cup C) = (A \cup B) \cup C$ $A \cap (B \cap C) = (A \cap B) \cap C$	Associative laws
$A \cup (B \cap C) = (A \cup B) \cap (A \cup C)$ $A \cap (B \cup C) = (A \cap B) \cup (A \cap C)$	Distributive laws
$\overline{A \cap B} = \overline{A} \cup \overline{B}$ $\overline{A \cup B} = \overline{A} \cap \overline{B}$	De Morgan's laws
$A \cup (A \cap B) = A$ $A \cap (A \cup B) = A$	Absorption laws
$A \cup \overline{A} = U$ $A \cap \overline{A} = \emptyset$	Complement laws

EXAMPLE 10   Prove that $\overline{A \cap B} = \overline{A} \cup \overline{B}$.

This identity says that the complement of the intersection of two sets is the union of their complements.

Solution: We will prove that the two sets $\overline{A \cap B}$ and $\overline{A} \cup \overline{B}$ are equal by showing that each set is a subset of the other.

First, we will show that $\overline{A \cap B} \subseteq \overline{A} \cup \overline{B}$. We do this by showing that if x is in $\overline{A \cap B}$, then it must also be in $\overline{A} \cup \overline{B}$. Now suppose that $x \in \overline{A \cap B}$. By the definition of complement, $x \notin A \cap B$. Using the definition of intersection, we see that the proposition $\neg((x \in A) \wedge (x \in B))$ is true.

By applying De Morgan's law for propositions, we see that $\neg(x \in A)$ or $\neg(x \in B)$. Using the definition of negation of propositions, we have $x \notin A$ or $x \notin B$. Using the definition of the complement of a set, we see that this implies that $x \in \overline{A}$ or $x \in \overline{B}$. Consequently, by the definition of union, we see that $x \in \overline{A} \cup \overline{B}$. We have now shown that $\overline{A \cap B} \subseteq \overline{A} \cup \overline{B}$.

Extra Examples

Next, we will show that $\overline{A} \cup \overline{B} \subseteq \overline{A \cap B}$. We do this by showing that if x is in $\overline{A} \cup \overline{B}$, then it must also be in $\overline{A \cap B}$. Now suppose that $x \in \overline{A} \cup \overline{B}$. By the definition of union, we know that $x \in \overline{A}$ or $x \in \overline{B}$. Using the definition of complement, we see that $x \notin A$ or $x \notin B$. Consequently, the proposition $\neg(x \in A) \vee \neg(x \in B)$ is true.

By De Morgan's law for propositions, we conclude that $\neg((x \in A) \wedge (x \in B))$ is true. By the definition of intersection, it follows that $\neg(x \in A \cap B)$. We now use the definition of complement to conclude that $x \in \overline{A \cap B}$. This shows that $\overline{A} \cup \overline{B} \subseteq \overline{A \cap B}$.

Because we have shown that each set is a subset of the other, the two sets are equal, and the identity is proved.   ◄

We can more succinctly express the reasoning used in Example 10 using set builder notation, as Example 11 illustrates.

EXAMPLE 11   Use set builder notation and logical equivalences to establish the first De Morgan law $\overline{A \cap B} = \overline{A} \cup \overline{B}$.

Solution: We can prove this identity with the following steps.

$$
\begin{aligned}
\overline{A \cap B} &= \{x \mid x \notin A \cap B\} && \text{by definition of complement}\\
&= \{x \mid \neg(x \in (A \cap B))\} && \text{by definition of does not belong symbol}\\
&= \{x \mid \neg(x \in A \wedge x \in B)\} && \text{by definition of intersection}\\
&= \{x \mid \neg(x \in A) \vee \neg(x \in B)\} && \text{by the first De Morgan law for logical equivalences}\\
&= \{x \mid x \notin A \vee x \notin B\} && \text{by definition of does not belong symbol}\\
&= \{x \mid x \in \overline{A} \vee x \in \overline{B}\} && \text{by definition of complement}\\
&= \{x \mid x \in \overline{A} \cup \overline{B}\} && \text{by definition of union}\\
&= \overline{A} \cup \overline{B} && \text{by meaning of set builder notation}
\end{aligned}
$$

Note that besides the definitions of complement, union, set membership, and set builder notation, this proof uses the second De Morgan law for logical equivalences.   ◀

Proving a set identity involving more than two sets by showing each side of the identity is a subset of the other often requires that we keep track of different cases, as illustrated by the proof in Example 12 of one of the distributive laws for sets.

EXAMPLE 12   Prove the second distributive law from Table 1, which states that $A \cap (B \cup C) = (A \cap B) \cup (A \cap C)$ for all sets A, B, and C.

Solution: We will prove this identity by showing that each side is a subset of the other side.
   Suppose that $x \in A \cap (B \cup C)$. Then $x \in A$ and $x \in B \cup C$. By the definition of union, it follows that $x \in A$, and $x \in B$ or $x \in C$ (or both). In other words, we know that the compound proposition $(x \in A) \wedge ((x \in B) \vee (x \in C))$ is true. By the distributive law for conjunction over disjunction, it follows that $((x \in A) \wedge (x \in B)) \vee ((x \in A) \wedge (x \in C))$. We conclude that either $x \in A$ and $x \in B$, or $x \in A$ and $x \in C$. By the definition of intersection, it follows that $x \in A \cap B$ or $x \in A \cap C$. Using the definition of union, we conclude that $x \in (A \cap B) \cup (A \cap C)$. We conclude that $A \cap (B \cup C) \subseteq (A \cap B) \cup (A \cap C)$.
   Now suppose that $x \in (A \cap B) \cup (A \cap C)$. Then, by the definition of union, $x \in A \cap B$ or $x \in A \cap C$. By the definition of intersection, it follows that $x \in A$ and $x \in B$ or that $x \in A$ and $x \in C$. From this we see that $x \in A$, and $x \in B$ or $x \in C$. Consequently, by the definition of union we see that $x \in A$ and $x \in B \cup C$. Furthermore, by the definition of intersection, it follows that $x \in A \cap (B \cup C)$. We conclude that $(A \cap B) \cup (A \cap C) \subseteq A \cap (B \cup C)$. This completes the proof of the identity.   ◀

Set identities can also be proved using **membership tables**. We consider each combination of sets that an element can belong to and verify that elements in the same combinations of sets belong to both the sets in the identity. To indicate that an element is in a set, a 1 is used; to indicate that an element is not in a set, a 0 is used. (The reader should note the similarity between membership tables and truth tables.)

EXAMPLE 13   Use a membership table to show that $A \cap (B \cup C) = (A \cap B) \cup (A \cap C)$.

Solution: The membership table for these combinations of sets is shown in Table 2. This table has eight rows. Because the columns for $A \cap (B \cup C)$ and $(A \cap B) \cup (A \cap C)$ are the same, the identity is valid.   ◀

Additional set identities can be established using those that we have already proved. Consider Example 14.

TABLE 2 **A Membership Table for the Distributive Property.**							
A	B	C	$B \cup C$	$A \cap (B \cup C)$	$A \cap B$	$A \cap C$	$(A \cap B) \cup (A \cap C)$
1	1	1	1	1	1	1	1
1	1	0	1	1	1	0	1
1	0	1	1	1	0	1	1
1	0	0	0	0	0	0	0
0	1	1	1	0	0	0	0
0	1	0	1	0	0	0	0
0	0	1	1	0	0	0	0
0	0	0	0	0	0	0	0

EXAMPLE 14   Let A, B, and C be sets. Show that

$$\overline{A \cup (B \cap C)} = (\overline{C} \cup \overline{B}) \cap \overline{A}.$$

Solution: We have

$$
\begin{aligned}
\overline{A \cup (B \cap C)} &= \overline{A} \cap \overline{(B \cap C)} \quad &\text{by the first De Morgan law}\\
&= \overline{A} \cap (\overline{B} \cup \overline{C}) \quad &\text{by the second De Morgan law}\\
&= (\overline{B} \cup \overline{C}) \cap \overline{A} \quad &\text{by the commutative law for intersections}\\
&= (\overline{C} \cup \overline{B}) \cap \overline{A} \quad &\text{by the commutative law for unions.}
\end{aligned}
$$

Generalized Unions and Intersections

Because unions and intersections of sets satisfy associative laws, the sets $A \cup B \cup C$ and $A \cap B \cap C$ are well defined; that is, the meaning of this notation is unambiguous when A, B, and C are sets. That is, we do not have to use parentheses to indicate which operation comes first because $A \cup (B \cup C) = (A \cup B) \cup C$ and $A \cap (B \cap C) = (A \cap B) \cap C$. Note that $A \cup B \cup C$ contains those elements that are in at least one of the sets A, B, and C, and that $A \cap B \cap C$ contains those elements that are in all of A, B, and C. These combinations of the three sets, A, B, and C, are shown in Figure 5.

EXAMPLE 15   Let $A = \{0, 2, 4, 6, 8\}$, $B = \{0, 1, 2, 3, 4\}$, and $C = \{0, 3, 6, 9\}$. What are $A \cup B \cup C$ and $A \cap B \cap C$?

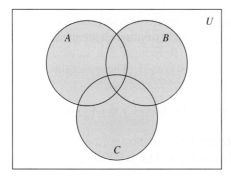
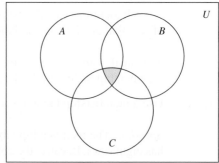

(a) $A \cup B \cup C$ is shaded.   (b) $A \cap B \cap C$ is shaded.

FIGURE 5   **The Union and Intersection of A, B, and C.**

Solution: The set $A \cup B \cup C$ contains those elements in at least one of A, B, and C. Hence,

$$A \cup B \cup C = \{0, 1, 2, 3, 4, 6, 8, 9\}.$$

The set $A \cap B \cap C$ contains those elements in all three of A, B, and C. Thus,

$$A \cap B \cap C = \{0\}. \qquad \blacktriangleleft$$

We can also consider unions and intersections of an arbitrary number of sets. We introduce these definitions.

DEFINITION 6   The *union* of a collection of sets is the set that contains those elements that are members of at least one set in the collection.

We use the notation

$$A_1 \cup A_2 \cup \cdots \cup A_n = \bigcup_{i=1}^{n} A_i$$

to denote the union of the sets $A_1, A_2, \ldots, A_n$.

DEFINITION 7   The *intersection* of a collection of sets is the set that contains those elements that are members of all the sets in the collection.

We use the notation

$$A_1 \cap A_2 \cap \cdots \cap A_n = \bigcap_{i=1}^{n} A_i$$

to denote the intersection of the sets $A_1, A_2, \ldots, A_n$. We illustrate generalized unions and intersections with Example 16.

EXAMPLE 16   For $i = 1, 2, \ldots$, let $A_i = \{i, i+1, i+2, \ldots\}$. Then,

$$\bigcup_{i=1}^{n} A_i = \bigcup_{i=1}^{n} \{i, i+1, i+2, \ldots\} = \{1, 2, 3, \ldots\},$$

and

$$\bigcap_{i=1}^{n} A_i = \bigcap_{i=1}^{n} \{i, i+1, i+2, \ldots\} = \{n, n+1, n+2, \ldots\} = A_n. \qquad \blacktriangleleft$$

We can extend the notation we have introduced for unions and intersections to other families of sets. In particular, we use the notation

$$A_1 \cup A_2 \cup \cdots \cup A_n \cup \cdots = \bigcup_{i=1}^{\infty} A_i$$

to denote the union of the sets $A_1, A_2, \ldots, A_n, \ldots$. Similarly, the intersection of these sets is denoted by

$$A_1 \cap A_2 \cap \cdots \cap A_n \cap \cdots = \bigcap_{i=1}^{\infty} A_i.$$

More generally, when I is a set, the notations $\bigcap_{i \in I} A_i$ and $\bigcup_{i \in I} A_i$ are used to denote the intersection and union of the sets A_i for $i \in I$, respectively. Note that we have $\bigcap_{i \in I} A_i = \{x \mid \forall i \in I \ (x \in A_i)\}$ and $\bigcup_{i \in I} A_i = \{x \mid \exists i \in I \ (x \in A_i)\}$.

EXAMPLE 17   Suppose that $A_i = \{1, 2, 3, \ldots, i\}$ for $i = 1, 2, 3, \ldots$. Then,

$$\bigcup_{i=1}^{\infty} A_i = \bigcup_{i=1}^{\infty} \{1, 2, 3, \ldots, i\} = \{1, 2, 3, \ldots\} = \mathbf{Z}^+$$

and

$$\bigcap_{i=1}^{\infty} A_i = \bigcap_{i=1}^{\infty} \{1, 2, 3, \ldots, i\} = \{1\}.$$

To see that the union of these sets is the set of positive integers, note that every positive integer n is in at least one of the sets, because it belongs to $A_n = \{1, 2, \ldots, n\}$, and every element of the sets in the union is a positive integer. To see that the intersection of these sets is the set $\{1\}$, note that the only element that belongs to all the sets $A_1, A_2, \ldots$ is 1. To see this note that $A_1 = \{1\}$ and $1 \in A_i$ for $i = 1, 2, \ldots$.   ◀

Computer Representation of Sets

There are various ways to represent sets using a computer. One method is to store the elements of the set in an unordered fashion. However, if this is done, the operations of computing the union, intersection, or difference of two sets would be time-consuming, because each of these operations would require a large amount of searching for elements. We will present a method for storing elements using an arbitrary ordering of the elements of the universal set. This method of representing sets makes computing combinations of sets easy.

Assume that the universal set U is finite (and of reasonable size so that the number of elements of U is not larger than the memory size of the computer being used). First, specify an arbitrary ordering of the elements of U, for instance $a_1, a_2, \ldots, a_n$. Represent a subset A of U with the bit string of length n, where the ith bit in this string is 1 if a_i belongs to A and is 0 if a_i does not belong to A. Example 18 illustrates this technique.

EXAMPLE 18   Let $U = \{1, 2, 3, 4, 5, 6, 7, 8, 9, 10\}$, and the ordering of elements of U has the elements in increasing order; that is, $a_i = i$. What bit strings represent the subset of all odd integers in U, the subset of all even integers in U, and the subset of integers not exceeding 5 in U?

Solution: The bit string that represents the set of odd integers in U, namely, $\{1, 3, 5, 7, 9\}$, has a one bit in the first, third, fifth, seventh, and ninth positions, and a zero elsewhere. It is

   10 1010 1010.

(We have split this bit string of length ten into blocks of length four for easy reading.) Similarly, we represent the subset of all even integers in U, namely, $\{2, 4, 6, 8, 10\}$, by the string

   01 0101 0101.

The set of all integers in U that do not exceed 5, namely, $\{1, 2, 3, 4, 5\}$, is represented by the string

   11 1110 0000.   ◀

   Using bit strings to represent sets, it is easy to find complements of sets and unions, intersections, and differences of sets. To find the bit string for the complement of a set from the bit string for that set, we simply change each 1 to a 0 and each 0 to 1, because $x \in A$ if and only if $x \notin \overline{A}$. Note that this operation corresponds to taking the negation of each bit when we associate a bit with a truth value—with 1 representing true and 0 representing false.

EXAMPLE 19 We have seen that the bit string for the set $\{1, 3, 5, 7, 9\}$ (with universal set $\{1, 2, 3, 4, 5, 6, 7, 8, 9, 10\}$) is

   10 1010 1010.

What is the bit string for the complement of this set?

Solution: The bit string for the complement of this set is obtained by replacing 0s with 1s and vice versa. This yields the string

   01 0101 0101,

which corresponds to the set $\{2, 4, 6, 8, 10\}$.   ◀

   To obtain the bit string for the union and intersection of two sets we perform bitwise Boolean operations on the bit strings representing the two sets. The bit in the ith position of the bit string of the union is 1 if either of the bits in the ith position in the two strings is 1 (or both are 1), and is 0 when both bits are 0. Hence, the bit string for the union is the bitwise *OR* of the bit strings for the two sets. The bit in the ith position of the bit string of the intersection is 1 when the bits in the corresponding position in the two strings are both 1, and is 0 when either of the two bits is 0 (or both are). Hence, the bit string for the intersection is the bitwise *AND* of the bit strings for the two sets.

EXAMPLE 20 The bit strings for the sets $\{1, 2, 3, 4, 5\}$ and $\{1, 3, 5, 7, 9\}$ are 11 1110 0000 and 10 1010 1010, respectively. Use bit strings to find the union and intersection of these sets.

Solution: The bit string for the union of these sets is

   $11\ 1110\ 0000 \vee 10\ 1010\ 1010 = 11\ 1110\ 1010,$

which corresponds to the set $\{1, 2, 3, 4, 5, 7, 9\}$. The bit string for the intersection of these sets is

   $11\ 1110\ 0000 \wedge 10\ 1010\ 1010 = 10\ 1010\ 0000,$

which corresponds to the set $\{1, 3, 5\}$.   ◀

Exercises

1. Let A be the set of students who live within one mile of school and let B be the set of students who walk to classes. Describe the students in each of these sets.
   a) $A \cap B$
   b) $A \cup B$
   c) $A - B$
   d) $B - A$

2. Let $A = \{a, b, c, d, e\}$ and $B = \{a, b, c, d, e, f, g, h\}$. Find
   a) $A \cup B$.
   b) $A \cap B$.
   c) $A - B$.
   d) $B - A$.

In Exercises 3–4 assume that A is a subset of some underlying universal set U.

3. Prove the complementation law in Table 1 by showing that $\overline{\overline{A}} = A$.

4. Prove the complement laws in Table 1 by showing that
   a) $A \cup \overline{A} = U$.
   b) $A \cap \overline{A} = \emptyset$.

5. Prove the first absorption law from Table 1 by showing that if A and B are sets, then $A \cup (A \cap B) = A$.

6. Find the sets A and B if $A - B = \{1, 5, 7, 8\}$, $B - A = \{2, 10\}$, and $A \cap B = \{3, 6, 9\}$.

7. Prove the second De Morgan law in Table 1 by showing that if A and B are sets, then $\overline{A \cup B} = \overline{A} \cap \overline{B}$
   a) by showing each side is a subset of the other side.
   b) using a membership table.

8. Let A and B be sets. Show that
   a) $(A \cap B) \subseteq A$.
   b) $A \subseteq (A \cup B)$.
   c) $A - B \subseteq A$.
   d) $A \cap (B - A) = \emptyset$.
   e) $A \cup (B - A) = A \cup B$.

9. Show that if A, B, and C are sets, then $\overline{A \cap B \cap C} = \overline{A} \cup \overline{B} \cup \overline{C}$
   a) by showing each side is a subset of the other side.
   b) using a membership table.

10. Let A, B, and C be sets. Show that
    a) $(A \cup B) \subseteq (A \cup B \cup C)$.
    b) $(A \cap B \cap C) \subseteq (A \cap B)$.
    c) $(A - B) - C \subseteq A - C$.
    d) $(A - C) \cap (C - B) = \emptyset$.
    e) $(B - A) \cup (C - A) = (B \cup C) - A$.

11. Show that if A and B are sets, then
    a) $A - B = A \cap \overline{B}$.
    b) $(A \cap B) \cup (A \cap \overline{B}) = A$.

12. Show that if A and B are sets with $A \subseteq B$, then
    a) $A \cup B = B$.
    b) $A \cap B = A$.

13. Let $A = \{0, 2, 4, 6, 8, 10\}$, $B = \{0, 1, 2, 3, 4, 5, 6\}$, and $C = \{4, 5, 6, 7, 8, 9, 10\}$. Find
    a) $A \cap B \cap C$.
    b) $A \cup B \cup C$.
    c) $(A \cup B) \cap C$.
    d) $(A \cap B) \cup C$.

14. Draw the Venn diagrams for each of these combinations of the sets A, B, and C.
    a) $A \cap (B \cup C)$
    b) $\overline{A} \cap \overline{B} \cap \overline{C}$
    c) $(A - B) \cup (A - C) \cup (B - C)$

15. What can you say about the sets A and B if we know that
    a) $A \cup B = A$?
    b) $A \cap B = A$?
    c) $A - B = A$?
    d) $A \cap B = B \cap A$?
    e) $A - B = B - A$?

16. Can you conclude that $A = B$ if A, B, and C are sets such that
    a) $A \cup C = B \cup C$?
    b) $A \cap C = B \cap C$?
    c) $A \cup C = B \cup C$ and $A \cap C = B \cap C$?

17. Let A and B be subsets of a universal set U. Show that $A \subseteq B$ if and only if $\overline{B} \subseteq \overline{A}$.

The **symmetric difference** of A and B, denoted by $A \oplus B$, is the set containing those elements in either A or B, but not in both A and B.

18. Find the symmetric difference of $\{1, 3, 5\}$ and $\{1, 2, 3\}$.

19. Find the symmetric difference of the set of computer science majors at a school and the set of mathematics majors at this school.

20. Draw a Venn diagram for the symmetric difference of the sets A and B.

21. What can you say about the sets A and B if $A \oplus B = A$?

*22. Determine whether the symmetric difference is associative; that is, if A, B, and C are sets, does it follow that $A \oplus (B \oplus C) = (A \oplus B) \oplus C$?

*23. Suppose that A, B, and C are sets such that $A \oplus C = B \oplus C$. Must it be the case that $A = B$?

24. Show that if A and B are finite sets, then $A \cup B$ is a finite set.

25. Show that if A is an infinite set, then whenever B is a set, $A \cup B$ is also an infinite set.

*26. Show that if A, B, and C are sets, then

$$|A \cup B \cup C| = |A| + |B| + |C| - |A \cap B|$$
$$- |A \cap C| - |B \cap C| + |A \cap B \cap C|.$$

(This is a special case of the inclusion–exclusion principle, which will be studied in Chapter 8.)

27. Let $A_i = \{1, 2, 3, \ldots, i\}$ for $i = 1, 2, 3, \ldots$. Find
    a) $\bigcup_{i=1}^{n} A_i$.
    b) $\bigcap_{i=1}^{n} A_i$.

28. Let $A_i = \{\ldots, -2, -1, 0, 1, \ldots, i\}$. Find
    a) $\bigcup_{i=1}^{n} A_i$.
    b) $\bigcap_{i=1}^{n} A_i$.

29. Let A_i be the set of all nonempty bit strings (that is, bit strings of length at least one) of length not exceeding i. Find

a) $\bigcup_{i=1}^{n} A_i$.   **b)** $\bigcap_{i=1}^{n} A_i$.

30. Find $\bigcup_{i=1}^{\infty} A_i$ and $\bigcap_{i=1}^{\infty} A_i$ if for every positive integer i,

a) $A_i = \{i, i+1, i+2, \ldots\}$.

b) $A_i = \{0, i\}$.

c) $A_i = (0, i)$, that is, the set of real numbers x with $0 < x < i$.

d) $A_i = (i, \infty)$, that is, the set of real numbers x with $x > i$.

31. Find $\bigcup_{i=1}^{\infty} A_i$ and $\bigcap_{i=1}^{\infty} A_i$ if for every positive integer i,

a) $A_i = \{-i, -i+1, \ldots, -1, 0, 1, \ldots, i-1, i\}$.

b) $A_i = \{-i, i\}$.

c) $A_i = [-i, i]$, that is, the set of real numbers x with $-i \le x \le i$.

d) $A_i = [i, \infty)$, that is, the set of real numbers x with $x \ge i$.

32. Suppose that the universal set is $U = \{1, 2, 3, 4, 5, 6, 7, 8, 9, 10\}$. Express each of these sets with bit strings where the ith bit in the string is 1 if i is in the set and 0 otherwise.

a) $\{3, 4, 5\}$

b) $\{1, 3, 6, 10\}$

c) $\{2, 3, 4, 7, 8, 9\}$

33. Using the same universal set as in the last problem, find the set specified by each of these bit strings.

a) 11 1100 1111

b) 01 0111 1000

c) 10 0000 0001

34. What is the bit string corresponding to the symmetric difference of two sets?

35. Show how bitwise operations on bit strings can be used to find these combinations of $A = \{a, b, c, d, e\}$, $B = \{b, c, d, g, p, t, v\}$, $C = \{c, e, i, o, u, x, y, z\}$, and $D = \{d, e, h, i, n, o, t, u, x, y\}$.

a) $A \cup B$ **b)** $A \cap B$

c) $(A \cup D) \cap (B \cup C)$ **d)** $A \cup B \cup C \cup D$

36. How can the union and intersection of n sets that all are subsets of the universal set U be found using bit strings?

The **successor** of the set A is the set $A \cup \{A\}$.

37. Find the successors of the following sets.

a) $\{1, 2, 3\}$ **b)** $\emptyset$

c) $\{\emptyset\}$ **d)** $\{\emptyset, \{\emptyset\}\}$

38. How many elements does the successor of a set with n elements have?

Sometimes the number of times that an element occurs in an unordered collection matters. **Multisets** are unordered collections of elements where an element can occur as a member more than once. The notation $\{m_1 \cdot a_1, m_2 \cdot a_2, \ldots, m_r \cdot a_r\}$ denotes the multiset with element a_1 occurring m_1 times, element a_2 occurring m_2 times, and so on. The numbers m_i, $i = 1, 2, \ldots, r$ are called the **multiplicities** of the elements $a_i, i = 1, 2, \ldots, r$.

Let P and Q be multisets. The **union** of the multisets P and Q is the multiset where the multiplicity of an element is the maximum of its multiplicities in P and Q. The **intersection** of P and Q is the multiset where the multiplicity of an element is the minimum of its multiplicities in P and Q. The **difference** of P and Q is the multiset where the multiplicity of an element is the multiplicity of the element in P less its multiplicity in Q unless this difference is negative, in which case the multiplicity is 0. The **sum** of P and Q is the multiset where the multiplicity of an element is the sum of multiplicities in P and Q. The union, intersection, and difference of P and Q are denoted by $P \cup Q$, $P \cap Q$, and $P - Q$, respectively (where these operations should not be confused with the analogous operations for sets). The sum of P and Q is denoted by $P + Q$.

39. Let A and B be the multisets $\{3 \cdot a, 2 \cdot b, 1 \cdot c\}$ and $\{2 \cdot a, 3 \cdot b, 4 \cdot d\}$, respectively. Find

a) $A \cup B$.   **b)** $A \cap B$.   **c)** $A - B$.
d) $B - A$.   **e)** $A + B$.

Fuzzy sets are used in artificial intelligence. Each element in the universal set U has a **degree of membership**, which is a real number between 0 and 1 (including 0 and 1), in a fuzzy set S. The fuzzy set S is denoted by listing the elements with their degrees of membership (elements with 0 degree of membership are not listed). For instance, we write {0.6 Alice, 0.9 Brian, 0.4 Fred, 0.1 Oscar, 0.5 Rita} for the set F (of famous people) to indicate that Alice has a 0.6 degree of membership in F, Brian has a 0.9 degree of membership in F, Fred has a 0.4 degree of membership in F, Oscar has a 0.1 degree of membership in F, and Rita has a 0.5 degree of membership in F (so that Brian is the most famous and Oscar is the least famous of these people). Also suppose that R is the set of rich people with $R = \{0.4$ Alice, 0.8 Brian, 0.2 Fred, 0.9 Oscar, 0.7 Rita}.

40. The **complement** of a fuzzy set S is the set $\overline{S}$, with the degree of the membership of an element in $\overline{S}$ equal to 1 minus the degree of membership of this element in S. Find $\overline{F}$ (the fuzzy set of people who are not famous) and $\overline{R}$ (the fuzzy set of people who are not rich).

41. The **union** of two fuzzy sets S and T is the fuzzy set $S \cup T$, where the degree of membership of an element in $S \cup T$ is the maximum of the degrees of membership of this element in S and in T. Find the fuzzy set $F \cup R$ of rich or famous people.

42. The **intersection** of two fuzzy sets S and T is the fuzzy set $S \cap T$, where the degree of membership of an element in $S \cap T$ is the minimum of the degrees of membership of this element in S and in T. Find the fuzzy set $F \cap R$ of rich and famous people.

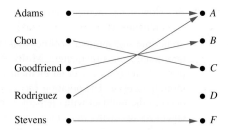

FIGURE 1   Assignment of Grades in a Discrete Mathematics Class.

2.3   Functions

Introduction

In many instances we assign to each element of a set a particular element of a second set (which may be the same as the first). For example, suppose that each student in a discrete mathematics class is assigned a letter grade from the set $\{A, B, C, D, F\}$. And suppose that the grades are A for Adams, C for Chou, B for Goodfriend, A for Rodriguez, and F for Stevens. This assignment of grades is illustrated in Figure 1.

This assignment is an example of a function. The concept of a function is extremely important in mathematics and computer science. For example, in discrete mathematics functions are used in the definition of such discrete structures as sequences and strings. Functions are also used to represent how long it takes a computer to solve problems of a given size. Many computer programs and subroutines are designed to calculate values of functions. Recursive functions, which are functions defined in terms of themselves, are used throughout computer science; they will be studied in Chapter 5. This section reviews the basic concepts involving functions needed in discrete mathematics.

DEFINITION 1

Let A and B be nonempty sets. A *function* f from A to B is an assignment of exactly one element of B to each element of A. We write $f(a) = b$ if b is the unique element of B assigned by the function f to the element a of A. If f is a function from A to B, we write $f : A \rightarrow B$.

Remark: Functions are sometimes also called **mappings** or **transformations**.

Assessment

Functions are specified in many different ways. Sometimes we explicitly state the assignments, as in Figure 1. Often we give a formula, such as $f(x) = x + 1$, to define a function. Other times we use a computer program to specify a function.

A function $f : A \rightarrow B$ can also be defined in terms of a relation from A to B. Recall from Section 2.1 that a relation from A to B is just a subset of $A \times B$. A relation from A to B that contains one, and only one, ordered pair (a, b) for every element $a \in A$, defines a function f from A to B. This function is defined by the assignment $f(a) = b$, where (a, b) is the unique ordered pair in the relation that has a as its first element.

DEFINITION 2

If f is a function from A to B, we say that A is the *domain* of f and B is the *codomain* of f. If $f(a) = b$, we say that b is the *image* of a and a is a *preimage* of b. The *range*, or *image*, of f is the set of all images of elements of A. Also, if f is a function from A to B, we say that f *maps* A to B.

Figure 2 represents a function f from A to B.

When we define a function we specify its domain, its codomain, and the mapping of elements of the domain to elements in the codomain. Two functions are **equal** when they have the same domain, have the same codomain, and map each element of their common domain to the same element in their common codomain. Note that if we change either the domain or the codomain of a function, then we obtain a different function. If we change the mapping of elements, then we also obtain a different function.

Examples 1–5 provide examples of functions. In each case, we describe the domain, the codomain, the range, and the assignment of values to elements of the domain.

EXAMPLE 1   What are the domain, codomain, and range of the function that assigns grades to students described in the first paragraph of the introduction of this section?

Solution: Let G be the function that assigns a grade to a student in our discrete mathematics class. Note that $G(\text{Adams}) = A$, for instance. The domain of G is the set {Adams, Chou, Goodfriend, Rodriguez, Stevens}, and the codomain is the set {A, B, C, D, F}. The range of G is the set {A, B, C, F}, because each grade except D is assigned to some student. ◄

EXAMPLE 2   Let R be the relation with ordered pairs (Abdul, 22), (Brenda, 24), (Carla, 21), (Desire, 22), (Eddie, 24), and (Felicia, 22). Here each pair consists of a graduate student and this student's age. Specify a function determined by this relation.

Solution: If f is a function specified by R, then $f(\text{Abdul}) = 22$, $f(\text{Brenda}) = 24$, $f(\text{Carla}) = 21$, $f(\text{Desire}) = 22$, $f(\text{Eddie}) = 24$, and $f(\text{Felicia}) = 22$. (Here, $f(x)$ is the age of x, where x is a student.) For the domain, we take the set {Abdul, Brenda, Carla, Desire, Eddie, Felicia}. We also need to specify a codomain, which needs to contain all possible ages of students. Because it is highly likely that all students are less than 100 years old, we can take the set of positive integers less than 100 as the codomain. (Note that we could choose a different codomain, such as the set of all positive integers or the set of positive integers between 10 and 90, but that would change the function. Using this codomain will also allow us to extend the function by adding the names and ages of more students later.) The range of the function we have specified is the set of different ages of these students, which is the set {21, 22, 24}. ◄

EXAMPLE 3   Let f be the function that assigns the last two bits of a bit string of length 2 or greater to that string. For example, $f(11010) = 10$. Then, the domain of f is the set of all bit strings of length 2 or greater, and both the codomain and range are the set {00, 01, 10, 11}. ◄

EXAMPLE 4   Let $f: \mathbf{Z} \to \mathbf{Z}$ assign the square of an integer to this integer. Then, $f(x) = x^2$, where the domain of f is the set of all integers, the codomain of f is the set of all integers, and the range of f is the set of all integers that are perfect squares, namely, {0, 1, 4, 9, ...}. ◄

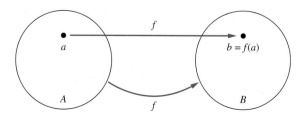

FIGURE 2   **The Function f Maps A to B.**

EXAMPLE 5   The domain and codomain of functions are often specified in programming languages. For instance, the Java statement

> int **floor**(float real){...}

and the C++ function statement

> int **function** (float x){...}

both tell us that the domain of the floor function is the set of real numbers (represented by floating point numbers) and its codomain is the set of integers.   ◀

A function is called **real-valued** if its codomain is the set of real numbers, and it is called **integer-valued** if its codomain is the set of integers. Two real-valued functions or two integer-valued functions with the same domain can be added, as well as multiplied.

DEFINITION 3   Let f_1 and f_2 be functions from A to $\mathbf{R}$. Then $f_1 + f_2$ and $f_1 f_2$ are also functions from A to $\mathbf{R}$ defined for all $x \in A$ by

$$(f_1 + f_2)(x) = f_1(x) + f_2(x),$$
$$(f_1 f_2)(x) = f_1(x) f_2(x).$$

Note that the functions $f_1 + f_2$ and $f_1 f_2$ have been defined by specifying their values at x in terms of the values of f_1 and f_2 at x.

EXAMPLE 6   Let f_1 and f_2 be functions from $\mathbf{R}$ to $\mathbf{R}$ such that $f_1(x) = x^2$ and $f_2(x) = x - x^2$. What are the functions $f_1 + f_2$ and $f_1 f_2$?

Solution: From the definition of the sum and product of functions, it follows that

$$(f_1 + f_2)(x) = f_1(x) + f_2(x) = x^2 + (x - x^2) = x$$

and

$$(f_1 f_2)(x) = x^2(x - x^2) = x^3 - x^4.$$ ◀

When f is a function from A to B, the image of a subset of A can also be defined.

DEFINITION 4   Let f be a function from A to B and let S be a subset of A. The *image* of S under the function f is the subset of B that consists of the images of the elements of S. We denote the image of S by $f(S)$, so

$$f(S) = \{t \mid \exists s \in S \, (t = f(s))\}.$$

We also use the shorthand $\{f(s) \mid s \in S\}$ to denote this set.

Remark: The notation $f(S)$ for the image of the set S under the function f is potentially ambiguous. Here, $f(S)$ denotes a set, and not the value of the function f for the set S.

EXAMPLE 7   Let $A = \{a, b, c, d, e\}$ and $B = \{1, 2, 3, 4\}$ with $f(a) = 2$, $f(b) = 1$, $f(c) = 4$, $f(d) = 1$, and $f(e) = 1$. The image of the subset $S = \{b, c, d\}$ is the set $f(S) = \{1, 4\}$.   ◀

FIGURE 3   A One-to-One Function.

One-to-One and Onto Functions

Some functions never assign the same value to two different domain elements. These functions are said to be **one-to-one**.

DEFINITION 5

A function f is said to be *one-to-one*, or an *injunction*, if and only if $f(a) = f(b)$ implies that $a = b$ for all a and b in the domain of f. A function is said to be *injective* if it is one-to-one.

Note that a function f is one-to-one if and only if $f(a) \neq f(b)$ whenever $a \neq b$. This way of expressing that f is one-to-one is obtained by taking the contrapositive of the implication in the definition.

Remark: We can express that f is one-to-one using quantifiers as $\forall a \forall b (f(a) = f(b) \rightarrow a = b)$ or equivalently $\forall a \forall b (a \neq b \rightarrow f(a) \neq f(b))$, where the universe of discourse is the domain of the function.

We illustrate this concept by giving examples of functions that are one-to-one and other functions that are not one-to-one.

EXAMPLE 8

Determine whether the function f from $\{a, b, c, d\}$ to $\{1, 2, 3, 4, 5\}$ with $f(a) = 4$, $f(b) = 5$, $f(c) = 1$, and $f(d) = 3$ is one-to-one.

Solution: The function f is one-to-one because f takes on different values at the four elements of its domain. This is illustrated in Figure 3.   ◀

EXAMPLE 9

Determine whether the function $f(x) = x^2$ from the set of integers to the set of integers is one-to-one.

Solution: The function $f(x) = x^2$ is not one-to-one because, for instance, $f(1) = f(-1) = 1$, but $1 \neq -1$.
   Note that the function $f(x) = x^2$ with its domain restricted to $\mathbf{Z}^+$ is one-to-one. (Technically, when we restrict the domain of a function, we obtain a new function whose values agree with those of the original function for the elements of the restricted domain. The restricted function is not defined for elements of the original domain outside of the restricted domain.)   ◀

EXAMPLE 10

Determine whether the function $f(x) = x + 1$ from the set of real numbers to itself is one-to-one.

Solution: The function $f(x) = x + 1$ is a one-to-one function. To demonstrate this, note that $x + 1 \neq y + 1$ when $x \neq y$.   ◀

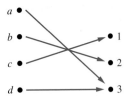

FIGURE 4   An Onto Function.

EXAMPLE 11   Suppose that each worker in a group of employees is assigned a job from a set of possible jobs, each to be done by a single worker. In this situation, the function f that assigns a job to each worker is one-to-one. To see this, note that if x and y are two different workers, then $f(x) \neq f(y)$ because the two workers x and y must be assigned different jobs.   ◄

We now give some conditions that guarantee that a function is one-to-one.

DEFINITION 6   A function f whose domain and codomain are subsets of the set of real numbers is called *increasing* if $f(x) \leq f(y)$, and *strictly increasing* if $f(x) < f(y)$, whenever $x < y$ and x and y are in the domain of f. Similarly, f is called *decreasing* if $f(x) \geq f(y)$, and *strictly decreasing* if $f(x) > f(y)$, whenever $x < y$ and x and y are in the domain of f. (The word *strictly* in this definition indicates a strict inequality.)

Remark: A function f is increasing if $\forall x \forall y (x < y \rightarrow f(x) \leq f(y))$, strictly increasing if $\forall x \forall y (x < y \rightarrow f(x) < f(y))$, decreasing if $\forall x \forall y (x < y \rightarrow f(x) \geq f(y))$, and strictly decreasing if $\forall x \forall y (x < y \rightarrow f(x) > f(y))$, where the universe of discourse is the domain of f.

From these definitions, it can be shown that a function that is either strictly increasing or strictly decreasing must be one-to-one. However, a function that is increasing, but not strictly increasing, or decreasing, but not strictly decreasing, is not one-to-one.

For some functions the range and the codomain are equal. That is, every member of the codomain is the image of some element of the domain. Functions with this property are called **onto** functions.

DEFINITION 7   A function f from A to B is called *onto*, or a *surjection*, if and only if for every element $b \in B$ there is an element $a \in A$ with $f(a) = b$. A function f is called *surjective* if it is onto.

Remark: A function f is onto if $\forall y \exists x (f(x) = y)$, where the domain for x is the domain of the function and the domain for y is the codomain of the function.

We now give examples of onto functions and functions that are not onto.

EXAMPLE 12   Let f be the function from $\{a, b, c, d\}$ to $\{1, 2, 3\}$ defined by $f(a) = 3$, $f(b) = 2$, $f(c) = 1$, and $f(d) = 3$. Is f an onto function?

Solution: Because all three elements of the codomain are images of elements in the domain, we see that f is onto. This is illustrated in Figure 4. Note that if the codomain were $\{1, 2, 3, 4\}$, then f would not be onto.   ◄

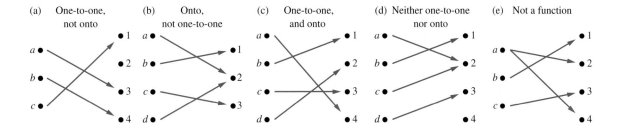

FIGURE 5 **Examples of Different Types of Correspondences.**

EXAMPLE 13   Is the function $f(x) = x^2$ from the set of integers to the set of integers onto?

Solution: The function f is not onto because there is no integer x with $x^2 = -1$, for instance. ◀

EXAMPLE 14   Is the function $f(x) = x + 1$ from the set of integers to the set of integers onto?

Solution: This function is onto, because for every integer y there is an integer x such that $f(x) = y$. To see this, note that $f(x) = y$ if and only if $x + 1 = y$, which holds if and only if $x = y - 1$. ◀

EXAMPLE 15   Consider the function f in Example 11 that assigns jobs to workers. The function f is onto if for every job there is a worker assigned this job. The function f is not onto when there is at least one job that has no worker assigned it. ◀

DEFINITION 8   The function f is a *one-to-one correspondence,* or a *bijection,* if it is both one-to-one and onto. We also say that such a function is *bijective.*

Examples 16 and 17 illustrate the concept of a bijection.

EXAMPLE 16   Let f be the function from $\{a, b, c, d\}$ to $\{1, 2, 3, 4\}$ with $f(a) = 4$, $f(b) = 2$, $f(c) = 1$, and $f(d) = 3$. Is f a bijection?

Solution: The function f is one-to-one and onto. It is one-to-one because no two values in the domain are assigned the same function value. It is onto because all four elements of the codomain are images of elements in the domain. Hence, f is a bijection. ◀

Figure 5 displays four functions where the first is one-to-one but not onto, the second is onto but not one-to-one, the third is both one-to-one and onto, and the fourth is neither one-to-one nor onto. The fifth correspondence in Figure 5 is not a function, because it sends an element to two different elements.

Suppose that f is a function from a set A to itself. If A is finite, then f is one-to-one if and only if it is onto. (This follows from the result in Exercise 48.) This is not necessarily the case if A is infinite (as will be shown in Section 2.5).

EXAMPLE 17   Let A be a set. The *identity function* on A is the function $\iota_A : A \to A$, where

$$\iota_A(x) = x$$

for all $x \in A$. In other words, the identity function ι_A is the function that assigns each element to itself. The function ι_A is one-to-one and onto, so it is a bijection. (Note that ι is the Greek letter iota.) ◀

For future reference, we summarize what needs be to shown to establish whether a function is one-to-one and whether it is onto. It is instructive to review Examples 8–17 in light of this summary.

To show that f is injective Show that if $f(x) = f(y)$ for arbitrary $x, y \in A$ with $x \neq y$, then $x = y$.

To show that f is not injective Find particular elements $x, y \in A$ such that $x \neq y$ and $f(x) = f(y)$.

To show that f is surjective Consider an arbitrary element $y \in B$ and find an element $x \in A$ such that $f(x) = y$.

To show that f is not surjective Find a particular $y \in B$ such that $f(x) \neq y$ for all $x \in A$.

Inverse Functions and Compositions of Functions

Now consider a one-to-one correspondence f from the set A to the set B. Because f is an onto function, every element of B is the image of some element in A. Furthermore, because f is also a one-to-one function, every element of B is the image of a *unique* element of A. Consequently, we can define a new function from B to A that reverses the correspondence given by f. This leads to Definition 9.

DEFINITION 9   Let f be a one-to-one correspondence from the set A to the set B. The *inverse function* of f is the function that assigns to an element b belonging to B the unique element a in A such that $f(a) = b$. The inverse function of f is denoted by f^{-1}. Hence, $f^{-1}(b) = a$ when $f(a) = b$.

Remark: Be sure not to confuse the function f^{-1} with the function $1/f$, which is the function that assigns to each x in the domain the value $1/f(x)$. Notice that the latter makes sense only when $f(x)$ is a non-zero real number.

Figure 6 illustrates the concept of an inverse function.

If a function f is not a one-to-one correspondence, we cannot define an inverse function of f. When f is not a one-to-one correspondence, either it is not one-to-one or it is not onto. If

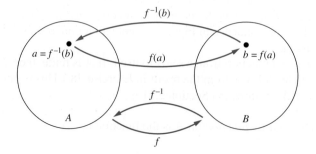

FIGURE 6   The Function f^{-1} Is the Inverse of Function f.

f is not one-to-one, some element b in the codomain is the image of more than one element in the domain. If f is not onto, for some element b in the codomain, no element a in the domain exists for which $f(a) = b$. Consequently, if f is not a one-to-one correspondence, we cannot assign to each element b in the codomain a unique element a in the domain such that $f(a) = b$ (because for some b there is either more than one such a or no such a).

A one-to-one correspondence is called **invertible** because we can define an inverse of this function. A function is **not invertible** if it is not a one-to-one correspondence, because the inverse of such a function does not exist.

EXAMPLE 18   Let f be the function from $\{a, b, c\}$ to $\{1, 2, 3\}$ such that $f(a) = 2$, $f(b) = 3$, and $f(c) = 1$. Is f invertible, and if it is, what is its inverse?

Solution: The function f is invertible because it is a one-to-one correspondence. The inverse function f^{-1} reverses the correspondence given by f, so $f^{-1}(1) = c$, $f^{-1}(2) = a$, and $f^{-1}(3) = b$.   ◄

EXAMPLE 19   Let $f : \mathbf{Z} \to \mathbf{Z}$ be such that $f(x) = x + 1$. Is f invertible, and if it is, what is its inverse?

Solution: The function f has an inverse because it is a one-to-one correspondence, as follows from Examples 10 and 14. To reverse the correspondence, suppose that y is the image of x, so that $y = x + 1$. Then $x = y - 1$. This means that $y - 1$ is the unique element of $\mathbf{Z}$ that is sent to y by f. Consequently, $f^{-1}(y) = y - 1$.   ◄

EXAMPLE 20   Let f be the function from $\mathbf{R}$ to $\mathbf{R}$ with $f(x) = x^2$. Is f invertible?

Solution: Because $f(-2) = f(2) = 4$, f is not one-to-one. If an inverse function were defined, it would have to assign two elements to 4. Hence, f is not invertible. (Note we can also show that f is not invertible because it is not onto.)   ◄

Sometimes we can restrict the domain or the codomain of a function, or both, to obtain an invertible function, as Example 21 illustrates.

EXAMPLE 21   Show that if we restrict the function $f(x) = x^2$ in Example 20 to a function from the set of all nonnegative real numbers to the set of all nonnegative real numbers, then f is invertible.

Solution: The function $f(x) = x^2$ from the set of nonnegative real numbers to the set of non-negative real numbers is one-to-one. To see this, note that if $f(x) = f(y)$, then $x^2 = y^2$, so $x^2 - y^2 = (x + y)(x - y) = 0$. This means that $x + y = 0$ or $x - y = 0$, so $x = -y$ or $x = y$. Because both x and y are nonnegative, we must have $x = y$. So, this function is one-to-one. Furthermore, $f(x) = x^2$ is onto when the codomain is the set of all nonnegative real numbers, because each nonnegative real number has a square root. That is, if y is a nonnegative real number, there exists a nonnegative real number x such that $x = \sqrt{y}$, which means that $x^2 = y$. Because the function $f(x) = x^2$ from the set of nonnegative real numbers to the set of non-negative real numbers is one-to-one and onto, it is invertible. Its inverse is given by the rule $f^{-1}(y) = \sqrt{y}$.   ◄

DEFINITION 10   Let g be a function from the set A to the set B and let f be a function from the set B to the set C. The *composition* of the functions f and g, denoted for all $a \in A$ by $f \circ g$, is defined by

$$(f \circ g)(a) = f(g(a)).$$

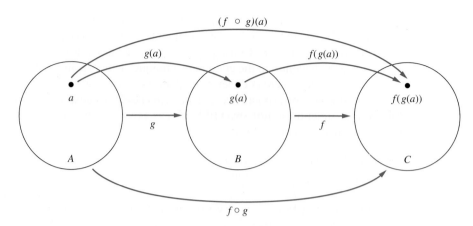

FIGURE 7 **The Composition of the Functions f and g.**

In other words, $f \circ g$ is the function that assigns to the element a of A the element assigned by f to $g(a)$. That is, to find $(f \circ g)(a)$ we first apply the function g to a to obtain $g(a)$ and then we apply the function f to the result $g(a)$ to obtain $(f \circ g)(a) = f(g(a))$. Note that the composition $f \circ g$ cannot be defined unless the range of g is a subset of the domain of f. In Figure 7 the composition of functions is shown.

EXAMPLE 22 Let g be the function from the set $\{a, b, c\}$ to itself such that $g(a) = b$, $g(b) = c$, and $g(c) = a$. Let f be the function from the set $\{a, b, c\}$ to the set $\{1, 2, 3\}$ such that $f(a) = 3$, $f(b) = 2$, and $f(c) = 1$. What is the composition of f and g, and what is the composition of g and f?

Solution: The composition $f \circ g$ is defined by $(f \circ g)(a) = f(g(a)) = f(b) = 2$, $(f \circ g)(b) = f(g(b)) = f(c) = 1$, and $(f \circ g)(c) = f(g(c)) = f(a) = 3$.

Note that $g \circ f$ is not defined, because the range of f is not a subset of the domain of g. ◀

EXAMPLE 23 Let f and g be the functions from the set of integers to the set of integers defined by $f(x) = 2x + 3$ and $g(x) = 3x + 2$. What is the composition of f and g? What is the composition of g and f?

Solution: Both the compositions $f \circ g$ and $g \circ f$ are defined. Moreover,

$$(f \circ g)(x) = f(g(x)) = f(3x + 2) = 2(3x + 2) + 3 = 6x + 7$$

and

$$(g \circ f)(x) = g(f(x)) = g(2x + 3) = 3(2x + 3) + 2 = 6x + 11.$$ ◀

Remark: Note that even though $f \circ g$ and $g \circ f$ are defined for the functions f and g in Example 23, $f \circ g$ and $g \circ f$ are not equal. In other words, the commutative law does not hold for the composition of functions.

When the composition of a function and its inverse is formed, in either order, an identity function is obtained. To see this, suppose that f is a one-to-one correspondence from the set A to the set B. Then the inverse function f^{-1} exists and is a one-to-one correspondence from B to A. The inverse function reverses the correspondence of the original function, so $f^{-1}(b) = a$ when $f(a) = b$, and $f(a) = b$ when $f^{-1}(b) = a$. Hence,

$$(f^{-1} \circ f)(a) = f^{-1}(f(a)) = f^{-1}(b) = a,$$

and

$$(f \circ f^{-1})(b) = f(f^{-1}(b)) = f(a) = b.$$

Consequently $f^{-1} \circ f = \iota_A$ and $f \circ f^{-1} = \iota_B$, where ι_A and ι_B are the identity functions on the sets A and B, respectively. That is, $(f^{-1})^{-1} = f$.

The Graphs of Functions

We can associate a set of pairs in $A \times B$ to each function from A to B. This set of pairs is called the **graph** of the function and is often displayed pictorially to aid in understanding the behavior of the function.

DEFINITION 11

Let f be a function from the set A to the set B. The *graph* of the function f is the set of ordered pairs $\{(a, b) \mid a \in A \text{ and } f(a) = b\}$.

From the definition, the graph of a function f from A to B is the subset of $A \times B$ containing the ordered pairs with the second entry equal to the element of B assigned by f to the first entry. Also, note that the graph of a function f from A to B is the same as the relation from A to B determined by the function f, as described on page 139.

EXAMPLE 24 Display the graph of the function $f(n) = 2n + 1$ from the set of integers to the set of integers.

Solution: The graph of f is the set of ordered pairs of the form $(n, 2n + 1)$, where n is an integer. This graph is displayed in Figure 8. ◀

EXAMPLE 25 Display the graph of the function $f(x) = x^2$ from the set of integers to the set of integers.

Solution: The graph of f is the set of ordered pairs of the form $(x, f(x)) = (x, x^2)$, where x is an integer. This graph is displayed in Figure 9. ◀

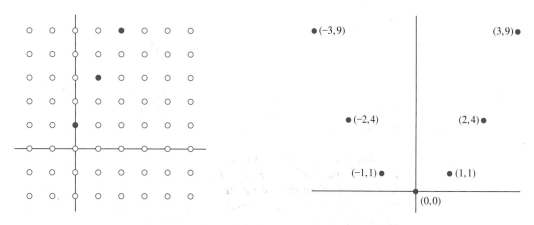

FIGURE 8   The Graph of
$f(n) = 2n + 1$ **from Z to Z.**

FIGURE 9   The Graph of
$f(x) = x^2$ **from Z to Z.**

Some Important Functions

Next, we introduce two important functions in discrete mathematics, namely, the floor and ceiling functions. Let x be a real number. The floor function rounds x down to the closest integer less than or equal to x, and the ceiling function rounds x up to the closest integer greater than or equal to x. These functions are often used when objects are counted. They play an important role in the analysis of the number of steps used by procedures to solve problems of a particular size.

DEFINITION 12   The *floor function* assigns to the real number x the largest integer that is less than or equal to x. The value of the floor function at x is denoted by $\lfloor x \rfloor$. The *ceiling function* assigns to the real number x the smallest integer that is greater than or equal to x. The value of the ceiling function at x is denoted by $\lceil x \rceil$.

Remark: The floor function is often also called the *greatest integer function*. It is often denoted by $[x]$.

EXAMPLE 26   These are some values of the floor and ceiling functions:

$$\lfloor \tfrac{1}{2} \rfloor = 0, \lceil \tfrac{1}{2} \rceil = 1, \lfloor -\tfrac{1}{2} \rfloor = -1, \lceil -\tfrac{1}{2} \rceil = 0, \lfloor 3.1 \rfloor = 3, \lceil 3.1 \rceil = 4, \lfloor 7 \rfloor = 7, \lceil 7 \rceil = 7. \quad \blacktriangleleft$$

We display the graphs of the floor and ceiling functions in Figure 10. In Figure 10(a) we display the graph of the floor function $\lfloor x \rfloor$. Note that this function has the same value throughout the interval $[n, n + 1)$, namely n, and then it jumps up to $n + 1$ when $x = n + 1$. In Figure 10(b) we display the graph of the ceiling function $\lceil x \rceil$. Note that this function has the same value throughout the interval $(n, n + 1]$, namely $n + 1$, and then jumps to $n + 2$ when x is a little larger than $n + 1$.

The floor and ceiling functions are useful in a wide variety of applications, including those involving data storage and data transmission. Consider Examples 27 and 28, typical of basic calculations done when database and data communications problems are studied.

EXAMPLE 27   Data stored on a computer disk or transmitted over a data network are usually represented as a string of bytes. Each byte is made up of 8 bits. How many bytes are required to encode 100 bits of data?

Solution: To determine the number of bytes needed, we determine the smallest integer that is at least as large as the quotient when 100 is divided by 8, the number of bits in a byte. Consequently, $\lceil 100/8 \rceil = \lceil 12.5 \rceil = 13$ bytes are required. $\quad \blacktriangleleft$

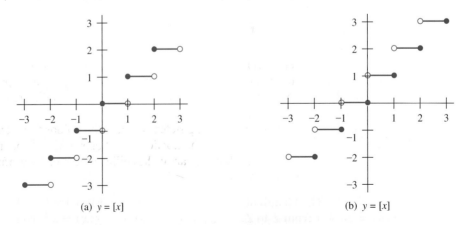

(a) $y = \lfloor x \rfloor$ (b) $y = \lceil x \rceil$

FIGURE 10   Graphs of the (a) Floor and (b) Ceiling Functions.

> **TABLE 1 Useful Properties of the Floor and Ceiling Functions.**
> (n is an integer, x is a real number)
>
> (1a) $\lfloor x \rfloor = n$ if and only if $n \le x < n + 1$
> (1b) $\lceil x \rceil = n$ if and only if $n - 1 < x \le n$
> (1c) $\lfloor x \rfloor = n$ if and only if $x - 1 < n \le x$
> (1d) $\lceil x \rceil = n$ if and only if $x \le n < x + 1$
>
> (2) $\quad x - 1 < \lfloor x \rfloor \le x \le \lceil x \rceil < x + 1$
>
> (3a) $\lfloor -x \rfloor = -\lceil x \rceil$
> (3b) $\lceil -x \rceil = -\lfloor x \rfloor$
>
> (4a) $\lfloor x + n \rfloor = \lfloor x \rfloor + n$
> (4b) $\lceil x + n \rceil = \lceil x \rceil + n$

EXAMPLE 28   In asynchronous transfer mode (ATM) (a communications protocol used on backbone networks), data are organized into cells of 53 bytes. How many ATM cells can be transmitted in 1 minute over a connection that transmits data at the rate of 500 kilobits per second?

Solution: In 1 minute, this connection can transmit $500{,}000 \cdot 60 = 30{,}000{,}000$ bits. Each ATM cell is 53 bytes long, which means that it is $53 \cdot 8 = 424$ bits long. To determine the number of cells that can be transmitted in 1 minute, we determine the largest integer not exceeding the quotient when $30{,}000{,}000$ is divided by 424. Consequently, $\lfloor 30{,}000{,}000/424 \rfloor = 70{,}754$ ATM cells can be transmitted in 1 minute over a 500 kilobit per second connection.   ◄

Table 1, with x denoting a real number, displays some simple but important properties of the floor and ceiling functions. Because these functions appear so frequently in discrete mathematics, it is useful to look over these identities. Each property in this table can be established using the definitions of the floor and ceiling functions. Properties (1a), (1b), (1c), and (1d) follow directly from these definitions. For example, (1a) states that $\lfloor x \rfloor = n$ if and only if the integer n is less than or equal to x and $n + 1$ is larger than x. This is precisely what it means for n to be the greatest integer not exceeding x, which is the definition of $\lfloor x \rfloor = n$. Properties (1b), (1c), and (1d) can be established similarly. We will prove property (4a) using a direct proof.

Proof: Suppose that $\lfloor x \rfloor = m$, where m is a positive integer. By property (1a), it follows that $m \le x < m + 1$. Adding n to all three quantities in this chain of two inequalities shows that $m + n \le x + n < m + n + 1$. Using property (1a) again, we see that $\lfloor x + n \rfloor = m + n = \lfloor x \rfloor + n$. This completes the proof. Proofs of the other properties are left as exercises.   ◁

The floor and ceiling functions enjoy many other useful properties besides those displayed in Table 1. There are also many statements about these functions that may appear to be correct, but actually are not. We will consider statements about the floor and ceiling functions in Examples 29 and 30.

A useful approach for considering statements about the floor function is to let $x = n + \epsilon$, where $n = \lfloor x \rfloor$ is an integer, and ϵ, the fractional part of x, satisfies the inequality $0 \le \epsilon < 1$. Similarly, when considering statements about the ceiling function, it is useful to write $x = n - \epsilon$, where $n = \lceil x \rceil$ is an integer and $0 \le \epsilon < 1$.

EXAMPLE 29   Prove that if x is a real number, then $\lfloor 2x \rfloor = \lfloor x \rfloor + \lfloor x + \frac{1}{2} \rfloor$.

Solution: To prove this statement we let $x = n + \epsilon$, where n is an integer and $0 \le \epsilon < 1$. There are two cases to consider, depending on whether ϵ is less than, or greater than or equal to $\frac{1}{2}$. (The reason we choose these two cases will be made clear in the proof.)

We first consider the case when $0 \le \epsilon < \frac{1}{2}$. In this case, $2x = 2n + 2\epsilon$ and $\lfloor 2x \rfloor = 2n$ because $0 \le 2\epsilon < 1$. Similarly, $x + \frac{1}{2} = n + (\frac{1}{2} + \epsilon)$, so $\lfloor x + \frac{1}{2} \rfloor = n$, because $0 < \frac{1}{2} + \epsilon < 1$. Consequently, $\lfloor 2x \rfloor = 2n$ and $\lfloor x \rfloor + \lfloor x + \frac{1}{2} \rfloor = n + n = 2n$.

Next, we consider the case when $\frac{1}{2} \le \epsilon < 1$. In this case, $2x = 2n + 2\epsilon = (2n + 1) + (2\epsilon - 1)$. Because $0 \le 2\epsilon - 1 < 1$, it follows that $\lfloor 2x \rfloor = 2n + 1$. Because $\lfloor x + \frac{1}{2} \rfloor = \lfloor n + (\frac{1}{2} + \epsilon) \rfloor = \lfloor n + 1 + (\epsilon - \frac{1}{2}) \rfloor$ and $0 \le \epsilon - \frac{1}{2} < 1$, it follows that $\lfloor x + \frac{1}{2} \rfloor = n + 1$. Consequently, $\lfloor 2x \rfloor = 2n + 1$ and $\lfloor x \rfloor + \lfloor x + \frac{1}{2} \rfloor = n + (n + 1) = 2n + 1$. This concludes the proof. ◀

EXAMPLE 30   Prove or disprove that $\lceil x + y \rceil = \lceil x \rceil + \lceil y \rceil$ for all real numbers x and y.

Solution: Although this statement may appear reasonable, it is false. A counterexample is supplied by $x = \frac{1}{2}$ and $y = \frac{1}{2}$. With these values we find that $\lceil x + y \rceil = \lceil \frac{1}{2} + \frac{1}{2} \rceil = \lceil 1 \rceil = 1$, but $\lceil x \rceil + \lceil y \rceil = \lceil \frac{1}{2} \rceil + \lceil \frac{1}{2} \rceil = 1 + 1 = 2$. ◀

There are certain types of functions that will be used throughout the text. These include polynomial, logarithmic, and exponential functions. A brief review of the properties of these functions needed in this text is given in Appendix 2. In this book the notation $\log x$ will be used to denote the logarithm to the base 2 of x, because 2 is the base that we will usually use for logarithms. We will denote logarithms to the base b, where b is any real number greater than 1, by $\log_b x$, and the natural logarithm by $\ln x$.

Another function we will use throughout this text is the **factorial function** $f : \mathbf{N} \to \mathbf{Z}^+$, denoted by $f(n) = n!$. The value of $f(n) = n!$ is the product of the first n positive integers, so $f(n) = 1 \cdot 2 \cdots (n - 1) \cdot n$ [and $f(0) = 0! = 1$].

EXAMPLE 31   We have $f(1) = 1! = 1$, $f(2) = 2! = 1 \cdot 2 = 2$, $f(6) = 6! = 1 \cdot 2 \cdot 3 \cdot 4 \cdot 5 \cdot 6 = 720$, and $f(20) = 1 \cdot 2 \cdot 3 \cdot 4 \cdot 5 \cdot 6 \cdot 7 \cdot 8 \cdot 9 \cdot 10 \cdot 11 \cdot 12 \cdot 13 \cdot 14 \cdot 15 \cdot 16 \cdot 17 \cdot 18 \cdot 19 \cdot 20 = 2,432,902,008,176,640,000$. ◀

Example 31 illustrates that the factorial function grows extremely rapidly as n grows. The rapid growth of the factorial function is made clearer by Stirling's formula, a result from

JAMES STIRLING (1692–1770)   James Stirling was born near the town of Stirling, Scotland. His family strongly supported the Jacobite cause of the Stuarts as an alternative to the British crown. The first information known about James is that he entered Balliol College, Oxford, on a scholarship in 1711. However, he later lost his scholarship when he refused to pledge his allegiance to the British crown. The first Jacobean rebellion took place in 1715, and Stirling was accused of communicating with rebels. He was charged with cursing King George, but he was acquitted of these charges. Even though he could not graduate from Oxford because of his politics, he remained there for several years. Stirling published his first work, which extended Newton's work on plane curves, in 1717. He traveled to Venice, where a chair of mathematics had been promised to him, an appointment that unfortunately fell through. Nevertheless, Stirling stayed in Venice, continuing his mathematical work. He attended the University of Padua in 1721, and in 1722 he returned to Glasgow. Stirling apparently fled Italy after learning the secrets of the Italian glass industry, avoiding the efforts of Italian glass makers to assassinate him to protect their secrets.

In late 1724 Stirling moved to London, staying there 10 years teaching mathematics and actively engaging in research. In 1730 he published *Methodus Differentialis*, his most important work, presenting results on infinite series, summations, interpolation, and quadrature. It is in this book that his asymptotic formula for $n!$ appears. Stirling also worked on gravitation and the shape of the earth; he stated, but did not prove, that the earth is an oblate spheroid. Stirling returned to Scotland in 1735, when he was appointed manager of a Scottish mining company. He was very successful in this role and even published a paper on the ventilation of mine shafts. He continued his mathematical research, but at a reduced pace, during his years in the mining industry. Stirling is also noted for surveying the River Clyde with the goal of creating a series of locks to make it navigable. In 1752 the citizens of Glasgow presented him with a silver teakettle as a reward for this work.

higher mathematics that tell us that $n! \sim \sqrt{2\pi n}(n/e)^n$. Here, we have used the notation $f(n) \sim g(n)$, which means that the ratio $f(n)/g(n)$ approaches 1 as n grows without bound (that is, $\lim_{n\to\infty} f(n)/g(n) = 1$). The symbol $\sim$ is read "is asymptotic to." Stirling's formula is named after James Stirling, a Scottish mathematician of the eighteenth century.

Partial Functions

A program designed to evaluate a function may not produce the correct value of the function for all elements in the domain of this function. For example, a program may not produce a correct value because evaluating the function may lead to an infinite loop or an overflow. Similarly, in abstract mathematics, we often want to discuss functions that are defined only for a subset of the real numbers, such as $1/x$, $\sqrt{x}$, and $\arcsin(x)$. Also, we may want to use such notions as the "youngest child" function, which is undefined for a couple having no children, or the "time of sunrise," which is undefined for some days above the Arctic Circle. To study such situations, we use the concept of a partial function.

DEFINITION 13

A *partial function* f from a set A to a set B is an assignment to each element a in a subset of A, called the *domain of definition* of f, of a unique element b in B. The sets A and B are called the *domain* and *codomain* of f, respectively. We say that f is *undefined* for elements in A that are not in the domain of definition of f. When the domain of definition of f equals A, we say that f is a *total function*.

Remark: We write $f : A \to B$ to denote that f is a partial function from A to B. Note that this is the same notation as is used for functions. The context in which the notation is used determines whether f is a partial function or a total function.

EXAMPLE 32   The function $f : \mathbf{Z} \to \mathbf{R}$ where $f(n) = \sqrt{n}$ is a partial function from $\mathbf{Z}$ to $\mathbf{R}$ where the domain of definition is the set of nonnegative integers. Note that f is undefined for negative integers. ◀

Exercises

1. Why is f not a function from $\mathbf{R}$ to $\mathbf{R}$ if
  a) $f(x) = 1/x$?
  b) $f(x) = \sqrt{x}$?
  c) $f(x) = \pm\sqrt{(x^2 + 1)}$?

2. Determine whether f is a function from $\mathbf{Z}$ to $\mathbf{R}$ if
  a) $f(n) = \pm n$.
  b) $f(n) = \sqrt{n^2 + 1}$.
  c) $f(n) = 1/(n^2 - 4)$.

3. Determine whether f is a function from the set of all bit strings to the set of integers if
  a) $f(S)$ is the position of a 0 bit in S.
  b) $f(S)$ is the number of 1 bits in S.
  c) $f(S)$ is the smallest integer i such that the ith bit of S is 1 and $f(S) = 0$ when S is the empty string, the string with no bits.

4. Find the domain and range of these functions. Note that in each case, to find the domain, determine the set of elements assigned values by the function.
  a) the function that assigns to each nonnegative integer its last digit

  b) the function that assigns the next largest integer to a positive integer
  c) the function that assigns to a bit string the number of one bits in the string
  d) the function that assigns to a bit string the number of bits in the string

5. Find these values.
  a) $\lceil \frac{3}{4} \rceil$
  b) $\lfloor \frac{7}{8} \rfloor$
  c) $\lceil -\frac{3}{4} \rceil$
  d) $\lfloor -\frac{7}{8} \rfloor$
  e) $\lceil 3 \rceil$
  f) $\lfloor -1 \rfloor$
  g) $\lfloor \frac{1}{2} + \lceil \frac{3}{2} \rceil \rfloor$
  h) $\lfloor \frac{1}{2} \cdot \lfloor \frac{5}{2} \rfloor \rfloor$

6. Determine whether each of these functions from $\{a, b, c, d\}$ to itself is one-to-one.
  a) $f(a) = b, f(b) = a, f(c) = c, f(d) = d$
  b) $f(a) = b, f(b) = b, f(c) = d, f(d) = c$
  c) $f(a) = d, f(b) = b, f(c) = c, f(d) = d$

7. Which functions in Exercise 6 are onto?

8. Determine whether each of these functions from $\mathbf{Z}$ to $\mathbf{Z}$ is one-to-one.
  a) $f(n) = n - 1$
  b) $f(n) = n^2 + 1$
  c) $f(n) = n^3$
  d) $f(n) = \lceil n/2 \rceil$

9. Which functions in Exercise 8 are onto?

10. Consider these functions from the set of students in a discrete mathematics class. Under what conditions is the function one-to-one if it assigns to a student his or her

   a) mobile phone number.

   b) student identification number.

   c) final grade in the class.

   d) home town.

11. Consider these functions from the set of teachers in a school. Under what conditions is the function one-to-one if it assigns to a teacher his or her

   a) office.

   b) assigned bus to chaperone in a group of buses taking students on a field trip.

   c) salary.

   d) social security number.

12. Specify a codomain for each of the functions in Exercise 16. Under what conditions is each of these functions with the codomain you specified onto?

13. Specify a codomain for each of the functions in Exercise 11. Under what conditions is each of the functions with the codomain you specified onto?

14. Give an example of a function from **N** to **N** that is

   a) one-to-one but not onto.

   b) onto but not one-to-one.

   c) both onto and one-to-one (but different from the identity function).

   d) neither one-to-one nor onto.

15. Determine whether each of these functions is a bijection from **R** to **R**.

   a) $f(x) = 2x + 1$

   b) $f(x) = x^2 + 1$

   c) $f(x) = x^3$

   d) $f(x) = (x^2 + 1)/(x^2 + 2)$

16. Show that the function $f(x) = e^x$ from the set of real numbers to the set of real numbers is not invertible, but if the codomain is restricted to the set of positive real numbers, the resulting function is invertible.

17. Show that the function $f(x) = |x|$ from the set of real numbers to the set of nonnegative real numbers is not invertible, but if the domain is restricted to the set of non-negative real numbers, the resulting function is invertible.

18. Let $S = \{-1, 0, 2, 4, 7\}$. Find $f(S)$ if

   a) $f(x) = 1$.   **b)** $f(x) = 2x + 1$.

   c) $f(x) = \lceil x/5 \rceil$.   **d)** $f(x) = \lfloor (x^2 + 1)/3 \rfloor$.

19. Suppose that g is a function from A to B and f is a function from B to C.

   a) Show that if both f and g are one-to-one functions, then $f \circ g$ is also one-to-one.

   b) Show that if both f and g are onto functions, then $f \circ g$ is also onto.

*20. If f and $f \circ g$ are one-to-one, does it follow that g is one-to-one? Justify your answer.

*21. If f and $f \circ g$ are onto, does it follow that g is onto? Justify your answer.

22. Find $f \circ g$ and $g \circ f$, where $f(x) = x^2 + 1$ and $g(x) = x + 2$, are functions from **R** to **R**.

23. Find $f + g$ and fg for the functions f and g given in Exercise 22.

24. Let $f(x) = ax + b$ and $g(x) = cx + d$, where $a, b, c,$ and d are constants. Determine necessary and sufficient conditions on the constants $a, b, c,$ and d so that $f \circ g = g \circ f$.

25. Show that the function $f(x) = ax + b$ from **R** to **R** is invertible, where a and b are constants, with $a \neq 0$, and find the inverse of f.

26. Let f be a function from the set A to the set B. Let S and T be subsets of A. Show that

   a) $f(S \cup T) = f(S) \cup f(T)$.

   b) $f(S \cap T) \subseteq f(S) \cap f(T)$.

27. a) Give an example to show that the inclusion in part (b) in Exercise 26 may be proper.

   b) Show that if f is one-to-one, the inclusion in part (b) in Exercise 26 is an equality.

Let f be a function from the set A to the set B. Let S be a subset of B. We define the **inverse image** of S to be the subset of A whose elements are precisely all pre-images of all elements of S. We denote the inverse image of S by $f^{-1}(S)$, so $f^{-1}(S) = \{a \in A \mid f(a) \in S\}$. (*Beware:* The notation f^{-1} is used in two different ways. Do not confuse the notation introduced here with the notation $f^{-1}(y)$ for the value at y of the inverse of the invertible function f. Notice also that $f^{-1}(S)$, the inverse image of the set S, makes sense for all functions f, not just invertible functions.)

28. Let f be the function from **R** to **R** defined by $f(x) = x^2$. Find

   a) $f^{-1}(\{1\})$.   **b)** $f^{-1}(\{x \mid 0 < x < 1\})$.

   c) $f^{-1}(\{x \mid x > 4\})$.

29. Show that $\lceil x - \frac{1}{2} \rceil$ is the closest integer to the number x, except when x is midway between two integers, when it is the smaller of these two integers.

30. Show that if x is a real number, then $\lceil x \rceil - \lfloor x \rfloor = 1$ if x is not an integer and $\lceil x \rceil - \lfloor x \rfloor = 0$ if x is an integer.

31. Show that if x is a real number, then $x - 1 < \lfloor x \rfloor \leq x \leq \lceil x \rceil < x + 1$.

32. Show that if x is a real number and m is an integer, then $\lceil x + m \rceil = \lceil x \rceil + m$.

33. Show that if x is a real number and n is an integer, then

   a) $x < n$ if and only if $\lfloor x \rfloor < n$.

   b) $n < x$ if and only if $n < \lceil x \rceil$.

34. Show that if x is a real number and n is an integer, then

   a) $x \leq n$ if and only if $\lceil x \rceil \leq n$.

   b) $n \leq x$ if and only if $n \leq \lfloor x \rfloor$.

35. Prove that if n is an integer, then $\lfloor n/2 \rfloor = n/2$ if n is even and $(n - 1)/2$ if n is odd.

36. Prove that if x is a real number, then $\lfloor -x \rfloor = -\lceil x \rceil$ and $\lceil -x \rceil = -\lfloor x \rfloor$.

37. The function INT is found on some calculators, where INT$(x) = \lfloor x \rfloor$ when x is a nonnegative real number and INT$(x) = \lceil x \rceil$ when x is a negative real number. Show that this INT function satisfies the identity INT$(-x) = -$INT(x).

38. Let a and b be real numbers with $a < b$. Use the floor and/or ceiling functions to express the number of integers n that satisfy the inequality $a \leq n \leq b$.

39. Let a and b be real numbers with $a < b$. Use the floor and/or ceiling functions to express the number of integers n that satisfy the inequality $a < n < b$.

40. How many bytes are required to encode n bits of data where n equals

    a) 4?   **b)** 10?   **c)** 500?   **d)** 3000?

41. How many bytes are required to encode n bits of data where n equals

    a) 7?   **b)** 17?   **c)** 1001?   **d)** 28,800?

42. Draw the graph of the function $f(n) = 1 - n^2$ from $\mathbf{Z}$ to $\mathbf{Z}$.

43. Draw the graph of the function $f(x) = \lfloor 2x \rfloor$ from $\mathbf{R}$ to $\mathbf{R}$.

44. Draw graphs of each of these functions.

    a) $f(x) = \lceil 3x - 2 \rceil$ **b)** $f(x) = \lceil 0.2x \rceil$
    c) $f(x) = \lfloor -1/x \rfloor$ **d)** $f(x) = \lfloor x^2 \rfloor$
    e) $f(x) = \lceil x/2 \rceil \lfloor x/2 \rfloor$ **f)** $f(x) = \lfloor x/2 \rfloor + \lceil x/2 \rceil$
    g) $f(x) = \lfloor 2 \lceil x/2 \rceil + \frac{1}{2} \rfloor$

45. Find the inverse function of $f(x) = x^3 + 1$.

46. Suppose that f is an invertible function from Y to Z and g is an invertible function from X to Y. Show that the inverse of the composition $f \circ g$ is given by $(f \circ g)^{-1} = g^{-1} \circ f^{-1}$.

47. Let S be a subset of a universal set U. The **characteristic function** f_S of S is the function from U to the set $\{0, 1\}$ such that $f_S(x) = 1$ if x belongs to S and $f_S(x) = 0$ if x does not belong to S. Let A and B be sets. Show that for all $x \in U$,

    a) $f_{A \cap B}(x) = f_A(x) \cdot f_B(x)$
    b) $f_{A \cup B}(x) = f_A(x) + f_B(x) - f_A(x) \cdot f_B(x)$
    c) $f_{\overline{A}}(x) = 1 - f_A(x)$
    d) $f_{A \oplus B}(x) = f_A(x) + f_B(x) - 2f_A(x)f_B(x)$

48. Suppose that f is a function from A to B, where A and B are finite sets with $|A| = |B|$. Show that f is one-to-one if and only if it is onto.

49. Prove or disprove each of these statements about the floor and ceiling functions.

    a) $\lfloor \lceil x \rceil \rfloor = \lceil x \rceil$ for all real numbers x.
    b) $\lfloor x + y \rfloor = \lfloor x \rfloor + \lfloor y \rfloor$ for all real numbers x and y.
    c) $\lceil \lceil x/2 \rceil /2 \rceil = \lceil x/4 \rceil$ for all real numbers x.
    d) $\lfloor \sqrt{\lceil x \rceil} \rfloor = \lfloor \sqrt{x} \rfloor$ for all positive real numbers x.
    e) $\lfloor x \rfloor + \lfloor y \rfloor + \lfloor x + y \rfloor \leq \lfloor 2x \rfloor + \lfloor 2y \rfloor$ for all real numbers x and y.

50. Prove that if x is a positive real number, then

    a) $\lfloor \sqrt{\lfloor x \rfloor} \rfloor = \lfloor \sqrt{x} \rfloor$.
    b) $\lceil \sqrt{\lceil x \rceil} \rceil = \lceil \sqrt{x} \rceil$.

51. For each of these partial functions, determine its domain, codomain, domain of definition, and the set of values for which it is undefined. Also, determine whether it is a total function.

    a) $f : \mathbf{Z} \rightarrow \mathbf{R}$, $f(n) = 1/n$
    b) $f : \mathbf{Z} \rightarrow \mathbf{Z}$, $f(n) = \lceil n/2 \rceil$
    c) $f : \mathbf{Z} \times \mathbf{Z} \rightarrow \mathbf{Q}$, $f(m, n) = m/n$
    d) $f : \mathbf{Z} \times \mathbf{Z} \rightarrow \mathbf{Z}$, $f(m, n) = mn$
    e) $f : \mathbf{Z} \times \mathbf{Z} \rightarrow \mathbf{Z}$, $f(m, n) = m - n$ if $m > n$

52. a) Show that a partial function from A to B can be viewed as a function f^* from A to $B \cup \{u\}$, where u is not an element of B and

$$f^*(a) = \begin{cases} f(a) & \text{if } a \text{ belongs to the domain} \\ & \text{of definition of } f \\ u & \text{if } f \text{ is undefined at } a. \end{cases}$$

    b) Using the construction in (a), find the function f^* corresponding to each partial function in Exercise 77.

53. a) Show that if a set S has cardinality m, where m is a positive integer, then there is a one-to-one correspondence between S and the set $\{1, 2, \ldots, m\}$.

    b) Show that if S and T are two sets each with m elements, where m is a positive integer, then there is a one-to-one correspondence between S and T.

2.4   Sequences and Summations

Introduction

Sequences are ordered lists of elements, used in discrete mathematics in many ways. For example, they can be used to represent solutions to certain counting problems, as we will see in Chapter 8. They are also an important data structure in computer science. We will often need to work with sums of terms of sequences in our study of discrete mathematics. This section reviews the use of summation notation, basic properties of summations, and formulas for the sums of terms of some particular types of sequences.

    The terms of a sequence can be specified by providing a formula for each term of the sequence. In this section we describe another way to specify the terms of a sequence using a recurrence relation, which expresses each term as a combination of the previous terms. We will introduce one method, known as iteration, for finding a closed formula for the terms of a sequence specified via a recurrence relation. Identifying a sequence when the first few terms are provided is a useful skill when solving problems in discrete mathematics. We will provide some tips, including a useful tool on the Web, for doing so.

Sequences

A sequence is a discrete structure used to represent an ordered list. For example, 1, 2, 3, 5, 8 is a sequence with five terms and $1, 3, 9, 27, 81, \ldots, 3^n, \ldots$ is an infinite sequence.

DEFINITION 1   A *sequence* is a function from a subset of the set of integers (usually either the set $\{0, 1, 2, \ldots\}$ or the set $\{1, 2, 3, \ldots\}$) to a set S. We use the notation a_n to denote the image of the integer n. We call a_n a *term* of the sequence.

We use the notation $\{a_n\}$ to describe the sequence. (Note that a_n represents an individual term of the sequence $\{a_n\}$. Be aware that the notation $\{a_n\}$ for a sequence conflicts with the notation for a set. However, the context in which we use this notation will always make it clear when we are dealing with sets and when we are dealing with sequences. Moreover, although we have used the letter a in the notation for a sequence, other letters or expressions may be used depending on the sequence under consideration. That is, the choice of the letter a is arbitrary.)

We describe sequences by listing the terms of the sequence in order of increasing subscripts.

EXAMPLE 1   Consider the sequence $\{a_n\}$, where

$$a_n = \frac{1}{n}.$$

The list of the terms of this sequence, beginning with a_1, namely,

$$a_1, a_2, a_3, a_4, \ldots,$$

starts with

$$1, \frac{1}{2}, \frac{1}{3}, \frac{1}{4}, \ldots. \qquad \blacktriangleleft$$

DEFINITION 2   A *geometric progression* is a sequence of the form

$$a, ar, ar^2, \ldots, ar^n, \ldots$$

where the *initial term a* and the *common ratio r* are real numbers.

Remark: A geometric progression is a discrete analogue of the exponential function $f(x) = ar^x$.

EXAMPLE 2   The sequences $\{b_n\}$ with $b_n = (-1)^n$, $\{c_n\}$ with $c_n = 2 \cdot 5^n$, and $\{d_n\}$ with $d_n = 6 \cdot (1/3)^n$ are geometric progressions with initial term and common ratio equal to 1 and -1; 2 and 5; and 6 and 1/3, respectively, if we start at $n = 0$. The list of terms $b_0, b_1, b_2, b_3, b_4, \ldots$ begins with

$$1, -1, 1, -1, 1, \ldots;$$

the list of terms $c_0, c_1, c_2, c_3, c_4, \ldots$ begins with

$$2, 10, 50, 250, 1250, \ldots ;$$

and the list of terms $d_0, d_1, d_2, d_3, d_4, \ldots$ begins with

$$6, 2, \frac{2}{3}, \frac{2}{9}, \frac{2}{27}, \ldots .$$ ◀

DEFINITION 3
An *arithmetic progression* is a sequence of the form

$$a, a + d, a + 2d, \ldots, a + nd, \ldots$$

where the *initial term a* and the *common difference d* are real numbers.

Remark: An arithmetic progression is a discrete analogue of the linear function $f(x) = dx + a$.

EXAMPLE 3
The sequences $\{s_n\}$ with $s_n = -1 + 4n$ and $\{t_n\}$ with $t_n = 7 - 3n$ are both arithmetic progressions with initial terms and common differences equal to -1 and 4, and 7 and -3, respectively, if we start at $n = 0$. The list of terms $s_0, s_1, s_2, s_3, \ldots$ begins with

$$-1, 3, 7, 11, \ldots ,$$

and the list of terms $t_0, t_1, t_2, t_3, \ldots$ begins with

$$7, 4, 1, -2, \ldots .$$ ◀

Sequences of the form $a_1, a_2, \ldots, a_n$ are often used in computer science. These finite sequences are also called **strings**. This string is also denoted by $a_1 a_2 \ldots a_n$. (Recall that bit strings, which are finite sequences of bits, were introduced in Section 1.1.) The **length** of a string is the number of terms in this string. The **empty string**, denoted by λ, is the string that has no terms. The empty string has length zero.

EXAMPLE 4
The string *abcd* is a string of length four. ◀

Recurrence Relations

In Examples 1–3 we specified sequences by providing explicit formulas for their terms. There are many other ways to specify a sequence. For example, another way to specify a sequence is to provide one or more initial terms together with a rule for determining subsequent terms from those that precede them.

DEFINITION 4
A *recurrence relation* for the sequence $\{a_n\}$ is an equation that expresses a_n in terms of one or more of the previous terms of the sequence, namely, $a_0, a_1, \ldots, a_{n-1}$, for all integers n with $n \geq n_0$, where n_0 is a nonnegative integer. A sequence is called a *solution* of a recurrence relation if its terms satisfy the recurrence relation. (A recurrence relation is said to *recursively define* a sequence. We will explain this alternative terminology in Chapter 5.)

EXAMPLE 5
Let $\{a_n\}$ be a sequence that satisfies the recurrence relation $a_n = a_{n-1} + 3$ for $n = 1, 2, 3, \ldots$, and suppose that $a_0 = 2$. What are a_1, a_2, and a_3?

Solution: We see from the recurrence relation that $a_1 = a_0 + 3 = 2 + 3 = 5$. It then follows that $a_2 = 5 + 3 = 8$ and $a_3 = 8 + 3 = 11$. ◀

EXAMPLE 6   Let $\{a_n\}$ be a sequence that satisfies the recurrence relation $a_n = a_{n-1} - a_{n-2}$ for $n = 2, 3, 4, \ldots$, and suppose that $a_0 = 3$ and $a_1 = 5$. What are a_2 and a_3?

Solution: We see from the recurrence relation that $a_2 = a_1 - a_0 = 5 - 3 = 2$ and $a_3 = a_2 - a_1 = 2 - 5 = -3$. We can find a_4, a_5, and each successive term in a similar way. ◀

The **initial conditions** for a recursively defined sequence specify the terms that precede the first term where the recurrence relation takes effect. For instance, the initial condition in Example 5 is $a_0 = 2$, and the initial conditions in Example 6 are $a_0 = 3$ and $a_1 = 5$. Using mathematical induction, a proof technique introduced in Chapter 5, it can be shown that a recurrence relation together with its initial conditions determines a unique solution.

Next, we define a particularly useful sequence defined by a recurrence relation, known as the **Fibonacci sequence**, after the Italian mathematician Fibonacci who was born in the 12th century (see Chapter 5 for his biography). We will study this sequence in depth in Chapters 5 and 8, where we will see why it is important for many applications, including modeling the population growth of rabbits.

Hop along to Chapter 8 to learn how to find a formula for the Fibonacci numbers.

DEFINITION 5

Links

The *Fibonacci sequence*, $f_0, f_1, f_2, \ldots$, is defined by the initial conditions $f_0 = 0$, $f_1 = 1$, and the recurrence relation

$$f_n = f_{n-1} + f_{n-2}$$

for $n = 2, 3, 4, \ldots$.

EXAMPLE 7   Find the Fibonacci numbers f_2, f_3, f_4, f_5, and f_6.

Solution: The recurrence relation for the Fibonacci sequence tells us that we find successive terms by adding the previous two terms. Because the initial conditions tell us that $f_0 = 0$ and $f_1 = 1$, using the recurrence relation in the definition we find that

$$f_2 = f_1 + f_0 = 1 + 0 = 1,$$
$$f_3 = f_2 + f_1 = 1 + 1 = 2,$$
$$f_4 = f_3 + f_2 = 2 + 1 = 3,$$
$$f_5 = f_4 + f_3 = 3 + 2 = 5,$$
$$f_6 = f_5 + f_4 = 5 + 3 = 8.$$

◀

EXAMPLE 8   Suppose that $\{a_n\}$ is the sequence of integers defined by $a_n = n!$, the value of the factorial function at the integer n, where $n = 1, 2, 3, \ldots$. Because $n! = n((n-1)(n-2)\ldots 2 \cdot 1) = n(n-1)! = n a_{n-1}$, we see that the sequence of factorials satisfies the recurrence relation $a_n = n a_{n-1}$, together with the initial condition $a_1 = 1$. ◀

We say that we have solved the recurrence relation together with the initial conditions when we find an explicit formula, called a **closed formula**, for the terms of the sequence.

EXAMPLE 9   Determine whether the sequence $\{a_n\}$, where $a_n = 3n$ for every nonnegative integer n, is a solution of the recurrence relation $a_n = 2a_{n-1} - a_{n-2}$ for $n = 2, 3, 4, \ldots$. Answer the same question where $a_n = 2^n$ and where $a_n = 5$.

Solution: Suppose that $a_n = 3n$ for every nonnegative integer n. Then, for $n \geq 2$, we see that $2a_{n-1} - a_{n-2} = 2(3(n-1)) - 3(n-2) = 3n = a_n$. Therefore, $\{a_n\}$, where $a_n = 3n$, is a solution of the recurrence relation.

Suppose that $a_n = 2^n$ for every nonnegative integer n. Note that $a_0 = 1, a_1 = 2$, and $a_2 = 4$. Because $2a_1 - a_0 = 2 \cdot 2 - 1 = 3 \neq a_2$, we see that $\{a_n\}$, where $a_n = 2^n$, is not a solution of the recurrence relation.

Suppose that $a_n = 5$ for every nonnegative integer n. Then for $n \geq 2$, we see that $a_n = 2a_{n-1} - a_{n-2} = 2 \cdot 5 - 5 = 5 = a_n$. Therefore, $\{a_n\}$, where $a_n = 5$, is a solution of the recurrence relation. ◀

Many methods have been developed for solving recurrence relations. Here, we will introduce a straightforward method known as iteration via several examples. In Chapter 8 we will study recurrence relations in depth. In that chapter we will show how recurrence relations can be used to solve counting problems and we will introduce several powerful methods that can be used to solve many different recurrence relations.

EXAMPLE 10 Solve the recurrence relation and initial condition in Example 5.

Solution: We can successively apply the recurrence relation in Example 5, starting with the initial condition $a_1 = 2$, and working upward until we reach a_n to deduce a closed formula for the sequence. We see that

$$a_2 = 2 + 3$$
$$a_3 = (2 + 3) + 3 = 2 + 3 \cdot 2$$
$$a_4 = (2 + 2 \cdot 3) + 3 = 2 + 3 \cdot 3$$
$$\vdots$$
$$a_n = a_{n-1} + 3 = (2 + 3 \cdot (n-2)) + 3 = 2 + 3(n-1).$$

We can also successively apply the recurrence relation in Example 5, starting with the term a_n and working downward until we reach the initial condition $a_1 = 2$ to deduce this same formula. The steps are

$$a_n = a_{n-1} + 3$$
$$= (a_{n-2} + 3) + 3 = a_{n-2} + 3 \cdot 2$$
$$= (a_{n-3} + 3) + 3 \cdot 2 = a_{n-3} + 3 \cdot 3$$
$$\vdots$$
$$= a_2 + 3(n-2) = (a_1 + 3) + 3(n-2) = 2 + 3(n-1).$$

At each iteration of the recurrence relation, we obtain the next term in the sequence by adding 3 to the previous term. We obtain the nth term after $n - 1$ iterations of the recurrence relation. Hence, we have added $3(n-1)$ to the initial term $a_0 = 2$ to obtain a_n. This gives us the closed formula $a_n = 2 + 3(n-1)$. Note that this sequence is an arithmetic progression. ◀

The technique used in Example 10 is called **iteration**. We have iterated, or repeatedly used, the recurrence relation. The first approach is called **forward substitution** – we found successive terms beginning with the initial condition and ending with a_n. The second approach is called **backward substitution**, because we began with a_n and iterated to express it in terms of falling terms of the sequence until we found it in terms of a_1. Note that when we use iteration, we essential guess a formula for the terms of the sequence. To prove that our guess is correct, we need to use mathematical induction, a technique we discuss in Chapter 5.

In Chapter 8 we will show that recurrence relations can be used to model a wide variety of problems. We provide one such example here, showing how to use a recurrence relation to find compound interest.

EXAMPLE 11   **Compound Interest**   Suppose that a person deposits $10,000 in a savings account at a bank yielding 11% per year with interest compounded annually. How much will be in the account after 30 years?

Solution: To solve this problem, let P_n denote the amount in the account after n years. Because the amount in the account after n years equals the amount in the account after $n - 1$ years plus interest for the nth year, we see that the sequence $\{P_n\}$ satisfies the recurrence relation

$$P_n = P_{n-1} + 0.11 P_{n-1} = (1.11) P_{n-1}.$$

The initial condition is $P_0 = 10,000$.

We can use an iterative approach to find a formula for P_n. Note that

$$P_1 = (1.11) P_0$$
$$P_2 = (1.11) P_1 = (1.11)^2 P_0$$
$$P_3 = (1.11) P_2 = (1.11)^3 P_0$$
$$\vdots$$
$$P_n = (1.11) P_{n-1} = (1.11)^n P_0.$$

When we insert the initial condition $P_0 = 10,000$, the formula $P_n = (1.11)^n 10,000$ is obtained.

Inserting $n = 30$ into the formula $P_n = (1.11)^n 10,000$ shows that after 30 years the account contains

$$P_{30} = (1.11)^{30} 10,000 = \$228,922.97. \qquad \blacktriangleleft$$

Special Integer Sequences

A common problem in discrete mathematics is finding a closed formula, a recurrence relation, or some other type of general rule for constructing the terms of a sequence. Sometimes only a few terms of a sequence solving a problem are known; the goal is to identify the sequence. Even though the initial terms of a sequence do not determine the entire sequence (after all, there are infinitely many different sequences that start with any finite set of initial terms), knowing the first few terms may help you make an educated conjecture about the identity of your sequence. Once you have made this conjecture, you can try to verify that you have the correct sequence.

When trying to deduce a possible formula, recurrence relation, or some other type of rule for the terms of a sequence when given the initial terms, try to find a pattern in these terms. You might also see whether you can determine how a term might have been produced from those preceding it. There are many questions you could ask, but some of the more useful are:

- Are there runs of the same value? That is, does the same value occur many times in a row?
- Are terms obtained from previous terms by adding the same amount or an amount that depends on the position in the sequence?
- Are terms obtained from previous terms by multiplying by a particular amount?
- Are terms obtained by combining previous terms in a certain way?
- Are there cycles among the terms?

EXAMPLE 12   Find formulae for the sequences with the following first five terms: (a) 1, 1/2, 1/4, 1/8, 1/16 (b) 1, 3, 5, 7, 9   (c) 1, −1, 1, −1, 1.

Extra Examples

Solution: (a) We recognize that the denominators are powers of 2. The sequence with $a_n = 1/2^n$, $n = 0, 1, 2, \ldots$ is a possible match. This proposed sequence is a geometric progression with $a = 1$ and $r = 1/2$.

(b) We note that each term is obtained by adding 2 to the previous term. The sequence with $a_n = 2n + 1$, $n = 0, 1, 2, \ldots$ is a possible match. This proposed sequence is an arithmetic progression with $a = 1$ and $d = 2$.

(c) The terms alternate between 1 and −1. The sequence with $a_n = (-1)^n$, $n = 0, 1, 2 \ldots$ is a possible match. This proposed sequence is a geometric progression with $a = 1$ and $r = -1$. ◀

Examples 13–15 illustrate how we can analyze sequences to find how the terms are constructed.

EXAMPLE 13   How can we produce the terms of a sequence if the first 10 terms are 1, 2, 2, 3, 3, 3, 4, 4, 4, 4?

Solution: In this sequence, the integer 1 appears once, the integer 2 appears twice, the integer 3 appears three times, and the integer 4 appears four times. A reasonable rule for generating this sequence is that the integer n appears exactly n times, so the next five terms of the sequence would all be 5, the following six terms would all be 6, and so on. The sequence generated this way is a possible match. ◀

EXAMPLE 14   How can we produce the terms of a sequence if the first 10 terms are 5, 11, 17, 23, 29, 35, 41, 47, 53, 59?

Solution: Note that each of the first 10 terms of this sequence after the first is obtained by adding 6 to the previous term. (We could see this by noticing that the difference between consecutive terms is 6.) Consequently, the nth term could be produced by starting with 5 and adding 6 a total of $n - 1$ times; that is, a reasonable guess is that the nth term is $5 + 6(n - 1) = 6n - 1$. (This is an arithmetic progression with $a = 5$ and $d = 6$.) ◀

EXAMPLE 15   How can we produce the terms of a sequence if the first 10 terms are 1, 3, 4, 7, 11, 18, 29, 47, 76, 123?

Solution: Observe that each successive term of this sequence, starting with the third term, is the sum of the two previous terms. That is, $4 = 3 + 1$, $7 = 4 + 3$, $11 = 7 + 4$, and so on. Consequently, if L_n is the nth term of this sequence, we guess that the sequence is determined by the recurrence relation $L_n = L_{n-1} + L_{n-2}$ with initial conditions $L_1 = 1$ and $L_2 = 3$ (the same recurrence relation as the Fibonacci sequence, but with different initial conditions). This sequence is known as the **Lucas sequence**, after the French mathematician François Édouard Lucas. Lucas studied this sequence and the Fibonacci sequence in the nineteenth century. ◀

Another useful technique for finding a rule for generating the terms of a sequence is to compare the terms of a sequence of interest with the terms of a well-known integer sequence, such as terms of an arithmetic progression, terms of a geometric progression, perfect squares, perfect cubes, and so on. The first 10 terms of some sequences you may want to keep in mind are displayed in Table 1.

EXAMPLE 16   Conjecture a simple formula for a_n if the first 10 terms of the sequence $\{a_n\}$ are 1, 7, 25, 79, 241, 727, 2185, 6559, 19681, 59047.

TABLE 1 Some Useful Sequences.	
nth Term	*First 10 Terms*
n^2	$1, 4, 9, 16, 25, 36, 49, 64, 81, 100, \ldots$
n^3	$1, 8, 27, 64, 125, 216, 343, 512, 729, 1000, \ldots$
n^4	$1, 16, 81, 256, 625, 1296, 2401, 4096, 6561, 10000, \ldots$
2^n	$2, 4, 8, 16, 32, 64, 128, 256, 512, 1024, \ldots$
3^n	$3, 9, 27, 81, 243, 729, 2187, 6561, 19683, 59049, \ldots$
$n!$	$1, 2, 6, 24, 120, 720, 5040, 40320, 362880, 3628800, \ldots$
f_n	$1, 1, 2, 3, 5, 8, 13, 21, 34, 55, 89, \ldots$

Solution: To attack this problem, we begin by looking at the difference of consecutive terms, but we do not see a pattern. When we form the ratio of consecutive terms to see whether each term is a multiple of the previous term, we find that this ratio, although not a constant, is close to 3. So it is reasonable to suspect that the terms of this sequence are generated by a formula involving 3^n. Comparing these terms with the corresponding terms of the sequence $\{3^n\}$, we notice that the nth term is 2 less than the corresponding power of 3. We see that $a_n = 3^n - 2$ for $1 \leq n \leq 10$ and conjecture that this formula holds for all n.   ◀

We will see throughout this text that integer sequences appear in a wide range of contexts in discrete mathematics. Sequences we have encountered or will encounter include the sequence of prime numbers (Chapter 4), the number of ways to order n discrete objects (Chapter 6), the number of moves required to solve the famous Tower of Hanoi puzzle with n disks (Chapter 8), and the number of rabbits on an island after n months (Chapter 8).

Check out the puzzles at the OEIS site.

Links

Integer sequences appear in an amazingly wide range of subject areas besides discrete mathematics, including biology, engineering, chemistry, and physics, as well as in puzzles. An amazing database of over 200,000 different integer sequences can be found in the *On-Line Encyclopedia of Integer Sequences (OEIS)*. This database was originated by Neil Sloane in the 1960s. The last printed version of this database was published in 1995 ([SIPI95]); the current encyclopedia would occupy more than 750 volumes of the size of the 1995 book with more than 10,000 new submissions a year. There is also a program accessible via the Web that you can use to find sequences from the encyclopedia that match initial terms you provide.

Summations

Next, we consider the addition of the terms of a sequence. For this we introduce **summation notation**. We begin by describing the notation used to express the sum of the terms

$$a_m, a_{m+1}, \ldots, a_n$$

from the sequence $\{a_n\}$. We use the notation

$$\sum_{j=m}^{n} a_j, \qquad \sum_{j=m}^{n} a_j, \qquad \text{or} \qquad \sum_{m \leq j \leq n} a_j$$

(read as the sum from $j = m$ to $j = n$ of a_j) to represent

$$a_m + a_{m+1} + \cdots + a_n.$$

Here, the variable j is called the **index of summation**, and the choice of the letter j as the variable is arbitrary; that is, we could have used any other letter, such as i or k. Or, in notation,

$$\sum_{j=m}^{n} a_j = \sum_{i=m}^{n} a_i = \sum_{k=m}^{n} a_k.$$

Here, the index of summation runs through all integers starting with its **lower limit** m and ending with its **upper limit** n. A large uppercase Greek letter sigma, $\sum$, is used to denote summation.

The usual laws for arithmetic apply to summations. For example, when a and b are real numbers, we have $\sum_{j=1}^{n}(ax_j + by_j) = a\sum_{y=1}^{n} x_j + b\sum_{j=1}^{n} y_j$, where $x_1, x_2, \ldots, x_n$ and $y_1, y_2, \ldots, y_n$ are real numbers. (We do not present a formal proof of this identity here. Such a proof can be constructed using mathematical induction, a proof method we introduce in Chapter 5. The proof also uses the commutative and associative laws for addition and the distributive law of multiplication over addition.)

We give some examples of summation notation.

EXAMPLE 17   Use summation notation to express the sum of the first 100 terms of the sequence $\{a_j\}$, where $a_j = 1/j$ for $j = 1, 2, 3, \ldots$.

Solution: The lower limit for the index of summation is 1, and the upper limit is 100. We write this sum as

$$\sum_{j=1}^{100} \frac{1}{j}.$$
◀

EXAMPLE 18   What is the value of $\sum_{j=1}^{5} j^2$?

Solution: We have

$$\sum_{j=1}^{5} j^2 = 1^2 + 2^2 + 3^2 + 4^2 + 5^2$$
$$= 1 + 4 + 9 + 16 + 25$$
$$= 55.$$
◀

EXAMPLE 19   What is the value of $\sum_{k=4}^{8} (-1)^k$?

Solution: We have

$$\sum_{k=4}^{8} (-1)^k = (-1)^4 + (-1)^5 + (-1)^6 + (-1)^7 + (-1)^8$$
$$= 1 + (-1) + 1 + (-1) + 1$$
$$= 1.$$
◀

NEIL SLOANE (BORN 1939)   Neil Sloane studied mathematics and electrical engineering at the University of Melbourne on a scholarship from the Australian state telephone company. He mastered many telephone-related jobs, such as erecting telephone poles, in his summer work. After graduating, he designed minimal-cost telephone networks in Australia. In 1962 he came to the United States and studied electrical engineering at Cornell University. His Ph.D. thesis was on what are now called neural networks. He took a job at Bell Labs in 1969, working in many areas, including network design, coding theory, and sphere packing. He now works for AT&T Labs, moving there from Bell Labs when AT&T split up in 1996. One of his favorite problems is the **kissing problem** (a name he coined), which asks how many spheres can be arranged in n dimensions so that they all touch a central sphere of the same size. (In two dimensions the answer is 6, because 6 pennies can be placed so that they touch a central penny. In three dimensions, 12 billiard balls can be placed so that they touch a central billiard ball. Two billiard balls that just touch are said to "kiss," giving rise to the terminology "kissing problem" and "kissing number.") Sloane, together with Andrew Odlyzko, showed that in 8 and 24 dimensions, the optimal kissing numbers are, respectively, 240 and 196,560. The kissing number is known in dimensions 1, 2, 3, 4, 8, and 24, but not in any other dimensions. Sloane's books include *Sphere Packings, Lattices and Groups,* 3d ed., with John Conway; *The Theory of Error-Correcting Codes* with Jessie MacWilliams; *The Encyclopedia of Integer Sequences* with Simon Plouffe (which has grown into the famous OEIS website); and *The Rock-Climbing Guide to New Jersey Crags* with Paul Nick. The last book demonstrates his interest in rock climbing; it includes more than 50 climbing sites in New Jersey.

Sometimes it is useful to shift the index of summation in a sum. This is often done when two sums need to be added but their indices of summation do not match. When shifting an index of summation, it is important to make the appropriate changes in the corresponding summand. This is illustrated by Example 20.

EXAMPLE 20   Suppose we have the sum

$$\sum_{j=1}^{5} j^2$$

but want the index of summation to run between 0 and 4 rather than from 1 to 5. To do this, we let $k = j - 1$. Then the new summation index runs from 0 (because $k = 1 - 0 = 0$ when $j = 1$) to 4 (because $k = 5 - 1 = 4$ when $j = 5$), and the term j^2 becomes $(k + 1)^2$. Hence,

$$\sum_{j=1}^{5} j^2 = \sum_{k=0}^{4} (k + 1)^2.$$

It is easily checked that both sums are $1 + 4 + 9 + 16 + 25 = 55$.   ◀

Sums of terms of geometric progressions commonly arise (such sums are called **geometric series**). Theorem 1 gives us a formula for the sum of terms of a geometric progression.

THEOREM 1   If a and r are real numbers and $r \neq 0$, then

$$\sum_{j=0}^{n} ar^j = \begin{cases} \dfrac{ar^{n+1} - a}{r - 1} & \text{if } r \neq 1 \\[2mm] (n + 1)a & \text{if } r = 1. \end{cases}$$

Proof: Let

$$S_n = \sum_{j=0}^{n} ar^j.$$

To compute S, first multiply both sides of the equality by r and then manipulate the resulting sum as follows:

$$rS_n = r \sum_{j=0}^{n} ar^j \qquad\qquad \text{substituting summation formula for } S$$

$$= \sum_{j=0}^{n} ar^{j+1} \qquad\qquad \text{by the distributive property}$$

$$= \sum_{k=1}^{n+1} ar^k \qquad\qquad \text{shifting the index of summation, with } k = j + 1$$

$$= \left(\sum_{k=0}^{n} ar^k \right) + (ar^{n+1} - a) \qquad \text{removing } k = n + 1 \text{ term and adding } k = 0 \text{ term}$$

$$= S_n + (ar^{n+1} - a) \qquad\qquad \text{substituting } S \text{ for summation formula}$$

From these equalities, we see that

$$r S_n = S_n + (ar^{n+1} - a).$$

Solving for S_n shows that if $r \neq 1$, then

$$S_n = \frac{ar^{n+1} - a}{r - 1}.$$

If $r = 1$, then the $S_n = \sum_{j=0}^{n} ar^j = \sum_{j=0}^{n} a = (n+1)a$. ◁

EXAMPLE 21 Double summations arise in many contexts (as in the analysis of nested loops in computer programs). An example of a double summation is

$$\sum_{i=1}^{4} \sum_{j=1}^{3} ij.$$

To evaluate the double sum, first expand the inner summation and then continue by computing the outer summation:

$$\sum_{i=1}^{4} \sum_{j=1}^{3} ij = \sum_{i=1}^{4} (i + 2i + 3i)$$

$$= \sum_{i=1}^{4} 6i$$

$$= 6 + 12 + 18 + 24 = 60.$$ ◀

We can also use summation notation to add all values of a function, or terms of an indexed set, where the index of summation runs over all values in a set. That is, we write

$$\sum_{s \in S} f(s)$$

to represent the sum of the values $f(s)$, for all members s of S.

EXAMPLE 22 What is the value of $\sum_{s \in \{0,2,4\}} s$?

Solution: Because $\sum_{s \in \{0,2,4\}} s$ represents the sum of the values of s for all the members of the set $\{0, 2, 4\}$, it follows that

$$\sum_{s \in \{0,2,4\}} s = 0 + 2 + 4 = 6.$$ ◀

Certain sums arise repeatedly throughout discrete mathematics. Having a collection of formulae for such sums can be useful; Table 2 provides a small table of formulae for commonly occurring sums.

We derived the first formula in this table in Theorem 1. The next three formulae give us the sum of the first n positive integers, the sum of their squares, and the sum of their cubes. These three formulae can be derived in many different ways (for example, see Exercises 23 and 24). Also note that each of these formulae, once known, can easily be proved using mathematical induction, the subject of Section 5.1. The last two formulae in the table involve infinite series and will be discussed shortly.

Example 23 illustrates how the formulae in Table 2 can be useful.

TABLE 2 Some Useful Summation Formulae.	
Sum	**Closed Form**
$\displaystyle\sum_{k=0}^{n} ar^k \;(r \neq 0)$	$\dfrac{ar^{n+1} - a}{r - 1}, r \neq 1$
$\displaystyle\sum_{k=1}^{n} k$	$\dfrac{n(n + 1)}{2}$
$\displaystyle\sum_{k=1}^{n} k^2$	$\dfrac{n(n + 1)(2n + 1)}{6}$
$\displaystyle\sum_{k=1}^{n} k^3$	$\dfrac{n^2(n + 1)^2}{4}$
$\displaystyle\sum_{k=0}^{\infty} x^k, |x| < 1$	$\dfrac{1}{1 - x}$
$\displaystyle\sum_{k=1}^{\infty} kx^{k-1}, |x| < 1$	$\dfrac{1}{(1 - x)^2}$

EXAMPLE 23 Find $\sum_{k=50}^{100} k^2$.

Solution: First note that because $\sum_{k=1}^{100} k^2 = \sum_{k=1}^{49} k^2 + \sum_{k=50}^{100} k^2$, we have

$$\sum_{k=50}^{100} k^2 = \sum_{k=1}^{100} k^2 - \sum_{k=1}^{49} k^2.$$

Using the formula $\sum_{k=1}^{n} k^2 = n(n + 1)(2n + 1)/6$ from Table 2 (and proved in Exercise 38), we see that

$$\sum_{k=50}^{100} k^2 = \frac{100 \cdot 101 \cdot 201}{6} - \frac{49 \cdot 50 \cdot 99}{6} = 338{,}350 - 40{,}425 = 297{,}925. \qquad \blacktriangleleft$$

SOME INFINITE SERIES   Although most of the summations in this book are finite sums, infinite series are important in some parts of discrete mathematics. Infinite series are usually studied in a course in calculus and even the definition of these series requires the use of calculus, but sometimes they arise in discrete mathematics, because discrete mathematics deals with infinite collections of discrete elements. In particular, in our future studies in discrete mathematics, we will find the closed forms for the infinite series in Examples 24 and 25 to be quite useful.

EXAMPLE 24 (*Requires calculus*) Let x be a real number with $|x| < 1$. Find $\sum_{n=0}^{\infty} x^n$.

Solution: By Theorem 1 with $a = 1$ and $r = x$ we see that $\sum_{n=0}^{k} x^n = \dfrac{x^{k+1} - 1}{x - 1}$. Because $|x| < 1$, x^{k+1} approaches 0 as k approaches infinity. It follows that

$$\sum_{n=0}^{\infty} x^n = \lim_{k \to \infty} \frac{x^{k+1} - 1}{x - 1} = \frac{0 - 1}{x - 1} = \frac{1}{1 - x}. \qquad \blacktriangleleft$$

We can produce new summation formulae by differentiating or integrating existing formulae.

EXAMPLE 25 (*Requires calculus*) Differentiating both sides of the equation

$$\sum_{k=0}^{\infty} x^k = \frac{1}{1-x},$$

from Example 24 we find that

$$\sum_{k=1}^{\infty} k x^{k-1} = \frac{1}{(1-x)^2}.$$

(This differentiation is valid for $|x| < 1$ by a theorem about infinite series.) ◄

Exercises

1. Find these terms of the sequence $\{a_n\}$, where $a_n = 2 \cdot (-3)^n + 5^n$.

   a) a_0 **b)** a_1 **c)** a_4 **d)** a_5

2. What are the terms a_0, a_1, a_2, and a_3 of the sequence $\{a_n\}$, where a_n equals

   a) $2^n + 1$?   **b)** $(n+1)^{n+1}$?

   c) $\lfloor n/2 \rfloor$?   **d)** $\lfloor n/2 \rfloor + \lceil n/2 \rceil$?

3. List the first 10 terms of each of these sequences.

   a) the sequence that begins with 2 and in which each successive term is 3 more than the preceding term

   b) the sequence that lists each positive integer three times, in increasing order

   c) the sequence that lists the odd positive integers in increasing order, listing each odd integer twice

   d) the sequence whose nth term is $n! - 2^n$

   e) the sequence that begins with 3, where each succeeding term is twice the preceding term

   f) the sequence whose first term is 2, second term is 4, and each succeeding term is the sum of the two preceding terms

   g) the sequence whose nth term is the number of bits in the binary expansion of the number n (defined in Section 4.2)

   h) the sequence where the nth term is the number of letters in the English word for the index n

4. Find at least three different sequences beginning with the terms 1, 2, 4 whose terms are generated by a simple formula or rule.

5. Find at least three different sequences beginning with the terms 3, 5, 7 whose terms are generated by a simple formula or rule.

6. Find the first five terms of the sequence defined by each of these recurrence relations and initial conditions.

   a) $a_n = 6a_{n-1}, a_0 = 2$

   b) $a_n = a_{n-1}^2, a_1 = 2$

   c) $a_n = a_{n-1} + 3a_{n-2}, a_0 = 1, a_1 = 2$

   d) $a_n = na_{n-1} + n^2 a_{n-2}, a_0 = 1, a_1 = 1$

   e) $a_n = a_{n-1} + a_{n-3}, a_0 = 1, a_1 = 2, a_2 = 0$

7. Let $a_n = 2^n + 5 \cdot 3^n$ for $n = 0, 1, 2, \ldots$.

   a) Find a_0, a_1, a_2, a_3, and a_4.

   b) Show that $a_2 = 5a_1 - 6a_0, a_3 = 5a_2 - 6a_1$, and $a_4 = 5a_3 - 6a_2$.

   c) Show that $a_n = 5a_{n-1} - 6a_{n-2}$ for all integers n with $n \geq 2$.

8. Show that the sequence $\{a_n\}$ is a solution of the recurrence relation $a_n = -3a_{n-1} + 4a_{n-2}$ if

   a) $a_n = 0$.   **b)** $a_n = 1$.

   c) $a_n = (-4)^n$.   **d)** $a_n = 2(-4)^n + 3$.

9. Find the solution to each of these recurrence relations and initial conditions. Use an iterative approach such as that used in Example 10.

   a) $a_n = 3a_{n-1}, a_0 = 2$

   b) $a_n = a_{n-1} + 2, a_0 = 3$

   c) $a_n = a_{n-1} + n, a_0 = 1$

   d) $a_n = a_{n-1} + 2n + 3, a_0 = 4$

   e) $a_n = 2a_{n-1} - 1, a_0 = 1$

   f) $a_n = 3a_{n-1} + 1, a_0 = 1$

   g) $a_n = na_{n-1}, a_0 = 5$

   h) $a_n = 2na_{n-1}, a_0 = 1$

10. A person deposits $1000 in an account that yields 9% interest compounded annually.

   a) Set up a recurrence relation for the amount in the account at the end of n years.

   b) Find an explicit formula for the amount in the account at the end of n years.

   c) How much money will the account contain after 100 years?

11. Suppose that the number of bacteria in a colony triples every hour.

   a) Set up a recurrence relation for the number of bacteria after n hours have elapsed.

   b) If 100 bacteria are used to begin a new colony, how many bacteria will be in the colony in 10 hours?

12. An employee joined a company in 2009 with a starting salary of $50,000. Every year this employee receives a raise of $1000 plus 5% of the salary of the previous year.

   a) Set up a recurrence relation for the salary of this employee n years after 2009.

   b) What will the salary of this employee be in 2017?

   c) Find an explicit formula for the salary of this employee n years after 2009.

13. Find a recurrence relation for the balance $B(k)$ owed at the end of k months on a loan of \$5000 at a rate of 7% if a payment of \$100 is made each month. [*Hint:* Express $B(k)$ in terms of $B(k - 1)$; the monthly interest is $(0.07/12)B(k - 1)$.]

14. a) Find a recurrence relation for the balance $B(k)$ owed at the end of k months on a loan at a rate of r if a payment P is made on the loan each month. [*Hint:* Express $B(k)$ in terms of $B(k - 1)$ and note that the monthly interest rate is $r/12$.]

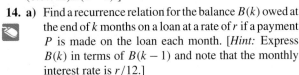

   b) Determine what the monthly payment P should be so that the loan is paid off after T months.

15. For each of these lists of integers, provide a simple formula or rule that generates the terms of an integer sequence that begins with the given list. Assuming that your formula or rule is correct, determine the next three terms of the sequence.

   a) $1, 0, 1, 1, 0, 0, 1, 1, 1, 0, 0, 0, 1, \ldots$
   b) $1, 2, 2, 3, 4, 4, 5, 6, 6, 7, 8, 8, \ldots$
   c) $1, 0, 2, 0, 4, 0, 8, 0, 16, 0, \ldots$
   d) $3, 6, 12, 24, 48, 96, 192, \ldots$
   e) $15, 8, 1, -6, -13, -20, -27, \ldots$
   f) $3, 5, 8, 12, 17, 23, 30, 38, 47, \ldots$
   g) $2, 16, 54, 128, 250, 432, 686, \ldots$
   h) $2, 3, 7, 25, 121, 721, 5041, 40321, \ldots$

**16. Show that if a_n denotes the nth positive integer that is not a perfect square, then $a_n = n + \{\sqrt{n}\}$, where $\{x\}$ denotes the integer closest to the real number x.

17. What are the values of these sums?

   a) $\displaystyle\sum_{k=1}^{5} (k + 1)$ **b)** $\displaystyle\sum_{j=0}^{4} (-2)^j$

   c) $\displaystyle\sum_{i=1}^{10} 3$ **d)** $\displaystyle\sum_{j=0}^{8} (2^{j+1} - 2^j)$

18. What are the values of these sums, where $S = \{1, 3, 5, 7\}$?

   a) $\displaystyle\sum_{j \in S} j$ **b)** $\displaystyle\sum_{j \in S} j^2$

   c) $\displaystyle\sum_{j \in S} (1/j)$ **d)** $\displaystyle\sum_{j \in S} 1$

19. What is the value of each of these sums of terms of a geometric progression?

   a) $\displaystyle\sum_{j=0}^{8} 3 \cdot 2^j$ **b)** $\displaystyle\sum_{j=1}^{8} 2^j$

   c) $\displaystyle\sum_{j=2}^{8} (-3)^j$ **d)** $\displaystyle\sum_{j=0}^{8} 2 \cdot (-3)^j$

20. Compute each of these double sums.

   a) $\displaystyle\sum_{i=1}^{2} \sum_{j=1}^{3} (i + j)$ **b)** $\displaystyle\sum_{i=0}^{2} \sum_{j=0}^{3} (2i + 3j)$

   c) $\displaystyle\sum_{i=1}^{3} \sum_{j=0}^{2} i$ **d)** $\displaystyle\sum_{i=0}^{2} \sum_{j=1}^{3} ij$

21. Show that $\sum_{j=1}^{n}(a_j - a_{j-1}) = a_n - a_0$, where $a_0, a_1, \ldots, a_n$ is a sequence of real numbers. This type of sum is called **telescoping**.

22. Use the identity $1/(k(k + 1)) = 1/k - 1/(k + 1)$ and Exercise 21 to compute $\sum_{k=1}^{n} 1/(k(k + 1))$.

23. Sum both sides of the identity $k^2 - (k - 1)^2 = 2k - 1$ from $k = 1$ to $k = n$ and use Exercise 35 to find

   a) a formula for $\sum_{k=1}^{n}(2k - 1)$ (the sum of the first n odd natural numbers).
   b) a formula for $\sum_{k=1}^{n} k$.

*24. Use the technique given in Exercise 35, together with the result of Exercise 37b, to derive the formula for $\sum_{k=1}^{n} k^2$ given in Table 2. [*Hint:* Take $a_k = k^3$ in the telescoping sum in Exercise 35.]

25. Find $\sum_{k=100}^{200} k$. (Use Table 2.)

26. Find $\sum_{k=99}^{200} k^3$. (Use Table 2.)

*27. Find a formula for $\sum_{k=0}^{m} \lfloor \sqrt{k} \rfloor$, when m is a positive integer.

*28. Find a formula for $\sum_{k=0}^{m} \lfloor \sqrt[3]{k} \rfloor$, when m is a positive integer.

There is also a special notation for products. The product of $a_m, a_{m+1}, \ldots, a_n$ is represented by $\prod_{j=m}^{n} a_j$, read as the product from $j = m$ to $j = n$ of a_j.

29. What are the values of the following products?

   a) $\prod_{i=0}^{10} i$ **b)** $\prod_{i=5}^{8} i$
   c) $\prod_{i=1}^{100} (-1)^i$ **d)** $\prod_{i=1}^{10} 2$

Recall that the value of the factorial function at a positive integer n, denoted by $n!$, is the product of the positive integers from 1 to n, inclusive. Also, we specify that $0! = 1$.

30. Express $n!$ using product notation.

31. Find $\sum_{j=0}^{4} j!$.

32. Find $\prod_{j=0}^{4} j!$.

2.5 Cardinality of Sets

Introduction

In Definition 4 of Section 2.1 we defined the cardinality of a finite set as the number of elements in the set. We use the cardinalities of finite sets to tell us when they have the same size, or when one is bigger than the other. In this section we extend this notion to infinite sets. That is, we will define what it means for two infinite sets to have the same cardinality, providing us with a way to measure the relative sizes of infinite sets.

We will be particularly interested in countably infinite sets, which are sets with the same cardinality as the set of positive integers. We will establish the surprising result that the set of rational numbers is countably infinite. We will also provide an example of an uncountable set when we show that the set of real numbers is not countable.

The concepts developed in this section have important applications to computer science. A function is called uncomputable if no computer program can be written to find all its values, even with unlimited time and memory. We will use the concepts in this section to explain why uncomputable functions exist.

We now define what it means for two sets to have the same size, or cardinality. In Section 2.1, we discussed the cardinality of finite sets and we defined the size, or cardinality, of such sets. In Exercise 53 of Section 2.3 we showed that there is a one-to-one correspondence between any two finite sets with the same number of elements. We use this observation to extend the concept of cardinality to all sets, both finite and infinite.

DEFINITION 1

A set S is finite with cardinality $n \in N$ if there is a bijection rom the set $\{0, 1, \ldots, n - 1\}$ to S. A set is infinite if it is not finte.

THEOREM 1

The set N of natural numbers is an infinite set.

Proof: Consider the injection $f : N \to N$ deined as $f(x) = 3x$. The range of f is a subset of the domain of f.

Some facts which could easily be seen are:

1. If S' is infinite and is a subset of S, S is infinite.
2. Every subset of a finite set is finite.
3. If $F : S \to T$ be an injection and S is infinite, then T is infinite.
4. If S is an infinite set $P(S)$ is infinite.
5. If S and T are infinite sets $S \cup T$ is infinite.
6. If S is infinite and $T \neq \emptyset$ the $S \times T$ is infinite.
7. If S is infinite and $T \neq \emptyset$ the set of functions from T to S is infinite.

DEFINITION 2

The sets A and B have the same *cardinality* if and only if there is a one-to-one correspondence from A to B. When A and B have the same cardinality, we write $|A| = |B|$.

For infinite sets the definition of cardinality provides a relative measure of the sizes of two sets, rather than a measure of the size of one particular set. We can also define what it means for one set to have a smaller cardinality than another set.

DEFINITION 3

If there is a one-to-one function from A to B, the cardinality of A is less than or the same as the cardinality of B and we write $|A| \leq |B|$. Moreover, when $|A| \leq |B|$ and A and B have different cardinality, we say that the cardinality of A is less than the cardinality of B and we write $|A| < |B|$.

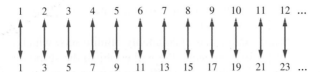

FIGURE 1   A One-to-One Correspondence Between $\mathbf{Z}^+$ and the Set of Odd Positive Integers.

Countable Sets

We will now split infinite sets into two groups, those with the same cardinality as the set of natural numbers and those with a different cardinality.

DEFINITION 4

A set that is either finite or has the same cardinality as the set of positive integers is called *countable*. A set that is not countable is called *uncountable*. When an infinite set S is countable, we denote the cardinality of S by $\aleph_0$ (where $\aleph$ is aleph, the first letter of the Hebrew alphabet). We write $|S| = \aleph_0$ and say that S has cardinality "aleph null."

We illustrate how to show a set is countable in the next example.

EXAMPLE 1   Show that the set of odd positive integers is a countable set.

Solution: To show that the set of odd positive integers is countable, we will exhibit a one-to-one correspondence between this set and the set of positive integers. Consider the function

$$f(n) = 2n - 1$$

from $\mathbf{Z}^+$ to the set of odd positive integers. We show that f is a one-to-one correspondence by showing that it is both one-to-one and onto. To see that it is one-to-one, suppose that $f(n) = f(m)$. Then $2n - 1 = 2m - 1$, so $n = m$. To see that it is onto, suppose that t is an odd positive integer. Then t is 1 less than an even integer $2k$, where k is a natural number. Hence $t = 2k - 1 = f(k)$. We display this one-to-one correspondence in Figure 1.   ◀

An infinite set is countable if and only if it is possible to list the elements of the set in a sequence (indexed by the positive integers). The reason for this is that a one-to-one correspondence f from the set of positive integers to a set S can be expressed in terms of a sequence $a_1, a_2, \ldots, a_n, \ldots$, where $a_1 = f(1)$, $a_2 = f(2), \ldots, a_n = f(n), \ldots$.

You can always get a room at Hilbert's Grand Hotel!

Links

HILBERT'S GRAND HOTEL   We now describe a paradox that shows that something impossible with finite sets may be possible with infinite sets. The famous mathematician David Hilbert invented the notion of the **Grand Hotel**, which has a countably infinite number of rooms, each occupied by a guest. When a new guest arrives at a hotel with a finite number of rooms, and all rooms are occupied, this guest cannot be accommodated without evicting a current guest. However, we can always accommodate a new guest at the Grand Hotel, even when all rooms are already occupied, as we show in Example 2. Exercises 4 and 6 ask you to show that we can accommodate a finite number of new guests and a countable number of new guests, respectively, at the fully occupied Grand Hotel.

FIGURE 2 **A New Guest Arrives at Hilbert's Grand Hotel.**

EXAMPLE 2 How can we accommodate a new guest arriving at the fully occupied Grand Hotel without removing any of the current guests?

Solution: Because the rooms of the Grand Hotel are countable, we can list them as Room 1, Room 2, Room 3, and so on. When a new guest arrives, we move the guest in Room 1 to Room 2, the guest in Room 2 to Room 3, and in general, the guest in Room n to Room $n + 1$, for all positive integers n. This frees up Room 1, which we assign to the new guest, and all the current guests still have rooms. We illustrate this situation in Figure 2. ◀

When there are finitely many room in a hotel, the notion that all rooms are occupied is equivalent to the notion that no new guests can be accommodated. However, Hilbert's paradox of the Grand Hotel can be explained by noting that this equivalence no longer holds when there are infinitely many room.

EXAMPLES OF COUNTABLE AND UNCOUNTABLE SETS We will now show that certain sets of numbers are countable. We begin with the set of all integers. Note that we can show that the set of all integers is countable by listing its members.

EXAMPLE 3 Show that the set of all integers is countable.

Solution: We can list all integers in a sequence by starting with 0 and alternating between positive and negative integers: $0, 1, -1, 2, -2, \ldots$. Alternatively, we could find a one-to-one correspondence between the set of positive integers and the set of all integers. We leave it to the

DAVID HILBERT (1862–1943) Hilbert, born in Königsberg, the city famous in mathematics for its seven bridges, was the son of a judge. During his tenure at Göttingen University, from 1892 to 1930, he made many fundamental contributions to a wide range of mathematical subjects. He almost always worked on one area of mathematics at a time, making important contributions, then moving to a new mathematical subject. Some areas in which Hilbert worked are the calculus of variations, geometry, algebra, number theory, logic, and mathematical physics. Besides his many outstanding original contributions, Hilbert is remembered for his famous list of 23 difficult problems. He described these problems at the 1900 International Congress of Mathematicians, as a challenge to mathematicians at the birth of the twentieth century. Since that time, they have spurred a tremendous amount and variety of research. Although many of these problems have now been solved, several remain open, including the Riemann hypothesis, which is part of Problem 8 on Hilbert's list. Hilbert was also the author of several important textbooks in number theory and geometry.

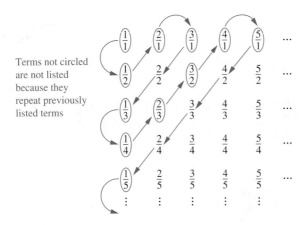

Terms not circled are not listed because they repeat previously listed terms

FIGURE 3   **The Positive Rational Numbers Are Countable.**

reader to show that the function $f(n) = n/2$ when n is even and $f(n) = -(n-1)/2$ when n is odd is such a function. Consequently, the set of all integers is countable.   ◄

It is not surprising that the set of odd integers and the set of all integers are both countable sets (as shown in Examples 1 and 3). Many people are amazed to learn that the set of rational numbers is countable, as Example 4 demonstrates.

EXAMPLE 4   Show that the set of positive rational numbers is countable.

Solution: It may seem surprising that the set of positive rational numbers is countable, but we will show how we can list the positive rational numbers as a sequence $r_1, r_2, \ldots, r_n, \ldots$. First, note that every positive rational number is the quotient p/q of two positive integers. We can arrange the positive rational numbers by listing those with denominator $q = 1$ in the first row, those with denominator $q = 2$ in the second row, and so on, as displayed in Figure 3.

The key to listing the rational numbers in a sequence is to first list the positive rational numbers p/q with $p + q = 2$, followed by those with $p + q = 3$, followed by those with $p + q = 4$, and so on, following the path shown in Figure 3. Whenever we encounter a number p/q that is already listed, we do not list it again. For example, when we come to $2/2 = 1$ we do not list it because we have already listed $1/1 = 1$. The initial terms in the list of positive rational numbers we have constructed are 1, 1/2, 2, 3, 1/3, 1/4, 2/3, 3/2, 4, 5, and so on. These numbers are shown circled; the uncircled numbers in the list are those we leave out because they are already listed. Because all positive rational numbers are listed once, as the reader can verify, we have shown that the set of positive rational numbers is countable.   ◄

An Uncountable Set

Not all infinite sets have the same size!

Links

We have seen that the set of positive rational numbers is a countable set. Do we have a promising candidate for an uncountable set? The first place we might look is the set of real numbers. In Example 5 we use an important proof method, introduced in 1879 by Georg Cantor and known as the **Cantor diagonalization argument**, to prove that the set of real numbers is not countable. This proof method is used extensively in mathematical logic and in the theory of computation.

EXAMPLE 5   Show that the set of real numbers is an uncountable set.

Extra
Examples

Solution: To show that the set of real numbers is uncountable, we suppose that the set of real numbers is countable and arrive at a contradiction. Then, the subset of all real numbers that fall between 0 and 1 would also be countable (because any subset of a countable set is also countable; see Exercise 13). Under this assumption, the real numbers between 0 and 1 can be listed in some order, say, $r_1, r_2, r_3, \ldots$. Let the decimal representation of these real numbers be

$$r_1 = 0.d_{11}d_{12}d_{13}d_{14} \ldots$$
$$r_2 = 0.d_{21}d_{22}d_{23}d_{24} \ldots$$
$$r_3 = 0.d_{31}d_{32}d_{33}d_{34} \ldots$$
$$r_4 = 0.d_{41}d_{42}d_{43}d_{44} \ldots$$
$$\vdots$$

where $d_{ij} \in \{0, 1, 2, 3, 4, 5, 6, 7, 8, 9\}$. (For example, if $r_1 = 0.23794102 \ldots$, we have $d_{11} = 2$, $d_{12} = 3$, $d_{13} = 7$, and so on.) Then, form a new real number with decimal expansion $r = 0.d_1 d_2 d_3 d_4 \ldots$, where the decimal digits are determined by the following rule:

$$d_i = \begin{cases} 4 \text{ if } d_{ii} \neq 4 \\ 5 \text{ if } d_{ii} = 4. \end{cases}$$

(As an example, suppose that $r_1 = 0.23794102 \ldots$, $r_2 = 0.44590138 \ldots$, $r_3 = 0.09118764 \ldots$, $r_4 = 0.80553900 \ldots$, and so on. Then we have $r = 0.d_1 d_2 d_3 d_4 \ldots = 0.4544 \ldots$, where $d_1 = 4$ because $d_{11} \neq 4$, $d_2 = 5$ because $d_{22} = 4$, $d_3 = 4$ because $d_{33} \neq 4$, $d_4 = 4$ because $d_{44} \neq 4$, and so on.)

A number with a decimal expansion that terminates has a second decimal expansion ending with an infinite sequence of 9s because $1 = 0.999 \ldots$.

Every real number has a unique decimal expansion (when the possibility that the expansion has a tail end that consists entirely of the digit 9 is excluded). Therefore, the real number r is not equal to any of $r_1, r_2, \ldots$ because the decimal expansion of r differs from the decimal expansion of r_i in the ith place to the right of the decimal point, for each i.

Because there is a real number r between 0 and 1 that is not in the list, the assumption that all the real numbers between 0 and 1 could be listed must be false. Therefore, all the real numbers between 0 and 1 cannot be listed, so the set of real numbers between 0 and 1 is uncountable. Any set with an uncountable subset is uncountable (see Exercise 13). Hence, the set of real numbers is uncountable.   ◄

RESULTS ABOUT CARDINALITY   We will now discuss some results about the cardinality of sets. First, we will prove that the union of two countable sets is also countable.

THEOREM 2   If A and B are countable sets, then $A \cup B$ is also countable.

This proof uses WLOG and cases.

Proof: Suppose that A and B are both countable sets. Without loss of generality, we can assume that A and B are disjoint. (If they are not, we can replace B by $B - A$, because $A \cap (B - A) = \emptyset$ and $A \cup (B - A) = A \cup B$.) Furthermore, without loss of generality, if one of the two sets is countably infinite and other finite, we can assume that B is the one that is finite.

There are three cases to consider: (i) A and B are both finite, (ii) A is infinite and B is finite, and (iii) A and B are both countably infinite.

Case (i): Note that when A and B are finite, $A \cup B$ is also finite, and therefore, countable.

Case (ii): Because A is countably infinite, its elements can be listed in an infinite sequence $a_1, a_2, a_3, \ldots, a_n, \ldots$ and because B is finite, its terms can be listed as $b_1, b_2, \ldots, b_m$ for some positive integer m. We can list the elements of $A \cup B$ as $b_1, b_2, \ldots, b_m, a_1, a_2, a_3, \ldots, a_n, \ldots$. This means that $A \cup B$ is countably infinite.

Case (iii): Because both A and B are countably infinite, we can list their elements as a_1, a_2, a_3, ..., a_n, ... and b_1, b_2, b_3, ..., b_n, ..., respectively. By alternating terms of these two sequences we can list the elements of $A \cup B$ in the infinite sequence $a_1, b_1, a_2, b_2, a_3, b_3, ..., a_n, b_n, ...$. This means $A \cup B$ must be countably infinite.

We have completed the proof, as we have shown that $A \cup B$ is countable in all three cases.   ◁

Because of its importance, we now state a key theorem in the study of cardinality.

THEOREM 3   **SCHRÖDER-BERNSTEIN THEOREM**   If A and B are sets with $|A| \le |B|$ and $|B| \le |A|$, then $|A| = |B|$. In other words, if there are one-to-one functions f from A to B and g from B to A, then there is a one-to-one correspondence between A and B.

Because Theorem 2 seems to be quite straightforward, we might expect that it has an easy proof. However, even though it can be proved without using advanced mathematics, no known proof is easy to explain. Consequently, we omit a proof here. We refer the interested reader to [AiZiHo09] and [Ve06] for a proof. This result is called the Schröder-Bernstein theorem after Ernst Schröder who published a flawed proof of it in 1898 and Felix Bernstein, a student of Georg Cantor, who presented a proof in 1897. However, a proof of this theorem was found in notes of Richard Dedekind dated 1887. Dedekind was a German mathematician who made important contributions to the foundations of mathematics, abstract algebra, and number theory.

We illustrate the use of Theorem 2 with an example.

EXAMPLE 6   Show that the $|(0, 1)| = |(0, 1]|$.

Solution: It is not at all obvious how to find a one-to-one correspondence between $(0, 1)$ and $(0, 1]$ to show that $|(0, 1)| = |(0, 1]|$. Fortunately, we can use the Schröder-Bernstein theorem instead. Finding a one-to-one function from $(0, 1)$ to $(0, 1]$ is simple. Because $(0, 1) \subset (0, 1]$, $f(x) = x$ is a one-to-one function from $(0, 1)$ to $(0, 1]$. Finding a one-to-one function from $(0, 1]$ to $(0, 1)$ is also not difficult. The function $g(x) = x/2$ is clearly one-to-one and maps $(0, 1]$ to $(0, 1/2] \subset (0, 1)$. As we have found one-to-one functions from $(0, 1)$ to $(0, 1]$ and from $(0, 1]$ to $(0, 1)$, the Schröder-Bernstein theorem tells us that $|(0, 1)| = |(0, 1]|$.   ◀

DEFINITION 5   Let S be a set. An enumeration of S is a surjective function f from an initial segment of N to S. If f is injective also (and hence bijective), then f is an enumeration without repetitions. If f is not injective, then f is an enumeration with repetitions.

EXAMPLE 7   1. If $S = \{\alpha, \beta, \gamma, \delta\}$, then $< \alpha, \beta, \gamma, \delta >$ is an enumeration. $< \alpha, \gamma, \beta, \beta, \delta, \alpha >$ is an enumeration with repetition. $< \gamma, \alpha, \delta, \beta >$ is an enumeration without repetition.

2. Let S be the set of natural numbers of the form $3n$, where $n \in N$. $< 0, 3, 6, 9, \ldots >$ is an enumeration defined by $f(n) = 3n$ $< 6, 3, 0, 15, 12, 9, \ldots >$ is also an enumeration defined by

$$f(n) = 3n + 6 \text{ if } n = 3k \text{ for some } k$$

$$f(n) = 3n \text{ if } n = 3k + 1 \text{ for some } k \in N$$

$$f(n) = 3n - 6 \text{ if } n = 3k + 2 \text{ for some } k \in N$$

It is straightforward to see that a set S is countable if and only if there exists an enumeration of S.

◄

EXAMPLE 8   Let $\Sigma = \{a, b\}$. The set of strings over Σ^* is a countably infinite set. An enumeration τ of Σ^* is possible. $\tau(0) = \lambda$, the empty string. Assuming a comes before b in Σ. $\tau(1) = a$ and $\tau(2) = b$. $\tau(3)$ to $\tau(6)$ are strings of length 3 and so on. What is the 49th string in the enumeration?

$$\tau(0) \text{ --string of length 0}$$

$$\tau(1) - \tau(2) \text{ --string of length 1}$$

$$\tau(3) - \tau(6) \text{ --string of length 2}$$

$$\tau(7) - \tau(14) \text{ --string of length 3}$$

$$\tau(15) - \tau(30) \text{ --string of length 4}$$

$$\tau(31) - \tau(62) \text{ --string of length 5}$$

Hence the 49th string is of length 5. Among these $\tau(31)$ to $\tau(46)$ begin with a and $\tau(47)$ to $\tau(62)$ begin with b.

$$\tau(47) = b\,a\,a\,a$$

$$\tau(48) = b\,a\,a\,b$$

$$\tau(49) = b\,a\,b\,a$$

Hence the 49th string in the enumeration is $b\,a\,b\,a$.

◄

THEOREM 4   Every infinite set contains a countably infinite subset.

Proof: Let S be an infinite set. Select a sequence of elements $< a_0, a_1, \ldots >$ from S as follows:

Select a_0 from S

Select a_1 from $S - \{a_0\}$

Select a_2 from $S - \{a_0, a_1\}$

and so on. a_{i+1} will be selected from $S - \{a_0, a_1, \ldots, a_i\}$. Each $S - \{a_0, \ldots, a_i\}$ is infinite. Otherwise $(S - \{a_0, \ldots, a_i\}) \cup \{a_0, \ldots, a_i\} = S$ will be finite. This set $\{a_0, \ldots, a_i, \ldots\}$ is countable infinite subset of S.

THEOREM 5   The union of a countable collection of countable sets is countable.

Proof: Let S_0, S_1, S_2, ... be a countable collection of countable sets and $S_i = <a_{i0}, a_{i1}, a_{i2}, ... >$

$$
\begin{array}{llll}
S_0 : & (a_{00} & a_{01} & a_{02} & a_{03}...) \\
S_1 : & (a_{10} & a_{11} & a_{12} & a_{13}...) \\
S_2 : & (a_{20} & a_{21} & a_{22} & a_{23}...) \\
S_3 : & (a_{30} & a_{31} & a_{32} & a_{33}...)
\end{array}
$$

Enumerate the elements as shown in the diagram.
Let b_0, b_1, ... be the sequence of elements.

$b_0 = a_{00}$

$b_1 = a_{01}$

$b_2 = a_{10}$

$b_3 = a_{02}$

$b_4 = a_{11}$

$b_5 = a_{20}$ and so on.

This is an enumeration of the elements of $\bigcup_{i=0}^{\infty} A_i$ and hence $\bigcup_{i=0}^{\infty} A_i$ is countable.

THEOREM 6   Let Σ be a finite alphabet and Σ^* the set of all strings over Σ. Then $\mathcal{P}(\Sigma^*)$ is uncountable.

Proof: Let $< x_0, x_1, x_2, ... >$ be an enumeration of strings in Σ^* and if possible let $< A_0, A_1, ... >$ be an enumeration of the sets in $\mathcal{P}(\Sigma^*)$. Construct a binary matrix M as follows.

	x_0	x_1	x_2	$\cdots$
A_0	a_{00}	a_{01}	a_{02}	$\cdots$
A_1	a_{10}	a_{11}	a_{12}	$\cdots$
A_2	a_{20}	a_{21}	a_{22}	$\cdots$
$\vdots$	$\vdots$	$\vdots$	$\vdots$	

$a_{ij} = 1$ if $x_j \in A_i$ and $a_{ij} = 0$ if $x_j \notin A_i$. Now we construct a set A as follows $x_i \in A$ if $a_{ii} = 0$ (i.e., $x_i \notin A_i$, $A = \{x_i | x_i \notin A_i, i \in N\}$).

Now we can see that A cannot be any A_j and hence cannot appear in the enumeration $< A_0, A_1, ... >$ even though $A \in \mathcal{P}(\Sigma^*)$. This is seen by the following contradiction. Suppose if possible let $A = A_j$.

A consists of strings x_j such that $a_{jj} = 0$. So if $x_j \in A$, then $a_{jj} = 0$ and by the way we constructed the matrix M, if $a_{jj} = 0$ $x_j \notin A_j$. Hence $A \neq A_j$. Hence we cannot have an enumeration of the sets in $\mathcal{P}(\Sigma^*)$ and so $\mathcal{P}(\Sigma^*)$ is an uncountable set.

Can we compare the cardinality of uncountable sets? We have seen that the set $[0, 1]$ (real numbers between 0 and 1) and $\mathcal{P}(\Sigma^*)$ are uncountable sets. Can we look at them as having the same cardinality? Are there uncountable sets X_1 and X_2 such that $|X_1| \leq |X_2|$?

DEFINITION 6

A set S is of cardinality c if there is a bijection from $[0, 1]$ to S. The label c is given to this because of the fact that the set $[0, 1]$ is often called a continuum.

DEFINITION 7

Let S and T be sets. Then S and T are equipotent or have the same cardinality, denoted by $|S| = |T|$, if there is a bijection from S to T.

The following result follows immediately.

THEOREM 7

Equipotence is an equivalence relation over any collection of sets.

DEFINITION 8

The cardinality of S is no greater than the cardinality of T, denoted as $|S| \leq |T|$, if there is an injection from S to T. The cardinality of S is less than the cardinality of T, written $|S| < |T|$, if there exists an injection but no bijection from S to T.

It is easy to see that if S and T are sets, then exactly one of the following three conditions will hold:

1. $|S| < |T|$
2. $|S| = |T|$
3. $|T| < |S|$

Also we can see that if $|S| \leq |T|$ and $|T| \leq |S|$, then $|T| = |S|$.

THEOREM 8

Let S be a finite set. Then $|S| < \aleph_0 < c$.

Proof: Let $|S| = n$. Let $Z_n = \{0, \ldots, n-1\}$, $|S| = |Z_n|$. Then there is an injection $f : Z_n \to N$ defined as $f(x) = x$. But we cannot have a bijection from $N \to S$. Hence $|S| < |N|$ and thus $|S| < \aleph_0$.

We have also seen that there cannot be a bijection from $[0, 1]$ to N. But we can have an injection from N to $[0, 1]$ defined as

$$f(n) = \frac{1}{n + 2}.$$

Hence $|N| < |[0, 1]|$ or $\aleph_0 < c$.

THEOREM 9   If S is an infinite set, then $\aleph_0 \leq |S|$.

Proof: We have already seen that if S is infinite, then S contain a countably infinite subset S'. Since the mapping f defined as $f : S' \to S$

$$f(x) = x \text{ for all } x \in S'$$

is an injection from A' to A, we conclude that $|S'| \leq |S|$. $|S'| = \aleph_0$. Hence $\aleph_0 \leq |S|$.

This leads us to the famous continuum hypothesis, which asserts that there is no cardinal number between $\aleph_0$ and c. In other words, the **continuum hypothesis** states that there is no set A such that $\aleph_0$, the cardinality of the set of positive integers, is less than $|A|$ and $|A|$ is less than c, the cardinality of real numbers.

THEOREM 10   Let S be a set. Then $|S| < |\mathcal{P}(S)|$.

Proof: Consider the injection $: S \to \mathcal{P}(S)$ defined as $f(a) = \{a\}$ for all $a \in S$. Since we have such an injection

$$|S| \leq |\mathcal{P}(S)|.$$

Next we show $|S| \neq |\mathcal{P}(S)|$. Let g be an arbitrary function from S to $\mathcal{P}(S)$. We can show that g is not surjective and hence not bijective.

The function g maps each element of S to a subset of S. An element x may or may not be in the subset defined by $g(x)$. Now let us define a subset A of S as follows:

$$A = \{x \mid x \notin g(x)\}$$

A is a subset of S and hence $A \in \mathcal{P}(S)$.
We show that for no a in S, $g(a) = A$.
Suppose for some a, $g(a) = A$, then

$$a \in A \leftrightarrow a \in \{x \mid x \notin g(x)\}$$
$$\leftrightarrow a \notin g(a)$$
$$\leftrightarrow a \notin A$$

and we arrive at a contradiction and the assumption that $g(a) = A$ for some a in S is not correct. It follows that g is not surjective and hence not bijective. We have chosen g arbitrarily and hence we can conclude that no bijection exists from S to $\mathcal{P}(S)$ and therefore $|S| \neq |\mathcal{P}(S)|$.

Hence $|S| < |\mathcal{P}(S)|$.

Using the above result, we see that we have an infinite hierarchy of uncountable sets (i.e., infinite sequence of cardinal numbers, each of which is smaller than the next in the sequence).

$$\aleph_0 = |N| < |\mathcal{P}(N)| < |\mathcal{P}(\mathcal{P}(N))| < \cdots$$

The **continuum hypothesis** says that there is no cardinal number X between X_0 and c. It was stated by Cantor in 1877. He labored unsuccessfully to prove it, becoming extremely dismayed that he could not. By 1900, settling the continuum hypothesis was considered to be among the most important unsolved problems in mathematics. It was the first problem posed by David Hilbert in his famous 1900 list of open problems in mathematics.

The continuum hypothesis is still an open question and remains an area for active research. However, it has been shown that it can be neither proved nor disproved under the standard set theory axioms in modern mathematics, the Zermelo-Fraenkel axioms. The Zermelo-Fraenkel axioms were formulated to avoid the paradoxes of naive set theory, such as Russell's paradox, but there is much controversy whether they should be replaced by some other set of axioms for set theory.

UNCOMPUTABLE FUNCTIONS   We will now describe an important application of the concepts of this section to computer science. In particular, we will show that there are functions whose values cannot be computed by any computer program.

DEFINITION 9   We say that a function is **computable** if there is a computer program in some programming language that finds the values of this function. If a function is not computable we say it is **uncomputable**.

To show that there are uncomputable functions, we need to establish two results. First, we need to show that the set of all computer programs in any particular programming language is countable. This can be proved by noting that a computer programs in a particular language can be thought of as a string of characters from a finite alphabet (see Exercise 29). Next, we show that there are uncountably many different functions from a particular countably infinite set to itself. In particular, Exercise 30 shows that the set of functions from the set of positive integers to itself is uncountable. This is a consequence of the uncountability of the real numbers between 0 and 1 (see Example 5). Putting these two results together shows that there are uncomputable functions.

Exercises

1. Determine whether each of these sets is finite, countably infinite, or uncountable. For those that are countably infinite, exhibit a one-to-one correspondence between the set of positive integers and that set.

a) the negative integers

b) the even integers

c) the integers less than 100

d) the real numbers between 0 and $\frac{1}{2}$

e) the positive integers less than 1,000,000,000

f) the integers that are multiples of 7

2. Determine whether each of these sets is finite, countably infinite, or uncountable. For those that are countably infinite, exhibit a one-to-one correspondence between the set of positive integers and that set.

a) the integers greater than 10

b) the odd negative integers

c) the integers with absolute value less than 1,000,000

d) the real numbers between 0 and 2

e) the set $A \times \mathbf{Z}^+$ where $A = \{2, 3\}$

f) the integers that are multiples of 10

3. Determine whether each of these sets is countable or uncountable. For those that are countably infinite, exhibit a one-to-one correspondence between the set of positive integers and that set.

a) all bit strings not containing the bit 0

b) all positive rational numbers that cannot be written with denominators less than 4

c) the real numbers not containing 0 in their decimal representation

d) the real numbers containing only a finite number of 1s in their decimal representation

4. Show that a finite group of guests arriving at Hilbert's fully occupied Grand Hotel can be given rooms without evicting any current guest.

5. Suppose that Hilbert's Grand Hotel is fully occupied on the day the hotel expands to a second building which also contains a countably infinite number of rooms. Show that the current guests can be spread out to fill every room of the two buildings of the hotel.

6. Show that a countably infinite number of guests arriving at Hilbert's fully occupied Grand Hotel can be given rooms without evicting any current guest.

*7. Suppose that a countably infinite number of buses, each containing a countably infinite number of guests, arrive at Hilbert's fully occupied Grand Hotel. Show that all the arriving guests can be accommodated without evicting any current guest.

8. Give an example of two uncountable sets A and B such that $A - B$ is

a) finite.

b) countably infinite.

c) uncountable.

9. Give an example of two uncountable sets A and B such that $A \cap B$ is

a) finite.

b) countably infinite.

c) uncountable.

10. Show that if A and B are sets and $A \subset B$ then $|A| \leq |B|$.

11. Explain why the set A is countable if and only if $|A| \leq |\mathbf{Z}^+|$.

12. Show that if A and B are sets with the same cardinality, then $|A| \leq |B|$ and $|B| \leq |A|$.

13. Show that if A and B are sets, A is uncountable, and $A \subseteq B$, then B is uncountable.

14. Show that a subset of a countable set is also countable.

15. If A is an uncountable set and B is a countable set, must $A - B$ be uncountable?

16. Show that if A and B are sets $|A| = |B|$, then $|\mathcal{P}(A)| = |\mathcal{P}(B)|$.

17. Show that if A is an infinite set, then it contains a countably infinite subset.

18. Show that there is no infinite set A such that $|A| < |\mathbf{Z}^+| = \aleph_0$.

19. Prove that if it is possible to label each element of an infinite set S with a finite string of keyboard characters, from a finite list characters, where no two elements of S have the same label, then S is a countably infinite set.

20. Use Exercise 19 to provide a proof different from that in the text that the set of rational numbers is countable. [*Hint:* Show that you can express a rational number as a string of digits with a slash and possibly a minus sign.]

*21. Show that the union of a countable number of countable sets is countable.

*22. Show that the set of real numbers that are solutions of quadratic equations $ax^2 + bx + c = 0$, where a, b, and c are integers, is countable.

*23. Show that $\mathbf{Z}^+ \times \mathbf{Z}^+$ is countable by showing that the polynomial function $f : \mathbf{Z}^+ \times \mathbf{Z}^+ \to \mathbf{Z}^+$ with $f(m, n) = (m + n - 2)(m + n - 1)/2 + m$ is one-to-one and onto.

*24. Show that when you substitute $(3n + 1)^2$ for each occurrence of n and $(3m + 1)^2$ for each occurrence of m in the right-hand side of the formula for the function $f(m, n)$ in Exercise 31, you obtain a one-to-one polynomial function $\mathbf{Z} \times \mathbf{Z} \to \mathbf{Z}$. It is an open question whether there is a one-to-one polynomial function $\mathbf{Q} \times \mathbf{Q} \to \mathbf{Q}$.

25. Use the Schröder-Bernstein theorem to show that $(0, 1)$ and $[0, 1]$ have the same cardinality

26. Show that $(0, 1)$ and $\mathbf{R}$ have the same cardinality. [*Hint:* Use the Schröder-Bernstein theorem.]

27. Show that there is no one-to-one correspondence from the set of positive integers to the power set of the set of positive integers. [*Hint:* Assume that there is such a one-to-one correspondence. Represent a subset of the set of positive integers as an infinite bit string with ith bit 1 if i belongs to the subset and 0 otherwise. Suppose that you can list these infinite strings in a sequence indexed by the positive integers. Construct a new bit string with its ith bit equal to the complement of the ith bit of the ith string in the list. Show that this new bit string cannot appear in the list.]

*28. Show that there is a one-to-one correspondence from the set of subsets of the positive integers to the set real numbers between 0 and 1. Use this result and Exercises 26 and 27 to conclude that $\aleph_0 < |\mathcal{P}(\mathbf{Z}^+)| = |\mathbf{R}|$. [*Hint:* Look at the first part of the hint for Exercise 27.]

*29. Show that the set of all computer programs in a particular programming language is countable. [*Hint:* A computer program written in a programming language can be thought of as a string of symbols from a finite alphabet.]

*30. Show that the set of functions from the positive integers to the set $\{0, 1, 2, 3, 4, 5, 6, 7, 8, 9\}$ is uncountable. [*Hint:* First set up a one-to-one correspondence between the set of real numbers between 0 and 1 and a subset of these functions. Do this by associating to the real number $0.d_1 d_2 \ldots d_n \ldots$ the function f with $f(n) = d_n$.]

2.6   Matrices

Introduction

Matrices are used throughout discrete mathematics to express relationships between elements in sets. In subsequent chapters we will use matrices in a wide variety of models. For instance, matrices will be used in models of communications networks and transportation systems. Many algorithms will be developed that use these matrix models. This section reviews matrix arithmetic that will be used in these algorithms.

DEFINITION 1   A *matrix* is a rectangular array of numbers. A matrix with m rows and n columns is called an $m \times n$ matrix. The plural of matrix is *matrices*. A matrix with the same number of rows as columns is called *square*. Two matrices are *equal* if they have the same number of rows and the same number of columns and the corresponding entries in every position are equal.

EXAMPLE 1   The matrix $\begin{bmatrix} 1 & 1 \\ 0 & 2 \\ 1 & 3 \end{bmatrix}$ is a 3×2 matrix.   ◀

We now introduce some terminology about matrices. Boldface uppercase letters will be used to represent matrices.

DEFINITION 2   Let m and n be positive integers and let

$$\mathbf{A} = \begin{bmatrix} a_{11} & a_{12} & \cdots & a_{1n} \\ a_{21} & a_{22} & \cdots & a_{2n} \\ \cdot & \cdot & & \cdot \\ \cdot & \cdot & & \cdot \\ \cdot & \cdot & & \cdot \\ a_{m1} & a_{m2} & \cdots & a_{mn} \end{bmatrix}.$$

The ith *row* of $\mathbf{A}$ is the $1 \times n$ matrix $[a_{i1}, a_{i2}, \ldots, a_{in}]$. The jth *column* of $\mathbf{A}$ is the $m \times 1$ matrix

$$\begin{bmatrix} a_{1j} \\ a_{2j} \\ \cdot \\ \cdot \\ \cdot \\ a_{mj} \end{bmatrix}.$$

The (i, j)th *element* or *entry* of $\mathbf{A}$ is the element a_{ij}, that is, the number in the ith row and jth column of $\mathbf{A}$. A convenient shorthand notation for expressing the matrix $\mathbf{A}$ is to write $\mathbf{A} = [a_{ij}]$, which indicates that $\mathbf{A}$ is the matrix with its (i, j)th element equal to a_{ij}.

Matrix Arithmetic

The basic operations of matrix arithmetic will now be discussed, beginning with a definition of matrix addition.

DEFINITION 3   Let $\mathbf{A} = [a_{ij}]$ and $\mathbf{B} = [b_{ij}]$ be $m \times n$ matrices. The *sum* of $\mathbf{A}$ and $\mathbf{B}$, denoted by $\mathbf{A} + \mathbf{B}$, is the $m \times n$ matrix that has $a_{ij} + b_{ij}$ as its (i, j)th element. In other words, $\mathbf{A} + \mathbf{B} = [a_{ij} + b_{ij}]$.

The sum of two matrices of the same size is obtained by adding elements in the corresponding positions. Matrices of different sizes cannot be added, because the sum of two matrices is defined only when both matrices have the same number of rows and the same number of columns.

EXAMPLE 2

We have $\begin{bmatrix} 1 & 0 & -1 \\ 2 & 2 & -3 \\ 3 & 4 & 0 \end{bmatrix} + \begin{bmatrix} 3 & 4 & -1 \\ 1 & -3 & 0 \\ -1 & 1 & 2 \end{bmatrix} = \begin{bmatrix} 4 & 4 & -2 \\ 3 & -1 & -3 \\ 2 & 5 & 2 \end{bmatrix}.$ ◀

We now discuss matrix products. A product of two matrices is defined only when the number of columns in the first matrix equals the number of rows of the second matrix.

DEFINITION 4   Let $\mathbf{A}$ be an $m \times k$ matrix and $\mathbf{B}$ be a $k \times n$ matrix. The *product* of $\mathbf{A}$ and $\mathbf{B}$, denoted by $\mathbf{AB}$, is the $m \times n$ matrix with its (i, j)th entry equal to the sum of the products of the corresponding elements from the ith row of $\mathbf{A}$ and the jth column of $\mathbf{B}$. In other words, if $\mathbf{AB} = [c_{ij}]$, then

$$c_{ij} = a_{i1}b_{1j} + a_{i2}b_{2j} + \cdots + a_{ik}b_{kj}.$$

In Figure 1 the colored row of $\mathbf{A}$ and the colored column of $\mathbf{B}$ are used to compute the element c_{ij} of $\mathbf{AB}$. The product of two matrices is not defined when the number of columns in the first matrix and the number of rows in the second matrix are not the same.

We now give some examples of matrix products.

EXAMPLE 3   Let

$$\mathbf{A} = \begin{bmatrix} 1 & 0 & 4 \\ 2 & 1 & 1 \\ 3 & 1 & 0 \\ 0 & 2 & 2 \end{bmatrix} \quad \text{and} \quad \mathbf{B} = \begin{bmatrix} 2 & 4 \\ 1 & 1 \\ 3 & 0 \end{bmatrix}.$$

Find $\mathbf{AB}$ if it is defined.

$$\begin{bmatrix} a_{11} & a_{12} & \cdots & a_{1k} \\ a_{21} & a_{22} & \cdots & a_{2k} \\ \vdots & \vdots & & \vdots \\ a_{i1} & a_{i2} & \cdots & a_{ik} \\ \vdots & \vdots & & \vdots \\ a_{m1} & a_{m2} & \cdots & a_{mk} \end{bmatrix} \begin{bmatrix} b_{11} & b_{12} & \cdots & b_{1j} & \cdots & b_{1n} \\ b_{21} & b_{22} & \cdots & b_{2j} & \cdots & b_{2n} \\ \vdots & \vdots & & \vdots & & \vdots \\ b_{k1} & b_{k2} & \cdots & b_{kj} & \cdots & b_{kn} \end{bmatrix} = \begin{bmatrix} c_{11} & c_{12} & \cdots & c_{1n} \\ c_{21} & c_{22} & \cdots & c_{2n} \\ \vdots & \vdots & c_{ij} & \vdots \\ c_{m1} & c_{m2} & \cdots & c_{mn} \end{bmatrix}$$

FIGURE 1   The Product of $\mathbf{A} = [a_{ij}]$ and $\mathbf{B} = [b_{ij}]$.

Extra Examples

Solution: Because **A** is a 4 × 3 matrix and **B** is a 3 × 2 matrix, the product **AB** is defined and is a 4 × 2 matrix. To find the elements of **AB**, the corresponding elements of the rows of **A** and the columns of **B** are first multiplied and then these products are added. For instance, the element in the (3, 1)th position of **AB** is the sum of the products of the corresponding elements of the third row of **A** and the first column of **B**; namely, $3 \cdot 2 + 1 \cdot 1 + 0 \cdot 3 = 7$. When all the elements of **AB** are computed, we see that

$$\mathbf{AB} = \begin{bmatrix} 14 & 4 \\ 8 & 9 \\ 7 & 13 \\ 8 & 2 \end{bmatrix}.$$

◀

Matrix multiplication is *not* commutative. That is, if **A** and **B** are two matrices, it is not necessarily true that **AB** and **BA** are the same. In fact, it may be that only one of these two products is defined. For instance, if **A** is 2 × 3 and **B** is 3 × 4, then **AB** is defined and is 2 × 4; however, **BA** is not defined, because it is impossible to multiply a 3 × 4 matrix and a 2 × 3 matrix.

In general, suppose that **A** is an $m \times n$ matrix and **B** is an $r \times s$ matrix. Then **AB** is defined only when $n = r$ and **BA** is defined only when $s = m$. Moreover, even when **AB** and **BA** are both defined, they will not be the same size unless $m = n = r = s$. Hence, if both **AB** and **BA** are defined and are the same size, then both **A** and **B** must be square and of the same size. Furthermore, even with **A** and **B** both $n \times n$ matrices, **AB** and **BA** are not necessarily equal, as Example 4 demonstrates.

EXAMPLE 4 Let

$$\mathbf{A} = \begin{bmatrix} 1 & 1 \\ 2 & 1 \end{bmatrix} \qquad \text{and} \qquad \mathbf{B} = \begin{bmatrix} 2 & 1 \\ 1 & 1 \end{bmatrix}.$$

Does **AB = BA**?

Solution: We find that

$$\mathbf{AB} = \begin{bmatrix} 3 & 2 \\ 5 & 3 \end{bmatrix} \qquad \text{and} \qquad \mathbf{BA} = \begin{bmatrix} 4 & 3 \\ 3 & 2 \end{bmatrix}.$$

Hence, **AB ≠ BA**.

◀

Transposes and Powers of Matrices

We now introduce an important matrix with entries that are zeros and ones.

DEFINITION 5   The *identity matrix of order n* is the $n \times n$ matrix $\mathbf{I}_n = [\delta_{ij}]$, where $\delta_{ij} = 1$ if $i = j$ and $\delta_{ij} = 0$ if $i \neq j$. Hence

$$\mathbf{I}_n = \begin{bmatrix} 1 & 0 & \dots & 0 \\ 0 & 1 & \dots & 0 \\ \cdot & \cdot & & \cdot \\ \cdot & \cdot & & \cdot \\ \cdot & \cdot & & \cdot \\ 0 & 0 & \dots & 1 \end{bmatrix}.$$

Multiplying a matrix by an appropriately sized identity matrix does not change this matrix. In other words, when $\mathbf{A}$ is an $m \times n$ matrix, we have

$$\mathbf{AI}_n = \mathbf{I}_m\mathbf{A} = \mathbf{A}.$$

Powers of square matrices can be defined. When $\mathbf{A}$ is an $n \times n$ matrix, we have

$$\mathbf{A}^0 = \mathbf{I}_n, \qquad \mathbf{A}^r = \underbrace{\mathbf{AAA}\cdots\mathbf{A}}_{r \text{ times}}.$$

The operation of interchanging the rows and columns of a square matrix arises in many contexts.

DEFINITION 6   Let $\mathbf{A} = [a_{ij}]$ be an $m \times n$ matrix. The *transpose* of $\mathbf{A}$, denoted by $\mathbf{A}^t$, is the $n \times m$ matrix obtained by interchanging the rows and columns of $\mathbf{A}$. In other words, if $\mathbf{A}^t = [b_{ij}]$, then $b_{ij} = a_{ji}$ for $i = 1, 2, \ldots, n$ and $j = 1, 2, \ldots, m$.

EXAMPLE 5   The transpose of the matrix $\begin{bmatrix} 1 & 2 & 3 \\ 4 & 5 & 6 \end{bmatrix}$ is the matrix $\begin{bmatrix} 1 & 4 \\ 2 & 5 \\ 3 & 6 \end{bmatrix}$.   ◀

Matrices that do not change when their rows and columns are interchanged are often important.

DEFINITION 7   A square matrix $\mathbf{A}$ is called *symmetric* if $\mathbf{A} = \mathbf{A}^t$. Thus $\mathbf{A} = [a_{ij}]$ is symmetric if $a_{ij} = a_{ji}$ for all i and j with $1 \le i \le n$ and $1 \le j \le n$.

Note that a matrix is symmetric if and only if it is square and it is symmetric with respect to its main diagonal (which consists of entries that are in the ith row and ith column for some i). This symmetry is displayed in Figure 2.

EXAMPLE 6   The matrix $\begin{bmatrix} 1 & 1 & 0 \\ 1 & 0 & 1 \\ 0 & 1 & 0 \end{bmatrix}$ is symmetric.   ◀

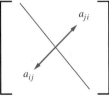

FIGURE 2   A Symmetric Matrix.

Zero–One Matrices

A matrix all of whose entries are either 0 or 1 is called a **zero–one matrix**. Zero–one matrices are often used to represent discrete structures, as we will see in Chapters 9 and 10. Algorithms using these structures are based on Boolean arithmetic with zero–one matrices. This arithmetic is based on the Boolean operations $\wedge$ and $\vee$, which operate on pairs of bits, defined by

$$b_1 \wedge b_2 = \begin{cases} 1 & \text{if } b_1 = b_2 = 1 \\ 0 & \text{otherwise,} \end{cases}$$

$$b_1 \vee b_2 = \begin{cases} 1 & \text{if } b_1 = 1 \text{ or } b_2 = 1 \\ 0 & \text{otherwise.} \end{cases}$$

DEFINITION 8   Let $\mathbf{A} = [a_{ij}]$ and $\mathbf{B} = [b_{ij}]$ be $m \times n$ zero–one matrices. Then the *join* of $\mathbf{A}$ and $\mathbf{B}$ is the zero–one matrix with (i, j)th entry $a_{ij} \vee b_{ij}$. The join of $\mathbf{A}$ and $\mathbf{B}$ is denoted by $\mathbf{A} \vee \mathbf{B}$. The *meet* of $\mathbf{A}$ and $\mathbf{B}$ is the zero–one matrix with (i, j)th entry $a_{ij} \wedge b_{ij}$. The meet of $\mathbf{A}$ and $\mathbf{B}$ is denoted by $\mathbf{A} \wedge \mathbf{B}$.

EXAMPLE 7   Find the join and meet of the zero–one matrices

$$\mathbf{A} = \begin{bmatrix} 1 & 0 & 1 \\ 0 & 1 & 0 \end{bmatrix}, \qquad \mathbf{B} = \begin{bmatrix} 0 & 1 & 0 \\ 1 & 1 & 0 \end{bmatrix}.$$

Solution: We find that the join of $\mathbf{A}$ and $\mathbf{B}$ is

$$\mathbf{A} \vee \mathbf{B} = \begin{bmatrix} 1 \vee 0 & 0 \vee 1 & 1 \vee 0 \\ 0 \vee 1 & 1 \vee 1 & 0 \vee 0 \end{bmatrix} = \begin{bmatrix} 1 & 1 & 1 \\ 1 & 1 & 0 \end{bmatrix}.$$

The meet of $\mathbf{A}$ and $\mathbf{B}$ is

$$\mathbf{A} \wedge \mathbf{B} = \begin{bmatrix} 1 \wedge 0 & 0 \wedge 1 & 1 \wedge 0 \\ 0 \wedge 1 & 1 \wedge 1 & 0 \wedge 0 \end{bmatrix} = \begin{bmatrix} 0 & 0 & 0 \\ 0 & 1 & 0 \end{bmatrix}.$$

◀

We now define the **Boolean product** of two matrices.

DEFINITION 9   Let $\mathbf{A} = [a_{ij}]$ be an $m \times k$ zero–one matrix and $\mathbf{B} = [b_{ij}]$ be a $k \times n$ zero–one matrix. Then the *Boolean product* of $\mathbf{A}$ and $\mathbf{B}$, denoted by $\mathbf{A} \odot \mathbf{B}$, is the $m \times n$ matrix with (i, j)th entry c_{ij} where

$$c_{ij} = (a_{i1} \wedge b_{1j}) \vee (a_{i2} \wedge b_{2j}) \vee \cdots \vee (a_{ik} \wedge b_{kj}).$$

Note that the Boolean product of $\mathbf{A}$ and $\mathbf{B}$ is obtained in an analogous way to the ordinary product of these matrices, but with addition replaced with the operation $\vee$ and with multiplication replaced with the operation $\wedge$. We give an example of the Boolean products of matrices.

EXAMPLE 8   Find the Boolean product of $\mathbf{A}$ and $\mathbf{B}$, where

$$\mathbf{A} = \begin{bmatrix} 1 & 0 \\ 0 & 1 \\ 1 & 0 \end{bmatrix}, \qquad \mathbf{B} = \begin{bmatrix} 1 & 1 & 0 \\ 0 & 1 & 1 \end{bmatrix}.$$

Solution: The Boolean product $\mathbf{A} \odot \mathbf{B}$ is given by

$$\mathbf{A} \odot \mathbf{B} = \begin{bmatrix} (1 \wedge 1) \vee (0 \wedge 0) & (1 \wedge 1) \vee (0 \wedge 1) & (1 \wedge 0) \vee (0 \wedge 1) \\ (0 \wedge 1) \vee (1 \wedge 0) & (0 \wedge 1) \vee (1 \wedge 1) & (0 \wedge 0) \vee (1 \wedge 1) \\ (1 \wedge 1) \vee (0 \wedge 0) & (1 \wedge 1) \vee (0 \wedge 1) & (1 \wedge 0) \vee (0 \wedge 1) \end{bmatrix}$$

$$= \begin{bmatrix} 1 \vee 0 & 1 \vee 0 & 0 \vee 0 \\ 0 \vee 0 & 0 \vee 1 & 0 \vee 1 \\ 1 \vee 0 & 1 \vee 0 & 0 \vee 0 \end{bmatrix}$$

$$= \begin{bmatrix} 1 & 1 & 0 \\ 0 & 1 & 1 \\ 1 & 1 & 0 \end{bmatrix}.$$

◀

We can also define the Boolean powers of a square zero–one matrix. These powers will be used in our subsequent studies of paths in graphs, which are used to model such things as communications paths in computer networks.

DEFINITION 10

Hence

$$\mathbf{A}^{[r]} = \underbrace{\mathbf{A} \odot \mathbf{A} \odot \mathbf{A} \odot \cdots \odot \mathbf{A}}_{r \text{ times}}.$$

(This is well defined because the Boolean product of matrices is associative.) We also define $\mathbf{A}^{[0]}$ to be $\mathbf{I}_n$.

EXAMPLE 9   Let $\mathbf{A} = \begin{bmatrix} 0 & 0 & 1 \\ 1 & 0 & 0 \\ 1 & 1 & 0 \end{bmatrix}$. Find $\mathbf{A}^{[n]}$ for all positive integers n.

Solution: We find that

$$\mathbf{A}^{[2]} = \mathbf{A} \odot \mathbf{A} = \begin{bmatrix} 1 & 1 & 0 \\ 0 & 0 & 1 \\ 1 & 0 & 1 \end{bmatrix}.$$

We also find that

$$\mathbf{A}^{[3]} = \mathbf{A}^{[2]} \odot \mathbf{A} = \begin{bmatrix} 1 & 0 & 1 \\ 1 & 1 & 0 \\ 1 & 1 & 1 \end{bmatrix}, \qquad \mathbf{A}^{[4]} = \mathbf{A}^{[3]} \odot \mathbf{A} = \begin{bmatrix} 1 & 1 & 1 \\ 1 & 0 & 1 \\ 1 & 1 & 1 \end{bmatrix}.$$

Additional computation shows that

$$\mathbf{A}^{[5]} = \begin{bmatrix} 1 & 1 & 1 \\ 1 & 1 & 1 \\ 1 & 1 & 1 \end{bmatrix}.$$

The reader can now see that $\mathbf{A}^{[n]} = \mathbf{A}^{[5]}$ for all positive integers n with $n \geq 5$. ◀

Exercises

1. Let $\mathbf{A} = \begin{bmatrix} 1 & 1 & 1 & 3 \\ 2 & 0 & 4 & 6 \\ 1 & 1 & 3 & 7 \end{bmatrix}$.

   a) What size is $\mathbf{A}$?
   b) What is the third column of $\mathbf{A}$?
   c) What is the second row of $\mathbf{A}$?
   d) What is the element of $\mathbf{A}$ in the (3, 2)th position?
   e) What is $\mathbf{A}^t$?

2. Find $\mathbf{A} + \mathbf{B}$, where

   a) $\mathbf{A} = \begin{bmatrix} 1 & 0 & 4 \\ -1 & 2 & 2 \\ 0 & -2 & -3 \end{bmatrix}$,

$\mathbf{B} = \begin{bmatrix} -1 & 3 & 5 \\ 2 & 2 & -3 \\ 2 & -3 & 0 \end{bmatrix}$.

   b) $\mathbf{A} = \begin{bmatrix} -1 & 0 & 5 & 6 \\ -4 & -3 & 5 & -2 \end{bmatrix}$,

$\mathbf{B} = \begin{bmatrix} -3 & 9 & -3 & 4 \\ 0 & -2 & -1 & 2 \end{bmatrix}$.

3. Find $\mathbf{AB}$ if

   a) $\mathbf{A} = \begin{bmatrix} 2 & 1 \\ 3 & 2 \end{bmatrix}$, $\mathbf{B} = \begin{bmatrix} 0 & 4 \\ 1 & 3 \end{bmatrix}$.

b) $\mathbf{A} = \begin{bmatrix} 1 & -1 \\ 0 & 1 \\ 2 & 3 \end{bmatrix}$, $\mathbf{B} = \begin{bmatrix} 3 & -2 & -1 \\ 1 & 0 & 2 \end{bmatrix}$.

c) $\mathbf{A} = \begin{bmatrix} 4 & -3 \\ 3 & -1 \\ 0 & -2 \\ -1 & 5 \end{bmatrix}$, $\mathbf{B} = \begin{bmatrix} -1 & 3 & 2 & -2 \\ 0 & -1 & 4 & -3 \end{bmatrix}$.

4. Find the product $\mathbf{AB}$, where

a) $\mathbf{A} = \begin{bmatrix} 1 & 0 & 1 \\ 0 & -1 & -1 \\ -1 & 1 & 0 \end{bmatrix}$, $\mathbf{B} = \begin{bmatrix} 0 & 1 & -1 \\ 1 & -1 & 0 \\ -1 & 0 & 1 \end{bmatrix}$.

b) $\mathbf{A} = \begin{bmatrix} 1 & -3 & 0 \\ 1 & 2 & 2 \\ 2 & 1 & -1 \end{bmatrix}$, $\mathbf{B} = \begin{bmatrix} 1 & -1 & 2 & 3 \\ -1 & 0 & 3 & -1 \\ -3 & -2 & 0 & 2 \end{bmatrix}$.

c) $\mathbf{A} = \begin{bmatrix} 0 & -1 \\ 7 & 2 \\ -4 & -3 \end{bmatrix}$, $\mathbf{B} = \begin{bmatrix} 4 & -1 & 2 & 3 & 0 \\ -2 & 0 & 3 & 4 & 1 \end{bmatrix}$.

5. Find a matrix $\mathbf{A}$ such that

$$\begin{bmatrix} 2 & 3 \\ 1 & 4 \end{bmatrix} \mathbf{A} = \begin{bmatrix} 3 & 0 \\ 1 & 2 \end{bmatrix}.$$

[*Hint:* Finding $\mathbf{A}$ requires that you solve systems of linear equations.]

6. Find a matrix $\mathbf{A}$ such that

$$\begin{bmatrix} 1 & 3 & 2 \\ 2 & 1 & 1 \\ 4 & 0 & 3 \end{bmatrix} \mathbf{A} = \begin{bmatrix} 7 & 1 & 3 \\ 1 & 0 & 3 \\ -1 & -3 & 7 \end{bmatrix}.$$

7. Let $\mathbf{A}$ be an $m \times n$ matrix and let $\mathbf{0}$ be the $m \times n$ matrix that has all entries equal to zero. Show that $\mathbf{A} = \mathbf{0} + \mathbf{A} = \mathbf{A} + \mathbf{0}$.

8. Show that matrix addition is commutative; that is, show that if $\mathbf{A}$ and $\mathbf{B}$ are both $m \times n$ matrices, then $\mathbf{A} + \mathbf{B} = \mathbf{B} + \mathbf{A}$.

9. Show that matrix addition is associative; that is, show that if $\mathbf{A}$, $\mathbf{B}$, and $\mathbf{C}$ are all $m \times n$ matrices, then $\mathbf{A} + (\mathbf{B} + \mathbf{C}) = (\mathbf{A} + \mathbf{B}) + \mathbf{C}$.

10. Let $\mathbf{A}$ be a 3×4 matrix, $\mathbf{B}$ be a 4×5 matrix, and $\mathbf{C}$ be a 4×4 matrix. Determine which of the following products are defined and find the size of those that are defined.

a) AB　　**b) BA**　　**c) AC**
d) CA　　**e) BC**　　**f) CB**

11. What do we know about the sizes of the matrices $\mathbf{A}$ and $\mathbf{B}$ if both of the products $\mathbf{AB}$ and $\mathbf{BA}$ are defined?

12. In this exercise we show that matrix multiplication is distributive over matrix addition.

a) Suppose that $\mathbf{A}$ and $\mathbf{B}$ are $m \times k$ matrices and that $\mathbf{C}$ is a $k \times n$ matrix. Show that $(\mathbf{A} + \mathbf{B})\mathbf{C} = \mathbf{AC} + \mathbf{BC}$.

b) Suppose that $\mathbf{C}$ is an $m \times k$ matrix and that $\mathbf{A}$ and $\mathbf{B}$ are $k \times n$ matrices. Show that $\mathbf{C}(\mathbf{A} + \mathbf{B}) = \mathbf{CA} + \mathbf{CB}$.

13. In this exercise we show that matrix multiplication is associative. Suppose that $\mathbf{A}$ is an $m \times p$ matrix, $\mathbf{B}$ is a $p \times k$ matrix, and $\mathbf{C}$ is a $k \times n$ matrix. Show that $\mathbf{A}(\mathbf{BC}) = (\mathbf{AB})\mathbf{C}$.

14. The $n \times n$ matrix $\mathbf{A} = [a_{ij}]$ is called a **diagonal matrix** if $a_{ij} = 0$ when $i \neq j$. Show that the product of two $n \times n$ diagonal matrices is again a diagonal matrix. Give a simple rule for determining this product.

15. Let

$$\mathbf{A} = \begin{bmatrix} 1 & 1 \\ 0 & 1 \end{bmatrix}.$$

Find a formula for $\mathbf{A}^n$, whenever n is a positive integer.

16. Show that $(\mathbf{A}^t)^t = \mathbf{A}$.

17. Let $\mathbf{A}$ and $\mathbf{B}$ be two $n \times n$ matrices. Show that

a) $(\mathbf{A} + \mathbf{B})^t = \mathbf{A}^t + \mathbf{B}^t$.

b) $(\mathbf{AB})^t = \mathbf{B}^t \mathbf{A}^t$.

If $\mathbf{A}$ and $\mathbf{B}$ are $n \times n$ matrices with $\mathbf{AB} = \mathbf{BA} = \mathbf{I}_n$, then $\mathbf{B}$ is called the **inverse** of $\mathbf{A}$ (this terminology is appropriate because such a matrix $\mathbf{B}$ is unique) and $\mathbf{A}$ is said to be **invertible**. The notation $\mathbf{B} = \mathbf{A}^{-1}$ denotes that $\mathbf{B}$ is the inverse of $\mathbf{A}$.

18. Show that

$$\begin{bmatrix} 2 & 3 & -1 \\ 1 & 2 & 1 \\ -1 & -1 & 3 \end{bmatrix}$$

is the inverse of

$$\begin{bmatrix} 7 & -8 & 5 \\ -4 & 5 & -3 \\ 1 & -1 & 1 \end{bmatrix}.$$

19. Let $\mathbf{A}$ be the 2×2 matrix

$$\mathbf{A} = \begin{bmatrix} a & b \\ c & d \end{bmatrix}.$$

Show that if $ad - bc \neq 0$, then

$$\mathbf{A}^{-1} = \begin{bmatrix} \dfrac{d}{ad - bc} & \dfrac{-b}{ad - bc} \\ \dfrac{-c}{ad - bc} & \dfrac{a}{ad - bc} \end{bmatrix}.$$

20. Let

$$\mathbf{A} = \begin{bmatrix} -1 & 2 \\ 1 & 3 \end{bmatrix}.$$

a) Find $\mathbf{A}^{-1}$. [*Hint:* Use Exercise 19.]

b) Find $\mathbf{A}^3$.

c) Find $(\mathbf{A}^{-1})^3$.

d) Use your answers to (b) and (c) to show that $(\mathbf{A}^{-1})^3$ is the inverse of $\mathbf{A}^3$.

21. Let $\mathbf{A}$ be an invertible matrix. Show that $(\mathbf{A}^n)^{-1} = (\mathbf{A}^{-1})^n$ whenever n is a positive integer.

22. Let $\mathbf{A}$ be a matrix. Show that the matrix $\mathbf{AA}^t$ is symmetric. [*Hint:* Show that this matrix equals its transpose with the help of Exercise 17b.]

23. Suppose that **A** is an $n \times n$ matrix where n is a positive integer. Show that $\mathbf{A} + \mathbf{A}^t$ is symmetric.

24. a) Show that the system of simultaneous linear equations

$$a_{11}x_1 + a_{12}x_2 + \cdots + a_{1n}x_n = b_1$$

$$a_{21}x_1 + a_{22}x_2 + \cdots + a_{2n}x_n = b_2$$

$$\vdots$$

$$a_{n1}x_1 + a_{n2}x_2 + \cdots + a_{nn}x_n = b_n.$$

in the variables $x_1, x_2, \ldots, x_n$ can be expressed as $\mathbf{AX} = \mathbf{B}$, where $\mathbf{A} = [a_{ij}]$, **X** is an $n \times 1$ matrix with x_i the entry in its ith row, and **B** is an $n \times 1$ matrix with b_i the entry in its ith row.

b) Show that if the matrix $\mathbf{A} = [a_{ij}]$ is invertible (as defined in the preamble to Exercise 18), then the solution of the system in part (a) can be found using the equation $\mathbf{X} = \mathbf{A}^{-1}\mathbf{B}$.

25. Use Exercises 18 and 24 to solve the system

$$7x_1 - 8x_2 + 5x_3 = 5$$

$$-4x_1 + 5x_2 - 3x_3 = -3$$

$$x_1 - x_2 + x_3 = 0$$

26. Let

$$\mathbf{A} = \begin{bmatrix} 1 & 1 \\ 0 & 1 \end{bmatrix} \quad \text{and} \quad \mathbf{B} = \begin{bmatrix} 0 & 1 \\ 1 & 0 \end{bmatrix}.$$

Find

a) $\mathbf{A} \vee \mathbf{B}$.   **b)** $\mathbf{A} \wedge \mathbf{B}$.   **c)** $\mathbf{A} \odot \mathbf{B}$.

27. Let

$$\mathbf{A} = \begin{bmatrix} 1 & 0 & 1 \\ 1 & 1 & 0 \\ 0 & 0 & 1 \end{bmatrix} \quad \text{and} \quad \mathbf{B} = \begin{bmatrix} 0 & 1 & 1 \\ 1 & 0 & 1 \\ 1 & 0 & 1 \end{bmatrix}.$$

Find

a) $\mathbf{A} \vee \mathbf{B}$.   **b)** $\mathbf{A} \wedge \mathbf{B}$.   **c)** $\mathbf{A} \odot \mathbf{B}$.

28. Find the Boolean product of **A** and **B**, where

$$\mathbf{A} = \begin{bmatrix} 1 & 0 & 0 & 1 \\ 0 & 1 & 0 & 1 \\ 1 & 1 & 1 & 1 \end{bmatrix} \quad \text{and} \quad \mathbf{B} = \begin{bmatrix} 1 & 0 \\ 0 & 1 \\ 1 & 1 \\ 1 & 0 \end{bmatrix}.$$

29. Let

$$\mathbf{A} = \begin{bmatrix} 1 & 0 & 0 \\ 1 & 0 & 1 \\ 0 & 1 & 0 \end{bmatrix}.$$

Find

a) $\mathbf{A}^{[2]}$.   **b)** $\mathbf{A}^{[3]}$.

c) $\mathbf{A} \vee \mathbf{A}^{[2]} \vee \mathbf{A}^{[3]}$.

30. Let **A** be a zero–one matrix. Show that

a) $\mathbf{A} \vee \mathbf{A} = \mathbf{A}$.   **b)** $\mathbf{A} \wedge \mathbf{A} = \mathbf{A}$.

31. In this exercise we show that the meet and join operations are commutative. Let **A** and **B** be $m \times n$ zero–one matrices. Show that

a) $\mathbf{A} \vee \mathbf{B} = \mathbf{B} \vee \mathbf{A}$.   **b)** $\mathbf{B} \wedge \mathbf{A} = \mathbf{A} \wedge \mathbf{B}$.

32. In this exercise we show that the meet and join operations are associative. Let **A**, **B**, and **C** be $m \times n$ zero–one matrices. Show that

a) $(\mathbf{A} \vee \mathbf{B}) \vee \mathbf{C} = \mathbf{A} \vee (\mathbf{B} \vee \mathbf{C})$.

b) $(\mathbf{A} \wedge \mathbf{B}) \wedge \mathbf{C} = \mathbf{A} \wedge (\mathbf{B} \wedge \mathbf{C})$.

33. We will establish distributive laws of the meet over the join operation in this exercise. Let **A**, **B**, and **C** be $m \times n$ zero–one matrices. Show that

a) $\mathbf{A} \vee (\mathbf{B} \wedge \mathbf{C}) = (\mathbf{A} \vee \mathbf{B}) \wedge (\mathbf{A} \vee \mathbf{C})$.

b) $\mathbf{A} \wedge (\mathbf{B} \vee \mathbf{C}) = (\mathbf{A} \wedge \mathbf{B}) \vee (\mathbf{A} \wedge \mathbf{C})$.

34. Let **A** be an $n \times n$ zero–one matrix. Let **I** be the $n \times n$ identity matrix. Show that $\mathbf{A} \odot \mathbf{I} = \mathbf{I} \odot \mathbf{A} = \mathbf{A}$.

35. In this exercise we will show that the Boolean product of zero–one matrices is associative. Assume that **A** is an $m \times p$ zero–one matrix, **B** is a $p \times k$ zero–one matrix, and **C** is a $k \times n$ zero–one matrix. Show that $\mathbf{A} \odot (\mathbf{B} \odot \mathbf{C}) = (\mathbf{A} \odot \mathbf{B}) \odot \mathbf{C}$.

Key Terms and Results

TERMS

set: a collection of distinct objects

axiom: a basic assumption of a theory

paradox: a logical inconsistency

element, member of a set: an object in a set

roster method: a method that describes a set by listing its elements

set builder notation: the notation that describes a set by stating a property an element must have to be a member

Ø (empty set, null set): the set with no members

universal set: the set containing all objects under consideration

Venn diagram: a graphical representation of a set or sets

S = T (set equality): S and T have the same elements

$S \subseteq T$ (S is a subset of T): every element of S is also an element of T

$S \subset T$ (S is a proper subset of T): S is a subset of T and $S \neq T$

finite set: a set with n elements, where n is a nonnegative integer

infinite set: a set that is not finite

$|S|$ (the cardinality of S): the number of elements in S

$P(S)$ (the power set of S): the set of all subsets of S

$A \cup B$ (the union of A and B): the set containing those elements that are in at least one of A and B

$A \cap B$ (the intersection of A and B): the set containing those elements that are in both A and B.

$A - B$ **(the difference of A and B):** the set containing those elements that are in A but not in B

$\bar{A}$ **(the complement of A):** the set of elements in the universal set that are not in A

$A \oplus B$ **(the symmetric difference of A and B):** the set containing those elements in exactly one of A and B

membership table: a table displaying the membership of elements in sets

function from A to B: an assignment of exactly one element of B to each element of A

domain of f: the set A, where f is a function from A to B

codomain of f: the set B, where f is a function from A to B

b is the image of a under f: $b = f(a)$

a is a pre-image of b under f: $f(a) = b$

range of f: the set of images of f

onto function, surjection: a function from A to B such that every element of B is the image of some element in A

one-to-one function, injection: a function such that the images of elements in its domain are distinct

one-to-one correspondence, bijection: a function that is both one-to-one and onto

inverse of f: the function that reverses the correspondence given by f (when f is a bijection)

$f \circ g$ (composition of f and g): the function that assigns $f(g(x))$ to x

$\lfloor x \rfloor$ (floor function): the largest integer not exceeding x

$\lceil x \rceil$ (ceiling function): the smallest integer greater than or equal to x

partial function: an assignment to each element in a subset of the domain a unique element in the codomain

sequence: a function with domain that is a subset of the set of integers

geometric progression: a sequence of the form $a, ar, ar^2, \ldots$, where a and r are real numbers

arithmetic progression: a sequence of the form $a, a + d, a + 2d, \ldots$, where a and d are real numbers

string: a finite sequence

empty string: a string of length zero

recurrence relation: a equation that expresses the nth term a_n of a sequence in terms of one or more of the previous terms of the sequence for all integers n greater than a particular integer

$\sum_{i=1}^{n} a_i$: the sum $a_1 + a_2 + \cdots + a_n$

$\prod_{i=1}^{n} a_i$: the product $a_1 a_2 \cdots a_n$

cardinality: two sets A and B have the same cardinality if there is a one-to-one correspondence from A to B

countable set: a set that either is finite or can be placed in one-to-one correspondence with the set of positive integers

uncountable set: a set that is not countable

$\aleph_0$ (aleph null): the cardinality of a countable set

$\mathfrak{c}$: the cardinality of the set of real numbers

Cantor diagonalization argument: a proof technique used to show that the set of real numbers is uncountable

computable function: a function for which there is a computer program in some programming language that finds its values

uncomputable function: a function for which no computer program in a programming language exists that finds its values

continuum hypothesis: the statement there no set A exists such that $\aleph_0 < |A| < \mathfrak{c}$

matrix: a rectangular array of numbers

matrix addition: see page 178

matrix multiplication: see page 179

I_n (identity matrix of order n): the $n \times n$ matrix that has entries equal to 1 on its diagonal and 0s elsewhere

A^t (transpose of A): the matrix obtained from A by interchanging the rows and columns

symmetric matrix: a matrix is symmetric if it equals its transpose

zero–one matrix: a matrix with each entry equal to either 0 or 1

$A \vee B$ (the join of A and B): see page 181

$A \wedge B$ (the meet of A and B): see page 181

$A \odot B$ (the Boolean product of A and B): see page 182

RESULTS

The set identities given in Table 1 in Section 2.2

The summation formulae in Table 2 in Section 2.4

The set of rational numbers is countable.

The set of real numbers is uncountable.

Review Questions

1. Explain what it means for one set to be a subset of another set. How do you prove that one set is a subset of another set?

2. What is the empty set? Show that the empty set is a subset of every set.

3. a) Define $|S|$, the cardinality of the set S.

   b) Give a formula for $|A \cup B|$, where A and B are sets.

4. a) Define the power set of a set S.

   b) When is the empty set in the power set of a set S?

   c) How many elements does the power set of a set S with n elements have?

5. a) Define the union, intersection, difference, and symmetric difference of two sets.

   b) What are the union, intersection, difference, and symmetric difference of the set of positive integers and the set of odd integers?

6. a) Explain what it means for two sets to be equal.

   b) Describe as many of the ways as you can to show that two sets are equal.

   c) Show in at least two different ways that the sets $A - (B \cap C)$ and $(A - B) \cup (A - C)$ are equal.

7. Explain the relationship between logical equivalences and set identities.

8. a) Define the domain, codomain, and range of a function.
   b) Let $f(n)$ be the function from the set of integers to the set of integers such that $f(n) = n^2 + 1$. What are the domain, codomain, and range of this function?

9. a) Define what it means for a function from the set of positive integers to the set of positive integers to be one-to-one.
   b) Define what it means for a function from the set of positive integers to the set of positive integers to be onto.
   c) Give an example of a function from the set of positive integers to the set of positive integers that is both one-to-one and onto.
   d) Give an example of a function from the set of positive integers to the set of positive integers that is one-to-one but not onto.
   e) Give an example of a function from the set of positive integers to the set of positive integers that is not one-to-one but is onto.
   f) Give an example of a function from the set of positive integers to the set of positive integers that is neither one-to-one nor onto.

10. a) Define the inverse of a function.
   b) When does a function have an inverse?
   c) Does the function $f(n) = 10 - n$ from the set of integers to the set of integers have an inverse? If so, what is it?

11. a) Define the floor and ceiling functions from the set of real numbers to the set of integers.
   b) For which real numbers x is it true that $\lfloor x \rfloor = \lceil x \rceil$?

12. Conjecture a formula for the terms of the sequence that begins 8, 14, 32, 86, 248 and find the next three terms of your sequence.

13. Suppose that $a_n = a_{n-1} - 5$ for $n = 1, 2, \ldots$. Find a formula for a_n.

14. What is the sum of the terms of the geometric progression $a + ar + \cdots + ar^n$ when $r \neq 1$?

15. Show that the set of odd integers is countable.

16. Give an example of an uncountable set.

17. Define the product of two matrices $\mathbf{A}$ and $\mathbf{B}$. When is this product defined?

18. Show that matrix multiplication is not commutative.

Supplementary Exercises

1. Let A be the set of English words that contain the letter x, and let B be the set of English words that contain the letter q. Express each of these sets as a combination of A and B.
   a) The set of English words that do not contain the letter x.
   b) The set of English words that contain both an x and a q.
   c) The set of English words that contain an x but not a q.
   d) The set of English words that do not contain either an x or a q.
   e) The set of English words that contain an x or a q, but not both.

2. Show that if A is a subset of B, then the power set of A is a subset of the power set of B.

3. Suppose that A and B are sets such that the power set of A is a subset of the power set of B. Does it follow that A is a subset of B?

4. Let $\mathbf{E}$ denote the set of even integers and $\mathbf{O}$ denote the set of odd integers. As usual, let $\mathbf{Z}$ denote the set of all integers. Determine each of these sets.
   a) $\mathbf{E} \cup \mathbf{O}$ **b)** $\mathbf{E} \cap \mathbf{O}$ **c)** $\mathbf{Z} - \mathbf{E}$ **d)** $\mathbf{Z} - \mathbf{O}$

5. Show that if A and B are finite sets, then $|A \cap B| \leq |A \cup B|$. Determine when this relationship is an equality.

6. Let A and B be sets in a finite universal set U. List the following in order of increasing size.
   a) $|A|, |A \cup B|, |A \cap B|, |U|, |\emptyset|$
   b) $|A - B|, |A \oplus B|, |A| + |B|, |A \cup B|, |\emptyset|$

7. Let f and g be functions from $\{1, 2, 3, 4\}$ to $\{a, b, c, d\}$ and from $\{a, b, c, d\}$ to $\{1, 2, 3, 4\}$, respectively, with $f(1) = d$, $f(2) = c$, $f(3) = a$, and $f(4) = b$, and $g(a) = 2, g(b) = 1, g(c) = 3$, and $g(d) = 2$.
   a) Is f one-to-one? Is g one-to-one?
   b) Is f onto? Is g onto?
   c) Does either f or g have an inverse? If so, find this inverse.

8. Suppose that f is a function from A to B where A and B are finite sets. Explain why $|f(S)| \leq |S|$ for all subsets S of A.

9. Suppose that f is a function from A to B where A and B are finite sets. Explain why $|f(S)| = |S|$ for all subsets S of A if and only if f is one-to-one.

Suppose that f is a function from A to B. We define the function S_f from $\mathcal{P}(A)$ to $\mathcal{P}(B)$ by the rule $S_f(X) = f(X)$ for each subset X of A. Similarly, we define the function $S_{f^{-1}}$ from $\mathcal{P}(B)$ to $\mathcal{P}(A)$ by the rule $S_{f^{-1}}(Y) = f^{-1}(Y)$ for each subset Y of B. Here, we are using Definition 4, and the definition of the inverse image of a set found in the preamble to Exercise 28, both in Section 2.3.

*10. Suppose that f is a function from the set A to the set B. Prove that
   a) if f is one-to-one, then S_f is a one-to-one function from $\mathcal{P}(A)$ to $\mathcal{P}(B)$.
   b) if f is onto function, then S_f is an onto function from $\mathcal{P}(A)$ to $\mathcal{P}(B)$.
   c) if f is onto function, then $S_{f^{-1}}$ is a one-to-one function from $\mathcal{P}(B)$ to $\mathcal{P}(A)$.

d) if f is one-to-one, then $S_{f^{-1}}$ is an onto function from $\mathcal{P}(B)$ to $\mathcal{P}(A)$.

e) if f is a one-to-one correspondence, then S_f is a one-to-one correspondence from $\mathcal{P}(A)$ to $\mathcal{P}(B)$ and $S_{f^{-1}}$ is a one-to-one correspondence from $\mathcal{P}(B)$ to $\mathcal{P}(A)$. [*Hint:* Use parts (a)-(d).]

11. Prove that if f and g are functions from A to B and $S_f = S_g$ (using the definition in the preamble to Exercise 10), then $f(x) = g(x)$ for all $x \in A$.

12. Show that if n is an integer, then $n = \lceil n/2 \rceil + \lfloor n/2 \rfloor$.

13. For which real numbers x and y is it true that $\lfloor x + y \rfloor = \lfloor x \rfloor + \lfloor y \rfloor$?

14. For which real numbers x and y is it true that $\lceil x + y \rceil = \lceil x \rceil + \lceil y \rceil$?

15. For which real numbers x and y is it true that $\lceil x + y \rceil = \lceil x \rceil + \lfloor y \rfloor$?

16. Prove that $\lfloor n/2 \rfloor \lceil n/2 \rceil = \lfloor n^2/4 \rfloor$ for all integers n.

17. Prove that if m is an integer, then $\lfloor x \rfloor + \lfloor m - x \rfloor = m - 1$, unless x is an integer, in which case, it equals m.

18. Prove that if x is a real number, then $\lfloor \lfloor x/2 \rfloor /2 \rfloor = \lfloor x/4 \rfloor$.

19. Prove that if n is an odd integer, then $\lceil n^2/4 \rceil = (n^2 + 3)/4$.

*20. We define the **Ulam numbers** by setting $u_1 = 1$ and $u_2 = 2$. Furthermore, after determining whether the integers less than n are Ulam numbers, we set n equal to the next Ulam number if it can be written uniquely as the sum of two different Ulam numbers. Note that $u_3 = 3$, $u_4 = 4$, $u_5 = 6$, and $u_6 = 8$.

a) Find the first 20 Ulam numbers.

b) Prove that there are infinitely many Ulam numbers.

21. Determine the value of $\prod_{k=1}^{100} \frac{k+1}{k}$. (The notation used here for products is defined in the preamble to Exercise 29 in Section 2.4.)

*22. Determine a rule for generating the terms of the sequence that begins 1, 3, 4, 8, 15, 27, 50, 92, ..., and find the next four terms of the sequence.

*23. Determine a rule for generating the terms of the sequence that begins 2, 3, 3, 5, 10, 13, 39, 43, 172, 177, 885, 891, ..., and find the next four terms of the sequence.

24. Show that the set of irrational numbers is an uncountable set.

25. Show that the set S is a countable set if there is a function f from S to the positive integers such that $f^{-1}(j)$ is countable whenever j is a positive integer.

26. Show that the set of all finite subsets of the set of positive integers is a countable set.

27. Find $\mathbf{A}^n$ if $\mathbf{A}$ is

$$\begin{bmatrix} 0 & 1 \\ -1 & 0 \end{bmatrix}.$$

28. Show that if $\mathbf{A} = c\mathbf{I}$, where c is a real number and $\mathbf{I}$ is the $n \times n$ identity matrix, then $\mathbf{AB} = \mathbf{BA}$ whenever $\mathbf{B}$ is an $n \times n$ matrix.

29. Show that if $\mathbf{A}$ is a 2×2 matrix such that $\mathbf{AB} = \mathbf{BA}$ whenever $\mathbf{B}$ is a 2×2 matrix, then $\mathbf{A} = c\mathbf{I}$, where c is a real number and $\mathbf{I}$ is the 2×2 identity matrix.

30. Show that if $\mathbf{A}$ and $\mathbf{B}$ are invertible matrices and $\mathbf{AB}$ exists, then $(\mathbf{AB})^{-1} = \mathbf{B}^{-1}\mathbf{A}^{-1}$.

31. Let $\mathbf{A}$ be an $n \times n$ matrix and let $\mathbf{0}$ be the $n \times n$ matrix all of whose entries are zero. Show that the following are true.

a) $\mathbf{A} \odot \mathbf{0} = \mathbf{0} \odot \mathbf{A} = \mathbf{0}$

b) $\mathbf{A} \vee \mathbf{0} = \mathbf{0} \vee \mathbf{A} = \mathbf{A}$

c) $\mathbf{A} \wedge \mathbf{0} = \mathbf{0} \wedge \mathbf{A} = \mathbf{0}$

Computer Projects

Write programs with the specified input and output.

1. Given subsets A and B of a set with n elements, use bit strings to find $\overline{A}$, $A \cup B$, $A \cap B$, $A - B$, and $A \oplus B$.

2. Given multisets A and B from the same universal set, find $A \cup B$, $A \cap B$, $A - B$, and $A + B$ (see preamble to Exercise 61 of Section 2.2).

3. Given fuzzy sets A and B, find $\overline{A}$, $A \cup B$, and $A \cap B$ (see preamble to Exercise 63 of Section 2.2).

4. Given a function f from $\{1, 2, \ldots, n\}$ to the set of integers, determine whether f is one-to-one.

5. Given a function f from $\{1, 2, \ldots, n\}$ to itself, determine whether f is onto.

6. Given a bijection f from the set $\{1, 2, \ldots, n\}$ to itself, find f^{-1}.

7. Given an $m \times k$ matrix $\mathbf{A}$ and a $k \times n$ matrix $\mathbf{B}$, find $\mathbf{AB}$.

8. Given a square matrix $\mathbf{A}$ and a positive integer n, find $\mathbf{A}^n$.

9. Given a square matrix, determine whether it is symmetric.

10. Given two $m \times n$ Boolean matrices, find their meet and join.

11. Given an $m \times k$ Boolean matrix $\mathbf{A}$ and a $k \times n$ Boolean matrix $\mathbf{B}$, find the Boolean product of $\mathbf{A}$ and $\mathbf{B}$.

12. Given a square Boolean matrix $\mathbf{A}$ and a positive integer n, find $\mathbf{A}^{[n]}$.

Computations and Explorations

Use a computational program or programs you have written to do these exercises.

1. Given two finite sets, list all elements in the Cartesian product of these two sets.

2. Given a finite set, list all elements of its power set.

3. Calculate the number of one-to-one functions from a set S to a set T, where S and T are finite sets of various sizes. Can you determine a formula for the number of such functions? (We will find such a formula in Chapter 6.)

4. Calculate the number of onto functions from a set S to a set T, where S and T are finite sets of various sizes. Can you determine a formula for the number of such functions? (We will find such a formula in Chapter 8.)

Writing Projects

Respond to these with essays using outside sources.

1. Discuss how an axiomatic set theory can be developed to avoid Russell's paradox. (See Exercise 46 of Section 2.1.)

2. Research where the concept of a function first arose, and describe how this concept was first used.

3. Explain the different ways in which the *Encyclopedia of Integer Sequences* has been found useful. Also, describe a few of the more unusual sequences in this encyclopedia and how they arise.

4. Define the recently invented EKG sequence and describe some of its properties and open questions about it.

5. Look up the definition of a transcendental number. Explain how to show that such numbers exist and how such numbers can be constructed. Which famous numbers can be shown to be transcendental and for which famous numbers is it still unknown whether they are transcendental?

3

Algorithms

Many problems can be solved by considering them as special cases of general problems. For instance, consider the problem of locating the largest integer in the sequence 101, 12, 144, 212, 98. This is a specific case of the problem of locating the largest integer in a sequence of integers. To solve this general problem we must give an algorithm, which specifies a sequence of steps used to solve this general problem. We will study algorithms for solving many different types of problems in this book. For example, in this chapter we will introduce algorithms for two of the most important problems in computer science, searching for an element in a list and sorting a list so its elements are in some prescribed order, such as increasing, decreasing, or alphabetic. Later in the book we will develop algorithms that find the greatest common divisor of two integers, that generate all the orderings of a finite set, that find the shortest path between nodes in a network, and for solving many other problems.

We will also introduce the notion of an algorithmic paradigm, which provides a general method for designing algorithms. In particular we will discuss brute-force algorithms, which find solutions using a straightforward approach without introducing any cleverness. We will also discuss greedy algorithms, a class of algorithms used to solve optimization problems. Proofs are important in the study of algorithms. In this chapter we illustrate this by proving that a particular greedy algorithm always finds an optimal solution.

One important consideration concerning an algorithm is its computational complexity, which measures the processing time and computer memory required by the algorithm to solve problems of a particular size. To measure the complexity of algorithms we use big-O and big-Theta notation, which we develop in this chapter. We will illustrate the analysis of the complexity of algorithms in this chapter, focusing on the time an algorithm takes to solve a problem. Furthermore, we will discuss what the time complexity of an algorithm means in practical and theoretical terms.

3.1 Algorithms

Introduction

There are many general classes of problems that arise in discrete mathematics. For instance: given a sequence of integers, find the largest one; given a set, list all its subsets; given a set of integers, put them in increasing order; given a network, find the shortest path between two vertices. When presented with such a problem, the first thing to do is to construct a model that translates the problem into a mathematical context. Discrete structures used in such models include sets, sequences, and functions—structures discussed in Chapter 2—as well as such other structures as permutations, relations, graphs, trees, networks, and finite state machines—concepts that will be discussed in later chapters.

Setting up the appropriate mathematical model is only part of the solution. To complete the solution, a method is needed that will solve the general problem using the model. Ideally, what is required is a procedure that follows a sequence of steps that leads to the desired answer. Such a sequence of steps is called an **algorithm**.

DEFINITION 1   An *algorithm* is a finite sequence of precise instructions for performing a computation or for solving a problem.

The term *algorithm* is a corruption of the name *al-Khowarizmi*, a mathematician of the ninth century, whose book on Hindu numerals is the basis of modern decimal notation. Originally, the word *algorism* was used for the rules for performing arithmetic using decimal notation. *Algorism* evolved into the word *algorithm* by the eighteenth century. With the growing interest in computing machines, the concept of an algorithm was given a more general meaning, to include all definite procedures for solving problems, not just the procedures for performing arithmetic. (We will discuss algorithms for performing arithmetic with integers in Chapter 4.)

In this book, we will discuss algorithms that solve a wide variety of problems. In this section we will use the problem of finding the largest integer in a finite sequence of integers to illustrate the concept of an algorithm and the properties algorithms have. Also, we will describe algorithms for locating a particular element in a finite set. In subsequent sections, procedures for finding the greatest common divisor of two integers, for finding the shortest path between two points in a network, for multiplying matrices, and so on, will be discussed.

EXAMPLE 1 Describe an algorithm for finding the maximum (largest) value in a finite sequence of integers.

 Extra Examples

Even though the problem of finding the maximum element in a sequence is relatively trivial, it provides a good illustration of the concept of an algorithm. Also, there are many instances where the largest integer in a finite sequence of integers is required. For instance, a university may need to find the highest score on a competitive exam taken by thousands of students. Or a sports organization may want to identify the member with the highest rating each month. We want to develop an algorithm that can be used whenever the problem of finding the largest element in a finite sequence of integers arises.

We can specify a procedure for solving this problem in several ways. One method is simply to use the English language to describe the sequence of steps used. We now provide such a solution.

Solution of Example 1: We perform the following steps.

1. Set the temporary maximum equal to the first integer in the sequence. (The temporary maximum will be the largest integer examined at any stage of the procedure.)
2. Compare the next integer in the sequence to the temporary maximum, and if it is larger than the temporary maximum, set the temporary maximum equal to this integer.
3. Repeat the previous step if there are more integers in the sequence.
4. Stop when there are no integers left in the sequence. The temporary maximum at this point is the largest integer in the sequence. ◀

An algorithm can also be described using a computer language. However, when that is done, only those instructions permitted in the language can be used. This often leads to a description of the algorithm that is complicated and difficult to understand. Furthermore, because many programming languages are in common use, it would be undesirable to choose one particular language. So, instead of using a particular computer language to specify algorithms, a form of **pseudocode**, described in Appendix 3, will be used in this book. (We will also describe algorithms using the English language.) Pseudocode provides an intermediate step between

 Links

ABU JA'FAR MOHAMMED IBN MUSA AL-KHOWARIZMI (C. 780–C. 850) al-Khowarizmi, an astronomer and mathematician, was a member of the House of Wisdom, an academy of scientists in Baghdad. The name al-Khowarizmi means "from the town of Kowarzizm," which was then part of Persia, but is now called *Khiva* and is part of Uzbekistan. al-Khowarizmi wrote books on mathematics, astronomy, and geography. Western Europeans first learned about algebra from his works. The word *algebra* comes from al-jabr, part of the title of his book *Kitab al-jabr w'al muquabala*. This book was translated into Latin and was a widely used textbook. His book on the use of Hindu numerals describes procedures for arithmetic operations using these numerals. European authors used a Latin corruption of his name, which later evolved to the word *algorithm*, to describe the subject of arithmetic with Hindu numerals.

an English language description of an algorithm and an implementation of this algorithm in a programming language. The steps of the algorithm are specified using instructions resembling those used in programming languages. However, in pseudocode, the instructions used can include any well-defined operations or statements. A computer program can be produced in any computer language using the pseudocode description as a starting point.

The pseudocode used in this book is designed to be easily understood. It can serve as an intermediate step in the construction of programs implementing algorithms in one of a variety of different programming languages. Although this pseudocode does not follow the syntax of Java, C, C++, or any other programming language, students familiar with a modern programming language will find it easy to follow. A key difference between this pseudocode and code in a programming language is that we can use any well-defined instruction even if it would take many lines of code to implement this instruction. The details of the pseudocode used in the text are given in Appendix 3. The reader should refer to this appendix whenever the need arises.

A pseudocode description of the algorithm for finding the maximum element in a finite sequence follows.

ALGORITHM 1 **Finding the Maximum Element in a Finite Sequence.**

procedure $max(a_1, a_2, \ldots, a_n$: integers)
$max := a_1$
for $i := 2$ **to** n
    if $max < a_i$ **then** $max := a_i$
return $max\{max$ is the largest element$\}$

This algorithm first assigns the initial term of the sequence, a_1, to the variable *max*. The "for" loop is used to successively examine terms of the sequence. If a term is greater than the current value of *max*, it is assigned to be the new value of *max*.

PROPERTIES OF ALGORITHMS   There are several properties that algorithms generally share. They are useful to keep in mind when algorithms are described. These properties are:

- *Input.* An algorithm has input values from a specified set.
- *Output.* From each set of input values an algorithm produces output values from a specified set. The output values are the solution to the problem.
- *Definiteness.* The steps of an algorithm must be defined precisely.
- *Correctness.* An algorithm should produce the correct output values for each set of input values.
- *Finiteness.* An algorithm should produce the desired output after a finite (but perhaps large) number of steps for any input in the set.
- *Effectiveness.* It must be possible to perform each step of an algorithm exactly and in a finite amount of time.
- *Generality.* The procedure should be applicable for all problems of the desired form, not just for a particular set of input values.

EXAMPLE 2   Show that Algorithm 1 for finding the maximum element in a finite sequence of integers has all the properties listed.

Solution: The input to Algorithm 1 is a sequence of integers. The output is the largest integer in the sequence. Each step of the algorithm is precisely defined, because only assignments, a finite loop, and conditional statements occur. To show that the algorithm is correct, we must show that when the algorithm terminates, the value of the variable *max* equals the maximum of the terms of the sequence. To see this, note that the initial value of *max* is the first term of the sequence; as successive terms of the sequence are examined, *max* is updated to the value of a term if the term exceeds the maximum of the terms previously examined. This (informal) argument shows that when all the terms have been examined, *max* equals the value of the largest term. (A rigorous proof of this requires techniques developed in Section 5.1.) The algorithm uses a finite number of steps, because it terminates after all the integers in the sequence have been examined. The algorithm can be carried out in a finite amount of time because each step is either a comparison or an assignment, there are a finite number of these steps, and each of these two operations takes a finite amount of time. Finally, Algorithm 1 is general, because it can be used to find the maximum of any finite sequence of integers. ◀

Searching Algorithms

The problem of locating an element in an ordered list occurs in many contexts. For instance, a program that checks the spelling of words searches for them in a dictionary, which is just an ordered list of words. Problems of this kind are called **searching problems**. We will discuss several algorithms for searching in this section. We will study the number of steps used by each of these algorithms in Section 3.3.

The general searching problem can be described as follows: Locate an element x in a list of distinct elements $a_1, a_2, \ldots, a_n$, or determine that it is not in the list. The solution to this search problem is the location of the term in the list that equals x (that is, i is the solution if $x = a_i$) and is 0 if x is not in the list.

THE LINEAR SEARCH   The first algorithm that we will present is called the **linear search**, or **sequential search**, algorithm. The linear search algorithm begins by comparing x and a_1. When $x = a_1$, the solution is the location of a_1, namely, 1. When $x \neq a_1$, compare x with a_2. If $x = a_2$, the solution is the location of a_2, namely, 2. When $x \neq a_2$, compare x with a_3. Continue this process, comparing x successively with each term of the list until a match is found, where the solution is the location of that term, unless no match occurs. If the entire list has been searched without locating x, the solution is 0. The pseudocode for the linear search algorithm is displayed as Algorithm 2.

ALGORITHM 2 **The Linear Search Algorithm.**

---

procedure *linear search*(x: integer, $a_1, a_2, \ldots, a_n$: distinct integers)
$i := 1$
while ($i \leq n$ and $x \neq a_i$)
  $i := i + 1$
if $i \leq n$ **then** *location* $:= i$
else *location* $:= 0$
return *location*{*location* is the subscript of the term that equals x, or is 0 if x is not found}

THE BINARY SEARCH   We will now consider another searching algorithm. This algorithm can be used when the list has terms occurring in order of increasing size (for instance: if the terms are numbers, they are listed from smallest to largest; if they are words, they are listed in lexicographic, or alphabetic, order). This second searching algorithm is called the **binary search algorithm**. It proceeds by comparing the element to be located to the middle term of the list. The list is then split into two smaller sublists of the same size, or where one of these smaller lists has one fewer term than the other. The search continues by restricting the search to the appropriate sublist based on the comparison of the element to be located and the middle term. In Section 3.3, it will be shown that the binary search algorithm is much more efficient than the linear search algorithm. Example 3 demonstrates how a binary search works.

EXAMPLE 3 To search for 19 in the list

$$1\ 2\ 3\ 5\ 6\ 7\ 8\ 10\ 12\ 13\ 15\ 16\ 18\ 19\ 20\ 22,$$

first split this list, which has 16 terms, into two smaller lists with eight terms each, namely,

$$1\ 2\ 3\ 5\ 6\ 7\ 8\ 10 \qquad 12\ 13\ 15\ 16\ 18\ 19\ 20\ 22.$$

Then, compare 19 and the largest term in the first list. Because $10 < 19$, the search for 19 can be restricted to the list containing the 9th through the 16th terms of the original list. Next, split this list, which has eight terms, into the two smaller lists of four terms each, namely,

$$12\ 13\ 15\ 16 \qquad 18\ 19\ 20\ 22.$$

Because $16 < 19$ (comparing 19 with the largest term of the first list) the search is restricted to the second of these lists, which contains the 13th through the 16th terms of the original list. The list 18 19 20 22 is split into two lists, namely,

$$18\ 19 \qquad 20\ 22.$$

Because 19 is not greater than the largest term of the first of these two lists, which is also 19, the search is restricted to the first list: 18 19, which contains the 13th and 14th terms of the original list. Next, this list of two terms is split into two lists of one term each: 18 and 19. Because $18 < 19$, the search is restricted to the second list: the list containing the 14th term of the list, which is 19. Now that the search has been narrowed down to one term, a comparison is made, and 19 is located as the 14th term in the original list. ◀

We now specify the steps of the binary search algorithm. To search for the integer x in the list $a_1, a_2, \ldots, a_n$, where $a_1 < a_2 < \cdots < a_n$, begin by comparing x with the middle term a_m of the list, where $m = \lfloor (n + 1)/2 \rfloor$. (Recall that $\lfloor x \rfloor$ is the greatest integer not exceeding x.) If $x > a_m$, the search for x is restricted to the second half of the list, which is $a_{m+1}, a_{m+2}, \ldots, a_n$. If x is not greater than a_m, the search for x is restricted to the first half of the list, which is $a_1, a_2, \ldots, a_m$.

The search has now been restricted to a list with no more than $\lceil n/2 \rceil$ elements. (Recall that $\lceil x \rceil$ is the smallest integer greater than or equal to x.) Using the same procedure, compare x to the middle term of the restricted list. Then restrict the search to the first or second half of the list. Repeat this process until a list with one term is obtained. Then determine whether this term is x. Pseudocode for the binary search algorithm is displayed as Algorithm 3.

ALGORITHM 3 **The Binary Search Algorithm.**

---

procedure *binary search* (x: integer, $a_1, a_2, \ldots, a_n$: increasing integers)
$i := 1\{i$ is left endpoint of search interval$\}$
$j := n$ $\{j$ is right endpoint of search interval$\}$
while $i < j$
    $m := \lfloor (i + j)/2 \rfloor$
    if $x > a_m$ **then** $i := m + 1$
    else $j := m$
if $x = a_i$ **then** *location* $:= i$
else *location* $:= 0$
return *location*$\{location$ is the subscript i of the term a_i equal to x, or 0 if x is not found$\}$

Algorithm 3 proceeds by successively narrowing down the part of the sequence being searched. At any given stage only the terms from a_i to a_j are under consideration. In other words, i and j are the smallest and largest subscripts of the remaining terms, respectively. Algorithm 3 continues narrowing the part of the sequence being searched until only one term of the sequence remains. When this is done, a comparison is made to see whether this term equals x.

Sorting

Demo

Ordering the elements of a list is a problem that occurs in many contexts. For example, to produce a telephone directory it is necessary to alphabetize the names of subscribers. Similarly, producing a directory of songs available for downloading requires that their titles be put in alphabetic order. Putting addresses in order in an e-mail mailing list can determine whether there are duplicated addresses. Creating a useful dictionary requires that words be put in alphabetical order. Similarly, generating a parts list requires that we order them according to increasing part number.

Suppose that we have a list of elements of a set. Furthermore, suppose that we have a way to order elements of the set. (The notion of ordering elements of sets will be discussed in detail in Section 9.6.) **Sorting** is putting these elements into a list in which the elements are in increasing order. For instance, sorting the list 7, 2, 1, 4, 5, 9 produces the list 1, 2, 4, 5, 7, 9. Sorting the list d, h, c, a, f (using alphabetical order) produces the list a, c, d, f, h.

An amazingly large percentage of computing resources is devoted to sorting one thing or another. Hence, much effort has been devoted to the development of sorting algorithms. A surprisingly large number of sorting algorithms have been devised using distinct strategies, with new ones introduced regularly. In his fundamental work, *The Art of Computer Programming,* Donald Knuth devotes close to 400 pages to sorting, covering around 15 different sorting algorithms in depth! More than 100 sorting algorithms have been devised, and it is surprising how often new sorting algorithms are developed. Among the newest sorting algorithms that have caught on is the the library sort, also known as the gapped insertion sort, invented as recently as 2006. There are many reasons why sorting algorithms interest computer scientists and mathematicians. Among these reasons are that some algorithms are easier to implement, some algorithms are more efficient (either in general, or when given input with certain characteristics, such as lists slightly out of order), some algorithms take advantage of particular computer architectures, and some algorithms are particularly clever. In this section we will introduce two sorting algorithms, the bubble sort and the insertion sort. Two other sorting algorithms, the selection sort and the binary insertion sort, are introduced in the exercises, and the shaker sort is introduced in the Supplementary Exercises. In Section 5.4 we will discuss the merge sort and introduce the quick sort in the exercises in that section; the tournament sort is introduced in the exercise set in Section 11.2. We cover sorting algorithms both because sorting is an important problem and because these algorithms can serve as examples for many important concepts.

> Sorting is thought to hold the record as the problem solved by the most fundamentally different algorithms!

THE BUBBLE SORT   The **bubble sort** is one of the simplest sorting algorithms, but not one of the most efficient. It puts a list into increasing order by successively comparing adjacent elements, interchanging them if they are in the wrong order. To carry out the bubble sort, we perform the basic operation, that is, interchanging a larger element with a smaller one following it, starting at the beginning of the list, for a full pass. We iterate this procedure until the sort is complete. Pseudocode for the bubble sort is given as Algorithm 4. We can imagine the elements in the list placed in a column. In the bubble sort, the smaller elements "bubble" to the top as they are interchanged with larger elements. The larger elements "sink" to the bottom. This is illustrated in Example 4.

Links

EXAMPLE 4   Use the bubble sort to put 3, 2, 4, 1, 5 into increasing order.

First pass   Second pass

Third pass   Fourth pass

⌒ : an interchange

(: pair in correct order

numbers in color guaranteed to be in correct order

FIGURE 1   **The Steps of a Bubble Sort.**

Solution: The steps of this algorithm are illustrated in Figure 1. Begin by comparing the first two elements, 3 and 2. Because 3 > 2, interchange 3 and 2, producing the list 2, 3, 4, 1, 5. Because 3 < 4, continue by comparing 4 and 1. Because 4 > 1, interchange 1 and 4, producing the list 2, 3, 1, 4, 5. Because 4 < 5, the first pass is complete. The first pass guarantees that the largest element, 5, is in the correct position.

The second pass begins by comparing 2 and 3. Because these are in the correct order, 3 and 1 are compared. Because 3 > 1, these numbers are interchanged, producing 2, 1, 3, 4, 5. Because 3 < 4, these numbers are in the correct order. It is not necessary to do any more comparisons for this pass because 5 is already in the correct position. The second pass guarantees that the two largest elements, 4 and 5, are in their correct positions.

The third pass begins by comparing 2 and 1. These are interchanged because 2 > 1, producing 1, 2, 3, 4, 5. Because 2 < 3, these two elements are in the correct order. It is not necessary to do any more comparisons for this pass because 4 and 5 are already in the correct positions. The third pass guarantees that the three largest elements, 3, 4, and 5, are in their correct positions.

The fourth pass consists of one comparison, namely, the comparison of 1 and 2. Because 1 < 2, these elements are in the correct order. This completes the bubble sort. ◀

ALGORITHM 4 The Bubble Sort.

procedure *bubblesort*($a_1, \ldots, a_n$: real numbers with $n \geq 2$)
for $i := 1$ **to** $n - 1$
    for $j := 1$ **to** $n - i$
        if $a_j > a_{j+1}$ **then** interchange a_j and a_{j+1}
{$a_1, \ldots, a_n$ is in increasing order}

THE INSERTION SORT   The **insertion sort** is a simple sorting algorithm, but it is usually not the most efficient. To sort a list with n elements, the insertion sort begins with the second element. The insertion sort compares this second element with the first element and inserts it before the first element if it does not exceed the first element and after the first element if it exceeds the first element. At this point, the first two elements are in the correct order. The third element is then compared with the first element, and if it is larger than the first element, it is compared with the second element; it is inserted into the correct position among the first three elements.

In general, in the jth step of the insertion sort, the jth element of the list is inserted into the correct position in the list of the previously sorted $j - 1$ elements. To insert the jth element in the list, a linear search technique is used (see Exercise 43); the jth element is successively compared with the already sorted $j - 1$ elements at the start of the list until the first element that is not less than this element is found or until it has been compared with all $j - 1$ elements; the jth element is inserted in the correct position so that the first j elements are sorted. The algorithm continues until the last element is placed in the correct position relative to the already sorted list of the first $n - 1$ elements. The insertion sort is described in pseudocode in Algorithm 5.

EXAMPLE 5   Use the insertion sort to put the elements of the list 3, 2, 4, 1, 5 in increasing order.

Solution: The insertion sort first compares 2 and 3. Because $3 > 2$, it places 2 in the first position, producing the list 2, 3, 4, 1, 5 (the sorted part of the list is shown in color). At this point, 2 and 3 are in the correct order. Next, it inserts the third element, 4, into the already sorted part of the list by making the comparisons $4 > 2$ and $4 > 3$. Because $4 > 3$, 4 remains in the third position. At this point, the list is 2, 3, 4, 1, 5 and we know that the ordering of the first three elements is correct. Next, we find the correct place for the fourth element, 1, among the already sorted elements, 2, 3, 4. Because $1 < 2$, we obtain the list 1, 2, 3, 4, 5. Finally, we insert 5 into the correct position by successively comparing it to 1, 2, 3, and 4. Because $5 > 4$, it stays at the end of the list, producing the correct order for the entire list.   ◄

ALGORITHM 5 The Insertion Sort.

procedure *insertion sort*($a_1, a_2, \ldots, a_n$: real numbers with $n \geq 2$)
for $j := 2$ **to** n
    $i := 1$
    while $a_j > a_i$
        $i := i + 1$
    $m := a_j$
    for $k := 0$ **to** $j - i - 1$
        $a_{j-k} := a_{j-k-1}$
    $a_i := m$
$\{a_1, \ldots, a_n$ is in increasing order$\}$

Greedy Algorithms

"Greed is good ... Greed is right, greed works. Greed clarifies ..." – spoken by the character Gordon Gecko in the film *Wall Street*.

Links

You have to prove that a greedy algorithm always finds an optimal solution.

Many algorithms we will study in this book are designed to solve **optimization problems**. The goal of such problems is to find a solution to the given problem that either minimizes or maximizes the value of some parameter. Optimization problems studied later in this text include finding a route between two cities with smallest total mileage, determining a way to encode messages using the fewest bits possible, and finding a set of fiber links between network nodes using the least amount of fiber.

Surprisingly, one of the simplest approaches often leads to a solution of an optimization problem. This approach selects the best choice at each step, instead of considering all sequences of steps that may lead to an optimal solution. Algorithms that make what seems to be the "best" choice at each step are called **greedy algorithms**. Once we know that a greedy algorithm finds a feasible solution, we need to determine whether it has found an optimal solution. (Note that we call the algoritm "greedy" whether or not it finds an optimal solution.) To do this, we either prove that the solution is optimal or we show that there is a counterexample where the algorithm yields a nonoptimal solution. To make these concepts more concrete, we will consider an algorithm that makes change using coins.

EXAMPLE 6   Consider the problem of making n cents change with quarters, dimes, nickels, and pennies, and using the least total number of coins. We can devise a greedy algorithm for making change for n cents by making a locally optimal choice at each step; that is, at each step we choose the coin of the largest denomination possible to add to the pile of change without exceeding n cents. For example, to make change for 67 cents, we first select a quarter (leaving 42 cents). We next select a second quarter (leaving 17 cents), followed by a dime (leaving 7 cents), followed by a nickel (leaving 2 cents), followed by a penny (leaving 1 cent), followed by a penny.   ◀

We display a greedy change-making algorithm for n cents, using any set of denominations of coins, as Algorithm 6.

ALGORITHM 6 Greedy Change-Making Algorithm.

---

procedure *change*($c_1, c_2, \ldots, c_r$: values of denominations of coins, where
   $c_1 > c_2 > \cdots > c_r$; n: a positive integer)
for $i := 1$ **to** r
   $d_i := 0$ {d_i counts the coins of denomination c_i used}
   while $n \geq c_i$
      $d_i := d_i + 1$ {add a coin of denomination c_i}
      $n := n - c_i$
{d_i is the number of coins of denomination c_i in the change for $i = 1, 2, \ldots, r$}

We have described a greedy algorithm for making change using any finite set of coins with denominations $c_1, c_2, \ldots, c_r$. In the particular case where the four denominations are quarters dimes, nickels, and pennies, we have $c_1 = 25$, $c_2 = 10$, $c_3 = 5$, and $c_4 = 1$. For this case, we will show that this algorithm leads to an optimal solution in the sense that it uses the fewest coins possible. Before we embark on our proof, we show that there are sets of coins for which the greedy algorithm (Algorithm 6) does not necessarily produce change using the fewest coins possible. For example, if we have only quarters, dimes, and pennies (and no nickels) to use, the greedy algorithm would make change for 30 cents using six coins—a quarter and five pennies—whereas we could have used three coins, namely, three dimes.

LEMMA 1   If n is a positive integer, then n cents in change using quarters, dimes, nickels, and pennies using the fewest coins possible has at most two dimes, at most one nickel, at most four pennies, and cannot have two dimes and a nickel. The amount of change in dimes, nickels, and pennies cannot exceed 24 cents.

Proof: We use a proof by contradiction. We will show that if we had more than the specified number of coins of each type, we could replace them using fewer coins that have the same value. We note that if we had three dimes we could replace them with a quarter and a nickel, if we had two nickels we could replace them with a dime, if we had five pennies we could replace them with a nickel, and if we had two dimes and a nickel we could replace them with a quarter. Because we can have at most two dimes, one nickel, and four pennies, but we cannot have two dimes and a nickel, it follows that 24 cents is the most money we can have in dimes, nickels, and pennies when we make change using the fewest number of coins for n cents.   ◁

THEOREM 1   The greedy algorithm (Algorithm 6) produces change using the fewest coins possible.

Proof: We will use a proof by contradiction. Suppose that there is a positive integer n such that there is a way to make change for n cents using quarters, dimes, nickels, and pennies that uses fewer coins than the greedy algorithm finds. We first note that q', the number of quarters used in this optimal way to make change for n cents, must be the same as q, the number of quarters used by the greedy algorithm. To show this, first note that the greedy algorithm uses the most quarters possible, so $q' \leq q$. However, it is also the case that q' cannot be less than q. If it were, we would need to make up at least 25 cents from dimes, nickels, and pennies in this optimal way to make change. But this is impossible by Lemma 1.

Because there must be the same number of quarters in the two ways to make change, the value of the dimes, nickels, and pennies in these two ways must be the same, and these coins are worth no more than 24 cents. There must be the same number of dimes, because the greedy algorithm used the most dimes possible and by Lemma 1, when change is made using the fewest coins possible, at most one nickel and at most four pennies are used, so that the most dimes possible are also used in the optimal way to make change. Similarly, we have the same number of nickels and, finally, the same number of pennies.   ◁

A greedy algorithm makes the best choice at each step according to a specified criterion. The next example shows that it can be difficult to determine which of many possible criteria to choose.

EXAMPLE 7   Suppose we have a group of proposed talks with preset start and end times. Devise a greedy algorithm to schedule as many of these talks as possible in a lecture hall, under the assumptions that once a talk starts, it continues until it ends, no two talks can proceed at the same time, and a talk can begin at the same time another one ends. Assume that talk j begins at time s_j (where s stands for *start*) and ends at time e_j (where e stands for *end*).

Solution: To use a greedy algorithm to schedule the most talks, that is, an optimal schedule, we need to decide how to choose which talk to add at each step. There are many criteria we could use to select a talk at each step, where we chose from the talks that do not overlap talks already selected. For example, we could add talks in order of earliest start time, we could add talks in order of shortest time, we could add talks in order of earliest finish time, or we could use some other criterion.

We now consider these possible criteria. Suppose we add the talk that starts earliest among the talks compatible with those already selected. We can construct a counterexample to see that the resulting algorithm does not always produce an optimal schedule. For instance, suppose that we have three talks: Talk 1 starts at 8 A.M. and ends at 12 noon, Talk 2 starts at 9 A.M. and ends at 10 A.M., and Talk 3 starts at 11 A.M. and ends at 12 noon. We first select the Talk 1 because it starts earliest. But once we have selected Talk 1 we cannot select either Talk 2 or Talk 3 because both overlap Talk 1. Hence, this greedy algorithm selects only one talk. This is not optimal because we could schedule Talk 2 and Talk 3, which do not overlap.

Now suppose we add the talk that is shortest among the talks that do not overlap any of those already selected. Again we can construct a counterexample to show that this greedy algorithm does not always produce an optimal schedule. So, suppose that we have three talks: Talk 1 starts at 8 A.M. and ends at 9:15 A.M., Talk 2 starts at 9 A.M. and ends at 10 A.M., and Talk 3 starts at 9:45 A.M. and ends at 11 A.M. We select Talk 2 because it is shortest, requiring one hour. Once we select Talk 2, we cannot select either Talk 1 or Talk 3 because neither is compatible with Talk 2. Hence, this greedy algorithm selects only one talk. However, it is possible to select two talks, Talk 1 and Talk 3, which are compatible.

However, it can be shown that we schedule the most talks possible if in each step we select the talk with the earliest ending time among the talks compatible with those already selected. We will prove this in Chapter 5 using the method of mathematical induction. The first step we will make is to sort the talks according to increasing finish time. After this sorting, we relabel the talks so that $e_1 \leq e_2 \leq \ldots \leq e_n$. The resulting greedy algorithm is given as Algorithm 7. ◀

ALGORITHM 7 Greedy Algorithm for Scheduling Talks.

procedure *schedule*($s_1 \le s_2 \le \cdots \le s_n$: start times of talks,
  $e_1 \le e_2 \le \cdots \le e_n$: ending times of talks)
sort talks by finish time and reorder so that $e_1 \le e_2 \le \ldots \le e_n$
$S := \emptyset$
for $j := 1$ **to** n
   if talk j is compatible with S **then**
      $S := S \cup \{\text{talk } j\}$
return $S\{S$ is the set of talks scheduled$\}$

The Halting Problem

We will now describe a proof of one of the most famous theorems in computer science. We will show that there is a problem that cannot be solved using any procedure. That is, we will show there are unsolvable problems. The problem we will study is the **halting problem**. It asks whether there is a procedure that does this: It takes as input a computer program and input to the program and determines whether the program will eventually stop when run with this input. It would be convenient to have such a procedure, if it existed. Certainly being able to test whether a program entered into an infinite loop would be helpful when writing and debugging programs. However, in 1936 Alan Turing showed that no such procedure exists (see his biography in Section 13.4).

Before we present a proof that the halting problem is unsolvable, first note that we cannot simply run a program and observe what it does to determine whether it terminates when run with the given input. If the program halts, we have our answer, but if it is still running after any fixed length of time has elapsed, we do not know whether it will never halt or we just did not wait long enough for it to terminate. After all, it is not hard to design a program that will stop only after more than a billion years has elapsed.

We will describe Turing's proof that the halting problem is unsolvable; it is a proof by contradiction. (The reader should note that our proof is not completely rigorous, because we have not explicitly defined what a procedure is. To remedy this, the concept of a Turing machine is needed.)

Proof: Assume there is a solution to the halting problem, a procedure called $H(P, I)$. The procedure $H(P, I)$ takes two inputs, one a program P and the other I, an input to the program P. $H(P, I)$ generates the string "halt" as output if H determines that P stops when given I as input. Otherwise, $H(P, I)$ generates the string "loops forever" as output. We will now derive a contradiction.

When a procedure is coded, it is expressed as a string of characters; this string can be interpreted as a sequence of bits. This means that a program itself can be used as data. Therefore a program can be thought of as input to another program, or even itself. Hence, H can take a program P as both of its inputs, which are a program and input to this program. H should be able to determine whether P will halt when it is given a copy of itself as input.

To show that no procedure H exists that solves the halting problem, we construct a simple procedure $K(P)$, which works as follows, making use of the output $H(P, P)$. If the output of $H(P, P)$ is "loops forever," which means that P loops forever when given a copy of itself as input, then $K(P)$ halts. If the output of $H(P, P)$ is "halt," which means that P halts when given a copy of itself as input, then $K(P)$ loops forever. That is, $K(P)$ does the opposite of what the output of $H(P, P)$ specifies. (See Figure 2.)

Now suppose we provide K as input to K. We note that if the output of $H(K, K)$ is "loops forever," then by the definition of K we see that $K(K)$ halts. Otherwise, if the output of $H(K, K)$

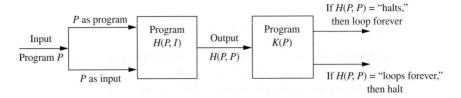

FIGURE 2   Showing that the Halting Problem is Unsolvable.

is "halt," then by the definition of K we see that $K(K)$ loops forever, in violation of what H tells us. In both cases, we have a contradiction.

Thus, H cannot always give the correct answers. Consequently, there is no procedure that solves the halting problem.   ◁

Exercises

1. List all the steps used by Algorithm 1 to find the maximum of the list 1, 8, 12, 9, 11, 2, 14, 5, 10, 4.

2. Determine which characteristics of an algorithm described in the text (after Algorithm 1) the following procedures have and which they lack.

   a) **procedure** *double*(*n*: positive integer)
      while $n > 0$
      $n := 2n$

   b) **procedure** *divide*(*n*: positive integer)
      while $n \geq 0$
      $m := 1/n$
      $n := n - 1$

   c) **procedure** *sum*(*n*: positive integer)
      $sum := 0$
      while $i < 10$
      $sum := sum + i$

   d) **procedure** *choose*(*a*, *b*: integers)
      $x := $ either a or b

3. Devise an algorithm that finds the sum of all the integers in a list.

4. Describe an algorithm that takes as input a list of n integers in nondecreasing order and produces the list of all values that occur more than once. (Recall that a list of integers is **nondecreasing** if each integer in the list is at least as large as the previous integer in the list.)

5. Describe an algorithm that takes as input a list of n integers and finds the number of negative integers in the list.

6. Describe an algorithm that takes as input a list of n integers and finds the location of the last even integer in the list or returns 0 if there are no even integers in the list.

7. A **palindrome** is a string that reads the same forward and backward. Describe an algorithm for determining whether a string of n characters is a palindrome.

8. Devise an algorithm to compute x^n, where x is a real number and n is an integer. [*Hint:* First give a procedure for computing x^n when n is nonnegative by successive multiplication by x, starting with 1. Then extend this procedure, and use the fact that $x^{-n} = 1/x^n$ to compute x^n when n is negative.]

9. Describe an algorithm that interchanges the values of the variables x and y, using only assignments. What is the minimum number of assignment statements needed to do this?

10. Describe an algorithm that uses only assignment statements that replaces the triple (x, y, z) with (y, z, x). What is the minimum number of assignment statements needed?

11. List all the steps used to search for 9 in the sequence 1, 3, 4, 5, 6, 8, 9, 11 using

    a) a linear search.   b) a binary search.

12. List all the steps used to search for 7 in the sequence given in Exercise 13 for both a linear search and a binary search.

13. Describe an algorithm that inserts an integer x in the appropriate position into the list $a_1, a_2, \ldots, a_n$ of integers that are in increasing order.

14. Describe an algorithm for finding the smallest integer in a finite sequence of natural numbers.

15. Describe an algorithm that locates the first occurrence of the largest element in a finite list of integers, where the integers in the list are not necessarily distinct.

16. Describe an algorithm that locates the last occurrence of the smallest element in a finite list of integers, where the integers in the list are not necessarily distinct.

17. Describe an algorithm that produces the maximum, median, mean, and minimum of a set of three integers. (The **median** of a set of integers is the middle element in the list when these integers are listed in order of increasing size. The **mean** of a set of integers is the sum of the integers divided by the number of integers in the set.)

18. Describe an algorithm for finding both the largest and the smallest integers in a finite sequence of integers.

19. Describe an algorithm that puts the first three terms of a sequence of integers of arbitrary length in increasing order.

20. Describe an algorithm that determines whether a function from a finite set of integers to another finite set of integers is onto.

21. Describe an algorithm that will count the number of 1s in a bit string by examining each bit of the string to determine whether it is a 1 bit.

22. Change Algorithm 3 so that the binary search procedure compares x to a_m at each stage of the algorithm, with the algorithm terminating if $x = a_m$. What advantage does this version of the algorithm have?

23. The **ternary search algorithm** locates an element in a list of increasing integers by successively splitting the list into three sublists of equal (or as close to equal as possible) size, and restricting the search to the appropriate piece. Specify the steps of this algorithm.

24. Specify the steps of an algorithm that locates an element in a list of increasing integers by successively splitting the list into four sublists of equal (or as close to equal as possible) size, and restricting the search to the appropriate piece.

In a list of elements the same element may appear several times. A **mode** of such a list is an element that occurs at least as often as each of the other elements; a list has more than one mode when more than one element appears the maximum number of times.

25. Devise an algorithm that finds a mode in a list of nondecreasing integers. (Recall that a list of integers is nondecreasing if each term is at least as large as the preceding term.)

26. Devise an algorithm that finds all modes. (Recall that a list of integers is nondecreasing if each term of the list is at least as large as the preceding term.)

27. Devise an algorithm that finds the first term of a sequence of integers that equals some previous term in the sequence.

28. Devise an algorithm that finds the first term of a sequence of positive integers that is less than the immediately preceding term of the sequence.

29. Use the bubble sort to sort 6, 2, 3, 1, 5, 4, showing the lists obtained at each step.

30. Use the bubble sort to sort d, f, k, m, a, b, showing the lists obtained at each step.

*31. Adapt the bubble sort algorithm so that it stops when no interchanges are required. Express this more efficient version of the algorithm in pseudocode.

32. Use the insertion sort to sort the list in Exercise 29, showing the lists obtained at each step.

33. Use the insertion sort to sort the list in Exercise 30, showing the lists obtained at each step.

The **selection sort** begins by finding the least element in the list. This element is moved to the front. Then the least element among the remaining elements is found and put into the second position. This procedure is repeated until the entire list has been sorted.

34. Write the selection sort algorithm in pseudocode.

35. Sort these lists using the selection sort.

a) 3, 5, 4, 1, 2   b) 5, 4, 3, 2, 1
c) 1, 2, 3, 4, 5

36. Describe an algorithm based on the binary search for determining the correct position in which to insert a new element in an already sorted list.

37. Describe an algorithm based on the linear search for determining the correct position in which to insert a new element in an already sorted list.

38. How many comparisons does the insertion sort use to sort the list $n, n - 1, \ldots, 2, 1$?

39. How many comparisons does the insertion sort use to sort the list $1, 2, \ldots, n$?

The **binary insertion sort** is a variation of the insertion sort that uses a binary search technique (see Exercise 36) rather than a linear search technique to insert the ith element in the correct place among the previously sorted elements.

40. Compare the number of comparisons used by the insertion sort and the binary insertion sort to sort the list 7, 4, 3, 8, 1, 5, 4, 2.

41. Show all the steps used by the binary insertion sort to sort the list 3, 2, 4, 5, 1, 6.

42. a) Devise a variation of the insertion sort that uses a linear search technique that inserts the jth element in the correct place by first comparing it with the $(j - 1)$st element, then the $(j - 2)$th element if necessary, and so on.

   b) Use your algorithm to sort 3, 2, 4, 5, 1, 6.

   c) Answer Exercise 38 using this algorithm.

   d) Answer Exercise 39 using this algorithm.

43. When a list of elements is in close to the correct order, would it be better to use an insertion sort or its variation described in Exercise 42?

44. Use the greedy algorithm to make change using quarters, dimes, nickels, and pennies for

   a) 51 cents.   b) 69 cents.
   c) 76 cents.   d) 60 cents.

45. Use the greedy algorithm to make change using quarters, dimes, and pennies (but no nickels) for each of the amounts given in Exercise 44. For which of these amounts does the greedy algorithm use the fewest coins of these denominations possible?

46. Show that if there were a coin worth 12 cents, the greedy algorithm using quarters, 12-cent coins, dimes, nickels, and pennies would not always produce change using the fewest coins possible.

47. Use Algorithm 7 to schedule the largest number of talks in a lecture hall from a proposed set of talks, if the starting and ending times of the talks are 9:00 A.M. and 9:45 A.M.; 9:30 A.M. and 10:00 A.M.; 9:50 A.M. and 10:15 A.M.; 10:00 A.M. and 10:30 A.M.; 10:10 A.M. and 10:25 A.M.; 10:30 A.M. and 10:55 A.M.; 10:15 A.M. and 10:45 A.M.; 10:30 A.M. and 11:00 A.M.; 10:45 A.M. and 11:30 A.M.; 10:55 A.M. and 11:25 A.M.; 11:00 A.M. and 11:15 A.M.

48. Show that a greedy algorithm that schedules talks in a lecture hall, as described in Example 7, by selecting at each step the talk that overlaps the fewest other talks, does not always produce an optimal schedule.

*49. a) Devise a greedy algorithm that determines the fewest lecture halls needed to accommodate n talks given the starting and ending time for each talk.
  b) Prove that your algorithm is optimal.

Suppose we have s men $m_1, m_2, \ldots, m_s$ and s women $w_1, w_2, \ldots, w_s$. We wish to match each person with a member of the opposite gender. Furthermore, suppose that each person ranks, in order of preference, with no ties, the people of the opposite gender. We say that a matching of people of opposite genders to form couples is **stable** if we cannot find a man m and a woman w who are not assigned to each other such that m prefers w over his assigned partner and w prefers m to her assigned partner.

50. Suppose we have three men m_1, m_2, and m_3 and three women w_1, w_2, and w_3. Furthermore, suppose that the preference rankings of the men for the three women, from highest to lowest, are m_1: w_3, w_1, w_2; m_2: w_1, w_2, w_3; m_3: w_2, w_3, w_1; and the preference rankings of the women for the three men, from highest to lowest, are w_1: m_1, m_2, m_3; w_2: m_2, m_1, m_3; w_3: m_3, m_2, m_1. For each of the six possible matchings of men and women to form three couples, determine whether this matching is stable.

3.2 The Growth of Functions

Introduction

In Section 3.1 we discussed the concept of an algorithm. We introduced algorithms that solve a variety of problems, including searching for an element in a list and sorting a list. In Section 3.3 we will study the number of operations used by these algorithms. In particular, we will estimate the number of comparisons used by the linear and binary search algorithms to find an element in a sequence of n elements. We will also estimate the number of comparisons used by the bubble sort and by the insertion sort to sort a list of n elements. The time required to solve a problem depends on more than only the number of operations it uses. The time also depends on the hardware and software used to run the program that implements the algorithm. However, when we change the hardware and software used to implement an algorithm, we can closely approximate the time required to solve a problem of size n by multiplying the previous time required by a constant. For example, on a supercomputer we might be able to solve a problem of size n a million times faster than we can on a PC. However, this factor of one million will not depend on n (except perhaps in some minor ways). One of the advantages of using **big-O notation**, which we introduce in this section, is that we can estimate the growth of a function without worrying about constant multipliers or smaller order terms. This means that, using big-O notation, we do not have to worry about the hardware and software used to implement an algorithm. Furthermore, using big-O notation, we can assume that the different operations used in an algorithm take the same time, which simplifies the analysis considerably.

Big-O notation is used extensively to estimate the number of operations an algorithm uses as its input grows. With the help of this notation, we can determine whether it is practical to use a particular algorithm to solve a problem as the size of the input increases. Furthermore, using big-O notation, we can compare two algorithms to determine which is more efficient as the size of the input grows. For instance, if we have two algorithms for solving a problem, one using $100n^2 + 17n + 4$ operations and the other using n^3 operations, big-O notation can help us see that the first algorithm uses far fewer operations when n is large, even though it uses more operations for small values of n, such as $n = 10$.

This section introduces big-O notation and the related big-Omega and big-Theta notations. We will explain how big-O, big-Omega, and big-Theta estimates are constructed and establish estimates for some important functions that are used in the analysis of algorithms.

Big-O Notation

The growth of functions is often described using a special notation. Definition 1 describes this notation.

DEFINITION 1   Let f and g be functions from the set of integers or the set of real numbers to the set of real numbers. We say that $f(x)$ is $O(g(x))$ if there are constants C and k such that

$$|f(x)| \leq C|g(x)|$$

whenever $x > k$. [This is read as "$f(x)$ is big-oh of $g(x)$."]

Remark: Intuitively, the definition that $f(x)$ is $O(g(x))$ says that $f(x)$ grows slower that some fixed multiple of $g(x)$ as x grows without bound.

Assessment

Links

The constants C and k in the definition of big-O notation are called **witnesses** to the relationship $f(x)$ is $O(g(x))$. To establish that $f(x)$ is $O(g(x))$ we need only one pair of witnesses to this relationship. That is, to show that $f(x)$ is $O(g(x))$, we need find only *one* pair of constants C and k, the witnesses, such that $|f(x)| \leq C|g(x)|$ whenever $x > k$.

Note that when there is one pair of witnesses to the relationship $f(x)$ is $O(g(x))$, there are *infinitely many* pairs of witnesses. To see this, note that if C and k are one pair of witnesses, then any pair C' and k', where $C < C'$ and $k < k'$, is also a pair of witnesses, because $|f(x)| \leq C|g(x)| \leq C'|g(x)|$ whenever $x > k' > k$.

THE HISTORY OF BIG-O NOTATION   Big-O notation has been used in mathematics for more than a century. In computer science it is widely used in the analysis of algorithms, as will be seen in Section 3.3. The German mathematician Paul Bachmann first introduced big-O notation in 1892 in an important book on number theory. The big-O symbol is sometimes called a **Landau symbol** after the German mathematician Edmund Landau, who used this notation throughout his work. The use of big-O notation in computer science was popularized by Donald Knuth, who also introduced the big-Ω and big-Θ notations defined later in this section.

WORKING WITH THE DEFINITION OF BIG-O NOTATION   A useful approach for finding a pair of witnesses is to first select a value of k for which the size of $|f(x)|$ can be readily

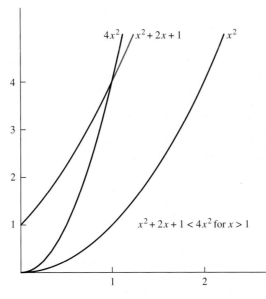

The part of the graph of $f(x) = x^2 + 2x + 1$ that satisfies $f(x) < 4x^2$ is shown in blue.

FIGURE 1   The Function $x^2 + 2x + 1$ is $O(x^2)$.

estimated when $x > k$ and to see whether we can use this estimate to find a value of C for which $|f(x)| \le C|g(x)|$ for $x > k$. This approach is illustrated in Example 1.

EXAMPLE 1   Show that $f(x) = x^2 + 2x + 1$ is $O(x^2)$.

Extra
Examples

Solution: We observe that we can readily estimate the size of $f(x)$ when $x > 1$ because $x < x^2$ and $1 < x^2$ when $x > 1$. It follows that

$$0 \le x^2 + 2x + 1 \le x^2 + 2x^2 + x^2 = 4x^2$$

whenever $x > 1$, as shown in Figure 1. Consequently, we can take $C = 4$ and $k = 1$ as witnesses to show that $f(x)$ is $O(x^2)$. That is, $f(x) = x^2 + 2x + 1 < 4x^2$ whenever $x > 1$. (Note that it is not necessary to use absolute values here because all functions in these equalities are positive when x is positive.)

Alternatively, we can estimate the size of $f(x)$ when $x > 2$. When $x > 2$, we have $2x \le x^2$ and $1 \le x^2$. Consequently, if $x > 2$, we have

$$0 \le x^2 + 2x + 1 \le x^2 + x^2 + x^2 = 3x^2.$$

It follows that $C = 3$ and $k = 2$ are also witnesses to the relation $f(x)$ is $O(x^2)$.

Observe that in the relationship "$f(x)$ is $O(x^2)$," x^2 can be replaced by any function with larger values than x^2. For example, $f(x)$ is $O(x^3)$, $f(x)$ is $O(x^2 + x + 7)$, and so on.

It is also true that x^2 is $O(x^2 + 2x + 1)$, because $x^2 < x^2 + 2x + 1$ whenever $x > 1$. This means that $C = 1$ and $k = 1$ are witnesses to the relationship x^2 is $O(x^2 + 2x + 1)$.   ◀

Note that in Example 1 we have two functions, $f(x) = x^2 + 2x + 1$ and $g(x) = x^2$, such that $f(x)$ is $O(g(x))$ and $g(x)$ is $O(f(x))$—the latter fact following from the inequality $x^2 \le x^2 + 2x + 1$, which holds for all nonnegative real numbers x. We say that two functions $f(x)$ and $g(x)$ that satisfy both of these big-O relationships are of the **same order**. We will return to this notion later in this section.

Remark: The fact that $f(x)$ is $O(g(x))$ is sometimes written $f(x) = O(g(x))$. However, the equals sign in this notation does *not* represent a genuine equality. Rather, this notation tells

Links

PAUL GUSTAV HEINRICH BACHMANN (1837–1920)   Paul Bachmann, the son of a Lutheran pastor, shared his father's pious lifestyle and love of music. His mathematical talent was discovered by one of his teachers, even though he had difficulties with some of his early mathematical studies. After recuperating from tuberculosis in Switzerland, Bachmann studied mathematics, first at the University of Berlin and later at Göttingen, where he attended lectures presented by the famous number theorist Dirichlet. He received his doctorate under the German number theorist Kummer in 1862; his thesis was on group theory. Bachmann was a professor at Breslau and later at Münster. After he retired from his professorship, he continued his mathematical writing, played the piano, and served as a music critic for newspapers. Bachmann's mathematical writings include a five-volume survey of results and methods in number theory, a two-volume work on elementary number theory, a book on irrational numbers, and a book on the famous conjecture known as Fermat's Last Theorem. He introduced big-O notation in his 1892 book *Analytische Zahlentheorie*.

Links

EDMUND LANDAU (1877–1938)   Edmund Landau, the son of a Berlin gynecologist, attended high school and university in Berlin. He received his doctorate in 1899, under the direction of Frobenius. Landau first taught at the University of Berlin and then moved to Göttingen, where he was a full professor until the Nazis forced him to stop teaching. Landau's main contributions to mathematics were in the field of analytic number theory. In particular, he established several important results concerning the distribution of primes. He authored a three-volume exposition on number theory as well as other books on number theory and mathematical analysis.

us that an inequality holds relating the values of the functions f and g for sufficiently large numbers in the domains of these functions. However, it is acceptable to write $f(x) \in O(g(x))$ because $O(g(x))$ represents the set of functions that are $O(g(x))$.

When $f(x)$ is $O(g(x))$, and $h(x)$ is a function that has larger absolute values than $g(x)$ does for sufficiently large values of x, it follows that $f(x)$ is $O(h(x))$. In other words, the function $g(x)$ in the relationship $f(x)$ is $O(g(x))$ can be replaced by a function with larger absolute values. To see this, note that if

$$|f(x)| \le C|g(x)| \qquad \text{if } x > k,$$

and if $|h(x)| > |g(x)|$ for all $x > k$, then

$$|f(x)| \le C|h(x)| \qquad \text{if } x > k.$$

Hence, $f(x)$ is $O(h(x))$.

When big-O notation is used, the function g in the relationship $f(x)$ is $O(g(x))$ is chosen to be as small as possible (sometimes from a set of reference functions, such as functions of the form x^n, where n is a positive integer).

In subsequent discussions, we will almost always deal with functions that take on only positive values. All references to absolute values can be dropped when working with big-O estimates for such functions. Figure 2 illustrates the relationship $f(x)$ is $O(g(x))$.

Example 2 illustrates how big-O notation is used to estimate the growth of functions.

 Links

 DONALD E. KNUTH (BORN 1938)   Knuth grew up in Milwaukee, where his father taught bookkeeping at a Lutheran high school and owned a small printing business. He was an excellent student, earning academic achievement awards. He applied his intelligence in unconventional ways, winning a contest when he was in the eighth grade by finding over 4500 words that could be formed from the letters in "Ziegler's Giant Bar." This won a television set for his school and a candy bar for everyone in his class.

Knuth had a difficult time choosing physics over music as his major at the Case Institute of Technology. He then switched from physics to mathematics, and in 1960 he received his bachelor of science degree, simultaneously receiving a master of science degree by a special award of the faculty who considered his work outstanding. At Case, he managed the basketball team and applied his talents by constructing a formula for the value of each player. This novel approach was covered by *Newsweek* and by Walter Cronkite on the CBS television network. Knuth began graduate work at the California Institute of Technology in 1960 and received his Ph.D. there in 1963. During this time he worked as a consultant, writing compilers for different computers.

Knuth joined the staff of the California Institute of Technology in 1963, where he remained until 1968, when he took a job as a full professor at Stanford University. He retired as Professor Emeritus in 1992 to concentrate on writing. He is especially interested in updating and completing new volumes of his series *The Art of Computer Programming,* a work that has had a profound influence on the development of computer science, which he began writing as a graduate student in 1962, focusing on compilers. In common jargon, "Knuth," referring to *The Art of Computer Programming,* has come to mean the reference that answers all questions about such topics as data structures and algorithms.

Knuth is the founder of the modern study of computational complexity. He has made fundamental contributions to the subject of compilers. His dissatisfaction with mathematics typography sparked him to invent the now widely used TeX and Metafont systems. TeX has become a standard language for computer typography. Two of the many awards Knuth has received are the 1974 Turing Award and the 1979 National Medal of Technology, awarded to him by President Carter.

Knuth has written for a wide range of professional journals in computer science and in mathematics. However, his first publication, in 1957, when he was a college freshman, was a parody of the metric system called "The Potrzebie Systems of Weights and Measures," which appeared in *MAD Magazine* and has been in reprint several times. He is a church organist, as his father was. He is also a composer of music for the organ. Knuth believes that writing computer programs can be an aesthetic experience, much like writing poetry or composing music.

Knuth pays \$2.56 for the first person to find each error in his books and \$0.32 for significant suggestions. If you send him a letter with an error (you will need to use regular mail, because he has given up reading e-mail), he will eventually inform you whether you were the first person to tell him about this error. Be prepared for a long wait, because he receives an overwhelming amount of mail. (The author received a letter years after sending an error report to Knuth, noting that this report arrived several months after the first report of this error.)

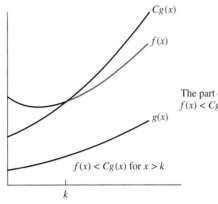

The part of the graph of $f(x)$ that satisfies $f(x) < Cg(x)$ is shown in color.

FIGURE 2 **The Function $f(x)$ is $O(g(x))$.**

EXAMPLE 2 Show that $7x^2$ is $O(x^3)$.

Solution: Note that when $x > 7$, we have $7x^2 < x^3$. (We can obtain this inequality by multiplying both sides of $x > 7$ by x^2.) Consequently, we can take $C = 1$ and $k = 7$ as witnesses to establish the relationship $7x^2$ is $O(x^3)$. Alternatively, when $x > 1$, we have $7x^2 < 7x^3$, so that $C = 7$ and $k = 1$ are also witnesses to the relationship $7x^2$ is $O(x^3)$. ◄

Example 3 illustrates how to show that a big-O relationship does not hold.

EXAMPLE 3 Show that n^2 is not $O(n)$.

Solution: To show that n^2 is not $O(n)$, we must show that no pair of witnesses C and k exist such that $n^2 \leq Cn$ whenever $n > k$. We will use a proof by contradiction to show this.

Suppose that there are constants C and k for which $n^2 \leq Cn$ whenever $n > k$. Observe that when $n > 0$ we can divide both sides of the inequality $n^2 \leq Cn$ by n to obtain the equivalent inequality $n \leq C$. However, no matter what C and k are, the inequality $n \leq C$ cannot hold for all n with $n > k$. In particular, once we set a value of k, we see that when n is larger than the maximum of k and C, it is not true that $n \leq C$ even though $n > k$. This contradiction shows that n^2 in not $O(n)$. ◄

EXAMPLE 4 Example 2 shows that $7x^2$ is $O(x^3)$. Is it also true that x^3 is $O(7x^2)$?

Solution: To determine whether x^3 is $O(7x^2)$, we need to determine whether witnesses C and k exist, so that $x^3 \leq C(7x^2)$ whenever $x > k$. We will show that no such witnesses exist using a proof by contradiction.

If C and k are witnesses, the inequality $x^3 \leq C(7x^2)$ holds for all $x > k$. Observe that the inequality $x^3 \leq C(7x^2)$ is equivalent to the inequality $x \leq 7C$, which follows by dividing both sides by the positive quantity x^2. However, no matter what C is, it is not the case that $x \leq 7C$ for all $x > k$ no matter what k is, because x can be made arbitrarily large. It follows that no witnesses C and k exist for this proposed big-O relationship. Hence, x^3 is *not* $O(7x^2)$. ◄

Big-O Estimates for Some Important Functions

Polynomials can often be used to estimate the growth of functions. Instead of analyzing the growth of polynomials each time they occur, we would like a result that can always be used to estimate the growth of a polynomial. Theorem 1 does this. It shows that the leading term of a polynomial dominates its growth by asserting that a polynomial of degree n or less is $O(x^n)$.

THEOREM 1   Let $f(x) = a_nx^n + a_{n-1}x^{n-1} + \cdots + a_1x + a_0$, where $a_0, a_1, \ldots, a_{n-1}, a_n$ are real numbers. Then $f(x)$ is $O(x^n)$.

Proof: Using the triangle inequality (see Exercise 7 in Section 1.8), if $x > 1$ we have

$$
\begin{aligned}
|f(x)| &= |a_nx^n + a_{n-1}x^{n-1} + \cdots + a_1x + a_0| \\
&\leq |a_n|x^n + |a_{n-1}|x^{n-1} + \cdots + |a_1|x + |a_0| \\
&= x^n \left(|a_n| + |a_{n-1}|/x + \cdots + |a_1|/x^{n-1} + |a_0|/x^n\right) \\
&\leq x^n \left(|a_n| + |a_{n-1}| + \cdots + |a_1| + |a_0|\right).
\end{aligned}
$$

This shows that

$$
|f(x)| \leq Cx^n,
$$

where $C = |a_n| + |a_{n-1}| + \cdots + |a_0|$ whenever $x > 1$. Hence, the witnesses $C = |a_n| + |a_{n-1}| + \cdots + |a_0|$ and $k = 1$ show that $f(x)$ is $O(x^n)$.   ◁

We now give some examples involving functions that have the set of positive integers as their domains.

EXAMPLE 5   How can big-O notation be used to estimate the sum of the first n positive integers?

Solution: Because each of the integers in the sum of the first n positive integers does not exceed n, it follows that

$$
1 + 2 + \cdots + n \leq n + n + \cdots + n = n^2.
$$

From this inequality it follows that $1 + 2 + 3 + \cdots + n$ is $O(n^2)$, taking $C = 1$ and $k = 1$ as witnesses. (In this example the domains of the functions in the big-O relationship are the set of positive integers.)   ◀

In Example 6 big-O estimates will be developed for the factorial function and its logarithm. These estimates will be important in the analysis of the number of steps used in sorting procedures.

EXAMPLE 6   Give big-O estimates for the factorial function and the logarithm of the factorial function, where the factorial function $f(n) = n!$ is defined by

$$
n! = 1 \cdot 2 \cdot 3 \cdot \cdots \cdot n
$$

whenever n is a positive integer, and $0! = 1$. For example,

$$
1! = 1, \quad 2! = 1 \cdot 2 = 2, \quad 3! = 1 \cdot 2 \cdot 3 = 6, \quad 4! = 1 \cdot 2 \cdot 3 \cdot 4 = 24.
$$

Note that the function $n!$ grows rapidly. For instance,

$$
20! = 2{,}432{,}902{,}008{,}176{,}640{,}000.
$$

Solution: A big-O estimate for $n!$ can be obtained by noting that each term in the product does not exceed n. Hence,

$$n! = 1 \cdot 2 \cdot 3 \cdot \cdots \cdot n$$
$$\leq n \cdot n \cdot n \cdot \cdots \cdot n$$
$$= n^n.$$

This inequality shows that $n!$ is $O(n^n)$, taking $C = 1$ and $k = 1$ as witnesses. Taking logarithms of both sides of the inequality established for $n!$, we obtain

$$\log n! \leq \log n^n = n \log n.$$

This implies that $\log n!$ is $O(n \log n)$, again taking $C = 1$ and $k = 1$ as witnesses. ◀

EXAMPLE 7   In Section 4.1, we will show that $n < 2^n$ whenever n is a positive integer. Show that this inequality implies that n is $O(2^n)$, and use this inequality to show that $\log n$ is $O(n)$.

Solution: Using the inequality $n < 2^n$, we quickly can conclude that n is $O(2^n)$ by taking $k = C = 1$ as witnesses. Note that because the logarithm function is increasing, taking logarithms (base 2) of both sides of this inequality shows that

$$\log n < n.$$

It follows that

$$\log n \text{ is } O(n).$$

(Again we take $C = k = 1$ as witnesses.)

If we have logarithms to a base b, where b is different from 2, we still have $\log_b n$ is $O(n)$ because

$$\log_b n = \frac{\log n}{\log b} < \frac{n}{\log b}$$

whenever n is a positive integer. We take $C = 1/\log b$ and $k = 1$ as witnesses. (We have used Theorem 3 in Appendix 2 to see that $\log_b n = \log n / \log b$.) ◀

As mentioned before, big-O notation is used to estimate the number of operations needed to solve a problem using a specified procedure or algorithm. The functions used in these estimates often include the following:

$$1, \ \log n, \ n, \ n \log n, \ n^2, \ 2^n, \ n!$$

Using calculus it can be shown that each function in the list is smaller than the succeeding function, in the sense that the ratio of a function and the succeeding function tends to zero as n grows without bound. Figure 3 displays the graphs of these functions, using a scale for the values of the functions that doubles for each successive marking on the graph. That is, the vertical scale in this graph is logarithmic.

USEFUL BIG-O ESTIMATES INVOLVING LOGARITHMS, POWERS, AND EXPONENTIAL FUNCTIONS   We now give some useful facts that help us determine whether big-O relationships hold between pairs of functions when each of the functions is a power of a logarithm, a power, or an exponential function of the form b^n where $b > 1$. Their proofs are left as Exercises 57–60 for readers skilled with calculus.

Theorem 1 shows that if $f(n)$ is a polynomial of degree d, then $f(n)$ is $O(n^d)$. Applying this theorem, we see that if $d > c > 1$, then n^c is $O(n^d)$. We leave it to the reader to show that the reverse of this relationship does not hold. Putting these facts together, we see that if $d > c > 1$, then

$$n^c \text{ is } O(n^d), \text{ but } n^d \text{ is not } O(n^c).$$

In Example 7 we showed that $\log_b n$ is $O(n)$ whenever $b > 1$. More generally, whenever $b > 1$ and c and d are positive, we have

$$(\log_b n)^c \text{ is } O(n^d), \text{ but } n^d \text{ is not } (O(\log_b n)^c).$$

This tells us that every positive power of the logarithm of n to the base b, where $b > 1$, is big-O of every positive power of n, but the reverse relationship never holds.

In Example 7, we also showed that n is $O(2^n)$. More generally, whenever d is positive and $b > 1$, we have

$$n^d \text{ is } O(b^n), \text{ but } b^n \text{ is not } O(n^d).$$

This tells us that every power of n is big-O of every exponential function of n with a base that is greater than one, but the reverse relationship never holds. Furthermore, we have when $c > b > 1$,

$$b^n \text{ is } O(c^n) \text{ but } c^n \text{ is not } O(b^n).$$

This tells us that if we have two exponential functions with different bases greater than one, one of these functions is big-O of the other if and only if its base is smaller or equal.

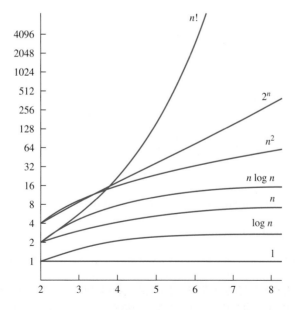

FIGURE 3   **A Display of the Growth of Functions Commonly Used in Big-O Estimates.**

The Growth of Combinations of Functions

Many algorithms are made up of two or more separate subprocedures. The number of steps used by a computer to solve a problem with input of a specified size using such an algorithm is the sum of the number of steps used by these subprocedures. To give a big-O estimate for the number of steps needed, it is necessary to find big-O estimates for the number of steps used by each subprocedure and then combine these estimates.

Big-O estimates of combinations of functions can be provided if care is taken when different big-O estimates are combined. In particular, it is often necessary to estimate the growth of the sum and the product of two functions. What can be said if big-O estimates for each of two functions are known? To see what sort of estimates hold for the sum and the product of two functions, suppose that $f_1(x)$ is $O(g_1(x))$ and $f_2(x)$ is $O(g_2(x))$.

From the definition of big-O notation, there are constants C_1, C_2, k_1, and k_2 such that

$$|f_1(x)| \le C_1|g_1(x)|$$

when $x > k_1$, and

$$|f_2(x)| \le C_2|g_2(x)|$$

when $x > k_2$. To estimate the sum of $f_1(x)$ and $f_2(x)$, note that

$$|(f_1 + f_2)(x)| = |f_1(x) + f_2(x)|$$
$$\le |f_1(x)| + |f_2(x)| \quad \text{using the triangle inequality } |a + b| \le |a| + |b|.$$

When x is greater than both k_1 and k_2, it follows from the inequalities for $|f_1(x)|$ and $|f_2(x)|$ that

$$|f_1(x)| + |f_2(x)| \le C_1|g_1(x)| + C_2|g_2(x)|$$
$$\le C_1|g(x)| + C_2|g(x)|$$
$$= (C_1 + C_2)|g(x)|$$
$$= C|g(x)|,$$

where $C = C_1 + C_2$ and $g(x) = \max(|g_1(x)|, |g_2(x)|)$. [Here $\max(a, b)$ denotes the maximum, or larger, of a and b.]

This inequality shows that $|(f_1 + f_2)(x)| \le C|g(x)|$ whenever $x > k$, where $k = \max(k_1, k_2)$. We state this useful result as Theorem 2.

THEOREM 2   Suppose that $f_1(x)$ is $O(g_1(x))$ and that $f_2(x)$ is $O(g_2(x))$. Then $(f_1 + f_2)(x)$ is $O(\max(|g_1(x)|, |g_2(x)|))$.

We often have big-O estimates for f_1 and f_2 in terms of the same function g. In this situation, Theorem 2 can be used to show that $(f_1 + f_2)(x)$ is also $O(g(x))$, because $\max(g(x), g(x)) = g(x)$. This result is stated in Corollary 1.

COROLLARY 1   Suppose that $f_1(x)$ and $f_2(x)$ are both $O(g(x))$. Then $(f_1 + f_2)(x)$ is $O(g(x))$.

In a similar way big-O estimates can be derived for the product of the functions f_1 and f_2. When x is greater than $\max(k_1, k_2)$ it follows that

$$|(f_1 f_2)(x)| = |f_1(x)||f_2(x)|$$
$$\leq C_1|g_1(x)|C_2|g_2(x)|$$
$$\leq C_1 C_2|(g_1 g_2)(x)|$$
$$\leq C|(g_1 g_2)(x)|,$$

where $C = C_1 C_2$. From this inequality, it follows that $f_1(x)f_2(x)$ is $O(g_1 g_2(x))$, because there are constants C and k, namely, $C = C_1 C_2$ and $k = \max(k_1, k_2)$, such that $|(f_1 f_2)(x)| \leq C|g_1(x)g_2(x)|$ whenever $x > k$. This result is stated in Theorem 3.

THEOREM 3 Suppose that $f_1(x)$ is $O(g_1(x))$ and $f_2(x)$ is $O(g_2(x))$. Then $(f_1 f_2)(x)$ is $O(g_1(x)g_2(x))$.

The goal in using big-O notation to estimate functions is to choose a function $g(x)$ as simple as possible, that grows relatively slowly so that $f(x)$ is $O(g(x))$. Examples 8 and 9 illustrate how to use Theorems 2 and 3 to do this. The type of analysis given in these examples is often used in the analysis of the time used to solve problems using computer programs.

EXAMPLE 8 Give a big-O estimate for $f(n) = 3n \log(n!) + (n^2 + 3) \log n$, where n is a positive integer.

Solution: First, the product $3n \log(n!)$ will be estimated. From Example 6 we know that $\log(n!)$ is $O(n \log n)$. Using this estimate and the fact that $3n$ is $O(n)$, Theorem 3 gives the estimate that $3n \log(n!)$ is $O(n^2 \log n)$.

Next, the product $(n^2 + 3) \log n$ will be estimated. Because $(n^2 + 3) < 2n^2$ when $n > 2$, it follows that $n^2 + 3$ is $O(n^2)$. Thus, from Theorem 3 it follows that $(n^2 + 3) \log n$ is $O(n^2 \log n)$. Using Theorem 2 to combine the two big-O estimates for the products shows that $f(n) = 3n \log(n!) + (n^2 + 3) \log n$ is $O(n^2 \log n)$. ◀

EXAMPLE 9 Give a big-O estimate for $f(x) = (x + 1) \log(x^2 + 1) + 3x^2$.

Solution: First, a big-O estimate for $(x + 1) \log(x^2 + 1)$ will be found. Note that $(x + 1)$ is $O(x)$. Furthermore, $x^2 + 1 \leq 2x^2$ when $x > 1$. Hence,

$$\log(x^2 + 1) \leq \log(2x^2) = \log 2 + \log x^2 = \log 2 + 2 \log x \leq 3 \log x,$$

if $x > 2$. This shows that $\log(x^2 + 1)$ is $O(\log x)$.

From Theorem 3 it follows that $(x + 1) \log(x^2 + 1)$ is $O(x \log x)$. Because $3x^2$ is $O(x^2)$, Theorem 2 tells us that $f(x)$ is $O(\max(x \log x, x^2))$. Because $x \log x \leq x^2$, for $x > 1$, it follows that $f(x)$ is $O(x^2)$. ◀

Big-Omega and Big-Theta Notation

Big-O notation is used extensively to describe the growth of functions, but it has limitations. In particular, when $f(x)$ is $O(g(x))$, we have an upper bound, in terms of $g(x)$, for the size of $f(x)$ for large values of x. However, big-O notation does not provide a lower bound for the size of $f(x)$ for large x. For this, we use **big-Omega (big-Ω) notation**. When we want to give both an upper and a lower bound on the size of a function $f(x)$, relative to a reference function $g(x)$, we use **big-Theta (big-Θ) notation**. Both big-Omega and big-Theta notation were introduced by Donald

Ω and Θ are the Greek uppercase letters omega and theta, respectively.

Knuth in the 1970s. His motivation for introducing these notations was the common misuse of big-O notation when both an upper and a lower bound on the size of a function are needed.

We now define big-Omega notation and illustrate its use. After doing so, we will do the same for big-Theta notation.

DEFINITION 2

Let f and g be functions from the set of integers or the set of real numbers to the set of real numbers. We say that $f(x)$ is $\Omega(g(x))$ if there are positive constants C and k such that

$$|f(x)| \geq C|g(x)|$$

whenever $x > k$. [This is read as "$f(x)$ is big-Omega of $g(x)$."]

There is a strong connection between big-O and big-Omega notation. In particular, $f(x)$ is $\Omega(g(x))$ if and only if $g(x)$ is $O(f(x))$. We leave the verification of this fact as a straightforward exercise for the reader.

EXAMPLE 10
The function $f(x) = 8x^3 + 5x^2 + 7$ is $\Omega(g(x))$, where $g(x)$ is the function $g(x) = x^3$. This is easy to see because $f(x) = 8x^3 + 5x^2 + 7 \geq 8x^3$ for all positive real numbers x. This is equivalent to saying that $g(x) = x^3$ is $O(8x^3 + 5x^2 + 7)$, which can be established directly by turning the inequality around. ◀

Often, it is important to know the order of growth of a function in terms of some relatively simple reference function such as x^n when n is a positive integer or c^x, where $c > 1$. Knowing the order of growth requires that we have both an upper bound and a lower bound for the size of the function. That is, given a function $f(x)$, we want a reference function $g(x)$ such that $f(x)$ is $O(g(x))$ and $f(x)$ is $\Omega(g(x))$. Big-Theta notation, defined as follows, is used to express both of these relationships, providing both an upper and a lower bound on the size of a function.

DEFINITION 3

Let f and g be functions from the set of integers or the set of real numbers to the set of real numbers. We say that $f(x)$ is $\Theta(g(x))$ if $f(x)$ is $O(g(x))$ and $f(x)$ is $\Omega(g(x))$. When $f(x)$ is $\Theta(g(x))$ we say that f is big-Theta of $g(x)$, that $f(x)$ is of *order* $g(x)$, and that $f(x)$ and $g(x)$ are of the *same order*.

When $f(x)$ is $\Theta(g(x))$, it is also the case that $g(x)$ is $\Theta(f(x))$. Also note that $f(x)$ is $\Theta(g(x))$ if and only if $f(x)$ is $O(g(x))$ and $g(x)$ is $O(f(x))$ (see Exercise 31). Furthermore, note that $f(x)$ is $\Theta(g(x))$ if and only if there are real numbers C_1 and C_2 and a positive real number k such that

$$C_1|g(x)| \leq |f(x)| \leq C_2|g(x)|$$

whenever $x > k$. The existence of the constants C_1, C_2, and k tells us that $f(x)$ is $\Omega(g(x))$ and that $f(x)$ is $O(g(x))$, respectively.

Usually, when big-Theta notation is used, the function $g(x)$ in $\Theta(g(x))$ is a relatively simple reference function, such as x^n, c^x, $\log x$, and so on, while $f(x)$ can be relatively complicated.

EXAMPLE 11
We showed (in Example 5) that the sum of the first n positive integers is $O(n^2)$. Is this sum of order n^2?

Solution: Let $f(n) = 1 + 2 + 3 + \cdots + n$. Because we already know that $f(n)$ is $O(n^2)$, to show that $f(n)$ is of order n^2 we need to find a positive constant C such that $f(n) > Cn^2$ for

Extra Examples

sufficiently large integers n. To obtain a lower bound for this sum, we can ignore the first half of the terms. Summing only the terms greater than $\lceil n/2 \rceil$, we find that

$$1 + 2 + \cdots + n \geq \lceil n/2 \rceil + (\lceil n/2 \rceil + 1) + \cdots + n$$

$$\geq \lceil n/2 \rceil + \lceil n/2 \rceil + \cdots + \lceil n/2 \rceil$$

$$= (n - \lceil n/2 \rceil + 1) \lceil n/2 \rceil$$

$$\geq (n/2)(n/2)$$

$$= n^2/4.$$

This shows that $f(n)$ is $\Omega(n^2)$. We conclude that $f(n)$ is of order n^2, or in symbols, $f(n)$ is $\Theta(n^2)$. ◄

EXAMPLE 12 Show that $3x^2 + 8x \log x$ is $\Theta(x^2)$.

Solution: Because $0 \leq 8x \log x \leq 8x^2$, it follows that $3x^2 + 8x \log x \leq 11x^2$ for $x > 1$. Consequently, $3x^2 + 8x \log x$ is $O(x^2)$. Clearly, x^2 is $O(3x^2 + 8x \log x)$. Consequently, $3x^2 + 8x \log x$ is $\Theta(x^2)$. ◄

One useful fact is that the leading term of a polynomial determines its order. For example, if $f(x) = 3x^5 + x^4 + 17x^3 + 2$, then $f(x)$ is of order x^5. This is stated in Theorem 4, whose proof is left as Exercise 36.

THEOREM 4 Let $f(x) = a_n x^n + a_{n-1} x^{n-1} + \cdots + a_1 x + a_0$, where $a_0, a_1, \ldots, a_n$ are real numbers with $a_n \neq 0$. Then $f(x)$ is of order x^n.

EXAMPLE 13 The polynomials $3x^8 + 10x^7 + 221x^2 + 1444$, $x^{19} - 18x^4 - 10{,}112$, and $-x^{99} + 40{,}001x^{98} + 100{,}003x$ are of orders x^8, x^{19}, and x^{99}, respectively. ◄

Unfortunately, as Knuth observed, big-O notation is often used by careless writers and speakers as if it had the same meaning as big-Theta notation. Keep this in mind when you see big-O notation used. The recent trend has been to use big-Theta notation whenever both upper and lower bounds on the size of a function are needed.

Exercises

In Exercises 1–12, to establish a big-O relationship, find witnesses C and k such that $|f(x)| \leq C|g(x)|$ whenever $x > k$.

1. Determine whether each of these functions is $O(x)$.
  a) $f(x) = 10$ **b)** $f(x) = 3x + 7$
  c) $f(x) = x^2 + x + 1$ **d)** $f(x) = 5 \log x$
  e) $f(x) = \lfloor x \rfloor$ **f)** $f(x) = \lceil x/2 \rceil$

2. Determine whether each of these functions is $O(x^2)$.
  a) $f(x) = 17x + 11$ **b)** $f(x) = x^2 + 1000$
  c) $f(x) = x \log x$ **d)** $f(x) = x^4/2$
  e) $f(x) = 2^x$ **f)** $f(x) = \lfloor x \rfloor \cdot \lceil x \rceil$

3. Use the definition of "$f(x)$ is $O(g(x))$" to show that $x^4 + 9x^3 + 4x + 7$ is $O(x^4)$.

4. Use the definition of "$f(x)$ is $O(g(x))$" to show that $2^x + 17$ is $O(3^x)$.

5. Show that $(x^2 + 1)/(x + 1)$ is $O(x)$.

6. Show that $(x^3 + 2x)/(2x + 1)$ is $O(x^2)$.

7. Find the least integer n such that $f(x)$ is $O(x^n)$ for each of these functions.
  a) $f(x) = 2x^3 + x^2 \log x$
  b) $f(x) = 3x^3 + (\log x)^4$
  c) $f(x) = (x^4 + x^2 + 1)/(x^3 + 1)$
  d) $f(x) = (x^4 + 5 \log x)/(x^4 + 1)$

8. Find the least integer n such that $f(x)$ is $O(x^n)$ for each of these functions.

a) $f(x) = 2x^2 + x^3 \log x$
b) $f(x) = 3x^5 + (\log x)^4$
c) $f(x) = (x^4 + x^2 + 1)/(x^4 + 1)$
d) $f(x) = (x^3 + 5 \log x)/(x^4 + 1)$

9. Show that $x^2 + 4x + 17$ is $O(x^3)$ but that x^3 is not $O(x^2 + 4x + 17)$.

10. Show that $x \log x$ is $O(x^2)$ but that x^2 is not $O(x \log x)$.

11. Determine whether x^3 is $O(g(x))$ for each of these functions $g(x)$.

a) $g(x) = x^2$ **b)** $g(x) = x^3$
c) $g(x) = x^2 + x^3$ **d)** $g(x) = x^2 + x^4$
e) $g(x) = 3^x$ **f)** $g(x) = x^3/2$

12. Explain what it means for a function to be $O(1)$.

13. Suppose that $f(x)$, $g(x)$, and $h(x)$ are functions such that $f(x)$ is $O(g(x))$ and $g(x)$ is $O(h(x))$. Show that $f(x)$ is $O(h(x))$.

14. Let k be a positive integer. Show that $1^k + 2^k + \cdots + n^k$ is $O(n^{k+1})$.

15. Give as good a big-O estimate as possible for each of these functions.

a) $(n^2 + 8)(n + 1)$ **b)** $(n \log n + n^2)(n^3 + 2)$
c) $(n! + 2^n)(n^3 + \log(n^2 + 1))$

16. Give a big-O estimate for each of these functions. For the function g in your estimate that $f(x)$ is $O(g(x))$, use a simple function g of the smallest order.

a) $n \log(n^2 + 1) + n^2 \log n$
b) $(n \log n + 1)^2 + (\log n + 1)(n^2 + 1)$
c) $n^{2^n} + n^{n^2}$

17. For each function in Exercise 2, determine whether that function is $\Omega(x^2)$ and whether it is $\Theta(x^2)$.

18. Show that each of these pairs of functions are of the same order.

a) $3x + 7$, x
b) $2x^2 + x - 7$, x^2
c) $\lfloor x + 1/2 \rfloor$, x
d) $\log(x^2 + 1)$, $\log_2 x$
e) $\log_{10} x$, $\log_2 x$

19. Show that $f(x)$ is $\Theta(g(x))$ if and only if $f(x)$ is $O(g(x))$ and $g(x)$ is $O(f(x))$.

20. Show that if $f(x)$ and $g(x)$ are functions from the set of real numbers to the set of real numbers, then $f(x)$ is $O(g(x))$ if and only if $g(x)$ is $\Omega(f(x))$.

21. Show that if $f(x)$ and $g(x)$ are functions from the set of real numbers to the set of real numbers, then $f(x)$ is $\Theta(g(x))$ if and only if there are positive constants k, C_1, and C_2 such that $C_1|g(x)| \leq |f(x)| \leq C_2|g(x)|$ whenever $x > k$.

22. a) Show that $3x^2 + x + 1$ is $\Theta(3x^2)$ by directly finding the constants k, C_1, and C_2 in Exercise 21.
   b) Express the relationship in part (a) using a picture showing the functions $3x^2 + x + 1$, $C_1 \cdot 3x^2$, and $C_2 \cdot 3x^2$, and the constant k on the x-axis, where C_1, C_2, and k are the constants you found in part (a) to show that $3x^2 + x + 1$ is $\Theta(3x^2)$.

23. Express the relationship $f(x)$ is $\Theta(g(x))$ using a picture. Show the graphs of the functions $f(x)$, $C_1|g(x)|$, and $C_2|g(x)|$, as well as the constant k on the x-axis.

24. Explain what it means for a function to be $\Omega(1)$.

25. Explain what it means for a function to be $\Theta(1)$.

26. Give a big-O estimate of the product of the first n odd positive integers.

27. Show that if f and g are real-valued functions such that $f(x)$ is $O(g(x))$, then for every positive integer n, $f^n(x)$ is $O(g^n(x))$. [Note that $f^n(x) = f(x)^n$.]

28. Suppose that $f(x)$ is $O(g(x))$ where f and g are increasing and unbounded functions. Show that $\log|f(x)|$ is $O(\log|g(x)|)$.

29. Suppose that $f(x)$ is $O(g(x))$. Does it follow that $2^{f(x)}$ is $O(2^{g(x)})$?

30. Let $f_1(x)$ and $f_2(x)$ be functions from the set of real numbers to the set of positive real numbers. Show that if $f_1(x)$ and $f_2(x)$ are both $\Theta(g(x))$, where $g(x)$ is a function from the set of real numbers to the set of positive real numbers, then $f_1(x) + f_2(x)$ is $\Theta(g(x))$. Is this still true if $f_1(x)$ and $f_2(x)$ can take negative values?

31. If $f_1(x)$ and $f_2(x)$ are functions from the set of positive integers to the set of positive real numbers and $f_1(x)$ and $f_2(x)$ are both $\Theta(g(x))$, is $(f_1 - f_2)(x)$ also $\Theta(g(x))$? Either prove that it is or give a counterexample.

32. Show that if $f_1(x)$ and $f_2(x)$ are functions from the set of positive integers to the set of real numbers and $f_1(x)$ is $\Theta(g_1(x))$ and $f_2(x)$ is $\Theta(g_2(x))$, then $(f_1 f_2)(x)$ is $\Theta((g_1 g_2)(x))$.

33. Find functions f and g from the set of positive integers to the set of real numbers such that $f(n)$ is not $O(g(n))$ and $g(n)$ is not $O(f(n))$.

34. Express the relationship $f(x)$ is $\Omega(g(x))$ using a picture. Show the graphs of the functions $f(x)$ and $Cg(x)$, as well as the constant k on the real axis.

35. Show that if $f_1(x)$ is $\Theta(g_1(x))$, $f_2(x)$ is $\Theta(g_2(x))$, and $f_2(x) \neq 0$ and $g_2(x) \neq 0$ for all real numbers $x > 0$, then $(f_1/f_2)(x)$ is $\Theta((g_1/g_2)(x))$.

36. Show that if $f(x) = a_n x^n + a_{n-1} x^{n-1} + \cdots + a_1 x + a_0$, where $a_0, a_1, \ldots, a_{n-1}$, and a_n are real numbers and $a_n \neq 0$, then $f(x)$ is $\Theta(x^n)$.

Big-O, big-Theta, and big-Omega notation can be extended to functions in more than one variable. For example, the statement $f(x, y)$ is $O(g(x, y))$ means that there exist constants C, k_1, and k_2 such that $|f(x, y)| \leq C|g(x, y)|$ whenever $x > k_1$ and $y > k_2$.

37. Define the statement $f(x, y)$ is $\Theta(g(x, y))$.

38. Define the statement $f(x, y)$ is $\Omega(g(x, y))$.

39. Show that $(x^2 + xy + x \log y)^3$ is $O(x^6 y^3)$.

40. Show that $x^5 y^3 + x^4 y^4 + x^3 y^5$ is $\Omega(x^3 y^3)$.

The following problems deal with another type of asymptotic notation, called **little-o** notation. Because little-o notation is based on the concept of limits, a knowledge of calculus is needed for these problems. We say that $f(x)$ is $o(g(x))$ [read $f(x)$ is "little-oh" of $g(x)$], when

$$\lim_{x \to \infty} \frac{f(x)}{g(x)} = 0.$$

41. (*Requires calculus*) Show that
   a) x^2 is $o(x^3)$.   **b)** $x \log x$ is $o(x^2)$.
   c) x^2 is $o(2^x)$.   **d)** $x^2 + x + 1$ is not $o(x^2)$.

42. (*Requires calculus*)
   a) Show that if $f(x)$ and $g(x)$ are functions such that $f(x)$ is $o(g(x))$ and c is a constant, then $cf(x)$ is $o(g(x))$, where $(cf)(x) = cf(x)$.
   b) Show that if $f_1(x)$, $f_2(x)$, and $g(x)$ are functions such that $f_1(x)$ is $o(g(x))$ and $f_2(x)$ is $o(g(x))$, then $(f_1 + f_2)(x)$ is $o(g(x))$, where $(f_1 + f_2)(x) = f_1(x) + f_2(x)$.

43. (*Requires calculus*) Represent pictorially that $x \log x$ is $o(x^2)$ by graphing $x \log x$, x^2, and $x \log x/x^2$. Explain how this picture shows that $x \log x$ is $o(x^2)$.

*44. (*Requires calculus*) Suppose that $f(x)$ is $o(g(x))$. Does it follow that $2^{f(x)}$ is $o(2^{g(x)})$?

45. (*Requires calculus*) The two parts of this exercise describe the relationship between little-o and big-O notation.

   a) Show that if $f(x)$ and $g(x)$ are functions such that $f(x)$ is $o(g(x))$, then $f(x)$ is $O(g(x))$.
   b) Show that if $f(x)$ and $g(x)$ are functions such that $f(x)$ is $O(g(x))$, then it does not necessarily follow that $f(x)$ is $o(g(x))$.

46. (*Requires calculus*) Show that if $f_1(x)$ is $O(g(x))$ and $f_2(x)$ is $o(g(x))$, then $f_1(x) + f_2(x)$ is $O(g(x))$.

*47. Show that $n \log n$ is $O(\log n!)$.

48. Determine whether $\log n!$ is $\Theta(n \log n)$. Justify your answer.

*49. Show that $\log n!$ is greater than $(n \log n)/4$ for $n > 4$. [*Hint:* Begin with the inequality $n! > n(n-1)(n-2) \cdots \lceil n/2 \rceil$.]

Let $f(x)$ and $g(x)$ be functions from the set of real numbers to the set of real numbers. We say that the functions f and g are **asymptotic** and write $f(x) \sim g(x)$ if $\lim_{x \to \infty} f(x)/g(x) = 1$.

50. (*Requires calculus*) For each of these pairs of functions, determine whether f and g are asymptotic.
   a) $f(x) = x^2 + 3x + 7$, $g(x) = x^2 + 10$
   b) $f(x) = x^2 \log x$, $g(x) = x^3$
   c) $f(x) = x^4 + \log(3x^8 + 7)$, $g(x) = (x^2 + 17x + 3)^2$
   d) $f(x) = (x^3 + x^2 + x + 1)^4$, $g(x) = (x^4 + x^3 + x^2 + x + 1)^3$.

3.3 Complexity of Algorithms

Introduction

When does an algorithm provide a satisfactory solution to a problem? First, it must always produce the correct answer. How this can be demonstrated will be discussed in Chapter 5. Second, it should be efficient. The efficiency of algorithms will be discussed in this section.

How can the efficiency of an algorithm be analyzed? One measure of efficiency is the time used by a computer to solve a problem using the algorithm, when input values are of a specified size. A second measure is the amount of computer memory required to implement the algorithm when input values are of a specified size.

Questions such as these involve the **computational complexity** of the algorithm. An analysis of the time required to solve a problem of a particular size involves the **time complexity** of the algorithm. An analysis of the computer memory required involves the **space complexity** of the algorithm. Considerations of the time and space complexity of an algorithm are essential when algorithms are implemented. It is obviously important to know whether an algorithm will produce an answer in a microsecond, a minute, or a billion years. Likewise, the required memory must be available to solve a problem, so that space complexity must be taken into account.

Considerations of space complexity are tied in with the particular data structures used to implement the algorithm. Because data structures are not dealt with in detail in this book, space complexity will not be considered. We will restrict our attention to time complexity.

Time Complexity

The time complexity of an algorithm can be expressed in terms of the number of operations used by the algorithm when the input has a particular size. The operations used to measure time complexity can be the comparison of integers, the addition of integers, the multiplication of integers, the division of integers, or any other basic operation.

Time complexity is described in terms of the number of operations required instead of actual computer time because of the difference in time needed for different computers to perform basic operations. Moreover, it is quite complicated to break all operations down to the basic bit operations that a computer uses. Furthermore, the fastest computers in existence can perform basic bit operations (for instance, adding, multiplying, comparing, or exchanging two bits) in 10^{-11} second (10 picoseconds), but personal computers may require 10^{-8} second (10 nanoseconds), which is 1000 times as long, to do the same operations.

We illustrate how to analyze the time complexity of an algorithm by considering Algorithm 1 of Section 3.1, which finds the maximum of a finite set of integers.

EXAMPLE 1 Describe the time complexity of Algorithm 1 of Section 3.1 for finding the maximum element in a finite set of integers.

Solution: The number of comparisons will be used as the measure of the time complexity of the algorithm, because comparisons are the basic operations used.

To find the maximum element of a set with n elements, listed in an arbitrary order, the temporary maximum is first set equal to the initial term in the list. Then, after a comparison $i \leq n$ has been done to determine that the end of the list has not yet been reached, the temporary maximum and second term are compared, updating the temporary maximum to the value of the second term if it is larger. This procedure is continued, using two additional comparisons for each term of the list—one $i \leq n$, to determine that the end of the list has not been reached and another $max < a_i$, to determine whether to update the temporary maximum. Because two comparisons are used for each of the second through the nth elements and one more comparison is used to exit the loop when $i = n + 1$, exactly $2(n - 1) + 1 = 2n - 1$ comparisons are used whenever this algorithm is applied. Hence, the algorithm for finding the maximum of a set of n elements has time complexity $\Theta(n)$, measured in terms of the number of comparisons used. Note that for this algorithm the number of comparisons is independent of particular input of n numbers. ◄

Next, we will analyze the time complexity of searching algorithms.

EXAMPLE 2 Describe the time complexity of the linear search algorithm (specified as Algortihm 2 in Section 3.1).

Solution: The number of comparisons used by Algorithm 2 in Section 3.1 will be taken as the measure of the time complexity. At each step of the loop in the algorithm, two comparisons are performed—one $i \leq n$, to see whether the end of the list has been reached and one $x \leq a_i$, to compare the element x with a term of the list. Finally, one more comparison $i \leq n$ is made outside the loop. Consequently, if $x = a_i$, $2i + 1$ comparisons are used. The most comparisons, $2n + 2$, are required when the element is not in the list. In this case, $2n$ comparisons are used to determine that x is not a_i, for $i = 1, 2, \ldots, n$, an additional comparison is used to exit the loop, and one comparison is made outside the loop. So when x is not in the list, a total of $2n + 2$ comparisons are used. Hence, a linear search requires $\Theta(n)$ comparisons in the worst case, because $2n + 2$ is $\Theta(n)$. ◄

WORST-CASE COMPLEXITY The type of complexity analysis done in Example 2 is a **worst-case** analysis. By the worst-case performance of an algorithm, we mean the largest number of operations needed to solve the given problem using this algorithm on input of specified size. Worst-case analysis tells us how many operations an algorithm requires to guarantee that it will produce a solution.

EXAMPLE 3 Describe the time complexity of the binary search algorithm (specified as Algorithm 3 in Section 3.1) in terms of the number of comparisons used (and ignoring the time required to compute $m = \lfloor (i + j)/2 \rfloor$ in each iteration of the loop in the algorithm).

Solution: For simplicity, assume there are $n = 2^k$ elements in the list $a_1, a_2, \ldots, a_n$, where k is a nonnegative integer. Note that $k = \log n$. (If n, the number of elements in the list, is not a power of 2, the list can be considered part of a larger list with 2^{k+1} elements, where $2^k < n < 2^{k+1}$. Here 2^{k+1} is the smallest power of 2 larger than n.)

At each stage of the algorithm, i and j, the locations of the first term and the last term of the restricted list at that stage, are compared to see whether the restricted list has more than one term. If $i < j$, a comparison is done to determine whether x is greater than the middle term of the restricted list.

At the first stage the search is restricted to a list with 2^{k-1} terms. So far, two comparisons have been used. This procedure is continued, using two comparisons at each stage to restrict the search to a list with half as many terms. In other words, two comparisons are used at the first stage of the algorithm when the list has 2^k elements, two more when the search has been reduced to a list with 2^{k-1} elements, two more when the search has been reduced to a list with 2^{k-2} elements, and so on, until two comparisons are used when the search has been reduced to a list with $2^1 = 2$ elements. Finally, when one term is left in the list, one comparison tells us that there are no additional terms left, and one more comparison is used to determine if this term is x.

Hence, at most $2k + 2 = 2 \log n + 2$ comparisons are required to perform a binary search when the list being searched has 2^k elements. (If n is not a power of 2, the original list is expanded to a list with 2^{k+1} terms, where $k = \lfloor \log n \rfloor$, and the search requires at most $2 \lceil \log n \rceil + 2$ comparisons.) It follows that in the worst case, binary search requires $O(\log n)$ comparisons. Note that in the worst case, $2 \log n + 2$ comparisons are used by the binary search. Hence, the binary search uses $\Theta(\log n)$ comparisons in the worst case, because $2 \log n + 2 = \Theta(\log n)$. From this analysis it follows that in the worst case, the binary search algorithm is more efficient than the linear search algorithm, because we know by Example 2 that the linear search algorithm has $\Theta(n)$ worst-case time complexity. ◀

AVERAGE-CASE COMPLEXITY   Another important type of complexity analysis, besides worst-case analysis, is called **average-case** analysis. The average number of operations used to solve the problem over all possible inputs of a given size is found in this type of analysis. Average-case time complexity analysis is usually much more complicated than worst-case analysis. However, the average-case analysis for the linear search algorithm can be done without difficulty, as shown in Example 4.

EXAMPLE 4   Describe the average-case performance of the linear search algorithm in terms of the average number of comparisons used, assuming that the integer x is in the list and it is equally likely that x is in any position.

Solution: By hypothesis, the integer x is one of the integers $a_1, a_2, \ldots, a_n$ in the list. If x is the first term a_1 of the list, three comparisons are needed, one $i \leq n$ to determine whether the end of the list has been reached, one $x \neq a_i$ to compare x and the first term, and one $i \leq n$ outside the loop. If x is the second term a_2 of the list, two more comparisons are needed, so that a total of five comparisons are used. In general, if x is the ith term of the list a_i, two comparisons will be used at each of the i steps of the loop, and one outside the loop, so that a total of $2i + 1$ comparisons are needed. Hence, the average number of comparisons used equals

$$\frac{3 + 5 + 7 + \cdots + (2n + 1)}{n} = \frac{2(1 + 2 + 3 + \cdots + n) + n}{n}.$$

Using the formula from line 2 of Table 2 in Section 2.4 (and see Exercise 37(b) of Section 2.4),

$$1 + 2 + 3 + \cdots + n = \frac{n(n + 1)}{2}.$$

Hence, the average number of comparisons used by the linear search algorithm (when x is known to be in the list) is

$$\frac{2[n(n+1)/2]}{n} + 1 = n + 2,$$

which is $\Theta(n)$. ◄

Remark: In the analysis in Example 4 we assumed that x is in the list being searched. It is also possible to do an average-case analysis of this algorithm when x may not be in the list (see Exercise 17).

Remark: Although we have counted the comparisons needed to determine whether we have reached the end of a loop, these comparisons are often not counted. From this point on we will ignore such comparisons.

WORST-CASE COMPLEXITY OF TWO SORTING ALGORITHMS We analyze the worst-case complexity of the bubble sort and the insertion sort in Examples 5 and 6.

EXAMPLE 5 What is the worst-case complexity of the bubble sort in terms of the number of comparisons made?

Solution: The bubble sort described before Example 4 in Section 3.1 sorts a list by performing a sequence of passes through the list. During each pass the bubble sort successively compares adjacent elements, interchanging them if necessary. When the ith pass begins, the $i-1$ largest elements are guaranteed to be in the correct positions. During this pass, $n-i$ comparisons are used. Consequently, the total number of comparisons used by the bubble sort to order a list of n elements is

$$(n-1) + (n-2) + \cdots + 2 + 1 = \frac{(n-1)n}{2}$$

using a summation formula from line 2 in Table 2 in Section 2.4 (and Exercise 23(b) in Section 2.4). Note that the bubble sort always uses this many comparisons, because it continues even if the list becomes completely sorted at some intermediate step. Consequently, the bubble sort uses $(n-1)n/2$ comparisons, so it has $\Theta(n^2)$ worst-case complexity in terms of the number of comparisons used. ◄

EXAMPLE 6 What is the worst-case complexity of the insertion sort in terms of the number of comparisons made?

Solution: The insertion sort (described in Section 3.1) inserts the jth element into the correct position among the first $j-1$ elements that have already been put into the correct order. It does this by using a linear search technique, successively comparing the jth element with successive terms until a term that is greater than or equal to it is found or it compares a_j with itself and stops because a_j is not less than itself. Consequently, in the worst case, j comparisons are required to insert the jth element into the correct position. Therefore, the total number of comparisons used by the insertion sort to sort a list of n elements is

$$2 + 3 + \cdots + n = \frac{n(n+1)}{2} - 1,$$

using the summation formula for the sum of consecutive integers in line 2 of Table 2 of Section 2.4 (and see Exercise 23(b) of Section 2.4), and noting that the first term, 1, is missing in this sum. Note that the insertion sort may use considerably fewer comparisons if the smaller elements started out at the end of the list. We conclude that the insertion sort has worst-case complexity $\Theta(n^2)$. ◀

In Examples 5 and 6 we showed that both the bubble sort and the insertion sort have worst-case time complexity $\Theta(n^2)$. However, the most efficient sorting algorithms can sort n items in $O(n \log n)$ time, as we will show in Sections 8.3 and 11.1 using techniques we develop in those sections. From this point on, we will assume that sorting n items can be done in $O(n \log n)$ time.

Complexity of Matrix Multiplication

The definition of the product of two matrices can be expressed as an algorithm for computing the product of two matrices. Suppose that $\mathbf{C} = [c_{ij}]$ is the $m \times n$ matrix that is the product of the $m \times k$ matrix $\mathbf{A} = [a_{ij}]$ and the $k \times n$ matrix $\mathbf{B} = [b_{ij}]$. The algorithm based on the definition of the matrix product is expressed in pseudocode in Algorithm 1.

ALGORITHM 1 Matrix Multiplication.

procedure *matrix multiplication*($\mathbf{A}$, $\mathbf{B}$: matrices)
for $i := 1$ **to** m
    for $j := 1$ **to** n
        $c_{ij} := 0$
        for $q := 1$ **to** k
            $c_{ij} := c_{ij} + a_{iq}b_{qj}$
return $\mathbf{C}$ {$\mathbf{C} = [c_{ij}]$ is the product of $\mathbf{A}$ and $\mathbf{B}$}

We can determine the complexity of this algorithm in terms of the number of additions and multiplications used.

EXAMPLE 7   How many additions of integers and multiplications of integers are used by Algorithm 1 to multiply two $n \times n$ matrices with integer entries?

Solution: There are n^2 entries in the product of $\mathbf{A}$ and $\mathbf{B}$. To find each entry requires a total of n multiplications and $n - 1$ additions. Hence, a total of n^3 multiplications and $n^2(n - 1)$ additions are used. ◀

Surprisingly, there are more efficient algorithms for matrix multiplication than that given in Algorithm 1. As Example 7 shows, multiplying two $n \times n$ matrices directly from the definition requires $O(n^3)$ multiplications and additions. Using other algorithms, two $n \times n$ matrices can be multiplied using $O(n^{\sqrt{7}})$ multiplications and additions. (Details of such algorithms can be found in [CoLeRiSt09].)

We can also analyze the complexity of the algorithm we described in Chapter 2 for computing the Boolean product of two matrices, which we display as Algorithm 2.

ALGORITHM 2 The Boolean Product of Zero-One Matrices.

---

procedure *Boolean product of Zero-One Matrices* (**A**, **B**: zero–one matrices)
for $i := 1$ **to** m
    for $j := 1$ **to** n
        $c_{ij} := 0$
        for $q := 1$ **to** k
            $c_{ij} := c_{ij} \vee (a_{iq} \wedge b_{qj})$
return C {**C** $= [c_{ij}]$ is the Boolean product of **A** and **B**}

The number of bit operations used to find the Boolean product of two $n \times n$ matrices can be easily determined.

EXAMPLE 8   How many bit operations are used to find **A** $\odot$ **B**, where **A** and **B** are $n \times n$ zero–one matrices?

Solution: There are n^2 entries in **A** $\odot$ **B**. Using Algorithm 2, a total of n *OR*s and n *AND*s are used to find an entry of **A** $\odot$ **B**. Hence, $2n$ bit operations are used to find each entry. Therefore, $2n^3$ bit operations are required to compute **A** $\odot$ **B** using Algorithm 2. ◀

MATRIX-CHAIN MULTIPLICATION   There is another important problem involving the complexity of the multiplication of matrices. How should the **matrix-chain** $\mathbf{A}_1 \mathbf{A}_2 \cdots \mathbf{A}_n$ be computed using the fewest multiplications of integers, where $\mathbf{A}_1, \mathbf{A}_2, \ldots, \mathbf{A}_n$ are $m_1 \times m_2, m_2 \times m_3, \ldots, m_n \times m_{n+1}$ matrices, respectively, and each has integers as entries? (Because matrix multiplication is associative, as shown in Exercise 13 in Section 2.6, the order of the multiplication used does not change the product.) Note that $m_1 m_2 m_3$ multiplications of integers are performed to multiply an $m_1 \times m_2$ matrix and an $m_2 \times m_3$ matrix using Algorithm 1. Example 9 illustrates this problem.

EXAMPLE 9   In which order should the matrices $\mathbf{A}_1$, $\mathbf{A}_2$, and $\mathbf{A}_3$—where $\mathbf{A}_1$ is 30×20, $\mathbf{A}_2$ is 20×40, and $\mathbf{A}_3$ is 40×10, all with integer entries—be multiplied to use the least number of multiplications of integers?

Solution: There are two possible ways to compute $\mathbf{A}_1 \mathbf{A}_2 \mathbf{A}_3$. These are $\mathbf{A}_1 (\mathbf{A}_2 \mathbf{A}_3)$ and $(\mathbf{A}_1 \mathbf{A}_2) \mathbf{A}_3$.
    If $\mathbf{A}_2$ and $\mathbf{A}_3$ are first multiplied, a total of $20 \cdot 40 \cdot 10 = 8000$ multiplications of integers are used to obtain the 20×10 matrix $\mathbf{A}_2 \mathbf{A}_3$. Then, to multiply $\mathbf{A}_1$ and $\mathbf{A}_2 \mathbf{A}_3$ requires $30 \cdot 20 \cdot 10 = 6000$ multiplications. Hence, a total of

$$8000 + 6000 = 14{,}000$$

multiplications are used. On the other hand, if $\mathbf{A}_1$ and $\mathbf{A}_2$ are first multiplied, then $30 \cdot 20 \cdot 40 = 24{,}000$ multiplications are used to obtain the 30×40 matrix $\mathbf{A}_1 \mathbf{A}_2$. Then, to multiply $\mathbf{A}_1 \mathbf{A}_2$ and $\mathbf{A}_3$ requires $30 \cdot 40 \cdot 10 = 12{,}000$ multiplications. Hence, a total of

$$24{,}000 + 12{,}000 = 36{,}000$$

multiplications are used.
    Clearly, the first method is more efficient. ◀

We will return to this problem in Exercise 57 in Section 8.1. Algorithms for determining the most efficient way to carry out matrix-chain multiplication are discussed in [CoLeRiSt09].

Algorithmic Paradigms

In Section 3.1 we introduced the basic notion of an algorithm. We provided examples of many different algorithms, including searching and sorting algorithms. We also introduced the concept of a greedy algorithm, giving examples of several problems that can be solved by greedy algorithms. Greedy algorithms provide an example of an **algorithmic paradigm**, that is, a general approach based on a particular concept that can be used to construct algorithms for solving a variety of problems.

In this book we will construct algorithms for solving many different problems based on a variety of algorithmic paradigms, including the most widely used algorithmic paradigms. These paradigms can serve as the basis for constructing efficient algorithms for solving a wide range of problems.

Some of the algorithms we have already studied are based on an algorithmic paradigm known as brute force, which we will describe in this section. Algorithmic paradigms, studied later in this book, include divide-and-conquer algorithms studied in Chapter 8, dynamic programming, also studied in Chapter 8, backtracking, studied in Chapter 10, and probabilistic algorithms, studied in Chapter 7. There are many important algorithmic paradigms besides those described in this book. Consult books on algorithm design such as [KlTa06] to learn more about them.

BRUTE-FORCE ALGORITHMS   Brute force is an important, and basic, algorithmic paradigm. In a **brute-force algorithm**, a problem is solved in the most straightforward manner based on the statement of the problem and the definitions of terms. Brute-force algorithms are designed to solve problems without regard to the computing resources required. For example, in some brute-force algorithms the solution to a problem is found by examining every possible solution, looking for the best possible. In general, brute-force algorithms are naive approaches for solving problems that do not take advantage of any special structure of the problem or clever ideas.

Note that Algorithm 1 in Section 3.1 for finding the maximum number in a sequence is a brute-force algorithm because it examines each of the n numbers in a sequence to find the maximum term. The algorithm for finding the sum of n numbers by adding one additional number at a time is also a brute-force algorithm, as is the algorithm for matrix multiplication based on its definition (Algorithm 1). The bubble, insertion, and selection sorts (described in Section 3.1 in Algorithms 4 and 5) are also considered to be brute-force algorithms; all three of these sorting algorithms are straightforward approaches much less efficient than other sorting algorithms such as the merge sort and the quick sort discussed in Chapters 5 and 8.

Although brute-force algorithms are often inefficient, they are often quite useful. A brute-force algorithm may be able to solve practical instances of problems, particularly when the input is not too large, even if it is impractical to use this algorithm for larger inputs. Furthermore, when designing new algorithms to solve a problem, the goal is often to find a new algorithm that is more efficient than a brute-force algorithm. One such problem of this type is described in Example 10.

EXAMPLE 10   Construct a brute-force algorithm for finding the closest pair of points in a set of n points in the plane and provide a worst-case big-O estimate for the number of bit operations used by the algorithm.

Solution: Suppose that we are given as input the points $(x_1, y_1), (x_2, y_2), \ldots, (x_n, y_n)$. Recall that the distance between (x_i, y_i) and (x_j, y_j) is $\sqrt{(x_j - x_i)^2 + (y_j - y_i)^2}$. A brute-force algorithm can find the closest pair of these points by computing the distances between all pairs of the n points and determining the smallest distance. (We can make one small simplification to make the computation easier; we can compute the square of the distance between pairs of points to find the closest pair, rather than the distance between these points. We can do this because

TABLE 1 Commonly Used Terminology for the Complexity of Algorithms.

Complexity	Terminology
$\Theta(1)$	Constant complexity
$\Theta(\log n)$	Logarithmic complexity
$\Theta(n)$	Linear complexity
$\Theta(n \log n)$	Linearithmic complexity
$\Theta(n^b)$	Polynomial complexity
$\Theta(b^n)$, where $b > 1$	Exponential complexity
$\Theta(n!)$	Factorial complexity

the square of the distance between a pair of points is smallest when the distance between these points is smallest.)

ALGORITHM 3 Brute-Force Algorithm for Closest Pair of Points.

procedure *closest-pair*($(x_1, y_1), (x_2, y_2), \ldots, (x_n, y_n)$): pairs of real numbers)
$min = \infty$
for $i := 2$ **to** n
    for $j := 1$ **to** $i - 1$
        if $(x_j - x_i)^2 + (y_j - y_i)^2 < min$ **then**
            $min := (x_j - x_i)^2 + (y_j - y_i)^2$
            closest pair $:= ((x_i, y_i), (x_j, y_j))$
return *closest pair*

To estimate the number of operations used by the algorithm, first note that there are $n(n-1)/2$ pairs of points $((x_i, y_i), (x_j, y_j))$ that we loop through (as the reader should verify). For each such pair we compute $(x_j - x_i)^2 + (y_j - y_i)^2$, compare it with the current value of *min*, and if it is smaller than *min* replace the current value of *min* by this new value. It follows that this algorithm uses $\Theta(n^2)$ operations, in terms of arithmetic operations and comparisons.

In Chapter 8 we will devise an algorithm that determines the closest pair of points when given n points in the plane as input that has $O(n \log n)$ worst-case complexity. The original discovery of such an algorithm, much more efficient than the brute-force approach, was considered quite surprising. ◄

Understanding the Complexity of Algorithms

Table 1 displays some common terminology used to describe the time complexity of algorithms. For example, an algorithm that finds the largest of the first 100 terms of a list of n elements by applying Algorithm 1 to the sequence of the first 100 terms, where n is an integer with $n \geq 100$, has **constant complexity** because it uses 99 comparisons no matter what n is (as the reader can verify). The linear search algorithm has **linear** (worst-case or average-case) **complexity** and the binary search algorithm has **logarithmic** (worst-case) **complexity.** Many important algorithms have $n \log n$, or **linearithmic** (worst-case) **complexity**, such as the merge sort, which we will introduce in Chapter 4. (The word *linearithmic* is a combination of the words *linear* and *logarithmic*.)

An algorithm has **polynomial complexity** if it has complexity $\Theta(n^b)$, where b is an integer with $b \geq 1$. For example, the bubble sort algorithm is a polynomial-time algorithm because it uses $\Theta(n^2)$ comparisons in the worst case. An algorithm has **exponential complexity** if it has time complexity $\Theta(b^n)$, where $b > 1$. The algorithm that determines whether a compound proposition in n variables is satisfiable by checking all possible assignments of truth variables is an algorithm with exponential complexity, because it uses $\Theta(2^n)$ operations. Finally, an algorithm has **factorial complexity** if it has $\Theta(n!)$ time complexity. The algorithm that finds all orders that a traveling salesperson could use to visit n cities has factorial complexity; we will discuss this algorithm in Chapter 9.

TRACTABILITY   A problem that is solvable using an algorithm with polynomial worst-case complexity is called **tractable**, because the expectation is that the algorithm will produce the solution to the problem for reasonably sized input in a relatively short time. However, if the polynomial in the big-Θ estimate has high degree (such as degree 100) or if the coefficients are extremely large, the algorithm may take an extremely long time to solve the problem. Consequently, that a problem can be solved using an algorithm with polynomial worst-case time complexity is no guarantee that the problem can be solved in a reasonable amount of time for even relatively small input values. Fortunately, in practice, the degree and coefficients of polynomials in such estimates are often small.

The situation is much worse for problems that cannot be solved using an algorithm with worst-case polynomial time complexity. Such problems are called **intractable**. Usually, but not always, an extremely large amount of time is required to solve the problem for the worst cases of even small input values. In practice, however, there are situations where an algorithm with a certain worst-case time complexity may be able to solve a problem much more quickly for most cases than for its worst case. When we are willing to allow that some, perhaps small, number of cases may not be solved in a reasonable amount of time, the average-case time complexity is a better measure of how long an algorithm takes to solve a problem. Many problems important in industry are thought to be intractable but can be practically solved for essentially all sets of input that arise in daily life. Another way that intractable problems are handled when they arise in practical applications is that instead of looking for exact solutions of a problem, approximate solutions are sought. It may be the case that fast algorithms exist for finding such approximate solutions, perhaps even with a guarantee that they do not differ by very much from an exact solution.

Some problems even exist for which it can be shown that no algorithm exists for solving them. Such problems are called **unsolvable** (as opposed to **solvable** problems that can be solved using an algorithm). The first proof that there are unsolvable problems was provided by the great English mathematician and computer scientist Alan Turing when he showed that the halting problem is unsolvable. Recall that we proved that the halting problem is unsolvable in Section 3.1.

P VERSUS NP   The study of the complexity of algorithms goes far beyond what we can describe here. Note, however, that many solvable problems are believed to have the property that no algorithm with polynomial worst-case time complexity solves them, but that a solution, if known, can be checked in polynomial time. Problems for which a solution can be checked in polynomial time are said to belong to the **class NP** (tractable problems are said to belong to **class P**). The abbreviation NP stands for *nondeterministic polynomial* time. The satisfiability problem, discussed in Section 1.3, is an example of an NP problem—we can quickly verify that an assignment of truth values to the variables of a compound proposition makes it true, but no polynomial time algorithm has been discovered for finding such an assignment of truth values. (For example, an exhaustive search of all possible truth values requires $\Omega(2^n)$ bit operations where n is the number of variables in the compound proposition.)

There is also an important class of problems, called **NP-complete problems**, with the property that if any of these problems can be solved by a polynomial worst-case time algorithm, then all problems in the class NP can be solved by polynomial worst-case time algorithms. The satisfiability problem, is also an example of an NP-complete problem. It is an NP problem and if a polynomial time algorithm for solving it were known, there would be polynomial time

Links

algorithms for all problems known to be in this class of problems (and there are many important problems in this class). This last statement follows from the fact that every problem in NP can be reduced in polynomial time to the satisfiability problem. Although more than 3000 NP-complete problems are now known, the satisfiability problem was the first problem shown to be NP-complete. The theorem that asserts this is known as the **Cook-Levin theorem** after Stephen Cook and Leonid Levin, who independently proved it in the early 1970s.

The **P versus NP problem** asks whether NP, the class of problems for which it is possible to check solutions in polynomial time, equals P, the class of tractable problems. If P≠NP, there would be some problems that cannot be solved in polynomial time, but whose solutions could be verified in polynomial time. The concept of NP-completeness is helpful in research aimed at solving the P versus NP problem, because NP-complete problems are the problems in NP considered most likely not to be in P, as every problem in NP can be reduced to an NP-complete problem in polynomial time. A large majority of theoretical computer scientists believe that $P \neq NP$, which would mean that no NP-complete problem can be solved in polynomial time. One reason for this belief is that despite extensive research, no one has succeeded in showing that $P = NP$. In particular, no one has been able to find an algorithm with worst-case polynomial time complexity that solves any NP-complete problem. The P versus NP problem is one of the most famous unsolved problems in the mathematical sciences (which include theoretical computer science). It is one of the seven famous Millennium Prize Problems, of which six remain unsolved. A prize of $1,000,000 is offered by the Clay Mathematics Institute for its solution.

For more information about the complexity of algorithms, consult the references, including [CoLeRiSt09], for this section listed at the end of this book. (Also, for a more formal discussion of computational complexity in terms of Turing machines, see Section 13.5.)

Links

PRACTICAL CONSIDERATIONS   Note that a big-Θ estimate of the time complexity of an algorithm expresses how the time required to solve the problem increases as the input grows in size. In practice, the best estimate (that is, with the smallest reference function) that can be shown is used. However, big-Θ estimates of time complexity cannot be directly translated into the actual amount of computer time used. One reason is that a big-Θ estimate $f(n)$ is $\Theta(g(n))$, where $f(n)$ is the time complexity of an algorithm and $g(n)$ is a reference function, means that $C_1 g(n) \leq f(n) \leq C_2 g(n)$ when $n > k$, where C_1, C_2, and k are constants. So without knowing the constants C_1, C_2, and k in the inequality, this estimate cannot be used to determine a lower bound and an upper bound on the number of operations used in the worst case. As remarked before, the time required for an operation depends on the type of operation and the computer being used. Often, instead of a big-Θ estimate on the worst-case time complexity of an algorithm, we have only a big-O estimate. Note that a big-O estimate on the time complexity

Links

STEPHEN COOK (BORN 1939)   Stephen Cook was born in Buffalo where his father worked as an industrial chemist and taught university courses. His mother taught English courses in a community college. While in high school Cook developed an interest in electronics through his work with a famous local inventor noted for inventing the first implantable cardiac pacemaker.

Cook was a mathematics major at the University of Michigan, graduating in 1961. He did graduate work at Harvard, receiving a master's degree in 1962 and a Ph.D. in 1966. Cook was appointed an assistant professor in the Mathematics Department at the University of California, Berkeley in 1966. He was not granted tenure there, possibly because the members of the Mathematics Department did not find his work on what is now considered to be one of the most important areas of theoretical computer science of sufficient interest. In 1970, he joined the University of Toronto as an assistant professor, holding a joint appointment in the Computer Science Department and the Mathematics Department. He has remained at the University of Toronto, where he was appointed a University Professor in 1985.

Cook is considered to be one of the founders of computational complexity theory. His 1971 paper "The Complexity of Theorem Proving Procedures" formalized the notions of NP-completeness and polynomial-time reduction, showed that NP-complete problems exist by showing that the satisfiability problem is such a problem, and introduced the notorious P versus NP problem.

Cook has received many awards, including the 1982 Turing Award. He is married and has two sons. Among his interests are playing the violin and racing sailboats.

TABLE 2 **The Computer Time Used by Algorithms.**						
Problem Size	*Bit Operations Used*					
n	$\log n$	*n*	$n \log n$	n^2	2^n	*n*!
10	3×10^{-11} s	10^{-10} s	3×10^{-10} s	10^{-9} s	10^{-8} s	3×10^{-7} s
10^2	7×10^{-11} s	10^{-9} s	7×10^{-9} s	10^{-7} s	4×10^{11} yr	*
10^3	1.0×10^{-10} s	10^{-8} s	1×10^{-7} s	10^{-5} s	*	*
10^4	1.3×10^{-10} s	10^{-7} s	1×10^{-6} s	10^{-3} s	*	*
10^5	1.7×10^{-10} s	10^{-6} s	2×10^{-5} s	0.1 s	*	*
10^6	2×10^{-10} s	10^{-5} s	2×10^{-4} s	0.17 min	*	*

of an algorithm provides an upper, but not a lower, bound on the worst-case time required for the algorithm as a function of the input size. Nevertheless, for simplicity, we will often use big-O estimates when describing the time complexity of algorithms, with the understanding that big-Θ estimates would provide more information.

Table 2 displays the time needed to solve problems of various sizes with an algorithm using the indicated number n of bit operations, assuming that each bit operation takes 10^{-11} seconds, a reasonable estimate of the time required for a bit operation using the fastest computers available today. Times of more than 10^{100} years are indicated with an asterisk. In the future, these times will decrease as faster computers are developed. We can use the times shown in Table 2 to see whether it is reasonable to expect a solution to a problem of a specified size using an algorithm with known worst-case time complexity when we run this algorithm on a modern computer. Note that we cannot determine the exact time a computer uses to solve a problem with input of a particular size because of a myriad of issues involving computer hardware and the particular software implementation of the algorithm.

It is important to have a reasonable estimate for how long it will take a computer to solve a problem. For instance, if an algorithm requires approximately 10 hours, it may be worthwhile to spend the computer time (and money) required to solve this problem. But, if an algorithm requires approximately 10 billion years to solve a problem, it would be unreasonable to use resources to implement this algorithm. One of the most interesting phenomena of modern technology is the tremendous increase in the speed and memory space of computers. Another important factor that decreases the time needed to solve problems on computers is **parallel processing**, which is the technique of performing sequences of operations simultaneously.

Efficient algorithms, including most algorithms with polynomial time complexity, benefit most from significant technology improvements. However, these technology improvements offer little help in overcoming the complexity of algorithms of exponential or factorial time complexity. Because of the increased speed of computation, increases in computer memory, and the use of algorithms that take advantage of parallel processing, many problems that were considered impossible to solve five years ago are now routinely solved, and certainly five years from now this statement will still be true. This is even true when the algorithms used are intractable.

Exercises

1. Give a big-O estimate for the number of operations (where an operation is an addition or a multiplication) used in this segment of an algorithm.

$t := 0$
for $i := 1$ **to** 3
  for $j := 1$ **to** 4
    $t := t + ij$

2. Give a big-O estimate for the number additions used in this segment of an algorithm.

$t := 0$
for $i := 1$ **to** n
  for $j := 1$ **to** n
    $t := t + i + j$

3. How many comparisons are used by the algorithm given in Exercise 14 of Section 3.1 to find the smallest natural number in a sequence of n natural numbers?

4. a) Use pseudocode to describe the algorithm that puts the first four terms of a list of real numbers of arbitrary length in increasing order using the insertion sort.

   b) Show that this algorithm has time complexity $O(1)$ in terms of the number of comparisons used.

5. Suppose that an element is known to be among the first four elements in a list of 32 elements. Would a linear search or a binary search locate this element more rapidly?

6. Given a real number x and a positive integer k, determine the number of multiplications used to find x^{2^k} starting with x and successively squaring (to find x^2, x^4, and so on). Is this a more efficient way to find x^{2^k} than by multiplying x by itself the appropriate number of times?

7. Give a big-O estimate for the number of comparisons used by the algorithm that determines the number of 1s in a bit string by examining each bit of the string to determine whether it is a 1 bit (see Exercise 21 of Section 3.1).

*8. a) Show that this algorithm determines the number of 1 bits in the bit string S:

> **procedure** *bit count*(S: bit string)
> *count* := 0
> **while** $S \neq 0$
> *count* := *count* + 1
> $S := S \wedge (S - 1)$
> **return** *count* {*count* is the number of 1s in S}

Here $S - 1$ is the bit string obtained by changing the rightmost 1 bit of S to a 0 and all the 0 bits to the right of this to 1s. [Recall that $S \wedge (S - 1)$ is the bitwise *AND* of S and $S - 1$.]

   b) How many bitwise *AND* operations are needed to find the number of 1 bits in a string S using the algorithm in part (a)?

9. The conventional algorithm for evaluating a polynomial $a_n x^n + a_{n-1} x^{n-1} + \cdots + a_1 x + a_0$ at $x = c$ can be expressed in pseudocode by

> **procedure** *polynomial*($c, a_0, a_1, \ldots, a_n$: real numbers)
> *power* := 1
> $y := a_0$
> **for** $i := 1$ **to** n
> *power* := *power* $* c$
> $y := y + a_i * power$
> **return** y {$y = a_n c^n + a_{n-1} c^{n-1} + \cdots + a_1 c + a_0$}

where the final value of y is the value of the polynomial at $x = c$.

a) Evaluate $3x^2 + x + 1$ at $x = 2$ by working through each step of the algorithm showing the values assigned at each assignment step.

b) Exactly how many multiplications and additions are used to evaluate a polynomial of degree n at $x = c$? (Do not count additions used to increment the loop variable.)

10. There is a more efficient algorithm (in terms of the number of multiplications and additions used) for evaluating polynomials than the conventional algorithm described in the previous exercise. It is called **Horner's method**. This pseudocode shows how to use this method to find the value of $a_n x^n + a_{n-1} x^{n-1} + \cdots + a_1 x + a_0$ at $x = c$.

> **procedure** *Horner*($c, a_0, a_1, a_2, \ldots, a_n$: real numbers)
> $y := a_n$
> **for** $i := 1$ **to** n
> $y := y * c + a_{n-i}$
> **return** y {$y = a_n c^n + a_{n-1} c^{n-1} + \cdots + a_1 c + a_0$}

a) Evaluate $3x^2 + x + 1$ at $x = 2$ by working through each step of the algorithm showing the values assigned at each assignment step.

b) Exactly how many multiplications and additions are used by this algorithm to evaluate a polynomial of degree n at $x = c$? (Do not count additions used to increment the loop variable.)

11. What is the largest n for which one can solve within one second a problem using an algorithm that requires $f(n)$ bit operations, where each bit operation is carried out in 10^{-9} seconds, with these functions $f(n)$?

  a) $\log n$ **b)** n **c)** $n \log n$
  d) n^2 **e)** 2^n **f)** $n!$

12. How much time does an algorithm take to solve a problem of size n if this algorithm uses $2n^2 + 2^n$ operations, each requiring 10^{-9} seconds, with these values of n?

  a) 10   **b)** 20   **c)** 50   **d)** 100

13. How much time does an algorithm using 2^{50} operations need if each operation takes these amounts of time?

  a) 10^{-6} s   **b)** 10^{-9} s   **c)** 10^{-12} s

14. What is the effect in the time required to solve a problem when you double the size of the input from n to $2n$, assuming that the number of milliseconds the algorithm uses to solve the problem with input size n is each of these function? [Express your answer in the simplest form possible, either as a ratio or a difference. Your answer may be a function of n or a constant.]

  a) $\log \log n$ **b)** $\log n$ **c)** $100n$
  d) $n \log n$ **e)** n^2 **f)** n^3
  g) 2^n

15. What is the effect in the time required to solve a problem when you increase the size of the input from n to $n + 1$, assuming that the number of milliseconds the algorithm uses to solve the problem with input size n is each of these function? [Express your answer in the simplest form possible, either as a ratio or a difference. Your answer may be a function of n or a constant.]

  a) $\log n$ **b)** $100n$ **c)** n^2
  d) n^3 **e)** 2^n **f)** 2^{n^2}
  g) $n!$

16. Determine the least number of comparisons, or best-case performance,

   a) required to find the maximum of a sequence of n integers, using Algorithm 1 of Section 3.1.

b) used to locate an element in a list of n terms with a linear search.

c) used to locate an element in a list of n terms using a binary search.

17. Analyze the average-case performance of the linear search algorithm, if exactly half the time the element x is not in the list and if x is in the list it is equally likely to be in any position.

18. An algorithm is called **optimal** for the solution of a problem with respect to a specified operation if there is no algorithm for solving this problem using fewer operations.

a) Show that Algorithm 1 in Section 3.1 is an optimal algorithm with respect to the number of comparisons of integers. [*Note:* Comparisons used for bookkeeping in the loop are not of concern here.]

b) Is the linear search algorithm optimal with respect to the number of comparisons of integers (not including comparisons used for bookkeeping in the loop)?

19. Describe the worst-case time complexity, measured in terms of comparisons, of the ternary search algorithm described in Exercise 23 of Section 3.1.

20. Analyze the worst-case time complexity of the algorithm you devised in Exercise 25 of Section 3.1 for locating a mode in a list of nondecreasing integers.

21. Analyze the worst-case time complexity of the algorithm you devised in Exercise 27 of Section 3.1 for finding the first term of a sequence of integers equal to some previous term.

22. Analyze the worst-case time complexity of the algorithm you devised in Exercise 28 of Section 3.1 for finding the first term of a sequence less than the immediately preceding term.

23. Determine the worst-case complexity in terms of comparisons of the algorithm from Exercise 7 in Section 3.1 for determining whether a string of n characters is a palindrome.

24. How many comparisons does the selection sort (see preamble to Exercise 34 in Section 3.1) use to sort n items? Use your answer to give a big-O estimate of the complexity of the selection sort in terms of number of comparisons for the selection sort.

25. Find a big-O estimate for the worst-case complexity in terms of number of comparisons used and the number of terms swapped by the binary insertion sort described in the preamble to Exercise 40 in Section 3.1.

26. Show that the greedy algorithm for making change for n cents using quarters, dimes, nickels, and pennies has $O(n)$ complexity measured in terms of comparisons needed.

27. Describe how the number of comparisons used in the worst case changes when these algorithms are used to search for an element of a list when the size of the list doubles from n to $2n$, where n is a positive integer.

a) linear search **b)** binary search

28. Describe how the number of comparisons used in the worst case changes when the size of the list to be sorted doubles from n to $2n$, where n is a positive integer when these sorting algorithms are used.

a) bubble sort **b)** insertion sort

c) selection sort (described in the preamble to Exercise 34 in Section 3.1)

d) binary insertion sort (described in the preamble to Exercise 40 in Section 3.1)

Key Terms and Results

TERMS

algorithm: a finite sequence of precise instructions for performing a computation or solving a problem

searching algorithm: the problem of locating an element in a list

linear search algorithm: a procedure for searching a list element by element

binary search algorithm: a procedure for searching an ordered list by successively splitting the list in half

sorting: the reordering of the elements of a list into prescribed order

$f(x)$ **is** $O(g(x))$**:** the fact that $|f(x)| \leq C|g(x)|$ for all $x > k$ for some constants C and k

witness to the relationship $f(x)$ **is** $O(g(x))$**:** a pair C and k such that $|f(x)| \leq C|g(x)|$ whenever $x > k$

$f(x)$ **is** $\Omega(g(x))$**:** the fact that $|f(x)| \geq C|g(x)|$ for all $x > k$ for some positive constants C and k

$f(x)$ **is** $\Theta(g(x))$**:** the fact that $f(x)$ is both $O(g(x))$ and $\Omega(g(x))$

time complexity: the amount of time required for an algorithm to solve a problem

space complexity: the amount of space in computer memory required for an algorithm to solve a problem

worst-case time complexity: the greatest amount of time required for an algorithm to solve a problem of a given size

average-case time complexity: the average amount of time required for an algorithm to solve a problem of a given size

algorithmic paradigm: a general approach for constructing algorithms based on a particular concept

brute force: the algorithmic paradigm based on constructing algorithms for solving problems in a naive manner from the statement of the problem and definitions

greedy algorithm: an algorithm that makes the best choice at each step according to some specified condition

tractable problem: a problem for which there is a worst-case polynomial-time algorithm that solves it

intractable problem: a problem for which no worst-case polynomial-time algorithm exists for solving it

solvable problem: a problem that can be solved by an algorithm

unsolvable problem: a problem that cannot be solved by an algorithm

RESULTS

linear and binary search algorithms: (given in Section 3.1)
bubble sort: a sorting that uses passes where successive items are interchanged if they in the wrong order
insertion sort: a sorting that at the jth step inserts the jth element into the correct position in in the list, when the first $j-1$ elements of the list are already sorted

The linear search has $O(n)$ worst case time complexity.
The binary search has $O(\log n)$ worst case time complexity.

The bubble and insertion sorts have $O(n^2)$ worst case time complexity.

$\log n!$ is $O(n \log n)$.

If $f_1(x)$ is $O(g_1(x))$ and $f_2(x)$ is $O(g_2(x))$, then $(f_1 + f_2)(x)$ is $O(\max(g_1(x), g_2(x)))$ and $(f_1 f_2)(x)$ is $O((g_1 g_2(x)))$.

If $a_0, a_1, \ldots, a_n$ are real numbers with $a_n \neq 0$, then $a_n x^n + a_{n-1}x^{n-1} + \cdots + a_1 x + a_0$ is $\Theta(x^n)$, and hence $O(n)$ and $\Omega(n)$.

Review Questions

1. **a)** Define the term *algorithm*.
   b) What are the different ways to describe algorithms?
   c) What is the difference between an algorithm for solving a problem and a computer program that solves this problem?

2. **a)** Describe, using English, an algorithm for finding the largest integer in a list of n integers.
   b) Express this algorithm in pseudocode.
   c) How many comparisons does the algorithm use?

3. **a)** State the definition of the fact that $f(n)$ is $O(g(n))$, where $f(n)$ and $g(n)$ are functions from the set of positive integers to the set of real numbers.
   b) Use the definition of the fact that $f(n)$ is $O(g(n))$ directly to prove or disprove that $n^2 + 18n + 107$ is $O(n^3)$.
   c) Use the definition of the fact that $f(n)$ is $O(g(n))$ directly to prove or disprove that n^3 is $O(n^2 + 18n + 107)$.

4. List these functions so that each function is big-O of the next function in the list: $(\log n)^3$, $n^3/1000000$, $\sqrt{n}$, $100n + 101$, 3^n, $n!$, $2^n n^2$.

5. **a)** How can you produce a big-O estimate for a function that is the sum of different terms where each term is the product of several functions?
   b) Give a big-O estimate for the function $f(n) = (n! + 1)(2^n + 1) + (n^{n-2} + 8n^{n-3})(n^3 + 2^n)$. For the function g in your estimate $f(x)$ is $O(g(x))$ use a simple function of smallest possible order.

6. **a)** Define what the worst-case time complexity, average-case time complexity, and best-case time complexity (in terms of comparisons) mean for an algorithm that finds the smallest integer in a list of n integers.
   b) What are the worst-case, average-case, and best-case time complexities, in terms of comparisons, of the algorithm that finds the smallest integer in a list of n integers by comparing each of the integers with the smallest integer found so far?

7. **a)** Describe the linear search and binary search algorithm for finding an integer in a list of integers in increasing order.
   b) Compare the worst-case time complexities of these two algorithms.
   c) Is one of these algorithms always faster than the other (measured in terms of comparisons)?

8. **a)** Describe the bubble sort algorithm.
   b) Use the bubble sort algorithm to sort the list 5, 2, 4, 1, 3.
   c) Give a big-O estimate for the number of comparisons used by the bubble sort.

9. **a)** Describe the insertion sort algorithm.
   b) Use the insertion sort algorithm to sort the list 2, 5, 1, 4, 3.
   c) Give a big-O estimate for the number of comparisons used by the insertion sort.

10. **a)** Explain the concept of a greedy algorithm.
    b) Provide an example of a greedy algorithm that produces an optimal solution and explain why it produces an optimal solution.
    c) Provide an example of a greedy algorithm that does not always produce an optimal solution and explain why it fails to do so.

11. Define what it means for a problem to be tractable and what it means for a problem to be solvable.

Supplementary Exercises

1. **a)** Describe an algorithm for locating the last occurrence of the largest number in a list of integers.
   b) Estimate the number of comparisons used.

2. **a)** Give an algorithm to determine whether a bit string contains a pair of consecutive zeros.
   b) How many comparisons does the algorithm use?

3. **a)** Adapt Algorithm 1 in Section 3.1 to find the maximum and the minimum of a sequence of n elements by employing a temporary maximum and a temporary minimum that is updated as each successive element is examined.
   b) Describe the algorithm from part (a) in pseudocode.

c) How many comparisons of elements in the sequence are carried out by this algorithm? (Do not count comparisons used to determine whether the end of the sequence has been reached.)

4. a) Describe in detail (and in English) the steps of an algorithm that finds the maximum and minimum of a sequence of n elements by examining pairs of successive elements, keeping track of a temporary maximum and a temporary minimum. If n is odd, both the temporary maximum and temporary minimum should initially equal the first term, and if n is even, the temporary minimum and temporary maximum should be found by comparing the initial two elements. The temporary maximum and temporary minimum should be updated by comparing them with the maximum and minimum of the pair of elements being examined.

b) Express the algorithm described in part (a) in pseudocode.

c) How many comparisons of elements of the sequence are carried out by this algorithm? (Do not count comparisons used to determine whether the end of the sequence has been reached.) How does this compare to the number of comparisons used by the algorithm in Exercise 3?

*5. Show that the worst-case complexity in terms of comparisons of an algorithm that finds the maximum and minimum of n elements is at least $\lceil 3n/2 \rceil - 2$.

6. Devise an efficient algorithm for finding the second largest element in a sequence of n elements and determine the worst-case complexity of your algorithm.

7. Devise an algorithm that finds all equal pairs of sums of two terms of a sequence of n numbers, and determine the worst-case complexity of your algorithm.

8. Devise an algorithm that finds the closest pair of integers in a sequence of n integers, and determine the worst-case complexity of your algorithm. [*Hint:* Sort the sequence. Use the fact that sorting can be done with worst-case time complexity $O(n \log n)$.]

The **shaker sort** (or **bidirectional bubble sort**) successively compares pairs of adjacent elements, exchanging them if they are out of order, and alternately passing through the list from the beginning to the end and then from the end to the beginning until no exchanges are needed.

9. Show the steps used by the shaker sort to sort the list 3, 5, 1, 4, 6, 2.

10. Show that the shaker sort has $O(n^2)$ complexity measured in terms of the number of comparisons it uses.

11. Explain why the shaker sort is efficient for sorting lists that are already in close to the correct order.

12. Show that $(n \log n + n^2)^3$ is $O(n^6)$.

13. Give a big-O estimate for $(x^2 + x(\log x)^3) \cdot (2^x + x^3)$.

*14. Show that $n!$ is not $O(2^n)$.

15. Find all pairs of functions of the same order in this list of functions: $n^2 + (\log n)^2$, $n^2 + n$, $n^2 + \log 2^n + 1$, $(n + 1)^3 - (n - 1)^3$, and $(n + \log n)^2$.

16. Find an integer n with $n > 2$ for which $n^{2^{100}} < 2^n$.

17. Find an integer n with $n > 2$ for which $(\log n)^{2^{100}} < \sqrt{n}$.

*18. Arrange the functions n^n, $(\log n)^2$, $n^{1.0001}$, $(1.0001)^n$, $2^{\sqrt{\log_2 n}}$, and $n(\log n)^{1001}$ in a list so that each function is big-O of the next function. [*Hint:* To determine the relative size of some of these functions, take logarithms.]

*19. Give an example of two increasing functions $f(n)$ and $g(n)$ from the set of positive integers to the set of positive integers such that neither $f(n)$ is $O(g(n))$ nor $g(n)$ is $O(f(n))$.

20. Show that if the denominations of coins are $c^0, c^1, \ldots, c^k$, where k is a positive integer and c is a positive integer, $c > 1$, the greedy algorithm always produces change using the fewest coins possible.

21. a) Use pseudocode to specify a brute-force algorithm that determines when given as input a sequence of n positive integers whether there are two distinct terms of the sequence that have as sum a third term. The algorithm should loop through all triples of terms of the sequence, checking whether the sum of the first two terms equals the third.

b) Give a big-O estimate for the complexity of the brute-force algorithm from part (a).

22. a) Devise a more efficient algorithm for solving the problem described in Exercise 21 that first sorts the input sequence and then checks for each pair of terms whether their difference is in the sequence.

b) Give a big-O estimate for the complexity of this algorithm. Is it more efficient than the brute-force algorithm from Exercise 21?

Suppose we have s men and s women each with their preference lists for the members of the opposite gender, as described in the preamble to Exercise 50 in Section 3.1. We say that a woman w is a **valid partner** for a man m if there is some stable matching in which they are paired. Similarly, a man m is a **valid partner** for a woman w if there is some stable matching in which they are paired. A matching in which each man is assigned his valid partner ranking highest on his preference list is called **male optimal**, and a matching in which each woman is assigned her valid partner ranking lowest on her preference list is called **female pessimal**.

23. Find all valid partners for each man and each woman if there are three men m_1, m_2, and m_3 and three women w_1, w_2, w_3 with these preference rankings of the men for the women, from highest to lowest: $m_1: w_3, w_1, w_2$; $m_2: w_3, w_2, w_1$; $m_3: w_2, w_3, w_1$; and with these preference rankings of the women for the men, from highest to lowest: $w_1: m_3, m_2, m_1$; $w_2: m_1, m_3, m_2$; $w_3: m_3, m_2, m_1$.

24. Define what it means for a matching to be female optimal and for a matching to be male pessimal.

Exercises 25–26 deal with the problem of scheduling n jobs on a single processor. To complete job j, the processor must run job j for time t_j without interruption. Each job has a deadline d_j. If we start job j at time s_j, it will be completed at time $e_j = s_j + t_j$. The **lateness** of the job measures how long it finishes after its deadline, that is, the lateness of job j is

$\max(0, e_j - d_j)$. We wish to devise a greedy algorithm that minimizes the maximum lateness of a job among the n jobs.

25. Suppose we have five jobs with specified required times and deadlines: $t_1 = 25, d_1 = 50$; $t_2 = 15, d_2 = 60$; $t_3 = 20, d_3 = 60$; $t_4 = 5, d_4 = 55$; $t_5 = 10, d_5 = 75$. Find the maximum lateness of any job when the jobs are scheduled in this order (and they start at time 0): Job 3, Job 1, Job 4, Job 2, Job 5. Answer the same question for the schedule Job 5, Job 4, Job 3, Job 1, Job 2.

26. The **slackness** of a job requiring time t and with deadline d is $d - t$, the difference between its deadline and the time it requires. Find an example that shows that scheduling jobs by increasing slackness does not always yield a schedule with the smallest possible maximum lateness.

27. Suppose that we have a knapsack with total capacity of W kg. We also have n items where item j has mass w_j. The **knapsack problem** asks for a subset of these n items with the largest possible total mass not exceeding W.

    a) Devise a brute-force algorithm for solving the knapsack problem.

    b) Solve the knapsack problem when the capacity of the knapsack is 18 kg and there are five items: a 5-kg sleeping bag, an 8-kg tent, a 7-kg food pack, a 4-kg container of water, and an 11-kg portable stove.

In Exercises 28–29 we will study the problem of load balancing. The input to the problem is a collection of p processors and n jobs, t_j is the time required to run job j, jobs run without interruption on a single machine until finished, and a processor can run only one job at a time. The **load** L_k of processor k is the sum over all jobs assigned to processor k of the times required to run these jobs. The **makespan** is the maximum load over all the p processors. The load balancing problem asks for an assignment of jobs to processors to minimize the makespan.

28. Suppose we have three processors and five jobs requiring times $t_1 = 3$, $t_2 = 5$, $t_3 = 4$, $t_4 = 7$, and $t_5 = 8$. Solve the load balancing problem for this input by finding the assignment of the five jobs to the three processors that minimizes the makespan.

29. Suppose that L^* is the minimum makespan when p processors are given n jobs, where t_j is the time required to run job j.

    a) Show that $L^* \geq \max_{j=1,2,\ldots,n} t_j$.

    b) Show that $L^* \geq \frac{1}{p} \sum_{j=1}^{n} t_j$.

Computer Projects

Write programs with these inputs and outputs.

1. Given a list of n integers, find the largest integer in the list.

2. Given a list of n integers, find the first and last occurrences of the largest integer in the list.

3. Given a list of n distinct integers, determine the position of an integer in the list using a linear search.

4. Given an ordered list of n distinct integers, determine the position of an integer in the list using a binary search.

5. Given a list of n integers, sort them using a bubble sort.

6. Given a list of n integers, sort them using an insertion sort.

7. Given an integer n, use the greedy algorithm to find the change for n cents using quarters, dimes, nickels, and pennies.

8. Given the starting and ending times of n talks, use the appropriate greedy algorithm to schedule the most talks possible in a single lecture hall.

9. Given an ordered list of n integers and an integer x in the list, find the number of comparisons used to determine the position of x in the list using a linear search and using a binary search.

10. Given a list of integers, determine the number of comparisons used by the bubble sort and by the insertion sort to sort this list.

Computations and Explorations

Use a computational program or programs you have written to do these exercises.

1. We know that n^b is $O(d^n)$ when b and d are positive numbers with $d \geq 2$. Give values of the constants C and k such that $n^b \leq Cd^n$ whenever $x > k$ for each of these sets of values: $b = 10, d = 2$; $b = 20, d = 3$; $b = 1000$, $d = 7$.

2. Compute the change for different values of n with coins of different denominations using the greedy algorithm and determine whether the smallest number of coins was used. Can you find conditions so that the greedy algorithm is guaranteed to use the fewest coins possible?

3. Using a generator of random orderings of the integers $1, 2, \ldots, n$, find the number of comparisons used by the bubble sort, insertion sort, binary insertion sort, and selection sort to sort these integers.

Writing Projects

Respond to these with essays using outside sources.

1. Examine the history of the word *algorithm* and describe the use of this word in early writings.

2. Explain how sorting algorithms can be classified into a taxonomy based on the underlying principle on which they are based.

3. Describe the radix sort algorithm.

4. Develop a detailed list of algorithmic paradigms and provide examples using each of these paradigms.

5. Explain what the Turing Award is and describe the criteria used to select winners. List six past winners of the award and why they received the award.

6. Describe what is meant by a parallel algorithm. Explain how the pseudocode used in this book can be extended to handle parallel algorithms.

7. Describe six different NP-complete problems.

8. Demonstrate how one of the many different NP-complete problems can be reduced to the satisfiability problem.

4

Number Theory and Cryptography

The part of mathematics devoted to the study of the set of integers and their properties is known as number theory. In this chapter we will develop some of the important concepts of number theory including many of those used in computer science. As we develop number theory, we will use the proof methods developed in Chapter 1 to prove many theorems.

We will first introduce the notion of divisibility of integers, which we use to introduce modular, or clock, arithmetic. Modular arithmetic operates with the remainders of integers when they are divided by a fixed positive integer, called the modulus. We will prove many important results about modular arithmetic which we will use extensively in this chapter.

Integers can be represented with any positive integer b greater than 1 as a base. In this chapter we discuss base b representations of integers and give an algorithm for finding them. In particular, we will discuss binary, octal, and hexadecimal (base 2, 8, and 16) representations. We will describe algorithms for carrying out arithmetic using these representations and study their complexity. These algorithms were the first procedures called algorithms.

We will discuss prime numbers, the positive integers that have only 1 and themselves as positive divisors. We will prove that there are infinitely many primes; the proof we give is considered to be one of the most beautiful proofs in mathematics. We will discuss the distribution of primes and many famous open questions concerning primes. We will introduce the concept of greatest common divisors and study the Euclidean algorithm for computing them. This algorithm was first described thousands of years ago. We will introduce the fundamental theorem of arithmetic, a key result which tells us that every positive integer has a unique factorization into primes.

We will explain how to solve linear congruences, as well as systems of linear congruences, which we solve using the famous Chinese remainder theorem. We will introduce the notion of pseudoprimes, which are composite integers masquerading as primes, and show how this notion can help us rapidly generate prime numbers.

This chapter introduces several important applications of number theory. In particular, we will use number theory to generate pseudorandom numbers, to assign memory locations to computer files, and to find check digits used to detect errors in various kinds of identification numbers. We also introduce the subject of cryptography. Number theory plays an essentially role both in classical cryptography, first used thousands of years ago, and modern cryptography, which plays an essential role in electronic communication. We will show how the ideas we develop can be used in cryptographical protocols, introducing protocols for sharing keys and for sending signed messages. Number theory, once considered the purest of subjects, has become an essential tool in providing computer and Internet security.

4.1 Divisibility and Modular Arithmetic

Introduction

The ideas that we will develop in this section are based on the notion of divisibility. Division of an integer by a positive integer produces a quotient and a remainder. Working with these remainders leads to modular arithmetic, which plays an important role in mathematics and which is used throughout computer science. We will discuss some important applications of modular arithmetic later in this chapter, including generating pseudorandom numbers, assigning computer memory locations to files, constructing check digits, and encrypting messages.

Division

When one integer is divided by a second nonzero integer, the quotient may or may not be an integer. For example, $12/3 = 4$ is an integer, whereas $11/4 = 2.75$ is not. This leads to Definition 1.

DEFINITION 1   If a and b are integers with $a \neq 0$, we say that a *divides* b if there is an integer c such that $b = ac$, or equivalently, if $\frac{b}{a}$ is an integer. When a divides b we say that a is a *factor* or *divisor* of b, and that b is a *multiple* of a. The notation $a \mid b$ denotes that a divides b. We write $a \nmid b$ when a does not divide b.

Remark: We can express $a \mid b$ using quantifiers as $\exists c(ac = b)$, where the universe of discourse is the set of integers.

In Figure 1 a number line indicates which integers are divisible by the positive integer d.

EXAMPLE 1   Determine whether $3 \mid 7$ and whether $3 \mid 12$.

Solution: We see that $3 \nmid 7$, because $7/3$ is not an integer. On the other hand, $3 \mid 12$ because $12/3 = 4$.   ◀

EXAMPLE 2   Let n and d be positive integers. How many positive integers not exceeding n are divisible by d?

Solution: The positive integers divisible by d are all the integers of the form dk, where k is a positive integer. Hence, the number of positive integers divisible by d that do not exceed n equals the number of integers k with $0 < dk \leq n$, or with $0 < k \leq n/d$. Therefore, there are $\lfloor n/d \rfloor$ positive integers not exceeding n that are divisible by d.   ◀

Some of the basic properties of divisibility of integers are given in Theorem 1.

THEOREM 1   Let a, b, and c be integers, where $a \neq 0$. Then

   (*i*) if $a \mid b$ and $a \mid c$, then $a \mid (b + c)$;
   (*ii*) if $a \mid b$, then $a \mid bc$ for all integers c;
   (*iii*) if $a \mid b$ and $b \mid c$, then $a \mid c$.

Proof: We will give a direct proof of (*i*). Suppose that $a \mid b$ and $a \mid c$. Then, from the definition of divisibility, it follows that there are integers s and t with $b = as$ and $c = at$. Hence,

$$b + c = as + at = a(s + t).$$

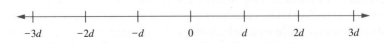

FIGURE 1   Integers Divisible by the Positive Integer d.

Therefore, a divides $b + c$. This establishes part (*i*) of the theorem. The proofs of parts (*ii*) and (*iii*) are left as Exercises 3 and 4.   ◁

Theorem 1 has this useful consequence.

COROLLARY 1

If a, b, and c are integers, where $a \neq 0$, such that $a \mid b$ and $a \mid c$, then $a \mid mb + nc$ whenever m and n are integers.

Proof: We will give a direct proof. By part (*ii*) of Theorem 1 we see that $a \mid mb$ and $a \mid nc$ whenever m and n are integers. By part (*i*) of Theorem 1 it follows that $a \mid mb + nc$.   ◁

The Division Algorithm

When an integer is divided by a positive integer, there is a quotient and a remainder, as the division algorithm shows.

THEOREM 2

THE DIVISION ALGORITHM   Let a be an integer and d a positive integer. Then there are unique integers q and r, with $0 \leq r < d$, such that $a = dq + r$.

We defer the proof of the division algorithm to Section 5.2. (See Example 5 and Exercise 29.)

Remark: Theorem 2 is not really an algorithm. (Why not?) Nevertheless, we use its traditional name.

DEFINITION 2

In the equality given in the division algorithm, d is called the *divisor*, a is called the *dividend*, q is called the *quotient*, and r is called the *remainder*. This notation is used to express the quotient and remainder:

$$q = a \textbf{ div } d, \quad r = a \textbf{ mod } d.$$

Remark: Note that both a **div** d and a **mod** d for a fixed d are functions on the set of integers. Furthermore, when a is an integer and d is a positive integer, we have a **div** $d = \lfloor a/d \rfloor$ and a **mod** $d = a - d$. (See exercise 18.)

Examples 3 and 4 illustrate the division algorithm.

EXAMPLE 3

What are the quotient and remainder when 101 is divided by 11?

Solution: We have

$$101 = 11 \cdot 9 + 2.$$

Hence, the quotient when 101 is divided by 11 is $9 = 101$ **div** 11, and the remainder is $2 = 101$ **mod** 11.   ◀

EXAMPLE 4   What are the quotient and remainder when -11 is divided by 3?

Solution: We have

$$-11 = 3(-4) + 1.$$

Extra
Examples

Hence, the quotient when -11 is divided by 3 is $-4 = -11$ **div** 3, and the remainder is $1 = -11$ **mod** 3.

Note that the remainder cannot be negative. Consequently, the remainder is *not* -2, even though

$$-11 = 3(-3) - 2,$$

because $r = -2$ does not satisfy $0 \le r < 3$.   ◀

Note that the integer a is divisible by the integer d if and only if the remainder is zero when a is divided by d.

Remark: A programming language may have one, or possibly two, operators for modular arithmetic, denoted by mod (in BASIC, Maple, Mathematica, EXCEL, and SQL), % (in C, C++, Java, and Python), rem (in Ada and Lisp), or something else. Be careful when using them, because for $a < 0$, some of these operators return $a - m\lceil a/m \rceil$ instead of a **mod** $m = a - m\lfloor a/m \rfloor$ (as shown in Exercise 15). Also, unlike a **mod** m, some of these operators are defined when $m < 0$, and even when $m = 0$.

Modular Arithmetic

In some situations we care only about the remainder of an integer when it is divided by some specified positive integer. For instance, when we ask what time it will be (on a 24-hour clock) 50 hours from now, we care only about the remainder when 50 plus the current hour is divided by 24. Because we are often interested only in remainders, we have special notations for them. We have already introduced the notation a **mod** m to represent the remainder when an integer a is divided by the positive integer m. We now introduce a different, but related, notation that indicates that two integers have the same remainder when they are divided by the positive integer m.

DEFINITION 3   If a and b are integers and m is a positive integer, then a is *congruent to b modulo m* if m divides $a - b$. We use the notation $a \equiv b \pmod{m}$ to indicate that a is congruent to b modulo m. We say that $a \equiv b \pmod{m}$ is a **congruence** and that m is its **modulus** (plural **moduli**). If a and b are not congruent modulo m, we write $a \not\equiv b \pmod{m}$.

Although both notations $a \equiv b \pmod{m}$ and a **mod** $m = b$ include "mod," they represent fundamentally different concepts. The first represents a relation on the set of integers, whereas the second represents a function. However, the relation $a \equiv b \pmod{m}$ and the **mod** m function are closely related, as described in Theorem 3.

THEOREM 3   Let a and b be integers, and let m be a positive integer. Then $a \equiv b \pmod{m}$ if and only if $a \bmod m = b \bmod m$.

The proof of Theorem 3 is left as Exercise 13. Recall that $a \bmod m$ and $b \bmod m$ are the remainders when a and b are divided by m, respectively. Consequently, Theorem 3 also says that $a \equiv b \pmod{m}$ if and only if a and b have the same remainder when divided by m.

EXAMPLE 5   Determine whether 17 is congruent to 5 modulo 6 and whether 24 and 14 are congruent modulo 6.

Solution: Because 6 divides $17 - 5 = 12$, we see that $17 \equiv 5 \pmod{6}$. However, because $24 - 14 = 10$ is not divisible by 6, we see that $24 \not\equiv 14 \pmod{6}$.   ◄

The great German mathematician Karl Friedrich Gauss developed the concept of congruences at the end of the eighteenth century. The notion of congruences has played an important role in the development of number theory.

Theorem 4 provides a useful way to work with congruences.

THEOREM 4   Let m be a positive integer. The integers a and b are congruent modulo m if and only if there is an integer k such that $a = b + km$.

Proof: If $a \equiv b \pmod{m}$, by the definition of congruence (Definition 3), we know that $m \mid (a - b)$. This means that there is an integer k such that $a - b = km$, so that $a = b + km$. Conversely, if there is an integer k such that $a = b + km$, then $km = a - b$. Hence, m divides $a - b$, so that $a \equiv b \pmod{m}$.   ◁

The set of all integers congruent to an integer a modulo m is called the **congruence class** of a modulo m. In Chapter 9 we will show that there are m pairwise disjoint equivalence classes modulo m and that the union of these equivalence classes is the set of integers.

Theorem 5 shows that additions and multiplications preserve congruences.

Links

KARL FRIEDRICH GAUSS (1777–1855)   Karl Friedrich Gauss, the son of a bricklayer, was a child prodigy. He demonstrated his potential at the age of 10, when he quickly solved a problem assigned by a teacher to keep the class busy. The teacher asked the students to find the sum of the first 100 positive integers. Gauss realized that this sum could be found by forming 50 pairs, each with the sum 101: $1 + 100, 2 + 99, \ldots, 50 + 51$. This brilliance attracted the sponsorship of patrons, including Duke Ferdinand of Brunswick, who made it possible for Gauss to attend Caroline College and the University of Göttingen. While a student, he invented the method of least squares, which is used to estimate the most likely value of a variable from experimental results. In 1796 Gauss made a fundamental discovery in geometry, advancing a subject that had not advanced since ancient times. He showed that a 17-sided regular polygon could be drawn using just a ruler and compass.

In 1799 Gauss presented the first rigorous proof of the fundamental theorem of algebra, which states that a polynomial of degree n has exactly n roots (counting multiplicities). Gauss achieved worldwide fame when he successfully calculated the orbit of the first asteroid discovered, Ceres, using scanty data.

Gauss was called the Prince of Mathematics by his contemporary mathematicians. Although Gauss is noted for his many discoveries in geometry, algebra, analysis, astronomy, and physics, he had a special interest in number theory, which can be seen from his statement "Mathematics is the queen of the sciences, and the theory of numbers is the queen of mathematics." Gauss laid the foundations for modern number theory with the publication of his book *Disquisitiones Arithmeticae* in 1801.

THEOREM 5   Let m be a positive integer. If $a \equiv b \pmod{m}$ and $c \equiv d \pmod{m}$, then

$$a + c \equiv b + d \pmod{m} \qquad \text{and} \qquad ac \equiv bd \pmod{m}.$$

Proof: We use a direct proof. Because $a \equiv b \pmod{m}$ and $c \equiv d \pmod{m}$, by Theorem 4 there are integers s and t with $b = a + sm$ and $d = c + tm$. Hence,

$$b + d = (a + sm) + (c + tm) = (a + c) + m(s + t)$$

and

$$bd = (a + sm)(c + tm) = ac + m(at + cs + stm).$$

Hence,

$$a + c \equiv b + d \pmod{m} \qquad \text{and} \qquad ac \equiv bd \pmod{m}. \qquad \triangleleft$$

EXAMPLE 6   Because $7 \equiv 2 \pmod{5}$ and $11 \equiv 1 \pmod{5}$, it follows from Theorem 5 that

$$18 = 7 + 11 \equiv 2 + 1 = 3 \pmod{5}$$

and that

$$77 = 7 \cdot 11 \equiv 2 \cdot 1 = 2 \pmod{5}. \qquad \blacktriangleleft$$

We must be careful working with congruences. Some properties we may expect to be true are not valid. For example, if $ac \equiv bc \pmod{m}$, the congruence $a \equiv b \pmod{m}$ may be false. Similarly, if $a \equiv b \pmod{m}$ and $c \equiv d \pmod{m}$, the congruence $a^c \equiv b^d \pmod{m}$ may be false. (See Exercise 27.)

You cannot always divide both sides of a congruence by the same number!

Corollary 2 shows how to find the values of the **mod** m function at the sum and product of two integers using the values of this function at each of these integers. We will use this result in Section 5.4.

COROLLARY 2   Let m be a positive integer and let a and b be integers. Then

$$(a + b) \bmod m = ((a \bmod m) + (b \bmod m)) \bmod m$$

and

$$ab \bmod m = ((a \bmod m)(b \bmod m)) \bmod m.$$

Proof: By the definitions of **mod** m and of congruence modulo m, we know that $a \equiv (a \bmod m) \pmod{m}$ and $b \equiv (b \bmod m) \pmod{m}$. Hence, Theorem 5 tells us that

$$a + b \equiv (a \bmod m) + (b \bmod m) \pmod{m}$$

and

$$ab \equiv (a \bmod m)(b \bmod m) \pmod{m}.$$

The equalities in this corollary follow from these last two congruences by Theorem 3. $\qquad \triangleleft$

Arithmetic Modulo m

We can define arithmetic operations on $\mathbf{Z}_m$, the set of nonnegative integers less than m, that is, the set $\{0, 1, \ldots, m - 1\}$. In particular, we define addition of these integers, denoted by $+_m$ by

$$a +_m b = (a + b) \bmod m,$$

where the addition on the right-hand side of this equation is the ordinary addition of integers, and we define multiplication of these integers, denoted by $\cdot_m$ by

$$a \cdot_m b = (a \cdot b) \bmod m,$$

where the multiplication on the right-hand side of this equation is the ordinary multiplication of integers. The operations $+_m$ and $\cdot_m$ are called addition and multiplication modulo m and when we use these operations, we are said to be doing **arithmetic modulo** m.

EXAMPLE 7   Use the definition of addition and multiplication in $\mathbf{Z}_m$ to find $7 +_{11} 9$ and $7 \cdot_{11} 9$.

Solution: Using the definition of addition modulo 11, we find that

$$7 +_{11} 9 = (7 + 9) \bmod 11 = 16 \bmod 11 = 5,$$

and

$$7 \cdot_{11} 9 = (7 \cdot 9) \bmod 11 = 63 \bmod 11 = 8.$$

Hence $7 +_{11} 9 = 5$ and $7 \cdot_{11} 9 = 8$.   ◀

The operations $+_m$ and $\cdot_m$ satisfy many of the same properties of ordinary addition and multiplication of integers. In particular, they satisfy these properties:

Closure If a and b belong to $\mathbf{Z}_m$, then $a +_m b$ and $a \cdot_m b$ belong to $\mathbf{Z}_m$.

Associativity If a, b, and c belong to $\mathbf{Z}_m$, then $(a +_m b) +_m c = a +_m (b +_m c)$ and $(a \cdot_m b) \cdot_m c = a \cdot_m (b \cdot_m c)$.

Commutativity If a and b belong to $\mathbf{Z}_m$, then $a +_m b = b +_m a$ and $a \cdot_m b = b \cdot_m a$.

Identity elements The elements 0 and 1 are identity elements for addition and multiplication modulo m, respectively. That is, if a belongs to $\mathbf{Z}_m$, then $a +_m 0 = 0 +_m a = a$ and $a \cdot_m 1 = 1 \cdot_m a = a$.

Additive inverses If $a \neq 0$ belongs to $\mathbf{Z}_m$, then $m - a$ is an additive inverse of a modulo m and 0 is its own additive inverse. That is $a +_m (m - a) = 0$ and $0 +_m 0 = 0$.

Distributivity If a, b, and c belong to $\mathbf{Z}_m$, then $a \cdot_m (b +_m c) = (a \cdot_m b) +_m (a \cdot_m c)$ and $(a +_m b) \cdot_m c = (a \cdot_m c) +_m (b \cdot_m c)$.

These properties follow from the properties we have developed for congruences and remainders modulo m, together with the properties of integers (See Exercise 32). Note that we have listed the property that every element of $\mathbf{Z}_m$ has an additive inverse, but no analogous property for multiplicative inverses has been included. This is because multiplicative inverses do not always exists modulo m. For instance, there is no multiplicative inverse of 2 modulo 6, as the reader can verify. We will return to the question of when an integer has a multiplicative inverse modulo m later in this chapter.

Remark: Because $\mathbf{Z}_m$ with the operations of addition and multiplication modulo m satisfies the properties listed, $\mathbf{Z}_m$ with modular addition is said to be a **commutative group** and $\mathbf{Z}_m$ with both of these operations is said to be a **commutative ring**. Note that the set of integers with ordinary addition and multiplication also forms a commutative ring. Groups and rings are studied in courses that cover abstract algebra.

Remark: In Exercise 30, and in later sections, we will use the notations $+$ and $\cdot$ for $+_m$ and $\cdot_m$ without the subscript m on the symbol for the operator whenever we work with $\mathbf{Z}_m$.

Exercises

1. Does 17 divide each of these numbers?
   a) 68 **b)** 84 **c)** 357 **d)** 1001

2. Prove that if a is an integer other than 0, then
   a) 1 divides a. **b)** a divides 0.

3. Prove that part (*ii*) of Theorem 1 is true.

4. Prove that part (*iii*) of Theorem 1 is true.

5. Show that if $a \mid b$ and $b \mid a$, where a and b are integers, then $a = b$ or $a = -b$.

6. Show that if $a, b, c,$ and d are integers, where $a \neq 0$, such that $a \mid c$ and $b \mid d$, then $ab \mid cd$.

7. Show that if $a, b,$ and c are integers, where $a \neq 0$ and $c \neq 0$, such that $ac \mid bc$, then $a \mid b$.

8. Prove or disprove that if $a \mid bc$, where $a, b,$ and c are positive integers and $a \neq 0$, then $a \mid b$ or $a \mid c$.

9. What are the quotient and remainder when
   a) 19 is divided by 7?
   b) -111 is divided by 11?
   c) 789 is divided by 23?
   d) 1001 is divided by 13?
   e) 0 is divided by 19?
   f) 3 is divided by 5?
   g) -1 is divided by 3?
   h) 4 is divided by 1?

10. What time does a 12-hour clock read
    a) 80 hours after it reads 11:00?
    b) 40 hours before it reads 12:00?
    c) 100 hours after it reads 6:00?

11. What time does a 24-hour clock read
    a) 100 hours after it reads 2:00?
    b) 45 hours before it reads 12:00?
    c) 168 hours after it reads 19:00?

12. Suppose that a and b are integers, $a \equiv 4 \pmod{13}$, and $b \equiv 9 \pmod{13}$. Find the integer c with $0 \le c \le 12$ such that
    a) $c \equiv 9a \pmod{13}$.
    b) $c \equiv 11b \pmod{13}$.
    c) $c \equiv a + b \pmod{13}$.
    d) $c \equiv 2a + 3b \pmod{13}$.
    e) $c \equiv a^2 + b^2 \pmod{13}$.
    f) $c \equiv a^3 - b^3 \pmod{13}$.

13. **a)** Let m be a positive integer. Show that $a \equiv b \pmod{m}$ if $a \bmod m = b \bmod m$.
    b) Let m be a positive integer. Show that $a \bmod m = b \bmod m$ if $a \equiv b \pmod{m}$.

14. Show that if n and k are positive integers, then $\lceil n/k \rceil = \lfloor (n-1)/k \rfloor + 1$.

15. Show that if a is an integer and d is an integer greater than 1, then the quotient and remainder obtained when a is divided by d are $\lfloor a/d \rfloor$ and $a - d\lfloor a/d \rfloor$, respectively.

16. Find a formula for the integer with smallest absolute value that is congruent to an integer a modulo m, where m is a positive integer.

17. Evaluate these quantities.
    a) $13 \bmod 3$ **b)** $-97 \bmod 11$
    c) $155 \bmod 19$ **d)** $-221 \bmod 23$

18. Find $a \textbf{ div } m$ and $a \bmod m$ when
    a) $a = 228, m = 119$.
    b) $a = 9009, m = 223$.
    c) $a = -10101, m = 333$.
    d) $a = -765432, m = 38271$.

19. Find the integer a such that
    a) $a \equiv -15 \pmod{27}$ and $-26 \le a \le 0$.
    b) $a \equiv 24 \pmod{31}$ and $-15 \le a \le 15$.
    c) $a \equiv 99 \pmod{41}$ and $100 \le a \le 140$.

20. List five integers that are congruent to 4 modulo 12.

21. List all integers between -100 and 100 that are congruent to -1 modulo 25.

22. Decide whether each of these integers is congruent to 5 modulo 17.
    a) 80 **b)** 103
    c) -29 **d)** -122

23. Find each of these values.
    a) $(-133 \bmod 23 + 261 \bmod 23) \bmod 23$
    b) $(457 \bmod 23 \cdot 182 \bmod 23) \bmod 23$

24. Find each of these values.
    a) $(99^2 \bmod 32)^3 \bmod 15$
    b) $(3^4 \bmod 17)^2 \bmod 11$
    c) $(19^3 \bmod 23)^2 \bmod 31$
    d) $(89^3 \bmod 79)^4 \bmod 26$

25. Show that if $n \mid m$, where n and m are integers greater than 1, and if $a \equiv b \pmod{m}$, where a and b are integers, then $a \equiv b \pmod{n}$.

26. Show that if a, b, c, and m are integers such that $m \geq 2$, $c > 0$, and $a \equiv b \pmod{m}$, then $ac \equiv bc \pmod{mc}$.

27. Find counterexamples to each of these statements about congruences.

    a) If $ac \equiv bc \pmod{m}$, where a, b, c, and m are integers with $m \geq 2$, then $a \equiv b \pmod{m}$.

    b) If $a \equiv b \pmod{m}$ and $c \equiv d \pmod{m}$, where a, b, c, d, and m are integers with c and d positive and $m \geq 2$, then $a^c \equiv b^d \pmod{m}$.

28. Show that if n is an integer then $n^2 \equiv 0$ or $1 \pmod 4$.

29. Use Exercise 28 to show that if m is a positive integer of the form $4k + 3$ for some nonnegative integer k, then m is not the sum of the squares of two integers.

30. Prove that if n is an odd positive integer, then $n^2 \equiv 1 \pmod 8$.

31. Show that if a, b, k, and m are integers such that $k \geq 1$, $m \geq 2$, and $a \equiv b \pmod{m}$, then $a^k \equiv b^k \pmod{m}$.

32. Show that $\mathbf{Z}_m$ with multiplication modulo m, where $m \geq 2$ is an integer, satisfies the closure, associative, and commutativity properties, and 1 is a multiplicative identity.

33. Write out the addition and multiplication tables for $\mathbf{Z}_5$ (where by addition and multiplication we mean $+_5$ and $\cdot_5$).

34. Determine whether each of the functions $f(a) = a \operatorname{\mathbf{div}} d$ and $g(a) = a \operatorname{\mathbf{mod}} d$, where d is a fixed positive integer, from the set of integers to the set of integers, is one-to-one, and determine whether each of these functions is onto.

4.2   Integer Representations and Algorithms

Introduction

Integers can be expressed using any integer greater than one as a base, as we will show in this section. Although we commonly use decimal (base 10), representations, binary (base 2), octal (base 8), and hexadecimal (base 16) representations are often used, especially in computer science. Given a base b and an integer n, we will show how to construct the base b representation of this integer. We will also explain how to quickly covert between binary and octal and between binary and hexadecimal notations.

As mentioned in Section 3.1, the term *algorithm* originally referred to procedures for performing arithmetic operations using the decimal representations of integers. These algorithms, adapted for use with binary representations, are the basis for computer arithmetic. They provide good illustrations of the concept of an algorithm and the complexity of algorithms. For these reasons, they will be discussed in this section.

We will also introduce an algorithm for finding $a \operatorname{\mathbf{div}} d$ and $a \operatorname{\mathbf{mod}} d$ where a and d are integers with $d > 1$. Finally, we will describe an efficient algorithm for modular exponentiation, which is a particularly important algorithm for cryptography, as we will see in Section 4.6.

Representations of Integers

In everyday life we use decimal notation to express integers. For example, 965 is used to denote $9 \cdot 10^2 + 6 \cdot 10 + 5$. However, it is often convenient to use bases other than 10. In particular, computers usually use binary notation (with 2 as the base) when carrying out arithmetic, and octal (base 8) or hexadecimal (base 16) notation when expressing characters, such as letters or digits. In fact, we can use any integer greater than 1 as the base when expressing integers. This is stated in Theorem 1.

THEOREM 1

> Let b be an integer greater than 1. Then if n is a positive integer, it can be expressed uniquely in the form
>
> $$n = a_k b^k + a_{k-1} b^{k-1} + \cdots + a_1 b + a_0,$$
>
> where k is a nonnegative integer, $a_0, a_1, \ldots, a_k$ are nonnegative integers less than b, and $a_k \neq 0$.

A proof of this theorem can be constructed using mathematical induction, a proof method that is discussed in Section 5.1. It can also be found in [Ro10]. The representation of n given in Theorem 1 is called the **base b expansion of n**. The base b expansion of n is denoted by $(a_k a_{k-1} \ldots a_1 a_0)_b$. For instance, $(245)_8$ represents $2 \cdot 8^2 + 4 \cdot 8 + 5 = 165$. Typically, the subscript 10 is omitted for base 10 expansions of integers because base 10, or **decimal expansions**, are commonly used to represent integers.

BINARY EXPANSIONS   Choosing 2 as the base gives **binary expansions** of integers. In binary notation each digit is either a 0 or a 1. In other words, the binary expansion of an integer is just a bit string. Binary expansions (and related expansions that are variants of binary expansions) are used by computers to represent and do arithmetic with integers.

EXAMPLE 1   What is the decimal expansion of the integer that has $(1\ 0101\ 1111)_2$ as its binary expansion?

Solution: We have

$$(1\ 0101\ 1111)_2 = 1 \cdot 2^8 + 0 \cdot 2^7 + 1 \cdot 2^6 + 0 \cdot 2^5 + 1 \cdot 2^4$$
$$+ 1 \cdot 2^3 + 1 \cdot 2^2 + 1 \cdot 2^1 + 1 \cdot 2^0 = 351. \qquad \blacktriangleleft$$

OCTAL AND HEXADECIMAL EXPANSIONS   Among the most important bases in computer science are base 2, base 8, and base 16. Base 8 expansions are called **octal** expansions and base 16 expansions are **hexadecimal** expansions.

EXAMPLE 2   What is the decimal expansion of the number with octal expansion $(7016)_8$?

Solution: Using the definition of a base b expansion with $b = 8$ tells us that

$$(7016)_8 = 7 \cdot 8^3 + 0 \cdot 8^2 + 1 \cdot 8 + 6 = 3598. \qquad \blacktriangleleft$$

Sixteen different digits are required for hexadecimal expansions. Usually, the hexadecimal digits used are 0, 1, 2, 3, 4, 5, 6, 7, 8, 9, A, B, C, D, E, and F, where the letters A through F represent the digits corresponding to the numbers 10 through 15 (in decimal notation).

EXAMPLE 3   What is the decimal expansion of the number with hexadecimal expansion $(2AE0B)_{16}$?

Solution: Using the definition of a base b expansion with $b = 16$ tells us that

$$(2AE0B)_{16} = 2 \cdot 16^4 + 10 \cdot 16^3 + 14 \cdot 16^2 + 0 \cdot 16 + 11 = 175627. \qquad \blacktriangleleft$$

Each hexadecimal digit can be represented using four bits. For instance, we see that $(1110\ 0101)_2 = (E5)_{16}$ because $(1110)_2 = (E)_{16}$ and $(0101)_2 = (5)_{16}$. **Bytes**, which are bit strings of length eight, can be represented by two hexadecimal digits.

BASE CONVERSION   We will now describe an algorithm for constructing the base b expansion of an integer n. First, divide n by b to obtain a quotient and remainder, that is,

$$n = bq_0 + a_0, \qquad 0 \le a_0 < b.$$

The remainder, a_0, is the rightmost digit in the base b expansion of n. Next, divide q_0 by b to obtain

$$q_0 = bq_1 + a_1, \qquad 0 \le a_1 < b.$$

We see that a_1 is the second digit from the right in the base b expansion of n. Continue this process, successively dividing the quotients by b, obtaining additional base b digits as the remainders. This process terminates when we obtain a quotient equal to zero. It produces the base b digits of n from the right to the left.

EXAMPLE 4 Find the octal expansion of $(12345)_{10}$.

Solution: First, divide 12345 by 8 to obtain

$$12345 = 8 \cdot 1543 + 1.$$

Successively dividing quotients by 8 gives

$$1543 = 8 \cdot 192 + 7,$$
$$192 = 8 \cdot 24 + 0,$$
$$24 = 8 \cdot 3 + 0,$$
$$3 = 8 \cdot 0 + 3.$$

The successive remainders that we have found, 1, 7, 0, 0, and 3, are the digits from the right to the left of 12345 in base 8. Hence,

$$(12345)_{10} = (30071)_8.$$ ◀

EXAMPLE 5 Find the hexadecimal expansion of $(177130)_{10}$.

Solution: First divide 177130 by 16 to obtain

$$177130 = 16 \cdot 11070 + 10.$$

Successively dividing quotients by 16 gives

$$11070 = 16 \cdot 691 + 14,$$
$$691 = 16 \cdot 43 + 3,$$
$$43 = 16 \cdot 2 + 11,$$
$$2 = 16 \cdot 0 + 2.$$

The successive remainders that we have found, 10, 14, 3, 11, 2, give us the digits from the right to the left of 177130 in the hexadecimal (base 16) expansion of $(177130)_{10}$. It follows that

$$(177130)_{10} = (2B3EA)_{16}.$$

(Recall that the integers 10, 11, and 14 correspond to the hexadecimal digits A, B, and E, respectively.) ◀

EXAMPLE 6 Find the binary expansion of $(241)_{10}$.

Solution: First divide 241 by 2 to obtain

$$241 = 2 \cdot 120 + 1.$$

Successively dividing quotients by 2 gives

$$120 = 2 \cdot 60 + 0,$$
$$60 = 2 \cdot 30 + 0,$$
$$30 = 2 \cdot 15 + 0,$$
$$15 = 2 \cdot 7 + 1,$$
$$7 = 2 \cdot 3 + 1,$$
$$3 = 2 \cdot 1 + 1,$$
$$1 = 2 \cdot 0 + 1.$$

The successive remainders that we have found, 1, 0, 0, 0, 1, 1, 1, 1, are the digits from the right to the left in the binary (base 2) expansion of $(241)_{10}$. Hence,

$$(241)_{10} = (1111\ 0001)_2. \qquad \blacktriangleleft$$

The pseudocode given in Algorithm 1 finds the base b expansion $(a_{k-1} \ldots a_1 a_0)_b$ of the integer n.

ALGORITHM 1 Constructing Base b Expansions.

---

procedure *base b expansion(n, b: positive integers with $b > 1$)*
$q := n$
$k := 0$
while $q \neq 0$
    $a_k := q \bmod b$
    $q := q \textbf{ div } b$
    $k := k + 1$
return $(a_{k-1}, \ldots, a_1, a_0)$ $\{(a_{k-1} \ldots a_1 a_0)_b$ is the base b expansion of $n\}$

In Algorithm 1, q represents the quotient obtained by successive divisions by b, starting with $q = n$. The digits in the base b expansion are the remainders of these divisions and are given by $q \bmod b$. The algorithm terminates when a quotient $q = 0$ is reached.

Remark: Note that Algorithm 1 can be thought of as a greedy algorithm, as the base b digits are taken as large as possible in each step.

CONVERSION BETWEEN BINARY, OCTAL, AND HEXADECIMAL EXPANSIONS
Conversion between binary and octal and between binary and hexadecimal expansions is extremely easy because each octal digit corresponds to a block of three binary digits and each hexadecimal digit corresponds to a block of four binary digits, with these correspondences shown in Table 1 without initial 0s shown. (We leave it as Exercises 7 and 8 to show that this is the case.) This conversion is illustrated in Example 7.

EXAMPLE 7   Find the octal and hexadecimal expansions of $(11\ 1110\ 1011\ 1100)_2$ and the binary expansions of $(765)_8$ and $(A8D)_{16}$.

Solution: To convert $(11\ 1110\ 1011\ 1100)_2$ into octal notation we group the binary digits into blocks of three, adding initial zeros at the start of the leftmost block if necessary. These blocks, from left to right, are 011, 111, 010, 111, and 100, corresponding to 3, 7, 2, 7, and 4, respectively. Consequently, $(11\ 1110\ 1011\ 1100)_2 = (37274)_8$. To convert $(11\ 1110\ 1011\ 1100)_2$ into hexadecimal notation we group the binary digits into blocks of four, adding initial

TABLE 1 Hexadecimal, Octal, and Binary Representation of the Integers 0 through 15.															
Decimal	0	1	2	3	4	5	6	7	8	9	10	11	12	13	14	15
Hexadecimal	0	1	2	3	4	5	6	7	8	9	A	B	C	D	E	F
Octal	0	1	2	3	4	5	6	7	10	11	12	13	14	15	16	17
Binary	0	1	10	11	100	101	110	111	1000	1001	1010	1011	1100	1101	1110	1111

zeros at the start of the leftmost block if necessary. These blocks, from left to right, are 0011, 1110, 1011, and 1100, corresponding to the hexadecimal digits 3, E, B, and C, respectively. Consequently, $(11\ 1110\ 1011\ 1100)_2 = (3EBC)_{16}$.

To convert $(765)_8$ into binary notation, we replace each octal digit by a block of three binary digits. These blocks are 111, 110, and 101. Hence, $(765)_8 = (1\ 1111\ 0101)_2$. To convert $(A8D)_{16}$ into binary notation, we replace each hexadecimal digit by a block of four binary digits. These blocks are 1010, 1000, and 1101. Hence, $(A8D)_{16} = (1010\ 1000\ 1101)_2$. ◄

Algorithms for Integer Operations

The algorithms for performing operations with integers using their binary expansions are extremely important in computer arithmetic. We will describe algorithms for the addition and the multiplication of two integers expressed in binary notation. We will also analyze the computational complexity of these algorithms, in terms of the actual number of bit operations used. Throughout this discussion, suppose that the binary expansions of a and b are

$$a = (a_{n-1}a_{n-2}\ldots a_1a_0)_2, \quad b = (b_{n-1}b_{n-2}\ldots b_1b_0)_2,$$

so that a and b each have n bits (putting bits equal to 0 at the beginning of one of these expansions if necessary).

We will measure the complexity of algorithms for integer arithmetic in terms of the number of bits in these numbers.

ADDITION ALGORITHM   Consider the problem of adding two integers in binary notation. A procedure to perform addition can be based on the usual method for adding numbers with pencil and paper. This method proceeds by adding pairs of binary digits together with carries, when they occur, to compute the sum of two integers. This procedure will now be specified in detail.

To add a and b, first add their rightmost bits. This gives

$$a_0 + b_0 = c_0 \cdot 2 + s_0,$$

where s_0 is the rightmost bit in the binary expansion of $a + b$ and c_0 is the **carry**, which is either 0 or 1. Then add the next pair of bits and the carry,

$$a_1 + b_1 + c_0 = c_1 \cdot 2 + s_1,$$

where s_1 is the next bit (from the right) in the binary expansion of $a + b$, and c_1 is the carry. Continue this process, adding the corresponding bits in the two binary expansions and the carry, to determine the next bit from the right in the binary expansion of $a + b$. At the last stage, add a_{n-1}, b_{n-1}, and c_{n-2} to obtain $c_{n-1} \cdot 2 + s_{n-1}$. The leading bit of the sum is $s_n = c_{n-1}$. This procedure produces the binary expansion of the sum, namely, $a + b = (s_n s_{n-1} s_{n-2} \ldots s_1 s_0)_2$.

EXAMPLE 8 Add $a = (1110)_2$ and $b = (1011)_2$.

Solution: Following the procedure specified in the algorithm, first note that

$$a_0 + b_0 = 0 + 1 = 0 \cdot 2 + 1,$$

so that $c_0 = 0$ and $s_0 = 1$. Then, because

$$a_1 + b_1 + c_0 = 1 + 1 + 0 = 1 \cdot 2 + 0,$$

it follows that $c_1 = 1$ and $s_1 = 0$. Continuing,

$$a_2 + b_2 + c_1 = 1 + 0 + 1 = 1 \cdot 2 + 0,$$

so that $c_2 = 1$ and $s_2 = 0$. Finally, because

$$a_3 + b_3 + c_2 = 1 + 1 + 1 = 1 \cdot 2 + 1,$$

1 1 1
```
   1 1 1 0
 + 1 0 1 1
 ─────────
 1 1 0 0 1
```

follows that $c_3 = 1$ and $s_3 = 1$. This means that $s_4 = c_3 = 1$. Therefore, $s = a + b = (1\,1001)_2$. This addition is displayed in Figure 1, where carries are shown in blue. ◀

FIGURE 1
Adding $(1110)_2$
and $(1011)_2$.

The algorithm for addition can be described using pseudocode as follows.

ALGORITHM 2 Addition of Integers.

procedure *add*(*a*, *b*: positive integers)
{the binary expansions of a and b are $(a_{n-1}a_{n-2} \ldots a_1 a_0)_2$
   and $(b_{n-1}b_{n-2} \ldots b_1 b_0)_2$, respectively}
$c := 0$
for $j := 0$ **to** $n - 1$
    $d := \lfloor (a_j + b_j + c)/2 \rfloor$
    $s_j := a_j + b_j + c - 2d$
    $c := d$
$s_n := c$
return $(s_0, s_1, \ldots, s_n)$ {the binary expansion of the sum is $(s_n s_{n-1} \ldots s_0)_2$}

Next, the number of additions of bits used by Algorithm 2 will be analyzed.

EXAMPLE 9 How many additions of bits are required to use Algorithm 2 to add two integers with n bits (or less) in their binary representations?

Solution: Two integers are added by successively adding pairs of bits and, when it occurs, a carry. Adding each pair of bits and the carry requires two additions of bits. Thus, the total number of additions of bits used is less than twice the number of bits in the expansion. Hence, the number of additions of bits used by Algorithm 2 to add two n-bit integers is $O(n)$. ◀

MULTIPLICATION ALGORITHM   Next, consider the multiplication of two n-bit integers a and b. The conventional algorithm (used when multiplying with pencil and paper) works as follows. Using the distributive law, we see that

$$ab = a(b_0 2^0 + b_1 2^1 + \cdots + b_{n-1} 2^{n-1})$$
$$= a(b_0 2^0) + a(b_1 2^1) + \cdots + a(b_{n-1} 2^{n-1}).$$

We can compute ab using this equation. We first note that $ab_j = a$ if $b_j = 1$ and $ab_j = 0$ if $b_j = 0$. Each time we multiply a term by 2, we shift its binary expansion one place to the left and add a zero at the tail end of the expansion. Consequently, we can obtain $(ab_j)2^j$ by **shifting** the binary expansion of ab_j j places to the left, adding j zero bits at the tail end of this binary expansion. Finally, we obtain ab by adding the n integers $ab_j 2^j$, $j = 0, 1, 2, \ldots, n - 1$.

Algorithm 3 displays this procedure for multiplication.

ALGORITHM 3 Multiplication of Integers.

procedure *multiply*(a, b: positive integers)
{the binary expansions of a and b are $(a_{n-1}a_{n-2} \ldots a_1 a_0)_2$
  and $(b_{n-1}b_{n-2} \ldots b_1 b_0)_2$, respectively}
for $j := 0$ **to** $n - 1$
    if $b_j = 1$ **then** $c_j := a$ shifted j places
    else $c_j := 0$
{$c_0, c_1, \ldots, c_{n-1}$ are the partial products}
$p := 0$
for $j := 0$ **to** $n - 1$
    $p := p + c_j$
return p {p is the value of ab}

Example 10 illustrates the use of this algorithm.

EXAMPLE 10 Find the product of $a = (110)_2$ and $b = (101)_2$.

Solution: First note that

$$ab_0 \cdot 2^0 = (110)_2 \cdot 1 \cdot 2^0 = (110)_2,$$
$$ab_1 \cdot 2^1 = (110)_2 \cdot 0 \cdot 2^1 = (0000)_2,$$

and

$$ab_2 \cdot 2^2 = (110)_2 \cdot 1 \cdot 2^2 = (11000)_2.$$

$$
\begin{array}{r}
1\ 1\ 0 \\
\times\ 1\ 0\ 1 \\
\hline
1\ 1\ 0 \\
0\ 0\ 0 \\
1\ 1\ 0 \\
\hline
1\ 1\ 1\ 1\ 0
\end{array}
$$

FIGURE 2
Multiplying
$(110)_2$ and $(101)_2$.

To find the product, add $(110)_2$, $(0000)_2$, and $(11000)_2$. Carrying out these additions (using Algorithm 2, including initial zero bits when necessary) shows that $ab = (1\ 1110)_2$. This multiplication is displayed in Figure 2. ◀

Next, we determine the number of additions of bits and shifts of bits used by Algorithm 3 to multiply two integers.

EXAMPLE 11   How many additions of bits and shifts of bits are used to multiply a and b using Algorithm 3?

Solution: Algorithm 3 computes the products of a and b by adding the partial products $c_0, c_1, c_2, \ldots,$ and c_{n-1}. When $b_j = 1$, we compute the partial product c_j by shifting the binary expansion of a by j bits. When $b_j = 0$, no shifts are required because $c_j = 0$. Hence, to find all n of the integers $ab_j 2^j$, $j = 0, 1, \ldots, n-1$, requires at most

$$0 + 1 + 2 + \cdots + n - 1$$

shifts. Hence, by Example 5 in Section 3.2 the number of shifts required is $O(n^2)$.

To add the integers ab_j from $j = 0$ to $j = n - 1$ requires the addition of an n-bit integer, an $(n + 1)$-bit integer, $\ldots,$ and a $(2n)$-bit integer. We know from Example 9 that each of these additions requires $O(n)$ additions of bits. Consequently, a total of $O(n^2)$ additions of bits are required for all n additions. ◀

Surprisingly, there are more efficient algorithms than the conventional algorithm for multiplying integers. One such algorithm, which uses $O(n^{1.585})$ bit operations to multiply n-bit numbers, will be described in Section 8.3.

ALGORITHM FOR div AND mod   Given integers a and d, $d > 0$, we can find $q = a$ **div** d and $r = a$ **mod** d using Algorithm 4. In this brute-force algorithm, when a is positive we subtract d from a as many times as necessary until what is left is less than d. The number of times we perform this subtraction is the quotient and what is left over after all these subtractions is the remainder. Algorithm 4 also covers the case where a is negative. This algorithm finds the quotient q and remainder r when $|a|$ is divided by d. Then, when $a < 0$ and $r > 0$, it uses these to find the quotient $-(q + 1)$ and remainder $d - r$ when a is divided by d. We leave it to the reader (Exercise 45) to show that, assuming that $a > d$, this algorithm uses $O(q \log a)$ bit operations.

ALGORITHM 4 **Computing div and mod.**

---

procedure *division algorithm*(a: integer, d: positive integer)
$q := 0$
$r := |a|$
while $r \geq d$
$\quad r := r - d$
$\quad q := q + 1$
if $a < 0$ and $r > 0$ **then**
$\quad r := d - r$
$\quad q := -(q + 1)$
return (q, r) {$q = a$ **div** d is the quotient, $r = a$ **mod** d is the remainder}

There are more efficient algorithms than Algorithm 4 for determining the quotient $q = a$ **div** d and the remainder $r = a$ **mod** d when a positive integer a is divided by a positive integer d (see [Kn98] for details). These algorithms require $O(\log a \cdot \log d)$ bit operations. If both of the binary expansions of a and d contain n or fewer bits, then we can replace $\log a \cdot \log d$ by n^2. This means that we need $O(n^2)$ bit operations to find the quotient and remainder when a is divided by d.

Modular Exponentiation

In cryptography it is important to be able to find b^n **mod** m efficiently, where b, n, and m are large integers. It is impractical to first compute b^n and then find its remainder when divided by m because b^n will be a huge number. Instead, we can use an algorithm that employs the binary expansion of the exponent n.

Before we present this algorithm, we illustrate its basic idea. We will explain how to use the binary expansion of n, say $n = (a_{k-1} \ldots a_1 a_0)_2$, to compute b^n. First, note that

$$b^n = b^{a_{k-1} \cdot 2^{k-1} + \cdots + a_1 \cdot 2 + a_0} = b^{a_{k-1} \cdot 2^{k-1}} \cdots b^{a_1 \cdot 2} \cdot b^{a_0}.$$

This shows that to compute b^n, we need only compute the values of b, b^2, $(b^2)^2 = b^4$, $(b^4)^2 = b^8, \ldots, b^{2^k}$. Once we have these values, we multiply the terms b^{2^j} in this list, where $a_j = 1$. (For efficiency, after multiplying by each term, we reduce the result modulo m.) This gives us b^n. For example, to compute 3^{11} we first note that $11 = (1011)_2$, so that $3^{11} = 3^8 3^2 3^1$. By successively squaring, we find that $3^2 = 9$, $3^4 = 9^2 = 81$, and $3^8 = (81)^2 = 6561$. Consequently, $3^{11} = 3^8 3^2 3^1 = 6561 \cdot 9 \cdot 3 = 177,147$.

The algorithm successively finds b **mod** m, b^2 **mod** m, b^4 **mod** $m, \ldots, b^{2^{k-1}}$ **mod** m and multiplies together those terms b^{2^j} **mod** m where $a_j = 1$, finding the remainder of the product when divided by m after each multiplication. Pseudocode for this algorithm is shown in Algorithm 5. Note that in Algorithm 5 we can use the most efficient algorithm available to compute values of the **mod** function, not necessarily Algorithm 4.

Be sure to reduce modulo m after each multiplication!

ALGORITHM 5 **Modular Exponentiation.**

procedure *modular exponentiation*(b: integer, $n = (a_{k-1} a_{k-2} \ldots a_1 a_0)_2$,
$\qquad$ m: positive integers)
$x := 1$
power $:= b$ **mod** m
for $i := 0$ **to** $k - 1$
$\qquad$ **if** $a_i = 1$ **then** $x := (x \cdot power)$ **mod** m
$\qquad$ *power* $:= (power \cdot power)$ **mod** m
return x {x equals b^n **mod** m}

We illustrate how Algorithm 5 works in Example 12.

EXAMPLE 12   Use Algorithm 5 to find 3^{644} **mod** 645.

Solution: Algorithm 5 initially sets $x = 1$ and *power* $= 3$ **mod** $645 = 3$. In the computation of 3^{644} **mod** 645, this algorithm determines 3^{2^j} **mod** 645 for $j = 1, 2, \ldots, 9$ by successively squaring and reducing modulo 645. If $a_j = 1$ (where a_j is the bit in the jth position in the binary expansion of 644, which is $(1010000100)_2$), it multiplies the current value of x by 3^{2^j} **mod** 645 and reduces the result modulo 645. Here are the steps used:

$i = 0$: Because $a_0 = 0$, we have $x = 1$ and $power = 3^2 \bmod 645 = 9 \bmod 645 = 9$;

$i = 1$: Because $a_1 = 0$, we have $x = 1$ and $power = 9^2 \bmod 645 = 81 \bmod 645 = 81$;

$i = 2$: Because $a_2 = 1$, we have $x = 1 \cdot 81 \bmod 645 = 81$ and $power = 81^2 \bmod 645 = 6561 \bmod 645 = 111$;

$i = 3$: Because $a_3 = 0$, we have $x = 81$ and $power = 111^2 \bmod 645 = 12{,}321 \bmod 645 = 66$;

$i = 4$: Because $a_4 = 0$, we have $x = 81$ and $power = 66^2 \bmod 645 = 4356 \bmod 645 = 486$;

$i = 5$: Because $a_5 = 0$, we have $x = 81$ and $power = 486^2 \bmod 645 = 236{,}196 \bmod 645 = 126$;

$i = 6$: Because $a_6 = 0$, we have $x = 81$ and $power = 126^2 \bmod 645 = 15{,}876 \bmod 645 = 396$;

$i = 7$: Because $a_7 = 1$, we find that $x = (81 \cdot 396) \bmod 645 = 471$ and $power = 396^2 \bmod 645 = 156{,}816 \bmod 645 = 81$;

$i = 8$: Because $a_8 = 0$, we have $x = 471$ and $power = 81^2 \bmod 645 = 6561 \bmod 645 = 111$;

$i = 9$: Because $a_9 = 1$, we find that $x = (471 \cdot 111) \bmod 645 = 36$.

This shows that following the steps of Algorithm 5 produces the result $3^{644} \bmod 645 = 36$. ◀

Algorithm 5 is quite efficient; it uses $O((\log m)^2 \log n)$ bit operations to find $b^n \bmod m$ (see Exercise 58).

Exercises

1. Convert the decimal expansion of each of these integers to a binary expansion.
   a) 231   **b)** 4532   **c)** 97644

2. Convert the binary expansion of each of these integers to a decimal expansion.
   a) $(1\ 1111)_2$ **b)** $(10\ 0000\ 0001)_2$
   c) $(1\ 0101\ 0101)_2$ **d)** $(110\ 1001\ 0001\ 0000)_2$

3. Convert the octal expansion of each of these integers to a binary expansion.
   a) $(572)_8$ **b)** $(1604)_8$
   c) $(423)_8$ **d)** $(2417)_8$

4. Convert the hexadecimal expansion of each of these integers to a binary expansion.
   a) $(80E)_{16}$ **b)** $(135AB)_{16}$
   c) $(ABBA)_{16}$ **d)** $(DEFACED)_{16}$

5. Convert $(ABCDEF)_{16}$ from its hexadecimal expansion to its binary expansion.

6. Convert $(1011\ 0111\ 1011)_2$ from its binary expansion to its hexadecimal expansion.

7. Show that the hexadecimal expansion of a positive integer can be obtained from its binary expansion by grouping together blocks of four binary digits, adding initial zeros if necessary, and translating each block of four binary digits into a single hexadecimal digit.

8. Show that the binary expansion of a positive integer can be obtained from its hexadecimal expansion by translating each hexadecimal digit into a block of four binary digits.

9. Convert $(7345321)_8$ to its binary expansion and $(10\ 1011\ 1011)_2$ to its octal expansion.

10. Give a procedure for converting from the octal expansion of an integer to its hexadecimal expansion using binary notation as an intermediate step.

11. Find the sum and the product of each of these pairs of numbers. Express your answers as a binary expansion.
    a) $(100\ 0111)_2$, $(111\ 0111)_2$
    b) $(1110\ 1111)_2$, $(1011\ 1101)_2$
    c) $(10\ 1010\ 1010)_2$, $(1\ 1111\ 0000)_2$
    d) $(10\ 0000\ 0001)_2$, $(11\ 1111\ 1111)_2$

12. Find the sum and product of each of these pairs of numbers. Express your answers as a hexadecimal expansion.
    a) $(1AE)_{16}$, $(BBC)_{16}$
    b) $(20CBA)_{16}$, $(A01)_{16}$
    c) $(ABCDE)_{16}$, $(1111)_{16}$
    d) $(E0000E)_{16}$, $(BAAA)_{16}$

13. Find the sum and product of each of these pairs of numbers. Express your answers as an octal expansion.
    a) $(763)_8$, $(147)_8$
    b) $(6001)_8$, $(272)_8$
    c) $(1111)_8$, $(777)_8$
    d) $(54321)_8$, $(3456)_8$

14. Use Algorithm 5 to find $7^{644} \bmod 645$.

15. Use Algorithm 5 to find $3^{2003} \bmod 99$.

16. Show that every positive integer can be represented uniquely as the sum of distinct powers of 2. [*Hint:* Consider binary expansions of integers.]

17. It can be shown that every integer can be uniquely represented in the form

$$e_k 3^k + e_{k-1} 3^{k-1} + \cdots + e_1 3 + e_0,$$

where $e_j = -1, 0$, or 1 for $j = 0, 1, 2, \ldots, k$. Expansions of this type are called **balanced ternary expansions**. Find the balanced ternary expansions of

a) 5.   **b)** 13.   **c)** 37.   **d)** 79.

18. Show that a positive integer is divisible by 11 if and only if the difference of the sum of its decimal digits in even-numbered positions and the sum of its decimal digits in odd-numbered positions is divisible by 11.

19. Show that a positive integer is divisible by 3 if and only if the sum of its decimal digits is divisible by 3.

One's complement representations of integers are used to simplify computer arithmetic. To represent positive and negative integers with absolute value less than 2^{n-1}, a total of n bits is used. The leftmost bit is used to represent the sign. A 0 bit in this position is used for positive integers, and a 1 bit in this position is used for negative integers. For positive integers, the remaining bits are identical to the binary expansion of the integer. For negative integers, the remaining bits are obtained by first finding the binary expansion of the absolute value of the integer, and then taking the complement of each of these bits, where the complement of a 1 is a 0 and the complement of a 0 is a 1.

20. Find the one's complement representations, using bit strings of length six, of the following integers.

a) 22   **b)** 31   **c)** −7   **d)** −19

21. What integer does each of the following one's complement representations of length five represent?

a) 11001   **b)** 01101
c) 10001   **d)** 11111

22. If m is a positive integer less than 2^{n-1}, how is the one's complement representation of $-m$ obtained from the one's complement of m, when bit strings of length n are used?

23. How is the one's complement representation of the sum of two integers obtained from the one's complement representations of these integers?

24. How is the one's complement representation of the difference of two integers obtained from the one's complement representations of these integers?

25. Show that the integer m with one's complement representation $(a_{n-1}a_{n-2}\ldots a_1a_0)$ can be found using the equation $m = -a_{n-1}(2^{n-1}-1) + a_{n-2}2^{n-2} + \cdots + a_1 \cdot 2 + a_0$.

Two's complement representations of integers are also used to simplify computer arithmetic and are used more commonly than one's complement representations. To represent an integer x with $-2^{n-1} \le x \le 2^{n-1} - 1$ for a specified positive integer n, a total of n bits is used. The leftmost bit is used to represent the sign. A 0 bit in this position is used for positive

integers, and a 1 bit in this position is used for negative integers, just as in one's complement expansions. For a positive integer, the remaining bits are identical to the binary expansion of the integer. For a negative integer, the remaining bits are the bits of the binary expansion of $2^{n-1} - |x|$. Two's complement expansions of integers are often used by computers because addition and subtraction of integers can be performed easily using these expansions, where these integers can be either positive or negative.

26. Answer Exercise 20, but this time find the two's complement expansion using bit strings of length six.

27. Answer Exercise 21 if each expansion is a two's complement expansion of length five.

28. Answer Exercise 22 for two's complement expansions.

29. Answer Exercise 23 for two's complement expansions.

30. Answer Exercise 24 for two's complement expansions.

31. Show that the integer m with two's complement representation $(a_{n-1}a_{n-2}\ldots a_1a_0)$ can be found using the equation $m = -a_{n-1} \cdot 2^{n-1} + a_{n-2}2^{n-2} + \cdots + a_1 \cdot 2 + a_0$.

32. Give a simple algorithm for forming the two's complement representation of an integer from its one's complement representation.

33. Sometimes integers are encoded by using four-digit binary expansions to represent each decimal digit. This produces the **binary coded decimal** form of the integer. For instance, 791 is encoded in this way by 011110010001. How many bits are required to represent a number with n decimal digits using this type of encoding?

A **Cantor expansion** is a sum of the form

$$a_n n! + a_{n-1}(n-1)! + \cdots + a_2 2! + a_1 1!,$$

where a_i is an integer with $0 \le a_i \le i$ for $i = 1, 2, \ldots, n$.

34. Find the Cantor expansions of

a) 2.   **b)** 7.
c) 19.   **d)** 87.
e) 1000.   **f)** 1,000,000.

*35. Describe an algorithm that finds the Cantor expansion of an integer.

*36. Describe an algorithm to add two integers from their Cantor expansions.

37. Add $(10111)_2$ and $(11010)_2$ by working through each step of the algorithm for addition given in the text.

38. Multiply $(1110)_2$ and $(1010)_2$ by working through each step of the algorithm for multiplication given in the text.

39. Describe an algorithm for finding the difference of two binary expansions.

40. Estimate the number of bit operations used to subtract two binary expansions.

41. Devise an algorithm that, given the binary expansions of the integers a and b, determines whether $a > b$, $a = b$, or $a < b$.

42. How many bit operations does the comparison algorithm from Exercise 41 use when the larger of a and b has n bits in its binary expansion?

43. Estimate the complexity of Algorithm 1 for finding the base b expansion of an integer n in terms of the number of divisions used.

*44. Show that Algorithm 5 uses $O((\log m)^2 \log n)$ bit operations to find $b^n \bmod m$.

45. Show that Algorithm 4 uses $O(q \log a)$ bit operations, assuming that $a > d$.

4.3   Primes and Greatest Common Divisors

Introduction

In Section 4.1 we studied the concept of divisibility of integers. One important concept based on divisibility is that of a prime number. A prime is an integer greater than 1 that is divisible by no positive integers other than 1 and itself. The study of prime numbers goes back to ancient times. Thousands of years ago it was known that there are infinitely many primes; the proof of this fact, found in the works of Euclid, is famous for its elegance and beauty.

We will discuss the distribution of primes among the integers. We will describe some of the results about primes found by mathematicians in the last 400 years. In particular, we will introduce an important theorem, the fundamental theorem of arithmetic. This theorem, which asserts that every positive integer can be written uniquely as the product of primes in nondecreasing order, has many interesting consequences. We will also discuss some of the many old conjectures about primes that remain unsettled today.

Primes have become essential in modern cryptographic systems, and we will develop some of their properties important in cryptography. For example, finding large primes is essential in modern cryptography. The length of time required to factor large integers into their prime factors is the basis for the strength of some important modern cryptographic systems.

In this section we will also study the greatest common divisor of two integers, as well as the least common multiple of two integers. We will develop an important algorithm for computing greatest common divisors, called the Euclidean algorithm.

Primes

Every integer greater than 1 is divisible by at least two integers, because a positive integer is divisible by 1 and by itself. Positive integers that have exactly two different positive integer factors are called **primes**.

DEFINITION 1   An integer p greater than 1 is called *prime* if the only positive factors of p are 1 and p. A positive integer that is greater than 1 and is not prime is called *composite*.

Remark: The integer n is composite if and only if there exists an integer a such that $a \mid n$ and $1 < a < n$.

EXAMPLE 1   The integer 7 is prime because its only positive factors are 1 and 7, whereas the integer 9 is composite because it is divisible by 3.   ◀

The primes are the building blocks of positive integers, as the fundamental theorem of arithmetic shows. The proof will be given in Section 5.2.

THEOREM 1	**THE FUNDAMENTAL THEOREM OF ARITHMETIC** Every integer greater than 1 can be written uniquely as a prime or as the product of two or more primes where the prime factors are written in order of nondecreasing size.

Example 2 gives some prime factorizations of integers.

EXAMPLE 2   The prime factorizations of 100, 641, 999, and 1024 are given by

$$100 = 2 \cdot 2 \cdot 5 \cdot 5 = 2^2 5^2,$$
$$641 = 641,$$
$$999 = 3 \cdot 3 \cdot 3 \cdot 37 = 3^3 \cdot 37,$$
$$1024 = 2 \cdot 2 \cdot 2 \cdot 2 \cdot 2 \cdot 2 \cdot 2 \cdot 2 \cdot 2 \cdot 2 = 2^{10}.$$
◀

Trial Division

It is often important to show that a given integer is prime. For instance, in cryptology, large primes are used in some methods for making messages secret. One procedure for showing that an integer is prime is based on the following observation.

THEOREM 2	If n is a composite integer, then n has a prime divisor less than or equal to $\sqrt{n}$.

Proof: If n is composite, by the definition of a composite integer, we know that it has a factor a with $1 < a < n$. Hence, by the definition of a factor of a positive integer, we have $n = ab$, where b is a positive integer greater than 1. We will show that $a \leq \sqrt{n}$ or $b \leq \sqrt{n}$. If $a > \sqrt{n}$ and $b > \sqrt{n}$, then $ab > \sqrt{n} \cdot \sqrt{n} = n$, which is a contradiction. Consequently, $a \leq \sqrt{n}$ or $b \leq \sqrt{n}$. Because both a and b are divisors of n, we see that n has a positive divisor not exceeding $\sqrt{n}$. This divisor is either prime or, by the fundamental theorem of arithmetic, has a prime divisor less than itself. In either case, n has a prime divisor less than or equal to $\sqrt{n}$.   ◁

From Theorem 2, it follows that an integer is prime if it is not divisible by any prime less than or equal to its square root. This leads to the brute-force algorithm known as **trial division**. To use trial division we divide n by all primes not exceeding $\sqrt{n}$ and conclude that n is prime if it is not divisible by any of these primes. In Example 3 we use trial division to show that 101 is prime.

EXAMPLE 3   Show that 101 is prime.

Solution: The only primes not exceeding $\sqrt{101}$ are 2, 3, 5, and 7. Because 101 is not divisible by 2, 3, 5, or 7 (the quotient of 101 and each of these integers is not an integer), it follows that 101 is prime.   ◀

Because every integer has a prime factorization, it would be useful to have a procedure for finding this prime factorization. Consider the problem of finding the prime factorization of n. Begin by dividing n by successive primes, starting with the smallest prime, 2. If n has a prime factor, then by Theorem 3 a prime factor p not exceeding $\sqrt{n}$ will be found. So, if no prime factor not exceeding $\sqrt{n}$ is found, then n is prime. Otherwise, if a prime factor p is found, continue by factoring n/p. Note that n/p has no prime factors less than p. Again, if n/p has no prime factor greater than or equal to p and not exceeding its square root, then it is prime.

Otherwise, if it has a prime factor q, continue by factoring $n/(pq)$. This procedure is continued until the factorization has been reduced to a prime. This procedure is illustrated in Example 4.

EXAMPLE 4   Find the prime factorization of 7007.

Solution: To find the prime factorization of 7007, first perform divisions of 7007 by successive primes, beginning with 2. None of the primes 2, 3, and 5 divides 7007. However, 7 divides 7007, with $7007/7 = 1001$. Next, divide 1001 by successive primes, beginning with 7. It is immediately seen that 7 also divides 1001, because $1001/7 = 143$. Continue by dividing 143 by successive primes, beginning with 7. Although 7 does not divide 143, 11 does divide 143, and $143/11 = 13$. Because 13 is prime, the procedure is completed. It follows that $7007 = 7 \cdot 1001 = 7 \cdot 7 \cdot 143 = 7 \cdot 7 \cdot 11 \cdot 13$. Consequently, the prime factorization of 7007 is $7 \cdot 7 \cdot 11 \cdot 13 = 7^2 \cdot 11 \cdot 13$.   ◀

Links

Prime numbers were studied in ancient times for philosophical reasons. Today, there are highly practical reasons for their study. In particular, large primes play a crucial role in cryptography, as we will see in Section 4.6.

The Sieve of Eratosthenes

Note that composite integers not exceeding 100 must have a prime factor not exceeding 10. Because the only primes less than 10 are 2, 3, 5, and 7, the primes not exceeding 100 are these four primes and those positive integers greater than 1 and not exceeding 100 that are divisible by none of 2, 3, 5, or 7.

Links

The **sieve of Eratosthenes** is used to find all primes not exceeding a specified positive integer. For instance, the following procedure is used to find the primes not exceeding 100. We begin with the list of all integers between 1 and 100. To begin the sieving process, the integers that are divisible by 2, other than 2, are deleted. Because 3 is the first integer greater than 2 that is left, all those integers divisible by 3, other than 3, are deleted. Because 5 is the next integer left after 3, those integers divisible by 5, other than 5, are deleted. The next integer left is 7, so those integers divisible by 7, other than 7, are deleted. Because all composite integers not exceeding 100 are divisible by 2, 3, 5, or 7, all remaining integers except 1 are prime. In Table 1, the panels display those integers deleted at each stage, where each integer divisible by 2, other than 2, is underlined in the first panel, each integer divisible by 3, other than 3, is underlined in the second panel, each integer divisible by 5, other than 5, is underlined in the third panel, and each integer divisible by 7, other than 7, is underlined in the fourth panel. The integers not underlined are the primes not exceeding 100. We conclude that the primes less than 100 are 2, 3, 5, 7, 11, 13, 17, 19, 23, 29, 31, 37, 41, 43, 47, 53, 59, 61, 67, 71, 73, 79, 83, 89, and 97.

THE INFINITUDE OF PRIMES   It has long been known that there are infinitely many primes. This means that whenever $p_1, p_2, \ldots, p_n$ are the n smallest primes, we know there is a larger prime not listed. We will prove this fact using a proof given by Euclid in his famous mathematics text, *The Elements*. This simple, yet elegant, proof is considered by many mathematicians to be among the most beautiful proofs in mathematics. It is the first proof presented in the book *Proofs*

Links

ERATOSTHENES (276 B.C.E.–194 B.C.E.)   It is known that Eratosthenes was born in Cyrene, a Greek colony west of Egypt, and spent time studying at Plato's Academy in Athens. We also know that King Ptolemy II invited Eratosthenes to Alexandria to tutor his son and that later Eratosthenes became chief librarian at the famous library at Alexandria, a central repository of ancient wisdom. Eratosthenes was an extremely versatile scholar, writing on mathematics, geography, astronomy, history, philosophy, and literary criticism. Besides his work in mathematics, he is most noted for his chronology of ancient history and for his famous measurement of the size of the earth.

TABLE 1 The Sieve of Eratosthenes.

Integers divisible by 2 other than 2 receive an underline.

1	2	3	4	5	6	7	8	9	10
11	12	13	14	15	16	17	18	19	20
21	22	23	24	25	26	27	28	29	30
31	32	33	34	35	36	37	38	39	40
41	42	43	44	45	46	47	48	49	50
51	52	53	54	55	56	57	58	59	60
61	62	63	64	65	66	67	68	69	70
71	72	73	74	75	76	77	78	79	80
81	82	83	84	85	86	87	88	89	90
91	92	93	94	95	96	97	98	99	100

Integers divisible by 3 other than 3 receive an underline.

1	2	3	4	5	6	7	8	9	10
11	12	13	14	15	16	17	18	19	20
21	22	23	24	25	26	27	28	29	30
31	32	33	34	35	36	37	38	39	40
41	42	43	44	45	46	47	48	49	50
51	52	53	54	55	56	57	58	59	60
61	62	63	64	65	66	67	68	69	70
71	72	73	74	75	76	77	78	79	80
81	82	83	84	85	86	87	88	89	90
91	92	93	94	95	96	97	98	99	100

Integers divisible by 5 other than 5 receive an underline.

1	2	3	4	5	6	7	8	9	10
11	12	13	14	15	16	17	18	19	20
21	22	23	24	25	26	27	28	29	30
31	32	33	34	35	36	37	38	39	40
41	42	43	44	45	46	47	48	49	50
51	52	53	54	55	56	57	58	59	60
61	62	63	64	65	66	67	68	69	70
71	72	73	74	75	76	77	78	79	80
81	82	83	84	85	86	87	88	89	90
91	92	93	94	95	96	97	98	99	100

Integers divisible by 7 other than 7 receive an underline; integers in color are prime.

1	2	3	4	5	6	7	8	9	10
11	12	13	14	15	16	17	18	19	20
21	22	23	24	25	26	27	28	29	30
31	32	33	34	35	36	37	38	39	40
41	42	43	44	45	46	47	48	49	50
51	52	53	54	55	56	57	58	59	60
61	62	63	64	65	66	67	68	69	70
71	72	73	74	75	76	77	78	79	80
81	82	83	84	85	86	87	88	89	90
91	92	93	94	95	96	97	98	99	100

from THE BOOK [AiZi10], where THE BOOK refers to the imagined collection of perfect proofs that the famous mathematician Paul Erdős claimed is maintained by God. By the way, there are a vast number of different proofs than there are an infinitude of primes, and new ones are published surprisingly frequently.

THEOREM 3   There are infinitely many primes.

Proof: We will prove this theorem using a proof by contradiction. We assume that there are only finitely many primes, $p_1, p_2, \ldots, p_n$. Let

$$Q = p_1 p_2 \cdots p_n + 1.$$

By the fundamental theorem of arithmetic, Q is prime or else it can be written as the product of two or more primes. However, none of the primes p_j divides Q, for if $p_j \mid Q$, then p_j divides $Q - p_1 p_2 \cdots p_n = 1$. Hence, there is a prime not in the list $p_1, p_2, \ldots, p_n$. This prime is either Q, if it is prime, or a prime factor of Q. This is a contradiction because we assumed that we have listed all the primes. Consequently, there are infinitely many primes. ◁

Remark: Note that in this proof we do *not* state that Q is prime! Furthermore, in this proof, we have given a nonconstructive existence proof that given any n primes, there is a prime not in this list. For this proof to be constructive, we would have had to explicitly give a prime not in our original list of n primes.

Because there are infinitely many primes, given any positive integer there are primes greater than this integer. There is an ongoing quest to discover larger and larger prime numbers; for almost all the last 300 years, the largest prime known has been an integer of the special form $2^p - 1$, where p is also prime. (Note that $2^n - 1$ cannot be prime when n is not prime; see Exercise 7.) Such primes are called **Mersenne primes**, after the French monk Marin Mersenne, who studied them in the seventeenth century. The reason that the largest known prime has usually been a Mersenne prime is that there is an extremely efficient test, known as the Lucas–Lehmer test, for determining whether $2^p - 1$ is prime. Furthermore, it is not currently possible to test numbers not of this or certain other special forms anywhere near as quickly to determine whether they are prime.

EXAMPLE 5   The numbers $2^2 - 1 = 3$, $2^3 - 1 = 7$, $2^5 - 1 = 31$ and $2^7 - 1 = 127$ are Mersenne primes, while $2^{11} - 1 = 2047$ is not a Mersenne prime because $2047 = 23 \cdot 89$.   ◄

Progress in finding Mersenne primes has been steady since computers were invented. As of early 2011, 47 Mersenne primes were known, with 16 found since 1990. The largest Mersenne prime known (again as of early 2011) is $2^{43,112,609} - 1$, a number with nearly 13 million decimal digits, which was shown to be prime in 2008. A communal effort, the Great Internet Mersenne Prime Search (GIMPS), is devoted to the search for new Mersenne primes. You can join this search, and if you are lucky, find a new Mersenne prime and possibly even win a cash prize. By the way, even the search for Mersenne primes has practical implications. One quality control test for supercomputers has been to replicate the Lucas–Lehmer test that establishes the primality of a large Mersenne prime. (See [Ro10] for more information about the quest for finding Mersenne primes.)

THE DISTRIBUTION OF PRIMES   Theorem 3 tells us that there are infinitely many primes. However, how many primes are less than a positive number x? This question interested mathematicians for many years; in the late eighteenth century, mathematicians produced large tables of prime numbers to gather evidence concerning the distribution of primes. Using this evidence, the great mathematicians of the day, including Gauss and Legendre, conjectured, but did not prove, Theorem 4.

MARIN MERSENNE (1588–1648)   Mersenne was born in Maine, France, into a family of laborers and attended the College of Mans and the Jesuit College at La Flèche. He continued his education at the Sorbonne, studying theology from 1609 to 1611. He joined the religious order of the Minims in 1611, a group whose name comes from the word *minimi* (the members of this group were extremely humble; they considered themselves the least of all religious orders). Besides prayer, the members of this group devoted their energy to scholarship and study. In 1612 he became a priest at the Place Royale in Paris; between 1614 and 1618 he taught philosophy at the Minim Convent at Nevers. He returned to Paris in 1619, where his cell in the Minims de l'Annociade became a place for meetings of French scientists, philosophers, and mathematicians, including Fermat and Pascal. Mersenne corresponded extensively with scholars throughout Europe, serving as a clearinghouse for mathematical and scientific knowledge, a function later served by mathematical journals (and today also by the Internet). Mersenne wrote books covering mechanics, mathematical physics, mathematics, music, and acoustics. He studied prime numbers and tried unsuccessfully to construct a formula representing all primes. In 1644 Mersenne claimed that $2^p - 1$ is prime for $p = 2, 3, 5, 7, 13, 17, 19, 31, 67, 127, 257$ but is composite for all other primes less than 257. It took over 300 years to determine that Mersenne's claim was wrong five times. Specifically, $2^p - 1$ is not prime for $p = 67$ and $p = 257$ but is prime for $p = 61$, $p = 87$, and $p = 107$. It is also noteworthy that Mersenne defended two of the most famous men of his time, Descartes and Galileo, from religious critics. He also helped expose alchemists and astrologers as frauds.

THEOREM 4   **THE PRIME NUMBER THEOREM**   The ratio of the number of primes not exceeding x and $x / \ln x$ approaches 1 as x grows without bound. (Here $\ln x$ is the natural logarithm of x.)

The prime number theorem was first proved in 1896 by the French mathematician Jacques Hadamard and the Belgian mathematician Charles-Jean-Gustave-Nicholas de la Vallée-Poussin using the theory of complex variables. Although proofs not using complex variables have been found, all known proofs of the prime number theorem are quite complicated.

We can use the prime number theorem to estimate the odds that a randomly chosen number is prime. The prime number theorem tells us that the number of primes not exceeding x can be approximated by $x / \ln x$. Consequently, the odds that a randomly selected positive integer less than n is prime are approximately $(n / \ln n)/n = 1/\ln n$. Sometimes we need to find a prime with a particular number of digits. We would like an estimate of how many integers with a particular number of digits we need to select before we encounter a prime. Using the prime number theorem and calculus, it can be shown that the probability that an integer n is prime is also approximately $1/\ln n$. For example, the odds that an integer near 10^{1000} is prime are approximately $1/\ln 10^{1000}$, which is approximately $1/2300$. (Of course, by choosing only odd numbers, we double our chances of finding a prime.)

Using trial division with Theorem 2 gives procedures for factoring and for primality testing. However, these procedures are not efficient algorithms; many much more practical and efficient algorithms for these tasks have been developed. Factoring and primality testing have become important in the applications of number theory to cryptography. This has led to a great interest in developing efficient algorithms for both tasks. Clever procedures have been devised in the last 30 years for efficiently generating large primes. Moreover, in 2002, an important theoretical discovery was made by Manindra Agrawal, Neeraj Kayal, and Nitin Saxena. They showed there is a polynomial-time algorithm in the number of bits in the binary expansion of an integer for determining whether a positive integer is prime. Algorithms based on their work use $O((\log n)^6)$ bit operations to determine whether a positive integer n is prime.

However, even though powerful new factorization methods have been developed in the same time frame, factoring large numbers remains extraordinarily more time-consuming than primality testing. No polynomial-time algorithm for factoring integers is known. Nevertheless, the challenge of factoring large numbers interests many people. There is a communal effort on the Internet to factor large numbers, especially those of the special form $k^n \pm 1$, where k is a small positive integer and n is a large positive integer (such numbers are called *Cunningham numbers*). At any given time, there is a list of the "Ten Most Wanted" large numbers of this type awaiting factorization.

PRIMES AND ARITHMETIC PROGRESSIONS   Every odd integer is in one of the two arithmetic progressions $4k + 1$ or $4k + 3, k = 1, 2, \ldots$. Because we know that there are infinitely many primes, we can ask whether there are infinitely many primes in both of these arithmetic progressions. The primes 5, 13, 17, 29, 37, 41, ... are in the arithmetic progression $4k + 1$; the primes 3, 7, 11, 19, 23, 31, 43, ... are in the arithmetic progression $4k + 3$. Looking at the evidence hints that there may be infinitely many primes in both progressions. What about other arithmetic progressions $ak + b, k = 1, 2, \ldots$, where no integer greater than one divides both a and b? Do they contain infinitely many primes? The answer was provided by the German mathematician G. Lejeune Dirichlet, who proved that every such arithmetic progression contains infinitely many primes. His proof, and all proofs found later, are beyond the scope of this book. However, it is possible to prove special cases of Dirichlet's theorem using the ideas developed in this book. For example, Exercises 54 and 55 ask for proofs that there are infinitely many primes in the arithmetic progressions $3k + 2$ and $4k + 3$, where k is a positive integer. (The hint for each of these exercises supplies the basic idea needed for the proof.)

We have explained that every arithmetic progression $ak + b$, $k = 1, 2, \ldots$, where a and b have no common factor greater than one, contains infinitely many primes. But are there long arithmetic progressions made up of just primes? For example, some exploration shows that 5, 11, 17, 23, 29 is an arithmetic progression of five primes and 199, 409, 619, 829, 1039, 1249, 1459, 1669, 1879, 2089 is an arithmetic progression of ten primes. In the 1930s, the famous mathematician Paul Erdős conjectured that for every positive integer n greater than two, there is an arithmetic progression of length n made up entirely of primes. In 2006, Ben Green and Terence Tao were able to prove this conjecture. Their proof, considered to be a mathematical tour de force, is a nonconstructive proof that combines powerful ideas from several advanced areas of mathematics.

Conjectures and Open Problems About Primes

Number theory is noted as a subject for which it is easy to formulate conjectures, some of which are difficult to prove and others that remained open problems for many years. We will describe some conjectures in number theory and discuss their status in Examples 6–9.

EXAMPLE 6

Extra Examples

It would be useful to have a function $f(n)$ such that $f(n)$ is prime for all positive integers n. If we had such a function, we could find large primes for use in cryptography and other applications. Looking for such a function, we might check out different polynomial functions, as some mathematicians did several hundred years ago. After a lot of computation we may encounter the polynomial $f(n) = n^2 - n + 41$. This polynomial has the interesting property that $f(n)$ is prime for all positive integers n not exceeding 40. [We have $f(1) = 41$, $f(2) = 43$, $f(3) = 47$, $f(4) = 53$, and so on.] This can lead us to the conjecture that $f(n)$ is prime for all positive integers n. Can we settle this conjecture?

Solution: Perhaps not surprisingly, this conjecture turns out to be false; we do not have to look far to find a positive integer n for which $f(n)$ is composite, because $f(41) = 41^2 - 41 + 41 = 41^2$. Because $f(n) = n^2 - n + 41$ is prime for all positive integers n with $1 \le n \le 40$, we might be tempted to find a different polynomial with the property that $f(n)$ is prime for *all* positive integers n. However, there is no such polynomial. It can be shown that for every polynomial $f(n)$ with integer coefficients, there is a positive integer y such that $f(y)$ is composite. (See Exercise 23 in the Supplementary Exercises.) ◀

Many famous problems about primes still await ultimate resolution by clever people. We describe a few of the most accessible and better known of these open problems in Examples 7–9.

TERENCE TAO (BORN 1975)   Tao was born in Australia. His father is a pediatrician and his mother taught mathematics at a Hong Kong secondary school. Tao was a child prodigy, teaching himself arithmetic at the age of two. At 10, he became the youngest contestant at the International Mathematical Olympiad (IMO); he won an IMO gold medal at 13. Tao received his bachelors and masters degrees when he was 17, and began graduate studies at Princeton, receiving his Ph.D. in three years. In 1996 he became a faculty member at UCLA, where he continues to work.

Tao is extremely versatile; he enjoys working on problems in diverse areas, including harmonic analysis, partial differential equations, number theory, and combinatorics. You can follow his work by reading his blog where he discusses progress on various problems. His most famous result is the Green-Tao theorem, which says that there are arbitrarily long arithmetic progressions of primes. Tao has made important contributions to the applications of mathematics, such as developing a method for reconstructing digital images using the least possible amount of information. Tao has an amazing reputation among mathematicians; he has become a Mr. Fix-It for researchers in mathematics. The well-known mathematician Charles Fefferman, himself a child prodigy, has said that "if you're stuck on a problem, then one way out is to interest Terence Tao." In 2006 Tao was awarded a Fields Medal, the most prestigious award for mathematicians under the age of 40. He was also awarded a MacArthur Fellowship in 2006, and in 2008, he received the Allan T. Waterman award, which came with a $500,000 cash prize to support research work of scientists early in their career. Tao's wife Laura is an engineer at the Jet Propulsion Laboratory.

Number theory is noted for its wealth of easy-to-understand conjectures that resist attack by all but the most sophisticated techniques, or simply resist all attacks. We present these conjectures to show that many questions that seem relatively simple remain unsettled even in the twenty-first century.

EXAMPLE 7

Links

Goldbach's Conjecture   In 1742, Christian Goldbach, in a letter to Leonhard Euler, conjectured that every odd integer n, $n > 5$, is the sum of three primes. Euler replied that this conjecture is equivalent to the conjecture that every even integer n, $n > 2$, is the sum of two primes (see Exercise 21 in the Supplementary Exercises). The conjecture that every even integer n, $n > 2$, is the sum of two primes is now called **Goldbach's conjecture**. We can check this conjecture for small even numbers. For example, $4 = 2 + 2$, $6 = 3 + 3$, $8 = 5 + 3$, $10 = 7 + 3$, $12 = 7 + 5$, and so on. Goldbach's conjecture was verified by hand calculations for numbers up to the millions prior to the advent of computers. With computers it can be checked for extremely large numbers. As of mid 2011, the conjecture has been checked for all positive even integers up to $1.6 \cdot 10^{18}$.

Although no proof of Goldbach's conjecture has been found, most mathematicians believe it is true. Several theorems have been proved, using complicated methods from analytic number theory far beyond the scope of this book, establishing results weaker than Goldbach's conjecture. Among these are the result that every even integer greater than 2 is the sum of at most six primes (proved in 1995 by O. Ramaré) and that every sufficiently large positive integer is the sum of a prime and a number that is either prime or the product of two primes (proved in 1966 by J. R. Chen). Perhaps Goldbach's conjecture will be settled in the not too distant future.   ◀

EXAMPLE 8

Links

There are many conjectures asserting that there are infinitely many primes of certain special forms. A conjecture of this sort is the conjecture that there are infinitely many primes of the form $n^2 + 1$, where n is a positive integer. For example, $5 = 2^2 + 1$, $17 = 4^2 + 1$, $37 = 6^2 + 1$, and so on. The best result currently known is that there are infinitely many positive integers n such that $n^2 + 1$ is prime or the product of at most two primes (proved by Henryk Iwaniec in 1973 using advanced techniques from analytic number theory, far beyond the scope of this book).   ◀

EXAMPLE 9

Links

The Twin Prime Conjecture   **Twin primes** are pairs of primes that differ by 2, such as 3 and 5, 5 and 7, 11 and 13, 17 and 19, and 4967 and 4969. The twin prime conjecture asserts that there are infinitely many twin primes. The strongest result proved concerning twin primes is that there are infinitely many pairs p and $p + 2$, where p is prime and $p + 2$ is prime or the product of two primes (proved by J. R. Chen in 1966). The world's record for twin primes, as of mid 2011, consists of the numbers $65,516,468,355 \cdot 2^{333,333} \pm 1$, which have $100,355$ decimal digits.   ◀

Greatest Common Divisors and Least Common Multiples

The largest integer that divides both of two integers is called the **greatest common divisor** of these integers.

DEFINITION 2

Links

Let a and b be integers, not both zero. The largest integer d such that $d \mid a$ and $d \mid b$ is called the *greatest common divisor* of a and b. The greatest common divisor of a and b is denoted by $\gcd(a, b)$.

CHRISTIAN GOLDBACH (1690–1764)   Christian Goldbach was born in Königsberg, Prussia, the city noted for its famous bridge problem (which will be studied in Section 10.5). He became professor of mathematics at the Academy in St. Petersburg in 1725. In 1728 Goldbach went to Moscow to tutor the son of the Tsar. He entered the world of politics when, in 1742, he became a staff member in the Russian Ministry of Foreign Affairs. Goldbach is best known for his correspondence with eminent mathematicians, including Euler and Bernoulli, for his famous conjectures in number theory, and for several contributions to analysis.

The greatest common divisor of two integers, not both zero, exists because the set of common divisors of these integers is nonempty and finite. One way to find the greatest common divisor of two integers is to find all the positive common divisors of both integers and then take the largest divisor. This is done in Examples 10 and 11. Later, a more efficient method of finding greatest common divisors will be given.

EXAMPLE 10   What is the greatest common divisor of 24 and 36?

Solution: The positive common divisors of 24 and 36 are 1, 2, 3, 4, 6, and 12. Hence, $\gcd(24, 36) = 12$. ◀

EXAMPLE 11   What is the greatest common divisor of 17 and 22?

Solution: The integers 17 and 22 have no positive common divisors other than 1, so that $\gcd(17, 22) = 1$. ◀

Because it is often important to specify that two integers have no common positive divisor other than 1, we have Definition 3.

DEFINITION 3   The integers a and b are *relatively prime* if their greatest common divisor is 1.

EXAMPLE 12   By Example 11 it follows that the integers 17 and 22 are relatively prime, because $\gcd(17, 22) = 1$. ◀

Because we often need to specify that no two integers in a set of integers have a common positive divisor greater than 1, we make Definition 4.

DEFINITION 4   The integers $a_1, a_2, \ldots, a_n$ are *pairwise relatively prime* if $\gcd(a_i, a_j) = 1$ whenever $1 \leq i < j \leq n$.

EXAMPLE 13   Determine whether the integers 10, 17, and 21 are pairwise relatively prime and whether the integers 10, 19, and 24 are pairwise relatively prime.

Solution: Because $\gcd(10, 17) = 1$, $\gcd(10, 21) = 1$, and $\gcd(17, 21) = 1$, we conclude that 10, 17, and 21 are pairwise relatively prime.
  Because $\gcd(10, 24) = 2 > 1$, we see that 10, 19, and 24 are not pairwise relatively prime. ◀

Another way to find the greatest common divisor of two positive integers is to use the prime factorizations of these integers. Suppose that the prime factorizations of the positive integers a and b are

$$a = p_1^{a_1} p_2^{a_2} \cdots p_n^{a_n}, \; b = p_1^{b_1} p_2^{b_2} \cdots p_n^{b_n},$$

where each exponent is a nonnegative integer, and where all primes occurring in the prime factorization of either a or b are included in both factorizations, with zero exponents if necessary. Then $\gcd(a, b)$ is given by

$$\gcd(a, b) = p_1^{\min(a_1, b_1)} p_2^{\min(a_2, b_2)} \cdots p_n^{\min(a_n, b_n)},$$

where $\min(x, y)$ represents the minimum of the two numbers x and y. To show that this formula for $\gcd(a, b)$ is valid, we must show that the integer on the right-hand side divides both a and b, and that no larger integer also does. This integer does divide both a and b, because the power of each prime in the factorization does not exceed the power of this prime in either the factorization of a or that of b. Further, no larger integer can divide both a and b, because the exponents of the primes in this factorization cannot be increased, and no other primes can be included.

EXAMPLE 14   Because the prime factorizations of 120 and 500 are $120 = 2^3 \cdot 3 \cdot 5$ and $500 = 2^2 \cdot 5^3$, the greatest common divisor is

$$\gcd(120, 500) = 2^{\min(3, 2)} 3^{\min(1, 0)} 5^{\min(1, 3)} = 2^2 3^0 5^1 = 20.$$ ◀

Prime factorizations can also be used to find the **least common multiple** of two integers.

DEFINITION 5   The *least common multiple* of the positive integers a and b is the smallest positive integer that is divisible by both a and b. The least common multiple of a and b is denoted by $\text{lcm}(a, b)$.

The least common multiple exists because the set of integers divisible by both a and b is nonempty (as ab belongs to this set, for instance), and every nonempty set of positive integers has a least element (by the well-ordering property, which will be discussed in Section 5.2). Suppose that the prime factorizations of a and b are as before. Then the least common multiple of a and b is given by

$$\text{lcm}(a, b) = p_1^{\max(a_1, b_1)} p_2^{\max(a_2, b_2)} \cdots p_n^{\max(a_n, b_n)},$$

where $\max(x, y)$ denotes the maximum of the two numbers x and y. This formula is valid because a common multiple of a and b has at least $\max(a_i, b_i)$ factors of p_i in its prime factorization, and the least common multiple has no other prime factors besides those in a and b.

EXAMPLE 15   What is the least common multiple of $2^3 3^5 7^2$ and $2^4 3^3$?

Solution: We have

$$\text{lcm}(2^3 3^5 7^2, 2^4 3^3) = 2^{\max(3, 4)} 3^{\max(5, 3)} 7^{\max(2, 0)} = 2^4 3^5 7^2.$$ ◀

Theorem 5 gives the relationship between the greatest common divisor and least common multiple of two integers. It can be proved using the formulae we have derived for these quantities. The proof of this theorem is left as an exercise.

THEOREM 5   Let a and b be positive integers. Then

$$ab = \gcd(a, b) \cdot \text{lcm}(a, b).$$

The Euclidean Algorithm

Links

Computing the greatest common divisor of two integers directly from the prime factorizations of these integers is inefficient. The reason is that it is time-consuming to find prime factorizations. We will give a more efficient method of finding the greatest common divisor, called the **Euclidean algorithm**. This algorithm has been known since ancient times. It is named after the ancient Greek mathematician Euclid, who included a description of this algorithm in his book *The Elements*.

Before describing the Euclidean algorithm, we will show how it is used to find $\gcd(91, 287)$. First, divide 287, the larger of the two integers, by 91, the smaller, to obtain

$$287 = 91 \cdot 3 + 14.$$

Any divisor of 91 and 287 must also be a divisor of $287 - 91 \cdot 3 = 14$. Also, any divisor of 91 and 14 must also be a divisor of $287 = 91 \cdot 3 + 14$. Hence, the greatest common divisor of 91 and 287 is the same as the greatest common divisor of 91 and 14. This means that the problem of finding $\gcd(91, 287)$ has been reduced to the problem of finding $\gcd(91, 14)$.

Next, divide 91 by 14 to obtain

$$91 = 14 \cdot 6 + 7.$$

Because any common divisor of 91 and 14 also divides $91 - 14 \cdot 6 = 7$ and any common divisor of 14 and 7 divides 91, it follows that $\gcd(91, 14) = \gcd(14, 7)$.

Continue by dividing 14 by 7, to obtain

$$14 = 7 \cdot 2.$$

Because 7 divides 14, it follows that $\gcd(14, 7) = 7$. Furthermore, because $\gcd(287, 91) = \gcd(91, 14) = \gcd(14, 7) = 7$, the original problem has been solved.

We now describe how the Euclidean algorithm works in generality. We will use successive divisions to reduce the problem of finding the greatest common divisor of two positive integers to the same problem with smaller integers, until one of the integers is zero.

The Euclidean algorithm is based on the following result about greatest common divisors and the division algorithm.

LEMMA 1   Let $a = bq + r$, where a, b, q, and r are integers. Then $\gcd(a, b) = \gcd(b, r)$.

Proof: If we can show that the common divisors of a and b are the same as the common divisors of b and r, we will have shown that $\gcd(a, b) = \gcd(b, r)$, because both pairs must have the same *greatest* common divisor.

So suppose that d divides both a and b. Then it follows that d also divides $a - bq = r$ (from Theorem 1 of Section 4.1). Hence, any common divisor of a and b is also a common divisor of b and r.

Links

EUCLID (325 B.C.E.–265 B.C.E.)   Euclid was the author of the most successful mathematics book ever written, *The Elements*, which appeared in over 1000 different editions from ancient to modern times. Little is known about Euclid's life, other than that he taught at the famous academy at Alexandria in Egypt. Apparently, Euclid did not stress applications. When a student asked what he would get by learning geometry, Euclid explained that knowledge was worth acquiring for its own sake and told his servant to give the student a coin "because he must make a profit from what he learns."

Likewise, suppose that d divides both b and r. Then d also divides $bq + r = a$. Hence, any common divisor of b and r is also a common divisor of a and b.

Consequently, $\gcd(a, b) = \gcd(b, r)$.   ◁

Suppose that a and b are positive integers with $a \geq b$. Let $r_0 = a$ and $r_1 = b$. When we successively apply the division algorithm, we obtain

$$
\begin{aligned}
r_0 &= r_1 q_1 + r_2 & 0 \leq r_2 < r_1, \\
r_1 &= r_2 q_2 + r_3 & 0 \leq r_3 < r_2, \\
&\ \ \vdots \\
r_{n-2} &= r_{n-1} q_{n-1} + r_n & 0 \leq r_n < r_{n-1}, \\
r_{n-1} &= r_n q_n.
\end{aligned}
$$

Eventually a remainder of zero occurs in this sequence of successive divisions, because the sequence of remainders $a = r_0 > r_1 > r_2 > \cdots \geq 0$ cannot contain more than a terms. Furthermore, it follows from Lemma 1 that

$$
\gcd(a, b) = \gcd(r_0, r_1) = \gcd(r_1, r_2) = \cdots = \gcd(r_{n-2}, r_{n-1})
$$
$$
= \gcd(r_{n-1}, r_n) = \gcd(r_n, 0) = r_n.
$$

Hence, the greatest common divisor is the last nonzero remainder in the sequence of divisions.

EXAMPLE 16 Find the greatest common divisor of 414 and 662 using the Euclidean algorithm.

Solution: Successive uses of the division algorithm give:

$$
\begin{aligned}
662 &= 414 \cdot 1 + 248 \\
414 &= 248 \cdot 1 + 166 \\
248 &= 166 \cdot 1 + 82 \\
166 &= 82 \cdot 2 + 2 \\
82 &= 2 \cdot 41.
\end{aligned}
$$

Hence, $\gcd(414, 662) = 2$, because 2 is the last nonzero remainder.   ◀

The Euclidean algorithm is expressed in pseudocode in Algorithm 1.

ALGORITHM 1 The Euclidean Algorithm.

---

procedure $gcd(a, b$: positive integers)
$x := a$
$y := b$
while $y \neq 0$
    $r := x \bmod y$
    $x := y$
    $y := r$
return $x\{\gcd(a, b)$ is $x\}$

In Algorithm 1, the initial values of x and y are a and b, respectively. At each stage of the procedure, x is replaced by y, and y is replaced by x **mod** y, which is the remainder when x is divided by y. This process is repeated as long as $y \neq 0$. The algorithm terminates when $y = 0$, and the value of x at that point, the last nonzero remainder in the procedure, is the greatest common divisor of a and b.

We will study the time complexity of the Euclidean algorithm in Section 5.3, where we will show that the number of divisions required to find the greatest common divisor of a and b, where $a \geq b$, is $O(\log b)$.

gcds as Linear Combinations

An important result we will use throughout the remainder of this section is that the greatest common divisor of two integers a and b can be expressed in the form

$$sa + tb,$$

where s and t are integers. In other words, $\gcd(a, b)$ can be expressed as a **linear combination** with integer coefficients of a and b. For example, $\gcd(6, 14) = 2$, and $2 = (-2) \cdot 6 + 1 \cdot 14$. We state this fact as Theorem 6.

THEOREM 6   **BÉZOUT'S THEOREM**   If a and b are positive integers, then there exist integers s and t such that $\gcd(a, b) = sa + tb$.

DEFINITION 6   If a and b are positive integers, then integers s and t such that $\gcd(a, b) = sa + tb$ are called *Bézout coefficients* of a and b (after Étienne Bézout, a French mathematician of the eighteenth century). Also, the equation $\gcd(a, b) = sa + tb$ is called *Bézout's identity*.

We will not give a formal proof of Theorem 6 here (see Exercise 28 in Section 5.2 and [Ro10] for proofs). We will provide an example of a general method that can be used to find a linear combination of two integers equal to their greatest common divisor. (In this section, we will assume that a linear combination has integer coefficients.) The method proceeds by working backward through the divisions of the Euclidean algorithm, so this method requires a forward pass and a backward pass through the steps of the Euclidean algorithm. (In the exercises we will describe an algorithm called the **extended Euclidean algorithm**, which can be used to express $\gcd(a, b)$ as a linear combination of a and b using a single pass through the steps of the Euclidean algorithm; see the preamble to Exercise 30.)

ÉTIENNE BÉZOUT (1730–1783)   Bézout was born in Nemours, France, where his father was a magistrate. Reading the writings of the great mathematician Leonhard Euler enticed him to become a mathematician. In 1758 he was appointed to a position at the Académie des Sciences in Paris; in 1763 he was appointed examiner of the Gardes de la Marine, where he was assigned the task of writing mathematics textbooks. This assignment led to a four-volume textbook completed in 1767. Bézout is well known for his six-volume comprehensive textbook on mathematics. His textbooks were extremely popular and were studied by many generations of students hoping to enter the École Polytechnique, the famous engineering and science school. His books were translated into English and used in North America, including at Harvard.

His most important original work was published in 1779 in the book *Théorie générale des équations algébriques*, where he introduced important methods for solving simultaneous polynomial equations in many unknowns. The most well-known result in this book is now called *Bézout's theorem*, which in its general form tells us that the number of common points on two plane algebraic curves equals the product of the degrees of these curves. Bézout is also credited with inventing the determinant (which was called the Bézoutian by the great English mathematician James Joseph Sylvester). He was considered to be a kind person with a warm heart, although he had a reserved and somber personality. He was happily married and a father.

EXAMPLE 17 Express $\gcd(252, 198) = 18$ as a linear combination of 252 and 198.

Solution: To show that $\gcd(252, 198) = 18$, the Euclidean algorithm uses these divisions:

$$252 = 1 \cdot 198 + 54$$
$$198 = 3 \cdot 54 + 36$$
$$54 = 1 \cdot 36 + 18$$
$$36 = 2 \cdot 18.$$

Using the next-to-last division (the third division), we can express $\gcd(252, 198) = 18$ as a linear combination of 54 and 36. We find that

$$18 = 54 - 1 \cdot 36.$$

The second division tells us that

$$36 = 198 - 3 \cdot 54.$$

Substituting this expression for 36 into the previous equation, we can express 18 as a linear combination of 54 and 198. We have

$$18 = 54 - 1 \cdot 36 = 54 - 1 \cdot (198 - 3 \cdot 54) = 4 \cdot 54 - 1 \cdot 198.$$

The first division tells us that

$$54 = 252 - 1 \cdot 198.$$

Substituting this expression for 54 into the previous equation, we can express 18 as a linear combination of 252 and 198. We conclude that

$$18 = 4 \cdot (252 - 1 \cdot 198) - 1 \cdot 198 = 4 \cdot 252 - 5 \cdot 198,$$

completing the solution. ◀

We will use Theorem 6 to develop several useful results. One of our goals will be to prove the part of the fundamental theorem of arithmetic asserting that a positive integer has at most one prime factorization. We will show that if a positive integer has a factorization into primes, where the primes are written in nondecreasing order, then this factorization is unique.

First, we need to develop some results about divisibility.

LEMMA 2 If a, b, and c are positive integers such that $\gcd(a, b) = 1$ and $a \mid bc$, then $a \mid c$.

Proof: Because $\gcd(a, b) = 1$, by Bézout's theorem there are integers s and t such that

$$sa + tb = 1.$$

Multiplying both sides of this equation by c, we obtain

$$sac + tbc = c.$$

We can now use Theorem 1 of Section 4.1 to show that $a \mid c$. By part *(ii)* of that theorem, $a \mid tbc$. Because $a \mid sac$ and $a \mid tbc$, by part *(i)* of that theorem, we conclude that a divides $sac + tbc$. Because $sac + tbc = c$, we conclude that $a \mid c$, completing the proof. ◁

We will use the following generalization of Lemma 2 in the proof of uniqueness of prime factorizations. (The proof of Lemma 3 is left as Exercise 52 in Section 5.1, because it can be most easily carried out using the method of mathematical induction, covered in that section.)

LEMMA 3   If p is a prime and $p \mid a_1 a_2 \cdots a_n$, where each a_i is an integer, then $p \mid a_i$ for some i.

We can now show that a factorization of an integer into primes is unique. That is, we will show that every integer can be written as the product of primes in nondecreasing order in at most one way. This is part of the fundamental theorem of arithmetic. We will prove the other part, that every integer has a factorization into primes, in Section 5.2.

Proof (of the uniqueness of the prime factorization of a positive integer): We will use a proof by contradiction. Suppose that the positive integer n can be written as the product of primes in two different ways, say, $n = p_1 p_2 \cdots p_s$ and $n = q_1 q_2 \cdots q_t$, each p_i and q_j are primes such that $p_1 \leq p_2 \leq \cdots \leq p_s$ and $q_1 \leq q_2 \leq \cdots \leq q_t$.

When we remove all common primes from the two factorizations, we have

$$p_{i_1} p_{i_2} \cdots p_{i_u} = q_{j_1} q_{j_2} \cdots q_{j_v},$$

where no prime occurs on both sides of this equation and u and v are positive integers. By Lemma 3 it follows that p_{i_1} divides q_{j_k} for some k. Because no prime divides another prime, this is impossible. Consequently, there can be at most one factorization of n into primes in nondecreasing order. ◁

Lemma 2 can also be used to prove a result about dividing both sides of a congruence by the same integer. We have shown (Theorem 5 in Section 4.1) that we can multiply both sides of a congruence by the same integer. However, dividing both sides of a congruence by an integer does not always produce a valid congruence, as Example 18 shows.

EXAMPLE 18   The congruence $14 \equiv 8 \pmod{6}$ holds, but both sides of this congruence cannot be divided by 2 to produce a valid congruence because $14/2 = 7$ and $8/2 = 4$, but $7 \not\equiv 4 \pmod{6}$. ◀

Although we cannot divide both sides of a congruence by any integer to produce a valid congruence, we can if this integer is relatively prime to the modulus. Theorem 7 establishes this important fact. We use Lemma 2 in the proof.

THEOREM 7   Let m be a positive integer and let a, b, and c be integers. If $ac \equiv bc \pmod{m}$ and $\gcd(c, m) = 1$, then $a \equiv b \pmod{m}$.

Proof: Because $ac \equiv bc \pmod{m}$, $m \mid ac - bc = c(a - b)$. By Lemma 2, because $\gcd(c, m) = 1$, it follows that $m \mid a - b$. We conclude that $a \equiv b \pmod{m}$. ◁

Exercises

1. Determine whether each of these integers is prime.
   a) 21
   b) 29
   c) 71
   d) 97
   e) 111
   f) 143

2. Find the prime factorization of each of these integers.
   a) 88
   b) 126
   c) 729
   d) 1001
   e) 1111
   f) 909,090

3. Find the prime factorization of 10!.

*4. How many zeros are there at the end of 100!?

5. Express in pseudocode the trial division algorithm for determining whether an integer is prime.

6. Express in pseudocode the algorithm described in the text for finding the prime factorization of an integer.

7. Show that if $a^m + 1$ is composite if a and m are integers greater than 1 and m is odd. [*Hint:* Show that $x + 1$ is a factor of the polynomial $x^m + 1$ if m is odd.]

8. Show that if $2^m + 1$ is an odd prime, then $m = 2^n$ for some nonnegative integer n. [*Hint:* First show that the polynomial identity $x^m + 1 = (x^k + 1)(x^{k(t-1)} - x^{k(t-2)} + \cdots - x^k + 1)$ holds, where $m = kt$ and t is odd.]

*9. Show that $\log_2 3$ is an irrational number. Recall that an irrational number is a real number x that cannot be written as the ratio of two integers.

10. Prove that for every positive integer n, there are n consecutive composite integers. [*Hint:* Consider the n consecutive integers starting with $(n + 1)! + 2$.]

*11. Prove or disprove that there are three consecutive odd positive integers that are primes, that is, odd primes of the form p, $p + 2$, and $p + 4$.

12. Which positive integers less than 30 are relatively prime to 30?

13. Determine whether the integers in each of these sets are pairwise relatively prime.
    a) 11, 15, 19
    b) 14, 15, 21
    c) 12, 17, 31, 37
    d) 7, 8, 9, 11

14. We call a positive integer **perfect** if it equals the sum of its positive divisors other than itself.
    a) Show that 6 and 28 are perfect.
    b) Show that $2^{p-1}(2^p - 1)$ is a perfect number when $2^p - 1$ is prime.

15. Show that if $2^n - 1$ is prime, then n is prime. [*Hint:* Use the identity $2^{ab} - 1 = (2^a - 1) \cdot (2^{a(b-1)} + 2^{a(b-2)} + \cdots + 2^a + 1)$.]

16. Determine whether each of these integers is prime, verifying some of Mersenne's claims.
    a) $2^7 - 1$
    b) $2^9 - 1$
    c) $2^{11} - 1$
    d) $2^{13} - 1$

The value of the **Euler ϕ-function** at the positive integer n is defined to be the number of positive integers less than or equal to n that are relatively prime to n. [*Note:* ϕ is the Greek letter phi.]

17. Find these values of the Euler ϕ-function.
    a) $\phi(4)$.
    b) $\phi(10)$.
    c) $\phi(13)$.

18. Show that n is prime if and only if $\phi(n) = n - 1$.

19. What is the value of $\phi(p^k)$ when p is prime and k is a positive integer?

20. What are the greatest common divisors of these pairs of integers?
    a) $3^7 \cdot 5^3 \cdot 7^3, 2^{11} \cdot 3^5 \cdot 5^9$
    b) $11 \cdot 13 \cdot 17, 2^9 \cdot 3^7 \cdot 5^5 \cdot 7^3$
    c) $23^{31}, 23^{17}$
    d) $41 \cdot 43 \cdot 53, 41 \cdot 43 \cdot 53$
    e) $3^{13} \cdot 5^{17}, 2^{12} \cdot 7^{21}$
    f) $1111, 0$

21. What is the least common multiple of each pair in Exercise 20?

22. Find $\gcd(1000, 625)$ and $\text{lcm}(1000, 625)$ and verify that $\gcd(1000, 625) \cdot \text{lcm}(1000, 625) = 1000 \cdot 625$.

23. If the product of two integers is $2^7 3^8 5^2 7^{11}$ and their greatest common divisor is $2^3 3^4 5$, what is their least common multiple?

24. Use the Euclidean algorithm to find
    a) $\gcd(12, 18)$.
    b) $\gcd(111, 201)$.
    c) $\gcd(1001, 1331)$.
    d) $\gcd(12345, 54321)$.
    e) $\gcd(1000, 5040)$.
    f) $\gcd(9888, 6060)$.

25. How many divisions are required to find $\gcd(34, 55)$ using the Euclidean algorithm?

*26. Show that if a and b are both positive integers, then $(2^a - 1) \bmod (2^b - 1) = 2^{a \bmod b} - 1$.

*27. Use Exercise 26 to show that if a and b are positive integers, then $\gcd(2^a - 1, 2^b - 1) = 2^{\gcd(a, b)} - 1$. [*Hint:* Show that the remainders obtained when the Euclidean algorithm is used to compute $\gcd(2^a - 1, 2^b - 1)$ are of the form $2^r - 1$, where r is a remainder arising when the Euclidean algorithm is used to find $\gcd(a, b)$.]

28. Use Exercise 27 to show that the integers $2^{35} - 1, 2^{34} - 1, 2^{33} - 1, 2^{31} - 1, 2^{29} - 1$, and $2^{23} - 1$ are pairwise relatively prime.

29. Using the method followed in Example 17, express the greatest common divisor of each of these pairs of integers as a linear combination of these integers.
    a) 10, 11
    b) 21, 44
    c) 36, 48
    d) 34, 55
    e) 117, 213
    f) 0, 223
    g) 123, 2347
    h) 3454, 4666
    i) 9999, 11111

The **extended Euclidean algorithm** can be used to express $\gcd(a, b)$ as a linear combination with integer coefficients of the integers a and b. We set $s_0 = 1, s_1 = 0, t_0 = 0$, and $t_1 = 1$ and let $s_j = s_{j-2} - q_{j-1}s_{j-1}$ and $t_j = t_{j-2} - q_{j-1}t_{j-1}$ for $j = 2, 3, \ldots, n$, where the q_j are the quotients in the divisions used when the Euclidean algorithm finds $\gcd(a, b)$, as shown in the text. It can be shown (see [Ro10]) that $\gcd(a, b) = s_n a + t_n b$. The main advantage of the extended Euclidean algorithm is that it uses one pass through the steps of the Euclidean algorithm to find Bézout coefficients of a and b, unlike the method in the text which uses two passes.

30. Use the extended Euclidean algorithm to express $\gcd(26, 91)$ as a linear combination of 26 and 91.

31. Use the extended Euclidean algorithm to express $\gcd(144, 89)$ as a linear combination of 144 and 89.

32. Find the smallest positive integer with exactly n different positive factors when n is

a) 3.   **b)** 4.   **c)** 5.

d) 6.   **e)** 10.

33. Can you find a formula or rule for the nth term of a sequence related to the prime numbers or prime factorizations so that the initial terms of the sequence have these values?

a) $0, 1, 1, 0, 1, 0, 1, 0, 0, 0, 1, 0, 1, \ldots$
b) $1, 2, 3, 2, 5, 2, 7, 2, 3, 2, 11, 2, 13, 2, \ldots$
c) $1, 2, 2, 3, 2, 4, 2, 4, 3, 4, 2, 6, 2, 4, \ldots$
d) $1, 1, 1, 0, 1, 1, 1, 1, 0, 0, 1, 1, 0, 1, 1, \ldots$
e) $1, 2, 3, 3, 5, 5, 7, 7, 7, 7, 11, 11, 13, 13, \ldots$
f) $1, 2, 6, 30, 210, 2310, 30030, 510510, 9699690,$
$223092870, \ldots$

34. Prove that the product of any three consecutive integers is divisible by 6.

∗35. Prove or disprove that $n^2 - 79n + 1601$ is prime whenever n is a positive integer.

36. Prove or disprove that $p_1 p_2 \cdots p_n + 1$ is prime for every positive integer n, where $p_1, p_2, \ldots, p_n$ are the n smallest prime numbers.

37. Show that there is a composite integer in every arithmetic progression $ak + b, k = 1, 2, \ldots$ where a and b are positive integers.

38. Adapt the proof in the text that there are infinitely many primes to prove that there are infinitely many primes of the form $4k + 3$, where k is a nonnegative integer. [*Hint:* Suppose that there are only finitely many such primes $q_1, q_2, \ldots, q_n$, and consider the number $4q_1 q_2 \cdots q_n - 1$.]

4.4   Solving Congruences

Introduction

Solving linear congruences, which have the form $ax \equiv b \pmod{m}$, is an essential task in the study of number theory and its applications, just as solving linear equations plays an important role in calculus and linear algebra. To solve linear congruences, we employ inverses modulo m. We explain how to work backwards through the steps of the Euclidean algorithm to find inverses modulo m. Once we have found an inverse of a modulo m, we solve the congruence $ax \equiv b \pmod{m}$ by multiplying both sides of the congruence by this inverse.

Simultaneous systems of linear congruence have been studied since ancient times. For example, the Chinese mathematician Sun-Tsu studied them in the first century. We will show how to solve systems of linear congruences modulo pairwise relatively prime moduli. The result we will prove is called the Chinese remainder theorem, and our proof will give a method to find all solutions of such systems of congruences. We will also show how to use the Chinese remainder theorem as a basis for performing arithmetic with large integers.

We will introduce a useful result of Fermat, known as Fermat's little theorem, which states that if p is prime and p does not divide a, then $a^{p-1} \equiv 1 \pmod{p}$. We will examine the converse of this statement, which will lead us to the concept of a pseudoprime. A pseudoprime m to the base a is a composite integer m that masquerades as a prime by satisfying the congruence $a^{m-1} \equiv 1 \pmod{m}$. We will also give an example of a Carmichael number, which is a composite integer that is a pseudoprime to all bases a relatively prime to it.

We also introduce the notion of discrete logarithms, which are analogous to ordinary logarithms. To define discrete logarithms we must first define primitive roots. A primitive root of a prime p is an integer r such that every integer not divisible by p is congruent to a power of r modulo p. If r is a primitive root of p and $r^e \equiv a \pmod{p}$, then e is the discrete logarithm of a modulo p to the base r. Finding discrete logarithms turns out to be an extremely difficult problem in general. The difficulty of this problem is the basis for the security of many cryptographic systems.

Linear Congruences

A congruence of the form

$$ax \equiv b \;(\mathrm{mod}\; m),$$

where m is a positive integer, a and b are integers, and x is a variable, is called a **linear congruence**. Such congruences arise throughout number theory and its applications.

How can we solve the linear congruence $ax \equiv b \;(\mathrm{mod}\; m)$, that is, how can we find all integers x that satisfy this congruence? One method that we will describe uses an integer $\overline{a}$ such that $\overline{a}a \equiv 1 \;(\mathrm{mod}\; m)$, if such an integer exists. Such an integer $\overline{a}$ is said to be an **inverse** of a modulo m. Theorem 1 guarantees that an inverse of a modulo m exists whenever a and m are relatively prime.

THEOREM 1   If a and m are relatively prime integers and $m > 1$, then an inverse of a modulo m exists. Furthermore, this inverse is unique modulo m. (That is, there is a unique positive integer $\overline{a}$ less than m that is an inverse of a modulo m and every other inverse of a modulo m is congruent to $\overline{a}$ modulo m.)

Proof: By Theorem 6 of Section 4.3, because $\gcd(a, m) = 1$, there are integers s and t such that

$$sa + tm = 1.$$

This implies that

$$sa + tm \equiv 1 \;(\mathrm{mod}\; m).$$

Because $tm \equiv 0 \;(\mathrm{mod}\; m)$, it follows that

$$sa \equiv 1 \;(\mathrm{mod}\; m).$$

Consequently, s is an inverse of a modulo m. That this inverse is unique modulo m is left as Exercise 5. ◁

Using inspection to find an inverse of a modulo m is easy when m is small. To find this inverse, we look for a multiple of a that exceeds a multiple of m by 1. For example, to find an inverse of 3 modulo 7, we can find $j \cdot 3$ for $j = 1, 2, \ldots, 6$, stopping when we find a multiple of 3 that is one more than a multiple of 7. We can speed this approach up if we note that $2 \cdot 3 \equiv -1 \;(\mathrm{mod}\; 7)$. This means that $(-2) \cdot 3 \equiv 1 \;(\mathrm{mod}\; 7)$. Hence, $5 \cdot 3 \equiv 1 \;(\mathrm{mod}\; 7)$, so 5 is an inverse of 3 modulo 7.

We can design a more efficient algorithm than brute force to find an inverse of a modulo m when $\gcd(a, m) = 1$ using the steps of the Euclidean algorithm. By reversing these steps as in Example 17 of Section 4.3, we can find a linear combination $sa + tm = 1$ where s and t are integers. Reducing both sides of this equation modulo m tells us that s is an inverse of a modulo m. We illustrate this procedure in Example 1.

EXAMPLE 1   Find an inverse of 3 modulo 7 by first finding Bézout coefficients of 3 and 7. (Note that we have already shown that 5 is an inverse of 3 modulo 7 by inspection.)

Solution: Because $\gcd(3, 7) = 1$, Theorem 1 tells us that an inverse of 3 modulo 7 exists. The Euclidean algorithm ends quickly when used to find the greatest common divisor of 3 and 7:

$$7 = 2 \cdot 3 + 1.$$

From this equation we see that

$$-2 \cdot 3 + 1 \cdot 7 = 1.$$

This shows that -2 and 1 are Bézout coefficients of 3 and 7. We see that -2 is an inverse of 3 modulo 7. Note that every integer congruent to -2 modulo 7 is also an inverse of 3, such as 5, -9, 12, and so on. ◀

EXAMPLE 2   Find an inverse of 101 modulo 4620.

Solution: For completeness, we present all steps used to compute an inverse of 101 modulo 4620. (Only the last step goes beyond methods developed in Section 4.3 and illustrated in Example 17 in that section.) First, we use the Euclidean algorithm to show that $\gcd(101, 4620) = 1$. Then we will reverse the steps to find Bézout coefficients a and b such that $101a + 4620b = 1$. It will then follow that a is an inverse of 101 modulo 4620. The steps used by the Euclidean algorithm to find $\gcd(101, 4620)$ are

$$4620 = 45 \cdot 101 + 75$$
$$101 = 1 \cdot 75 + 26$$
$$75 = 2 \cdot 26 + 23$$
$$26 = 1 \cdot 23 + 3$$
$$23 = 7 \cdot 3 + 2$$
$$3 = 1 \cdot 2 + 1$$
$$2 = 2 \cdot 1.$$

Because the last nonzero remainder is 1, we know that $\gcd(101, 4620) = 1$. We can now find the Bézout coefficients for 101 and 4620 by working backwards through these steps, expressing $\gcd(101, 4620) = 1$ in terms of each successive pair of remainders. In each step we eliminate the remainder by expressing it as a linear combination of the divisor and the dividend. We obtain

$$1 = 3 - 1 \cdot 2$$
$$= 3 - 1 \cdot (23 - 7 \cdot 3) = -1 \cdot 23 + 8 \cdot 3$$
$$= -1 \cdot 23 + 8 \cdot (26 - 1 \cdot 23) = 8 \cdot 26 - 9 \cdot 23$$
$$= 8 \cdot 26 - 9 \cdot (75 - 2 \cdot 26) = -9 \cdot 75 + 26 \cdot 26$$
$$= -9 \cdot 75 + 26 \cdot (101 - 1 \cdot 75) = 26 \cdot 101 - 35 \cdot 75$$
$$= 26 \cdot 101 - 35 \cdot (4620 - 45 \cdot 101) = -35 \cdot 4620 + 1601 \cdot 101.$$

That $-35 \cdot 4620 + 1601 \cdot 101 = 1$ tells us that -35 and 1601 are Bézout coefficients of 4620 and 101, and 1601 is an inverse of 101 modulo 4620. ◀

Once we have an inverse $\overline{a}$ of a modulo m, we can solve the congruence $ax \equiv b \pmod{m}$ by multiplying both sides of the linear congruence by $\overline{a}$, as Example 3 illustrates.

EXAMPLE 3   What are the solutions of the linear congruence $3x \equiv 4 \pmod{7}$?

Solution: By Example 1 we know that -2 is an inverse of 3 modulo 7. Multiplying both sides of the congruence by -2 shows that

$$-2 \cdot 3x \equiv -2 \cdot 4 \pmod{7}.$$

Because $-6 \equiv 1 \pmod 7$ and $-8 \equiv 6 \pmod 7$, it follows that if x is a solution, then $x \equiv -8 \equiv 6 \pmod 7$.

We need to determine whether every x with $x \equiv 6 \pmod 7$ is a solution. Assume that $x \equiv 6 \pmod 7$. Then, by Theorem 5 of Section 4.1, it follows that

$$3x \equiv 3 \cdot 6 = 18 \equiv 4 \pmod 7,$$

which shows that all such x satisfy the congruence. We conclude that the solutions to the congruence are the integers x such that $x \equiv 6 \pmod 7$, namely, $6, 13, 20, \ldots$ and $-1, -8, -15, \ldots$. ◀

The Chinese Remainder Theorem

Systems of linear congruences arise in many contexts. For example, as we will see later, they are the basis for a method that can be used to perform arithmetic with large integers. Such systems can even be found as word puzzles in the writings of ancient Chinese and Hindu mathematicians, such as that given in Example 4.

Links

EXAMPLE 4   In the first century, the Chinese mathematician Sun-Tsu asked:

> There are certain things whose number is unknown. When divided by 3, the remainder is 2; when divided by 5, the remainder is 3; and when divided by 7, the remainder is 2. What will be the number of things?

This puzzle can be translated into the following question: What are the solutions of the systems of congruences

$$x \equiv 2 \pmod 3,$$
$$x \equiv 3 \pmod 5,$$
$$x \equiv 2 \pmod 7?$$

We will solve this system, and with it Sun-Tsu's puzzle, later in this section. ◀

The *Chinese remainder theorem*, named after the Chinese heritage of problems involving systems of linear congruences, states that when the moduli of a system of linear congruences are pairwise relatively prime, there is a unique solution of the system modulo the product of the moduli.

THEOREM 2   **THE CHINESE REMAINDER THEOREM**   Let $m_1, m_2, \ldots, m_n$ be pairwise relatively prime positive integers greater than one and $a_1, a_2, \ldots, a_n$ arbitrary integers. Then the system

$$x \equiv a_1 \pmod{m_1},$$
$$x \equiv a_2 \pmod{m_2},$$
$$\vdots$$
$$x \equiv a_n \pmod{m_n}$$

has a unique solution modulo $m = m_1 m_2 \cdots m_n$. (That is, there is a solution x with $0 \le x < m$, and all other solutions are congruent modulo m to this solution.)

Proof: To establish this theorem, we need to show that a solution exists and that it is unique modulo m. We will show that a solution exists by describing a way to construct this solution; showing that the solution is unique modulo m is Exercise 22.

To construct a simultaneous solution, first let

$$M_k = m/m_k$$

for $k = 1, 2, \ldots, n$. That is, M_k is the product of the moduli except for m_k. Because m_i and m_k have no common factors greater than 1 when $i \neq k$, it follows that $\gcd(m_k, M_k) = 1$. Consequently, by Theorem 1, we know that there is an integer y_k, an inverse of M_k modulo m_k, such that

$$M_k y_k \equiv 1 \pmod{m_k}.$$

To construct a simultaneous solution, form the sum

$$x = a_1 M_1 y_1 + a_2 M_2 y_2 + \cdots + a_n M_n y_n.$$

We will now show that x is a simultaneous solution. First, note that because $M_j \equiv 0 \pmod{m_k}$ whenever $j \neq k$, all terms except the kth term in this sum are congruent to 0 modulo m_k. Because $M_k y_k \equiv 1 \pmod{m_k}$ we see that

$$x \equiv a_k M_k y_k \equiv a_k \pmod{m_k},$$

for $k = 1, 2, \ldots, n$. We have shown that x is a simultaneous solution to the n congruences. ◁

Example 5 illustrates how to use the construction given in our proof of the Chinese remainder theorem to solve a system of congruences. We will solve the system given in Example 4, arising in Sun-Tsu's puzzle.

EXAMPLE 5 To solve the system of congruences in Example 4, first let $m = 3 \cdot 5 \cdot 7 = 105$, $M_1 = m/3 = 35$, $M_2 = m/5 = 21$, and $M_3 = m/7 = 15$. We see that 2 is an inverse of $M_1 = 35$ modulo 3, because $35 \cdot 2 \equiv 2 \cdot 2 \equiv 1 \pmod{3}$; 1 is an inverse of $M_2 = 21$ modulo 5, because $21 \equiv 1 \pmod{5}$; and 1 is an inverse of $M_3 = 15 \pmod{7}$, because $15 \equiv 1 \pmod{7}$. The solutions to this system are those x such that

$$\begin{aligned} x &\equiv a_1 M_1 y_1 + a_2 M_2 y_2 + a_3 M_3 y_3 = 2 \cdot 35 \cdot 2 + 3 \cdot 21 \cdot 1 + 2 \cdot 15 \cdot 1 \\ &= 233 \equiv 23 \pmod{105}. \end{aligned}$$

It follows that 23 is the smallest positive integer that is a simultaneous solution. We conclude that 23 is the smallest positive integer that leaves a remainder of 2 when divided by 3, a remainder of 3 when divided by 5, and a remainder of 2 when divided by 7. ◀

Although the construction in Theorem 2 provides a general method for solving systems of linear congruences with pairwise relatively prime moduli, it can be easier to solve a system using a different method. Example 6 illustrates the use of a method known as **back substitution**.

EXAMPLE 6 Use the method of back substitution to find all integers x such that $x \equiv 1 \pmod{5}$, $x \equiv 2 \pmod{6}$, and $x \equiv 3 \pmod{7}$.

Solution: By Theorem 4 in Section 4.1, the first congruence can be rewritten as an equality, $x = 5t + 1$ where t is an integer. Substituting this expression for x into the second congruence tells us that

$$5t + 1 \equiv 2 \pmod{6},$$

which can be easily solved to show that $t \equiv 5 \pmod{6}$ (as the reader should verify). Using Theorem 4 in Section 4.1 again, we see that $t = 6u + 5$ where u is an integer. Substituting this expression for t back into the equation $x = 5t + 1$ tells us that $x = 5(6u + 5) + 1 = 30u + 26$. We insert this into the third equation to obtain

$$30u + 26 \equiv 3 \pmod 7.$$

Solving this congruence tells us that $u \equiv 6 \pmod 7$ (as the reader should verify). Hence, Theorem 4 in Section 4.1 tells us that $u = 7v + 6$ where v is an integer. Substituting this expression for u into the equation $x = 30u + 26$ tells us that $x = 30(7v + 6) + 26 = 210u + 206$. Translating this back into a congruence, we find the solution to the simultaneous congruences,

$$x \equiv 206 \pmod{210}. \qquad \blacktriangleleft$$

Computer Arithmetic with Large Integers

Suppose that $m_1, m_2, \ldots, m_n$ are pairwise relatively prime moduli and let m be their product. By the Chinese remainder theorem, we can show (see Exercise 20) that an integer a with $0 \le a < m$ can be uniquely represented by the n-tuple consisting of its remainders upon division by $m_i, i = 1, 2, \ldots, n$. That is, we can uniquely represent a by

$$(a \bmod m_1, a \bmod m_2, \ldots, a \bmod m_n).$$

EXAMPLE 7   What are the pairs used to represent the nonnegative integers less than 12 when they are represented by the ordered pair where the first component is the remainder of the integer upon division by 3 and the second component is the remainder of the integer upon division by 4?

Solution: We have the following representations, obtained by finding the remainder of each integer when it is divided by 3 and by 4:

$$
\begin{array}{lll}
0 = (0, 0) & 4 = (1, 0) & 8 = (2, 0) \\
1 = (1, 1) & 5 = (2, 1) & 9 = (0, 1) \\
2 = (2, 2) & 6 = (0, 2) & 10 = (1, 2) \\
3 = (0, 3) & 7 = (1, 3) & 11 = (2, 3).
\end{array}
$$

$\blacktriangleleft$

To perform arithmetic with large integers, we select moduli $m_1, m_2, \ldots, m_n$, where each m_i is an integer greater than 2, $\gcd(m_i, m_j) = 1$ whenever $i \neq j$, and $m = m_1 m_2 \cdots m_n$ is greater than the results of the arithmetic operations we want to carry out.

Once we have selected our moduli, we carry out arithmetic operations with large integers by performing componentwise operations on the n-tuples representing these integers using their remainders upon division by $m_i, i = 1, 2, \ldots, n$. Once we have computed the value of each component in the result, we recover its value by solving a system of n congruences modulo $m_i, i = 1, 2, \ldots, n$. This method of performing arithmetic with large integers has several valuable features. First, it can be used to perform arithmetic with integers larger than can ordinarily be carried out on a computer. Second, computations with respect to the different moduli can be done in parallel, speeding up the arithmetic.

EXAMPLE 8   Suppose that performing arithmetic with integers less than 100 on a certain processor is much quicker than doing arithmetic with larger integers. We can restrict almost all our computations to integers less than 100 if we represent integers using their remainders modulo pairwise relatively prime integers less than 100. For example, we can use the moduli of 99, 98, 97, and 95. (These integers are relatively prime pairwise, because no two have a common factor greater than 1.)

By the Chinese remainder theorem, every nonnegative integer less than $99 \cdot 98 \cdot 97 \cdot 95 = 89,403,930$ can be represented uniquely by its remainders when divided by these four moduli. For example, we represent 123,684 as (33, 8, 9, 89), because 123,684 **mod** 99 = 33; 123,684 **mod** 98 = 8; 123,684 **mod** 97 = 9; and 123,684 **mod** 95 = 89. Similarly, we represent 413,456 as (32, 92, 42, 16).

To find the sum of 123,684 and 413,456, we work with these 4-tuples instead of these two integers directly. We add the 4-tuples componentwise and reduce each component with respect to the appropriate modulus. This yields

$$(33, 8, 9, 89) + (32, 92, 42, 16)$$
$$= (65 \bmod 99, 100 \bmod 98, 51 \bmod 97, 105 \bmod 95)$$
$$= (65, 2, 51, 10).$$

To find the sum, that is, the integer represented by (65, 2, 51, 10), we need to solve the system of congruences

$$x \equiv 65 \pmod{99},$$
$$x \equiv 2 \pmod{98},$$
$$x \equiv 51 \pmod{97},$$
$$x \equiv 10 \pmod{95}.$$

It can be shown (see Exercise 35) that 537,140 is the unique nonnegative solution of this system less than 89,403,930. Consequently, 537,140 is the sum. Note that it is only when we have to recover the integer represented by (65, 2, 51, 10) that we have to do arithmetic with integers larger than 100. ◄

Particularly good choices for moduli for arithmetic with large integers are sets of integers of the form $2^k - 1$, where k is a positive integer, because it is easy to do binary arithmetic modulo such integers, and because it is easy to find sets of such integers that are pairwise relatively prime. [The second reason is a consequence of the fact that $\gcd(2^a - 1, 2^b - 1) = 2^{\gcd(a, b)} - 1$, as Exercise 27 in Section 4.3 shows.] Suppose, for instance, that we can do arithmetic with integers less than 2^{35} easily on our computer, but that working with larger integers requires special procedures. We can use pairwise relatively prime moduli less than 2^{35} to perform arithmetic with integers as large as their product. For example, as Exercise 38 in Section 4.3 shows, the integers $2^{35} - 1$, $2^{34} - 1$, $2^{33} - 1$, $2^{31} - 1$, $2^{29} - 1$, and $2^{23} - 1$ are pairwise relatively prime. Because the product of these six moduli exceeds 2^{184}, we can perform arithmetic with integers as large as 2^{184} (as long as the results do not exceed this number) by doing arithmetic modulo each of these six moduli, none of which exceeds 2^{35}.

Fermat's Little Theorem

The great French mathematician Pierre de Fermat made many important discoveries in number theory. One of the most useful of these states that p divides $a^{p-1} - 1$ whenever p is prime and a is an integer not divisible by p. Fermat announced this result in a letter to one of his correspondents. However, he did not include a proof in the letter, stating that he feared the proof would be too long. Although Fermat never published a proof of this fact, there is little doubt that he knew how to prove it, unlike the result known as Fermat's last theorem. The first published proof is credited to Leonhard Euler. We now state this theorem in terms of congruences.

THEOREM 3   **FERMAT'S LITTLE THEOREM**   If p is prime and a is an integer not divisible by p, then

$$a^{p-1} \equiv 1 \pmod{p}.$$

Furthermore, for every integer a we have

$$a^p \equiv a \pmod{p}.$$

Remark: Fermat's little theorem tells us that if $a \in \mathbf{Z}_p$, then $a^{p-1} = 1$ in $\mathbf{Z}_p$.

The proof of Theorem 3 is outlined in Exercise 19.

Fermat's little theorem is extremely useful in computing the remainders modulo p of large powers of integers, as Example 9 illustrates.

EXAMPLE 9   Find 7^{222} **mod** 11.

Solution: We can use Fermat's little theorem to evaluate 7^{222} **mod** 11 rather than using the fast modular exponentiation algorithm. By Fermat's little theorem we know that $7^{10} \equiv 1 \pmod{11}$, so $(7^{10})^k \equiv 1 \pmod{11}$ for every positive integer k. To take advantage of this last congruence, we divide the exponent 222 by 10, finding that $222 = 22 \cdot 10 + 2$. We now see that

$$7^{222} = 7^{22 \cdot 10 + 2} = (7^{10})^{22} 7^2 \equiv (1)^{22} \cdot 49 \equiv 5 \pmod{11}.$$

It follows that 7^{222} **mod** 11 = 5.   ◀

Example 9 illustrated how we can use Fermat's little theorem to compute a^n **mod** p, where p is prime and $p \nmid a$. First, we use the division algorithm to find the quotient q and remainder r when n is divided by $p - 1$, so that $n = q(p - 1) + r$ where $0 \leq r < p - 1$. It follows that $a^n = a^{q(p-1)+r} = (a^{p-1})^q a^r \equiv 1^q a^r \equiv a^r \pmod{p}$. Hence, to find a^n **mod** p, we only need to compute a^r **mod** p. We will take advantage of this simplification many times in our study of number theory.

Pseudoprimes

In Section 4.2 we showed that an integer n is prime when it is not divisible by any prime p with $p \leq \sqrt{n}$. Unfortunately, using this criterion to show that a given integer is prime is inefficient. It requires that we find all primes not exceeding $\sqrt{n}$ and that we carry out trial division by each such prime to see whether it divides n.

Are there more efficient ways to determine whether an integer is prime? According to some sources, ancient Chinese mathematicians believed that n was an odd prime if and only if

$$2^{n-1} \equiv 1 \pmod{n}.$$

If this were true, it would provide an efficient primality test. Why did they believe this congruence could be used to determine whether an integer $n > 2$ is prime? First, they observed that the congruence holds whenever n is an odd prime. For example, 5 is prime and

$$2^{5-1} = 2^4 = 16 \equiv 1 \pmod{5}.$$

By Fermat's little theorem, we know that this observation was correct, that is, $2^{n-1} \equiv 1 \pmod{n}$ whenever n is an odd prime. Second, they never found a composite integer n for which the congruence holds. However, the ancient Chinese were only partially correct. They were correct in thinking that the congruence holds whenever n is prime, but they were incorrect in concluding that n is necessarily prime if the congruence holds.

Unfortunately, there are composite integers n such that $2^{n-1} \equiv 1 \pmod{n}$. Such integers are called **pseudoprimes** to the base 2.

EXAMPLE 10   The integer 341 is a pseudoprime to the base 2 because it is composite ($341 = 11 \cdot 31$) and as Exercise 25 shows

$$2^{340} \equiv 1 \pmod{341}.$$ ◄

We can use an integer other than 2 as the base when we study pseudoprimes.

DEFINITION 1   Let b be a positive integer. If n is a composite positive integer, and $b^{n-1} \equiv 1 \pmod{n}$, then n is called a *pseudoprime to the base b*.

Given a positive integer n, determining whether $2^{n-1} \equiv 1 \pmod{n}$ is a useful test that provides some evidence concerning whether n is prime. In particular, if n satisfies this congruence, then it is either prime or a pseudoprime to the base 2; if n does not satisfy this congruence, it is composite. We can perform similar tests using bases b other than 2 and obtain more evidence as to whether n is prime. If n passes all such tests, it is either prime or a pseudoprime to all the bases b we have chosen. Furthermore, among the positive integers not exceeding x, where x is a positive real number, compared to primes there are relatively few pseudoprimes to the base b, where b is a positive integer. For example, among the positive integers less than 10^{10} there are 455,052,512 primes, but only 14,884 pseudoprimes to the base 2. Unfortunately, we cannot distinguish between primes and pseudoprimes just by choosing sufficiently many bases, because there are composite integers n that pass all tests with bases b such that $\gcd(b, n) = 1$. This leads to Definition 2.

DEFINITION 2   A composite integer n that satisfies the congruence $b^{n-1} \equiv 1 \pmod{n}$ for all positive integers b with $\gcd(b, n) = 1$ is called a *Carmichael number*. (These numbers are named after Robert Carmichael, who studied them in the early twentieth century.)

EXAMPLE 11   The integer 561 is a Carmichael number. To see this, first note that 561 is composite because $561 = 3 \cdot 11 \cdot 17$. Next, note that if $\gcd(b, 561) = 1$, then $\gcd(b, 3) = \gcd(b, 11) = \gcd(b, 17) = 1$.

PIERRE DE FERMAT (1601–1665)   Pierre de Fermat, one of the most important mathematicians of the seventeenth century, was a lawyer by profession. He is the most famous amateur mathematician in history. Fermat published little of his mathematical discoveries. It is through his correspondence with other mathematicians that we know of his work. Fermat was one of the inventors of analytic geometry and developed some of the fundamental ideas of calculus. Fermat, along with Pascal, gave probability theory a mathematical basis. Fermat formulated what was the most famous unsolved problem in mathematics. He asserted that the equation $x^n + y^n = z^n$ has no nontrivial positive integer solutions when n is an integer greater than 2. For more than 300 years, no proof (or counterexample) was found. In his copy of the works of the ancient Greek mathematician Diophantus, Fermat wrote that he had a proof but that it would not fit in the margin. Because the first proof, found by Andrew Wiles in 1994, relies on sophisticated, modern mathematics, most people think that Fermat thought he had a proof, but that the proof was incorrect. However, he may have been tempting others to look for a proof, not being able to find one himself.

Using Fermat's little theorem we find that

$$b^2 \equiv 1 \pmod 3, b^{10} \equiv 1 \pmod{11}, \text{ and } b^{16} \equiv 1 \pmod{17}.$$

It follows that

$$b^{560} = (b^2)^{280} \equiv 1 \pmod 3,$$
$$b^{560} = (b^{10})^{56} \equiv 1 \pmod{11},$$
$$b^{560} = (b^{16})^{35} \equiv 1 \pmod{17}.$$

By Exercise 21, it follows that $b^{560} \equiv 1 \pmod{561}$ for all positive integers b with $\gcd(b, 561) = 1$. Hence 561 is a Carmichael number. ◀

Although there are infinitely many Carmichael numbers, more delicate tests, described in the exercise set, can be devised that can be used as the basis for efficient probabilistic primality tests. Such tests can be used to quickly show that it is almost certainly the case that a given integer is prime. More precisely, if an integer is not prime, then the probability that it passes a series of tests is close to 0. We will describe such a test in Chapter 7 and discuss the notions from probability theory that this test relies on. These probabilistic primality tests can be used, and are used, to find large primes extremely rapidly on computers.

Primitive Roots and Discrete Logarithms

In the set of positive real numbers, if $b > 1$, and $x = b^y$, we say that y is the logarithm of x to the base b. Here, we will show that we can also define the concept of logarithms modulo p of positive integers where p is a prime. Before we do so, we need a definition.

DEFINITION 3

A *primitive root* modulo a prime p is an integer r in $\mathbf{Z}_p$ such that every nonzero element of $\mathbf{Z}_p$ is a power of r.

EXAMPLE 12

Determine whether 2 and 3 are primitive roots modulo 11.

Solution: When we compute the powers of 2 in $\mathbf{Z}_{11}$, we obtain $2^1 = 2, 2^2 = 4, 2^3 = 8, 2^4 = 5, 2^5 = 10, 2^6 = 9, 2^7 = 7, 2^8 = 3, 2^9 = 6, 2^{10} = 1$. Because every element of $\mathbf{Z}_{11}$ is a power of 2, 2 is a primitive root of 11.

When we compute the powers of 3 modulo 11, we obtain $3^1 = 3, 3^2 = 9, 3^3 = 5, 3^4 = 4, 3^5 = 1$. We note that this pattern repeats when we compute higher powers of 3. Because not all elements of $\mathbf{Z}_{11}$ are powers of 3, we conclude that 3 is not a primitive root of 11. ◀

Links

ROBERT DANIEL CARMICHAEL (1879–1967) Robert Daniel Carmichael was born in Alabama. He received his undergraduate degree from Lineville College in 1898 and his Ph.D. in 1911 from Princeton. Carmichael held positions at Indiana University from 1911 until 1915 and at the University of Illinois from 1915 until 1947. Carmichael was an active researcher in a wide variety of areas, including number theory, real analysis, differential equations, mathematical physics, and group theory. His Ph.D. thesis, written under the direction of G. D. Birkhoff, is considered the first significant American contribution to the subject of differential equations.

An important fact in number theory is that there is a primitive root modulo p for every prime p. We refer the reader to [Ro10] for a proof of this fact. Suppose that p is prime and r is a primitive root modulo p. If a is an integer between 1 and $p - 1$, that is, an element of $\mathbf{Z}_p$, we know that there is an unique exponent e such that $r^e = a$ in $\mathbf{Z}_p$, that is, $r^e \bmod p = a$.

DEFINITION 4   Suppose that p is a prime, r is a primitive root modulo p, and a is an integer between 1 and $p - 1$ inclusive. If $r^e \bmod p = a$ and $0 \le e \le p - 1$, we say that e is the *discrete logarithm* of a modulo p to the base r and we write $\log_r a = e$ (where the prime p is understood).

EXAMPLE 13   Find the discrete logarithms of 3 and 5 modulo 11 to the base 2.

Solution: When we computed the powers of 2 modulo 11 in Example 12, we found that $2^8 = 3$ and $2^4 = 5$ in $\mathbf{Z}_{11}$. Hence, the discrete logarithms of 3 and 5 modulo 11 to the base 2 are 8 and 4, respectively. (These are the powers of 2 that equal 3 and 5, respectively, in $\mathbf{Z}_{11}$.) We write $\log_2 3 = 8$ and $\log_2 5 = 4$ (where the modulus 11 is understood and not explicitly noted in the notation).   ◀

The discrete logarithm problem is hard!

The **discrete logarithm problem** takes as input a prime p, a primitive root r modulo p, and a positive integer $a \in \mathbf{Z}_p$; its output is the discrete logarithm of a modulo p to the base r. Although this problem might seem not to be that difficult, it turns out that no polynomial time algorithm is known for solving it. The difficulty of this problem plays an important role in cryptography, as we will see in Section 4.6

Exercises

1. Show that 15 is an inverse of 7 modulo 26.

2. Show that 937 is an inverse of 13 modulo 2436.

3. By inspection (as discussed prior to Example 1), find an inverse of 4 modulo 9.

4. Find an inverse of a modulo m for each of these pairs of relatively prime integers using the method followed in Example 2.
   a) $a = 4, m = 9$
   b) $a = 19, m = 141$
   c) $a = 55, m = 89$
   d) $a = 89, m = 232$

*5. Show that if a and m are relatively prime positive integers, then the inverse of a modulo m is unique modulo m. [*Hint:* Assume that there are two solutions b and c of the congruence $ax \equiv 1 \pmod{m}$. Use Theorem 7 of Section 4.3 to show that $b \equiv c \pmod{m}$.]

6. Show that an inverse of a modulo m, where a is an integer and $m > 2$ is a positive integer, does not exist if $\gcd(a, m) > 1$.

7. Solve the congruence $4x \equiv 5 \pmod{9}$ using the inverse of 4 modulo 9 found in part (a) of Exercise 4.

8. Solve each of these congruences using the modular inverses found in parts (b), (c), and (d) of Exercise 4.
   a) $19x \equiv 4 \pmod{141}$
   b) $55x \equiv 34 \pmod{89}$
   c) $89x \equiv 2 \pmod{232}$

9. Find the solutions of the congruence $15x^2 + 19x \equiv 5 \pmod{11}$. [*Hint:* Show the congruence is equivalent to the congruence $15x^2 + 19x + 6 \equiv 0 \pmod{11}$. Factor the left-hand side of the congruence; show that a solution of the quadratic congruence is a solution of one of the two different linear congruences.]

10. a) Show that the positive integers less than 11, except 1 and 10, can be split into pairs of integers such that each pair consists of integers that are inverses of each other modulo 11.
    b) Use part (a) to show that $10! \equiv -1 \pmod{11}$.

*11. Show that if m is an integer greater than 1 and $ac \equiv bc \pmod{m}$, then $a \equiv b \pmod{m/\gcd(c, m)}$.

12. Show that if p is prime, the only solutions of $x^2 \equiv 1 \pmod{p}$ are integers x such that $x \equiv 1 \pmod{p}$ or $x \equiv -1 \pmod{p}$.

*13. This exercise outlines a proof of Fermat's little theorem.
    a) Suppose that a is not divisible by the prime p. Show that no two of the integers $1 \cdot a, 2 \cdot a, \ldots, (p - 1)a$ are congruent modulo p.
    b) Conclude from part (a) that the product of $1, 2, \ldots, p - 1$ is congruent modulo p to the product of $a, 2a, \ldots, (p - 1)a$. Use this to show that
    $$(p - 1)! \equiv a^{p-1}(p - 1)! \pmod{p}.$$

c) Use Theorem 7 of Section 4.3 to show from part (b) that $a^{p-1} \equiv 1 \pmod{p}$ if $p \nmid a$. [*Hint:* Use Lemma 3 of Section 4.3 to show that p does not divide $(p-1)!$ and then use Theorem 7 of Section 4.3. Alternatively, use Wilson's theorem from Exercise 14(b).]

d) Use part (c) to show that $a^{p} \equiv a \pmod{p}$ for all integers a.

*14. a) Generalize the result in part (a) of Exercise 10; that is, show that if p is a prime, the positive integers less than p, except 1 and $p-1$, can be split into $(p-3)/2$ pairs of integers such that each pair consists of integers that are inverses of each other. [*Hint:* Use the result of Exercise 12.]

b) From part (a) conclude that $(p-1)! \equiv -1 \pmod{p}$ whenever p is prime. This result is known as **Wilson's theorem**.

c) What can we conclude if n is a positive integer such that $(n-1)! \not\equiv -1 \pmod{n}$?

15. Use the construction in the proof of the Chinese remainder theorem to find all solutions to the system of congruences $x \equiv 1 \pmod{2}$, $x \equiv 2 \pmod{3}$, $x \equiv 3 \pmod{5}$, and $x \equiv 4 \pmod{11}$.

16. Solve the system of congruence $x \equiv 3 \pmod{6}$ and $x \equiv 4 \pmod{7}$ using the method of back substitution.

17. Solve the system of congruences in Exercise 15 using the method of back substitution.

*18. Find all solutions, if any, to the system of congruences $x \equiv 5 \pmod{6}$, $x \equiv 3 \pmod{10}$, and $x \equiv 8 \pmod{15}$.

*19. Find all solutions, if any, to the system of congruences $x \equiv 7 \pmod{9}$, $x \equiv 4 \pmod{12}$, and $x \equiv 16 \pmod{21}$.

20. Use the Chinese remainder theorem to show that an integer a, with $0 \le a < m = m_1 m_2 \cdots m_n$, where the positive integers $m_1, m_2, \ldots, m_n$ are pairwise relatively prime, can be represented uniquely by the n-tuple $(a \bmod m_1, a \bmod m_2, \ldots, a \bmod m_n)$.

*21. Let $m_1, m_2, \ldots, m_n$ be pairwise relatively prime integers greater than or equal to 2. Show that if $a \equiv b \pmod{m_i}$ for $i = 1, 2, \ldots, n$, then $a \equiv b \pmod{m}$, where $m = m_1 m_2 \cdots m_n$. (This result will be used in Exercise 22 to prove the Chinese remainder theorem. Consequently, do not use the Chinese remainder theorem to prove it.)

*22. Complete the proof of the Chinese remainder theorem by showing that the simultaneous solution of a system of linear congruences modulo pairwise relatively prime moduli is unique modulo the product of these moduli. [*Hint:* Assume that x and y are two simultaneous solutions. Show that $m_i \mid x - y$ for all i. Using Exercise 21, conclude that $m = m_1 m_2 \cdots m_n \mid x - y$.]

23. Which integers leave a remainder of 1 when divided by 2 and also leave a remainder of 1 when divided by 3?

24. Use Fermat's little theorem to find $7^{121} \bmod 13$.

25. a) Show that $2^{340} \equiv 1 \pmod{11}$ by Fermat's little theorem and noting that $2^{340} = (2^{10})^{34}$.

b) Show that $2^{340} \equiv 1 \pmod{31}$ using the fact that $2^{340} = (2^5)^{68} = 32^{68}$.

c) Conclude from parts (a) and (b) that $2^{340} \equiv 1 \pmod{341}$.

26. a) Use Fermat's little theorem to compute $5^{2003} \bmod 7$, $5^{2003} \bmod 11$, and $5^{2003} \bmod 13$.

b) Use your results from part (a) and the Chinese remainder theorem to find $5^{2003} \bmod 1001$. (Note that $1001 = 7 \cdot 11 \cdot 13$.)

27. Show that if p is an odd prime, then every divisor of the Mersenne number $2^p - 1$ is of the form $2kp + 1$, where k is a nonnegative integer. [*Hint:* Use Fermat's little theorem and Exercise 27 of Section 4.3.]

28. Use Exercise 27 to determine whether $M_{13} = 2^{13} - 1 = 8191$ and $M_{23} = 2^{23} - 1 = 8,388,607$ are prime.

29. Use Exercise 27 to determine whether $M_{11} = 2^{11} - 1 = 2047$ and $M_{17} = 2^{17} - 1 = 131,071$ are prime.

Let n be a positive integer and let $n - 1 = 2^s t$, where s is a nonnegative integer and t is an odd positive integer. We say that n passes **Miller's test for the base b** if either $b^t \equiv 1 \pmod{n}$ or $b^{2^j t} \equiv -1 \pmod{n}$ for some j with $0 \le j \le s - 1$. It can be shown (see [Ro10]) that a composite integer n passes Miller's test for fewer than $n/4$ bases b with $1 < b < n$. A composite positive integer n that passes Miller's test to the base b is called a **strong pseudoprime to the base b**.

*30. Show that if n is prime and b is a positive integer with $n \nmid b$, then n passes Miller's test to the base b.

31. Show that 2047 is a strong pseudoprime to the base 2 by showing that it passes Miller's test to the base 2, but is composite.

32. Show that 2821 is a Carmichael number.

33. Express each nonnegative integer a less than 15 as a pair $(a \bmod 3, a \bmod 5)$.

34. Explain how to use the pairs found in Exercise 33 to add 4 and 7.

35. Solve the system of congruences that arises in Example 8.

36. Show that 2 is a primitive root of 19.

37. Find the discrete logarithms of 5 and 6 to the base 2 modulo 19.

If m is a positive integer, the integer a is a **quadratic residue** of m if $\gcd(a, m) = 1$ and the congruence $x^2 \equiv a \pmod{m}$ has a solution. In other words, a quadratic residue of m is an integer relatively prime to m that is a perfect square modulo m. If a is not a quadratic residue of m and $\gcd(a, m) = 1$, we say that it is a **quadratic nonresidue** of m. For example, 2 is a quadratic residue of 7 because $\gcd(2, 7) = 1$ and $3^2 \equiv 2 \pmod{7}$ and 3 is a quadratic nonresidue of 7 because $\gcd(3, 7) = 1$ and $x^2 \equiv 3 \pmod{7}$ has no solution.

38. Which integers are quadratic residues of 11?

If p is an odd prime and a is an integer not divisible by p, the **Legendre symbol** $\left(\dfrac{a}{p}\right)$ is defined to be 1 if a is a quadratic residue of p and -1 otherwise.

39. Show that if p is an odd prime and a and b are integers with $a \equiv b \pmod{p}$, then

$$\left(\frac{a}{p}\right) = \left(\frac{b}{p}\right).$$

40. Prove **Euler's criterion**, which states that if p is an odd prime and a is a positive integer not divisible by p, then

$$\left(\frac{a}{p}\right) \equiv a^{(p-1)/2} \pmod{p}.$$

[*Hint:* If a is a quadratic residue modulo p, apply Fermat's little theorem; otherwise, apply Wilson's theorem, given in Exercise 14(b).]

41. Use Exercise 40 to show that if p is an odd prime and a and b are integers not divisible by p, then

$$\left(\frac{ab}{p}\right) = \left(\frac{a}{p}\right)\left(\frac{b}{p}\right).$$

42. Show that if p is an odd prime, then -1 is a quadratic residue of p if $p \equiv 1 \pmod 4$, and -1 is not a quadratic residue of p if $p \equiv 3 \pmod 4$. [*Hint:* Use Exercise 40.]

43. Find all solutions of the congruence $x^2 \equiv 29 \pmod{35}$. [*Hint:* Find the solutions of this congruence modulo 5 and modulo 7, and then use the Chinese remainder theorem.]

44. Find all solutions of the congruence $x^2 \equiv 16 \pmod{105}$. [*Hint:* Find the solutions of this congruence modulo 3, modulo 5, and modulo 7, and then use the Chinese remainder theorem.]

4.5   Applications of Congruences

Congruences have many applications to discrete mathematics, computer science, and many other disciplines. We will introduce three applications in this section: the use of congruences to assign memory locations to computer files, the generation of pseudorandom numbers, and check digits.

Suppose that a customer identification number is ten digits long. To retrieve customer files quickly, we do not want to assign a memory location to a customer record using the ten-digit identification number. Instead, we want to use a smaller integer associated to the identification number. This can be done using what is known as a hashing function. In this section we will show how we can use modular arithmetic to do hashing.

Constructing sequences of random numbers is important for randomized algorithms, for simulations, and for many other purposes. Constructing a sequence of truly random numbers is extremely difficult, or perhaps impossible, because any method for generating what are supposed to be random numbers may generate numbers with hidden patterns. As a consequence, methods have been developed for finding sequences of numbers that have many desirable properties of random numbers, and which can be used for various purposes in place of random numbers. In this section we will show how to use congruences to generate sequences of pseudorandom numbers. The advantage is that the pseudorandom numbers so generated are constructed quickly; the disadvantage is that they have too much predictability to be used for many tasks.

Congruences also can be used to produce check digits for identification numbers of various kinds, such as code numbers used to identify retail products, numbers used to identify books, airline ticket numbers, and so on. We will explain how to construct check digits using congruences for a variety of types of identification numbers. We will show that these check digits can be used to detect certain kinds of common errors made when identification numbers are printed.

Hashing Functions

Links

The central computer at an insurance company maintains records for each of its customers. How can memory locations be assigned so that customer records can be retrieved quickly? The solution to this problem is to use a suitably chosen **hashing function**. Records are identified using a **key**, which uniquely identifies each customer's records. For instance, customer records are often identified using the Social Security number of the customer as the key. A hashing function h assigns memory location $h(k)$ to the record that has k as its key.

In practice, many different hashing functions are used. One of the most common is the function

$$h(k) = k \bmod m$$

where m is the number of available memory locations.

Hashing functions should be easily evaluated so that files can be quickly located. The hashing function $h(k) = k \bmod m$ meets this requirement; to find $h(k)$, we need only compute the remainder when k is divided by m. Furthermore, the hashing function should be onto, so that all memory locations are possible. The function $h(k) = k \bmod m$ also satisfies this property.

EXAMPLE 1   Find the memory locations assigned by the hashing function $h(k) = k \bmod 111$ to the records of customers with Social Security numbers 064212848 and 037149212.

Solution: The record of the customer with Social Security number 064212848 is assigned to memory location 14, because

$$h(064212848) = 064212848 \bmod 111 = 14.$$

Similarly, because

$$h(037149212) = 037149212 \bmod 111 = 65,$$

the record of the customer with Social Security number 037149212 is assigned to memory location 65.   ◀

Because a hashing function is not one-to-one (because there are more possible keys than memory locations), more than one file may be assigned to a memory location. When this happens, we say that a **collision** occurs. One way to resolve a collision is to assign the first free location following the occupied memory location assigned by the hashing function.

EXAMPLE 2   After making the assignments of records to memory locations in Example 1, assign a memory location to the record of the customer with Social Security number 107405723.

Solution: First note that the hashing function $h(k) = k \bmod 111$ maps the Social Security number 107405723 to location 14, because

$$h(107405723) = 107405723 \bmod 111 = 14.$$

However, this location is already occupied (by the file of the customer with Social Security number 064212848). But, because memory location 15, the first location following memory location 14, is free, we assign the record of the customer with Social Security number 107405723 to this location.   ◀

In Example 1 we used a **linear probing function**, namely $h(k, i) = h(k) + i \bmod m$, to look for the first free memory location, where i runs from 0 to $m - 1$. There are many other ways to resolve collisions that are discussed in the references on hashing functions given at the end of the book.

Pseudorandom Numbers

Randomly chosen numbers are often needed for computer simulations. Different methods have been devised for generating numbers that have properties of randomly chosen numbers. Because numbers generated by systematic methods are not truly random, they are called **pseudorandom numbers**.

The most commonly used procedure for generating pseudorandom numbers is the **linear congruential method**. We choose four integers: the **modulus** m, **multiplier** a, **increment** c, and **seed** x_0, with $2 \le a < m$, $0 \le c < m$, and $0 \le x_0 < m$. We generate a sequence of pseudorandom numbers $\{x_n\}$, with $0 \le x_n < m$ for all n, by successively using the recursively defined function

$$x_{n+1} = (ax_n + c) \bmod m.$$

(This is an example of a recursive definition, discussed in Section 5.3. In that section we will show that such sequences are well defined.)

Many computer experiments require the generation of pseudorandom numbers between 0 and 1. To generate such numbers, we divide numbers generated with a linear congruential generator by the modulus: that is, we use the numbers x_n/m.

EXAMPLE 3   Find the sequence of pseudorandom numbers generated by the linear congruential method with modulus $m = 9$, multiplier $a = 7$, increment $c = 4$, and seed $x_0 = 3$.

Solution: We compute the terms of this sequence by successively using the recursively defined function $x_{n+1} = (7x_n + 4) \bmod 9$, beginning by inserting the seed $x_0 = 3$ to find x_1. We find that

$$x_1 = 7x_0 + 4 \bmod 9 = 7 \cdot 3 + 4 \bmod 9 = 25 \bmod 9 = 7,$$
$$x_2 = 7x_1 + 4 \bmod 9 = 7 \cdot 7 + 4 \bmod 9 = 53 \bmod 9 = 8,$$
$$x_3 = 7x_2 + 4 \bmod 9 = 7 \cdot 8 + 4 \bmod 9 = 60 \bmod 9 = 6,$$
$$x_4 = 7x_3 + 4 \bmod 9 = 7 \cdot 6 + 4 \bmod 9 = 46 \bmod 9 = 1,$$
$$x_5 = 7x_4 + 4 \bmod 9 = 7 \cdot 1 + 4 \bmod 9 = 11 \bmod 9 = 2,$$
$$x_6 = 7x_5 + 4 \bmod 9 = 7 \cdot 2 + 4 \bmod 9 = 18 \bmod 9 = 0,$$
$$x_7 = 7x_6 + 4 \bmod 9 = 7 \cdot 0 + 4 \bmod 9 = 4 \bmod 9 = 4,$$
$$x_8 = 7x_7 + 4 \bmod 9 = 7 \cdot 4 + 4 \bmod 9 = 32 \bmod 9 = 5,$$
$$x_9 = 7x_8 + 4 \bmod 9 = 7 \cdot 5 + 4 \bmod 9 = 39 \bmod 9 = 3.$$

Because $x_9 = x_0$ and because each term depends only on the previous term, we see that the sequence

$$3, 7, 8, 6, 1, 2, 0, 4, 5, 3, 7, 8, 6, 1, 2, 0, 4, 5, 3, \dots$$

is generated. This sequence contains nine different numbers before repeating.   ◄

Most computers do use linear congruential generators to generate pseudorandom numbers. Often, a linear congruential generator with increment $c = 0$ is used. Such a generator is called a **pure multiplicative generator**. For example, the pure multiplicative generator with modulus $2^{31} - 1$ and multiplier $7^5 = 16,807$ is widely used. With these values, it can be shown that $2^{31} - 2$ numbers are generated before repetition begins.

Pseudorandom numbers generated by linear congruential generators have long been used for many tasks. Unfortunately, it has been shown that sequences of pseudorandom numbers generated in this way do not share some important statistical properties that true random numbers have. Because of this, it is not advisable to use them for some tasks, such as large simulations. For such sensitive tasks, other methods are used to produce sequences of pseudorandom numbers, either using some sort of algorithm or sampling numbers arising from a random physical phenomenon. For more details on pseudorandom number, see [Kn97] and [Re10].

Check Digits

Congruences are used to check for errors in digit strings. A common technique for detecting errors in such strings is to add an extra digit at the end of the string. This final digit, or check digit, is calculated using a particular function. Then, to determine whether a digit string is correct, a check is made to see whether this final digit has the correct value. We begin with an application of this idea for checking the correctness of bit strings.

EXAMPLE 4   **Parity Check Bits**   Digital information is represented by bit string, split into blocks of a specified size. Before each block is stored or transmitted, an extra bit, called a **parity check bit**, can be appended to each block. The parity check bit x_{n+1} for the bit string $x_1 x_2 \ldots x_n$ is defined by

$$x_{n+1} = x_1 + x_2 + \cdots + x_n \textbf{ mod } 2.$$

It follows that x_{n+1} is 0 if there are an even number of 1 bits in the block of n bits and it is 1 if there are an odd number of 1 bits in the block of n bits. When we examine a string that includes a parity check bit, we know that there is an error in it if the parity check bit is wrong. However, when the parity check bit is correct, there still may be an error. A parity check can detect an odd number of errors in the previous bits, but not an even number of errors. (See Exercise 14.)

Suppose we receive in a transmission the bit strings 01100101 and 11010110, each ending with a parity check bit. Should we accept these bit strings as correct?

Solution: Before accepting these strings as correct, we examine their parity check bits. The parity check bit of the first string is 1. Because $0 + 1 + 1 + 0 + 0 + 1 + 0 \equiv 1 \pmod 2$, the parity check bit is correct. The parity check bit of the second string is 0. We find that $1 + 1 + 0 + 1 + 0 + 1 + 1 \equiv 1 \pmod 2$, so the parity check is incorrect. We conclude that the first string may have been transmitted correctly and we know for certain that the second string was transmitted incorrectly. We accept the first string as correct (even though it still may contain an even number of errors), but we reject the second string.   ◀

Check bits computed using congruences are used extensively to verify the correctness of various kinds of identification numbers. Examples 5 and 6 show how check bits are computed for codes that identify products (Universal Product Codes) and books (International Standard Book Numbers). The preambles to Exercises 13, 19, and 21 introduce the use of congruences to find and use check digits in money order numbers, airline ticket numbers, and identification numbers for periodicals, respectively. Note that congruences are also used to compute check digits for bank account numbers, drivers license numbers, credit card numbers, and many other types of identification numbers.

EXAMPLE 5   **UPCs**   Retail products are identified by their **Universal Product Codes (UPCs)**. The most common form of a UPC has 12 decimal digits: the first digit identifies the product category, the next five digits identify the manufacturer, the following five identify the particular product, and the last digit is a check digit. The check digit is determined by the congruence

$$3x_1 + x_2 + 3x_3 + x_4 + 3x_5 + x_6 + 3x_7 + x_8 + 3x_9 + x_{10} + 3x_{11} + x_{12} \equiv 0 \pmod{10}.$$

Answer these questions:

(a) Suppose that the first 11 digits of a UPC are 79357343104. What is the check digit?

(b) Is 041331021641 a valid UPC?

Solution: (a) We insert the digits of 79357343104 into the congruence for UPC check digits. This gives $3 \cdot 7 + 9 + 3 \cdot 3 + 5 + 3 \cdot 7 + 3 + 3 \cdot 4 + 3 + 3 \cdot 1 + 0 + 3 \cdot 4 + x_{12} \equiv 0 \pmod{10}$. Simplifying, we have $21 + 9 + 9 + 5 + 21 + 3 + 12 + 3 + 3 + 0 + 12 + x_{12} \equiv 0 \pmod{10}$. Hence, $98 + x_{12} \equiv 0 \pmod{10}$. It follows that $x_{12} \equiv 2 \pmod{10}$, so the check digit is 2.

(b) To check whether 041331021641 is valid, we insert the digits into the congruence these digits must satisfy. This gives $3 \cdot 0 + 4 + 3 \cdot 1 + 3 + 3 \cdot 3 + 1 + 3 \cdot 0 + 2 + 3 \cdot 1 + 6 + 3 \cdot 4 + 1 \equiv 0 + 4 + 3 + 3 + 9 + 1 + 0 + 2 + 3 + 6 + 12 + 1 \equiv 4 \not\equiv 0 \pmod{10}$. Hence, 041331021641 is not a valid UPC. ◀

EXAMPLE 6

Remember that the check digit of an ISBN-10 can be an X!

ISBNs All books are identified by an **International Standard Book Number (ISBN-10)**, a 10-digit code $x_1 x_2 \ldots x_{10}$, assigned by the publisher. (Recently, a 13-digit code known as ISBN-13 was introduced to identify a larger number of published works; see the preamble to Exercise 31 in the Supplementary Exercises.) An ISBN-10 consists of blocks identifying the language, the publisher, the number assigned to the book by its publishing company, and finally, a check digit that is either a digit or the letter X (used to represent 10). This check digit is selected so that

$$x_{10} \equiv \sum_{i=1}^{9} i x_i \pmod{11},$$

or equivalently, so that

$$\sum_{i=1}^{10} i x_i \equiv 0 \pmod{11}.$$

Answer these questions about ISBN-10s:

(a) The first nine digits of the ISBN-10 of the sixth edition of this book are 007288008. What is the check digit?

(b) Is 084930149X a valid ISBN-10?

Solution: (a) The check digit is determined by the congruence $\sum_{i=1}^{10} i x_i \equiv 0 \pmod{11}$. Inserting the digits 007288008 gives $x_{10} \equiv 1 \cdot 0 + 2 \cdot 0 + 3 \cdot 7 + 4 \cdot 2 + 5 \cdot 8 + 6 \cdot 8 + 7 \cdot 0 + 8 \cdot 0 + 9 \cdot 8 \pmod{11}$. This means that $x_{10} \equiv 0 + 0 + 21 + 8 + 40 + 48 + 0 + 0 + 72 \pmod{11}$, so $x_{10} \equiv 189 \equiv 2 \pmod{11}$. Hence, $x_{10} = 2$.

(b) To see whether 084930149X is a valid ISBN-10, we see if $\sum_{i=1}^{10} i x_i \equiv 0 \pmod{11}$. We see that $1 \cdot 0 + 2 \cdot 8 + 3 \cdot 4 + 4 \cdot 9 + 5 \cdot 3 + 6 \cdot 0 + 7 \cdot 1 + 8 \cdot 4 + 9 \cdot 9 + 10 \cdot 10 = 0 + 16 + 12 + 36 + 15 + 0 + 7 + 32 + 81 + 100 = 299 \equiv 2 \not\equiv 0 \pmod{11}$. Hence, 084930149X is not a valid ISBN-10. ◀

Publishers sometimes do not calculate ISBNs correctly for their books, as was done for an earlier edition of this text.

Several kinds of errors often arise in identification numbers. A **single error**, an error in one digit of an identification number, is perhaps the most common type of error. Another common kind of error is a **transposition error**, which occurs when two digits are accidentally interchanged. For each type of identification number, including a check digit, we would like to be able to detect these common types of errors, as well as other types of errors. We will investigate whether the check digit for ISBNs can detect single errors and transposition errors. Whether check digits for UPCs can detect these kinds of errors is left as Exercise 18.

Suppose that $x_1 x_2 \dots x_{10}$ is a valid ISBN (so that $\sum_{i=1}^{10} x_i \equiv 0 \pmod{10}$). We will show that we can detect a single error and a transposition of two digits (where we include the possibility that one of the two digits is the check digit X, representing 10). Suppose that this ISBN has been printed with a single error as $y_1 y_2 \dots y_{10}$. If there is a single error, then, for some integer j, $y_i = x_i$ for $i \neq j$ and $y_j = x_j + a$ where $-10 \leq a \leq 10$ and $a \neq 0$. Note that $a = y_j - x_j$ is the error in the jth place. It then follows that

$$\sum_{i=1}^{10} i y_i = \left(\sum_{i=1}^{10} i x_i\right) + ja \equiv ja \not\equiv 0 \pmod{11}.$$

These last two congruences hold because $\sum_{i=1}^{10} x_i \equiv 0 \pmod{10}$ and $11 \nmid ja$, because $11 \nmid j$ and $11 \nmid a$. We conclude that $y_1 y_2 \dots y_{10}$ is not a valid ISBN. So, we have detected the single error.

Now suppose that two unequal digits have been transposed. It follows that there are distinct integers j and k such that $y_j = x_k$ and $y_k = x_j$, and $y_i = x_i$ for $i \neq j$ and $i \neq k$. Hence,

$$\sum_{i=1}^{10} i y_i = \left(\sum_{i=1}^{10} i x_i\right) + (j x_k - j x_j) + (k x_j - k x_k) \equiv (j - k)(x_k - x_j) \not\equiv 0 \pmod{11},$$

because $\sum_{i=1}^{10} x_i \equiv 0 \pmod{10}$ and $11 \nmid (j - k)$ and $11 \nmid (x_k - x_j)$. We see that $y_1 y_2 \dots y_{10}$ is not a valid ISBN. Thus, we can detect the interchange of two unequal digits.

Exercises

1. Which memory locations are assigned by the hashing function $h(k) = k \bmod 97$ to the records of insurance company customers with these Social Security numbers?

a) 034567981   **b)** 183211232
c) 220195744   **d)** 987255335

2. A parking lot has 31 visitor spaces, numbered from 0 to 30. Visitors are assigned parking spaces using the hashing function $h(k) = k \bmod 31$, where k is the number formed from the first three digits on a visitor's license plate.

a) Which spaces are assigned by the hashing function to cars that have these first three digits on their license plates: 317, 918, 007, 100, 111, 310?

b) Describe a procedure visitors should follow to find a free parking space, when the space they are assigned is occupied.

Another way to resolve collisions in hashing is to use *double hashing*. We use an initial hashing function $h(k) = k \bmod p$ where p is prime. We also use a second hashing function $g(k) = (k + 1) \bmod (p - 2)$. When a collision occurs, we use a *probing sequence* $h(k, i) = (h(k) + i \cdot g(k)) \bmod p$.

3. What sequence of pseudorandom numbers is generated using the linear congruential generator $x_{n+1} = (3x_n + 2) \bmod 13$ with seed $x_0 = 1$?

4. Use the double hashing procedure we have described with $p = 4969$ to assign memory locations to files for employees with social security numbers $k_1 = 132489971$, $k_2 = 509496993$, $k_3 = 546332190$, $k_4 = 034367980$,

$k_5 = 047900151$, $k_6 = 329938157$, $k_7 = 212228844$, $k_8 = 325510778$, $k_9 = 353354519$, $k_{10} = 053708912$.

5. What sequence of pseudorandom numbers is generated using the pure multiplicative generator $x_{n+1} = 3x_n \bmod 11$ with seed $x_0 = 2$?

The **middle-square method** for generating pseudorandom numbers begins with an n-digit integer. This number is squared, initial zeros are appended to ensure that the result has $2n$ digits, and its middle n digits are used to form the next number in the sequence. This process is repeated to generate additional terms.

6. Find the first eight terms of the sequence of four-digit pseudorandom numbers generated by the middle square method starting with 2357.

7. Explain why both 3792 and 2916 would be bad choices for the initial term of a sequence of four-digit pseudorandom numbers generated by the middle square method.

The **power generator** is a method for generating pseudorandom numbers. To use the power generator, parameters p and d are specified, where p is a prime, d is a positive integer such that $p \nmid d$, and a seed x_0 is specified. The pseudorandom numbers $x_1, x_2, \dots$ are generated using the recursive definition $x_{n+1} = x_n^d \bmod p$.

8. Find the sequence of pseudorandom numbers generated by the power generator with $p = 7$, $d = 3$, and seed $x_0 = 2$.

9. Suppose you received these bit strings over a communications link, where the last bit is a parity check bit. In which string are you sure there is an error?
   a) 00000111111
   b) 10101010101
   c) 11111100000
   d) 10111101111

10. Prove that a parity check bit can detect an error in a string if and only if the string contains an odd number of errors.

11. The first nine digits of the ISBN-10 of the European version of the fifth edition of this book are 0-07-119881. What is the check digit for that book?

12. Determine whether the check digit of the ISBN-10 for this textbook (the seventh edition of *Discrete Mathematics and its Applications*) was computed correctly by the publisher.

The United States Postal Service (USPS) sells money orders identified by an 11-digit number $x_1 x_2 \ldots x_{11}$. The first ten digits identify the money order; x_{11} is a check digit that satisfies $x_{11} = x_1 + x_2 + \cdots + x_{10}$ **mod** 9.

13. Determine whether each of these numbers is a valid USPS money order identification number.
    a) 74051489623
    b) 88382013445
    c) 56152240784
    d) 66606631178

14. One digit in each of these identification numbers of a postal money order is smudged. Can you recover the smudged digit, indicated by a Q, in each of these numbers?
    a) 493212Q0688
    b) 850Q9103858
    c) 2Q941007734
    d) 66687Q03201

15. Determine which single digit errors are detected by the USPS money order code.

16. Determine which transposition errors are detected by the USPS money order code.

17. Determine whether each of the strings of 12 digits is a valid UPC code.
    a) 036000291452
    b) 012345678903
    c) 782421843014
    d) 726412175425

18. Determine which transposition errors the check digit of a UPC code finds.

Some airline tickets have a 15-digit identification number $a_1 a_2 \ldots a_{15}$ where a_{15} is a check digit that equals $a_1 a_2 \ldots a_{14}$ **mod** 7.

19. Determine whether each of these 15-digit numbers is a valid airline ticket identification number.
    a) 101333341789013
    b) 007862342770445
    c) 113273438882531
    d) 000122347322871

*20. Can the accidental transposition of two consecutive digits in an airline ticket identification number be detected using the check digit?

Periodicals are identified using an **International Standard Serial Number (ISSN)**. An ISSN consists of two blocks of four digits. The last digit in the second block is a check digit. This check digit is determined by the congruence $d_8 \equiv 3d_1 + 4d_2 + 5d_3 + 6d_4 + 7d_5 + 8d_6 + 9d_7 \pmod{11}$. When $d_8 \equiv 10 \pmod{11}$, we use the letter X to represent d_8 in the code.

21. Are each of these eight-digit codes possible ISSNs? That is, do they end with a correct check digit?
    a) 1059-1027
    b) 0002-9890
    c) 1530-8669
    d) 1007-120X

22. Does the check digit of an ISSN detect every single error in an ISSN? Justify your answer with either a proof or a counterexample.

23. Does the check digit of an ISSN detect every error where two consecutive digits are accidentally interchanged? Justify your answer with either a proof or a counterexample.

4.6 Cryptography

Introduction

Number theory plays a key role in cryptography, the subject of transforming information so that it cannot be easily recovered without special knowledge. Number theory is the basis of many classical ciphers, first used thousands of years ago, and used extensively until the 20th century. These ciphers encrypt messages by changing each letter to a different letter, or each block of letters to a different block of letters. We will discuss some classical ciphers, including shift ciphers, which replace each letter by the letter a fixed number of positions later in the alphabet, wrapping around to the beginning of the alphabet when necessary. The classical ciphers we will discuss are examples of private key ciphers where knowing how to encrypt allows someone to also decrypt messages. With a private key cipher, two parties who wish to communicate in secret must share a secret key. The classical ciphers we will discuss are also vulnerable to cryptanalysis, which seeks to recover encrypted information without access to the secret information used to encrypt the message. We will show how to cryptanalyze messages sent using shift ciphers.

Number theory is also important in public key cryptography, a type of cryptography invented in the 1970s. In public key cryptography, knowing how to encrypt does not also tell someone how to decrypt. The most widely used public key system, called the RSA cryptosystem, encrypts messages using modular exponentiation, where the modulus is the product of two large primes. Knowing how to encrypt requires that someone know the modulus and an exponent. (It does not require that the two prime factors of the modulus be known.) As far as it is known, knowing how to decrypt requires someone to know how to invert the encryption function, which can only be done in a practical amount of time when someone knows these two large prime factors. In this chapter we will explain how the RSA cryptosystem works, including how to encrypt and decrypt messages.

The subject of cryptography also includes the subject of cryptographic protocols, which are exchanges of messages carried out by two or more parties to achieve a specific security goal. We will discuss two important protocols in this chapter. One allows two people to share a common secret key. The other can be used to send signed messages so that a recipient can be sure that they were sent by the purported sender.

Classical Cryptography

One of the earliest known uses of cryptography was by Julius Caesar. He made messages secret by shifting each letter three letters forward in the alphabet (sending the last three letters of the alphabet to the first three). For instance, using this scheme the letter B is sent to E and the letter X is sent to A. This is an example of **encryption**, that is, the process of making a message secret.

To express Caesar's encryption process mathematically, first replace each letter by an element of $\mathbf{Z}_{26}$, that is, an integer from 0 to 25 equal to one less than its position in the alphabet. For example, replace A by 0, K by 10, and Z by 25. Caesar's encryption method can be represented by the function f that assigns to the nonnegative integer p, $p \leq 25$, the integer $f(p)$ in the set $\{0, 1, 2, \ldots, 25\}$ with

$$f(p) = (p + 3) \bmod 26.$$

In the encrypted version of the message, the letter represented by p is replaced with the letter represented by $(p + 3) \bmod 26$.

EXAMPLE 1 What is the secret message produced from the message "MEET YOU IN THE PARK" using the Caesar cipher?

Solution: First replace the letters in the message with numbers. This produces

    12 4 4 19   24 14 20   8 13   19 7 4   15 0 17 10.

Now replace each of these numbers p by $f(p) = (p + 3) \bmod 26$. This gives

    15 7 7 22   1 17 23   11 16   22 10 7   18 3 20 13.

Translating this back to letters produces the encrypted message "PHHW BRX LQ WKH SDUN." ◀

To recover the original message from a secret message encrypted by the Caesar cipher, the function f^{-1}, the inverse of f, is used. Note that the function f^{-1} sends an integer p from

$\mathbf{Z}_{26}$, to $f^{-1}(p) = (p - 3)$ **mod** 26. In other words, to find the original message, each letter is shifted back three letters in the alphabet, with the first three letters sent to the last three letters of the alphabet. The process of determining the original message from the encrypted message is called **decryption**.

There are various ways to generalize the Caesar cipher. For example, instead of shifting the numerical equivalent of each letter by 3, we can shift the numerical equivalent of each letter by k, so that

$$f(p) = (p + k) \textbf{ mod } 26.$$

Such a cipher is called a *shift cipher*. Note that decryption can be carried out using

$$f^{-1}(p) = (p - k) \textbf{ mod } 26.$$

Here the integer k is called a **key**. We illustrate the use of a shift cipher in Examples 2 and 3.

EXAMPLE 2   Encrypt the plaintext message "STOP GLOBAL WARMING" using the shift cipher with shift $k = 11$.

Solution: To encrypt the message "STOP GLOBAL WARMING" we first translate each letter to the corresponding element of $\mathbf{Z}_{26}$. This produces the string

18 19 14 15   6 11 14 1 0 11   22 0 17 12 8 13 6.

We now apply the shift $f(p) = (p + 11) \textbf{ mod } 26$ to each number in this string. We obtain

3 4 25 0   17 22 25 12 11 22   7 11 2 23 19 24 17.

Translating this last string back to letters, we obtain the ciphertext "DEZA RWZMLW HLCX-TYR." ◀

EXAMPLE 3   Decrypt the ciphertext message "LEWLYPLUJL PZ H NYLHA ALHJOLY" that was encrypted with the shift cipher with shift $k = 7$.

Solution: To decrypt the ciphertext "LEWLYPLUJL PZ H NYLHA ALHJOLY" we first translate the letters back to elements of $\mathbf{Z}_{26}$. We obtain

11 4 22 11 24 15 11 20 9 11   15 25   7   13 24 11 7 0   0 11 7 9 14 11 24.

Next, we shift each of these numbers by $-k = -7$ modulo 26 to obtain

4 23 15 4 17 8 4 13 2 4   8 18   0   6 17 4 0 19   19 4 0 2 7 4 17.

Finally, we translate these numbers back to letters to obtain the plaintext. We obtain "EXPERIENCE IS A GREAT TEACHER." ◀

We can generalize shift ciphers further to slightly enhance security by using a function of the form

$$f(p) = (ap + b) \textbf{ mod } 26,$$

where a and b are integers, chosen so that f is a bijection. (The function $f(p) = (ap + b) \textbf{ mod } 26$ is a bijection if and only if $\gcd(a, 26) = 1$.) Such a mapping is called an *affine transformation*, and the resulting cipher is called an *affine cipher*.

EXAMPLE 4   What letter replaces the letter K when the function $f(p) = (7p + 3) \bmod 26$ is used for encryption?

Solution: First, note that 10 represents K. Then, using the encryption function specified, it follows that $f(10) = (7 \cdot 10 + 3) \bmod 26 = 21$. Because 21 represents V, K is replaced by V in the encrypted message.   ◀

We will now show how to decrypt messages encrypted using an affine cipher. Suppose that $c = (ap + b) \bmod 26$ with $\gcd(a, 26) = 1$. To decrypt we need to show how to express p in terms of c. To do this, we apply the encrypting congruence $c \equiv ap + b \pmod{26}$, and solve it for p. To do this, we first subtract b from both sides, to obtain $c - b \equiv ap \pmod{26}$. Because $\gcd(a, 26) = 1$, we know that there is an inverse $\bar{a}$ of a modulo 26. Multiplying both sides of the last equation by $\bar{a}$ gives us $\bar{a}(c - b) \equiv \bar{a}ap \pmod{26}$. Because $\bar{a}a \equiv 1 \pmod{26}$, this tells us that $p \equiv \bar{a}(c - b) \pmod{26}$. This determines p because p belongs to $\mathbf{Z}_{26}$.

CRYPTANALYSIS   The process of recovering plaintext from ciphertext without knowledge of both the encryption method and the key is known as **crytanalysis** or **breaking codes**. In general, cryptanalysis is a difficult process, especially when the encryption method is unknown. We will not discuss cryptanalysis in general, but we will explain how to break messages that were encrypted using a shift cipher.

If we know that a ciphertext message was produced by enciphering a message using a shift cipher, we can try to recover the message by shifting all characters of the ciphertext by each of the 26 possible shifts (including a shift of zero characters). One of these is guaranteed to be the plaintext. However, we can use a more intelligent approach, which we can build upon to cryptanalyze ciphertext resulting from other ciphers. The main tool for cryptanalyzing ciphertext encrypted using a shift cipher is the count of the frequency of letters in the ciphertext. The nine most common letters in English text and their approximate relative frequencies are E 13%, T 9%, A 8%, O 8%, I 7%, N 7%, S 7%, H 6%, and R 6%. To cryptanalyze ciphertext that we know was produced using a shift cipher, we first find the relative frequencies of letters in the ciphertext. We list the most common letters in the ciphertext in frequency order; we hypothesize that the most common letter in the ciphertext is produced by encrypting E. Then, we determine the value of the shift under this hypothesis, say k. If the message produced by shifting the ciphertext by $-k$ makes sense, we presume that our hypothesis is correct and that we have the correct value of k. If it does not make sense, we next consider the hypothesis that the most common letter in the ciphertext is produced by encrypting T, the second most common letter in English; we find k under this hypothesis, shift the letters of the message by $-k$, and see whether the resulting message makes sense. If it does not, we continue the process working our way through the letters from most common to least common.

Mathematicians make the best code breakers. Their work in World War II changed the course of the war.

EXAMPLE 5   Suppose that we intercepted the ciphertext message ZNK KGXRE HOXJ MKZY ZNK CUXS that we know was produced by a shift cipher. What was the original plaintext message?

Solution: Because we know that the intercepted ciphertext message was encrypted using a shift cipher, we begin by calculating the frequency of letters in the ciphertext. We find that the most common letter in the ciphertext is K. So, we hypothesize that the shift cipher sent the plaintext letter E to the ciphertext letter K. If this hypothesis is correct, we know that $10 = 4 + k \bmod 26$, so $k = 6$. Next, we shift the letters of the message by -6, obtaining THE EARLY BIRD GETS THE WORM. Because this message makes sense, we assume that the hypothesis that $k = 6$ is correct.   ◀

 Links

BLOCK CIPHERS   Shift ciphers and affine ciphers proceed by replacing each letter of the alphabet by another letter in the alphabet. Because of this, these ciphers are called **character** or **monoalphabetic ciphers**. Encryption methods of this kind are vulnerable to attacks based on the analysis of letter frequency in the ciphertext, as we just illustrated. We can make it harder to successfully attack ciphertext by replacing blocks of letters with other blocks of letters instead of replacing individual characters with individual characters; such ciphers are called **block ciphers**.

We will now introduce a simple type of block cipher, called the **transposition cipher**. As a key we use a permutation σ of the set $\{1, 2, \ldots, m\}$ for some positive integer m, that is, a one-to-one function from $\{1, 2, \ldots, m\}$ to itself. To encrypt a message we first split its letters into blocks of size m. (If the number of letters in the message is not divisible by m we add some random letters at the end to fill out the final block.) We encrypt the block $p_1 p_2 \ldots p_m$ as $c_1 c_2 \ldots c_m = p_{\sigma(1)} p_{\sigma(2)} \ldots, p_{\sigma(m)}$. To decryt a ciphertext block $c_1 c_2 \ldots c_m$, we transpose its letters using the permutation σ^{-1}, the inverse of σ. Example 6 illustrates encryption and decryption for a transposition cipher.

EXAMPLE 6   Using the transposition cipher based on the permutation σ of the set $\{1, 2, 3, 4\}$ with $\sigma(1) = 3$, $\sigma(2) = 1$, $\sigma(3) = 4$, and $\sigma(4) = 2$,

(a) Encrypt the plaintext message PIRATE ATTACK.

(b) Decrypt the ciphertext message SWUE TRAE OEHS, which was encrypted using this cipher.

Solution: (a) We first split the letters of the plaintext into blocks of four letters. We obtain PIRA TEAT TACK. To encrypt each block, we send the first letter to the third position, the second letter to the first position, the third letter to the fourth position, and the fourth letter to the second position. We obtain IAPR ETTA AKTC.

(b) We note that σ^{-1}, the inverse of σ, sends 1 to 2, sends 2 to 4, sends 3 to 1, and sends 4 to 3. Applying $\sigma^{-1}(m)$ to each block gives us the plaintext: USEW ATER HOSE. (Grouping together these letters to form common words, we surmise that the plaintext is USE WATER HOSE.) ◀

CRYPTOSYSTEMS   We have defined two families of ciphers: shift ciphers and affine ciphers. We now introduce the notion of a cryptosystem, which provides a general structure for defining new families of ciphers.

DEFINITION 1   A *cryptosystem* is a five-tuple $(\mathcal{P}, \mathcal{C}, \mathcal{K}, \mathcal{E}, \mathcal{D})$, where $\mathcal{P}$ is the set of plaintext strings, $\mathcal{C}$ is the set of ciphertext strings, $\mathcal{K}$ is the *keyspace* (the set of all possible keys), $\mathcal{E}$ is the set of encryption functions, and $\mathcal{D}$ is the set of decryption functions. We denote by E_k the encryption function in $\mathcal{E}$ corresponding to the key k and D_k the decryption function in $\mathcal{D}$ that decrypts ciphertext that was encrypted using E_k, that is $D_k(E_k(p)) = p$, for all plaintext strings p.

We now illustrate the use of the definition of a cryptosystem.

EXAMPLE 7   Describe the family of shift ciphers as a cryptosytem.

Solution: To encrypt a string of English letters with a shift cipher, we first translate each letter to an integer between 0 and 25, that is, to an element of $\mathbf{Z}_{26}$. We then shift each of these integers by a fixed integer modulo 26, and finally, we translate the integers back to letters. To apply the defintion of a cryptosystem to shift ciphers, we assume that our messages are already integers, that is, elements of $\mathbf{Z}_{26}$. That is, we assume that the translation between letters and integers is outside of the cryptosystem. Consequently, both the set of plaintext strings $\mathcal{P}$ and the set of ciphertext strings $\mathcal{C}$ are the set of strings of elements of $\mathbf{Z}_{26}$. The set of keys $\mathcal{K}$ is the set of possible shifts, so $\mathcal{K} = \mathbf{Z_{26}}$. The set $\mathcal{E}$ consists of functions of the form $E_k(p) = (p + k) \bmod 26$, and the set $\mathcal{D}$ of decryption functions is the same as the set of encrypting functions where $D_k(p) = (p - k) \bmod 26$. ◀

The concept of a cryptosystem is useful in the discussion of additional families of ciphers and is used extensively in cryptography.

Public Key Cryptography

All classical ciphers, including shift ciphers and affine ciphers, are examples of **private key cryptosystems**. In a private key cryptosystem, once you know an encryption key, you can quickly find the decryption key. So, knowing how to encrypt messages using a particular key allows you to decrypt messages that were encrypted using this key. For example, when a shift cipher is used with encryption key k, the plaintext integer p is sent to

$$c = (p + k) \bmod 26.$$

Decryption is carried out by shifting by $-k$; that is,

$$p = (c - k) \bmod 26.$$

So knowing how to encrypt with a shift cipher also tells you how to decrypt.

When a private key cryptosystem is used, two parties who wish to communicate in secret must share a secret key. Because anyone who knows this key can both encrypt and decrypt messages, two people who want to communicate securely need to securely exchange this key. (We will introduce a method for doing this later in this section.) The shift cipher and affine cipher cryptosystems are private key cryptosystems. They are quite simple and are extremely vulnerable to cryptanalysis. However, the same is not true of many modern private key cryptosystems. In particular, the current US government standard for private key cryptography, the Advanced Encryption Standard (AES), is extremely complex and is considered to be highly resistant to cryptanalysis. (See [St06] for details on AES and other modern private key cryptosystems.) AES is widely used in government and commercial communications. However, it still shares the property that for secure communications keys be shared. Furthermore, for extra security, a new key is used for each communication session between two parties, which requires a method for generating keys and securely sharing them.

To avoid the need for keys to be shared by every pair of parties that wish to communicate securely, in the 1970s cryptologists introduced the concept of **public key cryptosystems**. When such cryptosystems are used, knowing how to send an encrypted message does not help decrypt messages. In such a system, everyone can have a publicly known encryption key. Only the decryption keys are kept secret, and only the intended recipient of a message can decrypt it, because, as far as it is currently known, knowledge of the encryption key does not let someone recover the plaintext message without an extraordinary amount of work (such as billions of years of computer time).

The RSA Cryptosystem

M.I.T. is also known as the 'Tute.

In 1976, three researchers at the Massachusetts Institute of Technology—Ronald Rivest, Adi Shamir, and Leonard Adleman—introduced to the world a public key cryptosystem, known as the **RSA system**, from the initials of its inventors. As often happens with cryptographic discoveries, the RSA system had been discovered several years earlier in secret government research in the United Kingdom. Clifford Cocks, working in secrecy at the United Kingdom's Government Communications Headquarters (GCHQ), had discovered this cryptosystem in 1973. However, his invention was unknown to the outside world until the late 1990s, when he was allowed to share classified GCHQ documents from the early 1970s. (An excellent account of this earlier discovery, as well as the work of Rivest, Shamir, and Adleman, can be found in [Si99].)

Unfortunately, no one calls this the Cocks cryptosystem.

In the RSA cryptosystem, each individual has an encryption key (n, e) where $n = pq$, the modulus is the product of two large primes p and q, say with 200 digits each, and an exponent e that is relatively prime to $(p - 1)(q - 1)$. To produce a usable key, two large primes must be found. This can be done quickly on a computer using probabilistic primality tests, referred to earlier in this section. However, the product of these primes $n = pq$, with approximately 400 digits, cannot, as far as is currently known, be factored in a reasonable length of time. As we will see, this is an important reason why decryption cannot, as far as is currently known, be done quickly without a separate decryption key.

RSA Encryption

To encrypt messages using a particular key (n, e), we first translate a plaintext message M into sequences of integers. To do this, we first translate each plaintext letter into a two-digit number, using the same translation we employed for shift ciphers, with one key difference. That is, we include an initial zero for the letters A through J, so that A is translated into 00, B into 01, . . . , and J into 09. Then, we concatenate these two-digit numbers into strings of digits. Next, we divide this string into equally sized blocks of $2N$ digits, where $2N$ is the largest even number such that the number 2525 . . . 25 with $2N$ digits does not exceed n. (When necessary, we pad the plaintext message with dummy Xs to make the last block the same size as all other blocks.)

After these steps, we have translated the plaintext message M into a sequence of integers $m_1, m_2, \ldots, m_k$ for some integer k. Encryption proceeds by transforming each block m_i to a ciphertext block c_i. This is done using the function

$$C = M^e \bmod n.$$

(To perform the encryption, we use an algorithm for fast modular exponentiation, such as Algorithm 5 in Section 4.2.) We leave the encrypted message as blocks of numbers and send these to the intended recipient. Because the RSA cryptosystem encrypts blocks of characters into blocks of characters, it is a block cipher.

Example 8 illustrates how RSA encryption is performed. For practical reasons we use small primes p and q in this example, rather than primes with 200 or more digits. Although the cipher described in this example is not secure, it does illustrate the techniques used in the RSA cipher.

EXAMPLE 8   Encrypt the message STOP using the RSA cryptosystem with key (2537, 13). Note that $2537 = 43 \cdot 59$, $p = 43$ and $q = 59$ are primes, and

$$\gcd(e, (p - 1)(q - 1)) = \gcd(13, 42 \cdot 58) = 1.$$

CLIFFORD COCKS (BORN 1950)   Clifford Cocks, born in Cheshire, England, was a talented mathematics student. In 1968 he won a silver medal at the International Mathematical Olympiad. Cocks attended King's College, Cambridge, studying mathematics. He also spent a short time at Oxford University working in number theory. In 1973 he decided not to complete his graduate work, instead taking a mathematical job at the Government Communications Headquarters (GCHQ) of British intelligence. Two months after joining GCHQ, Cocks learned about public key cryptography from an internal GCHQ report written by James Ellis. Cocks used his number theory knowledge to invent what is now called the RSA cryptosystem. He quickly realized that a public key cryptosystem could be based on the difficulty of reversing the process of multiplying two large primes. In 1997 he was allowed to reveal declassified GCHQ internal documents describing his discovery. Cocks is also known for his invention of a secure identity based encryption scheme, which uses information about a user's identity as a public key. In 2001, Cocks became the Chief Mathematician at GCHQ. He has also set up the Heilbronn Institute for Mathematical Research, a partnership between GCHQ and the University of Bristol.

Solution: To encrypt, we first translate the letters in STOP into their numerical equivalents. We then group these numbers into blocks of four digits (because $2525 < 2537 < 252525$), to obtain

1819 1415.

We encrypt each block using the mapping

$$C = M^{13} \bmod 2537.$$

Computations using fast modular multiplication show that $1819^{13} \bmod 2537 = 2081$ and $1415^{13} \bmod 2537 = 2182$. The encrypted message is 2081 2182. ◄

RSA Decryption

The plaintext message can be quickly recovered from a ciphertext message when the decryption key d, an inverse of e modulo $(p-1)(q-1)$, is known. [Such an inverse exists because $\gcd(e, (p-1)(q-1)) = 1$.] To see this, note that if $de \equiv 1 \pmod{(p-1)(q-1)}$, there is an integer k such that $de = 1 + k(p-1)(q-1)$. It follows that

$$C^d \equiv (M^e)^d = M^{de} = M^{1+k(p-1)(q-1)} \pmod n.$$

By Fermat's little theorem [assuming that $\gcd(M, p) = \gcd(M, q) = 1$, which holds except in rare cases, which we cover in Exercise 28], it follows that $M^{p-1} \equiv 1 \pmod p$ and $M^{q-1} \equiv 1 \pmod q$. Consequently,

$$C^d \equiv M \cdot (M^{p-1})^{k(q-1)} \equiv M \cdot 1 = M \pmod p$$

and

$$C^d \equiv M \cdot (M^{q-1})^{k(p-1)} \equiv M \cdot 1 = M \pmod q.$$

RONALD RIVEST (BORN 1948)   Ronald Rivest received a B.A. from Yale in 1969 and his Ph.D. in computer science from Stanford in 1974. Rivest is a computer science professor at M.I.T. and was a cofounder of RSA Data Security, which held the patent on the RSA cryptosystem that he invented together with Adi Shamir and Leonard Adleman. Areas that Rivest has worked in besides cryptography include machine learning, VLSI design, and computer algorithms. He is a coauthor of a popular text on algorithms ([CoLeRiSt09]).

ADI SHAMIR (BORN 1952)   Adi Shamir was born in Tel Aviv, Israel. His undergraduate degree is from Tel Aviv University (1972) and his Ph.D. is from the Weizmann Institute of Science (1977). Shamir was a research assistant at the University of Warwick and an assistant professor at M.I.T. He is currently a professor in the Applied Mathematics Department at the Weizmann Institute and leads a group studying computer security. Shamir's contributions to cryptography, besides the RSA cryptosystem, include cracking knapsack cryptosystems, cryptanalysis of the Data Encryption Standard (DES), and the design of many cryptographic protocols.

LEONARD ADLEMAN (BORN 1945)   Leonard Adleman was born in San Francisco, California. He received a B.S. in mathematics (1968) and his Ph.D. in computer science (1976) from the University of California, Berkeley. Adleman was a member of the mathematics faculty at M.I.T. from 1976 until 1980, where he was a coinventor of the RSA cryptosystem, and in 1980 he took a position in the computer science department at the University of Southern California (USC). He was appointed to a chaired position at USC in 1985. Adleman has worked on computer security, computational complexity, immunology, and molecular biology. He invented the term "computer virus." Adleman's recent work on DNA computing has sparked great interest. He was a technical adviser for the movie *Sneakers*, in which computer security played an important role.

Because $\gcd(p, q) = 1$, it follows by the Chinese remainder theorem that

$$C^d \equiv M \ (\mathrm{mod} \ pq).$$

Example 9 illustrates how to decrypt messages sent using the RSA cryptosystem.

EXAMPLE 9   We receive the encrypted message 0981 0461. What is the decrypted message if it was encrypted using the RSA cipher from Example 8?

Solution: The message was encrypted using the RSA cryptosystem with $n = 43 \cdot 59$ and exponent 13. As Exercise 2 in Section 4.4 shows, $d = 937$ is an inverse of 13 modulo $42 \cdot 58 = 2436$. We use 937 as our decryption exponent. Consequently, to decrypt a block C, we compute

$$M = C^{937} \ \mathbf{mod} \ 2537.$$

To decrypt the message, we use the fast modular exponentiation algorithm to compute $0981^{937} \ \mathbf{mod} \ 2537 = 0704$ and $0461^{937} \ \mathbf{mod} \ 2537 = 1115$. Consequently, the numerical version of the original message is 0704 1115. Translating this back to English letters, we see that the message is HELP. ◀

RSA as a Public Key System

Links

Why is the RSA cryptosystem suitable for public key cryptography? First, it is possible to rapidly construct a public key by finding two large primes p and q, each with more than 200 digits, and to find an integer e relatively prime to $(p - 1)(q - 1)$. When we know the factorization of the modulus n, that is, when we know p and q, we can quickly find an inverse d of e modulo $(p - 1)(q - 1)$. [This is done by using the Euclidean algorithm to find Bézout coefficients s and t for d and $(p - 1)(q - 1)$, which shows that the inverse of d modulo $(p - 1)(q - 1)$ is $s \ \mathbf{mod} \ (p - 1)(q - 1)$.] Knowing d lets us decrypt messages sent using our key. However, no method is known to decrypt messages that is not based on finding a factorization of n, or that does not also lead to the factorization of n.

Factorization is believed to be a difficult problem, as opposed to finding large primes p and q, which can be done quickly. The most efficient factorization methods known (as of 2010) require billions of years to factor 400-digit integers. Consequently, when p and q are 200-digit primes, it is believed that messages encrypted using $n = pq$ as the modulus cannot be found in a reasonable time unless the primes p and q are known.

Although no polynomial-time algorithm is known for factoring large integers, active research is under way to find new ways to efficiently factor integers. Integers that were thought, as recently as several years ago, to be far too large to be factored in a reasonable amount of time can now be factored routinely. Integers with more than 150 digits, as well as some with more than 200 digits, have been factored using team efforts. When new factorization techniques are found, it will be necessary to use larger primes to ensure secrecy of messages. Unfortunately, messages that were considered secure earlier can be saved and subsequently decrypted by unintended recipients when it becomes feasible to factor the $n = pq$ in the key used for RSA encryption.

The RSA method is now widely used. However, the most commonly used cryptosystems are private key cryptosystems. The use of public key cryptography, via the RSA system, is growing. Nevertheless, there are applications that use both private key and public key systems. For example, a public key cryptosystem, such as RSA, can be used to distribute private keys to pairs of individuals when they wish to communicate. These people then use a private key system for encryption and decryption of messages.

Cryptographic Protocols

So far we have shown how cryptography can be used to make messages secure. However, there are many other important applications of cryptography. Among these applications are **cryptographic protocols**, which are exchanges of messages carried out by two or more parties to achieve a particular security goal. In particular, we will show how cryptography can be used to allow two people to exchange a secret key over an insecure communication channel. We will also show how cryptography can be used to send signed secret messages so that the recipient can be sure that the message came from the purported sender. We refer the reader to [St05] for thorough discussions of a variety of cryptographic protocols.

KEY EXCHANGE   We now discuss a protocol that two parties can use to exchange a secret key over an insecure communications channel without having shared any information in the past. Generating a key that two parties can share is important for many applications of cryptography. For example, for two people to send secure messages to each other using a private key cryptosystem they need to share a common key. The protocol we will describe is known as the **Diffie-Hellman key agreement protocol**, after Whitfield Diffie and Martin Hellman, who described it in 1976. However, this protocol was invented in 1974 by Malcolm Williamson in secret work at the British GCHQ. It was not until 1997 that his discovery was made public.

Suppose that Alice and Bob want to share a common key. The protocol follows these steps, where the computations are done in $\mathbf{Z}_p$.

(1) Alice and Bob agree to use a prime p and a primitive root a of p.

(2) Alice chooses a secret integer k_1 and sends $a^{k_1} \bmod p$ to Bob.

(3) Bob chooses a secret integer k_2 and sends $a^{k_2} \bmod p$ to Alice.

(4) Alice computes $(a^{k_2})^{k_1} \bmod p$.

(5) Bob computes $(a^{k_1})^{k_2} \bmod p$.

At the end of this protocol, Alice and Bob have computed their shared key, namely

$$(a^{k_2})^{k_1} \bmod p = (a^{k_1})^{k_2} \bmod p.$$

To analyze the security of this protocol, note that the messages sent in steps (1), (2), and (3) are not assumed to be sent securely. We can even assume that these communications were in the clear and that their contents are public information. So, p, a, $a^{k_1} \bmod p$, and $a^{k_2} \bmod p$ are assumed to be public information. The protocol ensures that k_1, k_2, and the common key $(a^{k_2})^{k_1} \bmod p = (a^{k_1})^{k_2} \bmod p$ are kept secret. To find the secret information from this public information requires that an adversary solves instances of the discrete logarithm problem, because the adversary would need to find k_1 and k_2 from $a^{k_1} \bmod p$ and $a^{k_2} \bmod p$, respectively. Furthermore, no other method is known for finding the shared key using just the public information. We have remarked that this is thought to be computationally infeasible when p and a are sufficiently large. With the computing power available now, this system is considered unbreakable when p has more than 300 decimal digits and k_1 and k_2 have more than 100 decimal digits each.

DIGITAL SIGNATURES   Not only can cryptography be used to secure the confidentiality of a message, but it also can be used so that the recipient of the message knows that it came from the person they think it came from. We first show how a message can be sent so that a recipient of the message will be sure that the message came from the purported sender of the message.

In particular, we can show how this can be accomplished using the RSA cryptosystem to apply a **digital signature** to a message.

Suppose that Alice's RSA public key is (n, e) and her private key is d. Alice encrypts a plain-text message x using the encryption function $E_{(n,e)}(x) = x^e \bmod n$. She decrypts a ciphertext message y using the decryption function $D_{(n,e)} = x^d \bmod n$. Alice wants to send the message M so that everyone who receives the message knows that it came from her. Just as in RSA encryption, she translates the letters into their numerical equivalents and splits the resulting string into blocks $m_1, m_2, \ldots, m_k$ such that each block is the same size which is as large as possible so that $0 \le m_i \le n$ for $i = 1, 2, \ldots, k$. She then applies her *decryption function* $D_{(n,e)}$ to each block, obtaining $D_{n,e}(m_i)$, $i = 1, 2, \ldots, k$. She sends the result to all intended recipients of the message.

When a recipient receives her message, they apply Alice's encryption function $E_{(n,e)}$ to each block, which everyone has available because Alice's key (n, e) is public information. The result is the original plaintext block because $E_{(n,e)}(D_{(n,e)}(x)) = x$. So, Alice can send her message to as many people as she wants and by signing it in this way, every recipient can be sure it came from Alice. Example 10 illustrates this protocol.

EXAMPLE 10 Suppose Alice's public RSA cryptosystem key is the same as in Example 8. That is, $n = 43 \cdot 59 = 2537$ and $e = 13$. Her decryption key is $d = 937$, as described in Example 9. She wants to send the message "MEET AT NOON" to her friends so that they are sure it came from her. What should she send?

Solution: Alice first translates the message into blocks of digits, obtaining 1204 0419 0019 1314 1413 (as the reader should verify). She then applies her decryption transformation $D_{(2537,13)}(x) = x^{937} \bmod 2537$ to each block. Using fast modular exponentiation (with the help of a computational aid), she finds that $1204^{937} \bmod 2537 = 817$, $419^{937} \bmod 2537 = 555$, $19^{937} \bmod 2537 = 1310$, $1314^{937} \bmod 2537 = 2173$, and $1413^{937} \bmod 2537 = 1026$.

So, the message she sends, split into blocks, is 0817 0555 1310 2173 1026. When one of her friends gets this message, they apply her encryption transformation $E_{(2537,13)}$ to each block. When they do this, they obtain the blocks of digits of the original message which they translate back to English letters. ◀

We have shown that signed messages can be sent using the RSA cryptosystem. We can extend this by sending signed secret messages. To do this, the sender applies RSA encryption using the publicly known encryption key of an intended recipient to each block that was encrypted using sender's decryption transformation. The recipient then first applies his private decryption transformation and then the sender's public encryption transformation. (Exercise 20 asks for this protocol to be carried out.)

Exercises

1. Encrypt the message DO NOT PASS GO by translating the letters into numbers, applying the given encryption function, and then translating the numbers back into letters.

a) $f(p) = (p + 3) \bmod 26$ (the Caesar cipher)
b) $f(p) = (p + 13) \bmod 26$
c) $f(p) = (3p + 7) \bmod 26$

2. Encrypt the message WATCH YOUR STEP by translating the letters into numbers, applying the given encryption function, and then translating the numbers back into letters.

a) $f(p) = (p + 14) \bmod 26$

b) $f(p) = (14p + 21) \bmod 26$

c) $f(p) = (-7p + 1) \bmod 26$

3. Decrypt these messages encrypted using the shift cipher $f(p) = (p + 10) \bmod 26$.

a) CEBBOXNOB XYG

b) LO WI PBSOXN

c) DSWO PYB PEX

4. Determine whether there is a key for which the enciphering function for the shift cipher is the same as the deciphering function.

5. What is the decryption function for an affine cipher if the encryption function is $c = (15p + 13) \bmod 26$?

*6. Find all pairs of integers keys (a, b) for affine ciphers for which the encryption function $c = (ap + b) \bmod 26$ is the same as the corresponding decryption function.

7. Suppose that the most common letter and the second most common letter in a long ciphertext produced by encrypting a plaintext using an affine cipher $f(p) = (ap + b) \bmod 26$ are Z and J, respectively. What are the most likely values of a and b?

8. Encrypt the message GRIZZLY BEARS using blocks of five letters and the transposition cipher based on the permutation of $\{1, 2, 3, 4, 5\}$ with $\sigma(1) = 3$, $\sigma(2) = 5$, $\sigma(3) = 1, \sigma(4) = 2$, and $\sigma(5) = 4$. For this exercise, use the letter X as many times as necessary to fill out the final block of fewer then five letters.

9. Decrypt the message EABW EFRO ATMR ASIN which is the ciphertext produced by encrypting a plaintext message using the transposition cipher with blocks of four letters and the permutation σ of $\{1, 2, 3, 4\}$ defined by $\sigma(1) = 3$, $\sigma(2) = 1$, $\sigma(3) = 4$, and $\sigma(4) = 2$.

The **Vigenère cipher** is a block cipher, with a key that is a string of letters with numerical equivalents $k_1 k_2 \ldots k_m$, where $k_i \in \mathbf{Z}_{26}$ for $i = 1, 2, \ldots, m$. Suppose that the numerical equivalents of the letters of a plaintext block are $p_1 p_2 \ldots p_m$. The corresponding numerical ciphertext block is $(p_1 + k_1) \bmod 26 \, (p_2 + k_2) \bmod 26 \ldots (p_m + k_m) \bmod 26$. Finally, we translate back to letters. For example, suppose that the key string is RED, with numerical equivalents 17 4 3. Then, the plaintext ORANGE, with numerical equivalents 14 17 00 13 06 04, is encrypted by first splitting it into two blocks 14 17 00 and 13 06 04. Then, in each block we shift the first letter by 17, the second by 4, and the third by 3. We obtain 5 21 03 and 04 10 07. The cipherext is FVDEKH.

10. Use the Vigenère cipher with key BLUE to encrypt the message SNOWFALL.

11. The ciphertext OIKYWVHBX was produced by encrypting a plaintext message using the Vigenère cipher with key HOT. What is the plaintext message?

To break a Vigenère cipher by recovering a plaintext message from the ciphertext message without having the key, the first step is to figure out the length of the key string. The second step is to figure out each character of the key string by determining the corresponding shift (see Exercise 12).

12. Suppose that when a long string of text is encrypted using a Vigenère cipher, the same string is found in the ciphertext starting at several different positions. Explain how this information can be used to help determine the length of the key.

*13. Show that we can easily factor n when we know that n is the product of two primes, p and q, and we know the value of $(p - 1)(q - 1)$.

In Exercises 14–15 first express your answers without computing modular exponentiations. Then use a computational aid to complete these computations.

14. Encrypt the message UPLOAD using the RSA system with $n = 53 \cdot 61$ and $e = 17$, translating each letter into integers and grouping together pairs of integers, as done in Example 8.

15. What is the original message encrypted using the RSA system with $n = 43 \cdot 59$ and $e = 13$ if the encrypted message is 0667 1947 0671? (To decrypt, first find the decryption exponent d which is the inverse of $e = 13$ modulo $42 \cdot 58$.)

*16. Suppose that (n, e) is an RSA encryption key, with $n = pq$ where p and q are large primes and $\gcd(e, (p - 1)(q - 1)) = 1$. Furthermore, suppose that d is an inverse of e modulo $(p - 1)(q - 1)$. Suppose that $C \equiv M^e \pmod{pq}$. In the text we showed that RSA decryption, that is, the congruence $C^d \equiv M \pmod{pq}$ holds when $\gcd(M, pq) = 1$. Show that this decryption congruence also holds when $\gcd(M, pq) > 1$. [*Hint:* Use congruences modulo p and modulo q and apply the Chinese remainder theorem.]

17. Describe the steps that Alice and Bob follow when they use the Diffie-Hellman key exchange protocol to generate a shared key. Assume that they use the prime $p = 23$ and take $a = 5$, which is a primitive root of 23, and that Alice selects $k_1 = 8$ and Bob selects $k_2 = 5$. (You may want to use some computational aid.)

18. Describe the steps that Alice and Bob follow when they use the Diffie-Hellman key exchange protocol to generate a shared key. Assume that they use the prime $p = 101$ and take $a = 2$, which is a primitive root of 101, and that Alice selects $k_1 = 7$ and Bob selects $k_2 = 9$. (You may want to use some computational aid.)

In Exercises 19–20 suppose that Alice and Bob have these public keys and corresponding private keys: $(n_{\text{Alice}}, e_{\text{Alice}}) = (2867, 7) = (61 \cdot 47, 7)$, $d_{\text{Alice}} = 1183$ and $(n_{\text{Bob}}, e_{\text{Bob}}) = (3127, 21) = (59 \cdot 53, 21)$, $d_{\text{Bob}} = 1149$. First express your answers without carrying out the calculations. Then, using a computational aid, if available, perform the calculation to get the numerical answers.

19. Alice wants to send to all her friends, including Bob, the message "SELL EVERYTHING" so that he knows that she sent it. What should she send to her friends, assuming she signs the message using the RSA cryptosystem.

20. Alice wants to send to Bob the message "BUY NOW" so that he knows that she sent it and so that only Bob can read it. What should she send to Bob, assuming she signs the message and then encrypts it using Bob's public key?

21. We describe a basic key exchange protocol using private key cryptography upon which more sophisticated protocols for key exchange are based. Encryption within the protocol is done using a private key cryptosystem (such as AES) that is considered secure. The protocol involves three parties, Alice and Bob, who wish to exchange a key, and a trusted third party Cathy. Assume that Alice has a secret key k_{Alice} that only she and Cathy know, and Bob has a secret key k_{Bob} which only he and Cathy know. The protocol has three steps:

(i) Alice sends the trusted third party Cathy the message "request a shared key with Bob" encrypted using Alice's key k_{Alice}.

(ii) Cathy sends back to Alice a key $k_{Alice, Bob}$, which she generates, encrypted using the key k_{Alice}, followed by this same key $k_{Alice, Bob}$, encrypted using Bob's key, k_{Bob}.

(iii) Alice sends to Bob the key $k_{Alice, Bob}$ encrypted using k_{Bob}, known only to Bob and to Cathy.

Explain why this protocol allows Alice and Bob to share the secret key $k_{Alice, Bob}$, known only to them and to Cathy.

Key Terms and Results

TERMS

$a \mid b$ (a divides b): there is an integer c such that $b = ac$

a and b are congruent modulo m: m divides $a - b$

modular arithmetic: arithmetic done modulo an integer $m \geq 2$

prime: an integer greater than 1 with exactly two positive integer divisors

composite: an integer greater than 1 that is not prime

Mersenne prime: a prime of the form $2^p - 1$, where p is prime

gcd(a, b) (greatest common divisor of a and b): the largest integer that divides both a and b

relatively prime integers: integers a and b such that $\gcd(a, b) = 1$

pairwise relatively prime integers: a set of integers with the property that every pair of these integers is relatively prime

lcm(a, b) (least common multiple of a and b): the smallest positive integer that is divisible by both a and b

a mod b: the remainder when the integer a is divided by the positive integer b

$a \equiv b$ (mod m) (a is congruent to b modulo m): $a - b$ is divisible by m

$n = (a_k a_{k-1} \ldots a_1 a_0)_b$: the base b representation of n

binary representation: the base 2 representation of an integer

octal representation: the base 8 representation of an integer

hexadecimal representation: the base 16 representation of an integer

linear combination of a and b with integer coefficients: an expression of the form $sa + tb$, where s and t are integers

Bézout coefficients of a and b: integers s and t such that the **Bézout identity** $sa + tb = \gcd(a, b)$ holds

inverse of a modulo m: an integer $\bar{a}$ such that $\bar{a}a \equiv 1$ (mod m)

linear congruence: a congruence of the form $ax \equiv b$ (mod m), where x is an integer variable

pseudoprime to the base b: a composite integer n such that $b^{n-1} \equiv 1$ (mod n)

Carmichael number: a composite integer n such that n is a pseudoprime to the base b for all positive integers b with $\gcd(b, n) = 1$

primitive root of a prime p: an integer r in $\mathbf{Z}_p$ such that every integer not divisible by p is congruent modulo p to a power of r

discrete logarithm of a to the base r modulo p: the integer e with $0 \leq e \leq p - 1$ such that $r^e \equiv a$ (mod p)

encryption: the process of making a message secret

decryption: the process of returning a secret message to its original form

encryption key: a value that determines which of a family of encryption functions is to be used

shift cipher: a cipher that encrypts the plaintext letter p as $(p + k)$ **mod** m for an integer k

affine cipher: a cipher that encrypts the plaintext letter p as $(ap + b)$ **mod** m for integers a and b with $\gcd(a, 26) = 1$

character cipher: a cipher that encrypts characters one by one

block cipher: a cipher that encrypts blocks of characters of a fixed size

crytanalysis: the process of recovering the plaintext from ciphertext without knowledge of the encryption method, or with knowledge of the encryption method, but not the key

cryptosystem: a five-tuple $(\mathcal{P}, \mathcal{C}, \mathcal{K}, \mathcal{E}, \mathcal{D})$ where $\mathcal{P}$ is the set of plaintext messages, $\mathcal{C}$ is the set of ciphertext messages, $\mathcal{K}$ is the set of keys, $\mathcal{E}$ is the set of encryption functions, and $\mathcal{D}$ is the set of decryption functions

private key encryption: encryption where both encryption keys and decryption keys must be kept secret

public key encryption: encryption where encryption keys are public knowledge, but decryption keys are kept secret

RSA cryptosystem: the cryptosystem where $\mathcal{P}$ and $\mathcal{C}$ are both $\mathbf{Z}_{26}$, $\mathcal{K}$ is the set of pairs $k = (n, e)$ where $n = pq$ where p and q are large primes and e is a positive integer, $E_k(p) = p^e$ **mod** n, and $D_k(c) = c^d$ **mod** n where d is the inverse of e modulo $(p - 1)(q - 1)$

key exchange protocol: a protocol used for two parties to generate a shared key

digital signature: a method that a recipient can use to determine that the purported sender of a message actually sent the message

RESULTS

division algorithm: Let a and d be integers with d positive. Then there are unique integers q and r with $0 \leq r < d$ such that $a = dq + r$.

Let b be an integer greater than 1. Then if n is a positive integer, it can be expressed uniquely in the form $n = a_k b^k + a_{k-1} b^{k-1} + \cdots + a_1 b + a_0$.

The algorithm for finding the base b expansion of an integer (see Algorithm 1 in Section 4.2)

The conventional algorithms for addition and multiplication of integers (given in Section 4.2)

The modular exponentiation algorithm (see Algorithm 5 in Section 4.2)

Euclidean algorithm: for finding greatest common divisors by successively using the division algorithm (see Algorithm 1 in Section 4.3)

Bézout's theorem: If a and b are positive integers, then $\gcd(a, b)$ is a linear combination of a and b.

sieve of Eratosthenes: A procedure for finding all primes not exceeding a specified number n, described in Section 4.3

fundamental theorem of arithmetic: Every positive integer can be written uniquely as the product of primes, where the prime factors are written in order of increasing size.

If a and b are positive integers, then $ab = \gcd(a, b) \cdot \text{lcm}(a, b)$.

If m is a positive integer and $\gcd(a, m) = 1$, then a has a unique inverse modulo m.

Chinese remainder theorem: A system of linear congruences modulo pairwise relatively prime integers has a unique solution modulo the product of these moduli.

Fermat's little theorem: If p is prime and $p \nmid a$, then $a^{p-1} \equiv 1 \pmod{p}$.

Review Questions

1. Find 210 **div** 17 and 210 **mod** 17.

2. **a)** Define what it means for a and b to be congruent modulo 7.
   b) Which pairs of the integers $-11, -8, -7, -1, 0, 3$, and 17 are congruent modulo 7?
   c) Show that if a and b are congruent modulo 7, then $10a + 13$ and $-4b + 20$ are also congruent modulo 7.

3. Show that if $a \equiv b \pmod{m}$ and $c \equiv d \pmod{m}$, then $a + c \equiv b + d \pmod{m}$.

4. Describe a procedure for converting decimal (base 10) expansions of integers into hexadecimal expansions.

5. Convert $(1101\ 1001\ 0101\ 1011)_2$ to octal and hexadecimal representations.

6. Convert $(7206)_8$ and $(A0EB)_{16}$ to a binary representation.

7. State the fundamental theorem of arithmetic.

8. **a)** Describe a procedure for finding the prime factorization of an integer.
   b) Use this procedure to find the prime factorization of 80,707.

9. **a)** Define the greatest common divisor of two integers.
   b) Describe at least three different ways to find the greatest common divisor of two integers. When does each method work best?
   c) Find the greatest common divisor of 1,234,567 and 7,654,321.
   d) Find the greatest common divisor of $2^3 3^5 5^7 7^9 11$ and $2^9 3^7 5^5 7^3 13$.

10. **a)** How can you find a linear combination (with integer coefficients) of two integers that equals their greatest common divisor?
    b) Express $\gcd(84, 119)$ as a linear combination of 84 and 119.

11. **a)** What does it mean for $\bar{a}$ to be an inverse of a modulo m?
    b) How can you find an inverse of a modulo m when m is a positive integer and $\gcd(a, m) = 1$?
    c) Find an inverse of 7 modulo 19.

12. **a)** How can an inverse of a modulo m be used to solve the congruence $ax \equiv b \pmod{m}$ when $\gcd(a, m) = 1$?
    b) Solve the linear congruence $7x \equiv 13 \pmod{19}$.

13. **a)** State the Chinese remainder theorem.
    b) Find the solutions to the system $x \equiv 1 \pmod{4}$, $x \equiv 2 \pmod{5}$, and $x \equiv 3 \pmod{7}$.

14. Suppose that $2^{n-1} \equiv 1 \pmod{n}$. Is n necessarily prime?

15. Use Fermat's little theorem to evaluate $9^{200} \bmod 19$.

16. Explain how the check digit is found for a 10-digit ISBN.

17. Encrypt the message APPLES AND ORANGES using a shift cipher with key $k = 13$.

18. **a)** What is the difference between a public key and a private key cryptosystem?
    b) Explain why using shift ciphers is a private key system.
    c) Explain why the RSA cryptosystem is a public key system.

19. Explain how encryption and decryption are done in the RSA cryptosystem.

20. Describe how two parties can share a secret key using the Diffie-Hellman key exchange protocol.

Supplementary Exercises

1. The odometer on a car goes to up 100,000 miles. The present owner of a car bought it when the odometer read 43,179 miles. He now wants to sell it; when you examine the car for possible purchase, you notice that the odometer reads 89,697 miles. What can you conclude about how many miles he drove the car, assuming that the odometer always worked correctly?

2. **a)** Explain why n **div** 7 equals the number of complete weeks in n days.
   b) Explain why n **div** 24 equals the number of complete days in n hours.

3. Find four numbers congruent to 5 modulo 17.

4. Show that if a and d are positive integers, then there are integers q and r such that $a = dq + r$ where $-d/2 < r \le d/2$.

*5. Show that if $ac \equiv bc \pmod{m}$, where a, b, c, and m are integers with $m > 2$, and $d = \gcd(m, c)$, then $a \equiv b \pmod{m/d}$.

6. Show that if $n^2 + 1$ is a perfect square, where n is an integer, then n is even.

7. Develop a test for divisibility of a positive integer n by 8 based on the binary expansion of n.

8. Devise an algorithm for guessing a number between 1 and $2^n - 1$ by successively guessing each bit in its binary expansion.

9. Show that an integer is divisible by 9 if and only if the sum of its decimal digits is divisible by 9.

10. Prove there are infinitely many primes by showing that $Q_n = n! + 1$ must have a prime factor greater than n whenever n is a positive integer.

11. Use Dirichlet's theorem, which states there are infinitely many primes in every arithmetic progression $ak + b$ where $\gcd(a, b) = 1$, to show that there are infinitely many primes that have a decimal expansion ending with a 1.

12. Prove that if n is a positive integer such that the sum of the divisors of n is $n + 1$, then n is prime.

*13. Show that every integer greater than 11 is the sum of two composite integers.

14. Find the five smallest consecutive composite integers.

15. Show that Goldbach's conjecture, which states that every even integer greater than 2 is the sum of two primes, is equivalent to the statement that every integer greater than 5 is the sum of three primes.

*16. Prove that if $f(x)$ is a nonconstant polynomial with integer coefficients, then there is an integer y such that $f(y)$ is composite. [*Hint:* Assume that $f(x_0) = p$ is prime. Show that p divides $f(x_0 + kp)$ for all integers k. Obtain a contradiction of the fact that a polynomial of degree n, where $n > 1$, takes on each value at most n times.]

17. Use the Euclidean algorithm to find the greatest common divisor of 10,223 and 33,341.

18. How many divisions are required to find $\gcd(144, 233)$ using the Euclidean algorithm?

19. Find $\gcd(2n + 1, 3n + 2)$, where n is a positive integer. [*Hint:* Use the Euclidean algorithm.]

20. a) Show that if a and b are positive integers with $a \geq b$, then $\gcd(a, b) = a$ if $a = b$, $\gcd(a, b) = 2 \gcd(a/2, b/2)$ if a and b are even, $\gcd(a, b) = \gcd(a/2, b)$ if a is even and b is odd, and $\gcd(a, b) = \gcd(a - b, b)$ if both a and b are odd.

b) Explain how to use (a) to construct an algorithm for computing the greatest common divisor of two positive integers that uses only comparisons, subtractions, and shifts of binary expansions, without using any divisions.

c) Find $\gcd(1202, 4848)$ using this algorithm.

21. Adapt the proof that there are infinitely many primes (Theorem 3 in Section 4.3) to show that are infinitely many primes in the arithmetic progression $6k + 5, k = 1, 2, \ldots$.

22. Explain why you cannot directly adapt the proof that there are infinitely many primes (Theorem 3 in Section 4.3) to show that there are infinitely many primes in the arithmetic progression $3k + 1, k = 1, 2, \ldots$.

23. Explain why you cannot directly adapt the proof that there are infinitely many primes (Theorem 3 in Section 4.3) to show that are infinitely many primes in the arithmetic progression $4k + 1, k = 1, 2, \ldots$.

24. Show that if the smallest prime factor p of the positive integer n is larger than $\sqrt[3]{n}$, then n/p is prime or equal to 1.

A set of integers is called **mutually relatively prime** if the greatest common divisor of these integers is 1.

25. Determine whether the integers in each of these sets are mutually relatively prime.

a) 8, 10, 12   **b)** 12, 15, 25

c) 15, 21, 28   **d)** 21, 24, 28, 32

*26. For which positive integers n is $n^4 + 4^n$ prime?

27. Show that the system of congruences $x \equiv 2 \pmod 6$ and $x \equiv 3 \pmod 9$ has no solutions.

28. Find all solutions of the system of congruences $x \equiv 4 \pmod 6$ and $x \equiv 13 \pmod{15}$.

29. Prove that 30 divides $n^9 - n$ for every nonnegative integer n.

30. Show that if p and q are distinct prime numbers, then $p^{q-1} + q^{p-1} \equiv 1 \pmod{pq}$.

The check digit a_{13} for an ISBN-13 with initial digits $a_1 a_2 \ldots a_{12}$ is determined by the congruence $(a_1 + a_3 + \cdots + a_{13}) + 3(a_2 + a_4 + \cdots + a_{12}) \equiv 0 \pmod{10}$.

31. Determine whether each of these 13-digit numbers is a valid ISBN-13.

a) 978-0-073-20679-1

b) 978-0-45424-521-1

c) 978-3-16-148410-0

d) 978-0-201-10179-9

32. Show that the check digit of an ISBN-13 can always detect a single error.

33. Show that there are transpositions of two digits that are not detected by an ISBN-13.

34. The encrypted version of a message is LJMKG MGMXF QEXMW. If it was encrypted using the affine cipher $f(p) = (7p + 10) \bmod 26$, what was the original message?

Autokey ciphers are ciphers where the nth letter of the plaintext is shifted by the numerical equivalent of the nth letter of a keystream. The keystream begins with a seed letter; its subsequent letters are constructed using either the plaintext or the ciphertext. When the plaintext is used, each character of the keystream, after the first, is the previous letter of the plaintext.

When the ciphertext is used, each subsequent character of the keystream, after the first, is the previous letter of the ciphertext computed so far. In both cases, plaintext letters are encrypted by shifting each character by the numerical equivalent of the corresponding keystream letter.

35. Use the autokey cipher to encrypt the message THE DREAM OF REASON (ignoring spaces) using

    a) the keystream with seed X followed by letters of the plaintext.

    b) the keystream with seed X followed by letters of the ciphertext.

c) Use the autokey cipher to encrypt the message NOW IS THE TIME TO DECIDE (ignoring spaces) using

    a.the keystream with seed X followed by letters of the plaintext.

    b.the keystream with seed X followed by letters of the ciphertext.

36. Use the autokey cipher to encrypt the message NOW IS THE TIME TO DECIDE (ignoring spaces) using

    a) the keystream with seed X followed by letters of the plaintext.

    b) the keystream with seed X followed by letters of the ciphertext.

Computer Projects

Write programs with these inputs and outputs.

1. Given integers n and b, each greater than 1, find the base b expansion of this integer.

2. Given the positive integers a, b, and m with $m > 1$, find $a^b \bmod m$.

3. Given a positive integer, find the Cantor expansion of this integer (see the preamble to Exercise 48 of Section 4.2).

4. Given a positive integer, determine whether it is prime using trial division.

5. Given a positive integer, find the prime factorization of this integer.

6. Given two positive integers, find their greatest common divisor using the Euclidean algorithm.

7. Given two positive integers, find their least common multiple.

8. Given positive integers a and b, find Bézout coefficients s and t of a and b.

9. Given relatively prime positive integers a and b, find an inverse of a modulo b.

10. Given a message and a positive integer k less than 26, encrypt this message using the shift cipher with key k; and given a message encrypted using a shift cipher with key k, decrypt this message.

11. Given a message and positive integers a and b less than 26 with $\gcd(a, 26)$, encrypt this message using an affine cipher with key (a, b); and given a message encrypted using the affine cipher with key (a, b), decrypt this message, by first finding the decryption key and then applying the appropriate decryption transformation.

12. Find the original plaintext message from the ciphertext message produced by encrypting the plaintext message using a shift cipher. Do this using a frequency count of letters in the ciphertext.

∗13. Construct a valid RSA encryption key by finding two primes p and q with 200 digits each and an integer $e > 1$ relatively prime to $(p-1)(q-1)$.

14. Given a message and an integer $n = pq$ where p and q are odd primes and an integer $e > 1$ relatively prime to $(p-1)(q-1)$, encrypt the message using the RSA cryptosystem with key (n, e).

15. Given a valid RSA key (n, e), and the primes p and q with $n = pq$, find the associated decryption key d.

16. Given a message encrypted using the RSA cryptosystem with key (n, e) and the associated decryption key d, decrypt this message.

Computations and Explorations

Use a computational program or programs you have written to do these exercises.

1. Determine whether $2^p - 1$ is prime for each of the primes not exceeding 100.

2. Test a range of large Mersenne numbers $2^p - 1$ to determine whether they are prime. (You may want to use software from the GIMPS project.)

3. Determine whether $Q_n = p_1 p_2 \cdots p_n + 1$ is prime where $p_1, p_2, \ldots, p_n$ are the n smallest primes, for as many positive integer n as possible.

4. Look for polynomials in one variables whose values at long runs of consecutive integers are all primes.

5. Find as many primes of the form $n^2 + 1$ where n is a positive integer as you can. It is not known whether there are infinitely many such primes.

Writing Projects

Respond to these with essays using outside sources.

1. Describe the Lucas–Lehmer test for determining whether a Mersenne number is prime. Discuss the progress of the GIMPS project in finding Mersenne primes using this test.

2. Explain how probabilistic primality tests are used in practice to produce extremely large numbers that are almost certainly prime. Do such tests have any potential drawbacks?

3. Show how a congruence can be used to tell the day of the week for any given date.

4. Describe how public key cryptography is being applied. Are the ways it is applied secure given the status of factoring algorithms? Will information kept secure using public key cryptography become insecure in the future?

5. Describe how public key cryptography can be used to produce signed secret messages so that the recipient is relatively sure the message was sent by the person expected to have sent it.

6. Describe the Rabin public key cryptosystem, explaining how to encrypt and how to decrypt messages and why it is suitable for use as a public key cryptosystem.

*7. Explain why it would not be suitable to use p, where p is a large prime, as the modulus for encryption in the RSA cryptosystem. That is, explain how someone could, without excessive computation, find a private key from the corresponding public key if the modulus were a large prime, rather than the product of two large primes.

8. Explain what is meant by a cryptographic hash function? What are the important properties such a function must have?

5 Induction and Recursion

Many mathematical statements assert that a property is true for all positive integers. Examples of such statements are that for every positive integer n: $n! \leq n^n$, $n^3 - n$ is divisible by 3; a set with n elements has 2^n subsets; and the sum of the first n positive integers is $n(n + 1)/2$. A major goal of this chapter, and the book, is to give the student a thorough understanding of mathematical induction, which is used to prove results of this kind.

Proofs using mathematical induction have two parts. First, they show that the statement holds for the positive integer 1. Second, they show that if the statement holds for a positive integer then it must also hold for the next larger integer. Mathematical induction is based on the rule of inference that tells us that if $P(1)$ and $\forall k(P(k) \rightarrow P(k + 1))$ are true for the domain of positive integers, then $\forall n P(n)$ is true. Mathematical induction can be used to prove a tremendous variety of results. Understanding how to read and construct proofs by mathematical induction is a key goal of learning discrete mathematics.

In Chapter 2 we explicitly defined sets and functions. That is, we described sets by listing their elements or by giving some property that characterizes these elements. We gave formulae for the values of functions. There is another important way to define such objects, based on mathematical induction. To define functions, some initial terms are specified, and a rule is given for finding subsequent values from values already known. (We briefly touched on this sort of definition in Chapter 2 when we showed how sequences can be defined using recurrence relations.) Sets can be defined by listing some of their elements and giving rules for constructing elements from those already known to be in the set. Such definitions, called *recursive definitions,* are used throughout discrete mathematics and computer science. Once we have defined a set recursively, we can use a proof method called structural induction to prove results about this set.

When a procedure is specified for solving a problem, this procedure must *always* solve the problem correctly. Just testing to see that the correct result is obtained for a set of input values does not show that the procedure always works correctly. The correctness of a procedure can be guaranteed only by proving that it always yields the correct result. The final section of this chapter contains an introduction to the techniques of program verification. This is a formal technique to verify that procedures are correct. Program verification serves as the basis for attempts under way to prove in a mechanical fashion that programs are correct.

5.1 Mathematical Induction

Introduction

Suppose that we have an infinite ladder, as shown in Figure 1, and we want to know whether we can reach every step on this ladder. We know two things:

1. We can reach the first rung of the ladder.
2. If we can reach a particular rung of the ladder, then we can reach the next rung.

Can we conclude that we can reach every rung? By (1), we know that we can reach the first rung of the ladder. Moreover, because we can reach the first rung, by (2), we can also reach the second rung; it is the next rung after the first rung. Applying (2) again, because we can reach the second rung, we can also reach the third rung. Continuing in this way, we can show that we

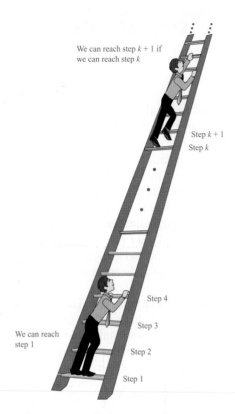

We can reach step $k + 1$ if we can reach step k

Step $k + 1$

Step k

Step 4

Step 3

Step 2

We can reach step 1

Step 1

FIGURE 1   Climbing an Infinite Ladder.

can reach the fourth rung, the fifth rung, and so on. For example, after 100 uses of (2), we know that we can reach the 101st rung. But can we conclude that we are able to reach every rung of this infinite ladder? The answer is yes, something we can verify using an important proof technique called **mathematical induction**. That is, we can show that $P(n)$ is true for every positive integer n, where $P(n)$ is the statement that we can reach the nth rung of the ladder.

Mathematical induction is an extremely important proof technique that can be used to prove assertions of this type. As we will see in this section and in subsequent sections of this chapter and later chapters, mathematical induction is used extensively to prove results about a large variety of discrete objects. For example, it is used to prove results about the complexity of algorithms, the correctness of certain types of computer programs, theorems about graphs and trees, as well as a wide range of identities and inequalities.

In this section, we will describe how mathematical induction can be used and why it is a valid proof technique. It is extremely important to note that mathematical induction can be used only to prove results obtained in some other way. It is *not* a tool for discovering formulae or theorems.

Mathematical Induction

Assessment

In general, mathematical induction * can be used to prove statements that assert that $P(n)$ is true for all positive integers n, where $P(n)$ is a propositional function. A proof by mathematical

---

*Unfortunately, using the terminology "mathematical induction" clashes with the terminology used to describe different types of reasoning. In logic, **deductive reasoning** uses rules of inference to draw conclusions from premises, whereas **inductive reasoning** makes conclusions only supported, but not ensured, by evidence. Mathematical proofs, including arguments that use mathematical induction, are deductive, not inductive.

induction has two parts, a **basis step**, where we show that $P(1)$ is true, and an **inductive step**, where we show that for all positive integers k, if $P(k)$ is true, then $P(k + 1)$ is true.

PRINCIPLE OF MATHEMATICAL INDUCTION   To prove that $P(n)$ is true for all positive integers n, where $P(n)$ is a propositional function, we complete two steps:

BASIS STEP: We verify that $P(1)$ is true.

INDUCTIVE STEP: We show that the conditional statement $P(k) \to P(k + 1)$ is true for all positive integers k.

To complete the inductive step of a proof using the principle of mathematical induction, we assume that $P(k)$ is true for an arbitrary positive integer k and show that under this assumption, $P(k + 1)$ must also be true. The assumption that $P(k)$ is true is called the **inductive hypothesis**. Once we complete both steps in a proof by mathematical induction, we have shown that $P(n)$ is true for all positive integers, that is, we have shown that $\forall n\, P(n)$ is true where the quantification is over the set of positive integers. In the inductive step, we show that $\forall k (P(k) \to P(k + 1))$ is true, where again, the domain is the set of positive integers.

Expressed as a rule of inference, this proof technique can be stated as

$$(P(1) \wedge \forall k (P(k) \to P(k + 1))) \to \forall n\, P(n),$$

when the domain is the set of positive integers. Because mathematical induction is such an important technique, it is worthwhile to explain in detail the steps of a proof using this technique. The first thing we do to prove that $P(n)$ is true for all positive integers n is to show that $P(1)$ is true. This amounts to showing that the particular statement obtained when n is replaced by 1 in $P(n)$ is true. Then we must show that $P(k) \to P(k + 1)$ is true for every positive integer k. To prove that this conditional statement is true for every positive integer k, we need to show that $P(k + 1)$ cannot be false when $P(k)$ is true. This can be accomplished by assuming that $P(k)$ is true and showing that *under this hypothesis $P(k + 1)$* must also be true.

Remark: In a proof by mathematical induction it is *not* assumed that $P(k)$ is true for all positive integers! It is only shown that *if it is assumed* that $P(k)$ is true, then $P(k + 1)$ is also true. Thus, a proof by mathematical induction is not a case of begging the question, or circular reasoning.

When we use mathematical induction to prove a theorem, we first show that $P(1)$ is true. Then we know that $P(2)$ is true, because $P(1)$ implies $P(2)$. Further, we know that $P(3)$ is true, because $P(2)$ implies $P(3)$. Continuing along these lines, we see that $P(n)$ is true for every positive integer n.

HISTORICAL NOTE   The first known use of mathematical induction is in the work of the sixteenth-century mathematician Francesco Maurolico (1494–1575). Maurolico wrote extensively on the works of classical mathematics and made many contributions to geometry and optics. In his book *Arithmeticorum Libri Duo,* Maurolico presented a variety of properties of the integers together with proofs of these properties. To prove some of these properties, he devised the method of mathematical induction. His first use of mathematical induction in this book was to prove that the sum of the first n odd positive integers equals n^2. Augustus De Morgan is credited with the first presentation in 1838 of formal proofs using mathematical induction, as well as introducing the terminology "mathematical induction." Maurolico's proofs were informal and he never used the word "induction." See [Gu11] to learn more about the history of the method of mathematical induction.

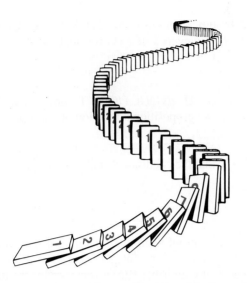

FIGURE 2 **Illustrating How Mathematical Induction Works Using Dominoes.**

WAYS TO REMEMBER HOW MATHEMATICAL INDUCTION WORKS Thinking of the infinite ladder and the rules for reaching steps can help you remember how mathematical induction works. Note that statements (1) and (2) for the infinite ladder are exactly the basis step and inductive step, respectively, of the proof that $P(n)$ is true for all positive integers n, where $P(n)$ is the statement that we can reach the nth rung of the ladder. Consequently, we can invoke mathematical induction to conclude that we can reach every rung.

Another way to illustrate the principle of mathematical induction is to consider an infinite row of dominoes, labeled $1, 2, 3, \ldots, n, \ldots$, where each domino is standing up. Let $P(n)$ be the proposition that domino n is knocked over. If the first domino is knocked over—i.e., if $P(1)$ is true—and if, whenever the kth domino is knocked over, it also knocks the $(k + 1)$st domino over—i.e., if $P(k) \to P(k + 1)$ is true for all positive integers k—then all the dominoes are knocked over. This is illustrated in Figure 2.

Why Mathematical Induction is Valid

Why is mathematical induction a valid proof technique? The reason comes from the well-ordering property, listed in Appendix 1, as an axiom for the set of positive integers, which states that every nonempty subset of the set of positive integers has a least element. So, suppose we know that $P(1)$ is true and that the proposition $P(k) \to P(k + 1)$ is true for all positive integers k. To show that $P(n)$ must be true for all positive integers n, assume that there is at least one positive integer for which $P(n)$ is false. Then the set S of positive integers for which $P(n)$ is false is nonempty. Thus, by the well-ordering property, S has a least element, which will be denoted by m. We know that m cannot be 1, because $P(1)$ is true. Because m is positive and greater than 1, $m - 1$ is a positive integer. Furthermore, because $m - 1$ is less than m, it is not in S, so $P(m - 1)$ must be true. Because the conditional statement $P(m - 1) \to P(m)$ is also true, it must be the case that $P(m)$ is true. This contradicts the choice of m. Hence, $P(n)$ must be true for every positive integer n.

The Good and the Bad of Mathematical Induction

An important point needs to be made about mathematical induction before we commence a study of its use. The good thing about mathematical induction is that it can be used to prove a conjecture once it is has been made (and is true). The bad thing about it is that it cannot be used to find new theorems. Mathematicians sometimes find proofs by mathematical induction unsatisfying because they do not provide insights as to why theorems are true. Many theorems can be proved in many ways, including by mathematical induction. Proofs of these theorems by methods other than mathematical induction are often preferred because of the insights they bring.

You can prove a theorem by mathematical induction even if you do not have the slightest idea why it is true!

Examples of Proofs by Mathematical Induction

Many theorems assert that $P(n)$ is true for all positive integers n, where $P(n)$ is a propositional function. Mathematical induction is a technique for proving theorems of this kind. In other words, mathematical induction can be used to prove statements of the form $\forall n \, P(n)$, where the domain is the set of positive integers. Mathematical induction can be used to prove an extremely wide variety of theorems, each of which is a statement of this form. (Remember, many mathematical assertions include an implicit universal quantifier. The statement "if n is a positive integer, then $n^3 - n$ is divisible by 3" is an example of this. Making the implicit universal quantifier explicit yields the statement "for every positive integer n, $n^3 - n$ is divisible by 3.)

We will use how theorems are proved using mathematical induction. The theorems we will prove include summation formulae, inequalities, identities for combinations of sets, divisibility results, theorems about algorithms, and some other creative results. In this section and in later sections, we will employ mathematical induction to prove many other types of results, including the correctness of computer programs and algorithms. Mathematical induction can be used to prove a wide variety of theorems, not just summation formulae, inequalities, and other types of examples we illustrate here. (For proofs by mathematical induction of many more interesting and diverse results, see the *Handbook of Mathematical Induction* by David Gunderson [Gu11]. This book is part of the extensive CRC Series in Discrete Mathematics, many of which may be of interest to readers. The author is the Series Editor of these books).

Note that there are many opportunities for errors in induction proofs. We will describe some incorrect proofs by mathematical induction at the end of this section and in the exercises. To avoid making errors in proofs by mathematical induction, try to follow the guidelines for such proofs given at the end of this section.

SEEING WHERE THE INDUCTIVE HYPOTHESIS IS USED   To help the reader understand each of the mathematical induction proofs in this section, we will note where the inductive hypothesis is used. We indicate this use in three different ways: by explicit mention in the text, by inserting the acronym IH (for inductive hypothesis) over an equals sign or a sign for an inequality, or by specifying the inductive hypothesis as the reason for a step in a multi-line display.

Look for the $\overset{\text{IH}}{=}$ symbol to see where the inductive hypothesis is used.

PROVING SUMMATION FORMULAE   We begin by using mathematical induction to prove several summation formulae. As we will see, mathematical induction is particularly well suited for proving that such formulae are valid. However, summation formulae can be proven in other ways. This is not surprising because there are often different ways to prove a theorem. The major disadvantage of using mathematical induction to prove a summation formula is that you cannot use it to derive this formula. That is, you must already have the formula before you attempt to prove it by mathematical induction.

Examples 1–4 illustrate how to use mathematical induction to prove summation formulae. The first summation formula we will prove by mathematical induction, in Example 1, is a closed formula for the sum of the smallest n positive integers.

EXAMPLE 1   Show that if n is a positive integer, then

$$1 + 2 + \cdots + n = \frac{n(n+1)}{2}.$$

Solution: Let $P(n)$ be the proposition that the sum of the first n positive integers, $1 + 2 + \cdots n = \frac{n(n+1)}{2}$, is $n(n+1)/2$. We must do two things to prove that $P(n)$ is true for $n = 1, 2, 3, \ldots$. Namely, we must show that $P(1)$ is true and that the conditional statement $P(k)$ implies $P(k+1)$ is true for $k = 1, 2, 3, \ldots$.

BASIS STEP: $P(1)$ is true, because $1 = \frac{1(1+1)}{2}$. (The left-hand side of this equation is 1 because 1 is the sum of the first positive integer. The right-hand side is found by substituting 1 for n in $n(n+1)/2$.)

INDUCTIVE STEP: For the inductive hypothesis we assume that $P(k)$ holds for an arbitrary positive integer k. That is, we assume that

$$1 + 2 + \cdots + k = \frac{k(k+1)}{2}.$$

If you are rusty simplifying algebraic expressions, this is the time to do some reviewing!

Under this assumption, it must be shown that $P(k+1)$ is true, namely, that

$$1 + 2 + \cdots + k + (k+1) = \frac{(k+1)[(k+1)+1]}{2} = \frac{(k+1)(k+2)}{2}$$

is also true. When we add $k+1$ to both sides of the equation in $P(k)$, we obtain

$$1 + 2 + \cdots + k + (k+1) \overset{\text{IH}}{=} \frac{k(k+1)}{2} + (k+1)$$

$$= \frac{k(k+1) + 2(k+1)}{2}$$

$$= \frac{(k+1)(k+2)}{2}.$$

This last equation shows that $P(k+1)$ is true under the assumption that $P(k)$ is true. This completes the inductive step.

We have completed the basis step and the inductive step, so by mathematical induction we know that $P(n)$ is true for all positive integers n. That is, we have proven that $1 + 2 + \cdots + n = n(n+1)/2$ for all positive integers n.   ◄

As we noted, mathematical induction is not a tool for finding theorems about all positive integers. Rather, it is a proof method for proving such results once they are conjectured. In Example 2, using mathematical induction to prove a summation formula, we will both formulate and then prove a conjecture.

EXAMPLE 2   Conjecture a formula for the sum of the first n positive odd integers. Then prove your conjecture using mathematical induction.

Solution: The sums of the first n positive odd integers for $n = 1, 2, 3, 4, 5$ are

$$1 = 1, \qquad 1 + 3 = 4, \qquad 1 + 3 + 5 = 9,$$
$$1 + 3 + 5 + 7 = 16, \quad 1 + 3 + 5 + 7 + 9 = 25.$$

From these values it is reasonable to conjecture that the sum of the first n positive odd integers is n^2, that is, $1 + 3 + 5 + \cdots + (2n - 1) = n^2$. We need a method to *prove* that this *conjecture* is correct, if in fact it is.

Let $P(n)$ denote the proposition that the sum of the first n odd positive integers is n^2. Our conjecture is that $P(n)$ is true for all positive integers. To use mathematical induction to prove this conjecture, we must first complete the basis step; that is, we must show that $P(1)$ is true. Then we must carry out the inductive step; that is, we must show that $P(k + 1)$ is true when $P(k)$ is assumed to be true. We now attempt to complete these two steps.

BASIS STEP: $P(1)$ states that the sum of the first one odd positive integer is 1^2. This is true because the sum of the first odd positive integer is 1. The basis step is complete.

INDUCTIVE STEP: To complete the inductive step we must show that the proposition $P(k) \rightarrow P(k + 1)$ is true for every positive integer k. To do this, we first assume the inductive hypothesis. The inductive hypothesis is the statement that $P(k)$ is true for an arbitrary positive integer k, that is,

$$1 + 3 + 5 + \cdots + (2k - 1) = k^2.$$

(Note that the kth odd positive integer is $(2k - 1)$, because this integer is obtained by adding 2 a total of $k - 1$ times to 1.) To show that $\forall k (P(k) \rightarrow P(k + 1))$ is true, we must show that if $P(k)$ is true (the inductive hypothesis), then $P(k + 1)$ is true. Note that $P(k + 1)$ is the statement that

$$1 + 3 + 5 + \cdots + (2k - 1) + (2k + 1) = (k + 1)^2.$$

So, assuming that $P(k)$ is true, it follows that

$$
\begin{aligned}
1 + 3 + 5 + \cdots + (2k - 1) + (2k + 1) &= [1 + 3 + \cdots + (2k - 1)] + (2k + 1) \\
&\stackrel{\text{IH}}{=} k^2 + (2k + 1) \\
&= k^2 + 2k + 1 \\
&= (k + 1)^2.
\end{aligned}
$$

This shows that $P(k + 1)$ follows from $P(k)$. Note that we used the inductive hypothesis $P(k)$ in the second equality to replace the sum of the first k odd positive integers by k^2.

We have now completed both the basis step and the inductive step. That is, we have shown that $P(1)$ is true and the conditional statement $P(k) \rightarrow P(k + 1)$ is true for all positive integers k. Consequently, by the principle of mathematical induction we can conclude that $P(n)$ is true for all positive integers n. That is, we know that $1 + 3 + 5 + \cdots + (2n - 1) = n^2$ for all positive integers n. ◄

Often, we will need to show that $P(n)$ is true for $n = b, \ b + 1, \ b + 2, \ldots$, where b is an integer other than 1. We can use mathematical induction to accomplish this, as long as we change the basis step by replacing $P(1)$ with $P(b)$. In other words, to use mathematical induction to show that $P(n)$ is true for $n = b, b + 1, b + 2, \ldots$, where b is an integer other than 1, we show that $P(b)$ is true in the basis step. In the inductive step, we show that the conditional statement $P(k) \rightarrow P(k + 1)$ is true for $k = b, b + 1, b + 2, \ldots$. Note that b can be negative, zero, or positive. Following the domino analogy we used earlier, imagine that we begin by knocking down the bth domino (the basis step), and as each domino falls, it knocks down the next domino (the inductive step). We leave it to the reader to show that this form of induction is valid (see Exercise 65).

We illustrate this notion in Example 3, which states that a summation formula is valid for all nonnegative integers. In this example, we need to prove that $P(n)$ is true for $n = 0, 1, 2, \ldots$. So, the basis step in Example 3 shows that $P(0)$ is true.

EXAMPLE 3   Use mathematical induction to show that

$$1 + 2 + 2^2 + \cdots + 2^n = 2^{n+1} - 1$$

for all nonnegative integers n.

Solution: Let $P(n)$ be the proposition that $1 + 2 + 2^2 + \cdots + 2^n = 2^{n+1} - 1$ for the integer n.

BASIS STEP: $P(0)$ is true because $2^0 = 1 = 2^1 - 1$. This completes the basis step.

INDUCTIVE STEP: For the inductive hypothesis, we assume that $P(k)$ is true for an arbitrary nonnegative integer k. That is, we assume that

$$1 + 2 + 2^2 + \cdots + 2^k = 2^{k+1} - 1.$$

To carry out the inductive step using this assumption, we must show that when we assume that $P(k)$ is true, then $P(k + 1)$ is also true. That is, we must show that

$$1 + 2 + 2^2 + \cdots + 2^k + 2^{k+1} = 2^{(k+1)+1} - 1 = 2^{k+2} - 1$$

assuming the inductive hypothesis $P(k)$. Under the assumption of $P(k)$, we see that

$$
\begin{aligned}
1 + 2 + 2^2 + \cdots + 2^k + 2^{k+1} &= (1 + 2 + 2^2 + \cdots + 2^k) + 2^{k+1} \\
&\overset{\text{IH}}{=} (2^{k+1} - 1) + 2^{k+1} \\
&= 2 \cdot 2^{k+1} - 1 \\
&= 2^{k+2} - 1.
\end{aligned}
$$

Note that we used the inductive hypothesis in the second equation in this string of equalities to replace $1 + 2 + 2^2 + \cdots + 2^k$ by $2^{k+1} - 1$. We have completed the inductive step.

Because we have completed the basis step and the inductive step, by mathematical induction we know that $P(n)$ is true for all nonnegative integers n. That is, $1 + 2 + \cdots + 2^n = 2^{n+1} - 1$ for all nonnegative integers n. ◀

The formula given in Example 3 is a special case of a general result for the sum of terms of a geometric progression (Theorem 1 in Section 2.4). We will use mathematical induction to provide an alternative proof of this formula.

EXAMPLE 4   **Sums of Geometric Progressions**   Use mathematical induction to prove this formula for the sum of a finite number of terms of a geometric progression with initial term a and common ratio r:

$$\sum_{j=0}^{n} ar^j = a + ar + ar^2 + \cdots + ar^n = \frac{ar^{n+1} - a}{r - 1} \qquad \text{when } r \neq 1,$$

where n is a nonnegative integer.

Solution: To prove this formula using mathematical induction, let $P(n)$ be the statement that the sum of the first $n + 1$ terms of a geometric progression in this formula is correct.

BASIS STEP: $P(0)$ is true, because

$$\frac{ar^{0+1} - a}{r - 1} = \frac{ar - a}{r - 1} = \frac{a(r - 1)}{r - 1} = a.$$

INDUCTIVE STEP: The inductive hypothesis is the statement that $P(k)$ is true, where k is an arbitrary nonnegative integer. That is, $P(k)$ is the statement that

$$a + ar + ar^2 + \cdots + ar^k = \frac{ar^{k+1} - a}{r - 1}.$$

To complete the inductive step we must show that if $P(k)$ is true, then $P(k + 1)$ is also true. To show that this is the case, we first add ar^{k+1} to both sides of the equality asserted by $P(k)$. We find that

$$a + ar + ar^2 + \cdots + ar^k + ar^{k+1} \overset{\text{IH}}{=} \frac{ar^{k+1} - a}{r - 1} + ar^{k+1}.$$

Rewriting the right-hand side of this equation shows that

$$\frac{ar^{k+1} - a}{r - 1} + ar^{k+1} = \frac{ar^{k+1} - a}{r - 1} + \frac{ar^{k+2} - ar^{k+1}}{r - 1}$$
$$= \frac{ar^{k+2} - a}{r - 1}.$$

Combining these last two equations gives

$$a + ar + ar^2 + \cdots + ar^k + ar^{k+1} = \frac{ar^{k+2} - a}{r - 1}.$$

This shows that if the inductive hypothesis $P(k)$ is true, then $P(k + 1)$ must also be true. This completes the inductive argument.

We have completed the basis step and the inductive step, so by mathematical induction $P(n)$ is true for all nonnegative integers n. This shows that the formula for the sum of the terms of a geometric series is correct. ◀

As previously mentioned, the formula in Example 3 is the case of the formula in Example 4 with $a = 1$ and $r = 2$. The reader should verify that putting these values for a and r into the general formula gives the same formula as in Example 3.

PROVING INEQUALITIES   Mathematical induction can be used to prove a variety of inequalities that hold for all positive integers greater than a particular positive integer, as Examples 5–7 illustrate.

EXAMPLE 5   Use mathematical induction to prove the inequality

$$n < 2^n$$

for all positive integers n.

Solution: Let $P(n)$ be the proposition that $n < 2^n$.

BASIS STEP: $P(1)$ is true, because $1 < 2^1 = 2$. This completes the basis step.

INDUCTIVE STEP: We first assume the inductive hypothesis that $P(k)$ is true for anarbitrary positive integer k. That is, the inductive hypothesis $P(k)$ is the statement that $k < 2^k$. To complete the inductive step, we need to show that if $P(k)$ is true, then $P(k + 1)$, which is the statement that $k + 1 < 2^{k+1}$, is true. That is, we need to show that if $k < 2^k$, then $k + 1 < 2^{k+1}$. To show that this conditional statement is true for the positive integer k, we first add 1 to both sides of $k < 2^k$, and then note that $1 \leq 2^k$. This tells us that

$$k + 1 \overset{\text{IH}}{<} 2^k + 1 \leq 2^k + 2^k = 2 \cdot 2^k = 2^{k+1}.$$

This shows that $P(k + 1)$ is true, namely, that $k + 1 < 2^{k+1}$, based on the assumption that $P(k)$ is true. The induction step is complete.

Therefore, because we have completed both the basis step and the inductive step, by the principle of mathematical induction we have shown that $n < 2^n$ is true for all positive integers n.   ◀

EXAMPLE 6   Use mathematical induction to prove that $2^n < n!$ for every integer n with $n \geq 4$. (Note that this inequality is false for $n = 1, 2$, and 3.)

Solution: Let $P(n)$ be the proposition that $2^n < n!$.

BASIS STEP: To prove the inequality for $n \geq 4$ requires that the basis step be $P(4)$. Note that $P(4)$ is true, because $2^4 = 16 < 24 = 4!$.

INDUCTIVE STEP: For the inductive step, we assume that $P(k)$ is true for an arbitrary integer k with $k \geq 4$. That is, we assume that $2^k < k!$ for the positive integer k with $k \geq 4$. We must show that under this hypothesis, $P(k + 1)$ is also true. That is, we must show that if $2^k < k!$ for an arbitrary positive integer k where $k \geq 4$, then $2^{k+1} < (k + 1)!$. We have

$$2^{k+1} = 2 \cdot 2^k \qquad \text{by definition of exponent}$$

$$< 2 \cdot k! \qquad \text{by the inductive hypothesis}$$

$$< (k + 1)k! \qquad \text{because } 2 < k + 1$$

$$= (k + 1)! \qquad \text{by definition of factorial function.}$$

This shows that $P(k + 1)$ is true when $P(k)$ is true. This completes the inductive step of the proof.

We have completed the basis step and the inductive step. Hence, by mathematical induction $P(n)$ is true for all integers n with $n \geq 4$. That is, we have proved that $2^n < n!$ is true for all integers n with $n \geq 4$.   ◀

An important inequality for the sum of the reciprocals of a set of positive integers will be proved in Example 7.

EXAMPLE 7   **An Inequality for Harmonic Numbers**   The **harmonic numbers** H_j, $j = 1, 2, 3, \ldots$, are defined by

$$H_j = 1 + \frac{1}{2} + \frac{1}{3} + \cdots + \frac{1}{j}.$$

For instance,

$$H_4 = 1 + \frac{1}{2} + \frac{1}{3} + \frac{1}{4} = \frac{25}{12}.$$

Use mathematical induction to show that

$$H_{2^n} \geq 1 + \frac{n}{2},$$

whenever n is a nonnegative integer.

Solution: To carry out the proof, let $P(n)$ be the proposition that $H_{2^n} \geq 1 + \dfrac{n}{2}$.

BASIS STEP: $P(0)$ is true, because $H_{2^0} = H_1 = 1 \geq 1 + \dfrac{0}{2}$.

INDUCTIVE STEP: The inductive hypothesis is the statement that $P(k)$ is true, that is, $H_{2^k} \geq 1 + \dfrac{k}{2}$, where k is an arbitrary nonnegative integer. We must show that if $P(k)$ is true, then $P(k+1)$, which states that $H_{2^{k+1}} \geq 1 + \dfrac{k+1}{2}$, is also true. So, assuming the inductive hypothesis, it follows that

$$
\begin{aligned}
H_{2^{k+1}} &= 1 + \frac{1}{2} + \frac{1}{3} + \cdots + \frac{1}{2^k} + \frac{1}{2^k + 1} + \cdots + \frac{1}{2^{k+1}} && \text{by the definition of harmonic number} \\
&= H_{2^k} + \frac{1}{2^k + 1} + \cdots + \frac{1}{2^{k+1}} && \text{by the definition of } 2^k \text{th harmonic number} \\
&\geq \left(1 + \frac{k}{2}\right) + \frac{1}{2^k + 1} + \cdots + \frac{1}{2^{k+1}} && \text{by the inductive hypothesis} \\
&\geq \left(1 + \frac{k}{2}\right) + 2^k \cdot \frac{1}{2^{k+1}} && \text{because there are } 2^k \text{ terms each} \geq 1/2^{k+1} \\
&\geq \left(1 + \frac{k}{2}\right) + \frac{1}{2} && \text{canceling a common factor of } 2^k \text{ in second term} \\
&= 1 + \frac{k+1}{2}.
\end{aligned}
$$

This establishes the inductive step of the proof.

We have completed the basis step and the inductive step. Thus, by mathematical induction $P(n)$ is true for all nonnegative integers n. That is, the inequality $H_{2^n} \geq 1 + \frac{n}{2}$ for the harmonic numbers holds for all nonnegative integers n. ◀

Remark: The inequality established here shows that the **harmonic series**

$$
1 + \frac{1}{2} + \frac{1}{3} + \cdots + \frac{1}{n} + \cdots
$$

is a divergent infinite series. This is an important example in the study of infinite series.

PROVING DIVISIBILITY RESULTS   Mathematical induction can be used to prove divisibility results about integers. Although such results are often easier to prove using basic results in number theory, it is instructive to see how to prove such results using mathematical induction, as Examples 8 and 9 illustrate.

EXAMPLE 8   Use mathematical induction to prove that $n^3 - n$ is divisible by 3 whenever n is a positive integer. (Note that this is the statement with $p = 3$ of Fermat's little theorem, which is Theorem 3 of Section 4.4.)

Extra Examples

Solution: To construct the proof, let $P(n)$ denote the proposition: "$n^3 - n$ is divisible by 3."

BASIS STEP: The statement $P(1)$ is true because $1^3 - 1 = 0$ is divisible by 3. This completes the basis step.

INDUCTIVE STEP: For the inductive hypothesis we assume that $P(k)$ is true; that is, we assume that $k^3 - k$ is divisible by 3 for an arbitrary positive integer k. To complete the inductive step, we must show that when we assume the inductive hypothesis, it follows that $P(k + 1)$, the statement that $(k + 1)^3 - (k + 1)$ is divisible by 3, is also true. That is, we must show that $(k + 1)^3 - (k + 1)$ is divisible by 3. Note that

$$(k + 1)^3 - (k + 1) = (k^3 + 3k^2 + 3k + 1) - (k + 1)$$
$$= (k^3 - k) + 3(k^2 + k).$$

Using the inductive hypothesis, we conclude that the first term $k^3 - k$ is divisible by 3. The second term is divisible by 3 because it is 3 times an integer. So, by part (i) of Theorem 1 in Section 4.1, we know that $(k + 1)^3 - (k + 1)$ is also divisible by 3. This completes the inductive step.

Because we have completed both the basis step and the inductive step, by the principle of mathematical induction we know that $n^3 - n$ is divisible by 3 whenever n is a positive integer. ◀

The next example presents a more challenging proof by mathematical induction of a divisibility result.

EXAMPLE 9 Use mathematical induction to prove that $7^{n+2} + 8^{2n+1}$ is divisible by 57 for every nonnegative integer n.

Solution: To construct the proof, let $P(n)$ denote the proposition: "$7^{n+2} + 8^{2n+1}$ is divisible by 57."

BASIS STEP: To complete the basis step, we must show that $P(0)$ is true, because we want to prove that $P(n)$ is true for every nonnegative integer. We see that $P(0)$ is true because $7^{0+2} + 8^{2·0+1} = 7^2 + 8^1 = 57$ is divisible by 57. This completes the basis step.

INDUCTIVE STEP: For the inductive hypothesis we assume that $P(k)$ is true for an arbitrary nonnegative integer k; that is, we assume that $7^{k+2} + 8^{2k+1}$ is divisible by 57. To complete the inductive step, we must show that when we assume that the inductive hypothesis $P(k)$ is true, then $P(k + 1)$, the statement that $7^{(k+1)+2} + 8^{2(k+1)+1}$ is divisible by 57, is also true.

The difficult part of the proof is to see how to use the inductive hypothesis. To take advantage of the inductive hypothesis, we use these steps:

$$7^{(k+1)+2} + 8^{2(k+1)+1} = 7^{k+3} + 8^{2k+3}$$
$$= 7 · 7^{k+2} + 8^2 · 8^{2k+1}$$
$$= 7 · 7^{k+2} + 64 · 8^{2k+1}$$
$$= 7(7^{k+2} + 8^{2k+1}) + 57 · 8^{2k+1}.$$

We can now use the inductive hypothesis, which states that $7^{k+2} + 8^{2k+1}$ is divisible by 57. We will use parts (i) and (ii) of Theorem 1 in Section 4.1. By part (ii) of this theorem, and the inductive hypothesis, we conclude that the first term in this last sum, $7(7^{k+2} + 8^{2k+1})$, is divisible by 57. By part (ii) of this theorem, the second term in this sum, $57 · 8^{2k+1}$, is divisible by 57. Hence, by part (i) of this theorem, we conclude that $7(7^{k+2} + 8^{2k+1}) + 57 · 8^{2k+1} = 7^{k+3} + 8^{2k+3}$ is divisible by 57. This completes the inductive step.

Because we have completed both the basis step and the inductive step, by the principle of mathematical induction we know that $7^{n+2} + 8^{2n+1}$ is divisible by 57 for every nonnegative integer n. ◀

PROVING RESULTS ABOUT SETS   Mathematical induction can be used to prove many results about sets. In particular, in Example 10 we prove a formula for the number of subsets of a finite set and in Example 11 we establish a set identity.

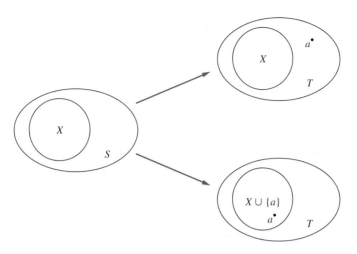

FIGURE 3 **Generating Subsets of a Set with $k + 1$ Elements. Here $T = S \cup \{a\}$.**

EXAMPLE 10   **The Number of Subsets of a Finite Set**   Use mathematical induction to show that if S is a finite set with n elements, where n is a nonnegative integer, then S has 2^n subsets. (We will prove this result directly in several ways in Chapter 6.)

Solution: Let $P(n)$ be the proposition that a set with n elements has 2^n subsets.

BASIS STEP: $P(0)$ is true, because a set with zero elements, the empty set, has exactly $2^0 = 1$ subset, namely, itself.

INDUCTIVE STEP: For the inductive hypothesis we assume that $P(k)$ is true for an arbitrary nonnegative integer k, that is, we assume that every set with k elements has 2^k subsets. It must be shown that under this assumption, $P(k + 1)$, which is the statement that every set with $k + 1$ elements has 2^{k+1} subsets, must also be true. To show this, let T be a set with $k + 1$ elements. Then, it is possible to write $T = S \cup \{a\}$, where a is one of the elements of T and $S = T - \{a\}$ (and hence $|S| = k$). The subsets of T can be obtained in the following way. For each subset X of S there are exactly two subsets of T, namely, X and $X \cup \{a\}$. (This is illustrated in Figure 3.) These constitute all the subsets of T and are all distinct. We now use the inductive hypothesis to conclude that S has 2^k subsets, because it has k elements. We also know that there are two subsets of T for each subset of S. Therefore, there are $2 \cdot 2^k = 2^{k+1}$ subsets of T. This finishes the inductive argument.

Because we have completed the basis step and the inductive step, by mathematical induction it follows that $P(n)$ is true for all nonnegative integers n. That is, we have proved that a set with n elements has 2^n subsets whenever n is a nonnegative integer.   ◀

EXAMPLE 11   Use mathematical induction to prove the following generalization of one of De Morgan's laws:

$$\overline{\bigcap_{j=1}^{n} A_j} = \bigcup_{j=1}^{n} \overline{A_j}$$

whenever $A_1, A_2, \ldots, A_n$ are subsets of a universal set U and $n \geq 2$.

Solution: Let $P(n)$ be the identity for n sets.

BASIS STEP: The statement $P(2)$ asserts that $\overline{A_1 \cap A_2} = \overline{A_1} \cup \overline{A_2}$. This is one of De Morgan's laws; it was proved in Example 11 of Section 2.2.

INDUCTIVE STEP: The inductive hypothesis is the statement that $P(k)$ is true, where k is an arbitrary integer with $k \geq 2$; that is, it is the statement that

$$\overline{\bigcap_{j=1}^{k} A_j} = \bigcup_{j=1}^{k} \overline{A_j}$$

whenever $A_1, A_2, \ldots, A_k$ are subsets of the universal set U. To carry out the inductive step, we need to show that this assumption implies that $P(k+1)$ is true. That is, we need to show that if this equality holds for every collection of k subsets of U, then it must also hold for every collection of $k+1$ subsets of U. Suppose that $A_1, A_2, \ldots, A_k, A_{k+1}$ are subsets of U. When the inductive hypothesis is assumed to hold, it follows that

$$\overline{\bigcap_{j=1}^{k+1} A_j} = \overline{\left(\bigcap_{j=1}^{k} A_j \right) \cap A_{k+1}} \qquad \text{by the definition of intersection}$$

$$= \overline{\left(\bigcap_{j=1}^{k} A_j \right)} \cup \overline{A_{k+1}} \qquad \text{by De Morgan's law (where the two sets are } \bigcap_{j=1}^{k} A_j \text{ and } A_{k+1})$$

$$= \left(\bigcup_{j=1}^{k} \overline{A_j} \right) \cup \overline{A_{k+1}} \qquad \text{by the inductive hypothesis}$$

$$= \bigcup_{j=1}^{k+1} \overline{A_j} \qquad \text{by the definition of union.}$$

This completes the inductive step.

Because we have completed both the basis step and the inductive step, by mathematical induction we know that $P(n)$ is true whenever n is a positive integer, $n \geq 2$. That is, we know that

$$\overline{\bigcap_{j=1}^{n} A_j} = \bigcup_{j=1}^{n} \overline{A_j}$$

whenever $A_1, A_2, \ldots, A_n$ are subsets of a universal set U and $n \geq 2$. ◀

PROVING RESULTS ABOUT ALGORITHMS   Next, we provide an example (somewhat more difficult than previous examples) that illustrates one of many ways mathematical induction is used in the study of algorithms. We will show how mathematical induction can be used to prove that a greedy algorithm we introduced in Section 3.1 always yields an optimal solution.

EXAMPLE 12   Recall the algorithm for scheduling talks discussed in Example 7 of Section 3.1. The input to this algorithm is a group of m proposed talks with preset starting and ending times. The goal is to schedule as many of these lectures as possible in the main lecture hall so that no two talks overlap. Suppose that talk t_j begins at time s_j and ends at time e_j. (No two lectures can proceed in the main lecture hall at the same time, but a lecture in this hall can begin at the same time another one ends.)

Without loss of generality, we assume that the talks are listed in order of nondecreasing ending time, so that $e_1 \leq e_2 \leq \cdots \leq e_m$. The greedy algorithm proceeds by selecting at each stage a talk with the earliest ending time among all those talks that begin no sooner than when the last talk scheduled in the main lecture hall has ended. Note that a talk with the earliest end time is always selected first by the algorithm. We will show that this greedy algorithm is optimal in the sense that it always schedules the most talks possible in the main lecture hall. To prove the optimality of this algorithm we use mathematical induction on the variable n, the number of talks scheduled by the algorithm. We let $P(n)$ be the proposition that if the greedy algorithm schedules n talks in the main lecture hall, then it is not possible to schedule more than n talks in this hall.

BASIS STEP: Suppose that the greedy algorithm managed to schedule just one talk, t_1, in the main lecture hall. This means that no other talk can start at or after e_1, the end time of t_1. Otherwise, the first such talk we come to as we go through the talks in order of nondecreasing end times could be added. Hence, at time e_1 each of the remaining talks needs to use the main lecture hall because they all start before e_1 and end after e_1. It follows that no two talks can be scheduled because both need to use the main lecture hall at time e_1. This shows that $P(1)$ is true and completes the basis step.

INDUCTIVE STEP: The inductive hypothesis is that $P(k)$ is true, where k is an arbitrary positive integer, that is, that the greedy algorithm always schedules the most possible talks when it selects k talks, where k is a positive integer, given any set of talks, no matter how many. We must show that $P(k + 1)$ follows from the assumption that $P(k)$ is true, that is, we must show that under the assumption of $P(k)$, the greedy algorithm always schedules the most possible talks when it selects $k + 1$ talks.

Now suppose that the greedy algorithm has selected $k + 1$ talks. Our first step in completing the inductive step is to show there is a schedule including the most talks possible that contains talk t_1, a talk with the earliest end time. This is easy to see because a schedule that begins with the talk t_i in the list, where $i > 1$, can be changed so that talk t_1 replaces talk t_i. To see this, note that because $e_1 \leq e_i$, all talks that were scheduled to follow talk t_i can still be scheduled.

Once we included talk t_1, scheduling the talks so that as many as possible are scheduled is reduced to scheduling as many talks as possible that begin at or after time e_1. So, if we have scheduled as many talks as possible, the schedule of talks other than talk t_1 is an optimal schedule of the original talks that begin once talk t_1 has ended. Because the greedy algorithm schedules k talks when it creates this schedule, we can apply the inductive hypothesis to conclude that it has scheduled the most possible talks. It follows that the greedy algorithm has scheduled the most possible talks, $k + 1$, when it produced a schedule with $k + 1$ talks, so $P(k + 1)$ is true. This completes the inductive step.

We have completed the basis step and the inductive step. So, by mathematical induction we know that $P(n)$ is true for all positive integers n. This completes the proof of optimality. That is, we have proved that when the greedy algorithm schedules n talks, when n is a positive integer, then it is not possible to schedule more than n talks. ◀

CREATIVE USES OF MATHEMATICAL INDUCTION   Mathematical induction can often be used in unexpected ways. We will illustrate two particularly clever uses of mathematical induction here, the first relating to survivors in a pie fight and the second relating to tilings with regular triominoes of checkerboards with one square missing.

EXAMPLE 13

Odd Pie Fights   An odd number of people stand in a yard at mutually distinct distances. At the same time each person throws a pie at their nearest neighbor, hitting this person. Use mathematical induction to show that there is at least one survivor, that is, at least one person who is not hit by a pie. (This problem was introduced by Carmony [Ca79]. Note that this result is false when there are an even number of people; see Exercise 62.)

Solution: Let $P(n)$ be the statement that there is a survivor whenever $2n + 1$ people stand in a yard at distinct mutual distances and each person throws a pie at their nearest neighbor. To prove this result, we will show that $P(n)$ is true for all positive integers n. This follows because as n runs through all positive integers, $2n + 1$ runs through all odd integers greater than or equal to 3. Note that one person cannot engage in a pie fight because there is no one else to throw the pie at.

BASIS STEP: When $n = 1$, there are $2n + 1 = 3$ people in the pie fight. Of the three people, suppose that the closest pair are A and B, and C is the third person. Because distances between pairs of people are different, the distance between A and C and the distance between B and C are both different from, and greater than, the distance between A and B. It follows that A and B throw pies at each other, while C throws a pie at either A or B, whichever is closer. Hence, C is not hit by a pie. This shows that at least one of the three people is not hit by a pie, completing the basis step.

INDUCTIVE STEP: For the inductive step, assume that $P(k)$ is true for an arbitrary odd integer k with $k \geq 3$. That is, assume that there is at least one survivor whenever $2k + 1$ people stand in a yard at distinct mutual distances and each throws a pie at their nearest neighbor. We must show that if the inductive hypothesis $P(k)$ is true, then $P(k + 1)$, the statement that there is at least one survivor whenever $2(k + 1) + 1 = 2k + 3$ people stand in a yard at distinct mutual distances and each throws a pie at their nearest neighbor, is also true.

So suppose that we have $2(k + 1) + 1 = 2k + 3$ people in a yard with distinct distances between pairs of people. Let A and B be the closest pair of people in this group of $2k + 3$ people. When each person throws a pie at the nearest person, A and B throw pies at each other. We have two cases to consider, *(i)* when someone else throws a pie at either A or B and *(ii)* when no one else throws a pie at either A or B.

Case (i): Because A and B throw pies at each other and someone else throws a pie at either A and B, at least three pies are thrown at A and B, and at most $(2k + 3) - 3 = 2k$ pies are thrown at the remaining $2k + 1$ people. This guarantees that at least one person is a survivor, for if each of these $2k + 1$ people was hit by at least one pie, a total of at least $2k + 1$ pies would have to be thrown at them. (The reasoning used in this last step is an example of the pigeonhole principle discussed further in Section 6.2.)

Case (ii): No one else throws a pie at either A and B. Besides A and B, there are $2k + 1$ people. Because the distances between pairs of these people are all different, we can use the inductive hypothesis to conclude that there is at least one survivor S when these $2k + 1$ people each throws a pie at their nearest neighbor. Furthermore, S is also not hit by either the pie thrown by A or the pie thrown by B because A and B throw their pies at each other, so S is a survivor because S is not hit by any of the pies thrown by these $2k + 3$ people.

We have completed both the basis step and the inductive step, using a proof by cases. So by mathematical induction it follows that $P(n)$ is true for all positive integers n. We conclude that whenever an odd number of people located in a yard at distinct mutual distances each throws a pie at their nearest neighbor, there is at least one survivor. ◀

Links

In Section 1.8 we discussed the tiling of checkerboards by polyominoes. Example 14 illustrates how mathematical induction can be used to prove a result about covering checkerboards with right triominoes, pieces shaped like the letter "L."

EXAMPLE 14 Let n be a positive integer. Show that every $2^n \times 2^n$ checkerboard with one square removed can be tiled using right triominoes, where these pieces cover three squares at a time, as shown in Figure 4.

Solution: Let $P(n)$ be the proposition that every $2^n \times 2^n$ checkerboard with one square removed can be tiled using right triominoes. We can use mathematical induction to prove that $P(n)$ is true for all positive integers n.

FIGURE 4 A Right Triomino.

BASIS STEP: $P(1)$ is true, because each of the four 2×2 checkerboards with one square removed can be tiled using one right triomino, as shown in Figure 5.

FIGURE 5   **Tiling 2 × 2 Checkerboards with One Square Removed.**

INDUCTIVE STEP: The inductive hypothesis is the assumption that $P(k)$ is true for the positive integer k; that is, it is the assumption that every $2^k \times 2^k$ checkerboard with one square removed can be tiled using right triominoes. It must be shown that under the assumption of the inductive hypothesis, $P(k + 1)$ must also be true; that is, any $2^{k+1} \times 2^{k+1}$ checkerboard with one square removed can be tiled using right triominoes.

To see this, consider a $2^{k+1} \times 2^{k+1}$ checkerboard with one square removed. Split this checkerboard into four checkerboards of size $2^k \times 2^k$, by dividing it in half in both directions. This is illustrated in Figure 6. No square has been removed from three of these four checkerboards. The fourth $2^k \times 2^k$ checkerboard has one square removed, so we now use the inductive hypothesis to conclude that it can be covered by right triominoes. Now temporarily remove the square from each of the other three $2^k \times 2^k$ checkerboards that has the center of the original, larger checkerboard as one of its corners, as shown in Figure 7. By the inductive hypothesis, each of these three $2^k \times 2^k$ checkerboards with a square removed can be tiled by right triominoes. Furthermore, the three squares that were temporarily removed can be covered by one right triomino. Hence, the entire $2^{k+1} \times 2^{k+1}$ checkerboard can be tiled with right triominoes.

We have completed the basis step and the inductive step. Therefore, by mathematical induction $P(n)$ is true for all positive integers n. This shows that we can tile every $2^n \times 2^n$ checkerboard, where n is a positive integer, with one square removed, using right triominoes. ◀

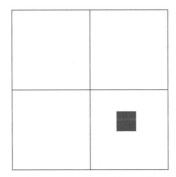

FIGURE 6   **Dividing a $2^{k+1} \times 2^{k+1}$ Checkerboard into Four $2^k \times 2^k$ Checkerboards.**

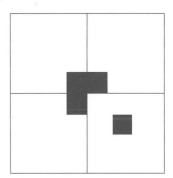

FIGURE 7   **Tiling the $2^{k+1} \times 2^{k+1}$ Checkerboard with One Square Removed.**

Mistaken Proofs By Mathematical Induction

Consult *Common Errors in Discrete Mathematics* on this book's website for more basic mistakes.

As with every proof method, there are many opportunities for making errors when using mathematical induction. Many well-known mistaken, and often entertaining, proofs by mathematical induction of clearly false statements have been devised, as exemplified by Example 15 and Exercises 39–41. Often, it is not easy to find where the error in reasoning occurs in such mistaken proofs.

To uncover errors in proofs by mathematical induction, remember that in every such proof, both the basis step and the inductive step must be done correctly. Not completing the basis step in a supposed proof by mathematical induction can lead to mistaken proofs of clearly ridiculous statements such as "$n = n + 1$ whenever n is a positive integer." (We leave it to the reader to show that it is easy to construct a correct inductive step in an attempted proof of this statement.) Locating the error in a faulty proof by mathematical induction, as Example 15 illustrates, can be quite tricky, especially when the error is hidden in the basis step.

EXAMPLE 15   Find the error in this "proof" of the clearly false claim that every set of lines in the plane, no two of which are parallel, meet in a common point.

"Proof:" Let $P(n)$ be the statement that every set of n lines in the plane, no two of which are parallel, meet in a common point. We will attempt to prove that $P(n)$ is true for all positive integers $n \geq 2$.

BASIS STEP: The statement $P(2)$ is true because any two lines in the plane that are not parallel meet in a common point (by the definition of parallel lines).

INDUCTIVE STEP: The inductive hypothesis is the statement that $P(k)$ is true for the positive integer k, that is, it is the assumption that every set of k lines in the plane, no two of which are parallel, meet in a common point. To complete the inductive step, we must show that if $P(k)$ is true, then $P(k + 1)$ must also be true. That is, we must show that if every set of k lines in the plane, no two of which are parallel, meet in a common point, then every set of $k + 1$ lines in the plane, no two of which are parallel, meet in a common point. So, consider a set of $k + 1$ distinct lines in the plane. By the inductive hypothesis, the first k of these lines meet in a common point p_1. Moreover, by the inductive hypothesis, the last k of these lines meet in a common point p_2. We will show that p_1 and p_2 must be the same point. If p_1 and p_2 were different points, all lines containing both of them must be the same line because two points determine a line. This contradicts our assumption that all these lines are distinct. Thus, p_1 and p_2 are the same point. We conclude that the point $p_1 = p_2$ lies on all $k + 1$ lines. We have shown that $P(k + 1)$ is true assuming that $P(k)$ is true. That is, we have shown that if we assume that every k, $k \geq 2$, distinct lines meet in a common point, then every $k + 1$ distinct lines meet in a common point. This completes the inductive step.

We have completed the basis step and the inductive step, and supposedly we have a correct proof by mathematical induction.

Solution: Examining this supposed proof by mathematical induction it appears that everything is in order. However, there is an error, as there must be. The error is rather subtle. Carefully looking at the inductive step shows that this step requires that $k \geq 3$. We cannot show that $P(2)$ implies $P(3)$. When $k = 2$, our goal is to show that every three distinct lines meet in a common point. The first two lines must meet in a common point p_1 and the last two lines must meet in a common point p_2. But in this case, p_1 and p_2 do not have to be the same, because only the second line is common to both sets of lines. Here is where the inductive step fails. ◄

Guidelines for Proofs by Mathematical Induction

Examples 1–14 illustrate proofs by mathematical induction of a diverse collection of theorems. Each of these examples includes all the elements needed in a proof by mathematical induction. We have provided an example of an invalid proof by mathematical induction. Summarizing what we have learned from these examples, we can provide some useful guidelines for constructing correct proofs by mathematical induction. We now present these guidelines.

Template for Proofs by Mathematical Induction

1. Express the statement that is to be proved in the form "for all $n \geq b$, $P(n)$" for a fixed integer b.
2. Write out the words "Basis Step." Then show that $P(b)$ is true, taking care that the correct value of b is used. This completes the first part of the proof.
3. Write out the words "Inductive Step."
4. State, and clearly identify, the inductive hypothesis, in the form "assume that $P(k)$ is true for an arbitrary fixed integer $k \geq b$."
5. State what needs to be proved under the assumption that the inductive hypothesis is true. That is, write out what $P(k + 1)$ says.
6. Prove the statement $P(k + 1)$ making use of the assumption $P(k)$. Be sure that your proof is valid for all integers k with $k \geq b$, taking care that the proof works for small values of k, including $k = b$.
7. Clearly identify the conclusion of the inductive step, such as by saying "this completes the inductive step."
8. After completing the basis step and the inductive step, state the conclusion, namely that by mathematical induction, $P(n)$ is true for all integers n with $n \geq b$.

It is worthwhile to revisit each of the mathematical induction proofs in Examples 1–14 to see how these steps are completed. It will be helpful to follow these guidelines in the solutions of the exercises that ask for proofs by mathematical induction. The guidelines that we presented can be adapted for each of the variants of mathematical induction that we introduce in the exercises and later in this chapter.

Exercises

1. There are infinitely many stations on a train route. Suppose that the train stops at the first station and suppose that if the train stops at a station, then it stops at the next station. Show that the train stops at all stations.

2. Suppose that you know that a golfer plays the first hole of a golf course with an infinite number of holes and that if this golfer plays one hole, then the golfer goes on to play the next hole. Prove that this golfer plays every hole on the course.

Use mathematical induction in Exercises 3–17 to prove summation formulae. Be sure to identify where you use the inductive hypothesis.

3. Let $P(n)$ be the statement that $1^2 + 2^2 + \cdots + n^2 = n(n + 1)(2n + 1)/6$ for the positive integer n.
   a) What is the statement $P(1)$?
   b) Show that $P(1)$ is true, completing the basis step of the proof.
   c) What is the inductive hypothesis?
   d) What do you need to prove in the inductive step?
   e) Complete the inductive step, identifying where you use the inductive hypothesis.
   f) Explain why these steps show that this formula is true whenever n is a positive integer.

4. Let $P(n)$ be the statement that $1^3 + 2^3 + \cdots + n^3 = (n(n + 1)/2)^2$ for the positive integer n.

   a) What is the statement $P(1)$?
   b) Show that $P(1)$ is true, completing the basis step of the proof.
   c) What is the inductive hypothesis?
   d) What do you need to prove in the inductive step?
   e) Complete the inductive step, identifying where you use the inductive hypothesis.
   f) Explain why these steps show that this formula is true whenever n is a positive integer.

5. Prove that $1^2 + 3^2 + 5^2 + \cdots + (2n + 1)^2 = (n + 1)(2n + 1)(2n + 3)/3$ whenever n is a nonnegative integer.

6. Prove that $1 \cdot 1! + 2 \cdot 2! + \cdots + n \cdot n! = (n + 1)! - 1$ whenever n is a positive integer.

7. Prove that $3 + 3 \cdot 5 + 3 \cdot 5^2 + \cdots + 3 \cdot 5^n = 3(5^{n+1} - 1)/4$ whenever n is a nonnegative integer.

8. Prove that $2 - 2 \cdot 7 + 2 \cdot 7^2 - \cdots + 2(-7)^n = (1 - (-7)^{n+1})/4$ whenever n is a nonnegative integer.

9. a) Find a formula for the sum of the first n even positive integers.

   b) Prove the formula that you conjectured in part (a).

10. a) Find a formula for

$$\frac{1}{1 \cdot 2} + \frac{1}{2 \cdot 3} + \cdots + \frac{1}{n(n+1)}$$

by examining the values of this expression for small values of n.

b) Prove the formula you conjectured in part (a).

11. a) Find a formula for

$$\frac{1}{2} + \frac{1}{4} + \frac{1}{8} + \cdots + \frac{1}{2^n}$$

by examining the values of this expression for small values of n.

b) Prove the formula you conjectured in part (a).

12. Prove that

$$\sum_{j=0}^{n} \left(-\frac{1}{2}\right)^j = \frac{2^{n+1} + (-1)^n}{3 \cdot 2^n}$$

whenever n is a nonnegative integer.

13. Prove that $1^2 - 2^2 + 3^2 - \cdots + (-1)^{n-1}n^2 = (-1)^{n-1} n(n+1)/2$ whenever n is a positive integer.

14. Prove that for every positive integer n, $\sum_{k=1}^{n} k2^k = (n-1)2^{n+1} + 2$.

15. Prove that for every positive integer n,

$$1 \cdot 2 + 2 \cdot 3 + \cdots + n(n+1) = n(n+1)(n+2)/3.$$

16. Prove that for every positive integer n,

$$1 \cdot 2 \cdot 3 + 2 \cdot 3 \cdot 4 + \cdots + n(n+1)(n+2)$$
$$= n(n+1)(n+2)(n+3)/4.$$

17. Prove that $\sum_{j=1}^{n} j^4 = n(n+1)(2n+1)(3n^2 + 3n - 1)/30$ whenever n is a positive integer.

Use mathematical induction to prove the inequalities in Exercises 18–26.

18. Let $P(n)$ be the statement that $n! < n^n$, where n is an integer greater than 1.

a) What is the statement $P(2)$?

b) Show that $P(2)$ is true, completing the basis step of the proof.

c) What is the inductive hypothesis?

d) What do you need to prove in the inductive step?

e) Complete the inductive step.

f) Explain why these steps show that this inequality is true whenever n is an integer greater than 1.

19. Let $P(n)$ be the statement that

$$1 + \frac{1}{4} + \frac{1}{9} + \cdots + \frac{1}{n^2} < 2 - \frac{1}{n},$$

where n is an integer greater than 1.

a) What is the statement $P(2)$?

b) Show that $P(2)$ is true, completing the basis step of the proof.

c) What is the inductive hypothesis?

d) What do you need to prove in the inductive step?

e) Complete the inductive step.

f) Explain why these steps show that this inequality is true whenever n is an integer greater than 1.

20. Prove that $3^n < n!$ if n is an integer greater than 6.

*21. Prove that if $h > -1$, then $1 + nh \leq (1+h)^n$ for all nonnegative integers n. This is called **Bernoulli's inequality**.

*22. Suppose that a and b are real numbers with $0 < b < a$. Prove that if n is a positive integer, then $a^n - b^n \leq na^{n-1}(a-b)$.

*23. Prove that for every positive integer n,

$$1 + \frac{1}{\sqrt{2}} + \frac{1}{\sqrt{3}} + \cdots + \frac{1}{\sqrt{n}} > 2(\sqrt{n+1} - 1).$$

24. Prove that $n^2 - 7n + 12$ is nonnegative whenever n is an integer with $n \geq 3$.

In Exercises 25 and 26, H_n denotes the nth harmonic number.

*25. Prove that $H_{2^n} \leq 1 + n$ whenever n is a nonnegative integer.

*26. Prove that

$$H_1 + H_2 + \cdots + H_n = (n+1)H_n - n.$$

Use mathematical induction in Exercises 27–30 to prove divisibility facts.

27. Prove that 2 divides $n^2 + n$ whenever n is a positive integer.

*28. Prove that $n^2 - 1$ is divisible by 8 whenever n is an odd positive integer.

*29. Prove that 21 divides $4^{n+1} + 5^{2n-1}$ whenever n is a positive integer.

*30. Prove that if n is a positive integer, then 133 divides $11^{n+1} + 12^{2n-1}$.

Use mathematical induction in Exercises 31–36 to prove results about sets.

31. Prove that if $A_1, A_2, \ldots, A_n$ and $B_1, B_2, \ldots, B_n$ are sets such that $A_j \subseteq B_j$ for $j = 1, 2, \ldots, n$, then

$$\bigcup_{j=1}^{n} A_j \subseteq \bigcup_{j=1}^{n} B_j.$$

32. Prove that if $A_1, A_2, \ldots, A_n$ and $B_1, B_2, \ldots, B_n$ are sets such that $A_j \subseteq B_j$ for $j = 1, 2, \ldots, n$, then

$$\bigcap_{j=1}^{n} A_j \subseteq \bigcap_{j=1}^{n} B_j.$$

33. Prove that if $A_1, A_2, \ldots, A_n$ and B are sets, then

$$(A_1 \cup A_2 \cup \cdots \cup A_n) \cap B$$
$$= (A_1 \cap B) \cup (A_2 \cap B) \cup \cdots \cup (A_n \cap B).$$

34. Prove that if $A_1, A_2, \ldots, A_n$ and B are sets, then

$$(A_1 \cap A_2 \cap \cdots \cap A_n) \cup B$$
$$= (A_1 \cup B) \cap (A_2 \cup B) \cap \cdots \cap (A_n \cup B).$$

35. Prove that a set with n elements has $n(n-1)/2$ subsets containing exactly two elements whenever n is an integer greater than or equal to 2.

*36. Prove that a set with n elements has $n(n-1)(n-2)/6$ subsets containing exactly three elements whenever n is an integer greater than or equal to 3.

In Exercises 37 and 38 we consider the problem of placing towers along a straight road, so that every building on the road receives cellular service. Assume that a building receives cellular service if it is within one mile of a tower.

37. Devise a greedy algorithm that uses the minimum number of towers possible to provide cell service to d buildings located at positions $x_1, x_2, \ldots, x_d$ from the start of the road. [*Hint:* At each step, go as far as possible along the road before adding a tower so as not to leave any buildings without coverage.]

*38. Use mathematical induction to prove that the algorithm you devised in Exercise 47 produces an optimal solution, that is, that it uses the fewest towers possible to provide cellular service to all buildings.

Exercises 39–41 present incorrect proofs using mathematical induction. You will need to identify an error in reasoning in each exercise.

39. What is wrong with this "proof" that all horses are the same color?

Let $P(n)$ be the proposition that all the horses in a set of n horses are the same color.

Basis Step: Clearly, $P(1)$ is true.

Inductive Step: Assume that $P(k)$ is true, so that all the horses in any set of k horses are the same color. Consider any $k+1$ horses; number these as horses $1, 2, 3, \ldots, k, k+1$. Now the first k of these horses all must have the same color, and the last k of these must also have the same color. Because the set of the first k horses and the set of the last k horses overlap, all $k+1$ must be the same color. This shows that $P(k+1)$ is true and finishes the proof by induction.

40. What is wrong with this "proof"?
"*Theorem*" For every positive integer n, $\sum_{i=1}^{n} i = (n+\frac{1}{2})^2/2$.

Basis Step: The formula is true for $n=1$.

Inductive Step: Suppose that $\sum_{i=1}^{n} i = (n+\frac{1}{2})^2/2$. Then $\sum_{i=1}^{n+1} i = (\sum_{i=1}^{n} i) + (n+1)$. By the inductive hypothesis, $\sum_{i=1}^{n+1} i = (n+\frac{1}{2})^2/2 + n + 1 = (n^2 + n + \frac{1}{4})/2 + n + 1 = (n^2 + 3n + \frac{9}{4})/2 = (n+\frac{3}{2})^2/2 = [(n+1) + \frac{1}{2}]^2/2$, completing the inductive step.

41. What is wrong with this "proof"?
"*Theorem*" For every positive integer n, if x and y are positive integers with $\max(x, y) = n$, then $x = y$.

Basis Step: Suppose that $n = 1$. If $\max(x, y) = 1$ and x and y are positive integers, we have $x = 1$ and $y = 1$.

Inductive Step: Let k be a positive integer. Assume that whenever $\max(x, y) = k$ and x and y are positive integers, then $x = y$. Now let $\max(x, y) = k+1$, where x and y are positive integers. Then $\max(x-1, y-1) = k$,

so by the inductive hypothesis, $x - 1 = y - 1$. It follows that $x = y$, completing the inductive step.

42. Use mathematical induction to show that given a set of $n+1$ positive integers, none exceeding $2n$, there is at least one integer in this set that divides another integer in the set.

*43. A knight on a chessboard can move one space horizontally (in either direction) and two spaces vertically (in either direction) or two spaces horizontally (in either direction) and one space vertically (in either direction). Suppose that we have an infinite chessboard, made up of all squares (m, n) where m and n are nonnegative integers that denote the row number and the column number of the square, respectively. Use mathematical induction to show that a knight starting at $(0, 0)$ can visit every square using a finite sequence of moves. [*Hint:* Use induction on the variable $s = m + n$.]

44. Suppose that
$$\mathbf{A} = \begin{bmatrix} a & 0 \\ 0 & b \end{bmatrix},$$
where a and b are real numbers. Show that
$$\mathbf{A}^n = \begin{bmatrix} a^n & 0 \\ 0 & b^n \end{bmatrix}$$
for every positive integer n.

45. (*Requires calculus*) Use mathematical induction to prove that the derivative of $f(x) = x^n$ equals nx^{n-1} whenever n is a positive integer. (For the inductive step, use the product rule for derivatives.)

46. Suppose that $\mathbf{A}$ and $\mathbf{B}$ are square matrices with the property $\mathbf{AB} = \mathbf{BA}$. Show that $\mathbf{AB}^n = \mathbf{B}^n\mathbf{A}$ for every positive integer n.

47. Suppose that m is a positive integer. Use mathematical induction to prove that if a and b are integers with $a \equiv b \pmod{m}$, then $a^k \equiv b^k \pmod{m}$ whenever k is a nonnegative integer.

48. Use mathematical induction to show that $\neg(p_1 \vee p_2 \vee \cdots \vee p_n)$ is equivalent to $\neg p_1 \wedge \neg p_2 \wedge \cdots \wedge \neg p_n$ whenever $p_1, p_2, \ldots, p_n$ are propositions.

*49. Show that
$$[(p_1 \rightarrow p_2) \wedge (p_2 \rightarrow p_3) \wedge \cdots \wedge (p_{n-1} \rightarrow p_n)]$$
$$\rightarrow [(p_1 \wedge p_2 \wedge \cdots \wedge p_{n-1}) \rightarrow p_n]$$
is a tautology whenever $p_1, p_2, \ldots, p_n$ are propositions, where $n \geq 2$.

*50. Show that n lines separate the plane into $(n^2 + n + 2)/2$ regions if no two of these lines are parallel and no three pass through a common point.

51. Let $a_1, a_2, \ldots, a_n$ be positive real numbers. The **arithmetic mean of these numbers is defined by
$$A = (a_1 + a_2 + \cdots + a_n)/n,$$
and the **geometric mean** of these numbers is defined by
$$G = (a_1 a_2 \cdots a_n)^{1/n}.$$
Use mathematical induction to prove that $A \geq G$.

52. Use mathematical induction to prove Lemma 3 of Section 4.3, which states that if p is a prime and $p \mid a_1 a_2 \cdots a_n$, where a_i is an integer for $i = 1, 2, 3, \ldots, n$, then $p \mid a_i$ for some integer i.

53. Show that if n is a positive integer, then

$$\sum_{\{a_1, \ldots, a_k\} \subseteq \{1, 2, \ldots, n\}} \frac{1}{a_1 a_2 \cdots a_k} = n.$$

(Here the sum is over all nonempty subsets of the set of the n smallest positive integers.)

*54. Use the well-ordering property to show that the following form of mathematical induction is a valid method to prove that $P(n)$ is true for all positive integers n.

Basis Step: $P(1)$ and $P(2)$ are true.

Inductive Step: For each positive integer k, if $P(k)$ and $P(k + 1)$ are both true, then $P(k + 2)$ is true.

55. Show that if $A_1, A_2, \ldots, A_n$ are sets where $n \geq 2$, and for all pairs of integers i and j with $1 \leq i < j \leq n$ either A_i is a subset of A_j or A_j is a subset of A_i, then there is an integer i, $1 \leq i \leq n$ such that A_i is a subset of A_j for all integers j with $1 \leq j \leq n$.

*56. A guest at a party is a **celebrity** if this person is known by every other guest, but knows none of them. There is at most one celebrity at a party, for if there were two, they would know each other. A particular party may have no celebrity. Your assignment is to find the celebrity, if one exists, at a party, by asking only one type of question—asking a guest whether they know a second guest. Everyone must answer your questions truthfully. That is, if Alice and Bob are two people at the party, you can ask Alice whether she knows Bob; she must answer correctly. Use mathematical induction to show that if there are n people at the party, then you can find the celebrity, if there is one, with $3(n - 1)$ questions. [*Hint:* First ask a question to eliminate one person as a celebrity. Then use the inductive hypothesis to identify a potential celebrity. Finally, ask two more questions to determine whether that person is actually a celebrity.]

Suppose there are n people in a group, each aware of a scandal no one else in the group knows about. These people communicate by telephone; when two people in the group talk, they share information about all scandals each knows about. For example, on the first call, two people share information, so by the end of the call, each of these people knows about two scandals. The **gossip problem** asks for $G(n)$, the minimum number of telephone calls that are needed for all n people to learn about all the scandals. Exercises 57–59 deal with the gossip problem.

57. Find $G(1)$, $G(2)$, $G(3)$, and $G(4)$.

58. Use mathematical induction to prove that $G(n) \leq 2n - 4$ for $n \geq 4$. [*Hint:* In the inductive step, have a new person call a particular person at the start and at the end.]

**59. Prove that $G(n) = 2n - 4$ for $n \geq 4$.

*60. Show that it is possible to arrange the numbers $1, 2, \ldots, n$ in a row so that the average of any two of these numbers never appears between them. [*Hint:* Show that it suffices to prove this fact when n is a power of 2. Then use mathematical induction to prove the result when n is a power of 2.]

*61. Show that if $I_1, I_2, \ldots, I_n$ is a collection of open intervals on the real number line, $n \geq 2$, and every pair of these intervals has a nonempty intersection, that is, $I_i \cap I_j \neq \emptyset$ whenever $1 \leq i \leq n$ and $1 \leq j \leq n$, then the intersection of all these sets is nonempty, that is, $I_1 \cap I_2 \cap \cdots \cap I_n \neq \emptyset$. (Recall that an **open interval** is the set of real numbers x with $a < x < b$, where a and b are real numbers with $a < b$.)

62. Let n be an even positive integer. Show that when n people stand in a yard at mutually distinct distances and each person throws a pie at their nearest neighbor, it is possible that everyone is hit by a pie.

63. Construct a tiling using right triominoes of the 8×8 checkerboard with the square in the upper left corner removed.

*64. Show that a three-dimensional $2^n \times 2^n \times 2^n$ checkerboard with one $1 \times 1 \times 1$ cube missing can be completely covered by $2 \times 2 \times 2$ cubes with one $1 \times 1 \times 1$ cube removed.

65. Use the principle of mathematical induction to show that $P(n)$ is true for $n = b, b + 1, b + 2, \ldots$, where b is an integer, if $P(b)$ is true and the conditional statement $P(k) \to P(k + 1)$ is true for all integers k with $k \geq b$.

5.2 Strong Induction and Well-Ordering

Introduction

In Section 5.1 we introduced mathematical induction and we showed how to use it to prove a variety of theorems. In this section we will introduce another form of mathematical induction, called **strong induction,** which can often be used when we cannot easily prove a result using mathematical induction. The basis step of a proof by strong induction is the same as a proof of the same result using mathematical induction. That is, in a strong induction proof that $P(n)$ is true for all positive integers n, the basis step shows that $P(1)$ is true. However, the inductive steps in these two proof methods are different. In a proof by mathematical induction, the inductive step shows that if the inductive hypothesis $P(k)$ is true, then $P(k + 1)$ is also true. In a proof by strong induction, the inductive step shows that if $P(j)$ is true for all positive integers not exceeding k, then $P(k + 1)$ is true. That is, for the inductive hypothesis we assume that $P(j)$ is true for $j = 1, 2, \ldots, k$.

The validity of both mathematical induction and strong induction follow from the well-ordering property in Appendix 1. In fact, mathematical induction, strong induction, and well-ordering are all equivalent principles (as shown in Exercises 31, 32, and 33). That is, the validity of each can be proved from either of the other two. This means that a proof using one of these two principles can be rewritten as a proof using either of the other two principles. Just as it is sometimes the case that it is much easier to see how to prove a result using strong induction rather than mathematical induction, it is sometimes easier to use well-ordering than one of the two forms of mathematical induction. In this section we will give some examples of how the well-ordering property can be used to prove theorems.

Strong Induction

Before we illustrate how to use strong induction, we state this principle again.

> *STRONG INDUCTION* To prove that $P(n)$ is true for all positive integers n, where $P(n)$ is a propositional function, we complete two steps:
>
> *BASIS STEP:* We verify that the proposition $P(1)$ is true.
>
> *INDUCTIVE STEP:* We show that the conditional statement $[P(1) \wedge P(2) \wedge \cdots \wedge P(k)] \rightarrow P(k + 1)$ is true for all positive integers k.

Note that when we use strong induction to prove that $P(n)$ is true for all positive integers n, our inductive hypothesis is the assumption that $P(j)$ is true for $j = 1, 2, \ldots, k$. That is, the inductive hypothesis includes all k statements $P(1), P(2), \ldots, P(k)$. Because we can use all k statements $P(1), P(2), \ldots, P(k)$ to prove $P(k + 1)$, rather than just the statement $P(k)$ as in a proof by mathematical induction, strong induction is a more flexible proof technique. Because of this, some mathematicians prefer to always use strong induction instead of mathematical induction, even when a proof by mathematical induction is easy to find.

You may be surprised that mathematical induction and strong induction are equivalent. That is, each can be shown to be a valid proof technique assuming that the other is valid. In particular, any proof using mathematical induction can also be considered to be a proof by strong induction because the inductive hypothesis of a proof by mathematical induction is part of the inductive hypothesis in a proof by strong induction. That is, if we can complete the inductive step of a proof using mathematical induction by showing that $P(k + 1)$ follows from $P(k)$ for every positive integer k, then it also follows that $P(k + 1)$ follows from all the statements $P(1)$, $P(2), \ldots, P(k)$, because we are assuming that not only $P(k)$ is true, but also more, namely, that the $k - 1$ statements $P(1), P(2), \ldots, P(k - 1)$ are true. However, it is much more awkward to convert a proof by strong induction into a proof using the principle of mathematical induction. (See Exercise 42.)

Strong induction is sometimes called the **second principle of mathematical induction** or **complete induction**. When the terminology "complete induction" is used, the principle of mathematical induction is called **incomplete induction**, a technical term that is a somewhat unfortunate choice because there is nothing incomplete about the principle of mathematical induction; after all, it is a valid proof technique.

STRONG INDUCTION AND THE INFINITE LADDER To better understand strong induction, consider the infinite ladder in Section 5.1. Strong induction tells us that we can reach all rungs if

1. we can reach the first rung, and
2. for every integer k, if we can reach all the first k rungs, then we can reach the $(k + 1)$st rung.

That is, if $P(n)$ is the statement that we can reach the nth rung of the ladder, by strong induction we know that $P(n)$ is true for all positive integers n, because (1) tells us $P(1)$ is true, completing the basis step and (2) tells us that $P(1) \land P(2) \land \cdots \land P(k)$ implies $P(k + 1)$, completing the inductive step.

Example 1 illustrates how strong induction can help us prove a result that cannot easily be proved using the principle of mathematical induction.

EXAMPLE 1   Suppose we can reach the first and second rungs of an infinite ladder, and we know that if we can reach a rung, then we can reach two rungs higher. Can we prove that we can reach every rung using the principle of mathematical induction? Can we prove that we can reach every rung using strong induction?

Solution: We first try to prove this result using the principle of mathematical induction.

BASIS STEP: The basis step of such a proof holds; here it simply verifies that we can reach the first rung.

ATTEMPTED INDUCTIVE STEP: The inductive hypothesis is the statement that we can reach the kth rung of the ladder. To complete the inductive step, we need to show that if we assume the inductive hypothesis for the positive integer k, namely, if we assume that we can reach the kth rung of the ladder, then we can show that we can reach the $(k + 1)$st rung of the ladder. However, there is no obvious way to complete this inductive step because we do not know from the given information that we can reach the $(k + 1)$st rung from the kth rung. After all, we only know that if we can reach a rung we can reach the rung two higher.

Now consider a proof using strong induction.

BASIS STEP: The basis step is the same as before; it simply verifies that we can reach the first rung.

INDUCTIVE STEP: The inductive hypothesis states that we can reach each of the first k rungs. To complete the inductive step, we need to show that if we assume that the inductive hypothesis is true, that is, if we can reach each of the first k rungs, then we can reach the $(k + 1)$st rung. We already know that we can reach the second rung. We can complete the inductive step by noting that as long as $k \geq 2$, we can reach the $(k + 1)$st rung from the $(k - 1)$st rung because we know we can climb two rungs from a rung we can already reach, and because $k - 1 \leq k$, by the inductive hypothesis we can reach the $(k - 1)$st rung. This completes the inductive step and finishes the proof by strong induction.

We have proved that if we can reach the first two rungs of an infinite ladder and for every positive integer k if we can reach all the first k rungs then we can reach the $(k + 1)$st rung, then we can reach all rungs of the ladder.   ◄

Examples of Proofs Using Strong Induction

Now that we have both mathematical induction and strong induction, how do we decide which method to apply in a particular situation? Although there is no cut-and-dried answer, we can supply some useful pointers. In practice, you should use mathematical induction when it is straightforward to prove that $P(k) \rightarrow P(k + 1)$ is true for all positive integers k. This is the case for all the proofs in the examples in Section 5.1. In general, you should restrict your use of the principle of mathematical induction to such scenarios. Unless you can clearly see that the inductive step of a proof by mathematical induction goes through, you should attempt a proof by strong induction. That is, use strong induction and not mathematical induction when you see how to prove that $P(k + 1)$ is true from the assumption that $P(j)$ is true for all positive integers j not exceeding k, but you cannot see how to prove that $P(k + 1)$ follows from just

$P(k)$. Keep this in mind as you examine the proofs in this section. For each of these proofs, consider why strong induction works better than mathematical induction.

We will illustrate how strong induction is employed in Examples 2–4. In these examples, we will prove a diverse collection of results. Pay particular attention to the inductive step in each of these examples, where we show that a result $P(k + 1)$ follows under the assumption that $P(j)$ holds for all positive integers j not exceeding k, where $P(n)$ is a propositional function.

We begin with one of the most prominent uses of strong induction, the part of the fundamental theorem of arithmetic that tells us that every positive integer can be written as the product of primes.

EXAMPLE 2 Show that if n is an integer greater than 1, then n can be written as the product of primes.

Solution: Let $P(n)$ be the proposition that n can be written as the product of primes.

BASIS STEP: $P(2)$ is true, because 2 can be written as the product of one prime, itself. (Note that $P(2)$ is the first case we need to establish.)

INDUCTIVE STEP: The inductive hypothesis is the assumption that $P(j)$ is true for all integers j with $2 \leq j \leq k$, that is, the assumption that j can be written as the product of primes whenever j is a positive integer at least 2 and not exceeding k. To complete the inductive step, it must be shown that $P(k + 1)$ is true under this assumption, that is, that $k + 1$ is the product of primes.

There are two cases to consider, namely, when $k + 1$ is prime and when $k + 1$ is composite. If $k + 1$ is prime, we immediately see that $P(k + 1)$ is true. Otherwise, $k + 1$ is composite and can be written as the product of two positive integers a and b with $2 \leq a \leq b < k + 1$. Because both a and b are integers at least 2 and not exceeding k, we can use the inductive hypothesis to write both a and b as the product of primes. Thus, if $k + 1$ is composite, it can be written as the product of primes, namely, those primes in the factorization of a and those in the factorization of b. ◀

Remark: Because 1 can be thought of as the *empty* product of no primes, we could have started the proof in Example 2 with $P(1)$ as the basis step. We chose not to do so because many people find this confusing.

Example 2 completes the proof of the fundamental theorem of arithmetic, which asserts that every nonnegative integer can be written uniquely as the product of primes in nondecreasing order. We showed in Section 4.3 that an integer has at most one such factorization into primes. Example 2 shows there is at least one such factorization.

Next, we show how strong induction can be used to prove that a player has a winning strategy in a game.

EXAMPLE 3 Consider a game in which two players take turns removing any positive number of matches they want from one of two piles of matches. The player who removes the last match wins the game. Show that if the two piles contain the same number of matches initially, the second player can always guarantee a win.

Solution: Let n be the number of matches in each pile. We will use strong induction to prove $P(n)$, the statement that the second player can win when there are initially n matches in each pile.

BASIS STEP: When $n = 1$, the first player has only one choice, removing one match from one of the piles, leaving a single pile with a single match, which the second player can remove to win the game.

INDUCTIVE STEP: The inductive hypothesis is the statement that $P(j)$ is true for all j with $1 \leq j \leq k$, that is, the assumption that the second player can always win whenever there are j

matches, where $1 \leq j \leq k$ in each of the two piles at the start of the game. We need to show that $P(k+1)$ is true, that is, that the second player can win when there are initially $k+1$ matches in each pile, under the assumption that $P(j)$ is true for $j = 1, 2, \ldots, k$. So suppose that there are $k+1$ matches in each of the two piles at the start of the game and suppose that the first player removes r matches $(1 \leq r \leq k)$ from one of the piles, leaving $k+1-r$ matches in this pile. By removing the same number of matches from the other pile, the second player creates the situation where there are two piles each with $k+1-r$ matches. Because $1 \leq k+1-r \leq k$, we can now use the inductive hypothesis to conclude that the second player can always win. We complete the proof by noting that if the first player removes all $k+1$ matches from one of the piles, the second player can win by removing all the remaining matches. ◀

Using the principle of mathematical induction, instead of strong induction, to prove the results in Examples 2 and 3 is difficult. However, as Example 4 shows, some results can be readily proved using either the principle of mathematical induction or strong induction.

Before we present Example 4, note that we can slightly modify strong induction to handle a wider variety of situations. In particular, we can adapt strong induction to handle cases where the inductive step is valid only for integers greater than a particular integer. Let b be a fixed integer and j a fixed positive integer. The form of strong induction we need tells us that $P(n)$ is true for all integers n with $n \geq b$ if we can complete these two steps:

BASIS STEP: We verify that the propositions $P(b), P(b+1), \ldots, P(b+j)$ are true.

INDUCTIVE STEP: We show that $[P(b) \wedge P(b+1) \wedge \cdots \wedge P(k)] \rightarrow P(k+1)$ is true for every integer $k \geq b+j$.

We will use this alternative form in the strong induction proof in Example 4.

EXAMPLE 4 Prove that every amount of postage of 12 cents or more can be formed using just 4-cent and 5-cent stamps.

Solution: We will prove this result using the principle of mathematical induction. Then we will present a proof using strong induction. Let $P(n)$ be the statement that postage of n cents can be formed using 4-cent and 5-cent stamps.

We begin by using the principle of mathematical induction.

BASIS STEP: Postage of 12 cents can be formed using three 4-cent stamps.

INDUCTIVE STEP: The inductive hypothesis is the statement that $P(k)$ is true. That is, under this hypothesis, postage of k cents can be formed using 4-cent and 5-cent stamps. To complete the inductive step, we need to show that when we assume $P(k)$ is true, then $P(k+1)$ is also true where $k \geq 12$. That is, we need to show that if we can form postage of k cents, then we can form postage of $k+1$ cents. So, assume the inductive hypothesis is true; that is, assume that we can form postage of k cents using 4-cent and 5-cent stamps. We consider two cases, when at least one 4-cent stamp has been used and when no 4-cent stamps have been used. First, suppose that at least one 4-cent stamp was used to form postage of k cents. Then we can replace this stamp with a 5-cent stamp to form postage of $k+1$ cents. But if no 4-cent stamps were used, we can form postage of k cents using only 5-cent stamps. Moreover, because $k \geq 12$, we needed at least three 5-cent stamps to form postage of k cents. So, we can replace three 5-cent stamps with four 4-cent stamps to form postage of $k+1$ cents. This completes the inductive step.

Because we have completed the basis step and the inductive step, we know that $P(n)$ is true for all $n \geq 12$. That is, we can form postage of n cents, where $n \geq 12$ using just 4-cent and 5-cent stamps. This completes the proof by mathematical induction.

Next, we will use strong induction to prove the same result. In this proof, in the basis step we show that $P(12), P(13), P(14)$, and $P(15)$ are true, that is, that postage of 12, 13, 14, or 15 cents can be formed using just 4-cent and 5-cent stamps. In the inductive step we show how to get postage of $k+1$ cents for $k \geq 15$ from postage of $k-3$ cents.

BASIS STEP: We can form postage of 12, 13, 14, and 15 cents using three 4-cent stamps, two 4-cent stamps and one 5-cent stamp, one 4-cent stamp and two 5-cent stamps, and three 5-cent stamps, respectively. This shows that $P(12)$, $P(13)$, $P(14)$, and $P(15)$ are true. This completes the basis step.

INDUCTIVE STEP: The inductive hypothesis is the statement that $P(j)$ is true for $12 \leq j \leq k$, where k is an integer with $k \geq 15$. To complete the inductive step, we assume that we can form postage of j cents, where $12 \leq j \leq k$. We need to show that under the assumption that $P(k + 1)$ is true, we can also form postage of $k + 1$ cents. Using the inductive hypothesis, we can assume that $P(k - 3)$ is true because $k - 3 \geq 12$, that is, we can form postage of $k - 3$ cents using just 4-cent and 5-cent stamps. To form postage of $k + 1$ cents, we need only add another 4-cent stamp to the stamps we used to form postage of $k - 3$ cents. That is, we have shown that if the inductive hypothesis is true, then $P(k + 1)$ is also true. This completes the inductive step.

Because we have completed the basis step and the inductive step of a strong induction proof, we know by strong induction that $P(n)$ is true for all integers n with $n \geq 12$. That is, we know that every postage of n cents, where n is at least 12, can be formed using 4-cent and 5-cent stamps. This finishes the proof by strong induction.

(There are other ways to approach this problem besides those described here. Can you find a solution that does not use mathematical induction?) ◀

Using Strong Induction in Computational Geometry

Our next example of strong induction will come from **computational geometry**, the part of discrete mathematics that studies computational problems involving geometric objects. Computational geometry is used extensively in computer graphics, computer games, robotics, scientific calculations, and a vast array of other areas. Before we can present this result, we introduce some terminology, possibly familiar from earlier studies in geometry.

A **polygon** is a closed geometric figure consisting of a sequence of line segments $s_1, s_2, \ldots, s_n$, called **sides**. Each pair of consecutive sides, s_i and s_{i+1}, $i = 1, 2, \ldots, n - 1$, as well as the last side s_n and the first side s_1, of the polygon meet at a common endpoint, called a **vertex**. A polygon is called **simple** if no two nonconsecutive sides intersect. Every simple polygon divides the plane into two regions: its **interior**, consisting of the points inside the curve, and its **exterior**, consisting of the points outside the curve. This last fact is surprisingly complicated to prove. It is a special case of the famous Jordan Curve Theorem, which tells us that every simple curve divides the plane into two regions; see [Or00], for example.

A polygon is called **convex** if every line segment connecting two points in the interior of the polygon lies entirely inside the polygon. (A polygon that is not convex is said to be **nonconvex**.) Figure 1 displays some polygons; polygons (a) and (b) are convex, but polygons (c) and (d) are not. A **diagonal** of a simple polygon is a line segment connecting two nonconsecutive vertices of the polygon, and a diagonal is called an **interior diagonal** if it lies entirely inside the polygon, except for its endpoints. For example, in polygon (d), the line segment connecting a and f is an interior diagonal, but the line segment connecting a and d is a diagonal that is not an interior diagonal.

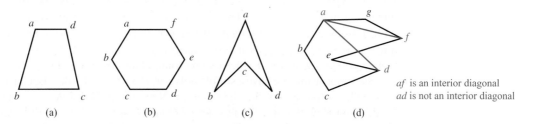

af is an interior diagonal
ad is not an interior diagonal

(a) (b) (c) (d)

FIGURE 1   **Convex and Nonconvex Polygons.**

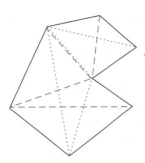

Two different triangulations of a simple polygon with seven sides into five triangles, shown with dotted lines and with dashed lines, respectively

FIGURE 2   Triangulations of a Polygon.

One of the most basic operations of computational geometry involves dividing a simple polygon into triangles by adding nonintersecting diagonals. This process is called **triangulation**. Note that a simple polygon can have many different triangulations, as shown in Figure 2. Perhaps the most basic fact in computational geometry is that it is possible to triangulate every simple polygon, as we state in Theorem 1. Furthermore, this theorem tells us that every triangulation of a simple polygon with n sides includes $n - 2$ triangles.

THEOREM 1   A simple polygon with n sides, where n is an integer with $n \geq 3$, can be triangulated into $n - 2$ triangles.

It seems obvious that we should be able to triangulate a simple polygon by successively adding interior diagonals. Consequently, a proof by strong induction seems promising. However, such a proof requires this crucial lemma.

LEMMA 1   Every simple polygon with at least four sides has an interior diagonal.

Although Lemma 1 seems particularly simple, it is surprisingly tricky to prove. In fact, as recently as 30 years ago, a variety of incorrect proofs thought to be correct were commonly seen in books and articles. We defer the proof of Lemma 1 until after we prove Theorem 1. It is not uncommon to prove a theorem pending the later proof of an important lemma.

Proof (of Theorem 1): We will prove this result using strong induction. Let $T(n)$ be the statement that every simple polygon with n sides can be triangulated into $n - 2$ triangles.

BASIS STEP: $T(3)$ is true because a simple polygon with three sides is a triangle. We do not need to add any diagonals to triangulate a triangle; it is already triangulated into one triangle, itself. Consequently, every simple polygon with $n = 3$ has can be triangulated into $n - 2 = 3 - 2 = 1$ triangle.

INDUCTIVE STEP: For the inductive hypothesis, we assume that $T(j)$ is true for all integers j with $3 \leq j \leq k$. That is, we assume that we can triangulate a simple polygon with j sides into $j - 2$ triangles whenever $3 \leq j \leq k$. To complete the inductive step, we must show that when we assume the inductive hypothesis, $P(k + 1)$ is true, that is, that every simple polygon with $k + 1$ sides can be triangulated into $(k + 1) - 2 = k - 1$ triangles.

So, suppose that we have a simple polygon P with $k + 1$ sides. Because $k + 1 \geq 4$, Lemma 1 tells us that P has an interior diagonal ab. Now, ab splits P into two simple polygons Q, with s sides, and R, with t sides. The sides of Q and R are the sides of P, together with the side ab, which is a side of both Q and R. Note that $3 \leq s \leq k$ and $3 \leq t \leq k$ because both Q and R have at least one fewer side than P does (after all, each of these is formed from P by deleting at least

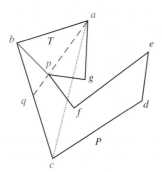

T is the triangle *abc*

p is the vertex of *P* inside *T* such that the $\angle bap$ is smallest

bp must be an interior diagonal of *P*

FIGURE 3   **Constructing an Interior Diagonal of a Simple Polygon.**

two sides and replacing these sides by the diagonal *ab*). Furthermore, the number of sides of *P* is two less than the sum of the numbers of sides of *Q* and the number of sides of *R*, because each side of *P* is a side of either *Q* or of *R*, but not both, and the diagonal *ab* is a side of both *Q* and *R*, but not *P*. That is, $k + 1 = s + t - 2$.

We now use the inductive hypothesis. Because both $3 \leq s \leq k$ and $3 \leq t \leq k$, by the inductive hypothesis we can triangulate *Q* and *R* into $s - 2$ and $t - 2$ triangles, respectively. Next, note that these triangulations together produce a triangulation of *P*. (Each diagonal added to triangulate one of these smaller polygons is also a diagonal of *P*.) Consequently, we can triangulate *P* into a total of $(s - 2) + (t - 2) = s + t - 4 = (k + 1) - 2$ triangles. This completes the proof by strong induction. That is, we have shown that every simple polygon with *n* sides, where $n \geq 3$, can be triangulated into $n - 2$ triangles.   ◁

We now return to our proof of Lemma 1. We present a proof published by Chung-Wu Ho [Ho75]. Note that although this proof may be omitted without loss of continuity, it does provide a correct proof of a result proved incorrectly by many mathematicians.

Proof: Suppose that *P* is a simple polygon drawn in the plane. Furthermore, suppose that *b* is the point of *P* or in the interior of *P* with the least *y*-coordinate among the vertices with the smallest *x*-coordinate. Then *b* must be a vertex of *P*, for if it is an interior point, there would have to be a vertex of *P* with a smaller *x*-coordinate. Two other vertices each share an edge with *b*, say *a* and *c*. It follows that the angle in the interior of *P* formed by *ab* and *bc* must be less than 180 degrees (otherwise, there would be points of *P* with smaller *x*-coordinates than *b*).

Now let *T* be the triangle $\triangle abc$. If there are no vertices of *P* on or inside *T*, we can connect *a* and *c* to obtain an interior diagonal. On the other hand, if there are vertices of *P* inside *T*, we will find a vertex *p* of *P* on or inside *T* such that *bp* is an interior diagonal. (This is the tricky part. Ho noted that in many published proofs of this lemma a vertex *p* was found such that *bp* was not necessarily an interior diagonal of *P*. See Exercise 16.) The key is to select a vertex *p* such that the angle $\angle bap$ is smallest. To see this, note that the ray starting at *a* and passing through *p* hits the line segment *bc* at a point, say *q*. It then follows that the triangle $\triangle baq$ cannot contain any vertices of *P* in its interior. Hence, we can connect *b* and *p* to produce an interior diagonal of *P*. Locating this vertex *p* is illustrated in Figure 3.   ◁

Proofs Using the Well-Ordering Property

The validity of both the principle of mathematical induction and strong induction follows from a fundamental axiom of the set of integers, the **well-ordering property** (see Appendix 1). The well-ordering property states that every nonempty set of nonnegative integers has a least element. We will show how the well-ordering property can be used directly in proofs. Furthermore, it can be shown (see Exercises 31, 32, and 33) that the well-ordering property, the principle of mathematical induction, and strong induction are all equivalent. That is, the validity of each of these three proof techniques implies the validity of the other two techniques. In Section 5.1 we showed that the principle of mathematical induction follows from the well-ordering property. The other parts of this equivalence are left as Exercises 22, 32, and 33.

THE WELL-ORDERING PROPERTY   Every nonempty set of nonnegative integers has a least element.

The well-ordering property can often be used directly in proofs.

EXAMPLE 5   Use the well-ordering property to prove the division algorithm. Recall that the division algorithm states that if a is an integer and d is a positive integer, then there are unique integers q and r with $0 \leq r < d$ and $a = dq + r$.

Solution: Let S be the set of nonnegative integers of the form $a - dq$, where q is an integer. This set is nonempty because $-dq$ can be made as large as desired (taking q to be a negative integer with large absolute value). By the well-ordering property, S has a least element $r = a - dq_0$.

The integer r is nonnegative. It is also the case that $r < d$. If it were not, then there would be a smaller nonnegative element in S, namely, $a - d(q_0 + 1)$. To see this, suppose that $r \geq d$. Because $a = dq_0 + r$, it follows that $a - d(q_0 + 1) = (a - dq_0) - d = r - d \geq 0$. Consequently, there are integers q and r with $0 \leq r < d$. The proof that q and r are unique is left as Exercise 29. ◀

EXAMPLE 6   In a round-robin tournament every player plays every other player exactly once and each match has a winner and a loser. We say that the players $p_1, p_2, \ldots, p_m$ form a *cycle* if p_1 beats p_2, p_2 beats $p_3, \ldots, p_{m-1}$ beats p_m, and p_m beats p_1. Use the well-ordering principle to show that if there is a cycle of length m ($m \geq 3$) among the players in a round-robin tournament, there must be a cycle of three of these players.

Solution: We assume that there is no cycle of three players. Because there is at least one cycle in the round-robin tournament, the set of all positive integers n for which there is a cycle of length n is nonempty. By the well-ordering property, this set of positive integers has a least element k, which by assumption must be greater than three. Consequently, there exists a cycle of players $p_1, p_2, p_3, \ldots, p_k$ and no shorter cycle exists.

Because there is no cycle of three players, we know that $k > 3$. Consider the first three elements of this cycle, p_1, p_2, and p_3. There are two possible outcomes of the match between p_1 and p_3. If p_3 beats p_1, it follows that p_1, p_2, p_3 is a cycle of length three, contradicting our assumption that there is no cycle of three players. Consequently, it must be the case that p_1 beats p_3. This means that we can omit p_2 from the cycle $p_1, p_2, p_3, \ldots, p_k$ to obtain the cycle $p_1, p_3, p_4, \ldots, p_k$ of length $k - 1$, contradicting the assumption that the smallest cycle has length k. We conclude that there must be a cycle of length three. ◀

Exercises

1. Use strong induction to show that if you can run one mile or two miles, and if you can always run two more miles once you have run a specified number of miles, then you can run any number of miles.

2. Use strong induction to show that all dominoes fall in an infinite arrangement of dominoes if you know that the first three dominoes fall, and that when a domino falls, the domino three farther down in the arrangement also falls.

3. Let $P(n)$ be the statement that a postage of n cents can be formed using just 3-cent stamps and 5-cent stamps. The

parts of this exercise outline a strong induction proof that $P(n)$ is true for $n \geq 8$.
a) Show that the statements $P(8)$, $P(9)$, and $P(10)$ are true, completing the basis step of the proof.
b) What is the inductive hypothesis of the proof?
c) What do you need to prove in the inductive step?
d) Complete the inductive step for $k \geq 10$.
e) Explain why these steps show that this statement is true whenever $n \geq 8$.

4. Let $P(n)$ be the statement that a postage of n cents can be formed using just 4-cent stamps and 7-cent stamps. The

parts of this exercise outline a strong induction proof that $P(n)$ is true for $n \geq 18$.

   a) Show statements $P(18)$, $P(19)$, $P(20)$, and $P(21)$ are true, completing the basis step of the proof.
   b) What is the inductive hypothesis of the proof?
   c) What do you need to prove in the inductive step?
   d) Complete the inductive step for $k \geq 21$.
   e) Explain why these steps show that this statement is true whenever $n \geq 18$.

*5. Use strong induction to prove that $\sqrt{2}$ is irrational. [*Hint:* Let $P(n)$ be the statement that $\sqrt{2} \neq n/b$ for any positive integer b.]

6. Assume that a chocolate bar consists of n squares arranged in a rectangular pattern. The entire bar, a smaller rectangular piece of the bar, can be broken along a vertical or a horizontal line separating the squares. Assuming that only one piece can be broken at a time, determine how many breaks you must successively make to break the bar into n separate squares. Use strong induction to prove your answer.

7. Consider this variation of the game of Nim. The game begins with n matches. Two players take turns removing matches, one, two, or three at a time. The player removing the last match loses. Use strong induction to show that if each player plays the best strategy possible, the first player wins if $n = 4j$, $4j + 2$, or $4j + 3$ for some nonnegative integer j and the second player wins in the remaining case when $n = 4j + 1$ for some nonnegative integer j.

8. Use strong induction to show that every positive integer n can be written as a sum of distinct powers of two, that is, as a sum of a subset of the integers $2^0 = 1, 2^1 = 2, 2^2 = 4$, and so on. [*Hint:* For the inductive step, separately consider the case where $k + 1$ is even and where it is odd. When it is even, note that $(k + 1)/2$ is an integer.]

*9. A jigsaw puzzle is put together by successively joining pieces that fit together into blocks. A move is made each time a piece is added to a block, or when two blocks are joined. Use strong induction to prove that no matter how the moves are carried out, exactly $n - 1$ moves are required to assemble a puzzle with n pieces.

10. Suppose you begin with a pile of n stones and split this pile into n piles of one stone each by successively splitting a pile of stones into two smaller piles. Each time you split a pile you multiply the number of stones in each of the two smaller piles you form, so that if these piles have r and s stones in them, respectively, you compute rs. Show that no matter how you split the piles, the sum of the products computed at each step equals $n(n - 1)/2$.

11. Prove that the first player has a winning strategy for the game of Chomp, introduced in Example 12 in Section 1.8, if the initial board is square. [*Hint:* Use strong induction to show that this strategy works. For the first move, the first player chomps all cookies except those in the left and top edges. On subsequent moves, after the second player has chomped cookies on either the top or left edge, the first player chomps cookies in the same relative positions in the left or top edge, respectively.]

*12. Prove that the first player has a winning strategy for the game of Chomp, introduced in Example 12 in Section 1.8, if the initial board is two squares wide, that is, a $2 \times n$ board. [*Hint:* Use strong induction. The first move of the first player should be to chomp the cookie in the bottom row at the far right.]

13. Use strong induction to show that if a simple polygon with at least four sides is triangulated, then at least two of the triangles in the triangulation have two sides that border the exterior of the polygon.

*14. Use strong induction to show that when a simple polygon P with consecutive vertices $v_1, v_2, \ldots, v_n$ is triangulated into $n - 2$ triangles, the $n - 2$ triangles can be numbered $1, 2, \ldots, n - 2$ so that v_i is a vertex of triangle i for $i = 1, 2, \ldots, n - 2$.

*15. **Pick's theorem** says that the area of a simple polygon P in the plane with vertices that are all lattice points (that is, points with integer coordinates) equals $I(P) + B(P)/2 - 1$, where $I(P)$ and $B(P)$ are the number of lattice points in the interior of P and on the boundary of P, respectively. Use strong induction on the number of vertices of P to prove Pick's theorem. [*Hint:* For the basis step, first prove the theorem for rectangles, then for right triangles, and finally for all triangles by noting that the area of a triangle is the area of a larger rectangle containing it with the areas of at most three triangles subtracted. For the inductive step, take advantage of Lemma 1.]

16. In the proof of Lemma 1 we mentioned that many incorrect methods for finding a vertex p such that the line segment bp is an interior diagonal of P have been published. This exercise presents some of the incorrect ways p has been chosen in these proofs. Show, by considering one of the polygons drawn here, that for each of these choices of p, the line segment bp is not necessarily an interior diagonal of P.

   a) p is the vertex of P such that the angle $\angle abp$ is smallest.

   b) p is the vertex of P with the least x-coordinate (other than b).

   c) p is the vertex of P that is closest to b.

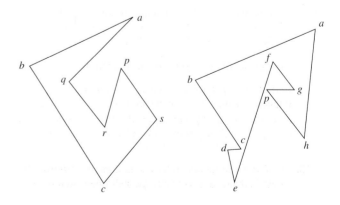

17. Let $E(n)$ be the statement that in a triangulation of a simple polygon with n sides, at least one of the triangles in the triangulation has two sides bordering the exterior of the polygon.

a) Explain where a proof using strong induction that $E(n)$ is true for all integers $n \geq 4$ runs into difficulties.

b) Show that we can prove that $E(n)$ is true for all integers $n \geq 4$ by proving by strong induction the stronger statement $T(n)$ for all integers $n \geq 4$, which states that in every triangulation of a simple polygon, at least two of the triangles in the triangulation have two sides bordering the exterior of the polygon.

*18. A stable assignment, defined in the preamble to Exercise 50 in Section 3.1, is called **optimal for suitors** if no stable assignment exists in which a suitor is paired with a suitee whom this suitor prefers to the person to whom this suitor is paired in this stable assignment. Use strong induction to show that the deferred acceptance algorithm produces a stable assignment that is optimal for suitors.

19. Suppose that $P(n)$ is a propositional function. Determine for which positive integers n the statement $P(n)$ must be true, and justify your answer, if

a) $P(1)$ is true; for all positive integers n, if $P(n)$ is true, then $P(n + 2)$ is true.

b) $P(1)$ and $P(2)$ are true; for all positive integers n, if $P(n)$ and $P(n + 1)$ are true, then $P(n + 2)$ is true.

c) $P(1)$ is true; for all positive integers n, if $P(n)$ is true, then $P(2n)$ is true.

d) $P(1)$ is true; for all positive integers n, if $P(n)$ is true, then $P(n + 1)$ is true.

20. Show that if the statement $P(n)$ is true for infinitely many positive integers n and $P(n + 1) \to P(n)$ is true for all positive integers n, then $P(n)$ is true for all positive integers n.

21. What is wrong with this "proof" by strong induction?

"Theorem"   For every nonnegative integer n, $5n = 0$.

Basis Step: $5 \cdot 0 = 0$.

Inductive Step: Suppose that $5j = 0$ for all nonnegative integers j with $0 \leq j \leq k$. Write $k + 1 = i + j$, where i and j are natural numbers less than $k + 1$. By the inductive hypothesis, $5(k + 1) = 5(i + j) = 5i + 5j = 0 + 0 = 0$.

*22. Find the flaw with the following "proof" that $a^n = 1$ for all nonnegative integers n, whenever a is a nonzero real number.

Basis Step: $a^0 = 1$ is true by the definition of a^0.

Inductive Step: Assume that $a^j = 1$ for all nonnegative integers j with $j \leq k$. Then note that

$$a^{k+1} = \frac{a^k \cdot a^k}{a^{k-1}} = \frac{1 \cdot 1}{1} = 1.$$

*23. Show that strong induction is a valid method of proof by showing that it follows from the well-ordering property.

24. Find the flaw with the following "proof" that every postage of three cents or more can be formed using just three-cent and four-cent stamps.

Basis Step: We can form postage of three cents with a single three-cent stamp and we can form postage of four cents using a single four-cent stamp.

Inductive Step: Assume that we can form postage of j cents for all nonnegative integers j with $j \leq k$ using just three-cent and four-cent stamps. We can then form postage of $k + 1$ cents by replacing one three-cent stamp with a four-cent stamp or by replacing two four-cent stamps by three three-cent stamps.

25. Show that we can prove that $P(n, k)$ is true for all pairs of positive integers n and k if we show

a) $P(1, 1)$ is true and $P(n, k) \to [P(n + 1, k) \wedge P(n, k + 1)]$ is true for all positive integers n and k.

b) $P(1, k)$ is true for all positive integers k, and $P(n, k) \to P(n + 1, k)$ is true for all positive integers n and k.

c) $P(n, 1)$ is true for all positive integers n, and $P(n, k) \to P(n, k + 1)$ is true for all positive integers n and k.

26. Prove that $\sum_{j=1}^{n} j(j + 1)(j + 2) \cdots (j + k - 1) = n(n + 1)(n + 2) \cdots (n + k)/(k + 1)$ for all positive integers k and n. [*Hint:* Use a technique from Exercise 25.]

*27. Show that if $a_1, a_2, \ldots, a_n$ are n distinct real numbers, exactly $n - 1$ multiplications are used to compute the product of these n numbers no matter how parentheses are inserted into their product. [*Hint:* Use strong induction and consider the last multiplication.]

*28. The well-ordering property can be used to show that there is a unique greatest common divisor of two positive integers. Let a and b be positive integers, and let S be the set of positive integers of the form $as + bt$, where s and t are integers.

a) Show that S is nonempty.

b) Use the well-ordering property to show that S has a smallest element c.

c) Show that if d is a common divisor of a and b, then d is a divisor of c.

d) Show that $c \mid a$ and $c \mid b$. [*Hint:* First, assume that $c \nmid a$. Then $a = qc + r$, where $0 < r < c$. Show that $r \in S$, contradicting the choice of c.]

e) Conclude from (c) and (d) that the greatest common divisor of a and b exists. Finish the proof by showing that this greatest common divisor is unique.

29. Let a be an integer and d be a positive integer. Show that the integers q and r with $a = dq + r$ and $0 \leq r < d$, which were shown to exist in Example 5, are unique.

30. Use mathematical induction to show that a rectangular checkerboard with an even number of cells and two squares missing, one white and one black, can be covered by dominoes.

*31. Show that the well-ordering property can be proved when the principle of mathematical induction is taken as an axiom.

*32. Show that the principle of mathematical induction and strong induction are equivalent; that is, each can be shown to be valid from the other.

*33. Show that we can prove the well-ordering property when we take strong induction as an axiom instead of taking the well-ordering property as an axiom.

5.3   Recursive Definitions and Structural Induction

Introduction

Sometimes it is difficult to define an object explicitly. However, it may be easy to define this object in terms of itself. This process is called **recursion**. For instance, the picture shown in Figure 1 is produced recursively. First, an original picture is given. Then a process of successively superimposing centered smaller pictures on top of the previous pictures is carried out.

We can use recursion to define sequences, functions, and sets. In Section 2.4, and in most beginning mathematics courses, the terms of a sequence are specified using an explicit formula. For instance, the sequence of powers of 2 is given by $a_n = 2^n$ for $n = 0, 1, 2, \ldots$. Recall from Section 2.4 that we can also define a sequence recursively by specifying how terms of the sequence are found from previous terms. The sequence of powers of 2 can also be defined by giving the first term of the sequence, namely, $a_0 = 1$, and a rule for finding a term of the sequence from the previous one, namely, $a_{n+1} = 2a_n$ for $n = 0, 1, 2, \ldots$. When we define a sequence recursively by specifying how terms of the sequence are found from previous terms, we can use induction to prove results about the sequence.

When we define a set recursively, we specify some initial elements in a basis step and provide a rule for constructing new elements from those we already have in the recursive step. To prove results about recursively defined sets we use a method called *structural induction*.

FIGURE 1   A Recursively Defined Picture.

Recursively Defined Functions

We use two steps to define a function with the set of nonnegative integers as its domain:

BASIS STEP: Specify the value of the function at zero.

RECURSIVE STEP: Give a rule for finding its value at an integer from its values at smaller integers.

Such a definition is called a **recursive** or **inductive definition**. Note that a function $f(n)$ from the set of nonnegative integers to the set of a real numbers is the same as a sequence $a_0, a_1, \ldots$ where a_i is a real number for every nonnegative integer i. So, defining a real-valued sequence $a_0, a_1, \ldots$ using a recurrence relation, as was done in Section 2.4, is the same as defining a function from the set of nonnegative integers to the set of real numbers.

EXAMPLE 1 Suppose that f is defined recursively by

$$f(0) = 3,$$
$$f(n + 1) = 2f(n) + 3.$$

Find $f(1)$, $f(2)$, $f(3)$, and $f(4)$.

Solution: From the recursive definition it follows that

$$f(1) = 2f(0) + 3 = 2 \cdot 3 + 3 = 9,$$
$$f(2) = 2f(1) + 3 = 2 \cdot 9 + 3 = 21,$$
$$f(3) = 2f(2) + 3 = 2 \cdot 21 + 3 = 45,$$
$$f(4) = 2f(3) + 3 = 2 \cdot 45 + 3 = 93.$$
◀

Recursively defined functions are **well defined**. That is, for every positive integer, the value of the function at this integer is determined in an unambiguous way. This means that given any positive integer, we can use the two parts of the definition to find the value of the function at that integer, and that we obtain the same value no matter how we apply the two parts of the definition. This is a consequence of the principle of mathematical induction. Additional examples of recursive definitions are given in Examples 2 and 3.

EXAMPLE 2 Give a recursive definition of a^n, where a is a nonzero real number and n is a nonnegative integer.

Solution: The recursive definition contains two parts. First a^0 is specified, namely, $a^0 = 1$. Then the rule for finding a^{n+1} from a^n, namely, $a^{n+1} = a \cdot a^n$, for $n = 0, 1, 2, 3, \ldots$, is given. These two equations uniquely define a^n for all nonnegative integers n. ◀

EXAMPLE 3 Give a recursive definition of

$$\sum_{k=0}^{n} a_k.$$

Solution: The first part of the recursive definition is

$$\sum_{k=0}^{0} a_k = a_0.$$

The second part is

$$\sum_{k=0}^{n+1} a_k = \left(\sum_{k=0}^{n} a_k\right) + a_{n+1}.$$

◀

In some recursive definitions of functions, the values of the function at the first k positive integers are specified, and a rule is given for determining the value of the function at larger integers from its values at some or all of the preceding k integers. That recursive definitions defined in this way produce well-defined functions follows from strong induction.

Recall from Section 2.4 that the Fibonacci numbers, $f_0, f_1, f_2, \ldots$, are defined by the equations $f_0 = 0$, $f_1 = 1$, and

$$f_n = f_{n-1} + f_{n-2}$$

Links

for $n = 2, 3, 4, \ldots$. [We can think of the Fibonacci number f_n either as the nth term of the sequence of Fibonacci numbers $f_0, f_1, \ldots$ or as the value at the integer n of a function $f(n)$.]

We can use the recursive definition of the Fibonacci numbers to prove many properties of these numbers. We give one such property in Example 4.

EXAMPLE 4   Show that whenever $n \geq 3$, $f_n > \alpha^{n-2}$, where $\alpha = (1 + \sqrt{5})/2$.

Extra Examples

Solution: We can use strong induction to prove this inequality. Let $P(n)$ be the statement $f_n > \alpha^{n-2}$. We want to show that $P(n)$ is true whenever n is an integer greater than or equal to 3.

BASIS STEP: First, note that

$$\alpha < 2 = f_3, \qquad \alpha^2 = (3 + \sqrt{5})/2 < 3 = f_4,$$

so $P(3)$ and $P(4)$ are true.

INDUCTIVE STEP: Assume that $P(j)$ is true, namely, that $f_j > \alpha^{j-2}$, for all integers j with $3 \leq j \leq k$, where $k \geq 4$. We must show that $P(k + 1)$ is true, that is, that $f_{k+1} > \alpha^{k-1}$. Because α is a solution of $x^2 - x - 1 = 0$ (as the quadratic formula shows), it follows that $\alpha^2 = \alpha + 1$. Therefore,

$$\alpha^{k-1} = \alpha^2 \cdot \alpha^{k-3} = (\alpha + 1)\alpha^{k-3} = \alpha \cdot \alpha^{k-3} + 1 \cdot \alpha^{k-3} = \alpha^{k-2} + \alpha^{k-3}.$$

By the inductive hypothesis, because $k \geq 4$, we have

$$f_{k-1} > \alpha^{k-3}, \qquad f_k > \alpha^{k-2}.$$

Therefore, it follows that

$$f_{k+1} = f_k + f_{k-1} > \alpha^{k-2} + \alpha^{k-3} = \alpha^{k-1}.$$

Hence, $P(k + 1)$ is true. This completes the proof.   ◀

Remark: The inductive step shows that whenever $k \geq 4$, $P(k + 1)$ follows from the assumption that $P(j)$ is true for $3 \leq j \leq k$. Hence, the inductive step does *not* show that $P(3) \rightarrow P(4)$. Therefore, we had to show that $P(4)$ is true separately.

We can now show that the Euclidean algorithm, introduced in Section 4.3, uses $O(\log b)$ divisions to find the greatest common divisor of the positive integers a and b, where $a \geq b$.

THEOREM 1   **LAMÉ'S THEOREM**   Let a and b be positive integers with $a \geq b$. Then the number of divisions used by the Euclidean algorithm to find $\gcd(a, b)$ is less than or equal to five times the number of decimal digits in b.

Proof: Recall that when the Euclidean algorithm is applied to find $\gcd(a, b)$ with $a \geq b$, this sequence of equations (where $a = r_0$ and $b = r_1$) is obtained.

$$r_0 = r_1 q_1 + r_2 \qquad 0 \leq r_2 < r_1,$$
$$r_1 = r_2 q_2 + r_3 \qquad 0 \leq r_3 < r_2,$$
$$\vdots$$
$$r_{n-2} = r_{n-1} q_{n-1} + r_n \qquad 0 \leq r_n < r_{n-1},$$
$$r_{n-1} = r_n q_n.$$

Here n divisions have been used to find $r_n = \gcd(a, b)$. Note that the quotients $q_1, q_2, \ldots, q_{n-1}$ are all at least 1. Moreover, $q_n \geq 2$, because $r_n < r_{n-1}$. This implies that

$$r_n \geq 1 = f_2,$$
$$r_{n-1} \geq 2r_n \geq 2f_2 = f_3,$$
$$r_{n-2} \geq r_{n-1} + r_n \geq f_3 + f_2 = f_4,$$
$$\vdots$$
$$r_2 \geq r_3 + r_4 \geq f_{n-1} + f_{n-2} = f_n,$$
$$b = r_1 \geq r_2 + r_3 \geq f_n + f_{n-1} = f_{n+1}.$$

It follows that if n divisions are used by the Euclidean algorithm to find $\gcd(a, b)$ with $a \geq b$, then $b \geq f_{n+1}$. By Example 4 we know that $f_{n+1} > \alpha^{n-1}$ for $n > 2$, where $\alpha = (1 + \sqrt{5})/2$. Therefore, it follows that $b > \alpha^{n-1}$. Furthermore, because $\log_{10} \alpha \approx 0.208 > 1/5$, we see that

$$\log_{10} b > (n - 1) \log_{10} \alpha > (n - 1)/5.$$

Hence, $n - 1 < 5 \cdot \log_{10} b$. Now suppose that b has k decimal digits. Then $b < 10^k$ and $\log_{10} b < k$. It follows that $n - 1 < 5k$, and because k is an integer, it follows that $n \leq 5k$. This finishes the proof. ◁

FIBONACCI (1170–1250)   Fibonacci (short for *filius Bonacci*, or "son of Bonacci") was also known as Leonardo of Pisa. He was born in the Italian commercial center of Pisa. Fibonacci was a merchant who traveled extensively throughout the Mideast, where he came into contact with Arabian mathematics. In his book *Liber Abaci*, Fibonacci introduced the European world to Arabic notation for numerals and algorithms for arithmetic. It was in this book that his famous rabbit problem (described in Section 8.1) appeared. Fibonacci also wrote books on geometry and trigonometry and on Diophantine equations, which involve finding integer solutions to equations.

Because the number of decimal digits in b, which equals $\lfloor \log_{10} b \rfloor + 1$, is less than or equal to $\log_{10} b + 1$, Theorem 1 tells us that the number of divisions required to find $\gcd(a, b)$ with $a > b$ is less than or equal to $5(\log_{10} b + 1)$. Because $5(\log_{10} b + 1)$ is $O(\log b)$, we see that $O(\log b)$ divisions are used by the Euclidean algorithm to find $\gcd(a, b)$ whenever $a > b$.

Recursively Defined Sets and Structures

We have explored how functions can be defined recursively. We now turn our attention to how sets can be defined recursively. Just as in the recursive definition of functions, recursive definitions of sets have two parts, a **basis step** and a **recursive step**. In the basis step, an initial collection of elements is specified. In the recursive step, rules for forming new elements in the set from those already known to be in the set are provided. Recursive definitions may also include an **exclusion rule**, which specifies that a recursively defined set contains nothing other than those elements specified in the basis step or generated by applications of the recursive step. In our discussions, we will always tacitly assume that the exclusion rule holds and no element belongs to a recursively defined set unless it is in the initial collection specified in the basis step or can be generated using the recursive step one or more times. Later we will see how we can use a technique known as structural induction to prove results about recursively defined sets.

Examples 5, 6, 8, and 9 illustrate the recursive definition of sets. In each example, we show those elements generated by the first few applications of the recursive step.

EXAMPLE 5 Consider the subset S of the set of integers recursively defined by

BASIS STEP: $3 \in S$.

RECURSIVE STEP: If $x \in S$ and $y \in S$, then $x + y \in S$.

The new elements found to be in S are 3 by the basis step, $3 + 3 = 6$ at the first application of the recursive step, $3 + 6 = 6 + 3 = 9$ and $6 + 6 = 12$ at the second application of the recursive step, and so on. We will show in Example 10 that S is the set of all positive multiples of 3. ◀

Recursive definitions play an important role in the study of strings. (See Chapter 13 for an introduction to the theory of formal languages, for example.) Recall from Section 2.4 that a string over an alphabet Σ is a finite sequence of symbols from Σ. We can define Σ^*, the set of strings over Σ, recursively, as Definition 1 shows.

DEFINITION 1 The set Σ^* of *strings* over the alphabet Σ is defined recursively by

BASIS STEP: $\lambda \in \Sigma^*$ (where λ is the empty string containing no symbols).

RECURSIVE STEP: If $w \in \Sigma^*$ and $x \in \Sigma$, then $wx \in \Sigma^*$.

The basis step of the recursive definition of strings says that the empty string belongs to Σ^*. The recursive step states that new strings are produced by adding a symbol from Σ to the end of strings in Σ^*. At each application of the recursive step, strings containing one additional symbol are generated.

EXAMPLE 6   If $\Sigma = \{0, 1\}$, the strings found to be in Σ^*, the set of all bit strings, are λ, specified to be in Σ^* in the basis step, 0 and 1 formed during the first application of the recursive step, 00, 01, 10, and 11 formed during the second application of the recursive step, and so on. ◀

Recursive definitions can be used to define operations or functions on the elements of recursively defined sets. This is illustrated in Definition 2 of the concatenation of two strings and Example 7 concerning the length of a string.

DEFINITION 2   Two strings can be combined via the operation of *concatenation*. Let Σ be a set of symbols and Σ^* the set of strings formed from symbols in Σ. We can define the concatenation of two strings, denoted by $\cdot$, recursively as follows.

BASIS STEP: If $w \in \Sigma^*$, then $w \cdot \lambda = w$, where λ is the empty string.

RECURSIVE STEP: If $w_1 \in \Sigma^*$ and $w_2 \in \Sigma^*$ and $x \in \Sigma$, then $w_1 \cdot (w_2 x) = (w_1 \cdot w_2)x$.

The concatenation of the strings w_1 and w_2 is often written as $w_1 w_2$ rather than $w_1 \cdot w_2$. By repeated application of the recursive definition, it follows that the concatenation of two strings w_1 and w_2 consists of the symbols in w_1 followed by the symbols in w_2. For instance, the concatenation of $w_1 = abra$ and $w_2 = cadabra$ is $w_1 w_2 = abracadabra$.

EXAMPLE 7   **Length of a String**   Give a recursive definition of $l(w)$, the length of the string w.

Solution: The length of a string can be recursively defined by

$$l(\lambda) = 0;$$
$$l(wx) = l(w) + 1 \text{ if } w \in \Sigma^* \text{ and } x \in \Sigma.$$ ◀

Another important use of recursive definitions is to define **well-formed formulae** of various types. This is illustrated in Examples 8 and 9.

EXAMPLE 8   **Well-Formed Formulae in Propositional Logic**   We can define the set of well-formed formulae in propositional logic involving **T, F**, propositional variables, and operators from the set $\{\neg, \wedge, \vee, \rightarrow, \leftrightarrow\}$.

BASIS STEP: **T**, **F**, and s, where s is a propositional variable, are well-formed formulae.

RECURSIVE STEP: If E and F are well-formed formulae, then $(\neg E)$, $(E \wedge F)$, $(E \vee F)$, $(E \rightarrow F)$, and $(E \leftrightarrow F)$ are well-formed formulae.

For example, by the basis step we know that **T**, **F**, p, and q are well-formed formulae, where p and q are propositional variables. From an initial application of the recursive step, we know that $(p \vee q)$, $(p \rightarrow \mathbf{F})$, $(\mathbf{F} \rightarrow q)$, and $(q \wedge \mathbf{F})$ are well-formed formulae. A second application of the recursive step shows that $((p \vee q) \rightarrow (q \wedge \mathbf{F}))$, $(q \vee (p \vee q))$, and $((p \rightarrow \mathbf{F}) \rightarrow \mathbf{T})$ are well-formed formulae. We leave it to the reader to show that $p\neg \wedge q$, $pq\wedge$, and $\neg \wedge pq$ are *not* well-formed formulae, by showing that none can be obtained using the basis step and one or more applications of the recursive step. ◀

EXAMPLE 9   **Well-Formed Formulae of Operators and Operands**   We can define the set of well-formed formulae consisting of variables, numerals, and operators from the set $\{+, -, *, /, \uparrow\}$ (where $*$ denotes multiplication and $\uparrow$ denotes exponentiation) recursively.

BASIS STEP: x is a well-formed formula if x is a numeral or a variable.

RECURSIVE STEP: If F and G are well-formed formulae, then $(F + G)$, $(F - G)$, $(F * G)$, (F/G), and $(F \uparrow G)$ are well-formed formulae.

For example, by the basis step we see that x, y, 0, and 3 are well-formed formulae (as is any variable or numeral). Well-formed formulae generated by applying the recursive step once include $(x + 3)$, $(3 + y)$, $(x - y)$, $(3 - 0)$, $(x * 3)$, $(3 * y)$, $(3/0)$, (x/y), $(3 \uparrow x)$, and $(0 \uparrow 3)$. Applying the recursive step twice shows that formulae such as $((x + 3) + 3)$ and $(x - (3 * y))$ are well-formed formulae. [Note that $(3/0)$ is a well-formed formula because we are concerned only with syntax matters here.] We leave it to the reader to show that each of the formulae $x3 +$, $y * + x$, and $* x/y$ is *not* a well-formed formula by showing that none of them can be obtained from the basis step and one or more applications of the recursive step. ◀

We will study trees extensively in Chapter 11. A tree is a special type of a graph; a graph is made up of vertices and edges connecting some pairs of vertices. We will study graphs in Chapter 10. We will briefly introduce them here to illustrate how they can be defined recursively.

DEFINITION 3

The set of *rooted trees,* where a rooted tree consists of a set of vertices containing a distinguished vertex called the *root,* and edges connecting these vertices, can be defined recursively by these steps:

BASIS STEP: A single vertex r is a rooted tree.

RECURSIVE STEP: Suppose that $T_1, T_2, \ldots, T_n$ are disjoint rooted trees with roots $r_1, r_2, \ldots, r_n$, respectively. Then the graph formed by starting with a root r, which is not in any of the rooted trees $T_1, T_2, \ldots, T_n$, and adding an edge from r to each of the vertices $r_1, r_2, \ldots, r_n$, is also a rooted tree.

In Figure 2 we illustrate some of the rooted trees formed starting with the basis step and applying the recursive step one time and two times. Note that infinitely many rooted trees are formed at each application of the recursive definition.

Binary trees are a special type of rooted trees. We will provide recursive definitions of two types of binary trees—full binary trees and extended binary trees. In the recursive step of the definition of each type of binary tree, two binary trees are combined to form a new tree with one of these trees designated the left subtree and the other the right subtree. In extended binary trees, the left subtree or the right subtree can be empty, but in full binary trees this is not

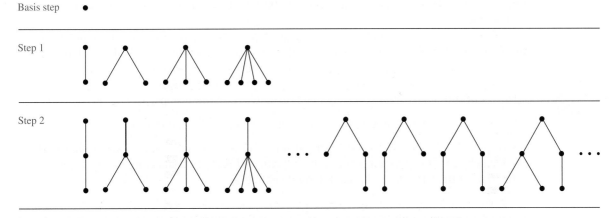

FIGURE 2 **Building Up Rooted Trees.**

Basis step $\emptyset$

Step 1   •

Step 2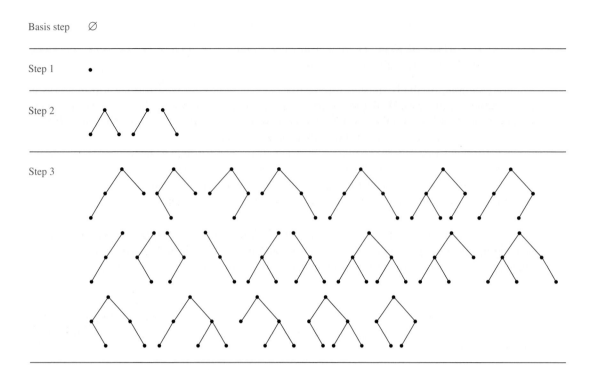

Step 3

FIGURE 3   Building Up Extended Binary Trees.

possible. Binary trees are one of the most important types of structures in computer science. In Chapter 11 we will see how they can be used in searching and sorting algorithms, in algorithms for compressing data, and in many other applications. We first define extended binary trees.

DEFINITION 4   The set of *extended binary trees* can be defined recursively by these steps:

BASIS STEP: The empty set is an extended binary tree.

RECURSIVE STEP: If T_1 and T_2 are disjoint extended binary trees, there is an extended binary tree, denoted by $T_1 \cdot T_2$, consisting of a root r together with edges connecting the root to each of the roots of the left subtree T_1 and the right subtree T_2 when these trees are nonempty.

Figure 3 shows how extended binary trees are built up by applying the recursive step from one to three times.

We now show how to define the set of full binary trees. Note that the difference between this recursive definition and that of extended binary trees lies entirely in the basis step.

DEFINITION 5   The set of full binary trees can be defined recursively by these steps:

BASIS STEP: There is a full binary tree consisting only of a single vertex r.

RECURSIVE STEP: If T_1 and T_2 are disjoint full binary trees, there is a full binary tree, denoted by $T_1 \cdot T_2$, consisting of a root r together with edges connecting the root to each of the roots of the left subtree T_1 and the right subtree T_2.

Figure 4 shows how full binary trees are built up by applying the recursive step one and two times.

Basis step

Step 1

Step 2

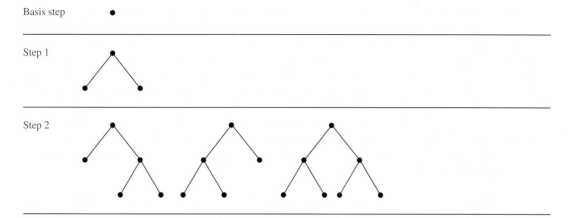

FIGURE 4   Building Up Full Binary Trees.

Structural Induction

To prove results about recursively defined sets, we generally use some form of mathematical induction. Example 10 illustrates the connection between recursively defined sets and mathematical induction.

EXAMPLE 10   Show that the set S defined in Example 5 by specifying that $3 \in S$ and that if $x \in S$ and $y \in S$, then $x + y \in S$, is the set of all positive integers that are multiples of 3.

Solution: Let A be the set of all positive integers divisible by 3. To prove that $A = S$, we must show that A is a subset of S and that S is a subset of A. To prove that A is a subset of S, we must show that every positive integer divisible by 3 is in S. We will use mathematical induction to prove this.

Let $P(n)$ be the statement that $3n$ belongs to S. The basis step holds because by the first part of the recursive definition of S, $3 \cdot 1 = 3$ is in S. To establish the inductive step, assume that $P(k)$ is true, namely, that $3k$ is in S. Because $3k$ is in S and because 3 is in S, it follows from the second part of the recursive definition of S that $3k + 3 = 3(k + 1)$ is also in S.

To prove that S is a subset of A, we use the recursive definition of S. First, the basis step of the definition specifies that 3 is in S. Because $3 = 3 \cdot 1$, all elements specified to be in S in this step are divisible by 3 and are therefore in A. To finish the proof, we must show that all integers in S generated using the second part of the recursive definition are in A. This consists of showing that $x + y$ is in A whenever x and y are elements of S also assumed to be in A. Now if x and y are both in A, it follows that $3 \mid x$ and $3 \mid y$. By part (i) of Theorem 1 of Section 4.1, it follows that $3 \mid x + y$, completing the proof.   ◀

In Example 10 we used mathematical induction over the set of positive integers and a recursive definition to prove a result about a recursively defined set. However, instead of using mathematical induction directly to prove results about recursively defined sets, we can use a more convenient form of induction known as **structural induction**. A proof by structural induction consists of two parts. These parts are

BASIS STEP: Show that the result holds for all elements specified in the basis step of the recursive definition to be in the set.

RECURSIVE STEP: Show that if the statement is true for each of the elements used to construct new elements in the recursive step of the definition, the result holds for these new elements.

The validity of structural induction follows from the principle of mathematical induction for the nonnegative integers. To see this, let $P(n)$ state that the claim is true for all elements of the set that are generated by n or fewer applications of the rules in the recursive step of a recursive definition. We will have established that the principle of mathematical induction implies the principle of structural induction if we can show that $P(n)$ is true whenever n is a positive integer. In the basis step of a proof by structural induction we show that $P(0)$ is true. That is, we show that the result is true of all elements specified to be in the set in the basis step of the definition. A consequence of the recursive step is that if we assume $P(k)$ is true, it follows that $P(k + 1)$ is true. When we have completed a proof using structural induction, we have shown that $P(0)$ is true and that $P(k)$ implies $P(k + 1)$. By mathematical induction it follows that $P(n)$ is true for all nonnegative integers n. This also shows that the result is true for all elements generated by the recursive definition, and shows that structural induction is a valid proof technique.

EXAMPLES OF PROOFS USING STRUCTURAL INDUCTION   Structural induction can be used to prove that all members of a set constructed recursively have a particular property. We will illustrate this idea by using structural induction to prove results about well-formed formulae, strings, and binary trees. For each proof, we have to carry out the appropriate basis step and the appropriate recursive step. For example, to use structural induction to prove a result about the set of well-formed formulae defined in Example 8, where we specify that $\mathbf{T}$, $\mathbf{F}$, and every propositional variable s are well-formed formulae and where we specify that if E and F are well-formed formulae, then $(\neg E)$, $(E \wedge F)$, $(E \vee F)$, $(E \rightarrow F)$, and $(E \leftrightarrow F)$ are well-formed formulae, we need to complete this basis step and this recursive step.

BASIS STEP: Show that the result is true for $\mathbf{T}$, $\mathbf{F}$, and s whenever s is a propositional variable.

RECURSIVE STEP: Show that if the result is true for the compound propositions p and q, it is also true for $(\neg p)$, $(p \vee q)$, $(p \wedge q)$, $(p \rightarrow q)$, and $(p \leftrightarrow q)$.

Example 11 illustrates how we can prove results about well-formed formulae using structural induction.

EXAMPLE 11   Show that every well-formed formula for compound propositions, as defined in Example 8, contains an equal number of left and right parentheses.

Solution:

BASIS STEP: Each of the formula $\mathbf{T}$, $\mathbf{F}$, and s contains no parentheses, so clearly they contain an equal number of left and right parentheses.

RECURSIVE STEP: Assume p and q are well-formed formulae each containing an equal number of left and right parentheses. That is, if l_p and l_q are the number of left parentheses in p and q, respectively, and r_p and r_q are the number of right parentheses in p and q, respectively, then $l_p = r_p$ and $l_q = r_q$. To complete the inductive step, we need to show that each of $(\neg p)$, $(p \vee q)$, $(p \wedge q)$, $(p \rightarrow q)$, and $(p \leftrightarrow q)$ also contains an equal number of left and right parentheses. The number of left parentheses in the first of these compound propositions equals $l_p + 1$ and in each of the other compound propositions equals $l_p + l_q + 1$. Similarly, the number of right parentheses in the first of these compound propositions equals $r_p + 1$ and in each of the other compound propositions equals $r_p + r_q + 1$. Because $l_p = r_p$ and $l_q = r_q$, it follows that each of these compound expressions contains the same number of left and right parentheses. This completes the proof by structural induction.   ◀

Suppose that $P(w)$ is a propositional function over the set of strings $w \in \Sigma^*$. To use structural induction to prove that $P(w)$ holds for all strings $w \in \Sigma^*$, we need to complete both a basis step and a recursive step. These steps are:

BASIS STEP: Show that $P(\lambda)$ is true.

RECURSIVE STEP: Assume that $P(w)$ is true, where $w \in \Sigma^*$. Show that if $x \in \Sigma$, then $P(wx)$ must also be true.

Example 12 illustrates how structural induction can be used in proofs about strings.

EXAMPLE 12   Use structural induction to prove that $l(xy) = l(x) + l(y)$, where x and y belong to Σ^*, the set of strings over the alphabet Σ.

Solution: We will base our proof on the recursive definition of the set Σ^* given in Definition 1 and the definition of the length of a string in Example 7, which specifies that $l(\lambda) = 0$ and $l(wx) = l(w) + 1$ when $w \in \Sigma^*$ and $x \in \Sigma$. Let $P(y)$ be the statement that $l(xy) = l(x) + l(y)$ whenever x belongs to Σ^*.

BASIS STEP: To complete the basis step, we must show that $P(\lambda)$ is true. That is, we must show that $l(x\lambda) = l(x) + l(\lambda)$ for all $x \in \Sigma^*$. Because $l(x\lambda) = l(x) = l(x) + 0 = l(x) + l(\lambda)$ for every string x, it follows that $P(\lambda)$ is true.

RECURSIVE STEP: To complete the inductive step, we assume that $P(y)$ is true and show that this implies that $P(ya)$ is true whenever $a \in \Sigma$. What we need to show is that $l(xya) = l(x) + l(ya)$ for every $a \in \Sigma$. To show this, note that by the recursive definition of $l(w)$ (given in Example 7), we have $l(xya) = l(xy) + 1$ and $l(ya) = l(y) + 1$. And, by the inductive hypothesis, $l(xy) = l(x) + l(y)$. We conclude that $l(xya) = l(x) + l(y) + 1 = l(x) + l(ya)$.◄

We can prove results about trees or special classes of trees using structural induction. For example, to prove a result about full binary trees using structural induction we need to complete this basis step and this recursive step.

BASIS STEP: Show that the result is true for the tree consisting of a single vertex.

RECURSIVE STEP: Show that if the result is true for the trees T_1 and T_2, then it is true for tree $T_1 \cdot T_2$ consisting of a root r, which has T_1 as its left subtree and T_2 as its right subtree.

Before we provide an example showing how structural induction can be used to prove a result about full binary trees, we need some definitions. We will recursively define the height $h(T)$ and the number of vertices $n(T)$ of a full binary tree T. We begin by defining the height of a full binary tree.

DEFINITION 6   We define the height $h(T)$ of a full binary tree T recursively.

BASIS STEP: The height of the full binary tree T consisting of only a root r is $h(T) = 0$.

RECURSIVE STEP: If T_1 and T_2 are full binary trees, then the full binary tree $T = T_1 \cdot T_2$ has height $h(T) = 1 + \max(h(T_1), h(T_2))$.

If we let $n(T)$ denote the number of vertices in a full binary tree, we observe that $n(T)$ satisfies the following recursive formula:

BASIS STEP: The number of vertices $n(T)$ of the full binary tree T consisting of only a root r is $n(T) = 1$.

RECURSIVE STEP: If T_1 and T_2 are full binary trees, then the number of vertices of the full binary tree $T = T_1 \cdot T_2$ is $n(T) = 1 + n(T_1) + n(T_2)$.

We now show how structural induction can be used to prove a result about full binary trees.

THEOREM 2   If T is a full binary tree T, then $n(T) \leq 2^{h(T)+1} - 1$.

Proof: We prove this inequality using structural induction.

BASIS STEP: For the full binary tree consisting of just the root r the result is true because $n(T) = 1$ and $h(T) = 0$, so that $n(T) = 1 \leq 2^{0+1} - 1 = 1$.

RECURSIVE STEP: For the inductive hypothesis we assume that $n(T_1) \leq 2^{h(T_1)+1} - 1$ and $n(T_2) \leq 2^{h(T_2)+1} - 1$ whenever T_1 and T_2 are full binary trees. By the recursive formulae for $n(T)$ and $h(T)$ we have $n(T) = 1 + n(T_1) + n(T_2)$ and $h(T) = 1 + \max(h(T_1), h(T_2))$.

We find that

$$
\begin{aligned}
n(T) &= 1 + n(T_1) + n(T_2) && \text{by the recursive formula for } n(T) \\
&\leq 1 + (2^{h(T_1)+1} - 1) + (2^{h(T_2)+1} - 1) && \text{by the inductive hypothesis} \\
&\leq 2 \cdot \max(2^{h(T_1)+1}, 2^{h(T_2)+1}) - 1 && \text{because the sum of two terms is at most 2} \\
& && \quad\text{times the larger} \\
&= 2 \cdot 2^{\max(h(T_1), h(T_2))+1} - 1 && \text{because } \max(2^x, 2^y) = 2^{\max(x,y)} \\
&= 2 \cdot 2^{h(T)} - 1 && \text{by the recursive definition of } h(T) \\
&= 2^{h(T)+1} - 1.
\end{aligned}
$$

This completes the recursive step.   ◁

Generalized Induction

We can extend mathematical induction to prove results about other sets that have the well-ordering property besides the set of integers. Although we will discuss this concept in detail in Section 9.6, we provide an example here to illustrate the usefulness of such an approach.

As an example, note that we can define an ordering on $\mathbf{N} \times \mathbf{N}$, the ordered pairs of non-negative integers, by specifying that (x_1, y_1) is less than or equal to (x_2, y_2) if either $x_1 < x_2$, or $x_1 = x_2$ and $y_1 < y_2$; this is called the **lexicographic ordering**. The set $\mathbf{N} \times \mathbf{N}$ with this ordering has the property that every subset of $\mathbf{N} \times \mathbf{N}$ has a least element (see Exercise 53 in Section 9.6). This implies that we can recursively define the terms $a_{m,n}$, with $m \in \mathbf{N}$ and $n \in \mathbf{N}$, and prove results about them using a variant of mathematical induction, as illustrated in Example 13.

EXAMPLE 13   Suppose that $a_{m,n}$ is defined recursively for $(m, n) \in \mathbf{N} \times \mathbf{N}$ by $a_{0,0} = 0$ and

$$
a_{m,n} = \begin{cases} a_{m-1,n} + 1 & \text{if } n = 0 \text{ and } m > 0 \\ a_{m,n-1} + n & \text{if } n > 0. \end{cases}
$$

Show that $a_{m,n} = m + n(n + 1)/2$ for all $(m, n) \in \mathbf{N} \times \mathbf{N}$, that is, for all pairs of nonnegative integers.

Solution: We can prove that $a_{m,n} = m + n(n + 1)/2$ using a generalized version of mathematical induction. The basis step requires that we show that this formula is valid when $(m, n) = (0, 0)$. The induction step requires that we show that if the formula holds for all pairs smaller than (m, n) in the lexicographic ordering of $\mathbf{N} \times \mathbf{N}$, then it also holds for (m, n).

BASIS STEP: Let $(m, n) = (0, 0)$. Then by the basis case of the recursive definition of $a_{m,n}$ we have $a_{0,0} = 0$. Furthermore, when $m = n = 0$, $m + n(n + 1)/2 = 0 + (0 \cdot 1)/2 = 0$. This completes the basis step.

INDUCTIVE STEP: Suppose that $a_{m',n'} = m' + n'(n' + 1)/2$ whenever (m', n') is less than (m, n) in the lexicographic ordering of $\mathbf{N} \times \mathbf{N}$. By the recursive definition, if $n = 0$, then $a_{m,n} = a_{m-1,n} + 1$. Because $(m - 1, n)$ is smaller than (m, n), the inductive hypothesis tells us that $a_{m-1,n} = m - 1 + n(n + 1)/2$, so that $a_{m,n} = m - 1 + n(n + 1)/2 + 1 =$

$m + n(n + 1)/2$, giving us the desired equality. Now suppose that $n > 0$, so $a_{m,n} = a_{m,n-1} + n$. Because $(m, n - 1)$ is smaller than (m, n), the inductive hypothesis tells us that $a_{m,n-1} = m + (n - 1)n/2$, so $a_{m,n} = m + (n - 1)n/2 + n = m + (n^2 - n + 2n)/2 = m + n(n + 1)/2$. This finishes the inductive step. ◀

As mentioned, we will justify this proof technique in Section 9.6.

Exercises

1. Find $f(1)$, $f(2)$, $f(3)$, and $f(4)$ if $f(n)$ is defined recursively by $f(0) = 1$ and for $n = 0, 1, 2, \ldots$
  a) $f(n + 1) = f(n) + 2$.
  b) $f(n + 1) = 3f(n)$.
  c) $f(n + 1) = 2^{f(n)}$.
  d) $f(n + 1) = f(n)^2 + f(n) + 1$.

2. Find $f(2)$, $f(3)$, $f(4)$, and $f(5)$ if f is defined recursively by $f(0) = -1$, $f(1) = 2$, and for $n = 1, 2, \ldots$
  a) $f(n + 1) = f(n) + 3f(n - 1)$.
  b) $f(n + 1) = f(n)^2 f(n - 1)$.
  c) $f(n + 1) = 3f(n)^2 - 4f(n - 1)^2$.
  d) $f(n + 1) = f(n - 1)/f(n)$.

3. Determine whether each of these proposed definitions is a valid recursive definition of a function f from the set of nonnegative integers to the set of integers. If f is well defined, find a formula for $f(n)$ when n is a nonnegative integer and prove that your formula is valid.
  a) $f(0) = 0$, $f(n) = 2f(n - 2)$ for $n \geq 1$
  b) $f(0) = 1$, $f(n) = f(n - 1) - 1$ for $n \geq 1$
  c) $f(0) = 2$, $f(1) = 3$, $f(n) = f(n - 1) - 1$ for $n \geq 2$
  d) $f(0) = 1$, $f(1) = 2$, $f(n) = 2f(n - 2)$ for $n \geq 2$
  e) $f(0) = 1$, $f(n) = 3f(n - 1)$ if n is odd and $n \geq 1$ and $f(n) = 9f(n - 2)$ if n is even and $n \geq 2$

4. Give a recursive definition of the sequence $\{a_n\}$, $n = 1, 2, 3, \ldots$ if
  a) $a_n = 6n$.   **b)** $a_n = 2n + 1$.
  c) $a_n = 10^n$.   **d)** $a_n = 5$.

5. Let F be the function such that $F(n)$ is the sum of the first n positive integers. Give a recursive definition of $F(n)$.

6. Give a recursive definition of $S_m(n)$, the sum of the integer m and the nonnegative integer n.

7. Give a recursive definition of $P_m(n)$, the product of the integer m and the nonnegative integer n.

In Exercises 8–13 f_n is the nth Fibonacci number.

8. Prove that $f_1^2 + f_2^2 + \cdots + f_n^2 = f_n f_{n+1}$ when n is a positive integer.

9. Prove that $f_1 + f_3 + \cdots + f_{2n-1} = f_{2n}$ when n is a positive integer.

*10. Show that $f_{n+1} f_{n-1} - f_n^2 = (-1)^n$ when n is a positive integer.

11. Determine the number of divisions used by the Euclidean algorithm to find the greatest common divisor of the Fibonacci numbers f_n and f_{n+1}, where n is a nonnegative integer. Verify your answer using mathematical induction.

12. Let
$$\mathbf{A} = \begin{bmatrix} 1 & 1 \\ 1 & 0 \end{bmatrix}.$$
Show that
$$\mathbf{A}^n = \begin{bmatrix} f_{n+1} & f_n \\ f_n & f_{n-1} \end{bmatrix}$$
when n is a positive integer.

13. By taking determinants of both sides of the equation in Exercise 18, prove the identity given in Exercise 14. (Recall that the determinant of the matrix $\begin{vmatrix} a & b \\ c & d \end{vmatrix}$ is $ad - bc$.)

*14. Give a recursive definition of the functions max and min so that $\max(a_1, a_2, \ldots, a_n)$ and $\min(a_1, a_2, \ldots, a_n)$ are the maximum and minimum of the n numbers $a_1, a_2, \ldots, a_n$, respectively.

15. Give a recursive definition of the set of positive integers that are multiples of 5.

16. Give a recursive definition of
  a) the set of odd positive integers.
  b) the set of positive integer powers of 3.
  c) the set of polynomials with integer coefficients.

17. Give a recursive definition of
  a) the set of even integers.
  b) the set of positive integers congruent to 2 modulo 3.
  c) the set of positive integers not divisible by 5.

18. Let S be the subset of the set of ordered pairs of integers defined recursively by

Basis step: $(0, 0) \in S$.

Recursive step: If $(a, b) \in S$, then $(a, b + 1) \in S$, $(a + 1, b + 1) \in S$, and $(a + 2, b + 1) \in S$.

  a) List the elements of S produced by the first four applications of the recursive definition.
  b) Use strong induction on the number of applications of the recursive step of the definition to show that $a \leq 2b$ whenever $(a, b) \in S$.
  c) Use structural induction to show that $a \leq 2b$ whenever $(a, b) \in S$.

19. Give a recursive definition of each of these sets of ordered pairs of positive integers. Use structural induction to prove that the recursive definition you found is correct. [*Hint:* To find a recursive definition, plot the points in the set in the plane and look for patterns.]
  a) $S = \{(a, b) \mid a \in \mathbf{Z}^+, b \in \mathbf{Z}^+, \text{ and } a + b \text{ is even}\}$
  b) $S = \{(a, b) \mid a \in \mathbf{Z}^+, b \in \mathbf{Z}^+, \text{ and } a \text{ or } b \text{ is odd}\}$
  c) $S = \{(a, b) \mid a \in \mathbf{Z}^+, b \in \mathbf{Z}^+, a + b \text{ is odd, and } 3 \mid b\}$

20. Prove that in a bit string, the string 01 occurs at most one more time than the string 10.

21. Define well-formed formulae of sets, variables representing sets, and operators from $\{^-, \cup, \cap, -\}$.

22. a) Give a recursive definition of the function $ones(s)$, which counts the number of ones in a bit string s.
   b) Use structural induction to prove that $ones(st) = ones(s) + ones(t)$.

23. a) Give a recursive definition of the function $m(s)$, which equals the smallest digit in a nonempty string of decimal digits.
   b) Use structural induction to prove that $m(st) = \min(m(s), m(t))$.

The **reversal** of a string is the string consisting of the symbols of the string in reverse order. The reversal of the string w is denoted by w^R.

24. Find the reversal of the following bit strings.
   a) 0101   **b)** 1 1011   **c)** 1000 1001 0111

25. Give a recursive definition of the reversal of a string. [*Hint:* First define the reversal of the empty string. Then write a string w of length $n + 1$ as xy, where x is a string of length n, and express the reversal of w in terms of x^R and y.]

*26. Use structural induction to prove that $(w_1 w_2)^R = w_2^R w_1^R$.

27. Give a recursive definition of w^i, where w is a string and i is a nonnegative integer. (Here w^i represents the concatenation of i copies of the string w.)

*28. Give a recursive definition of the set of bit strings that are palindromes.

29. When does a string belong to the set A of bit strings defined recursively by

$$\lambda \in A$$
$$0x1 \in A \text{ if } x \in A,$$

where λ is the empty string?

*30. Recursively define the set of bit strings that have more zeros than ones.

31. Use Exercise 27 and mathematical induction to show that $l(w^i) = i \cdot l(w)$, where w is a string and i is a nonnegative integer.

*32. Show that $(w^R)^i = (w^i)^R$ whenever w is a string and i is a nonnegative integer; that is, show that the ith power of the reversal of a string is the reversal of the ith power of the string.

33. Use structural induction to show that $n(T) \geq 2h(T) + 1$, where T is a full binary tree, $n(T)$ equals the number of vertices of T, and $h(T)$ is the height of T.

The set of leaves and the set of internal vertices of a full binary tree can be defined recursively.

Basis step: The root r is a leaf of the full binary tree with exactly one vertex r. This tree has no internal vertices.

Recursive step: The set of leaves of the tree $T = T_1 \cdot T_2$ is the union of the sets of leaves of T_1 and of T_2. The internal vertices of T are the root r of T and the union of the set of internal vertices of T_1 and the set of internal vertices of T_2.

34. Use structural induction to show that $l(T)$, the number of leaves of a full binary tree T, is 1 more than $i(T)$, the number of internal vertices of T.

35. Use generalized induction as was done in Example 13 to show that if $a_{m,n}$ is defined recursively by $a_{0,0} = 0$ and

$$a_{m,n} = \begin{cases} a_{m-1,n} + 1 & \text{if } n = 0 \text{ and } m > 0 \\ a_{m,n-1} + 1 & \text{if } n > 0, \end{cases}$$

then $a_{m,n} = m + n$ for all $(m, n) \in \mathbf{N} \times \mathbf{N}$.

36. Use generalized induction as was done in Example 13 to show that if $a_{m,n}$ is defined recursively by $a_{1,1} = 5$ and

$$a_{m,n} = \begin{cases} a_{m-1,n} + 2 & \text{if } n = 1 \text{ and } m > 1 \\ a_{m,n-1} + 2 & \text{if } n > 1, \end{cases}$$

then $a_{m,n} = 2(m + n) + 1$ for all $(m, n) \in \mathbf{Z}^+ \times \mathbf{Z}^+$.

*37. A **partition** of a positive integer n is a way to write n as a sum of positive integers where the order of terms in the sum does not matter. For instance, $7 = 3 + 2 + 1 + 1$ is a partition of 7. Let P_m equal the number of different partitions of m, and let $P_{m,n}$ be the number of different ways to express m as the sum of positive integers not exceeding n.
   a) Show that $P_{m,m} = P_m$.
   b) Show that the following recursive definition for $P_{m,n}$ is correct:

$$P_{m,n} = \begin{cases} 1 & \text{if } m = 1 \\ 1 & \text{if } n = 1 \\ P_{m,m} & \text{if } m < n \\ 1 + P_{m,m-1} & \text{if } m = n > 1 \\ P_{m,n-1} + P_{m-n,n} & \text{if } m > n > 1. \end{cases}$$

   c) Find the number of partitions of 5 and of 6 using this recursive definition.

Consider an inductive definition of a version of **Ackermann's function**. This function was named after Wilhelm Ackermann, a German mathematician who was a student of the great mathematician David Hilbert. Ackermann's function plays an important role in the theory of recursive functions and in the study of the complexity of certain algorithms involving set unions. (There are several different variants of this function. All are called Ackermann's function and have similar properties even though their values do not always agree.)

$$A(m, n) = \begin{cases} 2n & \text{if } m = 0 \\ 0 & \text{if } m \geq 1 \text{ and } n = 0 \\ 2 & \text{if } m \geq 1 \text{ and } n = 1 \\ A(m - 1, A(m, n - 1)) & \text{if } m \geq 1 \text{ and } n \geq 2 \end{cases}$$

Exercises 38–40 involve this version of Ackermann's function.

38. Find these values of Ackermann's function.
   a) $A(1, 0)$ **b)** $A(0, 1)$
   c) $A(1, 1)$ **d)** $A(2, 2)$

39. Show that $A(m, 2) = 4$ whenever $m \geq 1$.

40. Show that $A(1, n) = 2^n$ whenever $n \geq 1$.

41. Show that each of these proposed recursive definitions of a function on the set of positive integers does not produce a well-defined function.

a) $F(n) = 1 + F(\lfloor (n+1)/2 \rfloor)$ for $n \geq 1$ and $F(1) = 1$.

b) $F(n) = 1 + F(n-2)$ for $n \geq 2$ and $F(1) = 0$.

c) $F(n) = 1 + F(n/3)$ for $n \geq 3$, $F(1) = 1$, $F(2) = 2$, and $F(3) = 3$.

d) $F(n) = 1 + F(n/2)$ if n is even and $n \geq 2$, $F(n) = 1 + F(n-2)$ if n is odd, and $F(1) = 1$.

e) $F(n) = 1 + F(F(n-1))$ if $n \geq 2$ and $F(1) = 2$.

Exercises 42–44 deal with iterations of the logarithm function. Let $\log n$ denote the logarithm of n to the base 2, as usual. The function $\log^{(k)} n$ is defined recursively by

$$\log^{(k)} n = \begin{cases} n & \text{if } k = 0 \\ \log(\log^{(k-1)} n) & \text{if } \log^{(k-1)} n \text{ is defined} \\ & \text{and positive} \\ \text{undefined} & \text{otherwise.} \end{cases}$$

The **iterated logarithm** is the function $\log^* n$ whose value at n is the smallest nonnegative integer k such that $\log^{(k)} n \leq 1$.

42. Find these values.

a) $\log^{(2)} 16$
b) $\log^{(3)} 256$
c) $\log^{(3)} 2^{65536}$
d) $\log^{(4)} 2^{2^{65536}}$

43. Find the value of $\log^* n$ for these values of n.

a) 2 **b)** 4 **c)** 8 **d)** 16
e) 256 **f)** 65536 **g)** 2^{2048}

44. Find the largest integer n such that $\log^* n = 5$. Determine the number of decimal digits in this number.

Exercises 45–47 deal with values of iterated functions. Suppose that $f(n)$ is a function from the set of real numbers, or positive real numbers, or some other set of real numbers, to the set of real numbers such that $f(n)$ is monotonically increasing [that is, $f(n) < f(m)$ when $n < m$] and $f(n) < n$ for all n in the domain of f.] The function $f^{(k)}(n)$ is defined recursively by

$$f^{(k)}(n) = \begin{cases} n & \text{if } k = 0 \\ f(f^{(k-1)}(n)) & \text{if } k > 0. \end{cases}$$

Furthermore, let c be a positive real number. The **iterated function** f_c^* is the number of iterations of f required to reduce its argument to c or less, so $f_c^*(n)$ is the smallest nonnegative integer k such that $f^k(n) \leq c$.

45. Let $f(n) = n - a$, where a is a positive integer. Find a formula for $f^{(k)}(n)$. What is the value of $f_0^*(n)$ when n is a positive integer?

46. Let $f(n) = n/2$. Find a formula for $f^{(k)}(n)$. What is the value of $f_1^*(n)$ when n is a positive integer?

47. Let $f(n) = \sqrt{n}$. Find a formula for $f^{(k)}(n)$. What is the value of $f_2^*(n)$ when n is a positive integer?

5.4 Recursive Algorithms

Introduction

Sometimes we can reduce the solution to a problem with a particular set of input values to the solution of the same problem with smaller input values. For instance, the problem of finding the greatest common divisor of two positive integers a and b, where $b > a$, can be reduced to finding the greatest common divisor of a pair of smaller integers, namely, $b \bmod a$ and a, because $\gcd(b \bmod a, a) = \gcd(a, b)$. When such a reduction can be done, the solution to the original problem can be found with a sequence of reductions, until the problem has been reduced to some initial case for which the solution is known. For instance, for finding the greatest common divisor, the reduction continues until the smaller of the two numbers is zero, because $\gcd(a, 0) = a$ when $a > 0$.

We will see that algorithms that successively reduce a problem to the same problem with smaller input are used to solve a wide variety of problems.

Here's a famous humorous quote: "To understand recursion, you must first understand recursion."

DEFINITION 1   An algorithm is called *recursive* if it solves a problem by reducing it to an instance of the same problem with smaller input.

We will describe a variety of different recursive algorithms in this section.

EXAMPLE 1   Give a recursive algorithm for computing $n!$, where n is a nonnegative integer.

Solution: We can build a recursive algorithm that finds $n!$, where n is a nonnegative integer, based on the recursive definition of $n!$, which specifies that $n! = n \cdot (n-1)!$ when n is a positive integer, and that $0! = 1$. To find $n!$ for a particular integer, we use the recursive step n times, each time replacing a value of the factorial function with the value of the factorial function at the next smaller integer. At this last step, we insert the value of $0!$. The recursive algorithm we obtain is displayed as Algorithm 1.

To help understand how this algorithm works, we trace the steps used by the algorithm to compute $4!$. First, we use the recursive step to write $4! = 4 \cdot 3!$. We then use the recursive step repeatedly to write $3! = 3 \cdot 2!$, $2! = 2 \cdot 1!$, and $1! = 1 \cdot 0!$. Inserting the value of $0! = 1$, and working back through the steps, we see that $1! = 1 \cdot 1 = 1$, $2! = 2 \cdot 1! = 2$, $3! = 3 \cdot 2! = 3 \cdot 2 = 6$, and $4! = 4 \cdot 3! = 4 \cdot 6 = 24$. ◀

ALGORITHM 1 A Recursive Algorithm for Computing $n!$.

procedure *factorial*(n: nonnegative integer)
if $n = 0$ **then return** 1
else return $n \cdot factorial(n - 1)$
{output is $n!$}

Example 2 shows how a recursive algorithm can be constructed to evaluate a function from its recursive definition.

EXAMPLE 2   Give a recursive algorithm for computing a^n, where a is a nonzero real number and n is a nonnegative integer.

Solution: We can base a recursive algorithm on the recursive definition of a^n. This definition states that $a^{n+1} = a \cdot a^n$ for $n > 0$ and the initial condition $a^0 = 1$. To find a^n, successively use the recursive step to reduce the exponent until it becomes zero. We give this procedure in Algorithm 2. ◀

ALGORITHM 2 A Recursive Algorithm for Computing a^n.

procedure *power*(a: nonzero real number, n: nonnegative integer)
if $n = 0$ **then return** 1
else return $a \cdot power(a, n - 1)$
{output is a^n}

Next we give a recursive algorithm for finding greatest common divisors.

EXAMPLE 3   Give a recursive algorithm for computing the greatest common divisor of two nonnegative integers a and b with $a < b$.

Solution: We can base a recursive algorithm on the reduction $\gcd(a, b) = \gcd(b \bmod a, a)$ and the condition $\gcd(0, b) = b$ when $b > 0$. This produces the procedure in Algorithm 3, which is a recursive version of the Euclidean algorithm.

We illustrate the workings of Algorithm 3 with a trace when the input is $a = 5$, $b = 8$. With this input, the algorithm uses the "else" clause to find that $\gcd(5, 8) = \gcd(8 \bmod 5, 5) = \gcd(3, 5)$. It uses this clause again to find that $\gcd(3, 5) = \gcd(5 \bmod 3, 3) = \gcd(2, 3)$, then to get $\gcd(2, 3) = \gcd(3 \bmod 2, 2) = \gcd(1, 2)$, then to get $\gcd(1, 2) = \gcd(2 \bmod 1, 1) = \gcd(0, 1)$. Finally, to find $\gcd(0, 1)$ it uses the first step with $a = 0$ to find that $\gcd(0, 1) = 1$. Consequently, the algorithm finds that $\gcd(5, 8) = 1$. ◀

ALGORITHM 3 **A Recursive Algorithm for Computing gcd(a, b).**

procedure $gcd(a, b$: nonnegative integers with $a < b$)
if $a = 0$ **then return** b
else return $gcd(b \bmod a, a)$
{output is $gcd(a, b)$}

EXAMPLE 4   Devise a recursive algorithm for computing $b^n \bmod m$, where b, n, and m are integers with $m \geq 2, n \geq 0$, and $1 \leq b < m$.

Solution: We can base a recursive algorithm on the fact that

$$b^n \bmod m = (b \cdot (b^{n-1} \bmod m)) \bmod m,$$

which follows by Corollary 2 in Section 4.1, and the initial condition $b^0 \bmod m = 1$. We leave this as Exercise 8 for the reader.

However, we can devise a much more efficient recursive algorithm based on the observation that

$$b^n \bmod m = (b^{n/2} \bmod m)^2 \bmod m$$

when n is even and

$$b^n \bmod m = \left((b^{\lfloor n/2 \rfloor} \bmod m)^2 \bmod m \cdot b \bmod m\right) \bmod m$$

when n is odd, which we describe in pseudocode as Algorithm 4.

We trace the execution of Algorithm 4 with input $b = 2, n = 5$, and $m = 3$ to illustrate how it works. First, because $n = 5$ is odd we use the "else" clause to see that $mpower(2, 5, 3) = (mpower(2, 2, 3)^2 \bmod 3 \cdot 2 \bmod 3) \bmod 3$. We next use the "else if" clause to see that $mpower(2, 2, 3) = mpower(2, 1, 3)^2 \bmod 3$. Using the "else" clause again, we see that $mpower(2, 1, 3) = (mpower(2, 0, 3)^2 \bmod 3 \cdot 2 \bmod 3) \bmod 3$. Finally, using the "if" clause, we see that $mpower(2, 0, 3) = 1$. Working backwards, it follows that $mpower(2, 1, 3) = (1^2 \bmod 3 \cdot 2 \bmod 3) \bmod 3 = 2$, so $mpower(2, 2, 3) = 2^2 \bmod 3 = 1$, and finally $mpower(2, 5, 3) = (1^2 \bmod 3 \cdot 2 \bmod 3) \bmod 3 = 2$.   ◀

ALGORITHM 4 **Recursive Modular Exponentiation.**

procedure $mpower(b, n, m$: integers with $b > 0$ and $m \geq 2, n \geq 0$)
if $n = 0$ **then**
    return 1
else if n is even **then**
    return $mpower(b, n/2, m)^2 \bmod m$
else
    return $(mpower(b, \lfloor n/2 \rfloor, m)^2 \bmod m \cdot b \bmod m) \bmod m$
{output is $b^n \bmod m$}

We will now give recursive versions of searching algorithms that were introduced in Section 3.1.

EXAMPLE 5   Express the linear search algorithm as a recursive procedure.

Solution: To *search* for the first occurrence of x in the sequence $a_1, a_2, \ldots, a_n$, at the ith step of the algorithm, x and a_i are compared. If x equals a_i, then the algorithm returns i, the location of x in the sequence. Otherwise, the search for the first occurrence of x is reduced to a search in a sequence with one fewer element, namely, the sequence $a_{i+1}, \ldots, a_n$. The algorithm returns 0 when x is never found in the sequence after all terms have been examined. We can now give a recursive procedure, which is displayed as pseudocode in Algorithm 5.

Let *search* (i, j, x) be the procedure that searches for the first occurrence of x in the sequence $a_i, a_{i+1}, \ldots, a_j$. The input to the procedure consists of the triple $(1, n, x)$. The algorithm terminates at a step if the first term of the remaining sequence is x or if there is only one term of the sequence and this is not x. If x is not the first term and there are additional terms, the same procedure is carried out but with a search sequence of one fewer term, obtained by deleting the first term of the search sequence. If the algorithm terminates without x having been found, the algorithm returns the value 0. ◄

ALGORITHM 5 A Recursive Linear Search Algorithm.

---

procedure *search*(i, j, x: i, j, x integers, $1 \leq i \leq j \leq n$)
if $a_i = x$ **then**
    return i
else if $i = j$ **then**
    return 0
else
    return *search*($i + 1, j, x$)
{output is the location of x in $a_1, a_2, \ldots, a_n$ if it appears; otherwise it is 0}

EXAMPLE 6 Construct a recursive version of a binary search algorithm.

Solution: Suppose we want to locate x in the sequence $a_1, a_2, \ldots, a_n$ of integers in increasing order. To perform a binary search, we begin by comparing x with the middle term, $a_{\lfloor (n+1)/2 \rfloor}$. Our algorithm will terminate if x equals this term and return the location of this term in the sequence. Otherwise, we reduce the search to a smaller search sequence, namely, the first half of the sequence if x is smaller than the middle term of the original sequence, and the second half otherwise. We have reduced the solution of the search problem to the solution of the same problem with a sequence at most half as long. If have we never encountered the search term x, our algorithm returns the value 0. We express this recursive version of a binary search algorithm as Algorithm 6. ◄

ALGORITHM 6 A Recursive Binary Search Algorithm.

---

procedure *binary search*(i, j, x: i, j, x integers, $1 \leq i \leq j \leq n$)
$m := \lfloor (i + j)/2 \rfloor$
if $x = a_m$ **then**
    return m
else if ($x < a_m$ and $i < m$) **then**
    return *binary search*($i, m - 1, x$)
else if ($x > a_m$ and $j > m$) **then**
    return *binary search*($m + 1, j, x$)
else return 0
{output is location of x in $a_1, a_2, \ldots, a_n$ if it appears; otherwise it is 0}

Proving Recursive Algorithms Correct

Mathematical induction, and its variant strong induction, can be used to prove that a recursive algorithm is correct, that is, that it produces the desired output for all possible input values. Examples 7 and 8 illustrate how mathematical induction or strong induction can be used to prove that recursive algorithms are correct. First, we will show that Algorithm 2 is correct.

EXAMPLE 7 Prove that Algorithm 2, which computes powers of real numbers, is correct.

Solution: We use mathematical induction on the exponent n.

BASIS STEP: If $n = 0$, the first step of the algorithm tells us that $power(a, 0) = 1$. This is correct because $a^0 = 1$ for every nonzero real number a. This completes the basis step.

INDUCTIVE STEP: The inductive hypothesis is the statement that $power(a, k) = a^k$ for all $a \neq 0$ for an arbitrary nonnegative integer k. That is, the inductive hypothesis is the statement that the algorithm correctly computes a^k. To complete the inductive step, we show that if the inductive hypothesis is true, then the algorithm correctly computes a^{k+1}. Because $k + 1$ is a positive integer, when the algorithm computes a^{k+1}, the algorithm sets $power(a, k + 1) = a \cdot power(a, k)$. By the inductive hypothesis, we have $power(a, k) = a^k$, so $power(a, k + 1) = a \cdot power(a, k) = a \cdot a^k = a^{k+1}$. This completes the inductive step.

We have completed the basis step and the inductive step, so we can conclude that Algorithm 2 always computes a^n correctly when $a \neq 0$ and n is a nonnegative integer. ◀

Generally, we need to use strong induction to prove that recursive algorithms are correct, rather than just mathematical induction. Example 8 illustrates this; it shows how strong induction can be used to prove that Algorithm 4 is correct.

EXAMPLE 8 Prove that Algorithm 4, which computes modular powers, is correct.

Solution: We use strong induction on the exponent n.

BASIS STEP: Let b be an integer and m an integer with $m \geq 2$. When $n = 0$, the algorithm sets $mpower(b, n, m)$ equal to 1. This is correct because $b^0 \bmod m = 1$. The basis step is complete.

INDUCTIVE STEP: For the inductive hypothesis we assume that $mpower(b, j, m) = b^j \bmod m$ for all integers $0 \leq j < k$ whenever b is a positive integer and m is an integer with $m \geq 2$. To complete the inductive step, we show that if the inductive hypothesis is correct, then $mpower(b, k, m) = b^k \bmod m$. Because the recursive algorithm handles odd and even values of k differently, we split the inductive step into two cases.
When k is even, we have

$$mpower(b, k, m) = (mpower(b, k/2, m))^2 \bmod m = (b^{k/2} \bmod m)^2 \bmod m = b^k \bmod m,$$

where we have used the inductive hypothesis to replace $mpower(b, k/2, m)$ by $b^{k/2} \bmod m$.
When k is odd, we have

$$mpower(b, k, m) = ((mpower(b, \lfloor k/2 \rfloor, m))^2 \bmod m \cdot b \bmod m) \bmod m$$
$$= ((b^{\lfloor k/2 \rfloor} \bmod m)^2 \bmod m \cdot b \bmod m) \bmod m$$
$$= b^{2\lfloor k/2 \rfloor + 1} \bmod m = b^k \bmod m,$$

using Corollary 2 in Section 4.1, because $2\lfloor k/2 \rfloor + 1 = 2(k - 1)/2 + 1 = k$ when k is odd. Here we have used the inductive hypothesis to replace $mpower(b, \lfloor k/2 \rfloor, m)$ by $b^{\lfloor k/2 \rfloor} \bmod m$. This completes the inductive step.

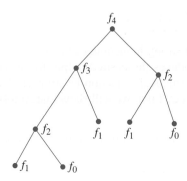

FIGURE 1   **Evaluating f_4 Recursively.**

We have completed the basis step and the inductive step, so by strong induction we know that Algorithm 4 is correct.   ◄

Recursion and Iteration

A recursive definition expresses the value of a function at a positive integer in terms of the values of the function at smaller integers. This means that we can devise a recursive algorithm to evaluate a recursively defined function at a positive integer. Instead of successively reducing the computation to the evaluation of the function at smaller integers, we can start with the value of the function at one or more integers, the base cases, and successively apply the recursive definition to find the values of the function at successive larger integers. Such a procedure is called **iterative**. Often an iterative approach for the evaluation of a recursively defined sequence requires much less computation than a procedure using recursion (unless special-purpose recursive machines are used). This is illustrated by the iterative and recursive procedures for finding the nth Fibonacci number. The recursive procedure is given first.

ALGORITHM 7 **A Recursive Algorithm for Fibonacci Numbers.**

procedure *fibonacci*(n: nonnegative integer)
if $n = 0$ **then return** 0
else if $n = 1$ **then return** 1
else return *fibonacci*($n - 1$) + *fibonacci*($n - 2$)
{output is *fibonacci*(n)}

When we use a recursive procedure to find f_n, we first express f_n as $f_{n-1} + f_{n-2}$. Then we replace both of these Fibonacci numbers by the sum of two previous Fibonacci numbers, and so on. When f_1 or f_0 arises, it is replaced by its value.

Note that at each stage of the recursion, until f_1 or f_0 is obtained, the number of Fibonacci numbers to be evaluated has doubled. For instance, when we find f_4 using this recursive algorithm, we must carry out all the computations illustrated in the tree diagram in Figure 1. This tree consists of a root labeled with f_4, and branches from the root to vertices labeled with the two Fibonacci numbers f_3 and f_2 that occur in the reduction of the computation of f_4. Each subsequent reduction produces two branches in the tree. This branching ends when f_0 and f_1 are reached. The reader can verify that this algorithm requires $f_{n+1} - 1$ additions to find f_n.

Now consider the amount of computation required to find f_n using the iterative approach in Algorithm 8.

ALGORITHM 8 An Iterative Algorithm for Computing Fibonacci Numbers.

procedure *iterative fibonacci*(n: nonnegative integer)
if $n = 0$ **then return** 0
else
    $x := 0$
    $y := 1$
    for $i := 1$ **to** $n - 1$
        $z := x + y$
        $x := y$
        $y := z$
    return y
{output is the nth Fibonacci number}

This procedure initializes x as $f_0 = 0$ and y as $f_1 = 1$. When the loop is traversed, the sum of x and y is assigned to the auxiliary variable z. Then x is assigned the value of y and y is assigned the value of the auxiliary variable z. Therefore, after going through the loop the first time, it follows that x equals f_1 and y equals $f_0 + f_1 = f_2$. Furthermore, after going through the loop $n - 1$ times, x equals f_{n-1} and y equals f_n (the reader should verify this statement). Only $n - 1$ additions have been used to find f_n with this iterative approach when $n > 1$. Consequently, this algorithm requires far less computation than does the recursive algorithm.

We have shown that a recursive algorithm may require far more computation than an iterative one when a recursively defined function is evaluated. It is sometimes preferable to use a recursive procedure even if it is less efficient than the iterative procedure. In particular, this is true when the recursive approach is easily implemented and the iterative approach is not. (Also, machines designed to handle recursion may be available that eliminate the advantage of using iteration.)

The Merge Sort

Links

We now describe a recursive sorting algorithm called the **merge sort** algorithm. We will demonstrate how the merge sort algorithm works with an example before describing it in generality.

EXAMPLE 9 Use the merge sort to put the terms of the list 8, 2, 4, 6, 9, 7, 10, 1, 5, 3 in increasing order.

Solution: A merge sort begins by splitting the list into individual elements by successively splitting lists in two. The progression of sublists for this example is represented with the balanced binary tree of height 4 shown in the upper half of Figure 2.

Sorting is done by successively merging pairs of lists. At the first stage, pairs of individual elements are merged into lists of length two in increasing order. Then successive merges of pairs of lists are performed until the entire list is put into increasing order. The succession of merged lists in increasing order is represented by the balanced binary tree of height 4 shown in the lower half of Figure 2 (note that this tree is displayed "upside down"). ◄

In general, a merge sort proceeds by iteratively splitting lists into two sublists of equal length (or where one sublist has one more element than the other) until each sublist contains one element. This succession of sublists can be represented by a balanced binary tree. The procedure continues by successively merging pairs of lists, where both lists are in increasing order, into a larger list with elements in increasing order, until the original list is put into increasing order. The succession of merged lists can be represented by a balanced binary tree.

Demo

We can also describe the merge sort recursively. To do a merge sort, we split a list into two sublists of equal, or approximately equal, size, sorting each sublist using the merge sort

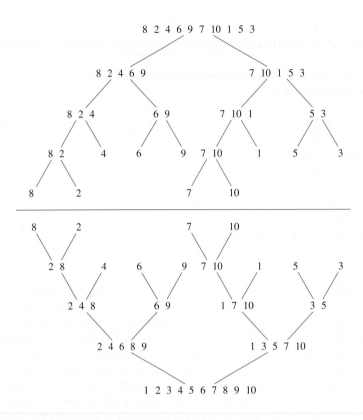

FIGURE 2   The Merge Sort of 8, 2, 4, 6, 9, 7, 10, 1, 5, 3.

algorithm, and then merging the two lists. The recursive version of the merge sort is given in Algorithm 9. This algorithm uses the subroutine *merge*, which is described in Algorithm 10.

ALGORITHM 9 A Recursive Merge Sort.

procedure *mergesort*($L = a_1, \ldots, a_n$)
if $n > 1$ **then**
    $m := \lfloor n/2 \rfloor$
    $L_1 := a_1, a_2, \ldots, a_m$
    $L_2 := a_{m+1}, a_{m+2}, \ldots, a_n$
    $L := merge(mergesort(L_1), \; mergesort(L_2))$
{L is now sorted into elements in nondecreasing order}

An efficient algorithm for merging two ordered lists into a larger ordered list is needed to implement the merge sort. We will now describe such a procedure.

EXAMPLE 10   Merge the two lists 2, 3, 5, 6 and 1, 4.

Solution: Table 1 illustrates the steps we use. First, compare the smallest elements in the two lists, 2 and 1, respectively. Because 1 is the smaller, put it at the beginning of the merged list and remove it from the second list. At this stage, the first list is 2, 3, 5, 6, the second is 4, and the combined list is 1.

Next, compare 2 and 4, the smallest elements of the two lists. Because 2 is the smaller, add it to the combined list and remove it from the first list. At this stage the first list is 3, 5, 6, the second is 4, and the combined list is 1, 2.

Continue by comparing 3 and 4, the smallest elements of their respective lists. Because 3 is the smaller of these two elements, add it to the combined list and remove it from the first list. At this stage the first list is 5, 6, and the second is 4. The combined list is 1, 2, 3.

Then compare 5 and 4, the smallest elements in the two lists. Because 4 is the smaller of these two elements, add it to the combined list and remove it from the second list. At this stage the first list is 5, 6, the second list is empty, and the combined list is 1, 2, 3, 4.

Finally, because the second list is empty, all elements of the first list can be appended to the end of the combined list in the order they occur in the first list. This produces the ordered list 1, 2, 3, 4, 5, 6. ◀

We will now consider the general problem of merging two ordered lists L_1 and L_2 into an ordered list L. We will describe an algorithm for solving this problem. Start with an empty list L. Compare the smallest elements of the two lists. Put the smaller of these two elements at the right end of L, and remove it from the list it was in. Next, if one of L_1 and L_2 is empty, append the other (nonempty) list to L, which completes the merging. If neither L_1 nor L_2 is empty, repeat this process. Algorithm 10 gives a pseudocode description of this procedure.

We will need estimates for the number of comparisons used to merge two ordered lists in the analysis of the merge sort. We can easily obtain such an estimate for Algorithm 10. Each time a comparison of an element from L_1 and an element from L_2 is made, an additional element is added to the merged list L. However, when either L_1 or L_2 is empty, no more comparisons are needed. Hence, Algorithm 10 is least efficient when $m + n - 2$ comparisons are carried out, where m and n are the number of elements in L_1 and L_2, respectively, leaving one element in each of L_1 and L_2. The next comparison will be the last one needed, because it will make one of these lists empty. Hence, Algorithm 10 uses no more than $m + n - 1$ comparisons. Lemma 1 summarizes this estimate.

ALGORITHM 10 **Merging Two Lists.**

procedure $merge(L_1, L_2:$ sorted lists)
$L :=$ empty list
while L_1 and L_2 are both nonempty
    remove smaller of first elements of L_1 and L_2 from its list; put it at the right end of L
    if this removal makes one list empty **then** remove all elements from the other list and
        append them to L
return $L\{L$ is the merged list with elements in increasing order}

TABLE 1 Merging the Two Sorted Lists 2, 3, 5, 6 and 1, 4.			
First List	*Second List*	*Merged List*	*Comparison*
2 3 5 6	1 4		$1 < 2$
2 3 5 6	4	1	$2 < 4$
3 5 6	4	1 2	$3 < 4$
5 6	4	1 2 3	$4 < 5$
5 6		1 2 3 4	
		1 2 3 4 5 6	

LEMMA 1 Two sorted lists with m elements and n elements can be merged into a sorted list using no more than $m + n - 1$ comparisons.

Sometimes two sorted lists of length m and n can be merged using far fewer than $m + n - 1$ comparisons. For instance, when $m = 1$, a binary search procedure can be applied to put the one element in the first list into the second list. This requires only $\lceil \log n \rceil$ comparisons, which is much smaller than $m + n - 1 = n$, for $m = 1$. On the other hand, for some values of m and n, Lemma 1 gives the best possible bound. That is, there are lists with m and n elements that cannot be merged using fewer than $m + n - 1$ comparisons. (See Exercise 31.)

We can now analyze the complexity of the merge sort. Instead of studying the general problem, we will assume that n, the number of elements in the list, is a power of 2, say 2^m. This will make the analysis less complicated, but when this is not the case, various modifications can be applied that will yield the same estimate.

At the first stage of the splitting procedure, the list is split into two sublists, of 2^{m-1} elements each, at level 1 of the tree generated by the splitting. This process continues, splitting the two sublists with 2^{m-1} elements into four sublists of 2^{m-2} elements each at level 2, and so on. In general, there are 2^{k-1} lists at level $k - 1$, each with 2^{m-k+1} elements. These lists at level $k - 1$ are split into 2^k lists at level k, each with 2^{m-k} elements. At the end of this process, we have 2^m lists each with one element at level m.

We start merging by combining pairs of the 2^m lists of one element into 2^{m-1} lists, at level $m - 1$, each with two elements. To do this, 2^{m-1} pairs of lists with one element each are merged. The merger of each pair requires exactly one comparison.

The procedure continues, so that at level k ($k = m, m - 1, m - 2, \ldots, 3, 2, 1$), 2^k lists each with 2^{m-k} elements are merged into 2^{k-1} lists, each with 2^{m-k+1} elements, at level $k - 1$. To do this a total of 2^{k-1} mergers of two lists, each with 2^{m-k} elements, are needed. But, by Lemma 1, each of these mergers can be carried out using at most $2^{m-k} + 2^{m-k} - 1 = 2^{m-k+1} - 1$ comparisons. Hence, going from level k to $k - 1$ can be accomplished using at most $2^{k-1}(2^{m-k+1} - 1)$ comparisons.

Summing all these estimates shows that the number of comparisons required for the merge sort is at most

$$\sum_{k=1}^{m} 2^{k-1}(2^{m-k+1} - 1) = \sum_{k=1}^{m} 2^m - \sum_{k=1}^{m} 2^{k-1} = m2^m - (2^m - 1) = n \log n - n + 1,$$

because $m = \log n$ and $n = 2^m$. (We evaluated $\sum_{k=1}^{m} 2^m$ by noting that it is the sum of m identical terms, each equal to 2^m. We evaluated $\sum_{k=1}^{m} 2^{k-1}$ using the formula for the sum of the terms of a geometric progression from Theorem 1 of Section 2.4.)

Theorem 1 summarizes what we have discovered about the worst-case complexity of the merge sort algorithm.

THEOREM 1 The number of comparisons needed to merge sort a list with n elements is $O(n \log n)$.

In Chapter 11 we will show that the fastest comparison-based sorting algorithm have $O(n \log n)$ time complexity. (A comparison-based sorting algorithm has the comparison of two elements as its basic operation.) Theorem 1 tells us that the merge sort achieves this best possible big-O estimate for the complexity of a sorting algorithm. We describe another efficient algorithm, the quick sort, in the preamble to Exercise 34.

Exercises

1. Trace Algorithm 1 when it is given $n = 5$ as input. That is, show all steps used by Algorithm 1 to find 5!, as is done in Example 1 to find 4!.

2. Trace Algorithm 3 when it finds gcd(8, 13). That is, show all the steps used by Algorithm 3 to find gcd(8, 13).

3. Trace Algorithm 4 when it is given $m = 5$, $n = 11$, and $b = 3$ as input. That is, show all the steps Algorithm 4 uses to find $3^{11} \bmod 5$.

4. Give a recursive algorithm for computing nx whenever n is a positive integer and x is an integer, using just addition.

5. Give a recursive algorithm for finding the sum of the first n odd positive integers.

6. Give a recursive algorithm for finding the maximum of a finite set of integers, making use of the fact that the maximum of n integers is the larger of the last integer in the list and the maximum of the first $n - 1$ integers in the list.

7. Give a recursive algorithm for finding the minimum of a finite set of integers, making use of the fact that the minimum of n integers is the smaller of the last integer in the list and the minimum of the first $n - 1$ integers in the list.

8. Devise a recursive algorithm for finding $x^n \bmod m$ whenever n, x, and m are positive integers based on the fact that $x^n \bmod m = (x^{n-1} \bmod m \cdot x \bmod m) \bmod m$.

9. Give a recursive algorithm for finding $n! \bmod m$ whenever n and m are positive integers.

10. Give a recursive algorithm for finding a **mode** of a list of integers. (A **mode** is an element in the list that occurs at least as often as every other element.)

11. Devise a recursive algorithm for computing the greatest common divisor of two nonnegative integers a and b with $a < b$ using the fact that $\gcd(a, b) = \gcd(a, b - a)$.

12. Describe a recursive algorithm for multiplying two nonnegative integers x and y based on the fact that $xy = 2(x \cdot (y/2))$ when y is even and $xy = 2(x \cdot \lfloor y/2 \rfloor) + x$ when y is odd, together with the initial condition $xy = 0$ when $y = 0$.

13. Prove that Algorithm 3 for computing $\gcd(a, b)$ when a and b are positive integers with $a < b$ is correct.

14. Prove that the algorithm you devised in Exercise 12 is correct.

15. Prove that the recursive algorithm that you found in Exercise 4 is correct.

16. Devise a recursive algorithm for computing n^2 where n is a nonnegative integer, using the fact that $(n + 1)^2 = n^2 + 2n + 1$. Then prove that this algorithm is correct.

17. **a)** Devise a recursive algorithm to find a^{2^n}, where a is a real number and n is a positive integer. [*Hint:* Use the equality $a^{2^{n+1}} = (a^{2^n})^2$.]
    b) How does the number of multiplications used by the algorithm in Exercise 17(a) compare to the number of multiplications used by Algorithm 2 to evaluate a^{2^n}?

18. How many additions are used by the recursive and iterative algorithms given in Algorithms 7 and 8, respectively, to find the Fibonacci number f_7?

19. Devise a recursive algorithm to find the nth term of the sequence defined by $a_0 = 1$, $a_1 = 2$, and $a_n = a_{n-1} \cdot a_{n-2}$, for $n = 2, 3, 4, \ldots$.

20. Devise an iterative algorithm to find the nth term of the sequence defined in Exercise 19.

21. Is the recursive or the iterative algorithm for finding the sequence in Exercise 19 more efficient?

22. Give iterative and recursive algorithms for finding the nth term of the sequence defined by $a_0 = 1$, $a_1 = 3$, $a_2 = 5$, and $a_n = a_{n-1} \cdot a_{n-2}^2 \cdot a_{n-3}^3$. Which is more efficient?

23. Give a recursive algorithm for finding the reversal of a bit string. (See the definition of the reversal of a bit string in the preamble of Exercise 24 in Section 5.3.)

24. Give a recursive algorithm to find the number of partitions of a positive integer based on the recursive definition given in Exercise 37 in Section 5.3.

25. Prove that the recursive algorithm for finding the reversal of a bit string that you gave in Exercise 23 is correct.

26. Give a recursive algorithm for finding the concatenation of i copies of a bit string and prove that it is correct.

*27. Give a recursive algorithm for tiling a $2^n \times 2^n$ checkerboard with one square missing using right triominoes.

28. Use a merge sort to sort 4, 3, 2, 5, 1, 8, 7, 6 into increasing order. Show all the steps used by the algorithm.

29. Use a merge sort to sort b, d, a, f, g, h, z, p, o, k into alphabetic order. Show all the steps used by the algorithm.

30. How many comparisons are required to merge these pairs of lists using Algorithm 10?
    a) 1, 3, 5, 7, 9; 2, 4, 6, 8, 10
    b) 1, 2, 3, 4, 5; 6, 7, 8, 9, 10
    c) 1, 5, 6, 7, 8; 2, 3, 4, 9, 10

31. Show that for all positive integers m and n there are sorted lists with m elements and n elements, respectively, such that Algorithm 10 uses $m + n - 1$ comparisons to merge them into one sorted list.

*32. What is the least number of comparisons needed to merge any two lists in increasing order into one list in increasing order when the number of elements in the two lists are
    a) 1, 4?   **b)** 2, 4?   **c)** 3, 4?   **d)** 4, 4?

*33. Prove that the merge sort algorithm is correct.

The **quick sort** is an efficient algorithm. To sort $a_1, a_2, \ldots, a_n$, this algorithm begins by taking the first element a_1 and forming two sublists, the first containing those elements that are less than a_1, in the order they arise, and the second containing those elements greater than a_1, in the order they arise. Then a_1 is put at the end of the first sublist. This procedure is repeated recursively for each sublist, until all sublists contain one item. The ordered list of n items is obtained by combining the sublists of one item in the order they occur.

34. Sort 3, 5, 7, 8, 1, 9, 2, 4, 6 using the quick sort.

35. Let $a_1, a_2, \ldots, a_n$ be a list of n distinct real numbers. How many comparisons are needed to form two sublists from this list, the first containing elements less than a_1 and the second containing elements greater than a_1?

36. Describe the quick sort algorithm using pseudocode.

37. What is the largest number of comparisons needed to order a list of four elements using the quick sort algorithm?

38. What is the least number of comparisons needed to order a list of four elements using the quick sort algorithm?

39. Determine the worst-case complexity of the quick sort algorithm in terms of the number of comparisons used.

5.5 Program Correctness

Introduction

Suppose that we have designed an algorithm to solve a problem and have written a program to implement it. How can we be sure that the program always produces the correct answer? After all the bugs have been removed so that the syntax is correct, we can test the program with sample input. It is not correct if an incorrect result is produced for any sample input. But even if the program gives the correct answer for all sample input, it may not always produce the correct answer (unless all possible input has been tested). We need a proof to show that the program *always* gives the correct output.

Program verification, the proof of correctness of programs, uses the rules of inference and proof techniques described in this chapter, including mathematical induction. Because an incorrect program can lead to disastrous results, a large amount of methodology has been constructed for verifying programs. Efforts have been devoted to automating program verification so that it can be carried out using a computer. However, only limited progress has been made toward this goal. Indeed, some mathematicians and theoretical computer scientists argue that it will never be realistic to mechanize the proof of correctness of complex programs.

Some of the concepts and methods used to prove that programs are correct will be introduced in this section. Many different methods have been devised for proving that programs are correct. We will discuss a widely used method for program verification introduced by Tony Hoare in this section; several other methods are also commonly used. Furthermore, we will not develop a complete methodology for program verification in this book. This section is meant to be a brief introduction to the area of program verification, which ties together the rules of logic, proof techniques, and the concept of an algorithm.

Program Verification

A program is said to be **correct** if it produces the correct output for every possible input. A proof that a program is correct consists of two parts. The first part shows that the correct answer is obtained if the program terminates. This part of the proof establishes the **partial correctness** of the program. The second part of the proof shows that the program always terminates.

To specify what it means for a program to produce the correct output, two propositions are used. The first is the **initial assertion**, which gives the properties that the input values must have. The second is the **final assertion**, which gives the properties that the output of the program should have, if the program did what was intended. The appropriate initial and final assertions must be provided when a program is checked.

DEFINITION 1   A program, or program segment, S is said to be *partially correct with respect to* the initial assertion p and the final assertion q if whenever p is true for the input values of S and S terminates, then q is true for the output values of S. The notation $p\{S\}q$ indicates that the program, or program segment, S is partially correct with respect to the initial assertion p and the final assertion q.

Note: The notation $p\{S\}q$ is known as a *Hoare triple*. Tony Hoare introduced the concept of partial correctness.

Links

Note that the notion of partial correctness has nothing to do with whether a program terminates; it focuses only on whether the program does what it is expected to do if it terminates.

A simple example illustrates the concepts of initial and final assertions.

EXAMPLE 1   Show that the program segment

$$y := 2$$
$$z := x + y$$

Extra Examples

is correct with respect to the initial assertion $p: x = 1$ and the final assertion $q: z = 3$.

Solution: Suppose that p is true, so that $x = 1$ as the program begins. Then y is assigned the value 2, and z is assigned the sum of the values of x and y, which is 3. Hence, S is correct with respect to the initial assertion p and the final assertion q. Thus, $p\{S\}q$ is true.   ◀

Rules of Inference

A useful rule of inference proves that a program is correct by splitting the program into a sequence of subprograms and then showing that each subprogram is correct.

Suppose that the program S is split into subprograms S_1 and S_2. Write $S = S_1; S_2$ to indicate that S is made up of S_1 followed by S_2. Suppose that the correctness of S_1 with respect to the initial assertion p and final assertion q, and the correctness of S_2 with respect to the initial assertion q and the final assertion r, have been established. It follows that if p is true and S_1 is executed and terminates, then q is true; and if q is true, and S_2 executes and terminates, then r is true. Thus, if p is true and $S = S_1; S_2$ is executed and terminates, then r is true. This rule of inference, called the **composition rule**, can be stated as

$$p\{S_1\}q$$
$$q\{S_2\}r$$
$$\therefore p\{S_1; S_2\}r.$$

This rule of inference will be used later in this section.

Next, some rules of inference for program segments involving conditional statements and loops will be given. Because programs can be split into segments for proofs of correctness, this will let us verify many different programs.

Conditional Statements

First, rules of inference for conditional statements will be given. Suppose that a program segment has the form

---

if *condition* **then**
    S

---

where S is a block of statements. Then S is executed if *condition* is true, and it is not executed when *condition* is false. To verify that this segment is correct with respect to the initial assertion p and final assertion q, two things must be done. First, it must be shown that when p is true and *condition* is also true, then q is true after S terminates. Second, it must be shown that when p is true and *condition* is false, then q is true (because in this case S does not execute).

This leads to the following rule of inference:

$$(p \wedge condition)\{S\}q$$
$$\underline{(p \wedge \neg condition) \rightarrow q}$$
$$\therefore p\{\textbf{if } condition \textbf{ then } S\}q.$$

Example 2 illustrates how this rule of inference is used.

EXAMPLE 2   Verify that the program segment

> **if** $x > y$ **then**
> $y := x$

is correct with respect to the initial assertion **T** and the final assertion $y \geq x$.

Solution: When the initial assertion is true and $x > y$, the assignment $y := x$ is carried out. Hence, the final assertion, which asserts that $y \geq x$, is true in this case. Moreover, when the initial assertion is true and $x > y$ is false, so that $x \leq y$, the final assertion is again true. Hence, using the rule of inference for program segments of this type, this program is correct with respect to the given initial and final assertions.   ◀

Similarly, suppose that a program has a statement of the form

> **if** *condition* **then**
> S_1
> **else**
> S_2

If *condition* is true, then S_1 executes; if *condition* is false, then S_2 executes. To verify that this program segment is correct with respect to the initial assertion p and the final assertion q, two

---

things must be done. First, it must be shown that when p is true and *condition* is true, then q is true after S_1 terminates. Second, it must be shown that when p is true and *condition* is false, then q is true after S_2 terminates. This leads to the following rule of inference:

$$(p \wedge condition)\{S_1\}q$$
$$(p \wedge \neg condition)\{S_2\}q$$
$$\overline{\therefore\ p\{\textbf{if } condition \textbf{ then } S_1 \textbf{ else } S_2\}q.}$$

Example 3 illustrates how this rule of inference is used.

EXAMPLE 3   Verify that the program segment

> **if** $x < 0$ **then**
> $abs := -x$
> **else**
> $abs := x$

is correct with respect to the initial assertion **T** and the final assertion $abs = |x|$.

Solution: Two things must be demonstrated. First, it must be shown that if the initial assertion is true and $x < 0$, then $abs = |x|$. This is correct, because when $x < 0$ the assignment statement $abs := -x$ sets $abs = -x$, which is $|x|$ by definition when $x < 0$. Second, it must be shown that if the initial assertion is true and $x < 0$ is false, so that $x \geq 0$, then $abs = |x|$. This is also correct, because in this case the program uses the assignment statement $abs := x$, and x is $|x|$ by definition when $x \geq 0$, so $abs := x$. Hence, using the rule of inference for program segments of this type, this segment is correct with respect to the given initial and final assertions.   ◀

Loop Invariants

Links

Next, proofs of correctness of **while** loops will be described. To develop a rule of inference for program segments of the type

> **while** *condition*
> S

note that S is repeatedly executed until *condition* becomes false. An assertion that remains true each time S is executed must be chosen. Such an assertion is called a **loop invariant**. In other words, p is a loop invariant if $(p \wedge condition)\{S\}p$ is true.

Suppose that p is a loop invariant. It follows that if p is true before the program segment is executed, p and $\neg condition$ are true after termination, if it occurs. This rule of inference is

$$(p \wedge condition)\{S\}p$$
$$\overline{\therefore\ p\{\textbf{while } condition\ S\}(\neg\ condition\ \wedge\ p).}$$

The use of a loop invariant is illustrated in Example 4.

EXAMPLE 4   A loop invariant is needed to verify that the program segment

$$i := 1$$
$$factorial := 1$$
while $i < n$
$\quad$ $i := i + 1$
$\quad$ $factorial := factorial \cdot i$

terminates with $factorial = n!$ when n is a positive integer.

Let p be the assertion "$factorial = i!$ and $i \leq n$." We first prove that p is a loop invariant. Suppose that, at the beginning of one execution of the **while** loop, p is true and the condition of the **while** loop holds; in other words, assume that $factorial = i!$ and that $i < n$. The new values i_{new} and $factorial_{new}$ of i and $factorial$ are $i_{new} = i + 1$ and $factorial_{new} = factorial \cdot (i + 1) = (i + 1)! = i_{new}!$. Because $i < n$, we also have $i_{new} = i + 1 \leq n$. Thus, p is true at the end of the execution of the loop. This shows that p is a loop invariant.

Now we consider the program segment. Just before entering the loop, $i = 1 \leq n$ and $factorial = 1 = 1! = i!$ both hold, so p is true. Because p is a loop invariant, the rule of inference just introduced implied that if the **while** loop terminates, it terminates with p true and with $i < n$ false. In this case, at the end, $factorial = i!$ and $i \leq n$ are true, but $i < n$ is false; in other words, $i = n$ and $factorial = i! = n!$, as desired.

Finally, we need to check that the **while** loop actually terminates. At the beginning of the program i is assigned the value 1, so after $n - 1$ traversals of the loop, the new value of i will be n, and the loop terminates at that point.   ◀

A final example will be given to show how the various rules of inference can be used to verify the correctness of a longer program.

EXAMPLE 5   We will outline how to verify the correctness of the program S for computing the product of two integers.

procedure $multiply(m, n:$ integers)

$S_1 \begin{cases} \textbf{if } n < 0 \textbf{ then } a := -n \\ \textbf{else } a := n \end{cases}$

$S_2 \begin{cases} k := 0 \\ x := 0 \end{cases}$

$S_3 \begin{cases} \textbf{while } k < a \\ \quad x := x + m \\ \quad k := k + 1 \end{cases}$

$S_4 \begin{cases} \textbf{if } n < 0 \textbf{ then } product := -x \\ \textbf{else } product := x \end{cases}$

return $product$
$\{product \text{ equals } mn\}$

The goal is to prove that after S is executed, $product$ has the value mn. The proof of correctness can be carried out by splitting S into four segments, with $S = S_1; S_2; S_3; S_4$, as shown in the listing of S. The rule of composition can be used to build the correctness proof. Here is how the argument proceeds. The details will be left as an exercise for the reader.

Let p be the initial assertion "m and n are integers." Then, it can be shown that $p\{S_1\}q$ is true, when q is the proposition $p \wedge (a = |n|)$. Next, let r be the proposition $q \wedge (k = 0) \wedge (x = 0)$. It is easily verified that $q\{S_2\}r$ is true. It can be shown that "$x = mk$ and $k \leq a$" is an invariant for the loop in S_3. Furthermore, it is easy to see that the loop terminates after a iterations, with $k = a$, so $x = ma$ at this point. Because r implies that $x = m \cdot 0$ and $0 \leq a$, the loop invariant is true before the loop is entered. Because the loop terminates with $k = a$, it follows that $r\{S_3\}s$ is true where s is the proposition "$x = ma$ and $a = |n|$." Finally, it can be shown that S_4 is correct with respect to the initial assertion s and final assertion t, where t is the proposition "$product = mn$."

Putting all this together, because $p\{S_1\}q$, $q\{S_2\}r$, $r\{S_3\}s$, and $s\{S_4\}t$ are all true, it follows from the rule of composition that $p\{S\}t$ is true. Furthermore, because all four segments terminate, S does terminate. This verifies the correctness of the program. ◀

Exercises

1. Prove that the program segment

$$y := 1$$
$$z := x + y$$

is correct with respect to the initial assertion $x = 0$ and the final assertion $z = 1$.

2. Verify that the program segment

if $x < 0$ **then** $x := 0$

is correct with respect to the initial assertion **T** and the final assertion $x \geq 0$.

3. Verify that the program segment

$$x := 2$$
$$z := x + y$$
if $y > 0$ **then**
$$z := z + 1$$
else
$$z := 0$$

is correct with respect to the initial assertion $y = 3$ and the final assertion $z = 6$.

4. Verify that the program segment

if $x < y$ **then**
$$min := x$$
else
$$min := y$$

is correct with respect to the initial assertion **T** and the final assertion $(x \leq y \wedge min = x) \vee (x > y \wedge min = y)$.

∗5. Devise a rule of inference for verification of partial correctness of statements of the form

if *condition* 1 **then**
$$S_1$$
else if *condition* 2 **then**
$$S_2$$
$$\vdots$$
else
$$S_n$$

where $S_1, S_2, \ldots, S_n$ are blocks.

6. Use the rule of inference developed in Exercise 5 to verify that the program

if $x < 0$ **then**
$$y := -2|x|/x$$
else if $x > 0$ **then**
$$y := 2|x|/x$$
else if $x = 0$ **then**
$$y := 2$$

is correct with respect to the initial assertion **T** and the final assertion $y = 2$.

7. Use a loop invariant to prove that the following program segment for computing the nth power, where n is a positive integer, of a real number x is correct.

$$power := 1$$
$$i := 1$$
while $i \leq n$
$$power := power * x$$
$$i := i + 1$$

8. Suppose that both the conditional statement $p_0 \rightarrow p_1$ and the program assertion $p_1\{S\}q$ are true. Show that $p_0\{S\}q$ also must be true.

9. Suppose that both the program assertion $p\{S\}q_0$ and the conditional statement $q_0 \rightarrow q_1$ are true. Show that $p\{S\}q_1$ also must be true.

10. This program computes quotients and remainders.

$$r := a$$
$$q := 0$$
while $r \geq d$
$$r := r - d$$
$$q := q + 1$$

Verify that it is partially correct with respect to the initial assertion "a and d are positive integers" and the final assertion "q and r are integers such that $a = dq + r$ and $0 \leq r < d$."

11. Use a loop invariant to verify that the Euclidean algorithm (Algorithm 1 in Section 4.3) is partially correct with respect to the initial assertion "a and b are positive integers" and the final assertion "$x = \gcd(a, b)$."

Key Terms and Results

TERMS

sequence: a function with domain that is a subset of the set of integers

geometric progression: a sequence of the form $a, ar, ar^2, \ldots,$ where a and r are real numbers

arithmetic progression: a sequence of the form $a, a + d, a + 2d, \ldots,$ where a and d are real numbers

the principle of mathematical induction: the statement $\forall n\, P(n)$ is true if $P(1)$ is true and $\forall k[P(k) \to P(k+1)]$ is true.

basis step: the proof of $P(1)$ in a proof by mathematical induction of $\forall n\, P(n)$

inductive step: the proof of $P(k) \to P(k+1)$ for all positive integers k in a proof by mathematical induction of $\forall n\, P(n)$

strong induction: the statement $\forall n\, P(n)$ is true if $P(1)$ is true and $\forall k[(P(1) \wedge \cdots \wedge P(k)) \to P(k+1)]$ is true

well-ordering property: Every nonempty set of nonnegative integers has a least element.

recursive definition of a function: a definition of a function that specifies an initial set of values and a rule for obtaining

values of this function at integers from its values at smaller integers

recursive definition of a set: a definition of a set that specifies an initial set of elements in the set and a rule for obtaining other elements from those in the set

structural induction: a technique for proving results about recursively defined sets

recursive algorithm: an algorithm that proceeds by reducing a problem to the same problem with smaller input

merge sort: a sorting algorithm that sorts a list by splitting it in two, sorting each of the two resulting lists, and merging the results into a sorted list

iteration: a procedure based on the repeated use of operations in a loop

program correctness: verification that a procedure always produces the correct result

loop invariant: a property that remains true during every traversal of a loop

initial assertion: the statement specifying the properties of the input values of a program

final assertion: the statement specifying the properties the output values should have if the program worked correctly

Review Questions

1. a) Can you use the principle of mathematical induction to find a formula for the sum of the first n terms of a sequence?

   b) Can you use the principle of mathematical induction to determine whether a given formula for the sum of the first n terms of a sequence is correct?

   c) Find a formula for the sum of the first n even positive integers, and prove it using mathematical induction.

2. a) For which positive integers n is $11n + 17 \le 2^n$?

   b) Prove the conjecture you made in part (a) using mathematical induction.

3. a) Which amounts of postage can be formed using only 5-cent and 9-cent stamps?

   b) Prove the conjecture you made using mathematical induction.

   c) Prove the conjecture you made using strong induction.

   d) Find a proof of your conjecture different from the ones you gave in (b) and (c).

4. Give two different examples of proofs that use strong induction.

5. a) State the well-ordering property for the set of positive integers.

   b) Use this property to show that every positive integer greater than one can be written as the product of primes.

6. a) Explain why a function f from the set of positive integers to the set of real numbers is well-defined if it is defined recursively by specifying $f(1)$ and a rule for finding $f(n)$ from $f(n-1)$.

   b) Provide a recursive definition of the function $f(n) = (n+1)!$.

7. a) Give a recursive definition of the Fibonacci numbers.

   b) Show that $f_n > \alpha^{n-2}$ whenever $n \ge 3$, where f_n is the nth term of the Fibonacci sequence and $\alpha = (1 + \sqrt{5})/2$.

8. a) Explain why a sequence a_n is well defined if it is defined recursively by specifying a_1 and a_2 and a rule for finding a_n from $a_1, a_2, \ldots, a_{n-1}$ for $n = 3, 4, 5, \ldots$.

   b) Find the value of a_n if $a_1 = 1$, $a_2 = 2$, and $a_n = a_{n-1} + a_{n-2} + \cdots + a_1$, for $n = 3, 4, 5, \ldots$.

9. Give two examples of how well-formed formulae are defined recursively for different sets of elements and operators.

10. a) Give a recursive definition of the length of a string.

    b) Use the recursive definition from part (a) and structural induction to prove that $l(xy) = l(x) + l(y)$.

11. a) What is a recursive algorithm?

    b) Describe a recursive algorithm for computing the sum of n numbers in a sequence.

12. Describe a recursive algorithm for computing the greatest common divisor of two positive integers.

13. a) Describe the merge sort algorithm.

    b) Use the merge sort algorithm to put the list 4, 10, 1, 5, 3, 8, 7, 2, 6, 9 in increasing order.

    c) Give a big-O estimate for the number of comparisons used by the merge sort.

14. a) Does testing a computer program to see whether it produces the correct output for certain input values verify that the program always produces the correct output?

   b) Does showing that a computer program is partially correct with respect to an initial assertion and a final

assertion verify that the program always produces the correct output? If not, what else is needed?

15. What techniques can you use to show that a long computer program is partially correct with respect to an initial assertion and a final assertion?

16. What is a loop invariant? How is a loop invariant used?

Supplementary Exercises

1. Use mathematical induction to show that $\frac{2}{3} + \frac{2}{9} + \frac{2}{27} + \cdots + \frac{2}{3^n} = 1 - \frac{1}{3^n}$ whenever n is a positive integer.

2. Use mathematical induction to show that $1 \cdot 2^0 + 2 \cdot 2^1 + 3 \cdot 2^2 + \cdots + n \cdot 2^{n-1} = (n-1) \cdot 2^n + 1$ whenever n is a positive integer.

3. Show that

$$\frac{1}{1 \cdot 4} + \frac{1}{4 \cdot 7} + \cdots + \frac{1}{(3n-2)(3n+1)} = \frac{n}{3n+1}$$

whenever n is a positive integer.

4. Use mathematical induction to show that $2^n > n^3$ whenever n is an integer greater than 9.

5. Use mathematical induction to prove that $a - b$ is a factor of $a^n - b^n$ whenever n is a positive integer.

6. Use mathematical induction to prove that 43 divides $6^{n+1} + 7^{2n-1}$ for every positive integer n.

7. Use mathematical induction to prove this formula for the sum of the terms of an arithmetic progression.

$$a + (a+d) + \cdots + (a+nd) = (n+1)(2a+nd)/2$$

8. Suppose that $a_j \equiv b_j \pmod{m}$ for $j = 1, 2, \ldots, n$. Use mathematical induction to prove that

   a) $\sum_{j=1}^{n} a_j \equiv \sum_{j=1}^{n} b_j \pmod{m}$.

   b) $\prod_{j=1}^{n} a_j \equiv \prod_{j=1}^{n} b_j \pmod{m}$.

9. Show that if n is a positive integer, then

$$\sum_{k=1}^{n} \frac{k+4}{k(k+1)(k+2)} = \frac{n(3n+7)}{2(n+1)(n+2)}.$$

10. For which positive integers n is $n + 6 < (n^2 - 8n)/16$? Prove your answer using mathematical induction.

＊11. Formulate a conjecture about which Fibonacci numbers are even, and use a form of mathematical induction to prove your conjecture.

Recall from Example 15 of Section 2.4 that the sequence of **Lucas numbers** is defined by $l_0 = 2$, $l_1 = 1$, and $l_n = l_{n-1} + l_{n-2}$ for $n = 2, 3, 4, \ldots$.

12. Show that $f_n + f_{n+2} = l_{n+1}$ whenever n is a positive integer, where f_i and l_i are the ith Fibonacci number and ith Lucas number, respectively.

13. Show that $l_0^2 + l_1^2 + \cdots + l_n^2 = l_n l_{n+1} + 2$ whenever n is a nonnegative integer and l_i is the ith Lucas number.

＊14. Use mathematical induction to show that the product of any n consecutive positive integers is divisible by $n!$. [*Hint:* Use the identity $m(m+1) \cdots (m+n-1)/n! = (m-1)m(m+1) \cdots (m+n-2)/n! + m(m+1) \cdots (m+n-2)/(n-1)!$.]

15. Use mathematical induction to show that $(\cos x + i \sin x)^n = \cos nx + i \sin nx$ whenever n is a positive integer. (Here i is the square root of -1.) [*Hint:* Use the identities $\cos(a+b) = \cos a \cos b - \sin a \sin b$ and $\sin(a+b) = \sin a \cos b + \cos a \sin b$.]

＊16. Use mathematical induction to show that $\sum_{j=1}^{n} \cos jx = \cos[(n+1)x/2] \sin(nx/2)/\sin(x/2)$ whenever n is a positive integer and $\sin(x/2) \neq 0$.

17. Use mathematical induction to prove that $\sum_{j=1}^{n} j^2 2^j = n^2 2^{n+1} - n2^{n+2} + 3 \cdot 2^{n+1} - 6$ for every positive integer n.

18. (*Requires calculus*) Suppose that the sequence $x_1, x_2, \ldots, x_n, \ldots$ is recursively defined by $x_1 = 0$ and $x_{n+1} = \sqrt{x_n + 6}$.

   a) Use mathematical induction to show that $x_1 < x_2 < \cdots < x_n < \cdots$, that is, the sequence $\{x_n\}$ is monotonically increasing.

   b) Use mathematical induction to prove that $x_n < 3$ for $n = 1, 2, \ldots$.

   c) Show that $\lim_{n \to \infty} x_n = 3$.

19. Show if n is a positive integer with $n \geq 2$, then

$$\sum_{j=2}^{n} \frac{1}{j^2 - 1} = \frac{(n-1)(3n+2)}{4n(n+1)}.$$

20. Use mathematical induction to prove Theorem 1 in Section 4.2, that is, show if b is an integer, where $b > 1$, and n is a positive integer, then n can be expressed uniquely in the form $n = a_k b^k + a_{k-1} b^{k-1} + \cdots + a_1 b + a_0$.

＊21. A **lattice point** in the plane is a point (x, y) where both x and y are integers. Use mathematical induction to show that at least $n + 1$ straight lines are needed to ensure that every lattice point (x, y) with $x \geq 0$, $y \geq 0$, and $x + y \leq n$ lies on one of these lines.

22. Use mathematical induction to show that if you draw lines in the plane you only need two colors to color the regions formed so that no two regions that have an edge in common have a common color.

23. Show that $n!$ can be represented as the sum of n of its distinct positive divisors whenever $n \geq 3$. [*Hint:* Use inductive loading. First try to prove this result using mathematical induction. By examining where your proof fails, find a stronger statement that you can easily prove using mathematical induction.]

24. Use mathematical induction to prove that if n people stand in a line, where n is a positive integer, and if the first person in the line is a woman and the last person in line is a man, then somewhere in the line there is a woman directly in front of a man.

25. Use mathematical induction to show that when n circles divide the plane into regions, these regions can be colored with two different colors such that no regions with a common boundary are colored the same.

*26. Suppose that among a group of cars on a circular track there is enough fuel for one car to complete a lap. Use mathematical induction to show that there is a car in the group that can complete a lap by obtaining gas from other cars as it travels around the track.

27. Show that if n is a positive integer, then

$$\sum_{j=1}^{n} (2j-1) \left(\sum_{k=j}^{n} 1/k \right) = n(n+1)/2.$$

28. Use mathematical induction to show that if a, b, and c are the lengths of the sides of a right triangle, where c is the length of the hypotenuse, then $a^n + b^n < c^n$ for all integers n with $n \geq 3$.

*29. Use mathematical induction to show that if n is a positive integers, the sequence $2 \bmod n$, $2^2 \bmod n$, $2^{2^2} \bmod n$, $2^{2^{2^2}} \bmod n, \dots$ is eventually constant (that is, all terms after a finite number of terms are all the same).

30. A **unit** or **Egyptian fraction** is a fraction of the form $1/n$, where n is a positive integer. In this exercise, we will use strong induction to show that a greedy algorithm can be used to express every rational number p/q with $0 < p/q < 1$ as the sum of distinct unit fractions. At each step of the algorithm, we find the smallest positive integer n such that $1/n$ can be added to the sum without exceeding p/q. For example, to express $5/7$ we first start the sum with $1/2$. Because $5/7 - 1/2 = 3/14$ we add $1/5$ to the sum because 5 is the smallest positive integer k such that $1/k < 3/14$. Because $3/14 - 1/5 = 1/70$, the algorithm terminates, showing that $5/7 = 1/2 + 1/5 + 1/70$. Let $T(p)$ be the statement that this algorithm terminates for all rational numbers p/q with $0 < p/q < 1$. We will prove that the algorithm always terminates by showing that $T(p)$ holds for all positive integers p.

a) Show that the basis step $T(1)$ holds.

b) Suppose that $T(k)$ holds for positive integers k with $k < p$. That is, assume that the algorithm terminates for all rational numbers k/r, where $1 \leq k < p$. Show that if we start with p/q and the fraction $1/n$ is selected in the first step of the algorithm, then $p/q = p'/q' + 1/n$, where $p' = np - q$ and $q' = nq$. After considering the case where $p/q = 1/n$, use the inductive hypothesis to show that the greedy algorithm terminates when it begins with p'/q' and complete the inductive step.

The **McCarthy 91 function** (defined by John McCarthy, one of the founders of artificial intelligence) is defined using the rule

$$M(n) = \begin{cases} n - 10 & \text{if } n > 100 \\ M(M(n+11)) & \text{if } n \leq 100 \end{cases}$$

for all positive integers n.

31. By successively using the defining rule for $M(n)$, find

   a) $M(102)$.   **b)** $M(101)$.   **c)** $M(99)$.

   d) $M(97)$.   **e)** $M(87)$.   **f)** $M(76)$.

Links

JOHN McCARTHY (BORN 1927)   John McCarthy was born in Boston. He grew up in Boston and in Los Angeles. He studied mathematics as both an undergraduate and a graduate student, receiving his B.S. in 1948 from the California Institute of Technology and his Ph.D. in 1951 from Princeton. After graduating from Princeton, McCarthy held positions at Princeton, Stanford, Dartmouth, and M.I.T. He held a position at Stanford from 1962 until 1994, and is now an emeritus professor there. At Stanford, he was the director of the Artificial Intelligence Laboratory, held a named chair in the School of Engineering, and was a senior fellow in the Hoover Institution.

McCarthy was a pioneer in the study of artificial intelligence, a term he coined in 1955. He worked on problems related to the reasoning and information needs required for intelligent computer behavior. McCarthy was among the first computer scientists to design time-sharing computer systems. He developed LISP, a programming language for computing using symbolic expressions. He played an important role in using logic to verify the correctness of computer programs. McCarthy has also worked on the social implications of computer technology. He is currently working on the problem of how people and computers make conjectures through assumptions that complications are absent from situations. McCarthy is an advocate of the sustainability of human progress and is an optimist about the future of humanity. He has also begun writing science fiction stories. Some of his recent writing explores the possibility that the world is a computer program written by some higher force.

Among the awards McCarthy has won are the Turing Award from the Association for Computing Machinery, the Research Excellence Award of the International Conference on Artificial Intelligence, the Kyoto Prize, and the National Medal of Science.

32. Is this proof that

$$\frac{1}{1 \cdot 2} + \frac{1}{2 \cdot 3} + \cdots + \frac{1}{(n-1)n} = \frac{3}{2} - \frac{1}{n},$$

whenever n is a positive integer, correct? Justify your answer.

Basis step: The result is true when $n = 1$ because

$$\frac{1}{1 \cdot 2} = \frac{3}{2} - \frac{1}{1}.$$

Inductive step: Assume that the result is true for n. Then

$$\frac{1}{1 \cdot 2} + \frac{1}{2 \cdot 3} + \cdots + \frac{1}{(n-1)n} + \frac{1}{n(n+1)}$$
$$= \frac{3}{2} - \frac{1}{n} + \left(\frac{1}{n} - \frac{1}{n+1}\right)$$
$$= \frac{3}{2} - \frac{1}{n+1}.$$

Hence, the result is true for $n + 1$ if it is true for n. This completes the proof.

∗33. Show that n circles divide the plane into $n^2 - n + 2$ regions if every two circles intersect in exactly two points and no three circles contain a common point.

∗34. Show that n planes divide three-dimensional space into $(n^3 + 5n + 6)/6$ regions if any three of these planes have exactly one point in common and no four contain a common point.

35. A set is **well ordered** if every nonempty subset of this set has a least element. Determine whether each of the following sets is well ordered.

a) the set of integers
b) the set of integers greater than -100
c) the set of positive rationals
d) the set of positive rationals with denominator less than 100

36. Find an explicit formula for $f(n)$ if $f(1) = 1$ and $f(n) = f(n-1) + 2n - 1$ for $n \geq 2$. Prove your result using mathematical induction.

37. Let S be the set of bit strings defined recursively by $\lambda \in S$ and $0x \in S$, $x1 \in S$ if $x \in S$, where λ is the empty string.

a) Find all strings in S of length not exceeding five.
b) Give an explicit description of the elements of S.

38. Let S be the set of strings defined recursively by $abc \in S$, $bac \in S$, and $acb \in S$, where a, b, and c are fixed letters; and for all $x \in S$, $abcx \in S$; $abxc \in S$, $axbc \in S$, and $xabc \in S$, where x is a variable representing a string of letters.

a) Find all elements of S of length eight or less.
b) Show that every element of S has a length divisible by three.

The set B of all **balanced strings of parentheses** is defined recursively by $\lambda \in B$, where λ is the empty string; $(x) \in B$, $xy \in B$ if $x, y \in B$.

39. Show that $(()())$ is a balanced string of parentheses and $(())$ is not a balanced string of parentheses.

40. Find all balanced strings of parentheses with exactly six symbols.

41. Find all balanced strings of parentheses with four or fewer symbols.

42. Use induction to show that if x is a balanced string of parentheses, then the number of left parentheses equals the number of right parentheses in x.

Define the function N on the set of strings of parentheses by

$$N(\lambda) = 0, N(() = 1, N()) = -1,$$
$$N(uv) = N(u) + N(v),$$

where λ is the empty string, and u and v are strings. It can be shown that N is well defined.

43. Find

a) $N(())$.
b) $N())()()((()$.
c) $N((()(())$.
d) $N()(((()))(())$.

∗44. Give a recursive algorithm for finding all balanced strings of parentheses containing n or fewer symbols.

45. Verify the program segment

> **if** $x > y$ **then**
> $\quad x := y$

with respect to the initial assertion **T** and the final assertion $x \leq y$.

∗46. Develop a rule of inference for verifying recursive programs and use it to verify the recursive algorithm for computing factorials given as Algorithm 1 in Section 5.4.

47. Devise a recursive algorithm that counts the number of times the integer 0 occurs in a list of integers.

Exercises 48–50 deal with some unusual sequences, informally called **self-generating sequences**, produced by simple recurrence relations or rules. In particular, Exercises 48–49 deal with the sequence $\{a(n)\}$ defined by $a(n) = n - a(a(n-1))$ for $n \geq 1$ and $a(0) = 0$. (This sequence is defined in Douglas Hofstader's fascinating book *Gödel, Escher, Bach* ([Ho99]).)

48. Find the first 10 terms of the sequence $\{a(n)\}$ defined in the preamble to this exercise.

∗49. Prove that this sequence is well defined. That is, show that $a(n)$ is uniquely defined for all nonnegative integers n.

Golomb's self-generating sequence is the unique nondecreasing sequence of positive integers $a_1, a_2, a_3, \ldots$ that has the property that it contains exactly a_k occurrences of k for each positive integer k.

50. Find the first 20 terms of Golomb's self-generating sequence.

Computer Projects

Write programs with these input and output.

1. Given a string, find its reversal.

2. Given a real number a and a nonnegative integer n, find a^n using recursion.

3. Given a real number a and a nonnegative integer n, find a^{2^n} using recursion.

4. Given two integers not both zero, find their greatest common divisor using recursion.

5. Given a list of integers and an element x, locate x in this list using a recursive implementation of a linear search.

6. Given a nonnegative integer n, find the nth Fibonacci number using iteration.

7. Given a nonnegative integer n, find the nth Fibonacci number using recursion.

8. Given a list of n integers, sort these integers using the merge sort.

Computations and Explorations

Use a computational program or programs you have written to do these exercises.

1. What are the largest values of n for which $n!$ has fewer than 100 decimal digits and fewer than 1000 decimal digits?

2. Determine which Fibonacci numbers are divisible by 5, which are divisible by 7, and which are divisible by 11. Prove that your conjectures are correct.

3. Construct tilings using right triominoes of various 16×16, 32×32, and 64×64 checkerboards with one square missing.

4. Explore which $m \times n$ checkerboards can be completely covered by right triominoes. Can you make a conjecture that answers this question?

5. Which values of Ackermann's function are small enough that you are able to compute them?

6. Compare either the number of operations or the time needed to compute Fibonacci numbers recursively versus that needed to compute them iteratively.

Writing Projects

Respond to these with essays using outside sources.

1. Describe the origins of mathematical induction. Who were the first people to use it and to which problems did they apply it?

2. Explain how to prove the Jordan curve theorem for simple polygons and describe an algorithm for determining whether a point is in the interior or exterior of a simple polygon.

3. Describe how the triangulation of simple polygons is used in some key algorithms in computational geometry.

4. Describe a variety of different applications of the Fibonacci numbers to the biological and the physical sciences.

5. Discuss the uses of Ackermann's function both in the theory of recursive definitions and in the analysis of the complexity of algorithms for set unions.

6. Discuss some of the various methodologies used to establish the correctness of programs and compare them to Hoare's methods described in Section 5.5.

7. Explain how the ideas and concepts of program correctness can be extended to prove that operating systems are secure.

6

Counting

Combinatorics, the study of arrangements of objects, is an important part of discrete mathematics. This subject was studied as long ago as the seventeenth century, when combinatorial questions arose in the study of gambling games. Enumeration, the counting of objects with certain properties, is an important part of combinatorics. We must count objects to solve many different types of problems. For instance, counting is used to determine the complexity of algorithms. Counting is also required to determine whether there are enough telephone numbers or Internet protocol addresses to meet demand. Recently, it has played a key role in mathematical biology, especially in sequencing DNA. Furthermore, counting techniques are used extensively when probabilities of events are computed.

The basic rules of counting, which we will study in Section 6.1, can solve a tremendous variety of problems. For instance, we can use these rules to enumerate the different telephone numbers possible in the United States, the allowable passwords on a computer system, and the different orders in which the runners in a race can finish. Another important combinatorial tool is the pigeonhole principle, which we will study in Section 6.2. This states that when objects are placed in boxes and there are more objects than boxes, then there is a box containing at least two objects. For instance, we can use this principle to show that among a set of 15 or more students, at least 3 were born on the same day of the week.

We can phrase many counting problems in terms of ordered or unordered arrangements of the objects of a set with or without repetitions. These arrangements, called permutations and combinations, are used in many counting problems. For instance, suppose the 100 top finishers on a competitive exam taken by 2000 students are invited to a banquet. We can count the possible sets of 100 students that will be invited, as well as the ways in which the top 10 prizes can be awarded.

Another problem in combinatorics involves generating all the arrangements of a specified kind. This is often important in computer simulations. We will devise algorithms to generate arrangements of various types.

6.1 The Basics of Counting

Introduction

Suppose that a password on a computer system consists of six, seven, or eight characters. Each of these characters must be a digit or a letter of the alphabet. Each password must contain at least one digit. How many such passwords are there? The techniques needed to answer this question and a wide variety of other counting problems will be introduced in this section.

Counting problems arise throughout mathematics and computer science. For example, we must count the successful outcomes of experiments and all the possible outcomes of these experiments to determine probabilities of discrete events. We need to count the number of operations used by an algorithm to study its time complexity.

We will introduce the basic techniques of counting in this section. These methods serve as the foundation for almost all counting techniques.

Basic Counting Principles

We first present two basic counting principles, the **product rule** and the **sum rule**. Then we will show how they can be used to solve many different counting problems.

The product rule applies when a procedure is made up of separate tasks.

> **THE PRODUCT RULE** Suppose that a procedure can be broken down into a sequence of two tasks. If there are n_1 ways to do the first task and for each of these ways of doing the first task, there are n_2 ways to do the second task, then there are $n_1 n_2$ ways to do the procedure.

Examples 1–10 show how the product rule is used.

EXAMPLE 1   A new company with just two employees, Sanchez and Patel, rents a floor of a building with 12 offices. How many ways are there to assign different offices to these two employees?

Solution: The procedure of assigning offices to these two employees consists of assigning an office to Sanchez, which can be done in 12 ways, then assigning an office to Patel different from the office assigned to Sanchez, which can be done in 11 ways. By the product rule, there are $12 \cdot 11 = 132$ ways to assign offices to these two employees.   ◄

EXAMPLE 2   The chairs of an auditorium are to be labeled with an uppercase English letter followed by a positive integer not exceeding 100. What is the largest number of chairs that can be labeled differently?

Solution: The procedure of labeling a chair consists of two tasks, namely, assigning to the seat one of the 26 uppercase English letters, and then assigning to it one of the 100 possible integers. The product rule shows that there are $26 \cdot 100 = 2600$ different ways that a chair can be labeled. Therefore, the largest number of chairs that can be labeled differently is 2600.   ◄

EXAMPLE 3   There are 32 microcomputers in a computer center. Each microcomputer has 24 ports. How many different ports to a microcomputer in the center are there?

Solution: The procedure of choosing a port consists of two tasks, first picking a microcomputer and then picking a port on this microcomputer. Because there are 32 ways to choose the microcomputer and 24 ways to choose the port no matter which microcomputer has been selected, the product rule shows that there are $32 \cdot 24 = 768$ ports.   ◄

An extended version of the product rule is often useful. Suppose that a procedure is carried out by performing the tasks $T_1, T_2, \ldots, T_m$ in sequence. If each task $T_i, i = 1, 2, \ldots, n$, can be done in n_i ways, regardless of how the previous tasks were done, then there are $n_1 \cdot n_2 \cdot \cdots \cdot n_m$ ways to carry out the procedure. This version of the product rule can be proved by mathematical induction from the product rule for two tasks (see Exercise 48).

EXAMPLE 4   How many different bit strings of length seven are there?

Solution: Each of the seven bits can be chosen in two ways, because each bit is either 0 or 1. Therefore, the product rule shows there are a total of $2^7 = 128$ different bit strings of length seven.   ◄

EXAMPLE 5 How many different license plates can be made if each plate contains a sequence of three uppercase English letters followed by three digits (and no sequences of letters are prohibited, even if they are obscene)?

$$\underbrace{\text{— — —}}_{\substack{\text{26 choices} \\ \text{for each} \\ \text{letter}}} \underbrace{\text{— — —}}_{\substack{\text{10 choices} \\ \text{for each} \\ \text{digit}}}$$

Solution: There are 26 choices for each of the three uppercase English letters and ten choices for each of the three digits. Hence, by the product rule there are a total of $26 \cdot 26 \cdot 26 \cdot 10 \cdot 10 \cdot 10 = 17{,}576{,}000$ possible license plates. ◀

EXAMPLE 6 **Counting Functions**   How many functions are there from a set with m elements to a set with n elements?

Solution: A function corresponds to a choice of one of the n elements in the codomain for each of the m elements in the domain. Hence, by the product rule there are $n \cdot n \cdot \cdots \cdot n = n^m$ functions from a set with m elements to one with n elements. For example, there are $5^3 = 125$ different functions from a set with three elements to a set with five elements. ◀

EXAMPLE 7 **Counting One-to-One Functions**   How many one-to-one functions are there from a set with m elements to one with n elements?

Counting the number of onto functions is harder. We'll do this in Chapter 8.

Solution: First note that when $m > n$ there are no one-to-one functions from a set with m elements to a set with n elements.

Now let $m \leq n$. Suppose the elements in the domain are $a_1, a_2, \ldots, a_m$. There are n ways to choose the value of the function at a_1. Because the function is one-to-one, the value of the function at a_2 can be picked in $n - 1$ ways (because the value used for a_1 cannot be used again). In general, the value of the function at a_k can be chosen in $n - k + 1$ ways. By the product rule, there are $n(n - 1)(n - 2) \cdots (n - m + 1)$ one-to-one functions from a set with m elements to one with n elements.

For example, there are $5 \cdot 4 \cdot 3 = 60$ one-to-one functions from a set with three elements to a set with five elements. ◀

EXAMPLE 8 **The Telephone Numbering Plan**   The *North American numbering plan (NANP)* specifies the format of telephone numbers in the U.S., Canada, and many other parts of North America. A telephone number in this plan consists of 10 digits, which are split into a three-digit area code, a three-digit office code, and a four-digit station code. Because of signaling considerations, there are certain restrictions on some of these digits. To specify the allowable format, let X denote a digit that can take any of the values 0 through 9, let N denote a digit that can take any of the values 2 through 9, and let Y denote a digit that must be a 0 or a 1. Two numbering plans, which will be called the old plan, and the new plan, will be discussed. (The old plan, in use in the 1960s, has been replaced by the new plan, but the recent rapid growth in demand for new numbers for mobile phones and devices will eventually make even this new plan obsolete. In this example, the letters used to represent digits follow the conventions of the *North American Numbering Plan*.) As will be shown, the new plan allows the use of more numbers.

Current projections are that by 2038, it will be necessary to add one or more digits to North American telephone numbers.

In the old plan, the formats of the area code, office code, and station code are *NYX*, *NNX*, and *XXXX*, respectively, so that telephone numbers had the form *NYX-NNX-XXXX*. In the new plan, the formats of these codes are *NXX*, *NXX*, and *XXXX*, respectively, so that telephone numbers have the form *NXX-NXX-XXXX*. How many different North American telephone numbers are possible under the old plan and under the new plan?

Solution: By the product rule, there are $8 \cdot 2 \cdot 10 = 160$ area codes with format *NYX* and $8 \cdot 10 \cdot 10 = 800$ area codes with format *NXX*. Similarly, by the product rule, there are $8 \cdot 8 \cdot 10 = 640$ office codes with format *NNX*. The product rule also shows that there are $10 \cdot 10 \cdot 10 \cdot 10 = 10{,}000$ station codes with format *XXXX*.

Note that we have ignored restrictions that rule out N11 station codes for most area codes.

Consequently, applying the product rule again, it follows that under the old plan there are

$$160 \cdot 640 \cdot 10{,}000 = 1{,}024{,}000{,}000$$

different numbers available in North America. Under the new plan, there are

$$800 \cdot 800 \cdot 10{,}000 = 6{,}400{,}000{,}000$$

different numbers available. ◀

EXAMPLE 9 What is the value of k after the following code, where $n_1, n_2, \ldots, n_m$ are positive integers, has been executed?

```
k := 0
for i₁ := 1 to n₁
    for i₂ := 1 to n₂
            .
            .
            .
        for iₘ := 1 to nₘ
            k := k + 1
```

Solution: The initial value of k is zero. Each time the nested loop is traversed, 1 is added to k. Let T_i be the task of traversing the ith loop. Then the number of times the loop is traversed is the number of ways to do the tasks $T_1, T_2, \ldots, T_m$. The number of ways to carry out the task T_j, $j = 1, 2, \ldots, m$, is n_j, because the jth loop is traversed once for each integer i_j with $1 \le i_j \le n_j$. By the product rule, it follows that the nested loop is traversed $n_1 n_2 \cdots n_m$ times. Hence, the final value of k is $n_1 n_2 \cdots n_m$. ◀

EXAMPLE 10 **Counting Subsets of a Finite Set**   Use the product rule to show that the number of different subsets of a finite set S is $2^{|S|}$.

Solution: Let S be a finite set. List the elements of S in arbitrary order. Recall from Section 2.2 that there is a one-to-one correspondence between subsets of S and bit strings of length $|S|$. Namely, a subset of S is associated with the bit string with a 1 in the ith position if the ith element in the list is in the subset, and a 0 in this position otherwise. By the product rule, there are $2^{|S|}$ bit strings of length $|S|$. Hence, $|P(S)| = 2^{|S|}$. (Recall that we used mathematical induction to prove this fact in Example 10 of Section 5.1.) ◀

The product rule is often phrased in terms of sets in this way: If $A_1, A_2, \ldots, A_m$ are finite sets, then the number of elements in the Cartesian product of these sets is the product of the number of elements in each set. To relate this to the product rule, note that the task of choosing an element in the Cartesian product $A_1 \times A_2 \times \cdots \times A_m$ is done by choosing an element in A_1, an element in $A_2, \ldots,$ and an element in A_m. By the product rule it follows that

$$|A_1 \times A_2 \times \cdots \times A_m| = |A_1| \cdot |A_2| \cdot \cdots \cdot |A_m|.$$

EXAMPLE 11 **DNA and Genomes**   The hereditary information of a living organism is encoded using deoxyribonucleic acid (DNA), or in certain viruses, ribonucleic acid (RNA). DNA and RNA are extremely complex molecules, with different molecules interacting in a vast variety of ways to enable living process. For our purposes, we give only the briefest description of how DNA and RNA encode genetic information.

DNA molecules consist of two strands consisting of blocks known as nucleotides. Each nucleotide contains subcomponents called **bases**, each of which is adenine (A), cytosine (C), guanine (G), or thymine (T). The two strands of DNA are held together by hydrogen bonds connecting different bases, with A bonding only with T, and C bonding only with G. Unlike DNA, RNA is single stranded, with uracil (U) replacing thymine as a base. So, in DNA the possible base pairs are A-T and C-G, while in RNA they are A-U, and C-G. The DNA of a living creature consists of multiple pieces of DNA forming separate chromosomes. A **gene** is a segment of a DNA molecule that encodes a particular protein. The entirety of genetic information of an organism is called its **genome**.

Sequences of bases in DNA and RNA encode long chains of proteins called amino acids. There are 22 essential amino acids for human beings. We can quickly see that a sequence of at least three bases are needed to encode these 22 different amino acid. First note, that because there are four possibilities for each base in DNA, A, C, G, and T, by the product rule there are $4^2 = 16 < 22$ different sequences of two bases. However, there are $4^3 = 64$ different sequences of three bases, which provide enough different sequences to encode the 22 different amino acids (even after taking into account that several different sequences of three bases encode the same amino acid).

The DNA of simple living creatures such as algae and bacteria have between 10^5 and 10^7 links, where each link is one of the four possible bases. More complex organisms, such as insects, birds, and mammals have between 10^8 and 10^{10} links in their DNA. So, by the product rule, there are at least 4^{10^5} different sequences of bases in the DNA of simple organisms and at least 4^{10^8} different sequences of bases in the DNA of more complex organisms. These are both incredibly huge numbers, which helps explain why there is such tremendous variability among living organisms. In the past several decades techniques have been developed for determining the genome of different organisms. The first step is to locate each gene in the DNA of an organism. The next task, called **gene sequencing**, is the determination of the sequence of links on each gene. (Of course, the specific sequence of kinks on these genes depends on the particular individual representative of a species whose DNA is analyzed.) For example, the human genome includes approximately 23,000 genes, each with 1,000 or more links. Gene sequencing techniques take advantage of many recently developed algorithms and are based on numerous new ideas in combinatorics. Many mathematicians and computer scientists work on problems involving genomes, taking part in the fast moving fields of bioinformatics and computational biology. ◀

Soon it won't be that costly to have your own genetic code found.

We now introduce the sum rule.

THE SUM RULE If a task can be done either in one of n_1 ways or in one of n_2 ways, where none of the set of n_1 ways is the same as any of the set of n_2 ways, then there are $n_1 + n_2$ ways to do the task.

Example 12 illustrates how the sum rule is used.

EXAMPLE 12 Suppose that either a member of the mathematics faculty or a student who is a mathematics major is chosen as a representative to a university committee. How many different choices are there for this representative if there are 37 members of the mathematics faculty and 83 mathematics majors and no one is both a faculty member and a student?

Solution: There are 37 ways to choose a member of the mathematics faculty and there are 83 ways to choose a student who is a mathematics major. Choosing a member of the mathematics faculty is never the same as choosing a student who is a mathematics major because no one is both a faculty member and a student. By the sum rule it follows that there are $37 + 83 = 120$ possible ways to pick this representative. ◀

We can extend the sum rule to more than two tasks. Suppose that a task can be done in one of n_1 ways, in one of n_2 ways, ... , or in one of n_m ways, where none of the set of n_i ways of doing the task is the same as any of the set of n_j ways, for all pairs i and j with $1 \leq i < j \leq m$. Then the number of ways to do the task is $n_1 + n_2 + \cdots + n_m$. This extended version of the sum rule is often useful in counting problems, as Examples 13 and 14 show. This version of the sum rule can be proved using mathematical induction from the sum rule for two sets. (This is Exercise 47.)

EXAMPLE 13   A student can choose a computer project from one of three lists. The three lists contain 23, 15, and 19 possible projects, respectively. No project is on more than one list. How many possible projects are there to choose from?

Solution: The student can choose a project by selecting a project from the first list, the second list, or the third list. Because no project is on more than one list, by the sum rule there are $23 + 15 + 19 = 57$ ways to choose a project. ◀

EXAMPLE 14   What is the value of k after the following code, where $n_1, n_2, \ldots, n_m$ are positive integers, has been executed?

```
k := 0
for i_1 := 1 to n_1
    k := k + 1
for i_2 := 1 to n_2
    k := k + 1
        .
        .
        .
for i_m := 1 to n_m
    k := k + 1
```

Solution: The initial value of k is zero. This block of code is made up of m different loops. Each time a loop is traversed, 1 is added to k. To determine the value of k after this code has been executed, we need to determine how many times we traverse a loop. Note that there are n_i ways to traverse the ith loop. Because we only traverse one loop at a time, the sum rule shows that the final value of k, which is the number of ways to traverse one of the m loops is $n_1 + n_2 + \cdots + n_m$. ◀

The sum rule can be phrased in terms of sets as: If $A_1, A_2, \ldots, A_m$ are pairwise disjoint finite sets, then the number of elements in the union of these sets is the sum of the numbers of elements in the sets. To relate this to our statement of the sum rule, note there are $|A_i|$ ways to choose an element from A_i for $i = 1, 2, \ldots, m$. Because the sets are pairwise disjoint, when we select an element from one of the sets A_i, we do not also select an element from a different set A_j. Consequently, by the sum rule, because we cannot select an element from two of these sets at the same time, the number of ways to choose an element from one of the sets, which is the number of elements in the union, is

$$|A_1 \cup A_2 \cup \cdots \cup A_m| = |A_1| + |A_2| + \cdots + |A_m| \text{ when } A_i \cap A_j = \emptyset \text{ for all } i, j.$$

This equality applies only when the sets in question are pairwise disjoint. The situation is much more complicated when these sets have elements in common. That situation will be briefly discussed later in this section and discussed in more depth in Chapter 8.

More Complex Counting Problems

Many counting problems cannot be solved using just the sum rule or just the product rule. However, many complicated counting problems can be solved using both of these rules in combination. We begin by counting the number of variable names in the programming language BASIC. (In the exercises, we consider the number of variable names in JAVA.) Then we will count the number of valid passwords subject to a particular set of restrictions.

EXAMPLE 15

Extra Examples

In a version of the computer language BASIC, the name of a variable is a string of one or two alphanumeric characters, where uppercase and lowercase letters are not distinguished. (An *alphanumeric* character is either one of the 26 English letters or one of the 10 digits.) Moreover, a variable name must begin with a letter and must be different from the five strings of two characters that are reserved for programming use. How many different variable names are there in this version of BASIC?

Solution: Let V equal the number of different variable names in this version of BASIC. Let V_1 be the number of these that are one character long and V_2 be the number of these that are two characters long. Then by the sum rule, $V = V_1 + V_2$. Note that $V_1 = 26$, because a one-character variable name must be a letter. Furthermore, by the product rule there are $26 \cdot 36$ strings of length two that begin with a letter and end with an alphanumeric character. However, five of these are excluded, so $V_2 = 26 \cdot 36 - 5 = 931$. Hence, there are $V = V_1 + V_2 = 26 + 931 = 957$ different names for variables in this version of BASIC. ◀

EXAMPLE 16

Each user on a computer system has a password, which is six to eight characters long, where each character is an uppercase letter or a digit. Each password must contain at least one digit. How many possible passwords are there?

Solution: Let P be the total number of possible passwords, and let P_6, P_7, and P_8 denote the number of possible passwords of length 6, 7, and 8, respectively. By the sum rule, $P = P_6 + P_7 + P_8$. We will now find P_6, P_7, and P_8. Finding P_6 directly is difficult. To find P_6 it is easier to find the number of strings of uppercase letters and digits that are six characters long, including those with no digits, and subtract from this the number of strings with no digits. By the product rule, the number of strings of six characters is 36^6, and the number of strings with no digits is 26^6. Hence,

$$P_6 = 36^6 - 26^6 = 2,176,782,336 - 308,915,776 = 1,867,866,560.$$

Similarly, we have

$$P_7 = 36^7 - 26^7 = 78,364,164,096 - 8,031,810,176 = 70,332,353,920$$

and

$$P_8 = 36^8 - 26^8 = 2,821,109,907,456 - 208,827,064,576$$
$$= 2,612,282,842,880.$$

Consequently,

$$P = P_6 + P_7 + P_8 = 2,684,483,063,360.$$

◀

EXAMPLE 17

Links

Counting Internet Addresses In the Internet, which is made up of interconnected physical networks of computers, each computer (or more precisely, each network connection of a computer) is assigned an *Internet address*. In Version 4 of the Internet Protocol (IPv4), now in use,

Bit Number	0	1	2	3	4		8	16	24	31

Class A	0	netid						hostid		
Class B	1	0	netid						hostid	
Class C	1	1	0	netid						hostid
Class D	1	1	1	0	Multicast Address					
Class E	1	1	1	1	0	Address				

FIGURE 1   Internet Addresses (IPv4).

an address is a string of 32 bits. It begins with a *network number* (*netid*). The netid is followed by a *host number* (*hostid*), which identifies a computer as a member of a particular network.

Three forms of addresses are used, with different numbers of bits used for netids and hostids. **Class A addresses**, used for the largest networks, consist of 0, followed by a 7-bit netid and a 24-bit hostid. **Class B addresses**, used for medium-sized networks, consist of 10, followed by a 14-bit netid and a 16-bit hostid. **Class C addresses**, used for the smallest networks, consist of 110, followed by a 21-bit netid and an 8-bit hostid. There are several restrictions on addresses because of special uses: 1111111 is not available as the netid of a Class A network, and the hostids consisting of all 0s and all 1s are not available for use in any network. A computer on the Internet has either a Class A, a Class B, or a Class C address. (Besides Class A, B, and C addresses, there are also Class D addresses, reserved for use in multicasting when multiple computers are addressed at a single time, consisting of 1110 followed by 28 bits, and Class E addresses, reserved for future use, consisting of 11110 followed by 27 bits. Neither Class D nor Class E addresses are assigned as the IPv4 address of a computer on the Internet.) Figure 1 illustrates IPv4 addressing. (Limitations on the number of Class A and Class B netids have made IPv4 addressing inadequate; IPv6, a new version of IP, uses 128-bit addresses to solve this problem.)

> The lack of available IPv4 address has become a crisis!

How many different IPv4 addresses are available for computers on the Internet?

Solution: Let x be the number of available addresses for computers on the Internet, and let x_A, x_B, and x_C denote the number of Class A, Class B, and Class C addresses available, respectively. By the sum rule, $x = x_A + x_B + x_C$.

To find x_A, note that there are $2^7 - 1 = 127$ Class A netids, recalling that the netid 1111111 is unavailable. For each netid, there are $2^{24} - 2 = 16,777,214$ hostids, recalling that the hostids consisting of all 0s and all 1s are unavailable. Consequently, $x_A = 127 \cdot 16,777,214 = 2,130,706,178$.

To find x_B and x_C, note that there are $2^{14} = 16,384$ Class B netids and $2^{21} = 2,097,152$ Class C netids. For each Class B netid, there are $2^{16} - 2 = 65,534$ hostids, and for each Class C netid, there are $2^8 - 2 = 254$ hostids, recalling that in each network the hostids consisting of all 0s and all 1s are unavailable. Consequently, $x_B = 1,073,709,056$ and $x_C = 532,676,608$.

We conclude that the total number of IPv4 addresses available is $x = x_A + x_B + x_C = 2,130,706,178 + 1,073,709,056 + 532,676,608 = 3,737,091,842$. ◀

The Subtraction Rule (Inclusion–Exclusion for Two Sets)

Suppose that a task can be done in one of two ways, but some of the ways to do it are common to both ways. In this situation, we cannot use the sum rule to count the number of ways to do the task. If we add the number of ways to do the tasks in these two ways, we get an overcount of the total number of ways to do it, because the ways to do the task that are common to the two ways are counted twice. To correctly count the number of ways to do the two tasks, we must subtract the number of ways that are counted twice. This leads us to an important counting rule.

> Overcounting is perhaps the most common enumeration error.

THE SUBTRACTION RULE If a task can be done in either n_1 ways or n_2 ways, then the number of ways to do the task is $n_1 + n_2$ minus the number of ways to do the task that are common to the two different ways.

The subtraction rule is also known as the **principle of inclusion–exclusion**, especially when it is used to count the number of elements in the union of two sets. Suppose that A_1 and A_2 are sets. Then, there are $|A_1|$ ways to select an element from A_1 and $|A_2|$ ways to select an element from A_2. The number of ways to select an element from A_1 or from A_2, that is, the number of ways to select an element from their union, is the sum of the number of ways to select an element from A_1 and the number of ways to select an element from A_2, minus the number of ways to select an element that is in both A_1 and A_2. Because there are $|A_1 \cup A_2|$ ways to select an element in either A_1 or in A_2, and $|A_1 \cap A_2|$ ways to select an element common to both sets, we have

$$|A_1 \cup A_2| = |A_1| + |A_2| - |A_1 \cap A_2|.$$

This is the formula given in Section 2.2 for the number of elements in the union of two sets.

Example 18 illustrates how we can solve counting problems using the subtraction principle.

EXAMPLE 18

How many bit strings of length eight either start with a 1 bit or end with the two bits 00?

Solution: We can construct a bit string of length eight that either starts with a 1 bit or ends with the two bits 00, by constructing a bit string of length eight beginning with a 1 bit or by constructing a bit string of length eight that ends with the two bits 00. We can construct a bit string of length eight that begins with a 1 in $2^7 = 128$ ways. This follows by the product rule, because the first bit can be chosen in only one way and each of the other seven bits can be chosen in two ways. Similarly, we can construct a bit string of length eight ending with the two bits 00, in $2^6 = 64$ ways. This follows by the product rule, because each of the first six bits can be chosen in two ways and the last two bits can be chosen in only one way.

Some of the ways to construct a bit string of length eight starting with a 1 are the same as the ways to construct a bit string of length eight that ends with the two bits 00. There are $2^5 = 32$ ways to construct such a string. This follows by the product rule, because the first bit can be chosen in only one way, each of the second through the sixth bits can be chosen in two ways, and the last two bits can be chosen in one way. Consequently, the number of bit strings of length eight that begin with a 1 or end with a 00, which equals the number of ways to construct a bit string of length eight that begins with a 1 or that ends with 00, equals $128 + 64 - 32 = 160$. ◄

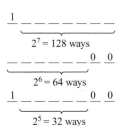

We present an example that illustrates how the formulation of the principle of inclusion–exclusion can be used to solve counting problems.

EXAMPLE 19

A computer company receives 350 applications from computer graduates for a job planning a line of new Web servers. Suppose that 220 of these applicants majored in computer science, 147 majored in business, and 51 majored both in computer science and in business. How many of these applicants majored neither in computer science nor in business?

Solution: To find the number of these applicants who majored neither in computer science nor in business, we can subtract the number of students who majored either in computer science or in business (or both) from the total number of applicants. Let A_1 be the set of students who majored in computer science and A_2 the set of students who majored in business. Then $A_1 \cup A_2$ is the set of students who majored in computer science or business (or both), and $A_1 \cap A_2$ is the set of students who majored both in computer science and in business. By the subtraction rule the number of students who majored either in computer science or in business (or both) equals

$$|A_1 \cup A_2| = |A_1| + |A_2| - |A_1 \cap A_2| = 220 + 147 - 51 = 316.$$

We conclude that $350 - 316 = 34$ of the applicants majored neither in computer science nor in business. ◀

The subtraction rule, or the principle of inclusion–exclusion, can be generalized to find the number of ways to do one of n different tasks or, equivalently, to find the number of elements in the union of n sets, whenever n is a positive integer. We will study the inclusion–exclusion principle and some of its many applications in Chapter 8.

The Division Rule

We have introduced the product, sum, and subtraction rules for counting. You may wonder whether there is also a division rule for counting. In fact, there is such a rule, which can be useful when solving certain types of enumeration problems.

> THE DIVISION RULE There are n/d ways to do a task if it can be done using a procedure that can be carried out in n ways, and for every way w, exactly d of the n ways correspond to way w.

We can restate the division rule in terms of sets: "If the finite set A is the union of n pairwise disjoint subsets each with d elements, then $n = |A|/d$."

We can also formulate the division rule in terms of functions: "If f is a function from A to B where A and B are finite sets, and that for every value $y \in B$ there are exactly d values $x \in A$ such that $f(x) = y$ (in which case, we say that f is d-to-one), then $|B| = |A|/d$."

We illustrate the use of the division rule for counting with an example.

EXAMPLE 20 How many different ways are there to seat four people around a circular table, where two seatings are considered the same when each person has the same left neighbor and the same right neighbor?

Solution: We arbitrarily select a seat at the table and label it seat 1. We number the rest of the seats in numerical order, proceeding clockwise around the table. Note that are four ways to select the person for seat 1, three ways to select the person for seat 2, two ways to select the person for seat 3, and one way to select the person for seat 4. Thus, there are $4! = 24$ ways to order the given four people for these seats. However, each of the four choices for seat 1 leads to the same arrangement, as we distinguish two arrangements only when one of the people has a different immediate left or immediate right neighbor. Because there are four ways to choose the person for seat 1, by the division rule there are $24/4 = 6$ different seating arrangements of four people around the circular table. ◀

Tree Diagrams

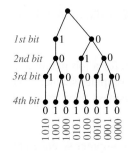

1st bit
2nd bit
3rd bit
4th bit

FIGURE 2 Bit Strings of Length Four without Consecutive 1s.

Counting problems can be solved using **tree diagrams**. A tree consists of a root, a number of branches leaving the root, and possible additional branches leaving the endpoints of other branches. (We will study trees in detail in Chapter 11.) To use trees in counting, we use a branch to represent each possible choice. We represent the possible outcomes by the leaves, which are the endpoints of branches not having other branches starting at them.

Note that when a tree diagram is used to solve a counting problem, the number of choices of which branch to follow to reach a leaf can vary (see Example 21, for example).

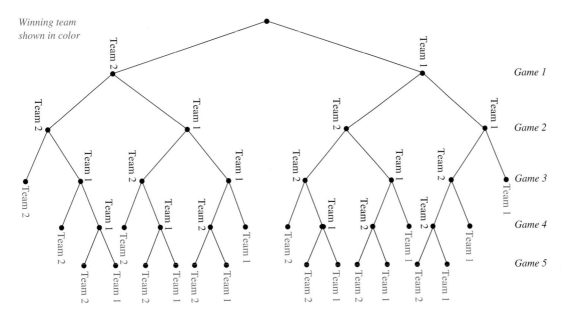

Winning team shown in color

FIGURE 3 **Best Three Games Out of Five Playoffs.**

EXAMPLE 21 How many bit strings of length four do not have two consecutive 1s?

Solution: The tree diagram in Figure 2 displays all bit strings of length four without two consecutive 1s. We see that there are eight bit strings of length four without two consecutive 1s. ◀

EXAMPLE 22 A playoff between two teams consists of at most five games. The first team that wins three games wins the playoff. In how many different ways can the playoff occur?

Solution: The tree diagram in Figure 3 displays all the ways the playoff can proceed, with the winner of each game shown. We see that there are 20 different ways for the playoff to occur. ◀

EXAMPLE 23 Suppose that "I Love New Jersey" T-shirts come in five different sizes: S, M, L, XL, and XXL. Further suppose that each size comes in four colors, white, red, green, and black, except for XL, which comes only in red, green, and black, and XXL, which comes only in green and black. How many different shirts does a souvenir shop have to stock to have at least one of each available size and color of the T-shirt?

Solution: The tree diagram in Figure 4 displays all possible size and color pairs. It follows that the souvenir shop owner needs to stock 17 different T-shirts. ◀

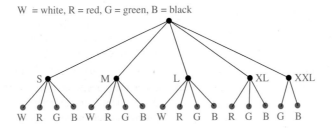

W = white, R = red, G = green, B = black

FIGURE 4 **Counting Varieties of T-Shirts.**

Exercises

1. There are 18 mathematics majors and 325 computer science majors at a college.

   a) In how many ways can two representatives be picked so that one is a mathematics major and the other is a computer science major?

   b) In how many ways can one representative be picked who is either a mathematics major or a computer science major?

2. A multiple-choice test contains 10 questions. There are four possible answers for each question.

   a) In how many ways can a student answer the questions on the test if the student answers every question?

   b) In how many ways can a student answer the questions on the test if the student can leave answers blank?

3. Six different airlines fly from New York to Denver and seven fly from Denver to San Francisco. How many different pairs of airlines can you choose on which to book a trip from New York to San Francisco via Denver, when you pick an airline for the flight to Denver and an airline for the continuation flight to San Francisco?

4. How many different three-letter initials can people have?

5. How many bit strings of length ten both begin and end with a 1?

6. How many bit strings with length not exceeding n, where n is a positive integer, consist entirely of 1s, not counting the empty string?

7. How many bit strings of length n, where n is a positive integer, start and end with 1s?

8. How many strings are there of lowercase letters of length four or less, not counting the empty string?

9. How many strings of five ASCII characters contain the character @ ("at" sign) at least once? [*Note:* There are 128 different ASCII characters.

10. How many 6-element RNA sequences

    a) do not contain U?

    b) end with GU?

    c) start with C?

    d) contain only A or U?

11. How many positive integers between 50 and 100

    a) are divisible by 7? Which integers are these?

    b) are divisible by 11? Which integers are these?

    c) are divisible by both 7 and 11? Which integers are these?

12. How many positive integers between 100 and 999 inclusive

    a) are divisible by 7?

    b) are odd?

    c) have the same three decimal digits?

    d) are not divisible by 4?

    e) are divisible by 3 or 4?

    f) are not divisible by either 3 or 4?

    g) are divisible by 3 but not by 4?

    h) are divisible by 3 and 4?

13. How many strings of three decimal digits

    a) do not contain the same digit three times?

    b) begin with an odd digit?

    c) have exactly two digits that are 4s?

14. A committee is formed consisting of one representative from each of the 50 states in the United States, where the representative from a state is either the governor or one of the two senators from that state. How many ways are there to form this committee?

15. How many license plates can be made using either two uppercase English letters followed by four digits or two digits followed by four uppercase English letters?

16. How many license plates can be made using either two or three uppercase English letters followed by either two or three digits?

17. How many strings of eight English letters are there

    a) that contain no vowels, if letters can be repeated?

    b) that contain no vowels, if letters cannot be repeated?

    c) that start with a vowel, if letters can be repeated?

    d) that start with a vowel, if letters cannot be repeated?

    e) that contain at least one vowel, if letters can be repeated?

    f) that contain exactly one vowel, if letters can be repeated?

    g) that start with X and contain at least one vowel, if letters can be repeated?

    h) that start and end with X and contain at least one vowel, if letters can be repeated?

18. How many different functions are there from a set with 10 elements to sets with the following numbers of elements?

    a) 2   b) 3   c) 4   d) 5

19. How many one-to-one functions are there from a set with five elements to sets with the following number of elements?

    a) 4   b) 5   c) 6   d) 7

20. How many functions are there from the set $\{1, 2, \ldots, n\}$, where n is a positive integer, to the set $\{0, 1\}$?

21. How many functions are there from the set $\{1, 2, \ldots, n\}$, where n is a positive integer, to the set $\{0, 1\}$

    a) that are one-to-one?

    b) that assign 0 to both 1 and n?

    c) that assign 1 to exactly one of the positive integers less than n?

22. How many partial functions (see Section 2.3) are there from a set with five elements to sets with each of these number of elements?

    a) 1   b) 2   c) 5   d) 9

23. How many partial functions (see Definition 13 of Section 2.3) are there from a set with m elements to a set with n elements, where m and n are positive integers?

24. How many subsets of a set with 100 elements have more than one element?

25. A **palindrome** is a string whose reversal is identical to the string. How many bit strings of length n are palindromes?

26. How many 4-element DNA sequences

    a) do not contain the base T?

    b) contain the sequence ACG?

    c) contain all four bases A, T, C, and G?

    d) contain exactly three of the four bases A, T, C, and G?

27. How many 4-element RNA sequences

    a) contain the base U?

    b) do not contain the sequence CUG?

    c) do not contain all four bases A, U, C, and G?

    d) contain exactly two of the four bases A, U, C, and G?

28. How many ways are there to seat four of a group of ten people around a circular table where two seatings are considered the same when everyone has the same immediate left and immediate right neighbor?

29. How many ways are there to seat six people around a circular table where two seatings are considered the same when everyone has the same two neighbors without regard to whether they are right or left neighbors?

30. In how many ways can a photographer at a wedding arrange 6 people in a row from a group of 10 people, where the bride and the groom are among these 10 people, if

    a) the bride must be in the picture?

    b) both the bride and groom must be in the picture?

    c) exactly one of the bride and the groom is in the picture?

31. In how many ways can a photographer at a wedding arrange six people in a row, including the bride and groom, if

    a) the bride must be next to the groom?

    b) the bride is not next to the groom?

    c) the bride is positioned somewhere to the left of the groom?

32. How many bit strings of length seven either begin with two 0s or end with three 1s?

33. How many bit strings of length 10 either begin with three 0s or end with two 0s?

*34. How many bit strings of length 10 contain either five consecutive 0s or five consecutive 1s?

**35. How many bit strings of length eight contain either three consecutive 0s or four consecutive 1s?

36. How many positive integers not exceeding 100 are divisible either by 4 or by 6?

37. Suppose that a password for a computer system must have at least 8, but no more than 12, characters, where each character in the password is a lowercase English letter, an uppercase English letter, a digit, or one of the six special characters $*$, $>$, $<$, $!$, $+$, and $=$.

    a) How many different passwords are available for this computer system?

    b) How many of these passwords contain at least one occurrence of at least one of the six special characters?

    c) Using your answer to part (a), determine how long it takes a hacker to try every possible password, assuming that it takes one nanosecond for a hacker to check each possible password.

38. The name of a variable in the C programming language is a string that can contain uppercase letters, lowercase letters, digits, or underscores. Further, the first character in the string must be a letter, either uppercase or lowercase, or an underscore. If the name of a variable is determined by its first eight characters, how many different variables can be named in C? (Note that the name of a variable may contain fewer than eight characters.)

39. The name of a variable in the JAVA programming language is a string of between 1 and 65,535 characters, inclusive, where each character can be an uppercase or a lowercase letter, a dollar sign, an underscore, or a digit, except that the first character must not be a digit. Determine the number of different variable names in JAVA.

40. The International Telecommunications Union (ITU) specifies that a telephone number must consist of a country code with between 1 and 3 digits, except that the code 0 is not available for use as a country code, followed by a number with at most 15 digits. How many available possible telephone numbers are there that satisfy these restrictions?

41. Suppose that at some future time every telephone in the world is assigned a number that contains a country code 1 to 3 digits long, that is, of the form X, XX, or XXX, followed by a 10-digit telephone number of the form *NXX-NXX-XXXX* (as described in Example 8). How many different telephone numbers would be available worldwide under this numbering plan?

42. A key in the Vigenère cryptosystem is a string of English letters, where the case of the letters does not matter. How many different keys for this cryptosystem are there with three, four, five, or six letters?

43. A wired equivalent privacy (WEP) key for a wireless fidelity (WiFi) network is a string of either 10, 26, or 58 hexadecimal digits. How many different WEP keys are there?

44. Suppose that p and q are prime numbers and that $n = pq$. Use the principle of inclusion–exclusion to find the number of positive integers not exceeding n that are relatively prime to n.

45. Use a tree diagram to determine the number of subsets of $\{3, 7, 9, 11, 24\}$ with the property that the sum of the elements in the subset is less than 28.

46. a) Suppose that a popular style of running shoe is available for both men and women. The woman's shoe comes in sizes 6, 7, 8, and 9, and the man's shoe comes in sizes 8, 9, 10, 11, and 12. The man's shoe comes in white and black, while the woman's shoe comes in white, red, and black. Use a tree diagram to determine the number of different shoes that a store has to stock to have at least one pair of this type of running shoe for all available sizes and colors for both men and women.

b) Answer the question in part (a) using counting rules.

47. Use mathematical induction to prove the sum rule for m tasks from the sum rule for two tasks.

48. Use mathematical induction to prove the product rule for m tasks from the product rule for two tasks.

49. How many diagonals does a convex polygon with n sides have? (Recall that a polygon is convex if every line segment connecting two points in the interior or boundary of the polygon lies entirely within this set and that a diagonal of a polygon is a line segment connecting two vertices that are not adjacent.)

50. Data are transmitted over the Internet in **datagrams**, which are structured blocks of bits. Each datagram contains header information organized into a maximum of 14 different fields (specifying many things, including the source and destination addresses) and a data area that contains the actual data that are transmitted. One of the 14 header fields is the **header length field** (denoted by HLEN), which is specified by the protocol to be 4 bits long and that specifies the header length in terms of 32-bit

blocks of bits. For example, if HLEN = 0110, the header is made up of six 32-bit blocks. Another of the 14 header fields is the 16-bit-long **total length field** (denoted by TOTAL LENGTH), which specifies the length in bits of the entire datagram, including both the header fields and the data area. The length of the data area is the total length of the datagram minus the length of the header.

a) The largest possible value of TOTAL LENGTH (which is 16 bits long) determines the maximum total length in octets (blocks of 8 bits) of an Internet datagram. What is this value?

b) The largest possible value of HLEN (which is 4 bits long) determines the maximum total header length in 32-bit blocks. What is this value? What is the maximum total header length in octets?

c) The minimum (and most common) header length is 20 octets. What is the maximum total length in octets of the data area of an Internet datagram?

d) How many different strings of octets in the data area can be transmitted if the header length is 20 octets and the total length is as long as possible?

6.2 The Pigeonhole Principle

Introduction

Links

Suppose that a flock of 20 pigeons flies into a set of 19 pigeonholes to roost. Because there are 20 pigeons but only 19 pigeonholes, a least one of these 19 pigeonholes must have at least two pigeons in it. To see why this is true, note that if each pigeonhole had at most one pigeon in it, at most 19 pigeons, one per hole, could be accommodated. This illustrates a general principle called the **pigeonhole principle**, which states that if there are more pigeons than pigeonholes, then there must be at least one pigeonhole with at least two pigeons in it (see Figure 1). Of course, this principle applies to other objects besides pigeons and pigeonholes.

THEOREM 1   **THE PIGEONHOLE PRINCIPLE**   If k is a positive integer and $k + 1$ or more objects are placed into k boxes, then there is at least one box containing two or more of the objects.

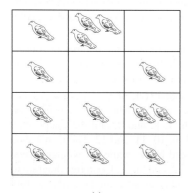
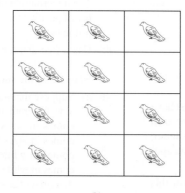
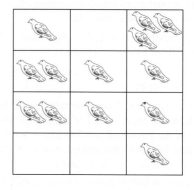

       (a)   (b)   (c)

FIGURE 1   **There Are More Pigeons Than Pigeonholes.**

Proof: We prove the pigeonhole principle using a proof by contraposition. Suppose that none of the k boxes contains more than one object. Then the total number of objects would be at most k. This is a contradiction, because there are at least $k + 1$ objects. ◁

The pigeonhole principle is also called the **Dirichlet drawer principle**, after the nineteenth-century German mathematician G. Lejeune Dirichlet, who often used this principle in his work. (Dirichlet was not the first person to use this principle; a demonstration that there were at least two Parisians with the same number of hairs on their heads dates back to the 17th century—see Exercise 26.) It is an important additional proof technique supplementing those we have developed in earlier chapters. We introduce it in this chapter because of its many important applications to combinatorics.

We will illustrate the usefulness of the pigeonhole principle. We first show that it can be used to prove a useful corollary about functions.

COROLLARY 1   A function f from a set with $k + 1$ or more elements to a set with k elements is not one-to-one.

Proof: Suppose that for each element y in the codomain of f we have a box that contains all elements x of the domain of f such that $f(x) = y$. Because the domain contains $k + 1$ or more elements and the codomain contains only k elements, the pigeonhole principle tells us that one of these boxes contains two or more elements x of the domain. This means that f cannot be one-to-one. ◁

Examples 1–3 show how the pigeonhole principle is used.

EXAMPLE 1   Among any group of 367 people, there must be at least two with the same birthday, because there are only 366 possible birthdays. ◀

EXAMPLE 2   In any group of 27 English words, there must be at least two that begin with the same letter, because there are 26 letters in the English alphabet. ◀

EXAMPLE 3   How many students must be in a class to guarantee that at least two students receive the same score on the final exam, if the exam is graded on a scale from 0 to 100 points?

Solution: There are 101 possible scores on the final. The pigeonhole principle shows that among any 102 students there must be at least 2 students with the same score. ◀

The pigeonhole principle is a useful tool in many proofs, including proofs of surprising results, such as that given in Example 4.

G. LEJEUNE DIRICHLET (1805–1859)   G. Lejeune Dirichlet was born into a Belgian family living near Cologne, Germany. His father was a postmaster. He became passionate about mathematics at a young age. He was spending all his spare money on mathematics books by the time he entered secondary school in Bonn at the age of 12. At 14 he entered the Jesuit College in Cologne, and at 16 he began his studies at the University of Paris. In 1825 he returned to Germany and was appointed to a position at the University of Breslau. In 1828 he moved to the University of Berlin. In 1855 he was chosen to succeed Gauss at the University of Göttingen. Dirichlet is said to be the first person to master Gauss's *Disquisitiones Arithmeticae*, which appeared 20 years earlier. He is said to have kept a copy at his side even when he traveled. Dirichlet made many important discoveries in number theory, including the theorem that there are infinitely many primes in arithmetical progressions $an + b$ when a and b are relatively prime. He proved the $n = 5$ case of Fermat's last theorem, that there are no nontrivial solutions in integers to $x^5 + y^5 = z^5$. Dirichlet also made many contributions to analysis. Dirichlet was considered to be an excellent teacher who could explain ideas with great clarity. He was married to Rebecca Mendelssohn, one of the sisters of the composer Frederick Mendelssohn.

EXAMPLE 4

Show that for every integer n there is a multiple of n that has only 0s and 1s in its decimal expansion.

Solution: Let n be a positive integer. Consider the $n + 1$ integers $1, 11, 111, \ldots, 11 \ldots 1$ (where the last integer in this list is the integer with $n + 1$ 1s in its decimal expansion). Note that there are n possible remainders when an integer is divided by n. Because there are $n + 1$ integers in this list, by the pigeonhole principle there must be two with the same remainder when divided by n. The larger of these integers less the smaller one is a multiple of n, which has a decimal expansion consisting entirely of 0s and 1s.   ◀

The Generalized Pigeonhole Principle

The pigeonhole principle states that there must be at least two objects in the same box when there are more objects than boxes. However, even more can be said when the number of objects exceeds a multiple of the number of boxes. For instance, among any set of 21 decimal digits there must be 3 that are the same. This follows because when 21 objects are distributed into 10 boxes, one box must have more than 2 objects.

THEOREM 2

THE GENERALIZED PIGEONHOLE PRINCIPLE   If N objects are placed into k boxes, then there is at least one box containing at least $\lceil N/k \rceil$ objects.

Proof: We will use a proof by contraposition. Suppose that none of the boxes contains more than $\lceil N/k \rceil - 1$ objects. Then, the total number of objects is at most

$$k \left(\left\lceil \frac{N}{k} \right\rceil - 1 \right) < k \left(\left(\frac{N}{k} + 1 \right) - 1 \right) = N,$$

where the inequality $\lceil N/k \rceil < (N/k) + 1$ has been used. This is a contradiction because there are a total of N objects.   ◁

A common type of problem asks for the minimum number of objects such that at least r of these objects must be in one of k boxes when these objects are distributed among the boxes. When we have N objects, the generalized pigeonhole principle tells us there must be at least r objects in one of the boxes as long as $\lceil N/k \rceil \geq r$. The smallest integer N with $N/k > r - 1$, namely, $N = k(r - 1) + 1$, is the smallest integer satisfying the inequality $\lceil N/k \rceil \geq r$. Could a smaller value of N suffice? The answer is no, because if we had $k(r - 1)$ objects, we could put $r - 1$ of them in each of the k boxes and no box would have at least r objects.

When thinking about problems of this type, it is useful to consider how you can avoid having at least r objects in one of the boxes as you add successive objects. To avoid adding a rth object to any box, you eventually end up with $r - 1$ objects in each box. There is no way to add the next object without putting an rth object in that box.

Examples 5–8 illustrate how the generalized pigeonhole principle is applied.

EXAMPLE 5

Among 100 people there are at least $\lceil 100/12 \rceil = 9$ who were born in the same month.   ◀

EXAMPLE 6

What is the minimum number of students required in a discrete mathematics class to be sure that at least six will receive the same grade, if there are five possible grades, A, B, C, D, and F?

Solution: The minimum number of students needed to ensure that at least six students receive the same grade is the smallest integer N such that $\lceil N/5 \rceil = 6$. The smallest such integer is

$N = 5 \cdot 5 + 1 = 26$. If you have only 25 students, it is possible for there to be five who have received each grade so that no six students have received the same grade. Thus, 26 is the minimum number of students needed to ensure that at least six students will receive the same grade. ◀

EXAMPLE 7   **a)** How many cards must be selected from a standard deck of 52 cards to guarantee that at least three cards of the same suit are chosen?
b) How many must be selected to guarantee that at least three hearts are selected?

A standard deck of 52 cards has 13 kinds of cards, with four cards of each of kind, one in each of the four suits, hearts, diamonds, spades, and clubs.

Solution: **a)** Suppose there are four boxes, one for each suit, and as cards are selected they are placed in the box reserved for cards of that suit. Using the generalized pigeonhole principle, we see that if N cards are selected, there is at least one box containing at least $\lceil N/4 \rceil$ cards. Consequently, we know that at least three cards of one suit are selected if $\lceil N/4 \rceil \geq 3$. The smallest integer N such that $\lceil N/4 \rceil \geq 3$ is $N = 2 \cdot 4 + 1 = 9$, so nine cards suffice. Note that if eight cards are selected, it is possible to have two cards of each suit, so more than eight cards are needed. Consequently, nine cards must be selected to guarantee that at least three cards of one suit are chosen. One good way to think about this is to note that after the eighth card is chosen, there is no way to avoid having a third card of some suit.

b) We do not use the generalized pigeonhole principle to answer this question, because we want to make sure that there are three hearts, not just three cards of one suit. Note that in the worst case, we can select all the clubs, diamonds, and spades, 39 cards in all, before we select a single heart. The next three cards will be all hearts, so we may need to select 42 cards to get three hearts. ◀

EXAMPLE 8   What is the least number of area codes needed to guarantee that the 25 million phones in a state can be assigned distinct 10-digit telephone numbers? (Assume that telephone numbers are of the form NXX-NXX-XXXX, where the first three digits form the area code, N represents a digit from 2 to 9 inclusive, and X represents any digit.)

Solution: There are eight million different phone numbers of the form NXX-XXXX (as shown in Example 8 of Section 6.1). Hence, by the generalized pigeonhole principle, among 25 million telephones, at least $\lceil 25,000,000/8,000,000 \rceil = 4$ of them must have identical phone numbers. Hence, at least four area codes are required to ensure that all 10-digit numbers are different. ◀

Example 9, although not an application of the generalized pigeonhole principle, makes use of similar principles.

EXAMPLE 9   Suppose that a computer science laboratory has 15 workstations and 10 servers. A cable can be used to directly connect a workstation to a server. For each server, only one direct connection to that server can be active at any time. We want to guarantee that at any time any set of 10 or fewer workstations can simultaneously access different servers via direct connections. Although we could do this by connecting every workstation directly to every server (using 150 connections), what is the minimum number of direct connections needed to achieve this goal?

Solution: Suppose that we label the workstations $W_1, W_2, \ldots, W_{15}$ and the servers $S_1, S_2, \ldots, S_{10}$. Furthermore, suppose that we connect W_k to S_k for $k = 1, 2, \ldots, 10$ and each of $W_{11}, W_{12}, W_{13}, W_{14}$, and W_{15} to all 10 servers. We have a total of 60 direct connections. Clearly any set of 10 or fewer workstations can simultaneously access different servers. We see this by noting that if workstation W_j is included with $1 \leq j \leq 10$, it can access server S_j, and for each workstation W_k with $k \geq 11$ included, there must be a corresponding workstation W_j with $1 \leq j \leq 10$ not included, so W_k can access server S_j. (This follows because there are at least as many available servers S_j as there are workstations W_j with $1 \leq j \leq 10$ not included.)

Now suppose there are fewer than 60 direct connections between workstations and servers. Then some server would be connected to at most $\lfloor 59/10 \rfloor = 5$ workstations. (If all servers were connected to at least six workstations, there would be at least $6 \cdot 10 = 60$ direct connections.) This means that the remaining nine servers are not enough to allow the other 10 workstations to simultaneously access different servers. Consequently, at least 60 direct connections are needed. It follows that 60 is the answer. ◄

Some Elegant Applications of the Pigeonhole Principle

In many interesting applications of the pigeonhole principle, the objects to be placed in boxes must be chosen in a clever way. A few such applications will be described here.

EXAMPLE 10   During a month with 30 days, a baseball team plays at least one game a day, but no more than 45 games. Show that there must be a period of some number of consecutive days during which the team must play exactly 14 games.

Solution: Let a_j be the number of games played on or before the jth day of the month. Then $a_1, a_2, \ldots, a_{30}$ is an increasing sequence of distinct positive integers, with $1 \leq a_j \leq 45$. Moreover, $a_1 + 14, a_2 + 14, \ldots, a_{30} + 14$ is also an increasing sequence of distinct positive integers, with $15 \leq a_j + 14 \leq 59$.

The 60 positive integers $a_1, a_2, \ldots, a_{30}, a_1 + 14, a_2 + 14, \ldots, a_{30} + 14$ are all less than or equal to 59. Hence, by the pigeonhole principle two of these integers are equal. Because the integers a_j, $j = 1, 2, \ldots, 30$ are all distinct and the integers $a_j + 14$, $j = 1, 2, \ldots, 30$ are all distinct, there must be indices i and j with $a_i = a_j + 14$. This means that exactly 14 games were played from day $j + 1$ to day i. ◄

EXAMPLE 11   Show that among any $n + 1$ positive integers not exceeding $2n$ there must be an integer that divides one of the other integers.

Solution: Write each of the $n + 1$ integers $a_1, a_2, \ldots, a_{n+1}$ as a power of 2 times an odd integer. In other words, let $a_j = 2^{k_j} q_j$ for $j = 1, 2, \ldots, n + 1$, where k_j is a nonnegative integer and q_j is odd. The integers $q_1, q_2, \ldots, q_{n+1}$ are all odd positive integers less than $2n$. Because there are only n odd positive integers less than $2n$, it follows from the pigeonhole principle that two of the integers $q_1, q_2, \ldots, q_{n+1}$ must be equal. Therefore, there are distinct integers i and j such that $q_i = q_j$. Let q be the common value of q_i and q_j. Then, $a_i = 2^{k_i} q$ and $a_j = 2^{k_j} q$. It follows that if $k_i < k_j$, then a_i divides a_j; while if $k_i > k_j$, then a_j divides a_i. ◄

A clever application of the pigeonhole principle shows the existence of an increasing or a decreasing subsequence of a certain length in a sequence of distinct integers. We review some definitions before this application is presented. Suppose that $a_1, a_2, \ldots, a_N$ is a sequence of real numbers. A **subsequence** of this sequence is a sequence of the form $a_{i_1}, a_{i_2}, \ldots, a_{i_m}$, where $1 \leq i_1 < i_2 < \cdots < i_m \leq N$. Hence, a subsequence is a sequence obtained from the original sequence by including some of the terms of the original sequence in their original order, and perhaps not including other terms. A sequence is called **strictly increasing** if each term is larger than the one that precedes it, and it is called **strictly decreasing** if each term is smaller than the one that precedes it.

THEOREM 3   Every sequence of $n^2 + 1$ distinct real numbers contains a subsequence of length $n + 1$ that is either strictly increasing or strictly decreasing.

We give an example before presenting the proof of Theorem 3.

EXAMPLE 12   The sequence 8, 11, 9, 1, 4, 6, 12, 10, 5, 7 contains 10 terms. Note that $10 = 3^2 + 1$. There are four strictly increasing subsequences of length four, namely, 1, 4, 6, 12; 1, 4, 6, 7; 1, 4, 6, 10; and 1, 4, 5, 7. There is also a strictly decreasing subsequence of length four, namely, 11, 9, 6, 5.   ◀

The proof of the theorem will now be given.

Proof: Let $a_1, a_2, \ldots, a_{n^2+1}$ be a sequence of $n^2 + 1$ distinct real numbers. Associate an ordered pair with each term of the sequence, namely, associate (i_k, d_k) to the term a_k, where i_k is the length of the longest increasing subsequence starting at a_k, and d_k is the length of the longest decreasing subsequence starting at a_k.

Suppose that there are no increasing or decreasing subsequences of length $n + 1$. Then i_k and d_k are both positive integers less than or equal to n, for $k = 1, 2, \ldots, n^2 + 1$. Hence, by the product rule there are n^2 possible ordered pairs for (i_k, d_k). By the pigeonhole principle, two of these $n^2 + 1$ ordered pairs are equal. In other words, there exist terms a_s and a_t, with $s < t$ such that $i_s = i_t$ and $d_s = d_t$. We will show that this is impossible. Because the terms of the sequence are distinct, either $a_s < a_t$ or $a_s > a_t$. If $a_s < a_t$, then, because $i_s = i_t$, an increasing subsequence of length $i_t + 1$ can be built starting at a_s, by taking a_s followed by an increasing subsequence of length i_t beginning at a_t. This is a contradiction. Similarly, if $a_s > a_t$, the same reasoning shows that d_s must be greater than d_t, which is a contradiction.   ◁

Links   The final example shows how the generalized pigeonhole principle can be applied to an important part of combinatorics called **Ramsey theory,** after the English mathematician F. P. Ramsey. In general, Ramsey theory deals with the distribution of subsets of elements of sets.

EXAMPLE 13   Assume that in a group of six people, each pair of individuals consists of two friends or two enemies. Show that there are either three mutual friends or three mutual enemies in the group.

Solution: Let A be one of the six people. Of the five other people in the group, there are either three or more who are friends of A, or three or more who are enemies of A. This follows from the generalized pigeonhole principle, because when five objects are divided into two sets, one of the sets has at least $\lceil 5/2 \rceil = 3$ elements. In the former case, suppose that B, C, and D are friends of A. If any two of these three individuals are friends, then these two and A form a group of three mutual friends. Otherwise, B, C, and D form a set of three mutual enemies. The proof in the latter case, when there are three or more enemies of A, proceeds in a similar manner.   ◀

The **Ramsey number** $R(m, n)$, where m and n are positive integers greater than or equal to 2, denotes the minimum number of people at a party such that there are either m mutual friends or n mutual enemies, assuming that every pair of people at the party are friends or enemies. Example 13 shows that $R(3, 3) \leq 6$. We conclude that $R(3, 3) = 6$ because in a group of five people where every two people are friends or enemies, there may not be three mutual friends or three mutual enemies (see Exercise 20).

It is possible to prove some useful properties about Ramsey numbers, but for the most part it is difficult to find their exact values. Note that by symmetry it can be shown that $R(m, n) = R(n, m)$ (see Exercise 24). We also have $R(2, n) = n$ for every positive integer $n \geq 2$ (see Exercise 23). The exact values of only nine Ramsey numbers $R(m, n)$ with $3 \leq m \leq n$ are known, including $R(4, 4) = 18$. Only bounds are known for many other Ramsey numbers, including $R(5, 5)$, which is known to satisfy $43 \leq R(5, 5) \leq 49$. The reader interested in learning more about Ramsey numbers should consult [MiRo91] or [GrRoSp90].

Exercises

1. Show that in any set of six classes, each meeting regularly once a week on a particular day of the week, there must be two that meet on the same day, assuming that no classes are held on weekends.

2. Show that if there are 30 students in a class, then at least two have last names that begin with the same letter.

3. A drawer contains a dozen brown socks and a dozen black socks, all unmatched. A man takes socks out at random in the dark.
   a) How many socks must he take out to be sure that he has at least two socks of the same color?
   b) How many socks must he take out to be sure that he has at least two black socks?

4. A bowl contains 10 red balls and 10 blue balls. A woman selects balls at random without looking at them.
   a) How many balls must she select to be sure of having at least three balls of the same color?
   b) How many balls must she select to be sure of having at least three blue balls?

5. Let n be a positive integer. Show that in any set of n consecutive integers there is exactly one divisible by n.

6. Show that if f is a function from S to T, where S and T are finite sets with $|S| > |T|$, then there are elements s_1 and s_2 in S such that $f(s_1) = f(s_2)$, or in other words, f is not one-to-one.

7. What is the minimum number of students, each of whom comes from one of the 50 states, who must be enrolled in a university to guarantee that there are at least 100 who come from the same state?

∗8. Let $(x_i, y_i), i = 1, 2, 3, 4, 5$, be a set of five distinct points with integer coordinates in the xy plane. Show that the midpoint of the line joining at least one pair of these points has integer coordinates.

∗9. Let $(x_i, y_i, z_i), i = 1, 2, 3, 4, 5, 6, 7, 8, 9$, be a set of nine distinct points with integer coordinates in xyz space. Show that the midpoint of at least one pair of these points has integer coordinates.

10. How many ordered pairs of integers (a, b) are needed to guarantee that there are two ordered pairs (a_1, b_1) and (a_2, b_2) such that $a_1 \bmod 5 = a_2 \bmod 5$ and $b_1 \bmod 5 = b_2 \bmod 5$?

11. a) Show that if five integers are selected from the first eight positive integers, there must be a pair of these integers with a sum equal to 9.
    b) Is the conclusion in part (a) true if four integers are selected rather than five?

12. a) Show that if seven integers are selected from the first 10 positive integers, there must be at least two pairs of these integers with the sum 11.
    b) Is the conclusion in part (a) true if six integers are selected rather than seven?

13. A company stores products in a warehouse. Storage bins in this warehouse are specified by their aisle, location in the aisle, and shelf. There are 50 aisles, 85 horizontal locations in each aisle, and 5 shelves throughout the warehouse. What is the least number of products the company can have so that at least two products must be stored in the same bin?

14. Suppose that there are nine students in a discrete mathematics class at a small college.
    a) Show that the class must have at least five male students or at least five female students.
    b) Show that the class must have at least three male students or at least seven female students.

15. Suppose that every student in a discrete mathematics class of 25 students is a freshman, a sophomore, or a junior.
    a) Show that there are at least nine freshmen, at least nine sophomores, or at least nine juniors in the class.
    b) Show that there are either at least three freshmen, at least 19 sophomores, or at least five juniors in the class.

16. Find an increasing subsequence of maximal length and a decreasing subsequence of maximal length in the sequence 22, 5, 7, 2, 23, 10, 15, 21, 3, 17.

17. Construct a sequence of 16 positive integers that has no increasing or decreasing subsequence of five terms.

18. Show that if there are 101 people of different heights standing in a line, it is possible to find 11 people in the order they are standing in the line with heights that are either increasing or decreasing.

∗19. Show that whenever 25 girls and 25 boys are seated around a circular table there is always a person both of whose neighbors are boys.

20. Show that in a group of five people (where any two people are either friends or enemies), there are not necessarily three mutual friends or three mutual enemies.

21. Show that in a group of 10 people (where any two people are either friends or enemies), there are either three mutual friends or four mutual enemies, and there are either three mutual enemies or four mutual friends.

22. Use Exercise 21 to show that among any group of 20 people (where any two people are either friends or enemies), there are either four mutual friends or four mutual enemies.

23. Show that if n is an integer with $n \geq 2$, then the Ramsey number $R(2, n)$ equals n. (Recall that Ramsey numbers were discussed after Example 13 in Section 6.2.)

24. Show that if m and n are integers with $m \geq 2$ and $n \geq 2$, then the Ramsey numbers $R(m, n)$ and $R(n, m)$ are equal. (Recall that Ramsey numbers were discussed after Example 13 in Section 6.2.)

25. Show that there are at least six people in California (population: 37 million) with the same three initials who were born on the same day of the year (but not necessarily in the same year). Assume that everyone has three initials.

26. In the 17th century, there were more than 800,000 inhabitants of Paris. At the time, it was believed that no one had more than 200,000 hairs on their head. Assuming these numbers are correct and that everyone has at least one hair on their head (that is, no one is completely bald), use the pigeonhole principle to show, as the French writer Pierre Nicole did, that there had to be two Parisians with the same number of hairs on their heads. Then use the generalized pigeonhole principle to show that there had to be at least five Parisians at that time with the same number of hairs on their heads.

27. There are 38 different time periods during which classes at a university can be scheduled. If there are 677 different classes, how many different rooms will be needed?

28. A computer network consists of six computers. Each computer is directly connected to at least one of the other computers. Show that there are at least two computers in the network that are directly connected to the same number of other computers.

29. A computer network consists of six computers. Each computer is directly connected to zero or more of the other computers. Show that there are at least two computers in the network that are directly connected to the same number of other computers. [*Hint:* It is impossible to have a computer linked to none of the others and a computer linked to all the others.]

30. Find the least number of cables required to connect eight computers to four printers to guarantee that for every choice of four of the eight computers, these four computers can directly access four different printers. Justify your answer.

31. Find the least number of cables required to connect 100 computers to 20 printers to guarantee that 2every subset of 20 computers can directly access 20 different printers. (Here, the assumptions about cables and computers are the same as in Example 9.) Justify your answer.

*32. Prove that at a party where there are at least two people, there are two people who know the same number of other people there.

33. An arm wrestler is the champion for a period of 75 hours. (Here, by an hour, we mean a period starting from an exact hour, such as 1 P.M., until the next hour.) The arm wrestler had at least one match an hour, but no more than 125 total matches. Show that there is a period of consecutive hours during which the arm wrestler had exactly 24 matches.

*34. Is the statement in Exercise 41 true if 24 is replaced by
   a) 2?   **b)** 23?   **c)** 25?   **d)** 30?

35. There are 51 houses on a street. Each house has an address between 1000 and 1099, inclusive. Show that at least two houses have addresses that are consecutive integers.

*36. An alternative proof of Theorem 3 based on the generalized pigeonhole principle is outlined in this exercise. The notation used is the same as that used in the proof in the text.

   a) Assume that $i_k \leq n$ for $k = 1, 2, \ldots, n^2 + 1$. Use the generalized pigeonhole principle to show that there are $n + 1$ terms $a_{k_1}, a_{k_2}, \ldots, a_{k_{n+1}}$ with $i_{k_1} = i_{k_2} = \cdots = i_{k_{n+1}}$, where $1 \leq k_1 < k_2 < \cdots < k_{n+1}$.

   b) Show that $a_{k_j} > a_{k_{j+1}}$ for $j = 1, 2, \ldots, n$. [*Hint:* Assume that $a_{k_j} < a_{k_{j+1}}$, and show that this implies that $i_{k_j} > i_{k_{j+1}}$, which is a contradiction.]

   c) Use parts (a) and (b) to show that if there is no increasing subsequence of length $n + 1$, then there must be a decreasing subsequence of this length.

6.3   Permutations and Combinations

Introduction

Many counting problems can be solved by finding the number of ways to arrange a specified number of distinct elements of a set of a particular size, where the order of these elements matters. Many other counting problems can be solved by finding the number of ways to select a particular number of elements from a set of a particular size, where the order of the elements selected does not matter. For example, in how many ways can we select three students from a group of five students to stand in line for a picture? How many different committees of three students can be formed from a group of four students? In this section we will develop methods to answer questions such as these.

Permutations

We begin by solving the first question posed in the introduction to this section, as well as related questions.

EXAMPLE 1   In how many ways can we select three students from a group of five students to stand in line for a picture? In how many ways can we arrange all five of these students in a line for a picture?

Solution: First, note that the order in which we select the students matters. There are five ways to select the first student to stand at the start of the line. Once this student has been selected, there are four ways to select the second student in the line. After the first and second students have been selected, there are three ways to select the third student in the line. By the product rule, there are $5 \cdot 4 \cdot 3 = 60$ ways to select three students from a group of five students to stand in line for a picture.

To arrange all five students in a line for a picture, we select the first student in five ways, the second in four ways, the third in three ways, the fourth in two ways, and the fifth in one way. Consequently, there are $5 \cdot 4 \cdot 3 \cdot 2 \cdot 1 = 120$ ways to arrange all five students in a line for a picture. ◄

Example 1 illustrates how ordered arrangements of distinct objects can be counted. This leads to some terminology.

A **permutation** of a set of distinct objects is an ordered arrangement of these objects. We also are interested in ordered arrangements of some of the elements of a set. An ordered arrangement of r elements of a set is called an ***r*-permutation**.

EXAMPLE 2   Let $S = \{1, 2, 3\}$. The ordered arrangement 3, 1, 2 is a permutation of S. The ordered arrangement 3, 2 is a 2-permutation of S. ◄

The number of r-permutations of a set with n elements is denoted by $P(n, r)$. We can find $P(n, r)$ using the product rule.

EXAMPLE 3   Let $S = \{a, b, c\}$. The 2-permutations of S are the ordered arrangements $a, b; a, c; b, a; b, c; c, a;$ and c, b. Consequently, there are six 2-permutations of this set with three elements. There are always six 2-permutations of a set with three elements. There are three ways to choose the first element of the arrangement. There are two ways to choose the second element of the arrangement, because it must be different from the first element. Hence, by the product rule, we see that $P(3, 2) = 3 \cdot 2 = 6$. the first element. By the product rule, it follows that $P(3, 2) = 3 \cdot 2 = 6$. ◄

We now use the product rule to find a formula for $P(n, r)$ whenever n and r are positive integers with $1 \le r \le n$.

THEOREM 1   If n is a positive integer and r is an integer with $1 \le r \le n$, then there are

$$P(n, r) = n(n - 1)(n - 2) \cdots (n - r + 1)$$

r-permutations of a set with n distinct elements.

Proof: We will use the product rule to prove that this formula is correct. The first element of the permutation can be chosen in n ways because there are n elements in the set. There are $n - 1$

ways to choose the second element of the permutation, because there are $n - 1$ elements left in the set after using the element picked for the first position. Similarly, there are $n - 2$ ways to choose the third element, and so on, until there are exactly $n - (r - 1) = n - r + 1$ ways to choose the rth element. Consequently, by the product rule, there are

$$n(n - 1)(n - 2) \cdots (n - r + 1)$$

r-permutations of the set.   ◁

Note that $P(n, 0) = 1$ whenever n is a nonnegative integer because there is exactly one way to order zero elements. That is, there is exactly one list with no elements in it, namely the empty list.

We now state a useful corollary of Theorem 1.

COROLLARY 1   If n and r are integers with $0 \leq r \leq n$, then $P(n, r) = \dfrac{n!}{(n - r)!}$.

Proof: When n and r are integers with $1 \leq r \leq n$, by Theorem 1 we have

$$P(n, r) = n(n - 1)(n - 2) \cdots (n - r + 1) = \frac{n!}{(n - r)!}$$

Because $\dfrac{n!}{(n - 0)!} = \dfrac{n!}{n!} = 1$ whenever n is a nonnegative integer, we see that the formula $P(n, r) = \dfrac{n!}{(n - r)!}$ also holds when $r = 0$.   ◁

By Theorem 1 we know that if n is a positive integer, then $P(n, n) = n!$. We will illustrate this result with some examples.

EXAMPLE 4   How many ways are there to select a first-prize winner, a second-prize winner, and a third-prize winner from 100 different people who have entered a contest?

Solution: Because it matters which person wins which prize, the number of ways to pick the three prize winners is the number of ordered selections of three elements from a set of 100 elements, that is, the number of 3-permutations of a set of 100 elements. Consequently, the answer is

$$P(100, 3) = 100 \cdot 99 \cdot 98 = 970,200.$$ ◀

EXAMPLE 5   Suppose that there are eight runners in a race. The winner receives a gold medal, the second-place finisher receives a silver medal, and the third-place finisher receives a bronze medal. How many different ways are there to award these medals, if all possible outcomes of the race can occur and there are no ties?

Solution: The number of different ways to award the medals is the number of 3-permutations of a set with eight elements. Hence, there are $P(8, 3) = 8 \cdot 7 \cdot 6 = 336$ possible ways to award the medals.   ◀

EXAMPLE 6   Suppose that a saleswoman has to visit eight different cities. She must begin her trip in a specified city, but she can visit the other seven cities in any order she wishes. How many possible orders can the saleswoman use when visiting these cities?

Solution: The number of possible paths between the cities is the number of permutations of seven elements, because the first city is determined, but the remaining seven can be ordered arbitrarily. Consequently, there are $7! = 7 \cdot 6 \cdot 5 \cdot 4 \cdot 3 \cdot 2 \cdot 1 = 5040$ ways for the saleswoman to choose her tour. If, for instance, the saleswoman wishes to find the path between the cities with minimum distance, and she computes the total distance for each possible path, she must consider a total of 5040 paths! ◄

EXAMPLE 7   How many permutations of the letters *ABCDEFGH* contain the string *ABC* ?

Solution: Because the letters *ABC* must occur as a block, we can find the answer by finding the number of permutations of six objects, namely, the block *ABC* and the individual letters *D*, *E*, *F*, *G*, and *H*. Because these six objects can occur in any order, there are $6! = 720$ permutations of the letters *ABCDEFGH* in which *ABC* occurs as a block. ◄

Combinations

We now turn our attention to counting unordered selections of objects. We begin by solving a question posed in the introduction to this section of the chapter.

EXAMPLE 8   How many different committees of three students can be formed from a group of four students?

Solution: To answer this question, we need only find the number of subsets with three elements from the set containing the four students. We see that there are four such subsets, one for each of the four students, because choosing three students is the same as choosing one of the four students to leave out of the group. This means that there are four ways to choose the three students for the committee, where the order in which these students are chosen does not matter. ◄

Example 8 illustrates that many counting problems can be solved by finding the number of subsets of a particular size of a set with *n* elements, where *n* is a positive integer.

An ***r*-combination** of elements of a set is an unordered selection of *r* elements from the set. Thus, an *r*-combination is simply a subset of the set with *r* elements.

EXAMPLE 9   Let *S* be the set $\{1, 2, 3, 4\}$. Then $\{1, 3, 4\}$ is a 3-combination from *S*. (Note that $\{4, 1, 3\}$ is the same 3-combination as $\{1, 3, 4\}$, because the order in which the elements of a set are listed does not matter.) ◄

The number of *r*-combinations of a set with *n* distinct elements is denoted by $C(n, r)$. Note that $C(n, r)$ is also denoted by $\binom{n}{r}$ and is called a **binomial coefficient**. We will learn where this terminology comes from in Section 6.4.

EXAMPLE 10   We see that $C(4, 2) = 6$, because the 2-combinations of $\{a, b, c, d\}$ are the six subsets $\{a, b\}$, $\{a, c\}$, $\{a, d\}$, $\{b, c\}$, $\{b, d\}$, and $\{c, d\}$. ◄

We can determine the number of *r*-combinations of a set with *n* elements using the formula for the number of *r*-permutations of a set. To do this, note that the *r*-permutations of a set can be obtained by first forming *r*-combinations and then ordering the elements in these combinations. The proof of Theorem 2, which gives the value of $C(n, r)$, is based on this observation.

THEOREM 2   The number of r-combinations of a set with n elements, where n is a nonnegative integer and r is an integer with $0 \leq r \leq n$, equals

$$C(n, r) = \frac{n!}{r!\,(n-r)!}.$$

Proof: The $P(n, r)$ r-permutations of the set can be obtained by forming the $C(n, r)$ r-combinations of the set, and then ordering the elements in each r-combination, which can be done in $P(r, r)$ ways. Consequently, by the product rule,

$$P(n, r) = C(n, r) \cdot P(r, r).$$

This implies that

$$C(n, r) = \frac{P(n, r)}{P(r, r)} = \frac{n!/(n-r)!}{r!/(r-r)!} = \frac{n!}{r!\,(n-r)!}.$$

We can also use the division rule for counting to construct a proof of this theorem. Because the order of elements in a combination does not matter and there are $P(r, r)$ ways to order r elements in an r-combination of n elements, each of the $C(n, r)$ r-combinations of a set with n elements corresponds to exactly $P(r, r)$ r-permutations. Hence, by the division rule, $C(n, r) = \frac{P(n,r)}{P(r,r)}$, which implies as before that $C(n, r) = \frac{n!}{r!\,(n-r)!}$.   ◁

The formula in Theorem 2, although explicit, is not helpful when $C(n, r)$ is computed for large values of n and r. The reasons are that it is practical to compute exact values of factorials exactly only for small integer values, and when floating point arithmetic is used, the formula in Theorem 2 may produce a value that is not an integer. When computing $C(n, r)$, first note that when we cancel out $(n - r)!$ from the numerator and denominator of the expression for $C(n, r)$ in Theorem 2, we obtain

$$C(n, r) = \frac{n!}{r!\,(n-r)!} = \frac{n(n-1)\cdots(n-r+1)}{r!}.$$

Consequently, to compute $C(n, r)$ you can cancel out all the terms in the larger factorial in the denominator from the numerator and denominator, then multiply all the terms that do not cancel in the numerator and finally divide by the smaller factorial in the denominator. [When doing this calculation by hand, instead of by machine, it is also worthwhile to factor out common factors in the numerator $n(n - 1) \cdots (n - r + 1)$ and in the denominator $r!$.] Note that many calculators have a built-in function for $C(n, r)$ that can be used for relatively small values of n and r and many computational programs can be used to find $C(n, r)$. [Such functions may be called *choose*(n, k) or *binom*(n, k)].

Example 11 illustrates how $C(n, k)$ is computed when k is relatively small compared to n and when k is close to n. It also illustrates a key identity enjoyed by the numbers $C(n, k)$.

EXAMPLE 11   How many poker hands of five cards can be dealt from a standard deck of 52 cards? Also, how many ways are there to select 47 cards from a standard deck of 52 cards?

Solution: Because the order in which the five cards are dealt from a deck of 52 cards does not matter, there are

$$C(52, 5) = \frac{52!}{5!\,47!}$$

different hands of five cards that can be dealt. To compute the value of $C(52, 5)$, first divide the numerator and denominator by 47! to obtain

$$C(52, 5) = \frac{52 \cdot 51 \cdot 50 \cdot 49 \cdot 48}{5 \cdot 4 \cdot 3 \cdot 2 \cdot 1}.$$

This expression can be simplified by first dividing the factor 5 in the denominator into the factor 50 in the numerator to obtain a factor 10 in the numerator, then dividing the factor 4 in the denominator into the factor 48 in the numerator to obtain a factor of 12 in the numerator, then dividing the factor 3 in the denominator into the factor 51 in the numerator to obtain a factor of 17 in the numerator, and finally, dividing the factor 2 in the denominator into the factor 52 in the numerator to obtain a factor of 26 in the numerator. We find that

$$C(52, 5) = 26 \cdot 17 \cdot 10 \cdot 49 \cdot 12 = 2{,}598{,}960.$$

Consequently, there are 2,598,960 different poker hands of five cards that can be dealt from a standard deck of 52 cards.

Note that there are

$$C(52, 47) = \frac{52!}{47!5!}$$

different ways to select 47 cards from a standard deck of 52 cards. We do not need to compute this value because $C(52, 47) = C(52, 5)$. (Only the order of the factors 5! and 47! is different in the denominators in the formulae for these quantities.) It follows that there are also 2,598,960 different ways to select 47 cards from a standard deck of 52 cards. ◀

In Example 11 we observed that $C(52, 5) = C(52, 47)$. This is a special case of the useful identity for the number of r-combinations of a set given in Corollary 2.

COROLLARY 2   Let n and r be nonnegative integers with $r \leq n$. Then $C(n, r) = C(n, n - r)$.

Proof: From Theorem 2 it follows that

$$C(n, r) = \frac{n!}{r! \, (n - r)!}$$

and

$$C(n, n - r) = \frac{n!}{(n - r)! \, [n - (n - r)]!} = \frac{n!}{(n - r)! \, r!}.$$

Hence, $C(n, r) = C(n, n - r)$. ◁

We can also prove Corollary 2 without relying on algebraic manipulation. Instead, we can use a combinatorial proof. We describe this important type of proof in Definition 1.

DEFINITION 1   A *combinatorial proof* of an identity is a proof that uses counting arguments to prove that both sides of the identity count the same objects but in different ways or a proof that is based on showing that there is a bijection between the sets of objects counted by the two sides of the identity. These two types of proofs are called *double counting proofs* and *bijective proofs*, respectively.

Many identities involving binomial coefficients can be proved using combinatorial proofs. We now show how to prove Corollary 2 using a combinatorial proof. We will provide both a double counting proof and a bijective proof, both based on the same basic idea.

Combinatorial proofs are almost always much shorter and provide more insights than proofs based on algebraic manipulation.

Proof: We will use a bijective proof to show that $C(n, r) = C(n, n - r)$ for all integers n and r with $0 \le r \le n$. Suppose that S is a set with n elements. The function that maps a subset A of S to $\overline{A}$ is a bijection between subsets of S with r elements and subsets with $n - r$ elements (as the reader should verify). The identity $C(n, r) = C(n, n - r)$ follows because when there is a bijection between two finite sets, the two sets must have the same number of elements.

Alternatively, we can reformulate this argument as a double counting proof. By definition, the number of subsets of S with r elements equals $C(n, r)$. But each subset A of S is also determined by specifying which elements are not in A, and so are in $\overline{A}$. Because the complement of a subset of S with r elements has $n - r$ elements, there are also $C(n, n - r)$ subsets of S with r elements. It follows that $C(n, r) = C(n, n - r)$. ◁

EXAMPLE 12

How many ways are there to select five players from a 10-member tennis team to make a trip to a match at another school?

Solution: The answer is given by the number of 5-combinations of a set with 10 elements. By Theorem 2, the number of such combinations is

$$C(10, 5) = \frac{10!}{5!\,5!} = 252.$$ ◀

EXAMPLE 13

A group of 30 people have been trained as astronauts to go on the first mission to Mars. How many ways are there to select a crew of six people to go on this mission (assuming that all crew members have the same job)?

Solution: The number of ways to select a crew of six from the pool of 30 people is the number of 6-combinations of a set with 30 elements, because the order in which these people are chosen does not matter. By Theorem 2, the number of such combinations is

$$C(30, 6) = \frac{30!}{6!\,24!} = \frac{30 \cdot 29 \cdot 28 \cdot 27 \cdot 26 \cdot 25}{6 \cdot 5 \cdot 4 \cdot 3 \cdot 2 \cdot 1} = 593,775.$$ ◀

EXAMPLE 14

How many bit strings of length n contain exactly r 1s?

Solution: The positions of r 1s in a bit string of length n form an r-combination of the set $\{1, 2, 3, \ldots, n\}$. Hence, there are $C(n, r)$ bit strings of length n that contain exactly r 1s. ◀

EXAMPLE 15

Suppose that there are 9 faculty members in the mathematics department and 11 in the computer science department. How many ways are there to select a committee to develop a discrete mathematics course at a school if the committee is to consist of three faculty members from the mathematics department and four from the computer science department?

Solution: By the product rule, the answer is the product of the number of 3-combinations of a set with nine elements and the number of 4-combinations of a set with 11 elements. By Theorem 2, the number of ways to select the committee is

$$C(9, 3) \cdot C(11, 4) = \frac{9!}{3!6!} \cdot \frac{11!}{4!7!} = 84 \cdot 330 = 27,720.$$ ◀

Exercises

1. List all the permutations of $\{a, b, c\}$.

2. How many different permutations are there of the set $\{a, b, c, d, e, f, g\}$?

3. How many permutations of $\{a, b, c, d, e, f, g\}$ end with a?

4. Let $S = \{1, 2, 3, 4, 5\}$.
   a) List all the 3-permutations of S.
   b) List all the 3-combinations of S.

5. Find the number of 5-permutations of a set with nine elements.

6. How many possibilities are there for the win, place, and show (first, second, and third) positions in a horse race with 12 horses if all orders of finish are possible?

7. How many bit strings of length 10 contain
   a) exactly four 1s?
   b) at most four 1s?
   c) at least four 1s?
   d) an equal number of 0s and 1s?

8. A group contains n men and n women. How many ways are there to arrange these people in a row if the men and women alternate?

9. In how many ways can a set of five letters be selected from the English alphabet?

10. How many subsets with an odd number of elements does a set with 10 elements have?

11. How many subsets with more than two elements does a set with 100 elements have?

12. A coin is flipped 10 times where each flip comes up either heads or tails. How many possible outcomes
    a) are there in total?
    b) contain exactly two heads?
    c) contain at most three tails?
    d) contain the same number of heads and tails?

13. How many permutations of the letters $ABCDEFG$ contain
    a) the string BCD?
    b) the string $CFGA$?
    c) the strings BA and GF?
    d) the strings ABC and DE?
    e) the strings ABC and CDE?
    f) the strings CBA and BED?

14. How many ways are there for eight men and five women to stand in a line so that no two women stand next to each other? [*Hint:* First position the men and then consider possible positions for the women.]

15. One hundred tickets, numbered 1, 2, 3, . . . , 100, are sold to 100 different people for a drawing. Four different prizes are awarded, including a grand prize (a trip to Tahiti). How many ways are there to award the prizes if
    a) there are no restrictions?
    b) the person holding ticket 47 wins the grand prize?
    c) the person holding ticket 47 wins one of the prizes?
    d) the person holding ticket 47 does not win a prize?

e) the people holding tickets 19 and 47 both win prizes?

f) the people holding tickets 19, 47, and 73 all win prizes?

g) the people holding tickets 19, 47, 73, and 97 all win prizes?

h) none of the people holding tickets 19, 47, 73, and 97 wins a prize?

i) the grand prize winner is a person holding ticket 19, 47, 73, or 97?

j) the people holding tickets 19 and 47 win prizes, but the people holding tickets 73 and 97 do not win prizes?

16. Thirteen people on a softball team show up for a game.
    a) How many ways are there to choose 10 players to take the field?
    b) How many ways are there to assign the 10 positions by selecting players from the 13 people who show up?
    c) Of the 13 people who show up, three are women. How many ways are there to choose 10 players to take the field if at least one of these players must be a woman?

17. A club has 25 members.
    a) How many ways are there to choose four members of the club to serve on an executive committee?
    b) How many ways are there to choose a president, vice president, secretary, and treasurer of the club, where no person can hold more than one office?

18. A professor writes 40 discrete mathematics true/false questions. Of the statements in these questions, 17 are true. If the questions can be positioned in any order, how many different answer keys are possible?

*19. How many 4-permutations of the positive integers not exceeding 100 contain three consecutive integers $k, k + 1, k + 2$, in the correct order
    a) where these consecutive integers can perhaps be separated by other integers in the permutation?
    b) where they are in consecutive positions in the permutation?

20. Seven women and nine men are on the faculty in the mathematics department at a school.
    a) How many ways are there to select a committee of five members of the department if at least one woman must be on the committee?
    b) How many ways are there to select a committee of five members of the department if at least one woman and at least one man must be on the committee?

21. The English alphabet contains 21 consonants and five vowels. How many strings of six lowercase letters of the English alphabet contain
    a) exactly one vowel?
    b) exactly two vowels?
    c) at least one vowel?
    d) at least two vowels?

22. How many strings of six lowercase letters from the English alphabet contain
    a) the letter a?

b) the letters a and b?

c) the letters a and b in consecutive positions with a preceding b, with all the letters distinct?

d) the letters a and b, where a is somewhere to the left of b in the string, with all the letters distinct?

23. Suppose that a department contains 10 men and 15 women. How many ways are there to form a committee with six members if it must have the same number of men and women?

24. Suppose that a department contains 10 men and 15 women. How many ways are there to form a committee with six members if it must have more women than men?

25. How many bit strings contain exactly eight 0s and 10 1s if every 0 must be immediately followed by a 1?

26. How many bit strings contain exactly five 0s and 14 1s if every 0 must be immediately followed by two 1s?

27. How many bit strings of length 10 contain at least three 1s and at least three 0s?

28. How many ways are there to select 12 countries in the United Nations to serve on a council if 3 are selected from a block of 45, 4 are selected from a block of 57, and the others are selected from the remaining 69 countries?

29. How many license plates consisting of three letters followed by three digits contain no letter or digit twice?

A **circular r-permutation of n** people is a seating of r of these n people around a circular table, where seatings are considered to be the same if they can be obtained from each other by rotating the table.

30. Find the number of circular 3-permutations of 5 people.

31. Find a formula for the number of circular r-permutations of n people.

32. Find a formula for the number of ways to seat r of n people around a circular table, where seatings are considered the same if every person has the same two neighbors without regard to which side these neighbors are sitting on.

33. How many ways are there for a horse race with three horses to finish if ties are possible? [*Note:* Two or three horses may tie.]

∗34. There are six runners in the 100-yard dash. How many ways are there for three medals to be awarded if ties are possible? (The runner or runners who finish with the fastest time receive gold medals, the runner or runners who finish with exactly one runner ahead receive silver medals, and the runner or runners who finish with exactly two runners ahead receive bronze medals.)

6.4   Binomial Coefficients and Identities

As we remarked in Section 6.3, the number of r-combinations from a set with n elements is often denoted by $\binom{n}{r}$. This number is also called a **binomial coefficient** because these numbers occur as coefficients in the expansion of powers of binomial expressions such as $(a + b)^n$. We will discuss the **binomial theorem**, which gives a power of a binomial expression as a sum of terms involving binomial coefficients. We will prove this theorem using a combinatorial proof. We will also show how combinatorial proofs can be used to establish some of the many different identities that express relationships among binomial coefficients.

The Binomial Theorem

Links

The binomial theorem gives the coefficients of the expansion of powers of binomial expressions. A **binomial** expression is simply the sum of two terms, such as $x + y$. (The terms can be products of constants and variables, but that does not concern us here.)

Example 1 illustrates how the coefficients in a typical expansion can be found and prepares us for the statement of the binomial theorem.

EXAMPLE 1   The expansion of $(x + y)^3$ can be found using combinatorial reasoning instead of multiplying the three terms out. When $(x + y)^3 = (x + y)(x + y)(x + y)$ is expanded, all products of a term in the first sum, a term in the second sum, and a term in the third sum are added. Terms of the form x^3, $x^2 y$, xy^2, and y^3 arise. To obtain a term of the form x^3, an x must be chosen in each of the sums, and this can be done in only one way. Thus, the x^3 term in the product has a coefficient of 1. To obtain a term of the form $x^2 y$, an x must be chosen in two of the three sums (and consequently a y in the other sum). Hence, the number of such terms is the number of 2-combinations of three objects, namely, $\binom{3}{2}$. Similarly, the number of terms of the form xy^2 is the number of ways to pick one of the three sums to obtain an x (and consequently take a y from each of the other two sums). This can be done in $\binom{3}{1}$ ways. Finally, the only way to obtain a y^3 term is to choose the y for each of the three sums in the product, and this can be done in exactly one way. Consequently, it follows that

$$
\begin{aligned}
(x+y)^3 &= (x+y)(x+y)(x+y) = (xx+xy+yx+yy)(x+y) \\
&= xxx + xxy + xyx + xyy + yxx + yxy + yyx + yyy \\
&= x^3 + 3x^2y + 3xy^2 + y^3.
\end{aligned}
$$

◀

We now state the binomial theorem.

THEOREM 1   **THE BINOMIAL THEOREM**   Let x and y be variables, and let n be a nonnegative integer. Then

$$
(x+y)^n = \sum_{j=0}^{n} \binom{n}{j} x^{n-j} y^j = \binom{n}{0} x^n + \binom{n}{1} x^{n-1} y + \cdots + \binom{n}{n-1} x y^{n-1} + \binom{n}{n} y^n.
$$

Proof: We use a combinatorial proof. The terms in the product when it is expanded are of the form $x^{n-j} y^j$ for $j = 0, 1, 2, \ldots, n$. To count the number of terms of the form $x^{n-j} y^j$, note that to obtain such a term it is necessary to choose $n - j$ xs from the n sums (so that the other j terms in the product are ys). Therefore, the coefficient of $x^{n-j} y^j$ is $\binom{n}{n-j}$, which is equal to $\binom{n}{j}$. This proves the theorem.   ◁

Some computational uses of the binomial theorem are illustrated in Examples 2–4.

EXAMPLE 2   What is the expansion of $(x+y)^4$?

Solution: From the binomial theorem it follows that

$$
(x+y)^4 = \sum_{j=0}^{4} \binom{4}{j} x^{4-j} y^j
$$

$$
= \binom{4}{0} x^4 + \binom{4}{1} x^3 y + \binom{4}{2} x^2 y^2 + \binom{4}{3} x y^3 + \binom{4}{4} y^4
$$

$$
= x^4 + 4x^3 y + 6x^2 y^2 + 4x y^3 + y^4.
$$

◀

EXAMPLE 3   What is the coefficient of $x^{12} y^{13}$ in the expansion of $(x+y)^{25}$?

Solution: From the binomial theorem it follows that this coefficient is

$$
\binom{25}{13} = \frac{25!}{13!\,12!} = 5{,}200{,}300.
$$

◀

EXAMPLE 4   What is the coefficient of $x^{12} y^{13}$ in the expansion of $(2x - 3y)^{25}$?

Solution: First, note that this expression equals $(2x + (-3y))^{25}$. By the binomial theorem, we have

$$
(2x + (-3y))^{25} = \sum_{j=0}^{25} \binom{25}{j} (2x)^{25-j} (-3y)^j.
$$

Consequently, the coefficient of $x^{12}y^{13}$ in the expansion is obtained when $j = 13$, namely,

$$\binom{25}{13}2^{12}(-3)^{13} = -\frac{25!}{13!\,12!}2^{12}3^{13}.$$

◀

We can prove some useful identities using the binomial theorem, as Corollaries 1, 2, and 3 demonstrate.

COROLLARY 1   Let n be a nonnegative integer. Then

$$\sum_{k=0}^{n}\binom{n}{k} = 2^n.$$

Proof: Using the binomial theorem with $x = 1$ and $y = 1$, we see that

$$2^n = (1+1)^n = \sum_{k=0}^{n}\binom{n}{k}1^k 1^{n-k} = \sum_{k=0}^{n}\binom{n}{k}.$$

This is the desired result.   ◁

There is also a nice combinatorial proof of Corollary 1, which we now present.

Proof: A set with n elements has a total of 2^n different subsets. Each subset has zero elements, one element, two elements, $\ldots$, or n elements in it. There are $\binom{n}{0}$ subsets with zero elements, $\binom{n}{1}$ subsets with one element, $\binom{n}{2}$ subsets with two elements, $\ldots$, and $\binom{n}{n}$ subsets with n elements. Therefore,

$$\sum_{k=0}^{n}\binom{n}{k}$$

counts the total number of subsets of a set with n elements. By equating the two formulas we have for the number of subsets of a set with n elements, we see that

$$\sum_{k=0}^{n}\binom{n}{k} = 2^n.$$

◁

COROLLARY 2   Let n be a positive integer. Then

$$\sum_{k=0}^{n}(-1)^k\binom{n}{k} = 0.$$

Proof: When we use the binomial theorem with $x = -1$ and $y = 1$, we see that

$$0 = 0^n = ((-1)+1)^n = \sum_{k=0}^{n}\binom{n}{k}(-1)^k 1^{n-k} = \sum_{k=0}^{n}\binom{n}{k}(-1)^k.$$

This proves the corollary.   ◁

Remark: Corollary 2 implies that

$$\binom{n}{0} + \binom{n}{2} + \binom{n}{4} + \cdots = \binom{n}{1} + \binom{n}{3} + \binom{n}{5} + \cdots .$$

COROLLARY 3

Let n be a nonnegative integer. Then

$$\sum_{k=0}^{n} 2^k \binom{n}{k} = 3^n.$$

Proof: We recognize that the left-hand side of this formula is the expansion of $(1 + 2)^n$ provided by the binomial theorem. Therefore, by the binomial theorem, we see that

$$(1 + 2)^n = \sum_{k=0}^{n} \binom{n}{k} 1^{n-k} 2^k = \sum_{k=0}^{n} \binom{n}{k} 2^k.$$

Hence

$$\sum_{k=0}^{n} 2^k \binom{n}{k} = 3^n.$$

◁

Pascal's Identity and Triangle

The binomial coefficients satisfy many different identities. We introduce one of the most important of these now.

THEOREM 2

PASCAL'S IDENTITY   Let n and k be positive integers with $n \geq k$. Then

$$\binom{n+1}{k} = \binom{n}{k-1} + \binom{n}{k}.$$

Proof: We will use a combinatorial proof. Suppose that T is a set containing $n + 1$ elements. Let a be an element in T, and let $S = T - \{a\}$. Note that there are $\binom{n+1}{k}$ subsets of T containing k elements. However, a subset of T with k elements either contains a together with $k - 1$ elements of S, or contains k elements of S and does not contain a. Because there are $\binom{n}{k-1}$ subsets of $k - 1$ elements of S, there are $\binom{n}{k-1}$ subsets of k elements of T that contain a. And there are $\binom{n}{k}$ subsets of k elements of T that do not contain a, because there are $\binom{n}{k}$ subsets of k elements of S. Consequently,

$$\binom{n+1}{k} = \binom{n}{k-1} + \binom{n}{k}.$$

◁

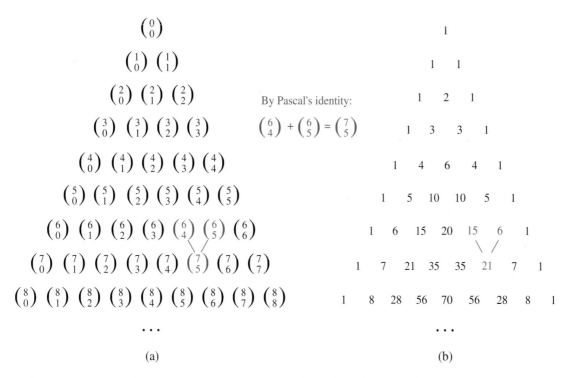

By Pascal's identity:

$$\binom{6}{4} + \binom{6}{5} = \binom{7}{5}$$

(a)   (b)

FIGURE 1   **Pascal's Triangle.**

Remark: It is also possible to prove this identity by algebraic manipulation from the formula for $\binom{n}{r}$ (see Exercise 11).

Remark: Pascal's identity, together with the initial conditions $\binom{n}{0} = \binom{n}{n} = 1$ for all integers n, can be used to recursively define binomial coefficients. This recursive definition is useful in the computation of binomial coefficients because only addition, and not multiplication, of integers is needed to use this recursive definition.

Pascal's identity is the basis for a geometric arrangement of the binomial coefficients in a triangle, as shown in Figure 1.

The nth row in the triangle consists of the binomial coefficients

$$\binom{n}{k}, \ k = 0, 1, \ldots, n.$$

This triangle is known as **Pascal's triangle**. Pascal's identity shows that when two adjacent binomial coefficients in this triangle are added, the binomial coefficient in the next row between these two coefficients is produced.

BLAISE PASCAL (1623–1662)   Blaise Pascal exhibited his talents at an early age, although his father, who had made discoveries in analytic geometry, kept mathematics books away from him to encourage other interests. At 16 Pascal discovered an important result concerning conic sections. At 18 he designed a calculating machine, which he built and sold. Pascal, along with Fermat, laid the foundations for the modern theory of probability. In this work, he made new discoveries concerning what is now called Pascal's triangle. In 1654, Pascal abandoned his mathematical pursuits to devote himself to theology. After this, he returned to mathematics only once. One night, distracted by a severe toothache, he sought comfort by studying the mathematical properties of the cycloid. Miraculously, his pain subsided, which he took as a sign of divine approval of the study of mathematics.

Other Identities Involving Binomial Coefficients

We conclude this section with combinatorial proofs of two of the many identities enjoyed by the binomial coefficients.

THEOREM 3   **VANDERMONDE'S IDENTITY**   Let m, n, and r be nonnegative integers with r not exceeding either m or n. Then

$$\binom{m+n}{r} = \sum_{k=0}^{r} \binom{m}{r-k}\binom{n}{k}.$$

Links

Remark: This identity was discovered by mathematician Alexandre-Théophile Vandermonde in the eighteenth century.

Proof: Suppose that there are m items in one set and n items in a second set. Then the total number of ways to pick r elements from the union of these sets is $\binom{m+n}{r}$.

Another way to pick r elements from the union is to pick k elements from the second set and then $r - k$ elements from the first set, where k is an integer with $0 \leq k \leq r$. Because there are $\binom{n}{k}$ ways to choose k elements from the second set and $\binom{m}{r-k}$ ways to choose $r - k$ elements from the first set, the product rule tells us that this can be done in $\binom{m}{r-k}\binom{n}{k}$ ways. Hence, the total number of ways to pick r elements from the union also equals $\sum_{k=0}^{r}\binom{m}{r-k}\binom{n}{k}$.

We have found two expressions for the number of ways to pick r elements from the union of a set with m items and a set with n items. Equating them gives us Vandermonde's identity. ◁

Corollary 4 follows from Vandermonde's identity.

COROLLARY 4   If n is a nonnegative integer, then

$$\binom{2n}{n} = \sum_{k=0}^{n} \binom{n}{k}^2.$$

Proof: We use Vandermonde's identity with $m = r = n$ to obtain

$$\binom{2n}{n} = \sum_{k=0}^{n} \binom{n}{n-k}\binom{n}{k} = \sum_{k=0}^{n} \binom{n}{k}^2.$$

Links

ALEXANDRE-THÉOPHILE VANDERMONDE (1735–1796)   Because Alexandre-Théophile Vandermonde was a sickly child, his physician father directed him to a career in music. However, he later developed an interest in mathematics. His complete mathematical work consists of four papers published in 1771–1772. These papers include fundamental contributions on the roots of equations, on the theory of determinants, and on the knight's tour problem (introduced in the exercises in Section 10.5). Vandermonde's interest in mathematics lasted for only 2 years. Afterward, he published papers on harmony, experiments with cold, and the manufacture of steel. He also became interested in politics, joining the cause of the French revolution and holding several different positions in government.

The last equality was obtained using the identity $\binom{n}{k} = \binom{n}{n-k}$. ◁

We can prove combinatorial identities by counting bit strings with different properties, as the proof of Theorem 4 will demonstrate.

THEOREM 4   Let n and r be nonnegative integers with $r \leq n$. Then

$$\binom{n+1}{r+1} = \sum_{j=r}^{n} \binom{j}{r}.$$

Proof: We use a combinatorial proof. By Example 14 in Section 6.3, the left-hand side, $\binom{n+1}{r+1}$, counts the bit strings of length $n+1$ containing $r+1$ ones.

We show that the right-hand side counts the same objects by considering the cases corresponding to the possible locations of the final 1 in a string with $r+1$ ones. This final one must occur at position $r+1, r+2, \ldots,$ or $n+1$. Furthermore, if the last one is the kth bit there must be r ones among the first $k-1$ positions. Consequently, by Example 14 in Section 6.3, there are $\binom{k-1}{r}$ such bit strings. Summing over k with $r+1 \leq k \leq n+1$, we find that there are

$$\sum_{k=r+1}^{n+1} \binom{k-1}{r} = \sum_{j=r}^{n} \binom{j}{r}$$

bit strings of length n containing exactly $r+1$ ones. (Note that the last step follows from the change of variables $j = k-1$.) Because the left-hand side and the right-hand side count the same objects, they are equal. This completes the proof. ◁

Exercises

1. Find the expansion of $(x+y)^4$
   a) using combinatorial reasoning, as in Example 1.
   b) using the binomial theorem.

2. Find the coefficient of $x^5 y^8$ in $(x+y)^{13}$.

3. How many terms are there in the expansion of $(x+y)^{100}$ after like terms are collected?

4. What is the coefficient of $x^{101} y^{99}$ in the expansion of $(2x - 3y)^{200}$?

∗5. Give a formula for the coefficient of x^k in the expansion of $(x^2 - 1/x)^{100}$, where k is an integer.

∗6. Give a formula for the coefficient of x^k in the expansion of $(x + 1/x)^{100}$, where k is an integer.

7. What is the row of Pascal's triangle containing the binomial coefficients $\binom{9}{k}$, $0 \leq k \leq 9$?

8. The row of Pascal's triangle containing the binomial coefficients $\binom{10}{k}$, $0 \leq k \leq 10$, is:

1 10 45 120 210 252 210 120 45 10 1

Use Pascal's identity to produce the row immediately following this row in Pascal's triangle.

9. Show that if n and k are integers with $1 \leq k \leq n$, then $\binom{n}{k} \leq n^k / 2^{k-1}$.

10. Suppose that b is an integer with $b \geq 7$. Use the binomial theorem and the appropriate row of Pascal's triangle to find the base-b expansion of $(11)_b^4$ [that is, the fourth power of the number $(11)_b$ in base-b notation].

11. Prove Pascal's identity, using the formula for $\binom{n}{r}$.

12. Suppose that k and n are integers with $1 \leq k < n$. Prove the **hexagon identity**

$$\binom{n-1}{k-1}\binom{n}{k+1}\binom{n+1}{k} = \binom{n-1}{k}\binom{n}{k-1}\binom{n+1}{k+1},$$

which relates terms in Pascal's triangle that form a hexagon.

13. Prove that if n and k are integers with $1 \leq k \leq n$, then $k\binom{n}{k} = n\binom{n-1}{k-1}$,
   a) using a combinatorial proof. [*Hint:* Show that the two sides of the identity count the number of ways to select a subset with k elements from a set with n elements and then an element of this subset.]
   b) using an algebraic proof based on the formula for $\binom{n}{r}$ given in Theorem 2 in Section 6.3.

14. Prove the identity $\binom{n}{r}\binom{r}{k} = \binom{n}{k}\binom{n-k}{r-k}$, whenever n, r, and k are nonnegative integers with $r \leq n$ and $k \leq r$,
   a) using a combinatorial argument.

b) using an argument based on the formula for the number of r-combinations of a set with n elements.

15. Show that if n and k are positive integers, then

$$\binom{n+1}{k} = (n+1)\binom{n}{k-1} / k.$$

Use this identity to construct an inductive definition of the binomial coefficients.

16. Show that if p is a prime and k is an integer such that $1 \leq k \leq p-1$, then p divides $\binom{p}{k}$.

17. Let n be a positive integer. Show that

$$\binom{2n}{n+1} + \binom{2n}{n} = \binom{2n+2}{n+1} / 2.$$

∗18. Let n and k be integers with $1 \leq k \leq n$. Show that

$$\sum_{k=1}^{n} \binom{n}{k}\binom{n}{k-1} = \binom{2n+2}{n+1} / 2 - \binom{2n}{n}.$$

∗19. Prove the **hockeystick identity**

$$\sum_{k=0}^{r} \binom{n+k}{k} = \binom{n+r+1}{r}$$

whenever n and r are positive integers,
a) using a combinatorial argument.
b) using Pascal's identity.

20. Show that if n is a positive integer, then $\binom{2n}{2} = 2\binom{n}{2} + n^2$
a) using a combinatorial argument.
b) by algebraic manipulation.

∗21. Give a combinatorial proof that $\sum_{k=1}^{n} k\binom{n}{k} = n2^{n-1}$.
[*Hint:* Count in two ways the number of ways to select a committee and to then select a leader of the committee.]

∗22. Give a combinatorial proof that $\sum_{k=1}^{n} k\binom{n}{k}^2 = n\binom{2n-1}{n-1}$.
[*Hint:* Count in two ways the number of ways to select a committee, with n members from a group of n mathematics professors and n computer science professors, such that the chairperson of the committee is a mathematics professor.]

23. In this exercise we will count the number of paths in the xy plane between the origin $(0, 0)$ and point (m, n), where m and n are nonnegative integers, such that each path is made up of a series of steps, where each step is a move one unit to the right or a move one unit upward. (No moves to the left or downward are allowed.) Two such paths from $(0, 0)$ to $(5, 3)$ are illustrated here.

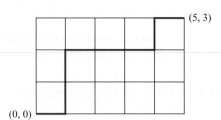

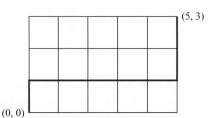

a) Show that each path of the type described can be represented by a bit string consisting of m 0s and n 1s, where a 0 represents a move one unit to the right and a 1 represents a move one unit upward.
b) Conclude from part (a) that there are $\binom{m+n}{n}$ paths of the desired type.

24. Use Exercise 23 to give an alternative proof of Corollary 2 in Section 6.3, which states that $\binom{n}{k} = \binom{n}{n-k}$ whenever k is an integer with $0 \leq k \leq n$. [*Hint:* Consider the number of paths of the type described in Exercise 23 from $(0, 0)$ to $(n-k, k)$ and from $(0, 0)$ to $(k, n-k)$.]

25. Use Exercise 23 to prove Theorem 4. [*Hint:* Count the number of paths with n steps of the type described in Exercise 23. Every such path must end at one of the points $(n-k, k)$ for $k = 0, 1, 2, \ldots, n$.]

26. Use Exercise 23 to prove Pascal's identity. [*Hint:* Show that a path of the type described in Exercise 23 from $(0, 0)$ to $(n+1-k, k)$ passes through either $(n+1-k, k-1)$ or $(n-k, k)$, but not through both.]

27. Use Exercise 23 to prove the hockeystick identity from Exercise 19. [*Hint:* First, note that the number of paths from $(0, 0)$ to $(n+1, r)$ equals $\binom{n+1+r}{r}$. Second, count the number of paths by summing the number of these paths that start by going k units upward for $k = 0, 1, 2, \ldots, r$.]

28. Give a combinatorial proof that if n is a positive integer then $\sum_{k=0}^{n} k^2\binom{n}{k} = n(n+1)2^{n-2}$. [*Hint:* Show that both sides count the ways to select a subset of a set of n elements together with two not necessarily distinct elements from this subset. Furthermore, express the right-hand side as $n(n-1)2^{n-2} + n2^{n-1}$.]

6.5   Generalized Permutations and Combinations

Introduction

In many counting problems, elements may be used repeatedly. For instance, a letter or digit may be used more than once on a license plate. When a dozen donuts are selected, each variety can be chosen repeatedly. This contrasts with the counting problems discussed earlier in the chapter where we considered only permutations and combinations in which each item could be used at most once. In this section we will show how to solve counting problems where elements may be used more than once.

Also, some counting problems involve indistinguishable elements. For instance, to count the number of ways the letters of the word *SUCCESS* can be rearranged, the placement of identical letters must be considered. This contrasts with the counting problems discussed earlier where all elements were considered distinguishable. In this section we will describe how to solve counting problems in which some elements are indistinguishable.

Moreover, in this section we will explain how to solve another important class of counting problems, problems involving counting the ways distinguishable elements can be placed in boxes. An example of this type of problem is the number of different ways poker hands can be dealt to four players.

Taken together, the methods described earlier in this chapter and the methods introduced in this section form a useful toolbox for solving a wide range of counting problems. When the additional methods discussed in Chapter 8 are added to this arsenal, you will be able to solve a large percentage of the counting problems that arise in a wide range of areas of study.

Permutations with Repetition

Counting permutations when repetition of elements is allowed can easily be done using the product rule, as Example 1 shows.

EXAMPLE 1   How many strings of length r can be formed from the uppercase letters of the English alphabet?

Solution: By the product rule, because there are 26 uppercase English letters, and because each letter can be used repeatedly, we see that there are 26^r strings of uppercase English letters of length r.   ◀

The number of r-permutations of a set with n elements when repetition is allowed is given in Theorem 1.

THEOREM 1   The number of r-permutations of a set of n objects with repetition allowed is n^r.

Proof: There are n ways to select an element of the set for each of the r positions in the r-permutation when repetition is allowed, because for each choice all n objects are available. Hence, by the product rule there are n^r r-permutations when repetition is allowed.   ◁

Combinations with Repetition

Consider these examples of combinations with repetition of elements allowed.

EXAMPLE 2   How many ways are there to select four pieces of fruit from a bowl containing apples, oranges, and pears if the order in which the pieces are selected does not matter, only the type of fruit and not the individual piece matters, and there are at least four pieces of each type of fruit in the bowl?

Solution: To solve this problem we list all the ways possible to select the fruit. There are 15 ways:

4 apples	4 oranges	4 pears
3 apples, 1 orange	3 apples, 1 pear	3 oranges, 1 apple
3 oranges, 1 pear	3 pears, 1 apple	3 pears, 1 orange
2 apples, 2 oranges	2 apples, 2 pears	2 oranges, 2 pears
2 apples, 1 orange, 1 pear	2 oranges, 1 apple, 1 pear	2 pears, 1 apple, 1 orange

The solution is the number of 4-combinations with repetition allowed from a three-element set, {*apple, orange, pear*}.   ◀

To solve more complex counting problems of this type, we need a general method for counting the *r*-combinations of an *n*-element set. In Example 3 we will illustrate such a method.

EXAMPLE 3   How many ways are there to select five bills from a cash box containing $1 bills, $2 bills, $5 bills, $10 bills, $20 bills, $50 bills, and $100 bills? Assume that the order in which the bills are chosen does not matter, that the bills of each denomination are indistinguishable, and that there are at least five bills of each type.

Solution: Because the order in which the bills are selected does not matter and seven different types of bills can be selected as many as five times, this problem involves counting 5-combinations with repetition allowed from a set with seven elements. Listing all possibilities would be tedious, because there are a large number of solutions. Instead, we will illustrate the use of a technique for counting combinations with repetition allowed.

Suppose that a cash box has seven compartments, one to hold each type of bill, as illustrated in Figure 1. These compartments are separated by six dividers, as shown in the picture. The choice of five bills corresponds to placing five markers in the compartments holding different types of bills. Figure 2 illustrates this correspondence for three different ways to select five bills, where the six dividers are represented by bars and the five bills by stars.

The number of ways to select five bills corresponds to the number of ways to arrange six bars and five stars in a row with a total of 11 positions. Consequently, the number of ways to select the five bills is the number of ways to select the positions of the five stars from the 11 positions. This corresponds to the number of unordered selections of 5 objects from a set of 11 objects, which can be done in $C(11, 5)$ ways. Consequently, there are

$$C(11, 5) = \frac{11!}{5!\,6!} = 462$$

ways to choose five bills from the cash box with seven types of bills.   ◀

Theorem 2 generalizes this discussion.

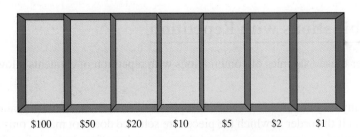

$100 $50 $20 $10 $5 $2 $1

FIGURE 1   Cash Box with Seven Types of Bills.

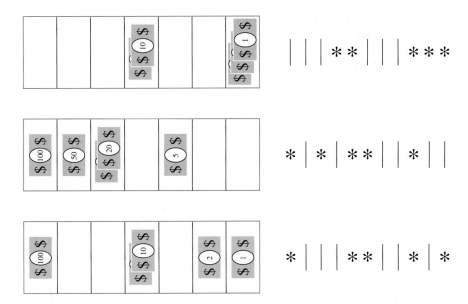

FIGURE 2   **Examples of Ways to Select Five Bills.**

THEOREM 2   There are $C(n + r - 1, r) = C(n + r - 1, n - 1)$ r-combinations from a set with n elements when repetition of elements is allowed.

Proof: Each r-combination of a set with n elements when repetition is allowed can be represented by a list of $n - 1$ bars and r stars. The $n - 1$ bars are used to mark off n different cells, with the ith cell containing a star for each time the ith element of the set occurs in the combination. For instance, a 6-combination of a set with four elements is represented with three bars and six stars. Here

$$** \mid * \mid \mid ***$$

represents the combination containing exactly two of the first element, one of the second element, none of the third element, and three of the fourth element of the set.

As we have seen, each different list containing $n - 1$ bars and r stars corresponds to an r-combination of the set with n elements, when repetition is allowed. The number of such lists is $C(n - 1 + r, r)$, because each list corresponds to a choice of the r positions to place the r stars from the $n - 1 + r$ positions that contain r stars and $n - 1$ bars. The number of such lists is also equal to $C(n - 1 + r, n - 1)$, because each list corresponds to a choice of the $n - 1$ positions to place the $n - 1$ bars.   ◁

Examples 4–6 show how Theorem 2 is applied.

EXAMPLE 4

Extra Examples

Suppose that a cookie shop has four different kinds of cookies. How many different ways can six cookies be chosen? Assume that only the type of cookie, and not the individual cookies or the order in which they are chosen, matters.

Solution: The number of ways to choose six cookies is the number of 6-combinations of a set with four elements. From Theorem 2 this equals $C(4 + 6 - 1, 6) = C(9, 6)$. Because

$$C(9, 6) = C(9, 3) = \frac{9 \cdot 8 \cdot 7}{1 \cdot 2 \cdot 3} = 84,$$

there are 84 different ways to choose the six cookies.   ◀

Theorem 2 can also be used to find the number of solutions of certain linear equations where the variables are integers subject to constraints. This is illustrated by Example 5.

EXAMPLE 5   How many solutions does the equation

$$x_1 + x_2 + x_3 = 11$$

have, where x_1, x_2, and x_3 are nonnegative integers?

Solution: To count the number of solutions, we note that a solution corresponds to a way of selecting 11 items from a set with three elements so that x_1 items of type one, x_2 items of type two, and x_3 items of type three are chosen. Hence, the number of solutions is equal to the number of 11-combinations with repetition allowed from a set with three elements. From Theorem 2 it follows that there are

$$C(3 + 11 - 1, 11) = C(13, 11) = C(13, 2) = \frac{13 \cdot 12}{1 \cdot 2} = 78$$

solutions.

The number of solutions of this equation can also be found when the variables are subject to constraints. For instance, we can find the number of solutions where the variables are integers with $x_1 \geq 1$, $x_2 \geq 2$, and $x_3 \geq 3$. A solution to the equation subject to these constraints corresponds to a selection of 11 items with x_1 items of type one, x_2 items of type two, and x_3 items of type three, where, in addition, there is at least one item of type one, two items of type two, and three items of type three. So, a solution corresponds to a choice of one item of type one, two of type two, and three of type three, together with a choice of five additional items of any type. By Theorem 2 this can be done in

$$C(3 + 5 - 1, 5) = C(7, 5) = C(7, 2) = \frac{7 \cdot 6}{1 \cdot 2} = 21$$

ways. Thus, there are 21 solutions of the equation subject to the given constraints.   ◀

Example 6 shows how counting the number of combinations with repetition allowed arises in determining the value of a variable that is incremented each time a certain type of nested loop is traversed.

EXAMPLE 6   What is the value of k after the following pseudocode has been executed?

```
k := 0
for i₁ := 1 to n
    for i₂ := 1 to i₁
        .
        .
        .
        for iₘ := 1 to iₘ₋₁
            k := k + 1
```

Solution: Note that the initial value of k is 0 and that 1 is added to k each time the nested loop is traversed with a sequence of integers $i_1, i_2, \ldots, i_m$ such that

$$1 \leq i_m \leq i_{m-1} \leq \cdots \leq i_1 \leq n.$$

TABLE 1 Combinations and Permutations With and Without Repetition.		
Type	*Repetition Allowed?*	*Formula*
r-permutations	No	$\dfrac{n!}{(n-r)!}$
r-combinations	No	$\dfrac{n!}{r!\,(n-r)!}$
r-permutations	Yes	n^r
r-combinations	Yes	$\dfrac{(n+r-1)!}{r!\,(n-1)!}$

The number of such sequences of integers is the number of ways to choose m integers from $\{1, 2, \ldots, n\}$, with repetition allowed. (To see this, note that once such a sequence has been selected, if we order the integers in the sequence in nondecreasing order, this uniquely defines an assignment of $i_m, i_{m-1}, \ldots, i_1$. Conversely, every such assignment corresponds to a unique unordered set.) Hence, from Theorem 2, it follows that $k = C(n + m - 1, m)$ after this code has been executed. ◀

The formulae for the numbers of ordered and unordered selections of r elements, chosen with and without repetition allowed from a set with n elements, are shown in Table 1.

Permutations with Indistinguishable Objects

Some elements may be indistinguishable in counting problems. When this is the case, care must be taken to avoid counting things more than once. Consider Example 7.

EXAMPLE 7   How many different strings can be made by reordering the letters of the word $SUCCESS$?

Solution: Because some of the letters of $SUCCESS$ are the same, the answer is *not* given by the number of permutations of seven letters. This word contains three Ss, two Cs, one U, and one E. To determine the number of different strings that can be made by reordering the letters, first note that the three Ss can be placed among the seven positions in $C(7, 3)$ different ways, leaving four positions free. Then the two Cs can be placed in $C(4, 2)$ ways, leaving two free positions. The U can be placed in $C(2, 1)$ ways, leaving just one position free. Hence E can be placed in $C(1, 1)$ way. Consequently, from the product rule, the number of different strings that can be made is

$$
\begin{aligned}
C(7, 3)C(4, 2)C(2, 1)C(1, 1) &= \frac{7!}{3!\,4!} \cdot \frac{4!}{2!\,2!} \cdot \frac{2!}{1!\,1!} \cdot \frac{1!}{1!\,0!} \\
&= \frac{7!}{3!\,2!\,1!\,1!} \\
&= 420.
\end{aligned}
$$

◀

We can prove Theorem 3 using the same sort of reasoning as in Example 7.

THEOREM 3

The number of different permutations of n objects, where there are n_1 indistinguishable objects of type 1, n_2 indistinguishable objects of type 2, ..., and n_k indistinguishable objects of type k, is

$$\frac{n!}{n_1! \, n_2! \cdots n_k!}.$$

Proof: To determine the number of permutations, first note that the n_1 objects of type one can be placed among the n positions in $C(n, n_1)$ ways, leaving $n - n_1$ positions free. Then the objects of type two can be placed in $C(n - n_1, n_2)$ ways, leaving $n - n_1 - n_2$ positions free. Continue placing the objects of type three, ..., type $k - 1$, until at the last stage, n_k objects of type k can be placed in $C(n - n_1 - n_2 - \cdots - n_{k-1}, n_k)$ ways. Hence, by the product rule, the total number of different permutations is

$$C(n, n_1) C(n - n_1, n_2) \cdots C(n - n_1 - \cdots - n_{k-1}, n_k)$$

$$= \frac{n!}{n_1! \, (n - n_1)!} \frac{(n - n_1)!}{n_2! \, (n - n_1 - n_2)!} \cdots \frac{(n - n_1 - \cdots - n_{k-1})!}{n_k! \, 0!}$$

$$= \frac{n!}{n_1! \, n_2! \cdots n_k!}.$$

◁

Distributing Objects into Boxes

Many counting problems can be solved by enumerating the ways objects can be placed into boxes (where the order these objects are placed into the boxes does not matter). The objects can be either *distinguishable*, that is, different from each other, or *indistinguishable*, that is, considered identical. Distinguishable objects are sometimes said to be *labeled*, whereas indistinguishable objects are said to be *unlabeled*. Similarly, boxes can be *distinguishable*, that is, different, or *indinguishable*, that is, identical. Distinguishable boxes are often said to be *labeled*, while indistinguishable boxes are said to be *unlabeled*. When you solve a counting problem using the model of distributing objects into boxes, you need to determine whether the objects are distinguishable and whether the boxes are distinguishable. Although the context of the counting problem makes these two decisions clear, counting problems are sometimes ambiguous and it may be unclear which model applies. In such a case it is best to state whatever assumptions you are making and explain why the particular model you choose conforms to your assumptions.

We will see that there are closed formulae for counting the ways to distribute objects, distinguishable or indistinguishable, into distinguishable boxes. We are not so lucky when we count the ways to distribute objects, distinguishable or indistinguishable, into indistinguishable boxes; there are no closed formulae to use in these cases.

DISTINGUISHABLE OBJECTS AND DISTINGUISHABLE BOXES   We first consider the case when distinguishable objects are placed into distinguishable boxes. Consider Example 8 in which the objects are cards and the boxes are hands of players.

EXAMPLE 8

How many ways are there to distribute hands of 5 cards to each of four players from the standard deck of 52 cards?

Solution: We will use the product rule to solve this problem. To begin, note that the first player can be dealt 5 cards in $C(52, 5)$ ways. The second player can be dealt 5 cards in $C(47, 5)$ ways, because only 47 cards are left. The third player can be dealt 5 cards in $C(42, 5)$ ways. Finally, the fourth player can be dealt 5 cards in $C(37, 5)$ ways. Hence, the total number of ways to deal four players 5 cards each is

$$C(52, 5)C(47, 5)C(42, 5)C(37, 5) = \frac{52!}{47!\,5!} \cdot \frac{47!}{42!\,5!} \cdot \frac{42!}{37!\,5!} \cdot \frac{37!}{32!\,5!}$$

$$= \frac{52!}{5!\,5!\,5!\,5!\,32!}.$$ ◀

Remark: The solution to Example 8 equals the number of permutations of 52 objects, with 5 indistinguishable objects of each of four different types, and 32 objects of a fifth type. This equality can be seen by defining a one-to-one correspondence between permutations of this type and distributions of cards to the players. To define this correspondence, first order the cards from 1 to 52. Then cards dealt to the first player correspond to the cards in the positions assigned to objects of the first type in the permutation. Similarly, cards dealt to the second, third, and fourth players, respectively, correspond to cards in the positions assigned to objects of the second, third, and fourth type, respectively. The cards not dealt to any player correspond to cards in the positions assigned to objects of the fifth type. The reader should verify that this is a one-to-one correspondence.

Example 8 is a typical problem that involves distributing distinguishable objects into distinguishable boxes. The distinguishable objects are the 52 cards, and the five distinguishable boxes are the hands of the four players and the rest of the deck. Counting problems that involve distributing distinguishable objects into boxes can be solved using Theorem 4.

THEOREM 4   The number of ways to distribute n distinguishable objects into k distinguishable boxes so that n_i objects are placed into box i, $i = 1, 2, \ldots, k$, equals

$$\frac{n!}{n_1!\,n_2! \cdots n_k!}.$$

Theorem 4 can be proved using the product rule. We leave the details as Exercise 47. It can also be proved (see Exercise 48) by setting up a one-to-one correspondence between the permutations counted by Theorem 3 and the ways to distribute objects counted by Theorem 4.

INDISTINGUISHABLE OBJECTS AND DISTINGUISHABLE BOXES   Counting the number of ways of placing n indistinguishable objects into k distinguishable boxes turns out to be the same as counting the number of n-combinations for a set with k elements when repetitions are allowed. The reason behind this is that there is a one-to-one correspondence between n-combinations from a set with k elements when repetition is allowed and the ways to place n indistinguishable balls into k distinguishable boxes. To set up this correspondence, we put a ball in the ith bin each time the ith element of the set is included in the n-combination.

EXAMPLE 9   How many ways are there to place 10 indistinguishable balls into eight distinguishable bins?

Solution: The number of ways to place 10 indistinguishable balls into eight bins equals the number of 10-combinations from a set with eight elements when repetition is allowed. Consequently, there are

$$C(8 + 10 - 1, 10) = C(17, 10) = \frac{17!}{10!7!} = 19{,}448.$$ ◀

This means that there are $C(n + r - 1, n - 1)$ ways to place r indistinguishable objects into n distinguishable boxes.

DISTINGUISHABLE OBJECTS AND INDISTINGUISHABLE BOXES   Counting the ways to place n distinguishable objects into k indistinguishable boxes is more difficult than counting the ways to place objects, distinguishable or indistinguishable objects, into distinguishable boxes. We illustrate this with an example.

EXAMPLE 10   How many ways are there to put four different employees into three indistinguishable offices, when each office can contain any number of employees?

Solution: We will solve this problem by enumerating all the ways these employees can be placed into the offices. We represent the four employees by A, B, C, and D. First, we note that we can distribute employees so that all four are put into one office, three are put into one office and a fourth is put into a second office, two employees are put into one office and two put into a second office, and finally, two are put into one office, and one each put into the other two offices. Each way to distribute these employees to these offices can be represented by a way to partition the elements A, B, C, and D into disjoint subsets.

We can put all four employees into one office in exactly one way, represented by $\{\{A, B, C, D\}\}$. We can put three employees into one office and the fourth employee into a different office in exactly four ways, represented by $\{\{A, B, C\}, \{D\}\}$, $\{\{A, B, D\}, \{C\}\}$, $\{\{A, C, D\}, \{B\}\}$, and $\{\{B, C, D\}, \{A\}\}$. We can put two employees into one office and two into a second office in exactly three ways, represented by $\{\{A, B\}, \{C, D\}\}$, $\{\{A, C\}, \{B, D\}\}$, and $\{\{A, D\}, \{B, C\}\}$. Finally, we can put two employees into one office, and one each into each of the remaining two offices in six ways, represented by $\{\{A, B\}, \{C\}, \{D\}\}$, $\{\{A, C\}, \{B\}, \{D\}\}$, $\{\{A, D\}, \{B\}, \{C\}\}$, $\{\{B, C\}, \{A\}, \{D\}\}$, $\{\{B, D\}, \{A\}, \{C\}\}$, and $\{\{C, D\}, \{A\}, \{B\}\}$.

Counting all the possibilities, we find that there are 14 ways to put four different employees into three indistinguishable offices. Another way to look at this problem is to look at the number of offices into which we put employees. Note that there are six ways to put four different employees into three indistinguishable offices so that no office is empty, seven ways to put four different employees into two indistinguishable offices so that no office is empty, and one way to put four employees into one office so that it is not empty.   ◄

There is no simple closed formula for the number of ways to distribute n distinguishable objects into j indistinguishable boxes. However, there is a formula involving a summation, which we will now describe. Let $S(n, j)$ denote the number of ways to distribute n distinguishable objects into j indistinguishable boxes so that no box is empty. The numbers $S(n, j)$ are called **Stirling numbers of the second kind**. For instance, Example 10 shows that $S(4, 3) = 6$, $S(4, 2) = 7$, and $S(4, 1) = 1$. We see that the number of ways to distribute n distinguishable objects into k indistinguishable boxes (where the number of boxes that are nonempty equals k, $k - 1, \ldots, 2$, or 1) equals $\sum_{j=1}^{k} S(n, j)$. For instance, following the reasoning in Example 10, the number of ways to distribute four distinguishable objects into three indistinguishable boxes equals $S(4, 1) + S(4, 2) + S(4, 3) = 1 + 7 + 6 = 14$. Using the inclusion–exclusion principle (see Section 8.6) it can be shown that

$$S(n, j) = \frac{1}{j!} \sum_{i=0}^{j-1} (-1)^i \binom{j}{i} (j - i)^n.$$

Consequently, the number of ways to distribute n distinguishable objects into k indistinguishable boxes equals

$$\sum_{j=1}^{k} S(n, j) = \sum_{j=1}^{k} \frac{1}{j!} \sum_{i=0}^{j-1} (-1)^i \binom{j}{i} (j - i)^n.$$

Remark: The reader may be curious about the Stirling numbers of the first kind. A combinatorial definition of the **signless Stirling numbers of the first kind**, the absolute values of the Stirling numbers of the first kind, can be found in the preamble to Exercise 37 in the Supplementary Exercises. For the definition of Stirling numbers of the first kind, for more information about Stirling numbers of the second kind, and to learn more about Stirling numbers of the first kind and the relationship between Stirling numbers of the first and second kind, see combinatorics textbooks such as [Bó07], [Br99], and [RoTe05], and Chapter 6 in [MiRo91].

INDISTINGUISHABLE OBJECTS AND INDISTINGUISHABLE BOXES   Some counting problems can be solved by determining the number of ways to distribute indistinguishable objects into indistinguishable boxes. We illustrate this principle with an example.

EXAMPLE 11   How many ways are there to pack six copies of the same book into four identical boxes, where a box can contain as many as six books?

Solution: We will enumerate all ways to pack the books. For each way to pack the books, we will list the number of books in the box with the largest number of books, followed by the numbers of books in each box containing at least one book, in order of decreasing number of books in a box. The ways we can pack the books are

> 6
> 5, 1
> 4, 2
> 4, 1, 1
> 3, 3
> 3, 2, 1
> 3, 1, 1, 1
> 2, 2, 2
> 2, 2, 1, 1.

For example, 4, 1, 1 indicates that one box contains four books, a second box contains a single book, and a third box contains a single book (and the fourth box is empty). We conclude that there are nine allowable ways to pack the books, because we have listed them all.   ◀

Observe that distributing n indistinguishable objects into k indistinguishable boxes is the same as writing n as the sum of at most k positive integers in nonincreasing order. If $a_1 + a_2 + \cdots + a_j = n$, where $a_1, a_2, \ldots, a_j$ are positive integers with $a_1 \geq a_2 \geq \cdots \geq a_j$, we say that $a_1, a_2, \ldots, a_j$ is a **partition** of the positive integer n into j positive integers. We see that if $p_k(n)$ is the number of partitions of n into at most k positive integers, then there are $p_k(n)$ ways to distribute n indistinguishable objects into k indistinguishable boxes. No simple closed formula exists for this number. For more information about partitions of positive integers, see [Ro11].

Exercises

1. In how many different ways can five elements be selected in order from a set with three elements when repetition is allowed?

2. How many strings of six letters are there?

3. How many ways are there to assign three jobs to five employees if each employee can be given more than one job?

4. How many ways are there to select five unordered elements from a set with three elements when repetition is allowed?

5. How many ways are there to select three unordered elements from a set with five elements when repetition is allowed?

6. A bagel shop has onion bagels, poppy seed bagels, egg bagels, salty bagels, pumpernickel bagels, sesame seed bagels, raisin bagels, and plain bagels. How many ways are there to choose

a) six bagels?

b) a dozen bagels?

c) two dozen bagels?

d) a dozen bagels with at least one of each kind?

e) a dozen bagels with at least three egg bagels and no more than two salty bagels?

7. How many ways are there to choose eight coins from a piggy bank containing 100 identical pennies and 80 identical nickels?

8. How many different combinations of pennies, nickels, dimes, quarters, and half dollars can a piggy bank contain if it has 20 coins in it?

9. A book publisher has 3000 copies of a discrete mathematics book. How many ways are there to store these books in their three warehouses if the copies of the book are indistinguishable?

10. How many solutions are there to the equation

$$x_1 + x_2 + x_3 + x_4 = 17,$$

where x_1, x_2, x_3, and x_4 are nonnegative integers?

11. How many solutions are there to the equation

$$x_1 + x_2 + x_3 + x_4 + x_5 = 21,$$

where x_i, $i = 1, 2, 3, 4, 5$, is a nonnegative integer such that

a) $x_1 \geq 1$?

b) $x_i \geq 2$ for $i = 1, 2, 3, 4, 5$?

c) $0 \leq x_1 \leq 10$?

d) $0 \leq x_1 \leq 3, 1 \leq x_2 < 4$, and $x_3 \geq 15$?

12. How many strings of 10 ternary digits (0, 1, or 2) are there that contain exactly two 0s, three 1s, and five 2s?

13. Suppose that a large family has 14 children, including two sets of identical triplets, three sets of identical twins, and two individual children. How many ways are there to seat these children in a row of chairs if the identical triplets or twins cannot be distinguished from one another?

14. How many solutions are there to the inequality

$$x_1 + x_2 + x_3 \leq 11,$$

where x_1, x_2, and x_3 are nonnegative integers? [*Hint:* Introduce an auxiliary variable x_4 such that $x_1 + x_2 + x_3 + x_4 = 11$.]

15. How many ways are there to distribute six indistinguishable balls into nine distinguishable bins?

16. How many ways are there to distribute 12 indistinguishable balls into six distinguishable bins?

17. How many ways are there to distribute 12 distinguishable objects into six distinguishable boxes so that two objects are placed in each box?

18. How many ways are there to distribute 15 distinguishable objects into five distinguishable boxes so that the boxes have one, two, three, four, and five objects in them, respectively.

19. How many positive integers less than 1,000,000 have the sum of their digits equal to 19?

20. How many positive integers less than 1,000,000 have exactly one digit equal to 9 and have a sum of digits equal to 13?

21. There are 10 questions on a discrete mathematics final exam. How many ways are there to assign scores to the problems if the sum of the scores is 100 and each question is worth at least 5 points?

22. Show that there are $C(n + r - q_1 - q_2 - \cdots - q_r - 1, n - q_1 - q_2 - \cdots - q_r)$ different unordered selections of n objects of r different types that include at least q_1 objects of type one, q_2 objects of type two, ..., and q_r objects of type r.

23. How many different bit strings can be transmitted if the string must begin with a 1 bit, must include three additional 1 bits (so that a total of four 1 bits is sent), must include a total of 12 0 bits, and must have at least two 0 bits following each 1 bit?

24. How many different strings can be made from the letters in *MISSISSIPPI*, using all the letters?

25. How many strings with seven or more characters can be formed from the letters in *EVERGREEN*?

26. How many different bit strings can be formed using six 1s and eight 0s?

27. A student has three mangos, two papayas, and two kiwi fruits. If the student eats one piece of fruit each day, and only the type of fruit matters, in how many different ways can these fruits be consumed?

28. A professor packs her collection of 40 issues of a mathematics journal in four boxes with 10 issues per box. How many ways can she distribute the journals if

a) each box is numbered, so that they are distinguishable?

b) the boxes are identical, so that they cannot be distinguished?

29. How many ways are there to travel in xyz space from the origin $(0, 0, 0)$ to the point $(4, 3, 5)$ by taking steps one unit in the positive x direction, one unit in the positive y direction, or one unit in the positive z direction? (Moving in the negative x, y, or z direction is prohibited, so that no backtracking is allowed.)

30. How many ways are there to deal hands of seven cards to each of five players from a standard deck of 52 cards?

31. How many ways are there to deal hands of five cards to each of six players from a deck containing 48 different cards?

32. In how many ways can a dozen books be placed on four distinguishable shelves

a) if the books are indistinguishable copies of the same title?

b) if no two books are the same, and the positions of the books on the shelves matter? [*Hint:* Break this into 12 tasks, placing each book separately. Start with the sequence 1, 2, 3, 4 to represent the shelves. Represent the books by b_i, $i = 1, 2, \ldots, 12$. Place b_1 to the right of one of the terms in 1, 2, 3, 4. Then successively place $b_2, b_3, \ldots$, and b_{12}.]

33. How many ways can n books be placed on k distinguishable shelves

    a) if the books are indistinguishable copies of the same title?

    b) if no two books are the same, and the positions of the books on the shelves matter?

34. A shelf holds 12 books in a row. How many ways are there to choose five books so that no two adjacent books are chosen? [*Hint:* Represent the books that are chosen by bars and the books not chosen by stars. Count the number of sequences of five bars and seven stars so that no two bars are adjacent.]

*35. Use the product rule to prove Theorem 4, by first placing objects in the first box, then placing objects in the second box, and so on.

*36. Prove Theorem 4 by first setting up a one-to-one correspondence between permutations of n objects with n_i indistinguishable objects of type $i, i = 1, 2, 3, \ldots, k$, and the distributions of n objects in k boxes such that n_i objects are placed in box $i, i = 1, 2, 3, \ldots, k$ and then applying Theorem 3.

*37. In this exercise we will prove Theorem 2 by setting up a one-to-one correspondence between the set of r-combinations with repetition allowed of $S = \{1, 2, 3, \ldots, n\}$ and the set of r-combinations of the set $T = \{1, 2, 3, \ldots, n + r - 1\}$.

    a) Arrange the elements in an r-combination, with repetition allowed, of S into an increasing sequence $x_1 \leq x_2 \leq \cdots \leq x_r$. Show that the sequence formed by adding $k - 1$ to the kth term is strictly increasing. Conclude that this sequence is made up of r distinct elements from T.

    b) Show that the procedure described in (a) defines a one-to-one correspondence between the set of r-combinations, with repetition allowed, of S and the r-combinations of T. [*Hint:* Show the correspondence can be reversed by associating to the r-combination $\{x_1, x_2, \ldots, x_r\}$ of T, with $1 \leq x_1 < x_2 < \cdots < x_r \leq n + r - 1$, the r-combination with repetition allowed from S, formed by subtracting $k - 1$ from the kth element.]

    c) Conclude that there are $C(n + r - 1, r)$ r-combinations with repetition allowed from a set with n elements.

38. How many ways are there to distribute six distinguishable objects into four indistinguishable boxes so that each of the boxes contains at least one object?

39. How many ways are there to put six temporary employees into four identical offices so that there is at least one temporary employee in each of these four offices?

40. How many ways are there to distribute five indistinguishable objects into three indistinguishable boxes?

41. How many ways are there to distribute six indistinguishable objects into four indistinguishable boxes so that each of the boxes contains at least one object?

42. How many ways are there to pack eight identical DVDs into five indistinguishable boxes so that each box contains at least one DVD?

43. How many ways are there to pack nine identical DVDs into three indistinguishable boxes so that each box contains at least two DVDs?

44. How many ways are there to distribute five balls into seven boxes if each box must have at most one ball in it if

    a) both the balls and boxes are labeled?

    b) the balls are labeled, but the boxes are unlabeled?

    c) the balls are unlabeled, but the boxes are labeled?

    d) both the balls and boxes are unlabeled?

45. How many ways are there to distribute five balls into three boxes if each box must have at least one ball in it if

    a) both the balls and boxes are labeled?

    b) the balls are labeled, but the boxes are unlabeled?

    c) the balls are unlabeled, but the boxes are labeled?

    d) both the balls and boxes are unlabeled?

46. Suppose that a basketball league has 32 teams, split into two conferences of 16 teams each. Each conference is split into three divisions. Suppose that the North Central Division has five teams. Each of the teams in the North Central Division plays four games against each of the other teams in this division, three games against each of the 11 remaining teams in the conference, and two games against each of the 16 teams in the other conference. In how many different orders can the games of one of the teams in the North Central Division be scheduled?

*47. Suppose that a weapons inspector must inspect each of five different sites twice, visiting one site per day. The inspector is free to select the order in which to visit these sites, but cannot visit site X, the most suspicious site, on two consecutive days. In how many different orders can the inspector visit these sites?

48. How many different terms are there in the expansion of $(x_1 + x_2 + \cdots + x_m)^n$ after all terms with identical sets of exponents are added?

*49. Prove the **Multinomial Theorem:** If n is a positive integer, then

$$(x_1 + x_2 + \cdots + x_m)^n$$

$$= \sum_{n_1 + n_2 + \cdots + n_m = n} C(n; n_1, n_2, \ldots, n_m) x_1^{n_1} x_2^{n_2} \cdots x_m^{n_m},$$

where

$$C(n; n_1, n_2, \ldots, n_m) = \frac{n!}{n_1! \, n_2! \cdots n_m!}$$

is a **multinomial coefficient**.

50. Find the expansion of $(x + y + z)^4$.

6.6   Generating Permutations and Combinations

Introduction

Methods for counting various types of permutations and combinations were described in the previous sections of this chapter, but sometimes permutations or combinations need to be generated, not just counted. Consider the following three problems. First, suppose that a salesperson must visit six different cities. In which order should these cities be visited to minimize total travel time? One way to determine the best order is to determine the travel time for each of the $6! = 720$ different orders in which the cities can be visited and choose the one with the smallest travel time. Second, suppose we are given a set of six positive integers and wish to find a subset of them that has 100 as their sum, if such a subset exists. One way to find these numbers is to generate all $2^6 = 64$ subsets and check the sum of their elements. Third, suppose a laboratory has 95 employees. A group of 12 of these employees with a particular set of 25 skills is needed for a project. (Each employee can have one or more of these skills.) One way to find such a set of employees is to generate all sets of 12 of these employees and check whether they have the desired skills. These examples show that it is often necessary to generate permutations and combinations to solve problems.

Generating Permutations

Any set with n elements can be placed in one-to-one correspondence with the set $\{1, 2, 3, \ldots, n\}$. We can list the permutations of any set of n elements by generating the permutations of the n smallest positive integers and then replacing these integers with the corresponding elements. Many different algorithms have been developed to generate the $n!$ permutations of this set. We will describe one of these that is based on the **lexicographic** (or **dictionary**) **ordering** of the set of permutations of $\{1, 2, 3, \ldots, n\}$. In this ordering, the permutation $a_1 a_2 \cdots a_n$ precedes the permutation of $b_1 b_2 \cdots b_n$, if for some k, with $1 \leq k \leq n$, $a_1 = b_1, a_2 = b_2, \ldots, a_{k-1} = b_{k-1}$, and $a_k < b_k$. In other words, a permutation of the set of the n smallest positive integers precedes (in lexicographic order) a second permutation if the number in this permutation in the first position where the two permutations disagree is smaller than the number in that position in the second permutation.

EXAMPLE 1   The permutation 23415 of the set $\{1, 2, 3, 4, 5\}$ precedes the permutation 23514, because these permutations agree in the first two positions, but the number in the third position in the first permutation, 4, is smaller than the number in the third position in the second permutation, 5. Similarly, the permutation 41532 precedes 52143.   ◀

An algorithm for generating the permutations of $\{1, 2, \ldots, n\}$ can be based on a procedure that constructs the next permutation in lexicographic order following a given permutation $a_1 a_2 \cdots a_n$. We will show how this can be done. First, suppose that $a_{n-1} < a_n$. Interchange a_{n-1} and a_n to obtain a larger permutation. No other permutation is both larger than the original permutation and smaller than the permutation obtained by interchanging a_{n-1} and a_n. For instance, the next larger permutation after 234156 is 234165. On the other hand, if $a_{n-1} > a_n$, then a larger permutation cannot be obtained by interchanging these last two terms in the permutation. Look at the last three integers in the permutation. If $a_{n-2} < a_{n-1}$, then the last three integers in the permutation can be rearranged to obtain the next largest permutation. Put the smaller of the two integers a_{n-1} and a_n that is greater than a_{n-2} in position $n - 2$. Then, place the remaining integer and a_{n-2} into the last two positions in increasing order. For instance, the next larger permutation after 234165 is 234516.

On the other hand, if $a_{n-2} > a_{n-1}$ (and $a_{n-1} > a_n$), then a larger permutation cannot be obtained by permuting the last three terms in the permutation. Based on these observations, a general method can be described for producing the next larger permutation in increasing order following a given permutation $a_1a_2 \cdots a_n$. First, find the integers a_j and a_{j+1} with $a_j < a_{j+1}$ and

$$a_{j+1} > a_{j+2} > \cdots > a_n,$$

that is, the last pair of adjacent integers in the permutation where the first integer in the pair is smaller than the second. Then, the next larger permutation in lexicographic order is obtained by putting in the jth position the least integer among $a_{j+1}, a_{j+2}, \ldots,$ and a_n that is greater than a_j and listing in increasing order the rest of the integers $a_j, a_{j+1}, \ldots, a_n$ in positions $j + 1$ to n. It is easy to see that there is no other permutation larger than the permutation $a_1a_2 \cdots a_n$ but smaller than the new permutation produced. (The verification of this fact is left as an exercise for the reader.)

EXAMPLE 2   What is the next permutation in lexicographic order after 362541?

Solution: The last pair of integers a_j and a_{j+1} where $a_j < a_{j+1}$ is $a_3 = 2$ and $a_4 = 5$. The least integer to the right of 2 that is greater than 2 in the permutation is $a_5 = 4$. Hence, 4 is placed in the third position. Then the integers 2, 5, and 1 are placed in order in the last three positions, giving 125 as the last three positions of the permutation. Hence, the next permutation is 364125.   ◀

To produce the $n!$ permutations of the integers $1, 2, 3, \ldots, n$, begin with the smallest permutation in lexicographic order, namely, $123 \cdots n$, and successively apply the procedure described for producing the next larger permutation of $n! - 1$ times. This yields all the permutations of the n smallest integers in lexicographic order.

EXAMPLE 3   Generate the permutations of the integers 1, 2, 3 in lexicographic order.

Solution: Begin with 123. The next permutation is obtained by interchanging 3 and 2 to obtain 132. Next, because $3 > 2$ and $1 < 3$, permute the three integers in 132. Put the smaller of 3 and 2 in the first position, and then put 1 and 3 in increasing order in positions 2 and 3 to obtain 213. This is followed by 231, obtained by interchanging 1 and 3, because $1 < 3$. The next larger permutation has 3 in the first position, followed by 1 and 2 in increasing order, namely, 312. Finally, interchange 1 and 2 to obtain the last permutation, 321. We have generated the permutations of 1, 2, 3 in lexicographic order. They are 123, 132, 213, 231, 312, and 321.   ◀

Algorithm 1 displays the procedure for finding the next permutation in lexicographic order after a permutation that is not $n\ n - 1\ n - 2\ \ldots\ 2\ 1$, which is the largest permutation.

ALGORITHM 1 Generating the Next Permutation in Lexicographic Order.

---

procedure *next permutation*($a_1a_2 \ldots a_n$: permutation of
$\quad$ $\{1, 2, \ldots, n\}$ not equal to $n\ n - 1\ \ldots\ 2\ 1$)
$j := n - 1$
while $a_j > a_{j+1}$
$\quad j := j - 1$
$\{j$ is the largest subscript with $a_j < a_{j+1}\}$
$k := n$

while $a_j > a_k$
   $k := k - 1$
{a_k is the smallest integer greater than a_j to the right of a_j}
interchange a_j and a_k
$r := n$
$s := j + 1$
while $r > s$
   interchange a_r and a_s
   $r := r - 1$
   $s := s + 1$
{this puts the tail end of the permutation after the jth position in increasing order}
{$a_1 a_2 \ldots a_n$ is now the next permutation}

Generating Combinations

Links

How can we generate all the combinations of the elements of a finite set? Because a combination is just a subset, we can use the correspondence between subsets of $\{a_1, a_2, \ldots, a_n\}$ and bit strings of length n.

Recall that the bit string corresponding to a subset has a 1 in position k if a_k is in the subset, and has a 0 in this position if a_k is not in the subset. If all the bit strings of length n can be listed, then by the correspondence between subsets and bit strings, a list of all the subsets is obtained.

Recall that a bit string of length n is also the binary expansion of an integer between 0 and $2^n - 1$. The 2^n bit strings can be listed in order of their increasing size as integers in their binary expansions. To produce all binary expansions of length n, start with the bit string $000 \ldots 00$, with n zeros. Then, successively find the next expansion until the bit string $111 \ldots 11$ is obtained. At each stage the next binary expansion is found by locating the first position from the right that is not a 1, then changing all the 1s to the right of this position to 0s and making this first 0 (from the right) a 1.

EXAMPLE 4   Find the next bit string after 10 0010 0111.

Solution: The first bit from the right that is not a 1 is the fourth bit from the right. Change this bit to a 1 and change all the following bits to 0s. This produces the next larger bit string, 10 0010 1000. ◄

The procedure for producing the next larger bit string after $b_{n-1} b_{n-2} \ldots b_1 b_0$ is given as Algorithm 2.

ALGORITHM 2 Generating the Next Larger Bit String.

procedure *next bit string*($b_{n-1} b_{n-2} \ldots b_1 b_0$: bit string not equal to $11 \ldots 11$)
$i := 0$
while $b_i = 1$
   $b_i := 0$
   $i := i + 1$
$b_i := 1$
{$b_{n-1} b_{n-2} \ldots b_1 b_0$ is now the next bit string}

Next, an algorithm for generating the r-combinations of the set $\{1, 2, 3, \ldots, n\}$ will be given. An r-combination can be represented by a sequence containing the elements in the subset in increasing order. The r-combinations can be listed using lexicographic order on these sequences. In this lexicographic ordering, the first r-combination is $\{1, 2, \ldots, r-1, r\}$ and the last r-combination is $\{n-r+1, n-r+2, \ldots, n-1, n\}$. The next r-combination after $a_1 a_2 \cdots a_r$ can be obtained in the following way: First, locate the last element a_i in the sequence such that $a_i \neq n - r + i$. Then, replace a_i with $a_i + 1$ and a_j with $a_i + j - i + 1$, for $j = i + 1, i + 2, \ldots, r$. It is left for the reader to show that this produces the next larger r-combination in lexicographic order. This procedure is illustrated with Example 5.

EXAMPLE 5   Find the next larger 4-combination of the set $\{1, 2, 3, 4, 5, 6\}$ after $\{1, 2, 5, 6\}$.

Solution: The last term among the terms a_i with $a_1 = 1$, $a_2 = 2$, $a_3 = 5$, and $a_4 = 6$ such that $a_i \neq 6 - 4 + i$ is $a_2 = 2$. To obtain the next larger 4-combination, increment a_2 by 1 to obtain $a_2 = 3$. Then set $a_3 = 3 + 1 = 4$ and $a_4 = 3 + 2 = 5$. Hence the next larger 4-combination is $\{1, 3, 4, 5\}$. ◀

Algorithm 3 displays pseudocode for this procedure.

ALGORITHM 3 Generating the Next r-Combination in Lexicographic Order.

procedure *next r-combination*($\{a_1, a_2, \ldots, a_r\}$: proper subset of
$\{1, 2, \ldots, n\}$ not equal to $\{n - r + 1, \ldots, n\}$ with
$\quad a_1 < a_2 < \cdots < a_r$)

$i := r$
while $a_i = n - r + i$
$\quad i := i - 1$
$a_i := a_i + 1$
for $j := i + 1$ **to** r
$\quad a_j := a_i + j - i$
$\{\{a_1, a_2, \ldots, a_r\}$ is now the next combination$\}$

Exercises

1. Place these permutations of $\{1, 2, 3, 4, 5\}$ in lexicographic order: 43521, 15432, 45321, 23451, 23514, 14532, 21345, 45213, 31452, 31542.

2. The name of a file in a computer directory consists of three uppercase letters followed by a digit, where each letter is either A, B, or C, and each digit is either 1 or 2. List the name of these files in lexicographic order, where we order letters using the usual alphabetic order of letters.

3. Find the next larger permutation in lexicographic order after each of these permutations.
   a) 1432   **b)** 54123   **c)** 12453
   d) 45231   **e)** 6714235   **f)** 31528764

4. Use Algorithm 1 to generate the 24 permutations of the first four positive integers in lexicographic order.

5. Use Algorithm 3 to list all the 3-combinations of $\{1, 2, 3, 4, 5\}$.

6. Show that Algorithm 1 produces the next larger permutation in lexicographic order.

7. Show that Algorithm 3 produces the next larger r-combination in lexicographic order after a given r-combination.

8. Develop an algorithm for generating the r-permutations of a set of n elements.

9. List all 3-permutations of $\{1, 2, 3, 4, 5\}$.

The remaining exercises in this section develop another algorithm for generating the permutations of $\{1, 2, 3, \ldots, n\}$. This algorithm is based on Cantor expansions of integers. Every nonnegative integer less than $n!$ has a unique Cantor expansion

$$a_1 1! + a_2 2! + \cdots + a_{n-1}(n-1)!$$

where a_i is a nonnegative integer not exceeding i, for $i = 1, 2, \ldots, n - 1$. The integers $a_1, a_2, \ldots, a_{n-1}$ are called the **Cantor digits** of this integer.

Given a permutation of $\{1, 2, \ldots, n\}$, let $a_{k-1}, k = 2, 3, \ldots, n$, be the number of integers less than k that follow k in the permutation. For instance, in the permutation 43215, a_1 is the number of integers less than 2 that follow 2, so $a_1 = 1$. Similarly, for this example $a_2 = 2$, $a_3 = 3$, and $a_4 = 0$. Consider the function from the set of permutations of $\{1, 2, 3, \ldots, n\}$ to the set of nonnegative integers less than $n!$ that sends a permutation to the integer that has $a_1, a_2, \ldots, a_{n-1}$, defined in this way, as its Cantor digits.

10. Find the Cantor digits $a_1, a_2, \ldots, a_{n-1}$ that correspond to these permutations.

    a) 246531   **b)** 12345   **c)** 654321

*11. Show that the correspondence described in the preamble is a bijection between the set of permutations of $\{1, 2, 3, \ldots, n\}$ and the nonnegative integers less than $n!$.

12. Find the permutations of $\{1, 2, 3, 4, 5\}$ that correspond to these integers with respect to the correspondence between Cantor expansions and permutations as described in the preamble to Exercise 14.

    a) 3   **b)** 89   **c)** 111

Key Terms and Results

TERMS

combinatorics: the study of arrangements of objects

enumeration: the counting of arrangements of objects

tree diagram: a diagram made up of a root, branches leaving the root, and other branches leaving some of the endpoints of branches

permutation: an ordered arrangement of the elements of a set

r-permutation: an ordered arrangement of r elements of a set

P(n,r): the number of r-permutations of a set with n elements

r-combination: an unordered selection of r elements of a set

C(n,r): the number of r-combinations of a set with n elements

binomial coefficient $\binom{n}{r}$: also the number of r-combinations of a set with n elements

combinatorial proof: a proof that uses counting arguments rather than algebraic manipulation to prove a result

Pascal's triangle: a representation of the binomial coefficients where the ith row of the triangle contains $\binom{i}{j}$ for $j = 0, 1, 2, \ldots, i$

S(n,j): the Stirling number of the second kind denoting the number of ways to distribute n distinguishable objects into j indistinguishable boxes so that no box is empty

RESULTS

product rule for counting: The number of ways to do a procedure that consists of two tasks is the product of the number of ways to do the first task and the number of ways to do the second task after the first task has been done.

product rule for sets: The number of elements in the Cartesian product of finite sets is the product of the number of elements in each set.

sum rule for counting: The number of ways to do a task in one of two ways is the sum of the number of ways to do these tasks if they cannot be done simultaneously.

sum rule for sets: The number of elements in the union of pairwise disjoint finite sets is the sum of the numbers of elements in these sets.

subtraction rule for counting or **inclusion–exclusion for sets:** If a task can be done in either n_1 ways or n_2 ways, then the number of ways to do the task is $n_1 + n_2$ minus the number of ways to do the task that are common to the two different ways.

subtraction rule or **inclusion–exclusion for sets:** The number of elements in the union of two sets is the sum of the number of elements in these sets minus the number of elements in their intersection.

division rule for counting: There are n/d ways to do a task if it can be done using a procedure that can be carried out in n ways, and for every way w, exactly d of the n ways correspond to way w.

division rule for sets: Suppose that a finite set A is the union of n disjoint subsets each with d elements. Then $n = |A|/d$.

the pigeonhole principle: When more than k objects are placed in k boxes, there must be a box containing more than one object.

the generalized pigeonhole principle: When N objects are placed in k boxes, there must be a box containing at least $\lceil N/k \rceil$ objects.

$$P(n, r) = \frac{n!}{(n-r)!}$$

$$C(n, r) = \binom{n}{r} = \frac{n!}{r!(n-r)!}$$

Pascal's identity: $\binom{n+1}{k} = \binom{n}{k-1} + \binom{n}{k}$

the binomial theorem: $(x + y)^n = \sum_{k=0}^{n} \binom{n}{k} x^{n-k} y^k$

There are n^r r-permutations of a set with n elements when repetition is allowed.

There are $C(n + r - 1, r)$ r-combinations of a set with n elements when repetition is allowed.

There are $n!/(n_1! n_2! \cdots n_k!)$ permutations of n objects of k types where there are n_i indistinguishable objects of type i for $i = 1, 2, 3, \ldots, k$.

the algorithm for generating the permutations of the set $\{1, 2, \ldots, n\}$

Review Questions

1. Explain how the sum and product rules can be used to find the number of bit strings with a length not exceeding 10.

2. Explain how to find the number of bit strings of length not exceeding 10 that have at least one 0 bit.

3. a) How can the product rule be used to find the number of functions from a set with m elements to a set with n elements?

    b) How many functions are there from a set with five elements to a set with 10 elements?

c) How can the product rule be used to find the number of one-to-one functions from a set with m elements to a set with n elements?

d) How many one-to-one functions are there from a set with five elements to a set with 10 elements?

e) How many onto functions are there from a set with five elements to a set with 10 elements?

4. How can you find the number of possible outcomes of a playoff between two teams where the first team that wins four games wins the playoff?

5. How can you find the number of bit strings of length ten that either begin with 101 or end with 010?

6. a) State the pigeonhole principle.

b) Explain how the pigeonhole principle can be used to show that among any 11 integers, at least two must have the same last digit.

7. a) State the generalized pigeonhole principle.

b) Explain how the generalized pigeonhole principle can be used to show that among any 91 integers, there are at least ten that end with the same digit.

8. a) What is the difference between an r-combination and an r-permutation of a set with n elements?

b) Derive an equation that relates the number of r-combinations and the number of r-permutations of a set with n elements.

c) How many ways are there to select six students from a class of 25 to serve on a committee?

d) How many ways are there to select six students from a class of 25 to hold six different executive positions on a committee?

9. a) What is Pascal's triangle?

b) How can a row of Pascal's triangle be produced from the one above it?

10. What is meant by a combinatorial proof of an identity? How is such a proof different from an algebraic one?

11. Explain how to prove Pascal's identity using a combinatorial argument.

12. a) State the binomial theorem.

b) Explain how to prove the binomial theorem using a combinatorial argument.

c) Find the coefficient of $x^{100}y^{101}$ in the expansion of $(2x + 5y)^{201}$.

13. a) Explain how to find a formula for the number of ways to select r objects from n objects when repetition is allowed and order does not matter.

b) How many ways are there to select a dozen objects from among objects of five different types if objects of the same type are indistinguishable?

c) How many ways are there to select a dozen objects from these five different types if there must be at least three objects of the first type?

d) How many ways are there to select a dozen objects from these five different types if there cannot be more than four objects of the first type?

e) How many ways are there to select a dozen objects from these five different types if there must be at least two objects of the first type, but no more than three objects of the second type?

14. a) Let n and r be positive integers. Explain why the number of solutions of the equation $x_1 + x_2 + \cdots + x_n = r$, where x_i is a nonnegative integer for $i = 1, 2, 3, \ldots, n$, equals the number of r-combinations of a set with n elements.

b) How many solutions in nonnegative integers are there to the equation $x_1 + x_2 + x_3 + x_4 = 17$?

c) How many solutions in positive integers are there to the equation in part (b)?

15. a) Derive a formula for the number of permutations of n objects of k different types, where there are n_1 indistinguishable objects of type one, n_2 indistinguishable objects of type two, $\ldots$, and n_k indistinguishable objects of type k.

b) How many ways are there to order the letters of the word *INDISCREETNESS*?

16. Describe an algorithm for generating all the permutations of the set of the n smallest positive integers.

17. a) How many ways are there to deal hands of five cards to six players from a standard 52-card deck?

b) How many ways are there to distribute n distinguishable objects into k distinguishable boxes so that n_i objects are placed in box i?

18. Describe an algorithm for generating all the combinations of the set of the n smallest positive integers.

Supplementary Exercises

1. How many ways are there to choose 6 items from 10 distinct items when

a) the items in the choices are ordered and repetition is not allowed?

b) the items in the choices are ordered and repetition is allowed?

c) the items in the choices are unordered and repetition is not allowed?

d) the items in the choices are unordered and repetition is allowed?

2. A test contains 100 true/false questions. How many different ways can a student answer the questions on the test, if answers may be left blank?

3. How many bit strings of length 10 over the alphabet $\{a, b, c\}$ have either exactly three as or exactly four bs?

4. An ice cream parlor has 28 different flavors, 8 different kinds of sauce, and 12 toppings.

a) In how many different ways can a dish of three scoops of ice cream be made where each flavor can be used more than once and the order of the scoops does not matter?

b) How many different kinds of small sundaes are there if a small sundae contains one scoop of ice cream, a sauce, and a topping?

c) How many different kinds of large sundaes are there if a large sundae contains three scoops of ice cream, where each flavor can be used more than once and the order of the scoops does not matter; two kinds of sauce, where each sauce can be used only once and the order of the sauces does not matter; and three toppings, where each topping can be used only once and the order of the toppings does not matter?

5. When the numbers from 1 to 1000 are written out in decimal notation, how many of each of these digits are used?

a) 0   **b)** 1   **c)** 2   **d)** 9

6. There are 12 signs of the zodiac. How many people are needed to guarantee that at least six of these people have the same sign?

7. A fortune cookie company makes 213 different fortunes. A student eats at a restaurant that uses fortunes from this company and gives each customer one fortune cookie at the end of a meal. What is the largest possible number of times that the student can eat at the restaurant without getting the same fortune four times?

8. How many people are needed to guarantee that at least two were born on the same day of the week and in the same month (perhaps in different years)?

9. Show that given any set of 10 positive integers not exceeding 50 there exist at least two different five-element subsets of this set that have the same sum.

10. A package of baseball cards contains 20 cards. How many packages must be purchased to ensure that two cards in these packages are identical if there are a total of 550 different cards?

11. a) How many cards must be chosen from a standard deck of 52 cards to guarantee that at least two of the four aces are chosen?

b) How many cards must be chosen from a standard deck of 52 cards to guarantee that at least two of the four aces and at least two of the 13 kinds are chosen?

c) How many cards must be chosen from a standard deck of 52 cards to guarantee that there are at least two cards of the same kind?

d) How many cards must be chosen from a standard deck of 52 cards to guarantee that there are at least two cards of each of two different kinds?

*12. Show that in any set of $n + 1$ positive integers not exceeding $2n$ there must be two that are relatively prime.

*13. Show that in a sequence of m integers there exists one or more consecutive terms with a sum divisible by m.

14. Show that if five points are picked in the interior of a square with a side length of 2, then at least two of these points are no farther than $\sqrt{2}$ apart.

15. Show that the decimal expansion of a rational number must repeat itself from some point onward.

16. Once a computer worm infects a personal computer via an infected e-mail message, it sends a copy of itself to 100 e-mail addresses it finds in the electronic message mailbox on this personal computer. What is the maximum number of different computers this one computer can infect in the time it takes for the infected message to be forwarded five times?

17. How many ways are there to choose a dozen donuts from 20 varieties

a) if there are no two donuts of the same variety?

b) if all donuts are of the same variety?

c) if there are no restrictions?

d) if there are at least two varieties among the dozen donuts chosen?

e) if there must be at least six blueberry-filled donuts?

f) if there can be no more than six blueberry-filled donuts?

18. Find n if

a) $P(n, 2) = 110$.   **b)** $P(n, n) = 5040$.

c) $P(n, 4) = 12P(n, 2)$.

19. Find n if

a) $C(n, 2) = 45$.   **b)** $C(n, 3) = P(n, 2)$.

c) $C(n, 5) = C(n, 2)$.

20. Show that if n and r are nonnegative integers and $n \geq r$, then

$$P(n + 1, r) = P(n, r)(n + 1)/(n + 1 - r).$$

*21. Suppose that S is a set with n elements. How many ordered pairs (A, B) are there such that A and B are subsets of S with $A \subseteq B$? [*Hint:* Show that each element of S belongs to A, $B - A$, or $S - B$.]

22. Give a combinatorial proof of Corollary 2 of Section 6.4 by setting up a correspondence between the subsets of a set with an even number of elements and the subsets of this set with an odd number of elements. [*Hint:* Take an element a in the set. Set up the correspondence by putting a in the subset if it is not already in it and taking it out if it is in the subset.]

23. Let n and r be integers with $1 \leq r < n$. Show that

$$C(n, r - 1) = C(n + 2, r + 1)$$
$$- 2C(n + 1, r + 1) + C(n, r + 1).$$

24. Prove using mathematical induction that $\sum_{j=2}^{n} C(j, 2) = C(n + 1, 3)$ whenever n is an integer greater than 1.

25. Show that if n is an integer then

$$\sum_{k=0}^{n} 3^k \binom{n}{k} = 4^n.$$

26. In this exercise we will derive a formula for the sum of the squares of the n smallest positive integers. We will count the number of triples (i, j, k) where i, j, and k are integers such that $0 \leq i < k$, $0 \leq j < k$, and $1 \leq k \leq n$ in two ways.

a) Show that there are k^2 such triples with a fixed k. Deduce that there are $\sum_{k=1}^{n} k^2$ such triples.

b) Show that the number of such triples with $0 \le i < j < k$ and the number of such triples with $0 \le j < i < k$ both equal $C(n + 1, 3)$.

c) Show that the number of such triples with $0 \le i = j < k$ equals $C(n + 1, 2)$.

d) Combining part (a) with parts (b) and (c), conclude that

$$\sum_{k=1}^{n} k^2 = 2C(n + 1, 3) + C(n + 1, 2)$$
$$= n(n + 1)(2n + 1)/6.$$

27. A professor writes 20 multiple-choice questions, each with the possible answer a, b, c, or d, for a discrete mathematics test. If the number of questions with a, b, c, and d as their answer is 8, 3, 4, and 5, respectively, how many different answer keys are possible, if the questions can be placed in any order?

28. How many different arrangements are there of eight people seated at a round table, where two arrangements are considered the same if one can be obtained from the other by a rotation?

29. How many ways are there to assign 24 students to five faculty advisors?

30. How many ways are there to choose a dozen apples from a bushel containing 20 indistinguishable Delicious apples, 20 indistinguishable Macintosh apples, and 20 indistinguishable Granny Smith apples, if at least three of each kind must be chosen?

31. How many subsets of a set with ten elements

a) have fewer than five elements?

b) have more than seven elements?

c) have an odd number of elements?

32. A witness to a hit-and-run accident tells the police that the license plate of the car in the accident, which contains three letters followed by three digits, starts with the letters AS and contains both the digits 1 and 2. How many different license plates can fit this description?

33. How many ways are there to put n identical objects into m distinct containers so that no container is empty?

34. How many ways are there to seat six boys and eight girls in a row of chairs so that no two boys are seated next to each other?

35. How many ways are there to distribute six objects to five boxes if

a) both the objects and boxes are labeled?

b) the objects are labeled, but the boxes are unlabeled?

c) the objects are unlabeled, but the boxes are labeled?

d) both the objects and the boxes are unlabeled?

36. How many ways are there to distribute five objects into six boxes if

a) both the objects and boxes are labeled?

b) the objects are labeled, but the boxes are unlabeled?

c) the objects are unlabeled, but the boxes are labeled?

d) both the objects and the boxes are unlabeled?

The **signless Stirling number of the first kind** $c(n, k)$, where k and n are integers with $1 \le k \le n$, equals the number of ways to arrange n people around k circular tables with at least one person seated at each table, where two seatings of m people around a circular table are considered the same if everyone has the same left neighbor and the same right neighbor.

37. Find these signless Stirling numbers of the first kind.

a) $c(3,2)$ **b)** $c(4,2)$

c) $c(4,3)$ **d)** $c(5,4)$

38. Show that if n is a positive integer, then $\sum_{j=1}^{n} c(n, j) = n!$.

39. Show that if n is a positive integer with $n \ge 3$, then $c(n, n - 2) = (3n - 1)C(n, 3)/4$.

40. Give a combinatorial proof that 2^n divides $n!$ whenever n is an even positive integer. [*Hint:* Use Theorem 3 in Section 6.5 to count the number of permutations of $2n$ objects where there are two indistinguishable objects of n different types.

41. How many 11-element RNA sequences consist of 4 As, 3Cs, 2Us, and 2Gs, and end with CAA?

42. Devise an algorithm for generating all the r-permutations of a finite set when repetition is allowed.

Computer Projects

Write programs with these input and output.

1. Given a positive integer n and a nonnegative integer not exceeding n, find the number of r-permutations and r-combinations of a set with n elements.

2. Given positive integers n and r, find the number of r-permutations when repetition is allowed and r-combinations when repetition is allowed of a set with n elements.

3. Given a sequence of positive integers, find the longest increasing and the longest decreasing subsequence of the sequence.

*4. Given an equation $x_1 + x_2 + \cdots + x_n = C$, where C is a constant, and $x_1, x_2, \ldots, x_n$ are nonnegative integers, list all the solutions.

5. Given a positive integer n, list all the permutations of the set $\{1, 2, 3, \ldots, n\}$ in lexicographic order.

6. Given a positive integer n and a nonnegative integer r not exceeding n, list all the r-combinations of the set $\{1, 2, 3, \ldots, n\}$ in lexicographic order.

7. Given a positive integer n and a nonnegative integer r not exceeding n, list all the r-permutations of the set $\{1, 2, 3, \ldots, n\}$ in lexicographic order.

8. Given a positive integer n, list all the combinations of the set $\{1, 2, 3, \ldots, n\}$.

9. Given positive integers n and r, list all the r-permutations, with repetition allowed, of the set $\{1, 2, 3, \ldots, n\}$.

10. Given positive integers n and r, list all the r-combinations, with repetition allowed, of the set $\{1, 2, 3, \ldots, n\}$.

Computations and Explorations

Use a computational program or programs you have written to do these exercises.

1. Find the number of possible outcomes in a two-team play-off when the winner is the first team to win 5 out of 9, 6 out of 11, 7 out of 13, and 8 out of 15.

2. Which binomial coefficients are odd? Can you formulate a conjecture based on numerical evidence?

3. Find as many odd integers n less than 200 as you can for which $C(n, \lfloor n/2 \rfloor)$ is not divisible by the square of a prime. Formulate a conjecture based on your evidence.

4. Generate all the permutations of a set with eight elements.

5. Generate all combinations of a set with eight elements.

Writing Projects

Respond to these with essays using outside sources.

1. Describe some of the earliest uses of the pigeonhole principle by Dirichlet and other mathematicians.

2. Discuss ways in which the current telephone numbering plan can be extended to accommodate the rapid demand for more telephone numbers. (See if you can find some of the proposals coming from the telecommunications industry.) For each new numbering plan you discuss, show how to find the number of different telephone numbers it supports.

3. Discuss the importance of combinatorial reasoning in gene sequencing and related problems involving genomes.

4. Many combinatorial identities are described in this book. Find some sources of such identities and describe important combinatorial identities besides those already introduced in this book. Give some representative proofs, including combinatorial ones, of some of these identities.

5. Describe the different models used to model the distribution of particles in statistical mechanics, including Maxwell–Boltzmann, Bose–Einstein, and Fermi–Dirac statistics. In each case, describe the counting techniques used in the model.

6. Describe at least one way to generate all the partitions of a positive integer n. (See Exercise 37 in Section 5.3.)

7

Discrete Probability

Combinatorics and probability theory share common origins. The theory of probability was first developed more than 300 years ago, when certain gambling games were analyzed. Although probability theory was originally invented to study gambling, it now plays an essential role in a wide variety of disciplines. For example, probability theory is extensively applied in the study of genetics, where it can be used to help understand the inheritance of traits. Of course, probability still remains an extremely popular part of mathematics because of its applicability to gambling, which continues to be an extremely popular human endeavor.

In computer science, probability theory plays an important role in the study of the complexity of algorithms. In particular, ideas and techniques from probability theory are used to determine the average-case complexity of algorithms. Probabilistic algorithms can be used to solve many problems that cannot be easily or practically solved by deterministic algorithms. In a probabilistic algorithm, instead of always following the same steps when given the same input, as a deterministic algorithm does, the algorithm makes one or more random choices, which may lead to different output. In combinatorics, probability theory can even be used to show that objects with certain properties exist. The probabilistic method, a technique in combinatorics introduced by Paul Erdős and Alfréd Rényi, shows that an object with a specified property exists by showing that there is a positive probability that a randomly constructed object has this property. Probability theory can help us answer questions that involve uncertainty, such as determining whether we should reject an incoming mail message as spam based on the words that appear in the message.

7.1 An Introduction to Discrete Probability

Introduction

Probability theory dates back to 1526 when the Italian mathematician, physician, and gambler Girolamo Cardano wrote the first known systematic treatment of the subject in his book *Liber de Ludo Aleae* (*Book on Games of Chance*). (This book was not published until 1663, which may have held back the development of probability theory.) In the seventeenth century the French mathematician Blaise Pascal determined the odds of winning some popular bets based on the outcome when a pair of dice is repeatedly rolled. In the eighteenth century, the French mathematician Laplace, who also studied gambling, defined the probability of an event as the number of successful outcomes divided by the number of possible outcomes. For instance, the probability that a die comes up an odd number when it is rolled is the number of successful outcomes—namely, the number of ways it can come up odd—divided by the number of possible outcomes—namely, the number of different ways the die can come up. There are a total of six possible outcomes—namely, 1, 2, 3, 4, 5, and 6—and exactly three of these are successful outcomes—namely, 1, 3, and 5. Hence, the probability that the die comes up an odd number is $3/6 = 1/2$. (Note that it has been assumed that all possible outcomes are equally likely, or, in other words, that the die is fair.)

In this section we will restrict ourselves to experiments that have finitely many, equally likely, outcomes. This permits us to use Laplace's definition of the probability of an event. We will continue our study of probability in Section 7.2, where we will study experiments with finitely many outcomes that are not necessarily equally likely. In Section 7.2 we will also introduce some key concepts in probability theory, including conditional probability, independence of events, and random variables. In Section 7.4 we will introduce the concepts of the expectation and variance of a random variable.

Finite Probability

An **experiment** is a procedure that yields one of a given set of possible outcomes. The **sample space** of the experiment is the set of possible outcomes. An **event** is a subset of the sample space. Laplace's definition of the probability of an event with finitely many possible outcomes will now be stated.

DEFINITION 1   If S is a finite nonempty sample space of equally likely outcomes, and E is an event, that is, a subset of S, then the *probability* of E is $p(E) = \dfrac{|E|}{|S|}$.

The probability of an event can never be negative or more than one!

According to Laplace's definition, the probability of an event is between 0 and 1. To see this, note that if E is an event from a finite sample space S, then $0 \le |E| \le |S|$, because $E \subseteq S$. Thus, $0 \le p(E) = |E|/|S| \le 1$.

Examples 1–7 illustrate how the probability of an event is found.

EXAMPLE 1   An urn contains four blue balls and five red balls. What is the probability that a ball chosen at random from the urn is blue?

Solution: To calculate the probability, note that there are nine possible outcomes, and four of these possible outcomes produce a blue ball. Hence, the probability that a blue ball is chosen is 4/9.   ◀

EXAMPLE 2   What is the probability that when two dice are rolled, the sum of the numbers on the two dice is 7?

Solution: There are a total of 36 equally likely possible outcomes when two dice are rolled. (The product rule can be used to see this; because each die has six possible outcomes, the total

GIROLAMO CARDANO (1501–1576)   Cardano, born in Pavia, Italy, was the illegitimate child of Fazio Cardano, a lawyer, mathematician, and friend of Leonardo da Vinci, and Chiara Micheria, a young widow. In spite of illness and poverty, Cardano was able to study at the universities of Pavia and Padua, from where he received his medical degree. Cardano was not accepted into Milan's College of Physicians because of his illegitimate birth, as well as his eccentricity and confrontational style. Nevertheless, his medical skills were highly regarded. One of his main accomplishments as a physician is the first description of typhoid fever.

Cardano published more than 100 books on a diverse range of subjects, including medicine, the natural sciences, mathematics, gambling, physical inventions and experiments, and astrology. He also wrote a fascinating autobiography. In mathematics, Cardano's book *Ars Magna*, published in 1545, established the foundations of abstract algebra. This was the most comprehensive book on abstract algebra for more than a century; it presents many novel ideas of Cardano and of others, including methods for solving cubic and quartic equations from their coefficients. Cardano also made several important contributions to cryptography. Cardano was an advocate of education for the deaf, believing, unlike his contemporaries, that deaf people could learn to read and write before learning to speak, and could use their minds just as well as hearing people.

Cardano was often short of money. However, he kept himself solvent through gambling and winning money by beating others at chess. His book about games of chance, *Liber de Ludo Aleae*, written in 1526 (but published in 1663), offers the first systematic treatment of probability; it also describes effective ways to cheat. Cardano was considered to be a man of dubious moral character; he was often described as a liar, gambler, lecher, and heretic.

number of outcomes when two dice are rolled is $6^2 = 36$.) There are six successful outcomes, namely, (1, 6), (2, 5), (3, 4), (4, 3), (5, 2), and (6, 1), where the values of the first and second dice are represented by an ordered pair. Hence, the probability that a seven comes up when two fair dice are rolled is $6/36 = 1/6$. ◀

Lotteries are extremely popular throughout the world. We can easily compute the odds of winning different types of lotteries, as illustrated in Examples 3 and 4. (The odd of winning the popular Mega Millions and Powerball lotteries are studied in the supplementary exercises.)

EXAMPLE 3   In a lottery, players win a large prize when they pick four digits that match, in the correct order, four digits selected by a random mechanical process. A smaller prize is won if only three digits are matched. What is the probability that a player wins the large prize? What is the probability that a player wins the small prize?

Solution: There is only one way to choose all four digits correctly. By the product rule, there are $10^4 = 10,000$ ways to choose four digits. Hence, the probability that a player wins the large prize is $1/10,000 = 0.0001$.

Players win the smaller prize when they correctly choose exactly three of the four digits. Exactly one digit must be wrong to get three digits correct, but not all four correct. By the sum rule, to find the number of ways to choose exactly three digits correctly, we add the number of ways to choose four digits matching the digits picked in all but the ith position, for $i = 1, 2, 3, 4$.

To count the number of successes with the first digit incorrect, note that there are nine possible choices for the first digit (all but the one correct digit), and one choice for each of the other digits, namely, the correct digits for these slots. Hence, there are nine ways to choose four digits where the first digit is incorrect, but the last three are correct. Similarly, there are nine ways to choose four digits where the second digit is incorrect, nine with the third digit incorrect, and nine with the fourth digit incorrect. Hence, there is a total of 36 ways to choose four digits with exactly three of the four digits correct. Thus, the probability that a player wins the smaller prize is $36/10,000 = 9/2500 = 0.0036$. ◀

EXAMPLE 4   There are many lotteries now that award enormous prizes to people who correctly choose a set of six numbers out of the first n positive integers, where n is usually between 30 and 60. What is the probability that a person picks the correct six numbers out of 40?

Solution: There is only one winning combination. The total number of ways to choose six numbers out of 40 is

$$C(40, 6) = \frac{40!}{34! \, 6!} = 3,838,380.$$

Consequently, the probability of picking a winning combination is $1/3,838,380 \approx 0.00000026$. (Here the symbol $\approx$ means approximately equal to.) ◀

PIERRE-SIMON LAPLACE (1749–1827)   Pierre-Simon Laplace came from humble origins in Normandy. In his childhood he was educated in a school run by the Benedictines. At 16 he entered the University of Caen intending to study theology. However, he soon realized his true interests were in mathematics. After completing his studies, he was named a provisional professor at Caen, and in 1769 he became professor of mathematics at the Paris Military School.

Laplace is best known for his contributions to celestial mechanics, the study of the motions of heavenly bodies. His *Traité de Mécanique Céleste* is considered one of the greatest scientific works of the early nineteenth century. Laplace was one of the founders of probability theory and made many contributions to mathematical statistics. His work in this area is documented in his book *Théorie Analytique des Probabilités*, in which he defined the probability of an event as the ratio of the number of favorable outcomes to the total number of outcomes of an experiment.

Laplace was famous for his political flexibility. He was loyal, in succession, to the French Republic, Napoleon, and King Louis XVIII. This flexibility permitted him to be productive before, during, and after the French Revolution.

Links

Poker, and other card games, are growing in popularity. To win at these games it helps to know the probability of different hands. We can find the probability of specific hands that arise in card games using the techniques developed so far. A deck of cards contains 52 cards. There are 13 different kinds of cards, with four cards of each kind. (Among the terms commonly used instead of "kind" are "rank," "face value," "denomination," and "value.") These kinds are twos, threes, fours, fives, sixes, sevens, eights, nines, tens, jacks, queens, kings, and aces. There are also four suits: spades, clubs, hearts, and diamonds, each containing 13 cards, with one card of each kind in a suit. In many poker games, a hand consists of five cards.

EXAMPLE 5 Find the probability that a hand of five cards in poker contains four cards of one kind.

Solution: By the product rule, the number of hands of five cards with four cards of one kind is the product of the number of ways to pick one kind, the number of ways to pick the four of this kind out of the four in the deck of this kind, and the number of ways to pick the fifth card. This is

$$C(13, 1)C(4, 4)C(48, 1).$$

By Example 11 in Section 6.3 there are $C(52, 5)$ different hands of five cards. Hence, the probability that a hand contains four cards of one kind is

$$\frac{C(13, 1)C(4, 4)C(48, 1)}{C(52, 5)} = \frac{13 \cdot 1 \cdot 48}{2,598,960} \approx 0.00024. \qquad \blacktriangleleft$$

EXAMPLE 6 What is the probability that a poker hand contains a full house, that is, three of one kind and two of another kind?

Solution: By the product rule, the number of hands containing a full house is the product of the number of ways to pick two kinds in order, the number of ways to pick three out of four for the first kind, and the number of ways to pick two out of four for the second kind. (Note that the order of the two kinds matters, because, for instance, three queens and two aces is different from three aces and two queens.) We see that the number of hands containing a full house is

$$P(13, 2)C(4, 3)C(4, 2) = 13 \cdot 12 \cdot 4 \cdot 6 = 3744.$$

Because there are $C(52, 5) = 2,598,960$ poker hands, the probability of a full house is

$$\frac{3744}{2,598,960} \approx 0.0014. \qquad \blacktriangleleft$$

EXAMPLE 7 What is the probability that the numbers 11, 4, 17, 39, and 23 are drawn in that order from a bin containing 50 balls labeled with the numbers 1, 2, . . . , 50 if (a) the ball selected is not returned to the bin before the next ball is selected and (b) the ball selected is returned to the bin before the next ball is selected?

Solution: (a) By the product rule, there are $50 \cdot 49 \cdot 48 \cdot 47 \cdot 46 = 254,251,200$ ways to select the balls because each time a ball is drawn there is one fewer ball to choose from. Consequently, the probability that 11, 4, 17, 39, and 23 are drawn in that order is $1/254,251,200$. This is an example of **sampling without replacement**.

(b) By the product rule, there are $50^5 = 312,500,000$ ways to select the balls because there are 50 possible balls to choose from each time a ball is drawn. Consequently, the probability that 11, 4, 17, 39, and 23 are drawn in that order is $1/312,500,000$. This is an example of **sampling with replacement**. $\qquad \blacktriangleleft$

Probabilities of Complements and Unions of Events

We can use counting techniques to find the probability of events derived from other events.

THEOREM 1   Let E be an event in a sample space S. The probability of the event $\overline{E} = S - E$, the complementary event of E, is given by

$$p(\overline{E}) = 1 - p(E).$$

Proof: To find the probability of the event $\overline{E} = S - E$, note that $|\overline{E}| = |S| - |E|$. Hence,

$$p(\overline{E}) = \frac{|S| - |E|}{|S|} = 1 - \frac{|E|}{|S|} = 1 - p(E).$$

◁

There is an alternative strategy for finding the probability of an event when a direct approach does not work well. Instead of determining the probability of the event, the probability of its complement can be found. This is often easier to do, as Example 8 shows.

EXAMPLE 8   A sequence of 10 bits is randomly generated. What is the probability that at least one of these bits is 0?

Solution: Let E be the event that at least one of the 10 bits is 0. Then $\overline{E}$ is the event that all the bits are 1s. Because the sample space S is the set of all bit strings of length 10, it follows that

$$p(E) = 1 - p(\overline{E}) = 1 - \frac{|\overline{E}|}{|S|} = 1 - \frac{1}{2^{10}}$$

$$= 1 - \frac{1}{1024} = \frac{1023}{1024}.$$

Hence, the probability that the bit string will contain at least one 0 bit is $1023/1024$. It is quite difficult to find this probability directly without using Theorem 1.   ◀

We can also find the probability of the union of two events.

THEOREM 2   Let E_1 and E_2 be events in the sample space S. Then

$$p(E_1 \cup E_2) = p(E_1) + p(E_2) - p(E_1 \cap E_2).$$

Proof: Using the formula given in Section 2.2 for the number of elements in the union of two sets, it follows that

$$|E_1 \cup E_2| = |E_1| + |E_2| - |E_1 \cap E_2|.$$

Hence,

$$p(E_1 \cup E_2) = \frac{|E_1 \cup E_2|}{|S|}$$

$$= \frac{|E_1| + |E_2| - |E_1 \cap E_2|}{|S|}$$

$$= \frac{|E_1|}{|S|} + \frac{|E_2|}{|S|} - \frac{|E_1 \cap E_2|}{|S|}$$

$$= p(E_1) + p(E_2) - p(E_1 \cap E_2).$$

◁

EXAMPLE 9

What is the probability that a positive integer selected at random from the set of positive integers not exceeding 100 is divisible by either 2 or 5?

Solution: Let E_1 be the event that the integer selected at random is divisible by 2, and let E_2 be the event that it is divisible by 5. Then $E_1 \cup E_2$ is the event that it is divisible by either 2 or 5. Also, $E_1 \cap E_2$ is the event that it is divisible by both 2 and 5, or equivalently, that it is divisible by 10. Because $|E_1| = 50$, $|E_2| = 20$, and $|E_1 \cap E_2| = 10$, it follows that

$$p(E_1 \cup E_2) = p(E_1) + p(E_2) - p(E_1 \cap E_2)$$

$$= \frac{50}{100} + \frac{20}{100} - \frac{10}{100} = \frac{3}{5}.$$

◀

Probabilistic Reasoning

A common problem is determining which of two events is more likely. Analyzing the probabilities of such events can be tricky. Example 10 describes a problem of this type. It discusses a famous problem originating with the television game show *Let's Make a Deal* and named after the host of the show, Monty Hall.

EXAMPLE 10

The Monty Hall Three-Door Puzzle   Suppose you are a game show contestant. You have a chance to win a large prize. You are asked to select one of three doors to open; the large prize is behind one of the three doors and the other two doors are losers. Once you select a door, the game show host, who knows what is behind each door, does the following. First, whether or not you selected the winning door, he opens one of the other two doors that he knows is a losing door (selecting at random if both are losing doors). Then he asks you whether you would like to switch doors. Which strategy should you use? Should you change doors or keep your original selection, or does it not matter?

Solution: The probability you select the correct door (before the host opens a door and asks you whether you want to change) is 1/3, because the three doors are equally likely to be the correct door. The probability this is the correct door does not change once the game show host opens one of the other doors, because he will always open a door that the prize is not behind.

   The probability that you selected incorrectly is the probability the prize is behind one of the two doors you did not select. Consequently, the probability that you selected incorrectly is 2/3. If you selected incorrectly, when the game show host opens a door to show you that the prize is not behind it, the prize is behind the other door. You will always win if your initial choice was incorrect and you change doors. So, by changing doors, the probability you win is 2/3. In other words, you should always change doors when given the chance to do so by the game show host. This doubles the probability that you will win. (A more rigorous treatment of this puzzle can be found in Exercise 9 of Section 7.3. For much more on this famous puzzle and its variations, see [Ro09].)

◀

Exercises

1. What is the probability that a card selected at random from a standard deck of 52 cards is an ace?

2. What is the probability that a fair die comes up six when it is rolled?

3. What is the probability that a randomly selected integer chosen from the first 100 positive integers is odd?

4. What is the probability that the sum of the numbers on two dice is even when they are rolled?

5. What is the probability that when a coin is flipped six times in a row, it lands heads up every time?

6. What is the probability that a five-card poker hand contains the ace of hearts?

7. What is the probability that a five-card poker hand does not contain the queen of hearts?

8. What is the probability that a five-card poker hand contains the two of diamonds and the three of spades?

9. What is the probability that a five-card poker hand contains the two of diamonds, the three of spades, the six of hearts, the ten of clubs, and the king of hearts?

10. What is the probability that a five-card poker hand contains exactly one ace?

11. What is the probability that a five-card poker hand contains at least one ace?

12. What is the probability that a five-card poker hand contains cards of five different kinds?

13. What is the probability that a five-card poker hand contains two pairs (that is, two of each of two different kinds and a fifth card of a third kind)?

14. What is the probability that a five-card poker hand contains a straight, that is, five cards that have consecutive kinds? (Note that an ace can be considered either the lowest card of an A-2-3-4-5 straight or the highest card of a 10-J-Q-K-A straight.)

*15. What is the probability that a five-card poker hand contains cards of five different kinds and does not contain a flush or a straight?

16. What is the probability that a five-card poker hand contains a royal flush, that is, the 10, jack, queen, king, and ace of one suit?

17. What is the probability that a fair die never comes up an even number when it is rolled six times?

18. What is the probability that a positive integer not exceeding 100 selected at random is divisible by 3?

19. What is the probability that a positive integer not exceeding 100 selected at random is divisible by 5 or 7?

20. Find the probability of winning a lottery by selecting the correct six integers, where the order in which these integers are selected does not matter, from the positive integers not exceeding
    a) 50.   b) 52.   c) 56.   d) 60.

21. Find the probability of selecting exactly one of the correct six integers in a lottery, where the order in which these integers are selected does not matter, from the positive integers not exceeding
    a) 40.   b) 48.   c) 56.   d) 64.

22. In a superlottery, players win a fortune if they choose the eight numbers selected by a computer from the positive integers not exceeding 100. What is the probability that a player wins this superlottery?

23. Suppose that 100 people enter a contest and that different winners are selected at random for first, second, and third prizes. What is the probability that Michelle wins one of these prizes if she is one of the contestants?

24. Suppose that 100 people enter a contest and that different winners are selected at random for first, second, and third prizes. What is the probability that Kumar, Janice, and Pedro each win a prize if each has entered the contest?

25. What is the probability that Abby, Barry, and Sylvia win the first, second, and third prizes, respectively, in a drawing if 200 people enter a contest and
    a) no one can win more than one prize.
    b) winning more than one prize is allowed.

26. What is the probability that Bo, Colleen, Jeff, and Rohini win the first, second, third, and fourth prizes, respectively, in a drawing if 50 people enter a contest and
    a) no one can win more than one prize.
    b) winning more than one prize is allowed.

27. In roulette, a wheel with 38 numbers is spun. Of these, 18 are red, and 18 are black. The other two numbers, which are neither black nor red, are 0 and 00. The probability that when the wheel is spun it lands on any particular number is 1/38.
    a) What is the probability that the wheel lands on a red number?
    b) What is the probability that the wheel lands on a black number twice in a row?
    c) What is the probability that the wheel lands on 0 or 00?
    d) What is the probability that in five spins the wheel never lands on either 0 or 00?
    e) What is the probability that the wheel lands on one of the first six integers on one spin, but does not land on any of them on the next spin?

28. Which is more likely: rolling a total of 8 when two dice are rolled or rolling a total of 8 when three dice are rolled?

29. Which is more likely: rolling a total of 9 when two dice are rolled or rolling a total of 9 when three dice are rolled?

30. Two events E_1 and E_2 are called **independent** if $p(E_1 \cap E_2) = p(E_1)p(E_2)$. For each of the following pairs of events, which are subsets of the set of all possible outcomes when a coin is tossed three times, determine whether or not they are independent.
    a) E_1: tails comes up with the coin is tossed the first time; E_2: heads comes up when the coin is tossed the second time.

b) E_1: the first coin comes up tails; E_2: two, and not three, heads come up in a row.

c) E_1: the second coin comes up tails; E_2: two, and not three, heads come up in a row.

(We will study independence of events in more depth in Section 7.2.)

31. Explain what is wrong with the statement that in the Monty Hall Three-Door Puzzle the probability that the prize is behind the first door you select and the probability that the prize is behind the other of the two doors that Monty does not open are both 1/2, because there are two doors left.

32. Suppose that instead of three doors, there are four doors in the Monty Hall puzzle. What is the probability that you win by not changing once the host, who knows what is behind each door, opens a losing door and gives you the chance to change doors? What is the probability that you win by changing the door you select to one of the two remaining doors among the three that you did not select?

33. This problem was posed by the Chevalier de Méré and was solved by Blaise Pascal and Pierre de Fermat.

a) Find the probability of rolling at least one six when a fair die is rolled four times.

b) Find the probability that a double six comes up at least once when a pair of dice is rolled 24 times. Answer the query the Chevalier de Méré made to Pascal asking whether this probability was greater than 1/2.

c) Is it more likely that a six comes up at least once when a fair die is rolled four times or that a double six comes up at least once when a pair of dice is rolled 24 times?

7.2 Probability Theory

Introduction

Links

In Section 7.1 we introduced the notion of the probability of an event. (Recall that an event is a subset of the possible outcomes of an experiment.) We defined the probability of an event E as Laplace did, that is,

$$p(E) = \frac{|E|}{|S|},$$

the number of outcomes in E divided by the total number of outcomes. This definition assumes that all outcomes are equally likely. However, many experiments have outcomes that are not equally likely. For instance, a coin may be biased so that it comes up heads twice as often as tails. Similarly, the likelihood that the input of a linear search is a particular element in a list, or is not in the list, depends on how the input is generated. How can we model the likelihood of events in such situations? In this section we will show how to define probabilities of outcomes to study probabilities of experiments where outcomes may not be equally likely.

Suppose that a fair coin is flipped four times, and the first time it comes up heads. Given this information, what is the probability that heads comes up three times? To answer this and similar questions, we will introduce the concept of *conditional probability*. Does knowing that the first flip comes up heads change the probability that heads comes up three times? If not, these two events are called *independent*, a concept studied later in this section.

Many questions address a particular numerical value associated with the outcome of an experiment. For instance, when we flip a coin 100 times, what is the probability that exactly 40 heads appear? How many heads should we expect to appear? In this section we will introduce *random variables*, which are functions that associate numerical values to the outcomes of experiments.

Links

HISTORICAL NOTE   The Chevalier de Méré was a French nobleman, a famous gambler, and a bon vivant. He was successful at making bets with odds slightly greater than 1/2 (such as having at least one six come up in four tosses of a fair die). His correspondence with Pascal asking about the probability of having at least one double six come up when a pair of dice is rolled 24 times led to the development of probability theory. According to one account, Pascal wrote to Fermat about the Chevalier saying something like "He's a good guy but, alas, he's no mathematician."

Assigning Probabilities

Let S be the sample space of an experiment with a finite or countable number of outcomes. We assign a probability $p(s)$ to each outcome s. We require that two conditions be met:

(*i*) $0 \leq p(s) \leq 1$ for each $s \in S$

and

(*ii*) $\displaystyle\sum_{s \in S} p(s) = 1.$

Condition (*i*) states that the probability of each outcome is a nonnegative real number no greater than 1. Condition (*ii*) states that the sum of the probabilities of all possible outcomes should be 1; that is, when we do the experiment, it is a certainty that one of these outcomes occurs. (Note that when the sample space is infinite, $\sum_{s \in S} p(s)$ is a convergent infinite series.) This is a generalization of Laplace's definition in which each of n outcomes is assigned a probability of $1/n$. Indeed, conditions (*i*) and (*ii*) are met when Laplace's definition of probabilities of equally likely outcomes is used and S is finite. (See Exercise 4.)

Note that when there are n possible outcomes, $x_1, x_2, \ldots, x_n$, the two conditions to be met are

(*i*) $0 \leq p(x_i) \leq 1$ for $i = 1, 2, \ldots, n$

and

(*ii*) $\displaystyle\sum_{i=1}^{n} p(x_i) = 1.$

The function p from the set of all outcomes of the sample space S is called a **probability distribution**.

To model an experiment, the probability $p(s)$ assigned to an outcome s should equal the limit of the number of times s occurs divided by the number of times the experiment is performed, as this number grows without bound. (We will assume that all experiments discussed have outcomes that are predictable on the average, so that this limit exists. We also assume that the outcomes of successive trials of an experiment do not depend on past results.)

Remark: We will not discuss probabilities of events when the set of outcomes is not finite or countable, such as when the outcome of an experiment can be any real number. In such cases, integral calculus is usually required for the study of the probabilities of events.

We can model experiments in which outcomes are either equally likely or not equally likely by choosing the appropriate function $p(s)$, as Example 1 illustrates.

EXAMPLE 1   What probabilities should we assign to the outcomes H (heads) and T (tails) when a fair coin is flipped? What probabilities should be assigned to these outcomes when the coin is biased so that heads comes up twice as often as tails?

Solution: For a fair coin, the probability that heads comes up when the coin is flipped equals the probability that tails comes up, so the outcomes are equally likely. Consequently, we assign the probability $1/2$ to each of the two possible outcomes, that is, $p(H) = p(T) = 1/2$.

For the biased coin we have

$$p(H) = 2p(T).$$

Because

$$p(H) + p(T) = 1,$$

it follows that

$$2p(T) + p(T) = 3p(T) = 1.$$

We conclude that $p(T) = 1/3$ and $p(H) = 2/3$. ◀

DEFINITION 1   Suppose that S is a set with n elements. The *uniform distribution* assigns the probability $1/n$ to each element of S.

We now define the probability of an event as the sum of the probabilities of the outcomes in this event.

DEFINITION 2   The *probability* of the event E is the sum of the probabilities of the outcomes in E. That is,

$$p(E) = \sum_{s \in E} p(s).$$

(Note that when E is an infinite set, $\sum_{s \in E} p(s)$ is a convergent infinite series.)

Note that when there are n outcomes in the event E, that is, if $E = \{a_1, a_2, \ldots, a_n\}$, then $p(E) = \sum_{i=1}^{n} p(a_i)$. Note also that the uniform distribution assigns the same probability to an event that Laplace's original definition of probability assigns to this event. The experiment of selecting an element from a sample space with a uniform distribution is called selecting an element of S **at random**.

EXAMPLE 2   Suppose that a die is biased (or loaded) so that 3 appears twice as often as each other number but that the other five outcomes are equally likely. What is the probability that an odd number appears when we roll this die?

Solution: We want to find the probability of the event $E = \{1, 3, 5\}$. By Exercise 2, we have

$$p(1) = p(2) = p(4) = p(5) = p(6) = 1/7; \ p(3) = 2/7.$$

It follows that

$$p(E) = p(1) + p(3) + p(5) = 1/7 + 2/7 + 1/7 = 4/7. \quad ◀$$

When possible outcomes are equally likely and there are a finite number of possible outcomes, the definition of the probability of an event given in this section (Definition 2) agrees with Laplace's definition (Definition 1 of Section 7.1). To see this, suppose that there are n equally likely outcomes; each possible outcome has probability $1/n$, because the sum of their probabilities is 1. Suppose the event E contains m outcomes. According to Definition 2,

$$p(E) = \sum_{i=1}^{m} \frac{1}{n} = \frac{m}{n}.$$

Because $|E| = m$ and $|S| = n$, it follows that

$$p(E) = \frac{m}{n} = \frac{|E|}{|S|}.$$

This is Laplace's definition of the probability of the event E.

Probabilities of Complements and Unions of Events

The formulae for probabilities of combinations of events in Section 7.1 continue to hold when we use Definition 2 to define the probability of an event. For example, Theorem 1 of Section 7.1 asserts that

$$p(\overline{E}) = 1 - p(E),$$

where $\overline{E}$ is the complementary event of the event E. This equality also holds when Definition 2 is used. To see this, note that because the sum of the probabilities of the n possible outcomes is 1, and each outcome is either in E or in $\overline{E}$, but not in both, we have

$$\sum_{s \in S} p(s) = 1 = p(E) + p(\overline{E}).$$

Hence, $p(\overline{E}) = 1 - p(E)$.

Under Laplace's definition, by Theorem 2 in Section 7.1, we have

$$p(E_1 \cup E_2) = p(E_1) + p(E_2) - p(E_1 \cap E_2)$$

whenever E_1 and E_2 are events in a sample space S. This also holds when we define the probability of an event as we do in this section. To see this, note that $p(E_1 \cup E_2)$ is the sum of the probabilities of the outcomes in $E_1 \cup E_2$. When an outcome x is in one, but not both, of E_1 and E_2, $p(x)$ occurs in exactly one of the sums for $p(E_1)$ and $p(E_2)$. When an outcome x is in both E_1 and E_2, $p(x)$ occurs in the sum for $p(E_1)$, in the sum for $p(E_2)$, and in the sum for $p(E_1 \cap E_2)$, so it occurs $1 + 1 - 1 = 1$ time on the right-hand side. Consequently, the left-hand side and right-hand side are equal.

Also, note that if the events E_1 and E_2 are disjoint, then $p(E_1 \cap E_2) = 0$, which implies that

$$p(E_1 \cup E_2) = p(E_1) + p(E_2) - p(E_1 \cap E_2) = p(E_1) + p(E_2).$$

Theorem 1 generalizes this last formula by providing a formula for the probability of the union of pairwise disjoint events.

THEOREM 1   If $E_1, E_2, \ldots$ is a sequence of pairwise disjoint events in a sample space S, then

$$p\left(\bigcup_i E_i\right) = \sum_i p(E_i).$$

(Note that this theorem applies when the sequence $E_1, E_2, \ldots$ consists of a finite number or a countably infinite number of pairwise disjoint events.)

We leave the proof of Theorem 1 to the reader (see Exercises 28 and 29).

Conditional Probability

Suppose that we flip a coin three times, and all eight possibilities are equally likely. Moreover, suppose we know that the event F, that the first flip comes up tails, occurs. Given this information, what is the probability of the event E, that an odd number of tails appears? Because the first flip comes up tails, there are only four possible outcomes: TTT, TTH, THT, and THH, where H

and T represent heads and tails, respectively. An odd number of tails appears only for the outcomes TTT and THH. Because the eight outcomes have equal probability, each of the four possible outcomes, given that F occurs, should also have an equal probability of $1/4$. This suggests that we should assign the probability of $2/4 = 1/2$ to E, given that F occurs. This probability is called the **conditional probability** of E given F.

In general, to find the conditional probability of E given F, we use F as the sample space. For an outcome from E to occur, this outcome must also belong to $E \cap F$. With this motivation, we make Definition 3.

DEFINITION 3   Let E and F be events with $p(F) > 0$. The *conditional probability* of E given F, denoted by $p(E \mid F)$, is defined as

$$p(E \mid F) = \frac{p(E \cap F)}{p(F)}.$$

EXAMPLE 3   A bit string of length four is generated at random so that each of the 16 bit strings of length four is equally likely. What is the probability that it contains at least two consecutive 0s, given that its first bit is a 0? (We assume that 0 bits and 1 bits are equally likely.)

Solution: Let E be the event that a bit string of length four contains at least two consecutive 0s, and let F be the event that the first bit of a bit string of length four is a 0. The probability that a bit string of length four has at least two consecutive 0s, given that its first bit is a 0, equals

$$p(E \mid F) = \frac{p(E \cap F)}{p(F)}.$$

Because $E \cap F = \{0000, 0001, 0010, 0011, 0100\}$, we see that $p(E \cap F) = 5/16$. Because there are eight bit strings of length four that start with a 0, we have $p(F) = 8/16 = 1/2$. Consequently,

$$p(E \mid F) = \frac{5/16}{1/2} = \frac{5}{8}.$$ ◀

EXAMPLE 4   What is the conditional probability that a family with two children has two boys, given they have at least one boy? Assume that each of the possibilities BB, BG, GB, and GG is equally likely, where B represents a boy and G represents a girl. (Note that BG represents a family with an older boy and a younger girl while GB represents a family with an older girl and a younger boy.)

Solution: Let E be the event that a family with two children has two boys, and let F be the event that a family with two children has at least one boy. It follows that $E = \{BB\}$, $F = \{BB, BG, GB\}$, and $E \cap F = \{BB\}$. Because the four possibilities are equally likely, it follows that $p(F) = 3/4$ and $p(E \cap F) = 1/4$. We conclude that

$$p(E \mid F) = \frac{p(E \cap F)}{p(F)} = \frac{1/4}{3/4} = \frac{1}{3}.$$ ◀

Independence

Links

Suppose a coin is flipped three times, as described in the introduction to our discussion of conditional probability. Does knowing that the first flip comes up tails (event F) alter the probability that tails comes up an odd number of times (event E)? In other words, is it the case that $p(E \mid F) = p(E)$? This equality is valid for the events E and F, because $p(E \mid F) = 1/2$ and $p(E) = 1/2$. Because this equality holds, we say that E and F are **independent events**. When two events are independent, the occurrence of one of the events gives no information about the probability that the other event occurs.

Because $p(E \mid F) = p(E \cap F)/p(F)$, asking whether $p(E \mid F) = p(E)$ is the same as asking whether $p(E \cap F) = p(E)p(F)$. This leads to Definition 4.

DEFINITION 4   The events E and F are *independent* if and only if $p(E \cap F) = p(E)p(F)$.

EXAMPLE 5   Suppose E is the event that a randomly generated bit string of length four begins with a 1 and F is the event that this bit string contains an even number of 1s. Are E and F independent, if the 16 bit strings of length four are equally likely?

Extra
Examples

Solution: There are eight bit strings of length four that begin with a one: 1000, 1001, 1010, 1011, 1100, 1101, 1110, and 1111. There are also eight bit strings of length four that contain an even number of ones: 0000, 0011, 0101, 0110, 1001, 1010, 1100, 1111. Because there are 16 bit strings of length four, it follows that

$$p(E) = p(F) = 8/16 = 1/2.$$

Because $E \cap F = \{1111, 1100, 1010, 1001\}$, we see that

$$p(E \cap F) = 4/16 = 1/4.$$

Because

$$p(E \cap F) = 1/4 = (1/2)(1/2) = p(E)p(F),$$

we conclude that E and F are independent.   ◄

Probability has many applications to genetics, as Examples 6 and 7 illustrate.

EXAMPLE 6   Assume, as in Example 4, that each of the four ways a family can have two children is equally likely. Are the events E, that a family with two children has two boys, and F, that a family with two children has at least one boy, independent?

Solution: Because $E = \{BB\}$, we have $p(E) = 1/4$. In Example 4 we showed that $p(F) = 3/4$ and that $p(E \cap F) = 1/4$. But $p(E)p(F) = \frac{1}{4} \cdot \frac{3}{4} = \frac{3}{16}$. Therefore $p(E \cap F) \neq p(E)p(F)$, so the events E and F are not independent.   ◄

EXAMPLE 7   Are the events E, that a family with three children has children of both sexes, and F, that this family has at most one boy, independent? Assume that the eight ways a family can have three children are equally likely.

Solution: By assumption, each of the eight ways a family can have three children, BBB, BBG, BGB, BGG, GBB, GBG, GGB, and GGG, has a probability of 1/8. Because $E = \{BBG, BGB, BGG, GBB, GBG, GGB\}$, $F = \{BGG, GBG, GGB, GGG\}$, and

$E \cap F = \{BGG, \, GBG, \, GGB\}$, it follows that $p(E) = 6/8 = 3/4$, $p(F) = 4/8 = 1/2$, and $p(E \cap F) = 3/8$. Because

$$p(E)p(F) = \frac{3}{4} \cdot \frac{1}{2} = \frac{3}{8},$$

it follows that $p(E \cap F) = p(E)p(F)$, so E and F are independent. (This conclusion may seem surprising. Indeed, if we change the number of children, the conclusion may no longer hold. See Exercise 21.)   ◄

PAIRWISE AND MUTUAL INDEPENDENCE   We can also define the independence of more than two events. However, there are two different types of independence, given in Definition 5.

DEFINITION 5

The events $E_1, E_2, \ldots, E_n$ are *pairwise independent* if and only if $p(E_i \cap E_j) = p(E_i)p(E_j)$ for all pairs of integers i and j with $1 \leq i < j \leq n$. These events are *mutually independent* if $p(E_{i_1} \cap E_{i_2} \cap \cdots \cap E_{i_m}) = p(E_{i_1})p(E_{i_2}) \cdots p(E_{i_m})$ whenever i_j, $j = 1, 2, \ldots, m$, are integers with $1 \leq i_1 < i_2 < \cdots < i_m \leq n$ and $m \geq 2$.

From Definition 5, we see that every set of n mutually independent events is also pairwise independent. However, n pairwise independent events are not necessarily mutually independent, as we see in Exercise 19 in the Supplementary Exercises. Many theorems about n events include the hypothesis that these events are mutually independent, and not just pairwise independent. We will introduce several such theorems later in this chapter.

Bernoulli Trials and the Binomial Distribution

Links

Suppose that an experiment can have only two possible outcomes. For instance, when a bit is generated at random, the possible outcomes are 0 and 1. When a coin is flipped, the possible outcomes are heads and tails. Each performance of an experiment with two possible outcomes is called a **Bernoulli trial**, after James Bernoulli, who made important contributions to probability theory. In general, a possible outcome of a Bernoulli trial is called a **success** or a **failure**. If p is the probability of a success and q is the probability of a failure, it follows that $p + q = 1$.

Many problems can be solved by determining the probability of k successes when an experiment consists of n mutually independent Bernoulli trials. (Bernoulli trials are **mutually independent** if the conditional probability of success on any given trial is p, given any information whatsoever about the outcomes of the other trials.) Consider Example 8.

EXAMPLE 8   A coin is biased so that the probability of heads is $2/3$. What is the probability that exactly four heads come up when the coin is flipped seven times, assuming that the flips are independent?

Solution: There are $2^7 = 128$ possible outcomes when a coin is flipped seven times. The number of ways four of the seven flips can be heads is $C(7, 4)$. Because the seven flips are independent, the probability of each of these outcomes (four heads and three tails) is $(2/3)^4(1/3)^3$. Consequently, the probability that exactly four heads appear is

$$C(7, 4)(2/3)^4(1/3)^3 = \frac{35 \cdot 16}{3^7} = \frac{560}{2187}.$$

◄

Following the same reasoning as was used in Example 8, we can find the probability of k successes in n independent Bernoulli trials.

THEOREM 2

The probability of exactly k successes in n independent Bernoulli trials, with probability of success p and probability of failure $q = 1 - p$, is

$$C(n, k) p^k q^{n-k}.$$

Proof: When n Bernoulli trials are carried out, the outcome is an n-tuple $(t_1, t_2, \ldots, t_n)$, where $t_i = S$ (for success) or $t_i = F$ (for failure) for $i = 1, 2, \ldots, n$. Because the n trials are independent, the probability of each outcome of n trials consisting of k successes and $n - k$ failures (in any order) is $p^k q^{n-k}$. Because there are $C(n, k)$ n-tuples of S's and F's that contain exactly k S's, the probability of exactly k successes is

$$C(n, k) p^k q^{n-k}.$$

$\triangleleft$

We denote by $b(k; n, p)$ the probability of k successes in n independent Bernoulli trials with probability of success p and probability of failure $q = 1 - p$. Considered as a function of k, we call this function the **binomial distribution**. Theorem 2 tells us that $b(k; n, p) = C(n, k) p^k q^{n-k}$.

EXAMPLE 9

Extra Examples

Suppose that the probability that a 0 bit is generated is 0.9, that the probability that a 1 bit is generated is 0.1, and that bits are generated independently. What is the probability that exactly eight 0 bits are generated when 10 bits are generated?

Solution: By Theorem 2, the probability that exactly eight 0 bits are generated is

$$b(8; 10, 0.9) = C(10, 8)(0.9)^8 (0.1)^2 = 0.1937102445.$$

$\blacktriangleleft$

Note that the sum of the probabilities that there are k successes when n independent Bernoulli trials are carried out, for $k = 0, 1, 2, \ldots, n$, equals

$$\sum_{k=0}^{n} C(n, k) p^k q^{n-k} = (p + q)^n = 1,$$

as should be the case. The first equality in this string of equalities is a consequence of the binomial theorem (see Section 6.4). The second equality follows because $q = 1 - p$.

Links

JAMES BERNOULLI (1654–1705) James Bernoulli (also known as Jacob I), was born in Basel, Switzerland. He is one of the eight prominent mathematicians in the Bernoulli family (see Section 10.1 for the Bernoulli family tree of mathematicians). Following his father's wish, James studied theology and entered the ministry. But contrary to the desires of his parents, he also studied mathematics and astronomy. He traveled throughout Europe from 1676 to 1682, learning about the latest discoveries in mathematics and the sciences. Upon returning to Basel in 1682, he founded a school for mathematics and the sciences. He was appointed professor of mathematics at the University of Basel in 1687, remaining in this position for the rest of his life.

James Bernoulli is best known for the work *Ars Conjectandi*, published eight years after his death. In this work, he described the known results in probability theory and in enumeration, often providing alternative proofs of known results. This work also includes the application of probability theory to games of chance and his introduction of the theorem known as the **law of large numbers**. This law states that if $\epsilon > 0$, as n becomes arbitrarily large the probability approaches 1 that the fraction of times an event E occurs during n trials is within ϵ of $p(E)$.

Random Variables

Many problems are concerned with a numerical value associated with the outcome of an experiment. For instance, we may be interested in the total number of one bits in a randomly generated string of 10 bits; or in the number of times tails come up when a coin is flipped 20 times. To study problems of this type we introduce the concept of a random variable.

DEFINITION 6   A *random variable* is a function from the sample space of an experiment to the set of real numbers. That is, a random variable assigns a real number to each possible outcome.

Remark: Note that a random variable is a function. It is not a variable, and it is not random! The name *random variable* (the translation of *variabile casuale*) was introduced by the Italian mathematician F. P. Cantelli in 1916. In the late 1940s, the mathematicians, W. Feller and J. L. Doob flipped a coin to see whether both would use "random variable" or the more fitting term "chance variable." Feller won; unfortunately "random varible" was used in both books and ever since.

EXAMPLE 10   Suppose that a coin is flipped three times. Let $X(t)$ be the random variable that equals the number of heads that appear when t is the outcome. Then $X(t)$ takes on the following values:

$$X(HHH) = 3,$$
$$X(HHT) = X(HTH) = X(THH) = 2,$$
$$X(TTH) = X(THT) = X(HTT) = 1,$$
$$X(TTT) = 0.$$
◀

DEFINITION 7   The *distribution* of a random variable X on a sample space S is the set of pairs $(r, p(X = r))$ for all $r \in X(S)$, where $p(X = r)$ is the probability that X takes the value r. (The set of pairs in this distribution is determined by the probabilities $p(X = r)$ for $r \in X(S)$.)

EXAMPLE 11   Each of the eight possible outcomes when a fair coin is flipped three times has probability $1/8$. So, the distribution of the random variable $X(t)$ in Example 10 is determined by the probabilities $P(X = 3) = 1/8$, $P(X = 2) = 3/8$, $P(X = 1) = 3/8$, and $P(X = 0) = 1/8$. Consequently, the distribution of $X(t)$ in Example 10 is the set of pairs $(3, 1/8)$, $(2, 3/8)$, $(1, 3/8)$, and $(0, 1/8)$.
◀

EXAMPLE 12   Let X be the sum of the numbers that appear when a pair of dice is rolled. What are the values of this random variable for the 36 possible outcomes (i, j), where i and j are the numbers that appear on the first die and the second die, respectively, when these two dice are rolled?

Solution: The random variable X takes on the following values:

$X((1, 1)) = 2,$

$X((1, 2)) = X((2, 1)) = 3,$

$X((1, 3)) = X((2, 2)) = X((3, 1)) = 4,$

$X((1, 4)) = X((2, 3)) = X((3, 2)) = X((4, 1)) = 5,$

$X((1, 5)) = X((2, 4)) = X((3, 3)) = X((4, 2)) = X((5, 1)) = 6,$

$X((1, 6)) = X((2, 5)) = X((3, 4)) = X((4, 3)) = X((5, 2)) = X((6, 1)) = 7,$

$X((2, 6)) = X((3, 5)) = X((4, 4)) = X((5, 3)) = X((6, 2)) = 8,$

$X((3, 6)) = X((4, 5)) = X((5, 4)) = X((6, 3)) = 9,$

$X((4, 6)) = X((5, 5)) = X((6, 4)) = 10,$

$X((5, 6)) = X((6, 5)) = 11,$

$X((6, 6)) = 12.$ ◀

We will continue our study of random variables in Section 7.4, where we will show how they can be used in a variety of applications.

The Birthday Problem

A famous puzzle asks for the smallest number of people needed in a room so that it is more likely than not that at least two of them have the same day of the year as their birthday. Most people find the answer, which we determine in Example 13, to be surprisingly small. After we solve this famous problem, we will show how similar reasoning can be adapted to solve a question about hashing functions.

EXAMPLE 13

The Birthday Problem   What is the minimum number of people who need to be in a room so that the probability that at least two of them have the same birthday is greater than $1/2$?

Solution: First, we state some assumptions. We assume that the birthdays of the people in the room are independent. Furthermore, we assume that each birthday is equally likely and that there are 366 days in the year. (In reality, more people are born on some days of the year than others, such as days nine months after some holidays including New Year's Eve, and only leap years have 366 days.)

To find the probability that at least two of n people in a room have the same birthday, we first calculate the probability p_n that these people all have different birthdays. Then, the probability that at least two people have the same birthday is $1- p_n$. To compute p_n, we consider the birthdays of the n people in some fixed order. Imagine them entering the room one at a time; we will compute the probability that each successive person entering the room has a birthday different from those of the people already in the room.

The birthday of the first person certainly does not match the birthday of someone already in the room. The probability that the birthday of the second person is different from that of the first person is $365/366$ because the second person has a different birthday when he or she was born on one of the 365 days of the year other than the day the first person was born. (The assumption that it is equally likely for someone to be born on any of the 366 days of the year enters into this and subsequent steps.)

The probability that the third person has a birthday different from both the birthdays of the first and second people given that these two people have different birthdays is $364/366$. In general, the probability that the jth person, with $2 \leq j \leq 366$, has a birthday different from the birthdays of the $j - 1$ people already in the room given that these $j - 1$ people have different birthdays is

$$\frac{366 - (j - 1)}{366} = \frac{367 - j}{366}.$$

Because we have assumed that the birthdays of the people in the room are independent, we can conclude that the probability that the n people in the room have different birthdays is

$$p_n = \frac{365}{366} \frac{364}{366} \frac{363}{366} \cdots \frac{367 - n}{366}.$$

It follows that the probability that among n people there are at least two people with the same birthday is

$$1 - p_n = 1 - \frac{365}{366} \frac{364}{366} \frac{363}{366} \cdots \frac{367 - n}{366}.$$

To determine the minimum number of people in the room so that the probability that at least two of them have the same birthday is greater than $1/2$, we use the formula we have found for $1 - p_n$ to compute it for increasing values of n until it becomes greater than $1/2$. (There are more sophisticated approaches using calculus that can eliminate this computation, but we will not use them here.) After considerable computation we find that for $n = 22$, $1 - p_n \approx 0.475$, while for $n = 23$, $1 - p_n \approx 0.506$. Consequently, the minimum number of people needed so that the probability that at least two people have the same birthday is greater than $1/2$ is 23. ◀

The solution to the birthday problem leads to the solution of the question in Example 14 about hashing functions.

EXAMPLE 14   **Probability of a Collision in Hashing Functions**   Recall from Section 4.5 that a hashing function $h(k)$ is a mapping of the keys (of the records that are to be stored in a database) to storage locations. Hashing functions map a large universe of keys (such as the approximately 300 million Social Security numbers in the United States) to a much smaller set of storage locations. A good hashing function yields few **collisions**, which are mappings of two different keys to the same memory location, when relatively few of the records are in play in a given application. What is the probability that no two keys are mapped to the same location by a hashing function, or, in other words, that there are no collisions?

Solution: To calculate this probability, we assume that the probability that a randomly selected key is mapped to a location is $1/m$, where m is the number of available locations, that is, the hashing function distributes keys uniformly. (In practice, hashing functions may not satisfy this assumption. However, for a good hashing function, this assumption should be close to correct.) Furthermore, we assume that the keys of the records selected have an equal probability to be any of the elements of the key universe and that these keys are independently selected.

Suppose that the keys are $k_1, k_2, \ldots, k_n$. When we add the second record, the probability that it is mapped to a location different from the location of the first record, that $h(k_2) \neq h(k_1)$, is $(m - 1)/m$ because there are $m - 1$ free locations after the first record has been placed. The probability that the third record is mapped to a free location after the first and second records have been placed without a collision is $(m - 2)/m$. In general, the probability that the jth record is mapped to a free location after the first $j - 1$ records have been mapped to locations $h(k_1)$, $h(k_2), \ldots, h(k_{j-1})$ without collisions is $(m - (j - 1))/m$ because $j - 1$ of the m locations are taken.

Because the keys are independent, the probability that all n keys are mapped to different locations is

$$p_n = \frac{m - 1}{m} \cdot \frac{m - 2}{m} \cdot \ldots \cdot \frac{m - n + 1}{m}.$$

It follows that the probability that there is at least one collision, that is, at least two keys are mapped to the same location, is

$$1 - p_n = 1 - \frac{m-1}{m} \cdot \frac{m-2}{m} \cdot \ldots \cdot \frac{m-n+1}{m}.$$

Techniques from calculus can be used to find the smallest value of n given a value of m such that the probability of a collision is greater than a particular threshold. It can be shown that the smallest integer n such that the probability of a collision is greater than $1/2$ is approximately $n = 1.177\sqrt{m}$. For example, when $m = 1,000,000$, the smallest integer n such that the probability of a collision is greater than $1/2$ is 1178. ◀

Monte Carlo Algorithms

The algorithms discussed so far in this book are all deterministic. That is, each algorithm always proceeds in the same way whenever given the same input. However, there are many situations where we would like an algorithm to make a random choice at one or more steps. Such a situation arises when a deterministic algorithm would have to go through a huge number, or even an unknown number, of possible cases. Algorithms that make random choices at one or more steps are called **probabilistic algorithms**. We will discuss a particular class of probabilistic algorithms in this section, namely, **Monte Carlo algorithms**, for decision problems. Monte Carlo algorithms always produce answers to problems, but a small probability remains that these answers may be incorrect. However, the probability that the answer is incorrect decreases rapidly when the algorithm carries out sufficient computation. Decision problems have either "true" or "false" as their answer. The designation "Monte Carlo" is a reference to the famous casino in Monaco; the use of randomness and the repetitive processes in these algorithms make them similar to some gambling games. This name was introduced by the inventors of Monte Carlo methods, including Stan Ulam, Enrico Fermi, and John von Neumann.

Monte Carlo methods were invented to help develop the first nuclear weapons.

A Monte Carlo algorithm for a decision problem uses a sequence of tests. The probability that the algorithm answers the decision problem correctly increases as more tests are carried out. At each step of the algorithm, possible responses are "true," which means that the answer is "true" and no additional iterations are needed, or "unknown," which means that the answer could be either "true" or "false." After running all the iterations in such an algorithm, the final answer produced is "true" if at least one iteration yields the answer "true," and the answer is "false" if every iteration yields the answer "unknown." If the correct answer is "false," then the algorithm answers "false," because every iteration will yield "unknown." However, if the correct answer is "true," then the algorithm could answer either "true" or "false," because it may be possible that each iteration produced the response "unknown" even though the correct response was "true." We will show that this possibility becomes extremely unlikely as the number of tests increases.

Suppose that p is the probability that the response of a test is "true," given that the answer is "true." It follows that $1 - p$ is the probability that the response is "unknown," given that the answer is "true." Because the algorithm answers "false" when all n iterations yield the answer "unknown" and the iterations perform independent tests, the probability of error is $(1-p)^n$. When $p \neq 0$, this probability approaches 0 as the number of tests increases. Consequently, the probability that the algorithm answers "true" when the answer is "true" approaches 1.

EXAMPLE 15   **Quality Control**   (This example is adapted from [AhUl95].) Suppose that a manufacturer orders processor chips in batches of size n, where n is a positive integer. The chip maker has tested only some of these batches to make sure that all the chips in the batch are good (replacing any bad chips found during testing with good ones). In previously untested batches, the probability that a particular chip is bad has been observed to be 0.1 when random testing is done. The PC manufacturer wants to decide whether all the chips in a batch are good. To do this, the PC manufacturer can test each chip in a batch to see whether it is good. However, this requires n tests. Assuming that each test can be carried out in constant time, these tests require $O(n)$ seconds. Can the PC manufacturer determine whether a batch of chips has been tested by the chip maker using less time?

Solution: We can use a Monte Carlo algorithm to determine whether a batch of chips has been tested by the chip maker as long as we are willing to accept some probability of error. The algorithm is set up to answer the question: "Has this batch of chips not been tested by the chip maker?" It proceeds by successively selecting chips at random from the batch and testing them one by one. When a bad chip is encountered, the algorithm answers "true" and stops. If a tested chip is good, the algorithm answers "unknown" and goes on to the next chip. After the algorithm has tested a specified number of chips, say k chips, without getting an answer of "true," the algorithm terminates with the answer "false"; that is, the algorithm concludes that the batch is good, that is, that the chip maker has tested all the chips in the batch.

The only way for this algorithm to answer incorrectly is for it to conclude that an untested batch of chips has been tested by the chip maker. The probability that a chip is good, but that it came from an untested batch, is $1 - 0.1 = 0.9$. Because the events of testing different chips from a batch are independent, the probability that all k steps of the algorithm produce the answer "unknown," given that the batch of chips is untested, is 0.9^k.

By taking k large enough, we can make this probability as small as we like. For example, by testing 66 chips, the probability that the algorithm decides a batch has been tested by the chip maker is 0.9^{66}, which is less than 0.001. That is, the probability is less than 1 in 1000 that the algorithm has answered incorrectly. Note that this probability is independent of n, the number of chips in a batch. That is, the Monte Carlo algorithm uses a constant number, or $O(1)$, tests and requires $O(1)$ seconds, no matter how many chips are in a batch. As long as the PC manufacturer can live with an error rate of less than 1 in 1000, the Monte Carlo algorithm will save the PC manufacturer a lot of testing. If a smaller error rate is needed, the PC manufacturer can test more chips in each batch; the reader can verify that 132 tests lower the error rate to less than 1 in 1,000,000. ◀

EXAMPLE 16 **Probabilistic Primality Testing** In Chapter 4 we remarked that a composite integer, that is, an integer greater than one that is not prime, passes Miller's test (see the preamble to Exercise 44 in Section 4.4) for fewer than $n/4$ bases b with $1 < b < n$. This observation is the basis for a Monte Carlo algorithm to determine whether an integer greater than one is prime. Because large primes play an essential role in public-key cryptography (see Section 4.6), being able to generate large primes quickly has become extremely important.

The goal of the algorithm is to decide the question "Is n composite?" Given an integer n greater than one, we select an integer b at random with $1 < b < n$ and determine whether n passes Miller's test to the base b. If n fails the test, the answer is "true" because n must be composite, and the algorithm ends. Otherwise, we perform the test k times, where k is a positive integer. Each time we select a random integer b and determine whether n passes Miller's test to the base b. If the answer is "unknown" at each step, the algorithm answers "false," that is, it says that n is not composite, so that it is prime. The only possibility for the algorithm to return an incorrect answer occurs when n is composite, and the answer "unknown" is the output at each of the k iterations. The probability that a composite integer n passes Miller's test for a randomly selected base b is less than $1/4$. Because the integer b with $1 < b < n$ is selected at random at each iteration and these iterations are independent, the probability that n is composite but the algorithm responds that n is prime is less than $(1/4)^k$. By taking k to be sufficiently large, we can make this probability extremely small. For example, with 10 iterations, the probability that the algorithm decides that n is prime when it really is composite is less than 1 in 1,000,000. With 30 iterations, this probability drops to less than 1 in 10^{18}, an extremely unlikely event.

A number that passes many iterations of a probabilistic primality test is called an *industrial strength prime*, even though it may be composite.

To generate large primes, say with 200 digits, we randomly choose an integer n with 200 digits and run this algorithm, with 30 iterations. If the algorithm decides that n is prime, we can use it as one of the two primes used in an encryption key for the RSA cryptosystem. If n is actually composite and is used as part of the key, the procedures used to decrypt messages will not produce the original encrypted message. The key is then discarded and two new possible primes are used. ◄

The Probabilistic Method

We discussed existence proofs in Chapter 1 and illustrated the difference between constructive existence proofs and nonconstructive existence proofs. The probabilistic method, introduced by Paul Erdős and Alfréd Rényi, is a powerful technique that can be used to create nonconstructive existence proofs. To use the probabilistic method to prove results about a set S, such as the existence of an element in S with a specified property, we assign probabilities to the elements of S. We then use methods from probability theory to prove results about the elements of S. In particular, we can show that an element with a specified property exists by showing that the probability an element $x \in S$ has this property is positive. The probabilistic method is based on the equivalent statement in Theorem 3.

THEOREM 3   **THE PROBABILISTIC METHOD**   If the probability that an element chosen at random from a S does not have a particular property is less than 1, there exists an element in S with this property.

An existence proof based on the probabilistic method is nonconstructive because it does not find a particular element with the desired property.

We illustrate the power of the probabilistic method by finding a lower bound for the Ramsey number $R(k, k)$. Recall from Section 6.2 that $R(k, k)$ equals the minimum number of people at a party needed to ensure that there are at least k mutual friends or k mutual enemies (assuming that any two people are friends or enemies).

THEOREM 4   If k is an integer with $k \geq 2$, then $R(k, k) \geq 2^{k/2}$.

Proof: We note that the theorem holds for $k = 2$ and $k = 3$ because $R(2, 2) = 2$ and $R(3, 3) = 6$, as was shown in Section 6.2. Now suppose that $k \geq 4$. We will use the probabilistic method to show that if there are fewer than $2^{k/2}$ people at a party, it is possible that no k of them are mutual friends or mutual enemies. This will show that $R(k, k)$ is at least $2^{k/2}$.

To use the probabilistic method, we assume that it is equally likely for two people to be friends or enemies. (Note that this assumption does not have to be realistic.) Suppose there are n people at the party. It follows that there are $\binom{n}{k}$ different sets of k people at this party, which we list as $S_1, S_2, \ldots, S_{\binom{n}{k}}$. Let E_i be the event that all k people in S_i are either mutual friends or mutual enemies. The probability that there are either k mutual friends or k mutual enemies among the n people equals $p(\bigcup_{i=1}^{\binom{n}{k}} E_i)$.

According to our assumption it is equally likely for two people to be friends or enemies. The probability that two people are friends equals the probability that they are enemies; both probabilities equal $1/2$. Furthermore, there are $\binom{k}{2} = k(k-1)/2$ pairs of people in S_i because there are k people in S_i. Hence, the probability that all k people in S_i are mutual friends and the probability that all k people in S_i are mutual enemies both equal $(1/2)^{k(k-1)/2}$. It follows that $p(E_i) = 2(1/2)^{k(k-1)/2}$.

The probability that there are either k mutual friends or k mutual enemies in the group of n people equals $p(\bigcup_{i=1}^{\binom{n}{k}} E_i)$. Using Boole's inequality (Exercise 12), it follows that

$$p\left(\bigcup_{i=1}^{\binom{n}{k}} E_i\right) \leq \sum_{i=1}^{\binom{n}{k}} p(E_i) = \binom{n}{k} \cdot 2\left(\frac{1}{2}\right)^{k(k-1)/2}.$$

By Exercise 17 in Section 6.4, we have $\binom{n}{k} \leq n^k/2^{k-1}$. Hence,

$$\binom{n}{k} 2\left(\frac{1}{2}\right)^{k(k-1)/2} \leq \frac{n^k}{2^{k-1}} 2\left(\frac{1}{2}\right)^{k(k-1)/2}.$$

Now if $n < 2^{k/2}$, we have

$$\frac{n^k}{2^{k-1}} 2\left(\frac{1}{2}\right)^{k(k-1)/2} < \frac{2^{k(k/2)}}{2^{k-1}} 2\left(\frac{1}{2}\right)^{k(k-1)/2} = 2^{2-(k/2)} \leq 1,$$

where the last step follows because $k \geq 4$.

We can now conclude that $p(\bigcup_{i=1}^{\binom{n}{k}} E_i) < 1$ when $k \geq 4$. Hence, the probability of the complementary event, that there is no set of either k mutual friends or mutual enemies at the party, is greater than 0. It follows that if $n < 2^{k/2}$, there is at least one set such that no subset of k people are mutual friends or mutual enemies.   ◁

Exercises

1. What probability should be assigned to the outcome of heads when a biased coin is tossed, if heads is three times as likely to come up as tails? What probability should be assigned to the outcome of tails?

2. Find the probability of each outcome when a loaded die is rolled, if a 3 is twice as likely to appear as each of the other five numbers on the die.

3. Find the probability of each outcome when a biased die is rolled, if rolling a 2 or rolling a 4 is three times as likely as rolling each of the other four numbers on the die and it is equally likely to roll a 2 or a 4.

4. Show that conditions (*i*) and (*ii*) are met under Laplace's definition of probability, when outcomes are equally likely.

5. A pair of dice is loaded. The probability that a 4 appears on the first die is 2/7, and the probability that a 3 appears on the second die is 2/7. Other outcomes for each die appear with probability 1/7. What is the probability of 7 appearing as the sum of the numbers when the two dice are rolled?

6. What is the probability of these events when we randomly select a permutation of $\{1, 2, 3, 4\}$?
   a) 1 precedes 4.
   b) 4 precedes 1.
   c) 4 precedes 1 and 4 precedes 2.
   d) 4 precedes 1, 4 precedes 2, and 4 precedes 3.
   e) 4 precedes 3 and 2 precedes 1.

7. What is the probability of these events when we randomly select a permutation of the 26 lowercase letters of the English alphabet?
   a) The permutation consists of the letters in reverse alphabetic order.
   b) z is the first letter of the permutation.
   c) z precedes a in the permutation.
   d) a immediately precedes z in the permutation.
   e) a immediately precedes m, which immediately precedes z in the permutation.
   f) m, n, and o are in their original places in the permutation.

8. What is the probability of these events when we randomly select a permutation of the 26 lowercase letters of the English alphabet?
   a) The first 13 letters of the permutation are in alphabetical order.
   b) a is the first letter of the permutation and z is the last letter.
   c) a and z are next to each other in the permutation.
   d) a and b are not next to each other in the permutation.
   e) a and z are separated by at least 23 letters in the permutation.
   f) z precedes both a and b in the permutation.

9. Suppose that E and F are events such that $p(E) = 0.7$ and $p(F) = 0.5$. Show that $p(E \cup F) \geq 0.7$ and $p(E \cap F) \geq 0.2$.

10. Suppose that E and F are events such that $p(E) = 0.8$ and $p(F) = 0.6$. Show that $p(E \cup F) \geq 0.8$ and $p(E \cap F) \geq 0.4$.

11. Show that if E and F are events, then $p(E \cap F) \geq p(E) + p(F) - 1$. This is known as **Bonferroni's inequality**.

12. Show that if $E_1, E_2, \ldots, E_n$ are events from a finite sample space, then
$$p(E_1 \cup E_2 \cup \cdots \cup E_n)$$
$$\leq p(E_1) + p(E_2) + \cdots + p(E_n).$$

This is known as **Boole's inequality**.

13. If E and F are independent events, prove or disprove that $\overline{E}$ and F are necessarily independent events.

In Exercises 14, 16, and 17 assume that the year has 366 days and all birthdays are equally likely. In Exercise 15 assume it is equally likely that a person is born in any given month of the year.

14. a) What is the probability that two people chosen at random were born on the same day of the week?

b) What is the probability that in a group of n people chosen at random, there are at least two born on the same day of the week?

c) How many people chosen at random are needed to make the probability greater than $1/2$ that there are at least two people born on the same day of the week?

15. a) What is the probability that two people chosen at random were born during the same month of the year?

b) What is the probability that in a group of n people chosen at random, there are at least two born in the same month of the year?

c) How many people chosen at random are needed to make the probability greater than $1/2$ that there are at least two people born in the same month of the year?

16. Find the smallest number of people you need to choose at random so that the probability that at least one of them has a birthday today exceeds $1/2$.

17. Find the smallest number of people you need to choose at random so that the probability that at least two of them were both born on April 1 exceeds $1/2$.

18. What is the conditional probability that exactly four heads appear when a fair coin is flipped five times, given that the first flip came up heads?

19. What is the conditional probability that a randomly generated bit string of length four contains at least two consecutive 0s, given that the first bit is a 1? (Assume the probabilities of a 0 and a 1 are the same.)

20. Let E be the event that a randomly generated bit string of length three contains an odd number of 1s, and let F be the event that the string starts with 1. Are E and F independent?

21. Let E and F be the events that a family of n children has children of both sexes and has at most one boy, respectively. Are E and F independent if

a) $n = 2$? **b)** $n = 4$? **c)** $n = 5$?

22. Assume that the probability a child is a boy is 0.51 and that the sexes of children born into a family are independent. What is the probability that a family of five children has

a) exactly three boys?
b) at least one boy?
c) at least one girl?
d) all children of the same sex?

23. A group of six people play the game of "odd person out" to determine who will buy refreshments. Each person flips a fair coin. If there is a person whose outcome is not the same as that of any other member of the group, this person has to buy the refreshments. What is the probability that there is an odd person out after the coins are flipped once?

24. Find the probability that a randomly generated bit string of length 10 does not contain a 0 if bits are independent and if

a) a 0 bit and a 1 bit are equally likely.
b) the probability that a bit is a 1 is 0.6.
c) the probability that the ith bit is a 1 is $1/2^i$ for $i = 1, 2, 3, \ldots, 10$.

25. Find the probability that a family with five children does not have a boy, if the sexes of children are independent and if

a) a boy and a girl are equally likely.
b) the probability of a boy is 0.51.
c) the probability that the ith child is a boy is $0.51 - (i/100)$.

26. Find the probability that the first child of a family with five children is a boy or that the last two children of the family are girls, for the same conditions as in parts (a), (b), and (c) of Exercise 25.

27. Find each of the following probabilities when n independent Bernoulli trials are carried out with probability of success p.

a) the probability of no failures
b) the probability of at least one failure
c) the probability of at most one failure
d) the probability of at least two failures

28. Use mathematical induction to prove that if $E_1, E_2, \ldots, E_n$ is a sequence of n pairwise disjoint events in a sample space S, where n is a positive integer, then $p(\bigcup_{i=1}^{n} E_i) = \sum_{i=1}^{n} p(E_i)$.

∗29. (*Requires calculus*) Show that if $E_1, E_2, \ldots$ is an infinite sequence of pairwise disjoint events in a sample space S, then $p(\bigcup_{i=1}^{\infty} E_i) = \sum_{i=1}^{\infty} p(E_i)$. [*Hint:* Use Exercise 28 and take limits.]

30. A pair of dice is rolled in a remote location and when you ask an honest observer whether at least one die came up six, this honest observer answers in the affirmative.

a) What is the probability that the sum of the numbers that came up on the two dice is seven, given the information provided by the honest observer?

b) Suppose that the honest observer tells us that at least one die came up five. What is the probability the sum of the numbers that came up on the dice is seven, given this information?

31. This exercise employs the probabilistic method to prove a result about round-robin tournaments. In a **round-robin tournament with m players, every two players play one game in which one player wins and the other loses.

We want to find conditions on positive integers m and k with $k < m$ such that it is possible for the outcomes of the tournament to have the property that for every set of k players, there is a player who beats every member in this set. So that we can use probabilistic reasoning to draw conclusions about round-robin tournaments, we assume that when two players compete it is equally likely that either player wins the game and we assume that the outcomes of different games are independent. Let E be the event that for every set S with k players, where k is a positive integer less than m, there is a player who has beaten all k players in S.

a) Show that $p(\overline{E}) \leq \sum_{j=1}^{\binom{m}{k}} p(F_j)$, where F_j is the event that there is no player who beats all k players from the jth set in a list of the $\binom{m}{k}$ sets of k players.

b) Show that the probability of F_j is $(1-2^{-k})^{m-k}$.

c) Conclude from parts (a) and (b) that $p(\overline{E}) \leq \binom{m}{k}(1 - 2^{-k})^{m-k}$ and, therefore, that there must be a tournament with the described property if $\binom{m}{k}(1-2^{-k})^{m-k} < 1$.

d) Use part (c) to find values of m such that there is a tournament with m players such that for every set S of two players, there is a player who has beaten both players in S. Repeat for sets of three players.

*32. Devise a Monte Carlo algorithm that determines whether a permutation of the integers 1 through n has already been sorted (that is, it is in increasing order), or instead, is a random permutation. A step of the algorithm should answer "true" if it determines the list is not sorted and "unknown" otherwise. After k steps, the algorithm decides that the integers are sorted if the answer is "unknown" in each step. Show that as the number of steps increases, the probability that the algorithm produces an incorrect answer is extremely small. [*Hint:* For each step, test whether certain elements are in the correct order. Make sure these tests are independent.]

7.3 Bayes' Theorem

Introduction

There are many times when we want to assess the probability that a particular event occurs on the basis of partial evidence. For example, suppose we know the percentage of people who have a particular disease for which there is a very accurate diagnostic test. People who test positive for this disease would like to know the likelihood that they actually have the disease. In this section we introduce a result that can be used to determine this probability, namely, the probability that a person has the disease given that this person tests positive for it. To use this result, we will need to know the percentage of people who do not have the disease but test positive for it and the percentage of people who have the disease but test negative for it.

Similarly, suppose we know the percentage of incoming e-mail messages that are spam. We will see that we can determine the likelihood that an incoming e-mail message is spam using the occurrence of words in the message. To determine this likelihood, we need to know the percentage of incoming messages that are spam, the percentage of spam messages in which each of these words occurs, and the percentage of messages that are not spam in which each of these words occurs.

The result that we can use to answer questions such as these is called Bayes' theorem and dates back to the eighteenth century. In the past two decades, Bayes' theorem has been extensively applied to estimate probabilities based on partial evidence in areas as diverse as medicine, law, machine learning, engineering, and software development.

Bayes' Theorem

We illustrate the idea behind Bayes' theorem with an example that shows that when extra information is available, we can derive a more realistic estimate that a particular event occurs. That is, suppose we know $p(F)$, the probability that an event F occurs, but we have knowledge that an event E occurs. Then the conditional probability that F occurs given that E occurs, $p(F \mid E)$, is a more realistic estimate than $p(F)$ that F occurs. In Example 1 we will see that we can find $p(F \mid E)$ when we know $p(F)$, $p(E \mid F)$, and $p(E \mid \overline{F})$.

EXAMPLE 1

Extra Examples

We have two boxes. The first contains two green balls and seven red balls; the second contains four green balls and three red balls. Bob selects a ball by first choosing one of the two boxes at random. He then selects one of the balls in this box at random. If Bob has selected a red ball, what is the probability that he selected a ball from the first box?

Solution: Let E be the event that Bob has chosen a red ball; $\overline{E}$ is the event that Bob has chosen a green ball. Let F be the event that Bob has chosen a ball from the first box; $\overline{F}$ is the event that Bob has chosen a ball from the second box. We want to find $p(F \mid E)$, the probability that the ball Bob selected came from the first box, given that it is red. By the definition of conditional probability, we have $p(F \mid E) = p(F \cap E)/p(E)$. Can we use the information provided to determine both $p(F \cap E)$ and $p(E)$ so that we can find $p(F \mid E)$?

First, note that because the first box contains seven red balls out of a total of nine balls, we know that $p(E \mid F) = 7/9$. Similarly, because the second box contains three red balls out of a total of seven balls, we know that $p(E \mid \overline{F}) = 3/7$. We assumed that Bob selects a box at random, so $p(F) = p(\overline{F}) = 1/2$. Because $p(E \mid F) = p(E \cap F)/p(F)$, it follows that $p(E \cap F) = p(E \mid F)p(F) = \frac{7}{9} \cdot \frac{1}{2} = \frac{7}{18}$ [as we remarked earlier, this is one of the quantities we need to find to determine $p(F \mid E)$]. Similarly, because $p(E \mid \overline{F}) = p(E \cap \overline{F})/p(\overline{F})$, it follows that $p(E \cap \overline{F}) = p(E \mid \overline{F})p(\overline{F}) = \frac{3}{7} \cdot \frac{1}{2} = \frac{3}{14}$.

We can now find $p(E)$. Note that $E = (E \cap F) \cup (E \cap \overline{F})$, where $E \cap F$ and $E \cap \overline{F}$ are disjoint sets. (If x belongs to both $E \cap F$ and $E \cap \overline{F}$, then x belongs to both F and $\overline{F}$, which is impossible.) It follows that

$$p(E) = p(E \cap F) + p(E \cap \overline{F}) = \frac{7}{18} + \frac{3}{14} = \frac{49}{126} + \frac{27}{126} = \frac{76}{126} = \frac{38}{63}.$$

We have now found both $p(F \cap E) = 7/18$ and $p(E) = 38/63$. We conclude that

$$p(F \mid E) = \frac{p(F \cap E)}{p(E)} = \frac{7/18}{38/63} = \frac{49}{76} \approx 0.645.$$

Before we had any extra information, we assumed that the probability that Bob selected the first box was $1/2$. However, with the extra information that the ball selected at random is red, this probability has increased to approximately 0.645. That is, the probability that Bob selected a ball from the first box increased from $1/2$, when no extra information was available, to 0.645 once we knew that the ball selected was red. ◀

Using the same type of reasoning as in Example 1, we can find the conditional probability that an event F occurs, given that an event E has occurred, when we know $p(E \mid F)$, $p(E \mid \overline{F})$, and $p(F)$. The result we can obtain is called **Bayes' theorem**; it is named after Thomas Bayes, an eighteenth-century British mathematician and minister who introduced this result.

THEOREM 1

BAYES' THEOREM   Suppose that E and F are events from a sample space S such that $p(E) \neq 0$ and $p(F) \neq 0$. Then

$$p(F \mid E) = \frac{p(E \mid F)p(F)}{p(E \mid F)p(F) + p(E \mid \overline{F})p(\overline{F})}.$$

Proof: The definition of conditional probability tells us that $p(F \mid E) = p(E \cap F)/p(E)$ and $p(E \mid F) = p(E \cap F)/p(F)$. Therefore, $p(E \cap F) = p(F \mid E)p(E)$ and $p(E \cap F) = p(E \mid F)p(F)$. Equating these two expressions for $p(E \cap F)$ shows that

$$p(F \mid E)p(E) = p(E \mid F)p(F).$$

Dividing both sides by $p(E)$, we find that

$$p(F \mid E) = \frac{p(E \mid F)p(F)}{p(E)}.$$

Next, we show that $p(E) = p(E \mid F)p(F) + p(E \mid \overline{F})p(\overline{F})$. To see this, first note that $E = E \cap S = E \cap (F \cup \overline{F}) = (E \cap F) \cup (E \cap \overline{F})$. Furthermore, $E \cap F$ and $E \cap \overline{F}$ are disjoint, because if $x \in E \cap F$ and $x \in E \cap \overline{F}$, then $x \in F \cap \overline{F} = \emptyset$. Consequently, $p(E) = p(E \cap F) + p(E \cap \overline{F})$. We have already shown that $p(E \cap F) = p(E \mid F)p(F)$. Moreover, we have $p(E \mid \overline{F}) = p(E \cap \overline{F})/p(\overline{F})$, which shows that $p(E \cap \overline{F}) = p(E \mid \overline{F})p(\overline{F})$. It now follows that

$$p(E) = p(E \cap F) + p(E \cap \overline{F}) = p(E \mid F)p(F) + p(E \mid \overline{F})p(\overline{F}).$$

To complete the proof we insert this expression for $p(E)$ into the equation $p(F \mid E) = p(E \mid F)p(F)/p(E)$. We have proved that

$$p(F \mid E) = \frac{p(E \mid F)p(F)}{p(E \mid F)p(F) + p(E \mid \overline{F})p(\overline{F})}.$$

◁

APPLYING BAYES' THEOREM   Bayes' theorem can be used to solve problems that arise in many disciplines. Next, we will discuss an application of Bayes' theorem to medicine. In particular, we will illustrate how Bayes' theorem can be used to assess the probability that someone testing positive for a disease actually has this disease. The results obtained from Bayes' theorem are often somewhat surprising, as Example 2 shows.

EXAMPLE 2   Suppose that one person in 100,000 has a particular rare disease for which there is a fairly accurate diagnostic test. This test is correct 99.0% of the time when given to a person selected at random who has the disease; it is correct 99.5% of the time when given to a person selected at random who does not have the disease. Given this information can we find

(a) the probability that a person who tests positive for the disease has the disease?
(b) the probability that a person who tests negative for the disease does not have the disease?

Should a person who tests positive be very concerned that he or she has the disease?

Solution: (a) Let F be the event that a person selected at random has the disease, and let E be the event that a person selected at random tests positive for the disease. We want to compute $p(F \mid E)$. To use Bayes' theorem to compute $p(F \mid E)$ we need to find $p(E \mid F)$, $p(E \mid \overline{F})$, $p(F)$, and $p(\overline{F})$.

We know that one person in 100,000 has this disease, so $p(F) = 1/100,000 = 0.00001$ and $p(\overline{F}) = 1 - 0.00001 = 0.99999$. Because a person who has the disease tests positive 99% of the time, we know that $p(E \mid F) = 0.99$; this is the probability of a true positive, that a person with the disease tests positive. It follows that $p(\overline{E} \mid F) = 1 - p(E \mid F) = 1 - 0.99 = 0.01$; this is the probability of a false negative, that a person who has the disease tests negative.

Furthermore, because a person who does not have the disease tests negative 99.5% of the time, we know that $p(\overline{E} \mid \overline{F}) = 0.995$. This is the probability of a true negative, that a person without the disease tests negative. Finally, we see that $p(E \mid \overline{F}) = 1 - p(\overline{E} \mid \overline{F}) = 1 - 0.995 = 0.005$; this is the probability of a false positive, that a person without the disease tests positive.

The probability that a person who tests positive for the disease actually has the disease is $p(F \mid E)$. By Bayes' theorem, we know that

$$p(F \mid E) = \frac{p(E \mid F)p(F)}{p(E \mid F)p(F) + p(E \mid \overline{F})p(\overline{F})}$$

$$= \frac{(0.99)(0.00001)}{(0.99)(0.00001) + (0.005)(0.99999)} \approx 0.002.$$

(b) The probability that someone who tests negative for the disease does not have the disease is $p(\overline{F} \mid \overline{E})$. By Bayes' theorem, we know that

$$p(\overline{F} \mid \overline{E}) = \frac{p(\overline{E} \mid \overline{F})p(\overline{F})}{p(\overline{E} \mid \overline{F})p(\overline{F}) + p(\overline{E} \mid F)p(F)}$$

$$= \frac{(0.995)(0.99999)}{(0.995)(0.99999) + (0.01)(0.00001)} \approx 0.9999999.$$

Consequently, 99.99999% of the people who test negative really do not have the disease.

In part (a) we showed that only 0.2% of people who test positive for the disease actually have the disease. Because the disease is extremely rare, the number of false positives on the diagnostic test is far greater than the number of true positives, making the percentage of people who test positive who actually have the disease extremely small. People who test positive for the diseases should not be overly concerned that they actually have the disease. ◄

GENERALIZING BAYES' THEOREM   Note that in the statement of Bayes' theorem, the events F and $\overline{F}$ are mutually exclusive and cover the entire sample space S (that is, $F \cup \overline{F} = S$). We can extend Bayes' theorem to any collection of mutually exclusive events that cover the entire sample space S, in the following way.

THEOREM 2   **GENERALIZED BAYES' THEOREM**   Suppose that E is an event from a sample space S and that $F_1, F_2, \ldots, F_n$ are mutually exclusive events such that $\bigcup_{i=1}^{n} F_i = S$. Assume that $p(E) \neq 0$ and $p(F_i) \neq 0$ for $i = 1, 2, \ldots, n$. Then

$$p(F_j \mid E) = \frac{p(E \mid F_j)p(F_j)}{\sum_{i=1}^{n} p(E \mid F_i)p(F_i)}.$$

We leave the proof of this generalized version of Bayes' theorem as Exercise 11.

Bayesian Spam Filters

Most electronic mailboxes receive a flood of unwanted and unsolicited messages, known as **spam**. Because spam threatens to overwhelm electronic mail systems, a tremendous amount of work has been devoted to filtering it out. Some of the first tools developed for eliminating spam were based on Bayes' theorem, such as **Bayesian spam filters**.

A Bayesian spam filter uses information about previously seen e-mail messages to guess whether an incoming e-mail message is spam. Bayesian spam filters look for occurrences of particular words in messages. For a particular word w, the probability that w appears in a spam e-mail message is estimated by determining the number of times w appears in a message from a large set of messages known to be spam and the number of times it appears in a large set of messages known not to be spam. When we examine e-mail messages to determine whether they might be spam, we look at words that might be indicators of spam, such as "offer," "special," or "opportunity," as well as words that might indicate that a message is not spam, such as "mom," "lunch," or "Jan" (where Jan is one of your friends). Unfortunately, spam filters sometimes fail to identify a spam message as spam; this is called a false negative. And they sometimes identify a message that is not spam as spam; this is called a false positive. When testing for spam, it is important to minimize false positives, because filtering out wanted e-mail is much worse than letting some spam through.

The use of the word *spam* for unsolicited e-mail comes from a Monty Python comedy sketch about a cafe where the food product Spam comes with everything regardless of whether customers want it.

We will develop some basic Bayesian spam filters. First, suppose we have a set B of messages known to be spam and a set G of messages known not to be spam. (For example, users could classify messages as spam when they examine them in their inboxes.) We next identify the words that occur in B and in G. We count the number of messages in the set containing each word to find $n_B(w)$ and $n_G(w)$, the number of messages containing the word w in the sets B and G, respectively. Then, the empirical probability that a spam message contains the word w is $p(w) = n_B(w)/|B|$, and the empirical probability that a message that is not spam contains the word w is $q(w) = n_G(w)/|G|$. We note that $p(w)$ and $q(w)$ estimate the probabilities that an incoming spam message, and an incoming message that is not spam, contain the word w, respectively.

THOMAS BAYES (1702–1761)   Thomas Bayes was the son a minister in a religious sect known as the Nonconformists. This sect was considered heretical in eighteenth-century Great Britain. Because of the secrecy of the Nonconformists, little is known of Thomas Bayes' life. When Thomas was young, his family moved to London. Thomas was likely educated privately; Nonconformist children generally did not attend school. In 1719 Bayes entered the University of Edinburgh, where he studied logic and theology. He was ordained as a Nonconformist minister like his father and began his work as a minister assisting his father. In 1733 he became minister of the Presbyterian Chapel in Tunbridge Wells, southeast of London, where he remained minister until 1752.

Bayes is best known for his essay on probability published in 1764, three years after his death. This essay was sent to the Royal Society by a friend who found it in the papers left behind when Bayes died. In the introduction to this essay, Bayes stated that his goal was to find a method that could measure the probability that an event happens, assuming that we know nothing about it, but that, under the same circumstances, it has happened a certain proportion of times. Bayes' conclusions were accepted by the great French mathematician Laplace but were later challenged by Boole, who questioned them in his book *Laws of Thought*. Since then Bayes' techniques have been subject to controversy.

Bayes also wrote an article that was published posthumously: "An Introduction to the Doctrine of Fluxions, and a Defense of the Mathematicians Against the Objections of the Author of The Analyst," which supported the logical foundations of calculus. Bayes was elected a Fellow of the Royal Society in 1742, with the support of important members of the Society, even though at that time he had no published mathematical works. Bayes' sole known publication during his lifetime was allegedly a mystical book entitled *Divine Benevolence*, discussing the original causation and ultimate purpose of the universe. Although the book is commonly attributed to Bayes, no author's name appeared on the title page, and the entire work is thought to be of dubious provenance. Evidence for Bayes' mathematical talents comes from a notebook that was almost certainly written by Bayes, which contains much mathematical work, including discussions of probability, trigonometry, geometry, solutions of equations, series, and differential calculus. There are also sections on natural philosophy, in which Bayes looks at topics that include electricity, optics, and celestial mechanics. Bayes is also the author of a mathematical publication on asymptotic series, which appeared after his death.

Now suppose we receive a new e-mail message containing the word w. Let S be the event that the message is spam. Let E be the event that the message contains the word w. The events S, that the message is spam, and $\overline{S}$, that the message is not spam, partition the set of all messages. Hence, by Bayes' theorem, the probability that the message is spam, given that it contains the word w, is

$$p(S \mid E) = \frac{p(E \mid S)p(S)}{p(E \mid S)p(S) + p(E \mid \overline{S})p(\overline{S})}.$$

To apply this formula, we first estimate $p(S)$, the probability that an incoming message is spam, as well as $p(\overline{S})$, the probability that the incoming message is not spam. Without prior knowledge about the likelihood that an incoming message is spam, for simplicity we assume that the message is equally likely to be spam as it is not to be spam. That is, we assume that $p(S) = p(\overline{S}) = 1/2$. Using this assumption, we find that the probability that a message is spam, given that it contains the word w, is

$$p(S \mid E) = \frac{p(E \mid S)}{p(E \mid S) + p(E \mid \overline{S})}.$$

(Note that if we have some empirical data about the ratio of spam messages to messages that are not spam, we can change this assumption to produce a better estimate for $p(S)$ and for $p(\overline{S})$; see Exercise 16.)

Next, we estimate $p(E \mid S)$, the conditional probability that the message contains the word w given that the message is spam, by $p(w)$. Similarly, we estimate $p(E \mid \overline{S})$, the conditional probability that the message contains the word w, given that the message is not spam, by $q(w)$. Inserting these estimates for $p(E \mid S)$ and $p(E \mid \overline{S})$ tells us that $p(S \mid E)$ can be estimated by

$$r(w) = \frac{p(w)}{p(w) + q(w)};$$

that is, $r(w)$ estimates the probability that the message is spam, given that it contains the word w. If $r(w)$ is greater than a threshold that we set, such as 0.9, then we classify the message as spam.

EXAMPLE 3   Suppose that we have found that the word "Rolex" occurs in 250 of 2000 messages known to be spam and in 5 of 1000 messages known not to be spam. Estimate the probability that an incoming message containing the word "Rolex" is spam, assuming that it is equally likely that an incoming message is spam or not spam. If our threshold for rejecting a message as spam is 0.9, will we reject such messages?

Solution: We use the counts that the word "Rolex" appears in spam messages and messages that are not spam to find that $p(\text{Rolex}) = 250/2000 = 0.125$ and $q(\text{Rolex}) = 5/1000 = 0.005$. Because we are assuming that it is equally likely for an incoming message to be spam as it is not to be spam, we can estimate the probability that an incoming message containing the word "Rolex" is spam by

$$r(\text{Rolex}) = \frac{p(\text{Rolex})}{p(\text{Rolex}) + q(\text{Rolex})} = \frac{0.125}{0.125 + 0.005} = \frac{0.125}{0.130} \approx 0.962.$$

Because $r(\text{Rolex})$ is greater than the threshold 0.9, we reject such messages as spam.   ◀

Detecting spam based on the presence of a single word can lead to excessive false positives and false negatives. Consequently, spam filters look at the presence of multiple words. For example, suppose that the message contains the words w_1 and w_2. Let E_1 and E_2 denote the events that the message contains the words w_1 and w_2, respectively. To make our computations simpler, we assume that E_1 and E_2 are independent events and that $E_1 \mid S$ and $E_2 \mid S$ are independent events and that we have no prior knowledge regarding whether or not the message is spam. (The assumptions that E_1 and E_2 are independent and that $E_1 \mid S$ and $E_2 \mid S$ are independent may introduce some error into our computations; we assume that this error is small.) Using Bayes' theorem and our assumptions, we can show (see Exercise 17) that $p(S \mid E_1 \cap E_2)$, the probability that the message is spam given that it contains both w_1 and w_2, is

$$p(S \mid E_1 \cap E_2) = \frac{p(E_1 \mid S)p(E_2 \mid S)}{p(E_1 \mid S)p(E_2 \mid S) + p(E_1 \mid \overline{S})p(E_2 \mid \overline{S})}.$$

We estimate the probability $p(S \mid E_1 \cap E_2)$ by

$$r(w_1, w_2) = \frac{p(w_1)p(w_2)}{p(w_1)p(w_2) + q(w_1)q(w_2)}.$$

That is, $r(w_1, w_2)$ estimates the probability that the message is spam, given that it contains the words w_1 and w_2. When $r(w_1, w_2)$ is greater than a preset threshold, such as 0.9, we determine that the message is likely spam.

EXAMPLE 4   Suppose that we train a Bayesian spam filter on a set of 2000 spam messages and 1000 messages that are not spam. The word "stock" appears in 400 spam messages and 60 messages that are not spam, and the word "undervalued" appears in 200 spam messages and 25 messages that are not spam. Estimate the probability that an incoming message containing both the words "stock" and "undervalued" is spam, assuming that we have no prior knowledge about whether it is spam. Will we reject such messages as spam when we set the threshold at 0.9?

Solution: Using the counts of each of these two words in messages known to be spam or known not to be spam, we obtain the following estimates: $p(\text{stock}) = 400/2000 = 0.2$, $q(\text{stock}) = 60/1000 = 0.06$, $p(\text{undervalued}) = 200/2000 = 0.1$, and $q(\text{undervalued}) = 25/1000 = 0.025$. Using these probabilities, we can estimate the probability that the message is spam by

$$r(\text{stock, undervalued}) = \frac{p(\text{stock})p(\text{undervalued})}{p(\text{stock})p(\text{undervalued}) + q(\text{stock})q(\text{undervalued})}$$

$$= \frac{(0.2)(0.1)}{(0.2)(0.1) + (0.06)(0.025)} \approx 0.930.$$

Because we have set the threshold for rejecting messages at 0.9, such messages will be rejected by the filter.   ◀

The more words we use to estimate the probability that an incoming mail message is spam, the better is our chance that we correctly determine whether it is spam. In general, if E_i is the event that the message contains word w_i, assuming that the number of incoming spam messages is approximately the same as the number of incoming messages that are not spam, and that the events $E_i \mid S$ are independent, then by Bayes' theorem the probability that a message containing all the words $w_1, w_2, \ldots, w_k$ is spam is

$$p\left(S \mid \bigcap_{i=1}^{k} E_i\right) = \frac{\prod_{i=1}^{k} p(E_i \mid S)}{\prod_{i=1}^{k} p(E_i \mid S) + \prod_{i=1}^{k} p(E_i \mid \overline{S})}.$$

We can estimate this probability by

$$r(w_1, w_2, \ldots, w_k) = \frac{\prod_{i=1}^{k} p(w_i)}{\prod_{i=1}^{k} p(w_i) + \prod_{i=1}^{k} q(w_i)}.$$

For the most effective spam filter, we choose words for which the probability that each of these words appears in spam is either very high or very low. When we compute this value for a particular message, we reject the message as spam if $r(w_1, w_2, \ldots, w_k)$ exceeds a preset threshold, such as 0.9.

Another way to improve the performance of a Bayesian spam filter is to look at the probabilities that particular pairs of words appear in spam and in messages that are not spam. We then treat appearances of these pairs of words as appearance of a single block, rather than as the appearance of two separate words. For example, the pair of words "enhance performance" most likely indicates spam, while "operatic performance" indicates a message that is not spam. Similarly, we can assess the likelihood that a message is spam by examining the structure of a message to determine where words appear in it. Also, spam filters look at appearances of certain types of strings of characters rather than just words. For example, a message with the valid e-mail address of one of your friends is less likely to be spam (if not sent by a worm) than one containing an e-mail address that came from a country known to originate a lot of spam. There is an ongoing war between people who create spam and those trying to filter their messages out. This leads to the introduction of many new techniques to defeat spam filters, including inserting into spam messages long strings of words that appear in messages that are not spam, as well as including words inside pictures. The techniques we have discussed here are only the first steps in fighting this war on spam.

Bayesian poisoning, the insertion of extra words to defeat spam filters, can use random or purposefully selected words.

Exercises

1. Suppose that E and F are events in a sample space and $p(E) = 1/3$, $p(F) = 1/2$, and $p(E \mid F) = 2/5$. Find $p(F \mid E)$.

2. Suppose that Frida selects a ball by first picking one of two boxes at random and then selecting a ball from this box at random. The first box contains two white balls and three blue balls, and the second box contains four white balls and one blue ball. What is the probability that Frida picked a ball from the first box if she has selected a blue ball?

3. Suppose that 8% of all bicycle racers use steroids, that a bicyclist who uses steroids tests positive for steroids 96% of the time, and that a bicyclist who does not use steroids tests positive for steroids 9% of the time. What is the probability that a randomly selected bicyclist who tests positive for steroids actually uses steroids?

4. Suppose that a test for opium use has a 2% false positive rate and a 5% false negative rate. That is, 2% of people who do not use opium test positive for opium, and 5% of opium users test negative for opium. Furthermore, suppose that 1% of people actually use opium.
   a) Find the probability that someone who tests negative for opium use does not use opium.
   b) Find the probability that someone who tests positive for opium use actually uses opium.

5. Suppose that 8% of the patients tested in a clinic are infected with HIV. Furthermore, suppose that when a blood test for HIV is given, 98% of the patients infected with HIV test positive and that 3% of the patients not infected with HIV test positive. What is the probability that
   a) a patient testing positive for HIV with this test is infected with it?
   b) a patient testing positive for HIV with this test is not infected with it?
   c) a patient testing negative for HIV with this test is infected with it?
   d) a patient testing negative for HIV with this test is not infected with it?

6. Suppose that 4% of the patients tested in a clinic are infected with avian influenza. Furthermore, suppose that when a blood test for avian influenza is given, 97% of the patients infected with avian influenza test positive and that 2% of the patients not infected with avian influenza test positive. What is the probability that
   a) a patient testing positive for avian influenza with this test is infected with it?
   b) a patient testing positive for avian influenza with this test is not infected with it?
   c) a patient testing negative for avian influenza with this test is infected with it?
   d) a patient testing negative for avian influenza with this test is not infected with it?

7. An electronics company is planning to introduce a new camera phone. The company commissions a marketing report for each new product that predicts either the success or the failure of the product. Of new products introduced by the company, 60% have been successes. Furthermore, 70% of their successful products were predicted to be successes, while 40% of failed products were predicted to be successes. Find the probability that this new camera phone will be successful if its success has been predicted.

8. Suppose that E, F_1, F_2, and F_3 are events from a sample space S and that F_1, F_2, and F_3 are pairwise disjoint and their union is S. Find $p(F_1 \mid E)$ if $p(E \mid F_1) = 1/8$, $p(E \mid F_2) = 1/4$, $p(E \mid F_3) = 1/6$, $p(F_1) = 1/4$, $p(F_2) = 1/4$, and $p(F_3) = 1/2$.

9. In this exercise we will use Bayes' theorem to solve the Monty Hall puzzle (Example 10 in Section 7.1). Recall that in this puzzle you are asked to select one of three doors to open. There is a large prize behind one of the three doors and the other two doors are losers. After you select a door, Monty Hall opens one of the two doors you did not select that he knows is a losing door, selecting at random if both are losing doors. Monty asks you whether you would like to switch doors. Suppose that the three doors in the puzzle are labeled 1, 2, and 3. Let W be the random variable whose value is the number of the winning door; assume that $p(W = k) = 1/3$ for $k = 1, 2, 3$. Let M denote the random variable whose value is the number of the door that Monty opens. Suppose you choose door i.

a) What is the probability that you will win the prize if the game ends without Monty asking you whether you want to change doors?

b) Find $p(M = j \mid W = k)$ for $j = 1, 2, 3$ and $k = 1, 2, 3$.

c) Use Bayes' theorem to find $p(W = j \mid M = k)$ where i and j and k are distinct values.

d) Explain why the answer to part (c) tells you whether you should change doors when Monty gives you the chance to do so.

10. Ramesh can get to work in three different ways: by bicycle, by car, or by bus. Because of commuter traffic, there is a 50% chance that he will be late when he drives his car. When he takes the bus, which uses a special lane reserved for buses, there is a 20% chance that he will be late. The probability that he is late when he rides his bicycle is only 5%. Ramesh arrives late one day. His boss wants to estimate the probability that he drove his car to work that day.

a) Suppose the boss assumes that there is a 1/3 chance that Ramesh takes each of the three ways he can get to work. What estimate for the probability that Ramesh drove his car does the boss obtain from Bayes' theorem under this assumption?

b) Suppose the boss knows that Ramesh drives 30% of the time, takes the bus only 10% of the time, and takes his bicycle 60% of the time. What estimate for the probability that Ramesh drove his car does the boss obtain from Bayes' theorem using this information?

∗11. Prove Theorem 2, the extended form of Bayes' theorem. That is, suppose that E is an event from a sample space S and that $F_1, F_2, \ldots, F_n$ are mutually exclusive events such that $\bigcup_{i=1}^{n} F_i = S$. Assume that $p(E) \neq 0$ and $p(F_i) \neq 0$ for $i = 1, 2, \ldots, n$. Show that

$$p(F_j \mid E) = \frac{p(E \mid F_j)p(F_j)}{\sum_{i=1}^{n} p(E \mid F_i)p(F_i)}.$$

[*Hint:* Use the fact that $E = \bigcup_{i=1}^{n}(E \cap F_i)$.]

12. Suppose that a Bayesian spam filter is trained on a set of 500 spam messages and 200 messages that are not spam. The word "exciting" appears in 40 spam messages and in 25 messages that are not spam. Would an incoming message be rejected as spam if it contains the word "exciting" and the threshold for rejecting spam is 0.9?

13. Suppose that a Bayesian spam filter is trained on a set of 1000 spam messages and 400 messages that are not spam. The word "opportunity" appears in 175 spam messages and 20 messages that are not spam. Would an incoming message be rejected as spam if it contains the word "opportunity" and the threshold for rejecting a message is 0.9?

14. Would we reject a message as spam in Example 4

a) using just the fact that the word "undervalued" occurs in the message?

b) using just the fact that the word "stock" occurs in the message?

15. Suppose that a Bayesian spam filter is trained on a set of 10,000 spam messages and 5000 messages that are not spam. The word "enhancement" appears in 1500 spam messages and 20 messages that are not spam, while the word "herbal" appears in 800 spam messages and 200 messages that are not spam. Estimate the probability that a received message containing both the words "enhancement" and "herbal" is spam. Will the message be rejected as spam if the threshold for rejecting spam is 0.9?

16. Suppose that we have prior information concerning whether a random incoming message is spam. In particular, suppose that over a time period, we find that s spam messages arrive and h messages arrive that are not spam.

a) Use this information to estimate $p(S)$, the probability that an incoming message is spam, and $p(\bar{S})$, the probability an incoming message is not spam.

b) Use Bayes' theorem and part (a) to estimate the probability that an incoming message containing the word w is spam, where $p(w)$ is the probability that w occurs in a spam message and $q(w)$ is the probability that w occurs in a message that is not spam.

17. Suppose that E_1 and E_2 are the events that an incoming mail message contains the words w_1 and w_2, respectively. Assuming that E_1 and E_2 are independent events and that $E_1 \mid S$ and $E_2 \mid S$ are independent events, where S is the event that an incoming message is spam, and that we have no prior knowledge regarding whether or not the message is spam, show that

$$p(S \mid E_1 \cap E_2)$$

$$= \frac{p(E_1 \mid S)p(E_2 \mid S)}{p(E_1 \mid S)p(E_2 \mid S) + p(E_1 \mid \overline{S})p(E_2 \mid \overline{S})}.$$

7.4   Expected Value and Variance

Introduction

The **expected value** of a random variable is the sum over all elements in a sample space of the product of the probability of the element and the value of the random variable at this element. Consequently, the expected value is a weighted average of the values of a random variable. The expected value of a random variable provides a central point for the distribution of values of this random variable. We can solve many problems using the notion of the expected value of a random variable, such as determining who has an advantage in gambling games and computing the average-case complexity of algorithms. Another useful measure of a random variable is its **variance**, which tells us how spread out the values of this random variable are. We can use the variance of a random variable to help us estimate the probability that a random variable takes values far removed from its expected value.

Expected Values

Links

Many questions can be formulated in terms of the value we expect a random variable to take, or more precisely, the average value of a random variable when an experiment is performed a large number of times. Questions of this kind include: How many heads are expected to appear when a coin is flipped 100 times? What is the expected number of comparisons used to find an element in a list using a linear search? To study such questions we introduce the concept of the expected value of a random variable.

DEFINITION 1   The *expected value*, also called the *expectation* or *mean*, of the random variable X on the sample space S is equal to

$$E(X) = \sum_{s \in S} p(s)X(s).$$

The *deviation* of X at $s \in S$ is $X(s) - E(X)$, the difference between the value of X and the mean of X.

Note that when the sample space S has n elements $S = \{x_1, x_2, \ldots, x_n\}$, $E(X) = \sum_{i=1}^{n} p(x_i)X(x_i)$.

Remark: When there are infinitely many elements of the sample space, the expectation is defined only when the infinite series in the definition is absolutely convergent. In particular, the expectation of a random variable on an infinite sample space is finite if it exists.

EXAMPLE 1   **Expected Value of a Die**   Let X be the number that comes up when a fair die is rolled. What is the expected value of X?

Solution: The random variable X takes the values 1, 2, 3, 4, 5, or 6, each with probability 1/6. It follows that

$$E(X) = \frac{1}{6} \cdot 1 + \frac{1}{6} \cdot 2 + \frac{1}{6} \cdot 3 + \frac{1}{6} \cdot 4 + \frac{1}{6} \cdot 5 + \frac{1}{6} \cdot 6 = \frac{21}{6} = \frac{7}{2}.$$ ◀

EXAMPLE 2   A fair coin is flipped three times. Let S be the sample space of the eight possible outcomes, and let X be the random variable that assigns to an outcome the number of heads in this outcome. What is the expected value of X?

Extra Examples

Solution: In Example 10 of Section 7.2 we listed the values of X for the eight possible outcomes when a coin is flipped three times. Because the coin is fair and the flips are independent, the probability of each outcome is 1/8. Consequently,

$$E(X) = \frac{1}{8}[X(HHH) + X(HHT) + X(HTH) + X(THH) + X(TTH)$$

$$+ X(THT) + X(HTT) + X(TTT)]$$

$$= \frac{1}{8}(3 + 2 + 2 + 2 + 1 + 1 + 1 + 0) = \frac{12}{8}$$

$$= \frac{3}{2}.$$

Consequently, the expected number of heads that come up when a fair coin is flipped three times is 3/2.   ◀

When an experiment has relatively few outcomes, we can compute the expected value of a random variable directly from its definition, as was done in Example 2. However, when an experiment has a large number of outcomes, it may be inconvenient to compute the expected value of a random variable directly from its definition. Instead, we can find the expected value of a random variable by grouping together all outcomes assigned the same value by the random variable, as Theorem 1 shows.

THEOREM 1   If X is a random variable and $p(X = r)$ is the probability that $X = r$, so that $p(X = r) = \sum_{s \in S, X(s) = r} p(s)$, then

$$E(X) = \sum_{r \in X(S)} p(X = r)r.$$

Proof: Suppose that X is a random variable with range $X(S)$, and let $p(X = r)$ be the probability that the random variable X takes the value r. Consequently, $p(X = r)$ is the sum of the probabilities of the outcomes s such that $X(s) = r$. It follows that

$$E(X) = \sum_{r \in X(S)} p(X = r)r.$$ ◁

Example 3 and the proof of Theorem 2 will illustrate the use of this formula. In Example 3 we will find the expected value of the sum of the numbers that appear on two fair dice when they are rolled. In Theorem 2 we will find the expected value of the number of successes when n Bernoulli trials are performed.

EXAMPLE 3   What is the expected value of the sum of the numbers that appear when a pair of fair dice is rolled?

Solution: Let X be the random variable equal to the sum of the numbers that appear when a pair of dice is rolled. In Example 12 of Section 7.2 we listed the value of X for the 36 outcomes of this experiment. The range of X is $\{2, 3, 4, 5, 6, 7, 8, 9, 10, 11, 12\}$. By Example 12 of Section 7.2 we see that

$$p(X = 2) = p(X = 12) = 1/36,$$

$$p(X = 3) = p(X = 11) = 2/36 = 1/18,$$

$$p(X = 4) = p(X = 10) = 3/36 = 1/12,$$

$$p(X = 5) = p(X = 9) = 4/36 = 1/9,$$

$$p(X = 6) = p(X = 8) = 5/36,$$

$$p(X = 7) = 6/36 = 1/6.$$

Substituting these values in the formula, we have

$$E(X) = 2 \cdot \frac{1}{36} + 3 \cdot \frac{1}{18} + 4 \cdot \frac{1}{12} + 5 \cdot \frac{1}{9} + 6 \cdot \frac{5}{36} + 7 \cdot \frac{1}{6}$$

$$+ 8 \cdot \frac{5}{36} + 9 \cdot \frac{1}{9} + 10 \cdot \frac{1}{12} + 11 \cdot \frac{1}{18} + 12 \cdot \frac{1}{36}$$

$$= 7.$$

◀

THEOREM 2   The expected number of successes when n mutually independent Bernoulli trials are performed, where p is the probability of success on each trial, is np.

Proof: Let X be the random variable equal to the number of successes in n trials. By Theorem 2 of Section 7.2 we see that $p(X = k) = C(n, k)p^k q^{n-k}$. Hence, we have

$$E(X) = \sum_{k=1}^{n} kp(X = k) \qquad\qquad \text{by Theorem 1}$$

$$= \sum_{k=1}^{n} kC(n, k)p^k q^{n-k} \qquad\qquad \text{by Theorem 2 in Section 7.2}$$

$$= \sum_{k=1}^{n} nC(n-1, k-1)p^k q^{n-k} \qquad\qquad \text{by Exercise 21 in Section 6.4}$$

$$= np \sum_{k=1}^{n} C(n-1, k-1)p^{k-1}q^{n-k} \qquad \text{factoring } np \text{ from each term}$$

$$= np \sum_{j=0}^{n-1} C(n-1, j)p^j q^{n-1-j} \qquad \text{shifting index of summation with } j = k - 1$$

$$= np(p + q)^{n-1} \qquad\qquad \text{by the binomial theorem}$$

$$= np. \qquad\qquad \text{because } p + q = 1$$

This completes the proof because it shows that the expected number of successes in n mutually independent Bernoulli trials is np. ◁

We will also show that the hypothesis that the Bernoulli trials are mutually independent in Theorem 2 is not necessary.

Linearity of Expectations

Theorem 3 tells us that expected values are linear. For example, the expected value of the sum of random variables is the sum of their expected values. We will find this property exceedingly useful.

THEOREM 3   If X_i, $i = 1, 2, \ldots, n$ with n a positive integer, are random variables on S, and if a and b are real numbers, then

(*i*) $E(X_1 + X_2 + \cdots + X_n) = E(X_1) + E(X_2) + \cdots + E(X_n)$
(*ii*) $E(aX + b) = aE(X) + b$.

Proof: Part (*i*) follows for $n = 2$ directly from the definition of expected value, because

$$E(X_1 + X_2) = \sum_{s \in S} p(s)(X_1(s) + X_2(s))$$

$$= \sum_{s \in S} p(s)X_1(s) + \sum_{s \in S} p(s)X_2(s)$$

$$= E(X_1) + E(X_2).$$

The case for n random variables follows easily by mathematical induction using the case of two random variables. (We leave it to the reader to complete the proof.)

To prove part (*ii*), note that

$$E(aX + b) = \sum_{s \in S} p(s)(aX(s) + b)$$
$$= a \sum_{s \in S} p(s)X(s) + b \sum_{s \in S} p(s)$$
$$= aE(X) + b \text{ because } \sum_{s \in S} p(s) = 1.$$ ◁

Examples 4 and 5 illustrate how to use Theorem 3.

EXAMPLE 4   Use Theorem 3 to find the expected value of the sum of the numbers that appear when a pair of fair dice is rolled. (This was done in Example 3 without the benefit of this theorem.)

Solution: Let X_1 and X_2 be the random variables with $X_1((i, j)) = i$ and $X_2((i, j)) = j$, so that X_1 is the number appearing on the first die and X_2 is the number appearing on the second die. It is easy to see that $E(X_1) = E(X_2) = 7/2$ because both equal $(1 + 2 + 3 + 4 + 5 + 6)/6 = 21/6 = 7/2$. The sum of the two numbers that appear when the two dice are rolled is the sum $X_1 + X_2$. By Theorem 3, the expected value of the sum is $E(X_1 + X_2) = E(X_1) + E(X_2) = 7/2 + 7/2 = 7$.   ◀

EXAMPLE 5   In the proof of Theorem 2 we found the expected value of the number of successes when n independent Bernoulli trials are performed, where p is the probability of success on each trial by direct computation. Show how Theorem 3 can be used to derive this result where the Bernoulli trials are not necessarily independent.

Solution: Let X_i be the random variable with $X_i((t_1, t_2, \ldots, t_n)) = 1$ if t_i is a success and $X_i((t_1, t_2, \ldots, t_n)) = 0$ if t_i is a failure. The expected value of X_i is $E(X_i) = 1 \cdot p + 0 \cdot (1 - p) = p$ for $i = 1, 2, \ldots, n$. Let $X = X_1 + X_2 + \cdots + X_n$, so that X counts the number of successes when these n Bernoulli trials are performed. Theorem 3, applied to the sum of n random variables, shows that $E(X) = E(X_1) + E(X_2) + \cdots + E(X_n) = np$. ◀

We can take advantage of the linearity of expectations to find the solutions of many seemingly difficult problems. The key step is to express a random variable whose expectation we wish to find as the sum of random variables whose expectations are easy to find. Examples 6 and 7 illustrate this technique.

EXAMPLE 6   **Expected Value in the Hatcheck Problem**   A new employee checks the hats of n people at a restaurant, forgetting to put claim check numbers on the hats. When customers return for their hats, the checker gives them back hats chosen at random from the remaining hats. What is the expected number of hats that are returned correctly?

Solution: Let X be the random variable that equals the number of people who receive the correct hat from the checker. Let X_i be the random variable with $X_i = 1$ if the ith person receives the correct hat and $X_i = 0$ otherwise. It follows that

$$X = X_1 + X_2 + \cdots + X_n.$$

Because it is equally likely that the checker returns any of the hats to this person, it follows that the probability that the ith person receives the correct hat is $1/n$. Consequently, by Theorem 1, for all i we have

$$E(X_i) = 1 \cdot p(X_i = 1) + 0 \cdot p(X_i = 0) = 1 \cdot 1/n + 0 = 1/n.$$

By the linearity of expectations (Theorem 3), it follows that

$$E(X) = E(X_1) + E(X_2) + \cdots + E(X_n) = n \cdot 1/n = 1.$$

Consequently, the average number of people who receive the correct hat is exactly 1. Note that this answer is independent of the number of people who have checked their hats! (We will find an explicit formula for the probability that no one receives the correct hat in Example 4 of Section 8.6.) ◀

EXAMPLE 7   **Expected Number of Inversions in a Permutation**   The ordered pair (i, j) is called an **inversion** in a permutation of the first n positive integers if $i < j$ but j precedes i in the permutation. For instance, there are six inversions in the permutation 3, 5, 1, 4, 2; these inversions are

$$(1, 3), (1, 5), (2, 3), (2, 4), (2, 5), (4, 5).$$

Let $I_{i,j}$ be the random variable on the set of all permutations of the first n positive integers with $I_{i,j} = 1$ if (i, j) is an inversion of the permutation and $I_{i,j} = 0$ otherwise. It follows that if X is the random variable equal to the number of inversions in the permutation, then

$$X = \sum_{1 \le i < j \le n} I_{i,j}.$$

Note that it is equally likely for i to precede j in a randomly chosen permutation as it is for j to precede i. (To see this, note that there are an equal number of permutations with each of these properties.) Consequently, for all pairs i and j we have

$$E(I_{i,j}) = 1 \cdot p(I_{i,j} = 1) + 0 \cdot p(I_{i,j} = 0) = 1 \cdot 1/2 + 0 = 1/2.$$

Because there are $\binom{n}{2}$ pairs i and j with $1 \leq i < j \leq n$ and by the linearity of expectations (Theorem 3), we have

$$E(X) = \sum_{1 \leq i < j \leq n} E(I_{i,j}) = \binom{n}{2} \cdot \frac{1}{2} = \frac{n(n-1)}{4}.$$

It follows that there are an average of $n(n-1)/4$ inversions in a permutation of the first n positive integers. ◀

Average-Case Computational Complexity

Links

Computing the average-case computational complexity of an algorithm can be interpreted as computing the expected value of a random variable. Let the sample space of an experiment be the set of possible inputs a_j, $j = 1, 2, \ldots, n$, and let X be the random variable that assigns to a_j the number of operations used by the algorithm when given a_j as input. Based on our knowledge of the input, we assign a probability $p(a_j)$ to each possible input value a_j. Then, the average-case complexity of the algorithm is

$$E(X) = \sum_{j=1}^{n} p(a_j)X(a_j).$$

This is the expected value of X.

Finding the average-case computational complexity of an algorithm is usually much more difficult than finding its worst-case computational complexity, and often involves the use of sophisticated methods. However, there are some algorithms for which the analysis required to find the average-case computational complexity is not difficult. For instance, in Example 8 we will illustrate how to find the average-case computational complexity of the linear search algorithm under different assumptions concerning the probability that the element for which we search is an element of the list.

EXAMPLE 8   **Average-Case Complexity of the Linear Search Algorithm**   We are given a real number x and a list of n distinct real numbers. The linear search algorithm, described in Section 3.1, locates x by successively comparing it to each element in the list, terminating when x is located or when all the elements have been examined and it has been determined that x is not in the list. What is the average-case computational complexity of the linear search algorithm if the probability that x is in the list is p and it is equally likely that x is any of the n elements in the list? (There are $n + 1$ possible types of input: one type for each of the n numbers in the list and a last type for numbers not in the list, which we treat as a single input.)

Solution: In Example 4 of Section 3.3 we showed that $2i + 1$ comparisons are used if x equals the ith element of the list and, in Example 2 of Section 3.3, we showed that $2n + 2$ comparisons are used if x is not in the list. The probability that x equals a_i, the ith element in the list, is p/n, and the probability that x is not in the list is $q = 1 - p$. It follows that the average-case computational complexity of the linear search algorithm is

$$E = \frac{3p}{n} + \frac{5p}{n} + \cdots + \frac{(2n+1)p}{n} + (2n+2)q$$

$$= \frac{p}{n}(3 + 5 + \cdots + (2n+1)) + (2n+2)q$$

$$= \frac{p}{n}((n+1)^2 - 1) + (2n+2)q$$

$$= p(n+2) + (2n+2)q.$$

(The third equality follows from Example 2 of Section 5.1.) For instance, when x is guaranteed to be in the list, we have $p = 1$ (so the probability that $x = a_i$ is $1/n$ for each i) and $q = 0$. Then $E = n + 2$, as we showed in Example 4 in Section 3.3.

When p, the probability that x is in the list, is $1/2$, it follows that $q = 1 - p = 1/2$, so $E = (n + 2)/2 + n + 1 = (3n + 4)/2$. Similarly, if the probability that x is in the list is $3/4$, we have $p = 3/4$ and $q = 1/4$, so $E = 3(n + 2)/4 + (n + 1)/2 = (5n + 8)/4$.

Finally, when x is guaranteed not to be in the list, we have $p = 0$ and $q = 1$. It follows that $E = 2n + 2$, which is not surprising because we have to search the entire list.   ◀

Example 9 illustrates how the linearity of expectations can help us find the average-case complexity of a sorting algorithm, the insertion sort.

EXAMPLE 9   **Average-Case Complexity of the Insertion Sort**   What is the average number of comparisons used by the insertion sort to sort n distinct elements?

Solution: We first suppose that X is the random variable equal to the number of comparisons used by the insertion sort (described in Section 3.1) to sort a list $a_1, a_2, \ldots, a_n$ of n distinct elements. Then $E(X)$ is the average number of comparisons used. (Recall that at step i for $i = 2, \ldots, n$, the insertion sort inserts the ith element in the original list into the correct position in the sorted list of the first $i - 1$ elements of the original list.)

We let X_i be the random variable equal to the number of comparisons used to insert a_i into the proper position after the first $i - 1$ elements $a_1, a_2, \ldots, a_{i-1}$ have been sorted. Because

$$X = X_2 + X_3 + \cdots + X_n,$$

we can use the linearity of expectations to conclude that

$$E(X) = E(X_2 + X_3 + \cdots + X_n) = E(X_2) + E(X_3) + \cdots + E(X_n).$$

To find $E(X_i)$ for $i = 2, 3, \ldots, n$, let $p_j(k)$ denote the probability that the largest of the first j elements in the list occurs at the kth position, that is, that $\max(a_1, a_2, \ldots, a_j) = a_k$, where $1 \leq k \leq j$. Because the elements of the list are randomly distributed, it is equally likely for the largest element among the first j elements to occur at any position. Consequently, $p_j(k) = 1/j$. If $X_i(k)$ equals the number of comparisons used by the insertion sort if a_i is inserted into the kth position in the list once $a_1, a_2, \ldots, a_{i-1}$ have been sorted, it follows that $X_i(k) = k$. Because it is possible that a_i is inserted in any of the first i positions, we find that

$$E(X_i) = \sum_{k=1}^{i} p_i(k) \cdot X_i(k) = \sum_{k=1}^{i} \frac{1}{i} \cdot k = \frac{1}{i} \cdot \sum_{k=1}^{i} k = \frac{1}{i} \cdot \frac{i(i+1)}{2} = \frac{i+1}{2}.$$

It follows that

$$E(X) = \sum_{i=2}^{n} E(X_i) = \sum_{i=2}^{n} \frac{i+1}{2} = \frac{1}{2} \sum_{j=3}^{n+1} j$$

$$= \frac{1}{2} \frac{(n+1)(n+2)}{2} - \frac{1}{2}(1+2) = \frac{n^2 + 3n - 4}{4}.$$

To obtain the third of these equalities we shifted the index of summation, setting $j = i + 1$. To obtain the fourth equality, we used the formula $\sum_{k=1}^{m} k = m(m + 1)/2$ (from Table 2 in Section 2.4) with $m = n + 1$, subtracting off the missing terms with $j = 1$ and $j = 2$. We conclude that the average number of comparisons used by the insertion sort to sort n elements equals $(n^2 + 3n - 4)/4$, which is $\Theta(n^2)$.   ◀

The Geometric Distribution

We now turn our attention to a random variable with infinitely many possible outcomes.

EXAMPLE 10   Suppose that the probability that a coin comes up tails is p. This coin is flipped repeatedly until it comes up tails. What is the expected number of flips until this coin comes up tails?

Links

Solution: We first note that the sample space consists of all sequences that begin with any number of heads, denoted by H, followed by a tail, denoted by T. Therefore, the sample space is the set $\{T, HT, HHT, HHHT, HHHHT, \ldots\}$. Note that this is an infinite sample space. We can determine the probability of an element of the sample space by noting that the coin flips are independent and that the probability of a head is $1 - p$. Therefore, $p(T) = p$, $p(HT) = (1 - p)p$, $p(HHT) = (1 - p)^2 p$, and in general the probability that the coin is flipped n times before a tail comes up, that is, that $n - 1$ heads come up followed by a tail, is $(1 - p)^{n-1} p$. (Exercise 12 asks for a verification that the sum of the probabilities of the points in the sample space is 1.)

Now let X be the random variable equal to the number of flips in an element in the sample space. That is, $X(T) = 1$, $X(HT) = 2$, $X(HHT) = 3$, and so on. Note that $p(X = j) = (1 - p)^{j-1} p$. The expected number of flips until the coin comes up tails equals $E(X)$.

Using Theorem 1, we find that

$$E(X) = \sum_{j=1}^{\infty} j \cdot p(X = j) = \sum_{j=1}^{\infty} j(1 - p)^{j-1} p = p \sum_{j=1}^{\infty} j(1 - p)^{j-1} = p \cdot \frac{1}{p^2} = \frac{1}{p}.$$

[The third equality in this chain follows from Table 2 in Section 2.4, which tells us that $\sum_{j=1}^{\infty} j(1 - p)^{j-1} = 1/(1 - (1 - p))^2 = 1/p^2$.] It follows that the expected number of times the coin is flipped until tails comes up is $1/p$. Note that when the coin is fair we have $p = 1/2$, so the expected number of flips until it comes up tails is $1/(1/2) = 2$. ◀

The random variable X that equals the number of flips expected before a coin comes up tails is an example of a random variable with a **geometric distribution**.

DEFINITION 2   A random variable X has a *geometric distribution with parameter p* if $p(X = k) = (1 - p)^{k-1} p$ for $k = 1, 2, 3, \ldots$, where p is a real number with $0 \leq p \leq 1$.

Geometric distributions arise in many applications because they are used to study the time required before a particular event happens, such as the time required before we find an object with a certain property, the number of attempts before an experiment succeeds, the number of times a product can be used before it fails, and so on.

When we computed the expected value of the number of flips required before a coin comes up tails, we proved Theorem 4.

THEOREM 4   If the random variable X has the geometric distribution with parameter p, then $E(X) = 1/p$.

Independent Random Variables

We have already discussed independent events. We will now define what it means for two random variables to be independent.

DEFINITION 3

The random variables X and Y on a sample space S are *independent* if

$$p(X = r_1 \text{ and } Y = r_2) = p(X = r_1) \cdot p(Y = r_2),$$

or in words, if the probability that $X = r_1$ and $Y = r_2$ equals the product of the probabilities that $X = r_1$ and $Y = r_2$, for all real numbers r_1 and r_2.

EXAMPLE 11   Are the random variables X_1 and X_2 from Example 4 independent?

Solution: Let $S = \{1, 2, 3, 4, 5, 6\}$, and let $i \in S$ and $j \in S$. Because there are 36 possible outcomes when the pair of dice is rolled and each is equally likely, we have

$$p(X_1 = i \text{ and } X_2 = j) = 1/36.$$

Furthermore, $p(X_1 = i) = 1/6$ and $p(X_2 = j) = 1/6$, because the probability that i appears on the first die and the probability that j appears on the second die are both $1/6$. It follows that

$$p(X_1 = i \text{ and } X_2 = j) = \frac{1}{36} \quad \text{and} \quad p(X_1 = i)p(X_2 = j) = \frac{1}{6} \cdot \frac{1}{6} = \frac{1}{36},$$

so X_1 and X_2 are independent. ◀

EXAMPLE 12   Show that the random variables X_1 and $X = X_1 + X_2$, where X_1 and X_2 are as defined in Example 4, are not independent.

Solution: Note that $p(X_1 = 1 \text{ and } X = 12) = 0$, because $X_1 = 1$ means the number appearing on the first die is 1, which implies that the sum of the numbers appearing on the two dice cannot equal 12. On the other hand, $p(X_1 = 1) = 1/6$ and $p(X = 12) = 1/36$. Hence $p(X_1 = 1 \text{ and } X = 12) \neq p(X_1 = 1) \cdot p(X = 12)$. This counterexample shows that X_1 and X are not independent. ◀

The expected value of the product of two independent random variables is the product of their expected values, as Theorem 5 shows.

THEOREM 5   If X and Y are independent random variables on a sample space S, then $E(XY) = E(X)E(Y)$.

Proof: To prove this formula, we use the key observation that the event $XY = r$ is the disjoint union of the events $X = r_1$ and $Y = r_2$ over all $r_1 \in X(S)$ and $r_2 \in Y(S)$ with $r = r_1 r_2$. We have

$$E(XY) = \sum_{r \in XY(S)} r \cdot p(XY = r) \qquad \text{by Theorem 1}$$

$$= \sum_{r_1 \in X(S), r_2 \in Y(S)} r_1 r_2 \cdot p(X = r_1 \text{ and } Y = r_2) \qquad \text{expressing } XY = r \text{ as a disjoint union}$$

$$= \sum_{r_1 \in X(S)} \sum_{r_2 \in Y(S)} r_1 r_2 \cdot p(X = r_1 \text{ and } Y = r_2) \qquad \text{using a double sum to order the terms}$$

$$= \sum_{r_1 \in X(S)} \sum_{r_2 \in Y(S)} r_1 r_2 \cdot p(X = r_1) \cdot p(Y = r_2) \qquad \text{by the independence of } X \text{ and } Y$$

$$= \sum_{r_1 \in X(S)} \left(r_1 \cdot p(X = r_1) \cdot \sum_{r_2 \in Y(S)} r_2 \cdot p(Y = r_2) \right) \qquad \text{by factoring out } r_1 \cdot p(X = r_1)$$

$$= \sum_{r_1 \in X(S)} r_1 \cdot p(X = r_1) \cdot E(Y) \qquad \text{by the definition of } E(Y)$$

$$= E(Y) \left(\sum_{r_1 \in X(S)} r_1 \cdot p(X = r_1) \right) \qquad \text{by factoring out } E(Y)$$

$$= E(Y)E(X) \qquad \text{by the definition of } E(X)$$

We complete the proof by noting that $E(Y)E(X) = E(X)E(Y)$, which is a consequence of the commutative law for multiplication. ◁

Note that when X and Y are random variables that are not independent, we cannot conclude that $E(XY) = E(X)E(Y)$, as Example 13 shows.

EXAMPLE 13 Let X and Y be random variables that count the number of heads and the number of tails when a coin is flipped twice. Because $p(X = 2) = 1/4$, $p(X = 1) = 1/2$, and $p(X = 0) = 1/4$, by Theorem 1 we have

$$E(X) = 2 \cdot \frac{1}{4} + 1 \cdot \frac{1}{2} + 0 \cdot \frac{1}{4} = 1.$$

A similar computation shows that $E(Y) = 1$. We note that $XY = 0$ when either two heads and no tails or two tails and no heads come up and that $XY = 1$ when one head and one tail come up. Hence,

$$E(XY) = 1 \cdot \frac{1}{2} + 0 \cdot \frac{1}{2} = \frac{1}{2}.$$

It follows that

$$E(XY) \neq E(X)E(Y).$$

This does not contradict Theorem 5 because X and Y are not independent, as the reader should verify (see Exercise 16). ◀

Variance

Links

The expected value of a random variable tells us its average value, but nothing about how widely its values are distributed. For example, if X and Y are the random variables on the set $S = \{1, 2, 3, 4, 5, 6\}$, with $X(s) = 0$ for all $s \in S$ and $Y(s) = -1$ if $s \in \{1, 2, 3\}$ and $Y(s) = 1$ if $s \in \{4, 5, 6\}$, then the expected values of X and Y are both zero. However, the random variable X never varies from 0, while the random variable Y always differs from 0 by 1. The variance of a random variable helps us characterize how widely a random variable is distributed. In particular, it provides a measure of how widely X is distributed about its expected value.

DEFINITION 4   Let X be a random variable on a sample space S. The *variance* of X, denoted by $V(X)$, is

$$V(X) = \sum_{s \in S}(X(s) - E(X))^2 p(s).$$

That is, $V(X)$ is the weighted average of the square of the deviation of X. The *standard deviation* of X, denoted $\sigma(X)$, is defined to be $\sqrt{V(X)}$.

Theorem 6 provides a useful simple expression for the variance of a random variable.

THEOREM 6   If X is a random variable on a sample space S, then $V(X) = E(X^2) - E(X)^2$.

Proof: Note that

$$V(X) = \sum_{s \in S}(X(s) - E(X))^2 p(s)$$
$$= \sum_{s \in S} X(s)^2 p(s) - 2E(X)\sum_{s \in S} X(s)p(s) + E(X)^2 \sum_{s \in S} p(s)$$
$$= E(X^2) - 2E(X)E(X) + E(X)^2$$
$$= E(X^2) - E(X)^2.$$

We have used the fact that $\sum_{s \in S} p(s) = 1$ in the next-to-last step.   ◁

We can use Theorems 3 and 6 to derive an alternative formula for $V(X)$ that provides some insight into the meaning of the variance of a random variable.

COROLLARY 1   If X is a random variable on a sample space S and $E(X) = \mu$, then $V(X) = E((X - \mu)^2)$.

μ is the Greek letter mu.

Proof: If X is a random variable with $E(X) = \mu$, then

$$E((X - \mu)^2) = E(X^2 - 2\mu X + \mu^2) \qquad \text{expanding } (X - \mu)^2$$
$$= E(X^2) - E(2\mu X) + E(\mu^2) \quad \text{by part } (i) \text{ of Theorem 3}$$
$$= E(X^2) - 2\mu E(X) + E(\mu^2) \quad \text{by part } (ii) \text{ of Theorem 3, noting that } \mu \text{ is a constant}$$
$$= E(X^2) - 2\mu E(X) + \mu^2 \qquad \text{as } E(\mu^2) = \mu^2, \text{ because } \mu^2 \text{ is a constant}$$
$$= E(X^2) - 2\mu^2 + \mu^2 \qquad \text{because } E(X) = \mu$$
$$= E(X^2) - \mu^2 \qquad \text{simplifying}$$
$$= V(X) \qquad \text{by Theorem 6 and noting that } E(X) = \mu.$$

This completes the proof.   ◁

Corollary 1 tells us that the variance of a random variable X is the expected value of the square of the difference between X and its own expected value. This is commonly expressed as saying that the variance of X is the mean of the square of its deviation. We also say that the standard deviation of X is the square root of the mean of the square of its deviation (often read as the "root mean square" of the deviation).

We now compute the variance of some random variables.

EXAMPLE 14   What is the variance of the random variable X with $X(t) = 1$ if a Bernoulli trial is a success and $X(t) = 0$ if it is a failure, where p is the probability of success and q is the probability of failure?

Extra Examples

Solution: Because X takes only the values 0 and 1, it follows that $X^2(t) = X(t)$. Hence,

$$V(X) = E(X^2) - E(X)^2 = p - p^2 = p(1 - p) = pq.$$ ◀

EXAMPLE 15   **Variance of the Value of a Die**   What is the variance of the random variable X, where X is the number that comes up when a fair die is rolled?

Solution: We have $V(X) = E(X^2) - E(X)^2$. By Example 1 we know that $E(X) = 7/2$. To find $E(X^2)$ note that X^2 takes the values i^2, $i = 1, 2, \ldots, 6$, each with probability $1/6$. It follows that

$$E(X^2) = \frac{1}{6}(1^2 + 2^2 + 3^2 + 4^2 + 5^2 + 6^2) = \frac{91}{6}.$$

We conclude that

$$V(X) = \frac{91}{6} - \left(\frac{7}{2}\right)^2 = \frac{35}{12}.$$ ◀

EXAMPLE 16   What is the variance of the random variable $X((i, j)) = 2i$, where i is the number appearing on the first die and j is the number appearing on the second die, when two fair dice are rolled?

Solution: We will use Theorem 6 to find the variance of X. To do so, we need to find the expected values of X and X^2. Note that because $p(X = k)$ is $1/6$ for $k = 2, 4, 6, 8, 10, 12$ and is 0 otherwise,

$$E(X) = (2 + 4 + 6 + 8 + 10 + 12)/6 = 7,$$

and

$$E(X^2) = (2^2 + 4^2 + 6^2 + 8^2 + 10^2 + 12^2)/6 = 182/3.$$

It follows from Theorem 6 that

$$V(X) = E(X^2) - E(X)^2 = 182/3 - 49 = 35/3.$$ ◀

Another useful property is that the variance of the sum of two or more independent random variables is the sum of their variances. The formula that expresses this property is known as **Bienaymé's formula**, after Irenée-Jules Bienaymé, the French mathematician who discovered it in 1853. Bienaymé's formula is useful for computing the variance of the result of n independent Bernoulli trials, for instance.

THEOREM 7   **BIENAYMÉ'S FORMULA**   If X and Y are two independent random variables on a sample space S, then $V(X + Y) = V(X) + V(Y)$. Furthermore, if $X_i, i = 1, 2, \ldots, n$, with n a positive integer, are pairwise independent random variables on S, then $V(X_1 + X_2 + \cdots + X_n) = V(X_1) + V(X_2) + \cdots + V(X_n)$.

Proof: From Theorem 6, we have

$$V(X + Y) = E((X + Y)^2) - E(X + Y)^2.$$

It follows that

$$
\begin{aligned}
V(X + Y) &= E(X^2 + 2XY + Y^2) - (E(X) + E(Y))^2 \\
&= E(X^2) + 2E(XY) + E(Y^2) - E(X)^2 - 2E(X)E(Y) - E(Y)^2.
\end{aligned}
$$

Because X and Y are independent, by Theorem 5 we have $E(XY) = E(X)E(Y)$. It follows that

$$
\begin{aligned}
V(X + Y) &= (E(X^2) - E(X)^2) + (E(Y^2) - E(Y)^2) \\
&= V(X) + V(Y).
\end{aligned}
$$

We leave the proof of the case for n pairwise independent random variables to the reader. Such a proof can be constructed by generalizing the proof we have given for the case for two random variables. Note that it is not possible to use mathematical induction in a straightforward way to prove the general case.   ◁

Links

IRENÉE-JULES BIENAYMÉ (1796–1878)   Bienaymé, born in Paris, moved with his family to Bruges in 1803 when his father became a government administrator. Bienaymé attended the Lycée impérial in Bruges, and when his family returned to Paris in 1811, the Lycée Louis-le-Grand. As a teenager, he helped defend Paris during the 1814 Napoleonic Wars; in 1815, he became a student at the École Polytechnique. In 1816 he joined the Ministry of Finances to help support his family. In 1819, he left the civil service, taking a job lecturing mathematics at the Académie militaire de Saint-Cyr. Unhappy with conditions there, he soon returned to the Ministry of Finances. He attained the position of inspector general, remaining until forced to retire in 1848 for political reasons. He was able to return as inspector general in 1850, but he retired a second time in 1852. In 1851 he briefly was professor at the Sorbonne and also served as an expert statistician for Napoleon III. Bienaymé was one of the founders of the Société Mathématique de France, and in 1875 was its president.

Bienaymé was noted for his ingenuity, but his papers frustrated readers by omitting important proofs. He published sparsely, often in obscure journals. However, he made important contributions to probability and statistics, and to their applications to the social sciences and to finance. Among his important contributions are the Bienaymé-Chebyshev inequality, which provides a simple proof of the law of large numbers, a generalization of Laplace's least square method, and Bienaymé's formula for the variance of a sum of random variables. He studied the extinction of aristocratic families, declining despite general population growth. Bienaymé was a skilled linguist; he translated the works of Chebyshev, a close friend, from Russian to French. It has been suggested that his relative obscurity results from his modesty, his lack of interest in asserting the priority of his discoveries, and the fact that his work was often ahead of its time. He and his brother married two sisters who were daughters of a family friend. Bienaymé and his wife had two sons and three daughters.

EXAMPLE 17   Find the variance and standard deviation of the random variable X whose value when two fair dice are rolled is $X((i, j)) = i + j$, where i is the number appearing on the first die and j is the number appearing on the second die.

Solution: Let X_1 and X_2 be the random variables defined by $X_1((i, j)) = i$ and $X_2((i, j)) = j$ for a roll of the dice. Then $X = X_1 + X_2$, and X_1 and X_2 are independent, as Example 11 showed. From Theorem 7 it follows that $V(X) = V(X_1) + V(X_2)$. A simple computation as in Example 16, together with Exercise 23 in the Supplementary Exercises, tells us that $V(X_1) = V(X_2) = 35/12$. Hence, $V(X) = 35/12 + 35/12 = 35/6$ and $\sigma(X) = \sqrt{35/6}$. ◀

We will now find the variance of the random variable that counts the number of successes when n independent Bernoulli trials are carried out.

EXAMPLE 18   What is the variance of the number of successes when n independent Bernoulli trials are performed, where, on each trial, p is the probability of success and q is the probability of failure?

Solution: Let X_i be the random variable with $X_i((t_1, t_2, \ldots, t_n)) = 1$ if trial t_i is a success and $X_i((t_1, t_2, \ldots, t_n)) = 0$ if trial t_i is a failure. Let $X = X_1 + X_2 + \cdots + X_n$. Then X counts the number of successes in the n trials. From Theorem 7 it follows that $V(X) = V(X_1) + V(X_2) + \cdots + V(X_n)$. Using Example 14 we have $V(X_i) = pq$ for $i = 1, 2, \ldots, n$. It follows that $V(X) = npq$. ◀

Chebyshev's Inequality

How likely is it that a random variable takes a value far from its expected value? Theorem 8, called Chebyshev's inequality, helps answer this question by providing an upper bound on the probability that the value of a random variable differs from the expected value of the random variable by more than a specified amount.

THEOREM 8   **CHEBYSHEV'S INEQUALITY**   Let X be a random variable on a sample space S with probability function p. If r is a positive real number, then

$$p(|X(s) - E(X)| \geq r) \leq V(X)/r^2.$$

Proof: Let A be the event

$$A = \{s \in S \mid |X(s) - E(X)| \geq r\}.$$

Links

PAFNUTY LVOVICH CHEBYSHEV (1821–1894)   Chebyshev was born into the gentry in Okatovo, Russia. His father was a retired army officer who had fought against Napoleon. In 1832 the family, with its nine children, moved to Moscow, where Pafnuty completed his high school education at home. He entered the Department of Physics and Mathematics at Moscow University. As a student, he developed a new method for approximating the roots of equations. He graduated from Moscow University in 1841 with a degree in mathematics, and he continued his studies, passing his master's exam in 1843 and completing his master's thesis in 1846.

Chebyshev was appointed in 1847 to a position as an assistant at the University of St. Petersburg. He wrote and defended a thesis in 1847. He became a professor at St. Petersburg in 1860, a position he held until 1882. His book on the theory of congruences written in 1849 was influential in the development of number theory. His work on the distribution of prime numbers was seminal. He proved Bertrand's conjecture that for every integer $n > 3$, there is a prime between n and $2n - 2$. Chebyshev helped develop ideas that were later used to prove the prime number theorem. Chebyshev's work on the approximation of functions using polynomials is used extensively when computers are used to find values of functions. Chebyshev was also interested in mechanics. He studied the conversion of rotary motion into rectilinear motion by mechanical coupling. The Chebyshev parallel motion is three linked bars approximating rectilinear motion.

What we want to prove is that $p(A) \le V(X)/r^2$. Note that

$$V(X) = \sum_{s \in S}(X(s) - E(X))^2 p(s)$$
$$= \sum_{s \in A}(X(s) - E(X))^2 p(s) + \sum_{s \notin A}(X(s) - E(X))^2 p(s).$$

The second sum in this expression is nonnegative, because each of its summands is nonnegative. Also, because for each element s in A, $(X(s) - E(X))^2 \ge r^2$, the first sum in this expression is at least $\sum_{s \in A} r^2 p(s)$. Hence, $V(X) \ge \sum_{s \in A} r^2 p(s) = r^2 p(A)$. It follows that $V(X)/r^2 \ge p(A)$, so $p(A) \le V(X)/r^2$, completing the proof.   ◁

EXAMPLE 19   **Deviation from the Mean when Counting Tails**   Suppose that X is the random variable that counts the number of tails when a fair coin is tossed n times. Note that X is the number of successes when n independent Bernoulli trials, each with probability of success $1/2$, are performed. It follows that $E(X) = n/2$ (by Theorem 2) and $V(X) = n/4$ (by Example 18). Applying Chebyshev's inequality with $r = \sqrt{n}$ shows that

$$p(|X(s) - n/2| \ge \sqrt{n}) \le (n/4)/(\sqrt{n})^2 = 1/4.$$

Consequently, the probability is no more than $1/4$ that the number of tails that come up when a fair coin is tossed n times deviates from the mean by more than $\sqrt{n}$.   ◀

Chebyshev's inequality, although applicable to any random variable, often fails to provide a practical estimate for the probability that the value of a random variable exceeds its mean by a large amount. This is illustrated by Example 20.

EXAMPLE 20   Let X be the random variable whose value is the number appearing when a fair die is rolled. We have $E(X) = 7/2$ (see Example 1) and $V(X) = 35/12$ (see Example 15). Because the only possible values of X are 1, 2, 3, 4, 5, and 6, X cannot take a value more than $5/2$ from its mean, $E(X) = 7/2$. Hence, $p(|X - 7/2| \ge r) = 0$ if $r > 5/2$. By Chebyshev's inequality we know that $p(|X - 7/2| \ge r) \le (35/12)/r^2$.

For example, when $r = 3$, Chebyshev's inequality tells us that $p(|X - 7/2| \ge 3) \le (35/12)/9 = 35/108 \approx 0.324$, which is a poor estimate, because $p(|X - 7/2| \ge 3) = 0$.   ◀

Exercises

1. What is the expected number of heads that come up when a fair coin is flipped five times?

2. What is the expected number of times a 6 appears when a fair die is rolled 10 times?

3. What is the expected sum of the numbers that appear on two dice, each biased so that a 3 comes up twice as often as each other number?

4. What is the expected value when a \$1 lottery ticket is bought in which the purchaser wins exactly \$10 million if the ticket contains the six winning numbers chosen from the set $\{1, 2, 3, \ldots, 50\}$ and the purchaser wins nothing otherwise?

5. The final exam of a discrete mathematics course consists of 50 true/false questions, each worth two points, and 25 multiple-choice questions, each worth four points.

The probability that Linda answers a true/false question correctly is 0.9, and the probability that she answers a multiple-choice question correctly is 0.8. What is her expected score on the final?

6. What is the expected sum of the numbers that appear when three fair dice are rolled?

7. Suppose that the probability that x is in a list of n distinct integers is $2/3$ and that it is equally likely that x equals any element in the list. Find the average number of comparisons used by the linear search algorithm to find x or to determine that it is not in the list.

8. Suppose that we flip a fair coin until either it comes up tails twice or we have flipped it six times. What is the expected number of times we flip the coin?

9. Suppose that we roll a fair die until a 6 comes up or we have rolled it 10 times. What is the expected number of times we roll the die?

10. Suppose that we roll a fair die until a 6 comes up.
   a) What is the probability that we roll the die n times?
   b) What is the expected number of times we roll the die?

11. Suppose that we roll a pair of fair dice until the sum of the numbers on the dice is seven. What is the expected number of times we roll the dice?

12. Show that the sum of the probabilities of a random variable with geometric distribution with parameter p, where $0 < p \leq 1$, equals 1.

13. Show that if the random variable X has the geometric distribution with parameter p, and j is a positive integer, then $p(X \geq j) = (1 - p)^{j-1}$.

14. Let X and Y be the random variables that count the number of heads and the number of tails that come up when two fair coins are flipped. Show that X and Y are not independent.

15. Estimate the expected number of integers with 1000 digits that need to be selected at random to find a prime, if the probability a number with 1000 digits is prime is approximately $1/2302$.

16. Suppose that X and Y are random variables and that X and Y are nonnegative for all points in a sample space S. Let Z be the random variable defined by $Z(s) = \max(X(s), Y(s))$ for all elements $s \in S$. Show that $E(Z) \leq E(X) + E(Y)$.

17. Let X be the number appearing on the first die when two fair dice are rolled and let Y be the sum of the numbers appearing on the two dice. Show that $E(X)E(Y) \neq E(XY)$.

18. Let A be an event. Then I_A, the **indicator random variable** of A, equals 1 if A occurs and equals 0 otherwise. Show that the expectation of the indicator random variable of A equals the probability of A, that is, $E(I_A) = p(A)$.

19. A **run** is a maximal sequence of successes in a sequence of Bernoulli trials. For example, in the sequence $S, S, S, F, S, S, F, F, S$, where S represents success and F represents failure, there are three runs consisting of three successes, two successes, and one success, respectively. Let R denote the random variable on the set of sequences of n independent Bernoulli trials that counts the number of runs in this sequence. Find $E(R)$. [*Hint:* Show that $R = \sum_{j=1}^{n} I_j$, where $I_j = 1$ if a run begins at the jth Bernoulli trial and $I_j = 0$ otherwise. Find $E(I_1)$ and then find $E(I_j)$, where $1 < j \leq n$.]

20. Let $X(s)$ be a random variable, where $X(s)$ is a nonnegative integer for all $s \in S$, and let A_k be the event that $X(s) \geq k$. Show that $E(X) = \sum_{k=1}^{\infty} p(A_k)$.

21. What is the variance of the number of heads that come up when a fair coin is flipped 10 times?

22. What is the variance of the number of times a 6 appears when a fair die is rolled 10 times?

23. Let X_n be the random variable that equals the number of tails minus the number of heads when n fair coins are flipped.

a) What is the expected value of X_n?
b) What is the variance of X_n?

24. Provide an example that shows that the variance of the sum of two random variables is not necessarily equal to the sum of their variances when the random variables are not independent.

25. Use Chebyshev's inequality to find an upper bound on the probability that the number of tails that come up when a fair coin is tossed n times deviates from the mean by more than $5\sqrt{n}$.

26. Use Chebyshev's inequality to find an upper bound on the probability that the number of tails that come up when a biased coin with probability of heads equal to 0.6 is tossed n times deviates from the mean by more than $\sqrt{n}$.

27. Let X be a random variable on a sample space S such that $X(s) \geq 0$ for all $s \in S$. Show that $p(X(s) \geq a) \leq E(X)/a$ for every positive real number a. This inequality is called **Markov's inequality**.

28. Suppose that the number of cans of soda pop filled in a day at a bottling plant is a random variable with an expected value of 10,000 and a variance of 1000.
   a) Use Markov's inequality (Exercise 27) to obtain an upper bound on the probability that the plant will fill more than 11,000 cans on a particular day.
   b) Use Chebyshev's inequality to obtain a lower bound on the probability that the plant will fill between 9000 and 11,000 cans on a particular day.

29. Suppose that the number of tin cans recycled in a day at a recycling center is a random variable with an expected value of 50,000 and a variance of 10,000.
   a) Use Markov's inequality (Exercise 27) to find an upper bound on the probability that the center will recycle more than 55,000 cans on a particular day.
   b) Use Chebyshev's inequality to provide a lower bound on the probability that the center will recycle 40,000 to 60,000 cans on a certain day.

*30. Suppose the probability that x is the ith element in a list of n distinct integers is $i/[n(n + 1)]$. Find the average number of comparisons used by the linear search algorithm to find x or to determine that it is not in the list.

*31. In this exercise we derive an estimate of the average-case complexity of the variant of the bubble sort algorithm that terminates once a pass has been made with no interchanges. Let X be the random variable on the set of permutations of a set of n distinct integers $\{a_1, a_2, \ldots, a_n\}$ with $a_1 < a_2 < \cdots < a_n$ such that $X(P)$ equals the number of comparisons used by the bubble sort to put these integers into increasing order.
   a) Show that, under the assumption that the input is equally likely to be any of the $n!$ permutations of these integers, the average number of comparisons used by the bubble sort equals $E(X)$.
   b) Use Example 5 in Section 3.3 to show that $E(X) \leq n(n - 1)/2$.
   c) Show that the sort makes at least one comparison for every inversion of two integers in the input.

d) Let $I(P)$ be the random variable that equals the number of inversions in the permutation P. Show that $E(X) \geq E(I)$.

e) Let $I_{j,k}$ be the random variable with $I_{j,k}(P) = 1$ if a_k precedes a_j in P and $I_{j,k} = 0$ otherwise. Show that $I(P) = \sum_k \sum_{j<k} I_{j,k}(P)$.

f) Show that $E(I) = \sum_k \sum_{j<k} E(I_{j,k})$.

g) Show that $E(I_{j,k}) = 1/2$. [*Hint:* Show that $E(I_{j,k}) =$ probability that a_k precedes a_j in a permutation P. Then show it is equally likely for a_k to precede a_j as it is for a_j to precede a_k in a permutation.]

h) Use parts (f) and (g) to show that $E(I) = n(n-1)/4$.

i) Conclude from parts (b), (d), and (h) that the average number of comparisons used to sort n integers is $\Theta(n^2)$.

*32. In this exercise we find the average-case complexity of the quick sort algorithm, described in the preamble to Exercise 34 in Section 5.4, assuming a uniform distribution on the set of permutations.

a) Let X be the number of comparisons used by the quick sort algorithm to sort a list of n distinct integers. Show that the average number of comparisons used by the quick sort algorithm is $E(X)$ (where the sample space is the set of all $n!$ permutations of n integers).

b) Let $I_{j,k}$ denote the random variable that equals 1 if the jth smallest element and the kth smallest element of the initial list are ever compared as the quick sort algorithm sorts the list and equals 0 otherwise. Show that $X = \sum_{k=2}^{n} \sum_{j=1}^{k-1} I_{j,k}$.

c) Show that $E(X) = \sum_{k=2}^{n} \sum_{j=1}^{k-1} p$(the jth smallest element and the kth smallest element are compared).

d) Show that p(the jth smallest element and the kth smallest element are compared), where $k > j$, equals $2/(k-j+1)$.

e) Use parts (c) and (d) to show that $E(X) = 2(n+1)(\sum_{i=2}^{n} 1/i) - 2(n-1)$.

f) Conclude from part (e) and the fact that $\sum_{j=1}^{n} 1/j \approx \ln n + \gamma$, where $\gamma = 0.57721\ldots$ is Euler's constant, that the average number of comparisons used by the quick sort algorithm is $\Theta(n \log n)$.

*33. What is the variance of the number of **fixed elements**, that is, elements left in the same position, of a randomly selected permutation of n elements? [*Hint:* Let X denote the number of fixed points of a random permutation. Write $X = X_1 + X_2 + \cdots + X_n$, where $X_i = 1$ if the permutation fixes the ith element and $X_i = 0$ otherwise.]

The **covariance** of two random variables X and Y on a sample space S, denoted by $\mathrm{Cov}(X, Y)$, is defined to be the expected value of the random variable $(X - E(X))(Y - E(Y))$. That is, $\mathrm{Cov}(X, Y) = E((X - E(X))(Y - E(Y)))$.

34. Show that $\mathrm{Cov}(X, Y) = E(XY) - E(X)E(Y)$, and use this result to conclude that $\mathrm{Cov}(X, Y) = 0$ if X and Y are independent random variables.

35. Show that $V(X + Y) = V(X) + V(Y) + 2\,\mathrm{Cov}(X, Y)$.

36. Find $\mathrm{Cov}(X, Y)$ if X and Y are the random variables with $X((i, j)) = 2i$ and $Y((i, j)) = i + j$, where i and j are the numbers that appear on the first and second of two dice when they are rolled.

37. When m balls are distributed into n bins uniformly at random, what is the probability that the first bin remains empty?

38. What is the expected number of balls that fall into the first bin when m balls are distributed into n bins uniformly at random?

39. What is the expected number of bins that remain empty when m balls are distributed into n bins uniformly at random?

Key Terms and Results

TERMS

sample space: the set of possible outcomes of an experiment

event: a subset of the sample space of an experiment

probability of an event (Laplace's definition): the number of successful outcomes of this event divided by the number of possible outcomes

probability distribution: a function p from the set of all outcomes of a sample space S for which $0 \leq p(x_i) \leq 1$ for $i = 1, 2, \ldots, n$ and $\sum_{i=1}^{n} p(x_i) = 1$, where $x_1, \ldots, x_n$ are the possible outcomes

probability of an event E: the sum of the probabilities of the outcomes in E

$p(E|F)$ (conditional probability of E given F): the ratio $p(E \cap F)/p(F)$

independent events: events E and F such that $p(E \cap F) = p(E)p(F)$

pairwise independent events: events $E_1, E_2, \ldots, E_n$ such that $p(E_i \cap E_j) = p(E_i)p(E_j)$ for all pairs of integers i and j with $1 \leq j < k \leq n$

mutually independent events: events $E_1, E_2, \ldots, E_n$ such that $p(E_{i_1} \cap E_{i_2} \cap \cdots \cap E_{i_m}) = p(E_{i_1})p(E_{i_2}) \cdots p(E_{i_m})$ whenever i_j, $j = 1, 2, \ldots, m$, are integers with $1 \leq i_1 < i_2 < \cdots < i_m \leq n$ and $m \geq 2$

random variable: a function that assigns a real number to each possible outcome of an experiment

distribution of a random variable X: the set of pairs $(r, p(X = r))$ for $r \in X(S)$

uniform distribution: the assignment of equal probabilities to the elements of a finite set

expected value of a random variable: the weighted average of a random variable, with values of the random variable weighted by the probability of outcomes, that is, $E(X) = \sum_{s \in S} p(s)X(s)$

geometric distribution: the distribution of a random variable X such that $p(X = k) = (1-p)^{k-1}p$ for $k = 1, 2, \ldots$ for some real number p with $0 \leq p \leq 1$.

independent random variables: random variables X and Y such that $p(X = r_1 \text{ and } Y = r_2) = p(X = r_1)p(Y = r_2)$ for all real numbers r_1 and r_2

variance of a random variable X: the weighted average of the square of the difference between the value of X and its expected value $E(X)$, with weights given by the probability of outcomes, that is, $V(X) = \sum_{s \in S}(X(s) - E(X))^2 p(s)$

standard deviation of a random variable X: the square root of the variance of X, that is, $\sigma(X) = \sqrt{V(X)}$

Bernoulli trial: an experiment with two possible outcomes

probabilistic (or Monte Carlo) algorithm: an algorithm in which random choices are made at one or more steps

probabilistic method: a technique for proving the existence of objects in a set with certain properties that proceeds by assigning probabilities to objects and showing that the probability that an object has these properties is positive

RESULTS

The probability of exactly k successes when n independent Bernoulli trials are carried out equals $C(n, k)p^k q^{n-k}$,

where p is the probability of success and $q = 1 - p$ is the probability of failure.

Bayes' theorem: If E and F are events from a sample space S such that $p(E) \neq 0$ and $p(F) \neq 0$, then

$$p(F \mid E) = \frac{p(E \mid F)p(F)}{p(E \mid F)p(F) + p(E \mid \overline{F})p(\overline{F})}.$$

$E(X) = \sum_{r \in X(S)} p(X = r)r.$

linearity of expectations: $E(X_1 + X_2 + \cdots + X_n) = E(X_1) + E(X_2) + \cdots + E(X_n)$ if $X_1, X_2, \ldots, X_n$ are random variables

If X and Y are independent random variables, then $E(XY) = E(X)E(Y)$.

Bienaymé's formula: If $X_1, X_2, \ldots, X_n$ are independent random variables, then $V(X_1 + X_2 + \cdots + X_n) = V(X_1) + V(X_2) + \cdots + V(X_n)$.

Chebyshev's inequality: $p(|X(s) - E(X)| \geq r) \leq V(X)/r^2$, where X is a random variable with probability function p and r is a positive real number.

Review Questions

1. a) Define the probability of an event when all outcomes are equally likely.
   b) What is the probability that you select the six winning numbers in a lottery if the six different winning numbers are selected from the first 50 positive integers?

2. a) What conditions should be met by the probabilities assigned to the outcomes from a finite sample space?
   b) What probabilities should be assigned to the outcome of heads and the outcome of tails if heads comes up three times as often as tails?

3. a) Define the conditional probability of an event E given an event F.
   b) Suppose E is the event that when a die is rolled it comes up an even number, and F is the event that when a die is rolled it comes up 1, 2, or 3. What is the probability of F given E?

4. a) When are two events E and F independent?
   b) Suppose E is the event that an even number appears when a fair die is rolled, and F is the event that a 5 or 6 comes up. Are E and F independent?

5. a) What is a random variable?
   b) What are the possible values assigned by the random variable X that assigns to a roll of two dice the larger number that appears on the two dice?

6. a) Define the expected value of a random variable X.
   b) What is the expected value of the random variable X that assigns to a roll of two dice the larger number that appears on the two dice?

7. a) Explain how the average-case computational complexity of an algorithm, with finitely many possible input values, can be interpreted as an expected value.

   b) What is the average-case computational complexity of the linear search algorithm, if the probability that the element for which we search is in the list is $1/3$, and it is equally likely that this element is any of the n elements in the list?

8. a) What is meant by a Bernoulli trial?
   b) What is the probability of k successes in n independent Bernoulli trials?
   c) What is the expected value of the number of successes in n independent Bernoulli trials?

9. a) What does the linearity of expectations of random variables mean?
   b) How can the linearity of expectations help us find the expected number of people who receive the correct hat when a hatcheck person returns hats at random?

10. a) How can probability be used to solve a decision problem, if a small probability of error is acceptable?
    b) How can we quickly determine whether a positive integer is prime, if we are willing to accept a small probability of making an error?

11. State Bayes' theorem and use it to find $p(F \mid E)$ if $p(E \mid F) = 1/3$, $p(E \mid \overline{F}) = 1/4$, and $p(F) = 2/3$, where E and F are events from a sample space S.

12. a) What does it mean to say that a random variable has a geometric distribution with parameter p?
    b) What is the mean of a geometric distribution with parameter p?

13. a) What is the variance of a random variable?
    b) What is the variance of a Bernoulli trial with probability p of success?

14. a) What is the variance of the sum of n independent random variables?

b) What is the variance of the number of successes when n independent Bernoulli trials, each with probability p of success, are carried out?

15. What does Chebyshev's inequality tell us about the probability that a random variable deviates from its mean by more than a specified amount?

Supplementary Exercises

1. What is the probability that six consecutive integers will be chosen as the winning numbers in a lottery where each number chosen is an integer between 1 and 40 (inclusive)?

2. What is the probability that a hand of 13 cards contains no pairs?

3. What is the probability that a 13-card bridge hand contains

    a) all 13 hearts?

    b) 13 cards of the same suit?

    c) seven spades and six clubs?

    d) seven cards of one suit and six cards of a second suit?

    e) four diamonds, six hearts, two spades, and one club?

    f) four cards of one suit, six cards of a second suit, two cards of a third suit, and one card of the fourth suit?

4. What is the probability that a seven-card poker hand contains

    a) four cards of one kind and three cards of a second kind?

    b) three cards of one kind and pairs of each of two different kinds?

    c) pairs of each of three different kinds and a single card of a fourth kind?

    d) pairs of each of two different kinds and three cards of a third, fourth, and fifth kind?

    e) cards of seven different kinds?

    f) a seven-card flush?

    g) a seven-card straight?

    h) a seven-card straight flush?

An **octahedral die** has eight faces that are numbered 1 through 8.

5. a) What is the expected value of the number that comes up when a fair octahedral die is rolled?

    b) What is the variance of the number that comes up when a fair octahedral die is rolled?

A **dodecahedral die** has 12 faces that are numbered 1 through 12.

6. a) What is the expected value of the number that comes up when a fair dodecahedral die is rolled?

    b) What is the variance of the number that comes up when a fair dodecahedral die is rolled?

7. Suppose that a fair standard (cubic) die and a fair octahedral die are rolled together.

    a) What is the expected value of the sum of the numbers that come up?

    b) What is the variance of the sum of the numbers that come up?

8. Suppose that a fair octahedral die and a fair dodecahedral die are rolled together.

    a) What is the expected value of the sum of the numbers that come up?

    b) What is the variance of the sum of the numbers that come up?

9. Suppose n people, $n \geq 3$, play "odd person out" to decide who will buy the next round of refreshments. The n people each flip a fair coin simultaneously. If all the coins but one come up the same, the person whose coin comes up different buys the refreshments. Otherwise, the people flip the coins again and continue until just one coin comes up different from all the others.

    a) What is the probability that the odd person out is decided in just one coin flip?

    b) What is the probability that the odd person out is decided with the kth flip?

    c) What is the expected number of flips needed to decide odd person out with n people?

10. Suppose that p and q are primes and $n = pq$. What is the probability that a randomly chosen positive integer less than n is not divisible by either p or q?

★11. Suppose that m and n are positive integers. What is the probability that a randomly chosen positive integer less than mn is not divisible by either m or n?

12. Suppose that $E_1, E_2, \ldots, E_n$ are n events with $p(E_i) > 0$ for $i = 1, 2, \ldots, n$. Show that

$$p(E_1 \cap E_2 \cap \cdots \cap E_n)$$
$$= p(E_1) p(E_2 \mid E_1) p(E_3 \mid E_1 \cap E_2)$$
$$\cdots p(E_n \mid E_1 \cap E_2 \cap \cdots \cap E_{n-1}).$$

13. There are three cards in a box. Both sides of one card are black, both sides of one card are red, and the third card has one black side and one red side. We pick a card at random and observe only one side.

    a) If the side is black, what is the probability that the other side is also black?

    b) What is the probability that the opposite side is the same color as the one we observed?

14. What is the probability that a randomly selected bit string of length 10 is a palindrome?

15. Consider the following game. A person flips a coin repeatedly until a head comes up. This person receives a payment of 2^n dollars if the first head comes up at the nth flip.

a) Let X be a random variable equal to the amount of money the person wins. Show that the expected value of X does not exist (that is, it is infinite). Show that a rational gambler, that is, someone willing to pay to play the game as long as the price to play is not more than the expected payoff, should be willing to wager any amount of money to play this game. (This is known as the **St. Petersburg paradox**. Why do you suppose it is called a paradox?)

b) Suppose that the person receives 2^n dollars if the first head comes up on the nth flip where $n < 8$ and $2^8 = 256$ dollars if the first head comes up on or after the eighth flip. What is the expected value of the amount of money the person wins? How much money should a person be willing to pay to play this game?

16. Suppose that n balls are tossed into b bins so that each ball is equally likely to fall into any of the bins and that the tosses are independent.

a) Find the probability that a particular ball lands in a specified bin.

b) What is the expected number of balls that land in a particular bin?

c) What is the expected number of balls tossed until a particular bin contains a ball?

*d) What is the expected number of balls tossed until all bins contain a ball? [*Hint:* Let X_i denote the number of tosses required to have a ball land in an ith bin once $i - 1$ bins contain a ball. Find $E(X_i)$ and use the linearity of expectations.]

17. Suppose that A and B are events with probabilities $p(A) = 3/4$ and $p(B) = 1/3$.

a) What is the largest $p(A \cap B)$ can be? What is the smallest it can be? Give examples to show that both extremes for $p(A \cap B)$ are possible.

b) What is the largest $p(A \cup B)$ can be? What is the smallest it can be? Give examples to show that both extremes for $p(A \cup B)$ are possible.

18. Suppose that A and B are events with probabilities $p(A) = 2/3$ and $p(B) = 1/2$.

a) What is the largest $p(A \cap B)$ can be? What is the smallest it can be? Give examples to show that both extremes for $p(A \cap B)$ are possible.

b) What is the largest $p(A \cup B)$ can be? What is the smallest it can be? Give examples to show that both extremes for $p(A \cup B)$ are possible.

19. Recall from Definition 5 in Section 7.2 that the events $E_1, E_2, \ldots, E_n$ are **mutually independent** if $p(E_{i_1} \cap E_{i_2} \cap \cdots \cap E_{i_m}) = p(E_{i_1})p(E_{i_2}) \cdots p(E_{i_m})$ whenever $i_j, j = 1, 2, \ldots, m$, are integers with $1 \leq i_1 < i_2 < \cdots < i_m \leq n$ and $m \geq 2$.

a) Write out the conditions required for three events E_1, E_2, and E_3 to be mutually independent.

b) Let E_1, E_2, and E_3 be the events that the first flip comes up heads, that the second flip comes up tails, and that the third flip comes up tails, respectively, when a fair coin is flipped three times. Are E_1, E_2, and E_3 mutually independent?

c) Let E_1, E_2, and E_3 be the events that the first flip comes up heads, that the third flip comes up heads, and that an even number of heads come up, respectively, when a fair coin is flipped three times. Are E_1, E_2, and E_3 pairwise independent? Are they mutually independent?

d) Let E_1, E_2, and E_3 be the events that the first flip comes up heads, that the third flip comes up heads, and that exactly one of the first flip and third flip come up heads, respectively, when a fair coin is flipped three times. Are E_1, E_2, and E_3 pairwise independent? Are they mutually independent?

e) How many conditions must be checked to show that n events are mutually independent?

20. Suppose that A and B are events from a sample space S such that $p(A) \neq 0$ and $p(B) \neq 0$. Show that if $p(B \mid A) < p(B)$, then $p(A \mid B) < p(A)$.

In Exercise 21 we consider the **two children problem**, introduced in 1959 by Martin Garnder is his Mathematical Games column in *Scientific American*. A version of the puzzle asks: "We meet Mr. Smith as he is walking down the street with a young child whom he introduces as his son. He also tells us that he has two children. What is the probability that his other child is a son?" We will show that this puzzle is ambiguous, leading to a paradox, by showing that there are two reasonable answers to this problem and we will describe how to make the puzzle unambiguous.

*21. a) Solve this puzzle in two different ways. First, answer the problem by considering the probability of the gender of the second child. Then, determine the probability differently, by considering the four different possibilities for a family of two children.

b) Show that the answer to the puzzle becomes unambiguous if we also know that Mr. Smith chose his walking companion at random from his two children.

c) Another variation of this puzzle asks "When we meet Mr. Smith, he tells us that he has two children and at least one is a son. What is the probability that his other child is a son?" Solve this variation of the puzzle, explaining why it is unambiguous.

22. In 2010, the puzzle designer Gary Foshee posed this problem: "Mr. Smith has two children, one of whom is a son born on a Tuesday. What is the probability that Mr. Smith has two sons?" Show that there are two different answers to this puzzle, depending on whether Mr. Smith specifically mentioned his son because he was born on a Tuesday or whether he randomly chose a child and reported its gender and birth day of the week. [*Hint:* For the first possibility, enumerate all the equally likely possibilities for the gender and birth day of the week of the other child. To do, this consider first the cases where the older child is a boy born on a Tuesday and then the case where the older child is not a boy born on a Tuesday.]

23. Let X be a random variable on a sample space S. Show that $V(aX + b) = a^2 V(X)$ whenever a and b are real numbers.

24. Use Chebyshev's inequality to show that the probability that more than 10 people get the correct hat back when a hatcheck person returns hats at random does not exceed $1/100$ no matter how many people check their hats. [*Hint:* Use Example 6 and Exercise 33 in Section 7.4.]

25. Suppose that at least one of the events E_j, $j = 1, 2, \ldots, m$, is guaranteed to occur and no more than two can occur. Show that if $p(E_j) = q$ for $j = 1, 2, \ldots, m$ and $p(E_j \cap E_k) = r$ for $1 \leq j < k \leq m$, then $q \geq 1/m$ and $r \leq 2/m$.

26. Show that if m is a positive integer, then the probability that the mth success occurs on the $(m + n)$th trial when independent Bernoulli trials, each with probability p of success, are run, is $\binom{n+m-1}{n}q^n p^m$.

27. There are n different types of collectible cards you can get as prizes when you buy a particular product. Suppose that every time you buy this product it is equally likely that you get any type of these cards. Let X be the random variable equal to the number of products that need to be purchased to obtain at least one of each type of card and let X_j be the random variable equal to the number of additional products that must be purchased after j different cards have been collected until a new card is obtained for $j = 0, 1, \ldots, n - 1$.

a) Show that $X = \sum_{j=0}^{n-1} X_j$.

b) Show that after j distinct types of cards have been obtained, the card obtained with the next purchase will be a card of a new type with probability $(n - j)/n$.

c) Show that X_j has a geometric distribution with parameter $(n - j)/n$.

d) Use parts (a) and (c) to show that $E(X) = n \sum_{j=1}^{n} 1/j$.

e) Use the approximation $\sum_{j=1}^{n} 1/j \approx \ln n + \gamma$, where $\gamma = 0.57721 \ldots$ is Euler's constant, to find the expected number of products that you need to buy to get one card of each type if there are 50 different types of cards.

28. The **maximum satisfiability problem** asks for an assignment of truth values to the variables in a compound proposition in conjunctive normal form (which expresses a compound proposition as the conjunction of clauses where each clause is the disjunction of two or more variables or their negations) that makes as many of these clauses true as possible. For example, three but not four of the clauses in

$$(p \vee q) \wedge (p \vee \neg q) \wedge (\neg p \vee r) \wedge (\neg p \vee \neg r)$$

can be made true by an assignment of truth values to p, q, and r. We will show that probabilistic methods can provide a lower bound for the number of clauses that can be made true by an assignment of truth values to the variables.

a) Suppose that there are n variables in a compound proposition in conjunctive normal form. If we pick a truth value for each variable randomly by flipping a coin and assigning true to the variable if the coin comes up heads and false if it comes up tails, what is the probability of each possible assignment of truth values to the n variables?

b) Assuming that each clause is the disjunction of exactly two distinct variables or their negations, what is the probability that a given clause is true, given the random assignment of truth values from part (a)?

c) Suppose that there are D clauses in the compound proposition. What is the expected number of these clauses that are true, given the random assignment of truth values of the variables?

d) Use part (c) to show that for every compound proposition in conjunctive normal form there is an assignment of truth values to the variables that makes at least 3/4 of the clauses true.

Computer Projects

Write programs with these input and output.

1. Given a real number p with $0 \leq p \leq 1$, generate random numbers taken from a Bernoulli distribution with probability p.

2. Given positive integers n and m, generate m random permutations of the first n positive integers. Sort each permutation using the insertion sort, counting the number of comparisons used. Determine the average number of comparisons used over all m permutations.

3. Given positive integers n and m, generate m random permutations of the first n positive integers. Sort each permutation using the version of the bubble sort that terminates when a pass has been made with no interchanges, counting

the number of comparisons used. Determine the average number of comparisons used over all m permutations.

4. Given a positive integer m, simulate the collection of cards that come with the purchase of products to find the number of products that must be purchased to obtain a full set of m different collector cards. (See Supplementary Exercise 27.)

5. Given positive integers m and n, simulate the placement of n keys, where a record with key k is placed at location $h(k) = k \bmod m$ and determine whether there is at least one collision.

6. Given a positive integer n, find the probability of selecting the six integers from the set $\{1, 2, \ldots, n\}$ that were mechanically selected in a lottery.

7. Simulate repeated trials of the Monty Hall Three-Door problem (Example 10 in Section 7.1) to calculate the probability of winning with each strategy.

8. Given a list of words and the empirical probabilities they occur in spam e-mails and in e-mails that are not spam, determine the probability that a new e-mail message is spam.

Computations and Explorations

Use a computational program or programs you have written to do these exercises.

1. Find the probabilities of each type of hand in five-card poker and rank the types of hands by their probability.

2. Find some conditions such that the expected value of buying a $1 lottery ticket in the New Jersey Pick-6 lottery has an expected value of more than $1. To win you have to select the six numbers drawn, where order does not matter, from the positive integers 1 to 49, inclusive. The winnings are split evenly among holders of winning tickets. Be sure to consider the total size of the pot going into the drawing and the number of people buying tickets.

3. Estimate the probability that two integers selected at random are relatively prime by testing a large number of randomly selected pairs of integers. Look up the theorem that gives this probability and compare your results with the correct probability.

4. Determine the number of people needed to ensure that the probability at least two of them have the same day of the year as their birthday is at least 70%, at least 80%, at least 90%, at least 95%, at least 98%, and at least 99%.

5. Generate a list of 100 randomly selected permutations of the set of the first 100 positive integers.

6. Given a collection of e-mail messages, each determined to be spam or not to be spam, develop a Bayesian filter based on the appearance of particular words in these messages.

7. Simulate the odd-person-out procedure (described in Exercise 13 of the Supplementary Exercises) for n people with $3 \leq n \leq 10$. Run a large number of trials for each value of n and use the results to estimate the expected number of flips needed to find the odd person out. Does your result agree with that found in Exercise 23 in Section 7.2? Vary the problem by supposing that exactly one person has a biased coin with probability of heads $p \neq 0.5$.

8. Given a positive integer n, simulate a hatcheck person randomly giving hats back to people. Determine the number of people who get the correct hat back.

Writing Projects

Respond to these with essays using outside sources.

1. Describe the origins of probability theory and the first uses of this theory, including those by Cardano, Pascal, and Laplace.

2. Describe the different bets you can make when you play roulette. Find the probability of each of these bets in the American version where the wheel contains the numbers 0 and 00. Which is the best bet and which is the worst for you?

3. Discuss the probability of winning when you play the game of blackjack versus a casino. Is there a winning strategy for the person playing against the house?

4. Investigate the game of craps and discuss the probability that the shooter wins and how close to a fair game it is.

5. Discuss issues involved in developing successful spam filters and the current situation in the war between spammers and people trying to filter spam out.

6. Discuss the history and solution of what is known as the Newton–Pepys problem, which asks which is most likely: rolling at least one six when six dice are rolled, rolling at least two sixes when 12 dice are rolled, or rolling at least three sixes when 18 dice are rolled.

7. Explain how Erdős and Rényi first used the probabilistic method and describe some other applications of this method.

8. Discuss the different types of probabilistic algorithms and describe some examples of each type.

8

Advanced Counting Techniques

Many counting problems cannot be solved easily using the methods discussed in Chapter 6. One such problem is: How many bit strings of length n do not contain two consecutive zeros? To solve this problem, let a_n be the number of such strings of length n. An argument can be given that shows that the sequence $\{a_n\}$ satisfies the recurrence relation $a_{n+1} = a_n + a_{n-1}$ and the initial conditions $a_1 = 2$ and $a_2 = 3$. This recurrence relation and the initial conditions determine the sequence $\{a_n\}$. Moreover, an explicit formula can be found for a_n from the equation relating the terms of the sequence. As we will see, a similar technique can be used to solve many different types of counting problems.

We will discuss two ways that recurrence relations play important roles in the study of algorithms. First, we will introduce an important algorithmic paradigm known as dynamic programming. Algorithms that follow this paradigm break down a problem into overlapping subproblems. The solution to the problem is then found from the solutions to the subproblems through the use of a recurrence relation. Second, we will study another important algorithmic paradigm, divide-and-conquer. Algorithms that follow this paradigm can be used to solve a problem by recursively breaking it into a fixed number of nonoverlapping subproblems until these problems can be solved directly. The complexity of such algorithms can be analyzed using a special type of recurrence relation. In this chapter we will discuss a variety of divide-and-conquer algorithms and analyze their complexity using recurrence relations.

We will also see that many counting problems can be solved using formal power series, called generating functions, where the coefficients of powers of x represent terms of the sequence we are interested in. Besides solving counting problems, we will also be able to use generating functions to solve recurrence relations and to prove combinatorial identities.

Many other kinds of counting problems cannot be solved using the techniques discussed in Chapter 6, such as: How many ways are there to assign seven jobs to three employees so that each employee is assigned at least one job? How many primes are there less than 1000? Both of these problems can be solved by counting the number of elements in the union of sets. We will develop a technique, called the principle of inclusion–exclusion, that counts the number of elements in a union of sets, and we will show how this principle can be used to solve counting problems.

The techniques studied in this chapter, together with the basic techniques of Chapter 6, can be used to solve many counting problems.

8.1 Applications of Recurrence Relations

Introduction

Recall from Chapter 2 that a recursive definition of a sequence specifies one or more initial terms and a rule for determining subsequent terms from those that precede them. Also, recall that a rule of the latter sort (whether or not it is part of a recursive definition) is called a **recurrence relation** and that a sequence is called a *solution* of a recurrence relation if its terms satisfy the recurrence relation.

In this section we will show that such relations can be used to study and to solve counting problems. For example, suppose that the number of bacteria in a colony doubles every hour. If a colony begins with five bacteria, how many will be present in n hours? To solve this problem, let a_n be the number of bacteria at the end of n hours. Because the number of bacteria doubles every hour, the relationship $a_n = 2a_{n-1}$ holds whenever n is a positive integer. This recurrence relation, together with the initial condition $a_0 = 5$, uniquely determines a_n for all nonnegative integers n. We can find a formula for a_n using the iterative approach followed in Chapter 2, namely that $a_n = 5 \cdot 2^n$ for all nonnegative integers n.

Some of the counting problems that cannot be solved using the techniques discussed in Chapter 6 can be solved by finding recurrence relations involving the terms of a sequence, as was done in the problem involving bacteria. In this section we will study a variety of counting problems that can be modeled using recurrence relations. In Chapter 2 we developed methods for solving certain recurrence relation. In Section 8.2 we will study methods for finding explicit formulae for the terms of sequences that satisfy certain types of recurrence relations.

We conclude this section by introducing the algorithmic paradigm of dynamic programming. After explaining how this paradigm works, we will illustrate its use with an example.

Modeling With Recurrence Relations

We can use recurrence relations to model a wide variety of problems, such as finding compound interest (see Example 11 in Section2.4), counting rabbits on an island, determining the number of moves in the Tower of Hanoi puzzle, and counting bit strings with certain properties.

Example 1 shows how the population of rabbits on an island can be modeled using a recurrence relation.

EXAMPLE 1 **Rabbits and the Fibonacci Numbers**   Consider this problem, which was originally posed by Leonardo Pisano, also known as Fibonacci, in the thirteenth century in his book *Liber abaci*. A young pair of rabbits (one of each sex) is placed on an island. A pair of rabbits does not breed until they are 2 months old. After they are 2 months old, each pair of rabbits produces another pair each month, as shown in Figure 1. Find a recurrence relation for the number of pairs of rabbits on the island after n months, assuming that no rabbits ever die.

Reproducing pairs (at least two months old)	Young pairs (less than two months old)	Month	Reproducing pairs	Young pairs	Total pairs
		1	0	1	1
		2	0	1	1
		3	1	1	2
		4	1	2	3
		5	2	3	5
		6	3	5	8

FIGURE 1   **Rabbits on an Island.**

The Fibonacci numbers appear in many other places in nature, including the number of petals on flowers and the number of spirals on seedheads.

Solution: Denote by f_n the number of pairs of rabbits after n months. We will show that f_n, $n = 1, 2, 3, \ldots$, are the terms of the Fibonacci sequence.

The rabbit population can be modeled using a recurrence relation. At the end of the first month, the number of pairs of rabbits on the island is $f_1 = 1$. Because this pair does not breed during the second month, $f_2 = 1$ also. To find the number of pairs after n months, add the number on the island the previous month, f_{n-1}, and the number of newborn pairs, which equals f_{n-2}, because each newborn pair comes from a pair at least 2 months old.

Consequently, the sequence $\{f_n\}$ satisfies the recurrence relation

$$f_n = f_{n-1} + f_{n-2}$$

for $n \geq 3$ together with the initial conditions $f_1 = 1$ and $f_2 = 1$. Because this recurrence relation and the initial conditions uniquely determine this sequence, the number of pairs of rabbits on the island after n months is given by the nth Fibonacci number. ◀

Demo

Example 2 involves a famous puzzle.

EXAMPLE 2

Links

The Tower of Hanoi   A popular puzzle of the late nineteenth century invented by the French mathematician Édouard Lucas, called the Tower of Hanoi, consists of three pegs mounted on a board together with disks of different sizes. Initially these disks are placed on the first peg in order of size, with the largest on the bottom (as shown in Figure 2). The rules of the puzzle allow disks to be moved one at a time from one peg to another as long as a disk is never placed on top of a smaller disk. The goal of the puzzle is to have all the disks on the second peg in order of size, with the largest on the bottom.

Let H_n denote the number of moves needed to solve the Tower of Hanoi problem with n disks. Set up a recurrence relation for the sequence $\{H_n\}$.

Solution: Begin with n disks on peg 1. We can transfer the top $n - 1$ disks, following the rules of the puzzle, to peg 3 using H_{n-1} moves (see Figure 3 for an illustration of the pegs and disks at this point). We keep the largest disk fixed during these moves. Then, we use one move to transfer the largest disk to the second peg. We can transfer the $n - 1$ disks on peg 3 to peg 2 using H_{n-1} additional moves, placing them on top of the largest disk, which always stays fixed on the bottom of peg 2. Moreover, it is easy to see that the puzzle cannot be solved using fewer steps. This shows that

Schemes for efficiently backing up computer files on multiple tapes or other media are based on the moves used to solve the Tower of Hanoi puzzle.

$$H_n = 2H_{n-1} + 1.$$

The initial condition is $H_1 = 1$, because one disk can be transferred from peg 1 to peg 2, according to the rules of the puzzle, in one move.

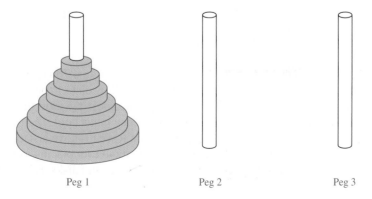

Peg 1   Peg 2   Peg 3

FIGURE 2   The Initial Position in the Tower of Hanoi.

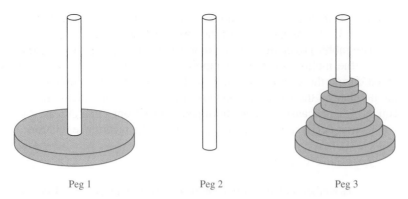

Peg 1   Peg 2   Peg 3

FIGURE 3   An Intermediate Position in the Tower of Hanoi.

We can use an iterative approach to solve this recurrence relation. Note that

$$
\begin{aligned}
H_n &= 2H_{n-1} + 1 \\
&= 2(2H_{n-2} + 1) + 1 = 2^2 H_{n-2} + 2 + 1 \\
&= 2^2(2H_{n-3} + 1) + 2 + 1 = 2^3 H_{n-3} + 2^2 + 2 + 1 \\
&\ \vdots \\
&= 2^{n-1} H_1 + 2^{n-2} + 2^{n-3} + \cdots + 2 + 1 \\
&= 2^{n-1} + 2^{n-2} + \cdots + 2 + 1 \\
&= 2^n - 1.
\end{aligned}
$$

We have used the recurrence relation repeatedly to express H_n in terms of previous terms of the sequence. In the next to last equality, the initial condition $H_1 = 1$ has been used. The last equality is based on the formula for the sum of the terms of a geometric series, which can be found in Theorem 1 in Section 2.4.

The iterative approach has produced the solution to the recurrence relation $H_n = 2H_{n-1} + 1$ with the initial condition $H_1 = 1$. This formula can be proved using mathematical induction. This is left for the reader as Exercise 1.

A myth created to accompany the puzzle tells of a tower in Hanoi where monks are transferring 64 gold disks from one peg to another, according to the rules of the puzzle. The myth says that the world will end when they finish the puzzle. How long after the monks started will the world end if the monks take one second to move a disk?

From the explicit formula, the monks require

$$
2^{64} - 1 = 18{,}446{,}744{,}073{,}709{,}551{,}615
$$

moves to transfer the disks. Making one move per second, it will take them more than 500 billion years to complete the transfer, so the world should survive a while longer than it already has. ◀

Links

Remark: Many people have studied variations of the original Tower of Hanoi puzzle discussed in Example 2. Some variations use more pegs, some allow disks to be of the same size, and some restrict the types of allowable disk moves. One of the oldest and most interesting variations is the **Reve's puzzle**,* proposed in 1907 by Henry Dudeney in his book *The Canterbury Puzzles*. The Reve's puzzle involves pilgrims challenged by the Reve to move a stack of cheeses of varying sizes from the first of four stools to another stool without ever placing a cheese on one of smaller diameter. The Reve's puzzle, expressed in terms of pegs and disks, follows the same rules as the

---

*Reve, more commonly spelled *reeve*, is an archaic word for *governor*.

Tower of Hanoi puzzle, except that four pegs are used. You may find it surprising that no one has been able to establish the minimum number of moves required to solve this puzzle for n disks. However, there is a conjecture, now more than 50 years old, that the minimum number of moves required equals the number of moves used by an algorithm invented by Frame and Stewart in 1939. (See Exercises 30–31 and [St94] for more information.)

Example 3 illustrates how recurrence relations can be used to count bit strings of a specified length that have a certain property.

EXAMPLE 3   Find a recurrence relation and give initial conditions for the number of bit strings of length n that do not have two consecutive 0s. How many such bit strings are there of length five?

Solution: Let a_n denote the number of bit strings of length n that do not have two consecutive 0s. To obtain a recurrence relation for $\{a_n\}$, note that by the sum rule, the number of bit strings of length n that do not have two consecutive 0s equals the number of such bit strings ending with a 0 plus the number of such bit strings ending with a 1. We will assume that $n \geq 3$, so that the bit string has at least three bits.

The bit strings of length n ending with 1 that do not have two consecutive 0s are precisely the bit strings of length $n - 1$ with no two consecutive 0s with a 1 added at the end. Consequently, there are a_{n-1} such bit strings.

Bit strings of length n ending with a 0 that do not have two consecutive 0s must have 1 as their $(n - 1)$st bit; otherwise they would end with a pair of 0s. It follows that the bit strings of length n ending with a 0 that have no two consecutive 0s are precisely the bit strings of length $n - 2$ with no two consecutive 0s with 10 added at the end. Consequently, there are a_{n-2} such bit strings.

We conclude, as illustrated in Figure 4, that $a_n = a_{n-1} + a_{n-2}$ for $n \geq 3$.

The initial conditions are $a_1 = 2$, because both bit strings of length one, 0 and 1 do not have consecutive 0s, and $a_2 = 3$, because the valid bit strings of length two are 01, 10, and 11. To obtain a_5, we use the recurrence relation three times to find that

$$a_3 = a_2 + a_1 = 3 + 2 = 5,$$
$$a_4 = a_3 + a_2 = 5 + 3 = 8,$$
$$a_5 = a_4 + a_3 = 8 + 5 = 13.$$

◀

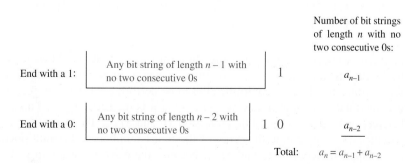

Number of bit strings of length n with no two consecutive 0s:

End with a 1:   Any bit string of length $n - 1$ with no two consecutive 0s   1 a_{n-1}

End with a 0:   Any bit string of length $n - 2$ with no two consecutive 0s   1   0 $\underline{a_{n-2}}$

Total: $a_n = a_{n-1} + a_{n-2}$

FIGURE 4   **Counting Bit Strings of Length n with No Two Consecutive 0s.**

> *Remark:* Note that $\{a_n\}$ satisfies the same recurrence relation as the Fibonacci sequence. Because $a_1 = f_3$ and $a_2 = f_4$ it follows that $a_n = f_{n+2}$.

Example 4 shows how a recurrence relation can be used to model the number of codewords that are allowable using certain validity checks.

EXAMPLE 4   **Codeword Enumeration**   A computer system considers a string of decimal digits a valid codeword if it contains an even number of 0 digits. For instance, 1230407869 is valid, whereas 120987045608 is not valid. Let a_n be the number of valid n-digit codewords. Find a recurrence relation for a_n.

Solution: Note that $a_1 = 9$ because there are 10 one-digit strings, and only one, namely, the string 0, is not valid. A recurrence relation can be derived for this sequence by considering how a valid n-digit string can be obtained from strings of $n - 1$ digits. There are two ways to form a valid string with n digits from a string with one fewer digit.

First, a valid string of n digits can be obtained by appending a valid string of $n - 1$ digits with a digit other than 0. This appending can be done in nine ways. Hence, a valid string with n digits can be formed in this manner in $9a_{n-1}$ ways.

Second, a valid string of n digits can be obtained by appending a 0 to a string of length $n - 1$ that is not valid. (This produces a string with an even number of 0 digits because the invalid string of length $n - 1$ has an odd number of 0 digits.) The number of ways that this can be done equals the number of invalid $(n - 1)$-digit strings. Because there are 10^{n-1} strings of length $n - 1$, and a_{n-1} are valid, there are $10^{n-1} - a_{n-1}$ valid n-digit strings obtained by appending an invalid string of length $n - 1$ with a 0.

Because all valid strings of length n are produced in one of these two ways, it follows that there are

$$a_n = 9a_{n-1} + (10^{n-1} - a_{n-1}) = 8a_{n-1} + 10^{n-1}$$

valid strings of length n.   ◀

Example 5 establishes a recurrence relation that appears in many different contexts.

EXAMPLE 5   Find a recurrence relation for C_n, the number of ways to parenthesize the product of $n + 1$ numbers, $x_0 \cdot x_1 \cdot x_2 \cdot \cdots \cdot x_n$, to specify the order of multiplication. For example, $C_3 = 5$ because there are five ways to parenthesize $x_0 \cdot x_1 \cdot x_2 \cdot x_3$ to determine the order of multiplication:

$$((x_0 \cdot x_1) \cdot x_2) \cdot x_3 \qquad (x_0 \cdot (x_1 \cdot x_2)) \cdot x_3 \qquad (x_0 \cdot x_1) \cdot (x_2 \cdot x_3)$$
$$x_0 \cdot ((x_1 \cdot x_2) \cdot x_3) \qquad x_0 \cdot (x_1 \cdot (x_2 \cdot x_3)).$$

Solution: To develop a recurrence relation for C_n, we note that however we insert parentheses in the product $x_0 \cdot x_1 \cdot x_2 \cdot \cdots \cdot x_n$, one "·" operator remains outside all parentheses, namely, the operator for the final multiplication to be performed. [For example, in $(x_0 \cdot (x_1 \cdot x_2)) \cdot x_3$, it is the final "·", while in $(x_0 \cdot x_1) \cdot (x_2 \cdot x_3)$ it is the second "·".] This final operator appears between two of the $n + 1$ numbers, say, x_k and x_{k+1}. There are $C_k C_{n-k-1}$ ways to insert parentheses to determine the order of the $n + 1$ numbers to be multiplied when the final operator appears between x_k and x_{k+1}, because there are C_k ways to insert parentheses in the product $x_0 \cdot x_1 \cdot \cdots \cdot x_k$ to determine the order in which these $k + 1$ numbers are to be multiplied and C_{n-k-1} ways to insert parentheses in the product $x_{k+1} \cdot x_{k+2} \cdot \cdots \cdot x_n$ to determine the order in which these $n - k$ numbers are to be multiplied. Because this final operator can appear between any two of the $n + 1$ numbers, it follows that

$$C_n = C_0 C_{n-1} + C_1 C_{n-2} + \cdots + C_{n-2} C_1 + C_{n-1} C_0$$

$$= \sum_{k=0}^{n-1} C_k C_{n-k-1}.$$

Note that the initial conditions are $C_0 = 1$ and $C_1 = 1$.   ◄

The recurrence relation in Example 5 can be solved using the method of generating functions, which will be discussed in Section 8.4. It can be shown that $C_n = C(2n, n)/(n + 1)$ (see Exercise 41 in Section 8.4) and that $C_n \sim \frac{4^n}{n^{3/2}\sqrt{\pi}}$ (see [GrKnPa94]). The sequence $\{C_n\}$ is the sequence of **Catalan numbers**, named after Eugène Charles Catalan. This sequence appears as the solution of many different counting problems besides the one considered here (see the chapter on Catalan numbers in [MiRo91] or [Ro84a] for details).

Algorithms and Recurrence Relations

Recurrence relations play an important role in many aspects of the study of algorithms and their complexity. In Section 8.3, we will show how recurrence relations can be used to analyze the complexity of divide-and-conquer algorithms, such as the merge sort algorithm introduced in Section 5.4. As we will see in Section 8.3, divide-and-conquer algorithms recursively divide a problem into a fixed number of non-overlapping subproblems until they become simple enough to solve directly. We conclude this section by introducing another algorithmic paradigm known as **dynamic programming**, which can be used to solve many optimization problems efficiently.

An algorithm follows the dynamic programming paradigm when it recursively breaks down a problem into simpler overlapping subproblems, and computes the solution using the solutions of the subproblems. Generally, recurrence relations are used to find the overall solution from the solutions of the subproblems. Dynamic programming has been used to solve important problems in such diverse areas as economics, computer vision, speech recognition, artificial intelligence, computer graphics, and bioinformatics. In this section we will illustrate the use of dynamic programming by constructing an algorithm for solving a scheduling problem. Before doing so, we will relate the amusing origin of the name *dynamic programming*, which was

EUGÈNE CHARLES CATALAN (1814–1894)   Eugène Catalan was born in Bruges, then part of France. His father became a successful architect in Paris while Eugène was a boy. Catalan attended a Parisian school for design hoping to follow in his father's footsteps. At 15, he won the job of teaching geometry to his design school classmates. After graduating, Catalan attended a school for the fine arts, but because of his mathematical aptitude his instructors recommended that he enter the École Polytechnique. He became a student there, but after his first year, he was expelled because of his politics. However, he was readmitted, and in 1835, he graduated and won a position at the Collège de Châlons sur Marne.

In 1838, Catalan returned to Paris where he founded a preparatory school with two other mathematicians, Sturm and Liouville. After teaching there for a short time, he was appointed to a position at the École Polytechnique. He received his doctorate from the École Polytechnique in 1841, but his political activity in favor of the French Republic hurt his career prospects. In 1846 Catalan held a position at the Collège de Charlemagne; he was appointed to the Lycée Saint Louis in 1849. However, when Catalan would not take a required oath of allegiance to the new Emperor Louis-Napoleon Bonaparte, he lost his job. For 13 years he held no permanent position. Finally, in 1865 he was appointed to a chair of mathematics at the University of Liège, Belgium, a position he held until his 1884 retirement.

Catalan made many contributions to number theory and to the related subject of continued fractions. He defined what are now known as the Catalan numbers when he solved the problem of dissecting a polygon into triangles using non-intersecting diagonals. Catalan is also well known for formulating what was known as the *Catalan conjecture*. This asserted that 8 and 9 are the only consecutive powers of integers, a conjecture not solved until 2003. Catalan wrote many textbooks, including several that became quite popular and appeared in as many as 12 editions. Perhaps this textbook will have a 12th edition someday!

introduced by the mathematician Richard Bellman in the 1950s. Bellman was working at the RAND Corporation on projects for the U.S. military, and at that time, the U.S. Secretary of Defense was hostile to mathematical research. Bellman decided that to ensure funding, he needed a name not containing the word mathematics for his method for solving scheduling and planning problems. He decided to use the adjective *dynamic* because, as he said "it's impossible to use the word dynamic in a pejorative sense" and he thought that dynamic programming was "something not even a Congressman could object to."

AN EXAMPLE OF DYNAMIC PROGRAMMING   The problem we use to illustrate dynamic programming is related to the problem studied in Example 7 in Section 3.1. In that problem our goal was to schedule as many talks as possible in a single lecture hall. These talks have preset start and end times; once a talk starts, it continues until it ends; no two talks can proceed at the same time; and a talk can begin at the same time another one ends. We developed a greedy algorithm that always produces an optimal schedule, as we proved in Example 12 in Section 5.1. Now suppose that our goal is not to schedule the most talks possible, but rather to have the largest possible combined attendance of the scheduled talks.

We formalize this problem by supposing that we have n talks, where talk j begins at time t_j, ends at time e_j, and will be attended by w_j students. We want a schedule that maximizes the total number of student attendees. That is, we wish to schedule a subset of talks to maximize the sum of w_j over all scheduled talks. (Note that when a student attends more than one talk, this student is counted according to the number of talks attended.) We denote by $T(j)$ the maximum number of total attendees for an optimal schedule from the first j talks, so $T(n)$ is the maximal number of total attendees for an optimal schedule for all n talks. We first sort the talks in order of increasing end time. After doing this, we renumber the talks so that $e_1 \leq e_2 \leq \cdots \leq e_n$. We say that two talks are **compatible** if they can be part of the same schedule, that is, if the times they are scheduled do not overlap (other than the possibility one ends and the other starts at the same time). We define $p(j)$ to be largest integer i, $i < j$, for which $e_i \leq s_j$, if such an integer exists, and $p(j) = 0$ otherwise. That is, talk $p(j)$ is the talk ending latest among talks compatible with talk j that end before talk j ends, if such a talk exists, and $p(j) = 0$ if there are no such talks.

Links

RICHARD BELLMAN (1920–1984)   Richard Bellman, born in Brooklyn, where his father was a grocer, spent many hours in the museums and libraries of New York as a child. After graduating high school, he studied mathematics at Brooklyn College and graduated in 1941. He began postgraduate work at Johns Hopkins University, but because of the war, left to teach electronics at the University of Wisconsin. He was able to continue his mathematics studies at Wisconsin, and in 1943 he received his masters degree there. Later, Bellman entered Princeton University, teaching in a special U.S. Army program. In late 1944, he was drafted into the army. He was assigned to the Manhattan Project at Los Alamos where he worked in theoretical physics. After the war, he returned to Princeton and received his Ph.D. in 1946.

After briefly teaching at Princeton, he moved to Stanford University, where he attained tenure. At Stanford he pursued his fascination with number theory. However, Bellman decided to focus on mathematical questions arising from real-world problems. In 1952, he joined the RAND Corporation, working on multistage decision processes, operations research problems, and applications to the social sciences and medicine. He worked on many military projects while at RAND. In 1965 he left RAND to become professor of mathematics, electrical and biomedical engineering and medicine at the University of Southern California.

In the 1950s Bellman pioneered the use of dynamic programming, a technique invented earlier, in a wide range of settings. He is also known for his work on stochastic control processes, in which he introduced what is now called the Bellman equation. He coined the term *curse of dimensionality* to describe problems caused by the exponential increase in volume associated with adding extra dimensions to a space. He wrote an amazing number of books and research papers with many coauthors, including many on industrial production and economic systems. His work led to the application of computing techniques in a wide variety of areas ranging from the design of guidance systems for space vehicles, to network optimization, and even to pest control.

Tragically, in 1973 Bellman was diagnosed with a brain tumor. Although it was removed successfully, complications left him severely disabled. Fortunately, he managed to continue his research and writing during his remaining ten years of life. Bellman received many prizes and awards, including the first Norbert Wiener Prize in Applied Mathematics and the IEEE Gold Medal of Honor. He was elected to the National Academy of Sciences. He was held in high regard for his achievements, courage, and admirable qualities. Bellman was the father of two children.

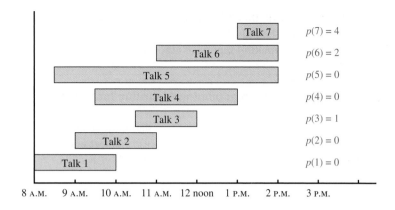

8 A.M.   9 A.M.   10 A.M.   11 A.M.   12 noon   1 P.M.   2 P.M.   3 P.M.

FIGURE 5 **A Schedule of Lectures with the Values of** $p(n)$ **Shown.**

EXAMPLE 6   Consider seven talks with these start times and end times, as illustrated in Figure 5.

Talk 1: start 8 A.M., end 10 A.M.   Talk 5: start 8:30 A.M., end 2 P.M.

Talk 2: start 9 A.M., end 11 A.M.   Talk 6: start 11 A.M., end 2 P.M.

Talk 3: start 10:30 A.M., end 12 noon   Talk 7: start 1 P.M., end 2 P.M.

Talk 4: start 9:30 A.M., end 1 P.M.

Find $p(j)$ for $j = 1, 2, \ldots, 7$.

Solution: We have $p(1) = 0$ and $p(2) = 0$, because no talks end before either of the first two talks begin. We have $p(3) = 1$ because talk 3 and talk 1 are compatible, but talk 3 and talk 2 are not compatible; $p(4) = 0$ because talk 4 is not compatible with any of talks 1, 2, and 3; $p(5) = 0$ because talk 5 is not compatible with any of talks 1, 2, 3, and 4; and $p(6) = 2$ because talk 6 and talk 2 are compatible, but talk 6 is not compatible with any of talks 3, 4, and 5. Finally, $p(7) = 4$, because talk 7 and talk 4 are compatible, but talk 7 is not compatible with either of talks 5 or 6.   ◀

To develop a dynamic programming algorithm for this problem, we first develop a key recurrence relation. To do this, first note that if $j \leq n$, there are two possibilities for an optimal schedule of the first j talks (recall that we are assuming that the n talks are ordered by increasing end time): *(i)* talk j belongs to the optimal schedule or *(ii)* it does not.

Case (i): We know that talks $p(j) + 1, \ldots, j - 1$ do not belong to this schedule, for none of these other talks are compatible with talk j. Furthermore, the other talks in this optimal schedule must comprise an optimal schedule for talks $1, 2, \ldots, p(j)$. For if there were a better schedule for talks $1, 2, \ldots, p(j)$, by adding talk j, we will have a schedule better than the overall optimal schedule. Consequently, in case *(i)*, we have $T(j) = w_j + T(p(j))$.

Case (ii): When talk j does not belong to an optimal schedule, it follows that an optimal schedule from talks $1, 2, \ldots, j$ is the same as an optimal schedule from talks $1, 2, \ldots, j - 1$. Consequently, in case *(ii)*, we have $T(j) = T(j - 1)$. Combining cases *(i)* and *(ii)* leads us to the recurrence relation

$$T(j) = \max(w_j + T(p(j)), T(j - 1)).$$

Now that we have developed this recurrence relation, we can construct an efficient algorithm, Algorithm 1, for computing the maximum total number of attendees. We ensure that the algorithm is efficient by storing the value of each $T(j)$ after we compute it. This allows us to compute $T(j)$

only once. If we did not do this, the algorithm would have exponential worst-case complexity. The process of storing the values as each is computed is known as **memoization** and is an important technique for making recursive algorithms efficient.

ALGORITHM 1 Dynamic Programming Algorithm for Scheduling Talks.

procedure *Maximum Attendees* ($s_1, s_2, \ldots, s_n$: start times of talks;
$\quad$ $e_1, e_2, \ldots, e_n$: end times of talks; $w_1, w_2, \ldots, w_n$: number of attendees to talks)
$\quad$ sort talks by end time and relabel so that $e_1 \le e_2 \le \cdots \le e_n$
for $j := 1$ **to** n
$\quad$ **if** no job i with $i < j$ is compatible with job j
$\quad\quad$ $p(j) = 0$
$\quad$ **else** $p(j) := \max\{i \mid i < j$ and job i is compatible with job $j\}$
$\quad$ $T(0) := 0$
for $j := 1$ **to** n
$\quad$ $T(j) := \max(w_j + T(p(j)), T(j-1))$
return $T(n)\{T(n)$ is the maximum number of attendees}

In Algorithm 1 we determine the maximum number of attendees that can be achieved by a schedule of talks, but we do not find a schedule that achieves this maximum. To find talks we need to schedule, we use the fact that talk j belongs to an optimal solution for the first j talks if and only if $w_j + T(p(j)) \ge T(j-1)$. We leave it as Exercise 39 to construct an algorithm based on this observation that determines which talks should be scheduled to achieve the maximum total number of attendees.

Algorithm 1 is a good example of dynamic programming as the maximum total attendance is found using the optimal solutions of the overlapping subproblems, each of which determines the maximum total attendance of the first j talks for some j with $1 \le j \le n-1$. See Exercises 42 and 43 and Supplementary Exercises 10 and 13 for other examples of dynamic programming.

Exercises

1. Use mathematical induction to verify the formula derived in Example 2 for the number of moves required to complete the Tower of Hanoi puzzle.

2. A vending machine dispensing books of stamps accepts only one-dollar coins, $1 bills, and $5 bills.

  a) Find a recurrence relation for the number of ways to deposit n dollars in the vending machine, where the order in which the coins and bills are deposited matters.

  b) What are the initial conditions?

  c) How many ways are there to deposit $10 for a book of stamps?

3. a) A country uses as currency coins with values of 1 peso, 2 pesos, 5 pesos, and 10 pesos and bills with values of 5 pesos, 10 pesos, 20 pesos, 50 pesos, and 100 pesos. Find a recurrence relation for the number of ways to pay a bill of n pesos if the order in which the coins and bills are paid matters.

  b) How many ways are there to pay a bill of 17 pesos using the currency described above, where the order in which coins and bills are paid matters?

*4. a) Find a recurrence relation for the number of strictly increasing sequences of positive integers that have 1 as their first term and n as their last term, where n is a positive integer. That is, sequences $a_1, a_2, \ldots, a_k$, where $a_1 = 1$, $a_k = n$, and $a_j < a_{j+1}$ for $j = 1, 2, \ldots, k-1$.

  b) What are the initial conditions?

  c) How many sequences of the type described in (a) are there when n is an integer with $n \ge 2$?

5. a) Find a recurrence relation for the number of bit strings of length n that contain a pair of consecutive 0s.

  b) What are the initial conditions?

  c) How many bit strings of length seven contain two consecutive 0s?

6. a) Find a recurrence relation for the number of bit strings of length n that contain three consecutive 0s.

  b) What are the initial conditions?

  c) How many bit strings of length seven contain three consecutive 0s?

7. a) Find a recurrence relation for the number of bit strings of length n that do not contain three consecutive 0s.

  b) What are the initial conditions?

c) How many bit strings of length seven do not contain three consecutive 0s?

8. a) Find a recurrence relation for the number of ways to climb n stairs if the person climbing the stairs can take one stair or two stairs at a time.

b) What are the initial conditions?

c) In how many ways can this person climb a flight of eight stairs?

A string that contains only 0s, 1s, and 2s is called a **ternary string**.

9. a) Find a recurrence relation for the number of ternary strings of length n that do not contain two consecutive 0s.

b) What are the initial conditions?

c) How many ternary strings of length six do not contain two consecutive 0s?

10. a) Find a recurrence relation for the number of ternary strings of length n that contain two consecutive 0s.

b) What are the initial conditions?

c) How many ternary strings of length six contain two consecutive 0s?

*11. a) Find a recurrence relation for the number of ternary strings of length n that do not contain two consecutive 0s or two consecutive 1s.

b) What are the initial conditions?

c) How many ternary strings of length six do not contain two consecutive 0s or two consecutive 1s?

*12. a) Find a recurrence relation for the number of ternary strings of length n that do not contain consecutive symbols that are the same.

b) What are the initial conditions?

c) How many ternary strings of length six do not contain consecutive symbols that are the same?

13. Messages are transmitted over a communications channel using two signals. The transmittal of one signal requires 1 microsecond, and the transmittal of the other signal requires 2 microseconds.

a) Find a recurrence relation for the number of different messages consisting of sequences of these two signals, where each signal in the message is immediately followed by the next signal, that can be sent in n microseconds.

b) What are the initial conditions?

c) How many different messages can be sent in 10 microseconds using these two signals?

14. A bus driver pays all tolls, using only nickels and dimes, by throwing one coin at a time into the mechanical toll collector.

a) Find a recurrence relation for the number of different ways the bus driver can pay a toll of n cents (where the order in which the coins are used matters).

b) In how many different ways can the driver pay a toll of 45 cents?

15. a) Find the recurrence relation satisfied by R_n, where R_n is the number of regions that a plane is divided into by n lines, if no two of the lines are parallel and no three of the lines go through the same point.

b) Find R_n using iteration.

*16. a) Find the recurrence relation satisfied by R_n, where R_n is the number of regions into which the surface of a sphere is divided by n great circles (which are the intersections of the sphere and planes passing through the center of the sphere), if no three of the great circles go through the same point.

b) Find R_n using iteration.

*17. a) Find the recurrence relation satisfied by S_n, where S_n is the number of regions into which three-dimensional space is divided by n planes if every three of the planes meet in one point, but no four of the planes go through the same point.

b) Find S_n using iteration.

18. a) Find a recurrence relation for the number of ways to completely cover a $2 \times n$ checkerboard with 1×2 dominoes. [*Hint:* Consider separately the coverings where the position in the top right corner of the checkerboard is covered by a domino positioned horizontally and where it is covered by a domino positioned vertically.]

b) What are the initial conditions for the recurrence relation in part (a)?

c) How many ways are there to completely cover a 2×17 checkerboard with 1×2 dominoes?

19. a) Find a recurrence relation for the number of ways to lay out a walkway with slate tiles if the tiles are red, green, or gray, so that no two red tiles are adjacent and tiles of the same color are considered indistinguishable.

b) What are the initial conditions for the recurrence relation in part (a)?

c) How many ways are there to lay out a path of seven tiles as described in part (a)?

20. Show that the Fibonacci numbers satisfy the recurrence relation $f_n = 5f_{n-4} + 3f_{n-5}$ for $n = 5, 6, 7, \ldots$, together with the initial conditions $f_0 = 0$, $f_1 = 1$, $f_2 = 1$, $f_3 = 2$, and $f_4 = 3$. Use this recurrence relation to show that f_{5n} is divisible by 5, for $n = 1, 2, 3, \ldots$.

*21. Let $S(m, n)$ denote the number of onto functions from a set with m elements to a set with n elements. Show that $S(m, n)$ satisfies the recurrence relation

$$S(m, n) = n^m - \sum_{k=1}^{n-1} C(n, k)S(m, k)$$

whenever $m \geq n$ and $n > 1$, with the initial condition $S(m, 1) = 1$.

22. a) Write out all the ways the product $x_0 \cdot x_1 \cdot x_2 \cdot x_3 \cdot x_4$ can be parenthesized to determine the order of multiplication.

b) Use the recurrence relation developed in Example 5 to calculate C_4, the number of ways to parenthesize the product of five numbers so as to determine the order of multiplication. Verify that you listed the correct number of ways in part (a).

c) Check your result in part (b) by finding C_4, using the closed formula for C_n mentioned in the solution of Example 5.

23. a) Use the recurrence relation developed in Example 5 to determine C_5, the number of ways to parenthesize the product of six numbers so as to determine the order of multiplication.

b) Check your result with the closed formula for C_5 mentioned in the solution of Example 5.

*24. In the Tower of Hanoi puzzle, suppose our goal is to transfer all n disks from peg 1 to peg 3, but we cannot move a disk directly between pegs 1 and 3. Each move of a disk must be a move involving peg 2. As usual, we cannot place a disk on top of a smaller disk.

a) Find a recurrence relation for the number of moves required to solve the puzzle for n disks with this added restriction.

b) Solve this recurrence relation to find a formula for the number of moves required to solve the puzzle for n disks.

c) How many different arrangements are there of the n disks on three pegs so that no disk is on top of a smaller disk?

d) Show that every allowable arrangement of the n disks occurs in the solution of this variation of the puzzle.

Exercises 25–29 deal with a variation of the **Josephus problem** described by Graham, Knuth, and Patashnik in [GrKnPa94]. This problem is based on an account by the historian Flavius Josephus, who was part of a band of 41 Jewish rebels trapped in a cave by the Romans during the Jewish-Roman war of the first century. The rebels preferred suicide to capture; they decided to form a circle and to repeatedly count off around the circle, killing every third rebel left alive. However, Josephus and another rebel did not want to be killed this way; they determined the positions where they should stand to be the last two rebels remaining alive. The variation we consider begins with n people, numbered 1 to n, standing around a circle. In each stage, every second person still left alive is eliminated until only one survives. We denote the number of the survivor by $J(n)$.

25. Determine the value of $J(n)$ for each integer n with $1 \le n \le 16$.

26. Use the values you found in Exercise 25 to conjecture a formula for $J(n)$. [*Hint:* Write $n = 2^m + k$, where m is a nonnegative integer and k is a nonnegative integer less than 2^m.]

27. Show that $J(n)$ satisfies the recurrence relation $J(2n) = 2J(n) - 1$ and $J(2n + 1) = 2J(n) + 1$, for $n \ge 1$, and $J(1) = 1$.

28. Use mathematical induction to prove the formula you conjectured in Exercise 26, making use of the recurrence relation from Exercise 27.

29. Determine $J(100)$, $J(1000)$, and $J(10,000)$ from your formula for $J(n)$.

Exercises 30–31 involve the Reve's puzzle, the variation of the Tower of Hanoi puzzle with four pegs and n disks. Before presenting these exercises, we describe the Frame–Stewart algorithm for moving the disks from peg 1 to peg 4 so that no disk is ever on top of a smaller one. This algorithm, given the number of disks n as input, depends on a choice of an integer k with $1 \le k \le n$. When there is only one disk, move it from peg 1 to peg 4 and stop. For $n > 1$, the algorithm proceeds recursively, using these three steps. Recursively move the stack of the $n - k$ smallest disks from peg 1 to peg 2, using all four pegs. Next move the stack of the k largest disks from peg 1 to peg 4, using the three-peg algorithm from the Tower of Hanoi puzzle without using the peg holding the $n - k$ smallest disks. Finally, recursively move the smallest $n - k$ disks to peg 4, using all four pegs. Frame and Stewart showed that to produce the fewest moves using their algorithm, k should be chosen to be the smallest integer such that n does not exceed $t_k = k(k + 1)/2$, the kth triangular number, that is, $t_{k-1} < n \le t_k$. The unsettled conjecture, known as **Frame's conjecture**, is that this algorithm uses the fewest number of moves required to solve the puzzle, no matter how the disks are moved.

30. Show that the Reve's puzzle with three disks can be solved using five, and no fewer, moves.

31. Show that the Reve's puzzle with four disks can be solved using nine, and no fewer, moves.

Let $\{a_n\}$ be a sequence of real numbers. The **backward differences** of this sequence are defined recursively as shown next. The **first difference** ∇a_n is

$$\nabla a_n = a_n - a_{n-1}.$$

The $(k + 1)$**st difference** $\nabla^{k+1} a_n$ is obtained from $\nabla^k a_n$ by

$$\nabla^{k+1} a_n = \nabla^k a_n - \nabla^k a_{n-1}.$$

32. Find ∇a_n for the sequence $\{a_n\}$, where
a) $a_n = 4$. **b)** $a_n = 2n$.
c) $a_n = n^2$. **d)** $a_n = 2^n$.

33. Find $\nabla^2 a_n$ for the sequences in Exercise 32.

34. Show that $a_{n-1} = a_n - \nabla a_n$.

35. Show that $a_{n-2} = a_n - 2\nabla a_n + \nabla^2 a_n$.

*36. Prove that a_{n-k} can be expressed in terms of a_n, ∇a_n, $\nabla^2 a_n$, $\ldots$, $\nabla^k a_n$.

37. Express the recurrence relation $a_n = a_{n-1} + a_{n-2}$ in terms of a_n, ∇a_n, and $\nabla^2 a_n$.

38. Show that any recurrence relation for the sequence $\{a_n\}$ can be written in terms of a_n, ∇a_n, $\nabla^2 a_n$, $\ldots$. The resulting equation involving the sequences and its differences is called a **difference equation**.

*39. Construct the algorithm described in the text after Algorithm 1 for determining which talks should be scheduled to maximize the total number of attendees and not just the maximum total number of attendees determined by Algorithm 1.

40. Use Algorithm 1 to determine the maximum number of total attendees in the talks in Example 6 if w_i, the number of attendees of talk i, $i = 1, 2, \ldots, 7$, is
   a) 20, 10, 50, 30, 15, 25, 40.
   b) 100, 5, 10, 20, 25, 40, 30.
   c) 2, 3, 8, 5, 4, 7, 10.
   d) 10, 8, 7, 25, 20, 30, 5.

41. For each part of Exercise 40, use your algorithm from Exercise 39 to find the optimal schedule for talks so that the total number of attendees is maximized.

42. In this exercise we will develop a dynamic programming algorithm for finding the maximum sum of consecutive terms of a sequence of real numbers. That is, given a sequence of real numbers $a_1, a_2, \ldots, a_n$, the algorithm computes the maximum sum $\sum_{i=j}^{k} a_i$ where $1 \le j \le k \le n$.
   a) Show that if all terms of the sequence are nonnegative, this problem is solved by taking the sum of all terms. Then, give an example where the maximum sum of consecutive terms is not the sum of all terms.
   b) Let $M(k)$ be the maximum of the sums of consecutive terms of the sequence ending at a_k. That is, $M(k) = \max_{1 \le j \le k} \sum_{i=j}^{k} a_i$. Explain why the recurrence relation $M(k) = \max(M(k-1) + a_k, a_k)$ holds for $k = 2, \ldots, n$.
   c) Use part (b) to develop a dynamic programming algorithm for solving this problem.
   d) Show each step your algorithm from part (c) uses to find the maximum sum of consecutive terms of the sequence 2, −3, 4, 1, −2, 3.
   e) Show that the worst-case complexity in terms of the number of additions and comparisons of your algorithm from part (c) is linear.

∗43. Dynamic programming can be used to develop an algorithm for solving the matrix-chain multiplication problem introduced in Section 3.3. This

is the problem of determining how the product $A_1A_2 \cdots A_n$ can be computed using the fewest integer multiplications, where $A_1, A_2, \ldots, A_n$ are $m_1 \times m_2, m_2 \times m_3, \ldots, m_n \times m_{n+1}$ matrices, respectively, and each matrix has integer entries. Recall that by the associative law, the product does not depend on the order in which the matrices are multiplied.
   a) Show that the brute-force method of determining the minimum number of integer multiplications needed to solve a matrix-chain multiplication problem has exponential worst-case complexity.
   b) Denote by A_{ij} the product $A_iA_{i+1} \ldots, A_j$, and $M(i, j)$ the minimum number of integer multiplications required to find A_{ij}. Show that if the least number of integer multiplications are used to compute A_{ij}, where $i < j$, by splitting the product into the product of A_i through A_k and the product of A_{k+1} through A_j, then the first k terms must be parenthesized so that A_{ik} is computed in the optimal way using $M(i, k)$ integer multiplications and $A_{k+1,j}$ must be parenthesized so that $A_{k+1,j}$ is computed in the optimal way using $M(k+1, j)$ integer multiplications.
   c) Explain why part (b) leads to the recurrence relation $M(i, j) = \min_{i \le k < j}(M(i, k) + M(k+1, j) + m_i m_{k+1} m_{j+1})$ if $1 \le i \le j < j \le n$.
   d) Use the recurrence relation in part (c) to construct an efficient algorithm for determining the order the n matrices should be multiplied to use the minimum number of integer multiplications. Store the partial results $M(i, j)$ as you find them so that your algorithm will not have exponential complexity.
   e) Show that your algorithm from part (d) has $O(n^3)$ worst-case complexity in terms of multiplications of integers.

8.2 Solving Linear Recurrence Relations

Introduction

Links

A wide variety of recurrence relations occur in models. Some of these recurrence relations can be solved using iteration or some other ad hoc technique. However, one important class of recurrence relations can be explicitly solved in a systematic way. These are recurrence relations that express the terms of a sequence as linear combinations of previous terms.

DEFINITION 1   A *linear homogeneous recurrence relation of degree k with constant coefficients* is a recurrence relation of the form

$$a_n = c_1 a_{n-1} + c_2 a_{n-2} + \cdots + c_k a_{n-k},$$

where $c_1, c_2, \ldots, c_k$ are real numbers, and $c_k \neq 0$.

The recurrence relation in the definition is **linear** because the right-hand side is a sum of previous terms of the sequence each multiplied by a function of n. The recurrence relation is **homogeneous** because no terms occur that are not multiples of the a_js. The coefficients of the terms of the sequence are all **constants**, rather than functions that depend on n. The **degree** is k because a_n is expressed in terms of the previous k terms of the sequence.

A consequence of the second principle of mathematical induction is that a sequence satisfying the recurrence relation in the definition is uniquely determined by this recurrence relation and the k initial conditions

$$a_0 = C_0, a_1 = C_1, \ldots, a_{k-1} = C_{k-1}.$$

EXAMPLE 1 The recurrence relation $P_n = (1.11)P_{n-1}$ is a linear homogeneous recurrence relation of degree one. The recurrence relation $f_n = f_{n-1} + f_{n-2}$ is a linear homogeneous recurrence relation of degree two. The recurrence relation $a_n = a_{n-5}$ is a linear homogeneous recurrence relation of degree five. ◄

Example 2 presents some examples of recurrence relations that are not linear homogeneous recurrence relations with constant coefficients.

EXAMPLE 2 The recurrence relation $a_n = a_{n-1} + a_{n-2}^2$ is not linear. The recurrence relation $H_n = 2H_{n-1} + 1$ is not homogeneous. The recurrence relation $B_n = nB_{n-1}$ does not have constant coefficients. ◄

Linear homogeneous recurrence relations are studied for two reasons. First, they often occur in modeling of problems. Second, they can be systematically solved.

Solving Linear Homogeneous Recurrence Relations with Constant Coefficients

The basic approach for solving linear homogeneous recurrence relations is to look for solutions of the form $a_n = r^n$, where r is a constant. Note that $a_n = r^n$ is a solution of the recurrence relation $a_n = c_1 a_{n-1} + c_2 a_{n-2} + \cdots + c_k a_{n-k}$ if and only if

$$r^n = c_1 r^{n-1} + c_2 r^{n-2} + \cdots + c_k r^{n-k}.$$

When both sides of this equation are divided by r^{n-k} and the right-hand side is subtracted from the left, we obtain the equation

$$r^k - c_1 r^{k-1} - c_2 r^{k-2} - \cdots - c_{k-1} r - c_k = 0.$$

Consequently, the sequence $\{a_n\}$ with $a_n = r^n$ is a solution if and only if r is a solution of this last equation. We call this the **characteristic equation** of the recurrence relation. The solutions of this equation are called the **characteristic roots** of the recurrence relation. As we will see, these characteristic roots can be used to give an explicit formula for all the solutions of the recurrence relation.

We will first develop results that deal with linear homogeneous recurrence relations with constant coefficients of degree two. Then corresponding general results when the degree may be greater than two will be stated. Because the proofs needed to establish the results in the general case are more complicated, they will not be given here.

We now turn our attention to linear homogeneous recurrence relations of degree two. First, consider the case when there are two distinct characteristic roots.

THEOREM 1

Let c_1 and c_2 be real numbers. Suppose that $r^2 - c_1 r - c_2 = 0$ has two distinct roots r_1 and r_2. Then the sequence $\{a_n\}$ is a solution of the recurrence relation $a_n = c_1 a_{n-1} + c_2 a_{n-2}$ if and only if $a_n = \alpha_1 r_1^n + \alpha_2 r_2^n$ for $n = 0, 1, 2, \ldots$, where α_1 and α_2 are constants.

Proof: We must do two things to prove the theorem. First, it must be shown that if r_1 and r_2 are the roots of the characteristic equation, and α_1 and α_2 are constants, then the sequence $\{a_n\}$ with $a_n = \alpha_1 r_1^n + \alpha_2 r_2^n$ is a solution of the recurrence relation. Second, it must be shown that if the sequence $\{a_n\}$ is a solution, then $a_n = \alpha_1 r_1^n + \alpha_2 r_2^n$ for some constants α_1 and α_2.

Now we will show that if $a_n = \alpha_1 r_1^n + \alpha_2 r_2^n$, then the sequence $\{a_n\}$ is a solution of the recurrence relation. Because r_1 and r_2 are roots of $r^2 - c_1 r - c_2 = 0$, it follows that $r_1^2 = c_1 r_1 + c_2$, $r_2^2 = c_1 r_2 + c_2$.

From these equations, we see that

$$
\begin{aligned}
c_1 a_{n-1} + c_2 a_{n-2} &= c_1 (\alpha_1 r_1^{n-1} + \alpha_2 r_2^{n-1}) + c_2 (\alpha_1 r_1^{n-2} + \alpha_2 r_2^{n-2}) \\
&= \alpha_1 r_1^{n-2} (c_1 r_1 + c_2) + \alpha_2 r_2^{n-2} (c_1 r_2 + c_2) \\
&= \alpha_1 r_1^{n-2} r_1^2 + \alpha_2 r_2^{n-2} r_2^2 \\
&= \alpha_1 r_1^n + \alpha_2 r_2^n \\
&= a_n.
\end{aligned}
$$

This shows that the sequence $\{a_n\}$ with $a_n = \alpha_1 r_1^n + \alpha_2 r_2^n$ is a solution of the recurrence relation.

To show that every solution $\{a_n\}$ of the recurrence relation $a_n = c_1 a_{n-1} + c_2 a_{n-2}$ has $a_n = \alpha_1 r_1^n + \alpha_2 r_2^n$ for $n = 0, 1, 2, \ldots$, for some constants α_1 and α_2, suppose that $\{a_n\}$ is a solution of the recurrence relation, and the initial conditions $a_0 = C_0$ and $a_1 = C_1$ hold. It will be shown that there are constants α_1 and α_2 such that the sequence $\{a_n\}$ with $a_n = \alpha_1 r_1^n + \alpha_2 r_2^n$ satisfies these same initial conditions. This requires that

$$
a_0 = C_0 = \alpha_1 + \alpha_2,
$$
$$
a_1 = C_1 = \alpha_1 r_1 + \alpha_2 r_2.
$$

We can solve these two equations for α_1 and α_2. From the first equation it follows that $\alpha_2 = C_0 - \alpha_1$. Inserting this expression into the second equation gives

$$
C_1 = \alpha_1 r_1 + (C_0 - \alpha_1) r_2.
$$

Hence,

$$
C_1 = \alpha_1 (r_1 - r_2) + C_0 r_2.
$$

This shows that

$$
\alpha_1 = \frac{C_1 - C_0 r_2}{r_1 - r_2}
$$

and

$$
\alpha_2 = C_0 - \alpha_1 = C_0 - \frac{C_1 - C_0 r_2}{r_1 - r_2} = \frac{C_0 r_1 - C_1}{r_1 - r_2},
$$

where these expressions for α_1 and α_2 depend on the fact that $r_1 \neq r_2$. (When $r_1 = r_2$, this theorem is not true.) Hence, with these values for α_1 and α_2, the sequence $\{a_n\}$ with $\alpha_1 r_1^n + \alpha_2 r_2^n$ satisfies the two initial conditions.

We know that $\{a_n\}$ and $\{\alpha_1 r_1^n + \alpha_2 r_2^n\}$ are both solutions of the recurrence relation $a_n = c_1 a_{n-1} + c_2 a_{n-2}$ and both satisfy the initial conditions when $n = 0$ and $n = 1$. Because there is a unique solution of a linear homogeneous recurrence relation of degree two with two initial conditions, it follows that the two solutions are the same, that is, $a_n = \alpha_1 r_1^n + \alpha_2 r_2^n$ for all nonnegative integers n. We have completed the proof by showing that a solution of the linear homogeneous recurrence relation with constant coefficients of degree two must be of the form $a_n = \alpha_1 r_1^n + \alpha_2 r_2^n$, where α_1 and α_2 are constants. $\triangleleft$

The characteristic roots of a linear homogeneous recurrence relation with constant coefficients may be complex numbers. Theorem 1 (and also subsequent theorems in this section) still applies in this case. Recurrence relations with complex characteristic roots will not be discussed in the text. Readers familiar with complex numbers may wish to solve Exercises 28 and 29.

Examples 3 and 4 show how to use Theorem 1 to solve recurrence relations.

EXAMPLE 3   What is the solution of the recurrence relation

$$a_n = a_{n-1} + 2a_{n-2}$$

with $a_0 = 2$ and $a_1 = 7$?

Solution: Theorem 1 can be used to solve this problem. The characteristic equation of the recurrence relation is $r^2 - r - 2 = 0$. Its roots are $r = 2$ and $r = -1$. Hence, the sequence $\{a_n\}$ is a solution to the recurrence relation if and only if

$$a_n = \alpha_1 2^n + \alpha_2 (-1)^n,$$

for some constants α_1 and α_2. From the initial conditions, it follows that

$$a_0 = 2 = \alpha_1 + \alpha_2, \qquad a_1 = 7 = \alpha_1 \cdot 2 + \alpha_2 \cdot (-1).$$

Solving these two equations shows that $\alpha_1 = 3$ and $\alpha_2 = -1$. Hence, the solution to the recurrence relation and initial conditions is the sequence $\{a_n\}$ with

$$a_n = 3 \cdot 2^n - (-1)^n. \qquad \blacktriangleleft$$

EXAMPLE 4   Find an explicit formula for the Fibonacci numbers.

Solution: Recall that the sequence of Fibonacci numbers satisfies the recurrence relation $f_n = f_{n-1} + f_{n-2}$ and also satisfies the initial conditions $f_0 = 0$ and $f_1 = 1$. The roots of the characteristic equation $r^2 - r - 1 = 0$ are $r_1 = (1 + \sqrt{5})/2$ and $r_2 = (1 - \sqrt{5})/2$. Therefore, from Theorem 1 it follows that the Fibonacci numbers are given by

$$f_n = \alpha_1 \left(\frac{1 + \sqrt{5}}{2} \right)^n + \alpha_2 \left(\frac{1 - \sqrt{5}}{2} \right)^n,$$

for some constants α_1 and α_2. The initial conditions $f_0 = 0$ and $f_1 = 1$ can be used to find these constants. We have

$$f_0 = \alpha_1 + \alpha_2 = 0,$$

$$f_1 = \alpha_1 \left(\frac{1 + \sqrt{5}}{2} \right) + \alpha_2 \left(\frac{1 - \sqrt{5}}{2} \right) = 1.$$

The solution to these simultaneous equations for α_1 and α_2 is

$$\alpha_1 = 1/\sqrt{5}, \qquad \alpha_2 = -1/\sqrt{5}.$$

Consequently, the Fibonacci numbers are given by

$$f_n = \frac{1}{\sqrt{5}} \left(\frac{1 + \sqrt{5}}{2} \right)^n - \frac{1}{\sqrt{5}} \left(\frac{1 - \sqrt{5}}{2} \right)^n. \qquad \blacktriangleleft$$

Theorem 1 does not apply when there is one characteristic root of multiplicity two. If this happens, then $a_n = nr_0^n$ is another solution of the recurrence relation when r_0 is a root of multiplicity two of the characteristic equation. Theorem 2 shows how to handle this case.

THEOREM 2   Let c_1 and c_2 be real numbers with $c_2 \neq 0$. Suppose that $r^2 - c_1 r - c_2 = 0$ has only one root r_0. A sequence $\{a_n\}$ is a solution of the recurrence relation $a_n = c_1 a_{n-1} + c_2 a_{n-2}$ if and only if $a_n = \alpha_1 r_0^n + \alpha_2 n r_0^n$, for $n = 0, 1, 2, \ldots$, where α_1 and α_2 are constants.

The proof of Theorem 2 is left as Exercise 8. Example 5 illustrates the use of this theorem.

EXAMPLE 5   What is the solution of the recurrence relation

$$a_n = 6a_{n-1} - 9a_{n-2}$$

with initial conditions $a_0 = 1$ and $a_1 = 6$?

Solution: The only root of $r^2 - 6r + 9 = 0$ is $r = 3$. Hence, the solution to this recurrence relation is

$$a_n = \alpha_1 3^n + \alpha_2 n 3^n$$

for some constants α_1 and α_2. Using the initial conditions, it follows that

$$a_0 = 1 = \alpha_1, \qquad a_1 = 6 = \alpha_1 \cdot 3 + \alpha_2 \cdot 3.$$

Solving these two equations shows that $\alpha_1 = 1$ and $\alpha_2 = 1$. Consequently, the solution to this recurrence relation and the initial conditions is

$$a_n = 3^n + n3^n. \qquad \blacktriangleleft$$

We will now state the general result about the solution of linear homogeneous recurrence relations with constant coefficients, where the degree may be greater than two, under the assumption that the characteristic equation has distinct roots. The proof of this result will be left as Exercise 12.

THEOREM 3   Let $c_1, c_2, \ldots, c_k$ be real numbers. Suppose that the characteristic equation

$$r^k - c_1 r^{k-1} - \cdots - c_k = 0$$

has k distinct roots $r_1, r_2, \ldots, r_k$. Then a sequence $\{a_n\}$ is a solution of the recurrence relation

$$a_n = c_1 a_{n-1} + c_2 a_{n-2} + \cdots + c_k a_{n-k}$$

if and only if

$$a_n = \alpha_1 r_1^n + \alpha_2 r_2^n + \cdots + \alpha_k r_k^n$$

for $n = 0, 1, 2, \ldots$, where $\alpha_1, \alpha_2, \ldots, \alpha_k$ are constants.

We illustrate the use of the theorem with Example 6.

EXAMPLE 6   Find the solution to the recurrence relation

$$a_n = 6a_{n-1} - 11a_{n-2} + 6a_{n-3}$$

with the initial conditions $a_0 = 2$, $a_1 = 5$, and $a_2 = 15$.

Solution: The characteristic polynomial of this recurrence relation is

$$r^3 - 6r^2 + 11r - 6.$$

The characteristic roots are $r = 1$, $r = 2$, and $r = 3$, because $r^3 - 6r^2 + 11r - 6 = (r-1)(r-2)(r-3)$. Hence, the solutions to this recurrence relation are of the form

$$a_n = \alpha_1 \cdot 1^n + \alpha_2 \cdot 2^n + \alpha_3 \cdot 3^n.$$

To find the constants α_1, α_2, and α_3, use the initial conditions. This gives

$$a_0 = 2 = \alpha_1 + \alpha_2 + \alpha_3,$$
$$a_1 = 5 = \alpha_1 + \alpha_2 \cdot 2 + \alpha_3 \cdot 3,$$
$$a_2 = 15 = \alpha_1 + \alpha_2 \cdot 4 + \alpha_3 \cdot 9.$$

When these three simultaneous equations are solved for α_1, α_2, and α_3, we find that $\alpha_1 = 1$, $\alpha_2 = -1$, and $\alpha_3 = 2$. Hence, the unique solution to this recurrence relation and the given initial conditions is the sequence $\{a_n\}$ with

$$a_n = 1 - 2^n + 2 \cdot 3^n.$$

◀

We now state the most general result about linear homogeneous recurrence relations with constant coefficients, allowing the characteristic equation to have multiple roots. The key point is that for each root r of the characteristic equation, the general solution has a summand of the form $P(n)r^n$, where $P(n)$ is a polynomial of degree $m - 1$, with m the multiplicity of this root. We leave the proof of this result as Exercise 39.

THEOREM 4   Let $c_1, c_2, \ldots, c_k$ be real numbers. Suppose that the characteristic equation

$$r^k - c_1 r^{k-1} - \cdots - c_k = 0$$

has t distinct roots $r_1, r_2, \ldots, r_t$ with multiplicities $m_1, m_2, \ldots, m_t$, respectively, so that $m_i \geq 1$ for $i = 1, 2, \ldots, t$ and $m_1 + m_2 + \cdots + m_t = k$. Then a sequence $\{a_n\}$ is a solution of the recurrence relation

$$a_n = c_1 a_{n-1} + c_2 a_{n-2} + \cdots + c_k a_{n-k}$$

if and only if

$$
\begin{aligned}
a_n = {} & (\alpha_{1,0} + \alpha_{1,1} n + \cdots + \alpha_{1,m_1-1} n^{m_1-1}) r_1^n \\
& + (\alpha_{2,0} + \alpha_{2,1} n + \cdots + \alpha_{2,m_2-1} n^{m_2-1}) r_2^n \\
& + \cdots + (\alpha_{t,0} + \alpha_{t,1} n + \cdots + \alpha_{t,m_t-1} n^{m_t-1}) r_t^n
\end{aligned}
$$

for $n = 0, 1, 2, \ldots$, where $\alpha_{i,j}$ are constants for $1 \leq i \leq t$ and $0 \leq j \leq m_i - 1$.

Example 7 illustrates how Theorem 4 is used to find the general form of a solution of a linear homogeneous recurrence relation when the characteristic equation has several repeated roots.

EXAMPLE 7   Suppose that the roots of the characteristic equation of a linear homogeneous recurrence relation are 2, 2, 2, 5, 5, and 9 (that is, there are three roots, the root 2 with multiplicity three, the root 5 with multiplicity two, and the root 9 with multiplicity one). What is the form of the general solution?

Solution: By Theorem 4, the general form of the solution is

$$(\alpha_{1,0} + \alpha_{1,1} n + \alpha_{1,2} n^2) 2^n + (\alpha_{2,0} + \alpha_{2,1} n) 5^n + \alpha_{3,0} 9^n.$$

◀

We now illustrate the use of Theorem 4 to solve a linear homogeneous recurrence relation with constant coefficients when the characteristic equation has a root of multiplicity three.

EXAMPLE 8   Find the solution to the recurrence relation

$$a_n = -3a_{n-1} - 3a_{n-2} - a_{n-3}$$

with initial conditions $a_0 = 1$, $a_1 = -2$, and $a_2 = -1$.

Solution: The characteristic equation of this recurrence relation is

$$r^3 + 3r^2 + 3r + 1 = 0.$$

Because $r^3 + 3r^2 + 3r + 1 = (r + 1)^3$, there is a single root $r = -1$ of multiplicity three of the characteristic equation. By Theorem 4 the solutions of this recurrence relation are of the form

$$a_n = \alpha_{1,0}(-1)^n + \alpha_{1,1}n(-1)^n + \alpha_{1,2}n^2(-1)^n.$$

To find the constants $\alpha_{1,0}, \alpha_{1,1}$, and $\alpha_{1,2}$, use the initial conditions. This gives

$$a_0 = 1 = \alpha_{1,0}, \quad a_1 = -2 = -\alpha_{1,0} - \alpha_{1,1} - \alpha_{1,2}, \quad a_2 = -1 = \alpha_{1,0} + 2\alpha_{1,1} + 4\alpha_{1,2}.$$

The simultaneous solution of these three equations is $\alpha_{1,0} = 1$, $\alpha_{1,1} = 3$, and $\alpha_{1,2} = -2$. Hence, the unique solution to this recurrence relation and the given initial conditions is the sequence $\{a_n\}$ with

$$a_n = (1 + 3n - 2n^2)(-1)^n.$$

◀

Linear Nonhomogeneous Recurrence Relations with Constant Coefficients

We have seen how to solve linear homogeneous recurrence relations with constant coefficients. Is there a relatively simple technique for solving a linear, but not homogeneous, recurrence relation with constant coefficients, such as $a_n = 3a_{n-1} + 2n$? We will see that the answer is yes for certain families of such recurrence relations.

The recurrence relation $a_n = 3a_{n-1} + 2n$ is an example of a **linear nonhomogeneous recurrence relation with constant coefficients**, that is, a recurrence relation of the form

$$a_n = c_1 a_{n-1} + c_2 a_{n-2} + \cdots + c_k a_{n-k} + F(n),$$

where $c_1, c_2, \ldots, c_k$ are real numbers and $F(n)$ is a function not identically zero depending only on n. The recurrence relation

$$a_n = c_1 a_{n-1} + c_2 a_{n-2} + \cdots + c_k a_{n-k}$$

is called the **associated homogeneous recurrence relation**. It plays an important role in the solution of the nonhomogeneous recurrence relation.

EXAMPLE 9   Each of the recurrence relations $a_n = a_{n-1} + 2^n$, $a_n = a_{n-1} + a_{n-2} + n^2 + n + 1$, $a_n = 3a_{n-1} + n3^n$, and $a_n = a_{n-1} + a_{n-2} + a_{n-3} + n!$ is a linear nonhomogeneous recurrence relation with constant coefficients. The associated linear homogeneous recurrence relations are $a_n = a_{n-1}, a_n = a_{n-1} + a_{n-2}, a_n = 3a_{n-1}$, and $a_n = a_{n-1} + a_{n-2} + a_{n-3}$, respectively.   ◀

The key fact about linear nonhomogeneous recurrence relations with constant coefficients is that every solution is the sum of a particular solution and a solution of the associated linear homogeneous recurrence relation, as Theorem 5 shows.

THEOREM 5   If $\{a_n^{(p)}\}$ is a particular solution of the nonhomogeneous linear recurrence relation with constant coefficients

$$a_n = c_1 a_{n-1} + c_2 a_{n-2} + \cdots + c_k a_{n-k} + F(n),$$

then every solution is of the form $\{a_n^{(p)} + a_n^{(h)}\}$, where $\{a_n^{(h)}\}$ is a solution of the associated homogeneous recurrence relation

$$a_n = c_1 a_{n-1} + c_2 a_{n-2} + \cdots + c_k a_{n-k}.$$

Proof: Because $\{a_n^{(p)}\}$ is a particular solution of the nonhomogeneous recurrence relation, we know that

$$a_n^{(p)} = c_1 a_{n-1}^{(p)} + c_2 a_{n-2}^{(p)} + \cdots + c_k a_{n-k}^{(p)} + F(n).$$

Now suppose that $\{b_n\}$ is a second solution of the nonhomogeneous recurrence relation, so that

$$b_n = c_1 b_{n-1} + c_2 b_{n-2} + \cdots + c_k b_{n-k} + F(n).$$

Subtracting the first of these two equations from the second shows that

$$b_n - a_n^{(p)} = c_1(b_{n-1} - a_{n-1}^{(p)}) + c_2(b_{n-2} - a_{n-2}^{(p)}) + \cdots + c_k(b_{n-k} - a_{n-k}^{(p)}).$$

It follows that $\{b_n - a_n^p\}$ is a solution of the associated homogeneous linear recurrence, say, $\{a_n^{(h)}\}$. Consequently, $b_n = a_n^{(p)} + a_n^{(h)}$ for all n.   ◁

By Theorem 5, we see that the key to solving nonhomogeneous recurrence relations with constant coefficients is finding a particular solution. Then every solution is a sum of this solution and a solution of the associated homogeneous recurrence relation. Although there is no general method for finding such a solution that works for every function $F(n)$, there are techniques that work for certain types of functions $F(n)$, such as polynomials and powers of constants. This is illustrated in Examples 10 and 11.

EXAMPLE 10   Find all solutions of the recurrence relation $a_n = 3a_{n-1} + 2n$. What is the solution with $a_1 = 3$?

Solution: To solve this linear nonhomogeneous recurrence relation with constant coefficients, we need to solve its associated linear homogeneous equation and to find a particular solution for the given nonhomogeneous equation. The associated linear homogeneous equation is $a_n = 3a_{n-1}$. Its solutions are $a_n^{(h)} = \alpha 3^n$, where α is a constant.

We now find a particular solution. Because $F(n) = 2n$ is a polynomial in n of degree one, a reasonable trial solution is a linear function in n, say, $p_n = cn + d$, where c and d are constants. To determine whether there are any solutions of this form, suppose that $p_n = cn + d$ is such a solution. Then the equation $a_n = 3a_{n-1} + 2n$ becomes $cn + d = 3(c(n - 1) + d) + 2n$. Simplifying and combining like terms gives $(2 + 2c)n + (2d - 3c) = 0$. It follows that $cn + d$ is a solution if and only if $2 + 2c = 0$ and $2d - 3c = 0$. This shows that $cn + d$ is a solution if and only if $c = -1$ and $d = -3/2$. Consequently, $a_n^{(p)} = -n - 3/2$ is a particular solution.

By Theorem 5 all solutions are of the form

$$a_n = a_n^{(p)} + a_n^{(h)} = -n - \frac{3}{2} + \alpha \cdot 3^n,$$

where α is a constant.

To find the solution with $a_1 = 3$, let $n = 1$ in the formula we obtained for the general solution. We find that $3 = -1 - 3/2 + 3\alpha$, which implies that $\alpha = 11/6$. The solution we seek is $a_n = -n - 3/2 + (11/6)3^n$.   ◀

EXAMPLE 11

Extra
Examples

Find all solutions of the recurrence relation

$$a_n = 5a_{n-1} - 6a_{n-2} + 7^n.$$

Solution: This is a linear nonhomogeneous recurrence relation. The solutions of its associated homogeneous recurrence relation

$$a_n = 5a_{n-1} - 6a_{n-2}$$

are $a_n^{(h)} = \alpha_1 \cdot 3^n + \alpha_2 \cdot 2^n$, where α_1 and α_2 are constants. Because $F(n) = 7^n$, a reasonable trial solution is $a_n^{(p)} = C \cdot 7^n$, where C is a constant. Substituting the terms of this sequence into the recurrence relation implies that $C \cdot 7^n = 5C \cdot 7^{n-1} - 6C \cdot 7^{n-2} + 7^n$. Factoring out 7^{n-2}, this equation becomes $49C = 35C - 6C + 49$, which implies that $20C = 49$, or that $C = 49/20$. Hence, $a_n^{(p)} = (49/20)7^n$ is a particular solution. By Theorem 5, all solutions are of the form

$$a_n = \alpha_1 \cdot 3^n + \alpha_2 \cdot 2^n + (49/20)7^n.$$

◀

In Examples 10 and 11, we made an educated guess that there are solutions of a particular form. In both cases we were able to find particular solutions. This was not an accident. Whenever $F(n)$ is the product of a polynomial in n and the nth power of a constant, we know exactly what form a particular solution has, as stated in Theorem 6. We leave the proof of Theorem 6 as Exercise 40.

THEOREM 6

Suppose that $\{a_n\}$ satisfies the linear nonhomogeneous recurrence relation

$$a_n = c_1 a_{n-1} + c_2 a_{n-2} + \cdots + c_k a_{n-k} + F(n),$$

where $c_1, c_2, \ldots, c_k$ are real numbers, and

$$F(n) = (b_t n^t + b_{t-1} n^{t-1} + \cdots + b_1 n + b_0)s^n,$$

where $b_0, b_1, \ldots, b_t$ and s are real numbers. When s is not a root of the characteristic equation of the associated linear homogeneous recurrence relation, there is a particular solution of the form

$$(p_t n^t + p_{t-1} n^{t-1} + \cdots + p_1 n + p_0)s^n.$$

When s is a root of this characteristic equation and its multiplicity is m, there is a particular solution of the form

$$n^m(p_t n^t + p_{t-1} n^{t-1} + \cdots + p_1 n + p_0)s^n.$$

Note that in the case when s is a root of multiplicity m of the characteristic equation of the associated linear homogeneous recurrence relation, the factor n^m ensures that the proposed particular solution will not already be a solution of the associated linear homogeneous recurrence relation. We next provide Example 12 to illustrate the form of a particular solution provided by Theorem 6.

EXAMPLE 12   What form does a particular solution of the linear nonhomogeneous recurrence relation $a_n = 6a_{n-1} - 9a_{n-2} + F(n)$ have when $F(n) = 3^n$, $F(n) = n3^n$, $F(n) = n^2 2^n$, and $F(n) = (n^2 + 1)3^n$?

Solution: The associated linear homogeneous recurrence relation is $a_n = 6a_{n-1} - 9a_{n-2}$. Its characteristic equation, $r^2 - 6r + 9 = (r - 3)^2 = 0$, has a single root, 3, of multiplicity two. To apply Theorem 6, with $F(n)$ of the form $P(n)s^n$, where $P(n)$ is a polynomial and s is a constant, we need to ask whether s is a root of this characteristic equation.

Because $s = 3$ is a root with multiplicity $m = 2$ but $s = 2$ is not a root, Theorem 6 tells us that a particular solution has the form $p_0 n^2 3^n$ if $F(n) = 3^n$, the form $n^2(p_1 n + p_0)3^n$ if $F(n) = n3^n$, the form $(p_2 n^2 + p_1 n + p_0)2^n$ if $F(n) = n^2 2^n$, and the form $n^2(p_2 n^2 + p_1 n + p_0)3^n$ if $F(n) = (n^2 + 1)3^n$.   ◀

Care must be taken when $s = 1$ when solving recurrence relations of the type covered by Theorem 6. In particular, to apply this theorem with $F(n) = b_t n_t + b_{t-1} n_{t-1} + \cdots + b_1 n + b_0$, the parameter s takes the value $s = 1$ (even though the term 1^n does not explicitly appear). By the theorem, the form of the solution then depends on whether 1 is a root of the characteristic equation of the associated linear homogeneous recurrence relation. This is illustrated in Example 13, which shows how Theorem 6 can be used to find a formula for the sum of the first n positive integers.

EXAMPLE 13   Let a_n be the sum of the first n positive integers, so that

$$a_n = \sum_{k=1}^{n} k.$$

Note that a_n satisfies the linear nonhomogeneous recurrence relation

$$a_n = a_{n-1} + n.$$

(To obtain a_n, the sum of the first n positive integers, from a_{n-1}, the sum of the first $n - 1$ positive integers, we add n.) Note that the initial condition is $a_1 = 1$.

The associated linear homogeneous recurrence relation for a_n is

$$a_n = a_{n-1}.$$

The solutions of this homogeneous recurrence relation are given by $a_n^{(h)} = c(1)^n = c$, where c is a constant. To find all solutions of $a_n = a_{n-1} + n$, we need find only a single particular solution. By Theorem 6, because $F(n) = n = n \cdot (1)^n$ and $s = 1$ is a root of degree one of the characteristic equation of the associated linear homogeneous recurrence relation, there is a particular solution of the form $n(p_1 n + p_0) = p_1 n^2 + p_0 n$.

Inserting this into the recurrence relation gives $p_1 n^2 + p_0 n = p_1 (n - 1)^2 + p_0 (n - 1) + n$. Simplifying, we see that $n(2p_1 - 1) + (p_0 - p_1) = 0$, which means that $2p_1 - 1 = 0$ and $p_0 - p_1 = 0$, so $p_0 = p_1 = 1/2$. Hence,

$$a_n^{(p)} = \frac{n^2}{2} + \frac{n}{2} = \frac{n(n + 1)}{2}$$

is a particular solution. Hence, all solutions of the original recurrence relation $a_n = a_{n-1} + n$ are given by $a_n = a_n^{(h)} + a_n^{(p)} = c + n(n+1)/2$. Because $a_1 = 1$, we have $1 = a_1 = c + 1 \cdot 2/2 = c + 1$, so $c = 0$. It follows that $a_n = n(n+1)/2$. (This is the same formula given in Table 2 in Section 2.4 and derived previously.) ◀

Using the idea of substitution helps in solving recurrence relations. The following two examples illustrate this.

EXAMPLE 14   Solve the recurrence relation $a_n^2 - 2a_{n-1}^2 = 1$ for $n \geq 1$ where $a_0 = 1$.

This is a nonlinear equation. This can be converted to a linear equation by using the substitution $b_n = a_n^2$. Now the equation becomes $b_n - 2b_{n-1} = 1$ with $b_0 = a_0^2 = 1^2 = 1$.

Solving $b_n - 2b_{n-1} = 1$ with $b_0 = 1$ the homogeneous solution is $a2^n$ and the particular solution is c. So $b_n = a2^n + c$ where a and c are constants. Using the fact that $b_0 = 1$ and $b_1 = 3$ we get $a + C = 1$ and $2a + c = 3$, which gives $a = 2$ and $b = -1$. Hence the solution is $b_n = 2 \cdot 2^n - 1 = 2^{n+1} - 1$.

Substituting back $a_n^2 = 2^{n+1} - 1$

$$a_n = \sqrt{2^{n+1} - 1}.$$

Even though $a_n = \pm\sqrt{b_n}$, since $a_0 = 1$ a_n cannot be $-\sqrt{b_n}$.

In a similar manner, divide and conquer recurrence relations can be solved using appropriate substitutions. Generally, a divide and conquer relation is of the form $a_n = ca_{n/d} + f(n)$ where n is usually taken at as d^k. Hence this equation can be looked at as $b_k = cb_{k-1} + f(d^k)$ using substitution $n = d^k$ and solved. Initial conditions have to be changed appropriately. For substituting back, we use $k = \log_d n$. ◀

EXAMPLE 15   Solve the divide and conquer relation $a_n = 3a_{n/2} + n$ where $n = 2^k$ for $k \geq 1$ and $a_1 = 1$.

Using change of variables, the equation becomes $b_k = 3b_{k-1} + 2^k$. The homogeneous solution $b_k^h = A3^k$ and the particular solution is $b_k^p = B2^k$ and we get $b_k = A3^k + B2^k$. Since $a_1 = 1$, $b_0 = 1$. So $A + B = 1$ $a_2 = b_1 = 5$. So we get $A = 3$ and $B = -2$ So the solution is $b_k = 3.3^k - 2.2^k$. Substituting back,

$$a_n = 3.3^{\log_2^n} - 2.2^{\log_2^n} = 3.3^{\log_2^n} - 2n$$

But $3^{\log_2^n} = n^{\log_2^3}$. So $a_n = 3n^{\log_2^3} - 2n$. ◀

Exercises

1. Determine which of these are linear homogeneous recurrence relations with constant coefficients. Also, find the degree of those that are.

a) $a_n = 3a_{n-1} + 4a_{n-2} + 5a_{n-3}$

b) $a_n = 2na_{n-1} + a_{n-2}$ **c)** $a_n = a_{n-1} + a_{n-4}$

d) $a_n = a_{n-1} + 2$ **e)** $a_n = a_{n-1}^2 + a_{n-2}$

f) $a_n = a_{n-2}$ **g)** $a_n = a_{n-1} + n$

2. Solve these recurrence relations together with the initial conditions given.

a) $a_n = 2a_{n-1}$ for $n \geq 1$, $a_0 = 3$

b) $a_n = a_{n-1}$ for $n \geq 1$, $a_0 = 2$

c) $a_n = 5a_{n-1} - 6a_{n-2}$ for $n \geq 2$, $a_0 = 1$, $a_1 = 0$

d) $a_n = 4a_{n-1} - 4a_{n-2}$ for $n \geq 2$, $a_0 = 6$, $a_1 = 8$

e) $a_n = -4a_{n-1} - 4a_{n-2}$ for $n \geq 2$, $a_0 = 0$, $a_1 = 1$

f) $a_n = 4a_{n-2}$ for $n \geq 2$, $a_0 = 0$, $a_1 = 4$

g) $a_n = a_{n-2}/4$ for $n \geq 2$, $a_0 = 1$, $a_1 = 0$

3. How many different messages can be transmitted in n microseconds using the two signals described in Exercise 13 in Section 8.1?

4. How many different messages can be transmitted in n microseconds using three different signals if one signal requires 1 microsecond for transmittal, the other two signals require 2 microseconds each for transmittal, and a signal in a message is followed immediately by the next signal?

5. In how many ways can a $2 \times n$ rectangular checkerboard be tiled using 1×2 and 2×2 pieces?

6. A model for the number of lobsters caught per year is based on the assumption that the number of lobsters caught in a year is the average of the number caught in the two previous years.
   a) Find a recurrence relation for $\{L_n\}$, where L_n is the number of lobsters caught in year n, under the assumption for this model.
   b) Find L_n if 100,000 lobsters were caught in year 1 and 300,000 were caught in year 2.

7. A deposit of $100,000 is made to an investment fund at the beginning of a year. On the last day of each year two dividends are awarded. The first dividend is 20% of the amount in the account during that year. The second dividend is 45% of the amount in the account in the previous year.
   a) Find a recurrence relation for $\{P_n\}$, where P_n is the amount in the account at the end of n years if no money is ever withdrawn.
   b) How much is in the account after n years if no money has been withdrawn?

∗8. Prove Theorem 2.

9. The **Lucas numbers** satisfy the recurrence relation

$$L_n = L_{n-1} + L_{n-2},$$

and the initial conditions $L_0 = 2$ and $L_1 = 1$.
   a) Show that $L_n = f_{n-1} + f_{n+1}$ for $n = 2, 3, \ldots$, where f_n is the nth Fibonacci number.
   b) Find an explicit formula for the Lucas numbers.

10. Find the solution to $a_n = 7a_{n-2} + 6a_{n-3}$ with $a_0 = 9$, $a_1 = 10$, and $a_2 = 32$.

11. Find the solution to $a_n = 2a_{n-1} + 5a_{n-2} - 6a_{n-3}$ with $a_0 = 7$, $a_1 = -4$, and $a_2 = 8$.

∗12. Prove Theorem 3.

13. Prove this identity relating the Fibonacci numbers and the binomial coefficients:

$$f_{n+1} = C(n, 0) + C(n - 1, 1) + \cdots + C(n - k, k),$$

where n is a positive integer and $k = \lfloor n/2 \rfloor$. [*Hint:* Let $a_n = C(n, 0) + C(n - 1, 1) + \cdots + C(n - k, k)$. Show that the sequence $\{a_n\}$ satisfies the same recurrence relation and initial conditions satisfied by the sequence of Fibonacci numbers.]

14. Solve the recurrence relation $a_n = 6a_{n-1} - 12a_{n-2} + 8a_{n-3}$ with $a_0 = -5$, $a_1 = 4$, and $a_2 = 88$.

15. Solve the recurrence relation $a_n = -3a_{n-1} - 3a_{n-2} - a_{n-3}$ with $a_0 = 5$, $a_1 = -9$, and $a_2 = 15$.

16. Find the general form of the solutions of the recurrence relation $a_n = 8a_{n-2} - 16a_{n-4}$.

17. What is the general form of the solutions of a linear homogeneous recurrence relation if its characteristic equation has roots $1, 1, 1, 1, -2, -2, -2, 3, 3, -4$?

18. What is the general form of the solutions of a linear homogeneous recurrence relation if its characteristic equation has the roots $-1, -1, -1, 2, 2, 5, 5, 7$?

19. Consider the nonhomogeneous linear recurrence relation $a_n = 3a_{n-1} + 2^n$.
   a) Show that $a_n = -2^{n+1}$ is a solution of this recurrence relation.
   b) Use Theorem 5 to find all solutions of this recurrence relation.
   c) Find the solution with $a_0 = 1$.

20. a) Determine values of the constants A and B such that $a_n = An + B$ is a solution of recurrence relation $a_n = 2a_{n-1} + n + 5$.
   b) Use Theorem 5 to find all solutions of this recurrence relation.
   c) Find the solution of this recurrence relation with $a_0 = 4$.

21. What is the general form of the particular solution guaranteed to exist by Theorem 6 of the linear nonhomogeneous recurrence relation $a_n = 8a_{n-2} - 16a_{n-4} + F(n)$ if
   a) $F(n) = n^3$?
   b) $F(n) = (-2)^n$?
   c) $F(n) = n2^n$?
   d) $F(n) = n^2 4^n$?
   e) $F(n) = (n^2 - 2)(-2)^n$?
   f) $F(n) = n^4 2^n$?
   g) $F(n) = 2$?

22. a) Find all solutions of the recurrence relation $a_n = 2a_{n-1} + 2n^2$.
   b) Find the solution of the recurrence relation in part (a) with initial condition $a_1 = 4$.

23. a) Find all solutions of the recurrence relation $a_n = 2a_{n-1} + 3^n$.
   b) Find the solution of the recurrence relation in part (a) with initial condition $a_1 = 5$.

24. a) Find all solutions of the recurrence relation $a_n = -5a_{n-1} - 6a_{n-2} + 42 \cdot 4^n$.
   b) Find the solution of this recurrence relation with $a_1 = 56$ and $a_2 = 278$.

25. Find the solution of the recurrence relation $a_n = 4a_{n-1} - 3a_{n-2} + 2^n + n + 3$ with $a_0 = 1$ and $a_1 = 4$.

26. Let a_n be the sum of the first n perfect squares, that is, $a_n = \sum_{k=1}^{n} k^2$. Show that the sequence $\{a_n\}$ satisfies the linear nonhomogeneous recurrence relation $a_n = a_{n-1} + n^2$ and the initial condition $a_1 = 1$. Use Theorem 6 to determine a formula for a_n by solving this recurrence relation.

27. Let a_n be the sum of the first n triangular numbers, that is, $a_n = \sum_{k=1}^{n} t_k$, where $t_k = k(k+1)/2$. Show that $\{a_n\}$ satisfies the linear nonhomogeneous recurrence relation $a_n = a_{n-1} + n(n+1)/2$ and the initial condition $a_1 = 1$. Use Theorem 6 to determine a formula for a_n by solving this recurrence relation.

28. a) Find the characteristic roots of the linear homogeneous recurrence relation $a_n = 2a_{n-1} - 2a_{n-2}$. [*Note:* These are complex numbers.]

    b) Find the solution of the recurrence relation in part (a) with $a_0 = 1$ and $a_1 = 2$.

*29. a) Find the characteristic roots of the linear homogeneous recurrence relation $a_n = a_{n-4}$. [*Note:* These include complex numbers.]

    b) Find the solution of the recurrence relation in part (a) with $a_0 = 1$, $a_1 = 0$, $a_2 = -1$, and $a_3 = 1$.

*30. Solve the simultaneous recurrence relations

$$a_n = 3a_{n-1} + 2b_{n-1}$$
$$b_n = a_{n-1} + 2b_{n-1}$$

with $a_0 = 1$ and $b_0 = 2$.

*31. a) Use the formula found in Example 4 for f_n, the nth Fibonacci number, to show that f_n is the integer closest to

$$\frac{1}{\sqrt{5}} \left(\frac{1 + \sqrt{5}}{2} \right)^n.$$

    b) Determine for which n f_n is greater than

$$\frac{1}{\sqrt{5}} \left(\frac{1 + \sqrt{5}}{2} \right)^n$$

and for which n f_n is less than

$$\frac{1}{\sqrt{5}} \left(\frac{1 + \sqrt{5}}{2} \right)^n.$$

32. Show that if $a_n = a_{n-1} + a_{n-2}$, $a_0 = s$ and $a_1 = t$, where s and t are constants, then $a_n = sf_{n-1} + tf_n$ for all positive integers n.

33. Express the solution of the linear nonhomogenous recurrence relation $a_n = a_{n-1} + a_{n-2} + 1$ for $n \geq 2$ where $a_0 = 0$ and $a_1 = 1$ in terms of the Fibonacci numbers. [*Hint:* Let $b_n = a_n + 1$ and apply Exercise 32 to the sequence b_n.]

34. Suppose that each pair of a genetically engineered species of rabbits left on an island produces two new pairs of rabbits at the age of 1 month and six new pairs of rabbits at the age of 2 months and every month afterward. None of the rabbits ever die or leave the island.

    a) Find a recurrence relation for the number of pairs of rabbits on the island n months after one newborn pair is left on the island.

    b) By solving the recurrence relation in (a) determine the number of pairs of rabbits on the island n months after one pair is left on the island.

35. A new employee at an exciting new software company starts with a salary of $50,000 and is promised that at the end of each year her salary will be double her salary of the previous year, with an extra increment of $10,000 for each year she has been with the company.

    a) Construct a recurrence relation for her salary for her nth year of employment.

    b) Solve this recurrence relation to find her salary for her nth year of employment.

Some linear recurrence relations that do not have constant coefficients can be systematically solved. This is the case for recurrence relations of the form $f(n)a_n = g(n)a_{n-1} + h(n)$. Exercises 36–38 illustrate this.

*36. a) Show that the recurrence relation

$$f(n)a_n = g(n)a_{n-1} + h(n),$$

for $n \geq 1$, and with $a_0 = C$, can be reduced to a recurrence relation of the form

$$b_n = b_{n-1} + Q(n)h(n),$$

where $b_n = g(n+1)Q(n+1)a_n$, with

$$Q(n) = (f(1)f(2) \cdots f(n-1))/(g(1)g(2) \cdots g(n)).$$

    b) Use part (a) to solve the original recurrence relation to obtain

$$a_n = \frac{C + \sum_{i=1}^{n} Q(i)h(i)}{g(n+1)Q(n+1)}.$$

*37. Use Exercise 36 to solve the recurrence relation $(n+1)a_n = (n+3)a_{n-1} + n$, for $n \geq 1$, with $a_0 = 1$.

38. It can be shown that C_n, the average number of comparisons made by the quick sort algorithm (described in preamble to Exercise 34 in Section 5.4), when sorting n elements in random order, satisfies the recurrence relation

$$C_n = n + 1 + \frac{2}{n} \sum_{k=0}^{n-1} C_k$$

for $n = 1, 2, \ldots$, with initial condition $C_0 = 0$.

    a) Show that $\{C_n\}$ also satisfies the recurrence relation $nC_n = (n+1)C_{n-1} + 2n$ for $n = 1, 2, \ldots$.

    b) Use Exercise 36 to solve the recurrence relation in part (a) to find an explicit formula for C_n.

**39. Prove Theorem 4.

**40. Prove Theorem 6.

41. Solve the recurrence relation $T(n) = nT^2(n/2)$ with initial condition $T(1) = 6$ when $n = 2^k$ for some integer k. [*Hint:* Let $n = 2^k$ and then make the substitution $a_k = \log T(2^k)$ to obtain a linear nonhomogeneous recurrence relation.]

8.3 Divide-and-Conquer Algorithms and Recurrence Relations

Introduction

Links

Many recursive algorithms take a problem with a given input and divide it into one or more smaller problems. This reduction is successively applied until the solutions of the smaller problems can be found quickly. For instance, we perform a binary search by reducing the search for an element in a list to the search for this element in a list half as long. We successively apply this reduction until one element is left. When we sort a list of integers using the merge sort, we split the list into two halves of equal size and sort each half separately. We then merge the two sorted halves. Another example of this type of recursive algorithm is a procedure for multiplying integers that reduces the problem of the multiplication of two integers to three multiplications of pairs of integers with half as many bits. This reduction is successively applied until integers with one bit are obtained. These procedures follow an important algorithmic paradigm known as **divide-and-conquer**, and are called **divide-and-conquer algorithms**, because they *divide* a problem into one or more instances of the same problem of smaller size and they *conquer* the problem by using the solutions of the smaller problems to find a solution of the original problem, perhaps with some additional work.

"Divide et impera" (translation: *"Divide and conquer"* - Julius Caesar)

In this section we will show how recurrence relations can be used to analyze the computational complexity of divide-and-conquer algorithms. We will use these recurrence relations to estimate the number of operations used by many different divide-and-conquer algorithms, including several that we introduce in this section.

Divide-and-Conquer Recurrence Relations

Suppose that a recursive algorithm divides a problem of size n into a subproblems, where each subproblem is of size n/b (for simplicity, assume that n is a multiple of b; in reality, the smaller problems are often of size equal to the nearest integers either less than or equal to, or greater than or equal to, n/b). Also, suppose that a total of $g(n)$ extra operations are required in the conquer step of the algorithm to combine the solutions of the subproblems into a solution of the original problem. Then, if $f(n)$ represents the number of operations required to solve the problem of size n, it follows that f satisfies the recurrence relation

$$f(n) = af(n/b) + g(n).$$

This is called a **divide-and-conquer recurrence relation**.

We will first set up the divide-and-conquer recurrence relations that can be used to study the complexity of some important algorithms. Then we will show how to use these divide-and-conquer recurrence relations to estimate the complexity of these algorithms.

EXAMPLE 1

Extra Examples

Binary Search   We introduced a binary search algorithm in Section 3.1. This binary search algorithm reduces the search for an element in a search sequence of size n to the binary search for this element in a search sequence of size $n/2$, when n is even. (Hence, the problem of size n has been reduced to *one* problem of size $n/2$.) Two comparisons are needed to implement this reduction (one to determine which half of the list to use and the other to determine whether any terms of the list remain). Hence, if $f(n)$ is the number of comparisons required to search for an element in a search sequence of size n, then

$$f(n) = f(n/2) + 2$$

when n is even.   ◀

EXAMPLE 2   **Finding the Maximum and Minimum of a Sequence**   Consider the following algorithm for locating the maximum and minimum elements of a sequence $a_1, a_2, \ldots, a_n$. If $n = 1$, then a_1 is the maximum and the minimum. If $n > 1$, split the sequence into two sequences, either where both have the same number of elements or where one of the sequences has one more element than the other. The problem is reduced to finding the maximum and minimum of each of the two smaller sequences. The solution to the original problem results from the comparison of the separate maxima and minima of the two smaller sequences to obtain the overall maximum and minimum.

Let $f(n)$ be the total number of comparisons needed to find the maximum and minimum elements of the sequence with n elements. We have shown that a problem of size n can be reduced into two problems of size $n/2$, when n is even, using two comparisons, one to compare the maxima of the two sequences and the other to compare the minima of the two sequences. This gives the recurrence relation

$$f(n) = 2f(n/2) + 2$$

when n is even. ◀

EXAMPLE 3   **Merge Sort**   The merge sort algorithm (introduced in Section 5.4) splits a list to be sorted with n items, where n is even, into two lists with $n/2$ elements each, and uses fewer than n comparisons to merge the two sorted lists of $n/2$ items each into one sorted list. Consequently, the number of comparisons used by the merge sort to sort a list of n elements is less than $M(n)$, where the function $M(n)$ satisfies the divide-and-conquer recurrence relation

$$M(n) = 2M(n/2) + n.$$
◀

EXAMPLE 4   **Fast Multiplication of Integers**   Surprisingly, there are more efficient algorithms than the conventional algorithm (described in Section 4.2) for multiplying integers. One of these algorithms, which uses a divide-and-conquer technique, will be described here. This fast multiplication algorithm proceeds by splitting each of two $2n$-bit integers into two blocks, each with n bits. Then, the original multiplication is reduced from the multiplication of two $2n$-bit integers to three multiplications of n-bit integers, plus shifts and additions.

Suppose that a and b are integers with binary expansions of length $2n$ (add initial bits of zero in these expansions if necessary to make them the same length). Let

$$a = (a_{2n-1}a_{2n-2}\cdots a_1a_0)_2 \qquad \text{and} \qquad b = (b_{2n-1}b_{2n-2}\cdots b_1b_0)_2.$$

Let

$$a = 2^n A_1 + A_0, \quad b = 2^n B_1 + B_0,$$

where

$$A_1 = (a_{2n-1}\cdots a_{n+1}a_n)_2, \qquad A_0 = (a_{n-1}\cdots a_1a_0)_2,$$
$$B_1 = (b_{2n-1}\cdots b_{n+1}b_n)_2, \qquad B_0 = (b_{n-1}\cdots b_1b_0)_2.$$

The algorithm for fast multiplication of integers is based on the fact that ab can be rewritten as

$$ab = (2^{2n} + 2^n)A_1 B_1 + 2^n(A_1 - A_0)(B_0 - B_1) + (2^n + 1)A_0 B_0.$$

The important fact about this identity is that it shows that the multiplication of two $2n$-bit integers can be carried out using three multiplications of n-bit integers, together with additions, subtractions, and shifts. This shows that if $f(n)$ is the total number of bit operations needed to multiply two n-bit integers, then

$$f(2n) = 3f(n) + Cn.$$

The reasoning behind this equation is as follows. The three multiplications of n-bit integers are carried out using $3f(n)$-bit operations. Each of the additions, subtractions, and shifts uses a constant multiple of n-bit operations, and Cn represents the total number of bit operations used by these operations.   ◀

EXAMPLE 5

Fast Matrix Multiplication   In Example 7 of Section 3.3 we showed that multiplying two $n \times n$ matrices using the definition of matrix multiplication required n^3 multiplications and $n^2(n-1)$ additions. Consequently, computing the product of two $n \times n$ matrices in this way requires $O(n^3)$ operations (multiplications and additions). Surprisingly, there are more efficient divide-and-conquer algorithms for multiplying two $n \times n$ matrices. Such an algorithm, invented by Volker Strassen in 1969, reduces the multiplication of two $n \times n$ matrices, when n is even, to seven multiplications of two $(n/2) \times (n/2)$ matrices and 15 additions of $(n/2) \times (n/2)$ matrices. (See [CoLeRiSt09] for the details of this algorithm.) Hence, if $f(n)$ is the number of operations (multiplications and additions) used, it follows that

$$f(n) = 7f(n/2) + 15n^2/4$$

when n is even.   ◀

As Examples 1–5 show, recurrence relations of the form $f(n) = af(n/b) + g(n)$ arise in many different situations. It is possible to derive estimates of the size of functions that satisfy such recurrence relations. Suppose that f satisfies this recurrence relation whenever n is divisible by b. Let $n = b^k$, where k is a positive integer. Then

$$\begin{aligned}
f(n) &= af(n/b) + g(n) \\
&= a^2 f(n/b^2) + ag(n/b) + g(n) \\
&= a^3 f(n/b^3) + a^2 g(n/b^2) + ag(n/b) + g(n) \\
&\ \ \vdots \\
&= a^k f(n/b^k) + \sum_{j=0}^{k-1} a^j g(n/b^j).
\end{aligned}$$

Because $n/b^k = 1$, it follows that

$$f(n) = a^k f(1) + \sum_{j=0}^{k-1} a^j g(n/b^j).$$

We can use this equation for $f(n)$ to estimate the size of functions that satisfy divide-and-conquer relations.

THEOREM 1   Let f be an increasing function that satisfies the recurrence relation

$$f(n) = af(n/b) + c$$

whenever n is divisible by b, where $a \geq 1$, b is an integer greater than 1, and c is a positive real number. Then

$$f(n) \text{ is } \begin{cases} O(n^{\log_b a}) & \text{if } a > 1, \\ O(\log n) & \text{if } a = 1. \end{cases}$$

Furthermore, when $n = b^k$ and $a \neq 1$, where k is a positive integer,

$$f(n) = C_1 n^{\log_b a} + C_2,$$

where $C_1 = f(1) + c/(a-1)$ and $C_2 = -c/(a-1)$.

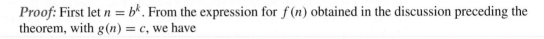

Proof: First let $n = b^k$. From the expression for $f(n)$ obtained in the discussion preceding the theorem, with $g(n) = c$, we have

$$f(n) = a^k f(1) + \sum_{j=0}^{k-1} a^j c = a^k f(1) + c \sum_{j=0}^{k-1} a^j.$$

When $a = 1$ we have

$$f(n) = f(1) + ck.$$

Because $n = b^k$, we have $k = \log_b n$. Hence,

$$f(n) = f(1) + c \log_b n.$$

When n is not a power of b, we have $b^k < n < b^{k+1}$, for a positive integer k. Because f is increasing, it follows that $f(n) \leq f(b^{k+1}) = f(1) + c(k+1) = (f(1) + c) + ck \leq (f(1) + c) + c \log_b n$. Therefore, in both cases, $f(n)$ is $O(\log n)$ when $a = 1$.

Now suppose that $a > 1$. First assume that $n = b^k$, where k is a positive integer. From the formula for the sum of terms of a geometric progression (Theorem 1 in Section 2.4), it follows that

$$f(n) = a^k f(1) + c(a^k - 1)/(a - 1)$$
$$= a^k [f(1) + c/(a-1)] - c/(a-1)$$
$$= C_1 n^{\log_b a} + C_2,$$

because $a^k = a^{\log_b n} = n^{\log_b a}$ (see Exercise 4 in Appendix 2), where $C_1 = f(1) + c/(a-1)$ and $C_2 = -c/(a-1)$.

Now suppose that n is not a power of b. Then $b^k < n < b^{k+1}$, where k is a nonnegative integer. Because f is increasing,

$$f(n) \le f(b^{k+1}) = C_1 a^{k+1} + C_2$$

$$\le (C_1 a)a^{\log_b n} + C_2$$

$$= (C_1 a)n^{\log_b a} + C_2,$$

because $k \le \log_b n < k + 1$.

Hence, we have $f(n)$ is $O(n^{\log_b a})$.   ◁

Examples 6–9 illustrate how Theorem 1 is used.

EXAMPLE 6   Let $f(n) = 5f(n/2) + 3$ and $f(1) = 7$. Find $f(2^k)$, where k is a positive integer. Also, estimate $f(n)$ if f is an increasing function.

Solution: From the proof of Theorem 1, with $a = 5$, $b = 2$, and $c = 3$, we see that if $n = 2^k$, then

$$f(n) = a^k[f(1) + c/(a-1)] + [-c/(a-1)]$$

$$= 5^k[7 + (3/4)] - 3/4$$

$$= 5^k(31/4) - 3/4.$$

Also, if $f(n)$ is increasing, Theorem 1 shows that $f(n)$ is $O(n^{\log_b a}) = O(n^{\log 5})$.   ◀

We can use Theorem 1 to estimate the computational complexity of the binary search algorithm and the algorithm given in Example 2 for locating the minimum and maximum of a sequence.

EXAMPLE 7   Give a big-O estimate for the number of comparisons used by a binary search.

Solution: In Example 1 it was shown that $f(n) = f(n/2) + 2$ when n is even, where f is the number of comparisons required to perform a binary search on a sequence of size n. Hence, from Theorem 1, it follows that $f(n)$ is $O(\log n)$.   ◀

EXAMPLE 8   Give a big-O estimate for the number of comparisons used to locate the maximum and minimum elements in a sequence using the algorithm given in Example 2.

Solution: In Example 2 we showed that $f(n) = 2f(n/2) + 2$, when n is even, where f is the number of comparisons needed by this algorithm. Hence, from Theorem 1, it follows that $f(n)$ is $O(n^{\log 2}) = O(n)$.   ◀

We now state a more general, and more complicated, theorem, which has Theorem 1 as a special case. This theorem (or more powerful versions, including big-Theta estimates) is sometimes known as the master theorem because it is useful in analyzing the complexity of many important divide-and-conquer algorithms.

THEOREM 2   **MASTER THEOREM**   Let f be an increasing function that satisfies the recurrence relation

$$f(n) = af(n/b) + cn^d$$

whenever $n = b^k$, where k is a positive integer, $a \geq 1$, b is an integer greater than 1, and c and d are real numbers with c positive and d nonnegative. Then

$$f(n) \text{ is } \begin{cases} O(n^d) & \text{if } a < b^d, \\ O(n^d \log n) & \text{if } a = b^d, \\ O(n^{\log_b a}) & \text{if } a > b^d. \end{cases}$$

The proof of Theorem 2 is left for the reader as Exercises 23–27.

EXAMPLE 9   **Complexity of Merge Sort**   In Example 3 we explained that the number of comparisons used by the merge sort to sort a list of n elements is less than $M(n)$, where $M(n) = 2M(n/2) + n$. By the master theorem (Theorem 2) we find that $M(n)$ is $O(n \log n)$, which agrees with the estimate found in Section 5.4.   ◄

EXAMPLE 10   Give a big-O estimate for the number of bit operations needed to multiply two n-bit integers using the fast multiplication algorithm described in Example 4.

Solution: Example 4 shows that $f(n) = 3f(n/2) + Cn$, when n is even, where $f(n)$ is the number of bit operations required to multiply two n-bit integers using the fast multiplication algorithm. Hence, from the master theorem (Theorem 2), it follows that $f(n)$ is $O(n^{\log 3})$. Note that $\log 3 \sim 1.6$. Because the conventional algorithm for multiplication uses $O(n^2)$ bit operations, the fast multiplication algorithm is a substantial improvement over the conventional algorithm in terms of time complexity for sufficiently large integers, including large integers that occur in practical applications.   ◄

EXAMPLE 11   Give a big-O estimate for the number of multiplications and additions required to multiply two $n \times n$ matrices using the matrix multiplication algorithm referred to in Example 5.

Solution: Let $f(n)$ denote the number of additions and multiplications used by the algorithm mentioned in Example 5 to multiply two $n \times n$ matrices. We have $f(n) = 7f(n/2) + 15n^2/4$, when n is even. Hence, from the master theorem (Theorem 2), it follows that $f(n)$ is $O(n^{\log 7})$. Note that $\log 7 \sim 2.8$. Because the conventional algorithm for multiplying two $n \times n$ matrices uses $O(n^3)$ additions and multiplications, it follows that for sufficiently large integers n, including those that occur in many practical applications, this algorithm is substantially more efficient in time complexity than the conventional algorithm.   ◄

THE CLOSEST-PAIR PROBLEM   We conclude this section by introducing a divide-and-conquer algorithm from computational geometry, the part of discrete mathematics devoted to algorithms that solve geometric problems.

EXAMPLE 12   **The Closest-Pair Problem**   Consider the problem of determining the closest pair of points in a set of n points $(x_1, y_1), \ldots, (x_n, y_n)$ in the plane, where the distance between two points (x_i, y_i) and (x_j, y_j) is the usual Euclidean distance $\sqrt{(x_i - x_j)^2 + (y_i - y_j)^2}$. This problem arises in many applications such as determining the closest pair of airplanes in the air space at a

Links

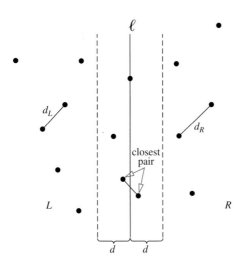

In this illustration the problem of finding the closest pair in a set of 16 points is reduced to two problems of finding the closest pair in a set of eight points *and* the problem of determining whether there are points closer than $d = \min(d_L, d_R)$ within the strip of width $2d$ centered at ℓ.

FIGURE 1   **The Recursive Step of the Algorithm for Solving the Closest-Pair Problem.**

particular altitude being managed by an air traffic controller. How can this closest pair of points be found in an efficient way?

Solution: To solve this problem we can first determine the distance between every pair of points and then find the smallest of these distances. However, this approach requires $O(n^2)$ computations of distances and comparisons because there are $C(n, 2) = n(n - 1)/2$ pairs of points. Surprisingly, there is an elegant divide-and-conquer algorithm that can solve the closest-pair problem for n points using $O(n \log n)$ computations of distances and comparisons. The algorithm we describe here is due to Michael Samos (see [PrSa85]).

It took researchers more than 10 year to find an algorithm with $O(n \log n)$ complexity that locates the closest pair of points among n points.

For simplicity, we assume that $n = 2^k$, where k is a positive integer. (We avoid some technical considerations that are needed when n is not a power of 2.) When $n = 2$, we have only one pair of points; the distance between these two points is the minimum distance. At the start of the algorithm we use the merge sort twice, once to sort the points in order of increasing x coordinates, and once to sort the points in order of increasing y coordinates. Each of these sorts requires $O(n \log n)$ operations. We will use these sorted lists in each recursive step.

The recursive part of the algorithm divides the problem into two subproblems, each involving half as many points. Using the sorted list of the points by their x coordinates, we construct a vertical line ℓ dividing the n points into two parts, a left part and a right part of equal size, each containing $n/2$ points, as shown in Figure 1. (If any points fall on the dividing line ℓ, we divide them among the two parts if necessary.) At subsequent steps of the recursion we need not sort on x coordinates again, because we can select the corresponding sorted subset of all the points. This selection is a task that can be done with $O(n)$ comparisons.

There are three possibilities concerning the positions of the closest points: (1) they are both in the left region L, (2) they are both in the right region R, or (3) one point is in the left region and the other is in the right region. Apply the algorithm recursively to compute d_L and d_R, where d_L is the minimum distance between points in the left region and d_R is the minimum distance between points in the right region. Let $d = \min(d_L, d_R)$. To successfully divide the problem of finding the closest two points in the original set into the two problems of finding the shortest distances between points in the two regions separately, we have to handle the conquer part of the algorithm, which requires that we consider the case where the closest points lie in different regions, that is, one point is in L and the other in R. Because there is a pair of points at distance d where both points lie in R or both points lie in L, for the closest points to lie in different regions requires that they must be a distance less than d apart.

For a point in the left region and a point in the right region to lie at a distance less than d apart, these points must lie in the vertical strip of width $2d$ that has the line ℓ as its center. (Otherwise,

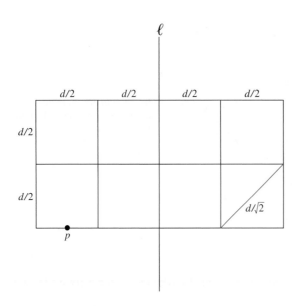

At most eight points, including p, can lie in or on the $2d \times d$ rectangle centered at ℓ because at most one point can lie in or on each of the eight $(d/2) \times (d/2)$ squares.

FIGURE 2   Showing That There Are at Most Seven Other Points to Consider for Each Point in the Strip.

the distance between these points is greater than the difference in their x coordinates, which exceeds d.) To examine the points within this strip, we sort the points so that they are listed in order of increasing y coordinates, using the sorted list of the points by their y coordinates. At each recursive step, we form a subset of the points in the region sorted by their y coordinates from the already sorted set of all points sorted by their y coordinates, which can be done with $O(n)$ comparisons.

Beginning with a point in the strip with the smallest y coordinate, we successively examine each point in the strip, computing the distance between this point and all other points in the strip that have larger y coordinates that could lie at a distance less than d from this point. Note that to examine a point p, we need only consider the distances between p and points in the set that lie within the rectangle of height d and width $2d$ with p on its base and with vertical sides at distance d from ℓ.

We can show that there are at most eight points from the set, including p, in or on this $2d \times d$ rectangle. To see this, note that there can be at most one point in each of the eight $d/2 \times d/2$ squares shown in Figure 2. This follows because the farthest apart points can be on or within one of these squares is the diagonal length $d/\sqrt{2}$ (which can be found using the Pythagorean theorem), which is less than d, and each of these $d/2 \times d/2$ squares lies entirely within the left region or the right region. This means that at this stage we need only compare at most seven distances, the distances between p and the seven or fewer other points in or on the rectangle, with d.

Because the total number of points in the strip of width $2d$ does not exceed n (the total number of points in the set), at most $7n$ distances need to be compared with d to find the minimum distance between points. That is, there are only $7n$ possible distances that could be less than d. Consequently, once the merge sort has been used to sort the pairs according to their x coordinates and according to their y coordinates, we find that the increasing function $f(n)$ satisfying the recurrence relation

$$f(n) = 2f(n/2) + 7n,$$

where $f(2) = 1$, exceeds the number of comparisons needed to solve the closest-pair problem for n points. By the master theorem (Theorem 2), it follows that $f(n)$ is $O(n \log n)$. The two sorts of points by their x coordinates and by their y coordinates each can be done using $O(n \log n)$

comparisons, by using the merge sort, and the sorted subsets of these coordinates at each of the $O(\log n)$ steps of the algorithm can be done using $O(n)$ comparisons each. Thus, we find that the closest-pair problem can be solved using $O(n \log n)$ comparisons. ◀

Exercises

1. How many comparisons are needed for a binary search in a set of 64 elements?

2. How many comparisons are needed to locate the maximum and minimum elements in a sequence with 128 elements using the algorithm in Example 2?

3. Multiply $(1110)_2$ and $(1010)_2$ using the fast multiplication algorithm.

4. Suppose that $f(n) = 2f(n/2) + 3$ when n is an even positive integer, and $f(1) = 5$. Find
   a) $f(2)$.   **b)** $f(8)$.   **c)** $f(64)$.   **d)** $f(1024)$.

5. Suppose that $f(n) = f(n/5) + 3n^2$ when n is a positive integer divisible by 5, and $f(1) = 4$. Find
   a) $f(5)$.   **b)** $f(125)$.   **c)** $f(3125)$.

6. Find $f(n)$ when $n = 2^k$, where f satisfies the recurrence relation $f(n) = f(n/2) + 1$ with $f(1) = 1$.

7. Give a big-O estimate for the function f in Exercise 6 if f is an increasing function.

8. Find $f(n)$ when $n = 3^k$, where f satisfies the recurrence relation $f(n) = 2f(n/3) + 4$ with $f(1) = 1$.

9. Give a big-O estimate for the function f in Exercise 8 if f is an increasing function.

10. Suppose that there are $n = 2^k$ teams in an elimination tournament, where there are $n/2$ games in the first round, with the $n/2 = 2^{k-1}$ winners playing in the second round, and so on. Develop a recurrence relation for the number of rounds in the tournament.

11. How many rounds are in the elimination tournament described in Exercise 10 when there are 32 teams?

12. Solve the recurrence relation for the number of rounds in the tournament described in Exercise 10.

13. Suppose that the votes of n people for different candidates (where there can be more than two candidates) for a particular office are the elements of a sequence. A person wins the election if this person receives a majority of the votes.
    a) Devise a divide-and-conquer algorithm that determines whether a candidate received a majority and, if so, determine who this candidate is. [*Hint:* Assume that n is even and split the sequence of votes into two sequences, each with $n/2$ elements. Note that a candidate could not have received a majority of votes without receiving a majority of votes in at least one of the two halves.]
    b) Use the master theorem to give a big-O estimate for the number of comparisons needed by the algorithm you devised in part (a).

14. Suppose that each person in a group of n people votes for exactly two people from a slate of candidates to fill two positions on a committee. The top two finishers both win positions as long as each receives more than $n/2$ votes.
    a) Devise a divide-and-conquer algorithm that determines whether the two candidates who received the most votes each received at least $n/2$ votes and, if so, determine who these two candidates are.
    b) Use the master theorem to give a big-O estimate for the number of comparisons needed by the algorithm you devised in part (a).

15. **a)** Set up a divide-and-conquer recurrence relation for the number of multiplications required to compute x^n, where x is a real number and n is a positive integer.
    b) Use the recurrence relation you found in part (a) to construct a big-O estimate for the number of multiplications used to compute x^n using the recursive algorithm.

16. **a)** Set up a divide-and-conquer recurrence relation for the number of modular multiplications required to compute $a^n \bmod m$, where a, m, and n are positive integers.
    b) Use the recurrence relation you found in part (a) to construct a big-O estimate for the number of modular multiplications used to compute $a^n \bmod m$ using the recursive algorithm.

17. Suppose that the function f satisfies the recurrence relation $f(n) = 2f(\sqrt{n}) + 1$ whenever n is a perfect square greater than 1 and $f(2) = 1$.
    a) Find $f(16)$.
    b) Give a big-O estimate for $f(n)$. [*Hint:* Make the substitution $m = \log n$.]

18. Suppose that the function f satisfies the recurrence relation $f(n) = 2f(\sqrt{n}) + \log n$ whenever n is a perfect square greater than 1 and $f(2) = 1$.
    a) Find $f(16)$.
    b) Find a big-O estimate for $f(n)$. [*Hint:* Make the substitution $m = \log n$.]

**19. This exercise deals with the problem of finding the largest sum of consecutive terms of a sequence of n real numbers. When all terms are positive, the sum of all terms provides the answer, but the situation is more complicated when some terms are negative. For example, the maximum sum of consecutive terms of the sequence $-2, 3, -1, 6, -7, 4$ is $3 + (-1) + 6 = 8$. (This exercise is based on [Be86].) Recall that in Exercise 42 in Section 8.1 we developed a dynamic programming algorithm for solving this problem. Here, we first look at the brute-force algorithm for solving this problem; then we develop a divide-and-conquer algorithm for solving it.

a) Use pseudocode to describe an algorithm that solves this problem by finding the sums of consecutive terms starting with the first term, the sums of consecutive terms starting with the second term, and so on, keeping track of the maximum sum found so far as the algorithm proceeds.

b) Determine the computational complexity of the algorithm in part (a) in terms of the number of sums computed and the number of comparisons made.

c) Devise a divide-and-conquer algorithm to solve this problem. [*Hint:* Assume that there are an even number of terms in the sequence and split the sequence into two halves. Explain how to handle the case when the maximum sum of consecutive terms includes terms in both halves.]

d) Use the algorithm from part (c) to find the maximum sum of consecutive terms of each of the sequences: $-2, 4, -1, 3, 5, -6, 1, 2$; $4, 1, -3, 7, -1, -5,\ \ 3, -2$; and $-1, 6, 3, -4, -5, 8, -1, 7$.

e) Find a recurrence relation for the number of sums and comparisons used by the divide-and-conquer algorithm from part (c).

f) Use the master theorem to estimate the computational complexity of the divide-and-conquer algorithm. How does it compare in terms of computational complexity with the algorithm from part (a)?

20. Apply the algorithm described in Example 12 for finding the closest pair of points, using the Euclidean distance between points, to find the closest pair of the points $(1, 3)$, $(1, 7)$, $(2, 4)$, $(2, 9)$, $(3, 1)$, $(3, 5)$, $(4, 3)$, and $(4, 7)$.

21. Construct a variation of the algorithm described in Example 12 along with justifications of the steps used by the algorithm to find the smallest distance between two points if the distance between two points is defined to be $d((x_i, y_i), (x_j, y_j)) = \max(|x_i - x_j|, |y_i - y_j|)$.

 *22. Suppose someone picks a number x from a set of n numbers. A second person tries to guess the number by successively selecting subsets of the n numbers and asking the first person whether x is in each set. The first person answers either "yes" or "no." When the first person answers each query truthfully, we can find x using $\log n$ queries by successively splitting the sets used in each query in half. Ulam's problem, proposed by

Stanislaw Ulam in 1976, asks for the number of queries required to find x, supposing that the first person is allowed to lie exactly once.

a) Show that by asking each question twice, given a number x and a set with n elements, and asking one more question when we find the lie, Ulam's problem can be solved using $2 \log n + 1$ queries.

b) Show that by dividing the initial set of n elements into four parts, each with $n/4$ elements, 1/4 of the elements can be eliminated using two queries. [*Hint:* Use two queries, where each of the queries asks whether the element is in the union of two of the subsets with $n/4$ elements and where one of the subsets of $n/4$ elements is used in both queries.]

c) Show from part (b) that if $f(n)$ equals the number of queries used to solve Ulam's problem using the method from part (b) and n is divisible by 4, then $f(n) = f(3n/4) + 2$.

d) Solve the recurrence relation in part (c) for $f(n)$.

e) Is the naive way to solve Ulam's problem by asking each question twice or the divide-and-conquer method based on part (b) more efficient? The most efficient way to solve Ulam's problem has been determined by A. Pelc [Pe87].

In Exercises 23–27, assume that f is an increasing function satisfying the recurrence relation $f(n) = af(n/b) + cn^d$, where $a \geq 1$, b is an integer greater than 1, and c and d are positive real numbers. These exercises supply a proof of Theorem 2.

*23. Show that if $a = b^d$ and n is a power of b, then $f(n) = f(1)n^d + cn^d \log_b n$.

24. Use Exercise 23 to show that if $a = b^d$, then $f(n)$ is $O(n^d \log n)$.

*25. Show that if $a \neq b^d$ and n is a power of b, then $f(n) = C_1 n^d + C_2 n^{\log_b a}$, where $C_1 = b^d c/(b^d - a)$ and $C_2 = f(1) + b^d c/(a - b^d)$.

26. Use Exercise 25 to show that if $a < b^d$, then $f(n)$ is $O(n^d)$.

27. Use Exercise 25 to show that if $a > b^d$, then $f(n)$ is $O(n^{\log_b a})$.

28. Find $f(n)$ when $n = 2^k$, where f satisfies the recurrence relation $f(n) = 8f(n/2) + n^2$ with $f(1) = 1$.

29. Give a big-O estimate for the function f in Exercise 28 if f is an increasing function.

8.4   Generating Functions

Introduction

Generating functions are used to represent sequences efficiently by coding the terms of a sequence as coefficients of powers of a variable x in a formal power series. Generating functions can be used to solve many types of counting problems, such as the number of ways to select or distribute objects of different kinds, subject to a variety of constraints, and the number of ways to make change for a dollar using coins of different denominations. Generating functions can be used to solve recurrence relations by translating a recurrence relation for the terms of a sequence into an equation involving a generating function. This equation can then be solved to find a closed form for the generating function. From this closed form, the coefficients of the power series for the generating function can be found, solving the original recurrence relation. Generating functions can also be used to prove combinatorial identities by taking advantage of relatively simple relationships between functions that can be translated into identities involving the terms of sequences. Generating functions are a helpful tool for studying many properties of sequences besides those described in this section, such as their use for establishing asymptotic formulae for the terms of a sequence.

We begin with the definition of the generating function for a sequence.

DEFINITION 1

The *generating function for the sequence* $a_0, a_1, \ldots, a_k, \ldots$ of real numbers is the infinite series

$$G(x) = a_0 + a_1 x + \cdots + a_k x^k + \cdots = \sum_{k=0}^{\infty} a_k x^k.$$

Remark: The generating function for $\{a_k\}$ given in Definition 1 is sometimes called the **ordinary generating function** of $\{a_k\}$ to distinguish it from other types of generating functions for this sequence.

EXAMPLE 1

The generating functions for the sequences $\{a_k\}$ with $a_k = 3, a_k = k + 1$, and $a_k = 2^k$ are $\sum_{k=0}^{\infty} 3x^k$, $\sum_{k=0}^{\infty} (k + 1)x^k$, and $\sum_{k=0}^{\infty} 2^k x^k$, respectively. ◀

We can define generating functions for finite sequences of real numbers by extending a finite sequence $a_0, a_1, \ldots, a_n$ into an infinite sequence by setting $a_{n+1} = 0, a_{n+2} = 0$, and so on. The generating function $G(x)$ of this infinite sequence $\{a_n\}$ is a polynomial of degree n because no terms of the form $a_j x^j$ with $j > n$ occur, that is,

$$G(x) = a_0 + a_1 x + \cdots + a_n x^n.$$

EXAMPLE 2

What is the generating function for the sequence 1, 1, 1, 1, 1, 1?

Solution: The generating function of 1, 1, 1, 1, 1, 1 is

$$1 + x + x^2 + x^3 + x^4 + x^5.$$

By Theorem 1 of Section 2.4 we have

$$(x^6 - 1)/(x - 1) = 1 + x + x^2 + x^3 + x^4 + x^5$$

when $x \neq 1$. Consequently, $G(x) = (x^6 - 1)/(x - 1)$ is the generating function of the sequence 1, 1, 1, 1, 1, 1. [Because the powers of x are only place holders for the terms of the sequence in a generating function, we do not need to worry that $G(1)$ is undefined.] ◀

EXAMPLE 3

Let m be a positive integer. Let $a_k = C(m, k)$, for $k = 0, 1, 2, \ldots, m$. What is the generating function for the sequence $a_0, a_1, \ldots, a_m$?

Solution: The generating function for this sequence is

$$G(x) = C(m, 0) + C(m, 1)x + C(m, 2)x^2 + \cdots + C(m, m)x^m.$$

The binomial theorem shows that $G(x) = (1 + x)^m$. ◀

Useful Facts About Power Series

When generating functions are used to solve counting problems, they are usually considered to be **formal power series**. Questions about the convergence of these series are ignored. However, to apply some results from calculus, it is sometimes important to consider for which x the power series converges. The fact that a function has a unique power series around $x = 0$ will also be important. Generally, however, we will not be concerned with questions of convergence or the uniqueness of power series in our discussions. Readers familiar with calculus can consult textbooks on this subject for details about power series, including the convergence of the series we consider here.

We will now state some important facts about infinite series used when working with generating functions. A discussion of these and related results can be found in calculus texts.

EXAMPLE 4   The function $f(x) = 1/(1 - x)$ is the generating function of the sequence $1, 1, 1, 1, \ldots$, because

$$1/(1 - x) = 1 + x + x^2 + \cdots$$

for $|x| < 1$. ◀

EXAMPLE 5   The function $f(x) = 1/(1 - ax)$ is the generating function of the sequence $1, a, a^2, a^3, \ldots$, because

$$1/(1 - ax) = 1 + ax + a^2 x^2 + \cdots$$

when $|ax| < 1$, or equivalently, for $|x| < 1/|a|$ for $a \neq 0$. ◀

We also will need some results on how to add and how to multiply two generating functions. Proofs of these results can be found in calculus texts.

THEOREM 1   Let $f(x) = \sum_{k=0}^{\infty} a_k x^k$ and $g(x) = \sum_{k=0}^{\infty} b_k x^k$. Then

$$f(x) + g(x) = \sum_{k=0}^{\infty} (a_k + b_k) x^k \quad \text{and} \quad f(x)g(x) = \sum_{k=0}^{\infty} \left(\sum_{j=0}^{k} a_j b_{k-j} \right) x^k.$$

Remark: Theorem 1 is valid only for power series that converge in an interval, as all series considered in this section do. However, the theory of generating functions is not limited to such series. In the case of series that do not converge, the statements in Theorem 1 can be taken as definitions of addition and multiplication of generating functions.

We will illustrate how Theorem 1 can be used with Example 6.

EXAMPLE 6   Let $f(x) = 1/(1 - x)^2$. Use Example 4 to find the coefficients $a_0, a_1, a_2, \ldots$ in the expansion $f(x) = \sum_{k=0}^{\infty} a_k x^k$.

Solution: From Example 4 we see that

$$1/(1 - x) = 1 + x + x^2 + x^3 + \cdots.$$

Hence, from Theorem 1, we have

$$1/(1 - x)^2 = \sum_{k=0}^{\infty} \left(\sum_{j=0}^{k} 1 \right) x^k = \sum_{k=0}^{\infty} (k + 1) x^k.$$

◀

Remark: This result also can be derived from Example 4 by differentiation. Taking derivatives is a useful technique for producing new identities from existing identities for generating functions.

To use generating functions to solve many important counting problems, we will need to apply the binomial theorem for exponents that are not positive integers. Before we state an extended version of the binomial theorem, we need to define extended binomial coefficients.

DEFINITION 2

Let u be a real number and k a nonnegative integer. Then the *extended binomial coefficient* $\binom{u}{k}$ is defined by

$$\binom{u}{k} = \begin{cases} u(u-1)\cdots(u-k+1)/k! & \text{if } k > 0, \\ 1 & \text{if } k = 0. \end{cases}$$

EXAMPLE 7

Find the values of the extended binomial coefficients $\binom{-2}{3}$ and $\binom{1/2}{3}$.

Solution: Taking $u = -2$ and $k = 3$ in Definition 2 gives us

$$\binom{-2}{3} = \frac{(-2)(-3)(-4)}{3!} = -4.$$

Similarly, taking $u = 1/2$ and $k = 3$ gives us

$$\binom{1/2}{3} = \frac{(1/2)(1/2 - 1)(1/2 - 2)}{3!} = (1/2)(-1/2)(-3/2)/6 = 1/16. \quad \blacktriangleleft$$

Example 8 provides a useful formula for extended binomial coefficients when the top parameter is a negative integer. It will be useful in our subsequent discussions.

EXAMPLE 8

When the top parameter is a negative integer, the extended binomial coefficient can be expressed in terms of an ordinary binomial coefficient. To see that this is the case, note that

$$\binom{-n}{r} = \frac{(-n)(-n-1)\cdots(-n-r+1)}{r!} \qquad \text{by definition of extended binomial coefficient}$$

$$= \frac{(-1)^r n(n+1)\cdots(n+r-1)}{r!} \qquad \text{factoring out –1 from each term in the numerator}$$

$$= \frac{(-1)^r (n+r-1)(n+r-2)\cdots n}{r!} \qquad \text{by the commutative law for multiplication}$$

$$= \frac{(-1)^r (n+r-1)!}{r!(n-1)!} \qquad \text{multiplying both the numerator and denominator by } (n-1)!$$

$$= (-1)^r \binom{n+r-1}{r} \qquad \text{by the definition of binomial coefficients}$$

$$= (-1)^r C(n+r-1, r). \qquad \text{using alternative notation for binomial coefficients} \quad \blacktriangleleft$$

We now state the extended binomial theorem.

THEOREM 2

THE EXTENDED BINOMIAL THEOREM Let x be a real number with $|x| < 1$ and let u be a real number. Then

$$(1+x)^u = \sum_{k=0}^{\infty} \binom{u}{k} x^k.$$

Theorem 2 can be proved using the theory of Maclaurin series. We leave its proof to the reader with a familiarity with this part of calculus.

Remark: When u is a positive integer, the extended binomial theorem reduces to the binomial theorem presented in Section 6.4, because in that case $\binom{u}{k} = 0$ if $k > u$.

Example 9 illustrates the use of Theorem 2 when the exponent is a negative integer.

EXAMPLE 9   Find the generating functions for $(1 + x)^{-n}$ and $(1 - x)^{-n}$, where n is a positive integer, using the extended binomial theorem.

Solution: By the extended binomial theorem, it follows that

$$(1 + x)^{-n} = \sum_{k=0}^{\infty} \binom{-n}{k} x^k.$$

Using Example 8, which provides a simple formula for $\binom{-n}{k}$, we obtain

$$(1 + x)^{-n} = \sum_{k=0}^{\infty} (-1)^k C(n + k - 1, k) x^k.$$

Replacing x by $-x$, we find that

$$(1 - x)^{-n} = \sum_{k=0}^{\infty} C(n + k - 1, k) x^k. \qquad \blacktriangleleft$$

Table 1 presents a useful summary of some generating functions that arise frequently.

Remark: Note that the second and third formulae in this table can be deduced from the first formula by substituting ax and x^r for x, respectively. Similarly, the sixth and seventh formulae can be deduced from the fifth formula using the same substitutions. The tenth and eleventh can be deduced from the ninth formula by substituting $-x$ and ax for x, respectively. Also, some of the formulae in this table can be derived from other formulae using methods from calculus (such as differentiation and integration). Students are encouraged to know the core formulae in this table (that is, formulae from which the others can be derived, perhaps the first, fourth, fifth, eighth, ninth, twelfth, and thirteenth formulae) and understand how to derive the other formulae from these core formulae.

Counting Problems and Generating Functions

Generating functions can be used to solve a wide variety of counting problems. In particular, they can be used to count the number of combinations of various types. In Chapter 6 we developed techniques to count the r-combinations from a set with n elements when repetition is allowed and additional constraints may exist. Such problems are equivalent to counting the solutions to equations of the form

$$e_1 + e_2 + \cdots + e_n = C,$$

where C is a constant and each e_i is a nonnegative integer that may be subject to a specified constraint. Generating functions can also be used to solve counting problems of this type, as Examples 10–12 show.

EXAMPLE 10   Find the number of solutions of $e_1 + e_2 + e_3 = 17$, where e_1, e_2, and e_3 are nonnegative integers with $2 \le e_1 \le 5$, $3 \le e_2 \le 6$, and $4 \le e_3 \le 7$.

Solution: The number of solutions with the indicated constraints is the coefficient of x^{17} in the expansion of $(x^2 + x^3 + x^4 + x^5)(x^3 + x^4 + x^5 + x^6)(x^4 + x^5 + x^6 + x^7)$. This follows because we obtain a term equal to x^{17} in the product by picking a term in the first sum x^{e_1}, a term in the second sum x^{e_2}, and a term in the third sum x^{e_3}, where the exponents e_1, e_2, and e_3 satisfy the equation $e_1 + e_2 + e_3 = 17$ and the given constraints.

It is not hard to see that the coefficient of x^{17} in this product is 3. Hence, there are three solutions. (Note that the calculating of this coefficient involves about as much work as enumerating all the solutions of the equation with the given constraints. However, the method that this illustrates often can be used to solve wide classes of counting problems with special formulae, as we will see. Furthermore, a computer algebra system can be used to do such computations.) ◀

EXAMPLE 11   In how many different ways can eight identical cookies be distributed among three distinct children if each child receives at least two cookies and no more than four cookies?

Solution: Because each child receives at least two but no more than four cookies, for each child there is a factor equal to $(x^2 + x^3 + x^4)$ in the generating function for the sequence $\{c_n\}$, where c_n is the number of ways to distribute n cookies. Because there are three children, this generating function is $(x^2 + x^3 + x^4)^3$. We need the coefficient of x^8 in this product. The reason is that the x^8 terms in the expansion correspond to the ways that three terms can be selected, with one from each factor, that have exponents adding up to 8. Furthermore, the exponents of the term from the first, second, and third factors are the numbers of cookies the first, second, and third children receive, respectively. Computation shows that this coefficient equals 6. Hence, there are six ways to distribute the cookies so that each child receives at least two, but no more than four, cookies. ◀

EXAMPLE 12   Use generating functions to determine the number of ways to insert tokens worth $1, $2, and $5 into a vending machine to pay for an item that costs r dollars in both the cases when the order in which the tokens are inserted does not matter and when the order does matter. (For example, there are two ways to pay for an item that costs $3 when the order in which the tokens are inserted does not matter: inserting three $1 tokens or one $1 token and a $2 token. When the order matters, there are three ways: inserting three $1 tokens, inserting a $1 token and then a $2 token, or inserting a $2 token and then a $1 token.)

Solution: Consider the case when the order in which the tokens are inserted does not matter. Here, all we care about is the number of each token used to produce a total of r dollars. Because we can use any number of $1 tokens, any number of $2 tokens, and any number of $5 tokens, the answer is the coefficient of x^r in the generating function

$$(1 + x + x^2 + x^3 + \cdots)(1 + x^2 + x^4 + x^6 + \cdots)(1 + x^5 + x^{10} + x^{15} + \cdots).$$

(The first factor in this product represents the $1 tokens used, the second the $2 tokens used, and the third the $5 tokens used.) For example, the number of ways to pay for an item costing $7 using $1, $2, and $5 tokens is given by the coefficient of x^7 in this expansion, which equals 6.

When the order in which the tokens are inserted matters, the number of ways to insert exactly n tokens to produce a total of r dollars is the coefficient of x^r in $(x + x^2 + x^5)^n$, because each of the r tokens may be a $1 token, a $2 token, or a $5 token. Because any number of tokens may be inserted, the number of ways to produce r dollars using $1, $2, or $5 tokens, when the order in which the tokens are inserted matters, is the coefficient of x^r in

$$1 + (x + x^2 + x^5) + (x + x^2 + x^5)^2 + \cdots = \frac{\cancel{1}}{1 - (x + x^2 + x^5)} = \frac{1}{1 - x - x^2 - x^5},$$

where we have added the number of ways to insert 0 tokens, 1 token, 2 tokens, 3 tokens, and so on, and where we have used the identity $1/(1 - x) = 1 + x + x^2 + \cdots$ with x replaced

TABLE 1 Useful Generating Functions.	
$G(x)$	a_k
$\displaystyle (1+x)^n = \sum_{k=0}^{n} C(n,k)x^k$ $= 1 + C(n,1)x + C(n,2)x^2 + \cdots + x^n$	$C(n,k)$
$\displaystyle (1+ax)^n = \sum_{k=0}^{n} C(n,k)a^k x^k$ $= 1 + C(n,1)ax + C(n,2)a^2 x^2 + \cdots + a^n x^n$	$C(n,k)a^k$
$\displaystyle (1+x^r)^n = \sum_{k=0}^{n} C(n,k)x^{rk}$ $= 1 + C(n,1)x^r + C(n,2)x^{2r} + \cdots + x^{rn}$	$C(n,k/r)$ if $r \mid k$; 0 otherwise
$\displaystyle \frac{1-x^{n+1}}{1-x} = \sum_{k=0}^{n} x^k = 1 + x + x^2 + \cdots + x^n$	1 if $k \le n$; 0 otherwise
$\displaystyle \frac{1}{1-x} = \sum_{k=0}^{\infty} x^k = 1 + x + x^2 + \cdots$	1
$\displaystyle \frac{1}{1-ax} = \sum_{k=0}^{\infty} a^k x^k = 1 + ax + a^2 x^2 + \cdots$	a^k
$\displaystyle \frac{1}{1-x^r} = \sum_{k=0}^{\infty} x^{rk} = 1 + x^r + x^{2r} + \cdots$	1 if $r \mid k$; 0 otherwise
$\displaystyle \frac{1}{(1-x)^2} = \sum_{k=0}^{\infty} (k+1)x^k = 1 + 2x + 3x^2 + \cdots$	$k+1$
$\displaystyle \frac{1}{(1-x)^n} = \sum_{k=0}^{\infty} C(n+k-1,k)x^k$ $= 1 + C(n,1)x + C(n+1,2)x^2 + \cdots$	$C(n+k-1,k) = C(n+k-1,n-1)$
$\displaystyle \frac{1}{(1+x)^n} = \sum_{k=0}^{\infty} C(n+k-1,k)(-1)^k x^k$ $= 1 - C(n,1)x + C(n+1,2)x^2 - \cdots$	$(-1)^k C(n+k-1,k) = (-1)^k C(n+k-1,n-1)$
$\displaystyle \frac{1}{(1-ax)^n} = \sum_{k=0}^{\infty} C(n+k-1,k)a^k x^k$ $= 1 + C(n,1)ax + C(n+1,2)a^2 x^2 + \cdots$	$C(n+k-1,k)a^k = C(n+k-1,n-1)a^k$
$\displaystyle e^x = \sum_{k=0}^{\infty} \frac{x^k}{k!} = 1 + x + \frac{x^2}{2!} + \frac{x^3}{3!} + \cdots$	$1/k!$
$\displaystyle \ln(1+x) = \sum_{k=1}^{\infty} \frac{(-1)^{k+1}}{k}x^k = x - \frac{x^2}{2} + \frac{x^3}{3} - \frac{x^4}{4} + \cdots$	$(-1)^{k+1}/k$

Note: The series for the last two generating functions can be found in most calculus books when power series are discussed.

with $x + x^2 + x^5$. For example, the number of ways to pay for an item costing $7 using $1, $2, and $5 tokens, when the order in which the tokens are used matters, is the coefficient of x^7 in this expansion, which equals 26. [*Hint:* To see that this coefficient equals 26 requires the addition of the coefficients of x^7 in the expansions $(x + x^2 + x^5)^k$ for $2 \le k \le 7$. This can be done by hand with considerable computation, or a computer algebra system can be used.] ◀

Example 13 shows the versatility of generating functions when used to solve problems with differing assumptions.

EXAMPLE 13   Use generating functions to find the number of k-combinations of a set with n elements. Assume that the binomial theorem has already been established.

Solution: Each of the n elements in the set contributes the term $(1 + x)$ to the generating function $f(x) = \sum_{k=0}^{n} a_k x^k$. Here $f(x)$ is the generating function for $\{a_k\}$, where a_k represents the number of k-combinations of a set with n elements. Hence, $f(x) = (1 + x)^n$. But by the binomial theorem, we have

$$f(x) = \sum_{k=0}^{n} \binom{n}{k} x^k, \quad \text{where} \quad \binom{n}{k} = \frac{n!}{k!(n-k)!}.$$

Hence, $C(n, k)$, the number of k-combinations of a set with n elements, is $\dfrac{n!}{k!(n-k)!}$. ◀

Remark: We proved the binomial theorem in Section 6.4 using the formula for the number of r-combinations of a set with n elements. This example shows that the binomial theorem, which can be proved by mathematical induction, can be used to derive the formula for the number of r-combinations of a set with n elements.

EXAMPLE 14   Use generating functions to find the number of r-combinations from a set with n elements when repetition of elements is allowed.

Solution: Let $G(x)$ be the generating function for the sequence $\{a_r\}$, where a_r equals the number of r-combinations of a set with n elements with repetitions allowed. That is, $G(x) = \sum_{r=0}^{\infty} a_r x^r$. Because we can select any number of a particular member of the set with n elements when we form an r-combination with repetition allowed, each of the n elements contributes $(1 + x + x^2 + x^3 + \cdots)$ to a product expansion for $G(x)$. Each element contributes this factor because it may be selected zero times, one time, two times, three times, and so on, when an r-combination is formed (with a total of r elements selected). Because there are n elements in the set and each contributes this same factor to $G(x)$, we have

$$G(x) = (1 + x + x^2 + \cdots)^n.$$

As long as $|x| < 1$, we have $1 + x + x^2 + \cdots = 1/(1 - x)$, so

$$G(x) = 1/(1 - x)^n = (1 - x)^{-n}.$$

Applying the extended binomial theorem (Theorem 2), it follows that

$$(1 - x)^{-n} = (1 + (-x))^{-n} = \sum_{r=0}^{\infty} \binom{-n}{r} (-x)^r.$$

The number of r-combinations of a set with n elements with repetitions allowed, when r is a positive integer, is the coefficient a_r of x^r in this sum. Consequently, using Example 8 we find that a_r equals

$$\binom{-n}{r} (-1)^r = (-1)^r C(n + r - 1, r) \cdot (-1)^r = C(n + r - 1, r).$$ ◀

Note that the result in Example 14 is the same result we stated as Theorem 2 in Section 6.5.

EXAMPLE 15   Use generating functions to find the number of ways to select r objects of n different kinds if we must select at least one object of each kind.

Solution: Because we need to select at least one object of each kind, each of the n kinds of objects contributes the factor $(x + x^2 + x^3 + \cdots)$ to the generating function $G(x)$ for the sequence $\{a_r\}$, where a_r is the number of ways to select r objects of n different kinds if we need at least one object of each kind. Hence,

$$G(x) = (x + x^2 + x^3 + \cdots)^n = x^n(1 + x + x^2 + \cdots)^n = x^n/(1 - x)^n.$$

Using the extended binomial theorem and Example 8, we have

$$G(x) = x^n/(1 - x)^n = x^n \cdot (1 - x)^{-n}$$

$$= x^n \sum_{r=0}^{\infty} \binom{-n}{r} (-x)^r = x^n \sum_{r=0}^{\infty} (-1)^r C(n + r - 1, r)(-1)^r x^r$$

$$= \sum_{r=0}^{\infty} C(n + r - 1, r)x^{n+r} = \sum_{t=n}^{\infty} C(t - 1, t - n)x^t = \sum_{r=n}^{\infty} C(r - 1, r - n)x^r.$$

We have shifted the summation in the next-to-last equality by setting $t = n + r$ so that $t = n$ when $r = 0$ and $n + r - 1 = t - 1$, and then we replaced t by r as the index of summation in the last equality to return to our original notation. Hence, there are $C(r - 1, r - n)$ ways to select r objects of n different kinds if we must select at least one object of each kind.   ◀

Exponential Generating Functions

Earlier we have seen that generating functions can be used for enumerating combinations. It is natural to see whether the idea can be extended to permutations. But the first obstacle we face is that in permutation abc is different from cba whereas selecting a, b, c from 2 set is the same in whatever order you select. Trying to use a power series for permutation of three elements a, b, c, we must get

$$1 + (a + b + c)x + (ab + ba + ac + bc + ca + cb)x^2$$
$$+ (abc + acb + bac + bca + cab + cba)x^3.$$

But this polynomial is equivalent to $1 + (a + b + c)x + 2(ab + bc + ca)x^2 + 6(abc)x^3$. We cannot distinguish between abc and bca. Since we do not want to discard commutative property in power series, the following idea is used for enumerating permutations.

A direct extension of the notion of the enumerators for combinations indicates that an enumerator for the permutation of n distinct objects would have the form

$$F(x) = P(n, 0)x^0 + P(n, 1)x + P(n, 2)x^2 + P(n, 3)x^3 + \cdots$$
$$+ P(n, r)x^r + \cdots + P(n, n)x^n$$

$$= 1 + \frac{n!}{(n-1)!}x + \frac{n!}{(n-2)!}x + \frac{n!}{(n-3)!}x + \cdots + + \frac{n!}{(n-r)!}x^r + \cdots + n!x^n$$

Unfortunately, there is no simple closed form expression for the above function. But we know

$$(1+x)^n = 1 + \binom{n}{1}x + \binom{n}{2}x^2 + \cdots + \binom{n}{n}x^n$$

$$= 1 + \frac{P(n,1)}{1!}x + \frac{P(n,2)}{2!}x^2 + \cdots + \frac{P(n,r)}{r!}x^r + \cdots + \frac{P(n,n)}{n!}x^n$$

This we call as an exponential generating function. Let $(a_0, a_1, a_2, \ldots, a_r, \ldots)$ be a sequence. The function

$$F(x) = a_0 + \frac{a_1}{1!}x + \frac{a_2}{2!}x^2 + \frac{a_3}{3!}x^3 + \cdots + \frac{a_r}{r!}x^r + \cdots$$

is called the exponential generating function of the sequence $(a_0, a_1, a_2, \ldots, a_r, \ldots)$

Thus $(1+x)^n$ is the exponential generating function of the $P(n,r)$ s., i.e. the permutations of r objects out of n objects.

Let us see how this idea can be used in solving problem related to permutations.

EXAMPLE 16   The exponential enumerator for the permutation of all p of p identical objects is $\dfrac{x^p}{p!}$ as there is only one way of doing so. Thus the exponential enumerator for the permutation of none, one, two, ..., p of p identical object is $1 + \dfrac{1}{1!}x + \dfrac{1}{2!}x^2 + \cdots + \dfrac{1}{p!}x^p$.

The exponential enumerator for the permutations of none, one, two, ..., $p + q$ of $p + q$ objects where p of them are of one kind and q of them are of another kind is

$$\left(1 + \frac{1}{1!}x + \frac{1}{2!}x^2 + \cdots + \frac{1}{p!}x^p\right)\left(1 + \frac{1}{1!}x + \frac{1}{2!}x^2 + \cdots + \frac{1}{q!}x^q\right).$$

To get the permutation of $p + q$ objects, where p of them are of one kind and q of them are of another kind, we see the factor $x^p x^q$, which is $\dfrac{x^p}{p!}\dfrac{x^q}{q!} = \dfrac{x^{p+q}}{p!q!}$. In the expression the answer we expect is given by

$$\frac{\alpha}{(p+q)!}x^{p+q}.$$

Hence we find

$$\alpha = \frac{(p+q)!}{p!q!}$$

which we know is correct.   ◀

EXAMPLE 17   Let the alphabet consist of $\{0, 1, 2\}$. Find the number of r-digit binary sequences that contain an even number of 0's.

Solution: The exponential enumerator for the permutation of digit 0 is

$$\left(1 + \frac{x^2}{2!} + \frac{x^4}{4!} + \frac{x^6}{6!} + \cdots\right) = \frac{1}{2}\left(e^x + e^{-x}\right).$$

The exponential enumerator for the permutations of each of the digits 1 and 2 is

$$\left(1 + \frac{x}{1!} + \frac{x^2}{2!} + \cdots\right) = e^x.$$

It follows that the exponential enumerator for the number of binary sequences containing an even number of 0's is

$$\frac{1}{2}\left(e^x + e^{-x}\right)e^x e^x = \frac{1}{2}\left(e^{3x} + e^+\right) = 1 + \sum_{r=1}^{\infty}\left(\frac{3^r + 1^r}{r!}\right)x^r.$$

Hence the number of r-digit ternary sequences that contain an even number of 0's is $\dfrac{3^r + 1}{2}$.

For example when $r = 1$ we have the possibilities 1 and 2 and $\dfrac{3^1 + 1}{2} = 2$. when $r = 2$ we have possibilities 00, 12, 21, 11, 22 and $\dfrac{3^2 + 1}{2} = 5$.   ◀

Using Generating Functions to Solve Recurrence Relations

We can find the solution to a recurrence relation and its initial conditions by finding an explicit formula for the associated generating function. This is illustrated in Examples 16 and 17.

EXAMPLE 18   Solve the recurrence relation $a_k = 3a_{k-1}$ for $k = 1, 2, 3, \ldots$ and initial condition $a_0 = 2$.

Solution: Let $G(x)$ be the generating function for the sequence $\{a_k\}$, that is, $G(x) = \sum_{k=0}^{\infty} a_k x^k$.

First note that

$$xG(x) = \sum_{k=0}^{\infty} a_k x^{k+1} = \sum_{k=1}^{\infty} a_{k-1}x^k.$$

Using the recurrence relation, we see that

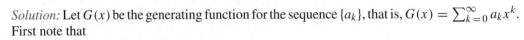
$$G(x) - 3xG(x) = \sum_{k=0}^{\infty} a_k x^k - 3\sum_{k=1}^{\infty} a_{k-1}x^k = a_0 + \sum_{k=1}^{\infty}(a_k - 3a_{k-1})x^k = 2,$$

because $a_0 = 2$ and $a_k = 3a_{k-1}$. Thus,

$$G(x) - 3xG(x) = (1 - 3x)G(x) = 2.$$

Solving for $G(x)$ shows that $G(x) = 2/(1 - 3x)$. Using the identity $1/(1 - ax) = \sum_{k=0}^{\infty} a^k x^k$, from Table 1, we have

$$G(x) = 2\sum_{k=0}^{\infty} 3^k x^k = \sum_{k=0}^{\infty} 2 \cdot 3^k x^k.$$

Consequently, $a_k = 2 \cdot 3^k$.   ◀

EXAMPLE 19   Suppose that a valid codeword is an n-digit number in decimal notation containing an even number of 0s. Let a_n denote the number of valid codewords of length n. In Example 4 of Section 8.1 we showed that the sequence $\{a_n\}$ satisfies the recurrence relation

$$a_n = 8a_{n-1} + 10^{n-1}$$

and the initial condition $a_1 = 9$. Use generating functions to find an explicit formula for a_n.

Solution: To make our work with generating functions simpler, we extend this sequence by setting $a_0 = 1$; when we assign this value to a_0 and use the recurrence relation, we have $a_1 = 8a_0 + 10^0 = 8 + 1 = 9$, which is consistent with our original initial condition. (It also makes sense because there is one code word of length 0—the empty string.)

We multiply both sides of the recurrence relation by x^n to obtain

$$a_n x^n = 8a_{n-1}x^n + 10^{n-1}x^n.$$

Let $G(x) = \sum_{n=0}^{\infty} a_n x^n$ be the generating function of the sequence $a_0, a_1, a_2, \ldots$. We sum both sides of the last equation starting with $n = 1$, to find that

$$
\begin{aligned}
G(x) - 1 = \sum_{n=1}^{\infty} a_n x^n &= \sum_{n=1}^{\infty}(8a_{n-1}x^n + 10^{n-1}x^n) \\
&= 8\sum_{n=1}^{\infty} a_{n-1}x^n + \sum_{n=1}^{\infty} 10^{n-1}x^n \\
&= 8x\sum_{n=1}^{\infty} a_{n-1}x^{n-1} + x\sum_{n=1}^{\infty} 10^{n-1}x^{n-1} \\
&= 8x\sum_{n=0}^{\infty} a_n x^n + x\sum_{n=0}^{\infty} 10^n x^n \\
&= 8xG(x) + x/(1 - 10x),
\end{aligned}
$$

where we have used Example 5 to evaluate the second summation. Therefore, we have

$$G(x) - 1 = 8xG(x) + x/(1 - 10x).$$

Solving for $G(x)$ shows that

$$G(x) = \frac{1 - 9x}{(1 - 8x)(1 - 10x)}.$$

Expanding the right-hand side of this equation into partial fractions (as is done in the integration of rational functions studied in calculus) gives

$$G(x) = \frac{1}{2}\left(\frac{1}{1 - 8x} + \frac{1}{1 - 10x}\right).$$

Using Example 5 twice (once with $a = 8$ and once with $a = 10$) gives

$$G(x) = \frac{1}{2}\left(\sum_{n=0}^{\infty} 8^n x^n + \sum_{n=0}^{\infty} 10^n x^n\right) = \sum_{n=0}^{\infty} \frac{1}{2}(8^n + 10^n)x^n.$$

Consequently, we have shown that

$$a_n = \frac{1}{2}(8^n + 10^n).$$

◀

Proving Identities via Generating Functions

In Chapter 6 we saw how combinatorial identities could be established using combinatorial proofs. Here we will show that such identities, as well as identities for extended binomial coefficients, can be proved using generating functions. Sometimes the generating function approach is simpler than other approaches, especially when it is simpler to work with the closed form of a generating function than with the terms of the sequence themselves. We illustrate how generating functions can be used to prove identities with Example 18.

EXAMPLE 20   Use generating functions to show that

$$\sum_{k=0}^{n} C(n,k)^2 = C(2n,n)$$

whenever n is a positive integer.

Solution: First note that by the binomial theorem $C(2n,n)$ is the coefficient of x^n in $(1+x)^{2n}$. However, we also have

$$(1+x)^{2n} = [(1+x)^n]^2$$
$$= [C(n,0) + C(n,1)x + C(n,2)x^2 + \cdots + C(n,n)x^n]^2.$$

The coefficient of x^n in this expression is

$$C(n,0)C(n,n) + C(n,1)C(n,n-1) + C(n,2)C(n,n-2) + \cdots + C(n,n)C(n,0).$$

This equals $\sum_{k=0}^{n} C(n,k)^2$, because $C(n,n-k) = C(n,k)$. Because both $C(2n,n)$ and $\sum_{k=0}^{n} C(n,k)^2$ represent the coefficient of x^n in $(1+x)^{2n}$, they must be equal.   ◀

Exercises 28 and 29 ask that Pascal's identity and Vandermonde's identity be proved using generating functions.

Exercises

1. Find the generating function for the finite sequence 2, 2, 2, 2, 2, 2.

2. Find the generating function for the finite sequence 1, 4, 16, 64, 256.

In Exercises 3–6, by a **closed form** we mean an algebraic expression not involving a summation over a range of values or the use of ellipses.

3. Find a closed form for the generating function for each of these sequences. (For each sequence, use the most obvious choice of a sequence that follows the pattern of the initial terms listed.)

a) 0, 2, 2, 2, 2, 2, 2, 0, 0, 0, 0, 0, ...

b) 0, 0, 0, 1, 1, 1, 1, 1, 1, ...

c) 0, 1, 0, 0, 1, 0, 0, 1, 0, 0, 1, ...

d) 2, 4, 8, 16, 32, 64, 128, 256, ...

e) $\binom{7}{0}, \binom{7}{1}, \binom{7}{2}, \ldots, \binom{7}{7}, 0, 0, 0, 0, 0, \ldots$

f) 2, −2, 2, −2, 2, −2, 2, −2, ...

g) 1, 1, 0, 1, 1, 1, 1, 1, 1, 1, ...

h) 0, 0, 0, 1, 2, 3, 4, ...

4. Find a closed form for the generating function for the sequence $\{a_n\}$, where

a) $a_n = 5$ for all $n = 0, 1, 2, \ldots$.

b) $a_n = 3^n$ for all $n = 0, 1, 2, \ldots$.

c) $a_n = 2$ for $n = 3, 4, 5, \ldots$ and $a_0 = a_1 = a_2 = 0$.

d) $a_n = 2n + 3$ for all $n = 0, 1, 2, \ldots$.

e) $a_n = \binom{8}{n}$ for all $n = 0, 1, 2, \ldots$.

f) $a_n = \binom{n+4}{n}$ for all $n = 0, 1, 2, \ldots$.

5. For each of these generating functions, provide a closed formula for the sequence it determines.

a) $(3x - 4)^3$ **b)** $(x^3 + 1)^3$

c) $1/(1 - 5x)$ **d)** $x^3/(1 + 3x)$

e) $x^2 + 3x + 7 + (1/(1 - x^2))$

f) $(x^4/(1 - x^4)) - x^3 - x^2 - x - 1$

g) $x^2/(1 - x)^2$ **h)** $2e^{2x}$

6. For each of these generating functions, provide a closed formula for the sequence it determines.

a) $(x^2 + 1)^3$ **b)** $(3x - 1)^3$

c) $1/(1 - 2x^2)$ **d)** $x^2/(1 - x)^3$

e) $x - 1 + (1/(1 - 3x))$ **f)** $(1 + x^3)/(1 + x)^3$

*g) $x/(1 + x + x^2)$ **h)** $e^{3x^2} - 1$

7. Find the coefficient of x^{10} in the power series of each of these functions.

a) $(1 + x^5 + x^{10} + x^{15} + \cdots)^3$

b) $(x^3 + x^4 + x^5 + x^6 + x^7 + \cdots)^3$

c) $(x^4 + x^5 + x^6)(x^3 + x^4 + x^5 + x^6 + x^7)(1 + x + x^2 + x^3 + x^4 + \cdots)$

d) $(x^2 + x^4 + x^6 + x^8 + \cdots)(x^3 + x^6 + x^9 + \cdots)(x^4 + x^8 + x^{12} + \cdots)$

e) $(1 + x^2 + x^4 + x^6 + x^8 + \cdots)(1 + x^4 + x^8 + x^{12} + \cdots)(1 + x^6 + x^{12} + x^{18} + \cdots)$

8. Find the coefficient of x^{10} in the power series of each of these functions.

a) $1/(1 - 2x)$ **b)** $1/(1 + x)^2$

c) $1/(1 - x)^3$ **d)** $1/(1 + 2x)^4$

e) $x^4/(1 - 3x)^3$

9. Use generating functions to determine the number of different ways 10 identical balloons can be given to four children if each child receives at least two balloons.

10. Use generating functions to find the number of ways to choose a dozen bagels from three varieties—egg, salty, and plain—if at least two bagels of each kind but no more than three salty bagels are chosen.

11. In how many ways can 25 identical donuts be distributed to four police officers so that each officer gets at least three but no more than seven donuts?

12. Use generating functions to find the number of ways to select 14 balls from a jar containing 100 red balls, 100 blue balls, and 100 green balls so that no fewer than 3 and no more than 10 blue balls are selected. Assume that the order in which the balls are drawn does not matter.

13. What is the generating function for the sequence $\{c_k\}$, where c_k is the number of ways to make change for k dollars using $1 bills, $2 bills, $5 bills, and $10 bills?

14. Give a combinatorial interpretation of the coefficient of x^4 in the expansion $(1 + x + x^2 + x^3 + \cdots)^3$. Use this interpretation to find this number.

15. a) What is the generating function for $\{a_k\}$, where a_k is the number of solutions of $x_1 + x_2 + x_3 = k$ when x_1, x_2, and x_3 are integers with $x_1 \geq 2$, $0 \leq x_2 \leq 3$, and $2 \leq x_3 \leq 5$?

b) Use your answer to part (a) to find a_6.

16. a) What is the generating function for $\{a_k\}$, where a_k is the number of solutions of $x_1 + x_2 + x_3 + x_4 = k$ when x_1, x_2, x_3, and x_4 are integers with $x_1 \geq 3$, $1 \leq x_2 \leq 5$, $0 \leq x_3 \leq 4$, and $x_4 \geq 1$?

b) Use your answer to part (a) to find a_7.

17. Explain how generating functions can be used to find the number of ways in which postage of r cents can be pasted on an envelope using 3-cent, 4-cent, and 20-cent stamps.

a) Assume that the order the stamps are pasted on does not matter.

b) Assume that the stamps are pasted in a row and the order in which they are pasted on matters.

c) Use your answer to part (a) to determine the number of ways 46 cents of postage can be pasted on an envelope using 3-cent, 4-cent, and 20-cent stamps when the order the stamps are pasted on does not matter. (Use of a computer algebra program is advised.)

d) Use your answer to part (b) to determine the number of ways 46 cents of postage can be pasted in a row on an envelope using 3-cent, 4-cent, and 20-cent stamps when the order in which the stamps are pasted on matters. (Use of a computer algebra program is advised.)

18. a) Show that $1/(1 - x - x^2 - x^3 - x^4 - x^5 - x^6)$ is the generating function for the number of ways that the sum n can be obtained when a die is rolled repeatedly and the order of the rolls matters.

b) Use part (a) to find the number of ways to roll a total of 8 when a die is rolled repeatedly, and the order of the rolls matters. (Use of a computer algebra package is advised.)

19. Use generating functions to find the number of ways to make change for $100 using

a) $10, $20, and $50 bills.

b) $5, $10, $20, and $50 bills.

c) $5, $10, $20, and $50 bills if at least one bill of each denomination is used.

d) $5, $10, and $20 bills if at least one and no more than four of each denomination is used.

20. If $G(x)$ is the generating function for the sequence $\{a_k\}$, what is the generating function for each of these sequences?

a) $2a_0, 2a_1, 2a_2, 2a_3, \ldots$

b) $0, a_0, a_1, a_2, a_3, \ldots$ (assuming that terms follow the pattern of all but the first term)

c) $0, 0, 0, 0, a_2, a_3, \ldots$ (assuming that terms follow the pattern of all but the first four terms)

d) $a_2, a_3, a_4, \ldots$

e) $a_1, 2a_2, 3a_3, 4a_4, \ldots$ [*Hint: Calculus required* here.]

f) $a_0^2, 2a_0a_1, a_1^2 + 2a_0a_2, 2a_0a_3 + 2a_1a_2, 2a_0a_4 + 2a_1a_3 + a_2^2, \ldots$

21. If $G(x)$ is the generating function for the sequence $\{a_k\}$, what is the generating function for each of these sequences?

a) $0, 0, 0, a_3, a_4, a_5, \ldots$ (assuming that terms follow the pattern of all but the first three terms)

b) $a_0, 0, a_1, 0, a_2, 0, \ldots$

c) $0, 0, 0, 0, a_0, a_1, a_2, \ldots$ (assuming that terms follow the pattern of all but the first four terms)

d) $a_0, 2a_1, 4a_2, 8a_3, 16a_4, \ldots$

e) $0, a_0, a_1/2, a_2/3, a_3/4, \ldots$ [*Hint: Calculus required here.*]

f) $a_0, a_0 + a_1, a_0 + a_1 + a_2, a_0 + a_1 + a_2 + a_3, \ldots$

22. Use generating functions to solve the recurrence relation $a_k = 7a_{k-1}$ with the initial condition $a_0 = 5$.

23. Use generating functions to solve the recurrence relation $a_k = 3a_{k-1} + 2$ with the initial condition $a_0 = 1$.

24. Use generating functions to solve the recurrence relation $a_k = 3a_{k-1} + 4^{k-1}$ with the initial condition $a_0 = 1$.

25. Use generating functions to solve the recurrence relation $a_k = 5a_{k-1} - 6a_{k-2}$ with initial conditions $a_0 = 6$ and $a_1 = 30$.

26. Use generating functions to solve the recurrence relation $a_k = 2a_{k-1} + 3a_{k-2} + 4^k + 6$ with initial conditions $a_0 = 20$, $a_1 = 60$.

27. Use generating functions to find an explicit formula for the Fibonacci numbers.

28. Use generating functions to prove Pascal's identity: $C(n, r) = C(n-1, r) + C(n-1, r-1)$ when n and r are positive integers with $r < n$. [*Hint:* Use the identity $(1+x)^n = (1+x)^{n-1} + x(1+x)^{n-1}$.]

29. Use generating functions to prove Vandermonde's identity: $C(m+n, r) = \sum_{k=0}^{r} C(m, r-k)C(n, k)$, whenever m, n, and r are nonnegative integers with r not exceeding either m or n. [*Hint:* Look at the coefficient of x^r in both sides of $(1+x)^{m+n} = (1+x)^m(1+x)^n$.]

30. This exercise shows how to use generating functions to derive a formula for the sum of the first n squares.

a) Show that $(x^2 + x)/(1-x)^4$ is the generating function for the sequence $\{a_n\}$, where $a_n = 1^2 + 2^2 + \cdots + n^2$.

b) Use part (a) to find an explicit formula for the sum $1^2 + 2^2 + \cdots + n^2$.

31. Find a closed form for the exponential generating function for the sequence $\{a_n\}$, where

a) $a_n = 2$. **b)** $a_n = (-1)^n$.

c) $a_n = 3^n$. **d)** $a_n = n + 1$.

e) $a_n = 1/(n+1)$.

32. Find the sequence with each of these functions as its exponential generating function.

a) $f(x) = e^{-x}$ **b)** $f(x) = 3x^{2x}$

c) $f(x) = e^{3x} - 3e^{2x}$ **d)** $f(x) = (1-x) + e^{-2x}$

e) $f(x) = e^{-2x} - (1/(1-x))$

f) $f(x) = e^{-3x} - (1+x) + (1/(1-2x))$

g) $f(x) = e^{x^2}$

33. A coding system encodes messages using strings of octal (base 8) digits. A codeword is considered valid if and only if it contains an even number of 7s.

a) Find a linear nonhomogeneous recurrence relation for the number of valid codewords of length n. What are the initial conditions?

b) Solve this recurrence relation using Theorem 6 in Section 8.2.

c) Solve this recurrence relation using generating functions.

*34. A coding system encodes messages using strings of base 4 digits (that is, digits from the set $\{0, 1, 2, 3\}$). A codeword is valid if and only if it contains an even number of 0s and an even number of 1s. Let a_n equal the number of valid codewords of length n. Furthermore, let b_n, c_n, and d_n equal the number of strings of base 4 digits of length n with an even number of 0s and an odd number of 1s, with an odd number of 0s and an even number of 1s, and with an odd number of 0s and an odd number of 1s, respectively.

a) Show that $d_n = 4^n - a_n - b_n - c_n$. Use this to show that $a_{n+1} = 2a_n + b_n + c_n$, $b_{n+1} = b_n - c_n + 4^n$, and $c_{n+1} = c_n - b_n + 4^n$.

b) What are a_1, b_1, c_1, and d_1?

c) Use parts (a) and (b) to find a_3, b_3, c_3, and d_3.

d) Use the recurrence relations in part (a), together with the initial conditions in part (b), to set up three equations relating the generating functions $A(x)$, $B(x)$, and $C(x)$ for the sequences $\{a_n\}$, $\{b_n\}$, and $\{c_n\}$, respectively.

e) Solve the system of equations from part (d) to get explicit formulae for $A(x)$, $B(x)$, and $C(x)$ and use these to get explicit formulae for a_n, b_n, c_n, and d_n.

Generating functions are useful in studying the number of different types of partitions of an integer n. A **partition** of a positive integer is a way to write this integer as the sum of positive integers where repetition is allowed and the order of the integers in the sum does not matter. For example, the partitions of 5 (with no restrictions) are $1+1+1+1+1$, $1+1+1+2$, $1+1+3$, $1+2+2$, $1+4$, $2+3$, and 5. Exercises 35–40 illustrate some of these uses.

35. Show that the coefficient $p(n)$ of x^n in the formal power series expansion of $1/((1-x)(1-x^2)(1-x^3) \cdots)$ equals the number of partitions of n.

36. Show that the coefficient $p_o(n)$ of x^n in the formal power series expansion of $1/((1-x)(1-x^3)(1-x^5) \cdots)$ equals the number of partitions of n into odd integers, that is, the number of partitions of n into odd positive integers, where the order does not matter and repetitions are allowed.

37. Show that the coefficient $p_d(n)$ of x^n in the formal power series expansion of $(1+x)(1+x^2)(1+x^3) \cdots$ equals the number of partitions of n into distinct parts, that is, the number of ways to write n as the sum of positive integers, where the order does not matter but no repetitions are allowed.

38. Find $p_o(n)$, the number of partitions of n into odd parts with repetitions allowed, and $p_d(n)$, the number of partitions of n into distinct parts, for $1 \leq n \leq 8$, by writing each partition of each type for each integer.

39. Show that if n is a positive integer, then the number of partitions of n into distinct parts equals the number of partitions of n into odd parts with repetitions allowed; that is, $p_o(n) = p_d(n)$. [*Hint:* Show that the generating functions for $p_o(n)$ and $p_d(n)$ are equal.]

**40. (*Requires calculus*) Use the generating function of $p(n)$ to show that $p(n) \leq e^{C\sqrt{n}}$ for some constant C. [Hardy and Ramanujan showed that $p(n) \sim e^{\pi\sqrt{2/3}\sqrt{n}}/(4\sqrt{3}n)$, which means that the ratio of $p(n)$ and the right-hand side approaches 1 as n approaches infinity.]

Suppose that X is a random variable on a sample space S such that $X(s)$ is a nonnegative integer for all $s \in S$. The **probability generating function** for X is

$$G_X(x) = \sum_{k=0}^{\infty} p(X(s) = k)x^k.$$

41. (*Requires calculus*) Show that if G_X is the probability generating function for a random variable X such that $X(s)$ is a nonnegative integer for all $s \in S$, then

a) $G_X(1) = 1.$ **b)** $E(X) = G_X'(1).$

c) $V(X) = G_X''(1) + G_X'(1) - G_X'(1)^2.$

42. Let X be the random variable whose value is n if the first success occurs on the nth trial when independent Bernoulli trials are performed, each with probability of success p.

a) Find a closed formula for the probability generating function G_X.

b) Find the expected value and the variance of X using Exercise 41 and the closed form for the probability generating function found in part (a).

43. Let m be a positive integer. Let X_m be the random variable whose value is n if the mth success occurs on the $(n + m)$th trial when independent Bernoulli trials are performed, each with probability of success p.

a) Using Exercise 26 in the Supplementary Exercises of Chapter 7, show that the probability generating function G_{X_m} is given by $G_{X_m}(x) = p^m/(1 - qx)^m$, where $q = 1 - p$.

b) Find the expected value and the variance of X_m using Exercise 41 and the closed form for the probability generating function in part (a).

44. Show that if X and Y are independent random variables on a sample space S such that $X(s)$ and $Y(s)$ are nonnegative integers for all $s \in S$, then $G_{X+Y}(x) = G_X(x)G_Y(x)$.

8.5   Inclusion–Exclusion

Introduction

A discrete mathematics class contains 30 women and 50 sophomores. How many students in the class are either women or sophomores? This question cannot be answered unless more information is provided. Adding the number of women in the class and the number of sophomores probably does not give the correct answer, because women sophomores are counted twice. This observation shows that the number of students in the class that are either sophomores or women is the sum of the number of women and the number of sophomores in the class minus the number of women sophomores. A technique for solving such counting problems was introduced in Section 6.1. In this section we will generalize the ideas introduced in that section to solve problems that require us to count the number of elements in the union of more than two sets.

The Principle of Inclusion–Exclusion

How many elements are in the union of two finite sets? In Section 2.2 we showed that the number of elements in the union of the two sets A and B is the sum of the numbers of elements in the sets minus the number of elements in their intersection. That is,

$$|A \cup B| = |A| + |B| - |A \cap B|.$$

As we showed in Section 6.1, the formula for the number of elements in the union of two sets is useful in counting problems. Examples 1–3 provide additional illustrations of the usefulness of this formula.

EXAMPLE 1   In a discrete mathematics class every student is a major in computer science or mathematics, or both. The number of students having computer science as a major (possibly along with mathematics) is 25; the number of students having mathematics as a major (possibly along with computer science) is 13; and the number of students majoring in both computer science and mathematics is 8. How many students are in this class?

Solution: Let A be the set of students in the class majoring in computer science and B be the set of students in the class majoring in mathematics. Then $A \cap B$ is the set of students in the class who are joint mathematics and computer science majors. Because every student in the class is majoring in either computer science or mathematics (or both), it follows that the number of students in the class is $|A \cup B|$. Therefore,

$$|A \cup B| = |A| + |B| - |A \cap B|$$
$$= 25 + 13 - 8 = 30.$$

Therefore, there are 30 students in the class. This computation is illustrated in Figure 1.   ◀

EXAMPLE 2   How many positive integers not exceeding 1000 are divisible by 7 or 11?

Solution: Let A be the set of positive integers not exceeding 1000 that are divisible by 7, and let B be the set of positive integers not exceeding 1000 that are divisible by 11. Then $A \cup B$ is the set of integers not exceeding 1000 that are divisible by either 7 or 11, and $A \cap B$ is the set of integers not exceeding 1000 that are divisible by both 7 and 11. From Example 2 of Section 4.1, we know that among the positive integers not exceeding 1000 there are $\lfloor 1000/7 \rfloor$ integers divisible by 7 and $\lfloor 1000/11 \rfloor$ divisible by 11. Because 7 and 11 are relatively prime, the integers divisible by both 7 and 11 are those divisible by $7 \cdot 11$. Consequently, there are $\lfloor 1000/(11 \cdot 7) \rfloor$ positive integers not exceeding 1000 that are divisible by both 7 and 11. It follows that there are

$$|A \cup B| = |A| + |B| - |A \cap B|$$
$$= \left\lfloor \frac{1000}{7} \right\rfloor + \left\lfloor \frac{1000}{11} \right\rfloor - \left\lfloor \frac{1000}{7 \cdot 11} \right\rfloor$$
$$= 142 + 90 - 12 = 220$$

positive integers not exceeding 1000 that are divisible by either 7 or 11. This computation is illustrated in Figure 2.   ◀

Example 3 shows how to find the number of elements in a finite universal set that are outside the union of two sets.

EXAMPLE 3   Suppose that there are 1807 freshmen at your school. Of these, 453 are taking a course in computer science, 567 are taking a course in mathematics, and 299 are taking courses in both computer science and mathematics. How many are not taking a course either in computer science or in mathematics?

Solution: To find the number of freshmen who are not taking a course in either mathematics or computer science, subtract the number that are taking a course in either of these subjects from the total number of freshmen. Let A be the set of all freshmen taking a course in computer science, and let B be the set of all freshmen taking a course in mathematics. It follows

$$|A \cup B|=|A|+|B|-|A \cap B| = 25 + 13 - 8 = 30$$

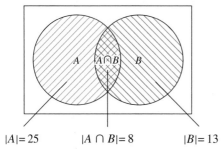

$|A|= 25$ $|A \cap B|= 8$ $|B|= 13$

FIGURE 1   The Set of Students in a Discrete Mathematics Class.

$$|A \cup B| = |A|+|B|-|A \cap B| = 142 + 90 - 12 = 220$$

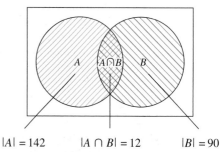

$|A| = 142$ $|A \cap B| = 12$ $|B| = 90$

FIGURE 2   The Set of Positive Integers Not Exceeding 1000 Divisible by Either 7 or 11.

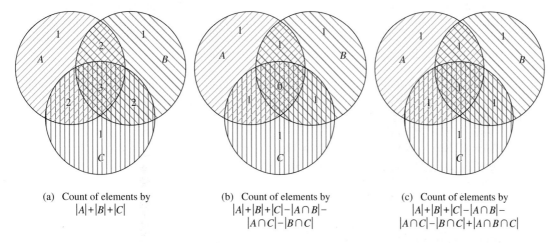

(a) Count of elements by $|A|+|B|+|C|$

(b) Count of elements by $|A|+|B|+|C|-|A \cap B|-$ $|A \cap C|-|B \cap C|$

(c) Count of elements by $|A|+|B|+|C|-|A \cap B|-$ $|A \cap C|-|B \cap C|+|A \cap B \cap C|$

FIGURE 3   Finding a Formula for the Number of Elements in the Union of Three Sets.

that $|A| = 453$, $|B| = 567$, and $|A \cap B| = 299$. The number of freshmen taking a course in either computer science or mathematics is

$$|A \cup B| = |A| + |B| - |A \cap B| = 453 + 567 - 299 = 721.$$

Consequently, there are $1807 - 721 = 1086$ freshmen who are not taking a course in computer science or mathematics. ◄

We will now begin our development of a formula for the number of elements in the union of a finite number of sets. The formula we will develop is called the **principle of inclusion–exclusion**. For concreteness, before we consider unions of n sets, where n is any positive integer, we will derive a formula for the number of elements in the union of three sets A, B, and C. To construct this formula, we note that $|A| + |B| + |C|$ counts each element that is in exactly one of the three sets once, elements that are in exactly two of the sets twice, and elements in all three sets three times. This is illustrated in the first panel in Figure 3.

To remove the overcount of elements in more than one of the sets, we subtract the number of elements in the intersections of all pairs of the three sets. We obtain

$$|A| + |B| + |C| - |A \cap B| - |A \cap C| - |B \cap C|.$$

This expression still counts elements that occur in exactly one of the sets once. An element that occurs in exactly two of the sets is also counted exactly once, because this element will occur in one of the three intersections of sets taken two at a time. However, those elements that occur in all three sets will be counted zero times by this expression, because they occur in all three intersections of sets taken two at a time. This is illustrated in the second panel in Figure 3.

To remedy this undercount, we add the number of elements in the intersection of all three sets. This final expression counts each element once, whether it is in one, two, or three of the sets. Thus,

$$|A \cup B \cup C| = |A| + |B| + |C| - |A \cap B| - |A \cap C| - |B \cap C| + |A \cap B \cap C|.$$

This formula is illustrated in the third panel of Figure 3.
Example 4 illustrates how this formula can be used.

EXAMPLE 4   A total of 1232 students have taken a course in Spanish, 879 have taken a course in French, and 114 have taken a course in Russian. Further, 103 have taken courses in both Spanish and French, 23 have taken courses in both Spanish and Russian, and 14 have taken courses in both French and Russian. If 2092 students have taken at least one of Spanish, French, and Russian, how many students have taken a course in all three languages?

Solution: Let S be the set of students who have taken a course in Spanish, F the set of students who have taken a course in French, and R the set of students who have taken a course in Russian. Then

$$|S| = 1232, \quad |F| = 879, \quad |R| = 114,$$
$$|S \cap F| = 103, |S \cap R| = 23, |F \cap R| = 14,$$

and

$$|S \cup F \cup R| = 2092.$$

When we insert these quantities into the equation

$$|S \cup F \cup R| = |S| + |F| + |R| - |S \cap F| - |S \cap R| - |F \cap R| + |S \cap F \cap R|$$

we obtain

$$2092 = 1232 + 879 + 114 - 103 - 23 - 14 + |S \cap F \cap R|.$$

We now solve for $|S \cap F \cap R|$. We find that $|S \cap F \cap R| = 7$. Therefore, there are seven students who have taken courses in Spanish, French, and Russian. This is illustrated in Figure 4.   ◀

We will now state and prove the inclusion–exclusion principle, which tells us how many elements are in the union of a finite number of finite sets.

THEOREM 1   **THE PRINCIPLE OF INCLUSION–EXCLUSION**   Let $A_1, A_2, \ldots, A_n$ be finite sets. Then

$$|A_1 \cup A_2 \cup \cdots \cup A_n| = \sum_{1 \le i \le n} |A_i| - \sum_{1 \le i < j \le n} |A_i \cap A_j|$$
$$+ \sum_{1 \le i < j < k \le n} |A_i \cap A_j \cap A_k| - \cdots + (-1)^{n+1} |A_1 \cap A_2 \cap \cdots \cap A_n|.$$

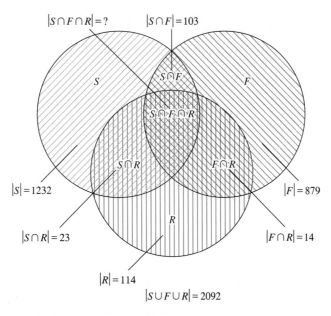

$|S \cap F \cap R| = ?$ $|S \cap F| = 103$

$S \cap F$

$S \cap F \cap R$

S F

$S \cap R$ $F \cap R$

$|S| = 1232$ $|F| = 879$

R

$|S \cap R| = 23$ $|F \cap R| = 14$

$|R| = 114$

$|S \cup F \cup R| = 2092$

FIGURE 4   **The Set of Students Who Have Taken Courses in Spanish, French, and Russian.**

Proof: We will prove the formula by showing that an element in the union is counted exactly once by the right-hand side of the equation. Suppose that a is a member of exactly r of the sets $A_1, A_2, \ldots, A_n$ where $1 \le r \le n$. This element is counted $C(r, 1)$ times by $\Sigma |A_i|$. It is counted $C(r, 2)$ times by $\Sigma |A_i \cap A_j|$. In general, it is counted $C(r, m)$ times by the summation involving m of the sets A_i. Thus, this element is counted exactly

$$C(r, 1) - C(r, 2) + C(r, 3) - \cdots + (-1)^{r+1} C(r, r)$$

times by the expression on the right-hand side of this equation. Our goal is to evaluate this quantity. By Corollary 2 of Section 6.4, we have

$$C(r, 0) - C(r, 1) + C(r, 2) - \cdots + (-1)^r C(r, r) = 0.$$

Hence,

$$1 = C(r, 0) = C(r, 1) - C(r, 2) + \cdots + (-1)^{r+1} C(r, r).$$

Therefore, each element in the union is counted exactly once by the expression on the right-hand side of the equation. This proves the principle of inclusion–exclusion. ◁

The inclusion–exclusion principle gives a formula for the number of elements in the union of n sets for every positive integer n. There are terms in this formula for the number of elements in the intersection of every nonempty subset of the collection of the n sets. Hence, there are $2^n - 1$ terms in this formula.

EXAMPLE 5   Give a formula for the number of elements in the union of four sets.

Solution: The inclusion–exclusion principle shows that

$$|A_1 \cup A_2 \cup A_3 \cup A_4| = |A_1| + |A_2| + |A_3| + |A_4|$$
$$- |A_1 \cap A_2| - |A_1 \cap A_3| - |A_1 \cap A_4| - |A_2 \cap A_3| - |A_2 \cap A_4|$$
$$- |A_3 \cap A_4| + |A_1 \cap A_2 \cap A_3| + |A_1 \cap A_2 \cap A_4| + |A_1 \cap A_3 \cap A_4|$$
$$+ |A_2 \cap A_3 \cap A_4| - |A_1 \cap A_2 \cap A_3 \cap A_4|.$$

Note that this formula contains 15 different terms, one for each nonempty subset of $\{A_1, A_2, A_3, A_4\}$.   ◀

Exercises

1. How many elements are in $A_1 \cup A_2$ if there are 12 elements in A_1, 18 elements in A_2, and
 a) $A_1 \cap A_2 = \emptyset$?   **b)** $|A_1 \cap A_2| = 1$?
 c) $|A_1 \cap A_2| = 6$?   **d)** $A_1 \subseteq A_2$?

2. There are 345 students at a college who have taken a course in calculus, 212 who have taken a course in discrete mathematics, and 188 who have taken courses in both calculus and discrete mathematics. How many students have taken a course in either calculus or discrete mathematics?

3. A survey of households in the United States reveals that 96% have at least one television set, 98% have telephone service, and 95% have telephone service and at least one television set. What percentage of households in the United States have neither telephone service nor a television set?

4. A marketing report concerning personal computers states that 650,000 owners will buy a printer for their machines next year and 1,250,000 will buy at least one software package. If the report states that 1,450,000 owners will buy either a printer or at least one software package, how many will buy both a printer and at least one software package?

5. Find the number of elements in $A_1 \cup A_2 \cup A_3$ if there are 100 elements in each set and if
 a) the sets are pairwise disjoint.
 b) there are 50 common elements in each pair of sets and no elements in all three sets.
 c) there are 50 common elements in each pair of sets and 25 elements in all three sets.
 d) the sets are equal.

6. In a survey of 270 college students, it is found that 64 like brussels sprouts, 94 like broccoli, 58 like cauliflower, 26 like both brussels sprouts and broccoli, 28 like both brussels sprouts and cauliflower, 22 like both broccoli and cauliflower, and 14 like all three vegetables. How many of the 270 students do not like any of these vegetables?

7. How many students are enrolled in a course either in calculus, discrete mathematics, data structures, or programming languages at a school if there are 507, 292, 312, and 344 students in these courses, respectively; 14 in both calculus and data structures; 213 in both calculus and programming languages; 211 in both discrete mathematics and data structures; 43 in both discrete mathematics and programming languages; and no student may take calculus and discrete mathematics, or data structures and programming languages, concurrently?

8. Find the number of positive integers not exceeding 100 that are not divisible by 5 or by 7.

9. Find the number of positive integers not exceeding 100 that are either odd or the square of an integer.

10. How many bit strings of length eight do not contain six consecutive 0s?

11. How many permutations of the 10 digits either begin with the 3 digits 987, contain the digits 45 in the fifth and sixth positions, or end with the 3 digits 123?

12. How many elements are in the union of four sets if each of the sets has 100 elements, each pair of the sets shares 50 elements, each three of the sets share 25 elements, and there are 5 elements in all four sets?

13. How many elements are in the union of four sets if the sets have 50, 60, 70, and 80 elements, respectively, each pair of the sets has 5 elements in common, each triple of the sets has 1 common element, and no element is in all four sets?

14. How many terms are there in the formula for the number of elements in the union of 10 sets given by the principle of inclusion–exclusion?

15. Write out the explicit formula given by the principle of inclusion–exclusion for the number of elements in the union of five sets.

16. Write out the explicit formula given by the principle of inclusion–exclusion for the number of elements in the union of six sets when it is known that no three of these sets have a common intersection.

17. Let E_1, E_2, and E_3 be three events from a sample space S. Find a formula for the probability of $E_1 \cup E_2 \cup E_3$.

18. Find the probability that when a fair coin is flipped five times tails comes up exactly three times, the first and last flips come up tails, or the second and fourth flips come up heads.

19. Find the probability that when four numbers from 1 to 100, inclusive, are picked at random with no repetitions allowed, either all are odd, all are divisible by 3, or all are divisible by 5.

20. Find a formula for the probability of the union of five events in a sample space if no four of them can occur at the same time.

21. Find a formula for the probability of the union of n events in a sample space.

8.6   Applications of Inclusion–Exclusion

Introduction

Many counting problems can be solved using the principle of inclusion–exclusion. For instance, we can use this principle to find the number of primes less than a positive integer. Many problems can be solved by counting the number of onto functions from one finite set to another. The inclusion–exclusion principle can be used to find the number of such functions. The famous hatcheck problem can be solved using the principle of inclusion–exclusion. This problem asks for the probability that no person is given the correct hat back by a hatcheck person who gives the hats back randomly.

An Alternative Form of Inclusion–Exclusion

There is an alternative form of the principle of inclusion–exclusion that is useful in counting problems. In particular, this form can be used to solve problems that ask for the number of elements in a set that have none of n properties $P_1, P_2, \ldots, P_n$.

Let A_i be the subset containing the elements that have property P_i. The number of elements with all the properties $P_{i_1}, P_{i_2}, \ldots, P_{i_k}$ will be denoted by $N(P_{i_1} P_{i_2} \ldots P_{i_k})$. Writing these quantities in terms of sets, we have

$$|A_{i_1} \cap A_{i_2} \cap \cdots \cap A_{i_k}| = N(P_{i_1} P_{i_2} \ldots P_{i_k}).$$

If the number of elements with none of the properties $P_1, P_2, \ldots, P_n$ is denoted by $N(P_1' P_2' \ldots P_n')$ and the number of elements in the set is denoted by N, it follows that

$$N(P_1' P_2' \ldots P_n') = N - |A_1 \cup A_2 \cup \cdots \cup A_n|.$$

From the inclusion–exclusion principle, we see that

$$N(P_1' P_2' \ldots P_n') = N - \sum_{1 \le i \le n} N(P_i) + \sum_{1 \le i < j \le n} N(P_i P_j)$$
$$- \sum_{1 \le i < j < k \le n} N(P_i P_j P_k) + \cdots + (-1)^n N(P_1 P_2 \ldots P_n).$$

Example 1 shows how the principle of inclusion–exclusion can be used to determine the number of solutions in integers of an equation with constraints.

EXAMPLE 1   How many solutions does

$$x_1 + x_2 + x_3 = 11$$

have, where x_1, x_2, and x_3 are nonnegative integers with $x_1 \leq 3$, $x_2 \leq 4$, and $x_3 \leq 6$?

Solution: To apply the principle of inclusion–exclusion, let a solution have property P_1 if $x_1 > 3$, property P_2 if $x_2 > 4$, and property P_3 if $x_3 > 6$. The number of solutions satisfying the inequalities $x_1 \leq 3$, $x_2 \leq 4$, and $x_3 \leq 6$ is

$$N(P_1' P_2' P_3') = N - N(P_1) - N(P_2) - N(P_3) + N(P_1 P_2)$$
$$+ N(P_1 P_3) + N(P_2 P_3) - N(P_1 P_2 P_3).$$

Using the same techniques as in Example 5 of Section 6.5, it follows that

- $N = $ total number of solutions $= C(3 + 11 - 1, 11) = 78,$
- $N(P_1) = $ (number of solutions with $x_1 \geq 4$) $= C(3 + 7 - 1, 7) = C(9, 7) = 36,$
- $N(P_2) = $ (number of solutions with $x_2 \geq 5$) $= C(3 + 6 - 1, 6) = C(8, 6) = 28,$
- $N(P_3) = $ (number of solutions with $x_3 \geq 7$) $= C(3 + 4 - 1, 4) = C(6, 4) = 15,$
- $N(P_1 P_2) = $ (number of solutions with $x_1 \geq 4$ and $x_2 \geq 5$) $= C(3 + 2 - 1, 2) = C(4, 2) = 6,$
- $N(P_1 P_3) = $ (number of solutions with $x_1 \geq 4$ and $x_3 \geq 7$) $= C(3 + 0 - 1, 0) = 1,$
- $N(P_2 P_3) = $ (number of solutions with $x_2 \geq 5$ and $x_3 \geq 7$) $= 0,$
- $N(P_1 P_2 P_3) = $ (number of solutions with $x_1 \geq 4$, $x_2 \geq 5$, and $x_3 \geq 7$) $= 0.$

Inserting these quantities into the formula for $N(P_1' P_2' P_3')$ shows that the number of solutions with $x_1 \leq 3$, $x_2 \leq 4$, and $x_3 \leq 6$ equals

$$N(P_1' P_2' P_3') = 78 - 36 - 28 - 15 + 6 + 1 + 0 - 0 = 6. \qquad \blacktriangleleft$$

The Sieve of Eratosthenes

In Section 4.3 we showed how to use the sieve of Eratosthenes to find all primes less than a specified positive integer n. Using the principle of inclusion–exclusion, we can find the number of primes not exceeding a specified positive integer with the same reasoning as is used in the sieve of Eratosthenes. Recall that a composite integer is divisible by a prime not exceeding its square root. So, to find the number of primes not exceeding 100, first note that composite integers not exceeding 100 must have a prime factor not exceeding 10. Because the only primes not exceeding 10 are 2, 3, 5, and 7, the primes not exceeding 100 are these four primes and those positive integers greater than 1 and not exceeding 100 that are divisible by none of 2, 3, 5, or 7. To apply the principle of inclusion–exclusion, let P_1 be the property that an integer is divisible by 2, let P_2 be the property that an integer is divisible by 3, let P_3 be the property that an integer is divisible by 5, and let P_4 be the property that an integer is divisible by 7. Thus, the number of primes not exceeding 100 is

$$4 + N(P_1' P_2' P_3' P_4').$$

Because there are 99 positive integers greater than 1 and not exceeding 100, the principle of inclusion–exclusion shows that

$$
\begin{aligned}
N(P_1'P_2'P_3'P_4') = {}& 99 - N(P_1) - N(P_2) - N(P_3) - N(P_4) \\
& + N(P_1P_2) + N(P_1P_3) + N(P_1P_4) + N(P_2P_3) + N(P_2P_4) + N(P_3P_4) \\
& - N(P_1P_2P_3) - N(P_1P_2P_4) - N(P_1P_3P_4) - N(P_2P_3P_4) \\
& + N(P_1P_2P_3P_4).
\end{aligned}
$$

The number of integers not exceeding 100 (and greater than 1) that are divisible by all the primes in a subset of $\{2, 3, 5, 7\}$ is $\lfloor 100/N \rfloor$, where N is the product of the primes in this subset. (This follows because any two of these primes have no common factor.) Consequently,

$$
\begin{aligned}
N(P_1'P_2'P_3'P_4') = {}& 99 - \left\lfloor \frac{100}{2} \right\rfloor - \left\lfloor \frac{100}{3} \right\rfloor - \left\lfloor \frac{100}{5} \right\rfloor - \left\lfloor \frac{100}{7} \right\rfloor \\
& + \left\lfloor \frac{100}{2 \cdot 3} \right\rfloor + \left\lfloor \frac{100}{2 \cdot 5} \right\rfloor + \left\lfloor \frac{100}{2 \cdot 7} \right\rfloor + \left\lfloor \frac{100}{3 \cdot 5} \right\rfloor + \left\lfloor \frac{100}{3 \cdot 7} \right\rfloor + \left\lfloor \frac{100}{5 \cdot 7} \right\rfloor \\
& - \left\lfloor \frac{100}{2 \cdot 3 \cdot 5} \right\rfloor - \left\lfloor \frac{100}{2 \cdot 3 \cdot 7} \right\rfloor - \left\lfloor \frac{100}{2 \cdot 5 \cdot 7} \right\rfloor - \left\lfloor \frac{100}{3 \cdot 5 \cdot 7} \right\rfloor + \left\lfloor \frac{100}{2 \cdot 3 \cdot 5 \cdot 7} \right\rfloor \\
= {}& 99 - 50 - 33 - 20 - 14 + 16 + 10 + 7 + 6 + 4 + 2 - 3 - 2 - 1 - 0 + 0 \\
= {}& 21.
\end{aligned}
$$

Hence, there are $4 + 21 = 25$ primes not exceeding 100.

The Number of Onto Functions

The principle of inclusion–exclusion can also be used to determine the number of onto functions from a set with m elements to a set with n elements. First consider Example 2.

EXAMPLE 2 How many onto functions are there from a set with six elements to a set with three elements?

Solution: Suppose that the elements in the codomain are b_1, b_2, and b_3. Let P_1, P_2, and P_3 be the properties that b_1, b_2, and b_3 are not in the range of the function, respectively. Note that a function is onto if and only if it has none of the properties P_1, P_2, or P_3. By the inclusion–exclusion principle it follows that the number of onto functions from a set with six elements to a set with three elements is

$$
\begin{aligned}
N(P_1'P_2'P_3') = {}& N - [N(P_1) + N(P_2) + N(P_3)] \\
& + [N(P_1P_2) + N(P_1P_3) + N(P_2P_3)] - N(P_1P_2P_3),
\end{aligned}
$$

where N is the total number of functions from a set with six elements to one with three elements. We will evaluate each of the terms on the right-hand side of this equation.

From Example 6 of Section 6.1, it follows that $N = 3^6$. Note that $N(P_i)$ is the number of functions that do not have b_i in their range. Hence, there are two choices for the value of the function at each element of the domain. Therefore, $N(P_i) = 2^6$. Furthermore, there are $C(3, 1)$ terms of this kind. Note that $N(P_iP_j)$ is the number of functions that do not have b_i and b_j in their range. Hence, there is only one choice for the value of the function at each element of the domain. Therefore, $N(P_iP_j) = 1^6 = 1$. Furthermore, there are $C(3, 2)$ terms of this kind.

Also, note that $N(P_1 P_2 P_3) = 0$, because this term is the number of functions that have none of b_1, b_2, and b_3 in their range. Clearly, there are no such functions. Therefore, the number of onto functions from a set with six elements to one with three elements is

$$3^6 - C(3, 1)2^6 + C(3, 2)1^6 = 729 - 192 + 3 = 540.$$ ◀

The general result that tells us how many onto functions there are from a set with m elements to one with n elements will now be stated. The proof of this result is left as an exercise for the reader.

THEOREM 1   Let m and n be positive integers with $m \geq n$. Then, there are

$$n^m - C(n, 1)(n - 1)^m + C(n, 2)(n - 2)^m - \cdots + (-1)^{n-1}C(n, n - 1) \cdot 1^m$$

onto functions from a set with m elements to a set with n elements.

An onto function from a set with m elements to a set with n elements corresponds to a way to distribute the m elements in the domain to n indistinguishable boxes so that no box is empty, and then to associate each of the n elements of the codomain to a box. This means that the number of onto functions from a set with m elements to a set with n elements is the number of ways to distribute m distinguishable objects to n indistinguishable boxes so that no box is empty multiplied by the number of permutations of a set with n elements. Consequently, the number of onto functions from a set with m elements to a set with n elements equals $n!S(m, n)$, where $S(m, n)$ is a *Stirling number of the second kind* defined in Section 6.5. This means that we can use Theorem 1 to deduce the formula given in Section 6.5 for $S(m, n)$. (See Chapter 6 of [MiRo91] for more details about Stirling numbers of the second kind.)

> Counting onto functions is much harder than counting one-to-one functions!

One of the many different applications of Theorem 1 will now be described.

EXAMPLE 3   How many ways are there to assign five different jobs to four different employees if every employee is assigned at least one job?

Solution: Consider the assignment of jobs as a function from the set of five jobs to the set of four employees. An assignment where every employee gets at least one job is the same as an onto function from the set of jobs to the set of employees. Hence, by Theorem 1 it follows that there are

$$4^5 - C(4, 1)3^5 + C(4, 2)2^5 - C(4, 3)1^5 = 1024 - 972 + 192 - 4 = 240$$

ways to assign the jobs so that each employee is assigned at least one job.   ◀

Derangements

The principle of inclusion–exclusion will be used to count the permutations of n objects that leave no objects in their original positions. Consider Example 4.

EXAMPLE 4   **The Hatcheck Problem**   A new employee checks the hats of n people at a restaurant, forgetting to put claim check numbers on the hats. When customers return for their hats, the checker gives them back hats chosen at random from the remaining hats. What is the probability that no one receives the correct hat?   ◀

Remark: The answer is the number of ways the hats can be arranged so that there is no hat in its original position divided by $n!$, the number of permutations of n hats. We will return to this example after we find the number of permutations of n objects that leave no objects in their original position.

Links

A **derangement** is a permutation of objects that leaves no object in its original position. To solve the problem posed in Example 4 we will need to determine the number of derangements of a set of n objects.

EXAMPLE 5   The permutation 21453 is a derangement of 12345 because no number is left in its original position. However, 21543 is not a derangement of 12345, because this permutation leaves 4 fixed.   ◄

Let D_n denote the number of derangements of n objects. For instance, $D_3 = 2$, because the derangements of 123 are 231 and 312. We will evaluate D_n, for all positive integers n, using the principle of inclusion–exclusion.

THEOREM 2   The number of derangements of a set with n elements is

$$D_n = n!\left[1 - \frac{1}{1!} + \frac{1}{2!} - \frac{1}{3!} + \cdots + (-1)^n\frac{1}{n!}\right].$$

Proof: Let a permutation have property P_i if it fixes element i. The number of derangements is the number of permutations having none of the properties P_i for $i = 1, 2, \ldots, n$. This means that

$$D_n = N(P_1' P_2' \ldots P_n').$$

Using the principle of inclusion–exclusion, it follows that

$$D_n = N - \sum_i N(P_i) + \sum_{i<j} N(P_i P_j) - \sum_{i<j<k} N(P_i P_j P_k) + \cdots + (-1)^n N(P_1 P_2 \ldots P_n),$$

where N is the number of permutations of n elements. This equation states that the number of permutations that fix no elements equals the total number of permutations, less the number that fix at least one element, plus the number that fix at least two elements, less the number that fix at least three elements, and so on. All the quantities that occur on the right-hand side of this equation will now be found.

First, note that $N = n!$, because N is simply the total number of permutations of n elements. Also, $N(P_i) = (n-1)!$. This follows from the product rule, because $N(P_i)$ is the number of

Links

HISTORICAL NOTE   In *rencontres* (matches), an old French card game, the 52 cards in a deck are laid out in a row. The cards of a second deck are laid out with one card of the second deck on top of each card of the first deck. The score is determined by counting the number of matching cards in the two decks. In 1708 Pierre Raymond de Montmort (1678–1719) posed *le problème de rencontres:* What is the probability that no matches take place in the game of rencontres? The solution to Montmort's problem is the probability that a randomly selected permutation of 52 objects is a derangement, namely, $D_{52}/52!$, which, as we will see, is approximately $1/e$.

permutations that fix element i, so the ith position of the permutation is determined, but each of the remaining positions can be filled arbitrarily. Similarly,

$$N(P_i P_j) = (n - 2)!,$$

because this is the number of permutations that fix elements i and j, but where the other $n - 2$ elements can be arranged arbitrarily. In general, note that

$$N(P_{i_1} P_{i_2} \ldots P_{i_m}) = (n - m)!,$$

because this is the number of permutations that fix elements $i_1, i_2, \ldots, i_m$, but where the other $n - m$ elements can be arranged arbitrarily. Because there are $C(n, m)$ ways to choose m elements from n, it follows that

$$\sum_{1 \leq i \leq n} N(P_i) = C(n, 1)(n - 1)!,$$

$$\sum_{1 \leq i < j \leq n} N(P_i P_j) = C(n, 2)(n - 2)!,$$

and in general,

$$\sum_{1 \leq i_1 < i_2 < \cdots < i_m \leq n} N(P_{i_1} P_{i_2} \ldots P_{i_m}) = C(n, m)(n - m)!.$$

Consequently, inserting these quantities into our formula for D_n gives

$$D_n = n! - C(n, 1)(n - 1)! + C(n, 2)(n - 2)! - \cdots + (-1)^n C(n, n)(n - n)!$$

$$= n! - \frac{n!}{1!(n - 1)!}(n - 1)! + \frac{n!}{2!(n - 2)!}(n - 2)! - \cdots + (-1)^n \frac{n!}{n! \, 0!} 0!.$$

Simplifying this expression gives

$$D_n = n! \left[1 - \frac{1}{1!} + \frac{1}{2!} - \cdots + (-1)^n \frac{1}{n!} \right]. \qquad \triangleleft$$

It is now simple to find D_n for a given positive integer n. For instance, using Theorem 2, it follows that

$$D_3 = 3! \left[1 - \frac{1}{1!} + \frac{1}{2!} - \frac{1}{3!} \right] = 6 \left(1 - 1 + \frac{1}{2} - \frac{1}{6} \right) = 2,$$

as we have previously remarked.

The solution of the problem in Example 4 can now be given.

Solution: The probability that no one receives the correct hat is $D_n / n!$. By Theorem 2, this probability is

$$\frac{D_n}{n!} = 1 - \frac{1}{1!} + \frac{1}{2!} - \cdots + (-1)^n \frac{1}{n!}.$$

The values of this probability for $2 \leq n \leq 7$ are displayed in Table 1.

TABLE 1 **The Probability of a Derangement.**						
n	2	3	4	5	6	7
$D_n/n!$	0.50000	0.33333	0.37500	0.36667	0.36806	0.36786

Using methods from calculus it can be shown that

$$e^{-1} = 1 - \frac{1}{1!} + \frac{1}{2!} - \cdots + (-1)^n \frac{1}{n!} + \cdots \approx 0.368.$$

Because this is an alternating series with terms tending to zero, it follows that as n grows without bound, the probability that no one receives the correct hat converges to $e^{-1} \approx 0.368$. In fact, this probability can be shown to be within $1/(n + 1)!$ of e^{-1}. ◀

Exercises

1. Suppose that in a bushel of 100 apples there are 20 that have worms in them and 15 that have bruises. Only those apples with neither worms nor bruises can be sold. If there are 10 bruised apples that have worms in them, how many of the 100 apples can be sold?

2. Of 1000 applicants for a mountain-climbing trip in the Himalayas, 450 get altitude sickness, 622 are not in good enough shape, and 30 have allergies. An applicant qualifies if and only if this applicant does not get altitude sickness, is in good shape, and does not have allergies. If there are 111 applicants who get altitude sickness and are not in good enough shape, 14 who get altitude sickness and have allergies, 18 who are not in good enough shape and have allergies, and 9 who get altitude sickness, are not in good enough shape, and have allergies, how many applicants qualify?

3. How many solutions does the equation $x_1 + x_2 + x_3 = 13$ have where $x_1, x_2,$ and x_3 are nonnegative integers less than 6?

4. An integer is called **squarefree** if it is not divisible by the square of a positive integer greater than 1. Find the number of squarefree positive integers less than 100.

5. How many positive integers less than 10,000 are not the second or higher power of an integer?

6. How many onto functions are there from a set with seven elements to one with five elements?

7. How many ways are there to distribute six different toys to three different children such that each child gets at least one toy?

8. In how many ways can eight distinct balls be distributed into three distinct urns if each urn must contain at least one ball?

9. In how many ways can seven different jobs be assigned to four different employees so that each employee is assigned at least one job and the most difficult job is assigned to the best employee?

10. List all the derangements of $\{1, 2, 3, 4\}$.

11. How many derangements are there of a set with seven elements?

12. What is the probability that none of 10 people receives the correct hat if a hatcheck person hands their hats back randomly?

13. A machine that inserts letters into envelopes goes haywire and inserts letters randomly into envelopes. What is the probability that in a group of 100 letters
    a) no letter is put into the correct envelope?
    b) exactly one letter is put into the correct envelope?
    c) exactly 98 letters are put into the correct envelopes?
    d) exactly 99 letters are put into the correct envelopes?
    e) all letters are put into the correct envelopes?

14. A group of n students is assigned seats for each of two classes in the same classroom. How many ways can these seats be assigned if no student is assigned the same seat for both classes?

*15. a) Use a combinatorial argument to show that the sequence $\{D_n\}$, where D_n denotes the number of derangements of n objects, satisfies the recurrence relation

$$D_n = (n - 1)(D_{n-1} + D_{n-2})$$

for $n \geq 2$. [*Hint:* Note that there are $n - 1$ choices for the first element k of a derangement. Consider separately the derangements that start with k that do and do not have 1 in the kth position.]

   b) Show that

$$D_n = nD_{n-1} + (-1)^n$$

for $n \geq 1$.

16. Use Exercise 15 to find an explicit formula for D_n.

17. For which positive integers n is D_n, the number of derangements of n objects, even?

18. Suppose that p and q are distinct primes. Use the principle of inclusion–exclusion to find $\phi(pq)$, the number of positive integers not exceeding pq that are relatively prime to pq.

19. How many derangements of $\{1, 2, 3, 4, 5, 6\}$ begin with the integers 1, 2, and 3, in some order?

20. How many derangements of $\{1, 2, 3, 4, 5, 6\}$ end with the integers 1, 2, and 3, in some order?

21. Prove Theorem 1.

Key Terms and Results

TERMS

recurrence relation: a formula expressing terms of a sequence, except for some initial terms, as a function of one or more previous terms of the sequence

initial conditions for a recurrence relation: the values of the terms of a sequence satisfying the recurrence relation before this relation takes effect

dynamic programming: an algorithmic paradigm that finds the solution to an optimization problem by recursively breaking down the problem into overlapping subproblems and combining their solutions with the help of a recurrence relation

linear homogeneous recurrence relation with constant coefficients: a recurrence relation that expresses the terms of a sequence, except initial terms, as a linear combination of previous terms

characteristic roots of a linear homogeneous recurrence relation with constant coefficients: the roots of the polynomial associated with a linear homogeneous recurrence relation with constant coefficients

linear nonhomogeneous recurrence relation with constant coefficients: a recurrence relation that expresses the terms of a sequence, except for initial terms, as a linear combination of previous terms plus a function that is not identically zero that depends only on the index

divide-and-conquer algorithm: an algorithm that solves a problem recursively by splitting it into a fixed number of smaller non-overlapping subproblems of the same type

generating function of a sequence: the formal series that has the nth term of the sequence as the coefficient of x^n

sieve of Eratosthenes: a procedure for finding the primes less than a specified positive integer

derangement: a permutation of objects such that no object is in its original place

RESULTS

the formula for the number of elements in the union of two finite sets:

$$|A \cup B| = |A| + |B| - |A \cap B|$$

the formula for the number of elements in the union of three finite sets:

$$|A \cup B \cup C| = |A| + |B| + |C| - |A \cap B| - |A \cap C| \\ - |B \cap C| + |A \cap B \cap C|$$

the principle of inclusion–exclusion:

$$|A_1 \cup A_2 \cup \cdots \cup A_n| = \sum_{1 \le i \le n} |A_i| - \sum_{1 \le i < j \le n} |A_i \cap A_j| \\ + \sum_{1 \le i < j < k \le n} |A_i \cap A_j \cap A_k| \\ - \cdots + (-1)^{n+1} |A_1 \cap A_2 \cap \cdots \cap A_n|$$

the number of onto functions from a set with m elements to a set with n elements:

$$n^m - C(n, 1)(n - 1)^m + C(n, 2)(n - 2)^m \\ - \cdots + (-1)^{n-1} C(n, n - 1) \cdot 1^m$$

the number of derangements of n objects:

$$D_n = n! \left[1 - \frac{1}{1!} + \frac{1}{2!} - \cdots + (-1)^n \frac{1}{n!} \right]$$

Review Questions

1. a) What is a recurrence relation?

    b) Find a recurrence relation for the amount of money that will be in an account after n years if \$1,000,000 is deposited in an account yielding 9% annually.

2. Explain how the Fibonacci numbers are used to solve Fibonacci's problem about rabbits.

3. a) Find a recurrence relation for the number of steps needed to solve the Tower of Hanoi puzzle.

    b) Show how this recurrence relation can be solved using iteration.

4. a) Explain how to find a recurrence relation for the number of bit strings of length n not containing two consecutive 1s.

b) Describe another counting problem that has a solution satisfying the same recurrence relation.

5. a) What is dynamic programming and how are recurrence relations used in algorithms that follow this paradigm?

b) Explain how dynamic programming can be used to schedule talks in a lecture hall from a set of possible talks to maximize overall attendance.

6. Define a linear homogeneous recurrence relation of degree k.

7. a) Explain how to solve linear homogeneous recurrence relations of degree 2.

b) Solve the recurrence relation $a_n = 13a_{n-1} - 22a_{n-2}$ for $n \geq 2$ if $a_0 = 3$ and $a_1 = 15$.

c) Solve the recurrence relation $a_n = 14a_{n-1} - 49a_{n-2}$ for $n \geq 2$ if $a_0 = 3$ and $a_1 = 35$.

8. a) Explain how to find $f(b^k)$ where k is a positive integer if $f(n)$ satisfies the divide-and-conquer recurrence relation $f(n) = af(n/b) + g(n)$ whenever b divides the positive integer n.

b) Find $f(256)$ if $f(n) = 3f(n/4) + 5n/4$ and $f(1) = 7$.

9. a) Derive a divide-and-conquer recurrence relation for the number of comparisons used to find a number in a list using a binary search.

b) Give a big-O estimate for the number of comparisons used by a binary search from the divide-and-conquer recurrence relation you gave in (a) using Theorem 1 in Section 8.3.

10. a) Give a formula for the number of elements in the union of three sets.

b) Explain why this formula is valid.

c) Explain how to use the formula from (a) to find the number of integers not exceeding 1000 that are divisible by 6, 10, or 15.

d) Explain how to use the formula from (a) to find the number of solutions in nonnegative integers to the equation $x_1 + x_2 + x_3 + x_4 = 22$ with $x_1 < 8$, $x_2 < 6$, and $x_3 < 5$.

11. a) Give a formula for the number of elements in the union of four sets and explain why it is valid.

b) Suppose the sets A_1, A_2, A_3, and A_4 each contain 25 elements, the intersection of any two of these sets contains 5 elements, the intersection of any three of these sets contains 2 elements, and 1 element is in all four of the sets. How many elements are in the union of the four sets?

12. a) State the principle of inclusion–exclusion.

b) Outline a proof of this principle.

13. Explain how the principle of inclusion–exclusion can be used to count the number of onto functions from a set with m elements to a set with n elements.

14. a) How can you count the number of ways to assign m jobs to n employees so that each employee is assigned at least one job?

b) How many ways are there to assign seven jobs to three employees so that each employee is assigned at least one job?

15. Explain how the inclusion–exclusion principle can be used to count the number of primes not exceeding the positive integer n.

16. a) Define a derangement.

b) Why is counting the number of ways a hatcheck person can return hats to n people, so that no one receives the correct hat, the same as counting the number of derangements of n objects?

c) Explain how to count the number of derangements of n objects.

Supplementary Exercises

1. A group of 10 people begin a chain letter, with each person sending the letter to four other people. Each of these people sends the letter to four additional people.

a) Find a recurrence relation for the number of letters sent at the nth stage of this chain letter, if no person ever receives more than one letter.

b) What are the initial conditions for the recurrence relation in part (a)?

c) How many letters are sent at the nth stage of the chain letter?

2. A nuclear reactor has created 18 grams of a particular radioactive isotope. Every hour 1% of this radioactive isotope decays.

a) Set up a recurrence relation for the amount of this isotope left n hours after its creation.

b) What are the initial conditions for the recurrence relation in part (a)?

c) Solve this recurrence relation.

3. Messages are sent over a communications channel using two different signals. One signal requires 2 microseconds for transmittal, and the other signal requires 3 microseconds for transmittal. Each signal of a message is followed immediately by the next signal.

a) Find a recurrence relation for the number of different signals that can be sent in n microseconds.

b) What are the initial conditions of the recurrence relation in part (a)?

c) How many different messages can be sent in 12 microseconds?

4. Find the solutions of the simultaneous system of recurrence relations

$$a_n = a_{n-1} + b_{n-1}$$
$$b_n = a_{n-1} - b_{n-1}$$

with $a_0 = 1$ and $b_0 = 2$.

5. Solve the recurrence relation $a_n = a_{n-1}^2/a_{n-2}$ if $a_0 = 1$ and $a_1 = 2$. [*Hint:* Take logarithms of both sides to obtain a recurrence relation for the sequence $\log a_n$, $n = 0, 1, 2, \ldots$.]

*6. Solve the recurrence relation $a_n = a_{n-1}^3 a_{n-2}^2$ if $a_0 = 2$ and $a_1 = 2$. (See the hint for Exercise 5.)

7. Find the solution of the recurrence relation $a_n = 3a_{n-1} - 3a_{n-2} + a_{n-3} + 1$ if $a_0 = 2$, $a_1 = 4$, and $a_2 = 8$.

8. Find the solution of the recurrence relation $a_n = 3a_{n-1} - 3a_{n-2} + a_{n-3}$ if $a_0 = 2$, $a_1 = 2$, and $a_2 = 4$.

*9. Suppose that in Example 1 of Section 8.1 a pair of rabbits leaves the island after reproducing twice. Find a recurrence relation for the number of rabbits on the island in the middle of the nth month.

*10. In this exercise we construct a dynamic programming algorithm for solving the problem of finding a subset S of items chosen from a set of n items where item i has a weight w_i, which is a positive integer, so that the total weight of the items in S is a maximum but does not exceed a fixed weight limit W. Let $M(j, w)$ denote the maximum total weight of the items in a subset of the first j items such that this total weight does not exceed w. This problem is known as the **knapsack problem**.

a) Show that if $w_j > w$, then $M(j, w) = M(j - 1, w)$.

b) Show that if $w_j \leq w$, then $M(j, w) = \max(M(j - 1, w), w_j + M(j - 1, w - w_j))$.

c) Use (a) and (b) to construct a dynamic programming algorithm for determining the maximum total weight of items so that this total weight does not exceed W. In your algorithm store the values $M(j, w)$ as they are found.

d) Explain how you can use the values $M(j, w)$ computed by the algorithm in part (c) to find a subset of items with maximum total weight not exceeding W.

In Exercises 11–13 we develop a dynamic programming algorithm for finding a longest common subsequence of two sequences $a_1, a_2, \ldots, a_m$ and $b_1, b_2, \ldots, b_n$, an important problem in the comparison of DNA of different organisms.

11. Suppose that $c_1, c_2, \ldots, c_p$ is a longest common subsequence of the sequences $a_1, a_2, \ldots, a_m$ and $b_1, b_2, \ldots, b_n$.

a) Show that if $a_m = b_n$, then $c_p = a_m = b_n$ and $c_1, c_2, \ldots, c_{p-1}$ is a longest common subsequence of $a_1, a_2, \ldots, a_{m-1}$ and $b_1, b_2, \ldots, b_{n-1}$ when $p > 1$.

b) Suppose that $a_m \neq b_n$. Show that if $c_p \neq a_m$, then $c_1, c_2, \ldots, c_p$ is a longest common subsequence of $a_1, a_2, \ldots, a_{m-1}$ and $b_1, b_2, \ldots, b_n$ and also show that if $c_p \neq b_n$, then $c_1, c_2, \ldots, c_p$ is a longest common subsequence of $a_1, a_2, \ldots, a_m$ and $b_1, b_2, \ldots, b_{n-1}$.

12. Let $L(i, j)$ denote the length of a longest common subsequence of $a_1, a_2, \ldots, a_i$ and $b_1, b_2, \ldots, b_j$, where $0 \leq i \leq m$ and $0 \leq j \leq n$. Use parts (a) and (b) of Exercise 11 to show that $L(i, j)$ satisfies the recurrence relation $L(i, j) = L(i - 1, j - 1) + 1$ if both i

and j are nonzero and $a_i = b_i$, and $L(i, j) = \max(L(i, j - 1), L(i - 1, j))$ if both i and j are nonzero and $a_i \neq b_i$, and the initial condition $L(i, j) = 0$ if $i = 0$ or $j = 0$.

13. Use Exercise 12 to construct a dynamic programming algorithm for computing the length of a longest common subsequence of two sequences $a_1, a_2, \ldots, a_m$ and $b_1, b_2, \ldots, b_n$, storing the values of $L(i, j)$ as they are found.

14. Find a recurrence relation that describes the number of comparisons used by the following algorithm: Find the largest and second largest elements of a sequence of n numbers recursively by splitting the sequence into two subsequences with an equal number of terms, or where there is one more term in one subsequence than in the other, at each stage. Stop when subsequences with two terms are reached.

15. Give a big-O estimate for the number of comparisons used by the algorithm described in Exercise 14.

Let $\{a_n\}$ be a sequence of real numbers. The **forward differences** of this sequence are defined recursively as follows: The **first forward difference** is $\Delta a_n = a_{n+1} - a_n$; the **$(k + 1)$st forward difference** $\Delta^{k+1} a_n$ is obtained from $\Delta^k a_n$ by $\Delta^{k+1} a_n = \Delta^k a_{n+1} - \Delta^k a_n$.

16. Find Δa_n, where

a) $a_n = 3$. **b)** $a_n = 4n + 7$. **c)** $a_n = n^2 + n + 1$.

17. Let $a_n = 3n^3 + n + 2$. Find $\Delta^k a_n$, where k equals

a) 2. **b)** 3. **c)** 4.

*18. Suppose that $a_n = P(n)$, where P is a polynomial of degree d. Prove that $\Delta^{d+1} a_n = 0$ for all nonnegative integers n.

19. Let $\{a_n\}$ and $\{b_n\}$ be sequences of real numbers. Show that

$$\Delta(a_n b_n) = a_{n+1}(\Delta b_n) + b_n(\Delta a_n).$$

20. Show that if $F(x)$ and $G(x)$ are the generating functions for the sequences $\{a_k\}$ and $\{b_k\}$, respectively, and c and d are real numbers, then $(cF + dG)(x)$ is the generating function for $\{ca_k + db_k\}$.

21. (*Requires calculus*) This exercise shows how generating functions can be used to solve the recurrence relation $(n + 1)a_{n+1} = a_n + (1/n!)$ for $n \geq 0$ with initial condition $a_0 = 1$.

a) Let $G(x)$ be the generating function for $\{a_n\}$. Show that $G'(x) = G(x) + e^x$ and $G(0) = 1$.

b) Show from part (a) that $(e^{-x}G(x))' = 1$, and conclude that $G(x) = xe^x + e^x$.

c) Use part (b) to find a closed form for a_n.

22. Suppose that 14 students receive an A on the first exam in a discrete mathematics class, and 18 receive an A on the second exam. If 22 students received an A on either the first exam or the second exam, how many students received an A on both exams?

23. There are 323 farms in Monmouth County that have at least one of horses, cows, and sheep. If 224 have horses, 85 have cows, 57 have sheep, and 18 farms have all three types of animals, how many farms have exactly two of these three types of animals?

24. Queries to a database of student records at a college produced the following data: There are 2175 students at the college, 1675 of these are not freshmen, 1074 students have taken a course in calculus, 444 students have taken a course in discrete mathematics, 607 students are not freshmen and have taken calculus, 350 students have taken calculus and discrete mathematics, 201 students are not freshmen and have taken discrete mathematics, and 143 students are not freshmen and have taken both calculus and discrete mathematics. Can all the responses to the queries be correct?

25. Students in the school of mathematics at a university major in one or more of the following four areas: applied mathematics (AM), pure mathematics (PM), operations research (OR), and computer science (CS). How many students are in this school if (including joint majors) there are 23 students majoring in AM; 17 in PM; 44 in OR; 63 in CS; 5 in AM and PM; 8 in AM and CS; 4 in AM and OR; 6 in PM and CS; 5 in PM and OR; 14 in OR and CS; 2 in PM, OR, and CS; 2 in AM, OR, and CS; 1 in PM, AM, and OR; 1 in PM, AM, and CS; and 1 in all four fields.

26. How many terms are needed when the inclusion–exclusion principle is used to express the number of elements in the union of seven sets if no more than five of these sets have a common element?

27. How many solutions in positive integers are there to the equation $x_1 + x_2 + x_3 = 20$ with $2 < x_1 < 6$, $6 < x_2 < 10$, and $0 < x_3 < 5$?

28. How many positive integers less than 1,000,000 are
   a) divisible by 2, 3, or 5?
   b) not divisible by 7, 11, or 13?
   c) divisible by 3 but not by 7?

29. How many positive integers less than 200 are
   a) second or higher powers of integers?
   b) either primes or second or higher powers of integers?
   c) not divisible by the square of an integer greater than 1?
   d) not divisible by the cube of an integer greater than 1?
   e) not divisible by three or more primes?

＊30. How many ways are there to assign six different jobs to three different employees if the hardest job is assigned to the most experienced employee and the easiest job is assigned to the least experienced employee?

31. What is the probability that exactly one person is given back the correct hat by a hatcheck person who gives n people their hats back at random?

32. How many bit strings of length six do not contain four consecutive 1s?

33. What is the probability that a bit string of length six chosen at random contains at least four 1s?

Computer Projects

Write programs with these input and output.

1. Given a positive integer n, list all the moves required in the Tower of Hanoi puzzle to move n disks from one peg to another according to the rules of the puzzle.

2. Given a positive integer n and an integer k with $1 \leq k \leq n$, list all the moves used by the Frame–Stewart algorithm (described in the preamble to Exercise 30 of Section 8.1) to move n disks from one peg to another using four pegs according to the rules of the puzzle.

3. Given a positive integer n, list all the bit sequences of length n that do not have a pair of consecutive 0s.

4. Given an integer n greater than 1, write out all ways to parenthesize the product of $n + 1$ variables.

5. Given a set of n talks, their start and end times, and the number of attendees at each talk, use dynamic programming to schedule a subset of these talks in a single lecture hall to maximize total attendance.

6. Given matrices $\mathbf{A}_1, \mathbf{A}_2, \ldots, \mathbf{A}_n$, with dimensions $m_1 \times m_2, m_2 \times m_3, \ldots, m_n \times m_{n+1}$, respectively, each with integer entries, use dynamic programming, as outlined in Exercise 43 in Section 8.1, to find the minimum number of multiplications of integers needed to compute $\mathbf{A}_1 \mathbf{A}_2 \cdots \mathbf{A}_n$.

7. Given a recurrence relation $a_n = c_1 a_{n-1} + c_2 a_{n-2}$, where c_1 and c_2 are real numbers, initial conditions $a_0 = C_0$ and $a_1 = C_1$, and a positive integer k, find a_k using iteration.

8. Given a recurrence relation $a_n = c_1 a_{n-1} + c_2 a_{n-2}$ and initial conditions $a_0 = C_0$ and $a_1 = C_1$, determine the unique solution.

9. Given a recurrence relation of the form $f(n) = af(n/b) + c$, where a is a real number, b is a positive integer, and c is a real number, and a positive integer k, find $f(b^k)$ using iteration.

10. Given the number of elements in the intersection of three sets, the number of elements in each pairwise intersection of these sets, and the number of elements in each set, find the number of elements in their union.

11. Given a positive integer n, produce the formula for the number of elements in the union of n sets.

12. Given positive integers m and n, find the number of onto functions from a set with m elements to a set with n elements.

13. Given a positive integer n, list all the derangements of the set $\{1, 2, 3, \ldots, n\}$.

Computations and Explorations

Use a computational program or programs you have written to do these exercises.

1. Find the exact value of f_{100}, f_{500}, and f_{1000}, where f_n is the nth Fibonacci number.

2. Write out all the moves required to solve the Tower of Hanoi puzzle with 10 disks.

3. Compute the number of operations required to multiply two integers with n bits for various integers n including 16, 64, 256, and 1024 using the fast multiplication described in Example 4 of Section 8.3 and the standard algorithm for multiplying integers (Algorithm 3 in Section 4.2).

4. Compute the number of operations required to multiply two $n \times n$ matrices for various integers n including 4, 16, 64, and 128 using the fast matrix multiplication described in Example 5 of Section 8.3 and the standard algorithm for multiplying matrices (Algorithm 1 in Section 3.3).

5. Find the number of primes not exceeding 10,000 using the method described in Section 8.6 to find the number of primes not exceeding 100.

6. List all the derangements of $\{1, 2, 3, 4, 5, 6, 7, 8\}$.

Writing Projects

Respond to these with essays using outside sources.

1. Find the original source where Fibonacci presented his puzzle about modeling rabbit populations. Discuss this problem and other problems posed by Fibonacci and give some information about Fibonacci himself.

2. Explain how the Fibonacci numbers arise in a variety of applications, such as in phyllotaxis, the study of arrangement of leaves in plants, in the study of reflections by mirrors, and so on.

3. Describe different variations of the Tower of Hanoi puzzle, including those with more than three pegs (including the Reve's puzzle discussed in the text and exercises), those where disk moves are restricted, and those where disks may have the same size. Include what is known about the number of moves required to solve each variation.

4. Discuss as many different problems as possible where the Catalan numbers arise.

5. Discuss some of the problems in which Richard Bellman first used dynamic programming.

6. Describe the role dynamic programming algorithms play in bioinformatics including for DNA sequence comparison, gene comparison, and RNA structure prediction.

7. Describe the use of dynamic programming in economics including its use to study optimal consumption and saving.

8. Explain how dynamic programming can be used to solve the egg-dropping puzzle which determines from which floors of a multistory building it is safe to drop eggs from without breaking.

9. Describe the solution of Ulam's problem (see Exercise 22 in Section 8.3) involving searching with one lie found by Andrzej Pelc.

10. Discuss variations of Ulam's problem (see Exercise 22 in Section 8.3) involving searching with more than one lie and what is known about this problem.

11. Define the convex hull of a set of points in the plane and describe three different algorithms, including a divide-and-conquer algorithm, for finding the convex hull of a set of points in the plane.

12. Describe how sieve methods are used in number theory. What kind of results have been established using such methods?

13. Look up the rules of the old French card game of *rencontres*. Describe these rules and describe the work of Pierre Raymond de Montmort on *le problème de rencontres*.

14. Describe how exponential generating functions can be used to solve a variety of counting problems.

15. Describe the Polyá theory of counting and the kind of counting problems that can be solved using this theory.

16. The *problème des ménages* (the problem of the households) asks for the number of ways to arrange n couples around a table so that the sexes alternate and no husband and wife are seated together. Explain the method used by E. Lucas to solve this problem.

17. Explain how *rook polynomials* can be used to solve counting problems.

Relations

Relationships between elements of sets occur in many contexts. Every day we deal with relationships such as those between a business and its telephone number, an employee and his or her salary, a person and a relative, and so on. In mathematics we study relationships such as those between a positive integer and one that it divides, an integer and one that it is congruent to modulo 5, a real number and one that is larger than it, a real number x and the value $f(x)$ where f is a function, and so on. Relationships such as that between a program and a variable it uses, and that between a computer language and a valid statement in this language often arise in computer science.

Relationships between elements of sets are represented using the structure called a relation, which is just a subset of the Cartesian product of the sets. Relations can be used to solve problems such as determining which pairs of cities are linked by airline flights in a network, finding a viable order for the different phases of a complicated project, or producing a useful way to store information in computer databases.

In some computer languages, only the first 31 characters of the name of a variable matter. The relation consisting of ordered pairs of strings where the first string has the same initial 31 characters as the second string is an example of a special type of relation, known as an equivalence relation. Equivalence relations arise throughout mathematics and computer science. We will study equivalence relations, and other special types of relations, in this chapter.

9.1 Relations and Their Properties

Introduction

The most direct way to express a relationship between elements of two sets is to use ordered pairs made up of two related elements. For this reason, sets of ordered pairs are called binary relations. In this section we introduce the basic terminology used to describe binary relations. Later in this chapter we will use relations to solve problems involving communications networks, project scheduling, and identifying elements in sets with common properties.

DEFINITION 1   Let A and B be sets. A *binary relation from A to B* is a subset of $A \times B$.

In other words, a binary relation from A to B is a set R of ordered pairs where the first element of each ordered pair comes from A and the second element comes from B. We use the notation $a\,R\,b$ to denote that $(a, b) \in R$ and $a\,\not\!R\,b$ to denote that $(a, b) \notin R$. Moreover, when (a, b) belongs to R, a is said to be **related to** b by R.

Binary relations represent relationships between the elements of two sets. We will introduce *n*-ary relations, which express relationships among elements of more than two sets, later in this chapter. We will omit the word *binary* when there is no danger of confusion.

Examples 1–3 illustrate the notion of a relation.

EXAMPLE 1   Let A be the set of students in your school, and let B be the set of courses. Let R be the relation that consists of those pairs (a, b), where a is a student enrolled in course b. For instance, if Jason Goodfriend and Deborah Sherman are enrolled in CS518, the pairs

(Jason Goodfriend, CS518) and (Deborah Sherman, CS518) belong to R. If Jason Goodfriend is also enrolled in CS510, then the pair (Jason Goodfriend, CS510) is also in R. However, if Deborah Sherman is not enrolled in CS510, then the pair (Deborah Sherman, CS510) is not in R.

Note that if a student is not currently enrolled in any courses there will be no pairs in R that have this student as the first element. Similarly, if a course is not currently being offered there will be no pairs in R that have this course as their second element. ◄

EXAMPLE 2 Let A be the set of cities in the U.S.A., and let B be the set of the 50 states in the U.S.A. Define the relation R by specifying that (a, b) belongs to R if a city with name a is in the state b. For instance, (Boulder, Colorado), (Bangor, Maine), (Ann Arbor, Michigan), (Middletown, New Jersey), (Middletown, New York), (Cupertino, California), and (Red Bank, New Jersey) are in R. ◄

EXAMPLE 3 Let $A = \{0, 1, 2\}$ and $B = \{a, b\}$. Then $\{(0, a), (0, b), (1, a), (2, b)\}$ is a relation from A to B. This means, for instance, that $0\,R\,a$, but that $1\,\not\!R\,b$. Relations can be represented graphically, as shown in Figure 1, using arrows to represent ordered pairs. Another way to represent this relation is to use a table, which is also done in Figure 1. We will discuss representations of relations in more detail in Section 9.3. ◄

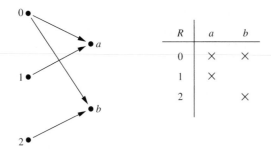

R	a	b
0	×	×
1	×	
2		×

FIGURE 1 **Displaying the Ordered Pairs in the Relation R from Example 3.**

Functions as Relations

Recall that a function f from a set A to a set B (as defined in Section 2.3) assigns exactly one element of B to each element of A. The graph of f is the set of ordered pairs (a, b) such that $b = f(a)$. Because the graph of f is a subset of $A \times B$, it is a relation from A to B. Moreover, the graph of a function has the property that every element of A is the first element of exactly one ordered pair of the graph.

Conversely, if R is a relation from A to B such that every element in A is the first element of exactly one ordered pair of R, then a function can be defined with R as its graph. This can be done by assigning to an element a of A the unique element $b \in B$ such that $(a, b) \in R$. (Note that the relation R in Example 2 is not the graph of a function because Middletown occurs more than once as the first element of an ordered pair in R.)

A relation can be used to express a one-to-many relationship between the elements of the sets A and B (as in Example 2), where an element of A may be related to more than one element of B. A function represents a relation where exactly one element of B is related to each element of A.

Relations are a generalization of graphs of functions; they can be used to express a much wider class of relationships between sets. (Recall that the graph of the function f from A to B is the set of ordered pairs $(a, f(a))$ for $a \in A$.)

Relations on a Set

Relations from a set A to itself are of special interest.

DEFINITION 2   A *relation on a set* A is a relation from A to A.

In other words, a relation on a set A is a subset of $A \times A$.

EXAMPLE 4   Let A be the set $\{1, 2, 3, 4\}$. Which ordered pairs are in the relation $R = \{(a, b) \mid a \text{ divides } b\}$?

Solution: Because (a, b) is in R if and only if a and b are positive integers not exceeding 4 such that a divides b, we see that

$$R = \{(1, 1), (1, 2), (1, 3), (1, 4), (2, 2), (2, 4), (3, 3), (4, 4)\}.$$

The pairs in this relation are displayed both graphically and in tabular form in Figure 2.   ◄

Next, some examples of relations on the set of integers will be given in Example 5.

EXAMPLE 5   Consider these relations on the set of integers:

$R_1 = \{(a, b) \mid a \le b\},$
$R_2 = \{(a, b) \mid a > b\},$
$R_3 = \{(a, b) \mid a = b \text{ or } a = -b\},$
$R_4 = \{(a, b) \mid a = b\},$
$R_5 = \{(a, b) \mid a = b + 1\},$
$R_6 = \{(a, b) \mid a + b \le 3\}.$

Which of these relations contain each of the pairs $(1, 1)$, $(1, 2)$, $(2, 1)$, $(1, -1)$, and $(2, 2)$?

Remark: Unlike the relations in Examples 1–4, these are relations on an infinite set.

Solution: The pair $(1, 1)$ is in R_1, R_3, R_4, and R_6; $(1, 2)$ is in R_1 and R_6; $(2, 1)$ is in R_2, R_5, and R_6; $(1, -1)$ is in R_2, R_3, and R_6; and finally, $(2, 2)$ is in R_1, R_3, and R_4.   ◄

It is not hard to determine the number of relations on a finite set, because a relation on a set A is simply a subset of $A \times A$.

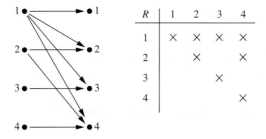

FIGURE 2   **Displaying the Ordered Pairs in the Relation R from Example 4.**

EXAMPLE 6   How many relations are there on a set with n elements?

Solution: A relation on a set A is a subset of $A \times A$. Because $A \times A$ has n^2 elements when A has n elements, and a set with m elements has 2^m subsets, there are 2^{n^2} subsets of $A \times A$. Thus, there are 2^{n^2} relations on a set with n elements. For example, there are $2^{3^2} = 2^9 = 512$ relations on the set $\{a, b, c\}$. ◀

Properties of Relations

There are several properties that are used to classify relations on a set. We will introduce the most important of these here.

In some relations an element is always related to itself. For instance, let R be the relation on the set of all people consisting of pairs (x, y) where x and y have the same mother and the same father. Then $x R x$ for every person x.

DEFINITION 3   A relation R on a set A is called *reflexive* if $(a, a) \in R$ for every element $a \in A$.

Remark: Using quantifiers we see that the relation R on the set A is reflexive if $\forall a((a, a) \in R)$, where the universe of discourse is the set of all elements in A.

We see that a relation on A is reflexive if every element of A is related to itself. Examples 7–9 illustrate the concept of a reflexive relation.

EXAMPLE 7   Consider the following relations on $\{1, 2, 3, 4\}$:

$R_1 = \{(1, 1), (1, 2), (2, 1), (2, 2), (3, 4), (4, 1), (4, 4)\}$,

$R_2 = \{(1, 1), (1, 2), (2, 1)\}$,

$R_3 = \{(1, 1), (1, 2), (1, 4), (2, 1), (2, 2), (3, 3), (4, 1), (4, 4)\}$,

$R_4 = \{(2, 1), (3, 1), (3, 2), (4, 1), (4, 2), (4, 3)\}$,

$R_5 = \{(1, 1), (1, 2), (1, 3), (1, 4), (2, 2), (2, 3), (2, 4), (3, 3), (3, 4), (4, 4)\}$,

$R_6 = \{(3, 4)\}$.

Which of these relations are reflexive?

Solution: The relations R_3 and R_5 are reflexive because they both contain all pairs of the form (a, a), namely, $(1, 1)$, $(2, 2)$, $(3, 3)$, and $(4, 4)$. The other relations are not reflexive because they do not contain all of these ordered pairs. In particular, R_1, R_2, R_4, and R_6 are not reflexive because $(3, 3)$ is not in any of these relations. ◀

EXAMPLE 8   Which of the relations from Example 5 are reflexive?

Solution: The reflexive relations from Example 5 are R_1 (because $a \le a$ for every integer a), R_3, and R_4. For each of the other relations in this example it is easy to find a pair of the form (a, a) that is not in the relation. (This is left as an exercise for the reader.) ◀

EXAMPLE 9   Is the "divides" relation on the set of positive integers reflexive?

Solution: Because $a \mid a$ whenever a is a positive integer, the "divides" relation is reflexive. (Note that if we replace the set of positive integers with the set of all integers the relation is not reflexive because by definition 0 does not divide 0.) ◀

In some relations an element is related to a second element if and only if the second element is also related to the first element. The relation consisting of pairs (x, y), where x and y are students at your school with at least one common class has this property. Other relations have the property that if an element is related to a second element, then this second element is not related to the first. The relation consisting of the pairs (x, y), where x and y are students at your school, where x has a higher grade point average than y has this property.

DEFINITION 4   A relation R on a set A is called *symmetric* if $(b, a) \in R$ whenever $(a, b) \in R$, for all $a, b \in A$. A relation R on a set A such that for all $a, b \in A$, if $(a, b) \in R$ and $(b, a) \in R$, then $a = b$ is called *antisymmetric*.

Remark: Using quantifiers, we see that the relation R on the set A is symmetric if $\forall a \forall b ((a, b) \in R \rightarrow (b, a) \in R)$. Similarly, the relation R on the set A is antisymmetric if $\forall a \forall b (((a, b) \in R \wedge (b, a) \in R) \rightarrow (a = b))$.

That is, a relation is symmetric if and only if a is related to b implies that b is related to a. A relation is antisymmetric if and only if there are no pairs of distinct elements a and b with a related to b and b related to a. That is, the only way to have a related to b and b related to a is for a and b to be the same element. The terms *symmetric* and *antisymmetric* are not opposites, because a relation can have both of these properties or may lack both of them (see Exercise 10). A relation cannot be both symmetric and antisymmetric if it contains some pair of the form (a, b), where $a \neq b$.

Remark: Although relatively few of the 2^{n^2} relations on a set with n elements are symmetric or antisymmetric, as counting arguments can show, many important relations have one of these properties. (See Exercise 33.)

EXAMPLE 10   Which of the relations from Example 7 are symmetric and which are antisymmetric?

Solution: The relations R_2 and R_3 are symmetric, because in each case (b, a) belongs to the relation whenever (a, b) does. For R_2, the only thing to check is that both $(2, 1)$ and $(1, 2)$ are in the relation. For R_3, it is necessary to check that both $(1, 2)$ and $(2, 1)$ belong to the relation, and $(1, 4)$ and $(4, 1)$ belong to the relation. The reader should verify that none of the other relations is symmetric. This is done by finding a pair (a, b) such that it is in the relation but (b, a) is not.

R_4, R_5, and R_6 are all antisymmetric. For each of these relations there is no pair of elements a and b with $a \neq b$ such that both (a, b) and (b, a) belong to the relation. The reader should verify that none of the other relations is antisymmetric. This is done by finding a pair (a, b) with $a \neq b$ such that (a, b) and (b, a) are both in the relation.   ◀

EXAMPLE 11   Which of the relations from Example 5 are symmetric and which are antisymmetric?

Solution: The relations R_3, R_4, and R_6 are symmetric. R_3 is symmetric, for if $a = b$ or $a = -b$, then $b = a$ or $b = -a$. R_4 is symmetric because $a = b$ implies that $b = a$. R_6 is symmetric because $a + b \leq 3$ implies that $b + a \leq 3$. The reader should verify that none of the other relations is symmetric.

The relations R_1, R_2, R_4, and R_5 are antisymmetric. R_1 is antisymmetric because the inequalities $a \leq b$ and $b \leq a$ imply that $a = b$. R_2 is antisymmetric because it is impossible that $a > b$ and $b > a$. R_4 is antisymmetric, because two elements are related with respect to R_4 if and only if they are equal. R_5 is antisymmetric because it is impossible that $a = b + 1$ and $b = a + 1$. The reader should verify that none of the other relations is antisymmetric.   ◀

EXAMPLE 12   Is the "divides" relation on the set of positive integers symmetric? Is it antisymmetric?

Solution: This relation is not symmetric because $1 \mid 2$, but $2 \nmid 1$. It is antisymmetric, for if a and b are positive integers with $a \mid b$ and $b \mid a$, then $a = b$ (the verification of this is left as an exercise for the reader).   ◀

Let R be the relation consisting of all pairs (x, y) of students at your school, where x has taken more credits than y. Suppose that x is related to y and y is related to z. This means that x has taken more credits than y and y has taken more credits than z. We can conclude that x has taken more credits than z, so that x is related to z. What we have shown is that R has the transitive property, which is defined as follows.

DEFINITION 5   A relation R on a set A is called *transitive* if whenever $(a, b) \in R$ and $(b, c) \in R$, then $(a, c) \in R$, for all $a, b, c \in A$.

Remark: Using quantifiers we see that the relation R on a set A is transitive if we have $\forall a \forall b \forall c (((a, b) \in R \wedge (b, c) \in R) \rightarrow (a, c) \in R)$.

EXAMPLE 13   Which of the relations in Example 7 are transitive?

Solution: R_4, R_5, and R_6 are transitive. For each of these relations, we can show that it is transitive by verifying that if (a, b) and (b, c) belong to this relation, then (a, c) also does. For instance, R_4 is transitive, because $(3, 2)$ and $(2, 1)$, $(4, 2)$ and $(2, 1)$, $(4, 3)$ and $(3, 1)$, and $(4, 3)$ and $(3, 2)$ are the only such sets of pairs, and $(3, 1)$, $(4, 1)$, and $(4, 2)$ belong to R_4. The reader should verify that R_5 and R_6 are transitive.

R_1 is not transitive because $(3, 4)$ and $(4, 1)$ belong to R_1, but $(3, 1)$ does not. R_2 is not transitive because $(2, 1)$ and $(1, 2)$ belong to R_2, but $(2, 2)$ does not. R_3 is not transitive because $(4, 1)$ and $(1, 2)$ belong to R_3, but $(4, 2)$ does not.   ◀

EXAMPLE 14   Which of the relations in Example 5 are transitive?

Solution: The relations R_1, R_2, R_3, and R_4 are transitive. R_1 is transitive because $a \leq b$ and $b \leq c$ imply that $a \leq c$. R_2 is transitive because $a > b$ and $b > c$ imply that $a > c$. R_3 is transitive because $a = \pm b$ and $b = \pm c$ imply that $a = \pm c$. R_4 is clearly transitive, as the reader should verify. R_5 is not transitive because $(2, 1)$ and $(1, 0)$ belong to R_5, but $(2, 0)$ does not. R_6 is not transitive because $(2, 1)$ and $(1, 2)$ belong to R_6, but $(2, 2)$ does not.   ◀

EXAMPLE 15   Is the "divides" relation on the set of positive integers transitive?

Solution: Suppose that a divides b and b divides c. Then there are positive integers k and l such that $b = ak$ and $c = bl$. Hence, $c = a(kl)$, so a divides c. It follows that this relation is transitive.   ◀

We can use counting techniques to determine the number of relations with specific properties. Finding the number of relations with a particular property provides information about how common this property is in the set of all relations on a set with n elements.

EXAMPLE 16   How many reflexive relations are there on a set with n elements?

Solution: A relation R on a set A is a subset of $A \times A$. Consequently, a relation is determined by specifying whether each of the n^2 ordered pairs in $A \times A$ is in R. However, if R is reflexive, each of the n ordered pairs (a, a) for $a \in A$ must be in R. Each of the other $n(n-1)$ ordered pairs of the form (a, b), where $a \neq b$, may or may not be in R. Hence, by the product rule for counting, there are $2^{n(n-1)}$ reflexive relations [this is the number of ways to choose whether each element (a, b), with $a \neq b$, belongs to R]. ◀

Formulas for the number of symmetric relations and the number of antisymmetric relations on a set with n elements can be found using reasoning similar to that in Example 16 (see Exercise 47). However, no general formula is known that counts the transitive relations on a set with n elements. Currently, $T(n)$, the number of transitive relations on a set with n elements, is known only for $n \leq 17$. For example, $T(4) = 3,994$, $T(5) = 154,303$, and $T(6) = 9,415,189$.

Combining Relations

Because relations from A to B are subsets of $A \times B$, two relations from A to B can be combined in any way two sets can be combined. Consider Examples 17–19.

EXAMPLE 17   Let $A = \{1, 2, 3\}$ and $B = \{1, 2, 3, 4\}$. The relations $R_1 = \{(1, 1), (2, 2), (3, 3)\}$ and $R_2 = \{(1, 1), (1, 2), (1, 3), (1, 4)\}$ can be combined to obtain

$$R_1 \cup R_2 = \{(1, 1), (1, 2), (1, 3), (1, 4), (2, 2), (3, 3)\},$$
$$R_1 \cap R_2 = \{(1, 1)\},$$
$$R_1 - R_2 = \{(2, 2), (3, 3)\},$$
$$R_2 - R_1 = \{(1, 2), (1, 3), (1, 4)\}.$$ ◀

EXAMPLE 18   Let A and B be the set of all students and the set of all courses at a school, respectively. Suppose that R_1 consists of all ordered pairs (a, b), where a is a student who has taken course b, and R_2 consists of all ordered pairs (a, b), where a is a student who requires course b to graduate. What are the relations $R_1 \cup R_2$, $R_1 \cap R_2$, $R_1 \oplus R_2$, $R_1 - R_2$, and $R_2 - R_1$?

Solution: The relation $R_1 \cup R_2$ consists of all ordered pairs (a, b), where a is a student who either has taken course b or needs course b to graduate, and $R_1 \cap R_2$ is the set of all ordered pairs (a, b), where a is a student who has taken course b and needs this course to graduate. Also, $R_1 \oplus R_2$ consists of all ordered pairs (a, b), where student a has taken course b but does not need it to graduate or needs course b to graduate but has not taken it. $R_1 - R_2$ is the set of ordered pairs (a, b), where a has taken course b but does not need it to graduate; that is, b is an elective course that a has taken. $R_2 - R_1$ is the set of all ordered pairs (a, b), where b is a course that a needs to graduate but has not taken. ◀

EXAMPLE 19   Let R_1 be the "less than" relation on the set of real numbers and let R_2 be the "greater than" relation on the set of real numbers, that is, $R_1 = \{(x, y) \mid x < y\}$ and $R_2 = \{(x, y) \mid x > y\}$. What are $R_1 \cup R_2$, $R_1 \cap R_2$, $R_1 - R_2$, $R_2 - R_1$, and $R_1 \oplus R_2$?

Solution: We note that $(x, y) \in R_1 \cup R_2$ if and only if $(x, y) \in R_1$ or $(x, y) \in R_2$. Hence, $(x, y) \in R_1 \cup R_2$ if and only if $x < y$ or $x > y$. Because the condition $x < y$ or $x > y$ is the same as the condition $x \neq y$, it follows that $R_1 \cup R_2 = \{(x, y) \mid x \neq y\}$. In other words, the union of the "less than" relation and the "greater than" relation is the "not equals" relation.

Next, note that it is impossible for a pair (x, y) to belong to both R_1 and R_2 because it is impossible that $x < y$ and $x > y$. It follows that $R_1 \cap R_2 = \emptyset$. We also see that $R_1 - R_2 = R_1$, $R_2 - R_1 = R_2$, and $R_1 \oplus R_2 = R_1 \cup R_2 - R_1 \cap R_2 = \{(x, y) \mid x \neq y\}$. ◀

There is another way that relations are combined that is analogous to the composition of functions.

DEFINITION 6   Let R be a relation from a set A to a set B and S a relation from B to a set C. The *composite* of R and S is the relation consisting of ordered pairs (a, c), where $a \in A$, $c \in C$, and for which there exists an element $b \in B$ such that $(a, b) \in R$ and $(b, c) \in S$. We denote the composite of R and S by $S \circ R$.

Computing the composite of two relations requires that we find elements that are the second element of ordered pairs in the first relation and the first element of ordered pairs in the second relation, as Examples 20 and 21 illustrate.

EXAMPLE 20   What is the composite of the relations R and S, where R is the relation from $\{1, 2, 3\}$ to $\{1, 2, 3, 4\}$ with $R = \{(1, 1), (1, 4), (2, 3), (3, 1), (3, 4)\}$ and S is the relation from $\{1, 2, 3, 4\}$ to $\{0, 1, 2\}$ with $S = \{(1, 0), (2, 0), (3, 1), (3, 2), (4, 1)\}$?

Solution: $S \circ R$ is constructed using all ordered pairs in R and ordered pairs in S, where the second element of the ordered pair in R agrees with the first element of the ordered pair in S. For example, the ordered pairs $(2, 3)$ in R and $(3, 1)$ in S produce the ordered pair $(2, 1)$ in $S \circ R$. Computing all the ordered pairs in the composite, we find

$$S \circ R = \{(1, 0), (1, 1), (2, 1), (2, 2), (3, 0), (3, 1)\}. \qquad \blacktriangleleft$$

EXAMPLE 21   **Composing the Parent Relation with Itself**   Let R be the relation on the set of all people such that $(a, b) \in R$ if person a is a parent of person b. Then $(a, c) \in R \circ R$ if and only if there is a person b such that $(a, b) \in R$ and $(b, c) \in R$, that is, if and only if there is a person b such that a is a parent of b and b is a parent of c. In other words, $(a, c) \in R \circ R$ if and only if a is a grandparent of c. $\qquad \blacktriangleleft$

The powers of a relation R can be recursively defined from the definition of a composite of two relations.

DEFINITION 7   Let R be a relation on the set A. The powers R^n, $n = 1, 2, 3, \ldots$, are defined recursively by

$$R^1 = R \qquad \text{and} \qquad R^{n+1} = R^n \circ R.$$

The definition shows that $R^2 = R \circ R$, $R^3 = R^2 \circ R = (R \circ R) \circ R$, and so on.

EXAMPLE 22   Let $R = \{(1, 1), (2, 1), (3, 2), (4, 3)\}$. Find the powers R^n, $n = 2, 3, 4, \ldots$.

Solution: Because $R^2 = R \circ R$, we find that $R^2 = \{(1, 1), (2, 1), (3, 1), (4, 2)\}$. Furthermore, because $R^3 = R^2 \circ R$, $R^3 = \{(1, 1), (2, 1), (3, 1), (4, 1)\}$. Additional computation shows that R^4 is the same as R^3, so $R^4 = \{(1, 1), (2, 1), (3, 1), (4, 1)\}$. It also follows that $R^n = R^3$ for $n = 5, 6, 7, \ldots$. The reader should verify this. $\qquad \blacktriangleleft$

The following theorem shows that the powers of a transitive relation are subsets of this relation. It will be used in Section 9.4.

THEOREM 1   The relation R on a set A is transitive if and only if $R^n \subseteq R$ for $n = 1, 2, 3, \ldots$.

Proof: We first prove the "if" part of the theorem. We suppose that $R^n \subseteq R$ for $n = 1$, $2, 3, \ldots$. In particular, $R^2 \subseteq R$. To see that this implies R is transitive, note that if $(a, b) \in R$ and $(b, c) \in R$, then by the definition of composition, $(a, c) \in R^2$. Because $R^2 \subseteq R$, this means that $(a, c) \in R$. Hence, R is transitive.

We will use mathematical induction to prove the only if part of the theorem. Note that this part of the theorem is trivially true for $n = 1$.

Assume that $R^n \subseteq R$, where n is a positive integer. This is the inductive hypothesis. To complete the inductive step we must show that this implies that R^{n+1} is also a subset of R. To show this, assume that $(a, b) \in R^{n+1}$. Then, because $R^{n+1} = R^n \circ R$, there is an element x with $x \in A$ such that $(a, x) \in R$ and $(x, b) \in R^n$. The inductive hypothesis, namely, that $R^n \subseteq R$, implies that $(x, b) \in R$. Furthermore, because R is transitive, and $(a, x) \in R$ and $(x, b) \in R$, it follows that $(a, b) \in R$. This shows that $R^{n+1} \subseteq R$, completing the proof. ◁

Exercises

1. List the ordered pairs in the relation R from $A = \{0, 1, 2, 3, 4\}$ to $B = \{0, 1, 2, 3\}$, where $(a, b) \in R$ if and only if

a) $a = b$. **b)** $a + b = 4$.

c) $a > b$. **d)** $a \mid b$.

e) $\gcd(a, b) = 1$. **f)** $\mathrm{lcm}(a, b) = 2$.

2. For each of these relations on the set $\{1, 2, 3, 4\}$, decide whether it is reflexive, whether it is symmetric, whether it is antisymmetric, and whether it is transitive.

a) $\{(2, 2), (2, 3), (2, 4), (3, 2), (3, 3), (3, 4)\}$

b) $\{(1, 1), (1, 2), (2, 1), (2, 2), (3, 3), (4, 4)\}$

c) $\{(2, 4), (4, 2)\}$

d) $\{(1, 2), (2, 3), (3, 4)\}$

e) $\{(1, 1), (2, 2), (3, 3), (4, 4)\}$

f) $\{(1, 3), (1, 4), (2, 3), (2, 4), (3, 1), (3, 4)\}$

3. Determine whether the relation R on the set of all Web pages is reflexive, symmetric, antisymmetric, and/or transitive, where $(a, b) \in R$ if and only if

a) everyone who has visited Web page a has also visited Web page b.

b) there are no common links found on both Web page a and Web page b.

c) there is at least one common link on Web page a and Web page b.

d) there is a Web page that includes links to both Web page a and Web page b.

4. Determine whether the relation R on the set of all integers is reflexive, symmetric, antisymmetric, and/or transitive, where $(x, y) \in R$ if and only if

a) $x \neq y$. **b)** $xy \geq 1$.

c) $x = y + 1$ or $x = y - 1$.

d) $x \equiv y \pmod{7}$. **e)** x is a multiple of y.

f) x and y are both negative or both nonnegative.

g) $x = y^2$. **h)** $x \geq y^2$.

5. Show that the relation $R = \emptyset$ on a nonempty set S is symmetric and transitive, but not reflexive.

6. Show that the relation $R = \emptyset$ on the empty set $S = \emptyset$ is reflexive, symmetric, and transitive.

A relation R on the set A is **irreflexive** if for every $a \in A$, $(a, a) \notin R$. That is, R is irreflexive if no element in A is related to itself.

7. Which relations in Exercise 2 are irreflexive?

8. Which relations in Exercise 3 are irreflexive?

9. Can a relation on a set be neither reflexive nor irreflexive?

10. Give an example of an irreflexive relation on the set of all people.

A relation R is called **asymmetric** if $(a, b) \in R$ implies that $(b, a) \notin R$. Exercises 11–14 explore the notion of an asymmetric relation.

11. Which relations in Exercise 2 are asymmetric?

12. Which relations in Exercise 3 are asymmetric?

13. Use quantifiers to express what it means for a relation to be asymmetric.

14. Give an example of an asymmetric relation on the set of all people.

15. How many different relations are there from a set with m elements to a set with n elements?

☞ Let R be a relation from a set A to a set B. The **inverse relation** from B to A, denoted by R^{-1}, is the set of ordered pairs $\{(b, a) \mid (a, b) \in R\}$. The **complementary relation** $\overline{R}$ is the set of ordered pairs $\{(a, b) \mid (a, b) \notin R\}$.

16. Let R be the relation $R = \{(a, b) \mid a < b\}$ on the set of integers. Find

a) R^{-1}. **b)** $\overline{R}$.

17. Let R be the relation $R = \{(a, b) \mid a \text{ divides } b\}$ on the set of positive integers. Find

a) R^{-1}. **b)** $\overline{R}$.

18. Let R be the relation on the set of all states in the United States consisting of pairs (a, b) where state a borders state b. Find

a) R^{-1}. **b)** $\overline{R}$.

19. Suppose that the function f from A to B is a one-to-one correspondence. Let R be the relation that equals the graph of f. That is, $R = \{(a, f(a)) \mid a \in A\}$. What is the inverse relation R^{-1}?

20. Let $R_1 = \{(1, 2), (2, 3), (3, 4)\}$ and $R_2 = \{(1, 1), (1, 2), (2, 1), (2, 2), (2, 3), (3, 1), (3, 2), (3, 3), (3, 4)\}$ be relations from $\{1, 2, 3\}$ to $\{1, 2, 3, 4\}$. Find
  a) $R_1 \cup R_2$.   **b)** $R_1 \cap R_2$.
  c) $R_1 - R_2$.   **d)** $R_2 - R_1$.

21. Let A be the set of students at your school and B the set of books in the school library. Let R_1 and R_2 be the relations consisting of all ordered pairs (a, b), where student a is required to read book b in a course, and where student a has read book b, respectively. Describe the ordered pairs in each of these relations.
  a) $R_1 \cup R_2$ **b)** $R_1 \cap R_2$
  c) $R_1 \oplus R_2$ **d)** $R_1 - R_2$
  e) $R_2 - R_1$

22. Let R be the relation $\{(1, 2), (1, 3), (2, 3), (2, 4), (3, 1)\}$, and let S be the relation $\{(2, 1), (3, 1), (3, 2), (4, 2)\}$. Find $S \circ R$.

23. Let R be the relation on the set of people consisting of pairs (a, b), where a is a parent of b. Let S be the relation on the set of people consisting of pairs (a, b), where a and b are siblings (brothers or sisters). What are $S \circ R$ and $R \circ S$?

Exercises 24–25 deal with these relations on the set of real numbers:

$R_1 = \{(a, b) \in \mathbf{R}^2 \mid a > b\}$, the "greater than" relation,
$R_2 = \{(a, b) \in \mathbf{R}^2 \mid a \geq b\}$, the "greater than or equal to" relation,
$R_3 = \{(a, b) \in \mathbf{R}^2 \mid a < b\}$, the "less than" relation,
$R_4 = \{(a, b) \in \mathbf{R}^2 \mid a \leq b\}$, the "less than or equal to" relation,
$R_5 = \{(a, b) \in \mathbf{R}^2 \mid a = b\}$, the "equal to" relation,
$R_6 = \{(a, b) \in \mathbf{R}^2 \mid a \neq b\}$, the "unequal to" relation.

24. Find
  a) $R_2 \cup R_4$.   **b)** $R_3 \cup R_6$.
  c) $R_3 \cap R_6$.   **d)** $R_4 \cap R_6$.
  e) $R_3 - R_6$.   **f)** $R_6 - R_3$.
  g) $R_2 \oplus R_6$.   **h)** $R_3 \oplus R_5$.

25. Find
  a) $R_2 \circ R_1$.   **b)** $R_2 \circ R_2$.
  c) $R_3 \circ R_5$.   **d)** $R_4 \circ R_1$.
  e) $R_5 \circ R_3$.   **f)** $R_3 \circ R_6$.
  g) $R_4 \circ R_6$.   **h)** $R_6 \circ R_6$.

26. Let R be the parent relation on the set of all people (see Example 21). When is an ordered pair in the relation R^3?

27. Let R be the relation on the set of people with doctorates such that $(a, b) \in R$ if and only if a was the thesis advisor of b. When is an ordered pair (a, b) in R^2? When is an ordered pair (a, b) in R^n, when n is a positive integer? (Assume that every person with a doctorate has a thesis advisor.)

28. Let R_1 and R_2 be the "divides" and "is a multiple of" relations on the set of all positive integers, respectively. That is, $R_1 = \{(a, b) \mid a \text{ divides } b\}$ and $R_2 = \{(a, b) \mid a \text{ is a multiple of } b\}$. Find
  a) $R_1 \cup R_2$.   **b)** $R_1 \cap R_2$.
  c) $R_1 - R_2$.   **d)** $R_2 - R_1$.
  e) $R_1 \oplus R_2$.

29. Let R_1 and R_2 be the "congruent modulo 3" and the "congruent modulo 4" relations, respectively, on the set of integers. That is, $R_1 = \{(a, b) \mid a \equiv b \pmod 3\}$ and $R_2 = \{(a, b) \mid a \equiv b \pmod 4\}$. Find
  a) $R_1 \cup R_2$.   **b)** $R_1 \cap R_2$.
  c) $R_1 - R_2$.   **d)** $R_2 - R_1$.
  e) $R_1 \oplus R_2$.

30. How many of the 16 different relations on $\{0, 1\}$ contain the pair $(0, 1)$?

31. a) How many relations are there on the set $\{a, b, c, d\}$?
  b) How many relations are there on the set $\{a, b, c, d\}$ that contain the pair (a, a)?

32. Let S be a set with n elements and let a and b be distinct elements of S. How many relations R are there on S such that
  a) $(a, b) \in R$?   **b)** $(a, b) \notin R$?
  c) no ordered pair in R has a as its first element?
  d) at least one ordered pair in R has a as its first element?
  e) no ordered pair in R has a as its first element or b as its second element?
  f) at least one ordered pair in R either has a as its first element or has b as its second element?

*33. How many relations are there on a set with n elements that are
  a) symmetric?   **b)** antisymmetric?
  c) asymmetric?   **d)** irreflexive?
  e) reflexive and symmetric?
  f) neither reflexive nor irreflexive?

*34. How many transitive relations are there on a set with n elements if
  a) $n = 1$?   **b)** $n = 2$?   **c)** $n = 3$?

35. Find the error in the "proof" of the following "theorem."

"*Theorem*": Let R be a relation on a set A that is symmetric and transitive. Then R is reflexive.

"*Proof*": Let $a \in A$. Take an element $b \in A$ such that $(a, b) \in R$. Because R is symmetric, we also have $(b, a) \in R$. Now using the transitive property, we can conclude that $(a, a) \in R$ because $(a, b) \in R$ and $(b, a) \in R$.

36. Suppose that R and S are reflexive relations on a set A. Prove or disprove each of these statements.
  a) $R \cup S$ is reflexive.
  b) $R \cap S$ is reflexive.
  c) $R \oplus S$ is irreflexive.
  d) $R - S$ is irreflexive.
  e) $S \circ R$ is reflexive.

37. Show that the relation R on a set A is symmetric if and only if $R = R^{-1}$, where R^{-1} is the inverse relation.

38. Show that the relation R on a set A is antisymmetric if and only if $R \cap R^{-1}$ is a subset of the diagonal relation $\Delta = \{(a, a) \mid a \in A\}$.

39. Show that the relation R on a set A is reflexive if and only if the inverse relation R^{-1} is reflexive.

40. Show that the relation R on a set A is reflexive if and only if the complementary relation $\overline{R}$ is irreflexive.

41. Let R be a relation that is reflexive and transitive. Prove that $R^n = R$ for all positive integers n.

42. Let R be the relation on the set $\{1, 2, 3, 4, 5\}$ containing the ordered pairs $(1, 1), (1, 2), (1, 3), (2, 3), (2, 4), (3, 1)$, $(3, 4)$, $(3, 5)$, $(4, 2)$, $(4, 5)$, $(5, 1)$, $(5, 2)$, and $(5, 4)$. Find

a) R^2.   **b)** R^3.   **c)** R^4.   **d)** R^5.

43. Let R be a reflexive relation on a set A. Show that R^n is reflexive for all positive integers n.

*44. Let R be a symmetric relation. Show that R^n is symmetric for all positive integers n.

45. Suppose that the relation R is irreflexive. Is R^2 necessarily irreflexive? Give a reason for your answer.

9.2 *n*-ary Relations and Their Applications

Introduction

Relationships among elements of more than two sets often arise. For instance, there is a relationship involving the name of a student, the student's major, and the student's grade point average. Similarly, there is a relationship involving the airline, flight number, starting point, destination, departure time, and arrival time of a flight. An example of such a relationship in mathematics involves three integers, where the first integer is larger than the second integer, which is larger than the third. Another example is the betweenness relationship involving points on a line, such that three points are related when the second point is between the first and the third.

We will study relationships among elements from more than two sets in this section. These relationships are called *n*-ary relations. These relations are used to represent computer databases. These representations help us answer queries about the information stored in databases, such as: Which flights land at O'Hare Airport between 3 A.M. and 4 A.M.? Which students at your school are sophomores majoring in mathematics or computer science and have greater than a 3.0 average? Which employees of a company have worked for the company less than 5 years and make more than $50,000?

n-ary Relations

We begin with the basic definition on which the theory of relational databases rests.

DEFINITION 1   Let $A_1, A_2, \ldots, A_n$ be sets. An *n-ary relation* on these sets is a subset of $A_1 \times A_2 \times \cdots \times A_n$. The sets $A_1, A_2, \ldots, A_n$ are called the *domains* of the relation, and n is called its *degree*.

EXAMPLE 1   Let R be the relation on $\mathbf{N} \times \mathbf{N} \times \mathbf{N}$ consisting of triples (a, b, c), where a, b, and c are integers with $a < b < c$. Then $(1, 2, 3) \in R$, but $(2, 4, 3) \notin R$. The degree of this relation is 3. Its domains are all equal to the set of natural numbers.   ◀

EXAMPLE 2   Let R be the relation on $\mathbf{Z} \times \mathbf{Z} \times \mathbf{Z}$ consisting of all triples of integers (a, b, c) in which a, b, and c form an arithmetic progression. That is, $(a, b, c) \in R$ if and only if there is an integer k such that $b = a + k$ and $c = a + 2k$, or equivalently, such that $b - a = k$ and $c - b = k$. Note that $(1, 3, 5) \in R$ because $3 = 1 + 2$ and $5 = 1 + 2 \cdot 2$, but $(2, 5, 9) \notin R$ because $5 - 2 = 3$ while $9 - 5 = 4$. This relation has degree 3 and its domains are all equal to the set of integers.   ◀

TABLE 1 Students.			
Student_name	*ID_number*	*Major*	*GPA*
Ackermann	231455	Computer Science	3.88
Adams	888323	Physics	3.45
Chou	102147	Computer Science	3.49
Goodfriend	453876	Mathematics	3.45
Rao	678543	Mathematics	3.90
Stevens	786576	Psychology	2.99

EXAMPLE 3 Let R be the relation on $\mathbf{Z} \times \mathbf{Z} \times \mathbf{Z}^+$ consisting of triples (a, b, m), where $a, b,$ and m are integers with $m \geq 1$ and $a \equiv b \pmod{m}$. Then $(8, 2, 3), (-1, 9, 5),$ and $(14, 0, 7)$ all belong to R, but $(7, 2, 3), (-2, -8, 5),$ and $(11, 0, 6)$ do not belong to R because $8 \equiv 2 \pmod 3, -1 \equiv 9 \pmod 5,$ and $14 \equiv 0 \pmod 7$, but $7 \not\equiv 2 \pmod 3, -2 \not\equiv -8 \pmod 5,$ and $11 \not\equiv 0 \pmod 6$. This relation has degree 3 and its first two domains are the set of all integers and its third domain is the set of positive integers. ◀

EXAMPLE 4 Let R be the relation consisting of 5-tuples (A, N, S, D, T) representing airplane flights, where A is the airline, N is the flight number, S is the starting point, D is the destination, and T is the departure time. For instance, if Nadir Express Airlines has flight 963 from Newark to Bangor at 15:00, then (Nadir, 963, Newark, Bangor, 15:00) belongs to R. The degree of this relation is 5, and its domains are the set of all airlines, the set of flight numbers, the set of cities, the set of cities (again), and the set of times. ◀

Databases and Relations

Links

The time required to manipulate information in a database depends on how this information is stored. The operations of adding and deleting records, updating records, searching for records, and combining records from overlapping databases are performed millions of times each day in a large database. Because of the importance of these operations, various methods for representing databases have been developed. We will discuss one of these methods, called the **relational data model**, based on the concept of a relation.

A database consists of **records**, which are n-tuples, made up of **fields**. The fields are the entries of the n-tuples. For instance, a database of student records may be made up of fields containing the name, student number, major, and grade point average of the student. The relational data model represents a database of records as an n-ary relation. Thus, student records are represented as 4-tuples of the form (*Student_name, ID_number, Major, GPA*). A sample database of six such records is

(Ackermann, 231455, Computer Science, 3.88)
(Adams, 888323, Physics, 3.45)
(Chou, 102147, Computer Science, 3.49)
(Goodfriend, 453876, Mathematics, 3.45)
(Rao, 678543, Mathematics, 3.90)
(Stevens, 786576, Psychology, 2.99).

Relations used to represent databases are also called **tables**, because these relations are often displayed as tables. Each column of the table corresponds to an *attribute* of the database. For instance, the same database of students is displayed in Table 1. The attributes of this database are Student Name, ID Number, Major, and GPA.

A domain of an *n*-ary relation is called a **primary key** when the value of the *n*-tuple from this domain determines the *n*-tuple. That is, a domain is a primary key when no two *n*-tuples in the relation have the same value from this domain.

Records are often added to or deleted from databases. Because of this, the property that a domain is a primary key is time-dependent. Consequently, a primary key should be chosen that remains one whenever the database is changed. The current collection of *n*-tuples in a relation is called the **extension** of the relation. The more permanent part of a database, including the name and attributes of the database, is called its **intension**. When selecting a primary key, the goal should be to select a key that can serve as a primary key for all possible extensions of the database. To do this, it is necessary to examine the intension of the database to understand the set of possible *n*-tuples that can occur in an extension.

EXAMPLE 5   Which domains are primary keys for the *n*-ary relation displayed in Table 1, assuming that no *n*-tuples will be added in the future?

Solution: Because there is only one 4-tuple in this table for each student name, the domain of student names is a primary key. Similarly, the ID numbers in this table are unique, so the domain of ID numbers is also a primary key. However, the domain of major fields of study is not a primary key, because more than one 4-tuple contains the same major field of study. The domain of grade point averages is also not a primary key, because there are two 4-tuples containing the same GPA.   ◀

Combinations of domains can also uniquely identify *n*-tuples in an *n*-ary relation. When the values of a set of domains determine an *n*-tuple in a relation, the Cartesian product of these domains is called a **composite key**.

EXAMPLE 6   Is the Cartesian product of the domain of major fields of study and the domain of GPAs a composite key for the *n*-ary relation from Table 1, assuming that no *n*-tuples are ever added?

Solution: Because no two 4-tuples from this table have both the same major and the same GPA, this Cartesian product is a composite key.   ◀

Because primary and composite keys are used to identify records uniquely in a database, it is important that keys remain valid when new records are added to the database. Hence, checks should be made to ensure that every new record has values that are different in the appropriate field, or fields, from all other records in this table. For instance, it makes sense to use the student identification number as a key for student records if no two students ever have the same student identification number. A university should not use the name field as a key, because two students may have the same name (such as John Smith).

Operations on *n*-ary Relations

There are a variety of operations on *n*-ary relations that can be used to form new *n*-ary relations. Applied together, these operations can answer queries on databases that ask for all *n*-tuples that satisfy certain conditions.

The most basic operation on an *n*-ary relation is determining all *n*-tuples in the *n*-ary relation that satisfy certain conditions. For example, we may want to find all the records of all computer science majors in a database of student records. We may want to find all students who have a grade point average above 3.5. We may want to find the records of all computer science majors who have a grade point average above 3.5. To perform such tasks we use the selection operator.

DEFINITION 2   Let R be an n-ary relation and C a condition that elements in R may satisfy. Then the *selection operator* s_C maps the n-ary relation R to the n-ary relation of all n-tuples from R that satisfy the condition C.

EXAMPLE 7   To find the records of computer science majors in the n-ary relation R shown in Table 1, we use the operator s_{C_1}, where C_1 is the condition Major = "Computer Science." The result is the two 4-tuples (Ackermann, 231455, Computer Science, 3.88) and (Chou, 102147, Computer Science, 3.49). Similarly, to find the records of students who have a grade point average above 3.5 in this database, we use the operator s_{C_2}, where C_2 is the condition GPA > 3.5. The result is the two 4-tuples (Ackermann, 231455, Computer Science, 3.88) and (Rao, 678543, Mathematics, 3.90). Finally, to find the records of computer science majors who have a GPA above 3.5, we use the operator s_{C_3}, where C_3 is the condition (Major = "Computer Science" $\wedge$ GPA > 3.5). The result consists of the single 4-tuple (Ackermann, 231455, Computer Science, 3.88).   ◀

TABLE 2 GPAs.	
Student_name	*GPA*
Ackermann	3.88
Adams	3.45
Chou	3.49
Goodfriend	3.45
Rao	3.90
Stevens	2.99

Projections are used to form new n-ary relations by deleting the same fields in every record of the relation.

DEFINITION 3   The *projection* $P_{i_1 i_2, \ldots, i_m}$ where $i_1 < i_2 < \cdots < i_m$, maps the n-tuple $(a_1, a_2, \ldots, a_n)$ to the m-tuple $(a_{i_1}, a_{i_2}, \ldots, a_{i_m})$, where $m \leq n$.

In other words, the projection $P_{i_1, i_2, \ldots, i_m}$ deletes $n - m$ of the components of an n-tuple, leaving the i_1th, i_2th, $\ldots$, and i_mth components.

EXAMPLE 8   What results when the projection $P_{1,3}$ is applied to the 4-tuples $(2, 3, 0, 4)$, (Jane Doe, 234111001, Geography, 3.14), and (a_1, a_2, a_3, a_4)?

Solution: The projection $P_{1,3}$ sends these 4-tuples to $(2, 0)$, (Jane Doe, Geography), and (a_1, a_3), respectively.   ◀

Example 9 illustrates how new relations are produced using projections.

EXAMPLE 9   What relation results when the projection $P_{1,4}$ is applied to the relation in Table 1?

Solution: When the projection $P_{1,4}$ is used, the second and third columns of the table are deleted, and pairs representing student names and grade point averages are obtained. Table 2 displays the results of this projection.   ◀

Fewer rows may result when a projection is applied to the table for a relation. This happens when some of the n-tuples in the relation have identical values in each of the m components of the projection, and only disagree in components deleted by the projection. For instance, consider the following example.

EXAMPLE 10   What is the table obtained when the projection $P_{1,2}$ is applied to the relation in Table 3?

Solution: Table 4 displays the relation obtained when $P_{1,2}$ is applied to Table 3. Note that there are fewer rows after this projection is applied.   ◀

The **join** operation is used to combine two tables into one when these tables share some identical fields. For instance, a table containing fields for airline, flight number, and gate, and

TABLE 3 Enrollments.		
Student	*Major*	*Course*
Glauser	Biology	BI 290
Glauser	Biology	MS 475
Glauser	Biology	PY 410
Marcus	Mathematics	MS 511
Marcus	Mathematics	MS 603
Marcus	Mathematics	CS 322
Miller	Computer Science	MS 575
Miller	Computer Science	CS 455

TABLE 4 Majors.	
Student	*Major*
Glauser	Biology
Marcus	Mathematics
Miller	Computer Science

another table containing fields for flight number, gate, and departure time can be combined into a table containing fields for airline, flight number, gate, and departure time.

DEFINITION 4

$(m + n − p)$-tuples $(a_1, a_2, \ldots, a_{m−p}, c_1, c_2, \ldots, c_p, b_1, b_2, \ldots, b_{n−p})$, where the m-tuple $(a_1, a_2, \ldots, a_{m−p}, c_1, c_2, \ldots, c_p)$ belongs to R and the n-tuple $(c_1, c_2, \ldots, c_p, b_1, b_2, \ldots, b_{n−p})$ belongs to S.

In other words, the join operator J_p produces a new relation from two relations by combining all m-tuples of the first relation with all n-tuples of the second relation, where the last p components of the m-tuples agree with the first p components of the n-tuples.

EXAMPLE 11 What relation results when the join operator J_2 is used to combine the relation displayed in Tables 5 and 6?

Solution: The join J_2 produces the relation shown in Table 7. ◄

There are other operators besides projections and joins that produce new relations from existing relations. A description of these operations can be found in books on database theory.

TABLE 5 Teaching_assignments.		
Professor	*Department*	*Course_number*
Cruz	Zoology	335
Cruz	Zoology	412
Farber	Psychology	501
Farber	Psychology	617
Grammer	Physics	544
Grammer	Physics	551
Rosen	Computer Science	518
Rosen	Mathematics	575

TABLE 6 Class_schedule.			
Department	*Course_number*	*Room*	*Time*
Computer Science	518	N521	2:00 P.M.
Mathematics	575	N502	3:00 P.M.
Mathematics	611	N521	4:00 P.M.
Physics	544	B505	4:00 P.M.
Psychology	501	A100	3:00 P.M.
Psychology	617	A110	11:00 A.M.
Zoology	335	A100	9:00 A.M.
Zoology	412	A100	8:00 A.M.

SQL

Links

The database query language SQL (short for Structured Query Language) can be used to carry out the operations we have described in this section. Example 12 illustrates how SQL commands are related to operations on *n*-ary relations.

EXAMPLE 12   We will illustrate how SQL is used to express queries by showing how SQL can be employed to make a query about airline flights using Table 8. The SQL statement

```
SELECT Departure_time
FROM Flights
WHERE Destination='Detroit'
```

is used to find the projection P_5 (on the Departure_time attribute) of the selection of 5-tuples in the Flights database that satisfy the condition: Destination = 'Detroit'. The output would be a list containing the times of flights that have Detroit as their destination, namely, 08:10, 08:47, and 09:44. SQL uses the FROM clause to identify the *n*-ary relation the query is applied to, the WHERE clause to specify the condition of the selection operation, and the SELECT clause to specify the projection operation that is to be applied. (*Beware:* SQL uses SELECT to represent a projection, rather than a selection operation. This is an unfortunate example of conflicting terminology.) ◄

Example 13 shows how SQL queries can be made involving more than one table.

EXAMPLE 13   The SQL statement

```
SELECT Professor, Time
FROM Teaching_assignments, Class_schedule
WHERE Department='Mathematics'
```

is used to find the projection $P_{1,5}$ of the 5-tuples in the database (shown in Table 7), which is the join J_2 of the Teaching_assignments and Class_schedule databases in Tables 5 and 6, respectively, which satisfy the condition: Department = Mathematics. The output would consist of the single 2-tuple (Rosen, 3:00 P.M.). The SQL FROM clause is used here to find the join of two different databases. ◄

We have only touched on the basic concepts of relational databases in this section. More information can be found in [AhUl95].

Professor	Department	Course_number	Room	Time
Cruz	Zoology	335	A100	9:00 A.M.
Cruz	Zoology	412	A100	8:00 A.M.
Farber	Psychology	501	A100	3:00 P.M.
Farber	Psychology	617	A110	11:00 A.M.
Grammer	Physics	544	B505	4:00 P.M.
Rosen	Computer Science	518	N521	2:00 P.M.
Rosen	Mathematics	575	N502	3:00 P.M.

TABLE 7 Teaching_schedule.

TABLE 8 Flights.				
Airline	*Flight_number*	*Gate*	*Destination*	*Departure_time*
Nadir	122	34	Detroit	08:10
Acme	221	22	Denver	08:17
Acme	122	33	Anchorage	08:22
Acme	323	34	Honolulu	08:30
Nadir	199	13	Detroit	08:47
Acme	222	22	Denver	09:10
Nadir	322	34	Detroit	09:44

Exercises

1. List the triples in the relation $\{(a, b, c) \mid a, b,$ and c are integers with $0 < a < b < c < 5\}$.

2. Assuming that no new *n*-tuples are added, find all the primary keys for the relations displayed in
   a) Table 3.   **b)** Table 5.
   c) Table 6.   **d)** Table 8.

3. Assuming that no new *n*-tuples are added, find a composite key with two fields containing the *Airline* field for the database in Table 8.

4. Assuming that no new *n*-tuples are added, find a composite key with two fields containing the Professor field for the database in Table 7.

5. The 3-tuples in a 3-ary relation represent the following attributes of a student database: student ID number, name, phone number.
   a) Is student ID number likely to be a primary key?
   b) Is name likely to be a primary key?
   c) Is phone number likely to be a primary key?

6. What do you obtain when you apply the selection operator s_C, where C is the condition Room = A100, to the database in Table 7?

7. What do you obtain when you apply the selection operator s_C, where C is the condition Destination = Detroit, to the database in Table 8?

8. What do you obtain when you apply the selection operator s_C, where C is the condition (Project = 2) $\wedge$ (Quantity $\geq$ 50), to the database in Table 10?

9. What do you obtain when you apply the selection operator s_C, where C is the condition (Airline = Nadir) $\vee$ (Destination = Denver), to the database in Table 8?

10. What do you obtain when you apply the projection $P_{2,3,5}$ to the 5-tuple (a, b, c, d, e)?

11. Display the table produced by applying the projection $P_{1,4}$ to Table 8.

12. How many components are there in the *n*-tuples in the table obtained by applying the join operator J_3 to two tables with 5-tuples and 8-tuples, respectively?

13. Construct the table obtained by applying the join operator J_2 to the relations in Tables 9 and 10.

14. Show that if C_1 and C_2 are conditions that elements of the *n*-ary relation R may satisfy, then $s_{C_1 \wedge C_2}(R) = s_{C_1}(s_{C_2}(R))$.

15. Show that if C_1 and C_2 are conditions that elements of the *n*-ary relation R may satisfy, then $s_{C_1}(s_{C_2}(R)) = s_{C_2}(s_{C_1}(R))$.

16. Show that if C is a condition that elements of the *n*-ary relations R and S may satisfy, then $s_C(R \cup S) = s_C(R) \cup s_C(S)$.

17. Show that if R and S are both *n*-ary relations, then $P_{i_1,i_2,\ldots,i_m}(R \cup S) = P_{i_1,i_2,\ldots,i_m}(R) \cup P_{i_1,i_2,\ldots,i_m}(S)$.

18. a) What are the operations that correspond to the query expressed using this SQL statement?

```
SELECT Supplier
FROM Part_needs
WHERE 1000 ≤ Part_number ≤ 5000
```

   b) What is the output of this query given the database in Table 9 as input?

19. a) What are the operations that correspond to the query expressed using this SQL statement?

```
SELECT Supplier, Project
FROM Part_needs, Parts_inventory
WHERE Quantity ≤ 10
```

   b) What is the output of this query given the databases in Tables 9 and 10 as input?

20. Determine whether there is a primary key for the relation in Example 2.

21. Determine whether there is a primary key for the relation in Example 3.

22. Show that an *n*-ary relation with a primary key can be thought of as the graph of a function that maps values of the primary key to $(n-1)$-tuples formed from values of the other domains.

TABLE 9 Part_needs.		
Supplier	*Part_number*	*Project*
23	1092	1
23	1101	3
23	9048	4
31	4975	3
31	3477	2
32	6984	4
32	9191	2
33	1001	1

TABLE 10 Parts_inventory.			
Part_number	*Project*	*Quantity*	*Color_code*
1001	1	14	8
1092	1	2	2
1101	3	1	1
3477	2	25	2
4975	3	6	2
6984	4	10	1
9048	4	12	2
9191	2	80	4

9.3   Representing Relations

Introduction

In this section, and in the remainder of this chapter, all relations we study will be binary relations. Because of this, in this section and in the rest of this chapter, the word relation will always refer to a binary relation. There are many ways to represent a relation between finite sets. As we have seen in Section 9.1, one way is to list its ordered pairs. Another way to represent a relation is to use a table, as we did in Example 3 in Section 9.1. In this section we will discuss two alternative methods for representing relations. One method uses zero–one matrices. The other method uses pictorial representations called directed graphs, which we will discuss later in this section.

Generally, matrices are appropriate for the representation of relations in computer programs. On the other hand, people often find the representation of relations using directed graphs useful for understanding the properties of these relations.

Representing Relations Using Matrices

A relation between finite sets can be represented using a zero–one matrix. Suppose that R is a relation from $A = \{a_1, a_2, \ldots, a_m\}$ to $B = \{b_1, b_2, \ldots, b_n\}$. (Here the elements of the sets A and B have been listed in a particular, but arbitrary, order. Furthermore, when $A = B$ we use the same ordering for A and B.) The relation R can be represented by the matrix $\mathbf{M}_R = [m_{ij}]$, where

$$m_{ij} = \begin{cases} 1 \text{ if } (a_i, b_j) \in R, \\ 0 \text{ if } (a_i, b_j) \notin R. \end{cases}$$

In other words, the zero–one matrix representing R has a 1 as its (i, j) entry when a_i is related to b_j, and a 0 in this position if a_i is not related to b_j. (Such a representation depends on the orderings used for A and B.)

The use of matrices to represent relations is illustrated in Examples 1–6.

EXAMPLE 1   Suppose that $A = \{1, 2, 3\}$ and $B = \{1, 2\}$. Let R be the relation from A to B containing (a, b) if $a \in A$, $b \in B$, and $a > b$. What is the matrix representing R if $a_1 = 1$, $a_2 = 2$, and $a_3 = 3$, and $b_1 = 1$ and $b_2 = 2$?

Solution: Because $R = \{(2, 1), (3, 1), (3, 2)\}$, the matrix for R is

$$\mathbf{M}_R = \begin{bmatrix} 0 & 0 \\ 1 & 0 \\ 1 & 1 \end{bmatrix}.$$

The 1s in $\mathbf{M}_R$ show that the pairs $(2, 1)$, $(3, 1)$, and $(3, 2)$ belong to R. The 0s show that no other pairs belong to R. ◄

EXAMPLE 2 Let $A = \{a_1, a_2, a_3\}$ and $B = \{b_1, b_2, b_3, b_4, b_5\}$. Which ordered pairs are in the relation R represented by the matrix

$$\mathbf{M}_R = \begin{bmatrix} 0 & 1 & 0 & 0 & 0 \\ 1 & 0 & 1 & 1 & 0 \\ 1 & 0 & 1 & 0 & 1 \end{bmatrix} ?$$

Solution: Because R consists of those ordered pairs (a_i, b_j) with $m_{ij} = 1$, it follows that

$$R = \{(a_1, b_2), (a_2, b_1), (a_2, b_3), (a_2, b_4), (a_3, b_1), (a_3, b_3), (a_3, b_5)\}. \quad ◄$$

The matrix of a relation on a set, which is a square matrix, can be used to determine whether the relation has certain properties. Recall that a relation R on A is reflexive if $(a, a) \in R$ whenever $a \in A$. Thus, R is reflexive if and only if $(a_i, a_i) \in R$ for $i = 1, 2, \ldots, n$. Hence, R is reflexive if and only if $m_{ii} = 1$, for $i = 1, 2, \ldots, n$. In other words, R is reflexive if all the elements on the main diagonal of $\mathbf{M}_R$ are equal to 1, as shown in Figure 1. Note that the elements off the main diagonal can be either 0 or 1.

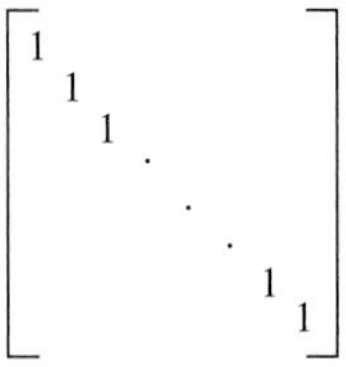

The relation R is symmetric if $(a, b) \in R$ implies that $(b, a) \in R$. Consequently, the relation R on the set $A = \{a_1, a_2, \ldots, a_n\}$ is symmetric if and only if $(a_j, a_i) \in R$ whenever $(a_i, a_j) \in R$. In terms of the entries of $\mathbf{M}_R$, R is symmetric if and only if $m_{ji} = 1$ whenever $m_{ij} = 1$. This also means $m_{ji} = 0$ whenever $m_{ij} = 0$. Consequently, R is symmetric if and only if $m_{ij} = m_{ji}$, for all pairs of integers i and j with $i = 1, 2, \ldots, n$ and $j = 1, 2, \ldots, n$. Recalling the definition of the transpose of a matrix from Section 2.6, we see that R is symmetric if and only if

$$\mathbf{M}_R = (\mathbf{M}_R)^t,$$

that is, if $\mathbf{M}_R$ is a symmetric matrix. The form of the matrix for a symmetric relation is illustrated in Figure 2(a).

The relation R is antisymmetric if and only if $(a, b) \in R$ and $(b, a) \in R$ imply that $a = b$. Consequently, the matrix of an antisymmetric relation has the property that if $m_{ij} = 1$ with $i \neq j$, then $m_{ji} = 0$. Or, in other words, either $m_{ij} = 0$ or $m_{ji} = 0$ when $i \neq j$. The form of the matrix for an antisymmetric relation is illustrated in Figure 2(b).

$$\begin{bmatrix} 1 & & & & \\ & 1 & & & \\ & & 1 & & \\ & & & \ddots & \\ & & & & 1 \\ & & & & & 1 \end{bmatrix}$$

FIGURE 1 The Zero–One Matrix for a Reflexive Relation. (Off Diagonal Elements Can Be 0 or 1.)

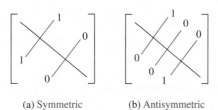

(a) Symmetric   (b) Antisymmetric

FIGURE 2 The Zero–One Matrices for Symmetric and Antisymmetric Relations.

EXAMPLE 3   Suppose that the relation R on a set is represented by the matrix

$$\mathbf{M}_R = \begin{bmatrix} 1 & 1 & 0 \\ 1 & 1 & 1 \\ 0 & 1 & 1 \end{bmatrix}.$$

Is R reflexive, symmetric, and/or antisymmetric?

Solution: Because all the diagonal elements of this matrix are equal to 1, R is reflexive. Moreover, because $\mathbf{M}_R$ is symmetric, it follows that R is symmetric. It is also easy to see that R is not antisymmetric.   ◀

The Boolean operations join and meet (discussed in Section 2.6) can be used to find the matrices representing the union and the intersection of two relations. Suppose that R_1 and R_2 are relations on a set A represented by the matrices $\mathbf{M}_{R_1}$ and $\mathbf{M}_{R_2}$, respectively. The matrix representing the union of these relations has a 1 in the positions where either $\mathbf{M}_{R_1}$ or $\mathbf{M}_{R_2}$ has a 1. The matrix representing the intersection of these relations has a 1 in the positions where both $\mathbf{M}_{R_1}$ and $\mathbf{M}_{R_2}$ have a 1. Thus, the matrices representing the union and intersection of these relations are

$$\mathbf{M}_{R_1 \cup R_2} = \mathbf{M}_{R_1} \vee \mathbf{M}_{R_2} \qquad \text{and} \qquad \mathbf{M}_{R_1 \cap R_2} = \mathbf{M}_{R_1} \wedge \mathbf{M}_{R_2}.$$

EXAMPLE 4   Suppose that the relations R_1 and R_2 on a set A are represented by the matrices

$$\mathbf{M}_{R_1} = \begin{bmatrix} 1 & 0 & 1 \\ 1 & 0 & 0 \\ 0 & 1 & 0 \end{bmatrix} \qquad \text{and} \qquad \mathbf{M}_{R_2} = \begin{bmatrix} 1 & 0 & 1 \\ 0 & 1 & 1 \\ 1 & 0 & 0 \end{bmatrix}.$$

What are the matrices representing $R_1 \cup R_2$ and $R_1 \cap R_2$?

Solution: The matrices of these relations are

$$\mathbf{M}_{R_1 \cup R_2} = \mathbf{M}_{R_1} \vee \mathbf{M}_{R_2} = \begin{bmatrix} 1 & 0 & 1 \\ 1 & 1 & 1 \\ 1 & 1 & 0 \end{bmatrix},$$

$$\mathbf{M}_{R_1 \cap R_2} = \mathbf{M}_{R_1} \wedge \mathbf{M}_{R_2} = \begin{bmatrix} 1 & 0 & 1 \\ 0 & 0 & 0 \\ 0 & 0 & 0 \end{bmatrix}.$$ ◀

We now turn our attention to determining the matrix for the composite of relations. This matrix can be found using the Boolean product of the matrices (discussed in Section 2.6) for these relations. In particular, suppose that R is a relation from A to B and S is a relation from B to C. Suppose that A, B, and C have m, n, and p elements, respectively. Let the zero–one matrices for $S \circ R$, R, and S be $\mathbf{M}_{S \circ R} = [t_{ij}]$, $\mathbf{M}_R = [r_{ij}]$, and $\mathbf{M}_S = [s_{ij}]$, respectively (these matrices have sizes $m \times p$, $m \times n$, and $n \times p$, respectively). The ordered pair (a_i, c_j) belongs to $S \circ R$ if and only if there is an element b_k such that (a_i, b_k) belongs to R and (b_k, c_j) belongs to S. It follows that $t_{ij} = 1$ if and only if $r_{ik} = s_{kj} = 1$ for some k. From the definition of the Boolean product, this means that

$$\mathbf{M}_{S \circ R} = \mathbf{M}_R \odot \mathbf{M}_S.$$

EXAMPLE 5 Find the matrix representing the relations $S \circ R$, where the matrices representing R and S are

$$\mathbf{M}_R = \begin{bmatrix} 1 & 0 & 1 \\ 1 & 1 & 0 \\ 0 & 0 & 0 \end{bmatrix} \quad \text{and} \quad \mathbf{M}_S = \begin{bmatrix} 0 & 1 & 0 \\ 0 & 0 & 1 \\ 1 & 0 & 1 \end{bmatrix}.$$

Solution: The matrix for $S \circ R$ is

$$\mathbf{M}_{S \circ R} = \mathbf{M}_R \odot \mathbf{M}_S = \begin{bmatrix} 1 & 1 & 1 \\ 0 & 1 & 1 \\ 0 & 0 & 0 \end{bmatrix}.$$

◀

The matrix representing the composite of two relations can be used to find the matrix for $\mathbf{M}_{R^n}$. In particular,

$$\mathbf{M}_{R^n} = \mathbf{M}_R^{[n]},$$

from the definition of Boolean powers. Exercise 35 asks for a proof of this formula.

EXAMPLE 6 Find the matrix representing the relation R^2, where the matrix representing R is

$$\mathbf{M}_R = \begin{bmatrix} 0 & 1 & 0 \\ 0 & 1 & 1 \\ 1 & 0 & 0 \end{bmatrix}.$$

Solution: The matrix for R^2 is

$$\mathbf{M}_{R^2} = \mathbf{M}_R^{[2]} = \begin{bmatrix} 0 & 1 & 1 \\ 1 & 1 & 1 \\ 0 & 1 & 0 \end{bmatrix}.$$

◀

Representing Relations Using Digraphs

We have shown that a relation can be represented by listing all of its ordered pairs or by using a zero–one matrix. There is another important way of representing a relation using a pictorial representation. Each element of the set is represented by a point, and each ordered pair is represented using an arc with its direction indicated by an arrow. We use such pictorial representations when we think of relations on a finite set as **directed graphs**, or **digraphs**.

DEFINITION 1 A *directed graph*, or *digraph*, consists of a set V of *vertices* (or *nodes*) together with a set E of ordered pairs of elements of V called *edges* (or *arcs*). The vertex a is called the *initial vertex* of the edge (a, b), and the vertex b is called the *terminal vertex* of this edge.

An edge of the form (a, a) is represented using an arc from the vertex a back to itself. Such an edge is called a **loop**.

EXAMPLE 7   The directed graph with vertices a, b, c, and d, and edges (a, b), (a, d), (b, b), (b, d), (c, a), (c, b), and (d, b) is displayed in Figure 3.   ◀

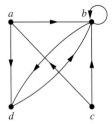

FIGURE 3
A Directed Graph.

The relation R on a set A is represented by the directed graph that has the elements of A as its vertices and the ordered pairs (a, b), where $(a, b) \in R$, as edges. This assignment sets up a one-to-one correspondence between the relations on a set A and the directed graphs with A as their set of vertices. Thus, every statement about relations corresponds to a statement about directed graphs, and vice versa. Directed graphs give a visual display of information about relations. As such, they are often used to study relations and their properties. (Note that relations from a set A to a set B can be represented by a directed graph where there is a vertex for each element of A and a vertex for each element of B, as shown in Section 9.1. However, when $A = B$, such representation provides much less insight than the digraph representations described here.) The use of directed graphs to represent relations on a set is illustrated in Examples 8–10.

EXAMPLE 8   The directed graph of the relation

$$R = \{(1, 1), (1, 3), (2, 1), (2, 3), (2, 4), (3, 1), (3, 2), (4, 1)\}$$

on the set $\{1, 2, 3, 4\}$ is shown in Figure 4.   ◀

EXAMPLE 9   What are the ordered pairs in the relation R represented by the directed graph shown in Figure 5?

Solution: The ordered pairs (x, y) in the relation are

$$R = \{(1, 3), (1, 4), (2, 1), (2, 2), (2, 3), (3, 1), (3, 3), (4, 1), (4, 3)\}.$$

Each of these pairs corresponds to an edge of the directed graph, with $(2, 2)$ and $(3, 3)$ corresponding to loops.   ◀

We will study directed graphs extensively in Chapter 10.

The directed graph representing a relation can be used to determine whether the relation has various properties. For instance, a relation is reflexive if and only if there is a loop at every vertex of the directed graph, so that every ordered pair of the form (x, x) occurs in the relation. A relation is symmetric if and only if for every edge between distinct vertices in its digraph there is an edge in the opposite direction, so that (y, x) is in the relation whenever (x, y) is in the relation. Similarly, a relation is antisymmetric if and only if there are never two edges in opposite directions between distinct vertices. Finally, a relation is transitive if and only if whenever there is an edge from a vertex x to a vertex y and an edge from a vertex y to a vertex z, there is an edge from x to z (completing a triangle where each side is a directed edge with the correct direction).

Remark: Note that a symmetric relation can be represented by an undirected graph, which is a graph where edges do not have directions. We will study undirected graphs in Chapter 10.

EXAMPLE 10   Determine whether the relations for the directed graphs shown in Figure 6 are reflexive, symmetric, antisymmetric, and/or transitive.

Solution: Because there are loops at every vertex of the directed graph of R, it is reflexive. R is neither symmetric nor antisymmetric because there is an edge from a to b but not one from b to a, but there are edges in both directions connecting b and c. Finally, R is not transitive because there is an edge from a to b and an edge from b to c, but no edge from a to c.

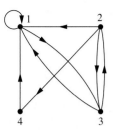

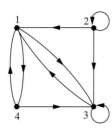

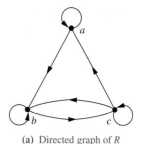

(a) Directed graph of R

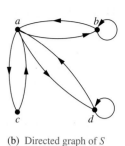

(b) Directed graph of S

FIGURE 4 The Directed Graph of the Relation R.

FIGURE 5 The Directed Graph of the Relation R.

FIGURE 6 The Directed Graphs of the Relations R and S.

Because loops are not present at all the vertices of the directed graph of S, this relation is not reflexive. It is symmetric and not antisymmetric, because every edge between distinct vertices is accompanied by an edge in the opposite direction. It is also not hard to see from the directed graph that S is not transitive, because (c, a) and (a, b) belong to S, but (c, b) does not belong to S.

Exercises

1. Represent each of these relations on $\{1, 2, 3\}$ with a matrix (with the elements of this set listed in increasing order).
   a) $\{(1, 1), (1, 2), (1, 3)\}$
   b) $\{(1, 2), (2, 1), (2, 2), (3, 3)\}$
   c) $\{(1, 1), (1, 2), (1, 3), (2, 2), (2, 3), (3, 3)\}$
   d) $\{(1, 3), (3, 1)\}$

2. List the ordered pairs in the relations on $\{1, 2, 3\}$ corresponding to these matrices (where the rows and columns correspond to the integers listed in increasing order).

   a) $\begin{bmatrix} 1 & 0 & 1 \\ 0 & 1 & 0 \\ 1 & 0 & 1 \end{bmatrix}$
   b) $\begin{bmatrix} 0 & 1 & 0 \\ 0 & 1 & 0 \\ 0 & 1 & 0 \end{bmatrix}$

   c) $\begin{bmatrix} 1 & 1 & 1 \\ 1 & 0 & 1 \\ 1 & 1 & 1 \end{bmatrix}$

3. How can the matrix representing a relation R on a set A be used to determine whether the relation is irreflexive?

4. Determine whether the relations represented by the matrices in Exercise 3 are reflexive, irreflexive, symmetric, antisymmetric, and/or transitive.

5. How many nonzero entries does the matrix representing the relation R on $A = \{1, 2, 3, \ldots, 100\}$ consisting of the first 100 positive integers have if R is
   a) $\{(a, b) \mid a > b\}$?
   b) $\{(a, b) \mid a \neq b\}$?
   c) $\{(a, b) \mid a = b + 1\}$?
   d) $\{(a, b) \mid a = 1\}$?
   e) $\{(a, b) \mid ab = 1\}$?

6. How many nonzero entries does the matrix representing the relation R on $A = \{1, 2, 3, \ldots, 1000\}$ consisting of the first 1000 positive integers have if R is

   a) $\{(a, b) \mid a \leq b\}$?
   b) $\{(a, b) \mid a = b \pm 1\}$?
   c) $\{(a, b) \mid a + b = 1000\}$?
   d) $\{(a, b) \mid a + b \leq 1001\}$?
   e) $\{(a, b) \mid a \neq 0\}$?

7. How can the matrix for $\overline{R}$, the complement of the relation R, be found from the matrix representing R, when R is a relation on a finite set A?

8. How can the matrix for R^{-1}, the inverse of the relation R, be found from the matrix representing R, when R is a relation on a finite set A?

9. Let R be the relation represented by the matrix

$$\mathbf{M}_R = \begin{bmatrix} 0 & 1 & 1 \\ 1 & 1 & 0 \\ 1 & 0 & 1 \end{bmatrix}.$$

   Find the matrix representing
   a) R^{-1}.
   b) $\overline{R}$.
   c) R^2.

10. Let R_1 and R_2 be relations on a set A represented by the matrices

$$\mathbf{M}_{R_1} = \begin{bmatrix} 0 & 1 & 0 \\ 1 & 1 & 1 \\ 1 & 0 & 0 \end{bmatrix} \quad \text{and} \quad \mathbf{M}_{R_2} = \begin{bmatrix} 0 & 1 & 0 \\ 0 & 1 & 1 \\ 1 & 1 & 1 \end{bmatrix}.$$

   Find the matrices that represent
   a) $R_1 \cup R_2$.
   b) $R_1 \cap R_2$.
   c) $R_2 \circ R_1$.
   d) $R_1 \circ R_1$.
   e) $R_1 \oplus R_2$.

11. Let R be the relation represented by the matrix

$$\mathbf{M}_R = \begin{bmatrix} 0 & 1 & 0 \\ 0 & 0 & 1 \\ 1 & 1 & 0 \end{bmatrix}.$$

Find the matrices that represent
a) R^2.   **b)** R^3.   **c)** R^4.

12. Let R be a relation on a set A with n elements. If there are k nonzero entries in $\mathbf{M}_R$, the matrix representing R, how many nonzero entries are there in $\mathbf{M}_{R^{-1}}$, the matrix representing R^{-1}, the inverse of R?

13. Let R be a relation on a set A with n elements. If there are k nonzero entries in $\mathbf{M}_R$, the matrix representing R, how many nonzero entries are there in $\mathbf{M}_{\overline{R}}$, the matrix representing $\overline{R}$, the complement of R?

14. Draw the directed graphs representing each of the relations from Exercise 1.

In Exercises 15–18 list the ordered pairs in the relations represented by the directed graphs.

15.

16.

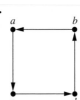

17.

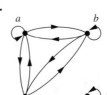

18.

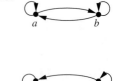

19. How can the directed graph of a relation R on a finite set A be used to determine whether a relation is asymmetric?

20. How can the directed graph of a relation R on a finite set A be used to determine whether a relation is irreflexive?

21. Determine whether the relations represented by the directed graphs shown in Exercises 15–16 are reflexive, irreflexive, symmetric, antisymmetric, and/or transitive.

22. Determine whether the relations represented by the directed graphs shown in Exercises 17–18 are reflexive, irreflexive, symmetric, antisymmetric, asymmetric, and/or transitive.

23. Let R be a relation on a set A. Explain how to use the directed graph representing R to obtain the directed graph representing the inverse relation R^{-1}.

24. Let R be a relation on a set A. Explain how to use the directed graph representing R to obtain the directed graph representing the complementary relation $\overline{R}$.

25. Show that if $\mathbf{M}_R$ is the matrix representing the relation R, then $\mathbf{M}_R^{[n]}$ is the matrix representing the relation R^n.

26. Given the directed graphs representing two relations, how can the directed graph of the union, intersection, symmetric difference, difference, and composition of these relations be found?

9.4   Closures of Relations

Introduction

A computer network has data centers in Boston, Chicago, Denver, Detroit, New York, and San Diego. There are direct, one-way telephone lines from Boston to Chicago, from Boston to Detroit, from Chicago to Detroit, from Detroit to Denver, and from New York to San Diego. Let R be the relation containing (a, b) if there is a telephone line from the data center in a to that in b. How can we determine if there is some (possibly indirect) link composed of one or more telephone lines from one center to another? Because not all links are direct, such as the link from Boston to Denver that goes through Detroit, R cannot be used directly to answer this. In the language of relations, R is not transitive, so it does not contain all the pairs that can be linked. As we will show in this section, we can find all pairs of data centers that have a link by constructing a transitive relation S containing R such that S is a subset of every transitive relation containing R. Here, S is the smallest transitive relation that contains R. This relation is called the **transitive closure** of R.

In general, let R be a relation on a set A. R may or may not have some property **P**, such as reflexivity, symmetry, or transitivity. If there is a relation S with property **P** containing R such that S is a subset of every relation with property **P** containing R, then S is called the **closure** of R with respect to **P**. (Note that the closure of a relation with respect to a property may not exist; see Exercises 15 and 35.) We will show how reflexive, symmetric, and transitive closures of relations can be found.

Closures

The relation $R = \{(1, 1), (1, 2), (2, 1), (3, 2)\}$ on the set $A = \{1, 2, 3\}$ is not reflexive. How can we produce a reflexive relation containing R that is as small as possible? This can be done by adding $(2, 2)$ and $(3, 3)$ to R, because these are the only pairs of the form (a, a) that are not in R. Clearly, this new relation contains R. Furthermore, *any* reflexive relation that contains R must also contain $(2, 2)$ and $(3, 3)$. Because this relation contains R, is reflexive, and is contained within every reflexive relation that contains R, it is called the **reflexive closure** of R.

As this example illustrates, given a relation R on a set A, the reflexive closure of R can be formed by adding to R all pairs of the form (a, a) with $a \in A$, not already in R. The addition of these pairs produces a new relation that is reflexive, contains R, and is contained within any reflexive relation containing R. We see that the reflexive closure of R equals $R \cup \Delta$, where $\Delta = \{(a, a) \mid a \in A\}$ is the **diagonal relation** on A. (The reader should verify this.)

EXAMPLE 1   What is the reflexive closure of the relation $R = \{(a, b) \mid a < b\}$ on the set of integers?

Solution: The reflexive closure of R is

$$R \cup \Delta = \{(a, b) \mid a < b\} \cup \{(a, a) \mid a \in \mathbf{Z}\} = \{(a, b) \mid a \leq b\}. \qquad \blacktriangleleft$$

The relation $\{(1, 1), (1, 2), (2, 2), (2, 3), (3, 1), (3, 2)\}$ on $\{1, 2, 3\}$ is not symmetric. How can we produce a symmetric relation that is as small as possible and contains R? To do this, we need only add $(2, 1)$ and $(1, 3)$, because these are the only pairs of the form (b, a) with $(a, b) \in R$ that are not in R. This new relation is symmetric and contains R. Furthermore, *any* symmetric relation that contains R must contain this new relation, because a symmetric relation that contains R must contain $(2, 1)$ and $(1, 3)$. Consequently, this new relation is called the **symmetric closure** of R.

As this example illustrates, the symmetric closure of a relation R can be constructed by adding all ordered pairs of the form (b, a), where (a, b) is in the relation, that are not already present in R. Adding these pairs produces a relation that is symmetric, that contains R, and that is contained in any symmetric relation that contains R. The symmetric closure of a relation can be constructed by taking the union of a relation with its inverse (defined in the preamble of Exercise 16 in Section 9.1); that is, $R \cup R^{-1}$ is the symmetric closure of R, where $R^{-1} = \{(b, a) \mid (a, b) \in R\}$. The reader should verify this statement.

EXAMPLE 2   What is the symmetric closure of the relation $R = \{(a, b) \mid a > b\}$ on the set of positive integers?

Solution: The symmetric closure of R is the relation

$$R \cup R^{-1} = \{(a, b) \mid a > b\} \cup \{(b, a) \mid a > b\} = \{(a, b) \mid a \neq b\}.$$

This last equality follows because R contains all ordered pairs of positive integers where the first element is greater than the second element and R^{-1} contains all ordered pairs of positive integers where the first element is less than the second. $\qquad \blacktriangleleft$

Suppose that a relation R is not transitive. How can we produce a transitive relation that contains R such that this new relation is contained within any transitive relation that contains R? Can the transitive closure of a relation R be produced by adding all the pairs of the form (a, c), where (a, b) and (b, c) are already in the relation? Consider the relation $R = \{(1, 3), (1, 4), (2, 1), (3, 2)\}$ on the set $\{1, 2, 3, 4\}$. This relation is not transitive because it does not contain all pairs of the form (a, c) where (a, b) and (b, c) are in R. The pairs of this form not in R are $(1, 2)$, $(2, 3)$, $(2, 4)$, and $(3, 1)$. Adding these pairs does *not* produce a transitive relation, because the resulting relation contains $(3, 1)$ and $(1, 4)$ but does not contain $(3, 4)$. This shows that constructing the transitive closure of a relation is more complicated than constructing either the reflexive or symmetric closure. The rest of this section develops algorithms for constructing transitive closures. As will be shown later in this section, the transitive closure of a relation can be found by adding new ordered pairs that must be present and then repeating this process until no new ordered pairs are needed.

Paths in Directed Graphs

We will see that representing relations by directed graphs helps in the construction of transitive closures. We now introduce some terminology that we will use for this purpose.

A path in a directed graph is obtained by traversing along edges (in the same direction as indicated by the arrow on the edge).

DEFINITION 1   A *path* from a to b in the directed graph G is a sequence of edges (x_0, x_1), (x_1, x_2), (x_2, x_3), ..., (x_{n-1}, x_n) in G, where n is a nonnegative integer, and $x_0 = a$ and $x_n = b$, that is, a sequence of edges where the terminal vertex of an edge is the same as the initial vertex in the next edge in the path. This path is denoted by $x_0, x_1, x_2, \ldots, x_{n-1}, x_n$ and has *length n*. We view the empty set of edges as a path of length zero from a to a. A path of length $n \geq 1$ that begins and ends at the same vertex is called a *circuit* or *cycle*.

A path in a directed graph can pass through a vertex more than once. Moreover, an edge in a directed graph can occur more than once in a path.

EXAMPLE 3   Which of the following are paths in the directed graph shown in Figure 1: a, b, e, d; a, e, c, d, b; b, a, c, b, a, a, b; d, c; c, b, a; e, b, a, b, a, b, e? What are the lengths of those that are paths? Which of the paths in this list are circuits?

Solution: Because each of (a, b), (b, e), and (e, d) is an edge, a, b, e, d is a path of length three. Because (c, d) is not an edge, a, e, c, d, b is not a path. Also, b, a, c, b, a, a, b is a path of length six because (b, a), (a, c), (c, b), (b, a), (a, a), and (a, b) are all edges. We see that d, c is a path of length one, because (d, c) is an edge. Also c, b, a is a path of length two, because (c, b) and (b, a) are edges. All of (e, b), (b, a), (a, b), (b, a), (a, b), and (b, e) are edges, so e, b, a, b, a, b, e is a path of length six.

The two paths b, a, c, b, a, a, b and e, b, a, b, a, b, e are circuits because they begin and end at the same vertex. The paths a, b, e, d; c, b, a; and d, c are not circuits.   ◀

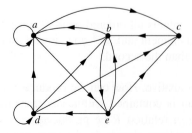

FIGURE 1   A Directed Graph.

The term *path* also applies to relations. Carrying over the definition from directed graphs to relations, there is a **path** from a to b in R if there is a sequence of elements $a, x_1, x_2, \ldots, x_{n-1}, b$ with $(a, x_1) \in R$, $(x_1, x_2) \in R, \ldots$, and $(x_{n-1}, b) \in R$. Theorem 1 can be obtained from the definition of a path in a relation.

THEOREM 1   Let R be a relation on a set A. There is a path of length n, where n is a positive integer, from a to b if and only if $(a, b) \in R^n$.

Proof: We will use mathematical induction. By definition, there is a path from a to b of length one if and only if $(a, b) \in R$, so the theorem is true when $n = 1$.

Assume that the theorem is true for the positive integer n. This is the inductive hypothesis. There is a path of length $n + 1$ from a to b if and only if there is an element $c \in A$ such that there is a path of length one from a to c, so $(a, c) \in R$, and a path of length n from c to b, that is, $(c, b) \in R^n$. Consequently, by the inductive hypothesis, there is a path of length $n + 1$ from a to b if and only if there is an element c with $(a, c) \in R$ and $(c, b) \in R^n$. But there is such an element if and only if $(a, b) \in R^{n+1}$. Therefore, there is a path of length $n + 1$ from a to b if and only if $(a, b) \in R^{n+1}$. This completes the proof.   ◁

Transitive Closures

We now show that finding the transitive closure of a relation is equivalent to determining which pairs of vertices in the associated directed graph are connected by a path. With this in mind, we define a new relation.

DEFINITION 2   Let R be a relation on a set A. The *connectivity relation* R^* consists of the pairs (a, b) such that there is a path of length at least one from a to b in R.

Because R^n consists of the pairs (a, b) such that there is a path of length n from a to b, it follows that R^* is the union of all the sets R^n. In other words,

$$R^* = \bigcup_{n=1}^{\infty} R^n.$$

The connectivity relation is useful in many models.

EXAMPLE 4   Let R be the relation on the set of all people in the world that contains (a, b) if a has met b. What is R^n, where n is a positive integer greater than one? What is R^*?

Solution: The relation R^2 contains (a, b) if there is a person c such that $(a, c) \in R$ and $(c, b) \in R$, that is, if there is a person c such that a has met c and c has met b. Similarly, R^n consists of those pairs (a, b) such that there are people $x_1, x_2, \ldots, x_{n-1}$ such that a has met x_1, x_1 has met $x_2, \ldots$, and x_{n-1} has met b.

The relation R^* contains (a, b) if there is a sequence of people, starting with a and ending with b, such that each person in the sequence has met the next person in the sequence. (There are many interesting conjectures about R^*. Do you think that this connectivity relation includes the pair with you as the first element and the president of Mongolia as the second element? We will use graphs to model this application in Chapter 10.)   ◀

EXAMPLE 5   Let R be the relation on the set of all subway stops in New York City that contains (a, b) if it is possible to travel from stop a to stop b without changing trains. What is R^n when n is a positive integer? What is R^*?

Solution: The relation R^n contains (a, b) if it is possible to travel from stop a to stop b by making at most $n - 1$ changes of trains. The relation R^* consists of the ordered pairs (a, b) where it is possible to travel from stop a to stop b making as many changes of trains as necessary. (The reader should verify these statements.) ◀

EXAMPLE 6   Let R be the relation on the set of all states in the United States that contains (a, b) if state a and state b have a common border. What is R^n, where n is a positive integer? What is R^*?

Solution: The relation R^n consists of the pairs (a, b), where it is possible to go from state a to state b by crossing exactly n state borders. R^* consists of the ordered pairs (a, b), where it is possible to go from state a to state b crossing as many borders as necessary. (The reader should verify these statements.) The only ordered pairs not in R^* are those containing states that are not connected to the continental United States (i.e., those pairs containing Alaska or Hawaii). ◀

Theorem 2 shows that the transitive closure of a relation and the associated connectivity relation are the same.

THEOREM 2   The transitive closure of a relation R equals the connectivity relation R^*.

Proof: Note that R^* contains R by definition. To show that R^* is the transitive closure of R we must also show that R^* is transitive and that $R^* \subseteq S$ whenever S is a transitive relation that contains R.

First, we show that R^* is transitive. If $(a, b) \in R^*$ and $(b, c) \in R^*$, then there are paths from a to b and from b to c in R. We obtain a path from a to c by starting with the path from a to b and following it with the path from b to c. Hence, $(a, c) \in R^*$. It follows that R^* is transitive.

Now suppose that S is a transitive relation containing R. Because S is transitive, S^n also is transitive (the reader should verify this) and $S^n \subseteq S$ (by Theorem 1 of Section 9.1). Furthermore, because

$$S^* = \bigcup_{k=1}^{\infty} S^k$$

and $S^k \subseteq S$, it follows that $S^* \subseteq S$. Now note that if $R \subseteq S$, then $R^* \subseteq S^*$, because any path in R is also a path in S. Consequently, $R^* \subseteq S^* \subseteq S$. Thus, any transitive relation that contains R must also contain R^*. Therefore, R^* is the transitive closure of R. ◁

Now that we know that the transitive closure equals the connectivity relation, we turn our attention to the problem of computing this relation. We do not need to examine arbitrarily long paths to determine whether there is a path between two vertices in a finite directed graph. As Lemma 1 shows, it is sufficient to examine paths containing no more than n edges, where n is the number of elements in the set.

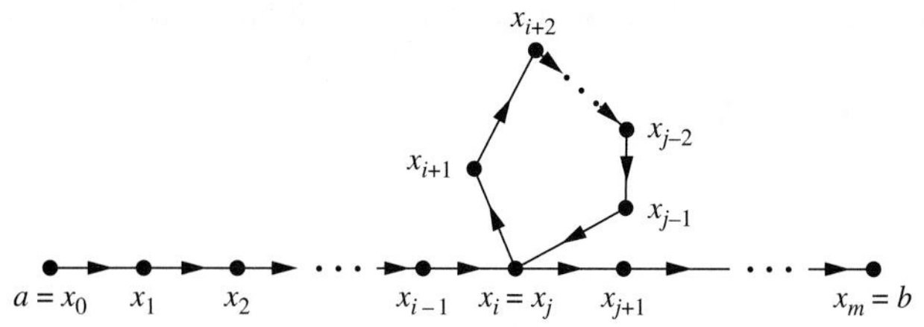

FIGURE 2 **Producing a Path with Length Not Exceeding n.**

LEMMA 1 Let A be a set with n elements, and let R be a relation on A. If there is a path of length at least one in R from a to b, then there is such a path with length not exceeding n. Moreover, when $a \neq b$, if there is a path of length at least one in R from a to b, then there is such a path with length not exceeding $n - 1$.

Proof: Suppose there is a path from a to b in R. Let m be the length of the shortest such path. Suppose that $x_0, x_1, x_2, \ldots, x_{m-1}, x_m$, where $x_0 = a$ and $x_m = b$, is such a path.

Suppose that $a = b$ and that $m > n$, so that $m \geq n + 1$. By the pigeonhole principle, because there are n vertices in A, among the m vertices $x_0, x_1, \ldots, x_{m-1}$, at least two are equal (see Figure 2).

Suppose that $x_i = x_j$ with $0 \leq i < j \leq m - 1$. Then the path contains a circuit from x_i to itself. This circuit can be deleted from the path from a to b, leaving a path, namely, $x_0, x_1, \ldots, x_i, x_{j+1}, \ldots, x_{m-1}, x_m$, from a to b of shorter length. Hence, the path of shortest length must have length less than or equal to n.

The case where $a \neq b$ is left as an exercise for the reader. ◁

From Lemma 1, we see that the transitive closure of R is the union of R, R^2, $R^3, \ldots,$ and R^n. This follows because there is a path in R^* between two vertices if and only if there is a path between these vertices in R^i, for some positive integer i with $i \leq n$. Because

$$R^* = R \cup R^2 \cup R^3 \cup \cdots \cup R^n$$

and the zero–one matrix representing a union of relations is the join of the zero–one matrices of these relations, the zero–one matrix for the transitive closure is the join of the zero–one matrices of the first n powers of the zero–one matrix of R.

THEOREM 3 Let $\mathbf{M}_R$ be the zero–one matrix of the relation R on a set with n elements. Then the zero–one matrix of the transitive closure R^* is

$$\mathbf{M}_{R^*} = \mathbf{M}_R \vee \mathbf{M}_R^{[2]} \vee \mathbf{M}_R^{[3]} \vee \cdots \vee \mathbf{M}_R^{[n]}.$$

EXAMPLE 7 Find the zero–one matrix of the transitive closure of the relation R where

$$\mathbf{M}_R = \begin{bmatrix} 1 & 0 & 1 \\ 0 & 1 & 0 \\ 1 & 1 & 0 \end{bmatrix}.$$

Solution: By Theorem 3, it follows that the zero–one matrix of R^* is

$$\mathbf{M}_{R^*} = \mathbf{M}_R \vee \mathbf{M}_R^{[2]} \vee \mathbf{M}_R^{[3]}.$$

Because

$$\mathbf{M}_R^{[2]} = \begin{bmatrix} 1 & 1 & 1 \\ 0 & 1 & 0 \\ 1 & 1 & 1 \end{bmatrix} \quad \text{and} \quad \mathbf{M}_R^{[3]} = \begin{bmatrix} 1 & 1 & 1 \\ 0 & 1 & 0 \\ 1 & 1 & 1 \end{bmatrix},$$

it follows that

$$\mathbf{M}_{R^*} = \begin{bmatrix} 1 & 0 & 1 \\ 0 & 1 & 0 \\ 1 & 1 & 0 \end{bmatrix} \vee \begin{bmatrix} 1 & 1 & 1 \\ 0 & 1 & 0 \\ 1 & 1 & 1 \end{bmatrix} \vee \begin{bmatrix} 1 & 1 & 1 \\ 0 & 1 & 0 \\ 1 & 1 & 1 \end{bmatrix} = \begin{bmatrix} 1 & 1 & 1 \\ 0 & 1 & 0 \\ 1 & 1 & 1 \end{bmatrix}. \qquad \blacktriangleleft$$

Links

Theorem 3 can be used as a basis for an algorithm for computing the matrix of the relation R^*. To find this matrix, the successive Boolean powers of $\mathbf{M}_R$, up to the nth power, are computed. As each power is calculated, its join with the join of all smaller powers is formed. When this is done with the nth power, the matrix for R^* has been found. This procedure is displayed as Algorithm 1.

ALGORITHM 1 A Procedure for Computing the Transitive Closure.

---

procedure *transitive closure* ($\mathbf{M}_R$: zero–one $n \times n$ matrix)
$\mathbf{A} := \mathbf{M}_R$
$\mathbf{B} := \mathbf{A}$
for $i := 2$ **to** n
    $\mathbf{A} := \mathbf{A} \odot \mathbf{M}_R$
    $\mathbf{B} := \mathbf{B} \vee \mathbf{A}$
return $\mathbf{B}$ {$\mathbf{B}$ is the zero–one matrix for R^*}

We can easily find the number of bit operations used by Algorithm 1 to determine the transitive closure of a relation. Computing the Boolean powers $\mathbf{M}_R, \mathbf{M}_R^{[2]}, \ldots, \mathbf{M}_R^{[n]}$ requires that $n - 1$ Boolean products of $n \times n$ zero–one matrices be found. Each of these Boolean products can be found using $n^2(2n - 1)$ bit operations. Hence, these products can be computed using $n^2(2n - 1)(n - 1)$ bit operations.

To find $\mathbf{M}_{R^*}$ from the n Boolean powers of $\mathbf{M}_R$, $n - 1$ joins of zero–one matrices need to be found. Computing each of these joins uses n^2 bit operations. Hence, $(n - 1)n^2$ bit operations are used in this part of the computation. Therefore, when Algorithm 1 is used, the matrix of the transitive closure of a relation on a set with n elements can be found using $n^2(2n - 1)(n - 1) + (n - 1)n^2 = 2n^3(n - 1)$, which is $O(n^4)$ bit operations. The remainder of this section describes a more efficient algorithm for finding transitive closures.

Warshall's Algorithm

Links

Warshall's algorithm, named after Stephen Warshall, who described this algorithm in 1960, is an efficient method for computing the transitive closure of a relation. Algorithm 1 can find the transitive closure of a relation on a set with n elements using $2n^3(n - 1)$ bit operations. However, the transitive closure can be found by Warshall's algorithm using only $2n^3$ bit operations.

Remark: Warshall's algorithm is sometimes called the Roy–Warshall algorithm, because Bernard Roy described this algorithm in 1959.

Suppose that R is a relation on a set with n elements. Let $v_1, v_2, \ldots, v_n$ be an arbitrary listing of these n elements. The concept of the **interior vertices** of a path is used in Warshall's algorithm. If $a, x_1, x_2, \ldots, x_{m-1}, b$ is a path, its interior vertices are $x_1, x_2, \ldots, x_{m-1}$, that is, all the vertices of the path that occur somewhere other than as the first and last vertices in the path. For instance, the interior vertices of a path a, c, d, f, g, h, b, j in a directed graph are $c, d, f, g, h,$ and b. The interior vertices of a, c, d, a, f, b are $c, d, a,$ and f. (Note that the first vertex in the path is not an interior vertex unless it is visited again by the path, except as the last vertex. Similarly, the last vertex in the path is not an interior vertex unless it was visited previously by the path, except as the first vertex.)

Warshall's algorithm is based on the construction of a sequence of zero–one matrices. These matrices are $\mathbf{W}_0, \mathbf{W}_1, \ldots, \mathbf{W}_n$, where $\mathbf{W}_0 = \mathbf{M}_R$ is the zero–one matrix of this relation, and $\mathbf{W}_k = [w_{ij}^{(k)}]$, where $w_{ij}^{(k)} = 1$ if there is a path from v_i to v_j such that all the interior vertices of this path are in the set $\{v_1, v_2, \ldots, v_k\}$ (the first k vertices in the list) and is 0 otherwise. (The first and last vertices in the path may be outside the set of the first k vertices in the list.) Note that $\mathbf{W}_n = \mathbf{M}_{R^*}$, because the (i, j)th entry of $\mathbf{M}_{R^*}$ is 1 if and only if there is a path from v_i to v_j, with all interior vertices in the set $\{v_1, v_2, \ldots, v_n\}$ (but these are the only vertices in the directed graph). Example 8 illustrates what the matrix $\mathbf{W}_k$ represents.

EXAMPLE 8   Let R be the relation with directed graph shown in Figure 3. Let a, b, c, d be a listing of the elements of the set. Find the matrices $\mathbf{W}_0, \mathbf{W}_1, \mathbf{W}_2, \mathbf{W}_3,$ and $\mathbf{W}_4$. The matrix $\mathbf{W}_4$ is the transitive closure of R.

Solution: Let $v_1 = a$, $v_2 = b$, $v_3 = c$, and $v_4 = d$. $\mathbf{W}_0$ is the matrix of the relation. Hence,

$$\mathbf{W}_0 = \begin{bmatrix} 0 & 0 & 0 & 1 \\ 1 & 0 & 1 & 0 \\ 1 & 0 & 0 & 1 \\ 0 & 0 & 1 & 0 \end{bmatrix}.$$

FIGURE 3
The Directed Graph of the Relation R.

$\mathbf{W}_1$ has 1 as its (i, j)th entry if there is a path from v_i to v_j that has only $v_1 = a$ as an interior vertex. Note that all paths of length one can still be used because they have no interior vertices.

Links

Also, there is now an allowable path from b to d, namely, b, a, d. Hence,

$$\mathbf{W}_1 = \begin{bmatrix} 0 & 0 & 0 & 1 \\ 1 & 0 & 1 & 1 \\ 1 & 0 & 0 & 1 \\ 0 & 0 & 1 & 0 \end{bmatrix}.$$

$\mathbf{W}_2$ has 1 as its (i, j)th entry if there is a path from v_i to v_j that has only $v_1 = a$ and/or $v_2 = b$ as its interior vertices, if any. Because there are no edges that have b as a terminal vertex, no new paths are obtained when we permit b to be an interior vertex. Hence, $\mathbf{W}_2 = \mathbf{W}_1$.

$\mathbf{W}_3$ has 1 as its (i, j)th entry if there is a path from v_i to v_j that has only $v_1 = a$, $v_2 = b$, and/or $v_3 = c$ as its interior vertices, if any. We now have paths from d to a, namely, d, c, a, and from d to d, namely, d, c, d. Hence,

$$\mathbf{W}_3 = \begin{bmatrix} 0 & 0 & 0 & 1 \\ 1 & 0 & 1 & 1 \\ 1 & 0 & 0 & 1 \\ 1 & 0 & 1 & 1 \end{bmatrix}.$$

Finally, $\mathbf{W}_4$ has 1 as its (i, j)th entry if there is a path from v_i to v_j that has $v_1 = a$, $v_2 = b$, $v_3 = c$, and/or $v_4 = d$ as interior vertices, if any. Because these are all the vertices of the graph, this entry is 1 if and only if there is a path from v_i to v_j. Hence,

$$\mathbf{W}_4 = \begin{bmatrix} 1 & 0 & 1 & 1 \\ 1 & 0 & 1 & 1 \\ 1 & 0 & 1 & 1 \\ 1 & 0 & 1 & 1 \end{bmatrix}.$$

This last matrix, $\mathbf{W}_4$, is the matrix of the transitive closure.   ◀

Warshall's algorithm computes $\mathbf{M}_{R*}$ by efficiently computing $\mathbf{W}_0 = \mathbf{M}_R, \mathbf{W}_1, \mathbf{W}_2, \ldots,$ $\mathbf{W}_n = \mathbf{M}_{R*}$. This observation shows that we can compute $\mathbf{W}_k$ directly from $\mathbf{W}_{k-1}$: There is a path from v_i to v_j with no vertices other than $v_1, v_2, \ldots, v_k$ as interior vertices if and only if either there is a path from v_i to v_j with its interior vertices among the first $k - 1$ vertices in the list, or there are paths from v_i to v_k and from v_k to v_j that have interior vertices only among the first $k - 1$ vertices in the list. That is, either a path from v_i to v_j already existed before v_k was permitted as an interior vertex, or allowing v_k as an interior vertex produces a path that goes from v_i to v_k and then from v_k to v_j. These two cases are shown in Figure 4.

The first type of path exists if and only if $w_{ij}^{[k-1]} = 1$, and the second type of path exists if and only if both $w_{ik}^{[k-1]}$ and $w_{kj}^{[k-1]}$ are 1. Hence, $w_{ij}^{[k]}$ is 1 if and only if either $w_{ij}^{[k-1]}$ is 1 or both $w_{ik}^{[k-1]}$ and $w_{kj}^{[k-1]}$ are 1. This gives us Lemma 2.

LEMMA 2   Let $\mathbf{W}_k = [w_{ij}^{[k]}]$ be the zero–one matrix that has a 1 in its (i, j)th position if and only if there is a path from v_i to v_j with interior vertices from the set $\{v_1, v_2, \ldots, v_k\}$. Then

$$w_{ij}^{[k]} = w_{ij}^{[k-1]} \vee (w_{ik}^{[k-1]} \wedge w_{kj}^{[k-1]}),$$

whenever i, j, and k are positive integers not exceeding n.

Lemma 2 gives us the means to compute efficiently the matrices $\mathbf{W}_k, k = 1, 2, \ldots, n$. We display the pseudocode for Warshall's algorithm, using Lemma 2, as Algorithm 2.

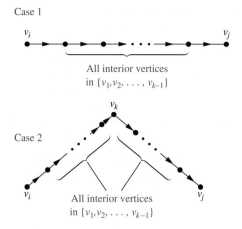

FIGURE 4   Adding v_k to the Set of Allowable Interior Vertices.

ALGORITHM 2 Warshall Algorithm.

procedure *Warshall* ($\mathbf{M}_R : n \times n$ zero–one matrix)
$\mathbf{W} := \mathbf{M}_R$
for $k := 1$ **to** n
    for $i := 1$ **to** n
        for $j := 1$ **to** n
            $w_{ij} := w_{ij} \lor (w_{ik} \land w_{kj})$
return $\mathbf{W}\{\mathbf{W} = [w_{ij}]$ is $\mathbf{M}_{R^*}\}$

The computational complexity of Warshall's algorithm can easily be computed in terms of bit operations. To find the entry $w_{ij}^{[k]}$ from the entries $w_{ij}^{[k-1]}$, $w_{ik}^{[k-1]}$, and $w_{kj}^{[k-1]}$ using Lemma 2 requires two bit operations. To find all n^2 entries of $\mathbf{W}_k$ from those of $\mathbf{W}_{k-1}$ requires $2n^2$ bit operations. Because Warshall's algorithm begins with $\mathbf{W}_0 = \mathbf{M}_R$ and computes the sequence of n zero–one matrices $\mathbf{W}_1, \mathbf{W}_2, \ldots, \mathbf{W}_n = \mathbf{M}_{R^*}$, the total number of bit operations used is $n \cdot 2n^2 = 2n^3$.

Exercises

1. Let R be the relation on the set $\{0, 1, 2, 3\}$ containing the ordered pairs $(0, 1)$, $(1, 1)$, $(1, 2)$, $(2, 0)$, $(2, 2)$, and $(3, 0)$. Find the
  a) reflexive closure of R.   **b)** symmetric closure of R.

2. Let R be the relation $\{(a, b) \mid a \neq b\}$ on the set of integers. What is the reflexive closure of R?

3. Let R be the relation $\{(a, b) \mid a$ divides $b\}$ on the set of integers. What is the symmetric closure of R?

4. How can the directed graph representing the reflexive closure of a relation on a finite set be constructed from the directed graph of the relation?

In Exercises 5–7 draw the directed graph of the reflexive closure of the relations with the directed graph shown.

5.

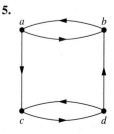

6.

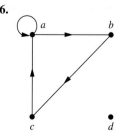

7.

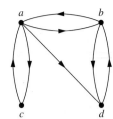

8. How can the directed graph representing the symmetric closure of a relation on a finite set be constructed from the directed graph for this relation?

9. Find the directed graphs of the symmetric closures of the relations with directed graphs shown in Exercises 5–7.

10. Find the smallest relation containing the relation in Example 2 that is both reflexive and symmetric.

11. Find the directed graph of the smallest relation that is both reflexive and symmetric that contains each of the relations with directed graphs shown in Exercises 5–7.

12. Suppose that the relation R on the finite set A is represented by the matrix $\mathbf{M}_R$. Show that the matrix that represents the reflexive closure of R is $\mathbf{M}_R \vee \mathbf{I}_n$.

13. Suppose that the relation R on the finite set A is represented by the matrix $\mathbf{M}_R$. Show that the matrix that represents the symmetric closure of R is $\mathbf{M}_R \vee \mathbf{M}_R^t$.

14. Show that the closure of a relation R with respect to a property **P**, if it exists, is the intersection of all the relations with property **P** that contain R.

15. When is it possible to define the "irreflexive closure" of a relation R, that is, a relation that contains R, is irreflexive, and is contained in every irreflexive relation that contains R?

16. Determine whether these sequences of vertices are paths in this directed graph.

a) a, b, c, e

b) b, e, c, b, e

c) a, a, b, e, d, e

d) b, c, e, d, a, a, b

e) b, c, c, b, e, d, e, d

f) $a, a, b, b, c, c, b, e, d$

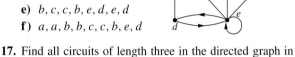

17. Find all circuits of length three in the directed graph in Exercise 16.

18. Determine whether there is a path in the directed graph in Exercise 16 beginning at the first vertex given and ending at the second vertex given.

a) a, b **b)** b, a **c)** b, b

d) a, e **e)** b, d **f)** c, d

g) d, d **h)** e, a **i)** e, c

19. Let R be the relation on the set $\{1, 2, 3, 4, 5\}$ containing the ordered pairs $(1, 3), (2, 4), (3, 1), (3, 5), (4, 3), (5, 1),$ $(5, 2),$ and $(5, 4)$. Find

a) R^2.   **b)** R^3.   **c)** R^4.

d) R^5.   **e)** R^6.   **f)** R^*.

20. Let R be the relation that contains the pair (a, b) if a and b are cities such that there is a direct non-stop airline flight from a to b. When is (a, b) in

a) R^2?   **b)** R^3?   **c)** R^*?

21. Let R be the relation on the set of all students containing the ordered pair (a, b) if a and b are in at least one common class and $a \neq b$. When is (a, b) in

a) R^2?   **b)** R^3?   **c)** R^*?

22. Suppose that the relation R is reflexive. Show that R^* is reflexive.

23. Suppose that the relation R is symmetric. Show that R^* is symmetric.

24. Suppose that the relation R is irreflexive. Is the relation R^2 necessarily irreflexive?

25. Use Algorithm 1 to find the transitive closures of these relations on $\{1, 2, 3, 4\}$.

a) $\{(1, 2), (2, 1), (2, 3), (3, 4), (4, 1)\}$

b) $\{(2, 1), (2, 3), (3, 1), (3, 4), (4, 1), (4, 3)\}$

c) $\{(1, 2), (1, 3), (1, 4), (2, 3), (2, 4), (3, 4)\}$

d) $\{(1, 1), (1, 4), (2, 1), (2, 3), (3, 1), (3, 2), (3, 4), (4, 2)\}$

26. Use Algorithm 1 to find the transitive closures of these relations on $\{a, b, c, d, e\}$.

a) $\{(a, c), (b, d), (c, a), (d, b), (e, d)\}$

b) $\{(b, c), (b, e), (c, e), (d, a), (e, b), (e, c)\}$

c) $\{(a, b), (a, c), (a, e), (b, a), (b, c), (c, a), (c, b), (d, a),$ $(e, d)\}$

d) $\{(a, e), (b, a), (b, d), (c, d), (d, a), (d, c), (e, a), (e, b),$ $(e, c), (e, e)\}$

27. Use Warshall's algorithm to find the transitive closures of the relations in Exercise 25.

28. Use Warshall's algorithm to find the transitive closures of the relations in Exercise 26.

29. Find the smallest relation containing the relation $\{(1, 2), (1, 4), (3, 3), (4, 1)\}$ that is

a) reflexive and transitive.

b) symmetric and transitive.

c) reflexive, symmetric, and transitive.

30. Finish the proof of the case when $a \neq b$ in Lemma 1.

31. Algorithms have been devised that use $O(n^{2.8})$ bit operations to compute the Boolean product of two $n \times n$ zero–one matrices. Assuming that these algorithms can be used, give big-O estimates for the number of bit operations using Algorithm 1 and using Warshall's algorithm to find the transitive closure of a relation on a set with n elements.

9.5 Equivalence Relations

Introduction

In some programming languages the names of variables can contain an unlimited number of characters. However, there is a limit on the number of characters that are checked when a compiler determines whether two variables are equal. For instance, in traditional C, only the first eight characters of a variable name are checked by the compiler. (These characters are uppercase or lowercase letters, digits, or underscores.) Consequently, the compiler considers strings longer than eight characters that agree in their first eight characters the same. Let R be the relation on the set of strings of characters such that $s R t$, where s and t are two strings, if s and t are at least eight characters long and the first eight characters of s and t agree, or $s = t$. It is easy to see that R is reflexive, symmetric, and transitive. Moreover, R divides the set of all strings into classes, where all strings in a particular class are considered the same by a compiler for traditional C.

The integers a and b are related by the "congruence modulo 4" relation when 4 divides $a - b$. We will show later that this relation is reflexive, symmetric, and transitive. It is not hard to see that a is related to b if and only if a and b have the same remainder when divided by 4. It follows that this relation splits the set of integers into four different classes. When we care only what remainder an integer leaves when it is divided by 4, we need only know which class it is in, not its particular value.

These two relations, R and congruence modulo 4, are examples of equivalence relations, namely, relations that are reflexive, symmetric, and transitive. In this section we will show that such relations split sets into disjoint classes of equivalent elements. Equivalence relations arise whenever we care only whether an element of a set is in a certain class of elements, instead of caring about its particular identity.

Equivalence Relations

Links

In this section we will study relations with a particular combination of properties that allows them to be used to relate objects that are similar in some way.

DEFINITION 1	A relation on a set A is called an *equivalence relation* if it is reflexive, symmetric, and transitive.

Equivalence relations are important in every branch of mathematics!

Equivalence relations are important throughout mathematics and computer science. One reason for this is that in an equivalence relation, when two elements are related it makes sense to say they are equivalent.

DEFINITION 2	Two elements a and b that are related by an equivalence relation are called *equivalent*. The notation $a \sim b$ is often used to denote that a and b are equivalent elements with respect to a particular equivalence relation.

For the notion of equivalent elements to make sense, every element should be equivalent to itself, as the reflexive property guarantees for an equivalence relation. It makes sense to say that a and b are related (not just that a is related to b) by an equivalence relation, because when a is related to b, by the symmetric property, b is related to a. Furthermore, because an equivalence relation is transitive, if a and b are equivalent and b and c are equivalent, it follows that a and c are equivalent.

Examples 1–5 illustrate the notion of an equivalence relation.

EXAMPLE 1 Let R be the relation on the set of integers such that aRb if and only if $a = b$ or $a = -b$. In Section 9.1 we showed that R is reflexive, symmetric, and transitive. It follows that R is an equivalence relation. ◄

EXAMPLE 2 Let R be the relation on the set of real numbers such that aRb if and only if $a - b$ is an integer. Is R an equivalence relation?

Solution: Because $a - a = 0$ is an integer for all real numbers a, aRa for all real numbers a. Hence, R is reflexive. Now suppose that aRb. Then $a - b$ is an integer, so $b - a$ is also an integer. Hence, bRa. It follows that R is symmetric. If aRb and bRc, then $a - b$ and $b - c$ are integers. Therefore, $a - c = (a - b) + (b - c)$ is also an integer. Hence, aRc. Thus, R is transitive. Consequently, R is an equivalence relation. ◄

One of the most widely used equivalence relations is congruence modulo m, where m is an integer greater than 1.

EXAMPLE 3 **Congruence Modulo m**   Let m be an integer with $m > 1$. Show that the relation

$$R = \{(a, b) \mid a \equiv b \pmod{m}\}$$

is an equivalence relation on the set of integers.

Solution: Recall from Section 4.1 that $a \equiv b \pmod{m}$ if and only if m divides $a - b$. Note that $a - a = 0$ is divisible by m, because $0 = 0 \cdot m$. Hence, $a \equiv a \pmod{m}$, so congruence modulo m is reflexive. Now suppose that $a \equiv b \pmod{m}$. Then $a - b$ is divisible by m, so $a - b = km$, where k is an integer. It follows that $b - a = (-k)m$, so $b \equiv a \pmod{m}$. Hence, congruence modulo m is symmetric. Next, suppose that $a \equiv b \pmod{m}$ and $b \equiv c \pmod{m}$. Then m divides both $a - b$ and $b - c$. Therefore, there are integers k and l with $a - b = km$ and $b - c = lm$. Adding these two equations shows that $a - c = (a - b) + (b - c) = km + lm = (k + l)m$. Thus, $a \equiv c \pmod{m}$. Therefore, congruence modulo m is transitive. It follows that congruence modulo m is an equivalence relation. ◄

EXAMPLE 4 Suppose that R is the relation on the set of strings of English letters such that aRb if and only if $l(a) = l(b)$, where $l(x)$ is the length of the string x. Is R an equivalence relation?

Solution: Because $l(a) = l(a)$, it follows that aRa whenever a is a string, so that R is reflexive. Next, suppose that aRb, so that $l(a) = l(b)$. Then bRa, because $l(b) = l(a)$. Hence, R is symmetric. Finally, suppose that aRb and bRc. Then $l(a) = l(b)$ and $l(b) = l(c)$. Hence, $l(a) = l(c)$, so aRc. Consequently, R is transitive. Because R is reflexive, symmetric, and transitive, it is an equivalence relation. ◄

EXAMPLE 5 Let n be a positive integer and S a set of strings. Suppose that R_n is the relation on S such that sR_nt if and only if $s = t$, or both s and t have at least n characters and the first n characters of s and t are the same. That is, a string of fewer than n characters is related only to itself; a string s with at least n characters is related to a string t if and only if t has at least n characters and t begins with the n characters at the start of s. For example, let $n = 3$ and let S be the set of all bit strings. Then sR_3t either when $s = t$ or both s and t are bit strings of length 3 or more that begin with the same three bits. For instance, $01R_301$ and $00111R_300101$, but $01\not R_3010$ and $01011\not R_301110$.

Show that for every set S of strings and every positive integer n, R_n is an equivalence relation on S.

Solution: The relation R_n is reflexive because $s = s$, so that $s R_n s$ whenever s is a string in S. If $s R_n t$, then either $s = t$ or s and t are both at least n characters long that begin with the same n characters. This means that $t R_n s$. We conclude that R_n is symmetric.

Now suppose that $s R_n t$ and $t R_n u$. Then either $s = t$ or s and t are at least n characters long and s and t begin with the same n characters, and either $t = u$ or t and u are at least n characters long and t and u begin with the same n characters. From this, we can deduce that either $s = u$ or both s and u are n characters long and s and u begin with the same n characters (because in this case we know that s, t, and u are all at least n characters long and both s and u begin with the same n characters as t does). Consequently, R_n is transitive. It follows that R_n is an equivalence relation. ◄

In Examples 6 and 7 we look at two relations that are not equivalence relations.

EXAMPLE 6   Show that the "divides" relation is the set of positive integers in not an equivalence relation.

Solution: By Examples 9 and 15 in Section 9.1, we know that the "divides" relation is reflexive and transitive. However, by Example 12 in Section 9.1, we know that this relation is not symmetric (for instance, $2 \mid 4$ but $4 \nmid 2$). We conclude that the "divides" relation on the set of positive integers is not an equivalence relation. ◄

EXAMPLE 7   Let R be the relation on the set of real numbers such that $x R y$ if and only if x and y are real numbers that differ by less than 1, that is $|x - y| < 1$. Show that R is not an equivalence relation.

Solution: R is reflexive because $|x - x| = 0 < 1$ whenever $x \in \mathbf{R}$. R is symmetric, for if $x R y$, where x and y are real numbers, then $|x - y| < 1$, which tells us that $|y - x| = |x - y| < 1$, so that $y R x$. However, R is not an equivalence relation because it is not transitive. Take $x = 2.8$, $y = 1.9$, and $z = 1.1$, so that $|x - y| = |2.8 - 1.9| = 0.9 < 1$, $|y - z| = |1.9 - 1.1| = 0.8 < 1$, but $|x - z| = |2.8 - 1.1| = 1.7 > 1$. That is, $2.8 \, R \, 1.9$, $1.9 \, R \, 1.1$, but $2.8 \, \cancel{R} \, 1.1$. ◄

Equivalence Classes

Let A be the set of all students in your school who graduated from high school. Consider the relation R on A that consists of all pairs (x, y), where x and y graduated from the same high school. Given a student x, we can form the set of all students equivalent to x with respect to R. This set consists of all students who graduated from the same high school as x did. This subset of A is called an equivalence class of the relation.

DEFINITION 3   Let R be an equivalence relation on a set A. The set of all elements that are related to an element a of A is called the *equivalence class* of a. The equivalence class of a with respect to R is denoted by $[a]_R$. When only one relation is under consideration, we can delete the subscript R and write $[a]$ for this equivalence class.

In other words, if R is an equivalence relation on a set A, the equivalence class of the element a is

$$[a]_R = \{s \mid (a, s) \in R\}.$$

If $b \in [a]_R$, then b is called a **representative** of this equivalence class. Any element of a class can be used as a representative of this class. That is, there is nothing special about the particular element chosen as the representative of the class.

EXAMPLE 8 What is the equivalence class of an integer for the equivalence relation of Example 1?

Solution: Because an integer is equivalent to itself and its negative in this equivalence relation, it follows that $[a] = \{-a, a\}$. This set contains two distinct integers unless $a = 0$. For instance, $[7] = \{-7, 7\}$, $[-5] = \{-5, 5\}$, and $[0] = \{0\}$. ◄

EXAMPLE 9 What are the equivalence classes of 0 and 1 for congruence modulo 4?

Solution: The equivalence class of 0 contains all integers a such that $a \equiv 0 \pmod 4$. The integers in this class are those divisible by 4. Hence, the equivalence class of 0 for this relation is

$$[0] = \{\ldots, -8, -4, 0, 4, 8, \ldots\}.$$

The equivalence class of 1 contains all the integers a such that $a \equiv 1 \pmod 4$. The integers in this class are those that have a remainder of 1 when divided by 4. Hence, the equivalence class of 1 for this relation is

$$[1] = \{\ldots, -7, -3, 1, 5, 9, \ldots\}. \quad ◄$$

In Example 9 the equivalence classes of 0 and 1 with respect to congruence modulo 4 were found. Example 9 can easily be generalized, replacing 4 with any positive integer m. The equivalence classes of the relation congruence modulo m are called the **congruence classes modulo** m. The congruence class of an integer a modulo m is denoted by $[a]_m$, so $[a]_m = \{\ldots, a - 2m, a - m, a, a + m, a + 2m, \ldots\}$. For instance, from Example 9 it follows that $[0]_4 = \{\ldots, -8, -4, 0, 4, 8, \ldots\}$ and $[1]_4 = \{\ldots, -7, -3, 1, 5, 9, \ldots\}$.

EXAMPLE 10 What is the equivalence class of the string 0111 with respect to the equivalence relation R_3 from Example 5 on the set of all bit strings? (Recall that $s R_3 t$ if and only if s and t are bit strings with $s = t$ or s and t are strings of at least three bits that start with the same three bits.)

Solution: The bit strings equivalent to 0111 are the bit strings with at least three bits that begin with 011. These are the bit strings 011, 0110, 0111, 01100, 01101, 01110, 01111, and so on. Consequently,

$$[011]_{R_3} = \{011, 0110, 0111, 01100, 01101, 01110, 01111, \ldots\}. \quad ◄$$

EXAMPLE 11 **Identifiers in the C Programming Language** In the C programming language, an **identifier** is the name of a variable, a function, or another type of entity. Each identifier is a nonempty string of characters where each character is a lowercase or an uppercase English letter, a digit, or an underscore, and the first character is a lowercase or an uppercase English letter. Identifiers can be any length. This allows developers to use as many characters as they want to name an entity, such as a variable. However, for compilers for some versions of C, there is a limit on the number of characters checked when two names are compared to see whether they refer to the same thing. For example, Standard C compilers consider two identifiers the same when they agree in their first 31 characters. Consequently, developers must be careful not to use identifiers with the same initial 31 characters for different things. We see that two identifiers are considered the same when they are related by the relation R_{31} in Example 5. Using Example 5, we know that R_{31}, on the set of all identifiers in Standard C, is an equivalence relation.

What are the equivalence classes of each of the identifiers Number_of_tropical_storms, Number_of_named_tropical_storms, and Number_of_named_tropical_storms_in_the_Atlantic_in_2005?

Solution: Note that when an identifier is less than 31 characters long, by the definition of R_{31}, its equivalence class contains only itself. Because the identifier Number_of_tropical_storms is 25 characters long, its equivalence class contains exactly one element, namely, itself.

The identifier Number_of_named_tropical_storms is exactly 31 characters long. An identifier is equivalent to it when it starts with these same 31 characters. Consequently, every identifier at least 31 characters long that starts with Number_of_named_tropical_storms is equivalent to this identifier. It follows that the equivalence class of Number_of_named_tropical_storms is the set of all identifiers that begin with the 31 characters Number_of_named_tropical_storms.

An identifier is equivalent to the Number_of_named_tropical_storms_in_the_Atlantic_in_2005 if and only if it begins with its first 31 characters. Because these characters are Number_of_named_tropical_storms, we see that an identifier is equivalent to Number_of_named_tropical_storms_in_the_Atlantic_in_2005 if and only if it is equivalent to Number_of_named_tropical_storms. It follows that these last two identifiers have the same equivalence class.   ◄

Equivalence Classes and Partitions

Let A be the set of students at your school who are majoring in exactly one subject, and let R be the relation on A consisting of pairs (x, y), where x and y are students with the same major. Then R is an equivalence relation, as the reader should verify. We can see that R splits all students in A into a collection of disjoint subsets, where each subset contains students with a specified major. For instance, one subset contains all students majoring (just) in computer science, and a second subset contains all students majoring in history. Furthermore, these subsets are equivalence classes of R. This example illustrates how the equivalence classes of an equivalence relation partition a set into disjoint, nonempty subsets. We will make these notions more precise in the following discussion.

Let R be a relation on the set A. Theorem 1 shows that the equivalence classes of two elements of A are either identical or disjoint.

THEOREM 1 Let R be an equivalence relation on a set A. These statements for elements a and b of A are equivalent:

    (i) aRb *(ii)* $[a] = [b]$ *(iii)* $[a] \cap [b] \neq \emptyset$

Proof: We first show that *(i)* implies *(ii)*. Assume that aRb. We will prove that $[a] = [b]$ by showing $[a] \subseteq [b]$ and $[b] \subseteq [a]$. Suppose $c \in [a]$. Then aRc. Because aRb and R is symmetric, we know that bRa. Furthermore, because R is transitive and bRa and aRc, it follows that bRc. Hence, $c \in [b]$. This shows that $[a] \subseteq [b]$. The proof that $[b] \subseteq [a]$ is similar; it is left as an exercise for the reader.

Second, we will show that *(ii)* implies *(iii)*. Assume that $[a] = [b]$. It follows that $[a] \cap [b] \neq \emptyset$ because $[a]$ is nonempty (because $a \in [a]$ because R is reflexive).

Next, we will show that *(iii)* implies *(i)*. Suppose that $[a] \cap [b] \neq \emptyset$. Then there is an element c with $c \in [a]$ and $c \in [b]$. In other words, aRc and bRc. By the symmetric property, cRb. Then by transitivity, because aRc and cRb, we have aRb.

Because *(i)* implies *(ii)*, *(ii)* implies *(iii)*, and *(iii)* implies *(i)*, the three statements, *(i)*, *(ii)*, and *(iii)*, are equivalent.   ◄

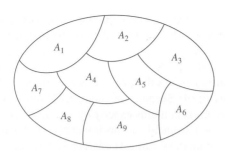

FIGURE 1   **A Partition of a Set.**

We are now in a position to show how an equivalence relation *partitions* a set. Let R be an equivalence relation on a set A. The union of the equivalence classes of R is all of A, because an element a of A is in its own equivalence class, namely, $[a]_R$. In other words,

$$\bigcup_{a \in A} [a]_R = A.$$

In addition, from Theorem 1, it follows that these equivalence classes are either equal or disjoint, so

$$[a]_R \cap [b]_R = \emptyset,$$

when $[a]_R \neq [b]_R$.

Recall that an *index set* is a set whose members label, or index, the elements of a set.

These two observations show that the equivalence classes form a partition of A, because they split A into disjoint subsets. More precisely, a **partition** of a set S is a collection of disjoint nonempty subsets of S that have S as their union. In other words, the collection of subsets A_i, $i \in I$ (where I is an index set) forms a partition of S if and only if

$$A_i \neq \emptyset \text{ for } i \in I,$$

$$A_i \cap A_j = \emptyset \text{ when } i \neq j,$$

and

$$\bigcup_{i \in I} A_i = S.$$

(Here the notation $\bigcup_{i \in I} A_i$ represents the union of the sets A_i for all $i \in I$.) Figure 1 illustrates the concept of a partition of a set.

EXAMPLE 12   Suppose that $S = \{1, 2, 3, 4, 5, 6\}$. The collection of sets $A_1 = \{1, 2, 3\}$, $A_2 = \{4, 5\}$, and $A_3 = \{6\}$ forms a partition of S, because these sets are disjoint and their union is S.   ◀

We have seen that the equivalence classes of an equivalence relation on a set form a partition of the set. The subsets in this partition are the equivalence classes. Conversely, every partition of a set can be used to form an equivalence relation. Two elements are equivalent with respect to this relation if and only if they are in the same subset of the partition.

To see this, assume that $\{A_i \mid i \in I\}$ is a partition on S. Let R be the relation on S consisting of the pairs (x, y), where x and y belong to the same subset A_i in the partition. To show that R is an equivalence relation we must show that R is reflexive, symmetric, and transitive.

We see that $(a, a) \in R$ for every $a \in S$, because a is in the same subset as itself. Hence, R is reflexive. If $(a, b) \in R$, then b and a are in the same subset of the partition, so that $(b, a) \in R$

as well. Hence, R is symmetric. If $(a, b) \in R$ and $(b, c) \in R$, then a and b are in the same subset X in the partition, and b and c are in the same subset Y of the partition. Because the subsets of the partition are disjoint and b belongs to X and Y, it follows that $X = Y$. Consequently, a and c belong to the same subset of the partition, so $(a, c) \in R$. Thus, R is transitive.

It follows that R is an equivalence relation. The equivalence classes of R consist of subsets of S containing related elements, and by the definition of R, these are the subsets of the partition. Theorem 2 summarizes the connections we have established between equivalence relations and partitions.

THEOREM 2   Let R be an equivalence relation on a set S. Then the equivalence classes of R form a partition of S. Conversely, given a partition $\{A_i \mid i \in I\}$ of the set S, there is an equivalence relation R that has the sets A_i, $i \in I$, as its equivalence classes.

Example 13 shows how to construct an equivalence relation from a partition.

EXAMPLE 13   List the ordered pairs in the equivalence relation R produced by the partition $A_1 = \{1, 2, 3\}$, $A_2 = \{4, 5\}$, and $A_3 = \{6\}$ of $S = \{1, 2, 3, 4, 5, 6\}$, given in Example 12.

Solution: The subsets in the partition are the equivalence classes of R. The pair $(a, b) \in R$ if and only if a and b are in the same subset of the partition. The pairs $(1, 1)$, $(1, 2)$, $(1, 3)$, $(2, 1)$, $(2, 2)$, $(2, 3)$, $(3, 1)$, $(3, 2)$, and $(3, 3)$ belong to R because $A_1 = \{1, 2, 3\}$ is an equivalence class; the pairs $(4, 4)$, $(4, 5)$, $(5, 4)$, and $(5, 5)$ belong to R because $A_2 = \{4, 5\}$ is an equivalence class; and finally the pair $(6, 6)$ belongs to R because $\{6\}$ is an equivalence class. No pair other than those listed belongs to R.   ◄

The congruence classes modulo m provide a useful illustration of Theorem 2. There are m different congruence classes modulo m, corresponding to the m different remainders possible when an integer is divided by m. These m congruence classes are denoted by $[0]_m, [1]_m, \ldots, [m-1]_m$. They form a partition of the set of integers.

EXAMPLE 14   What are the sets in the partition of the integers arising from congruence modulo 4?

Solution: There are four congruence classes, corresponding to $[0]_4$, $[1]_4$, $[2]_4$, and $[3]_4$. They are the sets

$$[0]_4 = \{\ldots, -8, -4, 0, 4, 8, \ldots\},$$
$$[1]_4 = \{\ldots, -7, -3, 1, 5, 9, \ldots\},$$
$$[2]_4 = \{\ldots, -6, -2, 2, 6, 10, \ldots\},$$
$$[3]_4 = \{\ldots, -5, -1, 3, 7, 11, \ldots\}.$$

These congruence classes are disjoint, and every integer is in exactly one of them. In other words, as Theorem 2 says, these congruence classes form a partition.   ◄

We now provide an example of a partition of the set of all strings arising from an equivalence relation on this set.

EXAMPLE 15   Let R_3 be the relation from Example 5. What are the sets in the partition of the set of all bit strings arising from the relation R_3 on the set of all bit strings? (Recall that $s R_3 t$, where s and t are bit strings, if $s = t$ or s and t are bit strings with at least three bits that agree in their first three bits.)

Solution: Note that every bit string of length less than three is equivalent only to itself. Hence $[\lambda]_{R_3} = \{\lambda\}$, $[0]_{R_3} = \{0\}$, $[1]_{R_3} = \{1\}$, $[00]_{R_3} = \{00\}$, $[01]_{R_3} = \{01\}$, $[10]_{R_3} = \{10\}$, and

$[11]_{R_3} = \{11\}$. Note that every bit string of length three or more is equivalent to one of the eight bit strings 000, 001, 010, 011, 100, 101, 110, and 111. We have

$$[000]_{R_3} = \{000, 0000, 0001, 00000, 00001, 00010, 00011, \ldots\},$$

$$[001]_{R_3} = \{001, 0010, 0011, 00100, 00101, 00110, 00111, \ldots\},$$

$$[010]_{R_3} = \{010, 0100, 0101, 01000, 01001, 01010, 01011, \ldots\},$$

$$[011]_{R_3} = \{011, 0110, 0111, 01100, 01101, 01110, 01111, \ldots\},$$

$$[100]_{R_3} = \{100, 1000, 1001, 10000, 10001, 10010, 10011, \ldots\},$$

$$[101]_{R_3} = \{101, 1010, 1011, 10100, 10101, 10110, 10111, \ldots\},$$

$$[110]_{R_3} = \{110, 1100, 1101, 11000, 11001, 11010, 11011, \ldots\},$$

$$[111]_{R_3} = \{111, 1110, 1111, 11100, 11101, 11110, 11111, \ldots\}.$$

These 15 equivalence classes are disjoint and every bit string is in exactly one of them. As Theorem 2 tells us, these equivalence classes partition the set of all bit strings. ◄

Exercises

1. Which of these relations on $\{0, 1, 2, 3\}$ are equivalence relations? Determine the properties of an equivalence relation that the others lack.
   a) $\{(0, 0), (1, 1), (2, 2), (3, 3)\}$
   b) $\{(0, 0), (0, 2), (2, 0), (2, 2), (2, 3), (3, 2), (3, 3)\}$
   c) $\{(0, 0), (1, 1), (1, 2), (2, 1), (2, 2), (3, 3)\}$
   d) $\{(0, 0), (1, 1), (1, 3), (2, 2), (2, 3), (3, 1), (3, 2), (3, 3)\}$
   e) $\{(0, 0), (0, 1), (0, 2), (1, 0), (1, 1), (1, 2), (2, 0), (2, 2), (3, 3)\}$

2. Which of these relations on the set of all people are equivalence relations? Determine the properties of an equivalence relation that the others lack.
   a) $\{(a, b) \mid a$ and b are the same age$\}$
   b) $\{(a, b) \mid a$ and b have the same parents$\}$
   c) $\{(a, b) \mid a$ and b share a common parent$\}$
   d) $\{(a, b) \mid a$ and b have met$\}$
   e) $\{(a, b) \mid a$ and b speak a common language$\}$

3. Which of these relations on the set of all functions from $\mathbf{Z}$ to $\mathbf{Z}$ are equivalence relations? Determine the properties of an equivalence relation that the others lack.
   a) $\{(f, g) \mid f(1) = g(1)\}$
   b) $\{(f, g) \mid f(0) = g(0)$ or $f(1) = g(1)\}$
   c) $\{(f, g) \mid f(x) - g(x) = 1$ for all $x \in \mathbf{Z}\}$

   d) $\{(f, g) \mid$ for some $C \in \mathbf{Z}$, for all $x \in \mathbf{Z}$, $f(x) - g(x) = C\}$
   e) $\{(f, g) \mid f(0) = g(1)$ and $f(1) = g(0)\}$

4. Define three equivalence relations on the set of students in your discrete mathematics class different from the relations discussed in the text. Determine the equivalence classes for each of these equivalence relations.

5. Define three equivalence relations on the set of buildings on a college campus. Determine the equivalence classes for each of these equivalence relations.

6. Show that the relation of logical equivalence on the set of all compound propositions is an equivalence relation. What are the equivalence classes of **F** and of **T**?

7. Suppose that A is a nonempty set, and f is a function that has A as its domain. Let R be the relation on A consisting of all ordered pairs (x, y) such that $f(x) = f(y)$.
   a) Show that R is an equivalence relation on A.
   b) What are the equivalence classes of R?

8. Suppose that A is a nonempty set and R is an equivalence relation on A. Show that there is a function f with A as its domain such that $(x, y) \in R$ if and only if $f(x) = f(y)$.

9. Show that the relation R consisting of all pairs (x, y) such that x and y are bit strings of length three or more that agree in their first three bits is an equivalence relation on the set of all bit strings of length three or more.

10. Show that the relation R consisting of all pairs (x, y) such that x and y are bit strings of length three or more that agree except perhaps in their first three bits is an equivalence relation on the set of all bit strings of length three or more.

11. Show that the relation R consisting of all pairs (x, y) such that x and y are bit strings that agree in their first and third bits is an equivalence relation on the set of all bit strings of length three or more.

12. Let R be the relation consisting of all pairs (x, y) such that x and y are strings of uppercase and lowercase English letters with the property that for every positive integer n, the nth characters in x and y are the same letter, either uppercase or lowercase. Show that R is an equivalence relation.

13. Let R be the relation on the set of ordered pairs of positive integers such that $((a, b), (c, d)) \in R$ if and only if $a + d = b + c$. Show that R is an equivalence relation.

14. Let R be the relation on the set of ordered pairs of positive integers such that $((a, b), (c, d)) \in R$ if and only if $ad = bc$. Show that R is an equivalence relation.

15. Let R be the relation on the set of all URLs (or Web addresses) such that $x \, R \, y$ if and only if the Web page at x is the same as the Web page at y. Show that R is an equivalence relation.

16. Let R be the relation on the set of all people who have visited a particular Web page such that $x \, R \, y$ if and only if person x and person y have followed the same set of links starting at this Web page (going from Web page to Web page until they stop using the Web). Show that R is an equivalence relation.

In Exercises 17–19 determine whether the relation with the directed graph shown is an equivalence relation.

17.   **18.**

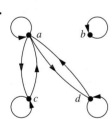

19.

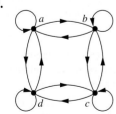

20. Determine whether the relations represented by these zero–one matrices are equivalence relations.

a) $\begin{bmatrix} 1 & 1 & 1 \\ 0 & 1 & 1 \\ 1 & 1 & 1 \end{bmatrix}$ **b)** $\begin{bmatrix} 1 & 0 & 1 & 0 \\ 0 & 1 & 0 & 1 \\ 1 & 0 & 1 & 0 \\ 0 & 1 & 0 & 1 \end{bmatrix}$ **c)** $\begin{bmatrix} 1 & 1 & 1 & 0 \\ 1 & 1 & 1 & 0 \\ 1 & 1 & 1 & 0 \\ 0 & 0 & 0 & 1 \end{bmatrix}$

21. Show that the relation R on the set of all bit strings such that $s \, R \, t$ if and only if s and t contain the same number of 1s is an equivalence relation.

22. What are the equivalence classes of the equivalence relations in Exercise 1?

23. What are the equivalence classes of the equivalence relations in Exercise 2?

24. What is the equivalence class of the bit string 011 for the equivalence relation in Exercise 21?

25. What are the equivalence classes of the bit strings?
  a) 010   **b)** 1011
  c) 11111   **d)** 01010101
  for the equivalence relations from Exercise 10.

26. What are the equivalence classes of the bit strings in Exercise 25 for the equivalence relation R_4 from Example 5 on the set of all bit strings? (Recall that bit strings s and t are equivalent under R_4 if and only if they are equal or they are both at least four bits long and agree in their first four bits.)

27. What is the congruence class $[n]_5$ (that is, the equivalence class of n with respect to congruence modulo 5) when n is
  a) 2? **b)** 3? **c)** 6? **d)** -3?

28. What is the congruence class $[4]_m$ when m is
  a) 2? **b)** 3? **c)** 6? **d)** 8?

29. Give a description of each of the congruence classes modulo 6.

30. What is the equivalence class of each of these strings with respect to the equivalence relation in Exercise 12?
  a) *No*   **b)** *Yes*   **c)** *Help*

31. a) What is the equivalence class of $(1, 2)$ with respect to the equivalence relation in Exercise 13?
  b) Give an interpretation of the equivalence classes for the equivalence relation R in Exercise 13. [*Hint:* Look at the difference $a - b$ corresponding to (a, b).]

32. Which of these collections of subsets are partitions of $\{1, 2, 3, 4, 5, 6\}$?
  a) $\{1, 2\}, \{2, 3, 4\}, \{4, 5, 6\}$ **b)** $\{1\}, \{2, 3, 6\}, \{4\}, \{5\}$
  c) $\{2, 4, 6\}, \{1, 3, 5\}$ **d)** $\{1, 4, 5\}, \{2, 6\}$

33. Which of these collections of subsets are partitions of the set of bit strings of length 8?
  a) the set of bit strings that begin with 1, the set of bit strings that begin with 00, and the set of bit strings that begin with 01
  b) the set of bit strings that contain the string 00, the set of bit strings that contain the string 01, the set of bit strings that contain the string 10, and the set of bit strings that contain the string 11
  c) the set of bit strings that end with 00, the set of bit strings that end with 01, the set of bit strings that end with 10, and the set of bit strings that end with 11
  d) the set of bit strings that end with 111, the set of bit strings that end with 011, and the set of bit strings that end with 00

e) the set of bit strings that contain $3k$ ones for some nonnegative integer k; the set of bit strings that contain $3k + 1$ ones for some nonnegative integer k; and the set of bit strings that contain $3k + 2$ ones for some nonnegative integer k.

34. Which of these collections of subsets are partitions of the set of integers?

a) the set of even integers and the set of odd integers

b) the set of positive integers and the set of negative integers

c) the set of integers divisible by 3, the set of integers leaving a remainder of 1 when divided by 3, and the set of integers leaving a remainder of 2 when divided by 3

d) the set of integers less than -100, the set of integers with absolute value not exceeding 100, and the set of integers greater than 100

e) the set of integers not divisible by 3, the set of even integers, and the set of integers that leave a remainder of 3 when divided by 6

35. Which of these are partitions of the set $\mathbf{Z} \times \mathbf{Z}$ of ordered pairs of integers?

a) the set of pairs (x, y), where x or y is odd; the set of pairs (x, y), where x is even; and the set of pairs (x, y), where y is even

b) the set of pairs (x, y), where both x and y are odd; the set of pairs (x, y), where exactly one of x and y is odd; and the set of pairs (x, y), where both x and y are even

c) the set of pairs (x, y), where x is positive; the set of pairs (x, y), where y is positive; and the set of pairs (x, y), where both x and y are negative

d) the set of pairs (x, y), where $3 \mid x$ and $3 \mid y$; the set of pairs (x, y), where $3 \mid x$ and $3 \nmid y$; the set of pairs (x, y), where $3 \nmid x$ and $3 \mid y$; and the set of pairs (x, y), where $3 \nmid x$ and $3 \nmid y$

e) the set of pairs (x, y), where $x > 0$ and $y > 0$; the set of pairs (x, y), where $x > 0$ and $y \le 0$; the set of pairs (x, y), where $x \le 0$ and $y > 0$; and the set of pairs (x, y), where $x \le 0$ and $y \le 0$

f) the set of pairs (x, y), where $x \ne 0$ and $y \ne 0$; the set of pairs (x, y), where $x = 0$ and $y \ne 0$; and the set of pairs (x, y), where $x \ne 0$ and $y = 0$

36. Which of these are partitions of the set of real numbers?

a) the negative real numbers, $\{0\}$, the positive real numbers

b) the set of irrational numbers, the set of rational numbers

c) the set of intervals $[k, k + 1]$, $k = \dots, -2, -1, 0, 1, 2, \dots$

d) the set of intervals $(k, k + 1)$, $k = \dots, -2, -1, 0, 1, 2, \dots$

e) the set of intervals $(k, k + 1]$, $k = \dots, -2, -1, 0, 1, 2, \dots$

f) the sets $\{x + n \mid n \in \mathbf{Z}\}$ for all $x \in [0, 1)$

37. List the ordered pairs in the equivalence relations produced by these partitions of $\{0, 1, 2, 3, 4, 5\}$.

a) $\{0\}, \{1, 2\}, \{3, 4, 5\}$

b) $\{0, 1\}, \{2, 3\}, \{4, 5\}$

c) $\{0, 1, 2\}, \{3, 4, 5\}$

d) $\{0\}, \{1\}, \{2\}, \{3\}, \{4\}, \{5\}$

A partition P_1 is called a **refinement** of the partition P_2 if every set in P_1 is a subset of one of the sets in P_2.

38. Show that the partition formed from congruence classes modulo 6 is a refinement of the partition formed from congruence classes modulo 3.

39. Show that the partition of the set of bit strings of length 16 formed by equivalence classes of bit strings that agree on the last eight bits is a refinement of the partition formed from the equivalence classes of bit strings that agree on the last four bits.

In Exercises 40 and 41, R_n refers to the family of equivalence relations defined in Example 5. Recall that $s \, R_n \, t$, where s and t are two strings if $s = t$ or s and t are strings with at least n characters that agree in their first n characters.

40. Show that the partition of the set of all bit strings formed by equivalence classes of bit strings with respect to the equivalence relation R_4 is a refinement of the partition formed by equivalence classes of bit strings with respect to the equivalence relation R_3.

41. Show that the partition of the set of all identifiers in C formed by the equivalence classes of identifiers with respect to the equivalence relation R_{31} is a refinement of the partition formed by the equivalence classes of identifiers with respect to the equivalence relation R_8. (Compilers for "old" C consider identifiers the same when their names agree in their first eight characters, while compilers in standard C consider identifiers the same when their names agree in their first 31 characters.)

42. Suppose that R_1 and R_2 are equivalence relations on a set A. Let P_1 and P_2 be the partitions that correspond to R_1 and R_2, respectively. Show that $R_1 \subseteq R_2$ if and only if P_1 is a refinement of P_2.

43. Find the smallest equivalence relation on the set $\{a, b, c, d, e\}$ containing the relation $\{(a, b), (a, c), (d, e)\}$.

44. Suppose that R_1 and R_2 are equivalence relations on the set S. Determine whether each of these combinations of R_1 and R_2 must be an equivalence relation.

a) $R_1 \cup R_2$ b) $R_1 \cap R_2$ c) $R_1 \oplus R_2$

45. Consider the equivalence relation from Example 2, namely, $R = \{(x, y) \mid x - y \text{ is an integer}\}$.

a) What is the equivalence class of 1 for this equivalence relation?

b) What is the equivalence class of 1/2 for this equivalence relation?

*46. Each bead on a bracelet with three beads is either red, white, or blue, as illustrated in the figure shown.

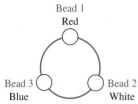

Bead 1
Red

Bead 3
Blue

Bead 2
White

Define the relation R between bracelets as: (B_1, B_2), where B_1 and B_2 are bracelets, belongs to R if and only if B_2 can be obtained from B_1 by rotating it or rotating it and then reflecting it.

a) Show that R is an equivalence relation.
b) What are the equivalence classes of R?

*47. Let R be the relation on the set of all colorings of the 2×2 checkerboard where each of the four squares is colored either red or blue so that (C_1, C_2), where C_1 and C_2 are 2×2 checkerboards with each of their four squares colored blue or red, belongs to R if and only if C_2 can be obtained from C_1 either by rotating the checkerboard or by rotating it and then reflecting it.

a) Show that R is an equivalence relation.

b) What are the equivalence classes of R?

48. Determine the number of different equivalence relations on a set with four elements by listing them.

*49. Do we necessarily get an equivalence relation when we form the transitive closure of the symmetric closure of the reflexive closure of a relation?

*50. Do we necessarily get an equivalence relation when we form the symmetric closure of the reflexive closure of the transitive closure of a relation?

51. Suppose we use Theorem 2 to form a partition P from an equivalence relation R. What is the equivalence relation R' that results if we use Theorem 2 again to form an equivalence relation from P?

52. Suppose we use Theorem 2 to form an equivalence relation R from a partition P. What is the partition P' that results if we use Theorem 2 again to form a partition from R?

53. Devise an algorithm to find the smallest equivalence relation containing a given relation.

9.6 Partial Orderings

Introduction

We often use relations to order some or all of the elements of sets. For instance, we order words using the relation containing pairs of words (x, y), where x comes before y in the dictionary. We schedule projects using the relation consisting of pairs (x, y), where x and y are tasks in a project such that x must be completed before y begins. We order the set of integers using the relation containing the pairs (x, y), where x is less than y. When we add all of the pairs of the form (x, x) to these relations, we obtain a relation that is reflexive, antisymmetric, and transitive. These are properties that characterize relations used to order the elements of sets.

Links

DEFINITION 1

A relation R on a set S is called a *partial ordering* or *partial order* if it is reflexive, antisymmetric, and transitive. A set S together with a partial ordering R is called a *partially ordered set*, or *poset*, and is denoted by (S, R). Members of S are called *elements* of the poset.

We give examples of posets in Examples 1–3.

EXAMPLE 1 Show that the "greater than or equal" relation ($\geq$) is a partial ordering on the set of integers.

Extra
Examples

Solution: Because $a \geq a$ for every integer a, $\geq$ is reflexive. If $a \geq b$ and $b \geq a$, then $a = b$. Hence, $\geq$ is antisymmetric. Finally, $\geq$ is transitive because $a \geq b$ and $b \geq c$ imply that $a \geq c$. It follows that $\geq$ is a partial ordering on the set of integers and $(\mathbf{Z}, \geq)$ is a poset. ◀

EXAMPLE 2 The divisibility relation $|$ is a partial ordering on the set of positive integers, because it is reflexive, antisymmetric, and transitive, as was shown in Section 9.1. We see that $(\mathbf{Z}^+, |)$ is a poset. Recall that ($\mathbf{Z}^+$ denotes the set of positive integers.) ◀

EXAMPLE 3　Show that the inclusion relation $\subseteq$ is a partial ordering on the power set of a set S.

Solution: Because $A \subseteq A$ whenever A is a subset of S, $\subseteq$ is reflexive. It is antisymmetric because $A \subseteq B$ and $B \subseteq A$ imply that $A = B$. Finally, $\subseteq$ is transitive, because $A \subseteq B$ and $B \subseteq C$ imply that $A \subseteq C$. Hence, $\subseteq$ is a partial ordering on $P(S)$, and $(P(S), \subseteq)$ is a poset. ◀

Example 4 illustrates a relation that is not a partial ordering.

EXAMPLE 4　Let R be the relation on the set of people such that $x R y$ if x and y are people and x is older than y. Show that R is not a partial ordering.

Solution: Note that R is antisymmetric because if a person x is older than a person y, then y is not older than x. That is, if $x R y$, then $y \not{R} x$. The relation R is transitive because if person x is older than person y and y is older than person z, then x is older than z. That is, if $x R y$ and $y R z$, then $x R z$. However, R is not reflexive, because no person is older than himself or herself. That is, $x \not{R} x$ for all people x. It follows that R is not a partial ordering. ◀

In different posets different symbols such as $\le$, $\subseteq$, and $|$, are used for a partial ordering. However, we need a symbol that we can use when we discuss the ordering relation in an arbitrary poset. Customarily, the notation $a \preccurlyeq b$ is used to denote that $(a, b) \in R$ in an arbitrary poset (S, R). This notation is used because the "less than or equal to" relation on the set of real numbers is the most familiar example of a partial ordering and the symbol $\preccurlyeq$ is similar to the $\le$ symbol. (Note that the symbol $\preccurlyeq$ is used to denote the relation in *any* poset, not just the "less than or equals" relation.) The notation $a \prec b$ denotes that $a \preccurlyeq b$, but $a \ne b$. Also, we say "a is less than b" or "b is greater than a" if $a \prec b$.

When a and b are elements of the poset $(S, \preccurlyeq)$, it is not necessary that either $a \preccurlyeq b$ or $b \preccurlyeq a$. For instance, in $(P(\mathbf{Z}), \subseteq)$, $\{1, 2\}$ is not related to $\{1, 3\}$, and vice versa, because neither set is contained within the other. Similarly, in $(\mathbf{Z}^+, |)$, 2 is not related to 3 and 3 is not related to 2, because $2 \nmid 3$ and $3 \nmid 2$. This leads to Definition 2.

DEFINITION 2　The elements a and b of a poset $(S, \preccurlyeq)$ are called *comparable* if either $a \preccurlyeq b$ or $b \preccurlyeq a$. When a and b are elements of S such that neither $a \preccurlyeq b$ nor $b \preccurlyeq a$, a and b are called *incomparable*.

EXAMPLE 5　In the poset $(\mathbf{Z}^+, |)$, are the integers 3 and 9 comparable? Are 5 and 7 comparable?

Solution: The integers 3 and 9 are comparable, because $3 \mid 9$. The integers 5 and 7 are incomparable, because $5 \nmid 7$ and $7 \nmid 5$. ◀

The adjective "partial" is used to describe partial orderings because pairs of elements may be incomparable. When every two elements in the set are comparable, the relation is called a **total ordering**.

DEFINITION 3　If $(S, \preccurlyeq)$ is a poset and every two elements of S are comparable, S is called a *totally ordered* or *linearly ordered set*, and $\preccurlyeq$ is called a *total order* or a *linear order*. A totally ordered set is also called a *chain*.

EXAMPLE 6　The poset $(\mathbf{Z}, \le)$ is totally ordered, because $a \le b$ or $b \le a$ whenever a and b are integers. ◀

EXAMPLE 7　The poset $(\mathbf{Z}^+, |)$ is not totally ordered because it contains elements that are incomparable, such as 5 and 7. ◀

In Chapter 6 we noted that $(\mathbf{Z}^+, \leq)$ is well-ordered, where $\leq$ is the usual "less than or equal to" relation. We now define well-ordered sets.

DEFINITION 4

$(S, \preccurlyeq)$ is a *well-ordered set* if it is a poset such that $\preccurlyeq$ is a total ordering and every nonempty subset of S has a least element.

EXAMPLE 8

The set of ordered pairs of positive integers, $\mathbf{Z}^+ \times \mathbf{Z}^+$, with $(a_1, a_2) \preccurlyeq (b_1, b_2)$ if $a_1 < b_1$, or if $a_1 = b_1$ and $a_2 \leq b_2$ (the lexicographic ordering), is a well-ordered set. The verification of this is left as Exercise 53. The set $\mathbf{Z}$, with the usual $\leq$ ordering, is not well-ordered because the set of negative integers, which is a subset of $\mathbf{Z}$, has no least element. ◄

At the end of Section 5.3 we showed how to use the principle of well-ordered induction (there called generalized induction) to prove results about a well-ordered set. We now state and prove that this proof technique is valid.

THEOREM 1

THE PRINCIPLE OF WELL-ORDERED INDUCTION   Suppose that S is a well-ordered set. Then $P(x)$ is true for all $x \in S$, if

INDUCTIVE STEP: For every $y \in S$, if $P(x)$ is true for all $x \in S$ with $x \prec y$, then $P(y)$ is true.

Proof: Suppose it is not the case that $P(x)$ is true for all $x \in S$. Then there is an element $y \in S$ such that $P(y)$ is false. Consequently, the set $A = \{x \in S \mid P(x) \text{ is false}\}$ is nonempty. Because S is well ordered, A has a least element a. By the choice of a as a least element of A, we know that $P(x)$ is true for all $x \in S$ with $x \prec a$. This implies by the inductive step $P(a)$ is true. This contradiction shows that $P(x)$ must be true for all $x \in S$. ◁

Remark: We do not need a basis step in a proof using the principle of well-ordered induction because if x_0 is the least element of a well ordered set, the inductive step tells us that $P(x_0)$ is true. This follows because there are no elements $x \in S$ with $x \prec x_0$, so we know (using a vacuous proof) that $P(x)$ is true for all $x \in S$ with $x \prec x_0$.

The principle of well-ordered induction is a versatile technique for proving results about well-ordered sets. Even when it is possible to use mathematical induction for the set of positive integers to prove a theorem, it may be simpler to use the principle of well-ordered induction, as we saw in Examples 5 and 6 in Section 6.2, where we proved a result about the well-ordered set $(\mathbf{N} \times \mathbf{N}, \preccurlyeq)$ where $\preccurlyeq$ is lexicographic ordering on $\mathbf{N} \times \mathbf{N}$.

Lexicographic Order

The words in a dictionary are listed in alphabetic, or lexicographic, order, which is based on the ordering of the letters in the alphabet. This is a special case of an ordering of strings on a set constructed from a partial ordering on the set. We will show how this construction works in any poset.

First, we will show how to construct a partial ordering on the Cartesian product of two posets, $(A_1, \preccurlyeq_1)$ and $(A_2, \preccurlyeq_2)$. The **lexicographic ordering** $\preccurlyeq$ on $A_1 \times A_2$ is defined by specifying that one pair is less than a second pair if the first entry of the first pair is less than (in A_1) the first entry of the second pair, or if the first entries are equal, but the second entry of this pair is less than (in A_2) the second entry of the second pair. In other words, (a_1, a_2) is less than (b_1, b_2), that is,

$$(a_1, a_2) \prec (b_1, b_2),$$

either if $a_1 \prec_1 b_1$ or if both $a_1 = b_1$ and $a_2 \prec_2 b_2$.

We obtain a partial ordering $\preccurlyeq$ by adding equality to the ordering $\prec$ on $A_1 \times A_2$. The verification of this is left as an exercise.

EXAMPLE 9   Determine whether $(3, 5) \prec (4, 8)$, whether $(3, 8) \prec (4, 5)$, and whether $(4, 9) \prec (4, 11)$ in the poset $(\mathbf{Z} \times \mathbf{Z}, \preccurlyeq)$, where $\preccurlyeq$ is the lexicographic ordering constructed from the usual $\leq$ relation on $\mathbf{Z}$.

Solution: Because $3 < 4$, it follows that $(3, 5) \prec (4, 8)$ and that $(3, 8) \prec (4, 5)$. We have $(4, 9) \prec (4, 11)$, because the first entries of $(4, 9)$ and $(4, 11)$ are the same but $9 < 11$.   ◀

In Figure 1 the ordered pairs in $\mathbf{Z}^+ \times \mathbf{Z}^+$ that are less than $(3, 4)$ are highlighted. A lexicographic ordering can be defined on the Cartesian product of n posets $(A_1, \preccurlyeq_1)$, $(A_2, \preccurlyeq_2), \ldots, (A_n, \preccurlyeq_n)$. Define the partial ordering $\preccurlyeq$ on $A_1 \times A_2 \times \cdots \times A_n$ by

$$(a_1, a_2, \ldots, a_n) \prec (b_1, b_2, \ldots, b_n)$$

if $a_1 \prec_1 b_1$, or if there is an integer $i > 0$ such that $a_1 = b_1, \ldots, a_i = b_i$, and $a_{i+1} \prec_{i+1} b_{i+1}$. In other words, one n-tuple is less than a second n-tuple if the entry of the first n-tuple in the first position where the two n-tuples disagree is less than the entry in that position in the second n-tuple.

EXAMPLE 10   Note that $(1, 2, 3, 5) \prec (1, 2, 4, 3)$, because the entries in the first two positions of these 4-tuples agree, but in the third position the entry in the first 4-tuple, 3, is less than that in the second 4-tuple, 4. (Here the ordering on 4-tuples is the lexicographic ordering that comes from the usual "less than or equals" relation on the set of integers.)   ◀

We can now define lexicographic ordering of strings. Consider the strings $a_1 a_2 \ldots a_m$ and $b_1 b_2 \ldots b_n$ on a partially ordered set S. Suppose these strings are not equal. Let t be the minimum

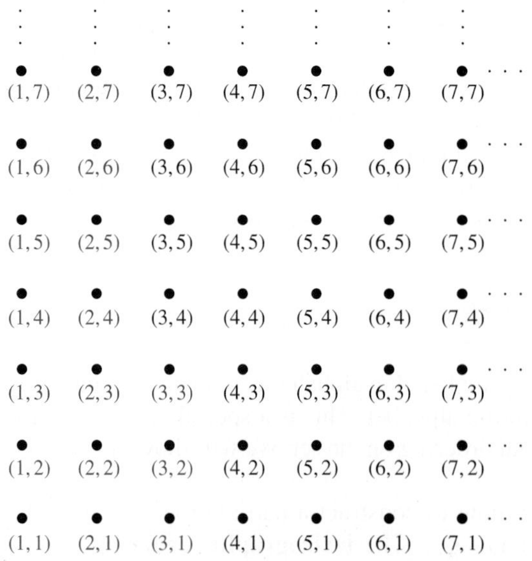

FIGURE 1   The Ordered Pairs Less Than (3, 4) in Lexicographic Order.

of m and n. The definition of lexicographic ordering is that the string $a_1 a_2 \ldots a_m$ is less than $b_1 b_2 \ldots b_n$ if and only if

$(a_1, a_2, \ldots, a_t) \prec (b_1, b_2, \ldots, b_t)$, or

$(a_1, a_2, \ldots, a_t) = (b_1, b_2, \ldots, b_t)$ and $m < n$,

where $\prec$ in this inequality represents the lexicographic ordering of S^t. In other words, to determine the ordering of two different strings, the longer string is truncated to the length of the shorter string, namely, to $t = \min(m, n)$ terms. Then the t-tuples made up of the first t terms of each string are compared using the lexicographic ordering on S^t. One string is less than another string if the t-tuple corresponding to the first string is less than the t-tuple of the second string, or if these two t-tuples are the same, but the second string is longer. The verification that this is a partial ordering is left as Exercise 38 for the reader.

EXAMPLE 11   Consider the set of strings of lowercase English letters. Using the ordering of letters in the alphabet, a lexicographic ordering on the set of strings can be constructed. A string is less than a second string if the letter in the first string in the first position where the strings differ comes before the letter in the second string in this position, or if the first string and the second string agree in all positions, but the second string has more letters. This ordering is the same as that used in dictionaries. For example,

   discreet $\prec$ *discrete*,

because these strings differ first in the seventh position, and $e \prec t$. Also,

   discreet $\prec$ *discreetness*,

because the first eight letters agree, but the second string is longer. Furthermore,

   discrete $\prec$ *discretion*,

because

   discrete $\prec$ *discreti*. $\blacktriangleleft$

Hasse Diagrams

Many edges in the directed graph for a finite poset do not have to be shown because they must be present. For instance, consider the directed graph for the partial ordering $\{(a, b) \mid a \leq b\}$ on the set $\{1, 2, 3, 4\}$, shown in Figure 2(a). Because this relation is a partial ordering, it is reflexive, and its directed graph has loops at all vertices. Consequently, we do not have to show these loops because they must be present; in Figure 2(b) loops are not shown. Because a partial ordering is transitive, we do not have to show those edges that must be present because of transitivity. For example, in Figure 2(c) the edges $(1, 3)$, $(1, 4)$, and $(2, 4)$ are not shown because they must be present. If we assume that all edges are pointed "upward" (as they are drawn in the figure), we do not have to show the directions of the edges; Figure 2(c) does not show directions.

In general, we can represent a finite poset $(S, \preccurlyeq)$ using this procedure: Start with the directed graph for this relation. Because a partial ordering is reflexive, a loop (a, a) is present at every vertex a. Remove these loops. Next, remove all edges that must be in the partial ordering because of the presence of other edges and transitivity. That is, remove all edges (x, y) for which there is an element $z \in S$ such that $x \prec z$ and $z \prec x$. Finally, arrange each edge so that

Links

HELMUT HASSE (1898–1979)   Helmut Hasse was born in Kassel, Germany. He served in the German navy after high school. He began his university studies at Göttingen University in 1918, moving in 1920 to Marburg University to study under the number theorist Kurt Hensel. During this time, Hasse made fundamental contributions to algebraic number theory. He became Hensel's successor at Marburg, later becoming director of the famous mathematical institute at Göttingen in 1934, and took a position at Hamburg University in 1950. Hasse served for 50 years as an editor of *Crelle's Journal*, a famous German mathematics periodical, taking over the job of chief editor in 1936 when the Nazis forced Hensel to resign. During World War II Hasse worked on applied mathematics research for the German navy. He was noted for the clarity and personal style of his lectures and was devoted both to number theory and to his students. (Hasse has been controversial for connections with the Nazi party. Investigations have shown he was a strong German nationalist but not an ardent Nazi.)

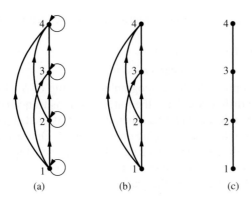

(a)   (b)   (c)

FIGURE 2 **Constructing the Hasse Diagram for ({1, 2, 3, 4}, ≤).**

its initial vertex is below its terminal vertex (as it is drawn on paper). Remove all the arrows on the directed edges, because all edges point "upward" toward their terminal vertex.

These steps are well defined, and only a finite number of steps need to be carried out for a finite poset. When all the steps have been taken, the resulting diagram contains sufficient information to find the partial ordering, as we will explain later. The resulting diagram is called the **Hasse diagram** of $(S, \preccurlyeq)$, named after the twentieth-century German mathematician Helmut Hasse who made extensive use of them.

Let $(S, \preccurlyeq)$ be a poset. We say that an element $y \in S$ **covers** an element $x \in S$ if $x \prec y$ and there is no element $z \in S$ such that $x \prec z \prec y$. The set of pairs (x, y) such that y covers x is called the **covering relation** of $(S, \preccurlyeq)$. From the description of the Hasse diagram of a poset, we see that the edges in the Hasse diagram of $(S, \preccurlyeq)$ are upwardly pointing edges corresponding to the pairs in the covering relation of $(S, \preccurlyeq)$. Furthermore, we can recover a poset from its covering relation, because it is the reflexive transitive closure of its covering relation. (Exercise 21 asks for a proof of this fact.) This tells us that we can construct a partial ordering from its Hasse diagram.

EXAMPLE 12   Draw the Hasse diagram representing the partial ordering $\{(a, b) \mid a \text{ divides } b\}$ on $\{1, 2, 3, 4, 6, 8, 12\}$.

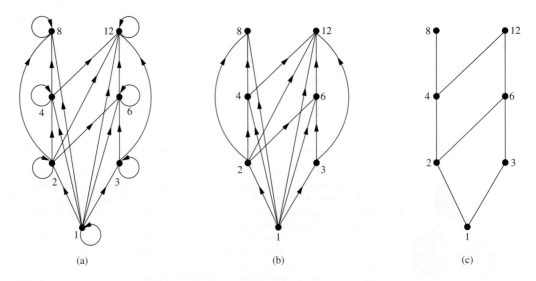

(a)   (b)   (c)

FIGURE 3 **Constructing the Hasse Diagram of ({1, 2, 3, 4, 6, 8, 12}, |).**

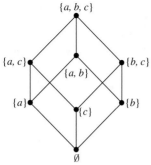

FIGURE 4 **The Hasse Diagram of $(P(\{a, b, c\}), \subseteq)$.**

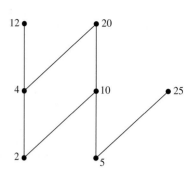

FIGURE 5 **The Hasse Diagram of a Poset.**

Solution: Begin with the digraph for this partial order, as shown in Figure 3(a). Remove all loops, as shown in Figure 3(b). Then delete all the edges implied by the transitive property. These are $(1, 4)$, $(1, 6)$, $(1, 8)$, $(1, 12)$, $(2, 8)$, $(2, 12)$, and $(3, 12)$. Arrange all edges to point upward, and delete all arrows to obtain the Hasse diagram. The resulting Hasse diagram is shown in Figure 3(c). ◀

EXAMPLE 13 Draw the Hasse diagram for the partial ordering $\{(A, B) \mid A \subseteq B\}$ on the power set $P(S)$ where $S = \{a, b, c\}$.

Solution: The Hasse diagram for this partial ordering is obtained from the associated digraph by deleting all the loops and all the edges that occur from transitivity, namely, $(\emptyset, \{a, b\})$, $(\emptyset, \{a, c\})$, $(\emptyset, \{b, c\})$, $(\emptyset, \{a, b, c\})$, $(\{a\}, \{a, b, c\})$, $(\{b\}, \{a, b, c\})$, and $(\{c\}, \{a, b, c\})$. Finally all edges point upward, and arrows are deleted. The resulting Hasse diagram is illustrated in Figure 4. ◀

Maximal and Minimal Elements

Elements of posets that have certain extremal properties are important for many applications. An element of a poset is called maximal if it is not less than any element of the poset. That is, a is **maximal** in the poset $(S, \preccurlyeq)$ if there is no $b \in S$ such that $a \prec b$. Similarly, an element of a poset is called minimal if it is not greater than any element of the poset. That is, a is **minimal** if there is no element $b \in S$ such that $b \prec a$. Maximal and minimal elements are easy to spot using a Hasse diagram. They are the "top" and "bottom" elements in the diagram.

EXAMPLE 14 Which elements of the poset $(\{2, 4, 5, 10, 12, 20, 25\}, |)$ are maximal, and which are minimal?

Solution: The Hasse diagram in Figure 5 for this poset shows that the maximal elements are 12, 20, and 25, and the minimal elements are 2 and 5. As this example shows, a poset can have more than one maximal element and more than one minimal element. ◀

Sometimes there is an element in a poset that is greater than every other element. Such an element is called the greatest element. That is, a is the **greatest element** of the poset $(S, \preccurlyeq)$ if $b \preccurlyeq a$ for all $b \in S$. The greatest element is unique when it exists [see Exercise 28(a)]. Likewise, an element is called the least element if it is less than all the other elements in the poset. That is, a is the **least element** of $(S, \preccurlyeq)$ if $a \preccurlyeq b$ for all $b \in S$. The least element is unique when it exists [see Exercise 28(b)].

EXAMPLE 15 Determine whether the posets represented by each of the Hasse diagrams in Figure 6 have a greatest element and a least element.

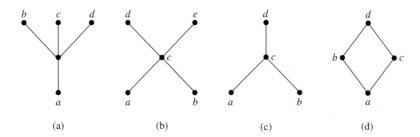

FIGURE 6 **Hasse Diagrams of Four Posets.**

Solution: The least element of the poset with Hasse diagram (a) is a. This poset has no greatest element. The poset with Hasse diagram (b) has neither a least nor a greatest element. The poset with Hasse diagram (c) has no least element. Its greatest element is d. The poset with Hasse diagram (d) has least element a and greatest element d. ◀

EXAMPLE 16 Let S be a set. Determine whether there is a greatest element and a least element in the poset $(P(S), \subseteq)$.

Solution: The least element is the empty set, because $\emptyset \subseteq T$ for any subset T of S. The set S is the greatest element in this poset, because $T \subseteq S$ whenever T is a subset of S. ◀

EXAMPLE 17 Is there a greatest element and a least element in the poset $(\mathbf{Z}^+, |)$?

Solution: The integer 1 is the least element because $1 | n$ whenever n is a positive integer. Because there is no integer that is divisible by all positive integers, there is no greatest element. ◀

Sometimes it is possible to find an element that is greater than or equal to all the elements in a subset A of a poset $(S, \preccurlyeq)$. If u is an element of S such that $a \preccurlyeq u$ for all elements $a \in A$, then u is called an **upper bound** of A. Likewise, there may be an element less than or equal to all the elements in A. If l is an element of S such that $l \preccurlyeq a$ for all elements $a \in A$, then l is called a **lower bound** of A.

EXAMPLE 18 Find the lower and upper bounds of the subsets $\{a, b, c\}$, $\{j, h\}$, and $\{a, c, d, f\}$ in the poset with the Hasse diagram shown in Figure 7.

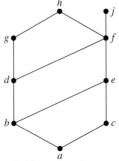

FIGURE 7 **The Hasse Diagram of a Poset.**

Solution: The upper bounds of $\{a, b, c\}$ are e, f, j, and h, and its only lower bound is a. There are no upper bounds of $\{j, h\}$, and its lower bounds are a, b, c, d, e, and f. The upper bounds of $\{a, c, d, f\}$ are f, h, and j, and its lower bound is a. ◀

The element x is called the **least upper bound** of the subset A if x is an upper bound that is less than every other upper bound of A. Because there is only one such element, if it exists, it makes sense to call this element *the* least upper bound [see Exercise 30(a)]. That is, x is the least upper bound of A if $a \preccurlyeq x$ whenever $a \in A$, and $x \preccurlyeq z$ whenever z is an upper bound of A. Similarly, the element y is called the **greatest lower bound** of A if y is a lower bound of A and $z \preccurlyeq y$ whenever z is a lower bound of A. The greatest lower bound of A is unique if it exists [see Exercise 30(b)]. The greatest lower bound and least upper bound of a subset A are denoted by $\mathrm{glb}(A)$ and $\mathrm{lub}(A)$, respectively.

EXAMPLE 19 Find the greatest lower bound and the least upper bound of $\{b, d, g\}$, if they exist, in the poset shown in Figure 7.

Solution: The upper bounds of $\{b, d, g\}$ are g and h. Because $g \prec h$, g is the least upper bound. The lower bounds of $\{b, d, g\}$ are a and b. Because $a \prec b$, b is the greatest lower bound. ◀

EXAMPLE 20 Find the greatest lower bound and the least upper bound of the sets $\{3, 9, 12\}$ and $\{1, 2, 4, 5, 10\}$, if they exist, in the poset $(\mathbf{Z}^+, |)$.

Solution: An integer is a lower bound of $\{3, 9, 12\}$ if 3, 9, and 12 are divisible by this integer. The only such integers are 1 and 3. Because $1 \mid 3$, 3 is the greatest lower bound of $\{3, 9, 12\}$. The only lower bound for the set $\{1, 2, 4, 5, 10\}$ with respect to $|$ is the element 1. Hence, 1 is the greatest lower bound for $\{1, 2, 4, 5, 10\}$.

An integer is an upper bound for $\{3, 9, 12\}$ if and only if it is divisible by 3, 9, and 12. The integers with this property are those divisible by the least common multiple of 3, 9, and 12, which is 36. Hence, 36 is the least upper bound of $\{3, 9, 12\}$. A positive integer is an upper bound for the set $\{1, 2, 4, 5, 10\}$ if and only if it is divisible by 1, 2, 4, 5, and 10. The integers with this property are those integers divisible by the least common multiple of these integers, which is 20. Hence, 20 is the least upper bound of $\{1, 2, 4, 5, 10\}$. ◀

Lattices

A partially ordered set in which every pair of elements has both a least upper bound and a greatest lower bound is called a **lattice**. Lattices have many special properties. Furthermore, lattices are used in many different applications such as models of information flow and play an important role in Boolean algebra.

EXAMPLE 21 Determine whether the posets represented by each of the Hasse diagrams in Figure 8 are lattices.

Solution: The posets represented by the Hasse diagrams in (a) and (c) are both lattices because in each poset every pair of elements has both a least upper bound and a greatest lower bound, as the reader should verify. On the other hand, the poset with the Hasse diagram shown in (b) is not a lattice, because the elements b and c have no least upper bound. To see this, note that each of the elements d, e, and f is an upper bound, but none of these three elements precedes the other two with respect to the ordering of this poset. ◀

EXAMPLE 22 Is the poset $(\mathbf{Z}^+, |)$ a lattice?

Solution: Let a and b be two positive integers. The least upper bound and greatest lower bound of these two integers are the least common multiple and the greatest common divisor of these integers, respectively, as the reader should verify. It follows that this poset is a lattice. ◀

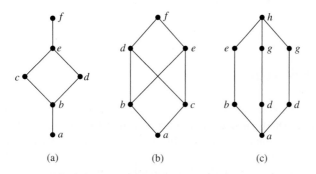

(a)   (b)   (c)

FIGURE 8 **Hasse Diagrams of Three Posets.**

EXAMPLE 23   Determine whether the posets ($\{1, 2, 3, 4, 5\}$, |) and ($\{1, 2, 4, 8, 16\}$, |) are lattices.

Solution: Because 2 and 3 have no upper bounds in ($\{1, 2, 3, 4, 5\}$, |), they certainly do not have a least upper bound. Hence, the first poset is not a lattice.

Every two elements of the second poset have both a least upper bound and a greatest lower bound. The least upper bound of two elements in this poset is the larger of the elements and the greatest lower bound of two elements is the smaller of the elements, as the reader should verify. Hence, this second poset is a lattice.   ◀

EXAMPLE 24   Determine whether ($P(S), \subseteq$) is a lattice where S is a set.

Solution: Let A and B be two subsets of S. The least upper bound and the greatest lower bound of A and B are $A \cup B$ and $A \cap B$, respectively, as the reader can show. Hence, ($P(S), \subseteq$) is a lattice.   ◀

EXAMPLE 25   **The Lattice Model of Information Flow**   In many settings the flow of information from one person or computer program to another is restricted via security clearances. We can use a lattice model to represent different information flow policies. For example, one common information flow policy is the *multilevel security policy* used in government and military systems. Each piece of information is assigned to a security class, and each security class is represented by a pair (A, C) where A is an *authority level* and C is a *category*. People and computer programs are then allowed access to information from a specific restricted set of security classes.

Links

The typical authority levels used in the U.S. government are unclassified (0), confidential (1), secret (2), and top secret (3). (Information is said to be classified if it is confidential, secret, or top secret.) Categories used in security classes are the subsets of a set of all *compartments* relevant to a particular area of interest. Each compartment represents a particular subject area. For example, if the set of compartments is {*spies, moles, double agents*}, then there are eight different categories, one for each of the eight subsets of the set of compartments, such as {*spies, moles*}.

There are billions of pages of classified U.S. government documents.

We can order security classes by specifying that $(A_1, C_1) \preccurlyeq (A_2, C_2)$ if and only if $A_1 \leq A_2$ and $C_1 \subseteq C_2$. Information is permitted to flow from security class (A_1, C_1) into security class (A_2, C_2) if and only if $(A_1, C_1) \preccurlyeq (A_2, C_2)$. For example, information is permitted to flow from the security class (*secret,* {*spies, moles*}) into the security class (*top secret,* {*spies, moles, double agents*}), whereas information is not allowed to flow from the security class (*top secret,* {*spies, moles*}) into either of the security classes (*secret,* {*spies, moles, double agents*}) or (*top secret,* {*spies*}).

We leave it to the reader (see Exercise 48) to show that the set of all security classes with the ordering defined in this example forms a lattice.   ◀

Topological Sorting

Suppose that a project is made up of 20 different tasks. Some tasks can be completed only after others have been finished. How can an order be found for these tasks? To model this problem we set up a partial order on the set of tasks so that $a \prec b$ if and only if a and b are tasks where b cannot be started until a has been completed. To produce a schedule for the project, we need to produce an order for all 20 tasks that is compatible with this partial order. We will show how this can be done.

Links

We begin with a definition. A total ordering $\preccurlyeq$ is said to be **compatible** with the partial ordering R if $a \preccurlyeq b$ whenever aRb. Constructing a compatible total ordering from a partial ordering is called **topological sorting**.* We will need to use Lemma 1.

LEMMA 1   Every finite nonempty poset $(S, \preccurlyeq)$ has at least one minimal element.

Proof: Choose an element a_0 of S. If a_0 is not minimal, then there is an element a_1 with $a_1 \prec a_0$. If a_1 is not minimal, there is an element a_2 with $a_2 \prec a_1$. Continue this process, so that if a_n is not minimal, there is an element a_{n+1} with $a_{n+1} \prec a_n$. Because there are only a finite number of elements in the poset, this process must end with a minimal element a_n. ◁

The topological sorting algorithm we will describe works for any finite nonempty poset. To define a total ordering on the poset $(A, \preccurlyeq)$, first choose a minimal element a_1; such an element exists by Lemma 1. Next, note that $(A - \{a_1\}, \preccurlyeq)$ is also a poset, as the reader should verify. (Here by $\preccurlyeq$ we mean the restriction of the original relation $\preccurlyeq$ on A to $A - \{a_1\}$.) If it is nonempty, choose a minimal element a_2 of this poset. Then remove a_2 as well, and if there are additional elements left, choose a minimal element a_3 in $A - \{a_1, a_2\}$. Continue this process by choosing a_{k+1} to be a minimal element in $A - \{a_1, a_2, \ldots, a_k\}$, as long as elements remain.

Because A is a finite set, this process must terminate. The end product is a sequence of elements $a_1, a_2, \ldots, a_n$. The desired total ordering $\preccurlyeq_t$ is defined by

$$a_1 \prec_t a_2 \prec_t \cdots \prec_t a_n.$$

This total ordering is compatible with the original partial ordering. To see this, note that if $b \prec c$ in the original partial ordering, c is chosen as the minimal element at a phase of the algorithm where b has already been removed, for otherwise c would not be a minimal element. Pseudocode for this topological sorting algorithm is shown in Algorithm 1.

ALGORITHM 1 **Topological Sorting.**

---

procedure *topological sort* $((S, \preccurlyeq)$: finite poset)
$k := 1$
while $S \neq \emptyset$
    $a_k := $ a minimal element of S {such an element exists by Lemma 1}
    $S := S - \{a_k\}$
    $k := k + 1$
return $a_1, a_2, \ldots, a_n$ {$a_1, a_2, \ldots, a_n$ is a compatible total ordering of S}

EXAMPLE 26   Find a compatible total ordering for the poset $(\{1, 2, 4, 5, 12, 20\}, |)$.

Solution: The first step is to choose a minimal element. This must be 1, because it is the only minimal element. Next, select a minimal element of $(\{2, 4, 5, 12, 20\}, |)$. There are two minimal elements in this poset, namely, 2 and 5. We select 5. The remaining elements are $\{2, 4, 12, 20\}$.

---

*"Topological sorting" is terminology used by computer scientists; mathematicians use the terminology "linearization of a partial ordering" for the same thing. In mathematics, topology is the branch of geometry dealing with properties of geometric figures that hold for all figures that can be transformed into one another by continuous bijections. In computer science, a topology is any arrangement of objects that can be connected with edges.

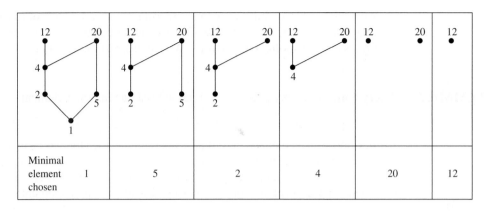

FIGURE 9   A Topological Sort of ({1, 2, 4, 5, 12, 20}, |).

The only minimal element at this stage is 2. Next, 4 is chosen because it is the only minimal element of ({4, 12, 20}, |). Because both 12 and 20 are minimal elements of ({12, 20}, |), either can be chosen next. We select 20, which leaves 12 as the last element left. This produces the total ordering

$$1 \prec 5 \prec 2 \prec 4 \prec 20 \prec 12.$$

The steps used by this sorting algorithm are displayed in Figure 9.   ◀

Topological sorting has an application to the scheduling of projects.

EXAMPLE 27   A development project at a computer company requires the completion of seven tasks. Some of these tasks can be started only after other tasks are finished. A partial ordering on tasks is set up by considering task $X \prec$ task Y if task Y cannot be started until task X has been completed. The Hasse diagram for the seven tasks, with respect to this partial ordering, is shown in Figure 10. Find an order in which these tasks can be carried out to complete the project.

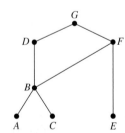

Solution: An ordering of the seven tasks can be obtained by performing a topological sort. The steps of a sort are illustrated in Figure 11. The result of this sort, $A \prec C \prec B \prec E \prec F \prec D \prec G$, gives one possible order for the tasks.   ◀

FIGURE 10   The Hasse Diagram for Seven Tasks.

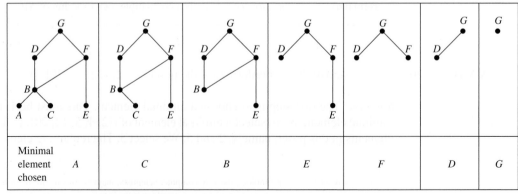

FIGURE 11   A Topological Sort of the Tasks.

Exercises

1. Which of these relations on $\{0, 1, 2, 3\}$ are partial order-ings? Determine the properties of a partial ordering that the others lack.
   a) $\{(0, 0), (1, 1), (2, 2), (3, 3)\}$
   b) $\{(0, 0), (1, 1), (2, 0), (2, 2), (2, 3), (3, 2), (3, 3)\}$
   c) $\{(0, 0), (1, 1), (1, 2), (2, 2), (3, 3)\}$
   d) $\{(0, 0), (1, 1), (1, 2), (1, 3), (2, 2), (2, 3), (3, 3)\}$
   e) $\{(0, 0), (0, 1), (0, 2), (1, 0), (1, 1), (1, 2), (2, 0),$ $(2, 2), (3, 3)\}$

2. Is (S, R) a poset if S is the set of all people in the world and $(a, b) \in R$, where a and b are people, if
   a) a is taller than b?
   b) a is not taller than b?
   c) $a = b$ or a is an ancestor of b?
   d) a and b have a common friend?

3. Which of these are posets?
   a) $(\mathbf{Z}, =)$ b) $(\mathbf{Z}, \neq)$ c) $(\mathbf{Z}, \geq)$ d) $(\mathbf{Z}, \nmid)$

4. Which of these are posets?
   a) $(\mathbf{R}, =)$ b) $(\mathbf{R}, <)$ c) $(\mathbf{R}, \leq)$ d) $(\mathbf{R}, \neq)$

5. Determine whether the relations represented by these zero–one matrices are partial orders.

   a) $\begin{bmatrix} 1 & 1 & 1 \\ 1 & 1 & 0 \\ 0 & 0 & 1 \end{bmatrix}$
   b) $\begin{bmatrix} 1 & 1 & 1 \\ 0 & 1 & 0 \\ 0 & 0 & 1 \end{bmatrix}$

   c) $\begin{bmatrix} 1 & 1 & 1 & 0 \\ 0 & 1 & 1 & 0 \\ 0 & 0 & 1 & 1 \\ 1 & 1 & 0 & 1 \end{bmatrix}$

In Exercises 6–7 determine whether the relation with the di-rected graph shown is a partial order.

6.

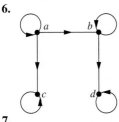

7.

8. Let (S, R) be a poset. Show that (S, R^{-1}) is also a poset, where R^{-1} is the inverse of R. The poset (S, R^{-1}) is called the **dual** of (S, R).

9. Find the duals of these posets.
   a) $(\{0, 1, 2\}, \leq)$ b) $(\mathbf{Z}, \geq)$
   c) $(P(\mathbf{Z}), \supseteq)$ d) $(\mathbf{Z}^+, |)$

10. Which of these pairs of elements are comparable in the poset $(\mathbf{Z}^+, |)$?
    a) $5, 15$ b) $6, 9$ c) $8, 16$ d) $7, 7$

11. Find two incomparable elements in these posets.
    a) $(P(\{0, 1, 2\}), \subseteq)$ b) $(\{1, 2, 4, 6, 8\}, |)$

12. Let $S = \{1, 2, 3, 4\}$. With respect to the lexicographic or-der based on the usual "less than" relation,
    a) find all pairs in $S \times S$ less than $(2, 3)$.
    b) find all pairs in $S \times S$ greater than $(3, 1)$.
    c) draw the Hasse diagram of the poset $(S \times S, \preceq)$.

13. Find the lexicographic ordering of these n-tuples:
    a) $(1, 1, 2), (1, 2, 1)$ b) $(0, 1, 2, 3), (0, 1, 3, 2)$
    c) $(1, 0, 1, 0, 1), (0, 1, 1, 1, 0)$

14. Find the lexicographic ordering of these strings of lower-case English letters:
    a) *quack, quick, quicksilver, quicksand, quacking*
    b) *open, opener, opera, operand, opened*
    c) *zoo, zero, zoom, zoology, zoological*

15. Find the lexicographic ordering of the bit strings 0, 01, 11, 001, 010, 011, 0001, and 0101 based on the ordering $0 < 1$.

16. Draw the Hasse diagram for the "less than or equal to" relation on $\{0, 2, 5, 10, 11, 15\}$.

In Exercises 17–18 list all ordered pairs in the partial ordering with the accompanying Hasse diagram.

17.

18.

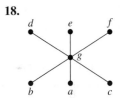

19. What is the covering relation of the partial order-ing $\{(A, B) \mid A \subseteq B\}$ on the power set of S, where $S = \{a, b, c\}$?

20. What is the covering relation of the partial ordering for the poset of security classes defined in Example 25?

21. Show that a finite poset can be reconstructed from its cov-ering relation. [*Hint:* Show that the poset is the reflexive transitive closure of its covering relation.]

22. Answer these questions for the partial order represented by this Hasse diagram.

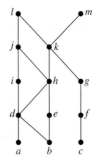

a) Find the maximal elements.

b) Find the minimal elements.

c) Is there a greatest element?

d) Is there a least element?

e) Find all upper bounds of $\{a, b, c\}$.

f) Find the least upper bound of $\{a, b, c\}$, if it exists.

g) Find all lower bounds of $\{f, g, h\}$.

h) Find the greatest lower bound of $\{f, g, h\}$, if it exists.

23. Answer these questions for the poset ($\{3, 5, 9, 15, 24, 45\}, |$).

a) Find the maximal elements.

b) Find the minimal elements.

c) Is there a greatest element?

d) Is there a least element?

e) Find all upper bounds of $\{3, 5\}$.

f) Find the least upper bound of $\{3, 5\}$, if it exists.

g) Find all lower bounds of $\{15, 45\}$.

h) Find the greatest lower bound of $\{15, 45\}$, if it exists.

24. Answer these questions for the poset ($\{\{1\}, \{2\}, \{4\}, \{1, 2\}, \{1, 4\}, \{2, 4\}, \{3, 4\}, \{1, 3, 4\}, \{2, 3, 4\}\}, \subseteq$).

a) Find the maximal elements.

b) Find the minimal elements.

c) Is there a greatest element?

d) Is there a least element?

e) Find all upper bounds of $\{\{2\}, \{4\}\}$.

f) Find the least upper bound of $\{\{2\}, \{4\}\}$, if it exists.

g) Find all lower bounds of $\{\{1, 3, 4\}, \{2, 3, 4\}\}$.

h) Find the greatest lower bound of $\{\{1, 3, 4\}, \{2, 3, 4\}\}$, if it exists.

25. Show that lexicographic order is a partial ordering on the Cartesian product of two posets.

26. Show that lexicographic order is a partial ordering on the set of strings from a poset.

27. Suppose that $(S, \preccurlyeq_1)$ and $(T, \preccurlyeq_2)$ are posets. Show that $(S \times T, \preccurlyeq)$ is a poset where $(s, t) \preccurlyeq (u, v)$ if and only if $s \preccurlyeq_1 u$ and $t \preccurlyeq_2 v$.

28. a) Show that there is exactly one greatest element of a poset, if such an element exists.

b) Show that there is exactly one least element of a poset, if such an element exists.

29. a) Show that there is exactly one maximal element in a poset with a greatest element.

b) Show that there is exactly one minimal element in a poset with a least element.

30. a) Show that the least upper bound of a set in a poset is unique if it exists.

b) Show that the greatest lower bound of a set in a poset is unique if it exists.

31. Determine whether the posets with these Hasse diagrams are lattices.

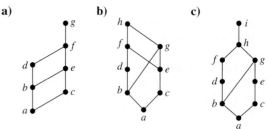

32. Determine whether these posets are lattices.

a) ($\{1, 3, 6, 9, 12\}, |$)   **b)** ($\{1, 5, 25, 125\}, |$)

c) ($\mathbf{Z}, \geq$)

d) ($P(S), \supseteq$), where $P(S)$ is the power set of a set S

33. Show that every nonempty finite subset of a lattice has a least upper bound and a greatest lower bound.

34. Show that if the poset (S, R) is a lattice then the dual poset (S, R^{-1}) is also a lattice.

35. In a company, the lattice model of information flow is used to control sensitive information with security classes represented by ordered pairs (A, C). Here A is an authority level, which may be nonproprietary (0), proprietary (1), restricted (2), or registered (3). A category C is a subset of the set of all projects {*Cheetah, Impala, Puma*}. (Names of animals are often used as code names for projects in companies.)

a) Is information permitted to flow from (*Proprietary*, {*Cheetah, Puma*}) into (*Restricted*, {*Puma*})?

b) Is information permitted to flow from (*Restricted*, {*Cheetah*}) into (*Registered*, {*Cheetah, Impala*})?

c) Into which classes is information from (*Proprietary*, {*Cheetah, Puma*}) permitted to flow?

d) From which classes is information permitted to flow into the security class (*Restricted*, {*Impala, Puma*})?

36. Show that the set S of security classes (A, C) is a lattice, where A is a positive integer representing an authority class and C is a subset of a finite set of compartments, with $(A_1, C_1) \preccurlyeq (A_2, C_2)$ if and only if $A_1 \leq A_2$ and $C_1 \subseteq C_2$. [*Hint:* First show that $(S, \preccurlyeq)$ is a poset and then show that the least upper bound and greatest lower bound of (A_1, C_1) and (A_2, C_2) are $(\max(A_1, A_2), C_1 \cup C_2)$ and $(\min(A_1, A_2), C_1 \cap C_2)$, respectively.]

∗37. Show that the set of all partitions of a set S with the relation $P_1 \preccurlyeq P_2$ if the partition P_1 is a refinement of the partition P_2 is a lattice. (See the preamble to Exercise 38 of Section 9.5.)

38. Show that every totally ordered set is a lattice.

39. Show that every finite lattice has a least element and a greatest element.

40. Give an example of an infinite lattice with

a) neither a least nor a greatest element.

b) a least but not a greatest element.

c) a greatest but not a least element.

d) both a least and a greatest element.

41. Verify that $(\mathbf{Z}^+ \times \mathbf{Z}^+, \preccurlyeq)$ is a well-ordered set, where $\preccurlyeq$ is lexicographic order, as claimed in Example 8.

42. Determine whether each of these posets is well-ordered.
  a) $(S, \leq)$, where $S = \{10, 11, 12, \ldots\}$
  b) $(\mathbf{Q} \cap [0, 1], \leq)$ (the set of rational numbers between 0 and 1 inclusive)
  c) $(S, \leq)$, where S is the set of positive rational numbers with denominators not exceeding 3
  d) $(\mathbf{Z}^-, \geq)$, where $\mathbf{Z}^-$ is the set of negative integers

A poset $(R, \preccurlyeq)$ is **well-founded** if there is no infinite decreasing sequence of elements in the poset, that is, elements $x_1, x_2, \ldots, x_n$ such that $\cdots \prec x_n \prec \cdots \prec x_2 \prec x_1$. A poset $(R, \preccurlyeq)$ is **dense** if for all $x \in S$ and $y \in S$ with $x \prec y$, there is an element $z \in R$ such that $x \prec z \prec y$.

43. Show that the poset $(\mathbf{Z}, \preccurlyeq)$, where $x \prec y$ if and only if $|x| < |y|$ is well-founded but is not a totally ordered set.

44. Show that a dense poset with at least two elements that are comparable is not well-founded.

45. Show that the poset of rational numbers with the usual "less than or equal to" relation, $(\mathbf{Q}, \leq)$, is a dense poset.

46. Show that a finite nonempty poset has a maximal element.

47. Find a compatible total order for the poset with the Hasse diagram shown in Exercise 22.

48. Find a compatible total order for the divisibility relation on the set $\{1, 2, 3, 6, 8, 12, 24, 36\}$.

49. Find all possible orders for completing the tasks in the development project in Example 27.

50. Schedule the tasks needed to build a house, by specifying their order, if the Hasse diagram representing these tasks is as shown in the figure.

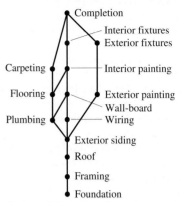

51. Find an ordering of the tasks of a software project if the Hasse diagram for the tasks of the project is as shown.

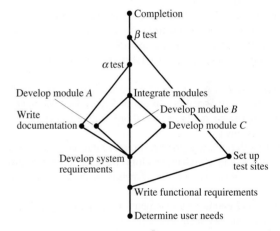

Key Terms and Results

TERMS

binary relation from A to B: a subset of $A \times B$

relation on A: a binary relation from A to itself (i.e., a subset of $A \times A$)

$S \circ R$: composite of R and S

R^{-1}: inverse relation of R

R^n: nth power of R

reflexive: a relation R on A is reflexive if $(a, a) \in R$ for all $a \in A$

symmetric: a relation R on A is symmetric if $(b, a) \in R$ whenever $(a, b) \in R$

antisymmetric: a relation R on A is antisymmetric if $a = b$ whenever $(a, b) \in R$ and $(b, a) \in R$

transitive: a relation R on A is transitive if $(a, b) \in R$ and $(b, c) \in R$ implies that $(a, c) \in R$

n-ary relation on $A_1, A_2, \ldots, A_n$: a subset of $A_1 \times A_2 \times \cdots \times A_n$

relational data model: a model for representing databases using n-ary relations

primary key: a domain of an n-ary relation such that an n-tuple is uniquely determined by its value for this domain

composite key: the Cartesian product of domains of an n-ary relation such that an n-tuple is uniquely determined by its values in these domains

selection operator: a function that selects the n-tuples in an n-ary relation that satisfy a specified condition

projection: a function that produces relations of smaller degree from an n-ary relation by deleting fields

join: a function that combines n-ary relations that agree on certain fields

directed graph or digraph: a set of elements called vertices and ordered pairs of these elements, called edges

loop: an edge of the form (a, a)

closure of a relation R with respect to a property P: the relation S (if it exists) that contains R, has property **P**, and is contained within any relation that contains R and has property **P**

path in a digraph: a sequence of edges $(a, x_1), (x_1, x_2), \ldots, (x_{n-2}, x_{n-1}), (x_{n-1}, b)$ such that the terminal vertex of each edge is the initial vertex of the succeeding edge in the sequence

circuit (or cycle) in a digraph: a path that begins and ends at the same vertex

R* (connectivity relation): the relation consisting of those ordered pairs (a, b) such that there is a path from a to b

equivalence relation: a reflexive, symmetric, and transitive relation

equivalent: if R is an equivalence relation, a is equivalent to b if $a R b$

$[a]_R$ (equivalence class of a with respect to R): the set of all elements of A that are equivalent to a

$[a]_m$ (congruence class modulo m): the set of integers congruent to a modulo m

partition of a set S: a collection of pairwise disjoint nonempty subsets that have S as their union

partial ordering: a relation that is reflexive, antisymmetric, and transitive

poset (S, R): a set S and a partial ordering R on this set

comparable: the elements a and b in the poset $(A, \preccurlyeq)$ are comparable if $a \preccurlyeq b$ or $b \preccurlyeq a$

incomparable: elements in a poset that are not comparable

total (or linear) ordering: a partial ordering for which every pair of elements are comparable

totally (or linearly) ordered set: a poset with a total (or linear) ordering

well-ordered set: a poset $(S, \preccurlyeq)$, where $\preccurlyeq$ is a total order and every nonempty subset of S has a least element

lexicographic order: a partial ordering of Cartesian products or strings

Hasse diagram: a graphical representation of a poset where loops and all edges resulting from the transitive property are not shown, and the direction of the edges is indicated by the position of the vertices

maximal element: an element of a poset that is not less than any other element of the poset

minimal element: an element of a poset that is not greater than any other element of the poset

greatest element: an element of a poset greater than all other elements in this set

least element: an element of a poset less than all other elements in this set

upper bound of a set: an element in a poset greater than all other elements in the set

lower bound of a set: an element in a poset less than all other elements in the set

least upper bound of a set: an upper bound of the set that is less than all other upper bounds

greatest lower bound of a set: a lower bound of the set that is greater than all other lower bounds

lattice: a partially ordered set in which every two elements have a greatest lower bound and a least upper bound

compatible total ordering for a partial ordering: a total ordering that contains the given partial ordering

topological sort: the construction of a total ordering compatible with a given partial ordering

RESULTS

The reflexive closure of a relation R on the set A equals $R \cup \Delta$, where $\Delta = \{(a, a) \mid a \in A\}$.

The symmetric closure of a relation R on the set A equals $R \cup R^{-1}$, where $R^{-1} = \{(b, a) \mid (a, b) \in R\}$.

The transitive closure of a relation equals the connectivity relation formed from this relation.

Warshall's algorithm for finding the transitive closure of a relation

Let R be an equivalence relation. Then the following three statements are equivalent: (1) $a R b$; (2) $[a]_R \cap [b]_R \neq \emptyset$; (3) $[a]_R = [b]_R$.

The equivalence classes of an equivalence relation on a set A form a partition of A. Conversely, an equivalence relation can be constructed from any partition so that the equivalence classes are the subsets in the partition.

The principle of well-ordered induction

The topological sorting algorithm

Review Questions

1. a) What is a relation on a set?
   b) How many relations are there on a set with n elements?

2. a) What is a reflexive relation?
   b) What is a symmetric relation?
   c) What is an antisymmetric relation?
   d) What is a transitive relation?

3. Give an example of a relation on the set $\{1, 2, 3, 4\}$ that is
   a) reflexive, symmetric, and not transitive.
   b) not reflexive, symmetric, and transitive.
   c) reflexive, antisymmetric, and not transitive.
   d) reflexive, symmetric, and transitive.
   e) reflexive, antisymmetric, and transitive.

4. a) How many reflexive relations are there on a set with n elements?
   b) How many symmetric relations are there on a set with n elements?
   c) How many antisymmetric relations are there on a set with n elements?

5. a) Explain how an n-ary relation can be used to represent information about students at a university.

   b) How can the 5-ary relation containing names of students, their addresses, telephone numbers, majors, and grade point averages be used to form a 3-ary relation containing the names of students, their majors, and their grade point averages?

   c) How can the 4-ary relation containing names of students, their addresses, telephone numbers, and majors and the 4-ary relation containing names of students, their student numbers, majors, and numbers of credit hours be combined into a single n-ary relation?

6. a) Explain how to use a zero–one matrix to represent a relation on a finite set.

   b) Explain how to use the zero–one matrix representing a relation to determine whether the relation is reflexive, symmetric, and/or antisymmetric.

7. a) Explain how to use a directed graph to represent a relation on a finite set.

b) Explain how to use the directed graph representing a relation to determine whether a relation is reflexive, symmetric, and/or antisymmetric.

8. a) Define the reflexive closure and the symmetric closure of a relation.

b) How can you construct the reflexive closure of a relation?

c) How can you construct the symmetric closure of a relation?

d) Find the reflexive closure and the symmetric closure of the relation $\{(1, 2), (2, 3), (2, 4), (3, 1)\}$ on the set $\{1, 2, 3, 4\}$.

9. a) Define the transitive closure of a relation.

b) Can the transitive closure of a relation be obtained by including all pairs (a, c) such that (a, b) and (b, c) belong to the relation?

c) Describe two algorithms for finding the transitive closure of a relation.

d) Find the transitive closure of the relation $\{(1,1), (1,3), (2,1), (2,3), (2,4), (3,2), (3,4), (4,1)\}$.

10. a) Define an equivalence relation.

b) Which relations on the set $\{a, b, c, d\}$ are equivalence relations and contain (a, b) and (b, d)?

11. a) Show that congruence modulo m is an equivalence relation whenever m is a positive integer.

b) Show that the relation $\{(a, b) \mid a \equiv \pm b \pmod 7\}$ is an equivalence relation on the set of integers.

12. a) What are the equivalence classes of an equivalence relation?

b) What are the equivalence classes of the "congruent modulo 5" relation?

c) What are the equivalence classes of the equivalence relation in Question 11(b)?

13. Explain the relationship between equivalence relations on a set and partitions of this set.

14. a) Define a partial ordering.

b) Show that the divisibility relation on the set of positive integers is a partial order.

15. Explain how partial orderings on the sets A_1 and A_2 can be used to define a partial ordering on the set $A_1 \times A_2$.

16. a) Explain how to construct the Hasse diagram of a partial order on a finite set.

b) Draw the Hasse diagram of the divisibility relation on the set $\{2, 3, 5, 9, 12, 15, 18\}$.

17. a) Define a maximal element of a poset and the greatest element of a poset.

b) Give an example of a poset that has three maximal elements.

c) Give an example of a poset with a greatest element.

18. a) Define a lattice.

b) Give an example of a poset with five elements that is a lattice and an example of a poset with five elements that is not a lattice.

19. a) Show that every finite subset of a lattice has a greatest lower bound and a least upper bound.

b) Show that every lattice with a finite number of elements has a least element and a greatest element.

20. a) Define a well-ordered set.

b) Describe an algorithm for producing a totally ordered set compatible with a given partially ordered set.

c) Explain how the algorithm from (b) can be used to order the tasks in a project if tasks are done one at a time and each task can be done only after one or more of the other tasks have been completed.

Supplementary Exercises

1. Let S be the set of all strings of English letters. Determine whether these relations are reflexive, irreflexive, symmetric, antisymmetric, and/or transitive.

a) $R_1 = \{(a, b) \mid a \text{ and } b \text{ have no letters in common}\}$
b) $R_2 = \{(a, b) \mid a \text{ and } b \text{ are not the same length}\}$
c) $R_3 = \{(a, b) \mid a \text{ is longer than } b\}$

2. Show that the relation R on $\mathbf{Z} \times \mathbf{Z}$ defined by $(a, b) \, R \, (c, d)$ if and only if $a + d = b + c$ is an equivalence relation.

3. Let R be a reflexive relation on a set A. Show that $R \subseteq R^2$.

4. Suppose that R is a symmetric relation on a set A. Is $\overline{R}$ also symmetric?

5. Let R_1 and R_2 be symmetric relations. Is $R_1 \cap R_2$ also symmetric? Is $R_1 \cup R_2$ also symmetric?

6. A relation R is called **circular** if $a R b$ and $b R c$ imply that $c R a$. Show that R is reflexive and circular if and only if it is an equivalence relation.

7. Show that a primary key in an n-ary relation is a primary key in any projection of this relation that contains this key as one of its fields.

8. Is the primary key in an n-ary relation also a primary key in a larger relation obtained by taking the join of this relation with a second relation?

9. Show that the reflexive closure of the symmetric closure of a relation is the same as the symmetric closure of its reflexive closure.

10. Let R be the relation on the set of all mathematicians that contains the ordered pair (a, b) if and only if a and b have written a published mathematical paper together.

a) Describe the relation R^2.

b) Describe the relation R^*.

c) The **Erdős number** of a mathematician is 1 if this mathematician wrote a paper with the prolific Hungarian mathematician Paul Erdős, it is 2 if this mathematician did not write a joint paper with Erdős but wrote a joint paper with someone who wrote a joint paper with Erdős, and so on (except that the Erdős number of Erdős himself is 0). Give a definition of the Erdős number in terms of paths in R.

11. a) Give an example to show that the transitive closure of the symmetric closure of a relation is not necessarily the same as the symmetric closure of the transitive closure of this relation.

b) Show, however, that the transitive closure of the symmetric closure of a relation must contain the symmetric closure of the transitive closure of this relation.

12. a) Let S be the set of subroutines of a computer program. Define the relation R by $\mathbf{P}\,R\,\mathbf{Q}$ if subroutine $\mathbf{P}$ calls subroutine $\mathbf{Q}$ during its execution. Describe the transitive closure of R.

b) For which subroutines $\mathbf{P}$ does $(\mathbf{P}, \mathbf{P})$ belong to the transitive closure of R?

c) Describe the reflexive closure of the transitive closure of R.

13. Suppose that R and S are relations on a set A with $R \subseteq S$ such that the closures of R and S with respect to a property $\mathbf{P}$ both exist. Show that the closure of R with respect to $\mathbf{P}$ is a subset of the closure of S with respect to $\mathbf{P}$.

14. Show that the symmetric closure of the union of two relations is the union of their symmetric closures.

*15. Devise an algorithm, based on the concept of interior vertices, that finds the length of the longest path between two vertices in a directed graph, or determines that there are arbitrarily long paths between these vertices.

16. Which of these are equivalence relations on the set of all people?

a) $\{(x, y) \mid x$ and y have the same sign of the zodiac$\}$

b) $\{(x, y) \mid x$ and y were born in the same year$\}$

c) $\{(x, y) \mid x$ and y have been in the same city$\}$

*17. How many different equivalence relations with exactly three different equivalence classes are there on a set with five elements?

18. Show that $\{(x, y) \mid x - y \in \mathbf{Q}\}$ is an equivalence relation on the set of real numbers, where $\mathbf{Q}$ denotes the set of rational numbers. What are $[1]$, $[\frac{1}{2}]$, and $[\pi]$?

19. Suppose that $P_1 = \{A_1, A_2, \ldots, A_m\}$ and $P_2 = \{B_1, B_2, \ldots, B_n\}$ are both partitions of the set S. Show that the collection of nonempty subsets of the form $A_i \cap B_j$ is a partition of S that is a refinement of both P_1 and P_2 (see the preamble to Exercise 38 of Section 9.5).

*20. Show that the transitive closure of the symmetric closure of the reflexive closure of a relation R is the smallest equivalence relation that contains R.

21. Let $\mathbf{R}(S)$ be the set of all relations on a set S. Define the relation $\preccurlyeq$ on $\mathbf{R}(S)$ by $R_1 \preccurlyeq R_2$ if $R_1 \subseteq R_2$, where R_1 and R_2 are relations on S. Show that $(\mathbf{R}(S), \preccurlyeq)$ is a poset.

22. Let $\mathbf{P}(S)$ be the set of all partitions of the set S. Define the relation $\preccurlyeq$ on $\mathbf{P}(S)$ by $P_1 \preccurlyeq P_2$ if P_1 is a refinement of P_2 (see Exercise 38 of Section 9.5). Show that $(\mathbf{P}(S), \preccurlyeq)$ is a poset.

Links

PAUL ERDŐS (1913–1996)   Paul Erdős, born in Budapest, Hungary, was the son of two high school mathematics teachers. He was a child prodigy; at age 3 he could multiply three-digit numbers in his head, and at 4 he discovered negative numbers on his own. Because his mother did not want to expose him to contagious diseases, he was mostly home-schooled. At 17 Erdős entered Eőtvős University, graduating four years later with a Ph.D. in mathematics. After graduating he spent four years at Manchester, England, on a postdoctoral fellowship. In 1938 he went to the United States because of the difficult political situation in Hungary, especially for Jews. He spent much of his time in the United States, except for 1954 to 1962, when he was banned as part of the paranoia of the McCarthy era. He also spent considerable time in Israel.

Erdős made many significant contributions to combinatorics and to number theory. One of the discoveries of which he was most proud is his elementary proof (in the sense that it does not use any complex analysis) of the prime number theorem, which provides an estimate for the number of primes not exceeding a fixed positive integer. He also participated in the modern development of the Ramsey theory.

Erdős traveled extensively throughout the world to work with other mathematicians, visiting conferences, universities, and research laboratories. He had no permanent home. He devoted himself almost entirely to mathematics, traveling from one mathematician to the next, proclaiming "My brain is open." Erdős was the author or coauthor of more than 1500 papers and had more than 500 coauthors. Copies of his articles are kept by Ron Graham, a famous discrete mathematician with whom he collaborated extensively and who took care of many of his worldly needs.

Erdős offered rewards, ranging from $10 to $10,000, for the solution of problems that he found particularly interesting, with the size of the reward depending on the difficulty of the problem. He paid out close to $4000. Erdős had his own special language, using such terms as "epsilon" (child), "boss" (woman), "slave" (man), "captured" (married), "liberated" (divorced), "Supreme Fascist" (God), "Sam" (United States), and "Joe" (Soviet Union). Although he was curious about many things, he concentrated almost all his energy on mathematical research. He had no hobbies and no full-time job. He never married and apparently remained celibate. Erdős was extremely generous, donating much of the money he collected from prizes, awards, and stipends for scholarships and to worthwhile causes. He traveled extremely lightly and did not like having many material possessions.

23. Schedule the tasks needed to cook a Chinese meal by specifying their order, if the Hasse diagram representing these tasks is as shown here.

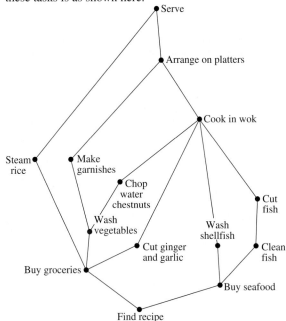

A subset of a poset such that every two elements of this subset are comparable is called a **chain**. A subset of a poset is called an **antichain** if every two elements of this subset are incomparable.

24. Find all chains in the posets with the Hasse diagrams shown in Exercises 17–18 in Section 9.6.

25. Find all antichains in the posets with the Hasse diagrams shown in Exercises 17–18 in Section 9.6.

26. Find an antichain with the greatest number of elements in the poset with the Hasse diagram of Exercise 22 in Section 9.6.

27. Show that every maximal chain in a finite poset $(S, \preccurlyeq)$ contains a minimal element of S. (A maximal chain is a chain that is not a subset of a larger chain.)

∗∗28. Show that every finite poset can be partitioned into k chains, where k is the largest number of elements in an antichain in this poset.

∗29. Show that in any group of $mn + 1$ people there is either a list of $m + 1$ people where a person in the list (except for the first person listed) is a descendant of the previous person on the list, or there are $n + 1$ people such that none of these people is a descendant of any of the other n people. [*Hint:* Use Exercise 28.]

Suppose that $(S, \preccurlyeq)$ is a well-founded partially ordered set. The *principle of well-founded induction* states that $P(x)$ is true for all $x \in S$ if $\forall x(\forall y(y \prec x \rightarrow P(y)) \rightarrow P(x))$.

30. Show that no separate basis case is needed for the principle of well-founded induction. That is, $P(u)$ is true for all minimal elements u in S if $\forall x(\forall y(y \prec x \rightarrow P(y)) \rightarrow P(x))$.

∗31. Show that the principle of well-founded induction is valid.

A relation R on a set A is a **quasi-ordering** on A if R is reflexive and transitive.

32. Let R be the relation on the set of all functions from $\mathbf{Z}^+$ to $\mathbf{Z}^+$ such that (f, g) belongs to R if and only if f is $O(g)$. Show that R is a quasi-ordering.

Let L be a lattice. Define the **meet** ($\wedge$) and **join** ($\vee$) operations by $x \wedge y = \text{glb}(x, y)$ and $x \vee y = \text{lub}(x, y)$.

33. Show that the following properties hold for all elements x, y, and z of a lattice L.
   a) $x \wedge y = y \wedge x$ and $x \vee y = y \vee x$ (**commutative laws**)
   b) $(x \wedge y) \wedge z = x \wedge (y \wedge z)$ and $(x \vee y) \vee z = x \vee (y \vee z)$ (**associative laws**)
   c) $x \wedge (x \vee y) = x$ and $x \vee (x \wedge y) = x$ (**absorption laws**)
   d) $x \wedge x = x$ and $x \vee x = x$ (**idempotent laws**)

34. Show that if x and y are elements of a lattice L, then $x \vee y = y$ if and only if $x \wedge y = x$.

A lattice L is **bounded** if it has both an **upper bound**, denoted by 1, such that $x \preccurlyeq 1$ for all $x \in L$ and a **lower bound**, denoted by 0, such that $0 \preccurlyeq x$ for all $x \in L$.

35. Show that if L is a bounded lattice with upper bound 1 and lower bound 0 then these properties hold for all elements $x \in L$.
   a) $x \vee 1 = 1$ b) $x \wedge 1 = x$
   c) $x \vee 0 = x$ d) $x \wedge 0 = 0$

A lattice is called **distributive** if $x \vee (y \wedge z) = (x \vee y) \wedge (x \vee z)$ and $x \wedge (y \vee z) = (x \wedge y) \vee (x \wedge z)$ for all x, y, and z in L.

36. Show that the lattice $(P(S), \subseteq)$ where $P(S)$ is the power set of a finite set S is distributive.

The **complement** of an element a of a bounded lattice L with upper bound 1 and lower bound 0 is an element b such that $a \vee b = 1$ and $a \wedge b = 0$. Such a lattice is **complemented** if every element of the lattice has a complement.

37. Show that the lattice $(P(S), \subseteq)$ where $P(S)$ is the power set of a finite set S is complemented.

∗38. Show that if L is a finite distributive lattice, then an element of L has at most one complement.

The game of Chomp, introduced in Example 12 in Section 1.8, can be generalized for play on any finite partially ordered set $(S, \preccurlyeq)$ with a least element a. In this game, a move consists of selecting an element x in S and removing x and all elements larger than it from S. The loser is the player who is forced to select the least element a.

39. Show that the game of Chomp with cookies arranged in an $m \times n$ rectangular grid, described in Example 12 in Section 1.8, is the same as the game of Chomp on the poset $(S, |)$, where S is the set of all positive integers that divide $p^{m-1}q^{n-1}$, where p and q are distinct primes.

40. Show that if $(S, \preccurlyeq)$ has a greatest element b, then a winning strategy for Chomp on this poset exists. [*Hint:* Generalize the argument in Example 12 in Section 1.8.]

Computer Projects

Write programs with these input and output.

1. Given the matrix representing a relation on a finite set, determine whether the relation is reflexive and/or irreflexive.
2. Given the matrix representing a relation on a finite set, determine whether the relation is symmetric and/or antisymmetric.
3. Given the matrix representing a relation on a finite set, determine whether the relation is transitive.
4. Given a positive integer n, display all the relations on a set with n elements.
5. Given an n-ary relation, find the projection of this relation when specified fields are deleted.
6. Given an m-ary relation and an n-ary relation, and a set of common fields, find the join of these relations with respect to these common fields.
7. Given the matrix representing a relation on a finite set, find the matrix representing the reflexive closure of this relation.

8. Given the matrix representing a relation on a finite set, find the matrix representing the symmetric closure of this relation.
9. Given the matrix representing a relation on a finite set, find the matrix representing the transitive closure of this relation by computing the join of the Boolean powers of the matrix representing the relation.
10. Given the matrix representing a relation on a finite set, find the matrix representing the transitive closure of this relation using Warshall's algorithm.
11. Given the matrix representing a relation on a finite set, find the matrix representing the smallest equivalence relation containing this relation.
12. Given a partial ordering on a finite set, find a total ordering compatible with it using topological sorting.

Computations and Explorations

Use a computational program or programs you have written to do these exercises.

1. Display all the different relations on a set with four elements.
2. Display all the different reflexive and symmetric relations on a set with six elements.
3. Display all the reflexive and transitive relations on a set with five elements.
4. Find the transitive closure of a relation of your choice on a set with at least 20 elements. Either use a relation that corresponds to direct links in a particular transportation or communications network or use a randomly generated relation.
5. Compute the number of different equivalence relations on a set with n elements for all positive integers n not exceeding 20.
6. Display all the equivalence relations on a set with seven elements.

Writing Projects

Respond to these with essays using outside sources.

1. Discuss the concept of a fuzzy relation. How are fuzzy relations used?
2. Describe the basic principles of relational databases, going beyond what was covered in Section 9.2. How widely used are relational databases as compared with other types of databases?
3. Look up the original papers by Warshall and by Roy (in French) in which they develop algorithms for finding transitive closures. Discuss their approaches. Why do you suppose that what we call Warshall's algorithm was discovered independently by more than one person?
4. Describe how equivalence classes can be used to define the rational numbers as classes of pairs of integers and how the basic arithmetic operations on rational numbers can be defined following this approach.
5. Explain how Helmut Hasse used what we now call Hasse diagrams.
6. Discuss the use of the Program Evaluation and Review Technique (PERT) to schedule the tasks of a large complicated project. How widely is PERT used?
7. Discuss the use of the Critical Path Method (CPM) to find the shortest time for the completion of a project. How widely is CPM used?
8. Discuss the concept of *duality* in a lattice. Explain how duality can be used to establish new results.

10

Graphs

Graphs are discrete structures consisting of vertices and edges that connect these vertices. There are different kinds of graphs, depending on whether edges have directions, whether multiple edges can connect the same pair of vertices, and whether loops are allowed. Problems in almost every conceivable discipline can be solved using graph models. We will give examples to illustrate how graphs are used as models in a variety of areas. For instance, we will show how graphs are used to represent the competition of different species in an ecological niche, how graphs are used to represent who influences whom in an organization, and how graphs are used to represent the outcomes of round-robin tournaments. We will describe how graphs can be used to model acquaintanceships between people, collaboration between researchers, telephone calls between telephone numbers, and links between websites. We will show how graphs can be used to model roadmaps and the assignment of jobs to employees of an organization.

Using graph models, we can determine whether it is possible to walk down all the streets in a city without going down a street twice, and we can find the number of colors needed to color the regions of a map. Graphs can be used to determine whether a circuit can be implemented on a planar circuit board. We can distinguish between two chemical compounds with the same molecular formula but different structures using graphs. We can determine whether two computers are connected by a communications link using graph models of computer networks. Graphs with weights assigned to their edges can be used to solve problems such as finding the shortest path between two cities in a transportation network. We can also use graphs to schedule exams and assign channels to television stations. This chapter will introduce the basic concepts of graph theory and present many different graph models. To solve the wide variety of problems that can be studied using graphs, we will introduce many different graph algorithms. We will also study the complexity of these algorithms.

10.1 Graphs and Graph Models

We begin with the definition of a graph.

DEFINITION 1   A *graph* $G = (V, E)$ consists of V, a nonempty set of *vertices* (or *nodes*) and E, a set of *edges*. Each edge has either one or two vertices associated with it, called its *endpoints*. An edge is said to *connect* its endpoints.

Remark: The set of vertices V of a graph G may be infinite. A graph with an infinite vertex set or an infinite number of edges is called an **infinite graph**, and in comparison, a graph with a finite vertex set and a finite edge set is called a **finite graph**. In this book we will usually consider only finite graphs.

Now suppose that a network is made up of data centers and communication links between computers. We can represent the location of each data center by a point and each communications link by a line segment, as shown in Figure 1.

This computer network can be modeled using a graph in which the vertices of the graph represent the data centers and the edges represent communication links. In general, we visualize

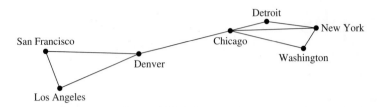

FIGURE 1   A Computer Network.

graphs by using points to represent vertices and line segments, possibly curved, to represent edges, where the endpoints of a line segment representing an edge are the points representing the endpoints of the edge. When we draw a graph, we generally try to draw edges so that they do not cross. However, this is not necessary because any depiction using points to represent vertices and any form of connection between vertices can be used. Indeed, there are some graphs that cannot be drawn in the plane without edges crossing (see Section 10.7). The key point is that the way we draw a graph is arbitrary, as long as the correct connections between vertices are depicted.

Note that each edge of the graph representing this computer network connects two different vertices. That is, no edge connects a vertex to itself. Furthermore, no two different edges connect the same pair of vertices. A graph in which each edge connects two different vertices and where no two edges connect the same pair of vertices is called a **simple graph**. Note that in a simple graph, each edge is associated to an unordered pair of vertices, and no other edge is associated to this same edge. Consequently, when there is an edge of a simple graph associated to $\{u, v\}$, we can also say, without possible confusion, that $\{u, v\}$ is an edge of the graph.

A computer network may contain multiple links between data centers, as shown in Figure 2. To model such networks we need graphs that have more than one edge connecting the same pair of vertices. Graphs that may have **multiple edges** connecting the same vertices are called **multigraphs**. When there are m different edges associated to the same unordered pair of vertices $\{u, v\}$, we also say that $\{u, v\}$ is an edge of multiplicity m. That is, we can think of this set of edges as m different copies of an edge $\{u, v\}$.

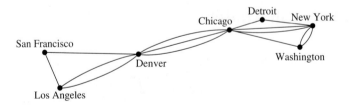

FIGURE 2   A Computer Network with Multiple Links between Data Centers.

Sometimes a communications link connects a data center with itself, perhaps a feedback loop for diagnostic purposes. Such a network is illustrated in Figure 3. To model this network we

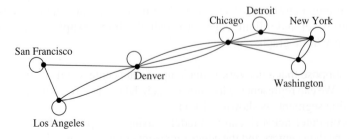

FIGURE 3   A Computer Network with Diagnostic Links.

FIGURE 4   **A Communications Network with One-Way Communications Links.**

need to include edges that connect a vertex to itself. Such edges are called **loops**, and sometimes we may even have more than one loop at a vertex. Graphs that may include loops, and possibly multiple edges connecting the same pair of vertices or a vertex to itself, are sometimes called **pseudographs**.

So far the graphs we have introduced are **undirected graphs**. Their edges are also said to be **undirected**. However, to construct a graph model, we may find it necessary to assign directions to the edges of a graph. For example, in a computer network, some links may operate in only one direction (such links are called single duplex lines). This may be the case if there is a large amount of traffic sent to some data centers, with little or no traffic going in the opposite direction. Such a network is shown in Figure 4.

To model such a computer network we use a directed graph. Each edge of a directed graph is associated to an ordered pair. The definition of directed graph we give here is more general than the one we used in Chapter 9, where we used directed graphs to represent relations.

DEFINITION 2   A *directed graph* (or *digraph*) (V, E) consists of a nonempty set of vertices V and a set of *directed edges* (or *arcs*) E. Each directed edge is associated with an ordered pair of vertices. The directed edge associated with the ordered pair (u, v) is said to *start* at u and *end* at v.

When we depict a directed graph with a line drawing, we use an arrow pointing from u to v to indicate the direction of an edge that starts at u and ends at v. A directed graph may contain loops and it may contain multiple directed edges that start and end at the same vertices. A directed graph may also contain directed edges that connect vertices u and v in both directions; that is, when a digraph contains an edge from u to v, it may also contain one or more edges from v to u. Note that we obtain a directed graph when we assign a direction to each edge in an undirected graph. When a directed graph has no loops and has no multiple directed edges, it is called a **simple directed graph**. Because a simple directed graph has at most one edge associated to each ordered pair of vertices (u, v), we call (u, v) an edge if there is an edge associated to it in the graph.

In some computer networks, multiple communication links between two data centers may be present, as illustrated in Figure 5. Directed graphs that may have **multiple directed edges** from a vertex to a second (possibly the same) vertex are used to model such networks. We called such graphs **directed multigraphs**. When there are m directed edges, each associated to an ordered pair of vertices (u, v), we say that (u, v) is an edge of **multiplicity** m.

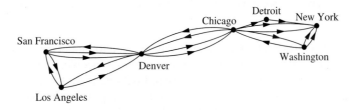

FIGURE 5   **A Computer Network with Multiple One-Way Links.**

TABLE 1 Graph Terminology.			
Type	*Edges*	*Multiple Edges Allowed?*	*Loops Allowed?*
Simple graph	Undirected	No	No
Multigraph	Undirected	Yes	No
Pseudograph	Undirected	Yes	Yes
Simple directed graph	Directed	No	No
Directed multigraph	Directed	Yes	Yes
Mixed graph	Directed and undirected	Yes	Yes

For some models we may need a graph where some edges are undirected, while others are directed. A graph with both directed and undirected edges is called a **mixed graph**. For example, a mixed graph might be used to model a computer network containing links that operate in both directions and other links that operate only in one direction.

This terminology for the various types of graphs is summarized in Table 1. We will sometimes use the term **graph** as a general term to describe graphs with directed or undirected edges (or both), with or without loops, and with or without multiple edges. At other times, when the context is clear, we will use the term graph to refer only to undirected graphs.

Links

Because of the relatively modern interest in graph theory, and because it has applications to a wide variety of disciplines, many different terminologies of graph theory have been introduced. The reader should determine how such terms are being used whenever they are encountered. The terminology used by mathematicians to describe graphs has been increasingly standardized, but the terminology used to discuss graphs when they are used in other disciplines is still quite varied. Although the terminology used to describe graphs may vary, three key questions can help us understand the structure of a graph:

- Are the edges of the graph undirected or directed (or both)?
- If the graph is undirected, are multiple edges present that connect the same pair of vertices? If the graph is directed, are multiple directed edges present?
- Are loops present?

Answering such questions helps us understand graphs. It is less important to remember the particular terminology used.

Graph Models

Links

Can you find a subject to which graph theory has not been applied?

Graphs are used in a wide variety of models. We began this section by describing how to construct graph models of communications networks linking data centers. We will complete this section by describing some diverse graph models for some interesting applications. We will return to many of these applications later in this chapter and in Chapter 11. We will introduce additional graph models in subsequent sections of this and later chapters. Also, recall that directed graph models for some applications were introduced in Chapter 9. When we build a graph model, we need to make sure that we have correctly answered the three key questions we posed about the structure of a graph.

SOCIAL NETWORKS   Graphs are extensively used to model social structures based on different kinds of relationships between people or groups of people. These social structures, and the graphs that represent them, are known as **social networks**. In these graph models, individuals or organizations are represented by vertices; relationships between individuals or organizations are represented by edges. The study of social networks is an extremely active multidisciplinary area, and many different types of relationships between people have been studied using them.

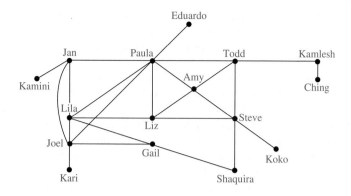

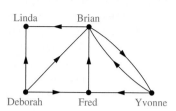

FIGURE 6   An Acquaintanceship Graph.   **FIGURE 7   An Influence Graph.**

We will introduce some of the most commonly studied social networks here. More information about social networks can be found in [Ne10] and [EaKl10].

EXAMPLE 1

Acquaintanceship and Friendship Graphs   We can use a simple graph to represent whether two people know each other, that is, whether they are acquainted, or whether they are friends (either in the real world in the virtual world via a social networking site such as Facebook). Each person in a particular group of people is represented by a vertex. An undirected edge is used to connect two people when these people know each other, when we are concerned only with acquaintanceship, or whether they are friends. No multiple edges and usually no loops are used. (If we want to include the notion of self-knowledge, we would include loops.) A small acquaintanceship graph is shown in Figure 6. The acquaintanceship graph of all people in the world has more than six billion vertices and probably more than one trillion edges! We will discuss this graph further in Section 10.4.   ◀

EXAMPLE 2   **Influence Graphs**   In studies of group behavior it is observed that certain people can influence the thinking of others. A directed graph called an **influence graph** can be used to model this behavior. Each person of the group is represented by a vertex. There is a directed edge from vertex a to vertex b when the person represented by vertex a can influence the person represented by vertex b. This graph does not contain loops and it does not contain multiple directed edges. An example of an influence graph for members of a group is shown in Figure 7. In the group modeled by this influence graph, Deborah cannot be influenced, but she can influence Brian, Fred, and Linda. Also, Yvonne and Brian can influence each other.   ◀

EXAMPLE 3

Collaboration Graphs   A **collaboration graph** is used to model social networks where two people are related by working together in a particular way. Collaboration graphs are simple graphs, as edges in these graphs are undirected and there are no multiple edges or loops. Vertices in these graphs represent people; two people are connected by an undirected edge when the people have collaborated. There are no loops nor multiple edges in these graphs. The **Hollywood graph** is a collaborator graph that represents actors by vertices and connects two actors with an edge if they have worked together on a movie or television show. The Hollywood graph is a huge graph with more than 1.5 million vertices (as of early 2011). We will discuss some aspects of the Hollywood graph later in Section 10.4.

In an **academic collaboration graph**, vertices represent people (perhaps restricted to members of a certain academic community), and edges link two people if they have jointly published a paper. The collaboration graph for people who have published research papers in mathematics was found in 2004 to have more than 400,000 vertices and 675,000 edges, and these numbers have grown considerably since then. We will have more to say about this graph in Section 10.4. Collaboration graphs have also been used in sports, where two professional athletes are considered to have collaborated if they have ever played on the same team during a regular season of their sport.   ◀

COMMUNICATION NETWORKS   We can model different communications networks using vertices to represent devices and edges to represent the particular type of communications links of interest. We have already modeled a data network in the first part of this section.

EXAMPLE 4

Call Graphs   Graphs can be used to model telephone calls made in a network, such as a long-distance telephone network. In particular, a directed multigraph can be used to model calls where each telephone number is represented by a vertex and each telephone call is represented by a directed edge. The edge representing a call starts at the telephone number from which the call was made and ends at the telephone number to which the call was made. We need directed edges because the direction in which the call is made matters. We need multiple directed edges because we want to represent each call made from a particular telephone number to a second number.

A small telephone call graph is displayed in Figure 8(a), representing seven telephone numbers. This graph shows, for instance, that three calls have been made from 732-555-1234 to 732-555-9876 and two in the other direction, but no calls have been made from 732-555-4444 to any of the other six numbers except 732-555-0011. When we care only whether there has been a call connecting two telephone numbers, we use an undirected graph with an edge connecting telephone numbers when there has been a call between these numbers. This version of the call graph is displayed in Figure 8(b).

Call graphs that model actual calling activities can be huge. For example, one call graph studied at AT&T, which models calls during 20 days, has about 290 million vertices and 4 billion edges. We will discuss call graphs further in Section 10.4.   ◀

INFORMATION NETWORKS   Graphs can be used to model various networks that link particular types of information. Here, we will describe how to model the World Wide Web using a graph. We will also describe how to use a graph to model the citations in different types of documents.

EXAMPLE 5

The Web Graph   The World Wide Web can be modeled as a directed graph where each Web page is represented by a vertex and where an edge starts at the Web page *a* and ends at the Web page *b* if there is a link on *a* pointing to *b*. Because new Web pages are created and others removed somewhere on the Web almost every second, the Web graph changes on an almost continual basis. Many people are studying the properties of the Web graph to better understand the nature of the Web. We will return to Web graphs in Section 10.4, and in Chapter 11 we will explain how the Web graph is used by the Web crawlers that search engines use to create indexes of Web pages.   ◀

EXAMPLE 6   **Citation Graphs**   Graphs can be used to represent citations in different types of documents, including academic papers, patents, and legal opinions. In such graphs, each document is represented by a vertex, and there is an edge from one document to a second document if the

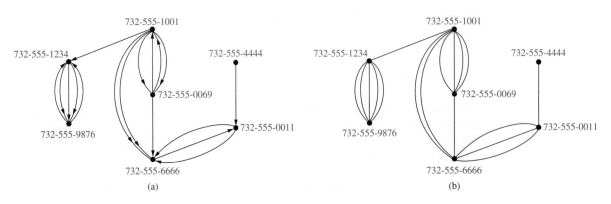

FIGURE 8   A Call Graph.

first document cites the second in its citation list. (In an academic paper, the citation list is the bibliography, or list of references; in a patent it is the list of previous patents that are cited; and in a legal opinion it is the list of previous opinions cited.) A citation graph is a directed graph without loops or multiple edges.   ◀

SOFTWARE DESIGN APPLICATIONS   Graph models are useful tools in the design of software. We will briefly describe two of these models here.

EXAMPLE 7   **Module Dependency Graphs**   One of the most important tasks in designing software is how to structure a program into different parts, or modules. Understanding how the different modules of a program interact is essential not only for program design, but also for testing and maintenance of the resulting software. A **module dependency graph** provides a useful tool for understanding how different modules of a program interact. In a program dependency graph, each module is represented by a vertex. There is a directed edge from a module to a second module if the second module depends on the first. An example of a program dependency graph for a web browser is shown in Figure 9.   ◀

EXAMPLE 8   **Precedence Graphs and Concurrent Processing**   Computer programs can be executed more rapidly by executing certain statements concurrently. It is important not to execute a statement that requires results of statements not yet executed. The dependence of statements on previous statements can be represented by a directed graph. Each statement is represented by a vertex, and there is an edge from one statement to a second statement if the second statement cannot be executed before the first statement. This resulting graph is called a **precedence graph**. A computer program and its graph are displayed in Figure 10. For instance, the graph shows that statement S_5 cannot be executed before statements S_1, S_2, and S_4 are executed.   ◀

TRANSPORTATION NETWORKS   We can use graphs to model many different types of transportation networks, including road, air, and rail networks, as well shipping networks.

EXAMPLE 9   **Airline Routes**   We can model airline networks by representing each airport by a vertex. In particular, we can model all the flights by a particular airline each day using a directed edge to represent each flight, going from the vertex representing the departure airport to the vertex representing the destination airport. The resulting graph will generally be a directed multigraph, as there may be multiple flights from one airport to some other airport during the same day.   ◀

EXAMPLE 10   **Road Networks**   Graphs can be used to model road networks. In such models, vertices represent intersections and edges represent roads. When all roads are two-way and there is at most one road connecting two intersections, we can use a simple undirected graph to model the road network. However, we will often want to model road networks when some roads are one-way and when there may be more than one road between two intersections. To build such models, we use undirected edges to represent two-way roads and we use directed edges to represent

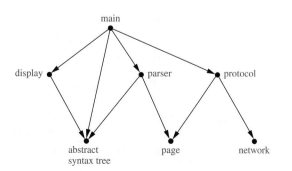

FIGURE 9   **A Module Dependency Graph.**

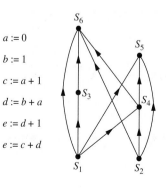

FIGURE 10   **A Precedence Graph.**

one-way roads. Multiple undirected edges represent multiple two-way roads connecting the same two intersections. Multiple directed edges represent multiple one-way roads that start at one intersection and end at a second intersection. Loops represent loop roads. Mixed graphs are needed to model road networks that include both one-way and two-way roads. ◄

BIOLOGICAL NETWORKS   Many aspects of the biological sciences can be modeled using graphs.

EXAMPLE 11

Links

Extra
Examples

Niche Overlap Graphs in Ecology   Graphs are used in many models involving the interaction of different species of animals. For instance, the competition between species in an ecosystem can be modeled using a **niche overlap graph**. Each species is represented by a vertex. An undirected edge connects two vertices if the two species represented by these vertices compete (that is, some of the food resources they use are the same). A niche overlap graph is a simple graph because no loops or multiple edges are needed in this model. The graph in Figure 11 models the ecosystem of a forest. We see from this graph that squirrels and raccoons compete but that crows and shrews do not. ◄

EXAMPLE 12   **Protein Interaction Graphs**   A protein interaction in a living cell occurs when two or more proteins in that cell bind to perform a biological function. Because protein interactions are crucial for most biological functions, many scientists work on discovering new proteins and understanding interactions between proteins. Protein interactions within a cell can be modeled using a **protein interaction graph** (also called a **protein–protein interaction network**), an undirected graph in which each protein is represented by a vertex, with an edge connecting the vertices representing each pair of proteins that interact. It is a challenging problem to determine genuine protein interactions in a cell, as experiments often produce false positives, which conclude that two proteins interact when they really do not. Protein interaction graphs can be used to deduce important biological information, such as by identifying the most important proteins for various functions and the functionality of newly discovered proteins.

Because there are thousands of different proteins in a typical cell, the protein interaction graph of a cell is extremely large and complex. For example, yeast cells have more than 6,000 proteins, and more than 80,000 interactions between them are known, and human cells have more than 100,000 proteins, with perhaps as many as 1,000,000 interactions between them. Additional vertices and edges are added to a protein interaction graph when new proteins and interactions between proteins are discovered. Because of the complexity of protein interaction graphs, they are often split into smaller graphs called modules that represent groups of proteins that are involved in a particular function of a cell. Figure 12 illustrates a module of the protein interaction graph described in [Bo04], comprising the complex of proteins that degrade RNA in human cells. To learn more about protein interaction graphs, see [Bo04], [Ne10], and [Hu07]. ◄

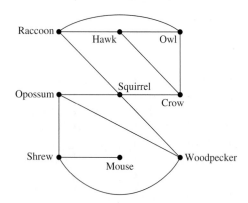

FIGURE 11   A Niche Overlap Graph.

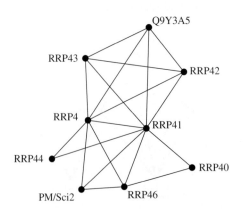

FIGURE 12   A Module of a Protein Interaction Graph.

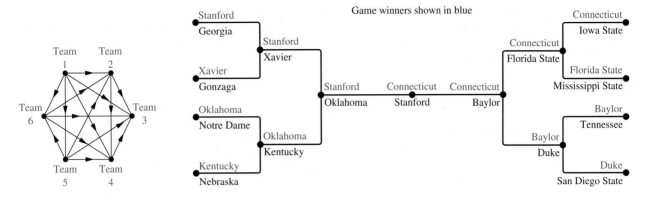

FIGURE 13 **A Graph Model of a Round-Robin Tournament.**

FIGURE 14 **A Single-Elimination Tournament.**

TOURNAMENTS   We now give some examples that show how graphs can also be used to model different kinds of tournaments.

EXAMPLE 13   **Round-Robin Tournaments**   A tournament where each team plays every other team exactly once and no ties are allowed is called a **round-robin tournament**. Such tournaments can be modeled using directed graphs where each team is represented by a vertex. Note that (a, b) is an edge if team a beats team b. This graph is a simple directed graph, containing no loops or multiple directed edges (because no two teams play each other more than once). Such a directed graph model is presented in Figure 13. We see that Team 1 is undefeated in this tournament, and Team 3 is winless.   ◄

EXAMPLE 14   **Single-Elimination Tournaments**   A tournament where each contestant is eliminated after one loss is called a **single-elimination tournament**. Single-elimination tournaments are often used in sports, including tennis championships and the yearly NCAA basketball championship. We can model such a tournament using a vertex to represent each game and a directed edge to connect a game to the next game the winner of this game played in. The graph in Figure 14 represents the games played by the final 16 teams in the 2010 NCAA women's basketball tournament.   ◄

Exercises

1. Draw graph models, stating the type of graph (from Table 1) used, to represent airline routes where every day there are four flights from Boston to Newark, two flights from Newark to Boston, three flights from Newark to Miami, two flights from Miami to Newark, one flight from Newark to Detroit, two flights from Detroit to Newark, three flights from Newark to Washington, two flights from Washington to Newark, and one flight from Washington to Miami, with

a) an edge between vertices representing cities that have a flight between them (in either direction).

b) an edge between vertices representing cities for each flight that operates between them (in either direction).

c) an edge between vertices representing cities for each flight that operates between them (in either direction), plus a loop for a special sightseeing trip that takes off and lands in Miami.

d) an edge from a vertex representing a city where a flight starts to the vertex representing the city where it ends.

e) an edge for each flight from a vertex representing a city where the flight begins to the vertex representing the city where the flight ends.

2. What kind of graph (from Table 1) can be used to model a highway system between major cities where

a) there is an edge between the vertices representing cities if there is an interstate highway between them?

b) there is an edge between the vertices representing cities for each interstate highway between them?

c) there is an edge between the vertices representing cities for each interstate highway between them, and there is a loop at the vertex representing a city if there is an interstate highway that circles this city?

For Exercises 3–5, determine whether the graph shown has directed or undirected edges, whether it has multiple edges, and whether it has one or more loops. Use your answers to determine the type of graph in Table 1 this graph is.

3. 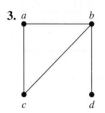   **4.**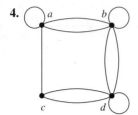

5.

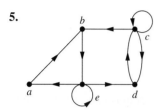

6. For each undirected graph in Exercises 3–5 that is not simple, find a set of edges to remove to make it simple.

7. Let G be a simple graph. Show that the relation R on the set of vertices of G such that uRv if and only if there is an edge associated to $\{u, v\}$ is a symmetric, irreflexive relation on G.

8. Let G be an undirected graph with a loop at every vertex. Show that the relation R on the set of vertices of G such that uRv if and only if there is an edge associated to $\{u, v\}$ is a symmetric, reflexive relation on G.

9. The **intersection graph** of a collection of sets A_1, $A_2, \ldots, A_n$ is the graph that has a vertex for each of these sets and has an edge connecting the vertices representing two sets if these sets have a nonempty intersection. Construct the intersection graph of these collections of sets.
 a) $A_1 = \{0, 2, 4, 6, 8\}$, $A_2 = \{0, 1, 2, 3, 4\}$,
  $A_3 = \{1, 3, 5, 7, 9\}$, $A_4 = \{5, 6, 7, 8, 9\}$,
  $A_5 = \{0, 1, 8, 9\}$

 b) $A_1 = \{\ldots, -4, -3, -2, -1, 0\}$,
  $A_2 = \{\ldots, -2, -1, 0, 1, 2, \ldots\}$,
  $A_3 = \{\ldots, -6, -4, -2, 0, 2, 4, 6, \ldots\}$,
  $A_4 = \{\ldots, -5, -3, -1, 1, 3, 5, \ldots\}$,
  $A_5 = \{\ldots, -6, -3, 0, 3, 6, \ldots\}$

 c) $A_1 = \{x \mid x < 0\}$,
  $A_2 = \{x \mid -1 < x < 0\}$,
  $A_3 = \{x \mid 0 < x < 1\}$,
  $A_4 = \{x \mid -1 < x < 1\}$,
  $A_5 = \{x \mid x > -1\}$,
  $A_6 = \mathbf{R}$

10. Use the niche overlap graph in Figure 11 to determine the species that compete with hawks.

11. Construct a niche overlap graph for six species of birds, where the hermit thrush competes with the robin and

with the blue jay, the robin also competes with the mockingbird, the mockingbird also competes with the blue jay, and the nuthatch competes with the hairy woodpecker.

12. Draw the acquaintanceship graph that represents that Tom and Patricia, Tom and Hope, Tom and Sandy, Tom and Amy, Tom and Marika, Jeff and Patricia, Jeff and Mary, Patricia and Hope, Amy and Hope, and Amy and Marika know each other, but none of the other pairs of people listed know each other.

13. We can use a graph to represent whether two people were alive at the same time. Draw such a graph to represent whether each pair of the mathematicians and computer scientists with biographies in the first five chapters of this book who died before 1900 were contemporaneous. (Assume two people lived at the same time if they were alive during the same year.)

14. Who can influence Fred and whom can Fred influence in the influence graph in Example 2?

15. In a round-robin tournament the Tigers beat the Blue Jays, the Tigers beat the Cardinals, the Tigers beat the Orioles, the Blue Jays beat the Cardinals, the Blue Jays beat the Orioles, and the Cardinals beat the Orioles. Model this outcome with a directed graph.

16. Construct the call graph for a set of seven telephone numbers 555-0011, 555-1221, 555-1333, 555-8888, 555-2222, 555-0091, and 555-1200 if there were three calls from 555-0011 to 555-8888 and two calls from 555-8888 to 555-0011, two calls from 555-2222 to 555-0091, two calls from 555-1221 to each of the other numbers, and one call from 555-1333 to each of 555-0011, 555-1221, and 555-1200.

17. Explain how the two telephone call graphs for calls made during the month of January and calls made during the month of February can be used to determine the new telephone numbers of people who have changed their telephone numbers.

18. a) Explain how graphs can be used to model electronic mail messages in a network. Should the edges be directed or undirected? Should multiple edges be allowed? Should loops be allowed?
 b) Describe a graph that models the electronic mail sent in a network in a particular week.

19. How can a graph that models e-mail messages sent in a network be used to find people who have recently changed their primary e-mail address?

20. Describe a graph model that represents whether each person at a party knows the name of each other person at the party. Should the edges be directed or undirected? Should multiple edges be allowed? Should loops be allowed?

21. Describe a graph model that represents traditional marriages between men and women. Does this graph have any special properties?

22. Which statements must be executed before S_6 is executed in the program in Example 8? (Use the precedence graph in Figure 10.)

23. Construct a precedence graph for the following program:

S_1: $x := 0$
S_2: $x := x + 1$
S_3: $y := 2$
S_4: $z := y$
S_5: $x := x + 2$
S_6: $y := x + z$
S_7: $z := 4$

24. Describe a discrete structure based on a graph that can be used to model airline routes and their flight times. [*Hint:* Add structure to a directed graph.]

25. Describe a discrete structure based on a graph that can be used to model relationships between pairs of individuals in a group, where each individual may either like, dislike, or be neutral about another individual, and the reverse relationship may be different. [*Hint:* Add structure to a directed graph. Treat separately the edges in opposite directions between vertices representing two individuals.]

26. Describe a graph model that can be used to represent all forms of electronic communication between two people in a single graph. What kind of graph is needed?

10.2 Graph Terminology and Special Types of Graphs

Introduction

Links

We introduce some of the basic vocabulary of graph theory in this section. We will use this vocabulary later in this chapter when we solve many different types of problems. One such problem involves determining whether a graph can be drawn in the plane so that no two of its edges cross. Another example is deciding whether there is a one-to-one correspondence between the vertices of two graphs that produces a one-to-one correspondence between the edges of the graphs. We will also introduce several important families of graphs often used as examples and in models. Several important applications will be described where these special types of graphs arise.

Basic Terminology

First, we give some terminology that describes the vertices and edges of undirected graphs.

DEFINITION 1   Two vertices u and v in an undirected graph G are called *adjacent* (or *neighbors*) in G if u and v are endpoints of an edge e of G. Such an edge e is called *incident with* the vertices u and v and e is said to *connect* u and v.

We will also find useful terminology describing the set of vertices adjacent to a particular vertex of a graph.

DEFINITION 2   The set of all neighbors of a vertex v of $G = (V, E)$, denoted by $N(v)$, is called the *neighborhood* of v. If A is a subset of V, we denote by $N(A)$ the set of all vertices in G that are adjacent to at least one vertex in A. So, $N(A) = \bigcup_{v \in A} N(v)$.

To keep track of how many edges are incident to a vertex, we make the following definition.

DEFINITION 3   The *degree of a vertex in an undirected graph* is the number of edges incident with it, except that a loop at a vertex contributes twice to the degree of that vertex. The degree of the vertex v is denoted by $\deg(v)$.

EXAMPLE 1   What are the degrees and what are the neighborhoods of the vertices in the graphs G and H displayed in Figure 1?

Solution: In G, $\deg(a) = 2$, $\deg(b) = \deg(c) = \deg(f) = 4$, $\deg(d) = 1$, $\deg(e) = 3$, and $\deg(g) = 0$. The neighborhoods of these vertices are $N(a) = \{b, f\}$, $N(b) = \{a, c, e, f\}$, $N(c) = \{b, d, e, f\}$, $N(d) = \{c\}$, $N(e) = \{b, c, f\}$, $N(f) = \{a, b, c, e\}$, and $N(g) = \emptyset$. In H, $\deg(a) = 4$, $\deg(b) = \deg(e) = 6$, $\deg(c) = 1$, and $\deg(d) = 5$. The neighborhoods of these vertices are $N(a) = \{b, d, e\}$, $N(b) = \{a, b, c, d, e\}$, $N(c) = \{b\}$, $N(d) = \{a, b, e\}$, and $N(e) = \{a, b, d\}$.   ◀

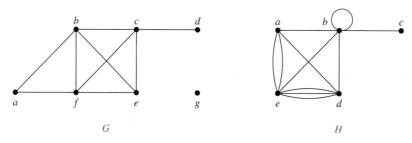

FIGURE 1   The Undirected Graphs G and H.

A vertex of degree zero is called **isolated**. It follows that an isolated vertex is not adjacent to any vertex. Vertex g in graph G in Example 1 is isolated. A vertex is **pendant** if and only if it has degree one. Consequently, a pendant vertex is adjacent to exactly one other vertex. Vertex d in graph G in Example 1 is pendant.

Examining the degrees of vertices in a graph model can provide useful information about the model, as Example 2 shows.

EXAMPLE 2   What does the degree of a vertex in a niche overlap graph (introduced in Example 11 in Section 10.1) represent? Which vertices in this graph are pendant and which are isolated? Use the niche overlap graph shown in Figure 11 of Section 10.1 to interpret your answers.

Solution: There is an edge between two vertices in a niche overlap graph if and only if the two species represented by these vertices compete. Hence, the degree of a vertex in a niche overlap graph is the number of species in the ecosystem that compete with the species represented by this vertex. A vertex is pendant if the species competes with exactly one other species in the ecosystem. Finally, the vertex representing a species is isolated if this species does not compete with any other species in the ecosystem.

For instance, the degree of the vertex representing the squirrel in the niche overlap graph in Figure 11 in Section 10.1 is four, because the squirrel competes with four other species: the crow, the opossum, the raccoon, and the woodpecker. In this niche overlap graph, the mouse is the only species represented by a pendant vertex, because the mouse competes only with the shrew and all other species compete with at least two other species. There are no isolated vertices in the graph in this niche overlap graph because every species in this ecosystem competes with at least one other species.   ◀

What do we get when we add the degrees of all the vertices of a graph $G = (V, E)$? Each edge contributes two to the sum of the degrees of the vertices because an edge is incident with exactly two (possibly equal) vertices. This means that the sum of the degrees of the vertices is twice the number of edges. We have the result in Theorem 1, which is sometimes called the handshaking theorem (and is also often known as the handshaking lemma), because of the analogy between an edge having two endpoints and a handshake involving two hands.

THEOREM 1 **THE HANDSHAKING THEOREM** Let $G = (V, E)$ be an undirected graph with m edges. Then

$$2m = \sum_{v \in V} \deg(v).$$

(Note that this applies even if multiple edges and loops are present.)

EXAMPLE 3 How many edges are there in a graph with 10 vertices each of degree six?

Solution: Because the sum of the degrees of the vertices is $6 \cdot 10 = 60$, it follows that $2m = 60$ where m is the number of edges. Therefore, $m = 30$. ◀

Theorem 1 shows that the sum of the degrees of the vertices of an undirected graph is even. This simple fact has many consequences, one of which is given as Theorem 2.

THEOREM 2 An undirected graph has an even number of vertices of odd degree.

Proof: Let V_1 and V_2 be the set of vertices of even degree and the set of vertices of odd degree, respectively, in an undirected graph $G = (V, E)$ with m edges. Then

$$2m = \sum_{v \in V} \deg(v) = \sum_{v \in V_1} \deg(v) + \sum_{v \in V_2} \deg(v).$$

Because $\deg(v)$ is even for $v \in V_1$, the first term in the right-hand side of the last equality is even. Furthermore, the sum of the two terms on the right-hand side of the last equality is even, because this sum is $2m$. Hence, the second term in the sum is also even. Because all the terms in this sum are odd, there must be an even number of such terms. Thus, there are an even number of vertices of odd degree. ◁

Terminology for graphs with directed edges reflects the fact that edges in directed graphs have directions.

DEFINITION 4 When (u, v) is an edge of the graph G with directed edges, u is said to be *adjacent to v* and v is said to be *adjacent from u*. The vertex u is called the *initial vertex* of (u, v), and v is called the *terminal* or *end vertex* of (u, v). The initial vertex and terminal vertex of a loop are the same.

Because the edges in graphs with directed edges are ordered pairs, the definition of the degree of a vertex can be refined to reflect the number of edges with this vertex as the initial vertex and as the terminal vertex.

DEFINITION 5 In a graph with directed edges the *in-degree of a vertex v*, denoted by $\deg^-(v)$, is the number of edges with v as their terminal vertex. The *out-degree of v*, denoted by $\deg^+(v)$, is the number of edges with v as their initial vertex. (Note that a loop at a vertex contributes 1 to both the in-degree and the out-degree of this vertex.)

EXAMPLE 4   Find the in-degree and out-degree of each vertex in the graph G with directed edges shown in Figure 2.

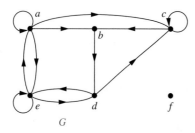

FIGURE 2   The Directed Graph G.

Solution: The in-degrees in G are $\deg^-(a) = 2$, $\deg^-(b) = 2$, $\deg^-(c) = 3$, $\deg^-(d) = 2$, $\deg^-(e) = 3$, and $\deg^-(f) = 0$. The out-degrees are $\deg^+(a) = 4$, $\deg^+(b) = 1$, $\deg^+(c) = 2$, $\deg^+(d) = 2$, $\deg^+(e) = 3$, and $\deg^+(f) = 0$.   ◀

Because each edge has an initial vertex and a terminal vertex, the sum of the in-degrees and the sum of the out-degrees of all vertices in a graph with directed edges are the same. Both of these sums are the number of edges in the graph. This result is stated as Theorem 3.

THEOREM 3   Let $G = (V, E)$ be a graph with directed edges. Then

$$\sum_{v \in V} \deg^-(v) = \sum_{v \in V} \deg^+(v) = |E|.$$

There are many properties of a graph with directed edges that do not depend on the direction of its edges. Consequently, it is often useful to ignore these directions. The undirected graph that results from ignoring directions of edges is called the **underlying undirected graph**. A graph with directed edges and its underlying undirected graph have the same number of edges.

Some Special Simple Graphs

We will now introduce several classes of simple graphs. These graphs are often used as examples and arise in many applications.

EXAMPLE 5   **Complete Graphs**   A **complete graph on n vertices**, denoted by K_n, is a simple graph that contains exactly one edge between each pair of distinct vertices. The graphs K_n, for $n = 1, 2, 3, 4, 5, 6$, are displayed in Figure 3. A simple graph for which there is at least one pair of distinct vertex not connected by an edge is called **noncomplete**.   ◀

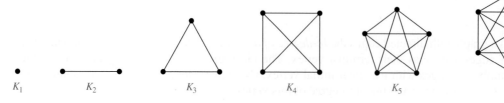

FIGURE 3   The Graphs K_n for $1 \leq n \leq 6$.

EXAMPLE 6 **Cycles** A **cycle** C_n, $n \geq 3$, consists of n vertices $v_1, v_2, \ldots, v_n$ and edges $\{v_1, v_2\}$, $\{v_2, v_3\}, \ldots, \{v_{n-1}, v_n\}$, and $\{v_n, v_1\}$. The cycles C_3, C_4, C_5, and C_6 are displayed in Figure 4. ◀

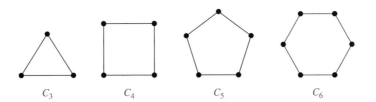

FIGURE 4 **The Cycles C_3, C_4, C_5, and C_6.**

EXAMPLE 7 **Wheels** We obtain a **wheel** W_n when we add an additional vertex to a cycle C_n, for $n \geq 3$, and connect this new vertex to each of the n vertices in C_n, by new edges. The wheels W_3, W_4, W_5, and W_6 are displayed in Figure 5. ◀

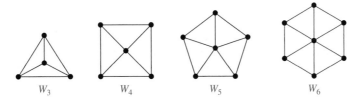

FIGURE 5 **The Wheels W_3, W_4, W_5, and W_6.**

EXAMPLE 8 **n-Cubes** An **n-dimensional hypercube**, or **n-cube**, denoted by Q_n, is a graph that has vertices representing the 2^n bit strings of length n. Two vertices are adjacent if and only if the bit strings that they represent differ in exactly one bit position. We display Q_1, Q_2, and Q_3 in Figure 6.

Note that you can construct the $(n + 1)$-cube Q_{n+1} from the n-cube Q_n by making two copies of Q_n, prefacing the labels on the vertices with a 0 in one copy of Q_n and with a 1 in the other copy of Q_n, and adding edges connecting two vertices that have labels differing only in the first bit. In Figure 6, Q_3 is constructed from Q_2 by drawing two copies of Q_2 as the top and bottom faces of Q_3, adding 0 at the beginning of the label of each vertex in the bottom face and 1 at the beginning of the label of each vertex in the top face. (Here, by *face* we mean a face of a cube in three-dimensional space. Think of drawing the graph Q_3 in three-dimensional space with copies of Q_2 as the top and bottom faces of a cube and then drawing the projection of the resulting depiction in the plane.) ◀

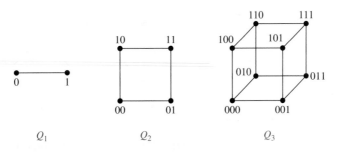

FIGURE 6 **The n-cube Q_n, $n = 1, 2, 3$.**

Bipartite Graphs

Sometimes a graph has the property that its vertex set can be divided into two disjoint subsets such that each edge connects a vertex in one of these subsets to a vertex in the other subset. For example, consider the graph representing marriages between men and women in a village, where each person is represented by a vertex and a marriage is represented by an edge. In this graph, each edge connects a vertex in the subset of vertices representing males and a vertex in the subset of vertices representing females. This leads us to Definition 6.

DEFINITION 6   A simple graph G is called *bipartite* if its vertex set V can be partitioned into two disjoint sets V_1 and V_2 such that every edge in the graph connects a vertex in V_1 and a vertex in V_2 (so that no edge in G connects either two vertices in V_1 or two vertices in V_2). When this condition holds, we call the pair (V_1, V_2) a *bipartition* of the vertex set V of G.

In Example 9 we will show that C_6 is bipartite, and in Example 10 we will show that K_3 is not bipartite.

EXAMPLE 9 C_6 is bipartite, as shown in Figure 7, because its vertex set can be partitioned into the two sets $V_1 = \{v_1, v_3, v_5\}$ and $V_2 = \{v_2, v_4, v_6\}$, and every edge of C_6 connects a vertex in V_1 and a vertex in V_2.   ◀

EXAMPLE 10 K_3 is not bipartite. To verify this, note that if we divide the vertex set of K_3 into two disjoint sets, one of the two sets must contain two vertices. If the graph were bipartite, these two vertices could not be connected by an edge, but in K_3 each vertex is connected to every other vertex by an edge.   ◀

EXAMPLE 11   Are the graphs G and H displayed in Figure 8 bipartite?

Solution: Graph G is bipartite because its vertex set is the union of two disjoint sets, $\{a, b, d\}$ and $\{c, e, f, g\}$, and each edge connects a vertex in one of these subsets to a vertex in the other subset. (Note that for G to be bipartite it is not necessary that every vertex in $\{a, b, d\}$ be adjacent to every vertex in $\{c, e, f, g\}$. For instance, b and g are not adjacent.)

Graph H is not bipartite because its vertex set cannot be partitioned into two subsets so that edges do not connect two vertices from the same subset. (The reader should verify this by considering the vertices a, b, and f.)   ◀

Theorem 4 provides a useful criterion for determining whether a graph is bipartite.

THEOREM 4   A simple graph is bipartite if and only if it is possible to assign one of two different colors to each vertex of the graph so that no two adjacent vertices are assigned the same color.

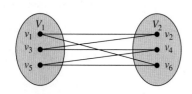

FIGURE 7   **Showing That C_6 Is Bipartite.**

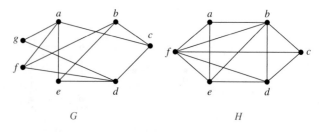

FIGURE 8   **The Undirected Graphs G and H.**

Proof: First, suppose that $G = (V, E)$ is a bipartite simple graph. Then $V = V_1 \cup V_2$, where V_1 and V_2 are disjoint sets and every edge in E connects a vertex in V_1 and a vertex in V_2. If we assign one color to each vertex in V_1 and a second color to each vertex in V_2, then no two adjacent vertices are assigned the same color.

Now suppose that it is possible to assign colors to the vertices of the graph using just two colors so that no two adjacent vertices are assigned the same color. Let V_1 be the set of vertices assigned one color and V_2 be the set of vertices assigned the other color. Then, V_1 and V_2 are disjoint and $V = V_1 \cup V_2$. Furthermore, every edge connects a vertex in V_1 and a vertex in V_2 because no two adjacent vertices are either both in V_1 or both in V_2. Consequently, G is bipartite. ◁

We illustrate how Theorem 4 can be used to determine whether a graph is bipartite in Example 12.

EXAMPLE 12   Use Theorem 4 to determine whether the graphs in Example 11 are bipartite.

Solution: We first consider the graph G. We will try to assign one of two colors, say red and blue, to each vertex in G so that no edge in G connects a red vertex and a blue vertex. Without loss of generality we begin by arbitrarily assigning red to a. Then, we must assign blue to c, e, f, and g, because each of these vertices is adjacent to a. To avoid having an edge with two blue endpoints, we must assign red to all the vertices adjacent to either c, e, f, or g. This means that we must assign red to both b and d (and means that a must be assigned red, which it already has been). We have now assigned colors to all vertices, with a, b, and d red and c, e, f, and g blue. Checking all edges, we see that every edge connects a red vertex and a blue vertex. Hence, by Theorem 4 the graph G is bipartite.

Next, we will try to assign either red or blue to each vertex in H so that no edge in H connects a red vertex and a blue vertex. Without loss of generality we arbitrarily assign red to a. Then, we must assign blue to b, e, and f, because each is adjacent to a. But this is not possible because e and f are adjacent, so both cannot be assigned blue. This argument shows that we cannot assign one of two colors to each of the vertices of H so that no adjacent vertices are assigned the same color. It follows by Theorem 4 that H is not bipartite. ◀

Theorem 4 is an example of a result in the part of graph theory known as graph colorings. Graph colorings is an important part of graph theory with important applications. We will study graph colorings further in Section 10.8.

Another useful criterion for determining whether a graph is bipartite is based on the notion of a path, a topic we study in Section 10.4. A graph is bipartite if and only if it is not possible to start at a vertex and return to this vertex by traversing an odd number of distinct edges. We will make this notion more precise when we discuss paths and circuits in graphs in Section 10.4 (see Exercise 49 in that section).

EXAMPLE 13   **Complete Bipartite Graphs**   A **complete bipartite graph** $K_{m,n}$ is a graph that has its vertex set partitioned into two subsets of m and n vertices, respectively with an edge between two vertices if and only if one vertex is in the first subset and the other vertex is in the second subset. The complete bipartite graphs $K_{2,3}$, $K_{3,3}$, $K_{3,5}$, and $K_{2,6}$ are displayed in Figure 9. ◀

Bipartite Graphs and Matchings

Bipartite graphs can be used to model many types of applications that involve matching the elements of one set to elements of another, as Example 14 illustrates.

EXAMPLE 14   **Job Assignments**   Suppose that there are m employees in a group and n different jobs that need to be done, where $m \geq n$. Each employee is trained to do one or more of these n jobs. We would like to assign an employee to each job. To help with this task, we can use a graph to model

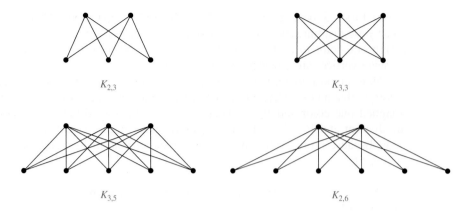

FIGURE 9   Some Complete Bipartite Graphs.

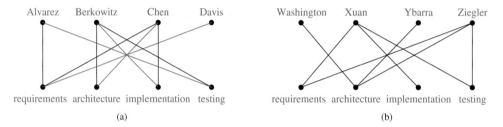

FIGURE 10   Modeling the Jobs for Which Employees Have Been Trained.

employee capabilities. We represent each employee by a vertex and each job by a vertex. For each employee, we include an edge from that employee to all jobs that the employee has been trained to do. Note that the vertex set of this graph can be partitioned into two disjoint sets, the set of employees and the set of jobs, and each edge connects an employee to a job. Consequently, this graph is bipartite, where the bipartition is (E, J) where E is the set of employees and J is the set of jobs. We now consider two different scenarios.

First, suppose that a group has four employees: Alvarez, Berkowitz, Chen, and Davis; and suppose that four jobs need to be done to complete Project 1: requirements, architecture, implementation, and testing. Suppose that Alvarez has been trained to do requirements and testing; Berkowitz has been trained to do architecture, implementation, and testing; Chen has been trained to do requirements, architecture, and implementation; and Davis has only been trained to do requirements. We model these employee capabilities using the bipartite graph in Figure 10(a).

Second, suppose that a group has second group also has four employees: Washington, Xuan, Ybarra, and Ziegler; and suppose that the same four jobs need to be done to complete Project 2 as are needed to complete Project 1. Suppose that Washington has been trained to do architecture; Xuan has been trained to do requirements, implementation, and testing; Ybarra has been trained to do architecture; and Ziegler has been trained to do requirements, architecture and testing. We model these employee capabilities using the bipartite graph in Figure 10(b).

To complete Project 1, we must assign an employee to each job so that every job has an employee assigned to it, and so that no employee is assigned more than one job. We can do this by assigning Alvarez to testing, Berkowitz to implementation, Chen to architecture, and Davis to requirements, as shown in Figure 10(a) (where blue lines show this assignment of jobs).

To complete Project 2, we must also assign an employee to each job so that every job has an employee assigned to it and no employee is assigned more than one job. However, this is impossible because there are only two employees, Xuan and Ziegler, who have been trained for at least one of the three jobs of requirements, implementation, and testing. Consequently, there is no way to assign three different employees to these three job so that each job is assigned an employee with the appropriate training. ◀

Finding an assignment of jobs to employees can be thought of as finding a matching in the graph model, where a **matching** M in a simple graph $G = (V, E)$ is a subset of the set E of edges of the graph such that no two edges are incident with the same vertex. In other words, a matching is a subset of edges such that if $\{s, t\}$ and $\{u, v\}$ are distinct edges of the matching, then $s, t, u,$ and v are distinct. A vertex that is the endpoint of an edge of a matching M is said to be **matched** in M; otherwise it is said to be **unmatched**. A **maximum matching** is a matching with the largest number of edges. We say that a matching M in a bipartite graph $G = (V, E)$ with bipartition (V_1, V_2) is a **complete matching from** V_1 **to** V_2 if every vertex in V_1 is the endpoint of an edge in the matching, or equivalently, if $|M| = |V_1|$. For example, to assign jobs to employees so that the largest number of jobs are assigned employees, we seek a maximum matching in the graph that models employee capabilities. To assign employees to all jobs we seek a complete matching from the set of jobs to the set of employees. In Example 14, we found a complete matching from the set of jobs to the set of employees for Project 1, and this matching is a maximun matching, and we showed that no complete matching exists from the set of jobs to the employees for Project 2.

We now give an example of how matchings can be used to model marriages.

EXAMPLE 15   **Marriages on an Island**   Suppose that there are m men and n women on an island. Each person has a list of members of the opposite gender acceptable as a spouse. We construct a bipartite graph $G = (V_1, V_2)$ where V_1 is the set of men and V_2 is the set of women so that there is an edge between a man and a woman if they find each other acceptable as a spouse. A matching in this graph consists of a set of edges, where each pair of endpoints of an edge is a husband-wife pair. A maximum matching is a largest possible set of married couples, and a complete matching of V_1 is a set of married couples where every man is married, but possibly not all women. ◀

NECESSARY AND SUFFICIENT CONDITIONS FOR COMPLETE MATCHINGS   We now turn our attention to the question of determining whether a complete matching from V_1 to V_2 exists when (V_1, V_2) is a bipartition of a bipartite graph $G = (V, E)$. We will introduce a theorem that provides a set of necessary and sufficient conditions for the existence of a complete matching. This theorem was proved by Philip Hall in 1935.

Hall's marriage theorem is an example of a theorem where obvious necessary conditions are sufficient too.

THEOREM 5   **HALL'S MARRIAGE THEOREM**   The bipartite graph $G = (V, E)$ with bipartition (V_1, V_2) has a complete matching from V_1 to V_2 if and only if $|N(A)| \geq |A|$ for all subsets A of V_1.

 Links

PHILIP HALL (1904–1982) Philip Hall grew up in London, where his mother was a dressmaker. He won a scholarship for board school reserved for needy children, and later a scholarship to King's College of Cambridge University. He received his bachelors degree in 1925. In 1926, unsure of his career goals, he took a civil service exam, but decided to continue his studies at Cambridge after failing.

In 1927 Hall was elected to a fellowship at King's College; soon after, he made his first important discovery in group theory. The results he proved are now known as Hall's theorems. In 1933 he was appointed as a Lecturer at Cambridge, where he remained until 1941. During World War II he worked as a cryptographer at Bletchley Park breaking Italian and Japanese codes. At the end of the war, Hall returned to King's College, and was soon promoted. In 1953 he was appointed to the Sadleirian Chair. His work during the 1950s proved to be extremely influential to the rapid development of group theory during the 1960s.

Hall loved poetry and recited it beautifully in Italian and Japanese, as well as English. He was interested in art, music, and botany. He was quite shy and disliked large groups of people. Hall had an incredibly broad and varied knowledge, and was respected for his integrity, intellectual standards, and judgement. He was beloved by his students.

Proof: We first prove the *only if* part of the theorem. To do so, suppose that there is a complete matching M from V_1 to V_2. Then, if $A \subseteq V_1$, for every vertex $v \in A$, there is an edge in M connecting v to a vertex in V_2. Consequently, there are at least as many vertices in V_2 that are neighbors of vertices in V_1 as there are vertices in V_1. It follows that $|N(A)| \geq |A|$.

To prove the *if* part of the theorem, the more difficult part, we need to show that if $|N(A)| \geq |A|$ for all $A \subseteq V_1$, then there is a complete matching M from V_1 to V_2. We will use strong induction on $|V_1|$ to prove this.

Basis step: If $|V_1| = 1$, then V_1 contains a single vertex v_0. Because $|N(\{v_0\})| \geq |\{v_0\}| = 1$, there is at least one edge connecting v_0 and a vertex $w_0 \in V_2$. Any such edge forms a complete matching from V_1 to V_2.

Inductive step: We first state the inductive hypothesis.

Inductive hypothesis: Let k be a positive integer. If $G = (V, E)$ is a bipartite graph with bipartition (V_1, V_2), and $|V_1| = j \leq k$, then there is a complete matching M from V_1 to V_2 whenever the condition that $|N(A)| \geq |A|$ for all $A \subseteq V_1$ is met.

Now suppose that $H = (W, F)$ is a bipartite graph with bipartition (W_1, W_2) and $|W_1| = k + 1$. We will prove that the inductive holds using a proof by cases, using two case. Case *(i)* applies when for all integers j with $1 \leq j \leq k$, the vertices in every set of j elements from W_1 are adjacent to at least $j + 1$ elements of W_2. Case *(ii)* applies when for some j with $1 \leq j \leq k$ there is a subset W_1' of j vertices such that there are exactly j neighbors of these vertices in W_2. Because either Case *(i)* or Case *(ii)* holds, we need only consider these cases to complete the inductive step.

Case (i): Suppose that for all integers j with $1 \leq j \leq k$, the vertices in every subset of j elements from W_1 are adjacent to at least $j + 1$ elements of W_2. Then, we select a vertex $v \in W_1$ and an element $w \in N(\{v\})$, which must exist by our assumption that $|N(\{v\}) \geq |\{v\}| = 1$. We delete v and w and all edges incident to them from H. This produces a bipartite graph H' with bipartition $(W_1 - \{v\}, W_2 - \{w\})$. Because $|W_1 - \{v\}| = k$, the inductive hypothesis tells us there is a complete matching from $W_1 - \{v\}$ to $W_2 - \{w\}$. Adding the edge from v to w to this complete matching produces a complete matching from W_1 to W_2.

Case (ii): Suppose that for some j with $1 \leq j \leq k$, there is a subset W_1' of j vertices such that there are exactly j neighbors of these vertices in W_2. Let W_2' be the set of these neighbors. Then, by the inductive hypothesis there is a complete matching from W_1' to W_2'. Remove these $2j$ vertices from W_1 and W_2 and all incident edges to produce a bipartite graph K with bipartition $(W_1 - W_1', W_2 - W_2')$.

We will show that the graph K satisfies the condition $|N(A)| \geq |A|$ for all subsets A of $W_1 - W_1'$. If not, there would be a subset of t vertices of $W_1 - W_1'$ where $1 \leq t \leq k + 1 - j$ such that the vertices in this subset have fewer than t vertices of $W_2 - W_2'$ as neighbors. Then, the set of $j + t$ vertices of W_1 consisting of these t vertices together with the j vertices we removed from W_1 has fewer than $j + t$ neighbors in W_2, contradicting the hypothesis that $|N(A)| \geq |A|$ for all $A \subseteq W_1$.

Hence, by the inductive hypothesis, the graph K has a complete matching. Combining this complete matching with the complete matching from W_1' to W_2', we obtain a complete matching from W_1 to W_2.

We have shown that in both cases there is a complete matching from W_1 to W_2. This completes the inductive step and completes the proof.　◁

We have used strong induction to prove Hall's marriage theorem. Although our proof is elegant, it does have some drawbacks. In particular, we cannot construct an algorithm based on this proof that finds a complete matching in a bipartite graph. For a constructive proof that can be used as the basis of an algorithm, see [Gi85].

Some Applications of Special Types of Graphs

We conclude this section by introducing some additional graph models that involve the special types of graph we have discussed in this section.

EXAMPLE 16

Local Area Networks   The various computers in a building, such as minicomputers and personal computers, as well as peripheral devices such as printers and plotters, can be connected using a *local area network*. Some of these networks are based on a *star topology*, where all devices are connected to a central control device. A local area network can be represented using a complete bipartite graph $K_{1,n}$, as shown in Figure 11(a). Messages are sent from device to device through the central control device.

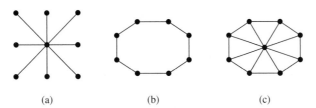

(a)   (b)   (c)

FIGURE 11   Star, Ring, and Hybrid Topologies for Local Area Networks.

Other local area networks are based on a *ring topology*, where each device is connected to exactly two others. Local area networks with a ring topology are modeled using *n*-cycles, C_n, as shown in Figure 11(b). Messages are sent from device to device around the cycle until the intended recipient of a message is reached.

Finally, some local area networks use a hybrid of these two topologies. Messages may be sent around the ring, or through a central device. This redundancy makes the network more reliable. Local area networks with this redundancy can be modeled using wheels W_n, as shown in Figure 11(c).   ◀

EXAMPLE 17   **Interconnection Networks for Parallel Computation**   For many years, computers executed programs one operation at a time. Consequently, the algorithms written to solve problems were designed to perform one step at a time; such algorithms are called **serial**. (Almost all algorithms described in this book are serial.) However, many computationally intense problems, such as weather simulations, medical imaging, and cryptanalysis, cannot be solved in a reasonable amount of time using serial operations, even on a supercomputer. Furthermore, there is a physical limit to how fast a computer can carry out basic operations, so there will always be problems that cannot be solved in a reasonable length of time using serial operations.

Parallel processing, which uses computers made up of many separate processors, each with its own memory, helps overcome the limitations of computers with a single processor. **Parallel algorithms**, which break a problem into a number of subproblems that can be solved concurrently, can then be devised to rapidly solve problems using a computer with multiple processors. In a parallel algorithm, a single instruction stream controls the execution of the algorithm, sending subproblems to different processors, and directs the input and output of these subproblems to the appropriate processors.

When parallel processing is used, one processor may need output generated by another processor. Consequently, these processors need to be interconnected. We can use the appropriate type of graph to represent the interconnection network of the processors in a computer with multiple processors. In the following discussion, we will describe the most commonly used types of interconnection networks for parallel processors. The type of interconnection network used to implement a particular parallel algorithm depends on the requirements for exchange of data between processors, the desired speed, and, of course, the available hardware.

The simplest, but most expensive, network-interconnecting processors include a two-way link between each pair of processors. This network can be represented by K_n, the complete

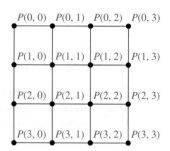

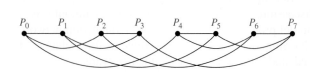

FIGURE 12   A Linear Array for Six Processors.

FIGURE 13   A Mesh Network for 16 Processors.

graph on n vertices, when there are n processors. However, there are serious problems with this type of interconnection network because the required number of connections is so large. In reality, the number of direct connections to a processor is limited, so when there are a large number of processors, a processor cannot be linked directly to all others. For example, when there are 64 processors, $C(64, 2) = 2016$ connections would be required, and each processor would have to be directly connected to 63 others.

On the other hand, perhaps the simplest way to interconnect n processors is to use an arrangement known as a **linear array**. Each processor P_i, other than P_1 and P_n, is connected to its neighbors P_{i-1} and P_{i+1} via a two-way link. P_1 is connected only to P_2, and P_n is connected only to P_{n-1}. The linear array for six processors is shown in Figure 12. The advantage of a linear array is that each processor has at most two direct connections to other processors. The disadvantage is that it is sometimes necessary to use a large number of intermediate links, called **hops**, for processors to share information.

The **mesh network** (or **two-dimensional array**) is a commonly used interconnection network. In such a network, the number of processors is a perfect square, say $n = m^2$. The n processors are labeled $P(i, j), 0 \le i \le m - 1, 0 \le j \le m - 1$. Two-way links connect processor $P(i, j)$ with its four neighbors, processors $P(i \pm 1, j)$ and $P(i, j \pm 1)$, as long as these are processors in the mesh. (Note that four processors, on the corners of the mesh, have only two adjacent processors, and other processors on the boundaries have only three neighbors. Sometimes a variant of a mesh network in which every processor has exactly four connections is used; see Exercise 54.) The mesh network limits the number of links for each processor. Communication between some pairs of processors requires $O(\sqrt{n}) = O(m)$ intermediate links. (See Exercise 55.) The graph representing the mesh network for 16 processors is shown in Figure 13.

One important type of interconnection network is the hypercube. For such a network, the number of processors is a power of 2, $n = 2^m$. The n processors are labeled $P_0, P_1, \ldots, P_{n-1}$. Each processor has two-way connections to m other processors. Processor P_i is linked to the processors with indices whose binary representations differ from the binary representation of i in exactly one bit. The hypercube network balances the number of direct connections for each processor and the number of intermediate connections required so that processors can communicate. Many computers have been built using a hypercube network, and many parallel

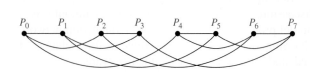

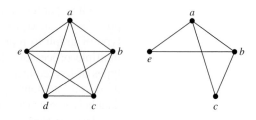

FIGURE 14   A Hypercube Network for Eight Processors.

FIGURE 15   A Subgraph of K_5.

algorithms have been devised that use a hypercube network. The graph Q_m, the m-cube, represents the hypercube network with $n = 2^m$ processors. Figure 14 displays the hypercube network for eight processors. (Figure 14 displays a different way to draw Q_3 than was shown in Figure 6.) ◀

New Graphs from Old

Sometimes we need only part of a graph to solve a problem. For instance, we may care only about the part of a large computer network that involves the computer centers in New York, Denver, Detroit, and Atlanta. Then we can ignore the other computer centers and all telephone lines not linking two of these specific four computer centers. In the graph model for the large network, we can remove the vertices corresponding to the computer centers other than the four of interest, and we can remove all edges incident with a vertex that was removed. When edges and vertices are removed from a graph, without removing endpoints of any remaining edges, a smaller graph is obtained. Such a graph is called a **subgraph** of the original graph.

DEFINITION 7   A *subgraph of a graph* $G = (V, E)$ is a graph $H = (W, F)$, where $W \subseteq V$ and $F \subseteq E$. A subgraph H of G is a *proper subgraph* of G if $H \neq G$.

Given a set of vertices of a graph, we can form a subgraph of this graph with these vertices and the edges of the graph that connect them.

DEFINITION 8   Let $G = (V, E)$ be a simple graph. The **subgraph induced** by a subset W of the vertex set V is the graph (W, F), where the edge set F contains an edge in E if and only if both endpoints of this edge are in W.

EXAMPLE 18   The graph G shown in Figure 15 is a subgraph of K_5. If we add the edge connecting c and e to G, we obtain the subgraph induced by $W = \{a, b, c, e\}$. ◀

REMOVING OR ADDING EDGES OF A GRAPH   Given a graph $G = (V, E)$ and an edge $e \in E$, we can produce a subgraph of G by removing the edge e. The resulting subgraph, denoted by $G - e$, has the same vertex set V as G. Its edge set is $E - e$. Hence,

$$G - e = (V, E - \{e\}).$$

Similarly, if E' is a subset of E, we can produce a subgraph of G by removing the edges in E' from the graph. The resulting subgraph has the same vertex set V as G. Its edge set is $E - E'$.

We can also add an edge e to a graph to produce a new larger graph when this edge connects two vertices already in G. We denote by $G + e$ the new graph produced by adding a new edge e, connecting two previously nonincident vertices, to the graph G Hence,

$$G + e = (V, E \cup \{e\}).$$

The vertex set of $G + e$ is the same as the vertex set of G and the edge set is the union of the edge set of G and the set $\{e\}$.

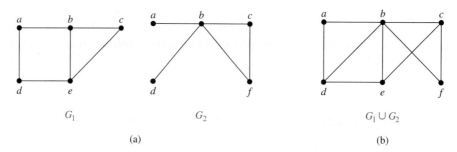

FIGURE 16 **(a) The Simple Graphs G_1 and G_2; (b) Their Union $G_1 \cup G_2$.**

EDGE CONTRACTIONS Sometimes when we remove an edge from a graph, we do not want to retain the endpoints of this edge as separate vertices in the resulting subgraph. In such a case we perform an **edge contraction** which removes an edge e with endpoints u and v and merges u and w into a new single vertex w, and for each edge with u or v as an endpoint replaces the edge with one with w as endpoint in place of u or v and with the same second endpoint. Hence, the contraction of the edge e with endpoints u and v in the graph $G = (V, E)$ produces a new graph $G' = (V', E')$ (which is not a subgraph of G), where $V' = V - \{u, v\} \cup \{w\}$ and E' contains the edges in E which do not have either u or v as endpoints and an edge connecting w to every neighbor of either u or v in V. For example, the contraction of the edge connecting the vertices e and c in the graph G_1 in Figure 16 produces a new graph G'_1 with vertices a, b, d, and w. As in G_1, there is an edge in G'_1 connecting a and b and an edge connecting a and d. There also is an edge in G'_1 that connects b and w that replaces the edges connecting b and c and connecting b and e in G_1 and an edge in G'_1 that connects d and w replacing the edge connecting d and e in G_1.

REMOVING VERTICES FROM A GRAPH When we remove a vertex v and all edges incident to it from $G = (V, E)$, we produce a subgraph, denoted by $G - v$. Observe that $G - v = (V - v, E')$, where E' is the set of edges of G not incident to v. Similarly, if V' is a subset of V, then the graph $G - V'$ is the subgraph $(V - V', E')$, where E' is the set of edges of G not incident to a vertex in V'.

GRAPH UNIONS Two or more graphs can be combined in various ways. The new graph that contains all the vertices and edges of these graphs is called the **union** of the graphs. We will give a more formal definition for the union of two simple graphs.

DEFINITION 9   The *union* of two simple graphs $G_1 = (V_1, E_1)$ and $G_2 = (V_2, E_2)$ is the simple graph with vertex set $V_1 \cup V_2$ and edge set $E_1 \cup E_2$. The union of G_1 and G_2 is denoted by $G_1 \cup G_2$.

EXAMPLE 19   Find the union of the graphs G_1 and G_2 shown in Figure 16(a).   ◀

Solution: The vertex set of the union $G_1 \cup G_2$ is the union of the two vertex sets, namely, $\{a, b, c, d, e, f\}$. The edge set of the union is the union of the two edge sets. The union is displayed in Figure 16(b).

Exercises

In Exercises 1–2 find the number of vertices, the number of edges, and the degree of each vertex in the given undirected graph. Identify all isolated and pendant vertices.

1.

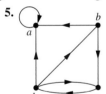

2.

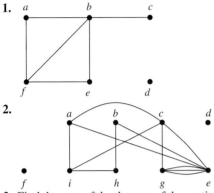

3. Find the sum of the degrees of the vertices of each graph in Exercises 1–2 and verify that it equals twice the number of edges in the graph.

4. Show that the sum, over the set of people at a party, of the number of people a person has shaken hands with, is even. Assume that no one shakes his or her own hand.

In Exercises 5–6 determine the number of vertices and edges and find the in-degree and out-degree of each vertex for the given directed multigraph.

5.

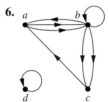

6.

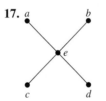

7. For each of the graphs in Exercises 5–6 determine the sum of the in-degrees of the vertices and the sum of the out-degrees of the vertices directly. Show that they are both equal to the number of edges in the graph.

8. What does the degree of a vertex represent in the acquaintanceship graph, where vertices represent all the people in the world? What does the neighborhood a vertex in this graph represent? What do isolated and pendant vertices in this graph represent? In one study it was estimated that the average degree of a vertex in this graph is 1000. What does this mean in terms of the model?

9. What does the degree of a vertex represent in an academic collaboration graph? What does the neighborhood of a vertex represent? What do isolated and pendant vertices represent?

10. What does the degree of a vertex in the Hollywood graph represent? What does the neighborhood of a vertex represent? What do the isolated and pendant vertices represent?

11. What do the in-degree and the out-degree of a vertex in a telephone call graph, as described in Example 4 of Section 10.1, represent? What does the degree of a vertex in the undirected version of this graph represent?

12. What do the in-degree and the out-degree of a vertex in the Web graph, as described in Example 5 of Section 10.1, represent?

13. What do the in-degree and the out-degree of a vertex in a directed graph modeling a round-robin tournament represent?

14. Show that in a simple graph with at least two vertices there must be two vertices that have the same degree.

15. Use Exercise 14 to show that in a group of people, there must be two people who are friends with the same number of other people in the group.

16. Draw these graphs.
   a) K_7 **b)** $K_{1,8}$ **c)** $K_{4,4}$
   d) C_7 **e)** W_7 **f)** Q_4

In Exercises 17–21 determine whether the graph is bipartite. You may find it useful to apply Theorem 4 and answer the question by determining whether it is possible to assign either red or blue to each vertex so that no two adjacent vertices are assigned the same color.

17.

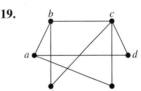

18.

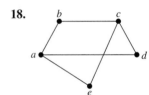

19.

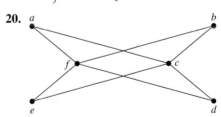

20.

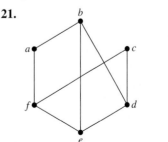

21.

22. For which values of n are these graphs bipartite?
   a) K_n **b)** C_n **c)** W_n **d)** Q_n

23. Suppose that a new company has five employees: Zamora, Agraharam, Smith, Chou, and Macintyre. Each employee will assume one of six responsiblities: planning, publicity, sales, marketing, development, and industry relations. Each employee is capable of doing one or more of these jobs: Zamora could do planning, sales, marketing, or industry relations; Agraharam could do planning or development; Smith could do publicity, sales, or industry relations; Chou could do planning, sales, or industry relations; and Macintyre could do planning, publicity, sales, or industry relations.

a) Model the capabilities of these employees using a bipartite graph.
b) Find an assignment of responsibilites such that each employee is assigned one responsibility.
c) Is the matching of responsibilities you found in part (b) a complete matching? Is it a maximum matching?

24. Suppose that there are five young women and six young men on an island. Each woman is willing to marry some of the men on the island and each man is willing to marry any woman who is willing to marry him. Suppose that Anna is willing to marry Jason, Larry, and Matt; Barbara is willing to marry Kevin and Larry; Carol is willing to marry Jason, Nick, and Oscar; Diane is willing to marry Jason, Larry, Nick, and Oscar; and Elizabeth is willing to marry Jason and Matt.

a) Model the possible marriages on the island using a bipartite graph.
b) Find a matching of the young women and the young men on the island such that each young woman is matched with a young man whom she is willing to marry.
c) Is the matching you found in part (b) a complete matching? Is it a maximum matching?

25. How many vertices and how many edges do these graphs have?

a) K_n b) C_n c) W_n
d) $K_{m,n}$ e) Q_n

The **degree sequence** of a graph is the sequence of the degrees of the vertices of the graph in nonincreasing order. For example, the degree sequence of the graph G in Example 1 is 4, 4, 4, 3, 2, 1, 0.

26. Find the degree sequences for each of the graphs in Exercises 17–21.

27. Find the degree sequence of each of the following graphs.

a) K_4 b) C_4 c) W_4
d) $K_{2,3}$ e) Q_3

28. What is the degree sequence of K_n, where n is a positive integer? Explain your answer.

29. How many edges does a graph have if its degree sequence is 5, 2, 2, 2, 2, 1? Draw such a graph.

A sequence $d_1, d_2, \ldots, d_n$ is called **graphic** if it is the degree sequence of a simple graph.

30. Determine whether each of these sequences is graphic. For those that are, draw a graph having the given degree sequence.

a) 5, 4, 3, 2, 1, 0 b) 6, 5, 4, 3, 2, 1 c) 2, 2, 2, 2, 2, 2
d) 3, 3, 3, 2, 2, 2 e) 3, 3, 2, 2, 2, 2 f) 1, 1, 1, 1, 1, 1
g) 5, 3, 3, 3, 3, 3 h) 5, 5, 4, 3, 2, 1

31. Determine whether each of these sequences is graphic. For those that are, draw a graph having the given degree sequence.

a) 3, 3, 3, 3, 2   b) 5, 4, 3, 2, 1   c) 4, 4, 3, 2, 1
d) 4, 4, 3, 3, 3   e) 3, 2, 2, 1, 0   f) 1, 1, 1, 1, 1

*32. Suppose that $d_1, d_2, \ldots, d_n$ is a graphic sequence. Show that there is a simple graph with vertices $v_1, v_2, \ldots, v_n$ such that $\deg(v_i) = d_i$ for $i = 1, 2, \ldots, n$ and v_1 is adjacent to $v_2, \ldots, v_{d_1+1}$.

33. Show that every nonincreasing sequence of nonnegative integers with an even sum of its terms is the degree sequence of a pseudograph, that is, an undirected graph where loops are allowed. [*Hint:* Construct such a graph by first adding as many loops as possible at each vertex. Then add additional edges connecting vertices of odd degree. Explain why this construction works.]

34. How many subgraphs with at least one vertex does K_2 have?

35. How many subgraphs with at least one vertex does K_3 have?

36. How many subgraphs with at least one vertex does W_3 have?

37. Draw all subgraphs of this graph.

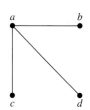

38. Let G be a graph with v vertices and e edges. Let M be the maximum degree of the vertices of G, and let m be the minimum degree of the vertices of G. Show that

a) $2e/v \geq m$.   b) $2e/v \leq M$.

A simple graph is called **regular** if every vertex of this graph has the same degree. A regular graph is called n-**regular** if every vertex in this graph has degree n.

39. For which values of n are these graphs regular?

a) K_n b) C_n c) W_n d) Q_n

40. How many vertices does a regular graph of degree four with 10 edges have?

In Exercises 41–42 find the union of the given pair of simple graphs. (Assume edges with the same endpoints are the same.)

41.

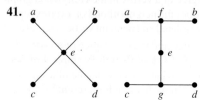

42.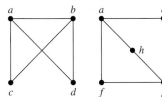

43. The **complementary graph** $\overline{G}$ of a simple graph G has the same vertices as G. Two vertices are adjacent in $\overline{G}$ if and only if they are not adjacent in G. Describe each of these graphs.

    a) $\overline{K_n}$ **b)** $\overline{K_{m,n}}$ **c)** $\overline{C_n}$ **d)** $\overline{Q_n}$

44. If G is a simple graph with 15 edges and $\overline{G}$ has 13 edges, how many vertices does G have?

45. If the simple graph G has v vertices and e edges, how many edges does $\overline{G}$ have?

46. If the degree sequence of the simple graph G is 4, 3, 3, 2, 2, what is the degree sequence of $\overline{G}$?

47. If the degree sequence of the simple graph G is $d_1, d_2, \ldots, d_n$, what is the degree sequence of $\overline{G}$?

48. Show that if G is a simple graph with n vertices, then the union of G and $\overline{G}$ is K_n.

The **converse** of a directed graph $G = (V, E)$, denoted by G^{conv}, is the directed graph (V, F), where the set F of edges of G^{conv} is obtained by reversing the direction of each edge in E.

49. Draw the converse of each of the graphs in Exercise 5 in Section 10.1.

50. Show that $(G^{conv})^{conv} = G$ whenever G is a directed graph.

51. Show that the graph G is its own converse if and only if the relation associated with G (see Section 9.3) is symmetric.

52. Show that if a bipartite graph $G = (V, E)$ is n-regular for some positive integer n (see the preamble to Exercise 39) and (V_1, V_2) is a bipartition of V, then $|V_1| = |V_2|$. That is, show that the two sets in a bipartition of the vertex set of an n-regular graph must contain the same number of vertices.

53. Draw the mesh network for interconnecting nine parallel processors.

54. In a variant of a mesh network for interconnecting $n = m^2$ processors, processor $P(i, j)$ is connected to the four processors $P((i \pm 1) \bmod m, j)$ and $P(i, (j \pm 1) \bmod m)$, so that connections wrap around the edges of the mesh. Draw this variant of the mesh network for 16 processors.

55. Show that every pair of processors in a mesh network of $n = m^2$ processors can communicate using $O(\sqrt{n}) = O(m)$ hops between directly connected processors.

10.3   Representing Graphs and Graph Isomorphism

Introduction

There are many useful ways to represent graphs. As we will see throughout this chapter, in working with a graph it is helpful to be able to choose its most convenient representation. In this section we will show how to represent graphs in several different ways.

    Sometimes, two graphs have exactly the same form, in the sense that there is a one-to-one correspondence between their vertex sets that preserves edges. In such a case, we say that the two graphs are **isomorphic**. Determining whether two graphs are isomorphic is an important problem of graph theory that we will study in this section.

Representing Graphs

One way to represent a graph without multiple edges is to list all the edges of this graph. Another way to represent a graph with no multiple edges is to use **adjacency lists**, which specify the vertices that are adjacent to each vertex of the graph.

EXAMPLE 1   Use adjacency lists to describe the simple graph given in Figure 1.

    Solution: Table 1 lists those vertices adjacent to each of the vertices of the graph.   ◄

EXAMPLE 2   Represent the directed graph shown in Figure 2 by listing all the vertices that are the terminal vertices of edges starting at each vertex of the graph.

    Solution: Table 2 represents the directed graph shown in Figure 2.   ◄

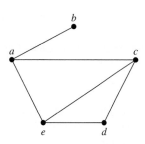

FIGURE 1 A Simple Graph.

| TABLE 1 An Adjacency List for a Simple Graph. ||
Vertex	*Adjacent Vertices*
a	b, c, e
b	a
c	a, d, e
d	c, e
e	a, c, d

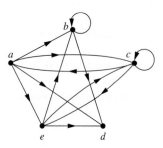

FIGURE 2 A Directed Graph.

| TABLE 2 An Adjacency List for a Directed Graph. ||
Initial Vertex	*Terminal Vertices*
a	b, c, d, e
b	b, d
c	a, c, e
d	
e	b, c, d

FIGURE 3
Simple Graph.

Adjacency Matrices

Carrying out graph algorithms using the representation of graphs by lists of edges, or by adjacency lists, can be cumbersome if there are many edges in the graph. To simplify computation, graphs can be represented using matrices. Two types of matrices commonly used to represent graphs will be presented here. One is based on the adjacency of vertices, and the other is based on incidence of vertices and edges.

Suppose that $G = (V, E)$ is a simple graph where $|V| = n$. Suppose that the vertices of G are listed arbitrarily as $v_1, v_2, \ldots, v_n$. The **adjacency matrix A** (or $\mathbf{A}_G$) of G, with respect to this listing of the vertices, is the $n \times n$ zero–one matrix with 1 as its (i, j)th entry when v_i and v_j are adjacent, and 0 as its (i, j)th entry when they are not adjacent. In other words, if its adjacency matrix is $\mathbf{A} = [a_{ij}]$, then

$$a_{ij} = \begin{cases} 1 & \text{if } \{v_i, v_j\} \text{ is an edge of } G, \\ 0 & \text{otherwise.} \end{cases}$$

EXAMPLE 3 Use an adjacency matrix to represent the graph shown in Figure 3.

Solution: We order the vertices as a, b, c, d. The matrix representing this graph is

$$\begin{bmatrix} 0 & 1 & 1 & 1 \\ 1 & 0 & 1 & 0 \\ 1 & 1 & 0 & 0 \\ 1 & 0 & 0 & 0 \end{bmatrix}.$$

EXAMPLE 4   Draw a graph with the adjacency matrix

FIGURE 4
**A Graph with the
Given Adjacency
Matrix.**

$$
\begin{bmatrix}
0 & 1 & 1 & 0 \\
1 & 0 & 0 & 1 \\
1 & 0 & 0 & 1 \\
0 & 1 & 1 & 0
\end{bmatrix}
$$

with respect to the ordering of vertices a, b, c, d.

Solution: A graph with this adjacency matrix is shown in Figure 4.   ◀

Note that an adjacency matrix of a graph is based on the ordering chosen for the vertices. Hence, there may be as many as $n!$ different adjacency matrices for a graph with n vertices, because there are $n!$ different orderings of n vertices.

The adjacency matrix of a simple graph is symmetric, that is, $a_{ij} = a_{ji}$, because both of these entries are 1 when v_i and v_j are adjacent, and both are 0 otherwise. Furthermore, because a simple graph has no loops, each entry $a_{ii}, i = 1, 2, 3, \ldots, n$, is 0.

Adjacency matrices can also be used to represent undirected graphs with loops and with multiple edges. A loop at the vertex v_i is represented by a 1 at the (i, i)th position of the adjacency matrix. When multiple edges connecting the same pair of vertices v_i and v_j, or multiple loops at the same vertex, are present, the adjacency matrix is no longer a zero–one matrix, because the (i, j)th entry of this matrix equals the number of edges that are associated to $\{v_i, v_j\}$. All undirected graphs, including multigraphs and pseudographs, have symmetric adjacency matrices.

EXAMPLE 5   Use an adjacency matrix to represent the pseudograph shown in Figure 5.

Solution: The adjacency matrix using the ordering of vertices a, b, c, d is

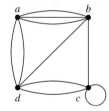

FIGURE 5
A Pseudograph.

$$
\begin{bmatrix}
0 & 3 & 0 & 2 \\
3 & 0 & 1 & 1 \\
0 & 1 & 1 & 2 \\
2 & 1 & 2 & 0
\end{bmatrix}.
$$

◀

We used zero–one matrices in Chapter 9 to represent directed graphs. The matrix for a directed graph $G = (V, E)$ has a 1 in its (i, j)th position if there is an edge from v_i to v_j, where $v_1, v_2, \ldots, v_n$ is an arbitrary listing of the vertices of the directed graph. In other words, if $\mathbf{A} = [a_{ij}]$ is the adjacency matrix for the directed graph with respect to this listing of the vertices, then

$$
a_{ij} =
\begin{cases}
1 & \text{if } (v_i, v_j) \text{ is an edge of } G, \\
0 & \text{otherwise.}
\end{cases}
$$

The adjacency matrix for a directed graph does not have to be symmetric, because there may not be an edge from v_j to v_i when there is an edge from v_i to v_j.

Adjacency matrices can also be used to represent directed multigraphs. Again, such matrices are not zero–one matrices when there are multiple edges in the same direction connecting two vertices. In the adjacency matrix for a directed multigraph, a_{ij} equals the number of edges that are associated to (v_i, v_j).

TRADE-OFFS BETWEEN ADJACENCY LISTS AND ADJACENCY MATRICES   When a simple graph contains relatively few edges, that is, when it is **sparse**, it is usually preferable to use adjacency lists rather than an adjacency matrix to represent the graph. For example, if each vertex has degree not exceeding c, where c is a constant much smaller than n, then each adjacency list contains c or fewer vertices. Hence, there are no more than cn items in all these adjacency lists. On the other hand, the adjacency matrix for the graph has n^2 entries. Note, however, that the adjacency matrix of a sparse graph is a **sparse matrix**, that is, a matrix with few nonzero entries, and there are special techniques for representing, and computing with, sparse matrices.

Now suppose that a simple graph is **dense**, that is, suppose that it contains many edges, such as a graph that contains more than half of all possible edges. In this case, using an adjacency matrix to represent the graph is usually preferable over using adjacency lists. To see why, we compare the complexity of determining whether the possible edge $\{v_i, v_j\}$ is present. Using an adjacency matrix, we can determine whether this edge is present by examining the (i, j)th entry in the matrix. This entry is 1 if the graph contains this edge and is 0 otherwise. Consequently, we need make only one comparison, namely, comparing this entry with 0, to determine whether this edge is present. On the other hand, when we use adjacency lists to represent the graph, we need to search the list of vertices adjacent to either v_i or v_j to determine whether this edge is present. This can require $\Theta(|V|)$ comparisons when many edges are present.

Incidence Matrices

Another common way to represent graphs is to use **incidence matrices**. Let $G = (V, E)$ be an undirected graph. Suppose that $v_1, v_2, \ldots, v_n$ are the vertices and $e_1, e_2, \ldots, e_m$ are the edges of G. Then the incidence matrix with respect to this ordering of V and E is the $n \times m$ matrix $\mathbf{M} = [m_{ij}]$, where

$$m_{ij} = \begin{cases} 1 & \text{when edge } e_j \text{ is incident with } v_i, \\ 0 & \text{otherwise.} \end{cases}$$

EXAMPLE 6   Represent the graph shown in Figure 6 with an incidence matrix.

Solution: The incidence matrix is

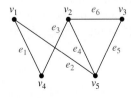

$$
\begin{array}{c}
\begin{array}{cccccc}
e_1 & e_2 & e_3 & e_4 & e_5 & e_6
\end{array} \\
\begin{array}{c} v_1 \\ v_2 \\ v_3 \\ v_4 \\ v_5 \end{array}
\left[
\begin{array}{cccccc}
1 & 1 & 0 & 0 & 0 & 0 \\
0 & 0 & 1 & 1 & 0 & 1 \\
0 & 0 & 0 & 0 & 1 & 1 \\
1 & 0 & 1 & 0 & 0 & 0 \\
0 & 1 & 0 & 1 & 1 & 0
\end{array}
\right].
\end{array}
$$

◀

FIGURE 6   An Undirected Graph.

Incidence matrices can also be used to represent multiple edges and loops. Multiple edges are represented in the incidence matrix using columns with identical entries, because these edges are incident with the same pair of vertices. Loops are represented using a column with exactly one entry equal to 1, corresponding to the vertex that is incident with this loop.

EXAMPLE 7   Represent the pseudograph shown in Figure 7 using an incidence matrix.

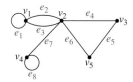

Solution: The incidence matrix for this graph is

$$
\begin{array}{c|cccccccc}
 & e_1 & e_2 & e_3 & e_4 & e_5 & e_6 & e_7 & e_8 \\
\hline
v_1 & 1 & 1 & 1 & 0 & 0 & 0 & 0 & 0 \\
v_2 & 0 & 1 & 1 & 1 & 0 & 1 & 1 & 0 \\
v_3 & 0 & 0 & 0 & 1 & 1 & 0 & 0 & 0 \\
v_4 & 0 & 0 & 0 & 0 & 0 & 0 & 1 & 1 \\
v_5 & 0 & 0 & 0 & 0 & 1 & 1 & 0 & 0 \\
\end{array}.
$$

FIGURE 7
A Pseudograph.

◀

Isomorphism of Graphs

We often need to know whether it is possible to draw two graphs in the same way. That is, do the graphs have the same structure when we ignore the identities of their vertices? For instance, in chemistry, graphs are used to model chemical compounds (in a way we will describe later). Different compounds can have the same molecular formula but can differ in structure. Such compounds can be represented by graphs that cannot be drawn in the same way. The graphs representing previously known compounds can be used to determine whether a supposedly new compound has been studied before.

There is a useful terminology for graphs with the same structure.

DEFINITION 1   The simple graphs $G_1 = (V_1, E_1)$ and $G_2 = (V_2, E_2)$ are *isomorphic* if there exists a one-to-one and onto function f from V_1 to V_2 with the property that a and b are adjacent in G_1 if and only if $f(a)$ and $f(b)$ are adjacent in G_2, for all a and b in V_1. Such a function f is called an *isomorphism.*[*] Two simple graphs that are not isomorphic are called *nonisomorphic*.

In other words, when two simple graphs are isomorphic, there is a one-to-one correspondence between vertices of the two graphs that preserves the adjacency relationship. Isomorphism of simple graphs is an equivalence relation. (We leave the verification of this as Exercise 29.)

EXAMPLE 8   Show that the graphs $G = (V, E)$ and $H = (W, F)$, displayed in Figure 8, are isomorphic.

Solution: The function f with $f(u_1) = v_1$, $f(u_2) = v_4$, $f(u_3) = v_3$, and $f(u_4) = v_2$ is a one-to-one correspondence between V and W. To see that this correspondence preserves adjacency, note that adjacent vertices in G are u_1 and u_2, u_1 and u_3, u_2 and u_4, and u_3 and u_4, and each of the pairs $f(u_1) = v_1$ and $f(u_2) = v_4$, $f(u_1) = v_1$ and $f(u_3) = v_3$, $f(u_2) = v_4$ and $f(u_4) = v_2$, and $f(u_3) = v_3$ and $f(u_4) = v_2$ consists of two adjacent vertices in H.   ◀

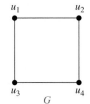

Determining whether Two Simple Graphs are Isomorphic

It is often difficult to determine whether two simple graphs are isomorphic. There are $n!$ possible one-to-one correspondences between the vertex sets of two simple graphs with n vertices. Testing each such correspondence to see whether it preserves adjacency and nonadjacency is impractical if n is at all large.

Sometimes it is not hard to show that two graphs are not isomorphic. In particular, we can show that two graphs are not isomorphic if we can find a property only one of the two graphs has, but that is preserved by isomorphism. A property preserved by isomorphism of graphs is

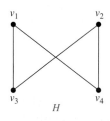

FIGURE 8   The Graphs G and H.

---

[*]The word *isomorphism* comes from the Greek roots *isos* for "equal" and *morphe* for "form."

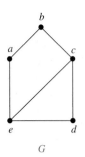

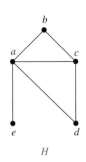

FIGURE 9 **The Graphs G and H.**

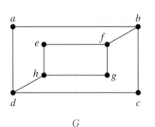
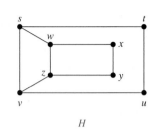

FIGURE 10 **The Graphs G and H.**

called a **graph invariant**. For instance, isomorphic simple graphs must have the same number of vertices, because there is a one-to-one correspondence between the sets of vertices of the graphs.

Isomorphic simple graphs also must have the same number of edges, because the one-to-one correspondence between vertices establishes a one-to-one correspondence between edges. In addition, the degrees of the vertices in isomorphic simple graphs must be the same. That is, a vertex v of degree d in G must correspond to a vertex $f(v)$ of degree d in H, because a vertex w in G is adjacent to v if and only if $f(v)$ and $f(w)$ are adjacent in H.

EXAMPLE 9 Show that the graphs displayed in Figure 9 are not isomorphic.

Solution: Both G and H have five vertices and six edges. However, H has a vertex of degree one, namely, e, whereas G has no vertices of degree one. It follows that G and H are not isomorphic. ◀

The number of vertices, the number of edges, and the number of vertices of each degree are all invariants under isomorphism. If any of these quantities differ in two simple graphs, these graphs cannot be isomorphic. However, when these invariants are the same, it does not necessarily mean that the two graphs are isomorphic. There are no useful sets of invariants currently known that can be used to determine whether simple graphs are isomorphic.

EXAMPLE 10 Determine whether the graphs shown in Figure 10 are isomorphic.

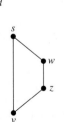

FIGURE 11 **The Subgraphs of G and H Made Up of Vertices of Degree Three and the Edges Connecting Them.**

Solution: The graphs G and H both have eight vertices and 10 edges. They also both have four vertices of degree two and four of degree three. Because these invariants all agree, it is still conceivable that these graphs are isomorphic.

However, G and H are not isomorphic. To see this, note that because $\deg(a) = 2$ in G, a must correspond to either t, u, x, or y in H, because these are the vertices of degree two in H. However, each of these four vertices in H is adjacent to another vertex of degree two in H, which is not true for a in G.

Another way to see that G and H are not isomorphic is to note that the subgraphs of G and H made up of vertices of degree three and the edges connecting them must be isomorphic if these two graphs are isomorphic (the reader should verify this). However, these subgraphs, shown in Figure 11, are not isomorphic. ◀

To show that a function f from the vertex set of a graph G to the vertex set of a graph H is an isomorphism, we need to show that f preserves the presence and absence of edges. One helpful way to do this is to use adjacency matrices. In particular, to show that f is an isomorphism, we can show that the adjacency matrix of G is the same as the adjacency matrix of H, when rows and columns are labeled to correspond to the images under f of the vertices in G that are the labels of these rows and columns in the adjacency matrix of G. We illustrate how this is done in Example 11.

EXAMPLE 11   Determine whether the graphs G and H displayed in Figure 12 are isomorphic.

Solution: Both G and H have six vertices and seven edges. Both have four vertices of degree two and two vertices of degree three. It is also easy to see that the subgraphs of G and H consisting of all vertices of degree two and the edges connecting them are isomorphic (as the reader should verify). Because G and H agree with respect to these invariants, it is reasonable to try to find an isomorphism f.

We now will define a function f and then determine whether it is an isomorphism. Because $\deg(u_1) = 2$ and because u_1 is not adjacent to any other vertex of degree two, the image of u_1 must be either v_4 or v_6, the only vertices of degree two in H not adjacent to a vertex of degree two. We arbitrarily set $f(u_1) = v_6$. [If we found that this choice did not lead to isomorphism, we would then try $f(u_1) = v_4$.] Because u_2 is adjacent to u_1, the possible images of u_2 are v_3 and v_5. We arbitrarily set $f(u_2) = v_3$. Continuing in this way, using adjacency of vertices and degrees as a guide, we set $f(u_3) = v_4$, $f(u_4) = v_5$, $f(u_5) = v_1$, and $f(u_6) = v_2$. We now have a one-to-one correspondence between the vertex set of G and the vertex set of H, namely, $f(u_1) = v_6$, $f(u_2) = v_3$, $f(u_3) = v_4$, $f(u_4) = v_5$, $f(u_5) = v_1$, $f(u_6) = v_2$. To see whether f preserves edges, we examine the adjacency matrix of G,

$$
\mathbf{A}_G = \begin{array}{c@{}c}
 & \begin{array}{cccccc} u_1 & u_2 & u_3 & u_4 & u_5 & u_6 \end{array} \\
\begin{array}{c} u_1 \\ u_2 \\ u_3 \\ u_4 \\ u_5 \\ u_6 \end{array} &
\left[\begin{array}{cccccc}
0 & 1 & 0 & 1 & 0 & 0 \\
1 & 0 & 1 & 0 & 0 & 1 \\
0 & 1 & 0 & 1 & 0 & 0 \\
1 & 0 & 1 & 0 & 1 & 0 \\
0 & 0 & 0 & 1 & 0 & 1 \\
0 & 1 & 0 & 0 & 1 & 0
\end{array}\right]
\end{array},
$$

and the adjacency matrix of H with the rows and columns labeled by the images of the corresponding vertices in G,

$$
\mathbf{A}_H = \begin{array}{c@{}c}
 & \begin{array}{cccccc} v_6 & v_3 & v_4 & v_5 & v_1 & v_2 \end{array} \\
\begin{array}{c} v_6 \\ v_3 \\ v_4 \\ v_5 \\ v_1 \\ v_2 \end{array} &
\left[\begin{array}{cccccc}
0 & 1 & 0 & 1 & 0 & 0 \\
1 & 0 & 1 & 0 & 0 & 1 \\
0 & 1 & 0 & 1 & 0 & 0 \\
1 & 0 & 1 & 0 & 1 & 0 \\
0 & 0 & 0 & 1 & 0 & 1 \\
0 & 1 & 0 & 0 & 1 & 0
\end{array}\right]
\end{array}.
$$

Because $\mathbf{A}_G = \mathbf{A}_H$, it follows that f preserves edges. We conclude that f is an isomorphism, so G and H are isomorphic. Note that if f turned out not to be an isomorphism, we would *not* have established that G and H are not isomorphic, because another correspondence of the vertices in G and H may be an isomorphism.   ◀

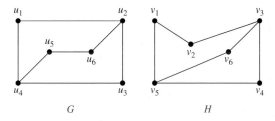

FIGURE 12   Graphs G and H.

ALGORITHMS FOR GRAPH ISOMORPHISM   The best algorithms known for determining whether two graphs are isomorphic have exponential worst-case time complexity (in the number of vertices of the graphs). However, linear average-case time complexity algorithms are known that solve this problem, and there is some hope, but also skepticism, that an algorithm with polynomial worst-case time complexity for determining whether two graphs are isomorphic can be found. The best practical general purpose software for isomorphism testing, called NAUTY, can be used to determine whether two graphs with as many as 100 vertices are isomorphic in less than a second on a modern PC. NAUTY software can be downloaded over the Internet and experimented with. Practical algorithms for determining whether two graphs are isomorphic exist for graphs that are restricted in various ways, such as when the maximum degree of vertices is small. The problem of determining whether any two graphs are isomorphic is of special interest because it is one of only a few NP problems (see Exercise 50) not known to be either tractable or NP-complete (see Section 3.3).

APPLICATIONS OF GRAPH ISOMORPHISMS   Graph isomorphisms, and functions that are almost graph isomorphisms, arise in applications of graph theory to chemistry and to the design of electronic circuits, and other areas including bioinformatics and computer vision. Chemists use multigraphs, known as molecular graphs, to model chemical compounds. In these graphs, vertices represent atoms and edges represent chemical bonds between these atoms. Two structural isomers, molecules with identical molecular formulas but with atoms bonded differently, have nonisomorphic molecular graphs. When a potentially new chemical compound is synthesized, a database of molecular graphs is checked to see whether the molecular graph of the compound is the same as one already known.

Electronic circuits are modeled using graphs in which vertices represent components and edges represent connections between them. Modern integrated circuits, known as chips, are miniaturized electronic circuits, often with millions of transistors and connections between them. Because of the complexity of modern chips, automation tools are used to design them. Graph isomorphism is the basis for the verification that a particular layout of a circuit produced by an automated tool corresponds to the original schematic of the design. Graph isomorphism can also be used to determine whether a chip from one vendor includes intellectual property from a different vendor. This can be done by looking for large isomorphic subgraphs in the graphs modeling these chips.

Exercises

In Exercises 1–2 use an adjacency list to represent the given graph.

1.

2.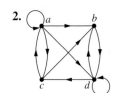

3. Represent the graph in Exercise 1 with an adjacency matrix.

4. Represent the graph in Exercise 2 with an adjacency matrix.

5. Represent each of these graphs with an adjacency matrix.
   a) K_4 **b)** $K_{1,4}$ **c)** $K_{2,3}$
   d) C_4 **e)** W_4 **f)** Q_3

In Exercises 6–7 draw a graph with the given adjacency matrix.

6. $\begin{bmatrix} 0 & 1 & 0 \\ 1 & 0 & 1 \\ 0 & 1 & 0 \end{bmatrix}$ **7.** $\begin{bmatrix} 0 & 0 & 1 & 1 \\ 0 & 0 & 1 & 0 \\ 1 & 1 & 0 & 1 \\ 1 & 1 & 1 & 0 \end{bmatrix}$

In Exercises 8–9 represent the given graph using an adjacency matrix.

8.

9.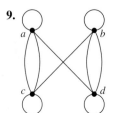

In Exercises 10–11 draw an undirected graph represented by the given adjacency matrix.

10. $\begin{bmatrix} 1 & 3 & 2 \\ 3 & 0 & 4 \\ 2 & 4 & 0 \end{bmatrix}$ **11.** $\begin{bmatrix} 1 & 2 & 0 & 1 \\ 2 & 0 & 3 & 0 \\ 0 & 3 & 1 & 1 \\ 1 & 0 & 1 & 0 \end{bmatrix}$

In Exercises 12–13 find the adjacency matrix of the given directed multigraph with respect to the vertices listed in alphabetic order.

12. 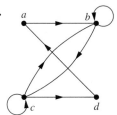 **13.**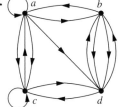

In Exercises 14–15 draw the graph represented by the given adjacency matrix.

14. $\begin{bmatrix} 1 & 0 & 1 \\ 0 & 0 & 1 \\ 1 & 1 & 1 \end{bmatrix}$ **15.** $\begin{bmatrix} 1 & 2 & 1 \\ 2 & 0 & 0 \\ 0 & 2 & 2 \end{bmatrix}$

16. Is every zero–one square matrix that is symmetric and has zeros on the diagonal the adjacency matrix of a simple graph?

17. Use an incidence matrix to represent the graphs in Exercises 8–9.

*18. What is the sum of the entries in a row of the adjacency matrix for an undirected graph? For a directed graph?

*19. What is the sum of the entries in a column of the adjacency matrix for an undirected graph? For a directed graph?

20. What is the sum of the entries in a row of the incidence matrix for an undirected graph?

21. What is the sum of the entries in a column of the incidence matrix for an undirected graph?

In Exercises 22–28 determine whether the given pair of graphs is isomorphic. Exhibit an isomorphism or provide a rigorous argument that none exists.

22.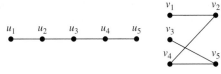

23.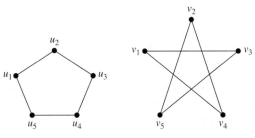

24.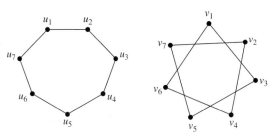

25.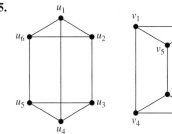

26.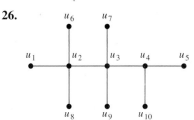

27.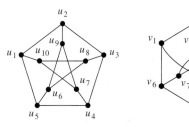

28.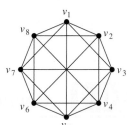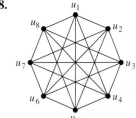

29. Show that isomorphism of simple graphs is an equivalence relation.

30. Suppose that G and H are isomorphic simple graphs. Show that their complementary graphs $\overline{G}$ and $\overline{H}$ are also isomorphic.

31. Describe the row and column of an adjacency matrix of a graph corresponding to an isolated vertex.

32. Describe the row of an incidence matrix of a graph corresponding to an isolated vertex.

33. Show that the vertices of a bipartite graph with two or more vertices can be ordered so that its adjacency matrix has the form

$$\begin{bmatrix} \mathbf{0} & \mathbf{A} \\ \mathbf{B} & \mathbf{0} \end{bmatrix},$$

where the four entries shown are rectangular blocks.

A simple graph G is called **self-complementary** if G and $\overline{G}$ are isomorphic.

34. Show that this graph is self-complementary.

35. Find a self-complementary simple graph with five vertices.

*36. Show that if G is a self-complementary simple graph with v vertices, then $v \equiv 0$ or $1 \pmod 4$.

37. For which integers n is C_n self-complementary?

38. How many nonisomorphic simple graphs are there with n vertices, when n is

a) 2?   **b)** 3?   **c)** 4?

39. How many nonisomorphic simple graphs are there with five vertices and three edges?

40. Are the simple graphs with the following adjacency matrices isomorphic?

a) $\begin{bmatrix} 0 & 0 & 1 \\ 0 & 0 & 1 \\ 1 & 1 & 0 \end{bmatrix}, \begin{bmatrix} 0 & 1 & 1 \\ 1 & 0 & 0 \\ 1 & 0 & 0 \end{bmatrix}$

b) $\begin{bmatrix} 0 & 1 & 0 & 1 \\ 1 & 0 & 0 & 1 \\ 0 & 0 & 0 & 1 \\ 1 & 1 & 1 & 0 \end{bmatrix}, \begin{bmatrix} 0 & 1 & 1 & 1 \\ 1 & 0 & 0 & 1 \\ 1 & 0 & 0 & 1 \\ 1 & 1 & 1 & 0 \end{bmatrix}$

c) $\begin{bmatrix} 0 & 1 & 1 & 0 \\ 1 & 0 & 0 & 1 \\ 1 & 0 & 0 & 1 \\ 0 & 1 & 1 & 0 \end{bmatrix}, \begin{bmatrix} 0 & 1 & 0 & 1 \\ 1 & 0 & 0 & 0 \\ 0 & 0 & 0 & 1 \\ 1 & 0 & 1 & 0 \end{bmatrix}$

41. Extend the definition of isomorphism of simple graphs to undirected graphs containing loops and multiple edges.

42. Define isomorphism of directed graphs.

In Exercises 43–46 determine whether the given pair of directed graphs are isomorphic. (See Exercise 42.)

43.

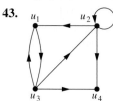 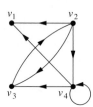

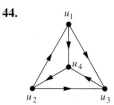

44.

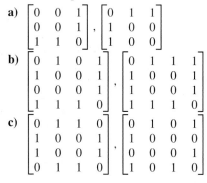

45. Show that if G and H are isomorphic directed graphs, then the converses of G and H (defined in the preamble of Exercise 49 of Section 10.2) are also isomorphic.

46. Show that the property that a graph is bipartite is an isomorphic invariant.

47. Find a pair of nonisomorphic graphs with the same degree sequence (defined in the preamble to Exercise 26 in Section 10.2) such that one graph is bipartite, but the other graph is not bipartite.

*48. How much storage is needed to represent a simple graph with n vertices and m edges using

a) adjacency lists?
b) an adjacency matrix?
c) an incidence matrix?

A **devil's pair** for a purported isomorphism test is a pair of nonisomorphic graphs that the test fails to show that they are not isomorphic.

49. Find a devil's pair for the test that checks the degree sequence (defined in the preamble to Exercise 26 in Section 10.2) in two graphs to make sure they agree.

50. Suppose that the function f from V_1 to V_2 is an isomorphism of the graphs $G_1 = (V_1, E_1)$ and $G_2 = (V_2, E_2)$. Show that it is possible to verify this fact in time polynomial in terms of the number of vertices of the graph, in terms of the number of comparisons needed.

10.4 Connectivity

Introduction

Many problems can be modeled with paths formed by traveling along the edges of graphs. For instance, the problem of determining whether a message can be sent between two computers using intermediate links can be studied with a graph model. Problems of efficiently planning routes for mail delivery, garbage pickup, diagnostics in computer networks, and so on can be solved using models that involve paths in graphs.

Paths

Informally, a **path** is a sequence of edges that begins at a vertex of a graph and travels from vertex to vertex along edges of the graph. As the path travels along its edges, it visits the vertices along this path, that is, the endpoints of these edges.

A formal definition of paths and related terminology is given in Definition 1.

DEFINITION 1

Let n be a nonnegative integer and G an undirected graph. A *path* of *length n* from u to v in G is a sequence of n edges $e_1, \ldots, e_n$ of G for which there exists a sequence $x_0 = u, x_1, \ldots, x_{n-1}, x_n = v$ of vertices such that e_i has, for $i = 1, \ldots, n$, the endpoints x_{i-1} and x_i. When the graph is simple, we denote this path by its vertex sequence $x_0, x_1, \ldots, x_n$ (because listing these vertices uniquely determines the path). The path is a *circuit* if it begins and ends at the same vertex, that is, if $u = v$, and has length greater than zero. The path or circuit is said to *pass through* the vertices $x_1, x_2, \ldots, x_{n-1}$ or *traverse* the edges $e_1, e_2, \ldots, e_n$. A path or circuit is *simple* if it does not contain the same edge more than once.

When it is not necessary to distinguish between multiple edges, we will denote a path $e_1, e_2, \ldots, e_n$, where e_i is associated with $\{x_{i-1}, x_i\}$ for $i = 1, 2, \ldots, n$ by its vertex sequence $x_0, x_1, \ldots, x_n$. This notation identifies a path only as far as which vertices it passes through. Consequently, it does not specify a unique path when there is more than one path that passes through this sequence of vertices, which will happen if and only if there are multiple edges between some successive vertices in the list. Note that a path of length zero consists of a single vertex.

Remark: There is considerable variation of terminology concerning the concepts defined in Definition 1. For instance, in some books, the term **walk** is used instead of *path*, where a walk is defined to be an alternating sequence of vertices and edges of a graph, $v_0, e_1, v_1, e_2, \ldots, v_{n-1}, e_n, v_n$, where v_{i-1} and v_i are the endpoints of e_i for $i = 1, 2, \ldots, n$. When this terminology is used, **closed walk** is used instead of *circuit* to indicate a walk that begins and ends at the same vertex, and **trail** is used to denote a walk that has no repeated edge (replacing the term *simple path*). When this terminology is used, the terminology **path** is often used for a trail with no repeated vertices, conflicting with the terminology in Definition 1. Because of this variation in terminology, you will need to make sure which set of definitions are used in a particular book or article when you read about traversing edges of a graph. The text [GrYe06] is a good reference for the alternative terminology described in this remark.

EXAMPLE 1

In the simple graph shown in Figure 1, a, d, c, f, e is a simple path of length 4, because $\{a, d\}$, $\{d, c\}$, $\{c, f\}$, and $\{f, e\}$ are all edges. However, d, e, c, a is not a path, because $\{e, c\}$ is not an edge. Note that b, c, f, e, b is a circuit of length 4 because $\{b, c\}$, $\{c, f\}$, $\{f, e\}$, and $\{e, b\}$ are edges, and this path begins and ends at b. The path a, b, e, d, a, b, which is of length 5, is not simple because it contains the edge $\{a, b\}$ twice. ◀

Paths and circuits in directed graphs were introduced in Chapter 9. We now provide more general definitions.

DEFINITION 2   Let n be a nonnegative integer and G a directed graph. A *path* of length n from u to v in G is a sequence of edges $e_1, e_2, \ldots, e_n$ of G such that e_1 is associated with (x_0, x_1), e_2 is associated with (x_1, x_2), and so on, with e_n associated with (x_{n-1}, x_n), where $x_0 = u$ and $x_n = v$. When there are no multiple edges in the directed graph, this path is denoted by its vertex sequence $x_0, x_1, x_2, \ldots, x_n$. A path of length greater than zero that begins and ends at the same vertex is called a *circuit* or *cycle*. A path or circuit is called *simple* if it does not contain the same edge more than once.

Remark: Terminology other than that given in Definition 2 is often used for the concepts defined there. In particular, the alternative terminology that uses *walk, closed walk, trail,* and *path* (described in the remarks following Definition 1) may be used for directed graphs. See [GrYe05] for details.

Note that the terminal vertex of an edge in a path is the initial vertex of the next edge in the path. When it is not necessary to distinguish between multiple edges, we will denote a path $e_1, e_2, \ldots, e_n$, where e_i is associated with (x_{i-1}, x_i) for $i = 1, 2, \ldots, n$, by its vertex sequence $x_0, x_1, \ldots, x_n$. The notation identifies a path only as far as which the vertices it passes through. There may be more than one path that passes through this sequence of vertices, which will happen if and only if there are multiple edges between two successive vertices in the list.

Paths represent useful information in many graph models, as Examples 2–4 demonstrate.

EXAMPLE 2   **Paths in Acquaintanceship Graphs**   In an acquaintanceship graph there is a path between two people if there is a chain of people linking these people, where two people adjacent in the chain know one another. For example, in Figure 6 in Section 10.1, there is a chain of six people linking Kamini and Ching. Many social scientists have conjectured that almost every pair of people in the world are linked by a small chain of people, perhaps containing just five or fewer people. This would mean that almost every pair of vertices in the acquaintanceship graph containing all people in the world is linked by a path of length not exceeding four. The play *Six Degrees of Separation* by John Guare is based on this notion. ◄

EXAMPLE 3   **Paths in Collaboration Graphs**   In a collaboration graph, two people a and b are connected by a path when there is a sequence of people starting with a and ending with b such that the endpoints of each edge in the path are people who have collaborated. We will consider two particular collaboration graphs here. First, in the academic collaboration graph of people who have written papers in mathematics, the **Erdős number** of a person m (defined in terms of relations in Supplementary Exercise 10 in Chapter 9) is the length of the shortest path between m and the extremely prolific mathematician Paul Erdős (who died in 1996). That is, the Erdős number of a mathematician is the length of the shortest chain of mathematicians that begins with Paul Erdős and ends with this mathematician, where each adjacent pair of mathematicians have written a joint paper. The number of mathematicians with each Erdős number as of early 2006, according to the Erdős Number Project, is shown in Table 1.

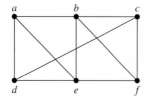

FIGURE 1   A Simple Graph.

Erdős Number	Number of People
0	1
1	504
2	6,593
3	33,605
4	83,642
5	87,760
6	40,014
7	11,591
8	3,146
9	819
10	244
11	68
12	23
13	5

TABLE 1 The Number of Mathematicians with a Given Erdős Number (as of early 2006).

Bacon Number	Number of People
0	1
1	2,367
2	242,407
3	785,389
4	200,602
5	14,048
6	1,277
7	114
8	16

TABLE 2 The Number of Actors with a Given Bacon Number (as of early 2011).

Links

Replace Kevin Bacon by your own favorite actor to invent a new party game

In the Hollywood graph (see Example 3 in Section 10.1) two actors a and b are linked when there is a chain of actors linking a and b, where every two actors adjacent in the chain have acted in the same movie. In the Hollywood graph, the **Bacon number** of an actor c is defined to be the length of the shortest path connecting c and the well-known actor Kevin Bacon. As new movies are made, including new ones with Kevin Bacon, the Bacon number of actors can change. In Table 2 we show the number of actors with each Bacon number as of early 2011 using data from the Oracle of Bacon website. The origins of the Bacon number of an actor dates back to the early 1990s, when Kevin Bacon remarked that he had worked with everyone in Hollywood or someone who worked with them. This lead some people to invent a party game where participants where challenged to find a sequence of movies leading from each actor named to Kevin Bacon. We can find a number similar to a Bacon number using any actor as the center of the acting universe. ◀

Connectedness in Undirected Graphs

When does a computer network have the property that every pair of computers can share information, if messages can be sent through one or more intermediate computers? When a graph is used to represent this computer network, where vertices represent the computers and edges represent the communication links, this question becomes: When is there always a path between two vertices in the graph?

DEFINITION 3   An undirected graph is called *connected* if there is a path between every pair of distinct vertices of the graph. An undirected graph that is not *connected* is called *disconnected*. We say that we *disconnect* a graph when we remove vertices or edges, or both, to produce a disconnected subgraph.

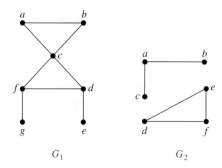

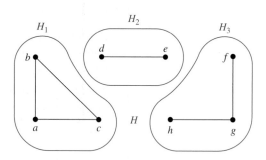

FIGURE 2   The Graphs G_1 and G_2.

FIGURE 3   The Graph H and Its Connected Components H_1, H_2, and H_3.

Thus, any two computers in the network can communicate if and only if the graph of this network is connected.

EXAMPLE 4   The graph G_1 in Figure 2 is connected, because for every pair of distinct vertices there is a path between them (the reader should verify this). However, the graph G_2 in Figure 2 is not connected. For instance, there is no path in G_2 between vertices a and d.   ◀

We will need the following theorem in Chapter 11.

THEOREM 1   There is a simple path between every pair of distinct vertices of a connected undirected graph.

Proof: Let u and v be two distinct vertices of the connected undirected graph $G = (V, E)$. Because G is connected, there is at least one path between u and v. Let $x_0, x_1, \ldots, x_n$, where $x_0 = u$ and $x_n = v$, be the vertex sequence of a path of least length. This path of least length is simple. To see this, suppose it is not simple. Then $x_i = x_j$ for some i and j with $0 \le i < j$. This means that there is a path from u to v of shorter length with vertex sequence $x_0, x_1, \ldots, x_{i-1}, x_j, \ldots, x_n$ obtained by deleting the edges corresponding to the vertex sequence $x_i, \ldots, x_{j-1}$.   ◁

CONNECTED COMPONENTS   A **connected component** of a graph G is a connected subgraph of G that is not a proper subgraph of another connected subgraph of G. That is, a connected component of a graph G is a maximal connected subgraph of G. A graph G that is not connected has two or more connected components that are disjoint and have G as their union.

EXAMPLE 5   What are the connected components of the graph H shown in Figure 3?

Solution: The graph H is the union of three disjoint connected subgraphs H_1, H_2, and H_3, shown in Figure 3. These three subgraphs are the connected components of H.   ◀

EXAMPLE 6   **Connected Components of Call Graphs**   Two vertices x and y are in the same component of a telephone call graph (see Example 4 in Section 10.1) when there is a sequence of telephone calls beginning at x and ending at y. When a call graph for telephone calls made during a particular day in the AT&T network was analyzed, this graph was found to have 53,767,087 vertices, more than 170 million edges, and more than 3.7 million connected components. Most of these components were small; approximately three-fourths consisted of two vertices representing pairs of telephone numbers that called only each other. This graph has one huge connected component with 44,989,297 vertices comprising more than 80% of the total. Furthermore, every vertex in this component can be linked to any other vertex by a chain of no more than 20 calls.   ◀

How Connected is a Graph?

Suppose that a graph represents a computer network. Knowing that this graph is connected tells us that any two computers on the network can communicate. However, we would also like to understand how reliable this network is. For instance, will it still be possible for all computers to communicate after a router or a communications link fails? To answer this and similar questions, we now develop some new concepts.

Sometimes the removal from a graph of a vertex and all incident edges produces a subgraph with more connected components. Such vertices are called **cut vertices** (or **articulation points**). The removal of a cut vertex from a connected graph produces a subgraph that is not connected. Analogously, an edge whose removal produces a graph with more connected components than in the original graph is called a **cut edge** or **bridge**. Note that in a graph representing a computer network, a cut vertex and a cut edge represent an essential router and an essential link that cannot fail for all computers to be able to communicate.

EXAMPLE 7   Find the cut vertices and cut edges in the graph G_1 shown in Figure 4.

Solution: The cut vertices of G_1 are b, c, and e. The removal of one of these vertices (and its adjacent edges) disconnects the graph. The cut edges are $\{a, b\}$ and $\{c, e\}$. Removing either one of these edges disconnects G_1.   ◀

VERTEX CONNECTIVITY   Not all graphs have cut vertices. For example, the complete graph K_n, where $n \geq 3$, has no cut vertices. When you remove a vertex from K_n and all edges

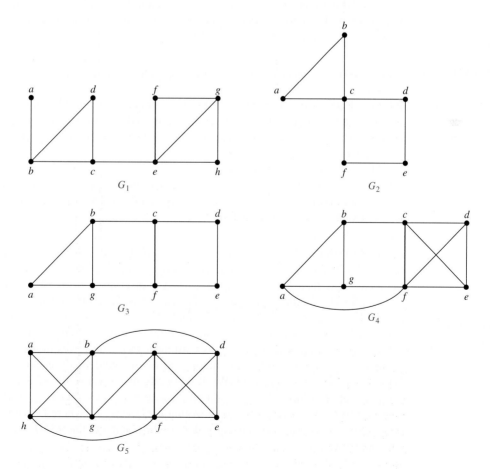

FIGURE 4   Some Connected Graphs

incident to it, the resulting subgraph is the complete graph K_{n-1}, a connected graph. Connected graphs without cut vertices are called **nonseparable graphs**, and can be thought of as more connected than those with a cut vertex. We can extend this notion by defining a more granulated measure of graph connectivity based on the minimum number of vertices that can be removed to disconnect a graph.

A subset V' of the vertex set V of $G = (V, E)$ is a **vertex cut**, or **separating set**, if $G - V'$ is disconnected. For instance, in the graph in Figure 1, the set $\{b, c, e\}$ is a vertex cut with three vertices, as the reader should verify. We leave it to the reader (Exercise 37) to show that every connected graph, except a complete graph, has a vertex cut. We define the **vertex connectivity** of a noncomplete graph G, denoted by $\kappa(G)$, as the minimum number of vertices in a vertex cut.

κ is the lowercase Greek letter kappa.

When G is a complete graph, it has no vertex cuts, because removing any subset of its vertices and all incident edges still leaves a complete graph. Consequently, we cannot define $\kappa(G)$ as the minimum number of vertices in a vertex cut when G is complete. Instead, we set $\kappa(K_n) = n - 1$, the number of vertices needed to be removed to produce a graph with a single vertex.

Consequently, for every graph G, $\kappa(G)$ is minimum number of vertices that can be removed from G to either disconnect G or produce a graph with a single vertex. We have $0 \leq \kappa(G) \leq n - 1$ if G has n vertices, $\kappa(G) = 0$ if and only if G is disconnected or $G = K_1$, and $\kappa(G) = n - 1$ if and only if G is complete [see Exercise 38(a)].

The larger $\kappa(G)$ is, the more connected we consider G to be. Disconnected graphs and K_1 have $\kappa(G) = 0$, connected graphs with cut vertices and K_2 have $\kappa(G) = 1$, graphs without cut vertices that can be disconnected by removing two vertices and K_3 have $\kappa(G) = 2$, and so on. We say that a graph is k-**connected** (or k-**vertex-connected**), if $\kappa(G) \geq k$. A graph G is 1-connected if it is connected and not a graph containing a single vertex; a graph is 2-connected, or **biconnected**, if it is nonseparable and has at least three vertices. Note that if G is a k-connected graph, then G is a j-connected graph for all j with $0 \leq j \leq k$.

EXAMPLE 8   Find the vertex connectivity for each of the graphs in Figure 4.

Solution: Each of the five graphs in Figure 4 is connected and has more than vertex, so each of these graphs has positive vertex connectivity. Because G_1 is a connected graph with a cut vertex, as shown in Example 7, we know that $\kappa(G_1) = 1$. Similarly, $\kappa(G_2) = 1$, because c is a cut vertex of G_2.

The reader should verify that G_3 has no cut vertices. but that $\{b, g\}$ is a vertex cut. Hence, $\kappa(G_3) = 2$. Similarly, because G_4 has a vertex cut of size two, $\{c, f\}$, but no cut vertices. It follows that $\kappa(G_4) = 2$. The reader can verify that G_5 has no vertex cut of size two, but $\{b, c, f\}$ is a vertex cut of G_5. Hence, $\kappa(G_5) = 3$. ◄

EDGE CONNECTIVITY   We can also measure the connectivity of a connected graph $G = (V, E)$ in terms of the minimum number of edges that we can remove to disconnect it. If a graph has a cut edge, then we need only remove it to disconnect G. If G does not have a cut edge, we look for the smallest set of edges that can be removed to disconnect it. A set of edges E' is called an **edge cut** of G if the subgraph $G - E'$ is disconnected. The **edge connectivity** of a graph G, denoted by $\lambda(G)$, is the minimum number of edges in an edge cut of G. This defines $\lambda(G)$ for all connected graphs with more than one vertex because it is always possible to disconnect such a graph by removing all edges incident to one of its vertices. Note that $\lambda(G) = 0$ if G is not connected. We also specify that $\lambda(G) = 0$ if G is a graph consisting of a single vertex. It follows that if G is a graph with n vertices, then $0 \leq \lambda(G) \leq n - 1$. We leave it to the reader [Exercise 38(b)] to show that $\lambda(G) = n - 1$ where G is a graph with n vertices if and only if $G = K_n$, which is equivalent to the statement that $\lambda(G) \leq n - 2$ when G is not a complete graph.

λ is the lowercase Greek letter lambda.

EXAMPLE 9   Find the edge connectivity of each of the graphs in Figure 4.

Solution: Each of the five graphs in Figure 4 is connected and has more than one vertex, so we know that all of them have positive edge connectivity. As we saw in Example 7, G_1 has a cut edge, so $\lambda(G_1) = 1$.

The graph G_2 has no cut edges, as the reader should verify, but the removal of the two edges $\{a, b\}$ and $\{a, c\}$ disconnects it. Hence, $\lambda(G_2) = 2$. Similarly, $\lambda(G_3) = 2$, because G_3 has no cut edges, but the removal of the two edges $\{b, c\}$ and $\{f, g\}$ disconnects it.

The reader should verify that the removal of no two edges disconnects G_4, but the removal of the three edges $\{b, c\}$, $\{a, f\}$, and $\{f, g\}$ disconnects it. Hence, $\lambda(G_4) = 3$. Finally, the reader should verify that $\lambda(G_5) = 3$, because the removal of any two of its edges does not disconnect it, but the removal of $\{a, b\}$, $\{a, g\}$, and $\{a, h\}$ does. ◀

AN INEQUALITY FOR VERTEX CONNECTIVITY AND EDGE CONNECTIVITY When $G = (V, E)$ is a noncomplete connected graph with at least three vertices, the minimum degree of a vertex of G is an upper bound for both the vertex connectivity of G and the edge connectivity of G. That is, $\kappa(G) \leq \min_{v \in V} \deg(v)$ and $\lambda(G) \leq \min_{v \in V} \deg(v)$. To see this, observe that deleting all the neighbors of a fixed vertex of minimum degree disconnects G, and deleting all the edges that have a fixed vertex of minimum degree as an endpoint disconnects G.

In Exercise 55, we ask the reader to show that $\kappa(G) \leq \lambda(G)$ when G is a connected non-complete graph. Note also that $\kappa(K_n) = \lambda(K_n) = \min_{v \in V} \deg(v) = n - 1$ when n is a positive integer and that $\kappa(G) = \lambda(G) = 0$ when G is a disconnected graph. Putting these facts together, establishes that for all graphs G,

$$\kappa(G) \leq \lambda(G) \leq \min_{v \in V} \deg(v).$$

APPLICATIONS OF VERTEX AND EDGE CONNECTIVITY   Graph connectivity plays an important role in many problems involving the reliability of networks. For instance, as we mentioned in our introduction of cut vertices and cut edges, we can model a data network using vertices to represent routers and edges to represent links between them. The vertex connectivity of the resulting graph equals the minimum number of routers that disconnect the network when they are out of service. If fewer routers are down, data transmission between every pair of routers is still possible. The edge connectivity represents the minimum number of fiber optic links that can be down to disconnect the network. If fewer links are down, it will still be possible for data to be transmitted between every pair of routers.

We can model a highway network, using vertices to represent highway intersections and edges to represent sections of roads running between intersections. The vertex connectivity of the resulting graph represents the minimum number of intersections that can be closed at a particular time that makes it impossible to travel between every two intersections. If fewer intersections are closed, travel between every pair of intersections is still possible. The edge connectivity represents the minimum number of roads that can be closed to disconnect the highway network. If fewer highways are closed, it will still be possible to travel between any two intersections. Clearly, it would be useful for the highway department to take this information into account when planning road repairs.

Connectedness in Directed Graphs

There are two notions of connectedness in directed graphs, depending on whether the directions of the edges are considered.

DEFINITION 4   A directed graph is *strongly connected* if there is a path from a to b and from b to a whenever a and b are vertices in the graph.

For a directed graph to be strongly connected there must be a sequence of directed edges from any vertex in the graph to any other vertex. A directed graph can fail to be strongly connected but still be in "one piece." Definition 5 makes this notion precise.

DEFINITION 5

A directed graph is *weakly connected* if there is a path between every two vertices in the underlying undirected graph.

That is, a directed graph is weakly connected if and only if there is always a path between two vertices when the directions of the edges are disregarded. Clearly, any strongly connected directed graph is also weakly connected.

EXAMPLE 10 Are the directed graphs G and H shown in Figure 5 strongly connected? Are they weakly connected?

Solution: G is strongly connected because there is a path between any two vertices in this directed graph (the reader should verify this). Hence, G is also weakly connected. The graph H is not strongly connected. There is no directed path from a to b in this graph. However, H is weakly connected, because there is a path between any two vertices in the underlying undirected graph of H (the reader should verify this). ◀

STRONG COMPONENTS OF A DIRECTED GRAPH   The subgraphs of a directed graph G that are strongly connected but not contained in larger strongly connected subgraphs, that is, the maximal strongly connected subgraphs, are called the **strongly connected components** or **strong components** of G. Note that if a and b are two vertices in a directed graph, their strong components are either the same or disjoint. (We leave the proof of this last fact as Exercise 11.)

EXAMPLE 11 The graph H in Figure 5 has three strongly connected components, consisting of the vertex a; the vertex e; and the subgraph consisting of the vertices b, c, and d and edges (b, c), (c, d), and (d, b). ◀

EXAMPLE 12 **The Strongly Connected Components of the Web Graph**   The Web graph introduced in Example 5 of Section 10.1 represents Web pages with vertices and links with directed edges. A snapshot of the Web in 1999 produced a Web graph with over 200 million vertices and over 1.5 billion edges (numbers that have now grown considerably). (See [Br00] for details.)

Links

In 2010 the Web graph was estimated to have at least 55 billion vertices and one trillion edges. This implies that more than 40 TB of computer memory would have been needed to represent its adjacency matrix.

The underlying undirected graph of this Web graph is not connected, but it has a connected component that includes approximately 90% of the vertices in the graph. The subgraph of the original directed graph corresponding to this connected component of the underlying undirected graph (that is, with the same vertices and all directed edges connecting vertices in this graph) has one very large strongly connected component and many small ones. The former is called the **giant strongly connected component (GSCC)** of the directed graph. A Web page in this component can be reached following links starting at any other page in this component. The

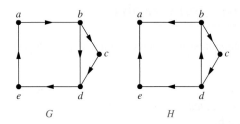

FIGURE 5   **The Directed Graphs G and H.**

GSCC in the Web graph produced by this study was found to have over 53 million vertices. The remaining vertices in the large connected component of the undirected graph represent three different types of Web pages: pages that can be reached from a page in the GSCC, but do not link back to these pages following a series of links; pages that link back to pages in the GSCC following a series of links, but cannot be reached by following links on pages in the GSCC; and pages that cannot reach pages in the GSCC and cannot be reached from pages in the GSCC following a series of links. In this study, each of these three other sets was found to have approximately 44 million vertices. (It is rather surprising that these three sets are close to the same size.) ◀

Paths and Isomorphism

There are several ways that paths and circuits can help determine whether two graphs are isomorphic. For example, the existence of a simple circuit of a particular length is a useful invariant that can be used to show that two graphs are not isomorphic. In addition, paths can be used to construct mappings that may be isomorphisms.

As we mentioned, a useful isomorphic invariant for simple graphs is the existence of a simple circuit of length k, where k is a positive integer greater than 2. (The proof that this is an invariant is left as Exercise 46.) Example 13 illustrates how this invariant can be used to show that two graphs are not isomorphic.

EXAMPLE 13   Determine whether the graphs G and H shown in Figure 6 are isomorphic.

Solution: Both G and H have six vertices and eight edges. Each has four vertices of degree three, and two vertices of degree two. So, the three invariants—number of vertices, number of edges, and degrees of vertices—all agree for the two graphs. However, H has a simple circuit of length three, namely, v_1, v_2, v_6, v_1, whereas G has no simple circuit of length three, as can be determined by inspection (all simple circuits in G have length at least four). Because the existence of a simple circuit of length three is an isomorphic invariant, G and H are not isomorphic. ◀

We have shown how the existence of a type of path, namely, a simple circuit of a particular length, can be used to show that two graphs are not isomorphic. We can also use paths to find mappings that are potential isomorphisms.

EXAMPLE 14   Determine whether the graphs G and H shown in Figure 7 are isomorphic.

Solution: Both G and H have five vertices and six edges, both have two vertices of degree three and three vertices of degree two, and both have a simple circuit of length three, a simple circuit

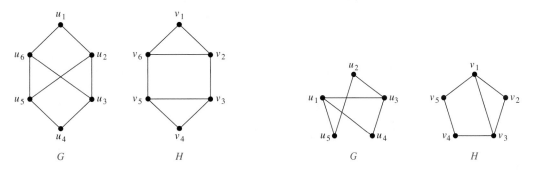

FIGURE 6   **The Graphs G and H.**   **FIGURE 7**   **The Graphs G and H.**

of length four, and a simple circuit of length five. Because all these isomorphic invariants agree, G and H may be isomorphic.

To find a possible isomorphism, we can follow paths that go through all vertices so that the corresponding vertices in the two graphs have the same degree. For example, the paths u_1, u_4, u_3, u_2, u_5 in G and v_3, v_2, v_1, v_5, v_4 in H both go through every vertex in the graph; start at a vertex of degree three; go through vertices of degrees two, three, and two, respectively; and end at a vertex of degree two. By following these paths through the graphs, we define the mapping f with $f(u_1) = v_3$, $f(u_4) = v_2$, $f(u_3) = v_1$, $f(u_2) = v_5$, and $f(u_5) = v_4$. The reader can show that f is an isomorphism, so G and H are isomorphic, either by showing that f preserves edges or by showing that with the appropriate orderings of vertices the adjacency matrices of G and H are the same. ◄

Counting Paths Between Vertices

The number of paths between two vertices in a graph can be determined using its adjacency matrix.

THEOREM 2   Let G be a graph with adjacency matrix $\mathbf{A}$ with respect to the ordering $v_1, v_2, \ldots, v_n$ of the vertices of the graph (with directed or undirected edges, with multiple edges and loops allowed). The number of different paths of length r from v_i to v_j, where r is a positive integer, equals the (i, j)th entry of $\mathbf{A}^r$.

Proof: The theorem will be proved using mathematical induction. Let G be a graph with adjacency matrix $\mathbf{A}$ (assuming an ordering $v_1, v_2, \ldots, v_n$ of the vertices of G). The number of paths from v_i to v_j of length 1 is the (i, j)th entry of $\mathbf{A}$, because this entry is the number of edges from v_i to v_j.

Assume that the (i, j)th entry of $\mathbf{A}^r$ is the number of different paths of length r from v_i to v_j. This is the inductive hypothesis. Because $\mathbf{A}^{r+1} = \mathbf{A}^r \mathbf{A}$, the (i, j)th entry of $\mathbf{A}^{r+1}$ equals

$$b_{i1}a_{1j} + b_{i2}a_{2j} + \cdots + b_{in}a_{nj},$$

where b_{ik} is the (i, k)th entry of $\mathbf{A}^r$. By the inductive hypothesis, b_{ik} is the number of paths of length r from v_i to v_k.

A path of length $r + 1$ from v_i to v_j is made up of a path of length r from v_i to some intermediate vertex v_k, and an edge from v_k to v_j. By the product rule for counting, the number of such paths is the product of the number of paths of length r from v_i to v_k, namely, b_{ik}, and the number of edges from v_k to v_j, namely, a_{kj}. When these products are added for all possible intermediate vertices v_k, the desired result follows by the sum rule for counting. ◄

EXAMPLE 15   How many paths of length four are there from a to d in the simple graph G in Figure 8?

Solution: The adjacency matrix of G (ordering the vertices as a, b, c, d) is

$$\mathbf{A} = \begin{bmatrix} 0 & 1 & 1 & 0 \\ 1 & 0 & 0 & 1 \\ 1 & 0 & 0 & 1 \\ 0 & 1 & 1 & 0 \end{bmatrix}.$$

FIGURE 8   **The Graph G.**

Hence, the number of paths of length four from a to d is the $(1, 4)$th entry of $\mathbf{A}^4$. Because

$$\mathbf{A}^4 = \begin{bmatrix} 8 & 0 & 0 & 8 \\ 0 & 8 & 8 & 0 \\ 0 & 8 & 8 & 0 \\ 8 & 0 & 0 & 8 \end{bmatrix},$$

Extra
Examples

there are exactly eight paths of length four from a to d. By inspection of the graph, we see that a, b, a, b, d; a, b, a, c, d; a, b, d, b, d; a, b, d, c, d; a, c, a, b, d; a, c, a, c, d; a, c, d, b, d; and a, c, d, c, d are the eight paths of length four from a to d.   ◀

Theorem 2 can be used to find the length of the shortest path between two vertices of a graph (see Exercise 42), and it can also be used to determine whether a graph is connected (see Exercises 47 and 48).

Exercises

1. Does each of these lists of vertices form a path in the following graph? Which paths are simple? Which are circuits? What are the lengths of those that are paths?

a) a, e, b, c, b **b)** a, e, a, d, b, c, a
c) e, b, a, d, b, e **d)** c, b, d, a, e, c

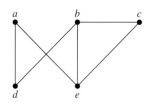

In Exercises 2–3 determine whether the given graph is connected.

2.

3.
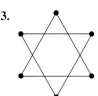

4. What do the connected components of acquaintanceship graphs represent?

5. Explain why in the collaboration graph of mathematicians (see Example 3 in Section 10.1) a vertex representing a mathematician is in the same connected component as the vertex representing Paul Erdős if and only if that mathematician has a finite Erdős number.

6. In the Hollywood graph (see Example 3 in Section 10.1), when is the vertex representing an actor in the same connected component as the vertex representing Kevin Bacon?

7. Determine whether each of these graphs is strongly connected and if not, whether it is weakly connected.

a)
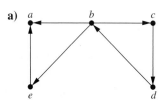

b)
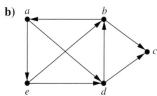

c)
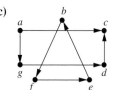

8. Find the strongly connected components of each of these graphs.

a)
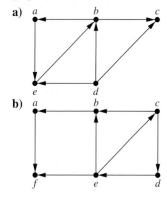

b)

c)

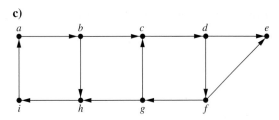

9. Find the strongly connected components of each of these graphs.

a)

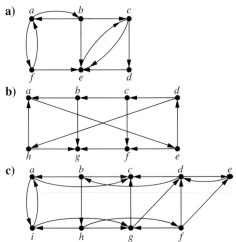

b)

c)

Suppose that $G = (V, E)$ is a directed graph. A vertex $w \in V$ is **reachable** from a vertex $v \in V$ if there is a directed path from v to w. The vertices v and w are **mutually reachable** if there are both a directed path from v to w and a directed path from w to v in G.

10. Show that if $G = (V, E)$ is a directed graph and u, v, and w are vertices in V for which u and v are mutually reachable and v and w are mutually reachable, then u and w are mutually reachable.

11. Show that if $G = (V, E)$ is a directed graph, then the strong components of two vertices u and v of V are either the same or disjoint. [*Hint:* Use Exercise 10.]

12. Find the number of paths of length n between two different vertices in K_4 if n is

a) 2.　　**b)** 3.　　**c)** 4.　　**d)** 5.

13. Use paths either to show that these graphs are not isomorphic or to find an isomorphism between them.

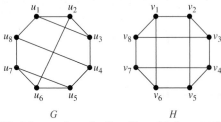

G　　　　　　H

14. Find the number of paths of length n between any two adjacent vertices in $K_{3,3}$ for the values of n in Exercise 12.

15. Find the number of paths of length n between any two nonadjacent vertices in $K_{3,3}$ for the values of n in Exercise 12.

16. Find the number of paths between c and d in the graph in Figure 1 of length

a) 2.　**b)** 3.　**c)** 4.　**d)** 5.　**e)** 6.　**f)** 7.

17. Find the number of paths from a to c in the directed graph in Exercise 7(b) of length

a) 2.　**b)** 3.　**c)** 4.　**d)** 5.　**e)** 6.　**f)** 7.

∗18. Show that every connected graph with n vertices has at least $n - 1$ edges.

19. Let $G = (V, E)$ be a simple graph. Let R be the relation on V consisting of pairs of vertices (u, v) such that there is a path from u to v or such that $u = v$. Show that R is an equivalence relation.

∗20. Show that in every simple graph there is a path from every vertex of odd degree to some other vertex of odd degree.

In Exercises 21–23 find all the cut vertices of the given graph.

21.　　　　　　　　　　**22.**

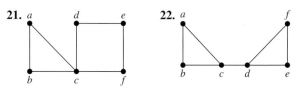

23.

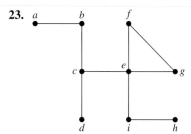

24. Find all the cut edges in the graphs in Exercises 21–23.

25. A communications link in a network should be provided with a backup link if its failure makes it impossible for some message to be sent. For each of the communications networks shown here in (a) and (b), determine those links that should be backed up.

a)

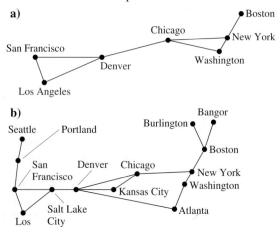

A **vertex basis** in a directed graph G is a minimal set B of vertices of G such that for each vertex v of G not in B there is a path to v from some vertex B.

26. Find a vertex basis for each of the directed graphs in Exercises 5–6 of Section 10.2.

27. What is the significance of a vertex basis in an influence graph (described in Example 2 of Section 10.1)? Find a vertex basis in the influence graph in that example.

28. Show that if a connected simple graph G is the union of the graphs G_1 and G_2, then G_1 and G_2 have at least one common vertex.

∗29. Show that if a simple graph G has k connected components and these components have $n_1, n_2, \ldots, n_k$ vertices, respectively, then the number of edges of G does not exceed

$$\sum_{i=1}^{k} C(n_i, 2).$$

∗30. Use Exercise 29 to show that a simple graph with n vertices and k connected components has at most $(n - k)(n - k + 1)/2$ edges. [*Hint:* First show that

$$\sum_{i=1}^{k} n_i^2 \le n^2 - (k - 1)(2n - k),$$

where n_i is the number of vertices in the ith connected component.]

∗31. Show that a simple graph G with n vertices is connected if it has more than $(n - 1)(n - 2)/2$ edges.

32. Describe the adjacency matrix of a graph with n connected components when the vertices of the graph are listed so that vertices in each connected component are listed successively.

33. How many nonisomorphic connected simple graphs are there with n vertices when n is

  a) 2?   **b)** 3?   **c)** 4?   **d)** 5?

34. Show that each of the following graphs has no cut vertices.

  a) C_n where $n \ge 3$
  b) W_n where $n \ge 3$
  c) $K_{m,n}$ where $m \ge 2$ and $n \ge 2$
  d) Q_n where $n \ge 2$

35. Show that each of the graphs in Exercise 34 has no cut edges.

36. For each of these graphs, find $\kappa(G)$, $\lambda(G)$, and $\min_{v \in V} \deg(v)$, and determine which of the two inequalities in $\kappa(G) \le \lambda(G) \le \min_{v \in V} \deg(v)$ are strict.

  a)
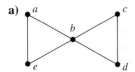

b)
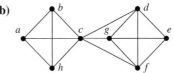

c) 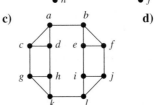   **d)**

37. Show that if G is a connected graph, then it is possible to remove vertices to disconnect G if and only if G is not a complete graph.

38. Show that if G is a connected graph with n vertices then
  a) $\kappa(G) = n - 1$ if and only if $G = K_n$.
  b) $\lambda(G) = n - 1$ if and only if $G = K_n$.

39. Find $\kappa(K_{m,n})$ and $\lambda(K_{m,n})$, where m and n are positive integers.

40. Construct a graph G with $\kappa(G) = 1$, $\lambda(G) = 2$, and $\min_{v \in V} \deg(v) = 3$.

∗41. Show that if G is a graph, then $\kappa(G) \le \lambda(G)$.

42. Explain how Theorem 2 can be used to find the length of the shortest path from a vertex v to a vertex w in a graph.

43. Use Theorem 2 to find the length of the shortest path between a and f in the graph in Figure 1.

44. Use Theorem 2 to find the length of the shortest path from a to c in the directed graph in Exercise 7(b).

45. Let P_1 and P_2 be two simple paths between the vertices u and v in the simple graph G that do not contain the same set of edges. Show that there is a simple circuit in G.

46. Show that the existence of a simple circuit of length k, where k is an integer greater than 2, is an invariant for graph isomorphism.

47. Explain how Theorem 2 can be used to determine whether a graph is connected.

48. Use Exercise 47 to show that the graph G_1 in Figure 2 is connected whereas the graph G_2 in that figure is not connected.

49. Show that a simple graph G is bipartite if and only if it has no circuits with an odd number of edges.

50. In an old puzzle attributed to Alcuin of York (735–804), a farmer needs to carry a wolf, a goat, and a cabbage across a river. The farmer only has a small boat, which can carry the farmer and only one object (an animal or a vegetable). He can cross the river repeatedly. However, if the farmer is on the other shore, the wolf will eat the goat, and, similarly, the goat will eat the cabbage. We can describe each state by listing what is on each shore. For example, we can use the pair (FG, WC) for the state where the farmer and goat are on the first shore and the wolf and cabbage are on the other shore. [The symbol $\emptyset$ is used when nothing is on a shore, so that $(FWGC, \emptyset)$ is the initial state.]

  a) Find all allowable states of the puzzle, where neither the wolf and the goat nor the goat and the cabbage are left on the same shore without the farmer.

b) Construct a graph such that each vertex of this graph represents an allowable state and the vertices representing two allowable states are connected by an edge if it is possible to move from one state to the other using one trip of the boat.

c) Explain why finding a path from the vertex representing ($FWGC$, Ø) to the vertex representing (Ø, $FWGC$) solves the puzzle.

d) Find two different solutions of the puzzle, each using seven crossings.

e) Suppose that the farmer must pay a toll of one dollar whenever he crosses the river with an animal. Which solution of the puzzle should the farmer use to pay the least total toll?

*51. Use a graph model and a path in your graph, as in Exercise 50, to solve the **jealous husbands problem**. Two married couples, each a husband and a wife, want to cross a river. They can only use a boat that can carry one or two people from one shore to the other shore. Each husband is extremely jealous and is not willing to leave his wife with the other husband, either in the boat or on shore. How can these four people reach the opposite shore?

52. Suppose that you have a three-gallon jug and a five-gallon jug. You may fill either jug with water, you may empty either jug, and you may transfer water from either jug into the other jug. Use a path in a directed graph to show that you can end up with a jug containing exactly one gallon. [*Hint:* Use an ordered pair (a, b) to indicate how much water is in each jug. Represent these ordered pairs by vertices. Add an edge for each allowable operation with the jugs.]

10.5   Euler and Hamilton Paths

Introduction

Can we travel along the edges of a graph starting at a vertex and returning to it by traversing each edge of the graph exactly once? Similarly, can we travel along the edges of a graph starting at a vertex and returning to it while visiting each vertex of the graph exactly once? Although these questions seem to be similar, the first question, which asks whether a graph has an *Euler circuit*, can be easily answered simply by examining the degrees of the vertices of the graph, while the second question, which asks whether a graph has a *Hamilton circuit*, is quite difficult to solve for most graphs. In this section we will study these questions and discuss the difficulty of solving them. Although both questions have many practical applications in many different areas, both arose in old puzzles. We will learn about these old puzzles as well as modern practical applications.

Euler Paths and Circuits

Links

Only five bridges connect Kaliningrad today. Of these, just two remain from Euler's day.

The town of Königsberg, Prussia (now called Kaliningrad and part of the Russian republic), was divided into four sections by the branches of the Pregel River. These four sections included the two regions on the banks of the Pregel, Kneiphof Island, and the region between the two branches of the Pregel. In the eighteenth century seven bridges connected these regions. Figure 1 depicts these regions and bridges.

The townspeople took long walks through town on Sundays. They wondered whether it was possible to start at some location in the town, travel across all the bridges once without crossing any bridge twice, and return to the starting point.

The Swiss mathematician Leonhard Euler solved this problem. His solution, published in 1736, may be the first use of graph theory. (For a translation of Euler's original paper see [BiLlWi99].) Euler studied this problem using the multigraph obtained when the four regions are represented by vertices and the bridges by edges. This multigraph is shown in Figure 2.

The problem of traveling across every bridge without crossing any bridge more than once can be rephrased in terms of this model. The question becomes: Is there a simple circuit in this multigraph that contains every edge?

DEFINITION 1   An *Euler circuit* in a graph G is a simple circuit containing every edge of G. An *Euler path* in G is a simple path containing every edge of G.

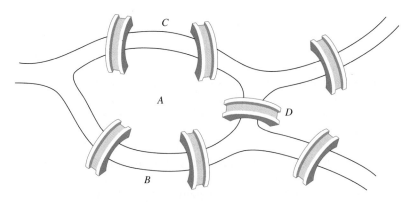

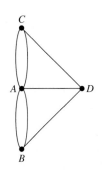

FIGURE 1 The Seven Bridges of Königsberg.

FIGURE 2 Multigraph Model of the Town of Königsberg.

Examples 1 and 2 illustrate the concept of Euler circuits and paths.

EXAMPLE 1 Which of the undirected graphs in Figure 3 have an Euler circuit? Of those that do not, which have an Euler path?

Solution: The graph G_1 has an Euler circuit, for example, a, e, c, d, e, b, a. Neither of the graphs G_2 or G_3 has an Euler circuit (the reader should verify this). However, G_3 has an Euler path, namely, a, c, d, e, b, d, a, b. G_2 does not have an Euler path (as the reader should verify). ◀

EXAMPLE 2 Which of the directed graphs in Figure 4 have an Euler circuit? Of those that do not, which have an Euler path?

Extra Examples

Solution: The graph H_2 has an Euler circuit, for example, $a, g, c, b, g, e, d, f, a$. Neither H_1 nor H_3 has an Euler circuit (as the reader should verify). H_3 has an Euler path, namely, c, a, b, c, d, b, but H_1 does not (as the reader should verify). ◀

NECESSARY AND SUFFICIENT CONDITIONS FOR EULER CIRCUITS AND PATHS
There are simple criteria for determining whether a multigraph has an Euler circuit or an Euler path. Euler discovered them when he solved the famous Königsberg bridge problem. We will assume that all graphs discussed in this section have a finite number of vertices and edges.

What can we say if a connected multigraph has an Euler circuit? What we can show is that every vertex must have even degree. To do this, first note that an Euler circuit begins with a vertex a and continues with an edge incident with a, say $\{a, b\}$. The edge $\{a, b\}$ contributes one to deg(a). Each time the circuit passes through a vertex it contributes two to the vertex's degree, because the circuit enters via an edge incident with this vertex and leaves via another such edge. Finally, the circuit terminates where it started, contributing one to deg(a). Therefore, deg(a)

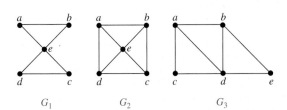

FIGURE 3 The Undirected Graphs G_1, G_2, and G_3.

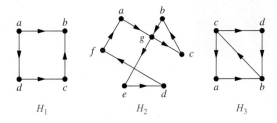

FIGURE 4 The Directed Graphs H_1, H_2, and H_3.

must be even, because the circuit contributes one when it begins, one when it ends, and two every time it passes through a (if it ever does). A vertex other than a has even degree because the circuit contributes two to its degree each time it passes through the vertex. We conclude that if a connected graph has an Euler circuit, then every vertex must have even degree.

Is this necessary condition for the existence of an Euler circuit also sufficient? That is, must an Euler circuit exist in a connected multigraph if all vertices have even degree? This question can be settled affirmatively with a construction.

Suppose that G is a connected multigraph with at least two vertices and the degree of every vertex of G is even. We will form a simple circuit that begins at an arbitrary vertex a of G, building it edge by edge. Let $x_0 = a$. First, we arbitrarily choose an edge $\{x_0, x_1\}$ incident with a which is possible because G is connected. We continue by building a simple path $\{x_0, x_1\}, \{x_1, x_2\}, \ldots, \{x_{n-1}, x_n\}$, successively adding edges one by one to the path until we cannot add another edge to the path. This happens when we reach a vertex for which we have already included all edges incident with that vertex in the path. For instance, in the graph G in Figure 5 we begin at a and choose in succession the edges $\{a, f\}$, $\{f, c\}$, $\{c, b\}$, and $\{b, a\}$.

The path we have constructed must terminate because the graph has a finite number of edges, so we are guaranteed to eventually reach a vertex for which no edges are available to add to the path. The path begins at a with an edge of the form $\{a, x\}$, and we now show that it must terminate at a with an edge of the form $\{y, a\}$. To see that the path must terminate at a, note that each time the path goes through a vertex with even degree, it uses only one edge to enter this vertex, so because the degree must be at least two, at least one edge remains for the path to leave the vertex. Furthermore, every time we enter and leave a vertex of even degree, there are an even number of edges incident with this vertex that we have not yet used in our path. Consequently, as we form the path, every time we enter a vertex other than a, we can leave it. This means that the path can end only at a. Next, note that the path we have constructed may use all the edges of the graph, or it may not if we have returned to a for the last time before using all the edges.

An Euler circuit has been constructed if all the edges have been used. Otherwise, consider the subgraph H obtained from G by deleting the edges already used and vertices that are not incident with any remaining edges. When we delete the circuit a, f, c, b, a from the graph in Figure 5, we obtain the subgraph labeled as H.

Because G is connected, H has at least one vertex in common with the circuit that has been deleted. Let w be such a vertex. (In our example, c is the vertex.)

Every vertex in H has even degree (because in G all vertices had even degree, and for each vertex, pairs of edges incident with this vertex have been deleted to form H). Note that H may not be connected. Beginning at w, construct a simple path in H by choosing edges as long as possible, as was done in G. This path must terminate at w. For instance, in Figure 5, c, d, e, c is a path in H. Next, form a circuit in G by splicing the circuit in H with the original circuit in G

LEONHARD EULER (1707–1783)   Leonhard Euler was the son of a Calvinist minister from the vicinity of Basel, Switzerland. At 13 he entered the University of Basel, pursuing a career in theology, as his father wished. At the university Euler was tutored by Johann Bernoulli of the famous Bernoulli family of mathematicians. His interest and skills led him to abandon his theological studies and take up mathematics. Euler obtained his master's degree in philosophy at the age of 16. In 1727 Peter the Great invited him to join the Academy at St. Petersburg. In 1741 he moved to the Berlin Academy, where he stayed until 1766. He then returned to St. Petersburg, where he remained for the rest of his life.

Euler was incredibly prolific, contributing to many areas of mathematics, including number theory, combinatorics, and analysis, as well as its applications to such areas as music and naval architecture. He wrote over 1100 books and papers and left so much unpublished work that it took 47 years after he died for all his work to be published. During his life his papers accumulated so quickly that he kept a large pile of articles awaiting publication. The Berlin Academy published the papers on top of this pile so later results were often published before results they depended on or superseded. Euler had 13 children and was able to continue his work while a child or two bounced on his knees. He was blind for the last 17 years of his life, but because of his fantastic memory this did not diminish his mathematical output. The project of publishing his collected works, undertaken by the Swiss Society of Natural Science, is ongoing and will require more than 75 volumes.

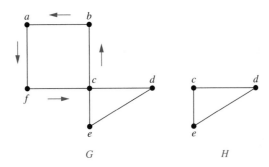

FIGURE 5   Constructing an Euler Circuit in G.

(this can be done because w is one of the vertices in this circuit). When this is done in the graph in Figure 5, we obtain the circuit a, f, c, d, e, c, b, a.

Continue this process until all edges have been used. (The process must terminate because there are only a finite number of edges in the graph.) This produces an Euler circuit. The construction shows that if the vertices of a connected multigraph all have even degree, then the graph has an Euler circuit.

We summarize these results in Theorem 1.

THEOREM 1   A connected multigraph with at least two vertices has an Euler circuit if and only if each of its vertices has even degree.

We can now solve the Königsberg bridge problem. Because the multigraph representing these bridges, shown in Figure 2, has four vertices of odd degree, it does not have an Euler circuit. There is no way to start at a given point, cross each bridge exactly once, and return to the starting point.

Algorithm 1 gives the constructive procedure for finding Euler circuits given in the discussion preceding Theorem 1. (Because the circuits in the procedure are chosen arbitrarily, there is some ambiguity. We will not bother to remove this ambiguity by specifying the steps of the procedure more precisely.)

ALGORITHM 1 Constructing Euler Circuits.

procedure *Euler*(*G*: connected multigraph with all vertices of
     even degree)
circuit := a circuit in *G* beginning at an arbitrarily chosen
     vertex with edges successively added to form a path that
     returns to this vertex
H := *G* with the edges of this circuit removed
while *H* has edges
     subcircuit := a circuit in *H* beginning at a vertex in *H* that
        also is an endpoint of an edge of *circuit*
     H := *H* with edges of *subcircuit* and all isolated vertices
       removed
     circuit := *circuit* with *subcircuit* inserted at the appropriate
       vertex
return *circuit* {*circuit* is an Euler circuit}

Algorithm 1 provides an efficient algorithm for finding Euler circuits in a connected multigraph G with all vertices of even degree. We leave it to the reader (Exercise 48) to show that the worst case complexity of this algorithm is $O(m)$, where m is the number of edges of G.

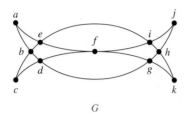

FIGURE 6   Mohammed's Scimitars.

Example 3 shows how Euler paths and circuits can be used to solve a type of puzzle.

EXAMPLE 3   Many puzzles ask you to draw a picture in a continuous motion without lifting a pencil so that no part of the picture is retraced. We can solve such puzzles using Euler circuits and paths. For example, can *Mohammed's scimitars*, shown in Figure 6, be drawn in this way, where the drawing begins and ends at the same point?

Solution: We can solve this problem because the graph G shown in Figure 6 has an Euler circuit. It has such a circuit because all its vertices have even degree. We will use Algorithm 1 to construct an Euler circuit. First, we form the circuit $a, b, d, c, b, e, i, f, e, a$. We obtain the subgraph H by deleting the edges in this circuit and all vertices that become isolated when these edges are removed. Then we form the circuit $d, g, h, j, i, h, k, g, f, d$ in H. After forming this circuit we have used all edges in G. Splicing this new circuit into the first circuit at the appropriate place produces the Euler circuit $a, b, d, g, h, j, i, h, k, g, f, d, c, b, e, i, f, e, a$. This circuit gives a way to draw the scimitars without lifting the pencil or retracing part of the picture.   ◀

Another algorithm for constructing Euler circuits, called Fleury's algorithm, is described in the premble to Exercise 32.

We will now show that a connected multigraph has an Euler path (and not an Euler circuit) if and only if it has exactly two vertices of odd degree. First, suppose that a connected multigraph does have an Euler path from a to b, but not an Euler circuit. The first edge of the path contributes one to the degree of a. A contribution of two to the degree of a is made every time the path passes through a. The last edge in the path contributes one to the degree of b. Every time the path goes through b there is a contribution of two to its degree. Consequently, both a and b have odd degree. Every other vertex has even degree, because the path contributes two to the degree of a vertex whenever it passes through it.

Now consider the converse. Suppose that a graph has exactly two vertices of odd degree, say a and b. Consider the larger graph made up of the original graph with the addition of an edge $\{a, b\}$. Every vertex of this larger graph has even degree, so there is an Euler circuit. The removal of the new edge produces an Euler path in the original graph. Theorem 2 summarizes these results.

THEOREM 2   A connected multigraph has an Euler path but not an Euler circuit if and only if it has exactly two vertices of odd degree.

EXAMPLE 4   Which graphs shown in Figure 7 have an Euler path?

Solution: G_1 contains exactly two vertices of odd degree, namely, b and d. Hence, it has an Euler path that must have b and d as its endpoints. One such Euler path is d, a, b, c, d, b. Similarly, G_2 has exactly two vertices of odd degree, namely, b and d. So it has an Euler path that must have b and d as endpoints. One such Euler path is $b, a, g, f, e, d, c, g, b, c, f, d$. G_3 has no Euler path because it has six vertices of odd degree.   ◀

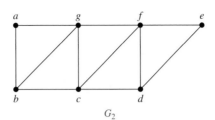

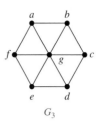

FIGURE 7 **Three Undirected Graphs.**

Returning to eighteenth-century Königsberg, is it possible to start at some point in the town, travel across all the bridges, and end up at some other point in town? This question can be answered by determining whether there is an Euler path in the multigraph representing the bridges in Königsberg. Because there are four vertices of odd degree in this multigraph, there is no Euler path, so such a trip is impossible.

Necessary and sufficient conditions for Euler paths and circuits in directed graphs are given in Exercises 10 and 11.

APPLICATIONS OF EULER PATHS AND CIRCUITS Euler paths and circuits can be used to solve many practical problems. For example, many applications ask for a path or circuit that traverses each street in a neighborhood, each road in a transportation network, each connection in a utility grid, or each link in a communications network exactly once. Finding an Euler path or circuit in the appropriate graph model can solve such problems. For example, if a postman can find an Euler path in the graph that represents the streets the postman needs to cover, this path produces a route that traverses each street of the route exactly once. If no Euler path exists, some streets will have to be traversed more than once. The problem of finding a circuit in a graph with the fewest edges that traverses every edge at least once is known as the *Chinese postman problem* in honor of Guan Meigu, who posed it in 1962. See [MiRo91] for more information on the solution of the Chinese postman problem when no Euler path exists.

Among the other areas where Euler circuits and paths are applied is in the layout of circuits, in network multicasting, and in molecular biology, where Euler paths are used in the sequencing of DNA.

Hamilton Paths and Circuits

We have developed necessary and sufficient conditions for the existence of paths and circuits that contain every edge of a multigraph exactly once. Can we do the same for simple paths and circuits that contain every vertex of the graph exactly once?

DEFINITION 2

A simple path in a graph G that passes through every vertex exactly once is called a *Hamilton path*, and a simple circuit in a graph G that passes through every vertex exactly once is called a *Hamilton circuit*. That is, the simple path $x_0, x_1, \ldots, x_{n-1}, x_n$ in the graph $G = (V, E)$ is a Hamilton path if $V = \{x_0, x_1, \ldots, x_{n-1}, x_n\}$ and $x_i \neq x_j$ for $0 \leq i < j \leq n$, and the simple circuit $x_0, x_1, \ldots, x_{n-1}, x_n, x_0$ (with $n > 0$) is a Hamilton circuit if $x_0, x_1, \ldots, x_{n-1}, x_n$ is a Hamilton path.

This terminology comes from a game, called the *Icosian puzzle*, invented in 1857 by the Irish mathematician Sir William Rowan Hamilton. It consisted of a wooden dodecahedron [a polyhedron with 12 regular pentagons as faces, as shown in Figure 8(a)], with a peg at each vertex of the dodecahedron, and string. The 20 vertices of the dodecahedron were labeled with different cities in the world. The object of the puzzle was to start at a city and travel along the

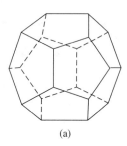 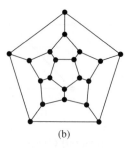 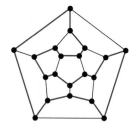

(a)   (b)

FIGURE 8   Hamilton's "A Voyage Round the World" Puzzle.

FIGURE 9   A Solution to the "A Voyage Round the World" Puzzle.

edges of the dodecahedron, visiting each of the other 19 cities exactly once, and end back at the first city. The circuit traveled was marked off using the string and pegs.

Because the author cannot supply each reader with a wooden solid with pegs and string, we will consider the equivalent question: Is there a circuit in the graph shown in Figure 8(b) that passes through each vertex exactly once? This solves the puzzle because this graph is isomorphic to the graph consisting of the vertices and edges of the dodecahedron. A solution of Hamilton's puzzle is shown in Figure 9.

EXAMPLE 5   Which of the simple graphs in Figure 10 have a Hamilton circuit or, if not, a Hamilton path?

Solution: G_1 has a Hamilton circuit: a, b, c, d, e, a. There is no Hamilton circuit in G_2 (this can be seen by noting that any circuit containing every vertex must contain the edge $\{a, b\}$ twice), but G_2 does have a Hamilton path, namely, a, b, c, d. G_3 has neither a Hamilton circuit nor a Hamilton path, because any path containing all vertices must contain one of the edges $\{a, b\}$, $\{e, f\}$, and $\{c, d\}$ more than once. ◄

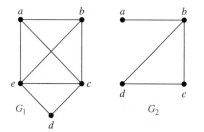 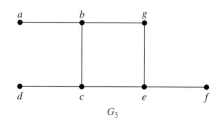

FIGURE 10   Three Simple Graphs.

CONDITIONS FOR THE EXISTENCE OF HAMILTON CIRCUITS   Is there a simple way to determine whether a graph has a Hamilton circuit or path? At first, it might seem that there should be an easy way to determine this, because there is a simple way to answer the similar question of whether a graph has an Euler circuit. Surprisingly, there are no known simple necessary and sufficient criteria for the existence of Hamilton circuits. However, many theorems are known that give sufficient conditions for the existence of Hamilton circuits. Also, certain properties can be used to show that a graph has no Hamilton circuit. For instance, a graph with a vertex of degree one cannot have a Hamilton circuit, because in a Hamilton circuit, each vertex is incident with two edges in the circuit. Moreover, if a vertex in the graph has degree two, then both edges that are incident with this vertex must be part of any Hamilton circuit. Also, note that when a Hamilton circuit is being constructed and this circuit has passed through a vertex, then all remaining edges incident with this vertex, other than the two used in the circuit, can be

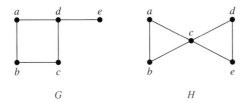

FIGURE 11 **Two Graphs That Do Not Have a Hamilton Circuit.**

removed from consideration. Furthermore, a Hamilton circuit cannot contain a smaller circuit within it.

EXAMPLE 6 Show that neither graph displayed in Figure 11 has a Hamilton circuit.

Solution: There is no Hamilton circuit in G because G has a vertex of degree one, namely, e. Now consider H. Because the degrees of the vertices a, b, d, and e are all two, every edge incident with these vertices must be part of any Hamilton circuit. It is now easy to see that no Hamilton circuit can exist in H, for any Hamilton circuit would have to contain four edges incident with c, which is impossible. ◀

EXAMPLE 7 Show that K_n has a Hamilton circuit whenever $n \geq 3$.

Solution: We can form a Hamilton circuit in K_n beginning at any vertex. Such a circuit can be built by visiting vertices in any order we choose, as long as the path begins and ends at the same vertex and visits each other vertex exactly once. This is possible because there are edges in K_n between any two vertices. ◀

Although no useful necessary and sufficient conditions for the existence of Hamilton circuits are known, quite a few sufficient conditions have been found. Note that the more edges a graph has, the more likely it is to have a Hamilton circuit. Furthermore, adding edges (but not vertices)

WILLIAM ROWAN HAMILTON (1805–1865)   William Rowan Hamilton, the most famous Irish scientist ever to have lived, was born in 1805 in Dublin. His father was a successful lawyer, his mother came from a family noted for their intelligence, and he was a child prodigy. By the age of 3 he was an excellent reader and had mastered advanced arithmetic. Because of his brilliance, he was sent off to live with his uncle James, a noted linguist. By age 8 Hamilton had learned Latin, Greek, and Hebrew; by 10 he had also learned Italian and French and he began his study of oriental languages, including Arabic, Sanskrit, and Persian. During this period he took pride in knowing as many languages as his age. At 17, no longer devoted to learning new languages and having mastered calculus and much mathematical astronomy, he began original work in optics, and he also found an important mistake in Laplace's work on celestial mechanics. Before entering Trinity College, Dublin, at 18, Hamilton had not attended school; rather, he received private tutoring. At Trinity, he was a superior student in both the sciences and the classics. Prior to receiving his degree, because of his brilliance he was appointed the Astronomer Royal of Ireland, beating out several famous astronomers for the post. He held this position until his death, living and working at Dunsink Observatory outside of Dublin. Hamilton made important contributions to optics, abstract algebra, and dynamics. Hamilton invented algebraic objects called quaternions as an example of a noncommutative system. He discovered the appropriate way to multiply quaternions while walking along a canal in Dublin. In his excitement, he carved the formula in the stone of a bridge crossing the canal, a spot marked today by a plaque. Later, Hamilton remained obsessed with quaternions, working to apply them to other areas of mathematics, instead of moving to new areas of research.

In 1857 Hamilton invented "The Icosian Game" based on his work in noncommutative algebra. He sold the idea for 25 pounds to a dealer in games and puzzles. (Because the game never sold well, this turned out to be a bad investment for the dealer.) The "Traveler's Dodecahedron," also called "A Voyage Round the World," the puzzle described in this section, is a variant of that game.

Hamilton married his third love in 1833, but his marriage worked out poorly, because his wife, a semi-invalid, was unable to cope with his household affairs. He suffered from alcoholism and lived reclusively for the last two decades of his life. He died from gout in 1865, leaving masses of papers containing unpublished research. Mixed in with these papers were a large number of dinner plates, many containing the remains of desiccated, uneaten chops.

to a graph with a Hamilton circuit produces a graph with the same Hamilton circuit. So as we add edges to a graph, especially when we make sure to add edges to each vertex, we make it increasingly likely that a Hamilton circuit exists in this graph. Consequently, we would expect there to be sufficient conditions for the existence of Hamilton circuits that depend on the degrees of vertices being sufficiently large. We state two of the most important sufficient conditions here. These conditions were found by Gabriel A. Dirac in 1952 and Øystein Ore in 1960.

THEOREM 3   **DIRAC'S THEOREM**   If G is a simple graph with n vertices with $n \geq 3$ such that the degree of every vertex in G is at least $n/2$, then G has a Hamilton circuit.

THEOREM 4   **ORE'S THEOREM**   If G is a simple graph with n vertices with $n \geq 3$ such that $\deg(u) + \deg(v) \geq n$ for every pair of nonadjacent vertices u and v in G, then G has a Hamilton circuit.

The proof of Ore's theorem is outlined in Exercise 65. Dirac's theorem can be proved as a corollary to Ore's theorem because the conditions of Dirac's theorem imply those of Ore's theorem.

Both Ore's theorem and Dirac's theorem provide sufficient conditions for a connected simple graph to have a Hamilton circuit. However, these theorems do not provide necessary conditions for the existence of a Hamilton circuit. For example, the graph C_5 has a Hamilton circuit but does not satisfy the hypotheses of either Ore's theorem or Dirac's theorem, as the reader can verify.

The best algorithms known for finding a Hamilton circuit in a graph or determining that no such circuit exists have exponential worst-case time complexity (in the number of vertices of the graph). Finding an algorithm that solves this problem with polynomial worst-case time

---

GABRIEL ANDREW DIRAC (1925–1984)   Gabriel Dirac was born in Budapest. He moved to England in 1937 when his mother married the famous physicist and Nobel Laureate Paul Adrien Maurice Dirac, who adopted him. Gabriel A. Dirac entered Cambridge University in 1942, but his studies were interrupted by wartime service in the aviation industry. He obtained his Ph.D. in mathematics in 1951 from the University of London. He held university positions in England, Canada, Austria, Germany, and Denmark, where he spent his last 14 years. Dirac became interested in graph theory early in his career and help raise its status as an important topic of research. He made important contributions to many aspects of graph theory, including graph coloring and Hamilton circuits. Dirac attracted many students to graph theory and was noted as an excellent lecturer.

Dirac was noted for his penetrating mind and held unconventional views on many topics, including politics and social life. Dirac was a man with many interests and held a great passion for fine art. He had a happy family life with his wife Rosemari and his four children.

ØYSTEIN ORE (1899–1968)   Ore was born in Kristiania (the old name for Oslo, Norway). In 1922 he received his bachelors degree and in 1925 his Ph.D. in mathematics from Kristiania University, after studies in Germany and in Sweden. In 1927 he was recruited to leave his junior position at Kristiania and join Yale University. He was promoted rapidly at Yale, becoming full professor in 1929 and Sterling Professor in 1931, a position he held until 1968.

Ore made many contributions to number theory, ring theory, lattice theory, graph theory, and probability theory. He was a prolific author of papers and books. His interest in the history of mathematics is reflected in his biographies of Abel and Cardano, and in his popular textbook *Number Theory and its History*. He wrote four books on graph theory in the 1960s.

During and after World War II Ore played a major role supporting his native Norway. In 1947 King Haakon VII of Norway gave him the Knight Order of St. Olaf to recognize these efforts. Ore possessed deep knowledge of painting and sculpture and was an ardent collector of ancient maps. He was married and had two children.

complexity would be a major accomplishment because it has been shown that this problem is NP-complete (see Section 3.3). Consequently, the existence of such an algorithm would imply that many other seemingly intractable problems could be solved using algorithms with polynomial worst-case time complexity.

Applications of Hamilton Circuits

Hamilton paths and circuits can be used to solve practical problems. For example, many applications ask for a path or circuit that visits each road intersection in a city, each place pipelines intersect in a utility grid, or each node in a communications network exactly once. Finding a Hamilton path or circuit in the appropriate graph model can solve such problems. The famous **traveling salesperson problem** or **TSP** (also known in older literature as the **traveling salesman problem**) asks for the shortest route a traveling salesperson should take to visit a set of cities. This problem reduces to finding a Hamilton circuit in a complete graph such that the total weight of its edges is as small as possible. We will return to this question in Section 10.6.

We now describe a less obvious application of Hamilton circuits to coding.

EXAMPLE 8   **Gray Codes**   The position of a rotating pointer can be represented in digital form. One way to do this is to split the circle into 2^n arcs of equal length and to assign a bit string of length n to each arc. Two ways to do this using bit strings of length three are shown in Figure 12.

The digital representation of the position of the pointer can be determined using a set of n contacts. Each contact is used to read one bit in the digital representation of the position. This is illustrated in Figure 13 for the two assignments from Figure 12.

When the pointer is near the boundary of two arcs, a mistake may be made in reading its position. This may result in a major error in the bit string read. For instance, in the coding scheme in Figure 12(a), if a small error is made in determining the position of the pointer, the bit string 100 is read instead of 011. All three bits are incorrect! To minimize the effect of an error in determining the position of the pointer, the assignment of the bit strings to the 2^n arcs should be made so that only one bit is different in the bit strings represented by adjacent arcs. This is exactly the situation in the coding scheme in Figure 12(b). An error in determining the position of the pointer gives the bit string 010 instead of 011. Only one bit is wrong.

A **Gray code** is a labeling of the arcs of the circle such that adjacent arcs are labeled with bit strings that differ in exactly one bit. The assignment in Figure 12(b) is a Gray code. We can find a Gray code by listing all bit strings of length n in such a way that each string differs in exactly one position from the preceding bit string, and the last string differs from the first in exactly one position. We can model this problem using the n-cube Q_n. What is needed to solve this problem is a Hamilton circuit in Q_n. Such Hamilton circuits are easily found. For instance, a Hamilton circuit for Q_3 is displayed in Figure 14. The sequence of bit strings differing in exactly one bit produced by this Hamilton circuit is 000, 001, 011, 010, 110, 111, 101, 100.

Gray codes are named after Frank Gray, who invented them in the 1940s at AT&T Bell Laboratories to minimize the effect of errors in transmitting digital signals. ◀

(a)   (b)

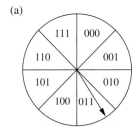 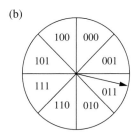

FIGURE 12   **Converting the Position of a Pointer into Digital Form.**

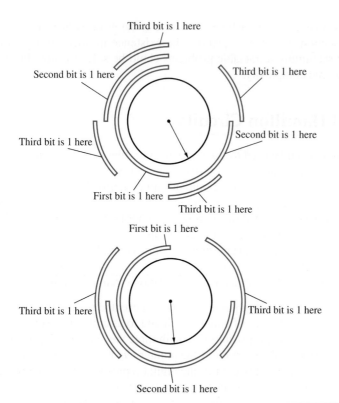

FIGURE 13   The Digital Representation of the Position of the Pointer.

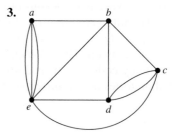

FIGURE 14   A Hamilton Circuit for Q_3.

Exercises

In Exercises 1–4 determine whether the given graph has an Euler circuit. Construct such a circuit when one exists. If no Euler circuit exists, determine whether the graph has an Euler path and construct such a path if one exists.

1.

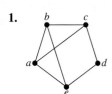

2.

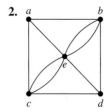

3.

4.

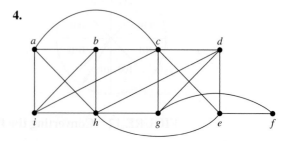

5. Suppose that in addition to the seven bridges of Königsberg (shown in Figure 1) there were two additional bridges, connecting regions B and C and regions B and D, respectively. Could someone cross all nine of these bridges exactly once and return to the starting point?

6. Devise a procedure, similar to Algorithm 1, for constructing Euler paths in multigraphs.

In Exercises 7–9 determine whether the picture shown can be drawn with a pencil in a continuous motion without lifting the pencil or retracing part of the picture.

7.

8.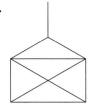

9.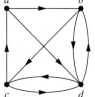

*10. Show that a directed multigraph having no isolated vertices has an Euler circuit if and only if the graph is weakly connected and the in-degree and out-degree of each vertex are equal.

*11. Show that a directed multigraph having no isolated vertices has an Euler path but not an Euler circuit if and only if the graph is weakly connected and the in-degree and out-degree of each vertex are equal for all but two vertices, one that has in-degree one larger than its out-degree and the other that has out-degree one larger than its in-degree.

In Exercises 12–14 determine whether the directed graph shown has an Euler circuit. Construct an Euler circuit if one exists. If no Euler circuit exists, determine whether the directed graph has an Euler path. Construct an Euler path if one exists.

12.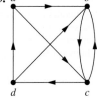

13.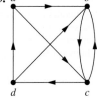

14.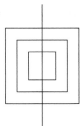

15. Devise an algorithm for constructing Euler paths in directed graphs.

16. For which values of n do these graphs have an Euler circuit?

   a) K_n **b)** C_n **c)** W_n **d)** Q_n

17. For which values of n do the graphs in Exercise 16 have an Euler path but no Euler circuit?

18. For which values of m and n does the complete bipartite graph $K_{m,n}$ have an

   a) Euler circuit?

   b) Euler path?

19. Find the least number of times it is necessary to lift a pencil from the paper when drawing each of the graphs in Exercises 1–4 without retracing any part of the graph.

In Exercises 20–22 determine whether the given graph has a Hamilton circuit. If it does, find such a circuit. If it does not, give an argument to show why no such circuit exists.

20.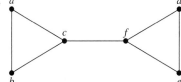

21.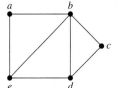

22.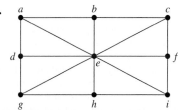

23. Does the graph in Exercise 20 have a Hamilton path? If so, find such a path. If it does not, give an argument to show why no such path exists.

24. Does the graph in Exercise 21 have a Hamilton path? If so, find such a path. If it does not, give an argument to show why no such path exists.

25. Does the graph in Exercise 22 have a Hamilton path? If so, find such a path. If it does not, give an argument to show why no such path exists.

26. For which values of n do the graphs in Exercise 16 have a Hamilton circuit?

27. For which values of m and n does the complete bipartite graph $K_{m,n}$ have a Hamilton circuit?

∗28. Show that the **Petersen graph**, shown here, does not have a Hamilton circuit, but that the subgraph obtained by deleting a vertex v, and all edges incident with v, does have a Hamilton circuit.

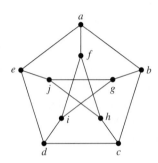

29. For each of these graphs, determine (i) whether Dirac's theorem can be used to show that the graph has a Hamilton circuit, (ii) whether Ore's theorem can be used to show that the graph has a Hamilton circuit, and (iii) whether the graph has a Hamilton circuit.

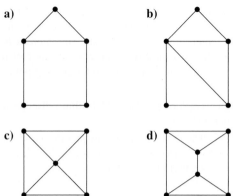

30. Can you find a simple graph with n vertices with $n \geq 3$ that does not have a Hamilton circuit, yet the degree of every vertex in the graph is at least $(n-1)/2$?

∗31. Show that there is a Gray code of order n whenever n is a positive integer, or equivalently, show that the n-cube $Q_n, n > 1$, always has a Hamilton circuit. [*Hint:* Use mathematical induction. Show how to produce a Gray code of order n from one of order $n-1$.]

 Fleury's algorithm, published in 1883, constructs Euler circuits by first choosing an arbitrary vertex of a connected multigraph, and then forming a circuit by choosing edges successively. Once an edge is chosen, it is removed. Edges are chosen successively so that each edge begins where the last edge ends, and so that this edge is not a cut edge unless there is no alternative.

32. Use Fleury's algorithm to find an Euler circuit in the graph G in Figure 5.

∗33. Express Fleury's algorithm in pseudocode.

Links

JULIUS PETER CHRISTIAN PETERSEN (1839–1910)   Julius Petersen was born in the Danish town of Sorø. His father was a dyer. In 1854 his parents were no longer able to pay for his schooling, so he became an apprentice in an uncle's grocery store. When this uncle died, he left Petersen enough money to return to school. After graduating, he began studying engineering at the Polytechnical School in Copenhagen, later deciding to concentrate on mathematics. He published his first textbook, a book on logarithms, in 1858. When his inheritance ran out, he had to teach to make a living. From 1859 until 1871 Petersen taught at a prestigious private high school in Copenhagen. While teaching high school he continued his studies, entering Copenhagen University in 1862. He married Laura Bertelsen in 1862; they had three children, two sons and a daughter.

Petersen obtained a mathematics degree from Copenhagen University in 1866 and finally obtained his doctorate in 1871 from that school. After receiving his doctorate, he taught at a polytechnic and military academy. In 1887 he was appointed to a professorship at the University of Copenhagen. Petersen was well known in Denmark as the author of a large series of textbooks for high schools and universities. One of his books, *Methods and Theories for the Solution of Problems of Geometrical Construction*, was translated into eight languages, with the English language version last reprinted in 1960 and the French version reprinted as recently as 1990, more than a century after the original publication date.

Petersen worked in a wide range of areas, including algebra, analysis, cryptography, geometry, mechanics, mathematical economics, and number theory. His contributions to graph theory, including results on regular graphs, are his best-known work. He was noted for his clarity of exposition, problem-solving skills, originality, sense of humor, vigor, and teaching. One interesting fact about Petersen was that he preferred not to read the writings of other mathematicians. This led him often to rediscover results already proved by others, often with embarrassing consequences. However, he was often angry when other mathematicians did not read his writings!

Petersen's death was front-page news in Copenhagen. A newspaper of the time described him as the Hans Christian Andersen of science—a child of the people who made good in the academic world.

**34. Prove that Fleury's algorithm always produces an Euler circuit.

*35. Give a variant of Fleury's algorithm to produce Euler paths.

36. A diagnostic message can be sent out over a computer network to perform tests over all links and in all devices. What sort of paths should be used to test all links? To test all devices?

37. Show that a bipartite graph with an odd number of vertices does not have a Hamilton circuit.

A **knight** is a chess piece that can move either two spaces horizontally and one space vertically or one space horizontally and two spaces vertically. That is, a knight on square (x, y) can move to any of the eight squares $(x \pm 2, y \pm 1)$, $(x \pm 1, y \pm 2)$, if these squares are on the chessboard, as illustrated here.

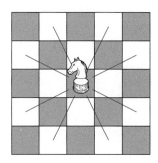

A **knight's tour** is a sequence of legal moves by a knight starting at some square and visiting each square exactly once. A knight's tour is called **reentrant** if there is a legal move that takes the knight from the last square of the tour back to where the tour began. We can model knight's tours using the graph that has a vertex for each square on the board, with an edge connecting two vertices if a knight can legally move between the squares represented by these vertices.

38. Draw the graph that represents the legal moves of a knight on a 3×3 chessboard.

39. Draw the graph that represents the legal moves of a knight on a 3×4 chessboard.

40. a) Show that finding a knight's tour on an $m \times n$ chessboard is equivalent to finding a Hamilton path on the graph representing the legal moves of a knight on that board.

   b) Show that finding a reentrant knight's tour on an $m \times n$ chessboard is equivalent to finding a Hamilton circuit on the corresponding graph.

*41. Show that there is a knight's tour on a 3×4 chessboard.

*42. Show that there is no knight's tour on a 3×3 chessboard.

*43. Show that there is no knight's tour on a 4×4 chessboard.

44. Show that the graph representing the legal moves of a knight on an $m \times n$ chessboard, whenever m and n are positive integers, is bipartite.

45. Show that there is no reentrant knight's tour on an $m \times n$ chessboard when m and n are both odd. [*Hint:* Use Exercises 37, 40(b), and 44.]

*46. Show that there is a knight's tour on an 8×8 chessboard. [*Hint:* You can construct a knight's tour using a method invented by H. C. Warnsdorff in 1823: Start in any square, and then always move to a square connected to the fewest number of unused squares. Although this method may not always produce a knight's tour, it often does.]

47. The parts of this exercise outline a proof of Ore's theorem. Suppose that G is a simple graph with n vertices, $n \geq 3$, and $\deg(x) + \deg(y) \geq n$ whenever x and y are nonadjacent vertices in G. Ore's theorem states that under these conditions, G has a Hamilton circuit.

   a) Show that if G does not have a Hamilton circuit, then there exists another graph H with the same vertices as G, which can be constructed by adding edges to G such that the addition of a single edge would produce a Hamilton circuit in H. [*Hint:* Add as many edges as possible at each successive vertex of G without producing a Hamilton circuit.]

   b) Show that there is a Hamilton path in H.

   c) Let $v_1, v_2, \ldots, v_n$ be a Hamilton path in H. Show that $\deg(v_1) + \deg(v_n) \geq n$ and that there are at most $\deg(v_1)$ vertices not adjacent to v_n (including v_n itself).

   d) Let S be the set of vertices preceding each vertex adjacent to v_1 in the Hamilton path. Show that S contains $\deg(v_1)$ vertices and $v_n \notin S$.

   e) Show that S contains a vertex v_k, which is adjacent to v_n, implying that there are edges connecting v_1 and v_{k+1} and v_k and v_n.

   f) Show that part (e) implies that $v_1, v_2, \ldots, v_{k-1}$, $v_k, v_n, v_{n-1}, \ldots, v_{k+1}, v_1$ is a Hamilton circuit in G. Conclude from this contradiction that Ore's theorem holds.

*48. Show that the worst case computational complexity of Algorithm 1 for finding Euler circuits in a connected graph with all vertices of even degree is $O(m)$, where m is the number of edges of G.

10.6   Shortest-Path Problems

Introduction

Many problems can be modeled using graphs with weights assigned to their edges. As an illustration, consider how an airline system can be modeled. We set up the basic graph model by representing cities by vertices and flights by edges. Problems involving distances can be

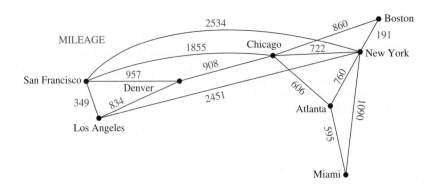

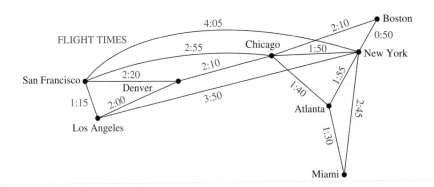

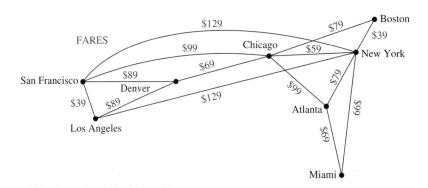

FIGURE 1 **Weighted Graphs Modeling an Airline System.**

modeled by assigning distances between cities to the edges. Problems involving flight time can be modeled by assigning flight times to edges. Problems involving fares can be modeled by assigning fares to the edges. Figure 1 displays three different assignments of weights to the edges of a graph representing distances, flight times, and fares, respectively.

Graphs that have a number assigned to each edge are called **weighted graphs**. Weighted graphs are used to model computer networks. Communications costs (such as the monthly cost of leasing a telephone line), the response times of the computers over these lines, or the distance between computers, can all be studied using weighted graphs. Figure 2 displays weighted graphs that represent three ways to assign weights to the edges of a graph of a computer network, corresponding to distance, response time, and cost.

Several types of problems involving weighted graphs arise frequently. Determining a path of least length between two vertices in a network is one such problem. To be more specific, let the **length** of a path in a weighted graph be the sum of the weights of the edges of this path. (The reader should note that this use of the term *length* is different from the use of *length* to denote the

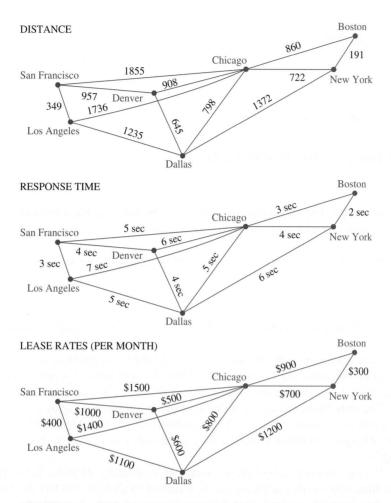

FIGURE 2   Weighted Graphs Modeling a Computer Network.

number of edges in a path in a graph without weights.) The question is: What is a shortest path, that is, a path of least length, between two given vertices? For instance, in the airline system represented by the weighted graph shown in Figure 1, what is a shortest path in air distance between Boston and Los Angeles? What combinations of flights has the smallest total flight time (that is, total time in the air, not including time between flights) between Boston and Los Angeles? What is the cheapest fare between these two cities? In the computer network shown in Figure 2, what is a least expensive set of telephone lines needed to connect the computers in San Francisco with those in New York? Which set of telephone lines gives a fastest response time for communications between San Francisco and New York? Which set of lines has a shortest overall distance?

Another important problem involving weighted graphs asks for a circuit of shortest total length that visits every vertex of a complete graph exactly once. This is the famous *traveling salesperson problem*, which asks for an order in which a salesperson should visit each of the cities on his route exactly once so that he travels the minimum total distance. We will discuss the traveling salesperson problem later in this section.

A Shortest-Path Algorithm

There are several different algorithms that find a shortest path between two vertices in a weighted graph. We will present a greedy algorithm discovered by the Dutch mathematician Edsger Di-

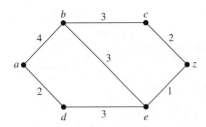

FIGURE 3 **A Weighted Simple Graph.**

jkstra in 1959. The version we will describe solves this problem in undirected weighted graphs where all the weights are positive. It is easy to adapt it to solve shortest-path problems in directed graphs.

Before giving a formal presentation of the algorithm, we will give an illustrative example.

EXAMPLE 1 What is the length of a shortest path between a and z in the weighted graph shown in Figure 3?

Solution: Although a shortest path is easily found by inspection, we will develop some ideas useful in understanding Dijkstra's algorithm. We will solve this problem by finding the length of a shortest path from a to successive vertices, until z is reached.

The only paths starting at a that contain no vertex other than a are formed by adding an edge that has a as one endpoint. These paths have only one edge. They are a, b of length 4 and a, d of length 2. It follows that d is the closest vertex to a, and the shortest path from a to d has length 2.

We can find the second closest vertex by examining all paths that begin with the shortest path from a to a vertex in the set $\{a, d\}$, followed by an edge that has one endpoint in $\{a, d\}$ and its other endpoint not in this set. There are two such paths to consider, a, d, e of length 7 and a, b of length 4. Hence, the second closest vertex to a is b and the shortest path from a to b has length 4.

To find the third closest vertex to a, we need examine only the paths that begin with the shortest path from a to a vertex in the set $\{a, d, b\}$, followed by an edge that has one endpoint in the set $\{a, d, b\}$ and its other endpoint not in this set. There are three such paths, a, b, c of length 7, a, b, e of length 7, and a, d, e of length 5. Because the shortest of these paths is a, d, e, the third closest vertex to a is e and the length of the shortest path from a to e is 5.

EDSGER WYBE DIJKSTRA (1930–2002)   Edsger Dijkstra, born in the Netherlands, began programming computers in the early 1950s while studying theoretical physics at the University of Leiden. In 1952, realizing that he was more interested in programming than in physics, he quickly completed the requirements for his physics degree and began his career as a programmer, even though programming was not a recognized profession. (In 1957, the authorities in Amsterdam refused to accept "programming" as his profession on his marriage license. However, they did accept "theoretical physicist" when he changed his entry to this.)

Dijkstra was one of the most forceful proponents of programming as a scientific discipline. He has made fundamental contributions to the areas of operating systems, including deadlock avoidance; programming languages, including the notion of structured programming; and algorithms. In 1972 Dijkstra received the Turing Award from the Association for Computing Machinery, one of the most prestigious awards in computer science. Dijkstra became a Burroughs Research Fellow in 1973, and in 1984 he was appointed to a chair in Computer Science at the University of Texas, Austin.

To find the fourth closest vertex to a, we need examine only the paths that begin with the shortest path from a to a vertex in the set $\{a, d, b, e\}$, followed by an edge that has one endpoint in the set $\{a, d, b, e\}$ and its other endpoint not in this set. There are two such paths, a, b, c of length 7 and a, d, e, z of length 6. Because the shorter of these paths is a, d, e, z, the fourth closest vertex to a is z and the length of the shortest path from a to z is 6. ◀

Example 1 illustrates the general principles used in Dijkstra's algorithm. Note that a shortest path from a to z could have been found by a brute force approach by examining the length of every path from a to z. However, this brute force approach is impractical for humans and even for computers for graphs with a large number of edges.

We will now consider the general problem of finding the length of a shortest path between a and z in an undirected connected simple weighted graph. Dijkstra's algorithm proceeds by finding the length of a shortest path from a to a first vertex, the length of a shortest path from a to a second vertex, and so on, until the length of a shortest path from a to z is found. As a side benefit, this algorithm is easily extended to find the length of the shortest path from a to all other vertices of the graph, and not just to z.

The algorithm relies on a series of iterations. A distinguished set of vertices is constructed by adding one vertex at each iteration. A labeling procedure is carried out at each iteration. In this labeling procedure, a vertex w is labeled with the length of a shortest path from a to w that contains only vertices already in the distinguished set. The vertex added to the distinguished set is one with a minimal label among those vertices not already in the set.

We now give the details of Dijkstra's algorithm. It begins by labeling a with 0 and the other vertices with ∞. We use the notation $L_0(a) = 0$ and $L_0(v) = \infty$ for these labels before any iterations have taken place (the subscript 0 stands for the "0th" iteration). These labels are the lengths of shortest paths from a to the vertices, where the paths contain only the vertex a. (Because no path from a to a vertex different from a exists, ∞ is the length of a shortest path between a and this vertex.)

Dijkstra's algorithm proceeds by forming a distinguished set of vertices. Let S_k denote this set after k iterations of the labeling procedure. We begin with $S_0 = \emptyset$. The set S_k is formed from S_{k-1} by adding a vertex u not in S_{k-1} with the smallest label.

Once u is added to S_k, we update the labels of all vertices not in S_k, so that $L_k(v)$, the label of the vertex v at the kth stage, is the length of a shortest path from a to v that contains vertices only in S_k (that is, vertices that were already in the distinguished set together with u). Note that the way we choose the vertex u to add to S_k at each step is an optimal choice at each step, making this a greedy algorithm. (We will prove shortly that this greedy algorithm always produces an optimal solution.)

Let v be a vertex not in S_k. To update the label of v, note that $L_k(v)$ is the length of a shortest path from a to v containing only vertices in S_k. The updating can be carried out efficiently when this observation is used: A shortest path from a to v containing only elements of S_k is either a shortest path from a to v that contains only elements of S_{k-1} (that is, the distinguished vertices not including u), or it is a shortest path from a to u at the $(k-1)$st stage with the edge $\{u, v\}$ added. In other words,

$$L_k(a, v) = \min\{L_{k-1}(a, v), L_{k-1}(a, u) + w(u, v)\},$$

where $w(u, v)$ is the length of the edge with u and v as endpoints. This procedure is iterated by successively adding vertices to the distinguished set until z is added. When z is added to the distinguished set, its label is the length of a shortest path from a to z.

Dijkstra's algorithm is given in Algorithm 1. Later we will give a proof that this algorithm is correct. Note that we can find the length of the shortest path from a to all other vertices of the graph if we continue this procedure until all vertices are added to the distinguished set.

Demo

ALGORITHM 1 Dijkstra's Algorithm.

procedure *Dijkstra*(G: weighted connected simple graph, with
    all weights positive)
{G has vertices $a = v_0, v_1, \ldots, v_n = z$ and lengths $w(v_i, v_j)$
    where $w(v_i, v_j) = \infty$ if $\{v_i, v_j\}$ is not an edge in G}
for $i := 1$ **to** n
    $L(v_i) := \infty$
$L(a) := 0$
$S := \emptyset$
{the labels are now initialized so that the label of a is 0 and all
    other labels are ∞, and S is the empty set}
while $z \notin S$
    $u :=$ a vertex not in S with $L(u)$ minimal
    $S := S \cup \{u\}$
    for all vertices v not in S
        if $L(u) + w(u, v) < L(v)$ **then** $L(v) := L(u) + w(u, v)$
        {this adds a vertex to S with minimal label and updates the
        labels of vertices not in S}
return $L(z)$ {$L(z) =$ length of a shortest path from a to z}

Example 2 illustrates how Dijkstra's algorithm works. Afterward, we will show that this algorithm always produces the length of a shortest path between two vertices in a weighted graph.

EXAMPLE 2   Use Dijkstra's algorithm to find the length of a shortest path between the vertices a and z in the weighted graph displayed in Figure 4(a).

Solution: The steps used by Dijkstra's algorithm to find a shortest path between a and z are shown in Figure 4. At each iteration of the algorithm the vertices of the set S_k are circled. A shortest path from a to each vertex containing only vertices in S_k is indicated for each iteration. The algorithm terminates when z is circled. We find that a shortest path from a to z is a, c, b, d, e, z, with length 13. ◀

Remark: In performing Dijkstra's algorithm it is sometimes more convenient to keep track of labels of vertices in each step using a table instead of redrawing the graph for each step.

Next, we use an inductive argument to show that Dijkstra's algorithm produces the length of a shortest path between two vertices a and z in an undirected connected weighted graph. Take as the inductive hypothesis the following assertion: At the kth iteration

(i) the label of every vertex v in S is the length of a shortest path from a to this vertex, and
(ii) the label of every vertex not in S is the length of a shortest path from a to this vertex that contains only (besides the vertex itself) vertices in S.

When $k = 0$, before any iterations are carried out, $S = \emptyset$, so the length of a shortest path from a to a vertex other than a is ∞. Hence, the basis case is true.

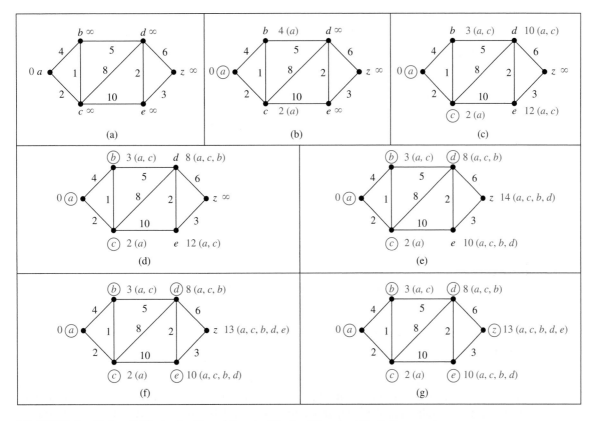

FIGURE 4 Using Dijkstra's Algorithm to Find a Shortest Path from a to z.

Assume that the inductive hypothesis holds for the kth iteration. Let v be the vertex added to S at the $(k + 1)$st iteration, so v is a vertex not in S at the end of the kth iteration with the smallest label (in the case of ties, any vertex with smallest label may be used).

From the inductive hypothesis we see that the vertices in S before the $(k + 1)$st iteration are labeled with the length of a shortest path from a. Also, v must be labeled with the length of a shortest path to it from a. If this were not the case, at the end of the kth iteration there would be a path of length less than $L_k(v)$ containing a vertex not in S [because $L_k(v)$ is the length of a shortest path from a to v containing only vertices in S after the kth iteration]. Let u be the first vertex not in S in such a path. There is a path with length less than $L_k(v)$ from a to u containing only vertices of S. This contradicts the choice of v. Hence, (i) holds at the end of the $(k + 1)$st iteration.

Let u be a vertex not in S after $k + 1$ iterations. A shortest path from a to u containing only elements of S either contains v or it does not. If it does not contain v, then by the inductive hypothesis its length is $L_k(u)$. If it does contain v, then it must be made up of a path from a to v of shortest possible length containing elements of S other than v, followed by the edge from v to u. In this case, its length would be $L_k(v) + w(v, u)$. This shows that (ii) is true, because $L_{k+1}(u) = \min\{L_k(u), L_k(v) + w(v, u)\}$.

We now state the thereom that we have proved.

THEOREM 1 Dijkstra's algorithm finds the length of a shortest path between two vertices in a connected simple undirected weighted graph.

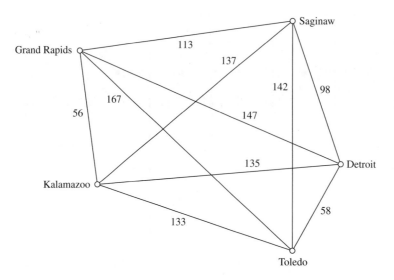

FIGURE 5   **The Graph Showing the Distances between Five Cities.**

We can now estimate the computational complexity of Dijkstra's algorithm (in terms of additions and comparisons). The algorithm uses no more than $n - 1$ iterations where n is the number of vertices in the graph, because one vertex is added to the distinguished set at each iteration. We are done if we can estimate the number of operations used for each iteration. We can identify the vertex not in S_k with the smallest label using no more than $n - 1$ comparisons. Then we use an addition and a comparison to update the label of each vertex not in S_k. It follows that no more than $2(n - 1)$ operations are used at each iteration, because there are no more than $n - 1$ labels to update at each iteration. Because we use no more than $n - 1$ iterations, each using no more than $2(n - 1)$ operations, we have Theorem 2.

THEOREM 2   Dijkstra's algorithm uses $O(n^2)$ operations (additions and comparisons) to find the length of a shortest path between two vertices in a connected simple undirected weighted graph with n vertices.

The Traveling Salesperson Problem

Links

We now discuss an important problem involving weighted graphs. Consider the following problem: A traveling salesperson wants to visit each of n cities exactly once and return to his starting point. For example, suppose that the salesperson wants to visit Detroit, Toledo, Saginaw, Grand Rapids, and Kalamazoo (see Figure 5). In which order should he visit these cities to travel the minimum total distance? To solve this problem we can assume the salesperson starts in Detroit (because this must be part of the circuit) and examine all possible ways for him to visit the other four cities and then return to Detroit (starting elsewhere will produce the same circuits). There are a total of 24 such circuits, but because we travel the same distance when we travel a circuit in reverse order, we need only consider 12 different circuits to find the minimum total distance he must travel. We list these 12 different circuits and the total distance traveled for each circuit. As can be seen from the list, the minimum total distance of 458 miles is traveled using the circuit Detroit–Toledo–Kalamazoo–Grand Rapids–Saginaw–Detroit (or its reverse).

Route	*Total Distance (miles)*
Detroit–Toledo–Grand Rapids–Saginaw–Kalamazoo–Detroit	610
Detroit–Toledo–Grand Rapids–Kalamazoo–Saginaw–Detroit	516
Detroit–Toledo–Kalamazoo–Saginaw–Grand Rapids–Detroit	588
Detroit–Toledo–Kalamazoo–Grand Rapids–Saginaw–Detroit	458
Detroit–Toledo–Saginaw–Kalamazoo–Grand Rapids–Detroit	540
Detroit–Toledo–Saginaw–Grand Rapids–Kalamazoo–Detroit	504
Detroit–Saginaw–Toledo–Grand Rapids–Kalamazoo–Detroit	598
Detroit–Saginaw–Toledo–Kalamazoo–Grand Rapids–Detroit	576
Detroit–Saginaw–Kalamazoo–Toledo–Grand Rapids–Detroit	682
Detroit–Saginaw–Grand Rapids–Toledo–Kalamazoo–Detroit	646
Detroit–Grand Rapids–Saginaw–Toledo–Kalamazoo–Detroit	670
Detroit–Grand Rapids–Toledo–Saginaw–Kalamazoo–Detroit	728

An 1832 handbook *Der Handlungsreisende* (The Traveling Salesman) mentions the traveling salesman problem, with sample tours through Germany and Switzerland.

We just described an instance of the **traveling salesperson problem**. The traveling salesperson problem asks for the circuit of minimum total weight in a weighted, complete, undirected graph that visits each vertex exactly once and returns to its starting point. This is equivalent to asking for a Hamilton circuit with minimum total weight in the complete graph, because each vertex is visited exactly once in the circuit.

The most straightforward way to solve an instance of the traveling salesperson problem is to examine all possible Hamilton circuits and select one of minimum total length. How many circuits do we have to examine to solve the problem if there are n vertices in the graph? Once a starting point is chosen, there are $(n - 1)!$ different Hamilton circuits to examine, because there are $n - 1$ choices for the second vertex, $n - 2$ choices for the third vertex, and so on. Because a Hamilton circuit can be traveled in reverse order, we need only examine $(n - 1)!/2$ circuits to find our answer. Note that $(n - 1)!/2$ grows extremely rapidly. Trying to solve a traveling salesperson problem in this way when there are only a few dozen vertices is impractical. For example, with 25 vertices, a total of $24!/2$ (approximately 3.1×10^{23}) different Hamilton circuits would have to be considered. If it took just one nanosecond (10^{-9} second) to examine each Hamilton circuit, a total of approximately ten million years would be required to find a minimum-length Hamilton circuit in this graph by exhaustive search techniques.

Because the traveling salesperson problem has both practical and theoretical importance, a great deal of effort has been devoted to devising efficient algorithms that solve it. However, no algorithm with polynomial worst-case time complexity is known for solving this problem. Furthermore, if a polynomial worst-case time complexity algorithm were discovered for the traveling salesperson problem, many other difficult problems would also be solvable using polynomial worst-case time complexity algorithms (such as determining whether a proposition in n variables is a tautology, discussed in Chapter 1). This follows from the theory of NP-completeness. (For more information about this, consult [GaJo79].)

A practical approach to the traveling salesperson problem when there are many vertices to visit is to use an **approximation algorithm**. These are algorithms that do not necessarily produce the exact solution to the problem but instead are guaranteed to produce a solution that is close to an exact solution. That is, they may produce a Hamilton circuit with total weight W' such that $W \leq W' \leq cW$, where W is the total length of an exact solution and c is a constant. For example, there is an algorithm with polynomial worst-case time complexity that works if the weighted graph satisfies the triangle inequality such that $c = 3/2$. For general weighted graphs for every positive real number k no algorithm is known that will always produce a solution at most k times a best solution. If such an algorithm existed, this would show that the class P would be the same as the class NP, perhaps the most famous open question about the complexity of algorithms (see Section 3.3).

In practice, algorithms have been developed that can solve traveling salesperson problems with as many as 1000 vertices within 2% of an exact solution using only a few minutes of computer time. For more information about the traveling salesperson problem, including history, applications, and algorithms, see the chapter on this topic in *Applications of Discrete Mathematics* [MiRo91] also available on the website for this book.

Exercises

1. For each of these problems about a subway system, describe a weighted graph model that can be used to solve the problem.

a) What is the least amount of time required to travel between two stops?

b) What is the minimum distance that can be traveled to reach a stop from another stop?

c) What is the least fare required to travel between two stops if fares between stops are added to give the total fare?

In Exercise 2 find the length of a shortest path between a and z in the given weighted graph.

2.

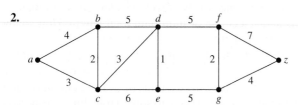

3. Find a shortest path between a and z in each of the weighted graphs in Exercise 2.

4. Find the length of a shortest path between these pairs of vertices in the weighted graph in Exercise 2.

a) a and d

b) a and f

c) c and f

d) b and z

5. Find a combination of flights with the least total air time between the pairs of flights. Use the flight times shown in Figure 1.

a) New York and Los Angeles

b) Boston and San Francisco

c) Miami and Denver

d) Miami and Los Angeles

6. Find a combination of flights with the least total air time between the pairs of cities in Exercise 5, using the flight times shown in Figure 1.

7. Find a shortest route (in distance) between computer centers in each of these pairs of cities in the communications network shown in Figure 2.

a) Boston and Los Angeles

b) New York and San Francisco

c) Dallas and San Francisco

d) Denver and New York

8. Find a route with the shortest response time between the pairs of computer centers in Exercise 7 using the response times given in Figure 2.

9. Find a least expensive route, in monthly lease charges, between the pairs of computer centers in Exercise 7 using the lease charges given in Figure 2.

10. Explain how to find a path with the least number of edges between two vertices in an undirected graph by considering it as a shortest path problem in a weighted graph.

11. Extend Dijkstra's algorithm for finding the length of a shortest path between two vertices in a weighted simple connected graph so that the length of a shortest path between the vertex a and every other vertex of the graph is found.

12. Extend Dijkstra's algorithm for finding the length of a shortest path between two vertices in a weighted simple connected graph so that a shortest path between these vertices is constructed.

13. The weighted graphs in the figures here show some major roads in New Jersey. Part (a) shows the distances between cities on these roads; part (b) shows the tolls.

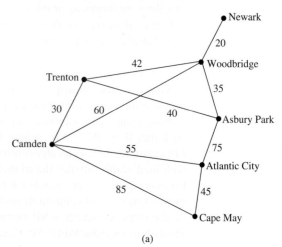

(a)

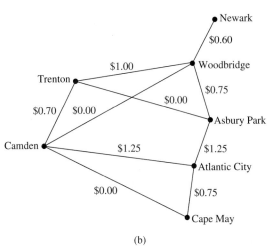

Newark

$0.60

$1.00   Woodbridge

Trenton

$0.75

$0.00

$0.70 $0.00

Asbury Park

Camden

$1.25 $1.25

Atlantic City

$0.00

$0.75

Cape May

(b)

a) Find a shortest route in distance between Newark and Camden, and between Newark and Cape May, using these roads.

b) Find a least expensive route in terms of total tolls using the roads in the graph between the pairs of cities in part (a) of this exercise.

14. Is a shortest path between two vertices in a weighted graph unique if the weights of edges are distinct?

15. What are some applications where it is necessary to find the length of a longest simple path between two vertices in a weighted graph?

16. What is the length of a longest simple path in the weighted graph in Figure 4 between a and z? Between c and z?

 Floyd's algorithm, displayed as Algorithm 2, can be used to find the length of a shortest path between all pairs of vertices in a weighted connected simple graph. However, this algorithm cannot be used to construct shortest paths. (We assign an infinite weight to any pair of vertices not connected by an edge in the graph.)

17. Use Floyd's algorithm to find the distance between all pairs of vertices in the weighted graph in Figure 4(a).

*18. Show that Dijkstra's algorithm may not work if edges can have negative weights.

19. Solve the traveling salesperson problem for this graph by finding the total weight of all Hamilton circuits and determining a circuit with minimum total weight.

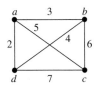

20. Find a route with the least total airfare that visits each of the cities in this graph, where the weight on an edge is the least price available for a flight between the two cities.

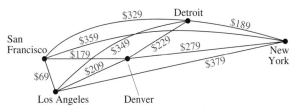

$329   Detroit
$189

San
Francisco $359
$179 $349 $229 $279   New
York

$69 $209
$379

Los Angeles   Denver

21. Construct a weighted undirected graph such that the total weight of a circuit that visits every vertex at least once is minimized for a circuit that visits some vertices more than once. [*Hint:* There are examples with three vertices.]

22. Show that the problem of finding a circuit of minimum total weight that visits every vertex of a weighted graph at least once can be reduced to the problem of finding a circuit of minimum total weight that visits each vertex of a weighted graph exactly once. Do so by constructing a new weighted graph with the same vertices and edges as the original graph but whose weight of the edge connecting the vertices u and v is equal to the minimum total weight of a path from u to v in the original graph.

*23. The **longest path problem** in a weighted directed graph with no simple circuits asks for a path in this graph such that the sum of its edge weights is a maximum. Devise an algorithm for solving the longest path problem. [*Hint*: First find a topological ordering of the vertices of the graph.]

ALGORITHM 2 Floyd's Algorithm.

procedure *Floyd*(G: weighted simple graph)
{G has vertices $v_1, v_2, \ldots, v_n$ and weights $w(v_i, v_j)$
   with $w(v_i, v_j) = \infty$ if $\{v_i, v_j\}$ is not an edge}
for $i := 1$ **to** n
   for $j := 1$ **to** n
      $d(v_i, v_j) := w(v_i, v_j)$
for $i := 1$ **to** n
   for $j := 1$ **to** n
      for $k := 1$ **to** n
         if $d(v_j, v_i) + d(v_i, v_k) < d(v_j, v_k)$
            then $d(v_j, v_k) := d(v_j, v_i) + d(v_i, v_k)$
return $\left[d(v_i, v_j)\right]$ {$d(v_i, v_j)$ is the length of a shortest path between v_i and v_j for $1 \le i \le n, 1 \le j \le n$}

10.7 Planar Graphs

Introduction

Consider the problem of joining three houses to each of three separate utilities, as shown in Figure 1. Is it possible to join these houses and utilities so that none of the connections cross?

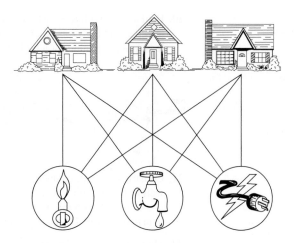

FIGURE 1 Three Houses and Three Utilities.

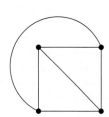

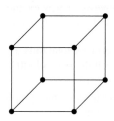

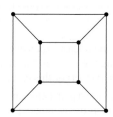

FIGURE 2 The Graph K_4.

FIGURE 3 K_4 Drawn with No Crossings.

FIGURE 4 The Graph Q_3.

FIGURE 5 A Planar Representation of Q_3.

This problem can be modeled using the complete bipartite graph $K_{3,3}$. The original question can be rephrased as: Can $K_{3,3}$ be drawn in the plane so that no two of its edges cross?

In this section we will study the question of whether a graph can be drawn in the plane without edges crossing. In particular, we will answer the houses-and-utilities problem.

There are always many ways to represent a graph. When is it possible to find at least one way to represent this graph in a plane without any edges crossing?

DEFINITION 1

A graph is called *planar* if it can be drawn in the plane without any edges crossing (where a crossing of edges is the intersection of the lines or arcs representing them at a point other than their common endpoint). Such a drawing is called a *planar representation* of the graph.

A graph may be planar even if it is usually drawn with crossings, because it may be possible to draw it in a different way without crossings.

EXAMPLE 1 Is K_4 (shown in Figure 2 with two edges crossing) planar?

Solution: K_4 is planar because it can be drawn without crossings, as shown in Figure 3. ◀

EXAMPLE 2 Is Q_3, shown in Figure 4, planar?

Solution: Q_3 is planar, because it can be drawn without any edges crossing, as shown in Figure 5. ◀

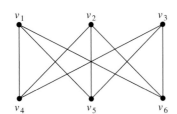

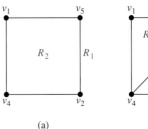

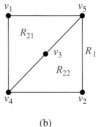

FIGURE 6   **The Graph $K_{3,3}$.**

(a)   (b)

FIGURE 7   **Showing that $K_{3,3}$ Is Nonplanar.**

We can show that a graph is planar by displaying a planar representation. It is harder to show that a graph is nonplanar. We will give an example to show how this can be done in an ad hoc fashion. Later we will develop some general results that can be used to do this.

EXAMPLE 3   Is $K_{3,3}$, shown in Figure 6, planar?

Solution: Any attempt to draw $K_{3,3}$ in the plane with no edges crossing is doomed. We now show why. In any planar representation of $K_{3,3}$, the vertices v_1 and v_2 must be connected to both v_4 and v_5. These four edges form a closed curve that splits the plane into two regions, R_1 and R_2, as shown in Figure 7(a). The vertex v_3 is in either R_1 or R_2. When v_3 is in R_2, the inside of the closed curve, the edges between v_3 and v_4 and between v_3 and v_5 separate R_2 into two subregions, R_{21} and R_{22}, as shown in Figure 7(b).

Next, note that there is no way to place the final vertex v_6 without forcing a crossing. For if v_6 is in R_1, then the edge between v_6 and v_3 cannot be drawn without a crossing. If v_6 is in R_{21}, then the edge between v_2 and v_6 cannot be drawn without a crossing. If v_6 is in R_{22}, then the edge between v_1 and v_6 cannot be drawn without a crossing.

A similar argument can be used when v_3 is in R_1. The completion of this argument is left for the reader (see Exercise 6). It follows that $K_{3,3}$ is not planar.   ◄

Example 3 solves the utilities-and-houses problem that was described at the beginning of this section. The three houses and three utilities cannot be connected in the plane without a crossing. A similar argument can be used to show that K_5 is nonplanar. (See Exercise 11.)

APPLICATIONS OF PLANAR GRAPHS   Planarity of graphs plays an important role in the design of electronic circuits. We can model a circuit with a graph by representing components of the circuit by vertices and connections between them by edges. We can print a circuit on a single board with no connections crossing if the graph representing the circuit is planar. When this graph is not planar, we must turn to more expensive options. For example, we can partition the vertices in the graph representing the circuit into planar subgraphs. We then construct the circuit using multiple layers. (See the preamble to Exercise 24 to learn about the thickness of a graph.) We can construct the circuit using insulated wires whenever connections cross. In this case, drawing the graph with the fewest possible crossings is important. (See the preamble to Exercise 22 to learn about the crossing number of a graph.)

The planarity of graphs is also useful in the design of road networks. Suppose we want to connect a group of cities by roads. We can model a road network connecting these cities using a simple graph with vertices representing the cities and edges representing the highways connecting them. We can built this road network without using underpasses or overpasses if the resulting graph is planar.

Euler's Formula

A planar representation of a graph splits the plane into **regions**, including an unbounded region. For instance, the planar representation of the graph shown in Figure 8 splits the plane into six

FIGURE 8   The Regions of the Planar Representation of a Graph.

regions. These are labeled in the figure. Euler showed that all planar representations of a graph split the plane into the same number of regions. He accomplished this by finding a relationship among the number of regions, the number of vertices, and the number of edges of a planar graph.

THEOREM 1   **EULER'S FORMULA**   Let G be a connected planar simple graph with e edges and v vertices. Let r be the number of regions in a planar representation of G. Then $r = e - v + 2$.

Proof: First, we specify a planar representation of G. We will prove the theorem by constructing a sequence of subgraphs $G_1, G_2, \ldots, G_e = G$, successively adding an edge at each stage. This is done using the following inductive definition. Arbitrarily pick one edge of G to obtain G_1. Obtain G_n from G_{n-1} by arbitrarily adding an edge that is incident with a vertex already in G_{n-1}, adding the other vertex incident with this edge if it is not already in G_{n-1}. This construction is possible because G is connected. G is obtained after e edges are added. Let r_n, e_n, and v_n represent the number of regions, edges, and vertices of the planar representation of G_n induced by the planar representation of G, respectively.

The proof will now proceed by induction. The relationship $r_1 = e_1 - v_1 + 2$ is true for G_1, because $e_1 = 1$, $v_1 = 2$, and $r_1 = 1$. This is shown in Figure 9.

Now assume that $r_k = e_k - v_k + 2$. Let $\{a_{k+1}, b_{k+1}\}$ be the edge that is added to G_k to obtain G_{k+1}. There are two possibilities to consider. In the first case, both a_{k+1} and b_{k+1} are already in G_k. These two vertices must be on the boundary of a common region R, or else it would be impossible to add the edge $\{a_{k+1}, b_{k+1}\}$ to G_k without two edges crossing (and G_{k+1} is planar). The addition of this new edge splits R into two regions. Consequently, in this case, $r_{k+1} = r_k + 1$, $e_{k+1} = e_k + 1$, and $v_{k+1} = v_k$. Thus, each side of the formula relating the number of regions, edges, and vertices increases by exactly one, so this formula is still true. In other words, $r_{k+1} = e_{k+1} - v_{k+1} + 2$. This case is illustrated in Figure 10(a).

In the second case, one of the two vertices of the new edge is not already in G_k. Suppose that a_{k+1} is in G_k but that b_{k+1} is not. Adding this new edge does not produce any new regions, because b_{k+1} must be in a region that has a_{k+1} on its boundary. Consequently, $r_{k+1} = r_k$. Moreover, $e_{k+1} = e_k + 1$ and $v_{k+1} = v_k + 1$. Each side of the formula relating the number of regions, edges, and vertices remains the same, so the formula is still true. In other words, $r_{k+1} = e_{k+1} - v_{k+1} + 2$. This case is illustrated in Figure 10(b).

We have completed the induction argument. Hence, $r_n = e_n - v_n + 2$ for all n. Because the original graph is the graph G_e, obtained after e edges have been added, the theorem is true. ◁

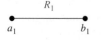

FIGURE 9   The Basis Case of the Proof of Euler's Formula.

Euler's formula is illustrated in Example 4.

EXAMPLE 4   Suppose that a connected planar simple graph has 20 vertices, each of degree 3. Into how many regions does a representation of this planar graph split the plane?

Solution: This graph has 20 vertices, each of degree 3, so $v = 20$. Because the sum of the degrees of the vertices, $3v = 3 \cdot 20 = 60$, is equal to twice the number of edges, $2e$, we have $2e = 60$, or $e = 30$. Consequently, from Euler's formula, the number of regions is

$$r = e - v + 2 = 30 - 20 + 2 = 12.$$

◀

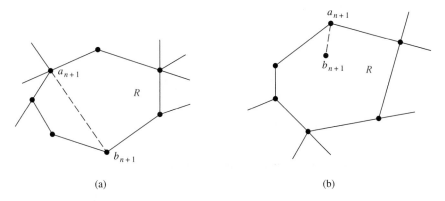

(a)   (b)

FIGURE 10   Adding an Edge to G_n to Produce G_{n+1}.

Euler's formula can be used to establish some inequalities that must be satisfied by planar graphs. One such inequality is given in Corollary 1.

COROLLARY 1   If G is a connected planar simple graph with e edges and v vertices, where $v \geq 3$, then $e \leq 3v - 6$.

Before we prove Corollary 1 we will use it to prove the following useful result.

COROLLARY 2   If G is a connected planar simple graph, then G has a vertex of degree not exceeding five.

Proof: If G has one or two vertices, the result is true. If G has at least three vertices, by Corollary 1 we know that $e \leq 3v - 6$, so $2e \leq 6v - 12$. If the degree of every vertex were at least six, then because $2e = \sum_{v \in V} \deg(v)$ (by the handshaking theorem), we would have $2e \geq 6v$. But this contradicts the inequality $2e \leq 6v - 12$. It follows that there must be a vertex with degree no greater than five.   ◁

The proof of Corollary 1 is based on the concept of the **degree** of a region, which is defined to be the number of edges on the boundary of this region. When an edge occurs twice on the boundary (so that it is traced out twice when the boundary is traced out), it contributes two to the degree. We denote the degree of a region R by $\deg(R)$. The degrees of the regions of the graph shown in Figure 11 are displayed in the figure.

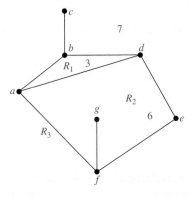

FIGURE 11   The Degrees of Regions.

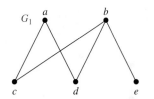
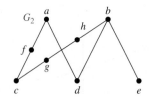
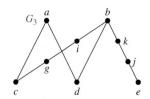

FIGURE 12   Homeomorphic Graphs.

The proof of Corollary 1 can now be given.

Proof: A connected planar simple graph drawn in the plane divides the plane into regions, say r of them. The degree of each region is at least three. (Because the graphs discussed here are simple graphs, no multiple edges that could produce regions of degree two, or loops that could produce regions of degree one, are permitted.) In particular, note that the degree of the unbounded region is at least three because there are at least three vertices in the graph.

Note that the sum of the degrees of the regions is exactly twice the number of edges in the graph, because each edge occurs on the boundary of a region exactly twice (either in two different regions, or twice in the same region). Because each region has degree greater than or equal to three, it follows that

$$2e = \sum_{\text{all regions } R} \deg(R) \geq 3r.$$

Hence,

$$(2/3)e \geq r.$$

Using $r = e - v + 2$ (Euler's formula), we obtain

$$e - v + 2 \leq (2/3)e.$$

It follows that $e/3 \leq v - 2$. This shows that $e \leq 3v - 6$.   ◁

This corollary can be used to demonstrate that K_5 is nonplanar.

EXAMPLE 5   Show that K_5 is nonplanar using Corollary 1.

Solution: The graph K_5 has five vertices and 10 edges. However, the inequality $e \leq 3v - 6$ is not satisfied for this graph because $e = 10$ and $3v - 6 = 9$. Therefore, K_5 is not planar.   ◀

It was previously shown that $K_{3,3}$ is not planar. Note, however, that this graph has six vertices and nine edges. This means that the inequality $e = 9 \leq 12 = 3 \cdot 6 - 6$ is satisfied. Consequently, the fact that the inequality $e \leq 3v - 6$ is satisfied does *not* imply that a graph is planar. However, the following corollary of Theorem 1 can be used to show that $K_{3,3}$ is nonplanar.

COROLLARY 3   If a connected planar simple graph has e edges and v vertices with $v \geq 3$ and no circuits of length three, then $e \leq 2v - 4$.

The proof of Corollary 3 is similar to that of Corollary 1, except that in this case the fact that there are no circuits of length three implies that the degree of a region must be at least four. The details of this proof are left for the reader (see Exercise 11).

EXAMPLE 6   Use Corollary 3 to show that $K_{3,3}$ is nonplanar.

Solution: Because $K_{3,3}$ has no circuits of length three (this is easy to see because it is bipartite), Corollary 3 can be used. $K_{3,3}$ has six vertices and nine edges. Because $e = 9$ and $2v - 4 = 8$, Corollary 3 shows that $K_{3,3}$ is nonplanar.   ◀

Kuratowski's Theorem

We have seen that $K_{3,3}$ and K_5 are not planar. Clearly, a graph is not planar if it contains either of these two graphs as a subgraph. Surprisingly, all nonplanar graphs must contain a subgraph that can be obtained from $K_{3,3}$ or K_5 using certain permitted operations.

If a graph is planar, so will be any graph obtained by removing an edge $\{u, v\}$ and adding a new vertex w together with edges $\{u, w\}$ and $\{w, v\}$. Such an operation is called an **elementary subdivision**. The graphs $G_1 = (V_1, E_1)$ and $G_2 = (V_2, E_2)$ are called **homeomorphic** if they can be obtained from the same graph by a sequence of elementary subdivisions.

EXAMPLE 7   Show that the graphs G_1, G_2, and G_3 displayed in Figure 12 are all homeomorphic.

Solution: These three graphs are homeomorphic because all three can be obtained from G_1 by elementary subdivisions. G_1 can be obtained from itself by an empty sequence of elementary subdivisions. To obtain G_2 from G_1 we can use this sequence of elementary subdivisions: (*i*) remove the edge $\{a, c\}$, add the vertex f, and add the edges $\{a, f\}$ and $\{f, c\}$; (*ii*) remove the edge $\{b, c\}$, add the vertex g, and add the edges $\{b, g\}$ and $\{g, c\}$; and (*iii*) remove the edge $\{b, g\}$, add the vertex h, and add the edges $\{g, h\}$ and $\{b, h\}$. We leave it to the reader to determine the sequence of elementary subdivisions needed to obtain G_3 from G_1.   ◀

The Polish mathematician Kazimierz Kuratowski established Theorem 2 in 1930, which characterizes planar graphs using the concept of graph homeomorphism.

THEOREM 2   A graph is nonplanar if and only if it contains a subgraph homeomorphic to $K_{3,3}$ or K_5.

It is clear that a graph containing a subgraph homeomorphic to $K_{3,3}$ or K_5 is nonplanar. However, the proof of the converse, namely that every nonplanar graph contains a subgraph homeomorphic to $K_{3,3}$ or K_5, is complicated and will not be given here. Examples 8 and 9 illustrate how Kuratowski's theorem is used.

EXAMPLE 8   Determine whether the graph G shown in Figure 13 is planar.

KAZIMIERZ KURATOWSKI (1896–1980)   Kazimierz Kuratowski, the son of a famous Warsaw lawyer, attended secondary school in Warsaw. He studied in Glasgow, Scotland, from 1913 to 1914 but could not return there after the outbreak of World War I. In 1915 he entered Warsaw University, where he was active in the Polish patriotic student movement. He published his first paper in 1919 and received his Ph.D. in 1921. He was an active member of the group known as the Warsaw School of Mathematics, working in the areas of the foundations of set theory and topology. He was appointed associate professor at the Lwów Polytechnical University, where he stayed for seven years, collaborating with the important Polish mathematicians Banach and Ulam. In 1930, while at Lwów, Kuratowski completed his work characterizing planar graphs.

In 1934 he returned to Warsaw University as a full professor. Until the start of World War II, he was active in research and teaching. During the war, because of the persecution of educated Poles, Kuratowski went into hiding under an assumed name and taught at the clandestine Warsaw University. After the war he helped revive Polish mathematics, serving as director of the Polish National Mathematics Institute. He wrote over 180 papers and three widely used textbooks.

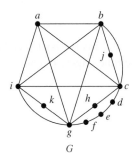

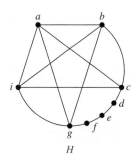

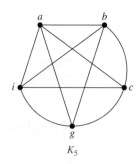

FIGURE 13   The Undirected Graph G, a Subgraph H Homeomorphic to K_5, and K_5.

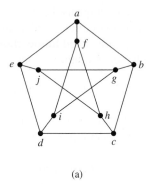

(a)

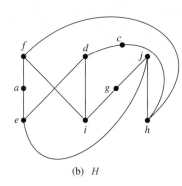

(b) H

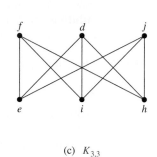

(c) $K_{3,3}$

FIGURE 14   (a) The Petersen Graph, (b) a Subgraph H Homeomorphic to $K_{3,3}$, and (c) $K_{3,3}$.

Extra Examples

Solution: G has a subgraph H homeomorphic to K_5. H is obtained by deleting h, j, and k and all edges incident with these vertices. H is homeomorphic to K_5 because it can be obtained from K_5 (with vertices a, b, c, g, and i) by a sequence of elementary subdivisions, adding the vertices d, e, and f. (The reader should construct such a sequence of elementary subdivisions.) Hence, G is nonplanar.   ◀

EXAMPLE 9   Is the Petersen graph, shown in Figure 14(a), planar? (The Danish mathematician Julius Petersen studied this graph in 1891; it is often used to illustrate various theoretical properties of graphs.)

Solution: The subgraph H of the Petersen graph obtained by deleting b and the three edges that have b as an endpoint, shown in Figure 14(b), is homeomorphic to $K_{3,3}$, with vertex sets $\{f, d, j\}$ and $\{e, i, h\}$, because it can be obtained by a sequence of elementary subdivisions, deleting $\{d, h\}$ and adding $\{c, h\}$ and $\{c, d\}$, deleting $\{e, f\}$ and adding $\{a, e\}$ and $\{a, f\}$, and deleting $\{i, j\}$ and adding $\{g, i\}$ and $\{g, j\}$. Hence, the Petersen graph is not planar.   ◀

Exercises

1. Can five houses be connected to two utilities without connections crossing?

In Exercise 2 draw the given planar graph without any crossings.

2.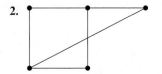

In Exercises 3–5 determine whether the given graph is planar. If so, draw it so that no edges cross.

3.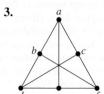

4.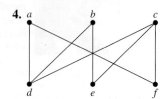

5.

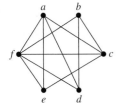

6. Complete the argument in Example 3.

7. Show that K_5 is nonplanar using an argument similar to that given in Example 3.

8. Suppose that a connected planar graph has eight vertices, each of degree three. Into how many regions is the plane divided by a planar representation of this graph?

9. Suppose that a connected planar graph has six vertices, each of degree four. Into how many regions is the plane divided by a planar representation of this graph?

10. Suppose that a connected planar graph has 30 edges. If a planar representation of this graph divides the plane into 20 regions, how many vertices does this graph have?

11. Prove Corollary 3.

12. Suppose that a connected bipartite planar simple graph has e edges and v vertices. Show that $e \leq 2v - 4$ if $v \geq 3$.

*13. Suppose that a connected planar simple graph with e edges and v vertices contains no simple circuits of length 4 or less. Show that $e \leq (5/3)v - (10/3)$ if $v \geq 4$.

14. Suppose that a planar graph has k connected components, e edges, and v vertices. Also suppose that the plane is divided into r regions by a planar representation of the graph. Find a formula for r in terms of e, v, and k.

15. Which of these nonplanar graphs have the property that the removal of any vertex and all edges incident with that vertex produces a planar graph?

  a) K_5 **b)** K_6 **c)** $K_{3,3}$ **d)** $K_{3,4}$

In Exercises 16–18 determine whether the given graph is homeomorphic to $K_{3,3}$.

16.

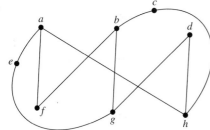

17.

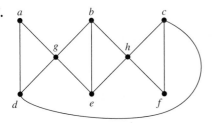

18.

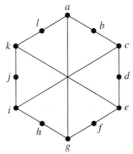

In Exercises 19–21 use Kuratowski's theorem to determine whether the given graph is planar.

19.

20.

21.

The **crossing number** of a simple graph is the minimum number of crossings that can occur when this graph is drawn in the plane where no three arcs representing edges are permitted to cross at the same point.

22. Show that $K_{3,3}$ has 1 as its crossing number.

**23. Find the crossing numbers of each of these nonplanar graphs.

  a) K_5 **b)** K_6 **c)** K_7
  d) $K_{3,4}$ **e)** $K_{4,4}$ **f)** $K_{5,5}$

The **thickness** of a simple graph G is the smallest number of planar subgraphs of G that have G as their union.

24. Show that $K_{3,3}$ has 2 as its thickness.

*25. Find the thickness of the graphs in Exercise 23.

26. Show that if G is a connected simple graph with v vertices and e edges, where $v \geq 3$, then the thickness of G is at least $\lceil e/(3v - 6) \rceil$.

*27. Use Exercise 26 to show that the thickness of K_n is at least $\lfloor (n + 7)/6 \rfloor$ whenever n is a positive integer.

10.8 Graph Coloring

Introduction

Problems related to the coloring of maps of regions, such as maps of parts of the world, have generated many results in graph theory. When a map* is colored, two regions with a common border are customarily assigned different colors. One way to ensure that two adjacent regions never have the same color is to use a different color for each region. However, this is inefficient, and on maps with many regions it would be hard to distinguish similar colors. Instead, a small number of colors should be used whenever possible. Consider the problem of determining the least number of colors that can be used to color a map so that adjacent regions never have the same color. For instance, for the map shown on the left in Figure 1, four colors suffice, but three colors are not enough. (The reader should check this.) In the map on the right in Figure 1, three colors are sufficient (but two are not).

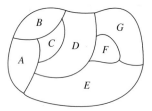 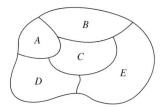

FIGURE 1   Two Maps.

Each map in the plane can be represented by a graph. To set up this correspondence, each region of the map is represented by a vertex. Edges connect two vertices if the regions represented by these vertices have a common border. Two regions that touch at only one point are not considered adjacent. The resulting graph is called the **dual graph** of the map. By the way in which dual graphs of maps are constructed, it is clear that any map in the plane has a planar dual graph. Figure 2 displays the dual graphs that correspond to the maps shown in Figure 1.

The problem of coloring the regions of a map is equivalent to the problem of coloring the vertices of the dual graph so that no two adjacent vertices in this graph have the same color. We now define a graph coloring.

DEFINITION 1   A *coloring* of a simple graph is the assignment of a color to each vertex of the graph so that no two adjacent vertices are assigned the same color.

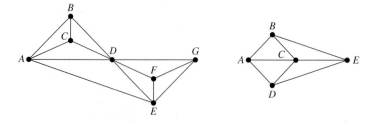

FIGURE 2   Dual Graphs of the Maps in Figure 1.

---

*We will assume that all regions in a map are connected. This eliminates any problems presented by such geographical entities as Michigan.

A graph can be colored by assigning a different color to each of its vertices. However, for most graphs a coloring can be found that uses fewer colors than the number of vertices in the graph. What is the least number of colors necessary?

DEFINITION 2 The *chromatic number* of a graph is the least number of colors needed for a coloring of this graph. The chromatic number of a graph G is denoted by $\chi(G)$. (Here χ is the Greek letter *chi*.)

Note that asking for the chromatic number of a planar graph is the same as asking for the minimum number of colors required to color a planar map so that no two adjacent regions are assigned the same color. This question has been studied for more than 100 years. The answer is provided by one of the most famous theorems in mathematics.

THEOREM 1 **THE FOUR COLOR THEOREM** The chromatic number of a planar graph is no greater than four.

The four color theorem was originally posed as a conjecture in the 1850s. It was finally proved by the American mathematicians Kenneth Appel and Wolfgang Haken in 1976. Prior to 1976, many incorrect proofs were published, often with hard-to-find errors. In addition, many futile attempts were made to construct counterexamples by drawing maps that require more than four colors. (Proving the five color theorem is not that difficult; see Exercise 26.)

Perhaps the most notorious fallacious proof in all of mathematics is the incorrect proof of the four color theorem published in 1879 by a London barrister and amateur mathematician, Alfred Kempe. Mathematicians accepted his proof as correct until 1890, when Percy Heawood found an error that made Kempe's argument incomplete. However, Kempe's line of reasoning turned out to be the basis of the successful proof given by Appel and Haken. Their proof relies on a careful case-by-case analysis carried out by computer. They showed that if the four color theorem were false, there would have to be a counterexample of one of approximately 2000 different types, and they then showed that none of these types exists. They used over 1000 hours of computer time in their proof. This proof generated a large amount of controversy, because computers played such an important role in it. For example, could there be an error in a computer program that led to incorrect results? Was their argument really a proof if it depended on what could be unreliable computer output? Since their proof appeared, simpler proofs that rely on checking fewer types of possible counterexamples have been found and a proof using an automated proof system has been created. However, no proof not relying on a computer has yet been found.

Note that the four color theorem applies only to planar graphs. Nonplanar graphs can have arbitrarily large chromatic numbers, as will be shown in Example 2.

Two things are required to show that the chromatic number of a graph is k. First, we must show that the graph can be colored with k colors. This can be done by constructing such a coloring. Second, we must show that the graph cannot be colored using fewer than k colors. Examples 1–4 illustrate how chromatic numbers can be found.

ALFRED BRAY KEMPE (1849–1922) Kempe was a barrister and a leading authority on ecclesiastical law. However, having studied mathematics at Cambridge University, he retained his interest in it, and later in life he devoted considerable time to mathematical research. Kempe made contributions to kinematics, the branch of mathematics dealing with motion, and to mathematical logic. However, Kempe is best remembered for his fallacious proof of the four color theorem.

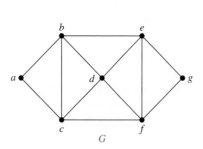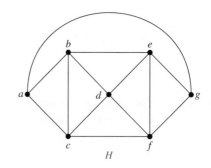

FIGURE 3 **The Simple Graphs G and H.**

EXAMPLE 1 What are the chromatic numbers of the graphs G and H shown in Figure 3?

Solution: The chromatic number of G is at least three, because the vertices a, b, and c must be assigned different colors. To see if G can be colored with three colors, assign red to a, blue to b, and green to c. Then, d can (and must) be colored red because it is adjacent to b and c. Furthermore, e can (and must) be colored green because it is adjacent only to vertices colored red and blue, and f can (and must) be colored blue because it is adjacent only to vertices colored red and green. Finally, g can (and must) be colored red because it is adjacent only to vertices colored blue and green. This produces a coloring of G using exactly three colors. Figure 4 displays such a coloring.

The graph H is made up of the graph G with an edge connecting a and g. Any attempt to color H using three colors must follow the same reasoning as that used to color G, except at the last stage, when all vertices other than g have been colored. Then, because g is adjacent (in H) to vertices colored red, blue, and green, a fourth color, say brown, needs to be used. Hence, H has a chromatic number equal to 4. A coloring of H is shown in Figure 4. ◀

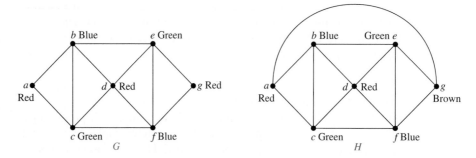

FIGURE 4 **Colorings of the Graphs G and H.**

HISTORICAL NOTE   In 1852, an ex-student of Augustus De Morgan, Francis Guthrie, noticed that the counties in England could be colored using four colors so that no adjacent counties were assigned the same color. On this evidence, he conjectured that the four color theorem was true. Francis told his brother Frederick, at that time a student of De Morgan, about this problem. Frederick in turn asked his teacher De Morgan about his brother's conjecture. De Morgan was extremely interested in this problem and publicized it throughout the mathematical community. In fact, the first written reference to the conjecture can be found in a letter from De Morgan to Sir William Rowan Hamilton. Although De Morgan thought Hamilton would be interested in this problem, Hamilton apparently was not interested in it, because it had nothing to do with quaternions.

HISTORICAL NOTE   Although a simpler proof of the four color theorem was found by Robertson, Sanders, Seymour, and Thomas in 1996, reducing the computational part of the proof to examining 633 configurations, no proof that does not rely on extensive computation has yet been found.

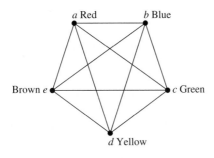

FIGURE 5 **A Coloring of K_5.**

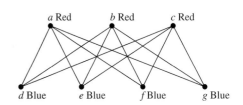

FIGURE 6 **A Coloring of $K_{3,4}$.**

EXAMPLE 2 What is the chromatic number of K_n?

Solution: A coloring of K_n can be constructed using n colors by assigning a different color to each vertex. Is there a coloring using fewer colors? The answer is no. No two vertices can be assigned the same color, because every two vertices of this graph are adjacent. Hence, the chromatic number of K_n is n. That is, $\chi(K_n) = n$. (Recall that K_n is not planar when $n \geq 5$, so this result does not contradict the four color theorem.) A coloring of K_5 using five colors is shown in Figure 5. ◄

EXAMPLE 3 What is the chromatic number of the complete bipartite graph $K_{m,n}$, where m and n are positive integers?

Solution: The number of colors needed may seem to depend on m and n. However, as Theorem 4 in Section 10.2 tells us, only two colors are needed, because $K_{m,n}$ is a bipartite graph. Hence, $\chi(K_{m,n}) = 2$. This means that we can color the set of m vertices with one color and the set of n vertices with a second color. Because edges connect only a vertex from the set of m vertices and a vertex from the set of n vertices, no two adjacent vertices have the same color. A coloring of $K_{3,4}$ with two colors is displayed in Figure 6. ◄

EXAMPLE 4 What is the chromatic number of the graph C_n, where $n \geq 3$? (Recall that C_n is the cycle with n vertices.)

Solution: We will first consider some individual cases. To begin, let $n = 6$. Pick a vertex and color it red. Proceed clockwise in the planar depiction of C_6 shown in Figure 7. It is necessary to assign a second color, say blue, to the next vertex reached. Continue in the clockwise direction; the third vertex can be colored red, the fourth vertex blue, and the fifth vertex red. Finally, the sixth vertex, which is adjacent to the first, can be colored blue. Hence, the chromatic number of C_6 is 2. Figure 7 displays the coloring constructed here.

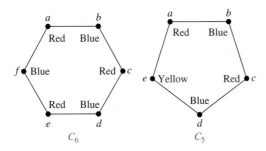

FIGURE 7 **Colorings of C_5 and C_6.**

Next, let $n = 5$ and consider C_5. Pick a vertex and color it red. Proceeding clockwise, it is necessary to assign a second color, say blue, to the next vertex reached. Continuing in the clockwise direction, the third vertex can be colored red, and the fourth vertex can be colored blue. The fifth vertex cannot be colored either red or blue, because it is adjacent to the fourth vertex and the first vertex. Consequently, a third color is required for this vertex. Note that we would have also needed three colors if we had colored vertices in the counterclockwise direction. Thus, the chromatic number of C_5 is 3. A coloring of C_5 using three colors is displayed in Figure 7.

In general, two colors are needed to color C_n when n is even. To construct such a coloring, simply pick a vertex and color it red. Proceed around the graph in a clockwise direction (using a planar representation of the graph) coloring the second vertex blue, the third vertex red, and so on. The nth vertex can be colored blue, because the two vertices adjacent to it, namely the $(n - 1)$st and the first vertices, are both colored red.

When n is odd and $n > 1$, the chromatic number of C_n is 3. To see this, pick an initial vertex. To use only two colors, it is necessary to alternate colors as the graph is traversed in a clockwise direction. However, the nth vertex reached is adjacent to two vertices of different colors, namely, the first and $(n - 1)$st. Hence, a third color must be used.

We have shown that $\chi(C_n) = 2$ if n is an even positive integer with $n \geq 4$ and $\chi(C_n) = 3$ if n is an odd positive integer with $n \geq 3$.   ◄

Links

The best algorithms known for finding the chromatic number of a graph have exponential worst-case time complexity (in the number of vertices of the graph). Even the problem of finding an approximation to the chromatic number of a graph is difficult. It has been shown that if there were an algorithm with polynomial worst-case time complexity that could approximate the chromatic number of a graph up to a factor of 2 (that is, construct a bound that was no more than double the chromatic number of the graph), then an algorithm with polynomial worst-case time complexity for finding the chromatic number of the graph would also exist.

Applications of Graph Colorings

Graph coloring has a variety of applications to problems involving scheduling and assignments. (Note that because no efficient algorithm is known for graph coloring, this does not lead to efficient algorithms for scheduling and assignments.) Examples of such applications will be given here. The first application deals with the scheduling of final exams.

EXAMPLE 5   **Scheduling Final Exams**   How can the final exams at a university be scheduled so that no student has two exams at the same time?

Solution: This scheduling problem can be solved using a graph model, with vertices representing courses and with an edge between two vertices if there is a common student in the courses they represent. Each time slot for a final exam is represented by a different color. A scheduling of the exams corresponds to a coloring of the associated graph.

For instance, suppose there are seven finals to be scheduled. Suppose the courses are numbered 1 through 7. Suppose that the following pairs of courses have common students: 1 and 2, 1 and 3, 1 and 4, 1 and 7, 2 and 3, 2 and 4, 2 and 5, 2 and 7, 3 and 4, 3 and 6, 3 and 7, 4 and 5, 4 and 6, 5 and 6, 5 and 7, and 6 and 7. In Figure 8 the graph associated with this set of classes is shown. A scheduling consists of a coloring of this graph.

Because the chromatic number of this graph is 4 (the reader should verify this), four time slots are needed. A coloring of the graph using four colors and the associated schedule are shown in Figure 9.   ◄

Now consider an application to the assignment of television channels.

EXAMPLE 6   **Frequency Assignments**   Television channels 2 through 13 are assigned to stations in North America so that no two stations within 150 miles can operate on the same channel. How can the assignment of channels be modeled by graph coloring?

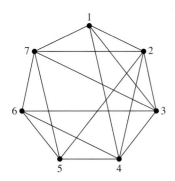

FIGURE 8 **The Graph Representing the Scheduling of Final Exams.**

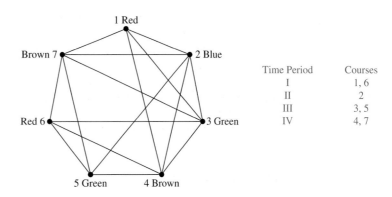

FIGURE 9 **Using a Coloring to Schedule Final Exams.**

Solution: Construct a graph by assigning a vertex to each station. Two vertices are connected by an edge if they are located within 150 miles of each other. An assignment of channels corresponds to a coloring of the graph, where each color represents a different channel. ◄

An application of graph coloring to compilers is considered in Example 7.

EXAMPLE 7 **Index Registers** In efficient compilers the execution of loops is speeded up when frequently used variables are stored temporarily in index registers in the central processing unit, instead of in regular memory. For a given loop, how many index registers are needed? This problem can be addressed using a graph coloring model. To set up the model, let each vertex of a graph represent a variable in the loop. There is an edge between two vertices if the variables they represent must be stored in index registers at the same time during the execution of the loop. Thus, the chromatic number of the graph gives the number of index registers needed, because different registers must be assigned to variables when the vertices representing these variables are adjacent in the graph. ◄

Exercises

In Exercises 1–2 construct the dual graph for the map shown. Then find the number of colors needed to color the map so that no two adjacent regions have the same color.

1.

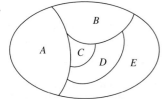

2.

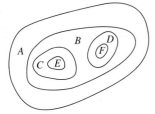

In Exercises 3–5 find the chromatic number of the given graph.

3.

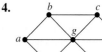

4.

5.

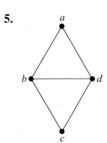

6. For the graphs in Exercises 3–5, decide whether it is possible to decrease the chromatic number by removing a single vertex and all edges incident with it.

7. Which graphs have a chromatic number of 1?

8. What is the least number of colors needed to color a map of the United States? Do not consider adjacent states that meet only at a corner. Suppose that Michigan is one region. Consider the vertices representing Alaska and Hawaii as isolated vertices.

9. What is the chromatic number of W_n?

10. Show that a simple graph that has a circuit with an odd number of vertices in it cannot be colored using two colors.

11. Schedule the final exams for Math 115, Math 116, Math 185, Math 195, CS 101, CS 102, CS 273, and CS 473, using the fewest number of different time slots, if there are no students taking both Math 115 and CS 473, both Math 116 and CS 473, both Math 195 and CS 101, both Math 195 and CS 102, both Math 115 and Math 116, both Math 115 and Math 185, and both Math 185 and Math 195, but there are students in every other pair of courses.

12. How many different channels are needed for six stations located at the distances shown in the table, if two stations cannot use the same channel when they are within 150 miles of each other?

	1	*2*	*3*	*4*	*5*	*6*
1	—	85	175	200	50	100
2	85	—	125	175	100	160
3	175	125	—	100	200	250
4	200	175	100	—	210	220
5	50	100	200	210	—	100
6	100	160	250	220	100	—

13. The mathematics department has six committees, each meeting once a month. How many different meeting times must be used to ensure that no member is scheduled to attend two meetings at the same time if the committees are $C_1 = \{$Arlinghaus, Brand, Zaslavsky$\}$, $C_2 = \{$Brand, Lee, Rosen$\}$, $C_3 = \{$Arlinghaus, Rosen, Zaslavsky$\}$, $C_4 = \{$Lee, Rosen, Zaslavsky$\}$, $C_5 = \{$Arlinghaus, Brand$\}$, and $C_6 = \{$Brand, Rosen, Zaslavsky$\}$?

14. A zoo wants to set up natural habitats in which to exhibit its animals. Unfortunately, some animals will eat some of the others when given the opportunity. How can a graph model and a coloring be used to determine the number of different habitats needed and the placement of the animals in these habitats?

An **edge coloring** of a graph is an assignment of colors to edges so that edges incident with a common vertex are assigned different colors. The **edge chromatic number** of a graph is the smallest number of colors that can be used in an edge coloring of the graph. The edge chromatic number of a graph G is denoted by $\chi'(G)$.

15. Find the edge chromatic number of each of the graphs in Exercises 3–5.

16. Find the edge chromatic numbers of
  a) C_n, where $n \geq 3$.
  b) W_n, where $n \geq 3$.

17. Show that if G is a graph with n vertices, then no more than $n/2$ edges can be colored the same in an edge coloring of G.

18. What can be said about the chromatic number of a graph that has K_n as a subgraph?

This algorithm can be used to color a simple graph: First, list the vertices $v_1, v_2, v_3, \ldots, v_n$ in order of decreasing degree so that $\deg(v_1) \geq \deg(v_2) \geq \cdots \geq \deg(v_n)$. Assign color 1 to v_1 and to the next vertex in the list not adjacent to v_1 (if one exists), and successively to each vertex in the list not adjacent to a vertex already assigned color 1. Then assign color 2 to the first vertex in the list not already colored. Successively assign color 2 to vertices in the list that have not already been colored and are not adjacent to vertices assigned color 2. If uncolored vertices remain, assign color 3 to the first vertex in the list not yet colored, and use color 3 to successively color those vertices not already colored and not adjacent to vertices assigned color 3. Continue this process until all vertices are colored.

19. Construct a coloring of the graph shown using this algorithm.

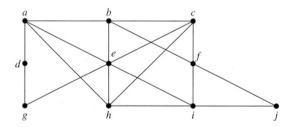

*20. Use pseudocode to describe this coloring algorithm.

*21. Show that the coloring produced by this algorithm may use more colors than are necessary to color a graph.

A connected graph G is called **chromatically k-critical** if the chromatic number of G is k, but for every edge of G, the chromatic number of the graph obtained by deleting this edge from G is $k - 1$.

22. Show that C_n is chromatically 3-critical whenever n is an odd positive integer, $n \geq 3$.

23. Show that W_n is chromatically 4-critical whenever n is an odd integer, $n \geq 3$.

24. Show that W_4 is not chromatically 3-critical.

25. Show that if G is a chromatically k-critical graph, then the degree of every vertex of G is at least $k - 1$.

A **k-tuple coloring** of a graph G is an assignment of a set of k different colors to each of the vertices of G such that no two adjacent vertices are assigned a common color. We denote by $\chi_k(G)$ the smallest positive integer n such that G has a k-tuple coloring using n colors. For example, $\chi_2(C_4) = 4$. To see this, note that using only four colors we can assign two colors to each vertex of C_4, as illustrated, so that no two adjacent vertices are assigned the same color. Furthermore, no fewer than four colors suffice because the vertices v_1 and v_2 each must be assigned two colors, and a common color cannot be assigned to both v_1 and v_2. (For more information about k-tuple coloring, see [MiRo91].)

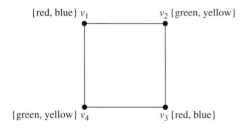

26. Find these values:
   a) $\chi_2(K_3)$ **b)** $\chi_2(K_4)$ **c)** $\chi_2(W_4)$
   d) $\chi_2(C_5)$ **e)** $\chi_2(K_{3,4})$ **f)** $\chi_3(K_5)$
   *g) $\chi_3(C_5)$ **h)** $\chi_3(K_{4,5})$

*27. Let G and H be the graphs displayed in Figure 3. Find
   a) $\chi_2(G)$.   **b)** $\chi_2(H)$.
   c) $\chi_3(G)$.   **d)** $\chi_3(H)$.

28. What is $\chi_k(G)$ if G is a bipartite graph and k is a positive integer?

29. Frequencies for mobile radio (or cellular) telephones are assigned by zones. Each zone is assigned a set of frequencies to be used by vehicles in that zone. The same frequency cannot be used in different zones when interference can occur between telephones in these zones. Explain how a k-tuple coloring can be used to assign k frequencies to each mobile radio zone in a region.

*30. Show that every planar graph G can be colored using six or fewer colors. [*Hint:* Use mathematical induction on the number of vertices of the graph. Apply Corollary 2 of Section 10.7 to find a vertex v with $\deg(v) \leq 5$. Consider the subgraph of G obtained by deleting v and all edges incident with it.]

**31. Show that every planar graph G can be colored using five or fewer colors. [*Hint:* Use the hint provided for Exercise 30.]

The famous Art Gallery Problem asks how many guards are needed to see all parts of an art gallery, where the gallery is the interior and boundary of a polygon with n sides. To state this problem more precisely, we need some terminology. A point x inside or on the boundary of a simple polygon P **covers** or **sees** a point y inside or on P if all points on the line segment xy are in the interior or on the boundary of P. We say that a set of points is a **guarding set** of a simple polygon P if for every point y inside P or on the boundary of P there is a point x in this guarding set that sees y. Denote by $G(P)$ the minimum number of points needed to guard the simple polygon P. The **art gallery problem** asks for the function $g(n)$, which is the maximum value of $G(P)$ over all simple polygons with n vertices. That is, $g(n)$ is the minimum positive integer for which it is guaranteed that a simple polygon with n vertices can be guarded with $g(n)$ or fewer guards.

32. Show that $g(3) = 1$ and $g(4) = 1$ by showing that all triangles and quadrilaterals can be guarded using one point.

*33. Show that $g(5) = 1$. That is, show that all pentagons can be guarded using one point. [*Hint:* Show that there are either 0, 1, or 2 vertices with an interior angle greater than 180 degrees and that in each case, one guard suffices.]

*34. Show that $g(6) = 2$ by first using Exercises 32 and 33 as well as Lemma 1 in Section 5.2 to show that $g(6) \leq 2$ and then find a simple hexagon for which two guards are needed.

*35. Show that $g(n) \geq \lfloor n/3 \rfloor$. [*Hint:* Consider the polygon with $3k$ vertices that resembles a comb with k prongs, such as the polygon with 15 sides shown here.]

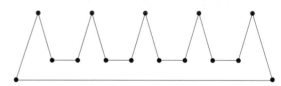

*36. Solve the art gallery problem by proving the **art gallery theorem**, which states that at most $\lfloor n/3 \rfloor$ guards are needed to guard the interior and boundary of a simple polygon with n vertices. [*Hint:* Use Theorem 1 in Section 5.2 to triangulate the simple polygon into $n - 2$ triangles. Then show that it is possible to color the vertices of the triangulated polygon using three colors so that no two adjacent vertices have the same color. Use induction and Exercise 17 in Section 5.2. Finally, put guards at all vertices that are colored red, where red is the color used least in the coloring of the vertices. Show that placing guards at these points is all that is needed.]

Key Terms and Results

TERMS

undirected edge: an edge associated to a set $\{u, v\}$, where u and v are vertices

directed edge: an edge associated to an ordered pair (u, v), where u and v are vertices

multiple edges: distinct edges connecting the same vertices

multiple directed edges: distinct directed edges associated with the same ordered pair (u, v), where u and v are vertices

loop: an edge connecting a vertex with itself

undirected graph: a set of vertices and a set of undirected edges each of which is associated with a set of one or two of these vertices

simple graph: an undirected graph with no multiple edges or loops

multigraph: an undirected graph that may contain multiple edges but no loops

pseudograph: an undirected graph that may contain multiple edges and loops

directed graph: a set of vertices together with a set of directed edges each of which is associated with an ordered pair of vertices

directed multigraph: a graph with directed edges that may contain multiple directed edges

simple directed graph: a directed graph without loops or multiple directed edges

adjacent: two vertices are adjacent if there is an edge between them

incident: an edge is incident with a vertex if the vertex is an endpoint of that edge

deg v (degree of the vertex v in an undirected graph): the number of edges incident with v with loops counted twice

$\deg^-(v)$ (the in-degree of the vertex v in a graph with directed edges): the number of edges with v as their terminal vertex

$\deg^+(v)$ (the out-degree of the vertex v in a graph with directed edges): the number of edges with v as their initial vertex

underlying undirected graph of a graph with directed edges: the undirected graph obtained by ignoring the directions of the edges

K_n (complete graph on n vertices): the undirected graph with n vertices where each pair of vertices is connected by an edge

bipartite graph: a graph with vertex set that can be partitioned into subsets V_1 and V_2 so that each edge connects a vertex in V_1 and a vertex in V_2. The pair (V_1, V_2) is called a **bipartition** of V.

$K_{m,n}$ (complete bipartite graph): the graph with vertex set partitioned into a subset of m elements and a subset of n elements with two vertices connected by an edge if and only if one is in the first subset and the other is in the second subset

C_n (cycle of size n), $n \geq 3$: the graph with n vertices $v_1, v_2, \ldots, v_n$ and edges $\{v_1, v_2\}, \{v_2, v_3\}, \ldots, \{v_{n-1}, v_n\}$, $\{v_n, v_1\}$

W_n (wheel of size n), $n \geq 3$: the graph obtained from C_n by adding a vertex and edges from this vertex to the original vertices in C_n

Q_n (n-cube), $n \geq 1$: the graph that has the 2^n bit strings of length n as its vertices and edges connecting every pair of bit strings that differ by exactly one bit

matching in a graph G: a set of edges such that no two edges have a common endpoint

complete matching M from V_1 to V_2: a matching such that every vertex in V_1 is an endpoint of an edge in M

maximum matching: a matching containing the most edges among all matchings in a graph

isolated vertex: a vertex of degree zero

pendant vertex: a vertex of degree one

regular graph: a graph where all vertices have the same degree

subgraph of a graph $G = (V, E)$: a graph (W, F), where W is a subset of V and F is a subset of E

$G_1 \cup G_2$ (union of G_1 and G_2): the graph $(V_1 \cup V_2, E_1 \cup E_2)$, where $G_1 = (V_1, E_1)$ and $G_2 = (V_2, E_2)$

adjacency matrix: a matrix representing a graph using the adjacency of vertices

incidence matrix: a matrix representing a graph using the incidence of edges and vertices

isomorphic simple graphs: the simple graphs $G_1 = (V_1, E_1)$ and $G_2 = (V_2, E_2)$ are isomorphic if there exists a one-to-one correspondence f from V_1 to V_2 such that $\{f(v_1), f(v_2)\} \in E_2$ if and only if $\{v_1, v_2\} \in E_1$ for all v_1 and v_2 in V_1

invariant for graph isomorphism: a property that isomorphic graphs either both have or both do not have

path from u to v in an undirected graph: a sequence of edges $e_1, e_2, \ldots, e_n$, where e_i is associated to $\{x_i, x_{i+1}\}$ for $i = 0, 1, \ldots, n$, where $x_0 = u$ and $x_{n+1} = v$

path from u to v in a graph with directed edges: a sequence of edges $e_1, e_2, \ldots, e_n$, where e_i is associated to (x_i, x_{i+1}) for $i = 0, 1, \ldots, n$, where $x_0 = u$ and $x_{n+1} = v$

simple path: a path that does not contain an edge more than once

circuit: a path of length $n \geq 1$ that begins and ends at the same vertex

connected graph: an undirected graph with the property that there is a path between every pair of vertices

cut vertex of *G*: a vertex *v* such that *G* − *v* is disconnected

cut edge of *G*: an edge *e* such that *G* − *e* is disconnected

nonseparable graph: a graph without a cut vertex

vertex cut of *G*: a subset *V′* of the set of vertices of *G* such that *G* − *V′* is disconnected

$\kappa(G)$ (the vertex connectivity of *G*): the size of a smallest vertex cut of *G*

k-connected graph: a graph that has a vertex connectivity no smaller than *k*

edge cut of *G*: a set of edges *E′* of *G* such that *G* − *E′* is disconnected

$\lambda(G)$ (the edge connectivity of *G*): the size of a smallest edge cut of *G*

connected component of a graph *G*: a maximal connected subgraph of *G*

strongly connected directed graph: a directed graph with the property that there is a directed path from every vertex to every vertex

strongly connected component of a directed graph *G*: a maximal strongly connected subgraph of *G*

Euler path: a path that contains every edge of a graph exactly once

Euler circuit: a circuit that contains every edge of a graph exactly once

Hamilton path: a path in a graph that passes through each vertex exactly once

Hamilton circuit: a circuit in a graph that passes through each vertex exactly once

weighted graph: a graph with numbers assigned to its edges

shortest-path problem: the problem of determining the path in a weighted graph such that the sum of the weights of the edges in this path is a minimum over all paths between specified vertices

traveling salesperson problem: the problem that asks for the circuit of shortest total length that visits every vertex of a weghted graph exactly once

planar graph: a graph that can be drawn in the plane with no crossings

regions of a representation of a planar graph: the regions the plane is divided into by the planar representation of the graph

elementary subdivision: the removal of an edge {*u*, *v*} of an undirected graph and the addition of a new vertex *w* together with edges {*u*, *w*} and {*w*, *v*}

homeomorphic: two undirected graphs are homeomorphic if they can be obtained from the same graph by a sequence of elementary subdivisions

graph coloring: an assignment of colors to the vertices of a graph so that no two adjacent vertices have the same color

chromatic number: the minimum number of colors needed in a coloring of a graph

RESULTS

The handshaking theorem: If $G = (V, E)$ be an undirected graph with *m* edges, then $2m = \sum_{v \in V} \deg(v)$.

Hall's marriage theorem: The bipartite graph $G = (V, E)$ with bipartition (V_1, V_2) has a complete matching from V_1 to V_2 if and only if $|N(A)| \geq |A|$ for all subsets *A* of V_1.

There is an Euler circuit in a connected multigraph if and only if every vertex has even degree.

There is an Euler path in a connected multigraph if and only if at most two vertices have odd degree.

Dijkstra's algorithm: a procedure for finding a shortest path between two vertices in a weighted graph (see Section 10.6).

Euler's formula: $r = e - v + 2$ where *r*, *e*, and *v* are the number of regions of a planar representation, the number of edges, and the number of vertices, respectively, of a connected planar graph.

Kuratowski's theorem: A graph is nonplanar if and only if it contains a subgraph homeomorphic to $K_{3,3}$ or K_5. (Proof beyond scope of this book.)

The four color theorem: Every planar graph can be colored using no more than four colors. (Proof far beyond the scope of this book!)

Review Questions

1. a) Define a simple graph, a multigraph, a pseudograph, a directed graph, and a directed multigraph.

  b) Use an example to show how each of the types of graph in part (a) can be used in modeling. For example, explain how to model different aspects of a computer network or airline routes.

2. Give at least four examples of how graphs are used in modeling.

3. What is the relationship between the sum of the degrees of the vertices in an undirected graph and the number of edges in this graph? Explain why this relationship holds.

4. Why must there be an even number of vertices of odd degree in an undirected graph?

5. What is the relationship between the sum of the in-degrees and the sum of the out-degrees of the vertices in a directed graph? Explain why this relationship holds.

6. Describe the following families of graphs.

  a) K_n, the complete graph on *n* vertices
  b) $K_{m,n}$, the complete bipartite graph on *m* and *n* vertices
  c) C_n, the cycle with *n* vertices
  d) W_n, the wheel of size *n*
  e) Q_n, the *n*-cube

7. How many vertices and how many edges are there in each of the graphs in the families in Question 6?

8. a) What is a bipartite graph?

  b) Which of the graphs K_n, C_n, and W_n are bipartite?

  c) How can you determine whether an undirected graph is bipartite?

9. a) Describe three different methods that can be used to represent a graph.

b) Draw a simple graph with at least five vertices and eight edges. Illustrate how it can be represented using the methods you described in part (a).

10. a) What does it mean for two simple graphs to be isomorphic?

b) What is meant by an invariant with respect to isomorphism for simple graphs? Give at least five examples of such invariants.

c) Give an example of two graphs that have the same numbers of vertices, edges, and degrees of vertices, but that are not isomorphic.

d) Is a set of invariants known that can be used to efficiently determine whether two simple graphs are isomorphic?

11. a) What does it mean for a graph to be connected?

b) What are the connected components of a graph?

12. a) Explain how an adjacency matrix can be used to represent a graph.

b) How can adjacency matrices be used to determine whether a function from the vertex set of a graph G to the vertex set of a graph H is an isomorphism?

c) How can the adjacency matrix of a graph be used to determine the number of paths of length r, where r is a positive integer, between two vertices of a graph?

13. a) Define an Euler circuit and an Euler path in an undirected graph.

b) Describe the famous Königsberg bridge problem and explain how to rephrase it in terms of an Euler circuit.

c) How can it be determined whether an undirected graph has an Euler path?

d) How can it be determined whether an undirected graph has an Euler circuit?

14. a) Define a Hamilton circuit in a simple graph.

b) Give some properties of a simple graph that imply that it does not have a Hamilton circuit.

15. Give examples of at least two problems that can be solved by finding the shortest path in a weighted graph.

16. a) Describe Dijkstra's algorithm for finding the shortest path in a weighted graph between two vertices.

b) Draw a weighted graph with at least 10 vertices and 20 edges. Use Dijkstra's algorithm to find the shortest path between two vertices of your choice in the graph.

17. a) What does it mean for a graph to be planar?

b) Give an example of a nonplanar graph.

18. a) What is Euler's formula for connected planar graphs?

b) How can Euler's formula for planar graphs be used to show that a simple graph is nonplanar?

19. State Kuratowski's theorem on the planarity of graphs and explain how it characterizes which graphs are planar.

20. a) Define the chromatic number of a graph.

b) What is the chromatic number of the graph K_n when n is a positive integer?

c) What is the chromatic number of the graph C_n when n is an integer greater than 2?

d) What is the chromatic number of the graph $K_{m,n}$ when m and n are positive integers?

21. State the four color theorem. Are there graphs that cannot be colored with four colors?

22. Explain how graph coloring can be used in modeling. Use at least two different examples.

Supplementary Exercises

1. How many edges does a 50-regular graph with 100 vertices have?

2. How many nonisomorphic subgraphs does K_3 have?

In Exercises 3–5 determine whether two given graphs are isomorphic.

3.

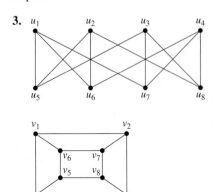

4.

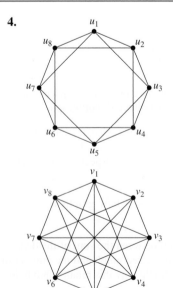

∗5.

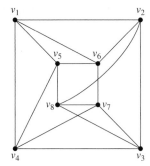

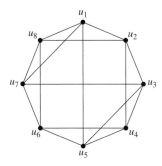

The **complete m-partite graph** $K_{n_1,n_2,\ldots,n_m}$ has vertices partitioned into m subsets of $n_1, n_2, \ldots, n_m$ elements each, and vertices are adjacent if and only if they are in different subsets in the partition.

6. Draw these graphs.

   a) $K_{1,2,3}$ **b)** $K_{2,2,2}$ **c)** $K_{1,2,2,3}$

∗7. How many vertices and how many edges does the complete m-partite graph $K_{n_1,n_2,\ldots,n_m}$ have?

8. Prove or disprove that there are always two vertices of the same degree in a finite multigraph having at least two vertices.

 A **clique** in a simple undirected graph is a complete subgraph that is not contained in any larger complete subgraph. In Exercises 9–11 find all cliques in the graph shown.

9.

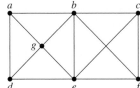

10.

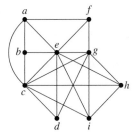

11.

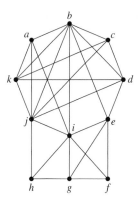

A **dominating set** of vertices in a simple graph is a set of vertices such that every other vertex is adjacent to at least one vertex of this set. A dominating set with the least number of vertices is called a **minimum dominating set**. In Exercises 12–13 find a minimum dominating set for the given graph.

12.   **13.**

 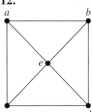 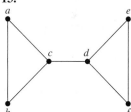

A simple graph can be used to determine the minimum number of queens on a chessboard that control the entire chessboard. An $n \times n$ chessboard has n^2 squares in an $n \times n$ configuration. A queen in a given position controls all squares in the same row, the same column, and on the two diagonals containing this square, as illustrated. The appropriate simple graph has n^2 vertices, one for each square, and two vertices are adjacent if a queen in the square represented by one of the vertices controls the square represented by the other vertex.

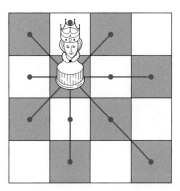

The Squares Controlled by a Queen

14. Construct the simple graph representing the $n \times n$ chessboard with edges representing the control of squares by queens for

   a) $n = 3$.   **b)** $n = 4$.

**15. Find the minimum number of queens controlling an $n \times n$ chessboard for

a) $n = 3$. **b)** $n = 4$. **c)** $n = 5$.

16. Suppose that G_1 and H_1 are isomorphic and that G_2 and H_2 are isomorphic. Prove or disprove that $G_1 \cup G_2$ and $H_1 \cup H_2$ are isomorphic.

17. Show that each of these properties is an invariant that isomorphic simple graphs either both have or both do not have.

a) connectedness

b) the existence of a Hamilton circuit

c) the existence of an Euler circuit

d) having crossing number C

e) having n isolated vertices

f) being bipartite

18. How can the adjacency matrix of $\overline{G}$ be found from the adjacency matrix of G, where G is a simple graph?

19. How many nonisomorphic connected bipartite simple graphs are there with four vertices?

*20. How many nonisomorphic simple connected graphs with five vertices are there

a) with no vertex of degree more than two?

b) with chromatic number equal to four?

c) that are nonplanar?

A directed graph is **self-converse** if it is isomorphic to its converse.

21. Determine whether the following graphs are self-converse.

a)

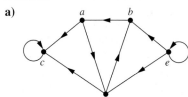

b)

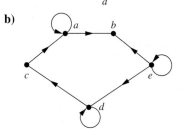

22. Show that if the directed graph G is self-converse and H is a directed graph isomorphic to G, then H is also self-converse.

An **orientation** of an undirected simple graph is an assignment of directions to its edges such that the resulting directed graph is strongly connected. When an orientation of

an undirected graph exists, this graph is called **orientable**. In Exercises 23–25 determine whether the given simple graph is orientable.

23.

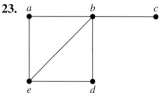

24.

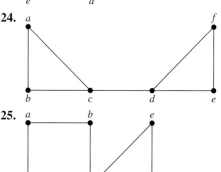

25.

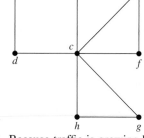

26. Because traffic is growing heavy in the central part of a city, traffic engineers are planning to change all the streets, which are currently two-way, into one-way streets. Explain how to model this problem.

*27. Show that a graph is not orientable if it has a cut edge.

A **tournament** is a simple directed graph such that if u and v are distinct vertices in the graph, exactly one of (u, v) and (v, u) is an edge of the graph.

28. How many different tournaments are there with n vertices?

29. What is the sum of the in-degree and out-degree of a vertex in a tournament?

*30. Show that every tournament has a Hamilton path.

31. Given two chickens in a flock, one of them is dominant. This defines the **pecking order** of the flock. How can a tournament be used to model pecking order?

32. Suppose that a connected graph G has n vertices and vertex connectivity $\kappa(G) = k$. Show that G must have at least $\lceil kn/2 \rceil$ edges.

A connected graph $G = (V, E)$ with n vertices and m edges is said to have **optimal connectivity** if $\kappa(G) = \lambda(G) = \min_{v \in V} \deg v = 2m/n$.

33. Show that a connected graph with optimal connectivity must be regular.

34. Show these graphs have optimal connectivity.

a) C_n for $n \geq 3$

b) K_n for $n \geq 3$

c) $K_{r,r}$ for $r \geq 2$

*35. Let G be a simple graph with n vertices. The **bandwidth** of G, denoted by $B(G)$, is the minimum, over all permutations $a_1, a_2, \ldots, a_n$ of the vertices of G, of $\max\{|i - j| \mid a_i$ and a_j are adjacent$\}$. That is, the bandwidth is the minimum over all listings of the vertices of the maximum difference in the indices assigned to adjacent vertices. Find the bandwidths of these graphs.

a) K_5 **b)** $K_{1,3}$ **c)** $K_{2,3}$
d) $K_{3,3}$ **e)** Q_3 **f)** C_5

*36. The **distance** between two distinct vertices v_1 and v_2 of a connected simple graph is the length (number of edges) of the shortest path between v_1 and v_2. The **radius** of a graph is the minimum over all vertices v of the maximum distance from v to another vertex. The **diameter** of a graph is the maximum distance between two distinct vertices. Find the radius and diameter of

a) K_6.   **b)** $K_{4,5}$.   **c)** Q_3.   **d)** C_6.

*37. a) Show that if the diameter of the simple graph G is at least four, then the diameter of its complement $\overline{G}$ is no more than two.

b) Show that if the diameter of the simple graph G is at least three, then the diameter of its complement $\overline{G}$ is no more than three.

*38. Suppose that a multigraph has $2m$ vertices of odd degree. Show that any circuit that contains every edge of the graph must contain at least m edges more than once.

39. Find the second shortest path between the vertices a and z in Figure 3 of Section 10.6.

40. Devise an algorithm for finding the second shortest path between two vertices in a simple connected weighted graph.

41. Find the shortest path between the vertices a and z that passes through the vertex f in the weighted graph in Exercise 2 in Section 10.6.

42. Devise an algorithm for finding the shortest path between two vertices in a simple connected weighted graph that passes through a specified third vertex.

*43. Show that if G is a simple graph with at least 11 vertices, then either G or $\overline{G}$, the complement of G, is nonplanar.

A set of vertices in a graph is called **independent** if no two vertices in the set are adjacent. The **independence number** of a graph is the maximum number of vertices in an independent set of vertices for the graph.

*44. What is the independence number of

a) K_n?   **b)** C_n?   **c)** Q_n?   **d)** $K_{m,n}$?

45. Show that the number of vertices in a simple graph is less than or equal to the product of the independence number and the chromatic number of the graph.

46. Show that the chromatic number of a graph is less than or equal to $n - i + 1$, where n is the number of vertices in the graph and i is the independence number of this graph.

47. Suppose that to generate a random simple graph with n vertices we first choose a real number p with $0 \le p \le 1$. For each of the $C(n, 2)$ pairs of distinct vertices we generate a random number x between 0 and 1. If $0 \le x \le p$, we connect these two vertices with an edge; otherwise these vertices are not connected.

a) What is the probability that a graph with m edges where $0 \le m \le C(n, 2)$ is generated?

b) What is the expected number of edges in a randomly generated graph with n vertices if each edge is included with probability p?

c) Show that if $p = 1/2$ then every simple graph with n vertices is equally likely to be generated.

A property retained whenever additional edges are added to a simple graph (without adding vertices) is called **monotone increasing**, and a property that is retained whenever edges are removed from a simple graph (without removing vertices) is called **monotone decreasing**.

48. For each of these properties, determine whether it is monotone increasing and determine whether it is monotone decreasing.

a) The graph G is connected.
b) The graph G is not connected.
c) The graph G has an Euler circuit.
d) The graph G has a Hamilton circuit.
e) The graph G is planar.
f) The graph G has chromatic number four.
g) The graph G has radius three.
h) The graph G has diameter three.

49. Show that the graph property P is monotone increasing if and only if the graph property Q is monotone decreasing where Q is the property of not having property P.

**50. Suppose that P is a monotone increasing property of simple graphs. Show that the probability a random graph with n vertices has property P is a monotonic nondecreasing function of p, the probability an edge is chosen to be in the graph.

Computer Projects

Write programs with these input and output.

1. Given the vertex pairs associated to the edges of an undirected graph, find the degree of each vertex.

2. Given the ordered pairs of vertices associated to the edges of a directed graph, determine the in-degree and out-degree of each vertex.

3. Given the list of edges of a simple graph, determine whether the graph is bipartite.

4. Given the vertex pairs associated to the edges of a graph, construct an adjacency matrix for the graph. (Produce a version that works when loops, multiple edges, or directed edges are present.)

5. Given an adjacency matrix of a graph, list the edges of this graph and give the number of times each edge appears.

6. Given the vertex pairs associated to the edges of an undirected graph and the number of times each edge appears, construct an incidence matrix for the graph.

7. Given an incidence matrix of an undirected graph, list its edges and give the number of times each edge appears.

8. Given a positive integer n, generate a simple graph with n vertices by producing an adjacency matrix for the graph so that all simple graphs with n vertices are equally likely to be generated.

9. Given a positive integer n, generate a simple directed graph with n vertices by producing an adjacency matrix for the graph so that all simple directed graphs with n vertices are equally likely to be generated.

10. Given the lists of edges of two simple graphs with no more than six vertices, determine whether the graphs are isomorphic.

11. Given an adjacency matrix of a graph and a positive integer n, find the number of paths of length n between two vertices. (Produce a version that works for directed and undirected graphs.)

12. Given the vertex pairs associated to the edges of a multigraph, determine whether it has an Euler circuit and, if not, whether it has an Euler path. Construct an Euler path or circuit if it exists.

13. Given the list of edges and weights of these edges of a weighted connected simple graph and two vertices in this graph, find the length of a shortest path between them using Dijkstra's algorithm. Also, find a shortest path.

14. Given the list of edges of an undirected graph, find a coloring of this graph using the algorithm given in the exercise set of Section 10.8.

15. Given a list of students and the courses that they are enrolled in, construct a schedule of final exams.

16. Given the distances between pairs of television stations and the minimum allowable distance between stations, assign frequencies to these stations.

Computations and Explorations

Use a computational program or programs you have written to do these exercises.

1. Display all simple graphs with four vertices.

2. Display a full set of nonisomorphic simple graphs with six vertices.

3. Display a full set of nonisomorphic directed graphs with four vertices.

4. Generate at random 10 different simple graphs each with 20 vertices so that each such graph is equally likely to be generated.

5. Construct a Gray code where the code words are bit strings of length six.

6. Construct knight's tours on chessboards of various sizes.

7. Determine whether each of the graphs you generated in Exercise 4 of this set is planar. If you can, determine the thickness of each of the graphs that are not planar.

8. Determine whether each of the graphs you generated in Exercise 4 of this set is connected. If a graph is not connected, determine the number of connected components of the graph.

9. Generate at random simple graphs with 10 vertices. Stop when you have constructed one with an Euler circuit. Display an Euler circuit in this graph.

10. Generate at random simple graphs with 10 vertices. Stop when you have constructed one with a Hamilton circuit. Display a Hamilton circuit in this graph.

Writing Projects

Respond to these with essays using outside sources.

1. Describe the origins and development of graph theory prior to the year 1900.

2. Discuss the applications of graph theory to the study of ecosystems.

3. Discuss the applications of graph theory to sociology and psychology.

4. Discuss what can be learned by investigating the properties of the Web graph.

5. Explain what community structure is in a graph representing a network, such as a social network, a computer network, an information network, or a biological network. Define what a community in such a graph is, and explain what communities represent in graphs representing the types of networks listed.

6. Describe some of the algorithms used to detect communities in graphs representing networks of the types listed in Question 5.

7. Describe algorithms for drawing a graph on paper or on a display given the vertices and edges of the graph. What considerations arise in drawing a graph so that it has the best appearance for understanding its properties?

8. Explain how graph theory can help uncover networks of criminals or terrorists by studying relevant social and communication networks.

9. What are some of the capabilities that a software tool for inputting, displaying, and manipulating graphs should have? Which of these capabilities do available tools have?

10. Describe some of the algorithms available for determining whether two graphs are isomorphic and the computational complexity of these algorithms. What is the most efficient such algorithm currently known?

11. What is the subgraph isomorphism problem and what are some of its important applications, including those to chemistry, bioinformatics, electronic circuit design, and computer vision?

12. Explain what the area of graph mining, an important area of data mining, is and describe some of the basic techniques used in graph mining.

13. Describe how Euler paths can be used to help determine DNA sequences.

14. Define *de Bruijn sequences* and discuss how they arise in applications. Explain how de Bruijn sequences can be constructed using Euler circuits.

15. Describe the *Chinese postman problem* and explain how to solve this problem.

16. Describe some of the different conditions that imply that a graph has a Hamilton circuit.

17. Describe some of the strategies and algorithms used to solve the traveling salesperson problem.

18. Describe several different algorithms for determining whether a graph is planar. What is the computational complexity of each of these algorithms?

19. In modeling, very large scale integration (VLSI) graphs are sometimes embedded in a book, with the vertices on the spine and the edges on pages. Define the *book number* of a graph and find the book number of various graphs including K_n for $n = 3, 4, 5$, and 6.

20. Discuss the history of the four color theorem.

21. Describe the role computers played in the proof of the four color theorem. How can we be sure that a proof that relies on a computer is correct?

22. Describe and compare several different algorithms for coloring a graph, in terms of whether they produce a coloring with the least number of colors possible and in terms of their complexity.

23. Explain how graph multicolorings can be used in a variety of different models.

24. Describe some of the applications of edge colorings.

25. Explain how the theory of random graphs can be used in nonconstructive existence proofs of graphs with certain properties.

11

Trees

A connected graph that contains no simple circuits is called a tree. Trees were used as long ago as 1857, when the English mathematician Arthur Cayley used them to count certain types of chemical compounds. Since that time, trees have been employed to solve problems in a wide variety of disciplines, as the examples in this chapter will show.

Trees are particularly useful in computer science, where they are employed in a wide range of algorithms. For instance, trees are used to construct efficient algorithms for locating items in a list. They can be used in algorithms, such as Huffman coding, that construct efficient codes saving costs in data transmission and storage. Trees can be used to study games such as checkers and chess and can help determine winning strategies for playing these games. Trees can be used to model procedures carried out using a sequence of decisions. Constructing these models can help determine the computational complexity of algorithms based on a sequence of decisions, such as sorting algorithms.

Procedures for building trees containing every vertex of a graph, including depth-first search and breadth-first search, can be used to systematically explore the vertices of a graph. Exploring the vertices of a graph via depth-first search, also known as backtracking, allows for the systematic search for solutions to a wide variety of problems, such as determining how eight queens can be placed on a chessboard so that no queen can attack another.

We can assign weights to the edges of a tree to model many problems. For example, using weighted trees we can develop algorithms to construct networks containing the least expensive set of telephone lines linking different network nodes.

11.1 Introduction to Trees

In Chapter 10 we showed how graphs can be used to model and solve many problems. In this chapter we will focus on a particular type of graph called a **tree**, so named because such graphs resemble trees. For example, *family trees* are graphs that represent genealogical charts. Family trees use vertices to represent the members of a family and edges to represent parent–child relationships. The family tree of the male members of the Bernoulli family of Swiss mathematicians is shown in Figure 1. The undirected graph representing a family tree (restricted to people of just one gender and with no inbreeding) is an example of a tree.

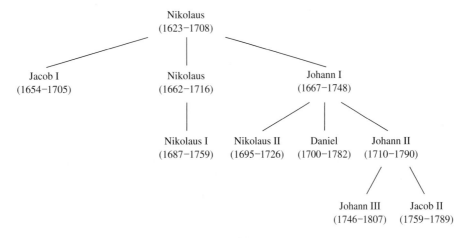

FIGURE 1 The Bernoulli Family of Mathematicians.

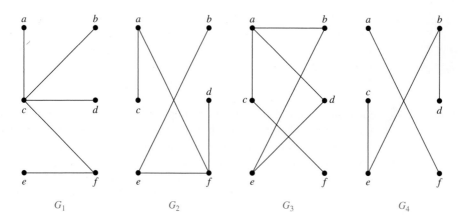

FIGURE 2   Examples of Trees and Graphs That Are Not Trees.

DEFINITION 1	A *tree* is a connected undirected graph with no simple circuits.

Because a tree cannot have a simple circuit, a tree cannot contain multiple edges or loops. Therefore any tree must be a simple graph.

EXAMPLE 1   Which of the graphs shown in Figure 2 are trees?

Solution: G_1 and G_2 are trees, because both are connected graphs with no simple circuits. G_3 is not a tree because e, b, a, d, e is a simple circuit in this graph. Finally, G_4 is not a tree because it is not connected.   ◀

Any connected graph that contains no simple circuits is a tree. What about graphs containing no simple circuits that are not necessarily connected? These graphs are called **forests** and have the property that each of their connected components is a tree. Figure 3 displays a forest.

Trees are often defined as undirected graphs with the property that there is a unique simple path between every pair of vertices. Theorem 1 shows that this alternative definition is equivalent to our definition.

THEOREM 1	An undirected graph is a tree if and only if there is a unique simple path between any two of its vertices.

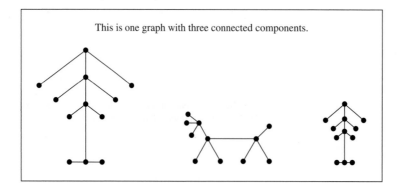

This is one graph with three connected components.

FIGURE 3   Example of a Forest.

Proof: First assume that T is a tree. Then T is a connected graph with no simple circuits. Let x and y be two vertices of T. Because T is connected, by Theorem 1 of Section 10.4 there is a simple path between x and y. Moreover, this path must be unique, for if there were a second such path, the path formed by combining the first path from x to y followed by the path from y to x obtained by reversing the order of the second path from x to y would form a circuit. This implies, using Exercise 45 of Section 10.4, that there is a simple circuit in T. Hence, there is a unique simple path between any two vertices of a tree.

Now assume that there is a unique simple path between any two vertices of a graph T. Then T is connected, because there is a path between any two of its vertices. Furthermore, T can have no simple circuits. To see that this is true, suppose T had a simple circuit that contained the vertices x and y. Then there would be two simple paths between x and y, because the simple circuit is made up of a simple path from x to y and a second simple path from y to x. Hence, a graph with a unique simple path between any two vertices is a tree.   ◁

Rooted Trees

In many applications of trees, a particular vertex of a tree is designated as the **root**. Once we specify a root, we can assign a direction to each edge as follows. Because there is a unique path from the root to each vertex of the graph (by Theorem 1), we direct each edge away from the root. Thus, a tree together with its root produces a directed graph called a **rooted tree**.

DEFINITION 2

> A *rooted tree* is a tree in which one vertex has been designated as the root and every edge is directed away from the root.

Rooted trees can also be defined recursively. Refer to Section 5.3 to see how this can be done. We can change an unrooted tree into a rooted tree by choosing any vertex as the root. Note that different choices of the root produce different rooted trees. For instance, Figure 4 displays the rooted trees formed by designating a to be the root and c to be the root, respectively, in the tree T. We usually draw a rooted tree with its root at the top of the graph. The arrows indicating the directions of the edges in a rooted tree can be omitted, because the choice of root determines the directions of the edges.

The terminology for trees has botanical and genealogical origins. Suppose that T is a rooted tree. If v is a vertex in T other than the root, the **parent** of v is the unique vertex u such that there is a directed edge from u to v (the reader should show that such a vertex is unique). When u is the parent of v, v is called a **child** of u. Vertices with the same parent are called **siblings**. The **ancestors** of a vertex other than the root are the vertices in the path from the root to this vertex, excluding the vertex itself and including the root (that is, its parent, its parent's parent, and so on, until the root is reached). The **descendants** of a vertex v are those vertices that have v as

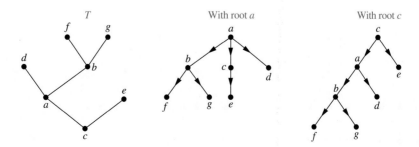

FIGURE 4   A Tree and Rooted Trees Formed by Designating Two Different Roots.

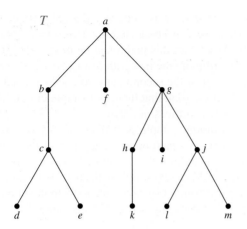

FIGURE 5 A Rooted Tree T.

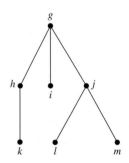

FIGURE 6 The Subtree Rooted at g.

an ancestor. A vertex of a rooted tree is called a **leaf** if it has no children. Vertices that have children are called **internal vertices**. The root is an internal vertex unless it is the only vertex in the graph, in which case it is a leaf.

If a is a vertex in a tree, the **subtree** with a as its root is the subgraph of the tree consisting of a and its descendants and all edges incident to these descendants.

EXAMPLE 2   In the rooted tree T (with root a) shown in Figure 5, find the parent of c, the children of g, the siblings of h, all ancestors of e, all descendants of b, all internal vertices, and all leaves. What is the subtree rooted at g?

Solution: The parent of c is b. The children of g are h, i, and j. The siblings of h are i and j. The ancestors of e are c, b, and a. The descendants of b are c, d, and e. The internal vertices are a, b, c, g, h, and j. The leaves are d, e, f, i, k, l, and m. The subtree rooted at g is shown in Figure 6.   ◀

Rooted trees with the property that all of their internal vertices have the same number of children are used in many different applications. Later in this chapter we will use such trees to study problems involving searching, sorting, and coding.

DEFINITION 3

A rooted tree is called an *m-ary tree* if every internal vertex has no more than m children. The tree is called a *full m-ary tree* if every internal vertex has exactly m children. An m-ary tree with $m = 2$ is called a *binary tree*.

EXAMPLE 3   Are the rooted trees in Figure 7 full m-ary trees for some positive integer m?

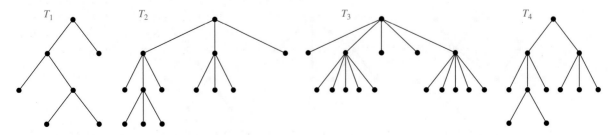

FIGURE 7 Four Rooted Trees.

Solution: T_1 is a full binary tree because each of its internal vertices has two children. T_2 is a full 3-ary tree because each of its internal vertices has three children. In T_3 each internal vertex has five children, so T_3 is a full 5-ary tree. T_4 is not a full m-ary tree for any m because some of its internal vertices have two children and others have three children. ◀

ORDERED ROOTED TREES   An **ordered rooted tree** is a rooted tree where the children of each internal vertex are ordered. Ordered rooted trees are drawn so that the children of each internal vertex are shown in order from left to right. Note that a representation of a rooted tree in the conventional way determines an ordering for its edges. We will use such orderings of edges in drawings without explicitly mentioning that we are considering a rooted tree to be ordered.

In an ordered binary tree (usually called just a **binary tree**), if an internal vertex has two children, the first child is called the **left child** and the second child is called the **right child**. The tree rooted at the left child of a vertex is called the **left subtree** of this vertex, and the tree rooted at the right child of a vertex is called the **right subtree** of the vertex. The reader should note that for some applications every vertex of a binary tree, other than the root, is designated as a right or a left child of its parent. This is done even when some vertices have only one child. We will make such designations whenever it is necessary, but not otherwise.

Ordered rooted trees can be defined recursively. Binary trees, a type of ordered rooted trees, were defined this way in Section 5.3.

EXAMPLE 4   What are the left and right children of d in the binary tree T shown in Figure 8(a) (where the order is that implied by the drawing)? What are the left and right subtrees of c?

Solution: The left child of d is f and the right child is g. We show the left and right subtrees of c in Figures 8(b) and 8(c), respectively. ◀

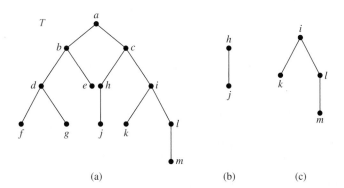

FIGURE 8   A Binary Tree T and Left and Right Subtrees of the Vertex c.

Just as in the case of graphs, there is no standard terminology used to describe trees, rooted trees, ordered rooted trees, and binary trees. This nonstandard terminology occurs because trees are used extensively throughout computer science, which is a relatively young field. The reader should carefully check meanings given to terms dealing with trees whenever they occur.

Trees as Models

Trees are used as models in such diverse areas as computer science, chemistry, geology, botany, and psychology. We will describe a variety of such models based on trees.

H
|
H—C—H
|
H—C—H
|
H—C—H
|
H—C—H
|
H

H   H   H
| | |
H—C——C——C—H
| | |
H   H
H—C—H
|
H

Butane   Isobutane

FIGURE 9   The Two Isomers of Butane.

EXAMPLE 5   **Saturated Hydrocarbons and Trees**   Graphs can be used to represent molecules, where atoms are represented by vertices and bonds between them by edges. The English mathematician Arthur Cayley discovered trees in 1857 when he was trying to enumerate the isomers of compounds of the form C_nH_{2n+2}, which are called *saturated hydrocarbons*.

In graph models of saturated hydrocarbons, each carbon atom is represented by a vertex of degree 4, and each hydrogen atom is represented by a vertex of degree 1. There are $3n + 2$ vertices in a graph representing a compound of the form C_nH_{2n+2}. The number of edges in such a graph is half the sum of the degrees of the vertices. Hence, there are $(4n + 2n + 2)/2 = 3n + 1$ edges in this graph. Because the graph is connected and the number of edges is one less than the number of vertices, it must be a tree (see Exercise 9).

The nonisomorphic trees with n vertices of degree 4 and $2n + 2$ of degree 1 represent the different isomers of C_nH_{2n+2}. For instance, when $n = 4$, there are exactly two nonisomorphic trees of this type (the reader should verify this). Hence, there are exactly two different isomers of C_4H_{10}. Their structures are displayed in Figure 9. These two isomers are called butane and isobutane.   ◀

EXAMPLE 6   **Representing Organizations**   The structure of a large organization can be modeled using a rooted tree. Each vertex in this tree represents a position in the organization. An edge from one vertex to another indicates that the person represented by the initial vertex is the (direct) boss of the person represented by the terminal vertex. The graph shown in Figure 10 displays such a tree. In the organization represented by this tree, the Director of Hardware Development works directly for the Vice President of R&D.   ◀

EXAMPLE 7   **Computer File Systems**   Files in computer memory can be organized into directories. A directory can contain both files and subdirectories. The root directory contains the entire file

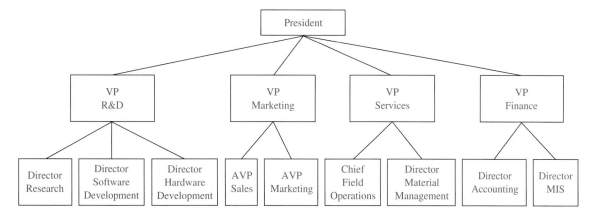

FIGURE 10   An Organizational Tree for a Computer Company.

system. Thus, a file system may be represented by a rooted tree, where the root represents the root directory, internal vertices represent subdirectories, and leaves represent ordinary files or empty directories. One such file system is shown in Figure 11. In this system, the file khr is in the directory rje. (Note that links to files where the same file may have more than one pathname can lead to circuits in computer file systems.)   ◄

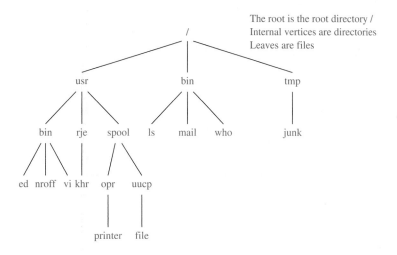

FIGURE 11   A Computer File System.

EXAMPLE 8

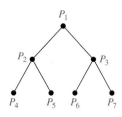

FIGURE 12   A Tree-Connected Network of Seven Processors.

Tree-Connected Parallel Processors   In Example 17 of Section 10.2 we described several interconnection networks for parallel processing. A **tree-connected network** is another important way to interconnect processors. The graph representing such a network is a complete binary tree, that is, a full binary tree where every root is at the same level. Such a network interconnects $n = 2^k - 1$ processors, where k is a positive integer. A processor represented by the vertex v that is not a root or a leaf has three two-way connections—one to the processor represented by the parent of v and two to the processors represented by the two children of v. The processor represented by the root has two two-way connections to the processors represented by its two children. A processor represented by a leaf v has a single two-way connection to the parent of v. We display a tree-connected network with seven processors in Figure 12.

We now illustrate how a tree-connected network can be used for parallel computation. In particular, we show how the processors in Figure 12 can be used to add eight numbers, using three steps. In the first step, we add x_1 and x_2 using P_4, x_3 and x_4 using P_5, x_5 and x_6 using P_6, and x_7 and x_8 using P_7. In the second step, we add $x_1 + x_2$ and $x_3 + x_4$ using P_2 and $x_5 + x_6$ and $x_7 + x_8$ using P_3. Finally, in the third step, we add $x_1 + x_2 + x_3 + x_4$ and $x_5 + x_6 + x_7 + x_8$ using P_1. The three steps used to add eight numbers compares favorably to the seven steps required to add eight numbers serially, where the steps are the addition of one number to the sum of the previous numbers in the list.   ◀

Properties of Trees

We will often need results relating the numbers of edges and vertices of various types in trees.

THEOREM 2	A tree with n vertices has $n - 1$ edges.

Links

Proof: We will use mathematical induction to prove this theorem. Note that for all the trees here we can choose a root and consider the tree rooted.

BASIS STEP: When $n = 1$, a tree with $n = 1$ vertex has no edges. It follows that the theorem is true for $n = 1$.

INDUCTIVE STEP: The inductive hypothesis states that every tree with k vertices has $k - 1$ edges, where k is a positive integer. Suppose that a tree T has $k + 1$ vertices and that v is a leaf of T (which must exist because the tree is finite), and let w be the parent of v. Removing from T the vertex v and the edge connecting w to v produces a tree T' with k vertices, because the resulting graph is still connected and has no simple circuits. By the inductive hypothesis, T' has $k - 1$ edges. It follows that T has k edges because it has one more edge than T', the edge connecting v and w. This completes the inductive step.   ◁

Recall that a tree is a connected undirected graph with no simple circuits. So, when G is an undirected graph with n vertices, Theorem 2 tells us that the two conditions *(i)* G is connected and *(ii)* G has no simple circuits, imply *(iii)* G has $n - 1$ edges. Also, when *(i)* and *(iii)* hold, then *(ii)* must also hold, and when *(ii)* and *(iii)* hold, *(i)* must also hold. That is, if G is connected and G has $n - 1$ edges, then G has no simple circuits, so that G is a tree (see Exercise 15(a)), and if G has no simple circuits and G has $n - 1$ edges, then G is connected, and so is a tree (see Exercise 15(b)). Consequently, when two of *(i)*, *(ii)*, and *(iii)* hold, the third condition must also hold, and G must be a tree.

COUNTING VERTICES IN FULL m-ARY TREES   The number of vertices in a full m-ary tree with a specified number of internal vertices is determined, as Theorem 3 shows. As in Theorem 2, we will use n to denote the number of vertices in a tree.

THEOREM 3	A full m-ary tree with i internal vertices contains $n = mi + 1$ vertices.

Proof: Every vertex, except the root, is the child of an internal vertex. Because each of the i internal vertices has m children, there are mi vertices in the tree other than the root. Therefore, the tree contains $n = mi + 1$ vertices.   ◁

Suppose that T is a full m-ary tree. Let i be the number of internal vertices and l the number of leaves in this tree. Once one of n, i, and l is known, the other two quantities are determined. Theorem 4 explains how to find the other two quantities from the one that is known.

THEOREM 4   A full *m*-ary tree with

    (*i*) *n* vertices has $i = (n-1)/m$ internal vertices and $l = [(m-1)n + 1]/m$ leaves,

    (*ii*) *i* internal vertices has $n = mi + 1$ vertices and $l = (m-1)i + 1$ leaves,

    (*iii*) *l* leaves has $n = (ml - 1)/(m-1)$ vertices and $i = (l-1)/(m-1)$ internal vertices.

Proof: Let *n* represent the number of vertices, *i* the number of internal vertices, and *l* the number of leaves. The three parts of the theorem can all be proved using the equality given in Theorem 3, that is, $n = mi + 1$, together with the equality $n = l + i$, which is true because each vertex is either a leaf or an internal vertex. We will prove part (*i*) here. The proofs of parts (*ii*) and (*iii*) are left as exercises for the reader.

Solving for *i* in $n = mi + 1$ gives $i = (n-1)/m$. Then inserting this expression for *i* into the equation $n = l + i$ shows that $l = n - i = n - (n-1)/m = [(m-1)n + 1]/m$.   ◁

Example 9 illustrates how Theorem 4 can be used.

EXAMPLE 9   Suppose that someone starts a chain letter. Each person who receives the letter is asked to send it on to four other people. Some people do this, but others do not send any letters. How many people have seen the letter, including the first person, if no one receives more than one letter and if the chain letter ends after there have been 100 people who read it but did not send it out? How many people sent out the letter?

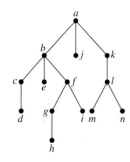

FIGURE 13   A Rooted Tree.

Solution: The chain letter can be represented using a 4-ary tree. The internal vertices correspond to people who sent out the letter, and the leaves correspond to people who did not send it out. Because 100 people did not send out the letter, the number of leaves in this rooted tree is $l = 100$. Hence, part (*iii*) of Theorem 4 shows that the number of people who have seen the letter is $n = (4 \cdot 100 - 1)/(4 - 1) = 133$. Also, the number of internal vertices is $133 - 100 = 33$, so 33 people sent out the letter.   ◀

BALANCED *m*-ARY TREES   It is often desirable to use rooted trees that are "balanced" so that the subtrees at each vertex contain paths of approximately the same length. Some definitions will make this concept clear. The **level** of a vertex *v* in a rooted tree is the length of the unique path from the root to this vertex. The level of the root is defined to be zero. The **height** of a rooted tree is the maximum of the levels of vertices. In other words, the height of a rooted tree is the length of the longest path from the root to any vertex.

EXAMPLE 10   Find the level of each vertex in the rooted tree shown in Figure 13. What is the height of this tree?

Solution: The root *a* is at level 0. Vertices *b*, *j*, and *k* are at level 1. Vertices *c*, *e*, *f*, and *l* are at level 2. Vertices *d*, *g*, *i*, *m*, and *n* are at level 3. Finally, vertex *h* is at level 4. Because the largest level of any vertex is 4, this tree has height 4.   ◀

A rooted *m*-ary tree of height *h* is **balanced** if all leaves are at levels *h* or $h - 1$.

EXAMPLE 11   Which of the rooted trees shown in Figure 14 are balanced?

Solution: T_1 is balanced, because all its leaves are at levels 3 and 4. However, T_2 is not balanced, because it has leaves at levels 2, 3, and 4. Finally, T_3 is balanced, because all its leaves are at level 3.   ◀

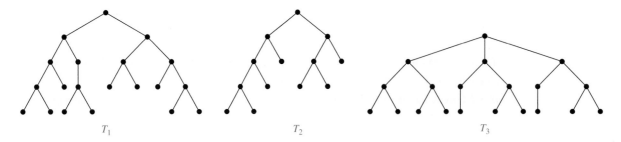

FIGURE 14 **Some Rooted Trees.**

A BOUND FOR THE NUMBER OF LEAVES IN AN m-ARY TREE   It is often useful to have an upper bound for the number of leaves in an m-ary tree. Theorem 5 provides such a bound in terms of the height of the m-ary tree.

THEOREM 5   There are at most m^h leaves in an m-ary tree of height h.

Proof: The proof uses mathematical induction on the height. First, consider m-ary trees of height 1. These trees consist of a root with no more than m children, each of which is a leaf. Hence, there are no more than $m^1 = m$ leaves in an m-ary tree of height 1. This is the basis step of the inductive argument.

Now assume that the result is true for all m-ary trees of height less than h; this is the inductive hypothesis. Let T be an m-ary tree of height h. The leaves of T are the leaves of the subtrees of T obtained by deleting the edges from the root to each of the vertices at level 1, as shown in Figure 15.

Each of these subtrees has height less than or equal to $h - 1$. So by the inductive hypothesis, each of these rooted trees has at most m^{h-1} leaves. Because there are at most m such subtrees, each with a maximum of m^{h-1} leaves, there are at most $m \cdot m^{h-1} = m^h$ leaves in the rooted tree. This finishes the inductive argument. ◁

COROLLARY 1   If an m-ary tree of height h has l leaves, then $h \geq \lceil \log_m l \rceil$. If the m-ary tree is full and balanced, then $h = \lceil \log_m l \rceil$. (We are using the ceiling function here. Recall that $\lceil x \rceil$ is the smallest integer greater than or equal to x.)

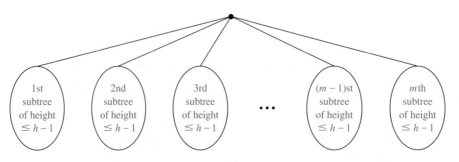

FIGURE 15   **The Inductive Step of the Proof.**

Proof: We know that $l \leq m^h$ from Theorem 5. Taking logarithms to the base m shows that $\log_m l \leq h$. Because h is an integer, we have $h \geq \lceil \log_m l \rceil$. Now suppose that the tree is balanced. Then each leaf is at level h or $h - 1$, and because the height is h, there is at least one leaf at level h. It follows that there must be more than m^{h-1} leaves (see Exercise 22). Because $l \leq m^h$, we have $m^{h-1} < l \leq m^h$. Taking logarithms to the base m in this inequality gives $h - 1 < \log_m l \leq h$. Hence, $h = \lceil \log_m l \rceil$. ◁

Exercises

1. Which of these graphs are trees?

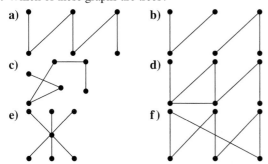

2. Answer these questions about the rooted tree illustrated.

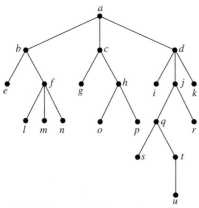

a) Which vertex is the root?

b) Which vertices are internal?

c) Which vertices are leaves?

d) Which vertices are children of j?

e) Which vertex is the parent of h?

f) Which vertices are siblings of o?

g) Which vertices are ancestors of m?

h) Which vertices are descendants of b?

3. Is the rooted tree in Exercise 2 a full m-ary tree for some positive integer m?

4. What is the level of each vertex of the rooted tree in Exercise 2?

5. Draw the subtree of the tree in Exercise 2 that is rooted at

  a) a.   **b)** c.   **c)** e.

6. a) How many nonisomorphic unrooted trees are there with three vertices?

b) How many nonisomorphic rooted trees are there with three vertices (using isomorphism for directed graphs)?

*7. a) How many nonisomorphic unrooted trees are there with four vertices?

b) How many nonisomorphic rooted trees are there with four vertices (using isomorphism for directed graphs)?

*8. Show that a simple graph is a tree if and only if it is connected but the deletion of any of its edges produces a graph that is not connected.

* 9. Let G be a simple graph with n vertices. Show that

  a) G is a tree if and only if it is connected and has $n - 1$ edges.

  b) G is a tree if and only if G has no simple circuits and has $n - 1$ edges. [*Hint:* To show that G is connected if it has no simple circuits and $n - 1$ edges, show that G cannot have more than one connected component.]

10. Which complete bipartite graphs $K_{m,n}$, where m and n are positive integers, are trees?

11. How many edges does a tree with 10,000 vertices have?

12. How many vertices does a full 5-ary tree with 100 internal vertices have?

13. How many edges does a full binary tree with 1000 internal vertices have?

14. How many leaves does a full 3-ary tree with 100 vertices have?

15. Suppose 1000 people enter a chess tournament. Use a rooted tree model of the tournament to determine how many games must be played to determine a champion, if a player is eliminated after one loss and games are played until only one entrant has not lost. (Assume there are no ties.)

16. A chain letter starts when a person sends a letter to five others. Each person who receives the letter either sends it to five other people who have never received it or does not send it to anyone. Suppose that 10,000 people send out the letter before the chain ends and that no one receives more than one letter. How many people receive the letter, and how many do not send it out?

*17. Either draw a full m-ary tree with 84 leaves and height 3, where m is a positive integer, or show that no such tree exists.

*18. A full m-ary tree T has 81 leaves and height 4.

  a) Give the upper and lower bounds for m.

b) What is m if T is also balanced?

A **complete m-ary tree** is a full m-ary tree in which every leaf is at the same level.

19. Construct a complete binary tree of height 4 and a complete 3-ary tree of height 3.

20. How many vertices and how many leaves does a complete m-ary tree of height h have?

21. Prove
   a) part (*ii*) of Theorem 4.
   b) part (*iii*) of Theorem 4.

22. Show that a full m-ary balanced tree of height h has more than m^{h-1} leaves.

23. How many edges are there in a forest of t trees containing a total of n vertices?

24. Explain how a tree can be used to represent the table of contents of a book organized into chapters, where each chapter is organized into sections, and each section is organized into subsections.

25. How many different isomers do these saturated hydrocarbons have?
   a) C_3H_8 **b)** C_5H_{12} **c)** C_6H_{14}

26. What does each of these represent in an organizational tree?
   a) the parent of a vertex
   b) a child of a vertex
   c) a sibling of a vertex
   d) the ancestors of a vertex
   e) the descendants of a vertex
   f) the level of a vertex
   g) the height of the tree

27. Let n be a power of 2. Show that n numbers can be added in $\log n$ steps using a tree-connected network of $n - 1$ processors.

28. A **labeled tree** is a tree where each vertex is assigned a label. Two labeled trees are considered isomorphic when there is an isomorphism between them that preserves the labels of vertices. How many nonisomorphic trees are there with three vertices labeled with different integers from the set $\{1, 2, 3\}$? How many nonisomorphic trees

are there with four vertices labeled with different integers from the set $\{1, 2, 3, 4\}$?

The **eccentricity** of a vertex in an unrooted tree is the length of the longest simple path beginning at this vertex. A vertex is called a **center** if no vertex in the tree has smaller eccentricity than this vertex. In Exercises 29–30 find every vertex that is a center in the given tree.

29.

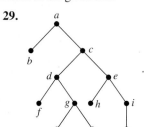

30.

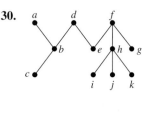

31. Show that a tree has either one center or two centers that are adjacent.

32. Show that every tree can be colored using two colors.

The **rooted Fibonacci trees** T_n are defined recursively in the following way. T_1 and T_2 are both the rooted tree consisting of a single vertex, and for $n = 3, 4, \ldots$, the rooted tree T_n is constructed from a root with T_{n-1} as its left subtree and T_{n-2} as its right subtree.

33. Draw the first seven rooted Fibonacci trees.

34. How many vertices, leaves, and internal vertices does the rooted Fibonacci tree T_n have, where n is a positive integer? What is its height?

35. What is wrong with the following "proof" using mathematical induction of the statement that every tree with n vertices has a path of length $n - 1$. *Basis step:* Every tree with one vertex clearly has a path of length 0. *Inductive step:* Assume that a tree with n vertices has a path of length $n - 1$, which has u as its terminal vertex. Add a vertex v and the edge from u to v. The resulting tree has $n + 1$ vertices and has a path of length n. This completes the inductive step.

36. Show that the average depth of a leaf in a binary tree with n vertices is $\Omega(\log n)$.

11.2   Applications of Trees

Introduction

We will discuss three problems that can be studied using trees. The first problem is: How should items in a list be stored so that an item can be easily located? The second problem is: What series of decisions should be made to find an object with a certain property in a collection of objects of a certain type? The third problem is: How should a set of characters be efficiently coded by bit strings?

Binary Search Trees

Searching for items in a list is one of the most important tasks that arises in computer science. Our primary goal is to implement a searching algorithm that finds items efficiently when the items are totally ordered. This can be accomplished through the use of a **binary search tree**, which is a binary tree in which each child of a vertex is designated as a right or left child, no vertex has more than one right child or left child, and each vertex is labeled with a key, which is one of the items. Furthermore, vertices are assigned keys so that the key of a vertex is both larger than the keys of all vertices in its left subtree and smaller than the keys of all vertices in its right subtree.

This recursive procedure is used to form the binary search tree for a list of items. Start with a tree containing just one vertex, namely, the root. The first item in the list is assigned as the key of the root. To add a new item, first compare it with the keys of vertices already in the tree, starting at the root and moving to the left if the item is less than the key of the respective vertex if this vertex has a left child, or moving to the right if the item is greater than the key of the respective vertex if this vertex has a right child. When the item is less than the respective vertex and this vertex has no left child, then a new vertex with this item as its key is inserted as a new left child. Similarly, when the item is greater than the respective vertex and this vertex has no right child, then a new vertex with this item as its key is inserted as a new right child. We illustrate this procedure with Example 1.

EXAMPLE 1   Form a binary search tree for the words *mathematics, physics, geography, zoology, meteorology, geology, psychology*, and *chemistry* (using alphabetical order).

Solution: Figure 1 displays the steps used to construct this binary search tree. The word *mathematics* is the key of the root. Because *physics* comes after *mathematics* (in alphabetical order), add a right child of the root with key *physics*. Because *geography* comes before *mathematics*, add a left child of the root with key *geography*. Next, add a right child of the vertex with key *physics*, and assign it the key *zoology*, because *zoology* comes after *mathematics* and after *physics*. Similarly, add a left child of the vertex with key *physics* and assign this new vertex the key *meteorology*. Add a right child of the vertex with key *geography* and assign this new vertex the key *geology*. Add a left child of the vertex with key *zoology* and assign it the key *psychology*.

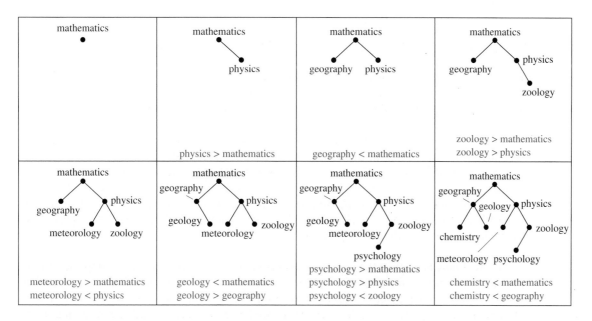

FIGURE 1   Constructing a Binary Search Tree.

Add a left child of the vertex with key *geography* and assign it the key *chemistry*. (The reader should work through all the comparisons needed at each step.) ◄

Once we have a binary search tree, we need a way to locate items in the binary search tree, as well as a way to add new items. Algorithm 1, an insertion algorithm, actually does both of these tasks, even though it may appear that it is only designed to add vertices to a binary search tree. That is, Algorithm 1 is a procedure that locates an item x in a binary search tree if it is present, and adds a new vertex with x as its key if x is not present. In the pseudocode, v is the vertex currently under examination and $label(v)$ represents the key of this vertex. The algorithm begins by examining the root. If x equals the key of v, then the algorithm has found the location of x and terminates; if x is less than the key of v, we move to the left child of v and repeat the procedure; and if x is greater than the key of v, we move to the right child of v and repeat the procedure. If at any step we attempt to move to a child that is not present, we know that x is not present in the tree, and we add a new vertex as this child with x as its key.

ALGORITHM 1 Locating an Item in or Adding an Item to a Binary Search Tree.

procedure *insertion*(T: binary search tree, x: item)
$v :=$ root of T
{a vertex not present in T has the value *null* }
while $v \neq null$ and $label(v) \neq x$
  if $x < label(v)$ **then**
    if left child of $v \neq null$ **then** $v :=$ left child of v
    else add *new vertex* as a left child of v and set $v := null$
  else
    if right child of $v \neq null$ **then** $v :=$ right child of v
    else add *new vertex* as a right child of v and set $v := null$
if root of $T = null$ **then** add a vertex v to the tree and label it with x
else if v is null or $label(v) \neq x$ **then** label *new vertex* with x and let v be this new vertex
return v {$v =$ location of x}

Example 2 illustrates the use of Algorithm 1 to insert a new item into a binary search tree.

EXAMPLE 2   Use Algorithm 1 to insert the word *oceanography* into the binary search tree in Example 1.

Solution: Algorithm 1 begins with v, the vertex under examination, equal to the root of T, so $label(v) = mathematics$. Because $v \neq null$ and $label(v) = mathematics < oceanography$, we next examine the right child of the root. This right child exists, so we set v, the vertex under examination, to be this right child. At this step we have $v \neq null$ and $label(v) = physics > oceanography$, so we examine the left child of v. This left child exists, so we set v, the vertex under examination, to this left child. At this step, we also have $v \neq null$ and $label(v) = metereology < oceanography$, so we try to examine the right child of v. However, this right child does not exist, so we add a new vertex as the right child of v (which at this point is the vertex with the key *metereology*) and we set $v := null$. We now exit the **while** loop because $v = null$. Because the root of T is not *null* and $v = null$, we use the **else if** statement at the end of the algorithm to label our new vertex with the key *oceanography*. ◄

We will now determine the computational complexity of this procedure. Suppose we have a binary search tree T for a list of n items. We can form a full binary tree U from T by adding unlabeled vertices whenever necessary so that every vertex with a key has two children. This is illustrated in Figure 2. Once we have done this, we can easily locate or add a new item as a key without adding a vertex.

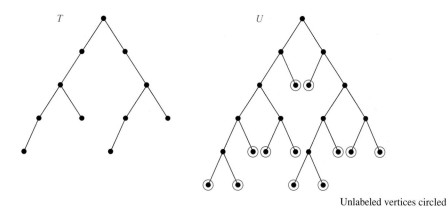

Unlabeled vertices circled

FIGURE 2   Adding Unlabeled Vertices to Make a Binary Search Tree Full.

The most comparisons needed to add a new item is the length of the longest path in U from the root to a leaf. The internal vertices of U are the vertices of T. It follows that U has n internal vertices. We can now use part (*ii*) of Theorem 4 in Section 11.1 to conclude that U has $n + 1$ leaves. Using Corollary 1 of Section 11.1, we see that the height of U is greater than or equal to $h = \lceil \log(n + 1) \rceil$. Consequently, it is necessary to perform at least $\lceil \log(n + 1) \rceil$ comparisons to add some item. Note that if U is balanced, its height is $\lceil \log(n + 1) \rceil$ (by Corollary 1 of Section 11.1). Thus, if a binary search tree is balanced, locating or adding an item requires no more than $\lceil \log(n + 1) \rceil$ comparisons. A binary search tree can become unbalanced as items are added to it. Because balanced binary search trees give optimal worst-case complexity for binary searching, algorithms have been devised that rebalance binary search trees as items are added. The interested reader can consult references on data structures for the description of such algorithms.

Decision Trees

Links

Rooted trees can be used to model problems in which a series of decisions leads to a solution. For instance, a binary search tree can be used to locate items based on a series of comparisons, where each comparison tells us whether we have located the item, or whether we should go right or left in a subtree. A rooted tree in which each internal vertex corresponds to a decision, with a subtree at these vertices for each possible outcome of the decision, is called a **decision tree**. The possible solutions of the problem correspond to the paths to the leaves of this rooted tree. Example 3 illustrates an application of decision trees.

EXAMPLE 3   Suppose there are seven coins, all with the same weight, and a counterfeit coin that weighs less than the others. How many weighings are necessary using a balance scale to determine which of the eight coins is the counterfeit one? Give an algorithm for finding this counterfeit coin.

Extra Examples

Solution: There are three possibilities for each weighing on a balance scale. The two pans can have equal weight, the first pan can be heavier, or the second pan can be heavier. Consequently, the decision tree for the sequence of weighings is a 3-ary tree. There are at least eight leaves in the decision tree because there are eight possible outcomes (because each of the eight coins can be the counterfeit lighter coin), and each possible outcome must be represented by at least one leaf. The largest number of weighings needed to determine the counterfeit coin is the height of the decision tree. From Corollary 1 of Section 11.1 it follows that the height of the decision tree is at least $\lceil \log_3 8 \rceil = 2$. Hence, at least two weighings are needed.

It is possible to determine the counterfeit coin using two weighings. The decision tree that illustrates how this is done is shown in Figure 3.   ◀

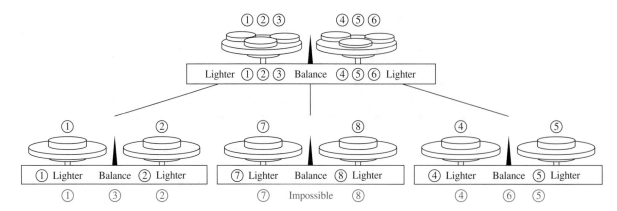

FIGURE 3   **A Decision Tree for Locating a Counterfeit Coin. The counterfeit coin is shown in color below each final weighing.**

THE COMPLEXITY OF COMPARISON-BASED SORTING ALGORITHMS   Many different sorting algorithms have been developed. To decide whether a particular sorting algorithm is efficient, its complexity is determined. Using decision trees as models, a lower bound for the worst-case complexity of sorting algorithms that are based on binary comparisons can be found.

We can use decision trees to model sorting algorithms and to determine an estimate for the worst-case complexity of these algorithms. Note that given n elements, there are $n!$ possible orderings of these elements, because each of the $n!$ permutations of these elements can be the correct order. The sorting algorithms studied in this book, and most commonly used sorting algorithms, are based on binary comparisons, that is, the comparison of two elements at a time. The result of each such comparison narrows down the set of possible orderings. Thus, a sorting algorithm based on binary comparisons can be represented by a binary decision tree in which each internal vertex represents a comparison of two elements. Each leaf represents one of the $n!$ permutations of n elements.

EXAMPLE 4   We display in Figure 4 a decision tree that orders the elements of the list a, b, c.   ◀

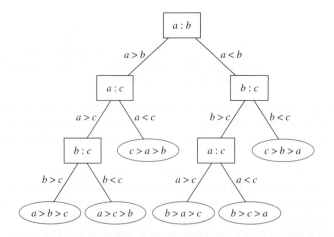

FIGURE 4   **A Decision Tree for Sorting Three Distinct Elements.**

The complexity of a sort based on binary comparisons is measured in terms of the number of such comparisons used. The largest number of binary comparisons ever needed to sort a list with n elements gives the worst-case performance of the algorithm. The most comparisons used equals the longest path length in the decision tree representing the sorting procedure. In other words, the largest number of comparisons ever needed is equal to the height of the decision tree. Because the height of a binary tree with $n!$ leaves is at least $\lceil \log n! \rceil$ (using Corollary 1 in Section 11.1), at least $\lceil \log n! \rceil$ comparisons are needed, as stated in Theorem 1.

THEOREM 1 A sorting algorithm based on binary comparisons requires at least $\lceil \log n! \rceil$ comparisons.

We can use Theorem 1 to provide a big-Omega estimate for the number of comparisons used by a sorting algorithm based on binary comparison. We need only note that by Exercise 48 in Section 3.2 we know that $\lceil \log n! \rceil$ is $\Theta(n \log n)$, one of the commonly used reference functions for the computational complexity of algorithms. Corollary 1 is a consequence of this estimate.

COROLLARY 1 The number of comparisons used by a sorting algorithm to sort n elements based on binary comparisons is $\Omega(n \log n)$.

A consequence of Corollary 1 is that a sorting algorithm based on binary comparisons that uses $\Theta(n \log n)$ comparisons, in the worst case, to sort n elements is optimal, in the sense that no other such algorithm has better worst-case complexity. Note that by Theorem 1 in Section 5.4 we see that the merge sort algorithm is optimal in this sense.

We can also establish a similar result for the average-case complexity of sorting algorithms. The average number of comparisons used by a sorting algorithm based on binary comparisons is the average depth of a leaf in the decision tree representing the sorting algorithm. By Exercise 36 in Section 11.1 we know that the average depth of a leaf in a binary tree with N vertices is $\Omega(\log N)$. We obtain the following estimate when we let $N = n!$ and note that a function that is $\Omega(\log n!)$ is also $\Omega(n \log n)$ because $\log n!$ is $\Theta(n \log n)$.

THEOREM 2 The average number of comparisons used by a sorting algorithm to sort n elements based on binary comparisons is $\Omega(n \log n)$.

Prefix Codes

Consider the problem of using bit strings to encode the letters of the English alphabet (where no distinction is made between lowercase and uppercase letters). We can represent each letter with a bit string of length five, because there are only 26 letters and there are 32 bit strings of length five. The total number of bits used to encode data is five times the number of characters in the text when each character is encoded with five bits. Is it possible to find a coding scheme of these letters such that, when data are coded, fewer bits are used? We can save memory and reduce transmittal time if this can be done.

Consider using bit strings of different lengths to encode letters. Letters that occur more frequently should be encoded using short bit strings, and longer bit strings should be used to encode rarely occurring letters. When letters are encoded using varying numbers of bits, some method must be used to determine where the bits for each character start and end. For instance, if e were encoded with 0, a with 1, and t with 01, then the bit string 0101 could correspond to *eat, tea, eaea,* or *tt.*

One way to ensure that no bit string corresponds to more than one sequence of letters is to encode letters so that the bit string for a letter never occurs as the first part of the bit string for another letter. Codes with this property are called **prefix codes**. For instance, the encoding of e as 0, a as 10, and t as 11 is a prefix code. A word can be recovered from the unique bit string that encodes its letters. For example, the string 10110 is the encoding of *ate.* To see this,

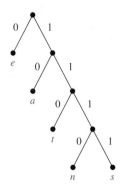

FIGURE 5   A Binary Tree with a Prefix Code.

note that the initial 1 does not represent a character, but 10 does represent *a* (and could not be the first part of the bit string of another letter). Then, the next 1 does not represent a character, but 11 does represent *t*. The final bit, 0, represents *e*.

A prefix code can be represented using a binary tree, where the characters are the labels of the leaves in the tree. The edges of the tree are labeled so that an edge leading to a left child is assigned a 0 and an edge leading to a right child is assigned a 1. The bit string used to encode a character is the sequence of labels of the edges in the unique path from the root to the leaf that has this character as its label. For instance, the tree in Figure 5 represents the encoding of *e* by 0, *a* by 10, *t* by 110, *n* by 1110, and *s* by 1111.

The tree representing a code can be used to decode a bit string. For instance, consider the word encoded by 11111011100 using the code in Figure 5. This bit string can be decoded by starting at the root, using the sequence of bits to form a path that stops when a leaf is reached. Each 0 bit takes the path down the edge leading to the left child of the last vertex in the path, and each 1 bit corresponds to the right child of this vertex. Consequently, the initial 1111 corresponds to the path starting at the root, going right four times, leading to a leaf in the graph that has *s* as its label, because the string 1111 is the code for *s*. Continuing with the fifth bit, we reach a leaf next after going right then left, when the vertex labeled with *a*, which is encoded by 10, is visited. Starting with the seventh bit, we reach a leaf next after going right three times and then left, when the vertex labeled with *n*, which is encoded by 1110, is visited. Finally, the last bit, 0, leads to the leaf that is labeled with *e*. Therefore, the original word is *sane*.

We can construct a prefix code from any binary tree where the left edge at each internal vertex is labeled by 0 and the right edge by a 1 and where the leaves are labeled by characters. Characters are encoded with the bit string constructed using the labels of the edges in the unique path from the root to the leaves.

HUFFMAN CODING   We now introduce an algorithm that takes as input the frequencies (which are the probabilities of occurrences) of symbols in a string and produces as output a prefix code that encodes the string using the fewest possible bits, among all possible binary prefix codes for these symbols. This algorithm, known as **Huffman coding**, was developed by David Huffman in a term paper he wrote in 1951 while a graduate student at MIT. (Note that this algorithm assumes that we already know how many times each symbol occurs in the string, so we can compute the frequency of each symbol by dividing the number of times this symbol occurs by the length of the string.) Huffman coding is a fundamental algorithm in *data compression*, the subject devoted to reducing the number of bits required to represent information. Huffman coding is extensively used to compress bit strings representing text and it also plays an important role in compressing audio and image files.

Algorithm 2 presents the Huffman coding algorithm. Given symbols and their frequencies, our goal is to construct a rooted binary tree where the symbols are the labels of the leaves. The algorithm begins with a forest of trees each consisting of one vertex, where each vertex has

DAVID A. HUFFMAN (1925–1999)   David Huffman grew up in Ohio. At the age of 18 he received his B.S. in electrical engineering from The Ohio State University. Afterward he served in the U.S. Navy as a radar maintenance officer on a destroyer that had the mission of clearing mines in Asian waters after World War II. Later, he earned his M.S. from Ohio State and his Ph.D. in electrical engineering from MIT. Huffman joined the MIT faculty in 1953, where he remained until 1967 when he became the founding member of the computer science department at the University of California at Santa Cruz. He played an important role in developing this department and spent the remainder of his career there, retiring in 1994.

Huffman is noted for his contributions to information theory and coding, signal designs for radar and for communications, and design procedures for asynchronous logical circuits. His work on surfaces with zero curvature led him to develop original techniques for folding paper and vinyl into unusual shapes considered works of art by many and publicly displayed in several exhibits. However, Huffman is best known for his development of what is now called Huffman coding, a result of a term paper he wrote during his graduate work at MIT.

Huffman enjoyed exploring the outdoors, hiking, and traveling extensively. He became certified as a scuba diver when he was in his late 60s. He kept poisonous snakes as pets.

a symbol as its label and where the weight of this vertex equals the frequency of the symbol that is its label. At each step, we combine two trees having the least total weight into a single tree by introducing a new root and placing the tree with larger weight as its left subtree and the tree with smaller weight as its right subtree. Furthermore, we assign the sum of the weights of the two subtrees of this tree as the total weight of the tree. (Although procedures for breaking ties by choosing between trees with equal weights can be specified, we will not specify such procedures here.) The algorithm is finished when it has constructed a tree, that is, when the forest is reduced to a single tree.

Demo

ALGORITHM 2 Huffman Coding.

procedure *Huffman*(C: symbols a_i with frequencies w_i, $i = 1, \ldots, n$)
$F :=$ forest of n rooted trees, each consisting of the single vertex a_i and assigned weight w_i
while F is not a tree
    Replace the rooted trees T and T' of least weights from F with $w(T) \geq w(T')$ with a tree
    having a new root that has T as its left subtree and T' as its right subtree. Label the new
    edge to T with 0 and the new edge to T' with 1.
    Assign $w(T) + w(T')$ as the weight of the new tree.
{the Huffman coding for the symbol a_i is the concatenation of the labels of the edges in the
unique path from the root to the vertex a_i }

Example 5 illustrates how Algorithm 2 is used to encode a set of five symbols.

EXAMPLE 5

Extra
Examples

Use Huffman coding to encode the following symbols with the frequencies listed: A: 0.08, B: 0.10, C: 0.12, D: 0.15, E: 0.20, F: 0.35. What is the average number of bits used to encode a character?

Solution: Figure 6 displays the steps used to encode these symbols. The encoding produced encodes A by 111, B by 110, C by 011, D by 010, E by 10, and F by 00. The average number of bits used to encode a symbol using this encoding is

◄

$$3 \cdot 0.08 + 3 \cdot 0.10 + 3 \cdot 0.12 + 3 \cdot 0.15 + 2 \cdot 0.20 + 2 \cdot 0.35 = 2.45.$$

Note that Huffman coding is a greedy algorithm. Replacing the two subtrees with the smallest weight at each step leads to an optimal code in the sense that no binary prefix code for these symbols can encode these symbols using fewer bits. We leave the proof that Huffman codes are optimal as Exercise 24.

Huffman coding is used in JPEG image coding

There are many variations of Huffman coding. For example, instead of encoding single symbols, we can encode blocks of symbols of a specified length, such as blocks of two symbols. Doing so may reduce the number of bits required to encode the string (see Exercise 22). We can also use more than two symbols to encode the original symbols in the string (see the preamble to Exercise 20). Furthermore, a variation known as adaptive Huffman coding (see [Sa00]) can be used when the frequency of each symbol in a string is not known in advance, so that encoding is done at the same time the string is being read.

Game Trees

Links

Trees can be used to analyze certain types of games such as tic-tac-toe, nim, checkers, and chess. In each of these games, two players take turns making moves. Each player knows the moves made by the other player and no element of chance enters into the game. We model such games using **game trees**; the vertices of these trees represent the positions that a game can be in as it

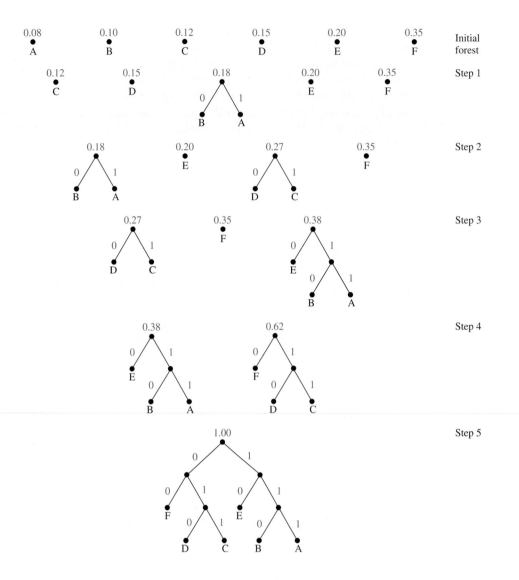

FIGURE 6 **Huffman Coding of Symbols in Example 4.**

progresses; the edges represent legal moves between these positions. Because game trees are usually large, we simplify game trees by representing all symmetric positions of a game by the same vertex. However, the same position of a game may be represented by different vertices if different sequences of moves lead to this position. The root represents the starting position. The usual convention is to represent vertices at even levels by boxes and vertices at odd levels by circles. When the game is in a position represented by a vertex at an even level, it is the first player's move; when the game is in a position represented by a vertex at an odd level, it is the second player's move. Game trees may be infinite when the games they represent never end, such as games that can enter infinite loops, but for most games there are rules that lead to finite game trees.

The leaves of a game tree represent the final positions of a game. We assign a value to each leaf indicating the payoff to the first player if the game terminates in the position represented by this leaf. For games that are win–lose, we label a terminal vertex represented by a circle with a 1 to indicate a win by the first player and we label a terminal vertex represented by a box with a −1 to indicate a win by the second player. For games where draws are allowed, we label a terminal vertex corresponding to a draw position with a 0. Note that for win–lose games, we

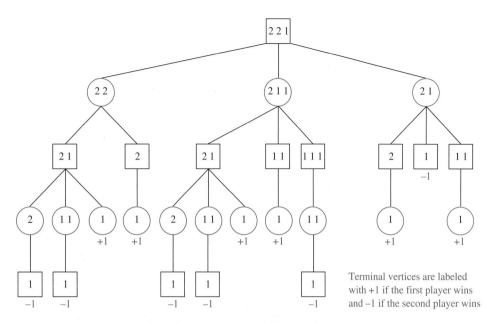

Terminal vertices are labeled
with +1 if the first player wins
and −1 if the second player wins

FIGURE 7 **The Game Tree for a Game of Nim.**

have assigned values to terminal vertices so that the larger the value, the better the outcome for
the first player.

In Example 6 we display a game tree for a well-known and well-studied game.

EXAMPLE 6

Although nim is an
ancient game, Charles
Bouton coined its modern
name in 1901 after an
archaic English word
meaning "to steal."

Nim In a version of the game of **nim**, at the start of a game there are a number of piles of
stones. Two players take turns making moves; a legal move consists of removing one or more
stones from one of the piles, without removing all the stones left. A player without a legal move
loses. (Another way to look at this is that the player removing the last stone loses because the
position with no piles of stones is not allowed.) The game tree shown in Figure 7 represents this
version of nim given the starting position where there are three piles of stones containing two,
two, and one stone each, respectively. We represent each position with an unordered list of the
number of stones in the different piles (the order of the piles does not matter). The initial move
by the first player can lead to three possible positions because this player can remove one stone
from a pile with two stones (leaving three piles containing one, one, and two stones); two stones
from a pile containing two stones (leaving two piles containing two stones and one stone); or
one stone from the pile containing one stone (leaving two piles of two stones). When only one
pile with one stone is left, no legal moves are possible, so such positions are terminal positions.
Because nim is a win–lose game, we label the terminal vertices with +1 when they represent
wins for the first player and −1 when they represent wins for the second player. ◀

EXAMPLE 7

Tic-tac-toe The game tree for tic-tac-toe is extremely large and cannot be drawn here, although
a computer could easily build such a tree. We show a portion of the game tic-tac-toe in Figure 8(a).
Note that by considering symmetric positions equivalent, we need only consider three possible
initial moves, as shown in Figure 8(a). We also show a subtree of this game tree leading to
terminal positions in Figure 8(b), where a player who can win makes a winning move. ◀

We can recursively define the values of all vertices in a game tree in a way that enables
us to determine the outcome of this game when both players follow optimal strategies. By a
strategy we mean a set of rules that tells a player how to select moves to win the game. An
optimal strategy for the first player is a strategy that maximizes the payoff to this player and for
the second player is a strategy that minimizes this payoff. We now recursively define the value
of a vertex.

(a)   (b)

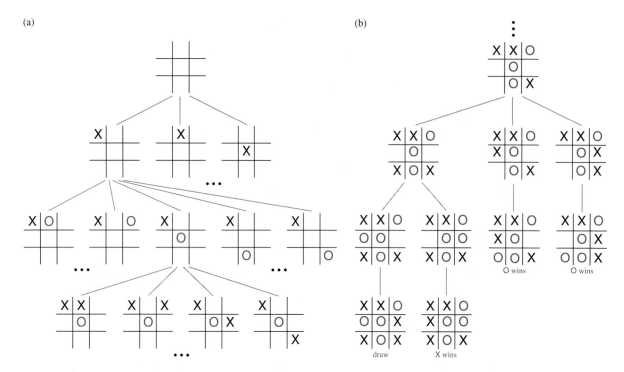

FIGURE 8   Some of the Game Tree for Tic-Tac-Toe.

DEFINITION 1   The *value of a vertex in a game tree* is defined recursively as:

  (*i*) the value of a leaf is the payoff to the first player when the game terminates in the position represented by this leaf.
  (*ii*) the value of an internal vertex at an even level is the maximum of the values of its children, and the value of an internal vertex at an odd level is the minimum of the values of its children.

The strategy where the first player moves to a position represented by a child with maximum value and the second player moves to a position of a child with minimum value is called the **minmax strategy**. We can determine who will win the game when both players follow the minmax strategy by calculating the value of the root of the tree; this value is called the **value** of the tree. This is a consequence of Theorem 3.

THEOREM 3   The value of a vertex of a game tree tells us the payoff to the first player if both players follow the minmax strategy and play starts from the position represented by this vertex.

Proof: We will use induction to prove this theorem.

BASIS STEP: If the vertex is a leaf, by definition the value assigned to this vertex is the payoff to the first player.

INDUCTIVE STEP: The inductive hypothesis is the assumption that the values of the children of a vertex are the payoffs to the first player, assuming that play starts at each of the positions represented by these vertices. We need to consider two cases, when it is the first player's turn and when it is the second player's turn.

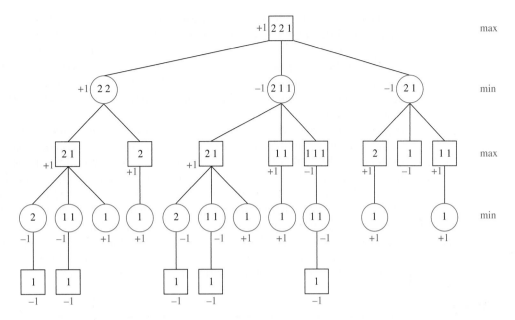

FIGURE 9 **Showing the Values of Vertices in the Game of Nim.**

When it is the first player's turn, this player follows the minmax strategy and moves to the position represented by the child with the largest value. By the inductive hypothesis, this value is the payoff to the first player when play starts at the position represented by this child and follows the minmax strategy. By the recursive step in the definition of the value of an internal vertex at an even level (as the maximum value of its children), the value of this vertex is the payoff when play begins at the position represented by this vertex.

When it is the second player's turn, this player follows the minmax strategy and moves to the position represented by the child with the least value. By the inductive hypothesis, this value is the payoff to the first player when play starts at the position represented by this child and both players follow the minmax strategy. By the recursive definition of the value of an internal vertex at an odd level as the minimum value of its children, the value of this vertex is the payoff when play begins at the position represented by this vertex. ◁

Remark: By extending the proof of Theorem 3, it can be shown that the minmax strategy is the optimal strategy for both players.

Example 8 illustrates how the minmax procedure works. It displays the values assigned to the internal vertices in the game tree from Example 6. Note that we can shorten the computation required by noting that for win–lose games, once a child of a square vertex with value +1 is found, the value of the square vertex is also +1 because +1 is the largest possible payoff. Similarly, once a child of a circle vertex with value −1 is found, this is the value of the circle vertex also.

EXAMPLE 8   In Example 6 we constructed the game tree for nim with a starting position where there are three piles containing two, two, and one stones. In Figure 9 we show the values of the vertices of this game tree. The values of the vertices are computed using the values of the leaves and working one level up at a time. In the right margin of this figure we indicate whether we use the maximum or minimum of the values of the children to find the value of an internal vertex at each level. For example, once we have found the values of the three children of the root, which are 1, −1, and −1, we find the value of the root by computing max(1, −1, −1) = 1. Because the value of the root is 1, it follows that the first player wins when both players follow a minmax strategy. ◀

Game trees for some well-known games can be extraordinarily large, because these games have many different possible moves. For example, the game tree for chess has been estimated to have as many as 10^{100} vertices! It may be impossible to use Theorem 3 directly to study a game because of the size of the game tree. Therefore, various approaches have been devised to help determine good strategies and to determine the outcome of such games. One useful technique, called *alpha-beta pruning*, eliminates much computation by pruning portions of the game tree that cannot affect the values of ancestor vertices. (For information about alpha-beta pruning, consult [Gr90].) Another useful approach is to use *evaluation functions*, which estimate the value of internal vertices in the game tree when it is not feasible to compute these values exactly. For example, in the game of tic-tac-toe, as an evaluation function for a position, we may use the number of files (rows, columns, and diagonals) containing no Os (used to indicate moves of the second player) minus the number of files containing no Xs (used to indicate moves of the first player). This evaluation function provides some indication of which player has the advantage in the game. Once the values of an evaluation function are inserted, the value of the game can be computed following the rules used for the minmax strategy. Computer programs created to play chess, such as the famous Deep Blue program, are based on sophisticated evaluation functions. For more information about how computers play chess see [Le91].

Chess programs on smartphones can now play at the grandmaster level.

Links

Exercises

1. Build a binary search tree for the words *banana, peach, apple, pear, coconut, mango,* and *papaya* using alphabetical order.

2. How many comparisons are needed to locate or to add each of these words in the search tree for Exercise 1, starting fresh each time?

   a) *pear*   **b)** *banana*
   c) *kumquat*   **d)** *orange*

3. Using alphabetical order, construct a binary search tree for the words in the sentence *"The quick brown fox jumps over the lazy dog."*

4. How many weighings of a balance scale are needed to find a lighter counterfeit coin among four coins? Describe an algorithm to find the lighter coin using this number of weighings.

5. How many weighings of a balance scale are needed to find a counterfeit coin among four coins if the counterfeit coin may be either heavier or lighter than the others? Describe an algorithm to find the counterfeit coin using this number of weighings.

*6. One of four coins may be counterfeit. If it is counterfeit, it may be lighter or heavier than the others. How many weighings are needed, using a balance scale, to determine whether there is a counterfeit coin, and if there is, whether it is lighter or heavier than the others? Describe an algorithm to find the counterfeit coin and determine whether it is lighter or heavier using this number of weighings.

7. Find the least number of comparisons needed to sort four elements and devise an algorithm that sorts these elements using this number of comparisons.

*8. Find the least number of comparisons needed to sort five elements and devise an algorithm that sorts these elements using this number of comparisons.

The **tournament sort** is a sorting algorithm that works by building an ordered binary tree. We represent the elements to be sorted by vertices that will become the leaves. We build up

the tree one level at a time as we would construct the tree representing the winners of matches in a tournament. Working left to right, we compare pairs of consecutive elements, adding a parent vertex labeled with the larger of the two elements under comparison. We make similar comparisons between labels of vertices at each level until we reach the root of the tree that is labeled with the largest element. The tree constructed by the tournament sort of 22, 8, 14, 17, 3, 9, 27, 11 is illustrated in part (a) of the figure. Once the largest element has been determined, the leaf with this label is relabeled by $-\infty$, which is defined to be less than every element. The labels of all vertices on the path from this vertex up to the root of the tree are recalculated, as shown in part (b) of the figure. This produces the second largest element. This process continues until the entire list has been sorted.

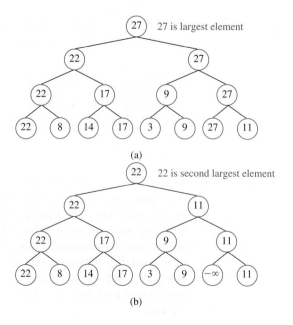

9. Complete the tournament sort of the list 22, 8, 14, 17, 3, 9, 27, 11. Show the labels of the vertices at each step.

10. Describe the tournament sort using pseudocode.

11. How many comparisons does the tournament sort use to find the second largest, the third largest, and so on, up to the $(n-1)$st largest (or second smallest) element?

12. Show that the tournament sort requires $\Theta(n \log n)$ comparisons to sort a list of n elements. [*Hint:* By inserting the appropriate number of dummy elements defined to be smaller than all integers, such as $-\infty$, assume that $n = 2^k$ for some positive integer k.]

13. Which of these codes are prefix codes?
    a) a: 11, e: 00, t: 10, s: 01
    b) a: 0, e: 1, t: 01, s: 001
    c) a: 101, e: 11, t: 001, s: 011, n: 010
    d) a: 010, e: 11, t: 011, s: 1011, n: 1001, i: 10101

14. Construct the binary tree with prefix codes representing these coding schemes.
    a) a: 11, e: 0, t: 101, s: 100
    b) a: 1, e: 01, t: 001, s: 0001, n: 00001
    c) a: 1010, e: 0, t: 11, s: 1011, n: 1001, i: 10001

15. What are the codes for a, e, i, k, o, p, and u if the coding scheme is represented by this tree?

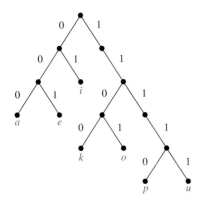

16. Given the coding scheme a: 001, b: 0001, e: 1, r: 0000, s: 0100, t: 011, x: 01010, find the word represented by
    a) 01110100011.   b) 0001110000.
    c) 0100101010.   d) 01100101010.

17. Use Huffman coding to encode these symbols with given frequencies: a: 0.20, b: 0.10, c: 0.15, d: 0.25, e: 0.30. What is the average number of bits required to encode a character?

18. Construct two different Huffman codes for these symbols and frequencies: t: 0.2, u: 0.3, v: 0.2, w: 0.3.

19. Construct a Huffman code for the letters of the English alphabet where the frequencies of letters in typical English text are as shown in this table.

Letter	Frequency	Letter	Frequency
A	0.0817	N	0.0662
B	0.0145	O	0.0781
C	0.0248	P	0.0156
D	0.0431	Q	0.0009
E	0.1232	R	0.0572
F	0.0209	S	0.0628
G	0.0182	T	0.0905
H	0.0668	U	0.0304
I	0.0689	V	0.0102
J	0.0010	W	0.0264
K	0.0080	X	0.0015
L	0.0397	Y	0.0211
M	0.0277	Z	0.0005

Suppose that m is a positive integer with $m \geq 2$. An m-ary Huffman code for a set of N symbols can be constructed analogously to the construction of a binary Huffman code. At the initial step, $((N-1) \bmod (m-1)) + 1$ trees consisting of a single vertex with least weights are combined into a rooted tree with these vertices as leaves. At each subsequent step, the m trees of least weight are combined into an m-ary tree.

20. Describe the m-ary Huffman coding algorithm in pseudocode.

21. Using the symbols 0, 1, and 2 use ternary ($m = 3$) Huffman coding to encode these letters with the given frequencies: A: 0.25, E: 0.30, N: 0.10, R: 0.05, T: 0.12, Z: 0.18.

22. Consider the three symbols A, B, and C with frequencies A: 0.80, B: 0.19, C: 0.01.
    a) Construct a Huffman code for these three symbols.
    b) Form a new set of nine symbols by grouping together blocks of two symbols, AA, AB, AC, BA, BB, BC, CA, CB, and CC. Construct a Huffman code for these nine symbols, assuming that the occurrences of symbols in the original text are independent.
    c) Compare the average number of bits required to encode text using the Huffman code for the three symbols in part (a) and the Huffman code for the nine blocks of two symbols constructed in part (b). Which is more efficient?

23. Given $n+1$ symbols $x_1, x_2, \ldots, x_n, x_{n+1}$ appearing 1, $f_1, f_2, \ldots, f_n$ times in a symbol string, respectively, where f_j is the jth Fibonacci number, what is the maximum number of bits used to encode a symbol when all possible tie-breaking selections are considered at each stage of the Huffman coding algorithm?

*24. Show that Huffman codes are optimal in the sense that they represent a string of symbols using the fewest bits among all binary prefix codes.

25. Draw a game tree for nim if the starting position consists of two piles with two and three stones, respectively. When drawing the tree represent by the same vertex symmetric

positions that result from the same move. Find the value of each vertex of the game tree. Who wins the game if both players follow an optimal strategy?

26. Suppose that in a variation of the game of nim we allow a player to either remove one or more stones from a pile or merge the stones from two piles into one pile as long as at least one stone remains. Draw the game tree for this variation of nim if the starting position consists of three piles containing two, two, and one stone, respectively. Find the values of each vertex in the game tree and determine the winner if both players follow an optimal strategy.

27. Draw the subtree of the game tree for tic-tac-toe beginning at each of these positions. Determine the value of each of these subtrees.

28. Suppose that the first four moves of a tic-tac-toe game are as shown. Does the first player (whose moves are marked by Xs) have a strategy that will always win?

29. Show that if a game of nim begins with two piles containing the same number of stones, as long as this number is at least two, then the second player wins when both players follow optimal strategies.

30. Show that if a game of nim begins with two piles containing different numbers of stones, the first player wins when both players follow optimal strategies.

31. How many children does the root of the game tree for checkers have? How many grandchildren does it have?

32. How many children does the root of the game tree for nim have and how many grandchildren does it have if the starting position is

a) piles with four and five stones, respectively.

b) piles with two, three, and four stones, respectively.

c) piles with one, two, three, and four stones, respectively.

d) piles with two, two, three, three, and five stones, respectively.

33. Draw the game tree for the game of tic-tac-toe for the levels corresponding to the first two moves. Assign the value of the evaluation function mentioned in the text that assigns to a position the number of files containing no Os minus the number of files containing no Xs as the value of each vertex at this level and compute the value of the tree for vertices as if the evaluation function gave the correct values for these vertices.

34. Use pseudocode to describe an algorithm for determining the value of a game tree when both players follow a minmax strategy.

11.3 Tree Traversal

Introduction

Links

Ordered rooted trees are often used to store information. We need procedures for visiting each vertex of an ordered rooted tree to access data. We will describe several important algorithms for visiting all the vertices of an ordered rooted tree. Ordered rooted trees can also be used to represent various types of expressions, such as arithmetic expressions involving numbers, variables, and operations. The different listings of the vertices of ordered rooted trees used to represent expressions are useful in the evaluation of these expressions.

Universal Address Systems

Procedures for traversing all vertices of an ordered rooted tree rely on the orderings of children. In ordered rooted trees, the children of an internal vertex are shown from left to right in the drawings representing these directed graphs.

We will describe one way we can totally order the vertices of an ordered rooted tree. To produce this ordering, we must first label all the vertices. We do this recursively:

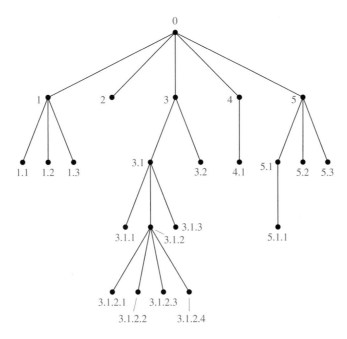

FIGURE 1   **The Universal Address System of an Ordered Rooted Tree.**

1. Label the root with the integer 0. Then label its k children (at level 1) from left to right with $1, 2, 3, \ldots, k$.
2. For each vertex v at level n with label A, label its k_v children, as they are drawn from left to right, with $A.1, A.2, \ldots, A.k_v$.

Following this procedure, a vertex v at level n, for $n \geq 1$, is labeled $x_1.x_2.\ldots.x_n$, where the unique path from the root to v goes through the x_1st vertex at level 1, the x_2nd vertex at level 2, and so on. This labeling is called the **universal address system** of the ordered rooted tree.

We can totally order the vertices using the lexicographic ordering of their labels in the universal address system. The vertex labeled $x_1.x_2.\ldots.x_n$ is less than the vertex labeled $y_1.y_2.\ldots.y_m$ if there is an i, $0 \leq i \leq n$, with $x_1 = y_1, x_2 = y_2, \ldots, x_{i-1} = y_{i-1}$, and $x_i < y_i$; or if $n < m$ and $x_i = y_i$ for $i = 1, 2, \ldots, n$.

EXAMPLE 1

We display the labelings of the universal address system next to the vertices in the ordered rooted tree shown in Figure 1. The lexicographic ordering of the labelings is

$$0 < 1 < 1.1 < 1.2 < 1.3 < 2 < 3 < 3.1 < 3.1.1 < 3.1.2 < 3.1.2.1 < 3.1.2.2$$
$$< 3.1.2.3 < 3.1.2.4 < 3.1.3 < 3.2 < 4 < 4.1 < 5 < 5.1 < 5.1.1 < 5.2 < 5.3$$ ◀

Traversal Algorithms

Procedures for systematically visiting every vertex of an ordered rooted tree are called **traversal algorithms**. We will describe three of the most commonly used such algorithms, **preorder traversal**, **inorder traversal**, and **postorder traversal**. Each of these algorithms can be defined recursively. We first define preorder traversal.

DEFINITION 1   Let T be an ordered rooted tree with root r. If T consists only of r, then r is the *preorder traversal* of T. Otherwise, suppose that $T_1, T_2, \ldots, T_n$ are the subtrees at r from left to right in T. The *preorder traversal* begins by visiting r. It continues by traversing T_1 in preorder, then T_2 in preorder, and so on, until T_n is traversed in preorder.

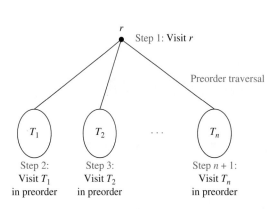

FIGURE 2   **Preorder Traversal.**

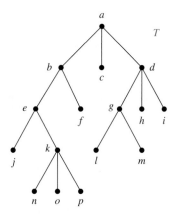

FIGURE 3   **The Ordered Rooted Tree T.**

The reader should verify that the preorder traversal of an ordered rooted tree gives the same ordering of the vertices as the ordering obtained using a universal address system. Figure 2 indicates how a preorder traversal is carried out.

Example 2 illustrates preorder traversal.

EXAMPLE 2   In which order does a preorder traversal visit the vertices in the ordered rooted tree T shown in Figure 3?

Solution: The steps of the preorder traversal of T are shown in Figure 4. We traverse T in preorder by first listing the root a, followed by the preorder list of the subtree with root b, the preorder list of the subtree with root c (which is just c) and the preorder list of the subtree with root d.

The preorder list of the subtree with root b begins by listing b, then the vertices of the subtree with root e in preorder, and then the subtree with root f in preorder (which is just f). The preorder list of the subtree with root d begins by listing d, followed by the preorder list of the subtree with root g, followed by the subtree with root h (which is just h), followed by the subtree with root i (which is just i).

The preorder list of the subtree with root e begins by listing e, followed by the preorder listing of the subtree with root j (which is just j), followed by the preorder listing of the subtree with root k. The preorder listing of the subtree with root g is g followed by l, followed by m. The preorder listing of the subtree with root k is k, n, o, p. Consequently, the preorder traversal of T is $a, b, e, j, k, n, o, p, f, c, d, g, l, m, h, i$.   ◀

We will now define inorder traversal.

DEFINITION 2   Let T be an ordered rooted tree with root r. If T consists only of r, then r is the *inorder traversal* of T. Otherwise, suppose that $T_1, T_2, \ldots, T_n$ are the subtrees at r from left to right. The *inorder traversal* begins by traversing T_1 in inorder, then visiting r. It continues by traversing T_2 in inorder, then T_3 in inorder, $\ldots$, and finally T_n in inorder.

Figure 5 indicates how inorder traversal is carried out. Example 3 illustrates how inorder traversal is carried out for a particular tree.

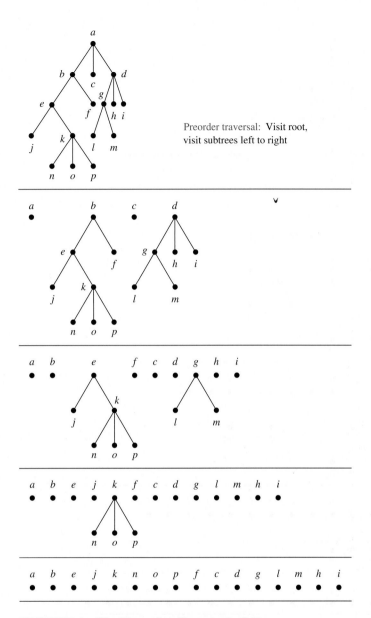

Preorder traversal: Visit root,
visit subtrees left to right

FIGURE 4   The Preorder Traversal of T.

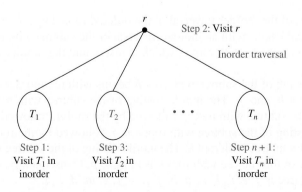

Step 2: Visit r

Inorder traversal

Step 1:
Visit T_1 in
inorder

Step 3:
Visit T_2 in
inorder

Step $n + 1$:
Visit T_n in
inorder

FIGURE 5   Inorder Traversal.

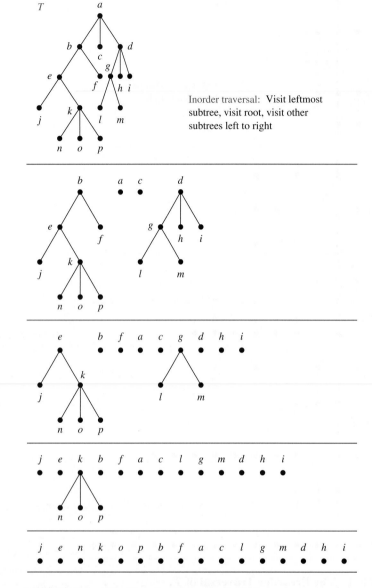

Inorder traversal: Visit leftmost subtree, visit root, visit other subtrees left to right

FIGURE 6 **The Inorder Traversal of *T*.**

EXAMPLE 3 In which order does an inorder traversal visit the vertices of the ordered rooted tree *T* in Figure 3?

Extra Examples

Solution: The steps of the inorder traversal of the ordered rooted tree *T* are shown in Figure 6. The inorder traversal begins with an inorder traversal of the subtree with root *b*, the root *a*, the inorder listing of the subtree with root *c*, which is just *c*, and the inorder listing of the subtree with root *d*.

The inorder listing of the subtree with root *b* begins with the inorder listing of the subtree with root *e*, the root *b*, and *f*. The inorder listing of the subtree with root *d* begins with the inorder listing of the subtree with root *g*, followed by the root *d*, followed by *h*, followed by *i*.

The inorder listing of the subtree with root *e* is *j*, followed by the root *e*, followed by the inorder listing of the subtree with root *k*. The inorder listing of the subtree with root *g* is *l*, *g*, *m*. The inorder listing of the subtree with root *k* is *n*, *k*, *o*, *p*. Consequently, the inorder listing of the ordered rooted tree is *j*, *e*, *n*, *k*, *o*, *p*, *b*, *f*, *a*, *c*, *l*, *g*, *m*, *d*, *h*, *i*. ◀

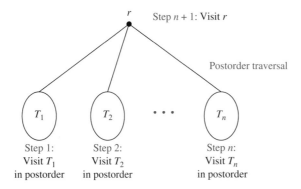

r Step *n* + 1: Visit *r*

Postorder traversal

T_1 T_2 $\cdots$ T_n

Step 1:   Step 2:   Step *n*:
Visit T_1 Visit T_2 Visit T_n
in postorder   in postorder   in postorder

FIGURE 7   Postorder Traversal.

We now define postorder traversal.

DEFINITION 3   Let T be an ordered rooted tree with root r. If T consists only of r, then r is the *postorder traversal* of T. Otherwise, suppose that $T_1, T_2, \ldots, T_n$ are the subtrees at r from left to right. The *postorder traversal* begins by traversing T_1 in postorder, then T_2 in postorder, $\ldots$, then T_n in postorder, and ends by visiting r.

Figure 7 illustrates how postorder traversal is done. Example 4 illustrates how postorder traversal works.

EXAMPLE 4   In which order does a postorder traversal visit the vertices of the ordered rooted tree T shown in Figure 3?

Solution: The steps of the postorder traversal of the ordered rooted tree T are shown in Figure 8. The postorder traversal begins with the postorder traversal of the subtree with root b, the postorder traversal of the subtree with root c, which is just c, the postorder traversal of the subtree with root d, followed by the root a.

The postorder traversal of the subtree with root b begins with the postorder traversal of the subtree with root e, followed by f, followed by the root b. The postorder traversal of the rooted tree with root d begins with the postorder traversal of the subtree with root g, followed by h, followed by i, followed by the root d.

The postorder traversal of the subtree with root e begins with j, followed by the postorder traversal of the subtree with root k, followed by the root e. The postorder traversal of the subtree with root g is l, m, g. The postorder traversal of the subtree with root k is n, o, p, k. Therefore, the postorder traversal of T is $j, n, o, p, k, e, f, b, c, l, m, g, h, i, d, a$.   ◄

There are easy ways to list the vertices of an ordered rooted tree in preorder, inorder, and postorder. To do this, first draw a curve around the ordered rooted tree starting at the root, moving along the edges, as shown in the example in Figure 9. We can list the vertices in preorder by listing each vertex the first time this curve passes it. We can list the vertices in inorder by listing a leaf the first time the curve passes it and listing each internal vertex the second time the curve passes it. We can list the vertices in postorder by listing a vertex the last time it is passed on the way back up to its parent. When this is done in the rooted tree in Figure 9, it follows that the preorder traversal gives $a, b, d, h, e, i, j, c, f, g, k$, the inorder traversal gives $h, d, b, i, e, j, a, f, c, k, g$; and the postorder traversal gives $h, d, i, j, e, b, f, k, g, c, a$.

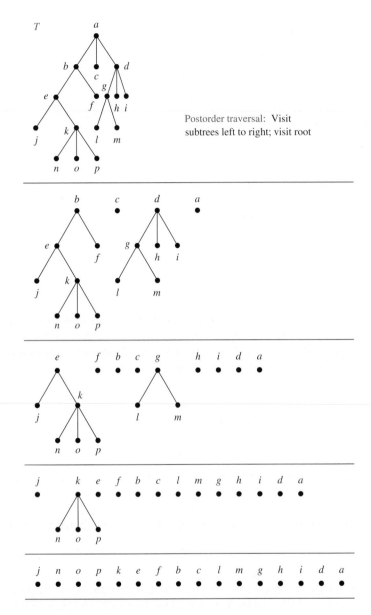

Postorder traversal: Visit
subtrees left to right; visit root

FIGURE 8   **The Postorder Traversal of T.**

Algorithms for traversing ordered rooted trees in preorder, inorder, or postorder are most easily expressed recursively.

ALGORITHM 1 Preorder Traversal.

procedure *preorder*(T: ordered rooted tree)
r := root of T
list r
for each child c of r from left to right
    $T(c)$:= subtree with c as its root
    preorder($T(c)$)

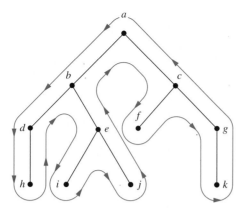

FIGURE 9 A Shortcut for Traversing an Ordered Rooted Tree in Preorder, Inorder, and Postorder.

ALGORITHM 2 Inorder Traversal.

procedure *inorder*(T: ordered rooted tree)
r := root of T
if r is a leaf **then** list r
else
   l := first child of r from left to right
   $T(l)$:= subtree with l as its root
   inorder($T(l)$)
   list r
   for each child c of r except for l from left to right
      $T(c)$:= subtree with c as its root
      inorder($T(c)$)

ALGORITHM 3 Postorder Traversal.

procedure *postorder*(T: ordered rooted tree)
r := root of T
for each child c of r from left to right
   $T(c)$:= subtree with c as its root
   postorder($T(c)$)
list r

Note that both the preorder traversal and the postorder traversal encode the structure of an ordered rooted tree when the number of children of each vertex is specified. That is, an ordered rooted tree is uniquely determined when we specify a list of vertices generated by a preorder traversal or by a postorder traversal of the tree, together with the number of children of each vertex (see Exercises 16 and 17). In particular, both a preorder traversal and a postorder traversal encode the structure of a full ordered m-ary tree. However, when the number of children of vertices is not specified, neither a preorder traversal nor a postorder traversal encodes the structure of an ordered rooted tree (see Exercises 18 and 19).

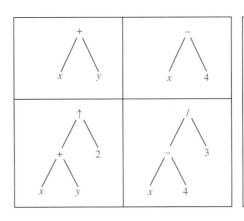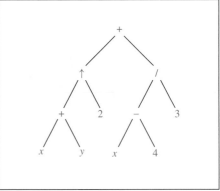

FIGURE 10   A Binary Tree Representing $((x + y) \uparrow 2) + ((x - 4)/3)$.

Infix, Prefix, and Postfix Notation

We can represent complicated expressions, such as compound propositions, combinations of sets, and arithmetic expressions using ordered rooted trees. For instance, consider the representation of an arithmetic expression involving the operators $+$ (addition), $-$ (subtraction), $*$ (multiplication), $/$ (division), and $\uparrow$ (exponentiation). We will use parentheses to indicate the order of the operations. An ordered rooted tree can be used to represent such expressions, where the internal vertices represent operations, and the leaves represent the variables or numbers. Each operation operates on its left and right subtrees (in that order).

EXAMPLE 5   What is the ordered rooted tree that represents the expression $((x + y) \uparrow 2) + ((x - 4)/3)$?

Solution: The binary tree for this expression can be built from the bottom up. First, a subtree for the expression $x + y$ is constructed. Then this is incorporated as part of the larger subtree representing $(x + y) \uparrow 2$. Also, a subtree for $x - 4$ is constructed, and then this is incorporated into a subtree representing $(x - 4)/3$. Finally the subtrees representing $(x + y) \uparrow 2$ and $(x - 4)/3$ are combined to form the ordered rooted tree representing $((x + y) \uparrow 2) + ((x - 4)/3)$. These steps are shown in Figure 10.   ◀

An inorder traversal of the binary tree representing an expression produces the original expression with the elements and operations in the same order as they originally occurred, except for unary operations, which instead immediately follow their operands. For instance, inorder traversals of the binary trees in Figure 11, which represent the expressions $(x + y)/(x + 3)$, $(x + (y/x)) + 3$, and $x + (y/(x + 3))$, all lead to the infix expression $x + y/x + 3$. To make such expressions unambiguous it is necessary to include parentheses in the inorder traversal whenever we encounter an operation. The fully parenthesized expression obtained in this way is said to be in **infix form**.

We obtain the **prefix form** of an expression when we traverse its rooted tree in preorder. Expressions written in prefix form are said to be in **Polish notation**, which is named after the Polish logician Jan Łukasiewicz. An expression in prefix notation (where each operation has a specified number of operands), is unambiguous, so no parentheses are needed in such an expression. The verification of this is left as an exercise for the reader.

EXAMPLE 6   What is the prefix form for $((x + y) \uparrow 2) + ((x - 4)/3)$?

Solution: We obtain the prefix form for this expression by traversing the binary tree that represents it in preorder, shown in Figure 10. This produces $+ \uparrow + x\ y\ 2 / - x\ 4\ 3$.   ◀

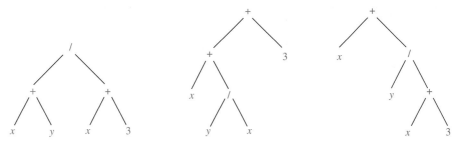

FIGURE 11 **Rooted Trees Representing** $(x + y)/(x + 3)$, $(x + (y/x)) + 3$, **and**
$x + (y/(x + 3))$.

In the prefix form of an expression, a binary operator, such as $+$, precedes its two operands. Hence, we can evaluate an expression in prefix form by working from right to left. When we encounter an operator, we perform the corresponding operation with the two operands immediately to the right of this operand. Also, whenever an operation is performed, we consider the result a new operand.

EXAMPLE 7 What is the value of the prefix expression $+ - * 2 3 5/ \uparrow 2 3 4$?

Solution: The steps used to evaluate this expression by working right to left, and performing operations using the operands on the right, are shown in Figure 12. The value of this expression is 3. ◄

Reverse polish notation was first proposed in 1954 by Burks, Warren, and Wright.

We obtain the **postfix form** of an expression by traversing its binary tree in postorder. Expressions written in postfix form are said to be in **reverse Polish notation**. Expressions in reverse Polish notation are unambiguous, so parentheses are not needed. The verification of this is left to the reader. Reverse polish notation was extensively used in electronic calculators in the 1970s and 1980s.

EXAMPLE 8 What is the postfix form of the expression $((x + y) \uparrow 2) + ((x - 4)/3)$?

Solution: The postfix form of the expression is obtained by carrying out a postorder traversal of the binary tree for this expression, shown in Figure 10. This produces the postfix expression: x $y + 2 \uparrow x 4 - 3 / +$. ◄

In the postfix form of an expression, a binary operator follows its two operands. So, to evaluate an expression from its postfix form, work from left to right, carrying out operations whenever an operator follows two operands. After an operation is carried out, the result of this operation becomes a new operand.

EXAMPLE 9 What is the value of the postfix expression $7 2 3 * - 4 \uparrow 9 3/+$?

Solution: The steps used to evaluate this expression by starting at the left and carrying out operations when two operands are followed by an operator are shown in Figure 13. The value of this expression is 4. ◄

+	−	*	2	3	5	/	↑	2	3	4

$2 \uparrow 3 = 8$

+	−	*	2	3	5	/	8	4

$8 / 4 = 2$

+	−	*	2	3	5	2

$2 * 3 = 6$

+	−	6	5	2

$6 - 5 = 1$

+	1	2

$1 + 2 = 3$

Value of expression: 3

FIGURE 12   Evaluating a Prefix Expression.

7	2	3	*	−	4	↑	9	3	/	+

$2 * 3 = 6$

7	6	−	4	↑	9	3	/	+

$7 - 6 = 1$

1	4	↑	9	3	/	+

$1^4 = 1$

1	9	3	/	+

$9 / 3 = 3$

1	3	+

$1 + 3 = 4$

Value of expression: 4

FIGURE 13   Evaluating a Postfix Expression.

Rooted trees can be used to represent other types of expressions, such as those representing compound propositions and combinations of sets. In these examples unary operators, such as the negation of a proposition, occur. To represent such operators and their operands, a vertex representing the operator and a child of this vertex representing the operand are used.

EXAMPLE 10   Find the ordered rooted tree representing the compound proposition $(\neg(p \wedge q)) \leftrightarrow (\neg p \vee \neg q)$. Then use this rooted tree to find the prefix, postfix, and infix forms of this expression.

Solution: The rooted tree for this compound proposition is constructed from the bottom up. First, subtrees for $\neg p$ and $\neg q$ are formed (where $\neg$ is considered a unary operator). Also, a subtree for $p \wedge q$ is formed. Then subtrees for $\neg(p \wedge q)$ and $(\neg p) \vee (\neg q)$ are constructed. Finally, these two subtrees are used to form the final rooted tree. The steps of this procedure are shown in Figure 14.

The prefix, postfix, and infix forms of this expression are found by traversing this rooted tree in preorder, postorder, and inorder (including parentheses), respectively. These traversals give $\leftrightarrow \neg \wedge pq \vee \neg p \neg q$, $pq \wedge \neg p \neg q \neg \vee \leftrightarrow$, and $(\neg(p \wedge q)) \leftrightarrow ((\neg p) \vee (\neg q))$, respectively. ◀

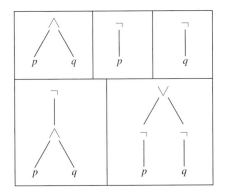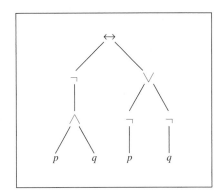

FIGURE 14 **Constructing the Rooted Tree for a Compound Proposition.**

Because prefix and postfix expressions are unambiguous and because they can be evaluated easily without scanning back and forth, they are used extensively in computer science. Such expressions are especially useful in the construction of compilers.

Exercises

In Exercise 1 construct the universal address system for the given ordered rooted tree. Then use this to order its vertices using the lexicographic order of their labels.

1.

2. Suppose that the address of the vertex v in the ordered rooted tree T is 3.4.5.2.4.

    a) At what level is v?

    b) What is the address of the parent of v?

    c) What is the least number of siblings v can have?

    d) What is the smallest possible number of vertices in T if v has this address?

    e) Find the other addresses that must occur.

3. Suppose that the vertex with the largest address in an ordered rooted tree T has address 2.3.4.3.1. Is it possible to determine the number of vertices in T?

4. Can the leaves of an ordered rooted tree have the following list of universal addresses? If so, construct such an ordered rooted tree.

    a) 1.1.1, 1.1.2, 1.2, 2.1.1.1, 2.1.2, 2.1.3, 2.2, 3.1.1, 3.1.2.1, 3.1.2.2, 3.2

    b) 1.1, 1.2.1, 1.2.2, 1.2.3, 2.1, 2.2.1, 2.3.1, 2.3.2, 2.4.2.1, 2.4.2.2, 3.1, 3.2.1, 3.2.2

    c) 1.1, 1.2.1, 1.2.2, 1.2.2.1, 1.3, 1.4, 2, 3.1, 3.2, 4.1.1.1

In Exercise 5 determine the order in which a preorder traversal visits the vertices of the given ordered rooted tree.

5.

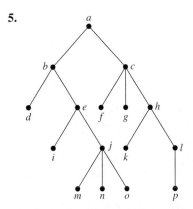

6. In which order are the vertices of the ordered rooted tree in Exercise 5 visited using an inorder traversal?

7. In which order are the vertices of the ordered rooted tree in Exercise 5 visited using a postorder traversal?

8. a) Represent the expression $((x + 2) \uparrow 3) *$ $(y - (3 + x)) - 5$ using a binary tree.
Write this expression in
b) prefix notation.
c) postfix notation.
d) infix notation.

9. a) Represent the expressions $(x + xy) + (x/y)$ and $x + ((xy + x)/y)$ using binary trees.
Write these expressions in
b) prefix notation.
c) postfix notation.
d) infix notation.

10. a) Represent the compound propositions $\neg(p \wedge q) \leftrightarrow (\neg p \vee \neg q)$ and $(\neg p \wedge (q \leftrightarrow \neg p)) \vee \neg q$ using ordered rooted trees.
Write these expressions in
b) prefix notation.
c) postfix notation.
d) infix notation.

11. a) Represent $(A \cap B) - (A \cup (B - A))$ using an ordered rooted tree.
Write this expression in
b) prefix notation.
c) postfix notation.
d) infix notation.

12. Draw the ordered rooted tree corresponding to each of these arithmetic expressions written in prefix notation. Then write each expression using infix notation.
a) $+ * + - 5\ 3\ 2\ 1\ 4$
b) $\uparrow + 2\ 3 - 5\ 1$
c) $* / 9\ 3 + * 2\ 4 - 7\ 6$

13. What is the value of each of these prefix expressions?
a) $- * 2 / 8\ 4\ 3$
b) $\uparrow - * 3\ 3 * 4\ 2\ 5$
c) $+ - \uparrow 3\ 2 \uparrow 2\ 3 / 6 - 4\ 2$
d) $* + 3 + 3 \uparrow 3 + 3\ 3\ 3$

14. What is the value of each of these postfix expressions?
a) $5\ 2\ 1 - - 3\ 1\ 4 + + *$
b) $9\ 3 / 5 + 7\ 2 - *$
c) $3\ 2 * 2 \uparrow 5\ 3 - 8\ 4 / * -$

15. Construct the ordered rooted tree whose preorder traversal is $a, b, f, c, g, h, i, d, e, j, k, l$, where a has four children, c has three children, j has two children, b and e have one child each, and all other vertices are leaves.

*16. Show that an ordered rooted tree is uniquely determined when a list of vertices generated by a preorder traversal of the tree and the number of children of each vertex are specified.

*17. Show that an ordered rooted tree is uniquely determined when a list of vertices generated by a postorder traversal of the tree and the number of children of each vertex are specified.

18. Show that preorder traversals of the two ordered rooted trees displayed below produce the same list of vertices. Note that this does not contradict the statement in Exercise 16, because the numbers of children of internal vertices in the two ordered rooted trees differ.

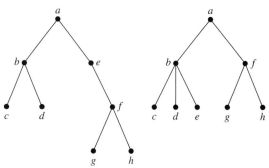

19. Show that postorder traversals of these two ordered rooted trees produce the same list of vertices. Note that this does not contradict the statement in Exercise 17, because the numbers of children of internal vertices in the two ordered rooted trees differ.

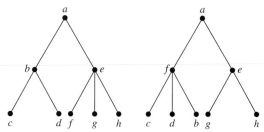

Well-formed formulae in prefix notation over a set of symbols and a set of binary operators are defined recursively by these rules:

(*i*) if x is a symbol, then x is a well-formed formula in prefix notation;

(*ii*) if X and Y are well-formed formulae and $*$ is an operator, then $* X Y$ is a well-formed formula.

20. Which of these are well-formed formulae over the symbols $\{x, y, z\}$ and the set of binary operators $\{\times, +, \circ\}$?
a) $\times + + x\ y\ x$
b) $\circ x\ y \times x\ z$
c) $\times \circ x\ z \times \times x\ y$
d) $\times + \circ x\ x \circ x\ x\ x$

*21. Show that any well-formed formula in prefix notation over a set of symbols and a set of binary operators contains exactly one more symbol than the number of operators.

22. Give a definition of well-formed formulae in postfix notation over a set of symbols and a set of binary operators.

23. Give six examples of well-formed formulae with three or more operators in postfix notation over the set of symbols $\{x, y, z\}$ and the set of operators $\{+, \times, \circ\}$.

24. Extend the definition of well-formed formulae in prefix notation to sets of symbols and operators where the operators may not be binary.

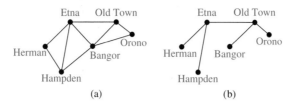

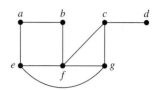

FIGURE 1 (a) A Road System and (b) a Set of Roads to Plow.

FIGURE 2 The Simple Graph *G*.

11.4 Spanning Trees

Introduction

Consider the system of roads in Maine represented by the simple graph shown in Figure 1(a). The only way the roads can be kept open in the winter is by frequently plowing them. The highway department wants to plow the fewest roads so that there will always be cleared roads connecting any two towns. How can this be done?

At least five roads must be plowed to ensure that there is a path between any two towns. Figure 1(b) shows one such set of roads. Note that the subgraph representing these roads is a tree, because it is connected and contains six vertices and five edges.

This problem was solved with a connected subgraph with the minimum number of edges containing all vertices of the original simple graph. Such a graph must be a tree.

DEFINITION 1   Let *G* be a simple graph. A *spanning tree* of *G* is a subgraph of *G* that is a tree containing every vertex of *G*.

A simple graph with a spanning tree must be connected, because there is a path in the spanning tree between any two vertices. The converse is also true; that is, every connected simple graph has a spanning tree. We will give an example before proving this result.

EXAMPLE 1   Find a spanning tree of the simple graph *G* shown in Figure 2.

Solution: The graph *G* is connected, but it is not a tree because it contains simple circuits. Remove the edge $\{a, e\}$. This eliminates one simple circuit, and the resulting subgraph is still connected and still contains every vertex of *G*. Next remove the edge $\{e, f\}$ to eliminate a second simple circuit. Finally, remove edge $\{c, g\}$ to produce a simple graph with no simple circuits. This subgraph is a spanning tree, because it is a tree that contains every vertex of *G*. The sequence of edge removals used to produce the spanning tree is illustrated in Figure 3.

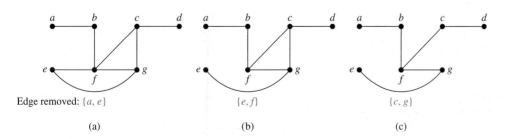

FIGURE 3   Producing a Spanning Tree for *G* by Removing Edges That Form Simple Circuits.

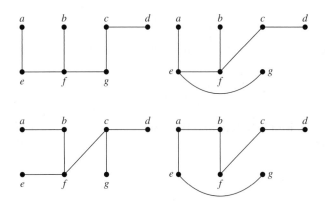

FIGURE 4   Spanning Trees of _G_.

The tree shown in Figure 3 is not the only spanning tree of _G_. For instance, each of the trees shown in Figure 4 is a spanning tree of _G_.   ◄

THEOREM 1   A simple graph is connected if and only if it has a spanning tree.

Proof: First, suppose that a simple graph _G_ has a spanning tree _T_. _T_ contains every vertex of _G_. Furthermore, there is a path in _T_ between any two of its vertices. Because _T_ is a subgraph of _G_, there is a path in _G_ between any two of its vertices. Hence, _G_ is connected.

Now suppose that _G_ is connected. If _G_ is not a tree, it must contain a simple circuit. Remove an edge from one of these simple circuits. The resulting subgraph has one fewer edge but still contains all the vertices of _G_ and is connected. This subgraph is still connected because when two vertices are connected by a path containing the removed edge, they are connected by a path not containing this edge. We can construct such a path by inserting into the original path, at the point where the removed edge once was, the simple circuit with this edge removed. If this subgraph is not a tree, it has a simple circuit; so as before, remove an edge that is in a simple circuit. Repeat this process until no simple circuits remain. This is possible because there are only a finite number of edges in the graph. The process terminates when no simple circuits remain. A tree is produced because the graph stays connected as edges are removed. This tree is a spanning tree because it contains every vertex of _G_.   ◁

Spanning trees are important in data networking, as Example 2 shows.

EXAMPLE 2

Links

IP Multicasting   Spanning trees play an important role in multicasting over Internet Protocol (IP) networks. To send data from a source computer to multiple receiving computers, each of which is a subnetwork, data could be sent separately to each computer. This type of networking, called unicasting, is inefficient, because many copies of the same data are transmitted over the network. To make the transmission of data to multiple receiving computers more efficient, IP multicasting is used. With IP multicasting, a computer sends a single copy of data over the network, and as data reaches intermediate routers, the data are forwarded to one or more other routers so that ultimately all receiving computers in their various subnetworks receive these data. (Routers are computers that are dedicated to forwarding IP datagrams between subnetworks in a network. In multicasting, routers use Class D addresses, each representing a session that receiving computers may join; see Example 17 in Section 6.1.)

For data to reach receiving computers as quickly as possible, there should be no loops (which in graph theory terminology are circuits or cycles) in the path that data take through the network. That is, once data have reached a particular router, data should never return to this

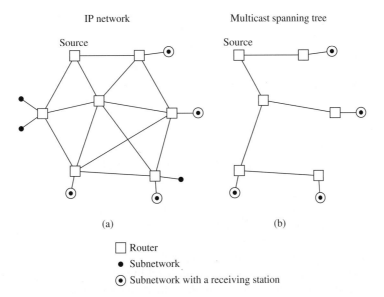

IP network   Multicast spanning tree

(a)   (b)

☐ Router
● Subnetwork
◉ Subnetwork with a receiving station

FIGURE 5 **A Multicast Spanning Tree.**

router. To avoid loops, the multicast routers use network algorithms to construct a spanning tree in the graph that has the multicast source, the routers, and the subnetworks containing receiving computers as vertices, with edges representing the links between computers and/or routers. The root of this spanning tree is the multicast source. The subnetworks containing receiving computers are leaves of the tree. (Note that subnetworks not containing receiving stations are not included in the graph.) This is illustrated in Figure 5. ◀

Depth-First Search

Links

Demo

The proof of Theorem 1 gives an algorithm for finding spanning trees by removing edges from simple circuits. This algorithm is inefficient, because it requires that simple circuits be identified. Instead of constructing spanning trees by removing edges, spanning trees can be built up by successively adding edges. Two algorithms based on this principle will be presented here.

We can build a spanning tree for a connected simple graph using **depth-first search**. We will form a rooted tree, and the spanning tree will be the underlying undirected graph of this rooted tree. Arbitrarily choose a vertex of the graph as the root. Form a path starting at this vertex by successively adding vertices and edges, where each new edge is incident with the last vertex in the path and a vertex not already in the path. Continue adding vertices and edges to this path as long as possible. If the path goes through all vertices of the graph, the tree consisting of this path is a spanning tree. However, if the path does not go through all vertices, more vertices and edges must be added. Move back to the next to last vertex in the path, and, if possible, form a new path starting at this vertex passing through vertices that were not already visited. If this cannot be done, move back another vertex in the path, that is, two vertices back in the path, and try again.

Repeat this procedure, beginning at the last vertex visited, moving back up the path one vertex at a time, forming new paths that are as long as possible until no more edges can be added. Because the graph has a finite number of edges and is connected, this process ends with the production of a spanning tree. Each vertex that ends a path at a stage of the algorithm will be a leaf in the rooted tree, and each vertex where a path is constructed starting at this vertex will be an internal vertex.

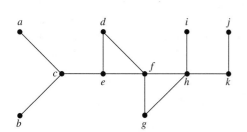

FIGURE 6   The Graph G.

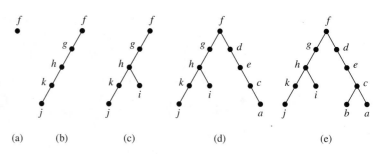

FIGURE 7   Depth-First Search of G.

The reader should note the recursive nature of this procedure. Also, note that if the vertices in the graph are ordered, the choices of edges at each stage of the procedure are all determined when we always choose the first vertex in the ordering that is available. However, we will not always explicitly order the vertices of a graph.

Depth-first search is also called **backtracking**, because the algorithm returns to vertices previously visited to add paths. Example 3 illustrates backtracking.

EXAMPLE 3   Use depth-first search to find a spanning tree for the graph G shown in Figure 6.

Solution: The steps used by depth-first search to produce a spanning tree of G are shown in Figure 7. We arbitrarily start with the vertex f. A path is built by successively adding edges incident with vertices not already in the path, as long as this is possible. This produces a path f, g, h, k, j (note that other paths could have been built). Next, backtrack to k. There is no path beginning at k containing vertices not already visited. So we backtrack to h. Form the path h, i. Then backtrack to h, and then to f. From f build the path f, d, e, c, a. Then backtrack to c and form the path c, b. This produces the spanning tree.   ◀

The edges selected by depth-first search of a graph are called **tree edges**. All other edges of the graph must connect a vertex to an ancestor or descendant of this vertex in the tree. These edges are called **back edges**. (Exercise 43 asks for a proof of this fact.)

EXAMPLE 4   In Figure 8 we highlight the tree edges found by depth-first search starting at vertex f by showing them with heavy colored lines. The back edges (e, f) and (f, h) are shown with thinner black lines.   ◀

We have explained how to find a spanning tree of a graph using depth-first search. However, our discussion so far has not brought out the recursive nature of depth-first search. To help make the recursive nature of the algorithm clear, we need a little terminology. We say that we

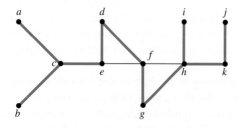

**FIGURE 8   The Tree Edges and Back Edges
of the Depth-First Search in Example 4.**

explore from a vertex v when we carry out the steps of depth-first search beginning when v is added to the tree and ending when we have backtracked back to v for the last time. The key observation needed to understand the recursive nature of the algorithm is that when we add an edge connecting a vertex v to a vertex w, we finish exploring from w before we return to v to complete exploring from v.

In Algorithm 1 we construct the spanning tree of a graph G with vertices $v_1, \ldots, v_n$ by first selecting the vertex v_1 to be the root. We initially set T to be the tree with just this one vertex. At each step we add a new vertex to the tree T together with an edge from a vertex already in T to this new vertex and we explore from this new vertex. Note that at the completion of the algorithm, T contains no simple circuits because no edge is ever added that connects two vertices in the tree. Moreover, T remains connected as it is built. (These last two observations can be easily proved via mathematical induction.) Because G is connected, every vertex in G is visited by the algorithm and is added to the tree (as the reader should verify). It follows that T is a spanning tree of G.

ALGORITHM 1 Depth-First Search.

---

procedure $DFS(G$: connected graph with vertices $v_1, v_2, \ldots, v_n)$
$T :=$ tree consisting only of the vertex v_1
$visit(v_1)$

procedure $visit(v$: vertex of $G)$
for each vertex w adjacent to v and not yet in T
    add vertex w and edge $\{v, w\}$ to T
    $visit(w)$

We now analyze the computational complexity of the depth-first search algorithm. The key observation is that for each vertex v, the procedure $visit(v)$ is called when the vertex v is first encountered in the search and it is not called again. Assuming that the adjacency lists for G are available (see Section 10.3), no computations are required to find the vertices adjacent to v. As we follow the steps of the algorithm, we examine each edge at most twice to determine whether to add this edge and one of its endpoints to the tree. Consequently, the procedure DFS constructs a spanning tree using $O(e)$, or $O(n^2)$, steps where e and n are the number of edges and vertices in G, respectively. [Note that a step involves examining a vertex to see whether it is already in the spanning tree as it is being built and adding this vertex and the corresponding edge if the vertex is not already in the tree. We have also made use of the inequality $e \leq n(n - 1)/2$, which holds for any simple graph.]

Depth-first search can be used as the basis for algorithms that solve many different problems. For example, it can be used to find paths and circuits in a graph, it can be used to determine the connected components of a graph, and it can be used to find the cut vertices of a connected graph. As we will see, depth-first search is the basis of backtracking techniques used to search for solutions of computationally difficult problems. (See [GrYe05], [Ma89], and [CoLeRiSt09] for a discussion of algorithms based on depth-first search.)

Breadth-First Search

Demo

We can also produce a spanning tree of a simple graph by the use of **breadth-first search**. Again, a rooted tree will be constructed, and the underlying undirected graph of this rooted tree forms the spanning tree. Arbitrarily choose a root from the vertices of the graph. Then add all

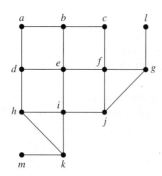

FIGURE 9   A Graph G.

Links

edges incident to this vertex. The new vertices added at this stage become the vertices at level 1 in the spanning tree. Arbitrarily order them. Next, for each vertex at level 1, visited in order, add each edge incident to this vertex to the tree as long as it does not produce a simple circuit. Arbitrarily order the children of each vertex at level 1. This produces the vertices at level 2 in the tree. Follow the same procedure until all the vertices in the tree have been added. The procedure ends because there are only a finite number of edges in the graph. A spanning tree is produced because we have produced a tree containing every vertex of the graph. An example of breadth-first search is given in Example 5.

EXAMPLE 5   Use breadth-first search to find a spanning tree for the graph shown in Figure 9.

Extra
Examples

Solution: The steps of the breadth-first search procedure are shown in Figure 10. We choose the vertex e to be the root. Then we add edges incident with all vertices adjacent to e, so edges from e to b, d, f, and i are added. These four vertices are at level 1 in the tree. Next, add the edges from these vertices at level 1 to adjacent vertices not already in the tree. Hence, the edges from b to a and c are added, as are edges from d to h, from f to j and g, and from i to k. The new vertices a, c, h, j, g, and k are at level 2. Next, add edges from these vertices to adjacent vertices not already in the graph. This adds edges from g to l and from k to m. ◀

We describe breadth-first search in pseudocode as Algorithm 2. In this algorithm, we assume the vertices of the connected graph G are ordered as $v_1, v_2, \ldots, v_n$. In the algorithm we use the term "process" to describe the procedure of adding new vertices, and corresponding edges, to the tree adjacent to the current vertex being processed as long as a simple circuit is not produced.

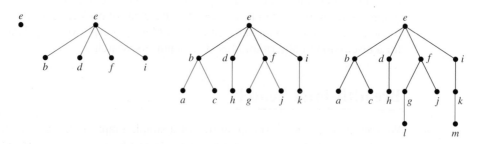

FIGURE 10   Breadth-First Search of G.

ALGORITHM 2 Breadth-First Search.

---

procedure *BFS* (*G*: connected graph with vertices $v_1, v_2, \ldots, v_n$)
T := tree consisting only of vertex v_1
L := empty list
put v_1 in the list L of unprocessed vertices
while L is not empty
   remove the first vertex, v, from L
    for each neighbor w of v
      if w is not in L and not in T **then**
        add w to the end of the list L
        add w and edge $\{v, w\}$ to T

We now analyze the computational complexity of breadth-first search. For each vertex v in the graph we examine all vertices adjacent to v and we add each vertex not yet visited to the tree T. Assuming we have the adjacency lists for the graph available, no computation is required to determine which vertices are adjacent to a given vertex. As in the analysis of the depth-first search algorithm, we see that we examine each edge at most twice to determine whether we should add this edge and its endpoint not already in the tree. It follows that the breadth-first search algorithm uses $O(e)$ or $O(n^2)$ steps.

Breadth-first search is one of the most useful algorithms in graph theory. In particular, it can serve as the basis for algorithms that solve a wide variety of problems. For example, algorithms that find the connected components of a graph, that determine whether a graph is bipartite, and that find the path with the fewest edges between two vertices in a graph can all be built using breadth-first search.

Backtracking Applications

There are problems that can be solved only by performing an exhaustive search of all possible solutions. One way to search systematically for a solution is to use a decision tree, where each internal vertex represents a decision and each leaf a possible solution. To find a solution via backtracking, first make a sequence of decisions in an attempt to reach a solution as long as this is possible. The sequence of decisions can be represented by a path in the decision tree. Once it is known that no solution can result from any further sequence of decisions, backtrack to the parent of the current vertex and work toward a solution with another series of decisions, if this is possible. The procedure continues until a solution is found, or it is established that no solution exists. Examples 6 to 8 illustrate the usefulness of backtracking.

EXAMPLE 6   **Graph Colorings**   How can backtracking be used to decide whether a graph can be colored using n colors?

Solution: We can solve this problem using backtracking in the following way. First pick some vertex a and assign it color 1. Then pick a second vertex b, and if b is not adjacent to a, assign it color 1. Otherwise, assign color 2 to b. Then go on to a third vertex c. Use color 1, if possible, for c. Otherwise use color 2, if this is possible. Only if neither color 1 nor color 2 can be used should color 3 be used. Continue this process as long as it is possible to assign one of the n colors to each additional vertex, always using the first allowable color in the list. If a vertex is reached that cannot be colored by any of the n colors, backtrack to the last assignment made and change the coloring of the last vertex colored, if possible, using the next allowable color in the list. If it is not possible to change this coloring, backtrack farther to previous assignments, one step back at a time, until it is possible to change a coloring of a vertex. Then continue assigning

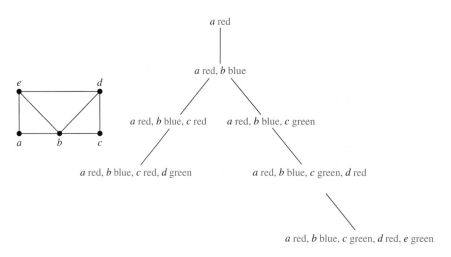

FIGURE 11   Coloring a Graph Using Backtracking.

colors of additional vertices as long as possible. If a coloring using n colors exists, backtracking will produce it. (Unfortunately this procedure can be extremely inefficient.)

In particular, consider the problem of coloring the graph shown in Figure 11 with three colors. The tree shown in Figure 11 illustrates how backtracking can be used to construct a 3-coloring. In this procedure, red is used first, then blue, and finally green. This simple example can obviously be done without backtracking, but it is a good illustration of the technique.

In this tree, the initial path from the root, which represents the assignment of red to a, leads to a coloring with a red, b blue, c red, and d green. It is impossible to color e using any of the three colors when a, b, c, and d are colored in this way. So, backtrack to the parent of the vertex representing this coloring. Because no other color can be used for d, backtrack one more level. Then change the color of c to green. We obtain a coloring of the graph by then assigning red to d and green to e. ◀

EXAMPLE 7

Links

The n-Queens Problem   The n-queens problem asks how n queens can be placed on an $n \times n$ chessboard so that no two queens can attack one another. How can backtracking be used to solve the n-queens problem?

Solution: To solve this problem we must find n positions on an $n \times n$ chessboard so that no two of these positions are in the same row, same column, or in the same diagonal [a diagonal consists of all positions (i, j) with $i + j = m$ for some m, or $i - j = m$ for some m]. We will use backtracking to solve the n-queens problem. We start with an empty chessboard. At stage $k + 1$ we attempt putting an additional queen on the board in the $(k + 1)$st column, where there are already queens in the first k columns. We examine squares in the $(k + 1)$st column starting with the square in the first row, looking for a position to place this queen so that it is not in the same row or on the same diagonal as a queen already on the board. (We already know it is not in the same column.) If it is impossible to find a position to place the queen in the $(k + 1)$st column, backtrack to the placement of the queen in the kth column, and place this queen in the next allowable row in this column, if such a row exists. If no such row exists, backtrack further.

In particular, Figure 12 displays a backtracking solution to the four-queens problem. In this solution, we place a queen in the first row and column. Then we put a queen in the third row of the second column. However, this makes it impossible to place a queen in the third column. So we backtrack and put a queen in the fourth row of the second column. When we do this, we can place a queen in the second row of the third column. But there is no way to add a queen to the fourth column. This shows that no solution results when a queen is placed in the first row and column. We backtrack to the empty chessboard, and place a queen in the second row of the first column. This leads to a solution as shown in Figure 12. ◀

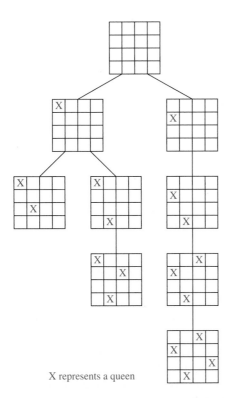

X represents a queen

FIGURE 12   A Backtracking Solution of the Four-Queens Problem.

EXAMPLE 8   **Sums of Subsets**   Consider this problem. Given a set of positive integers $x_1, x_2, \ldots, x_n$, find a subset of this set of integers that has M as its sum. How can backtracking be used to solve this problem?

Solution: We start with a sum with no terms. We build up the sum by successively adding terms. An integer in the sequence is included if the sum remains less than M when this integer is added to the sum. If a sum is reached such that the addition of any term is greater than M, backtrack by dropping the last term of the sum.

Figure 13 displays a backtracking solution to the problem of finding a subset of $\{31, 27, 15, 11, 7, 5\}$ with the sum equal to 39. ◄

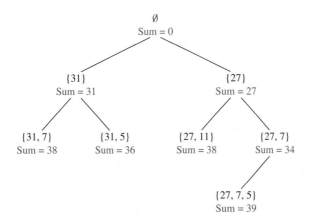

FIGURE 13   Find a Sum Equal to 39 Using Backtracking.

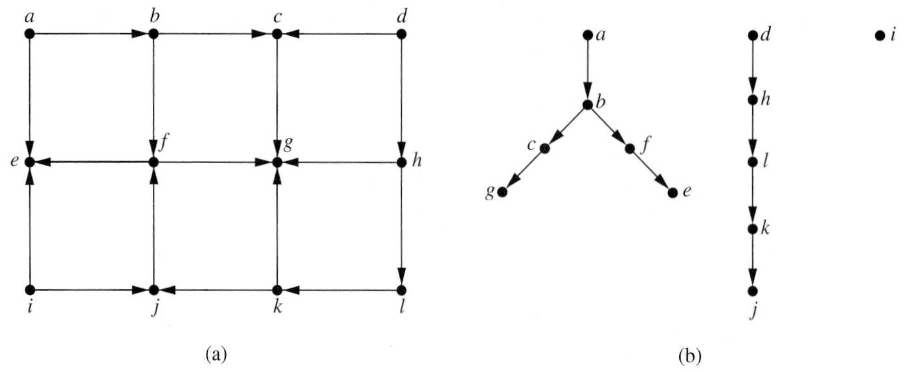

FIGURE 14 Depth-First Search of a Directed Graph.

Depth-First Search in Directed Graphs

We can easily modify both depth-first search and breadth-first search so that they can run given a directed graph as input. However, the output will not necessarily be a spanning tree, but rather a spanning forest. In both algorithms we can add an edge only when it is directed away from the vertex that is being visited and to a vertex not yet added. If at a stage of either algorithm we find that no edge exists starting at a vertex already added to one not yet added, the next vertex added by the algorithm becomes the root of a new tree in the spanning forest. This is illustrated in Example 9.

EXAMPLE 9 What is the output of depth-first search given the graph G shown in Figure 14(a) as input?

Solution: We begin the depth-first search at vertex a and add vertices b, c, and g and the corresponding edges where we are blocked. We backtrack to c but we are still blocked, and then backtrack to b, where we add vertices f and e and the corresponding edges. Backtracking takes us all the way back to a. We then start a new tree at d and add vertices h, l, k, and j and the corresponding edges. We backtrack to k, then l, then h, and back to d. Finally, we start a new tree at i, completing the depth-first search. The output is shown in Figure 14(b). ◀

Depth-first search in directed graphs is the basis of many algorithms (see [GrYe05], [Ma89], and [CoLeRiSt09]). It can be used to determine whether a directed graph has a circuit, it can be used to carry out a topological sort of a graph, and it can also be used to find the strongly connected components of a directed graph.

We conclude this section with an application of depth-first search and breadth-first search to search engines on the Web.

EXAMPLE 10 **Web Spiders** To index websites, search engines such as Google and Yahoo systematically explore the Web starting at known sites. These search engines use programs called Web spiders (or crawlers or bots) to visit websites and analyze their contents. Web spiders use both depth-first searching and breadth-first searching to create indices. As described in Example 5 in Section 10.1, Web pages and links between them can be modeled by a directed graph called the Web graph. Web pages are represented by vertices and links are represented by directed edges. Using depth-first search, an initial Web page is selected, a link is followed to a second Web page (if there is such a link), a link on the second Web page is followed to a third Web page, if there is such a link, and so on, until a page with no new links is found. Backtracking is then used to examine links at the previous level to look for new links, and so on. (Because of practical limitations, Web spiders have limits to the depth they search in depth-first search.) Using breadth-first search, an initial Web page is selected and a link on this page is followed to a second Web page, then a second link on the initial page is followed (if it exists), and so on, until all links of the initial page have been followed. Then links on the pages one level down are followed, page by page, and so on. ◀

Exercises

1. How many edges must be removed from a connected graph with n vertices and m edges to produce a spanning tree?

In Exercises 2–3 find a spanning tree for the graph shown by removing edges in simple circuits.

2.

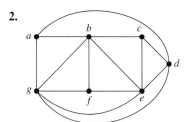

3.

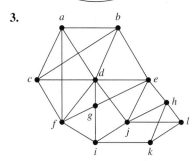

4. Find a spanning tree for each of these graphs.

   a) K_5 **b)** $K_{4,4}$ **c)** $K_{1,6}$

   d) Q_3 **e)** C_5 **f)** W_5

In Exercises 5–6 draw all the spanning trees of the given simple graphs.

5.

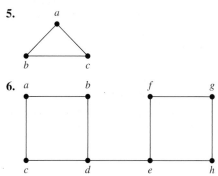

6.

*7. How many different spanning trees does each of these simple graphs have?

   a) K_3 **b)** K_4 **c)** $K_{2,2}$ **d)** C_5

*8. How many nonisomorphic spanning trees does each of these simple graphs have?

   a) K_3 **b)** K_4 **c)** K_5

9. Use depth-first search to produce a spanning tree for the given simple graph. Choose a as the root of this spanning tree and assume that the vertices are ordered alphabetically.

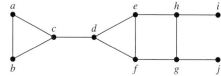

10. Use breadth-first search to produce a spanning tree for each of the simple graph in Exercise 9. Choose a as the root of each spanning tree.

11. Use depth-first search to find a spanning tree of each of these graphs.

   a) W_6 (see Example 7 of Section 10.2), starting at the vertex of degree 6

   b) K_5

   c) $K_{3,4}$, starting at a vertex of degree 3

   d) Q_3

12. Use breadth-first search to find a spanning tree of each of the graphs in Exercise 11.

13. Describe the trees produced by breadth-first search and depth-first search of the wheel graph W_n, starting at the vertex of degree n, where n is an integer with $n \geq 3$. (See Example 7 of Section 10.2.) Justify your answers.

14. Describe the trees produced by breadth-first search and depth-first search of the complete graph K_n, where n is a positive integer. Justify your answers.

*15. Show that the length of the shortest path between vertices v and u in a connected simple graph equals the level number of u in the breadth-first spanning tree of G with root v.

16. Use backtracking to try to find a coloring of the graph in Exercise 5 of Section 10.8 using three colors.

17. Use backtracking to solve the n-queens problem for these values of n.

   a) $n = 3$ **b)** $n = 5$ **c)** $n = 6$

18. Use backtracking to find a subset, if it exists, of the set $\{27, 24, 19, 14, 11, 8\}$ with sum

   a) 20.   **b)** 41.   **c)** 60.

19. Explain how backtracking can be used to find a Hamilton path or circuit in a graph.

20. a) Explain how backtracking can be used to find the way out of a maze, given a starting position and the exit position. Consider the maze divided into positions, where at each position the set of available moves includes one to four possibilities (up, down, right, left).

b) Find a path from the starting position marked by X to the exit in this maze.

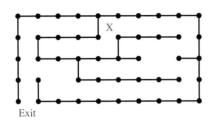

Exit

A **spanning forest** of a graph G is a forest that contains every vertex of G such that two vertices are in the same tree of the forest when there is a path in G between these two vertices.

21. Show that every finite simple graph has a spanning forest.

22. How many trees are in the spanning forest of a graph?

23. How many edges must be removed to produce the spanning forest of a graph with n vertices, m edges, and c connected components?

24. Let G be a connected graph. Show that if T is a spanning tree of G constructed using breadth-first search, then an edge of G not in T must connect vertices at the same level or at levels that differ by 1 in this spanning tree.

25. Explain how to use breadth-first search to find the length of a shortest path between two vertices in an undirected graph.

26. Devise an algorithm based on breadth-first search that determines whether a graph has a simple circuit, and if so, finds one.

27. Devise an algorithm based on breadth-first search for finding the connected components of a graph.

28. Explain how breadth-first search and how depth-first search can be used to determine whether a graph is bipartite.

29. Which connected simple graphs have exactly one spanning tree?

30. When must an edge of a connected simple graph be in every spanning tree for this graph?

31. For which graphs do depth-first search and breadth-first search produce identical spanning trees no matter which vertex is selected as the root of the tree? Justify your answer.

32. Prove that if G is a connected, simple graph with n vertices and G does not contain a simple path of length k then it contains at most $(k - 1)n$ edges. [*Hint:* If T is a spanning tree of a connected graph G constructed using depth-first search, then an edge of G not in T must be a back edge, that is, it must connect a vertex to one of its ancestors or one of its descendants in T.]

33. Use mathematical induction to prove that breadth-first search visits vertices in order of their level in the resulting spanning tree.

34. Use pseudocode to describe a variation of depth-first search that assigns the integer n to the nth vertex visited in the search. Show that this numbering corresponds to the numbering of the vertices created by a preorder traversal of the spanning tree.

35. Use pseudocode to describe a variation of breadth-first search that assigns the integer m to the mth vertex visited in the search.

36. Show that if G is a directed graph and T is a spanning tree constructed using depth-first search, then every edge not in the spanning tree is a **forward edge** connecting an ancestor to a descendant, a **back edge** connecting a descendant to an ancestor, or a **cross edge** connecting a vertex to a vertex in a previously visited subtree.

Let T_1 and T_2 be spanning trees of a graph. The **distance** between T_1 and T_2 is the number of edges in T_1 and T_2 that are not common to T_1 and T_2.

37. Find the distance between each pair of spanning trees shown in Figures 3(c) and 4 of the graph G shown in Figure 2.

∗38. Show that it is possible to find a sequence of spanning trees leading from any spanning tree to any other by successively removing one edge and adding another.

A **rooted spanning tree** of a directed graph is a rooted tree containing edges of the graph such that every vertex of the graph is an endpoint of one of the edges in the tree.

39. For each of the directed graphs in Exercises 12–14 of Section 10.5 either find a rooted spanning tree of the graph or determine that no such tree exists.

∗40. Show that a connected directed graph in which each vertex has the same in-degree and out-degree has a rooted spanning tree. [*Hint:* Use an Euler circuit.]

∗41. Give an algorithm to build a rooted spanning tree for connected directed graphs in which each vertex has the same in-degree and out-degree.

∗42. Show that if G is a directed graph and T is a spanning tree constructed using depth-first search, then G contains a circuit if and only if G contains a back edge (see Exercise 36) relative to the spanning tree T.

∗43. Use Exercise 42 to construct an algorithm for determining whether a directed graph contains a circuit.

11.5   Minimum Spanning Trees

Introduction

Links

A company plans to build a communications network connecting its five computer centers. Any pair of these centers can be linked with a leased telephone line. Which links should be made to

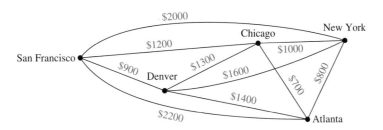

FIGURE 1   A Weighted Graph Showing Monthly Lease Costs for Lines in a Computer Network.

ensure that there is a path between any two computer centers so that the total cost of the network is minimized? We can model this problem using the weighted graph shown in Figure 1, where vertices represent computer centers, edges represent possible leased lines, and the weights on edges are the monthly lease rates of the lines represented by the edges. We can solve this problem by finding a spanning tree so that the sum of the weights of the edges of the tree is minimized. Such a spanning tree is called a **minimum spanning tree**.

Algorithms for Minimum Spanning Trees

A wide variety of problems are solved by finding a spanning tree in a weighted graph such that the sum of the weights of the edges in the tree is a minimum.

DEFINITION 1

A *minimum spanning tree* in a connected weighted graph is a spanning tree that has the smallest possible sum of weights of its edges.

Demo

We will present two algorithms for constructing minimum spanning trees. Both proceed by successively adding edges of smallest weight from those edges with a specified property that have not already been used. Both are greedy algorithms. Recall from Section 3.1 that a greedy algorithm is a procedure that makes an optimal choice at each of its steps. Optimizing at each step does not guarantee that the optimal overall solution is produced. However, the two algorithms presented in this section for constructing minimum spanning trees are greedy algorithms that do produce optimal solutions.

Links

The first algorithm that we will discuss was originally discovered by the Czech mathematician Vojtech Jarník in 1930, who described it in a paper in an obscure Czech journal. The algorithm became well known when it was rediscovered in 1957 by Robert Prim. Because of this, it is known as **Prim's algorithm** (and sometimes as the **Prim-Jarník algorithm**). Begin by choosing any edge with smallest weight, putting it into the spanning tree. Successively add to the tree edges of minimum weight that are incident to a vertex already in the tree, never forming a simple circuit with those edges already in the tree. Stop when $n - 1$ edges have been added.

Later in this section, we will prove that this algorithm produces a minimum spanning tree for any connected weighted graph. Algorithm 1 gives a pseudocode description of Prim's algorithm.

ROBERT CLAY PRIM (BORN 1921)   Robert Prim, born in Sweetwater, Texas, received his B.S. in electrical engineering in 1941 and his Ph.D. in mathematics from Princeton University in 1949. He was an engineer at the General Electric Company from 1941 until 1944, an engineer and mathematician at the United States Naval Ordnance Lab from 1944 until 1949, and a research associate at Princeton University from 1948 until 1949. Among the other positions he has held are director of mathematics and mechanics research at Bell Telephone Laboratories from 1958 until 1961 and vice president of research at Sandia Corporation. He is currently retired.

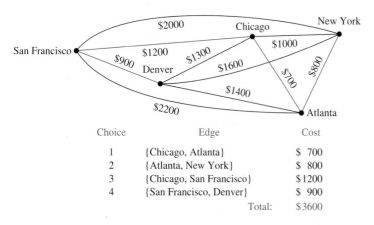

Choice	Edge	Cost
1	{Chicago, Atlanta}	$ 700
2	{Atlanta, New York}	$ 800
3	{Chicago, San Francisco}	$1200
4	{San Francisco, Denver}	$ 900
	Total:	$3600

FIGURE 2   A Minimum Spanning Tree for the Weighted Graph in Figure 1.

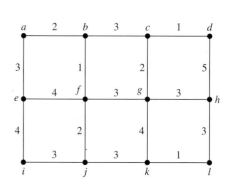

FIGURE 3   A Weighted Graph.

ALGORITHM 1 Prim's Algorithm.

procedure *Prim*(*G*: weighted connected undirected graph with *n* vertices)
T := a minimum-weight edge
for i := 1 **to** $n - 2$
    e := an edge of minimum weight incident to a vertex in T and not forming a
        simple circuit in T if added to T
    T := T with e added
return T {T is a minimum spanning tree of G}

Note that the choice of an edge to add at a stage of the algorithm is not determined when there is more than one edge with the same weight that satisfies the appropriate criteria. We need to order the edges to make the choices deterministic. We will not worry about this in the remainder of the section. Also note that there may be more than one minimum spanning tree for a given connected weighted simple graph. (See Exercise 5.) Examples 1 and 2 illustrate how Prim's algorithm is used.

EXAMPLE 1   Use Prim's algorithm to design a minimum-cost communications network connecting all the computers represented by the graph in Figure 1.

Solution: We solve this problem by finding a minimum spanning tree in the graph in Figure 1. Prim's algorithm is carried out by choosing an initial edge of minimum weight and successively adding edges of minimum weight that are incident to a vertex in the tree and that do not form simple circuits. The edges in color in Figure 2 show a minimum spanning tree produced by Prim's algorithm, with the choice made at each step displayed.   ◀

EXAMPLE 2   Use Prim's algorithm to find a minimum spanning tree in the graph shown in Figure 3.

Solution: A minimum spanning tree constructed using Prim's algorithm is shown in Figure 4. The successive edges chosen are displayed.   ◀

The second algorithm we will discuss was discovered by Joseph Kruskal in 1956, although the basic ideas it uses were described much earlier. To carry out **Kruskal's algorithm**, choose an edge in the graph with minimum weight.

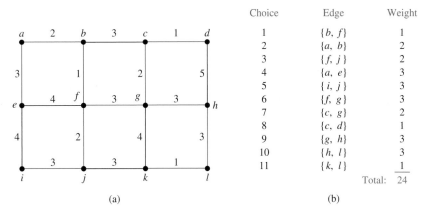

Choice	Edge	Weight
1	$\{b, f\}$	1
2	$\{a, b\}$	2
3	$\{f, j\}$	2
4	$\{a, e\}$	3
5	$\{i, j\}$	3
6	$\{f, g\}$	3
7	$\{c, g\}$	2
8	$\{c, d\}$	1
9	$\{g, h\}$	3
10	$\{h, l\}$	3
11	$\{k, l\}$	1
	Total:	24

(a)   (b)

FIGURE 4   A Minimum Spanning Tree Produced Using Prim's Algorithm.

Successively add edges with minimum weight that do not form a simple circuit with those edges already chosen. Stop after $n - 1$ edges have been selected.

The proof that Kruskal's algorithm produces a minimum spanning tree for every connected weighted graph is left as an exercise. Pseudocode for Kruskal's algorithm is given in Algorithm 2.

ALGORITHM 2   Kruskal's Algorithm.

procedure *Kruskal*(*G*: weighted connected undirected graph with *n* vertices)
$T :=$ empty graph
for $i := 1$ **to** $n - 1$
   $e :=$ any edge in G with smallest weight that does not form a simple circuit
     when added to T
   $T := T$ with e added
return T {T is a minimum spanning tree of G}

JOSEPH BERNARD KRUSKAL (1928–2010)   Joseph Kruskal was born in New York City, where his father was a fur dealer and his mother promoted the art of origami on early television. Kruskal attended the University of Chicago and received his Ph.D. from Princeton University in 1954. He was an instructor in mathematics at Princeton and at the University of Wisconsin, and later he was an assistant professor at the University of Michigan. In 1959 he became a member of the technical staff at Bell Laboratories, where he worked until his retirement in the late 1990s. Kruskal discovered his algorithm for producing minimum spanning trees when he was a second-year graduate student. He was not sure his $2\frac{1}{2}$-page paper on this subject was worthy of publication, but was convinced by others to submit it. His research interests included statistical linguistics and psychometrics. Besides his work on minimum spanning trees, Kruskal is also known for contributions to multidimensional scaling. It is noteworthy that Joseph Kruskal's two brothers, Martin and William, also were well known mathematicians.

HISTORICAL NOTE   Joseph Kruskal and Robert Prim developed their algorithms for constructing minimum spanning trees in the mid-1950s. However, they were not the first people to discover such algorithms. For example, the work of the anthropologist Jan Czekanowski, in 1909, contains many of the ideas required to find minimum spanning trees. In 1926, Otakar Boruvka described methods for constructing minimum spanning trees in work relating to the construction of electric power networks, and as mentioned in the text what is now called Prim's algorithm was discovered by Vojtěch Jarník in 1930.

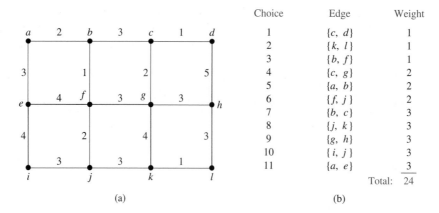

Choice	Edge	Weight
1	$\{c, d\}$	1
2	$\{k, l\}$	1
3	$\{b, f\}$	1
4	$\{c, g\}$	2
5	$\{a, b\}$	2
6	$\{f, j\}$	2
7	$\{b, c\}$	3
8	$\{j, k\}$	3
9	$\{g, h\}$	3
10	$\{i, j\}$	3
11	$\{a, e\}$	3
	Total:	24

(a) (b)

FIGURE 5   A Minimum Spanning Tree Produced by Kruskal's Algorithm.

The reader should note the difference between Prim's and Kruskal's algorithms. In Prim's algorithm edges of minimum weight that are incident to a vertex already in the tree, and not forming a circuit, are chosen; whereas in Kruskal's algorithm edges of minimum weight that are not necessarily incident to a vertex already in the tree, and that do not form a circuit, are chosen. Note that as in Prim's algorithm, if the edges are not ordered, there may be more than one choice for the edge to add at a stage of this procedure. Consequently, the edges need to be ordered for the procedure to be deterministic. Example 3 illustrates how Kruskal's algorithm is used.

EXAMPLE 3 Use Kruskal's algorithm to find a minimum spanning tree in the weighted graph shown in Figure 3.

Solution: A minimum spanning tree and the choices of edges at each stage of Kruskal's algorithm are shown in Figure 5. ◀

We will now prove that Prim's algorithm produces a minimum spanning tree of a connected weighted graph.

Proof: Let G be a connected weighted graph. Suppose that the successive edges chosen by Prim's algorithm are $e_1, e_2, \ldots, e_{n-1}$. Let S be the tree with $e_1, e_2, \ldots, e_{n-1}$ as its edges, and let S_k be the tree with $e_1, e_2, \ldots, e_k$ as its edges. Let T be a minimum spanning tree of G containing the edges $e_1, e_2, \ldots, e_k$, where k is the maximum integer with the property that a minimum spanning tree exists containing the first k edges chosen by Prim's algorithm. The theorem follows if we can show that $S = T$.

Suppose that $S \neq T$, so that $k < n - 1$. Consequently, T contains $e_1, e_2, \ldots, e_k$, but not e_{k+1}. Consider the graph made up of T together with e_{k+1}. Because this graph is connected and has n edges, too many edges to be a tree, it must contain a simple circuit. This simple circuit must contain e_{k+1} because there was no simple circuit in T. Furthermore, there must be an edge in the simple circuit that does not belong to S_{k+1} because S_{k+1} is a tree. By starting at an endpoint of e_{k+1} that is also an endpoint of one of the edges $e_1, \ldots, e_k$, and following the circuit until it reaches an edge not in S_{k+1}, we can find an edge e not in S_{k+1} that has an endpoint that is also an endpoint of one of the edges $e_1, e_2, \ldots, e_k$.

By deleting e from T and adding e_{k+1}, we obtain a tree T' with $n - 1$ edges (it is a tree because it has no simple circuits). Note that the tree T' contains $e_1, e_2, \ldots, e_k, e_{k+1}$. Furthermore, because e_{k+1} was chosen by Prim's algorithm at the $(k + 1)$st step, and e was also available at that step, the weight of e_{k+1} is less than or equal to the weight of e. From this observation, it follows that T' is also a minimum spanning tree, because the sum of the weights of its edges does not exceed the sum of the weights of the edges of T. This contradicts the choice of k as the maximum integer such that a minimum spanning tree exists containing $e_1, \ldots, e_k$. Hence, $k = n - 1$, and $S = T$. It follows that Prim's algorithm produces a minimum spanning tree. ◁

It can be shown (see [CoLeRiSt09]) that to find a minimum spanning tree of a graph with m edges and n vertices, Kruskal's algorithm can be carried out using $O(m \log m)$ operations and Prim's algorithm can be carried out using $O(m \log n)$ operations. Consequently, it is preferable to use Kruskal's algorithm for graphs that are **sparse**, that is, where m is very small compared to $C(n, 2) = n(n - 1)/2$, the total number of possible edges in an undirected graph with n vertices. Otherwise, there is little difference in the complexity of these two algorithms.

Exercises

1. The roads represented by this graph are all unpaved. The lengths of the roads between pairs of towns are represented by edge weights. Which roads should be paved so that there is a path of paved roads between each pair of towns so that a minimum road length is paved? (*Note:* These towns are in Nevada.)

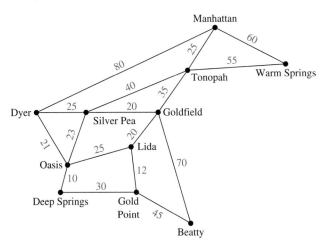

2. Use Prim's algorithm to find a minimum spanning tree for the given weighted graph.

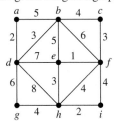

3. Use Kruskal's algorithm to design the communications network described at the beginning of the section.

4. Use Kruskal's algorithm to find a minimum spanning tree for the weighted graph in Exercise 2.

5. Find a connected weighted simple graph with the fewest edges possible that has more than one minimum spanning tree.

6. A **minimum spanning forest** in a weighted graph is a spanning forest with minimal weight. Explain how Prim's and Kruskal's algorithms can be adapted to construct minimum spanning forests.

A **maximum spanning tree** of a connected weighted undirected graph is a spanning tree with the largest possible weight.

7. Devise an algorithm similar to Prim's algorithm for constructing a maximum spanning tree of a connected weighted graph.

8. Devise an algorithm similar to Kruskal's algorithm for constructing a maximum spanning tree of a connected weighted graph.

9. Find a maximum spanning tree for the weighted graph in Exercise 2.

10. Find the second least expensive communications network connecting the five computer centers in the problem posed at the beginning of the section.

∗11. Devise an algorithm for finding the second shortest spanning tree in a connected weighted graph.

∗12. Show that an edge with smallest weight in a connected weighted graph must be part of any minimum spanning tree.

13. Show that there is a unique minimum spanning tree in a connected weighted graph if the weights of the edges are all different.

14. Suppose that the computer network connecting the cities in Figure 1 must contain a direct link between New York and Denver. What other links should be included so that there is a link between every two computer centers and the cost is minimized?

15. Find a spanning tree with minimal total weight containing the edges $\{e, i\}$ and $\{g, k\}$ in the weighted graph in Figure 3.

16. Describe an algorithm for finding a spanning tree with minimal weight containing a specified set of edges in a connected weighted undirected simple graph.

17. Express the algorithm devised in Exercise 16 in pseudocode.

Sollin's algorithm produces a minimum spanning tree from a connected weighted simple graph $G = (V, E)$ by successively adding groups of edges. Suppose that the vertices in V are ordered. This produces an ordering of the edges where $\{u_0, v_0\}$ precedes $\{u_1, v_1\}$ if u_0 precedes u_1 or if $u_0 = u_1$ and v_0 precedes v_1. The algorithm begins by simultaneously choosing the edge of least weight incident to each vertex. The first edge in the ordering is taken in the case of ties. This produces a graph with no simple circuits, that is, a forest of trees (Exercise 18 asks for a proof of this fact). Next, simultaneously choose for each tree in the forest the shortest edge between a vertex in this tree and a vertex in a different tree. Again the first edge in the ordering is chosen in the case of ties. (This produces a graph with no simple circuits containing fewer trees than were present before this step; see Exercise 18.) Continue the process of simultaneously adding edges connecting trees until $n - 1$ edges have been chosen. At this stage a minimum spanning tree has been constructed.

*18. Show that the addition of edges at each stage of Sollin's algorithm produces a forest.

19. Use Sollin's algorithm to produce a minimum spanning tree for the weighted graph shown in
   a) Figure 1.
   b) Figure 3.

*20. Show that the first step of Sollin's algorithm produces a forest containing at least $\lceil n/2 \rceil$ edges when the input is an undirected graph with n vertices.

*21. Show that if there are r trees in the forest at some intermediate step of Sollin's algorithm, then at least $\lceil r/2 \rceil$ edges are added by the next iteration of the algorithm.

*22. Show that when given as input an undirected graph with n vertices, no more than $\lfloor n/2^k \rfloor$ trees remain after the first step of Sollin's algorithm has been carried out and the second step of the algorithm has been carried out $k - 1$ times.

*23. Show that Sollin's algorithm requires at most $\log n$ iterations to produce a minimum spanning tree from a connected undirected weighted graph with n vertices.

24. Prove that Kruskal's algorithm produces minimum spanning trees.

Key Terms and Results

TERMS

tree: a connected undirected graph with no simple circuits

forest: an undirected graph with no simple circuits

rooted tree: a directed graph with a specified vertex, called the root, such that there is a unique path to every other vertex from this root

subtree: a subgraph of a tree that is also a tree

parent of v in a rooted tree: the vertex u such that (u, v) is an edge of the rooted tree

child of a vertex v in a rooted tree: any vertex with v as its parent

sibling of a vertex v in a rooted tree: a vertex with the same parent as v

ancestor of a vertex v in a rooted tree: any vertex on the path from the root to v

descendant of a vertex v in a rooted tree: any vertex that has v as an ancestor

internal vertex: a vertex that has children

leaf: a vertex with no children

level of a vertex: the length of the path from the root to this vertex

height of a tree: the largest level of the vertices of a tree

m-ary tree: a tree with the property that every internal vertex has no more than m children

full m-ary tree: a tree with the property that every internal vertex has exactly m children

binary tree: an m-ary tree with $m = 2$ (each child may be designated as a left or a right child of its parent)

ordered tree: a tree in which the children of each internal vertex are linearly ordered

balanced tree: a tree in which every leaf is at level h or $h - 1$, where h is the height of the tree

binary search tree: a binary tree in which the vertices are labeled with items so that a label of a vertex is greater than the labels of all vertices in the left subtree of this vertex and is less than the labels of all vertices in the right subtree of this vertex

decision tree: a rooted tree where each vertex represents a possible outcome of a decision and the leaves represent the possible solutions of a problem

game tree: a rooted tree where vertices represents the possible positions of a game as it progresses and edges represent legal moves between these positions

prefix code: a code that has the property that the code of a character is never a prefix of the code of another character

minmax strategy: the strategy where the first player and second player move to positions represented by a child with maximum and minimum value, respectively

value of a vertex in a game tree: for a leaf, the payoff to the first player when the game terminates in the position represented by this leaf; for an internal vertex, the maximum or minimum of the values of its children, for an internal vertex at an even or odd level, respectively

tree traversal: a listing of the vertices of a tree

preorder traversal: a listing of the vertices of an ordered rooted tree defined recursively—the root is listed, followed by the first subtree, followed by the other subtrees in the order they occur from left to right

inorder traversal: a listing of the vertices of an ordered rooted tree defined recursively—the first subtree is listed, followed by the root, followed by the other subtrees in the order they occur from left to right

postorder traversal: a listing of the vertices of an ordered rooted tree defined recursively—the subtrees are listed in the order they occur from left to right, followed by the root

infix notation: the form of an expression (including a full set of parentheses) obtained from an inorder traversal of the binary tree representing this expression

prefix (or Polish) notation: the form of an expression obtained from a preorder traversal of the tree representing this expression

postfix (or reverse Polish) notation: the form of an expression obtained from a postorder traversal of the tree representing this expression

spanning tree: a tree containing all vertices of a graph

minimum spanning tree: a spanning tree with smallest possible sum of weights of its edges

RESULTS

A graph is a tree if and only if there is a unique simple path between every pair of its vertices.

A tree with n vertices has $n - 1$ edges.

A full m-ary tree with i internal vertices has $mi + 1$ vertices.

The relationships among the numbers of vertices, leaves, and internal vertices in a full m-ary tree (see Theorem 4 in Section 11.1)

There are at most m^h leaves in an m-ary tree of height h.

If an m-ary tree has l leaves, its height h is at least $\lceil \log_m l \rceil$. If the tree is also full and balanced, then its height is $\lceil \log_m l \rceil$.

Huffman coding: a procedure for constructing an optimal binary code for a set of symbols, given the frequencies of these symbols

depth-first search, or backtracking: a procedure for constructing a spanning tree by adding edges that form a path until this is not possible, and then moving back up the path until a vertex is found where a new path can be formed

breadth-first search: a procedure for constructing a spanning tree that successively adds all edges incident to the last set of edges added, unless a simple circuit is formed

Prim's algorithm: a procedure for producing a minimum spanning tree in a weighted graph that successively adds edges with minimal weight among all edges incident to a vertex already in the tree so that no edge produces a simple circuit when it is added

Kruskal's algorithm: a procedure for producing a minimum spanning tree in a weighted graph that successively adds edges of least weight that are not already in the tree such that no edge produces a simple circuit when it is added

Review Questions

1. **a)** Define a tree.   **b)** Define a forest.

2. Can there be two different simple paths between the vertices of a tree?

3. Give at least three examples of how trees are used in modeling.

4. **a)** Define a rooted tree and the root of such a tree.
   b) Define the parent of a vertex and a child of a vertex in a rooted tree.
   c) What are an internal vertex, a leaf, and a subtree in a rooted tree?
   d) Draw a rooted tree with at least 10 vertices, where the degree of each vertex does not exceed 3. Identify the root, the parent of each vertex, the children of each vertex, the internal vertices, and the leaves.

5. **a)** How many edges does a tree with n vertices have?
   b) What do you need to know to determine the number of edges in a forest with n vertices?

6. **a)** Define a full m-ary tree.
   b) How many vertices does a full m-ary tree have if it has i internal vertices? How many leaves does the tree have?

7. **a)** What is the height of a rooted tree?
   b) What is a balanced tree?
   c) How many leaves can an m-ary tree of height h have?

8. **a)** What is a binary search tree?
   b) Describe an algorithm for constructing a binary search tree.
   c) Form a binary search tree for the words *vireo, warbler, egret, grosbeak, nuthatch,* and *kingfisher.*

9. **a)** What is a prefix code?
   b) How can a prefix code be represented by a binary tree?

10. **a)** Define preorder, inorder, and postorder tree traversal.
    b) Give an example of preorder, postorder, and inorder traversal of a binary tree of your choice with at least 12 vertices.

11. **a)** Explain how to use preorder, inorder, and postorder traversals to find the prefix, infix, and postfix forms of an arithmetic expression.
    b) Draw the ordered rooted tree that represents $((x - 3) + ((x/4) + (x - y) \uparrow 3))$.
    c) Find the prefix and postfix forms of the expression in part (b).

12. Show that the number of comparisons used by a sorting algorithm to sort a list of n elements is at least $\lceil \log n! \rceil$.

13. **a)** Describe the Huffman coding algorithm for constructing an optimal code for a set of symbols, given the frequency of these symbols.
    b) Use Huffman coding to find an optimal code for these symbols and frequencies: A: 0.2, B: 0.1, C: 0.3, D: 0.4.

14. Draw the game tree for nim if the starting position consists of two piles with one and four stones, respectively. Who wins the game if both players follow an optimal strategy?

15. **a)** What is a spanning tree of a simple graph?
    b) Which simple graphs have spanning trees?
    c) Describe at least two different applications that require that a spanning tree of a simple graph be found.

16. **a)** Describe two different algorithms for finding a spanning tree in a simple graph.
    b) Illustrate how the two algorithms you described in part (a) can be used to find the spanning tree of a simple graph of your choice with at least eight vertices and 15 edges.

17. a) Explain how backtracking can be used to determine whether a simple graph can be colored using n colors.

b) Show, with an example, how backtracking can be used to show that a graph with a chromatic number equal to 4 cannot be colored with three colors, but can be colored with four colors.

18. a) What is a minimum spanning tree of a connected weighted graph?

b) Describe at least two different applications that require that a minimum spanning tree of a connected weighted graph be found.

19. a) Describe Kruskal's algorithm and Prim's algorithm for finding minimum spanning trees.

b) Illustrate how Kruskal's algorithm and Prim's algorithm are used to find a minimum spanning tree, using a weighted graph with at least eight vertices and 15 edges.

Supplementary Exercises

*1. Show that a simple graph is a tree if and only if it contains no simple circuits and the addition of an edge connecting two nonadjacent vertices produces a new graph that has exactly one simple circuit (where circuits that contain the same edges are not considered different).

*2. How many nonisomorphic rooted trees are there with six vertices?

3. Show that every tree with at least one edge must have at least two pendant vertices.

4. Show that a tree with n vertices that has $n - 1$ pendant vertices must be isomorphic to $K_{1,n-1}$.

5. What is the sum of the degrees of the vertices of a tree with n vertices?

*6. Suppose that $d_1, d_2, \ldots, d_n$ are n positive integers with sum $2n - 2$. Show that there is a tree that has n vertices such that the degrees of these vertices are $d_1, d_2, \ldots, d_n$.

7. Show that every tree is a planar graph.

8. Show that every forest can be colored using two colors.

A **B-tree of degree k** is a rooted tree such that all its leaves are at the same level, its root has at least two and at most k children unless it is a leaf, and every internal vertex other than the root has at least $\lceil k/2 \rceil$, but no more than k, children. Computer files can be accessed efficiently when B-trees are used to represent them.

*9. Give an upper bound and a lower bound for the number of leaves in a B-tree of degree k with height h.

*10. Give an upper bound and a lower bound for the height of a B-tree of degree k with n leaves.

The **binomial trees** B_i, $i = 0, 1, 2, \ldots$, are ordered rooted trees defined recursively:

Basis step: The binomial tree B_0 is the tree with a single vertex.

Recursive step: Let k be a nonnegative integer. To construct the binomial tree B_{k+1}, add a copy of B_k to a second copy of B_k by adding an edge that makes the root of the first copy of B_k the leftmost child of the root of the second copy of B_k.

11. Draw B_k for $k = 0, 1, 2, 3, 4$.

12. How many vertices does B_k have? Prove that your answer is correct.

13. Find the height of B_k. Prove that your answer is correct.

14. How many vertices are there in B_k at depth j, where $0 \le j \le k$? Justify your answer.

15. What is the degree of the root of B_k? Prove that your answer is correct.

16. Show that the vertex of largest degree in B_k is the root.

A rooted tree T is called an **S_k-tree** if it satisfies this recursive definition. It is an S_0-tree if it has one vertex. For $k > 0$, T is an S_k-tree if it can be built from two S_{k-1}-trees by making the root of one the root of the S_k-tree and making the root of the other the child of the root of the first S_{k-1}-tree.

17. Draw an S_k-tree for $k = 0, 1, 2, 3, 4$.

18. Show that an S_k-tree has 2^k vertices and a unique vertex at level k. This vertex at level k is called the **handle**.

*19. Suppose that T is an S_k-tree with handle v. Show that T can be obtained from disjoint trees $T_0, T_1, \ldots, T_{k-1}$, with roots $r_0, r_1, \ldots, r_{k-1}$, respectively, where v is not in any of these trees, where T_i is an S_i-tree for $i = 0, 1, \ldots, k - 1$, by connecting v to r_0 and r_i to r_{i+1} for $i = 0, 1, \ldots, k - 2$.

The listing of the vertices of an ordered rooted tree in **level order** begins with the root, followed by the vertices at level 1 from left to right, followed by the vertices at level 2 from left to right, and so on.

20. List the vertices of the ordered rooted trees in Figures 3 and 9 of Section 11.3 in level order.

21. Devise an algorithm for listing the vertices of an ordered rooted tree in level order.

*22. Devise an algorithm for determining if a set of universal addresses can be the addresses of the leaves of a rooted tree.

23. Devise an algorithm for constructing a rooted tree from the universal addresses of its leaves.

A **cut set** of a graph is a set of edges such that the removal of these edges produces a subgraph with more connected components than in the original graph, but no proper subset of this set of edges has this property.

24. Show that a cut set of a graph must have at least one edge in common with any spanning tree of this graph.

A **cactus** is a connected graph in which no edge is in more than one simple circuit not passing through any vertex other than its initial vertex more than once or its initial vertex other than at its terminal vertex (where two circuits that contain the same edges are not considered different).

25. Which of these graphs are cacti?

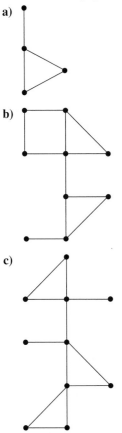

26. Is a tree necessarily a cactus?

27. Show that a cactus is formed if we add a circuit containing new edges beginning and ending at a vertex of a tree.

*28. Show that if every circuit not passing through any vertex other than its initial vertex more than once in a connected graph contains an odd number of edges, then this graph must be a cactus.

A **degree-constrained spanning tree** of a simple graph G is a spanning tree with the property that the degree of a vertex in this tree cannot exceed some specified bound. Degree-constrained spanning trees are useful in models of transportation systems where the number of roads at an intersection is limited, models of communications networks where the number of links entering a node is limited, and so on.

In Exercises 29–31 find a degree-constrained spanning tree of the given graph where each vertex has degree less than or equal to 3, or show that such a spanning tree does not exist.

29. 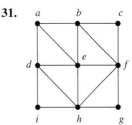 **30.**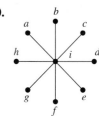

31.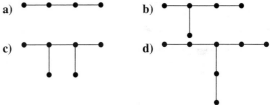

32. Show that a degree-constrained spanning tree of a simple graph in which each vertex has degree not exceeding 2 consists of a single Hamilton path in the graph.

33. A tree with n vertices is called **graceful** if its vertices can be labeled with the integers $1, 2, \ldots, n$ such that the absolute values of the difference of the labels of adjacent vertices are all different. Show that these trees are graceful.

a) 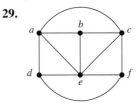 **b)**

c) **d)**

A **caterpillar** is a tree that contains a simple path such that every vertex not contained in this path is adjacent to a vertex in the path.

34. Which of the graphs in Exercise 33 are caterpillars?

35. How many nonisomorphic caterpillars are there with six vertices?

36. Suppose that in a long bit string the frequency of occurrence of a 0 bit is 0.9 and the frequency of a 1 bit is 0.1 and bits occur independently.

a) Construct a Huffman code for the four blocks of two bits, 00, 01, 10, and 11. What is the average number of bits required to encode a bit string using this code?

b) Construct a Huffman code for the eight blocks of three bits. What is the average number of bits required to encode a bit string using this code?

*37. Suppose that e is an edge in a weighted graph that is incident to a vertex v such that the weight of e does not exceed the weight of any other edge incident to v. Show that there exists a minimum spanning tree containing this edge.

38. Three couples arrive at the bank of a river. Each of the wives is jealous and does not trust her husband when he is with one of the other wives (and perhaps with other people), but not with her. How can six people cross to the other side of the river using a boat that can hold no more than two people so that no husband is alone with a woman other than his wife? Use a graph theory model.

*39. Show that if no two edges in a weighted graph have the same weight, then the edge with least weight incident to a vertex v is included in every minimum spanning tree.

40. Find a minimum spanning tree of each of these graphs where the degree of each vertex in the spanning tree does not exceed 2.

a)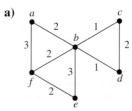

b)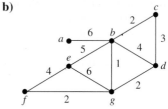

Computer Projects

Write programs with these input and output.

1. Given the adjacency matrix of an undirected simple graph, determine whether the graph is a tree.

2. Given the adjacency matrix of a rooted tree and a vertex in the tree, find the parent, children, ancestors, descendants, and level of this vertex.

3. Given the list of edges of a rooted tree and a vertex in the tree, find the parent, children, ancestors, descendants, and level of this vertex.

4. Given a list of items, construct a binary search tree containing these items.

5. Given a binary search tree and an item, locate or add this item to the binary search tree.

6. Given the ordered list of edges of an ordered rooted tree, find the universal addresses of its vertices.

7. Given the ordered list of edges of an ordered rooted tree, list its vertices in preorder, inorder, and postorder.

8. Given an arithmetic expression in prefix form, find its value.

9. Given an arithmetic expression in postfix form, find its value.

10. Given the frequency of symbols, use Huffman coding to find an optimal code for these symbols.

11. Given an initial position in the game of nim, determine an optimal strategy for the first player.

12. Given the adjacency matrix of a connected undirected simple graph, find a spanning tree for this graph using depth-first search.

13. Given the adjacency matrix of a connected undirected simple graph, find a spanning tree for this graph using breadth-first search.

14. Given a set of positive integers and a positive integer N, use backtracking to find a subset of these integers that have N as their sum.

15. Given the list of edges and their weights of a weighted undirected connected graph, use Prim's algorithm to find a minimum spanning tree of this graph.

16. Given the list of edges and their weights of a weighted undirected connected graph, use Kruskal's algorithm to find a minimum spanning tree of this graph.

Computations and Explorations

Use a computational program or programs you have written to do these exercises.

1. Display all trees with six vertices.

2. Display a full set of nonisomorphic trees with seven vertices.

3. Compute the number of different spanning trees of K_n for $n = 1, 2, 3, 4, 5, 6$. Conjecture a formula for the number of such spanning trees whenever n is a positive integer.

4. Compare the number of comparisons needed to sort lists of n elements for $n = 100, 1000$, and $10,000$ from the set of positive integers less than $1,000,000$, where the elements

are randomly selected positive integers, using the selection sort, the insertion sort, the merge sort, and the quick sort.

5. Compute the number of different ways n queens can be arranged on an $n \times n$ chessboard so that no two queens can attack each other for all positive integers n not exceeding 10.

6. Draw the complete game tree for a game of checkers on a 4×4 board.

Writing Projects

Respond to these with essays using outside sources.

1. Explain how Cayley used trees to enumerate the number of certain types of hydrocarbons.

2. Explain how trees are used to represent ancestral relations in the study of evolution.

3. Discuss hierarchical cluster trees and how they are used.

4. Define *AVL-trees* (sometimes also known as *height-balanced trees*). Describe how and why AVL-trees are used in a variety of different algorithms.

5. Define *quad trees* and explain how images can be represented using them. Describe how images can be rotated, scaled, and translated by manipulating the corresponding quad tree.

6. Define a *heap* and explain how trees can be turned into heaps. Why are heaps useful in sorting?

7. Describe dynamic algorithms for data compression based on letter frequencies as they change as characters are successively read, such as adaptive Huffman coding.

8. Explain how *alpha-beta pruning* can be used to simplify the computation of the value of a game tree.

9. Describe the techniques used by chess-playing programs such as Deep Blue.

10. Define the type of graph known as a *mesh of trees*. Explain how this graph is used in applications to very large system integration and parallel computing.

11. Discuss the algorithms used in IP multicasting to avoid loops between routers.

12. Describe an algorithm based on depth-first search for finding the articulation points of a graph.

13. Describe an algorithm based on depth-first search to find the strongly connected components of a directed graph.

14. Describe the search techniques used by the crawlers and spiders in different search engines on the Web.

15. Describe an algorithm for finding the minimum spanning tree of a graph such that the maximum degree of any vertex in the spanning tree does not exceed a fixed constant k.

16. Compare and contrast some of the most important sorting algorithms in terms of their complexity and when they are used.

17. Discuss the history and origins of algorithms for constructing minimum spanning trees.

18. Describe algorithms for producing random trees.

12 Algebraic Structures and Coding Theory

Generally, to study a phenomenon or process of a real world, we construct a suitable mathematical model to represent it and study the properties of the model to understand the phenomenon. Often the mathematical structure of a model is presented implicitly. But in this chapter, we specify in detail some mathematical structures and develop a few basic properties of these structures, emphasizing those properties which are useful for the models under consideration. The mathematical structures are called algebras or algebraic structures. The structures mainly considered are semigroups, monoids, groups, rings, and fields.

Semigroups are the simplest algebraic structures which satisfy the properties of closure and associativity. They are very important in the theory of sequential machines, formal languages, and in certain applications relating to computer arithmetic.

A monoid, in addition to being a semigroup, also satisfies the identity property. Monoids are used in a number of applications, but most particularly in the area of syntactic analysis and formal languages.

Groups are monoids which also possess inverse property. The application of group theory is important in the design of fast adders and error-correcting codes.

Rings and fields are algebraic systems with two binary operations.

In this chapter, we also study some useful and important concepts like isomorphism and homomorphism. The concept of isomorphism shows that two algebraic systems which are isomorphic to one another are structurally indistinguishable and that the results of operations in one system can be obtained from those of the other by simply renaming the elements and symbols for operations.

Another important concept studied here is that of homomorphism and congruence classes.

We also study coding theory in this chapter. When bits are transmitted through a communications channel, it is quite possible that some bits are erroneously transmitted due to noise or some fluctuations. In order to avoid erroneous transmission of bits, sequences of bits is divided into blocks and each block is encoded into a larger binary string and transmitted. The decoding scheme is defined in such a way that if a small error occurs, the bits transferred erroneously are found out, and corrected, and the correct original block of strings recovered. Various ways of encoding and decoding are studied here. The use of algebraic structure group is made use of in some encoding schemes. This is studied in detail.

In the last section, rings whose elements are polynomials and their use in defining cyclic codes is discussed.

12.1 The Structure of Algebras

In this section, we try to give a general introduction of an algebra which describes the concept and also we give some examples.

An algebra has the following components:

1. An underlying set S (sometimes it is called the carrier of the algebra).
2. Operations defined on this set.
3. Special elements of the underlying set possessing specific properties. These are called constants of the algebra.

The underlying set could be something like the set of integers, real numbers, or set of strings over an alphabet. An operation is a map from $S^p \to S$. Here p is called the 'arity' of the operation. For example, if the underlying set is the set of real numbers, unary minus is a unary operator mapping x to $-x$. Addition is a binary operator mapping x and y into $x + y$. Algebras are specified by specifying the underlying set, operations on the set, and the constants of the set in that order.

EXAMPLE 1   The underlying set is the set of real numbers R and operation is binary $+$. Here $+(a, b) = a + b$. Constant is 0.

$$a + 0 = a \text{ for all } a \text{ in } R$$
$$= 0 + a$$

The operation maps $R^2 \to R$.
This algebra can be specified as $(R, +, 0)$. ◄

EXAMPLE 2   The underlying set is the set of all strings over an alphabet Σ, denoted as Σ^*; the operation is concatenation.

$$\text{If } x = a_1 \ldots a_n$$
$$y = b_1 \ldots b_m$$
$$x \cdot y = xy = a_1 \ldots a_n b_1 \ldots b_m$$

It maps $\Sigma^* \times \Sigma^* \to \Sigma^*$ and is a binary operation. ◄

The constant is λ, the empty string with specific property $x \cdot \lambda = \lambda \cdot x = x$ for all $x \in \Sigma^*$. This can be denoted as $(\Sigma^*, \cdot, \lambda)$.

EXAMPLE 3 $(S, \oplus, \odot, 0, 1)$
The underlying set is the set of integers $S = \{0, 1, \ldots, p - 1\}$ where p is a prime.
Two operations are defined:

$\oplus S^2 \to S -$ mod p addition
$\quad a \oplus b = a + b$ if $a + b < p$
$\quad\quad\quad\quad a + b - p$ if $(a + b) \geq p$.
$\odot S^2 \to S -$ mod p multiplication
$\quad a \odot b = ab$ mod p.

Both are binary operators.
0 is a constant. $a \oplus 0 = 0 \oplus a = a$ for all $a \in S$
1 is another constant with the specific property that
$a \odot 1 = 1 \odot a = a$ for all $a \in S$.

This algebra can be specified as $(S, \oplus, \odot, 0, 1)$. ◄

Usually we would like to specify a class of algebras possessing some common properties rather than a single algebra.

We define a signature or species of an algebra first. Two algebras are of the same signature (or of the same species) if they have the same number of operations, same number of constants, and also the corresponding operations are of the same arity.

EXAMPLE 4 $(I, +, 0)$ and $(\sum^*, \cdot, \lambda)$ are of the same species.

   $(R, \cdot, 1)$ and $(I, -, 0)$ have the same signature.

   They have one binary operation and one constant. The $-$ here is a binary operation.

   It maps $I^2 \rightarrow I$. i.e.,

   It maps (a, b) to $a - b$.

   Similarly $\cdot$ is a multiplication operator mapping $R^2 \rightarrow R$.

   It maps (a, b) into ab.

   $\quad a \cdot 1 = 1 \cdot a = a$ for all a in R

   But $a - 0 \neq 0 - a$

   $\quad a - 0 = a$ for all a in I

   But $0 - a = -a$

   So they have different properties.

   Two algebras can have the same signature but may have different properties.   ◀

So we have to consider additional properties to define algebras of similar type. We define these properties and call them as axioms. Each axiom is an equation written in terms of the elements of the underlying set and the operations in the set. A set of axioms, together with a signature, specifies a class of algebras called a variety. Algebras which have the same signature and which obey the same set of axioms belong to the same variety. Examples are groups, rings, monoids, etc. Usually we explore and study results of algebras of particular varieties. The theorems are proved based on the axioms of the variety and the results hold for all algebras of the given variety.

DEFINITION 1   Let S be a set and let $*$ be a binary operation on S

   1. The operation $*$ is commutative over S, if $a * b = b * a$.
   2. The operation $*$ is associative over S, if $a * (b * c) = (a * b) * c$, for $a, b, c, \varepsilon\ S$.

EXAMPLE 5   Consider the variety of algebras with an underlying set, one binary operation, and one constant similar to $(I, +, \cdot)$ with the following axioms.

   (*i*) $x + y = y + x$

   (*ii*) $(x + y) + z = x + (y + z)$

   (*iii*) $x + 0 = x$

   Then $(R, +, 0)$, $(\sum^*, \cdot, \lambda)$, $(P(S), \cup, \phi)$, $(P(S), \cap, S)$, and $(I, \cdot, 1)$ satisfy these axioms and belong to the same variety. Any result proved for this variety will hold for all these algebras.

EXAMPLE 6   Consider the variety of algebras with the same signatures as $(R, +, \cdot, -, 0, 1)$ where $+$ and $\cdot$ are binary operations of addition and multiplication, respectively, and $-$ is a unary operator denoting unary minus. These operations satisfy the following axioms.

   (*i*) $x + y = y + x$

   (*ii*) $x \cdot y = y \cdot x$

   (*iii*) $(x + y) + z = x + (y + z)$

   (*iv*) $(x \cdot y) \cdot z = x \cdot (y \cdot z)$

   (*v*) $x \cdot (y + z) = x \cdot y + x \cdot z$

   (*vi*) $x + (-x) = 0$

(*vii*) $x + 0 = x$

(*viii*) $x \cdot 1 = x$

Then $(I, +, \cdot, -, 0, 1)$ and $(Q, +, \cdot, -, 0, 1)$ where Q is the set of rational numbers are of the same variety. But $(P(S), \cup, \cap, \bar{r}, \phi, S)$ where $\bar{r}$ denotes set complementation, is not of the same variety because axiom (vi) does not hold for this algebra. ◀

Let us denote an algebra by (S, O, C) where S is the underlying set, O is the set of operations, and C is the set of constants.

DEFINITION 2   Let S be a set and S' a subset of S. Let $\square$ be a binary operation of S and $\triangle$ a unary operation. S' is closed with respect to $\square$ if for all $a, b \in S'$, and $a \square b \in S'$ S' is closed with respect to $\triangle$, if for all $a \in S'$, and $\triangle a \in S'$.

If A is an algebra specified by (S, O, C), a subalgebra of A is an algebra with the same signature which is contained in A.

DEFINITION 3   Let $A = (S, O, C)$ be an algebra with $O = \{o_1, o_2, \ldots, o_n\}$ and $C = \{c_1, c_2, \ldots, c_k\}$. Then $A' = (S', O', C')$ is a subalgebra of A if

(*i*) $S' \subset S$

(*ii*) Each o_i is same as o_i restricted to S'

(*iii*) $C' = C$

If A' is a subalgebra of A, then A' has the same signature as A and obeys the same set of axioms. Moreover, the underlying set A' is a subset of the set A and A' is closed under all operations of A. The largest possible subalgebra of A is A itself.

If the set of constants of A is closed under the operations of A, then the algebra with this underlying set is the smallest subalgebra of A.

EXAMPLE 7   Let E be the set of even integers and I the set of integers.
Then $(E, +, 0)$ is a subalgebra of $(I, +, 0)$. ◀

EXAMPLE 8   Let $\cdot$ denote multiplication.
Then $([0, 1], \cdot, 1)$ is a subalgebra of $(R, \cdot, 1)$, where R is the set of real numbers. ◀

DEFINITION 4   Let $\square$ be a binary operation on a set T. An element $e \in S$ is an identity element (or unit element) for the operation $\square$ if for every $x \in T$

$$e \square x = x \square e = x.$$

An element $\mathbf{0} \in T$ is a zero for the operation $\square$, if for every $x \in T$,

$$\mathbf{0} \square x = x \square \mathbf{0} = \mathbf{0}.$$

EXAMPLE 9   Consider the set of integers. If addition is the operation, 0 is an identity element. If multiplication is the operation, 1 is the identity element and 0 is the zero element. ◀

DEFINITION 5

Let $\square$ be a binary operation an the set T. An element e_ℓ is a left identity for the operation $\square$ if for every $x \in T$, $e_\ell \,\square\, x = x$.

An element $\mathbf{0}_\ell$ is a left zero for the operation $\square$ if for every $x \in T$,

$$\mathbf{0}_\ell \,\square\, x = \mathbf{0}_\ell.$$

A right identity and right zero can be defined in a similar manner.

EXAMPLE 10

$\square$	a	b	c	d
a	a	c	d	a
b	a	b	c	d
c	a	b	a	c
d	a	b	b	b

Let $\{a, b, c, d\}$ be the underlying set. The binary operation is given by the above table. The operation is not commutative as

$a \,\square\, b = c$

$b \,\square\, a = a$

and they are not equal

The operation is not associative as

$a \,\square\, (b \,\square\, c) = a \,\square\, c = d$

$(a \,\square\, b) \,\square\, c = c \,\square\, c = a$

and they are not equal.

a is a right zero for the operation and b is a left identity. ◀

THEOREM 1

Let $\square$ be a binary operation on a set T with left identity e_ℓ and right identity e_r. Then $e_\ell = e_r$, and this element is a two-sided identity.

Proof: Since e_ℓ and e_r are left and right identities, $e_r = e_\ell \,\square\, e_r = e_r$.

THEOREM 2

Let $\square$ be a binary operation on a set T with left zero $\mathbf{0}_\ell$ and right zero $\mathbf{0}_r$. Then $\mathbf{0}_\ell = \mathbf{0}_r$, and this element is a two-sided zero.

Proof: Since $\mathbf{0}_\ell$ is a left zero,

$$\mathbf{0}_\ell \cdot \mathbf{0}_r = \mathbf{0}_\ell$$

Similarly, $\mathbf{0}_\ell \cdot \mathbf{0}_r = \mathbf{0}_r$ as $\mathbf{0}_r$ is the right zero. Therefore $\mathbf{0}_\ell = \mathbf{0}_r$.

COROLLARY 1

A two-sided identity (or zero) for a binary operation is unique.

Proof: If possible, let e_1 and e_2 be two identities.
Then $e_1 \square e_2 = e_1$ and also $e_1 \square e_2 = e_2$
Hence $e_1 = e_2$.
Similar proof can be given for the zero element.

DEFINITION 6   Let $\square$ be a binary operation on T and e an identity element for the operation $\square$. If $x \square y = e$, then x is the left inverse of y, and y is the right inverse of x with respect to the operation $\square$. If both $x \square y = e$ and $y \square x = e$, then x is the inverse of y (or a two-sided inverse of y) with respect to the operation $\square$.

EXAMPLE 11   The algebra $(I, +, 0)$ has an identity 0 and for each x in I, $-x$ is the inverse of x as $x + (-x) = (-x) + x = 0$. ◀

EXAMPLE 12   Let N_k be the first k natural numbers, where $k > 0$

$$N_k = \{0, 1, 2, \ldots, k - 1\}.$$

Define $\oplus$ as mod k addition, i.e., for every $x, y \in N_k$.

$$x \oplus y = x + y \text{ if } x + y < k$$
$$= x + y - k \text{ if } x + y \geq k$$

$\oplus$ is an associative binary operation with identity 0. Every element has an inverse. 0 is its own inverse. For other elements, the inverse of x is $k - x$. ◀

THEOREM 3   If an element has both a left inverse and a right inverse with respect to an associative operation, then the left and right inverse elements are equal.

Proof: Let e be an identity element for the operation $\square$. Let x be an element with y its left inverse and z its right inverse. Then we have to show $y = z$.
Since y is the left inverse, $y \square x = e$.
Since z is the right inverse, $x \square z = e$.

$$y = y \square e = y \square (x \square z) = (y \square x) \square z \text{ (associativity)}$$
$$= e \square z = z.$$

Exercises

1. Let $(A, \square)$ be an algebraic system where $\square$ is a binary operation such that, for any a and b in A, $a \square b = a$.

  a) Show that $\square$ is an associative operation
  b) Can $\square$ ever be a commutative operation?

2. Let N be the set of all natural numbers. For each of the following, determine whether $*$ is an associative operation.

  a) $a * b = \max(a, b)$
  b) $a * b = \min(a, b + 2)$
  c) $a * b = a + b + 3$
  d) $a * b = a + 2b$
  e) $a * b = \begin{cases} \min(a, b) & \text{if } \min(a, b) < 10 \\ \max(a, b) & \text{if } \min(a, b) \geq 10 \end{cases}$

12.2 Semigroups, Monoids, and Groups

Many specific algebraic varieties are useful in various applications in computer science and other areas. In this section, we study about some properties of semigroups, monoids, and groups.

DEFINITION 1

Let A be an algebra with an underlying set T and $\square$ a binary operation on T. $(T, \square)$ is called a semigroup if the following two conditions are satisfied.

1. T is closed with respect to $\square$.
2. $\square$ is an associative operation.

EXAMPLE 1

Let $(E, +)$ be a system.
   E is closed with respect to $+$ and $+$ is an associative operation.
   $\therefore (E, +)$ is a semigroup. ◄

EXAMPLE 2

Consider $(\sum^*, \text{concatenation})$ where $\sum$ is an alphabet.
   $\sum^*$ is closed with respect to concatenation and concatenation is an associative operation. Hence $(\sum^*, \text{concatenation})$ is a semigroup. ◄

DEFINITION 2

Let $(T, \square)$ be an algebraic system, where $\square$ is a binary operation on T. $(T, \square)$ is called a monoid if the following conditions are satisfied.

1. T is closed with respect to $\square$.
2. $\square$ is an associative operation.
3. There exists an identity element $e \in T$ for the operation $\square$.

Therefore, for any $x \in T$, $e \square x = x \square e = x$.

In the above examples both $(E, +)$ and $(\sum^*, \text{concatenation})$ are monoids.
For $(E, +)$, 0 is the identity element.
For $(\sum^*, \text{concatenation})$, λ, which is the empty word (sometimes also denoted as ε), is the identity element.

DEFINITION 3

Let $(T, \square)$ be an algebraic system, where $\square$ is a binary operation on T. Then $(T, \square)$ is called a group if the following conditions are satisfied.

1. T is closed with respect to $\square$.
2. $\square$ is an associative operation.
3. There exists an identity element $e \in T$ for the operation $\square$.
4. Each element $x \in T$ has an inverse element $x^{-1} \in T$ with respect to $\square$, Therefore,

$$x \square x^{-1} = x^{-1} \square x = e$$

In the examples considered previously, $(E, +)$ is a group with $-x$ as the inverse of x for every $x \in E$. $(\sum^*, \text{concatenation})$ is not a group, as inverse of a string x with respect to concatenation does not exist.

The algebraic system over $\mathbf{Z}_n$: modular arithmetic: If n is a positive integer, we define the operation $+_n$ (mod n addition) and $\times_n$ (mod n multiplication) as follows

$a +_n b =$ reminder after $a + b$ is divided by n
$a \times_n b =$ reminder after $a \times b$ is divided by n

Properties of modular arithmetic.

1. The result of doing arithmetic modulo n is always an integer between 0 and $n - 1$.
2. Addition modulo n is always commutative and associative. 0 is the identity of $+_n$.
3. $(\mathbf{Z}_n, +_n)$ is a group.
4. Multiplication modulo n is always commutative and associative. 1 is the identity of $\times_n$.
5. Multiplication modulo n is distributive over addition modulo n.
6. If $a \in \mathbf{Z}_n, a \neq 0$, then $-a = n - a$.

EXAMPLE 3   If $Z_n = \{0, 1, \ldots, n - 1\}$ and $\oplus$ is mod n addition operation (addition modulo n), then we can easily check that $(Z_n, \oplus)$ is a group.   ◀

See also Section 4.1 for more details on modular arithemetic.

EXAMPLE 4   Let $R = \{r_0, r_{60}, r_{120}, r_{180}, r_{240}, r_{300}\}$ where r_θ denotes rotation of geometric figures drawn on a plane by θ degrees. Let $\square$ be the operation defined as $r_{\theta_1} \square r_{\theta_2} = r_{\theta_1 + \theta_2}$. Then $(R, \square)$ is a group. Closure and associativity can easily be checked. r_0 is the identity element and $r_{360-\theta}$ is the inverse of r_θ.   ◀

A group $(A, \square)$ is called a commutative group or abelian group if $\square$ is a commutative operation. For example $(Z_n, \oplus)$ is a commutative group.

A group $(A, \square)$ is said to be finite if A is a finite set, and infinite if A is an infinite set. The size of A is often referred to as the order of the group. If A is a finite set $\{a_1, \ldots, a_n\}$ with n elements and the binary operation of the group is denoted by $\square$, the effect of this operation on pairs of elements of A can be given by a $n \times n$ matrix as given in the table below.

$\square$	a_1	$\ldots$	a_n
a_1	c_{11}	$\ldots$	c_{1n}
$\ldots$		$\ldots$	
a_n	c_{n1}	$\ldots$	c_{nn}

$c_{ij} = a_i \square a_j$. Because of the property that each element has an inverse, two elements in a row cannot be the same. Suppose $c_{ij} = c_{ik}, a_i \square a_j = a_i \square a_k$.

$$a_i^{-1} \square a_i \square a_j = a_i^{-1} \square a_i \square a_k$$

$$e \square a_j = e \square a_k$$

$$a_j = a_k$$

Similarly, two elements in a column cannot be the same. Hence each row of the previous table is a permutation of $a_1, \ldots, a_n$ and each column is also a permutation of $a_1, \ldots, a_n$.

If A has two elements $\{a, b\}$ with a as the identity element, the table is of the form

$\square$	a	b
a	a	b
b	b	a

a is its own inverse; similarly b is also its own inverse.

If A has 3 elements $\{a, b, c\}$ with a as identity, the table has the form

$\square$	a	b	c
a	a	b	c
b	b	c	a
c	c	a	b

a is its own inverse; b and c are inverses of each other. Note that these are abelian groups.

If A has 4 elements $\{a, b, c, d\}$ with a as identity, two possibilities exist:

$\square$	a	b	c	d
a	a	b	c	d
b	b	a	d	c
c	c	d	a	b
d	d	c	b	a

$\square$	a	b	c	d
a	a	b	c	d
b	b	c	d	a
c	c	d	a	b
d	d	a	b	c

Both are abelian groups. In the first case, each element is its own inverse. In the second case, a and c are their own inverses and b and d are inverses of each other. (Interchange of two rows and corresponding columns does not give rise to another group.)

Subgroups

Let $G = (T, \square)$ be a group and T' be a subset of T. $G' = (T', \square)$ is a subgroup of G if it satisfies the conditions of a group. For example $(E, +)$ is a subgroup of $(I, +)$. If $R' = \{r_0, r_{120}, r_{240}\}$, $(R', \square)$ is a subgroup of $(R, \square)$, as considered earlier.

In order to test whether $(T', \square)$ is a subgroup of $(T, \square)$, we have to check the following.

1. T' is closed with respect to $\square$.
2. Associative property will hold and need not be checked.
3. The identity element e of $(T, \square)$ should also be the identity for $(T', \square)$. Hence T' should contain e.
4. For each element $a \in T'$, the inverse of a also should be in T'.

THEOREM 1   Let $(T, \square)$ be a group and T' be a subset of T. If T' is a finite set, then $(T', \square)$ is a subgroup of $(T, \square)$ if T' is closed under $\square$.

What this result says is that it is enough to check the closure property alone, as the other properties will be satisfied if the closure property is satisfied if T' is a finite set.

Proof: Already we noted that the associative property will hold for $\square$ on T'. It is given that T' is closed with respect to $\square$. Let a be an element of T'. Hence $a^2, a^3, a^4, \ldots$ are all in T'. Because T' is a finite set, by the pigeonhole principle for some i and j, $i < j$, $a^i = a^j$. Therefore, $a^i = a^i \square a^{j-i}$.

Hence a^{j-i} is the identity of the operation $\square$ on T'. The identity is in T' if $j - i > 1$. Also $a^{j-i} = a \square a^{j-i-1}$. Hence a^{j-i-1} is the inverse of a and is in T'. If $j - i = 1$, we have $a^i = a^i \square a$. Hence a must be the identity element and hence its own inverse. Thus, we see that if T' is closed with respect to $\square$, the other properties of group follow, and $(T', \square)$ is a group.

Generators for a Group

Let $(T, \square)$ be an algebraic system where $\square$ is a closed operation. Let $S = \{a_1, a_2, \ldots\}$ be a subset of T. Let S_1 denote the subset of T which contains S as well as all elements $a_i \square a_j$ for a_i, a_j in S. S_1 is called the set generated directly by S. Similarly, let S_2 denote the set generated directly by $S_1, \ldots$ and S_{i+1} denote the set directly generated by S_i. Let S^* denote the union of $S, S_1,$

$S_2, \ldots$. The algebraic system $(S^*, \square)$ is called the subsystem generated by S, and an element is said to be generated by S if it is in S^*. Note that $\square$ is a closed operation on S^*. Thus for a group $(T, \square)$, if S^* is finite, then $(S^*, \square)$ is a subgroup. If $S^* = T$, S is called a generating set or a set of generators of the algebraic system $(T, \square)$. In the example of rotation of geometric figures, $\{60°\}$ is a generating, set and $\{120°, 180°\}$ is also a generating set.

A group that has a generating set consisting of a single element is known as a cyclic group. We considered two groups with four elements. The second one is a cyclic group with generating set $\{b\}$. $\{d\}$ is also a generating set for that group. The first one is not a cyclic group.

Let $(T, \square)$ be a cyclic group and $\{a\}$ a generating set of $(T, \square)$. Clearly elements of T can be expressed as $a, a^2, a^3, a^4, \ldots$ because of associating $a^i \square a^j = a^j \square a^i$ with a^{i+j}. Hence any cyclic group is a commutative group. Note that the group of four elements given in the left table is commutative but not cyclic.

Let $G = (T, \square)$ be a group and let $a \in T$. Here a^m is defined as $a \square a \square \ldots \square a$ (m factors), $a^0 = e$, and $a^{-m} = (a^{-1})^m$, where a^{-1} is the inverse of a.

LEMMA 1   If $G = (T, \square)$ is a group and $a \in T$, then
$$a^r \square a^s = a^{r+s} \qquad (a^r)^s = a^{rs}$$

Proof: For $r, s \in N$ (the set of nonnegative integers) the result is obvious.
  If r and s are negative integers,

$$r = -m \quad s = -n, \quad m, n > 0$$
$$a^r \square a^s = a^{-m} \square a^{-n} = (a^{-1})^m \square (a^{-1})^n$$
$$= (a^{-1})^{m+n} = a^{-(m+n)} = a^{(-m)+(-n)} = a^{r+s}.$$
$$(a^r)^s = (a^{-m})^{-n} = ((a^m)^{-1})^{-n}$$
$$= (((a^m)^{-1})^{-1})^n = (a^m)^n = a^{mn} = a^{(-m)(-n)} = a^{rs}.$$

The case where one of r and s is nonnegative and the other negative can similarly be proved.

THEOREM 2   In any group $G = (T, \square)$, the powers of any fixed element $a \in T$ constitute a subgroup of G.

Proof: Consider $G' = (T', \square)$ where T' consists of all powers of an element a. Closure under $\square$ is proved by previous lemma, and the associative property holds because all elements are of the form a^m. Here $a^0 = e$ is the identity element, and the inverse of a^r is a^{-r}.

THEOREM 3   Let $G = (T, \square)$ be a finite cyclic group generated by an element $a \in T$. If G is of order n (i.e., $|T| = n$) $a^n = e$, so that $T = \{a, a^2, a^3, \ldots, a^n = e\}$. Moreover, n is the least positive integer for which $a^n = e$.

Proof: If possible, let $a^m = e$ for some positive integer $m < n$. Since G is generated by a, any element of T can be written as a^k for some integer k. Here k can be written as $mq + r$, where q is some integer and $0 \le r < m$. This leads to

$$a^k = a^{mq+r} = (a^{mq}) \square a^r = (a^m)^q \square a^r = (e)^q \square a^r = e \square a^r = a^r$$

so that every element of T can be expressed as a^r for some $r, 0 \le r < m$. This means that T has at most m distinct elements and the order of G is $m < n$. Thus, we arrive at a contradiction. Hence, $a^m = e$ for $m < n$ is not possible.

We also note that all of the elements $a, a^2, \ldots, a^n$ are distinct and $a^n = e$. This can be seen as follows. Suppose if possible let $a^i = a^j, i < j \le n$. This means $a^{j-i} = e$, where $j < n$, and this is a contradiction.

Cosets and Lagrange's Theorem

Let $(T, \square)$ be an algebraic system, where $\square$ is a binary operation. Let a be an element in T and H be a subset of T. The left coset of H with respect to a, which is denoted by $a \square H$, is the set of elements $\{a \square x | x \in H\}$. Similarly, the right coset of H with respect to a is denoted as $H \square a$ and consists of elements $\{x \square a | x \in H\}$.

The cosets of groups have some interesting properties. Let $(T, \square)$ be a group and $(H, \square)$ be a subgroup of $(T, \square)$. If H has r elements say, $a \square H$ also has r elements. Since any element of T cannot occur twice in a row or in a column of the group table, no two elements of $a \square H$ can be identical.

THEOREM 4   Let $a \square H$ and $b \square H$ be two cosets of H. Then either $a \square H$ and $b \square H$ are disjoint or they are identical.

Proof: Suppose $a \square H$ and $b \square H$ are not disjoint. Let c be a common element of both. Therefore there exist elements h_1 and h_2 in H such that $c = a \square h_1 = b \square h_2$, which means $a = b \square h_2 \square h_1^{-1}$. Let $x \in a \square H$. Then $x = a \square h_3$. Hence, $x = (b \square h_2 \square h_1^{-1}) \square h_3 = b(h_2 \square h_1^{-1} \square h_3)$. But $h_2 \square h_1^{-1} \square h_3$ is an element of H, and hence, $x = b \square h_4$ for $h_4 = h_2 \square h_1^{-1} \square h_3$. Therefore, $x \in b \square H$. Similarly, if $y \in b \square H$, we can show $y \in a \square H$ too. Thus, $a \square H$ and $b \square H$ are identical. So if they are not disjoint, they are identical.

Let $(T, \square)$ be a group and $(H, \square)$ be a subgroup of $(T, \square)$. Because $(T, \square)$ is a group, no two elements in a column or no two elements in a row of the group table are the same. Hence it follows that for any $a \in T$ and h_1 and h_2 in H, $a \square h_1 \ne a \square h_2$. It follows that the size of any coset of H is the same as that of H. Here H contains the identity of the group. Hence if we compute all the left cosets of H, we would have exhausted all the elements in T. Consequently, we can conclude that the left cosets of H form a partition of T, in which all blocks are of the same size. Thus the size of T is the product of the size of H and the number of distinct cosets of H. Hence we have the following theorem.

THEOREM 5   **(Lagrange's Theorem)** The order of any subgroup of a finite group divides the order of the group.

From the above theorem, we can conclude that if a group is of prime order, it cannot have nontrivial subgroups. Trivial subgroups are the entire group itself and the one having just the identity element alone.

THEOREM 6   Any group of prime order is cyclic and any element other than the identity is a generator. It also follows that it is abelian.

Proof: Let $G = (T, \square)$ be a group of prime order and let $a \in T$ and $a \ne e$. The powers of a form a group. This should be G itself, as G has no nontrivial subgroup and hence any element a is a generator of G and G is cyclic.

Isomorphisms and Automorphisms

Let $(T, \square)$ be the algebraic system where $T = \{a, b, c\}$ and $(S, *)$ be an algebraic system with $S = \{\alpha, \beta, \gamma\}$. The tables for the operations are given below.

$\square$	a	b	c
a	a	b	c
b	b	c	a
c	c	a	b

$*$	α	β	γ
α	α	β	γ
β	β	γ	α
γ	γ	α	β

One can easily see the similarity between the two systems. In essence, they are the same, except for renaming of the elements and symbols used for operation. In this case, we say that $(T, \square)$ is isomorphic to $(S, *)$. We say two systems $(T, \square)$ and $(S, *)$ are isomorphic if there is a bijection f from T to S such that for any a_1, a_2 in T. So,

$$f(a_1 \square a_2) = f(a_1) * f(a_2)$$

The function f is called an isomorphism from $(T, \square)$ to $(S, *)$. $(S, *)$ is called an isomorphic image of $(T, \square)$. In the previous example, f is a function such that

$$f(a) = \alpha \qquad f(b) = \beta \qquad f(c) = \gamma$$

Note that a function g defined as follows is also an isomorphism from $(T, \square)$ to $(S, *)$.

$$g(a) = \alpha \qquad g(b) = \gamma \qquad g(c) = \beta$$

An isomorphism from an algebraic system $(T, \square)$ to $(T, \square)$ is called an automorphism. For example, the function

$$f(a) = a \qquad f(b) = c \qquad f(c) = b$$

is an automorphism.

EXAMPLE 5   Let $G = (T, \square)$ be a group of order p where p is a prime. Any group of order p is isomorphic to G.

Solution: We saw that for any integer n, $(Z_n, \oplus)$ is a group. Any group of order p is isomorphic to $(Z_p, \oplus)$. This can be seen as follows.

Let $G = (T, \square)$ be a group of prime order p. Then G is cyclic and elements of T are of the form $a, a^2, \ldots, a^p = e$ for any a in T that is not an identity. Define a mapping $\theta(a^i) = i$ for $1 \leq i \leq p$. Then θ is an isomorphism from $(T, \square)$ to $(Z_p, \oplus)$. This can be easily checked. Consequently, we can conclude that any group of order p is isomorphic to $(Z_p, \oplus)$. Let $G' = (T', \square')$ be another group of prime order. G' is isomorphic to $(Z_p, \oplus)$, and let θ' be the mapping defining the isomorphism. Define a mapping f from T' to T as follows: $f(x') = x$ if $\theta(x) = i$ and $\theta'(x') = i$, $1 \leq i \leq p$. It is straightforward to see that f is a bijection, and hence G' is isomorphic to G.   ◄

Extra Example

Links

JOSEPH LOUIS LAGRANGE (1736–1813)   Joseph Louis Lagrange was born in Italy on January 25, 1736. He lived part of his life in Prussia and part in France. He has contributed a lot to the areas of analysis, number theory, and classical and celestial mechanics. He had served as the director of mathematics at the Prussian Academy of Sciences in Berlin for 20 years. Then he moved over to France and was a member of the French academy till his death in 1813. Napolean named Lagrange to the Legion of Honour and made him a count of the empire in 1808. His treatise on analytical mechanics published in 1788 was considered to be the best treatment of classical mechanics in those days.

Permutation Groups

Let $S = \{a_1, a_2, a_3\}$ be a set and let p denote a permutation of S. That is, p is a bijective mapping $p : S \to S$.

Suppose $p(a_1) = a_2$, $p(a_2) = a_3$, $p(a_3) = a_1$. This may be represented as $\begin{pmatrix} a_1 & a_2 & a_3 \\ a_2 & a_3 & a_1 \end{pmatrix}$ or even as $\begin{pmatrix} 1 & 2 & 3 \\ 2 & 3 & 1 \end{pmatrix}$. The image of a_1 is a_2 and is written below it in this representation. Here p

can also be represented as $\begin{pmatrix} a_1 & a_3 & a_3 \\ a_2 & a_1 & a_3 \end{pmatrix}$ or as

Generally, we assume an ordering $a_1, a_2, \ldots$ among the elements and p is represented as $\begin{pmatrix} a_1 & a_2 & a_3 \\ a_2 & a_3 & a_1 \end{pmatrix}$ rather than $\begin{pmatrix} a_1 & a_3 & a_2 \\ a_2 & a_1 & a_3 \end{pmatrix}$.

If p_1 and p_2 are permutations, $p_1 \circ p_2$ is a permutation, where $p_1 \circ p_2$ is the composition of functions.

Suppose $p_1 = \begin{pmatrix} a_1 & a_2 & a_3 \\ a_2 & a_3 & a_1 \end{pmatrix}$, $p_2 = \begin{pmatrix} a_1 & a_2 & a_3 \\ a_3 & a_2 & a_1 \end{pmatrix}$.

$$p_1 \circ p_2 = \begin{pmatrix} a_1 & a_2 & a_3 \\ a_1 & a_3 & a_2 \end{pmatrix}$$

$$p_1 \circ p_2(a_1) = p_1(p_2(a_1)) = p_1(a_3) = a_1$$

$$p_1 \circ p_2(a_2) = p_1(p_2(a_2)) = p_1(a_2) = a_3$$

$$p_1 \circ p_2(a_3) = p_1(p_2(a_3)) = p_1(a_1) = a_2$$

If p_1, p_2, p_3 are permutations, we can easily see that the associative property holds. Here $p_1 \circ (p_2 \circ p_3) = (p_1 \circ p_2) \circ p_3$, and $\begin{pmatrix} a_1 & a_2 & a_3 \\ a_1 & a_2 & a_3 \end{pmatrix}$ is the identity permutation.

If $P = \{p_1, p_2, \ldots\}$ is the set of all permutation of elements of a set $S = \{a_1, \ldots, a_n\}$, it is not difficult to see $(P, \circ)$ is a group. The order of this group is $n!$. If $S = \{a_1, a_2\}$, there are only two permutations: $p_1 = \begin{pmatrix} a_1 & a_2 \\ a_1 & a_2 \end{pmatrix}$, $p_2 = \begin{pmatrix} a_1 & a_2 \\ a_2 & a_1 \end{pmatrix}$. The group table is

	p_1	p_2
p_1	p_1	p_2
p_1	p_2	p_1

If $S = \{a_1, a_2, a_3\}$, there are six permutations:

$$p_1 = \begin{pmatrix} a_1 & a_2 & a_3 \\ a_1 & a_2 & a_3 \end{pmatrix} \quad p_2 = \begin{pmatrix} a_1 & a_2 & a_3 \\ a_2 & a_1 & a_3 \end{pmatrix} \quad p_3 = \begin{pmatrix} a_1 & a_2 & a_3 \\ a_3 & a_2 & a_1 \end{pmatrix}$$

$$p_4 = \begin{pmatrix} a_1 & a_2 & a_3 \\ a_1 & a_3 & a_2 \end{pmatrix} \quad p_5 = \begin{pmatrix} a_1 & a_2 & a_3 \\ a_2 & a_3 & a_1 \end{pmatrix} \quad p_6 = \begin{pmatrix} a_1 & a_2 & a_3 \\ a_3 & a_1 & a_2 \end{pmatrix}$$

and the group table is given by

	p_1	p_2	p_3	p_4	p_5	p_6
p_1	p_1	p_2	p_3	p_4	p_5	p_6
p_2	p_2	p_1	p_6	p_5	p_4	p_3
p_3	p_3	p_5	p_1	p_6	p_2	p_4
p_4	p_4	p_6	p_5	p_1	p_3	p_2
p_5	p_5	p_3	p_4	p_2	p_6	p_1
p_6	p_6	p_4	p_2	p_3	p_1	p_5

This is called a permutation group. A permutation group is a group made up of elements which are permutations of a set. Note that this is not a commutative group. The order of this group is 6. We say that the degree of a permutation group is the cardinality of the set on which the permutations are defined. The degree of the above group is 3. In general, the set S_n of all permutation of n elements is a permutation group $(S_n, \circ)$. This is called the symmetric group. The $(S_n, \circ)$ is of order $n!$ and degree n. Note that $(\{p_1, p_4\}, \circ)$ is a subgroup of $(S_3, \circ)$ and has order 2 but degree 3. The group $(S_4, \circ)$ is of order 24 and degree 4. Consider a set of permutations:

$$p_1 = \begin{pmatrix} a_1 \ a_2 \ a_3 \ a_4 \\ a_1 \ a_2 \ a_3 \ a_4 \end{pmatrix} \qquad p_2 = \begin{pmatrix} a_1 \ a_2 \ a_3 \ a_4 \\ a_2 \ a_1 \ a_3 \ a_4 \end{pmatrix}$$

$$p_3 = \begin{pmatrix} a_1 \ a_2 \ a_3 \ a_4 \\ a_1 \ a_2 \ a_4 \ a_3 \end{pmatrix} \qquad p_4 = \begin{pmatrix} a_1 \ a_2 \ a_3 \ a_4 \\ a_2 \ a_1 \ a_4 \ a_3 \end{pmatrix}$$

They form a subgroup of $(S_4, \circ)$ with the following table.

	p_1	p_2	p_3	p_4
p_1	p_1	p_2	p_3	p_4
p_2	p_2	p_1	p_4	p_3
p_3	p_3	p_4	p_1	p_2
p_4	p_4	p_3	p_2	p_1

This permutation group is of order 4 and degree 4. If $p \begin{pmatrix} a_1 \ \cdots \ a_n \\ a_1' \ \cdots \ a_n' \end{pmatrix}$ is a permutation, an element is invariant under the permutation if $p(a_i) = a_i$ (i.e., $a_i' = a_i$). Sometimes it is of interest to count the number of invariant elements in a permutation. In the example considered above in p_1, the number of invariant elements is 4. In p_2, p_3 it is 2, and in p_4 it is 0.

By considering the symmetries of regular polygons, we obtain another class of groups. These are called dihedral groups. Let us consider the simplest regular polygon: the equilateral triangle. Let the vertices be named a_1, a_2, a_3. Now consider all possible rotations and reflections of the triangle that leave the final position of the triangle unchanged from its original position except for the renaming of the vertices.

For example,

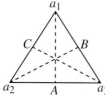 is the original position

 is obtained by rotating anticlockwise through $120°$

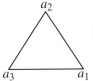 is obtained by rotating anticlockwise through 240°

If we rotate through 360°, we get the original triangle itself.

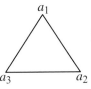 is obtained by reflecting about $a_1 A$

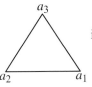 is obtained by reflecting about $a_2 B$

 is obtained by reflecting about $a_3 C$

We can easily see that these six variations of the triangle correspond to permutation p_1, p_5, p_6, p_4, p_3, p_2, respectively. This group of rotations and reflections of an equilateral triangle is called a dihedral group $(D_3, \diamond)$, where the operation $\diamond$ corresponds to performing one after another one of the two operations rotation and reflection.

For example, if rotation through 120° is performed and then reflection about the vertical,

 from first and

 from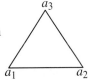

The first operation is represented by p_6, and the second by p_2, performing the first and second operations, which leads to a transformation represented by p_3, and we know $p_2 \cdot p_6 = p_3$.

Consider the next size regular polygon, which is a square.

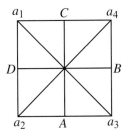

Rotation through 90°, 180°, and 270° leaves the square in the same position except for the renaming of the vertices. These are denoted as r_1, r_2, r_3. Reflection about AC and BD leave the

square in position, and they are denoted f_1 and f_2, respectively. Reflections about a_1, a_3 and a_2, a_4 also keep the square in position, and they are represented as f_3, f_4 respectively. Denoting by e the identity, these operations on the square give the following group table.

$\diamondsuit$	e	r_1	r_2	r_3	f_1	f_2	f_3	f_4
e	e	r_1	r_2	r_3	f_1	f_2	f_3	f_4
r_1	r_1	r_2	r_3	e	f_4	f_3	f_1	f_2
r_2	r_2	r_3	e	r_1	f_2	f_1	f_4	f_3
r_3	r_3	e	r_1	r_2	f_3	f_4	f_2	f_1
f_1	f_1	f_3	f_2	f_4	e	r_2	r_1	r_3
f_2	f_2	f_4	f_1	f_3	r_2	e	r_3	r_1
f_3	f_3	f_2	f_4	f_1	r_3	r_1	e	r_2
f_4	f_4	f_1	f_3	f_2	r_1	r_3	r_2	e

This is the dihedral group $(D_4, \diamondsuit)$

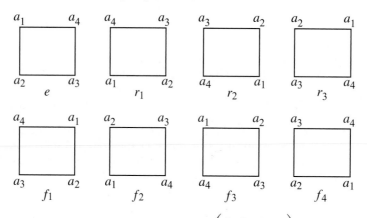

e represents the identity permutation $\begin{pmatrix} a_1\ a_2\ a_3\ a_4 \\ a_1\ a_2\ a_3\ a_4 \end{pmatrix}$

r_1 is represented by the permutation $\begin{pmatrix} a_1\ a_2\ a_3\ a_4 \\ a_4\ a_1\ a_2\ a_3 \end{pmatrix}$

r_2 is represented by the permutation $\begin{pmatrix} a_1\ a_2\ a_3\ a_4 \\ a_3\ a_4\ a_1\ a_2 \end{pmatrix}$

r_3 is represented by the permutation $\begin{pmatrix} a_1\ a_2\ a_3\ a_4 \\ a_2\ a_3\ a_4\ a_1 \end{pmatrix}$

f_1 is represented by the permutation $\begin{pmatrix} a_1\ a_2\ a_3\ a_4 \\ a_4\ a_3\ a_2\ a_1 \end{pmatrix}$

f_2 is represented by the permutation $\begin{pmatrix} a_1\ a_2\ a_3\ a_4 \\ a_2\ a_1\ a_4\ a_3 \end{pmatrix}$

f_3 is represented by the permutation $\begin{pmatrix} a_1\ a_2\ a_3\ a_4 \\ a_1\ a_4\ a_3\ a_2 \end{pmatrix}$

f_4 is represented by the permutation $\begin{pmatrix} a_1\ a_2\ a_3\ a_4 \\ a_3\ a_2\ a_1\ a_4 \end{pmatrix}$

Performing a rotation through 90° and performing reflection about vertical gives

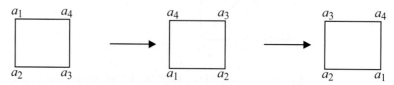

This is represented by $r_1 \diamond f_1 = f_4$. The corresponding operation on the respective permutation is

$$\begin{pmatrix} a_1\ a_2\ a_3\ a_4 \\ a_4\ a_1\ a_2\ a_3 \end{pmatrix} \circ \begin{pmatrix} a_1\ a_2\ a_3\ a_4 \\ a_4\ a_3\ a_2\ a_1 \end{pmatrix} = \begin{pmatrix} a_1\ a_2\ a_3\ a_4 \\ a_3\ a_2\ a_1\ a_4 \end{pmatrix}$$

This dihedral group $(D_4, \diamond)$ has eight elements, and $(S_4, \circ)$ has 24 elements. The permutations corresponding to the operations of rotation and reflection of a square form a subgroup of $(S_4, \circ)$. In general, $(S_n, \circ)$ is of order $n!$ and degree n. $(D_n, \circ)$ is of order $2n$. Here n of them correspond to rotations through angles $0, \dfrac{360}{n}, \dfrac{360}{n} \times 2, \cdots, 360 \times \dfrac{(n-1)}{n}$. The other n correspond to reflections. If n is odd, reflections are about lines joining a vertex to the midpoint of the opposite side. If n is even, reflections are about lines joining opposite vertices (diagonals) or about lines joining the midpoints of opposite sides. Thus we find $(D_n, \diamond)$ is isomorphic to a subgroup of $(S_n, \circ)$. When $n = 3$, the subgroup is the group itself.

The following example supplies an illustration for non-abelian group.

EXAMPLE 6 A rook matrix is a matrix that has only 0's and 1's as entities such that each row has exactly one 1 and each column has exactly one 1. The term rook matrix is derived from the fact that each rook matrix represents the placement of n rooks on an $n \times n$ chessboard such that none of the rooks can attack one another. A rook in chess can only move only vertically or horizontally, but not diagonally. Let R_n be the set of $n \times n$ rook matrices. There are six 3×3 rook matrices.

$$e = \begin{bmatrix} 1\ 0\ 0 \\ 0\ 1\ 0 \\ 0\ 0\ 1 \end{bmatrix} a = \begin{bmatrix} 0\ 1\ 0 \\ 0\ 0\ 1 \\ 1\ 0\ 0 \end{bmatrix} b = \begin{bmatrix} 0\ 0\ 1 \\ 1\ 0\ 0 \\ 0\ 1\ 0 \end{bmatrix} c = \begin{bmatrix} 1\ 0\ 0 \\ 0\ 0\ 1 \\ 0\ 1\ 0 \end{bmatrix} d = \begin{bmatrix} 0\ 0\ 1 \\ 0\ 1\ 0 \\ 1\ 0\ 0 \end{bmatrix} f = \begin{bmatrix} 0\ 1\ 0 \\ 1\ 0\ 0 \\ 0\ 0\ 1 \end{bmatrix}$$

The 2×2 rook matrices are $\begin{bmatrix} 1\ 0 \\ 0\ 1 \end{bmatrix}$ and $\begin{bmatrix} 0\ 1 \\ 1\ 0 \end{bmatrix}$. They form an abelian group under matrix multiplication.

If we write out the multiplication table for R_3, we could see the similarity between this and the table for $(S_3, \circ)$. This is not abelian as $a \times c \neq c \times a$.

The following can be taken as exercises.

How many 4×4 rook matrices are there? How many $n \times n$ rook matrices are there? Compare R_n with D_n and S_n. ◄

EXAMPLE 7 Imagine letters entering a conveyor belt to be postmarked. They are placed on the conveyor belt at random so that two sides are parallel to the belt. Suppose that a postmaster can recognize a stamp in the top right corner of the envelop on the side facing up. In the figure a sequence of operations is shown. This sequence will recognize a stamp on any letter, no matter what position the letter starts in. The letter P stands for postmaster. R and F stand for rotation through 90 degrees and flipping.

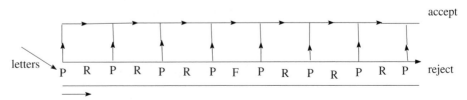

The arrows pointing up indicates that if a letter is postmarked, it is taken off the conveyor belt for delivering. If a letter reaches the end, it means it does not have a stamp. Compare the operations here with the dihedral group operation D_4. Can you achieve the same thing with more flips and less rotations than the way shown above? ◄

Even and Odd Permutations

Let the elements be $\{1, \ldots, n\}$. A permutation $\{a_1, \ldots, a_n\}$ of $\{1, 2, \ldots, n\}$ is called even or odd according to the number of pairs (a_i, a_j) where $a_i > a_j$ and $i < j$ is even or odd.

$\begin{pmatrix} 1\ 2\ 3\ 4 \\ 1\ 2\ 4\ 3 \end{pmatrix}$ is an odd permutation.

$\begin{pmatrix} 1\ 2\ 3\ 4 \\ 1\ 4\ 2\ 3 \end{pmatrix}$ is an even permutation.

The set of even permutations form a subgroup of $(S_n, \circ)$ called the alternating subgroup.

EXAMPLE 8   Consider the sliding-tile puzzle given in the following figure. Each numbered square is a tile and the dark square is a gap. Any tile that is adjacent to the gap can slide into the gap. In most versions of this puzzle, the tiles are locked into a frame so that they can be moved only in the manner described above. The object of the puzzle is to arrange the tile as they appear in (a). Figures (b) and (c) show typical starting point. We can show that the puzzle can be solved starting with (b) but not with (c).

1	2	3	4
5	6	7	8
9	10	11	12
13	14	15	

(a)

7	6	5	8
3	4	2	1
10	9	14	11
12	13	15	

(b)

7	6	5	8
3	4	1	2
10	9	14	11
12	13	15	

(c)

We can associate any configuration of the puzzle with a permutation in S_{16}. The gap is taken as the tile with number 16. If f is a configuration, the identity permutation represents (a). (b) and (c) are represented by f_1 and f_2 and we have

$$f_1 = \begin{pmatrix} 1\ 2\ 3\ 4\ 5\ 6\ 7\ 8\ 9 & 10\ 11\ 12\ 13\ 14\ 15\ 16 \\ 7\ 6\ 5\ 8\ 3\ 4\ 2\ 1\ 10\ 9 & 14\ 11\ 12\ 13\ 15\ 16 \end{pmatrix}$$

$$f_2 = \begin{pmatrix} 1\ 2\ 3\ 4\ 5\ 6\ 7\ 8\ 9 & 10\ 11\ 12\ 13\ 14\ 15\ 16 \\ 7\ 6\ 5\ 8\ 3\ 4\ 1\ 2\ 10\ 9 & 14\ 11\ 12\ 13\ 15\ 16 \end{pmatrix}$$

Consider what happens when the tile with number 12 of configuration (a) slides into the gap. The result is a configuration that we would interpret as $\begin{pmatrix} 12\ 16 \\ 16\ 12 \end{pmatrix}$. Now if we slide tile number 8 into old tile number 12 position, the result can be represented by $\begin{pmatrix} 8 & 12\ 16 \\ 16\ 8 & 12 \end{pmatrix}$. Every time you slide a tile into the gap, one exchange has occurred. To start with configuration (a), and end up in a configuration where the gap is in place 16, an even number of exchanges must have been made. Thus any permutation that leaves 16 fixed, cannot be solved if it is odd. Note that (f_2) represents an odd permutation and the puzzle cannot be solved starting with (c). But it can can be solved starting with (b). The proof that any puzzle with starting configuration representing an even permutation can be solved is left as an exercise.   ◀

THEOREM 7   Every finite group of order n is isomorphic to a permutation group of degree n.

Proof: Let $G = (T, \square)$ be a group of order n, i.e., T consists of n elements and the group table is a $n \times n$ array. Let $T = \{a_1, \ldots, a_n\}$ and let $e = a_1$. Each row and each column in the table is a permutation of elements in T. For each $a \in T$, we denote by p_a the permutation given by the column under a in the table. Thus $p_a(b) = b \square a$ for all $b \in T$. Each column represents a permutation of elements of T. Let them be denoted as p_{a1}, p_{a2}, ..., p_{an} and $P = \{p_{a1}, p_{a2}, \ldots, p_{an}\}$. Here P has n elements. We show $G_p = (P, \circ)$ as a group where $\circ$ denotes the composition (right) of permutations. The column under the identity element $a_1(= e)$ represents the identity permutation and is in P. Also $P_e \circ P_a = P_a \circ P_e = P_a$. If a_i and a_j are inverse elements in G

$$P_{ai} \circ P_{aj} = P_{aj} \circ P_{ai} = P_e$$

and

$$P_{ai} \circ P_{aj} = P_{ai \square aj}$$

This can be seen as

$$(P_{ai} \circ P_{aj})(b) = (b \square ai) \square aj = b(ai \square aj) = P_{ai \square aj}(b)$$

Thus we find that $G_p = (P, \circ)$ is a group isomorphic to $G = (T, \square)$ where the isomorphism is defined by the bijective mapping $f : T \to P$ as $f(a) = P_a$ for all $a \in T$.

We shall consider some problem solving techniques in this topic in Examples 9–12.

EXAMPLE 9   Show that any semigroup S can be extended to a monoid by adjoining an identity element.

Solution: Let $(A, *)$ be a semigroup. Add an element e to A and extend the operation $*$ to $A \cup \{e\}$ by defining $a * e = e * a = a$ for all a in $A \cup \{e\}$. For $(A \cup \{e\}, *)$, closure and associative properties hold and e by definition is the identity element. Hence $(A \cup \{e\}, *)$ is a monoid. ◀

EXAMPLE 10   Show that if z is a left zero of a semigroup $(S, *)$, then so are all its left multiples xz ($x \in S$).

Solution: Let z be a left zero for $(S, *)$. Then $zy = z$ for all y in S. So,

$$(xz)y = x(zy) \text{ (by associativity)}$$

$$= xz \text{ for any } y \text{ in } S$$

Hence, (xz) is also a left zero. ◀

EXAMPLE 11   a) Show that if $a^2 = e$ for all a in a group $G = (A, *)$, then G is commutative.
b) Show that the same is true in any monoid.

Solution: Let $G = (A, *)$ be a group with e as identity.
For any a in A, $a^2 = e$. So,

$$(a * a) * (b * b) = e * e = e$$

$$(a * b) * (a * b) = e$$

Hence, using associativity, we get

$$a * (a * b) * b = a * (b * a) * b$$

It follows

$$a^{-1} * a * (a * b) * b * b^{-1} = a^{-1} * a * (b * a) * b * b^{-1}$$

That is, $e * (a * b) * e = e * (b * a) * e$.
Hence, $a * b = b * a$. Therefore, G is commutative.
For a monoid,

$$(a * a) * (b * b) = e * e = e$$

$$(a * b) * (a * b) = e$$

Hence, using associativity, we get

$$a * (a * b) * b = a * (b * a) * b$$

It follows that

$$a * a * (a * b) * b * b = a * a * (b * a) * b * b$$

That is, $e * (a * b) * e = e * (b * a) * e$ ◀

Hence, $a * b = b * a$. Therefore, the monoid is commutative.

EXAMPLE 12 Let $(A, *)$ be a semi group. Furthermore for every a and b in A if $a \neq b$, then $a * b \neq b * a$. That is, if $a * b = b * a$, then $b = a$.

a) Show that for every a in A

$$a * a = a$$

b) Show that for every a, b in A

$$a * b * a = a$$

c) Show that for every a, b, c in A

$$a * b * c = a * c$$

Solution:

a) $a * (a * a) = (a * a) * a$
   Hence $a = a * a$
b) $(a * b * a) * a = a * b * (a * a) = a * b * a$
   $a * (a * b * a) = (a * a) * b * a = a * b * a$
   as $a * a = a$ by part a
   Hence $a * b * a = a$.
c) $(a * b * c) * (a * c) = a * b * (c * a * c) = a * b * c$
   $(a * c) * (a * b * c) = (a * c * a) * b * c = a * b * c$

   Hence $a * b * c = a * c$.   ◀

Exercises

1. Find the zeros of the semigroups $(P(x), \cap)$ and $(P(X), \cup)$ where X is any given set and $P(X)$ is its power set. Are these monoids? If so, what are the identities?

2. Let the alphabet $V = \{a, b\}$ and A be the set including λ of all sequences on V beginning with a. Show that $(A, \circ, \lambda)$ is a monoid.

3. Let $S = \{a, b\}$. Show that the semigroup $(S^S, \circ)$ is not commutative, where S^S denotes the set of all functions $S \to S$.

4. Let Z_n denote the set of integers $\{0, 1, 2, \ldots, n - 1\}$. Let $\odot$ be binary operation on Z_n such that $a \odot b =$ the remainder of ab divided by n.

   a) Construct the table for the operation $\odot$ for $n = 7$.
   b) Show that $(Z_n, \odot)$ is a semigroup for any n.

5. Let $(A, *)$ be a semigroup. Let a be an element in A. Consider a binary operation $\square$ on A such that, for every x and y in A,

   $$x \square y = x * a * y$$

Show that $\square$ is an associative operation.

6. An element $a \in S$, where $(S, *)$ is a semigroup, is called a left-cancellable element if for all x, $y \in S$, $a * x = a * y \rightarrow x = y$. Show that if a and b are left-cancellable, then $a * b$ is also left-cancellable.

7. Let $(A, \square)$ be a semigroup. Show that, for a, b, c in A, if $a \square c = c \square a$ and $b \square c = c \square b$, then $(a \square b)c = c \square (a \square b)$.

8. Let $(\{a, b\}, \square)$ be a semigroup where $a \square a = b$. Show the following.
   a) $a \square b = b \square a$
   b) $b \square b = b$

9. Let $(A, \square)$ be a commutative semigroup. Show that if $a \square a = a$ and $b \square b = b$, then $(a \square b) \square (a \square b) = a \square b$.

10. Show that every finite semigroup has an idempotent.

11. Show that a semigroup with more than one idempotent cannot be a group. Give an example of a semigroup which is not a group.

12. Show that the set of all the invertible elements of a monoid form a group under the same operation as that of the monoid.

13. In a monoid, show that the set of left-invertibles (right-invertibles) form a submonoid.

14. Let $(A, \square)$ be a semigroup. Furthermore, let there be an element a in A such that for every x in A there exist u and v in A satisfying the relation

$$a \square u = v \square a = x$$

Show that there is an identity element in A.

15. For $P = \{p_1, p_2, \ldots, p_5\}$ and $Q = \{q_1, q_2, \ldots, q_5\}$ explain why $(P, *)$ and $(Q, \square)$ are not groups. The operations $*$ and $\square$ are given in the following table:

$*$	p_1	p_2	p_3	p_4	p_5	$\square$	q_1	q_2	q_3	q_4	q_5
p_1	p_1	p_2	p_3	p_4	p_5		q_1	q_4	q_1	q_5	q_3	q_2
p_2	p_2	p_1	p_4	p_2	p_3		q_2	q_3	q_5	q_2	q_1	q_4
p_3	p_3	p_5	p_1	p_2	p_4		q_3	q_1	q_2	q_3	q_4	q_5
p_4	p_4	p_3	p_5	p_1	p_2		q_4	q_2	q_4	q_1	q_5	q_3
p_5	p_5	p_4	p_2	p_3	p_4		q_5	q_5	q_3	q_4	q_2	q_1

16. Consider a computer which uses words of k bits to represent nonnegative integers in binary notation. The only operation is addition. When overflow occurs, the high order bits are lost.
   a) What algebraic variety would be most appropriate to model addition in the machine? How big is the carrier?
   b) Suppose overflow causes the result to be set to the largest representable number. What algebraic variety would best model addition in this case?

17. a) Show that every group containing exactly two elements is isomorphic to $(Z_2, \oplus)$.
   b) Show that every group containing exactly three elements is isomorphic to $(Z_3, \oplus)$.
   c) How many nonisomorphic groups that contain exactly four elements are there?

d) What can you say about a group with five elements?

18. Let $(A, \blacklozenge)$ and $(B, *)$ be two algebraic systems. The Cartesian product of $(A, \blacklozenge)$ and $(B, *)$ is an algebraic system $(A \times B, \square)$, where $\square$ is a binary operation such that for any (a_1, b_1) and (a_2, b_2) in $A \times B$

$$(a_1, b_1) \square (a_2, b_2) = (a_1 \blacklozenge a_2, b_1 * b_2)$$

Show that the Cartesian product of two groups is a group.

19. Let $(A, \cdot)$ be a group. Show the following.
   a) $(ab)^{-1} = b^{-1}a^{-1}$
   b) $(a_1 a_2 \ldots a_{r-1} a_r)^{-1} = a_r^{-1} a_{r-1}^{-1} \ldots a_2^{-1} a_1^{-1}$
   c) $(a^i b^j)^{-1} = b^{-j} a^{-i}$ [b^{-j} denotes $(b^{-1})^j$ and a^{-i} denotes $(a^{-1})^i$]

20. Let $(A, *)$ be a monoid such that for every x in A, $x * x = e$, where e is the identity element. Show that $(A, *)$ is an abelian group.

21. Let $(G, \blacklozenge)$ be a group and H be a nonempty subset of G. Show that $(H, \blacklozenge)$ is a subgroup if for any a and b in H, where $a \blacklozenge b^{-1}$ is also in H.

22. Show that any subgroup of a cyclic group is cyclic.

23. Let $(H, \cdot)$ be a subgroup of a group $(G, \cdot)$. Let $N = \{x | x \in G, xHx^{-1} = H\}$. Show that $(N, \cdot)$ is a subgroup of $(G, \cdot)$.

24. Let $(H, \cdot)$ and $(K, \cdot)$ be subgroups of a group $(G, \cdot)$. Let $HK = \{h \cdot k | h \in H, k \in K\}$. Show that $(HK, \cdot)$ is a subgroup of $(G, \cdot)$ if and only if $HK = KH$.

25. a) Let $(A, \blacklozenge)$ be a group. Show that $(A, \blacklozenge)$ is an abelian group if and only if $a^2 \blacklozenge b^2 = (a \blacklozenge b)^2$ for all a and b in A.
   b) For all $a, b \in A$, show that $(a \blacklozenge b)^n = a^n \blacklozenge b^n$.

26. Let $(A, \blacklozenge)$ be a group. Show that $(A, \blacklozenge)$ is an abelian group if $a^3 \blacklozenge b^3 = (a \blacklozenge b)^3$, $a^4 \blacklozenge b^4 = (a \blacklozenge b)^4$, and $a^5 \blacklozenge b^5 = (a \blacklozenge b)^5$ for all a and b in A.

27. Let $(G, \blacklozenge)$ be a group of even order. Let $(H, \blacklozenge)$ be a subgroup of $(G, \blacklozenge)$ where $|H| = |G|/2$. Show that $(H, \blacklozenge)$ is a normal subgroup.

28. Let $(H, \blacklozenge)$ be a subgroup of a group $(G, \blacklozenge)$. Show that $(H, \blacklozenge)$ is a normal subgroup if and only if $a \blacklozenge H \blacklozenge a^{-1} \subseteq H$ for every $a \in G$.

29. Let $(G, \blacklozenge)$ be a group. Let $H = \{a | a \in G$ and $a \blacklozenge b = b \blacklozenge a$ for all $b \in G\}$. Show that H is a normal subgroup.

30. a) Let $(H, \blacklozenge)$ and $(K, \blacklozenge)$ be subgroups of a group $(G, \blacklozenge)$. Show that $(H \cap K, \blacklozenge)$ is also a subgroup.
   b) Show that, if $(H, \blacklozenge)$ and $(K, \blacklozenge)$ are normal subgroups, then $(H \cap K, \blacklozenge)$ is also a normal subgroup.

31. Show that if every element in a group is its own inverse then the group must be abelian.

32. Show that the set of all polynomials in x under the operation of addition is a group.

33. Show that the groups $(G, *)$ and $(S, \triangle)$ given by the following table are isomorphic.

$*$	p_1	p_2	p_3	p_4	$\triangle$	q_1	q_2	q_3	q_4
p_1	p_1	p_2	p_3	p_4	q_1	q_3	q_4	q_1	q_2
p_2	p_2	p_1	p_4	p_3	q_2	q_4	q_3	q_2	q_1
p_3	p_3	p_4	p_1	p_2	q_3	q_1	q_2	q_3	q_4
p_4	p_4	p_3	p_2	p_1	q_4	q_2	q_1	q_4	q_3

34. Find the left cosets of $\{p_1, p_5, p_6\}$ in the group $(S, \diamondsuit)$ given in the following table

$\diamondsuit$	p_1	p_2	p_3	p_4	p_5	p_6
p_1	p_1	p_2	p_3	p_4	p_5	p_6
p_2	p_2	p_1	p_5	p_6	p_3	p_4
p_3	p_3	p_6	p_1	p_5	p_4	p_2
p_4	p_4	p_5	p_6	p_1	p_2	p_3
p_5	p_5	p_4	p_2	p_3	p_6	p_1
p_6	p_6	p_3	p_4	p_2	p_1	p_5

35. Show that, if a group $(G, *)$ is of order n and $a \in G$ is such that $a^m = e$ for some integer $m \leq n$, then m must divide n.

36. Show that, if a group $(G, *)$ is of even order, then there must be an element $a \in G$ such that $a \neq e$ and $a * a = e$.

37. If an abelian group has subgroups of orders m and n, then show that it has a subgroup whose order is the least common multiple of m and n.

38. Show that among the cosets determined by a subgroup S in a group $(G, *)$, only one of the cosets is a subgroup.

39. Let

$$P_1 = \begin{pmatrix} 1\,2\,3\,4\,5 \\ 2\,3\,1\,4\,5 \end{pmatrix}, \quad P_2 = \begin{pmatrix} 1\,2\,3\,4\,5 \\ 1\,2\,3\,5\,4 \end{pmatrix},$$

$$P_3 = \begin{pmatrix} 1\,2\,3\,4\,5 \\ 5\,4\,3\,1\,2 \end{pmatrix}, \quad P_4 = \begin{pmatrix} 1\,2\,3\,4\,5 \\ 3\,2\,1\,5\,4 \end{pmatrix}.$$

Find $P_1 \circ P_2$, $P_2 \circ P_1$, $P_1 \circ P_1$, and $P_1 \circ P_2 \circ P_3$. Solve the equation $P_1 \circ x = P_2$.

40. Show that the set of permutations given below form a group.

$$\left\{ \begin{pmatrix} a\,b\,c\,d \\ a\,b\,c\,d \end{pmatrix}, \begin{pmatrix} a\,b\,c\,d \\ b\,a\,c\,d \end{pmatrix}, \begin{pmatrix} a\,b\,c\,d \\ a\,b\,d\,c \end{pmatrix}, \begin{pmatrix} a\,b\,c\,d \\ b\,a\,d\,c \end{pmatrix} \right\}$$

Draw the group table.

41. Show that $(s_4, \circ)$ is generated by

$$\begin{pmatrix} 1\,2\,3\,4 \\ 2\,1\,3\,4 \end{pmatrix} \begin{pmatrix} 1\,2\,3\,4 \\ 1\,4\,2\,3 \end{pmatrix} \begin{pmatrix} 1\,2\,3\,4 \\ 1\,3\,4\,2 \end{pmatrix}$$

12.3   Homomorphisms, Normal Subgroups, and Congruence Relations

In the case of isomorphism, two algebraic systems are structurally similar and also have the same order if they are finite. Two algebras may be very similar, but they may not be of the same order even though their orders are finite. In order to study such structures, we consider homomorphism (i.e., the function f defined in the case of homomorphism) need not be bijective, but other conditions will be satisfied.

The following figure explains the concept of homomorphism.

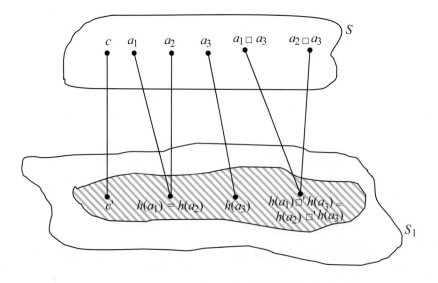

FIGURE 1   Representation of a homomorphism.

DEFINITION 1   Let $A = (S, O, C)$ be an algebra, $A' = (S', O', C')$ be another algebra with the same signature, and h be a function such that

$$h: S \to S'$$
$$h(a \,\square\, b) = h(a) \,\square'\, h(b)$$

where $\square$ and $\square'$ are corresponding operations in A and A', respectively.
    Here $h(c) = c'$ with c and c' being corresponding constants.

Alternatively, we can characterize a homomorphism from $A = (S, O, C)$ to $A' = (S', O', C')$ as a map h such that $h(c) = c'$, where c and c' are corresponding constants for the algebras. We have the following commuting diagram.

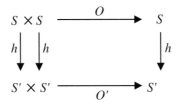

It should be noted that isomorphism is a particular case of homomorphism.

EXAMPLE 1 $f_k: I \to I$, where $f_k(x) = kx$ for an integer k is a homomorphism from $(I, +, 0)$ to $(I, +, 0)$.   ◄

EXAMPLE 2   Let $\sum$ be a finite non-empty alphabet, and let $|x|$ denote the length of a string $x \in \sum^*$. Then the function h is defined by $h: \sum^* \to N$.
    $h(x) = |x|$ is a homomorphism from $(\sum^*, \text{concatenation}, \lambda)$ to $(N, +, 0)$.

Question: Under what conditions is the above-mentioned homomorphism an isomorphism?   ◄

THEOREM 1   Let h be a homomorphism from $A = (S, O, C)$ to $A' = (S', O', C')$. Then $(h(S), O', C')$ is a subalgebra of A', called the homomorphic image of A under h.

Proof: To show that $(h(S), O', C')$ is a subalgebra, the following conditions must be satisfied.

1. $h(S) \subset S'$.
    This follows from the fact that $h: S \to S'$.
2. The constant k' is an element of $h(S)$. By definition of homomorphism, $h(k) = k'$. Since $k \in S$, it follows that

    $$k' = h(k) \in h(S).$$

3. The set $h(S)$ is closed under the operations O', That is, if $a, b \in h(S)$, then $a \circ' b \in h(S)$. Suppose $a, b \in h(S)$. Then there exist elements x, y in S such that $h(x) = a$ and $h(y) = b$ for

    $$h(x \circ y) = a \circ' b \in h(S)$$

The homomorphic image of an algebra is of the same variety as A. If A is a semigroup, the homomorphic image of A is a semigroup. If A is a monoid, the homomorphic image of A is a monoid. If A is a group, the homomorphic image of A is a group. This can be proved by verifying the conditions that have to be satisfied by that algebra.

A homomorphism is called a epimorphism if it is surjective and a monomorphism if it is injective. A homomorphism where the codomain and domain are the same is called an endomorphism. An isomorphism (homomorphism which is bijective) of an algebraic system with itself is called an automorphism.

THEOREM 2

Let $S = (T, \square)$ be a semigroup. The automorphisms of S form a group called the automorphism group of S.

Proof: Let $\mathcal{A}$ be the set of automorphisms of S. Let $A_1, A_2 \in \mathcal{A}$. The composition of A_1 and A_2 is defined as $A_1 * A_2 : A_1 * A_2(x) = A_1(A_2(x))$. It can be easily seen that if A_1 and A_2 are bijective $A_1 * A_2$ is bijective and hence $A_1 * A_2$ is an automorphism of S. The identity mapping 1_T defined as $1_T(x) = x$ for all $x \in T$ is an automorphism, which can be considered as an identity element for the algebraic system $\mathcal{G} = (\mathcal{A}, *)$. The associative property of $\mathcal{A}$ under $*$ is also obvious. If A is an automorphism, $A: T \rightarrow T$, A^{-1} can be defined as $A^{-1}(y) = x$ if $A(x) = y$. Hence $A^{-1}: T \rightarrow T$ is a bijection and A^{-1} can be taken as the inverse of A. Hence $\mathcal{G} = (\mathcal{A}, *)$ forms a group.

Congruence Relations

There is an alternative way to look at the notion of a homomorphism from one algebraic system to another. This is in terms of congruence relations.

A congruence relation is an equivalence relation defined on the carrier of an algebra such that the equivalence classes of the relation are "preserved" by the operations of the algebra. Recall that an equivalence relation is a relation which is reflexive, symmetric and transitive. An equivalence relation which is both right invariant and left invariant under the operations of the algebra is called a congruence relation.

Let us look at the example where ordered pairs of integers are associated with rational numbers. We define a fraction as an ordered pair of integers $< p, q >$ (written $\frac{p}{q}$) where $q \neq 0$, and let F be the set of all fractions.

The binary operation of $+, -$ and $\cdot$ and unary $-$ can be defined on F. Using the corresponding operations on integers as

$$\left(\frac{p}{q}\right) + \left(\frac{r}{s}\right) = \frac{(ps + rq)}{(qs)}$$

$$\left(\frac{p}{q}\right) - \left(\frac{r}{s}\right) = \frac{(ps - rq)}{(qs)}$$

$$\left(\frac{p}{q}\right) \cdot \left(\frac{r}{s}\right) = \frac{(pr)}{(qs)}$$

$$-\left(\frac{p}{q}\right) = \frac{(-p)}{q}$$

When we consider fractions as pairs of integers $\frac{1}{2}$ and $\frac{2}{4}$ are not equal. They represent different pairs $< 1, 2 >$ and $< 2, 4 >$. However we would like to treat them as equal. This is done by establishing an equivalence relation $\sim$ over F as

$$\frac{p}{q} \sim \frac{r}{s} \text{ iff } ps = rq$$

Each equivalence class of F corresponds to a rational number. Thus we find $\frac{1}{3}, \frac{2}{6}, \frac{3}{9}, \frac{-1}{-3}, \frac{-3}{-9}$ all belong to one equivalence class.

If a and b are fractions in the same equivalence class, $a + c$ and $b + c$ are in the same equivalence class for a fraction c. This can be seen as follows.

Suppose

$$a = \frac{p}{q} \quad b = \frac{r}{s} \quad c = \frac{t}{u}$$

$$a \sim b \quad \text{or} \quad ps = qr$$

$$a + c = \frac{p}{q} + \frac{t}{u} = \frac{pu + qt}{uq}$$

$$b + c = \frac{r}{s} + \frac{t}{u} = \frac{ru + st}{us}$$

$$p = \frac{qr}{s}$$

Hence

$$\frac{pu + qt}{uq} = \frac{\frac{qr}{s}u + qt}{uq} = \frac{ru + st}{su}$$

Hence $a + c \sim b + c$. Similarly,

$$c + a \sim c + b$$

$$c - a \sim c - b$$

$$a - c \sim b - c$$

$$a \cdot c \sim b \cdot c$$

$$c \cdot a \sim b \cdot a$$

$$-a \sim -b$$

The equivalence relation $\sim$ is right invariant and left invariant over the operations $+, -, \cdot$. We say that $\sim$ is a congruence relation with respect to $+, -, \cdot$ and unary $-$.

Let us consider an algebra with carrier T and binary operation $\square$. Let $\sim$ be an equivalence relation on $\sim$. Then $\sim$ is a congruence relation on T if and only if for all $a, b, c \in T$, if $a \sim b$, $a \square c \sim b \square c$ and $c \square a \sim b \square a$.

Informally, we say that a relation $\sim$ on a set T is a congruence relation with respect to an operation $\square$ if $\sim$ is a congruence relation on the algebra, $(T, \square)$. A relation $\sim$ is a congruence relation on an algebra A with underlying set T if and only if $\sim$ is a congruence relation on T with respect to each of the operations of A.

EXAMPLE 3   Equality relation is a congruence relation on any algebra.

For an arbitrary monoid $A = (S, \circ, 1)$ show that the equality and the universal relation $S \times S$ are both congruence relations on A.

Consider the equality relation

$$a R b \text{ iff } a = b$$

$$a \cdot x = b \cdot x \text{ if } a = b \text{ for any } a, b, x \text{ in } A$$

Hence $a \cdot x R b \cdot x$
Similarly, $x \cdot a R x \cdot b$

Hence R is both right invariant and left invariant. Equality is an equivalence relation and hence it is a congruence relation.

Each element of A is in one congruence class.

Consider the universality relation aRb for any $a, b \in A$.

Hence $a \cdot x R b \cdot x$ and $x \cdot a R x \cdot b$ as $a \cdot x$, $b \cdot x$, $x \cdot a$, and $x \cdot b$ are elements of A. All elements are in the same equivalence or congruence class, and the operation is both left invariant and right invariant. ◀

EXAMPLE 4   Consider the set of integers together with the operation of multiplication. The equivalence relation $\sim$ of "equivalence mod k" for some positive integer k is a congruence relation on the algebra $(I, \cdot)$.

$x \sim y$ if and only if $x \equiv y \bmod k$

Now to show $\sim$ is a congruence relation we have to show that, for any $z \in I$,

$x \cdot z = y \cdot z$ and

$z \cdot x = z \cdot y$

$\quad x = kn_1 + r$

$\quad y = kn_2 + r \qquad 0 \le r < k, \qquad n_1, n_2 \in I$

$z \cdot x = z(kn_1 + r) = zkn_1 + zr$

$z \cdot y = z(kn_2 + r) = zkn_2 + zr$

Let

$zr = n_3 k + r' \qquad 0 \le r' < k$

Then

$z \cdot x = zk(n_1) + n_3 k + r' = k(2n_1 + n_3) + r'$

$z \cdot y = zk(n_2) + n_3 k + r' = k(2n_2 + n_3) + r'$

Hence, $z \cdot x \sim z \cdot y$.

By commutativity of multiplication of integers, $x \cdot z = y \cdot z$.

Thus $\sim$ is a congruence relation over $(I, \cdot)$ ◀

EXAMPLE 5   Let I be the set of integers and $\square$ binary operation on I such that

$a \square b = \max(a, b)$

Consider $\sim$ as equivalence relation 'mod 7.'

That is, $a \sim b$ if $a \equiv b(\bmod 7)$.

Then $\sim$ is not a congruence relation for $\square$:

$5 \sim 12$

$5 \square 8 = 8$

$12 \square 8 = 12$

$8 \not\equiv 12(\bmod 7)$

$8 \not\sim 12$

Hence it is not a congruence relation over $\square$. ◀

THEOREM 3   The equivalence relation $\sim$ on a set T is a congruence relation with respect to the binary operation $\square$ if and only if, whenever $a \sim b$ and $c \sim d$, we have $a \square c \sim b \square d$.

Proof: **(a)** (only if). Let $\sim$ be a congruence relation with respect to the binary operation $\square$. Suppose $a \sim b$ and $c \sim d$, so $a \square c \sim b \square c$ and $b \square c \sim b \square d$. By transitivity of the equivalence relation $\sim$, we have $a \square c \sim b \square d$.

(b) (if) Suppose $\sim$ is an equivalence relation on T such that if $a \sim b$ and $c \sim d$, then $a \square c \sim b \square d$. We have to show $\sim$ is a congruence relation. We have $a \sim b$.

From $c \sim c$ we get $a \square c \sim b \square c$ and $c \square a \sim c \square b$. Hence $\square$ is a congruence relation.

A homomorphism h from an algebra A with carrier T to an algebra A' with carrier T' is a map from T to T', which preserves the operation of A. Any map induces a natural equivalence relation over its domain, and hence homomorphism also induces an equivalence relation over the underlying set. So $a \sim b$ if and only if $h(a) = h(b)$. We next show that the equivalence relation induced by h is in fact a congruence relation.

THEOREM 4   Let $A = (T, \square)$ be an algebra with T as the underlying set and $\square$ a binary operation in T. Let h be a homomorphism from $A = (T, \square)$ to $A' = (T', \square')$. Then the equivalence relation induced by h is a congruence relation on the algebra A.

Proof: For $a, b \in T$, $a \sim b$ if and only if $h(a) = h(b)$. In order to show that this is a congruence relation, it is enough to show that whenever $a \sim b$ and $c \sim d$, $a \square c \sim b \square d$ for all $a, b, c, d \in T$.

If $a \sim b$ and $c \sim d$,

$$h(a) = h(b) \text{ and } h(c) = h(d)$$

Hence $h(a) \square' h(c) = h(b) \square' h(d)$.

Since h is a homomorphism from $A = (T, \square)$ to $A' = (T, \square')$, $h(a \square c) = h(a) \square' h(c)$.

$$h(b \square d) = h(b) \square' h(d)$$

And it follows that

$$h(a \square c) = h(b \square d)$$

Therefore, $a \square c \sim b \square d$.

Thus $\sim$ is *a* congruence relation on T with respect to $\square$. That is, $\sim$ is a congruence relation on the algebra A.

EXAMPLE 6   Consider the algebraic system given below as a description of the interaction of six different kind of particles $\{a, b, c, d, e, f\}$. Suppose a, b, c are positively charged particles; d and e are neutral particles; and f is a negatively charged particle. If we let $\{+, 0, -\}$ denote the three kinds of particles, the table in part (b) shows how the three kinds of particles interact.

$\square$	a	b	c	d	e	f
a	a	b	c	a	c	d
b	b	c	a	b	c	e
c	a	c	a	b	c	e
d	a	b	b	d	e	f
e	c	c	c	e	e	f
f	d	e	e	f	f	f

(a)

$\square'$	$+$	0	$-$
$+$	$+$	$+$	0
0	$+$	0	$-$
$-$	0	$-$	$-$

(b)

The homomorphism

$$h(a) = h(b) = h(c) = +$$

$$h(d) = h(e) = 0$$

$$h(f) = -$$

maps the algebraic system in the table part (a) to the algebraic system given in the table part (b). ◀

Most computers represent numbers with binary sequence of fixed length. Only a finite set of numbers can be represented exactly, and "arithemetic overflow" occurs when the result of a computation is larger than any of the numbers which can be represented. Consider the following strategies for treating arithmetic overflow. For simplicity, we will treat only the natural numbers and the operation of addition. For each of the following function f, determine whether f is a homomorphism from $(N, +, 0)$ to the specified algebra $(S, \oplus,)$ where S is the set of binary sequences of length k. In each case, the operation $\oplus$ is based on binary addition and is described by means of examples. In the illustrative examples 7–8, we use $k = 3$.

EXAMPLE 7   The k bits represent the least significant digits of the k digit binary representation of each natural number. The operation $\oplus$ is the usual binary addition except that if overflow occurs, the leading digits are lost. This $f(4) = 100$, $f(5) = 101$, and $f(9) = f(4+5) = 100 \oplus 101 = 001 = f(8n+1)$ for all $n \in N$.

The function $F: N \rightarrow S$ is defined by

$$f(n) = n \bmod 2^k$$

and is a homomorphism since

$$f(a+b) = (a+b) \bmod 2^k = a \bmod 2^k \oplus a \bmod 2^k = f(a) \oplus f(b)$$

◀

EXAMPLE 8   If $n < 2^k$, then $f(n)$ is the k digit binary representation of n. If $n geq 2^k$, then $f(n)$ is represented by the k digit binary representation $2^k - 1$. Thus $f(4) = 100$, $f(f) = 101$, and $f(9) = f(4+5) = 100 \oplus 101 = 111 = f(x)$ for all $n \geq 7$.

It can be easily seen that if

$$a + b < 2^{k-1} \qquad f(a) \oplus f(b) = f(a+b)$$

$$\text{If } a + b \geq 2^{k-1} \qquad f(a) \oplus f(b) = f(a+b) = 2^k - 1$$

Hence f is a homomorphism. ◀

EXAMPLE 9 f is defined as follows:

One bit is reserved for an indication that overflow has occured (we will use 0 for n_0 overflow, 1 for overflow, and use the leftmost bit as the overflow indicator). For all numbers less than $2^{(k-1)}$, the numbers are represented in their $k - 1$ digit binary representation and the overflow is set to 0. If $n \geq 2^{(k-1)}$, the n $f(n)$ consists of the digit 1 followed by the $k - 1$ least significant digits y the binary representation of n e.g. if $= 3$ then $f(12) = 100$, $f(3) = 001$, $f(2) = 010$, and $f(312) = 011 \oplus 010 = 101 = f(4n+1)$ for all $n \in N$.

If we use $\cdot$ to dentote concatenation, the function f can be represented as

$$f(n) = 0 \cdot (n \bmod 2^{k-1}) = 0 \cdot n \text{ for } n < 2^{k-1}$$

$$= 1 \cdot (n \bmod 2^{k-1}) \text{ for } n \geq 2^{k-1}$$

If $a + b < 2^{k-1}$ then both a and b are less than 2^{k-1}.   ◀

$$f(a + b) = 0 \cdot (a + b)$$
$$= 0 \cdot a \oplus 0 \cdot b$$
$$= f(a) \oplus f(b)$$

If $a + b \geq 2^{k1-}$, then

$$f(a + b) = 1 \cdot (a + b) \bmod 2^{k-1}$$
$$= x \cdot (a \bmod 2^{k-1}) \oplus y \cdot (b \bmod 2^{k-1})$$

where $x = 0$ or $x = 1$ and $y = 0$ or $y = 1$.

In any case $f(a + b) = f(a) \oplus f(b)$ which establishes that f is a homomorphism.

Normal Subgroups

Let us now consider only groups. Given a group $G = (T, \square)$, what can we say about the homomorphic images of G? We have seen that a subgroup $H = (T', \square)$ of G induces a partition of T, which is determined by the cosets of the subgroup. Each coset is a block of the partition. This partition on T induces an equivalence relation. Is this equivalence relation a congruence relation? The answer is 'no'. But putting an additional restriction on the subgroup H will make this equivalence relation a congruence relation.

Let H be a subgroup of G. H is said to be a normal subgroup if, for any element a in G, the left coset $a \square H$ is equal to the right coset $H \square a$. It should be noted that if G is an abelian group, any subgroup of G is normal. Consider the following group G and its subgroup H.

G

$\square$	a	b	c	d	e	f
a	a	b	c	d	e	f
b	b	c	a	e	f	d
c	c	a	b	f	d	e
d	d	f	e	a	c	b
e	e	d	f	b	a	c
f	f	e	d	c	b	a

H

$\square$	a	b	c
a	a	b	c
b	b	c	a
c	c	a	b

H is a normal subgroup of G. For example,

$$e \square H = \{e \square a, e \square b, e \square c\}$$
$$= \{e, d, f\}$$
$$H \square e = \{a \square e, b \square e, c \square e\}$$
$$= \{e, f, d\}$$

We next show that the distinct left (right) cosets of a normal subgroup H are congruence classes of the underlying set of G. Let $a \square H$ and $b \square H$ be two cosets. We want to show that for all the elements a_1 in $a \square H$ and all the elements b_1 in $b \square H$, the elements $a_1 \square b_1$ are in one coset of H. Let

$$a_1 = a \square h_1$$
$$b_1 = b \square h_2$$

for some h_1 and h_2 in H.

We have

$$a_1 \,\square\, b_1 = (a \,\square\, h_1) \,\square\, (b \,\square\, h_2)$$

$$= (a \,\square\, h_1) \,\square\, (h_3 \,\square\, b) \quad \text{for some } h_3 \in H$$

$$= a \,\square\, h_1 \,\square\, h_3 \,\square\, b$$

$$= a \,\square\, h_4 \,\square\, b \qquad\qquad h_4 = h_1 \,\square\, h_3 \in H$$

$$= a \,\square\, b \,\square\, h_5 \qquad\qquad \text{for some } h_5 \in H$$

Thus $a_1 \,\square\, b_1$ is in the coset $(a \,\square\, b) \,\square\, H$.

For example, for the group $G = (\{a, b, c, d, e, f\}, \square)$ and the subgroup $H = (\{a, b, c\}, \square)$, the congruence classes are $\{a, b, c\}$ and $\{d, e, f\}$. Consequently we have the homomorphic image

$\square'$	$\{a, b, c\}$	$\{d, e, f\}$
$\{a, b, c\}$	$\{a, b, c\}$	$\{d, e, f\}$
$\{d, e, f\}$	$\{d, e, f\}$	$\{a, b, c\}$

Next we show that exhausting all normal subgroups of $(T, \square)$ would have exhausted all homomorphic images of $(T, \square)$.

Let f be a homomorphism from $(T, \square)$ to $(S, *)$. We want to show that f corresponds to a partition of T into congruence classes induced by a normal subgroup.

Let H be all the elements in T whose images under f are the identity of S, which we denote as e^*. This is called the kernel of the homomorphism.

We first show that $(H, \square)$ is a subgroup of $(T, \square)$.

1. Closure: To show $\square$ is closed on H.

   For any a and b in H, we have

   $$f(a \,\square\, b) = f(a) * f(b) = e^* \cdot e^* = e^*$$

   Hence $a \,\square\, b$ is in H.

2. We want to show that e, the identity of T is in H. For arbitrary element a in T, we have

   $$f(a \,\square\, e) = f(a) * f(e) \quad \text{or}$$

   $$f(a) = f(a) * f(e)$$

   Since $(S, *)$ is a group, $f(e)$ must be the identity of $(S, *)$, i.e., e^*. Consequently e is in H.

3. We want to show that a^{-1} is in H for any element a in H.

   Since $\qquad\qquad\qquad\qquad f(a \,\square\, a^{-1}) = f(a) * f(a^{-1})$

   or $\qquad\qquad\qquad\qquad\qquad f(e) = f(a) * f(a^{-1})$

   or $\qquad\qquad\qquad\qquad\qquad\; e^* = e^* * f(a^{-1})$

   or $\qquad\qquad\qquad\qquad\qquad f(a^{-1}) = e^*.$

Hence a^{-1} is in H.

Next we show that $(H, \square)$ is a normal subgroup. For any a in T and h in H.

$$f(a \square h \square a^{-1}) = f(a) * f(h) * f(a^{-1})$$
$$= f(a) * e^* * f(a^{-1})$$
$$= f(a) * f(a^{-1})$$
$$= f(a * a^{-1})$$
$$= f(e)$$
$$= e^*$$

Thus $a \square h \square a^{-1}$ is in H.

Let $a \square h \square a^{-1} = h_1$.

Then $a \square h = h_1 \square a$ for some h_1 in H. Hence every element of the right coset of H with respect to some element a is also in the left coset of H with respect to a.

Finally, we show that if a and b are in the same coset of H, then $f(a) = f(b)$. Since b can be written as $b = a \square h$ for some h in H, we have

$$f(b) = f(a) * f(\lambda) = f(a) \text{ as } f(h) = e$$

Conversely, if $f(a) = f(b)$, a and b are in the same coset of H.

Since $f(a^{-1} \square b) = f(a^{-1}) * f(b)$
$$= f(a^{-1}) * f(a)$$
$$= f(a^{-1} \square a)$$
$$= f(e)$$
$$= e*$$

$a^{-1} \square b$ is in H. That is, $a^{-1} \square b = h$ for some h in H, or $b = a \square h$ for some h in H. So a and b are in the same coset of H.

Note that $a = a \square e$ and $b = a \square h, e \in H$. So a and b are in the same coset $a \square H$.

Thus we find that f corresponds to a partition of T into congruence classes induced by the normal subgroup $(H, \square)$.

Exercises

1. Show that if $g: A \to B$ is a homomorphism of an algebraic system $(A, *)$ onto $(B, \square)$ and $(A_1, *)$ is a subalgebra of $(A, *)$, then the image of A_1 under g is a subalgebra of $(B, \square)$.

2. Show that the intersection of any two congruence relations on a set is also a congruence relation.

3. Show that the composition of two congruence relations on a set is not necessarily a congruence relation.

4. If $f: S \to T$ is a homomorphism from $(S, *)$ to (T, Δ) and $g: T \to P$ is also a homomorphism from (T, Δ) to (P, ∇), then $g \circ f: S \to P$ is a homomorphism from $(S, *)$ to (P, ∇).

5. Let $g: S \to T$ be an isomorphism of semigroups $(S, *)$ and (T, Δ). Show that, if z is a zero of S, then $g(z)$ must be a zero of (T, Δ).

6. Show that every monoid $(M, *, e)$ is isomorphic to a submonoid of $(M^M, \circ, \Delta)$ where Δ is the identity mapping of M and M^M is the set of all function from M to M.

7. Let f_1 and f_2 be homomorphisms from an algebraic system $(A, \blacklozenge)$ to another algebraic system $(B, *)$. Let g be a function from A to B such that

$$g(a) = f_1(a) * f_2(a)$$

for all a in A. Show that g is a homomorphism from $(A, \blacklozenge)$ to $(B, *)$ if $(B, *)$ is a commutative semigroup.

8. Let f and g be homomorphisms from a group $(G, \blacklozenge)$ to a group $(H, *)$. Show that $(C, \blacklozenge)$ is a subgroup of $(G, \blacklozenge)$, where

$$C = \{x \in G \,|\, f(x) = g(x)\}$$

12.4   Rings, Integral Domains, and Fields

Semigroups, monoids, and groups are algebraic systems with one binary operation. We now consider algebraic systems with two binary operations. Consider the set $S = \{a, b\}$ and two binary operations $\square$ and $*$ on S given by the following tables.

$\square$	a	b		$*$	a	b
a	a	b		a	a	a
b	b	a		b	a	b

$(S, \square)$ is a group and $(S, *)$ is a semigroup. Also for any α, β, γ in S,

$$\alpha * (\beta \square \gamma) = (\alpha * \beta) \square (\alpha * \gamma)$$

and $(\beta \square \gamma) * \alpha = (\beta * \alpha) \square (\gamma * \alpha)$

In this case, we say the operation $*$ is distributive the operation $\square$. In the above example, note that

$$b \square (a * b) = b \square a = b$$

and $(b \square a) * (b \square b) = b * a = a$

Hence $\square$ is not distributive over $*$. Now we consider the definitions of rings, integral domains and fields.

For convenience, we denote the two binary operation as $+$ and $\cdot$, respectively.

DEFINITION 1   An algebraic system $(S, +, \cdot)$ is called a ring if the following conditions are satisfied

1. $(S, +)$ is an abelian group.
2. $(S, \cdot)$ is a semigroup.
3. The operation $\cdot$ is distributive over $+$.

EXAMPLE 1   Let Z_n be the set of integers $\{0, 1, 2, \ldots, n - 1\}$. Let $\oplus$ denote the binary operation of mod n addition, That is,

$$a \oplus b = \begin{cases} a + b & \text{if } a + b < n \\ a + b - n & \text{if } a + b \geq n \end{cases}$$

Let $\odot$ denote mod n multiplication, that is,
$a \odot b =$ the remainder of ab divided by n.
    We have earlier seen that $(Z_n, \oplus)$ is an abelian group with 0 as identity and $n - i$ is the inverse of i. To see that $(Z_n, \odot)$ is a semigroup, we have to show that

1. Z_n is closed under.
2. $\odot$ is an associative operation.

It is easy to see that Z_n is closed under $\odot$. To see that $\odot$ is an associative operation, consider $a, b, c \in Z_n$.

$a \odot b =$ remainder when ab is divided by n
That is, $ab = kn + \alpha$ where $a \odot b = \alpha$, so
$(a \odot b) \odot c = \alpha \odot c =$ remainder when αc is divided by n
That is, $\alpha c = \ell n + \beta \qquad \alpha \odot c = \beta \quad 0 \le \beta < n$
$abc = (kn + \alpha)c = knc + \alpha c = knc + \ell n + b = n(kc + \ell) + \beta$
$\beta =$ remainder when abc is divided by n
$bc = k'n + \gamma$ where $b \odot c = \gamma$

Let $a\gamma = \ell'n + \delta$ so that $a \odot (b \odot c) = \delta \, 0 \le \delta < n$

$$abc = a(k'n + \gamma)$$
$$= ak'n + a\gamma$$
$$= ak'n + \ell'n + \delta$$
$$= n(ak' + \ell') + \delta$$

Hence we note $\beta = \delta$, that is, $(a \odot b) \odot c = a \odot (b \odot c)$
Hence $(Z_n, \odot)$ is a semigroup.
Let us now show that $\odot$ distributes over $\oplus$. This can also be checked easily.
That is, $a \odot (b \oplus c) = (a \odot b) \oplus (a \odot c), 0 \le a, b, c < n$

L.H.S.: $b \oplus c = b + c$
or $b + c - n$
$a \odot (b \oplus c) = a(b + c) \bmod n$
or $a(b + c - n) \bmod n$
$\qquad = a(b + c) \bmod n$
$\qquad = (ab + ac) \bmod n$
$\qquad = ab \bmod n + ac \bmod n$
or $(ab \bmod n + ac \bmod n) - n$

R.H.S.: $a \odot b = ab \bmod n = r$ (say)

$a \odot c = ac \bmod n = s$ (say)

rhs $= (r + s) \bmod n = r + s$

or $r + s - n$

$\qquad = (ab \bmod n + ac \bmod n)$

or $(ab \bmod n + ac \bmod n) - n$

Hence lhs = rhs. ◄

We shall hereafter call in this section the two operations $+$ and $\cdot$ as addition and multiplication respectively. We shall refer to $a + b$ as the sum and $a \cdot b$ as the product of a and b. The identity of the group $(S, +)$ will be called the additive identity and denoted by 0. The inverse of an element a in this group will be called additive inverse and denoted by $-a$. $a - b$ denotes the sum of a and the additive inverse of b.

It should be noted that for any element a in a ring $(S, +, \cdot), 0 \cdot a = a. 0 = 0$.
This follows from the fact that

$$0 \cdot a = (0 + 0) \cdot a = 0 \cdot a + 0 \cdot a$$

Hence $0 \cdot a = 0$. Similarly, it can be seen that $a \cdot 0 = 0$.

DEFINITION 2   Let $(S, +, \cdot)$ be an algebraic system with two binary operations. $< S, +, \cdot >$ is called an integral domain if

1. $(S, +)$ is an abelian group.
2. The operation $\cdot$ is commutative. Furthermore, if $c \neq 0$ and $c \cdot a = c \cdot b$, then $a = b$, where 0 is the additive identity.
3. The operation $\cdot$ is distributive over the operation $+$.

EXAMPLE 2   Consider the set of integers I with the operation of addition and multiplication. $(I, +)$ is a group. 0 is the identity of this group. $-a$ is the inverse of a for any integer a. Property 2 is also satisfied if $c \cdot a = c \cdot b$ and $c \neq 0$, dividing by c we get $a = b$. Multiplication of integers is distributive over addition of integers. Hence it follows that $(I, +, \cdot)$ is an integral domain.   ◀

DEFINITION 3   Let $(S, +, \cdot)$ be an algebraic system with two binary operations. $< S, +, \cdot >$ is called a field if:

1. $(S, +)$ is an abelian group.
2. $(S - \{0\}, \cdot)$ is an abelian group where 0 is the additive identity.
3. The operation $\cdot$ is distributive over the operation $+$.

EXAMPLE 3 $(Q, +, \cdot)$ and $(R, +, \cdot)$ are fields where Q is the set of rational numbers and R is the set of real numbers. $+$ and $\cdot$ are ordinary addition and multiplication respectively. 0 is the additive identity and 1 is the identity for multiplication. The properties can be easily checked. $(C, +, \cdot)$ is a field where C is the set of complex numbers, $+$ and $\cdot$ are addition and multiplication of complex numbers respectively. All the conditions can be checked. Here again 0 is the additive identity and 1 the identity for multiplication.   ◀

EXAMPLE 4   The algebraic system $(Z_n, \oplus, \odot)$ is a field if and only if n is a prime. It is clear that both $\oplus$ and $\odot$ are commutative operations. We have seen that $(Z_n, \oplus)$ is a group for any n. We shall now show that $(Z_n - \{0\}, \odot)$ is a group if and only if n is a prime. If n is not a prime, $n = ab$ for some a, b in $Z_n - \{0\}$. That is, $a \odot b = 0$. Since $\odot$ is not closed on the set $Z_n = \{0\}$, $(Z_n = \{0\}, \odot)$ cannot be a group. Consider the case when n is a prime. For any a and b in $Z_n = \{0\}$, $a \odot b$ is not equal to 0 and has a value in $Z_n = \{0\}$. So $Z_n = \{0\}$ is closed under $\odot$. It is straightforward to see $\odot$ is an associative operation and 1 is the identity for the operation. In a group table, we have seen that no two elements of a row are equal and no two elements of a column are equal. In $(Z_n, \oplus, \odot)$, we can show that for distinct b and c in $Z_n = \{0\}$. Here $a \odot b \neq a \odot c$ and $b \odot a \neq c \odot a$.

This is seen as follows,
Suppose $a \odot b = a \odot c$, so

$$ab = k_1 n + r$$

$$ac = k_2 n + r \qquad \text{for some } k_1, k_2, r$$

with $1 \leq r < n$. Assume $b > c$, and hence $k_1 > k_2$. So

$$ab - ac < (k_1 - k_2)n \qquad (12.1)$$

Since both a and $b - c$ are less than n, (1) is impossible when n is a prime.

Thus, we have a group table with column heading $1, \ldots, n - 1$ and row headings $1, \ldots, n - 1$ for the operation $\odot$. Since all elements of a row are different, 1 occurs in each row.

So for every a in $Z_n = \{0\}$, there is an element b in $Z_n = \{0\}$ such that $a \odot b = 1$. Here b is the inverse of a. Hence $(Z_n = \{0\}, \odot)$ is an abelian group. We have already seen $\odot$ is distributive over $\oplus$. So $(Z_n, \oplus, \odot)$ is a field when n is a prime. This is called as the field of integer modulo n.   ◀

Links

If p is prime, then $\mathbf{Z}_p$ is a field. i.e. $(\mathbf{Z}_p, \oplus, \odot)$ is a field. This is a finite field. $(\mathbf{Q}, +, .)$, $(\mathbf{R}, +, .), (\mathbf{C}, +, .)$ are examples of infinite fields. It is not necessary that the number of elements in a finite field should be prime. Denoting the additive identity by 0 and multiplicative identity by 1, a set $\{0, 1, a, b\}$ with two more elements and operations $+$ and . defined by the following tables forms a field. Note that the number of elements is four which is not a prime.

+	0	1	a	b
0	0	1	a	b
1	1	0	b	a
a	a	b	0	1
b	b	a	1	0

·	0	1	a	b
0	0	0	0	0
1	0	1	a	b
a	0	a	b	1
b	0	b	1	a

Ring Homomorphisms and Ideals

We have seen the definition of rings and fields in the previous section. Rings are of many types. Many rings are commutative in the sense that $a \cdot b = b \cdot a$. For all a in S where $(S, +, \cdot)$ is the ring, there exist noncommutative rings as well. Let $R = (S, +, \cdot)$ be a ring. Let $M_2(R)$ be the set of all 2×2 matrices $\begin{pmatrix} a & b \\ c & d \end{pmatrix}$ where $a, b, c, d \in S$.

These two by two matrices under the operation of matrix addition and matrix multiplication form a noncommutative ring. This can be checked in a straightforward manner.

We also note that the additive identity 0 of any ring also serves as the 'zero' for multiplication. It should be noted that if $a, b \in S$ where $(S, +, \cdot)$ is a ring $(-a)$ and $(-b)$ are the additive inverses of a and b, respectively, then $(-a)(-b) = ab$. This can be seen as

$$a + (-a) = 0$$
$$0 = 0 \cdot (-b) = (a + (-a)) \cdot (-b)$$
$$= a(-b) + (-a)(-b)$$

On the other hand

$$a(-b) + ab = a((-b) + b) = a \cdot 0 = 0$$

Hence $a(-b) + ab = a(-b) + (-a)(-b)$.
Cancelling $a(-b)$ on both sides,

$$ab = (-a)(-b)$$

DEFINITION 4   Let $\mathcal{R} = (A, +, \cdot)$ be a ring and $S = (A', +, \cdot)$ be an algebraic system with $A' \le A$. Here S is a subring of $\mathcal{R}$ if S is a ring by itself under $+$ and $\cdot$.

DEFINITION 5   A homomorphism of a ring $(A, +, \cdot)$ to $(B, +', \cdot')$ is a surjective function $\theta: A \to B$ such that for any x and y in A

$$\theta(x + y) = \theta(x) +' \theta(y)$$

$$\theta(x \cdot y) = \theta(x) \cdot' \theta(y)$$

Here $(B, +', \cdot')$ is called the homomorphic image of $(A, +, \cdot)$.

EXAMPLE 5   Let us consider $\mathcal{R} = (I, +, \cdot)$ where I is the set of integers and $+$ and $\cdot$ are ordinary addition and multiplication. Consider the following ring S over $\{even, odd\}$ and two operations $+'$ and $\cdot'$ defined by the following table.

$+'$	even	odd		$\cdot'$	even	odd
even	even	odd		even	even	even
odd	odd	even		odd	even	odd

$\mathcal{R}$ and S are rings.
Consider the mapping

$$\theta(n) = \begin{cases} even & \text{if } n \text{ is even} \\ odd & \text{if } n \text{ is odd} \end{cases}$$

Then

$$\theta(x + y) = \theta(x) +' \theta(y)$$

$$\theta(x \cdot y) = \theta(x) \cdot \theta(y)$$

Hence θ is a homomorphism from $\mathcal{R}$ to S.   ◄

We can also look at homomorphism in another way. Let R be a congruence relation on A with respect to both the operations $+$ and $\cdot$. That is, R is an equivalence relation on A and if $a_1 R a_2$ and $b_1 R b_2$, then $(a_1 + b_1) R (a_2 + b_2)$ and $(a_1 \cdot b_1) R (a_2 \cdot b_2)$. Let $B = \{c_1, \ldots, c_r\}$ be the set of congruence classes into which A is divided. We can define two binary operations $+'$ and $\cdot'$ on B such that $c_i +' c_j$ is equal to the congruence class to which $a_1 + a_2$ belongs to where $a_1 \in c_i$ and $a_2 \in c_j$. $c_i \cdot' c_j$ is equal to the congruence class to which $a_1 \cdot a_2$ belongs. It can be easily checked that $(B, +', \cdot')$ is a homomorphic image of $(A, +, \cdot)$.

We have seen earlier that, exhausting all the normal subgroups of a group, we get all homomorphic images of that group. Now we explore the possibility of a similar result for rings.

Let $(A, +, \cdot)$ be a ring and $(H, +)$ be a subgroup of $(A, +)$. Let R denote the equivalence relation that H induces on A. Because of the commutative property of $(A, +)$, R is a congruence relation on A with respect to 1. We want to find out under what condition R is also a congruence relation with respect to $\cdot$. Suppose aRb and cRd, so

$$a = b + h_1$$

and

$$c = d + h_2$$

for some h_1 and h_2 in H, because all left cosets of H exhaust A. It follows that

$$a \cdot c = (b + h_1) \cdot (d + h_2)$$

$$= (b \cdot d) + (b \cdot h_2) + (h_1 \cdot d) + (h_1 \cdot h_2)$$

We note that if both $b \cdot h_2$ and $h_1 \cdot d$ are in H, then

$$(b \cdot h_2) + (h_1 \cdot d) + (h_1 \cdot h_2)$$

is also in H and we shall have

$$(a \cdot c) R (b \cdot d)$$

This observation motivates the definition of an ideal in a ring (somewhat similar concept like normal subgroups). Let $(A, +, \cdot)$ be a ring and H be a subset of A. Here H is said to be an ideal if the following two conditions are satisfied.

1. $(H, +)$ is a subgroup of $(A, +)$.
2. For any a in A and h in H, $a \cdot h$ and $h \cdot a$ are both in H. We can say that an element a is swallowed or absorbed into the ideal when multiplied by an element of the ideal.

From the discussion we had, it follows immediately that an ideal H of a ring $(A, +, \cdot)$ induces a congruence relation on A with respect to the operation $+$ and $\cdot$.

On the other hand, can we say every homomorphism corresponds to an ideal of the ring? Let f be a homomorphism from a ring $(A, +, \cdot)$ to $(B, +', \cdot')$. Let H be the set of elements of A that are mapped into additive identity of $(B, +', \cdot')$ under f. We see that $(H, +)$ is a normal subgroup of $(A, +)$. To show that $(H, +', \cdot')$ is an ideal. We note that for any a in A and h in H.

$$f(a \cdot h) = f(a) \cdot' f(h) = f(a) \cdot' 0 = 0$$

where 0 is the additive identity of $(B, +', \cdot')$. Thus $a \cdot h$ is in H. In a similar manner, we can show that $h \cdot a$ is in H. Hence H is an ideal.

EXAMPLE 6 Consider the ring over the set $\{0, 1, \ldots, 7\}$ with $\oplus$ mod 8 addition $\odot$ mod 8 multiplication. The tables for the operations are given below.

$\oplus$	0	1	2	3	4	5	6	7
0	0	1	2	3	4	5	6	7
1	1	2	3	4	5	6	7	0
2	2	3	4	5	6	7	0	1
3	3	4	5	6	7	0	1	2
4	4	5	6	7	0	1	2	3
5	5	6	7	0	1	2	3	4
6	6	7	0	1	2	3	4	5
7	7	0	1	2	3	4	5	6

$\odot$	0	1	2	3	4	5	6	7
0	0	0	0	0	0	0	0	0
1	0	1	2	3	4	5	6	7
2	0	2	4	6	0	2	4	6
3	0	3	6	1	4	7	2	5
4	0	4	0	4	0	4	0	4
5	0	5	2	7	4	1	6	3
6	0	6	4	2	0	6	4	2
7	0	7	6	5	4	3	2	1

For this ring $\{0, 2, 4, 6\}$ is an ideal as $(\{0, 2, 4, 6\}, \oplus)$ is a normal subgroup of $(Z_8, \oplus)$. For any h in $\{0, 2, 4, 6\}$ and a in $\{0, 1, 2, 3, 4, 5, 6, 7\}$, $a \cdot h$ and $h \cdot a$ are in $\{0, 2, 4, 6\}$. The two congruence classes are $\{0, 2, 4, 6\}$ and $\{1, 3, 5, 7\}$, and if we denote them as J_1 and J_2, the homomorphic image is shown in the table.

$+'$	J_1	J_2
J_1	J_1	J_2
J_2	J_2	J_1

$\cdot'$	J_1	J_2
J_1	J_1	J_1
J_2	J_1	J_2

Here $(\{0, 4\}, \oplus)$ is a subgroup of $(Z_8, \oplus)$. This is also an ideal, as for any $h \in \{0, 4\}$ and $a \in \{0, \ldots, 7\}$, $a \cdot h$ and $h \cdot a$ are in $\{0, 4\}$. The corresponding congruence classes are $\{0, 4\}$, $\{1, 5\}$, $\{2, 6\}$ and $\{3, 7\}$. Denoting them as C_1, C_2, C_3, C_4 respectively the homomorphic image is given below.

$+'$	C_1	C_2	C_3	C_4
C_1	C_1	C_2	C_3	C_4
C_2	C_2	C_3	C_4	C_1
C_3	C_3	C_4	C_1	C_2
C_4	C_4	C_1	C_2	C_3

$\cdot'$	C_1	C_2	C_3	C_4
C_1	C_1	C_1	C_1	C_1
C_2	C_1	C_2	C_3	C_4
C_3	C_1	C_3	C_1	C_3
C_4	C_1	C_4	C_3	C_2

◀

Exercises

1. Let $(T, *)$ be a subsemigroup of $(S, *)$. T is called a left ideal of S if $S * T \subseteq T$. Similarly define a right ideal. If T is both a left and right ideal, then it is called an ideal. Show that in $(I, \cdot)$, where I is the set of integers under multiplication $\cdot$, the set of multiples of an integer n is an ideal.

2. Let $(A, \blacklozenge)$ and $(B, *)$ be two algebraic systems and f be a homomorphism from $(A, \blacklozenge)$ to $(B, *)$.

   a) Show that, if $\blacklozenge$ is an associative operation, so is $*$.

   b) Show that, if e is an identity element in $(A, \blacklozenge)$, then $f(e)$ is an identity element $n(B, *)$.

   c) Show that, if b is an inverse of a in $(A, \blacklozenge)$, then $f(b)$ is an inverse of $f(a)$ in $(B, *)$.

   d) Show that a homomorphic image of a ring is a ring.

3. Let $(A, \blacklozenge, *)$ be an algebraic system, where

   $$a \blacklozenge b = a$$

   for all a, b in A and $*$ is an arbitrary binary operation. Show that $*$ distributes over $\blacklozenge$.

4. Let $(A, \blacklozenge, *)$ be an algebraic system with e_1 and e_2 being the identity elements with respect to the operations $\blacklozenge$ and $*$, respectively. Given that the operations $\blacklozenge$ and $*$ distribute over each other show that $x \blacklozenge x = x$ and $x * x = x$ for all x in A.

5. Let $(A, +, \cdot)$ be a ring such that $a \cdot a = a$ for all a in A.

   a) Show that $a + a = 0$ for all a, where 0 is the additive identity.

   b) Show that the operation $\cdot$ is commutative.

6. A ring $(A, +, \cdot)$ is said to be commutative ring with unity if $(A, \cdot)$ is a commutative monoid.

   a) An ideal H of a ring is said to be a prime ideal if, for any two elements a and b that are not in H, $a \cdot b$ is not in H either. Show that the homomorphic image of a commutative ring with unity induced by an ideal is an integral domain if and only if the ideal is prime.

   b) An ideal H is said to be a maximal ideal if the only ideals of the ring that contain H are H and the ring itself. Show that the homomorphic image of a commutative ring with unity induced by an ideal is a field if and only if the ideal is maximal.

12.5   Quotient and Product Algebras

There are several ways of combining algebras to build new ones. Two such techniques are forming the quotient and products.

Quotient Algebras

Let us consider a set S and an equivalence relation $\sim$ on S. Here $[a]$ denotes the equivalence class to which a belongs for $a \in S$. Let us restrict ourselves to algebras with one binary operation $\square$, like semigroups, monoids, and groups. The concept can be extended to other algebras with more operations.

DEFINITION 1   Let $A = (T, \Box, k)$ be an algebra with T as the underlying set and $\Box$ as the binary operation on T. Here k denotes the constant(s) which have special properties. The third component can be ignored if no such constant exists. Let the equivalence relation $\sim$ be a congruence relation on A. The quotient algebra of A with respect to the relation $\sim$, denoted by $A/\sim$, is the algebra

$$(T/\sim, \Box', [k])$$

where

(*i*) $T/\sim$ is the quotient set of T under the relation $\sim$. That is, the elements of $T/\sim$ are equivalence classes of $\sim$.

(*ii*) For $[a]$ and $[b]$ in $T/\sim$,

$$[a] \Box' [b] = [a \Box b]$$

(*iii*) $[k]$ is the equivalence class to which k belongs.

We shall show that if $(T, \Box)$ is a semigroup, then $(T/\sim, \Box')$ is a semigroup. Closure follows from the definition. To see the associative property, we note that

$$[a] \Box' ([b] \Box' [c]) = [a] \Box' [b \Box c] = [a \Box (b \Box c)]$$

$$([a] \Box' [b]) \Box' [c] = [a \Box b] \Box' [c] = [(a \Box b) \Box c]$$

Since

$$a \Box (b \Box c) = (a \Box b) \Box c$$

we have $[a] \Box' ([b] \Box' [c]) = ([a] \Box' [b]) \Box' [c]$ and the associative property holds. Hence $(T/\sim, \Box')$ is a semigroup.

We also see that if $G = (T, \Box, e)$ is a group with identity e, then $G' = (T/\sim, \Box', [e])$ is a group. Closure, associativity follow as before. Since $[a] \Box' [e] = [a \Box e] = [a]$,

$[e]$ is the identity of $(T/\sim, \Box', [e])$.

Let a^{-1} be the inverse of a in G.

Then $[a^{-1}]$ is the inverse of $[a]$ in G', as $[a^{-1}] \Box' [a] = [a^{-1} \Box a] = [e]$.

Hence if G is a group, G' is a group. Similar results will hold for monoids also.

EXAMPLE 1   Let $\sum = \{a, b\}$. Consider the semigroup $(\sum^*, concatenation)$. Let $\sim$ be defined as $x \sim y$ if x and y have the same number of a's. Then $(\sum^* / \sim, \Box')$ is a semigroup where $[x] \Box' [y] = [xy]$.

We see that if x, y are strings in $\sum^*$ with m and n a's respectively, then xy will belong to the equivalence class containing strings having $m + n$ a's. Hence it follows that $[x] \Box' [y] = [xy]$.   ◀

EXAMPLE 2   We have seen that, if $G = (T, \Box)$ is a group and $H = (H, \Box)$ is a normal subgroup of G, then the cosets of H form congruence classes of T and $G' = (T, \Box')$, where T' is the set of congruence classes and $\Box'$ is defined as $[a] \Box' [b] = [a \Box b]$. G' is a group and a homomorphic image of G.   ◀

Product Algebras

Another way of constructing new algebras from old ones is by taking a 'direct product' of two algebras with the same signature.

DEFINITION 2   Let $A_1 = (T_1, \square_1)$ and $A_2 = (T_2, \square_2)$ be two algebras and k_1 and k_2 be their respective constants. The direct product of A_1 with A_2 is the algebra $A = A_1 \times A_2 = (T_1 \times T_2, \square)$ with constant $< k_1, k_2 >$, where $\square$ is defined as $< a, c > \square < b, d > = < a \square_1 b, c \square_2 d >$. A is called the product algebra of A_1 and A_2.

     If A_1 and A_2 are semigroups, it is not difficult to see $A_1 \times A_2$ is a semigroup. Similar results hold for monoids and groups.

EXAMPLE 3   Let $A_1 = (N, +, 0)$ and $A_2 = (N, +, 0)$. Then $A_1 \times A_2$ is $(N^2, +, < 0, 0 >)$. A_1 and A_2 are monoids and $A_1 \times A_2$ is also a monoid.   ◀

EXAMPLE 4   Let $A_1 = (Z_p, \oplus_p, 0)$ and $A_2 = (Z_q, \oplus_q, 0)$, when $\oplus_p$ is mod p addition. Then A_1 and A_2 are groups. $A = A_1 \times A_2 = (Z_p \times Z_q, \oplus, < 0, 0 >)$ is the product algebra. It can be seen that this is a group. Closure, associativity, and identity properties are easy to check. Inverse of $< a, b >$ is $< p - a, q - b >$.   ◀

Links

     We have considered quotient and product algebras in this section. These topics comes under what is known as **universal algebra**. It is possible to extend the concepts in many interesting and important ways. For example, **relational algebras** permit relations in the carrier to occur in the signature of the algebra. Another possibility would be to relax the requirement that the operations of an algebra be defined for all possible operands. It can be noted that as per our definition $(\mathbf{R}, /)$ is not an algebra as division by 0 is not defined. Permitting operations to be defined only for some of the possible operands gives another kind of mathematical structure called a **partial algebra**. Another extension is to **many-sorted algebra**. This is a mathematical structure in which elements from various sets (rather than a single carrier) can occur as operands, and not all operations be defined for all operands. Thus we could use one set to represent the integers, another set to represent the floating point numbers, and a third set to represent truth values. In such an algebra, arithmetic operations are defined on a set of numbers; Boolean operations on truth values. There can be operations which map elements of one set to another. Ceil and floor operations can transform real numbers to integers. Operators like $=, >, <$ can operate on real numbers to give boolean values. Such concepts have applications in programming languages. An operation like multiplication can be used on all sets and can be used carefully. This concept is similar to operator overloading in object oriented programming. Relational algebra has application in Database Management Systems.

Exercises

1. Let $S_k = \{x | x \in I \wedge x \geq k\}$, where $k \in N$. Let m and n be elements of N such that $nk \geq m$, and let h be the following homomorphism from $A = (S_m, +)$ to $A' = (S_k, +)$:

$$h: S_k \to S_m \qquad h(x) = nx$$

    Let $\sim$ be the congruence relation on A induced by h. Describe the quotient algebra $A/\sim$.

2. Let h be a homomorphism from $A = (S, O, C)$ to $A' = (S', O', C')$, and let $\sim$ be the equivalence relation induced on S by h:

$$x \sim y \Leftrightarrow h(x) = h(y)$$

    Show that $A/\sim$ is isomorphic to the subalgebra $(h(S), O', C')$ of A'.

3. Let $A = (S', o', e')$ and $A' = (S, o, e)$ be monoids. Show that the product algebra $A \times A'$ is a monoid.

4. Let $A = (\{1, 2, 3\}, \max, 1)$ and $A' = (\{5, 6\}, \min, 6)$. Specify the product algebra $A \times A'$ by constructing an operation table and identifying the constants.

5. Let A and A' be algebras with non-empty carriers and define the relation $\sim$ over a product algebra $A \times A'$ as

$$(w, x) \sim (y, z) \Leftrightarrow w = y$$

    a) Determine when $\sim$ is a congruence relation on $A \times A'$.

    b) Show that, if the relation $\sim$ defined above is a congruence relation, $(A \times A')/\sim$ is isomorphic to A.

6. Let $A_j = (N_j, +_j, 0)$ where $N_j = \{0, 1, 2, \ldots, j-1\}$ and $+_j$ denotes addition mod j.

    a) Show that $A_2 \times A_3$ is isomorphic to A_6.

    b) Describe the set of congruence relations on $A_2 \times A_3$.

    c) Describe the set of congruence relations on A_m, where m is a positive integer.

12.6 Coding Theory

Data communication consists of the transmission of characters from some finite alphabet through some communications channel. Whatever may be the communications channel, be it wires or cables or satellite communication systems, imperfections in the channel will cause a finite probability of error that a transmitted character will be incorrectly received by the receiver. For long messages, even a small error is not tolerable. Generally the characters are encoded in binary as the bits 0 and 1 and the symbols transmitted. Also when bits are transferred from one unit of a system (like memory) to another unit (CPU), errors may creep in and there should be a way of detecting and correcting errors.

Coding theory deals with the ways for improving the reliability of information transformation by systematic codes of various kinds. Most of these efficient codes are group codes, which are based on Lagrange's theorem.

The basic idea used in this is that of encoding and decoding. A sequence of characters to be transmitted is mapped into a longer sequence of the same characters by an encoding scheme. By inspecting the additional information in the extra characters, the receiver is then able to detect/correct errors which have been introduced by noise during transmission. The longer message received is then transferred to a sequence of characters of the original length by a proper method which is called decoding. This process can be pictorially represented as follows.

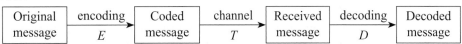

| Original message | encoding E | Coded message | channel T | Received message | decoding D | Decoded message |

DEFINITION 1   An (m, n)—code for a binary message consists of an encoding scheme $E: B^m \to B^n$ and a decoding scheme $D: B^n \to B^m$ $(m < n)$, which tries to make the transmission of m bit binary words error-free. Here B^n denotes the set of all binary strings of length n.

EXAMPLE 1   A simple scheme is to add an extra parity bit and have $E: B^m \to B^{m+1}$. If x is the m-bit original word and y is the $m + 1$ bit encoded word, the $(m + 1)$th bit added is 0 or 1 to make the number of 1's in the encoded word an even number. If $m = 4$ and $x = 0100$, y is 01001. For decoding, the last bit is omitted. If $y = 01100$, x is 0110. If the received word has odd number of 1's, error has occurred during transmission. ◀

 Links

RICHARD HAMMING (1915–1998)   Richard Wesley Hamming was born in Chicago, U.S. on February 11, 1915. This American mathematician's work had major impact in the field of telecommunications and computer science. The Hamming code which is used for error correction is due to him.

He received his bachelor's degree from the University of Chicago in 1937, a master's degree from the University of Nebraska in 1939, and PhD from University of Illinois at Urbana-Champaign in 1942. He was a professor at University of Louisville and later joined Bell Telephone Laboratories where he collaborated with C.E. Shannon. He worked on the Manhattan Project in 1945, where it was shown that the detonation of an atomic bomb would not ignite the atmosphere. This was instrumental for the U.S. to test the atomic bomb and later use it against Japan. He was also adjunct professor in the City College of New York and the Naval Postgraduate School. He was the recipient of the Turing award in 1968. He was also awarded the IEEE Emmanuel R. Piore award and Harold Pendere awards. He was a fellow of both the IEEE and ACM. For the ACM, he was a founder and president. His philosophy on scientific computing appears as preface to his 1962 book on numeric methods. He died in 1998.

See Examples 4–6 and Exercise 9–23 of Section 4.5.

EXAMPLE 2   Consider an encoding scheme $e: B^m \to B^{5m}$.

$$e(a_1 \ldots a_m) = a_1 \ldots a_m a_1 \ldots a_m a_1 \ldots a_m a_1 \ldots a_m a_1 \ldots a_m.$$

If $m = 2$, $e(10) = 1010101010$.

The corresponding decoding scheme is defined as follows:

$$d(b_1 \ldots \ldots b_{5m}) = a_1 \ldots a_m$$

$$a_i = 1 \text{ if 3 or more of } b_i, b_{i+m}, b_{i+2m}, b_{i+3m}, b_{i+4m} \text{ are 1's}$$

$$= 0 \text{ otherwise.}$$

Suppose, while transmitting 1010101010, the message received is 1011101010, the 4th bit is in error and while decoding it, since out of the bits 2, 4, 5, 6, and 10 only one is 1, the encoded character 2 is taken as 0.

Hence $d(1011101010) = 10$.

If a double error occurs also, it can be corrected.

If instead of receiving 1010101010, 1011011010 is received. Out of bits, 1, 3, 5, 7, and 9, four are 1's, and hence the encoded character is taken as 1.

Out of bits 2, 4, 6, 8, and 10, only one is a 1. Hence the encoded character is taken as 0. So $d(1011011010) = 10$.   ◄

The input is a binary sequence. It is divided into blocks of size m, and each block of size m is encoded as a binary string of length n and transmitted. Each such binary string used (of length n) is called a codeword. A coding scheme where every m bit string is encoded as a n bit string for the same n is called block code. Block codes are used to detect and correct errors (i.e., find the bits transmitted incorrectly and change them appropriately).

Let B denote the set of all binary sequences of length n. Let $\oplus$ be a binary operation on B such that for x and y in B, $x \oplus y$ is a bit string of length n that has 1's in those positions x and y differ and has 0's in those positions where x and y are the same. For example, if $x = 1001$, $y = 1100$, and $x \oplus y$ is 0101. We can show that $(B, \oplus)$ is a group. Closure and associativity can be easily shown, 0^n is the identity, and every n bit string is its own inverse.

DEFINITION 2   Let B denote the set of n bit binary strings and $x, y \in B$. The weight of x denoted $w(x)$ is the number of 1's in x. The distance between x and y denoted $d(x, y)$ is the weight of $x \oplus y$ [i.e., $w(x \oplus y)$]. It is the number of positions in which the two strings differ. It is also called the Hamming distance.

EXAMPLE 3   Let $x = 0\,1\,1\,0\,0\,1$ and $y = 1\,0\,1\,0\,0\,0$.

$$w(x) = 3$$

$$w(y) = 2$$

$$x \oplus y = 110001$$

$$w(x \oplus y) = 3$$

◄

We can easily verify the following properties of the distance function

(a) $d(x, y) = d(y, x)$
(b) $d(x, y) \geq 0$
(c) $d(x, y) = 0$ if and only if $x = y$
(d) $d(x, y) \leq d(x, z) + d(z, y)$

The last property can be seen as

$$w(\alpha \oplus \beta) \leq w(\alpha) + w(\beta)$$

$$w(x \oplus y) = w(x \oplus z + z \oplus y)$$

$$\leq w(x \oplus z) + w(z \oplus y) \quad \text{taking } \alpha = x \oplus z \text{ and } \beta = z \oplus y$$

[i.e., $d(x, y) \leq d(x, z) + d(z, y)$].

Minimum Distance, Error Correction, and Detection

Let G be a block code. The distance of G is the minimum between any pair of distinct codewords in G. The distance of a block code is related to its ability to detect and correct errors. An error occurs if a bit 1 is changed to 0 or 0 is changed to 1.

The relationship between the minimum distance of a code and the amount of error detection or correction possible is given by

$$M - 1 = C + D \quad \text{where } C \leq D$$

Here

$M =$ minimum distance of a code

$C =$ number of bits in error that can be corrected

$D =$ number of bits in error that can be detected.

No error can be corrected without detecting it. Hence $C \leq D$.

In a $B^m \to B^n$ code, 2^m strings of the total 2^n strings are chosen as codewords. Remaining $2^n - 2^m$ strings are not codewords. In a $(2, 5)$ code, out of 32 strings of length 5, only four are valid codewords. While transmitting, the length 2 string is coded into length 5 string and transmitted. If the received length 5 string is not a codeword, error has occurred. It should be detected and corrected.

An error-detection code is defined according to one less than the minimum error it will not always detect. Thus, if a code detects all single, double, and triple errors and some or no quadruple errors, it is called a triple-error detecting code. This would still be so even if the code detected all quintuple errors. An error correction code is defined in the same manner, i.e., according to one less than the minimum error it will not always correct.

The relationship between the minimum distance of a code and the amount of error correction or detection possible may be more graphically pictured if a "table-lookup" error detection system is considered. All the valid codewords in the code are stored in the table. Each codeword to be checked is compared with the codewords in the table. If a codeword in the table is found to match exactly, it is assumed that no error has occurred; if no codeword matches exactly, an error has been detected. Whether or not the error can be corrected depends upon the minimum distance of the code, as will be seen.

M	1	2	3		4		5			6		
C	0	0	0	1	0	1	0	1	2	0	1	2
D	0	1	2	1	3	2	4	3	2	5	4	3

In codes with a minimum distance of one, where two valid codewords may differ in only one bit position, a single error in a codeword could make that codeword appear like another valid codeword in the coding. This other valid codeword would be found in the table, and it would be falsely assumed that no error had occurred. Thus, in codes having a minimum distance of one, single errors can fail to be detected. Obviously, multiple errors cannot be detected.

In codes with a minimum distance of two, all codewords must differ in at least two bit positions. If there is a single bit error in a codeword, the codeword cannot possibly match any of those in the table; therefore, all single errors will be detected. Codes with a minimum distance of two are called single-error detecting codes. Errors in two or more bits might make the codeword match exactly some other valid codeword in the table, and therefore these errors would not be detected.

In codes with a minimum distance of three, all codewords must differ in at least three bit positions. A codeword with a single or double bit in error cannot match any codeword in the table; therefore, all single and double errors will be detected. Errors in three or more bits can result in another valid codeword, and therefore these errors cannot be detected.

Minimum distance three codes can be used for single-error correction. The key to error correction is that it must be possible to locate the bit or bits in error. If a single error occurs in a minimum distance three code, the resulting codeword will not match any codeword in the table, but it will come within one bit of matching the correct codeword. It will not come within one bit of matching any other codeword. To accomplish the correction, the one bit that does not match is changed.

In any code that can be used for correction, correction is "bought" at the expense of detection. If a minimum distance three code is used for correction and a double error occurs, the resulting codeword may come within a single bit of matching some other codeword in the table. Since there is no way of knowing that a double error has occurred, it would be assumed that the single bit was in error, and this bit would be erroneously "corrected." Thus, the error would be compounded, and an incorrect codeword will be taken. Minimum distance three codes thus will not detect double errors if they are used to correct single errors.

The codewords in minimum distance four codes differ in at least four bit positions. Single, double, and triple errors can be detected with these codes, since the resulting codeword cannot match any of those in the table. Errors in four or more bit positions can result in a codeword that matches some other valid codeword in the table, and thus these errors cannot be detected.

Instead of minimum distance four codes being used for triple-error detection, they can be used for single-error correction with double-error detection. If a single error occurs, the resulting codeword will not match exactly any codeword in the table, but it will come within one bit of matching the correct codeword. It will differ from all other codewords in the table by at least three bits. The correction is made by changing the one bit that does not match.

A codeword with a double error will come within two bits of matching the correct codeword of the table, but it may also come within two bits of matching with an incorrect codeword. There is, therefore, no way of knowing which bits are actually in error, and they cannot be corrected but only detected.

If a triple error occurs in a codeword, the resultant codeword will differ from the correct one in the table in three bit positions, but it may differ from some other codeword in the table in only one bit position. This one bit would thus be erroneously "corrected", and an incorrect codeword would result.

Minimum distance four codes are often referred to as single-error correcting, double-error detecting codes.

Maximum Likelihood Decoding and Minimum Distance Decoding

Let G be a block code where each codeword is of length n and $G = \{x_1, \ldots, x_N\}$ are codewords in G. Suppose a word x was transmitted and the word y was received. If $y \in G$, then we probably think it is the word transmitted. But this need not be so. By some errors, a word x_i might have been received as $x_j = y$. Let $P(x_i|y)$ denote the conditional probability that x_i was the transmitted word given that y was the received word. If $P(x_k|y)$ is the largest of all conditional probabilities computed, then we conclude that x_k was the transmitted word. Such a criteria for determining the transmitted word is known as maximum-likelihood decoding criterion.

It may not be easy to calculate the conditional probability $P(x_i|y)$. So generally another criterion is used which is the minimum distance decoding criterion. This is the one related to the error-detection explained in the previous section. For $i = 1, 2, \ldots, N$, we compute $d(x_i, y)$ and conclude that x_k was the transmitted word if $d(x_k, y)$ is the smallest among all distances computed. If it is assumed that the occurrence of errors in the positions are independent, and that the probability of the occurrence of an error is p, then $P(x_i|y) = (1-p)^{n-t}p^t$, where $t = d(x_i, y)$. For $p < \frac{1}{2}$, the smaller $d(x_i, y)$ is, the larger $p(x_i|y)$ will be. Thus we see both criteria are equivalent.

THEOREM 1   A code of distance $2t + 1$ can correct t or fewer transmission errors when the minimum-distance decoding criteria is considered.

Proof: Suppose a codeword x was sent and a word y was received. If no more than t errors have occurred during transmission, $d(x, y) \le t$.

Let x' be another codeword, $d(x, x') \ge 2t + 1$, $d(x, x') \le d(x, y) + d(y, x_1)$.

Hence $d(y, x_1) \ge t + 1$.

Hence minimum-distance decoding criterion will select x as the codeword transmitted.

A minimum distance $2t$ code cannot do it. Suppose x is the word sent and y is the word received. Suppose t bits are in error. Then $d(x, y) = t$. Let x' be another word such that $d(x', y) = t$, $d(x, x') \le d(x, y) + d(y', x') = 2t$ is satisfied. But x' may wrongly be chosen as the transmitted word.

Group Codes

In this subsection, we consider a class of block codes known as group codes. Let B be the set of all binary sequences of length n. A subset G of B is called a group code if $(G, \oplus)$ is a subgroup of $(B, \oplus)$.

EXAMPLE 4   Consider the following $(3, 6)$ block code.

It can be easily checked that it is a group code. f is the encoding function from $B^3 \to B^6$.

$$a^0 = f(000) = 0\,0\,0\,0\,0\,0$$

$$a^1 = f(001) = 0\,0\,1\,1\,1\,1$$

$$a^2 = f(010) = 0\,1\,0\,0\,1\,1$$

$$a^3 = f(011) = 0\,1\,1\,1\,0\,0$$

$$a^4 = f(100) = 1\,0\,0\,1\,1\,0$$

$$a^5 = f(101) = 1\,0\,1\,0\,0\,1$$

$$a^6 = f(110) = 1\,1\,0\,1\,0\,1$$

$$a^7 = f(111) = 1\,1\,1\,0\,1\,0$$

$(\{a^0, a^1, \ldots, a^7\}, \oplus)$ is a subgroup of $(B^6, \oplus)$. The minimum distance of the above code is 3.   ◄

THEOREM 2   Let $f: B^m \to B^n$ be a group code. The minimum distance of the group code f is the minimum weight of a nonzero codeword.

Proof: Let d be the minimum distance of the code. Let $d = d(x, y)$ where x, y are codewords. Let w be the minimum weight of a nonzero codeword, and suppose that $w = |z|$ for a code z. Since f is a group code, $x \oplus y$ is a nonzero codeword.

Thus $d = d(x, y) = |x \oplus y| \ge w$.

On the other hand, since 0 and z are distinct codewords, $w = |z| = |z \oplus 0| = d(z, 0) \geq d$. Hence $w = d$.

In the example considered above, $d = 3$ and w is also 3.

Now let us see how we can generate group codes. If $X = [x_{ij}]$ and $Y = [y_{ij}]$ are Boolean matrices of size $m \times n$, then $Z = X \oplus Y$ is a Boolean matrix of size $m \times n$, where $z_{ij} = x_{ij} \oplus y_{ij}$. If X is a $m \times n$ Boolean matrix and Y is a $n \times p$ Boolean matrix, Z is a $m \times p$ Boolean matrix, where $z_{ij} = x_{i1} \cdot y_{1j} \oplus x_{i2} \cdot y_{2j} \oplus \cdots \oplus x_{in} \cdot y_{nj}$, where $\cdot$ denotes $\wedge$ and $\oplus$ mod 2 addition, and $(X \oplus Y) * W = (X * W) \oplus (Y * W)$, where $*$ is Boolean matrix multiplication. Here X and Y are $m \times n$ matrices and W is a $n \times p$ matrix.

Let us consider a coding scheme $e \colon B^m \to B^n (m < n)$. We want the resulting codewords to form a subgroup of B^n. For this, consider a Boolean matrix E of size $m \times n$. The first $m \times m$ submatrix forms an identity matrix of size $m \times m$. For example, E can be chosen as

$$\begin{bmatrix} 1 & 0 & 0 & 1 & 1 & 0 \\ 0 & 1 & 0 & 0 & 1 & 1 \\ 0 & 0 & 1 & 1 & 1 & 1 \end{bmatrix}.$$

(In fact the identity matrix need not be in the first m columns. It can be anywhere.)

Then the coding of a word $x = x^1 x^2 x^3$ is given by $x * E$. So 000 will be coded as 000000.

THEOREM 3   The set of codewords defined by a coding scheme using a Boolean matrix form a group.

Proof: The order of the group is finite. Hence it is enough to prove the closure property given. We have also seen that o^m will be coded as o^n, which will serve as the identity. Suppose E is a $m \times n$ Boolean matrix and the coding scheme $e \colon B^m \to B^n$ is defined as $e(x) = x * E$, where $*$ denotes Boolean matrix multiplication. Let x' and $x'' \in B^m$ and $x = x' \oplus x''$. Hence $e(x) = e(x' \oplus x'')$.

Let $e(x) = y_1 y_2 \dots y_n$ and $E = [e_{ij}]$, so $y_j = \sum_{i=1}^{m} x_i e_{ij} = \sum_{i=1}^{m} (x_i' \oplus x_i'') e_{ij}$

The term $(x_i' \oplus x_i'') e_{ij} = 0$ if $e_{ij} = 0$. It is $x_i' \oplus x_i''$ if $e_{ij} = 1$. Thus, $x_i' e_{ij} = x_i'$ and $x_i'' e_{ij} = x_i''$ if $e_{ij} = 1$ Hence

$$\sum_{i=1}^{m} (x_i' \oplus x_i'') e_{ij} = \sum_{i=1}^{m} x_i' e_{ij} \oplus \sum_{i=1}^{m} x_i'' e_{ij}$$

Hence $E(x) = E(x') + E(x'')$. If $y' = x' * E$ and $y'' = x'' * E$, $e(x') = y'$ and $e(x'') = y''$
Hence it follows that

$$y = e(x) = e(x' \oplus x'') = e(x') \oplus e(x'') = y' \oplus y''$$

Thus closure is proved. So the set of codewords form a group.

Encoding, Decoding, and Error Correction using Binary Matrices.

Links

Consider an encoding scheme : $B^m \to B^n$. This could be achieved using a $m \times n$ boolean matrix G, having an identity matrix as a submatrix. G is called a generator matrix. This is a $m \times n$ matrix of the form $[I_m A]$, where I_m is the identity unit matrix. If w is the message of m bits, it is encoded as a codeword of n bits by wG. For example consider the encoding $e : B^2 \rightarrow B^5$ using the generator matrix

$$\begin{bmatrix} 1\ 0\ 0\ 1\ 1 \\ 0\ 1\ 1\ 0\ 1 \end{bmatrix}$$

This encoding gives

$$e(00) = \begin{bmatrix} 0\ 0 \end{bmatrix} \begin{bmatrix} 1\ 0\ 0\ 1\ 1 \\ 0\ 1\ 1\ 0\ 1 \end{bmatrix} = 00000 = a$$

$$e(01) = \begin{bmatrix} 0\ 1 \end{bmatrix} \begin{bmatrix} 1\ 0\ 0\ 1\ 1 \\ 0\ 1\ 1\ 0\ 1 \end{bmatrix} = 01101 = b$$

$$e(10) = \begin{bmatrix} 1\ 0 \end{bmatrix} \begin{bmatrix} 1\ 0\ 0\ 1\ 1 \\ 0\ 1\ 1\ 0\ 1 \end{bmatrix} = 10011 = c$$

$$e(11) = \begin{bmatrix} 1\ 1 \end{bmatrix} \begin{bmatrix} 1\ 0\ 0\ 1\ 1 \\ 0\ 1\ 1\ 0\ 1 \end{bmatrix} = 11110 = d$$

The four elements form a group

If a_1a_2 is the message word, the code word is $\begin{bmatrix} a_1\ a_2 \end{bmatrix} \begin{bmatrix} 1\ 0\ 0\ 1\ 1 \\ 0\ 1\ 1\ 0\ 1 \end{bmatrix} = a_1a_2a_2a_1a_1 + a_2$.

If $b_1b_2b_3b_4b_5$ is the codeword, $b_1 = a_1, b_2 = a_2, b_3 = a_2, b_4 = a_1, b_5 = a_1 + a_2 (mod\ 2)$. Since $-b_1$ is $+b_1$ in mod 2 addition we have

$$b_2 + b_3 \qquad = 0$$
$$b_1 \qquad + b_4 \qquad = 0$$
$$b_1 + b_2 \qquad + b_5 = 0$$

The coefficients corresponding to a matrix $[A^T\ I_3]$ is $\begin{bmatrix} 0\ 1\ 1\ 0\ 0 \\ 1\ 0\ 0\ 1\ 0 \\ 1\ 1\ 0\ 0\ 1 \end{bmatrix}$. Let us call this H. This is called a parity check matrix.

If c is a codeword $Hc = \begin{bmatrix} 0 \\ 0 \\ 0 \end{bmatrix}$.

If the received word is r and $Hr^T = \begin{bmatrix} 0 \\ 0 \\ 0 \end{bmatrix}$, this means that the received word is a code word and no error has occurred. If $Hr^T \neq \begin{bmatrix} 0 \\ 0 \\ 0 \end{bmatrix}$, an error has occurred. Note that the minimum weight of the codeword is 3 which is also the minimum distance. Hence it should be possible to correct single errors. Suppose it is meant to send the codeword 10011 but a single error occurs and the received word is 10111. The third bit is in error. We can find this as follows.

$$Hr^T = \begin{bmatrix} 0\ 1\ 1\ 0\ 0 \\ 1\ 0\ 0\ 1\ 0 \\ 1\ 1\ 0\ 0\ 1 \end{bmatrix} \begin{bmatrix} 1 \\ 0 \\ 1 \\ 1 \\ 1 \end{bmatrix} = \begin{bmatrix} 1 \\ 0 \\ 0 \end{bmatrix}.$$ The product of the multiplication gives $\begin{bmatrix} 1 \\ 0 \\ 0 \end{bmatrix}$ which is

the same as column 3 of H. Hence it is concluded that the third bit is in error and corrected as 10011. The first 2 bits are the message bits and decoding gives 10. If Hr^T is not $\begin{bmatrix} 0 \\ 0 \\ 0 \end{bmatrix}$ or any

column of H, more than one error has occurred. But it cannot be corrected.

Decoding in Group Codes

For group codes, there is an efficient way to determine the transmitted word corresponding to a received word. Let $(G, \oplus)$ be a group code. Let y be the received word. We know that $d(x_i, y) = w(x_i \oplus y)$. Hence the weights of the words in the coset $G \oplus y$ are the distances between codewords in G and y. If $\{x_0, x_1, \ldots, x_n\}$ are the words in G, we want to select x_j such that $x_j \oplus y$ is the smallest. Select the word with smallest weight in $G \oplus y$. Let this be x'. If more than one has the smallest weight, select any one of them. Let $x' = x_j \oplus y$ be for the transmitted word x_j, and hence $x_j = x' \oplus y$. So for a group code $(G, \oplus)$, the decoding procedure is as follows.

1. Determine all cosets of G
2. For each coset, pick the word of smallest weight (one of the smallest weight if there are more than one). This is called the leader of the coset.
3. For a received word y, $x' \oplus y$ is the sent word, where x' is the leader of the coset containing y.

EXAMPLE 5   Consider the group code $(G, \oplus)$ which is a subgroup of $(B^6, \oplus)$. Let

$$G = \{a^0, a^1, \ldots, a^7\}$$

as given in Example 4.

Consider the cosets of G. The following table gives the cosets of G.

000000	100110	010011	011100	001111	101001	110101	111010
100000	000110	110011	111100	101111	001001	010101	011010
010000	110110	000011	001100	011111	111001	100101	101010
001000	101110	011011	010100	000111	100001	111101	110010
000100	100010	010111	011000	001011	101101	110001	111110
000010	100100	010001	011110	001101	111011	110110	111000
000001	100111	010000	011101	001110	101000	110100	111011
000101	100011	010110	011001	001010	101100	110000	111111

The first row is G itself. The coset leaders are given in the first column. Except for the last row, coset leaders are uniquely chosen. In the last row, 000101 is chosen as coset leader. 001010 or 110000 could also be a leader. If the received word is 011110, locate it in the fourth column, sixth row of the table. The coset leader is 000010, and so the transmitted word is $011110 \oplus 000010 = 011100$. It is decoded as 011. A single error is found and corrected (minimum distance 3 code can correct single errors). If 111111 is the codeword received, a double error has occurred as the sent word could be 001111, 110101, or 111010. It is located in the last row last column. If 000101 is chosen as the coset leader, the sent word is $111111 \oplus 000101 = 111010$. If 001010 is chosen as the coset leader, the sent word is $111111 \oplus 001010 = 110101$. If 110000 is chosen as the coset leader, the sent word is $110000 \oplus 111111 = 001111$. A double error cannot be corrected. Decoding may take any one of 001111, 110101, or 111010 and give 001, 110, or 111.   ◄

Hamming Codes

An ingenious family of perfect codes which will correct all single errors was given by R.W. Hamming.

The Hamming codes are single-error-correcting codes (with minimum distance 3), which are perfect in the sense that for any r there exists a $(m = 2^r - 1 - r, n = 2^r - 1)$ code that corrects each single error which might occur, no other errors. We already noted that if minimum distance 3 code is suggested, it will correct all single errors but cannot detect double errors, and a double error may be taken as a wrong single error and corrected wrongly. If the decoding table in this case is constructed as in the earlier section, the coset leaders will consist of $00\ldots0$ and patterns where one of these 0's is replaced by 1. Hamming codes also provide a simple decoding scheme for locating the error and hence correct it. Though their length can be other than $2^r - 1$, in this section we assume the codewords to be of length $2^r - 1$.

The technique for constructing a Hamming code is as follows.

1. Choose an integer r. The message word is of length $2^r - 1 - r$ and the codeword is of length $2^r - 1$.
2. If the codeword is $b_1\,b_2 \ldots b_{2^r-1}$, the bits $b_1, b_2, b_4, b_8, \ldots, b_{2r} - 1$ are checkbits. If the message word is $a_1\,a_2 \ldots a_{2^r-1-r}$ then $b_3 = a_1, b_5 = a_2, b_6 = a_3, b_7 = a_4 \ldots$ and so on. Then

$$b_{2^r-1} = a_{2^r-1-r}$$

Suppose an $r = 3$ codeword is of length 7 and the message word is of length 4. Then

$$a = a_1\ a_2\ a_3\ a_4$$

is the message word and

$$b = b_1\ b_2\ b_3\ b_4\ b_5\ b_6\ b_7$$

is the codeword. Thus,
$$b_3 = a_1, b_5 = a_2, b_6 = a_3, \text{ and } b_7 = a_4, \text{ and } b_1, b_2, b_4 \text{ are checkbits.}$$
3. Form a matrix of $2^r - 1$ rows and r columns, where the row i is the binary number with value i. The matrix for $r = 3$ is

$$\begin{bmatrix} 0 & 0 & 1 \\ 0 & 1 & 0 \\ 0 & 1 & 1 \\ 1 & 0 & 0 \\ 1 & 0 & 1 \\ 1 & 1 & 0 \\ 1 & 1 & 1 \end{bmatrix}$$

4. Take $bM = 0$ where Boolean matrix multiplication is used. Suppose the codeword is $b = b_1 \ldots b_7$. We get

$$b_4 + b_5 + b_6 + b_7 = 0$$
$$b_2 + b_3 + b_6 + b_7 = 0$$
$$b_1 + b_3 + b_5 + b_7 = 0$$

If the message word is $a_1\ a_2\ a_3\ a_4$, we get

$$b_4 + a_2 + a_3 + a_4 = 0 (\text{mod } 2)$$
$$b_2 + a_1 + a_3 + a_4 = 0$$
$$b_1 + a_1 + a_2 + a_4 = 0$$

That is, b_1 is chosen so that it gets even parity with a_1, a_2, a_4. Then b_2 is chosen so that it gets even parity with a_1, a_3, a_4. Finally b_4 is chosen so that it gets even parity with a_2, a_3, a_4.

This procedure yields codewords which will have minimum weight 3, except for 0^{2^r-1}.

There is a simple way to decode these codes. Let a be the sent word and b' be the received word. If a single error has occurred, then b' differs from b in one bit position (say the ith bit). Then consider the error vector which has 1 in the ith position and 0 in other positions. Then $b' = b \oplus e$ so that $(b \oplus e)M = bM \oplus eM$. But $bM = 0$. Therefore $b'M = eM$. If the error vector is 0, the result is 0, as no error has occurred. If the error vector is of the form

$$\underbrace{000\ldots0}_{i-1}\ \underset{i}{1}\ \underbrace{000\ldots0}_{2-1-i}$$

then when this is multiplied by M, the 1 picks out the ith row of the matrix M, which is really the number i represented in binary. This shows that the ith bit is in error, and flipping it gives the corrected codeword.

EXAMPLE 6   Consider a (4, 7) Hamming code. Consider a codeword $b = 0\,0\,0\,1\,1\,1\,1$. Let $bM = 0\,0\,0$, and no error has occurred. Suppose in b, the fifth bit is changed to 0. Now the received word b' is $0\,0\,0\,1\,0\,1\,1$. This is obtained by adding $e = 0\,0\,0\,0\,1\,0\,0$ to b.

$$b'M = 0\,0\,0\,1\,0\,1\,1 \begin{bmatrix} 0\,0\,1 \\ 0\,1\,0 \\ 0\,1\,1 \\ 1\,0\,0 \\ 1\,0\,1 \\ 1\,1\,0 \\ 1\,1\,1 \end{bmatrix} = 1\,0\,1$$

This shows the fifth bit is in error, and it is changed from 0 to 1.

This cannot detect double errors, as the minimum distance is 3. Hamming code may be extended to perfect codes with minimum weight 4 by adding a parity bit to each codeword. This will give a single error-correcting, double error-detecting code.   ◀

Exercises

1. Find the weights of
   a) 10110
   b) 110110

2. Find the minimum distance between
   a) 11110 and 10001
   b) 1010101 and 0001100

3. Find the minimum distance of the following (2, 4) encoding scheme.
   a) $e(00) = 0000$
   b) $e(01) = 1011$
   c) $e(10) = 0110$
   d) $e(11) = 1100$

4. Let e be the encoding function $B^m \to B^{m+1}$ where the $(m+1)$th bit is added to get even parity. Let $d: B^{m+1} \to B^m$ be the corresponding decoding function:
   a) $m = 4$
      What is $e(1010)$, $e(1110)$?
      What is $d(11011)$, $d(10100)$?
   b) $m = 6$
      What is $e(110101)$, $e(101100)$?
      What is $d(1101100)$, $d(1101001)$, $d(1010100)$?

5. Let e be the encoding function $B^m \to B^{m+1}$ where the $(m+1)$th bit is added to get odd parity. Let $d: B^{m+1} \to B^m$ be the corresponding decoding function.
   a) $m = 4$
      What is $e(1010)$, $e(111)$?
      What is $d(11010)$, $d(10101)$, $d(11000)$?

   b) $m = 6$
      What is $e(110101)$, $e(101100)$?
      What is $d(1101101)$, $d(1101001)$?

6. Let e be an encoding scheme $B^m \to B^{3m}$ such that $e(a_1 \ldots a_m) = a_1 \ldots a_m a_1 \ldots a_m a_1 \ldots a_m$. How would you define a decoding scheme so that single errors can be corrected?

7. Consider the encoding scheme $e: B^3 \to B^{15}$ as discussed in Example 2.
   a) What is $e(101)$?

   b) What is $d(101001001101101)$, $d(111011110111111)$?

8. If minimum distance of a code is 5, how many errors can it correct/detect?

9. Consider the following code $e: B^2 \to B^5$ defined by

$$e(00) = 00000$$

$$e(01) = 01110$$

$$e(10) = 10101$$

$$e(11) = 11011$$

a) Show that it is a group code.

b) What is the minimum distance of the code?

c) Discuss the error detection/correction capability of this code.

10. Consider the $(2, 4)$ group encoding function $e: B^2 \to B^4$ defined by

$$e(00) = 0000 \qquad e(01) = 0111$$

$$e(10) = 1001 \qquad e(11) = 1111$$

Decode the following words using minimum distance decoding criterion.

a) 0011

b) 1011

c) 1111

11. Consider the $(3, 6)$ group encoding function $e: B^3 \to B^6$ defined by

$$e(000) = 000000 \quad e(100) = 100101$$

$$e(001) = 000110 \quad e(101) = 100011$$

$$e(010) = 010010 \quad e(110) = 110111$$

$$e(011) = 010100 \quad e(111) = 110001$$

Decode the following words using minimum distance decoding criterion.

a) 011110

b) 101011

c) 110010

12. Consider a (m, n) code and let $r = n - m$.

a) Given an encoding matrix G, show that there exits a parity-check matrix H such that

(i) If G is $m \times n$, H is $n \times n(n - m)$.

(ii) $GH = 0$.

(iii) Each column h_i, $1 \le i \le n - m$ of H is such that for no scalars θ_i, not all zero, does $\sum_{i=1}^{n-m} a_i h_i = 0$. That is, the columns of H are linearly independent.

H is called a parity-check matrix because, for any codeword c in the code generated by G, $cH = 0$.

b) Show that, given a parity-check matrix for a code, the minimum weight of a codeword (not the 0 codeword) is equal to the minimum number of rows of H which can be added together to give 0.

13. One Hamming code is described by the following matrix:

$$\begin{bmatrix} 1 & 1 & 1 & 0 & 0 & 0 & 0 \\ 1 & 0 & 0 & 1 & 1 & 0 & 0 \\ 0 & 1 & 0 & 1 & 0 & 1 & 0 \\ 1 & 1 & 0 & 1 & 0 & 0 & 1 \end{bmatrix} \qquad \begin{bmatrix} 0 & 0 & 1 \\ 0 & 1 & 0 \\ 0 & 1 & 1 \\ 1 & 0 & 0 \\ 1 & 0 & 1 \\ 1 & 1 & 0 \\ 1 & 1 & 1 \end{bmatrix}$$

(a) Generator matrix   (b) Parity–check matrix

Assuming that no more than one error occurs during transmission, what was the transmitted codeword vector when

a) 0111110 is received?

b) 0001111 is received?

14. Consider a $(3, 6)$ code with generator matrix

$$G = \begin{bmatrix} 1 & 0 & 0 & 0 & 1 & 1 \\ 0 & 1 & 0 & 1 & 0 & 1 \\ 0 & 0 & 1 & 1 & 1 & 1 \end{bmatrix}$$

What is the probability that a message of six digits will be received and accepted as correctly transmitted through a binary symmetric channel, when in fact at least one error has occurred?

15. Show that, if you try to use a 2×5 Boolean matrix which does not contain an identity matrix as a matrix for an encoding scheme $e: B^2 \to B^5$, you may not get proper encoding scheme.

12.7 Polynomial Rings and Polynomial Codes

Let $(A, +, \cdot)$ be a ring. An algebraic expression of the form $A(x) = a_0 + (a_1 \cdot x) + (a_2 \cdot x^2) + \cdots + (a_{n-1} \cdot x^{n-1}) + (a_n \cdot x^n) n \ge 0$ is called a polynomial. Here $a_1, a_2, \ldots, a_n$ are constants and x is the variable. We know that $+$ and $\cdot$ are associative operations. Hence $A(x)$ is defined unambiguously. The degree of the polynomial is the largest n for which a_n is not equal to 0.

Let us consider polynomials over a field F. Let

$$A(x) = a_0 + a_1 x + \cdots + a_n x^n$$

$$B(x) = b_0 + b_1 x + \cdots + b_n x^n$$

The addition of polynomials is defined as $C(x) = A(x) \diamondsuit B(x)$, $c_i = a_i + b_i$, and $0 \le i \le n$. The multiplication of polynomials is defined as $D(x) = A(x) \otimes B(x)$ where $d_i = a_0 b_i + a_1 b_{i-1} + a_2 b_{i-2} + \cdots + a_i b_0$. It can be easily checked that $(F[x], \diamondsuit)$ is an abelian group where $F[x]$ denotes all polynomials with coefficients in F. It can also be checked that $(F[x], \otimes)$ is a semigroup and $\otimes$ is commutative. The fact that $\otimes$ is distributive over $\diamondsuit$ can also be checked. Hence $(F[x], \diamondsuit, \otimes)$ forms a ring. It is called a polynomial ring.

The following fact is also well known.

Given polynomials $A(x)$ and $B(x)$ in $F[x]$ with $b(x) \neq 0$, there exists a quotient $q(x) \in F[x]$ and a remainder $r(x) \in F[x]$, where either $r(x) = 0$ or deg $r(x) <$ deg $B(x)$ such that $A(x) = B(x)q(x) + r(x)$.

The method of finding $q(x)$ and $r(x)$ is called the division algorithm.

A polynomial $A(x)$ in $F[x]$ is said to be reducible (over F) if $A(x) = a(x)b(x)$ for some non-constant $a(x)b(x) \in F[x]$. Otherwise $A(x)$ is said to be irreducible (over F).

Let $(F, +, \cdot)$ be a field. Let $F_n[x]$ denote the polynomials of degree less than n in $F[x]$. Let $m(x)$ be a polynomial of degree n in $F[x]$. Let $(F_n[x], \diamondsuit_+, \diamondsuit_\circ)$ be an algebraic system with two binary operations $\diamondsuit_+$ and $\diamondsuit_\circ$ defined as follows. If

$$a(x) = a_0 + a_1 x + \cdots + a_{n-1} x^{n-1} \qquad b(x) = b_0 + b_1 x + \cdots + b_{n-1} x^{n-1}$$

we have

$$a(x) \diamondsuit_+ b(x) = a(x) \diamondsuit b(x)$$

$$a(x) \diamondsuit_\circ b(x) = \text{ the remainder of } a(x) \otimes b(x) \text{ when divided by } m(x).$$

It is straightforward to check that $(F_n[x], \diamondsuit_+, \diamondsuit_\circ)$ is a ring. This is usually referred to as the ring of polynomials modulo $m(x)$.

For example, let $(F, +, \cdot)$ be a field of integers modulo 2. Let $F = \{0, 1\}$. There are four polynomials in $F_2[x]$. They are 0, 1, x, and $1 + x$. Let $m(x)$ be the polynomial $1 + x + x^2$. Then $\diamondsuit_+$ and $\diamondsuit_\circ$ are defined as given in the table.

$\diamondsuit_+$	0	1	x	$1 + x$		$\diamondsuit_\circ$	0	1	x	$1 + x$
0	0	1	x	$1 + x$		0	0	0	0	0
1	1	0	$1 + x$	x		1	0	1	x	$1 + x$
x	x	$1 + x$	0	1		x	0	x	$1 + x$	1
$1 + x$	$1 + x$	x	1	0		$1 + x$	0	$1 + x$	1	x

Next we consider a special class of group codes called polynomial codes. These are (m, n) codes. When an m bit message word is encoded as an n bit codeword, $k = n - m$ checkbits are added. Let $a_0 \ldots a_{m-1}$ be the message word. This is represented by the polynomial

$$a(x) = a_0 + a_1 x + \cdots + a_{m-1} x^{m-1}$$

Then if we have $m = 4$ and the message word is 1 1 0 1,

$$a(x) = 1 + 1 \cdot x + 0 \cdot x^2 + 1 \cdot x^3$$
$$= 1 + x + x^3$$

Let $(F_2, +, \cdot)$ be the finite field with mod 2 addition and multiplication and $F_2 = \{0, 1\}$. Let $F_2[x]$ denote the set of polynomials over F_2. Let $g(x) = g_0 + g_1 x + \cdots + g_k x^k$ be a polynomial of degree k. Then the codeword for message word $a_0 a_1 \ldots a_{m-1}$ is given by $b = b_0 \ldots b_{n-1}$, where $b(x) = b_0 + b_1 x + b_2 x^2 + \cdots + b_{n-1} x^{n-1}$ and $b(x) = a(x)g(x)$.

Here $g_0 \neq 0$ and $g_k \neq 0$. Otherwise the first and last bit of codewords will not be of use.

EXAMPLE 1   Consider a code $B^3 - B^6$ using encoding polynomial $1 + x^3$. Then message word $0\ 1\ 0$ is encoded as codeword by multiplying $0 + x + 0 \cdot x^2 = x$ by $(1 + x^3)$. That is, $x + x^4$ and the codeword is $0\ 1\ 0\ 0\ 1\ 0$.   ◀

For finding the codewords we can use a matrix of size $m \times (m + k)$.

$$
\begin{bmatrix}
g_0 & g_1 & \cdots & g_k & 0 & 0 & \cdots & 0 \\
0 & g_0 & g_1 & \cdots & g_k & 0 & \cdots & 0 \\
0 & 0 & g_0 & & \cdots & g_k & \cdots & 0 \\
\cdots & \cdots & \cdots & \cdots & \cdots & \cdots & \cdots & \\
& & & & & & & \\
\cdots & \cdots & \cdots & \cdots & \cdots & \cdots & \cdots & \\
0 & 0 & 0 & g_0 & g_1 & & & g_k
\end{bmatrix}
$$

EXAMPLE 2   The matrix or the encoding polynomial $1 + x^3$ will be

$$
\begin{bmatrix}
1 & 0 & 0 & 1 & 0 & 0 \\
0 & 1 & 0 & 0 & 1 & 0 \\
0 & 0 & 1 & 0 & 0 & 1
\end{bmatrix}
$$

and the encoding is given by

$$f(000) = 000000$$
$$f(001) = 001001$$
$$f(010) = 010010$$
$$f(011) = 011011$$
$$f(100) = 100100$$
$$f(101) = 101111$$
$$f(110) = 110110$$
$$f(111) = 111111$$

It is straightforward to see that the codewords form a subgroup of $(B^6, \oplus)$. As for group codes, we can see that the minimum distance of the coding scheme in this polynomial code with encoding polynomial $g(x)$ is the minimum weight of the production $a(x)g(x)$. If $(1 + x)$ is used as the encoding polynomial, each codeword has even number of 1's. This can be seen as follows. Let the message word be $a_0 + \cdots + a_m x^{m-1}$, and the codeword is given by

$$f(x) = (1 + x)a(x) \text{ where}$$
$$f(x) = f_0 \oplus f_1 x \oplus \cdots \oplus f_n x^{n-1}$$

Putting $x = 1$, we get $f(1) = f_0 \oplus f_1 \oplus \ldots \oplus f_n$ and we see $f(1) = (1 \oplus 1)a(x) = 0$. Here $f_0\ f_1 \ldots f_n$ has even number of 1's.   ◀

Let us see what are the advantages in using polynomial codes.

THEOREM 1   Let $e \colon B^m \to B^n$ be an encoding scheme defined by the encoding polynomial $g(x)$. If $g(x)$ divides no polynomial of the form $x^k - 1$ for $k < n$, then the minimum distance of the codewords is at least 3.

Proof: The codewords generated are represented by $g(x)\,a(x)$ with degree of $a(x)$ being $< m$. As the codewords form a group under the operation $\oplus$, we have seen by a result proved earlier that the minimum distance of the codeword is the minimum weight of the codewords (other than o^n).

If $b = b_0 \ldots b_n$ and $c = c_0 \ldots c_n$ are codewords and if $b(x) \oplus e(x) = c(x)$, then $e(x)$ must be a codeword. Then $e(x)$ represents the error bits. If the distance between two codewords is 2, then $e(x)$ has only two nonzero bits and can be represented by $x^i + x^j$ and $e(x) = g(x)e'(x)$ where $e'(x)$ is a polynomial representing a message word. Hence $g(x)$ divides $e(x)$. Similarly if $e(x)$ has only one 1, then it is represented by x^i and $g(x)$ divides x^i.

If $e(x) = x^i + x^j = x^i(1 + x^{j-i})$, $g(x)$ has $g(0) \neq 1$ and so it must divide $(1 + x^{j-i})$. However, $g(x)$ was selected so that it divides no polynomial of the form $(1 + x^k)$ for $k < n$. Note that $x^k - 1$ and $x^k + 1$ are the same when the underlying field is Z_2. So $e(x)$ cannot be of the form $x^i + x^j$.

If $e^x = x^i$ for some i, let $e(x) = g(x)e'(x)$ for message polynomial $e'(x)$. Hence $x^i = g(x)e'(x)$ and $g(x) = x^j$ for some j, $j \leq i$. We consider only $g(x)$ with $g(0) \neq 0$ and the only possibility is $g(0) = 1$, in which case $g(x)$ divides $x^i - 1$ (equivalent to $x^i + 1$ under Z_2). This is not possible and $e(x)$ cannot be of the form x^i.

Hence, we find that the minimum distance cannot be 1 or 2 but can be 3.

DEFINITION 1   A polynomial $g(x)$ of degree k over Z_2 is called primitive when $g(x)$ divides $x^m - 1$ for $m = 2^k - 1$ and for no smaller m.

EXAMPLE 3   The polynomial $1 + x^2 + x^3 \in Z_2[x]$ is a primitive polynomial of degree 3. It divides $x^7 - 1$ but not $x^j - 1$ for any $j < 7$. So $(x^4 + x^3 + x^2 + 1)(x^3 + x^2 + 1) = x^7 + 1$. (Note that $x^7 - 1$ and $x^7 + 1$ are equivalent in Z_2.) Let us consider the encoding scheme $B^4 \to B^7$ using encoding polynomial $1 + x^2 + x^3$.

$$f(0000) = 0000000$$

To get $f(0101)$:
Consider $(x + x^3)(1 + x^2 + x^3)$

$$= x + x^3$$
$$+ x^3 + x^5$$
$$+ x^4 + x^6$$
$$= x + x^4 + x^5 + x^6$$

Hence, $f(0101) = 0100111$. We find that

$f(0000) = 0000000$	$f(0110) = 0111010$	$f(1011) = 1000011$
$f(0001) = 0001011$	$f(0111) = 0110001$	$f(1100) = 1110100$
$f(0010) = 0010110$	$f(1000) = 1011000$	$f(1101) = 1111111$
$f(0011) = 0011101$	$f(1001) = 1010011$	$f(1110) = 1100010$
$f(0100) = 0101100$	$f(1010) = 1001110$	$f(1111) = 1101001$
$f(0101) = 0100111$		

It can be seen that this is a Hamming code with minimum distance 3 and can be used for single error correction.   ◄

THEOREM 2 The 'error polynomial' associated with any undetected error vector $e = e_0 \ldots e_{n-1}$ of a polynomial (m, n) code with encoding polynomial $g(x)$ is a nontrivial multiple of $g(x)$.

Proof: Let $a = a_0 \ldots a_m$ be the message word $b = b_0 \ldots b_{n-1}$ the codeword and $c = c_0 \ldots c_{n-1}$ the received word. If c has errors, $c = b + e$ and $e = e_0 \ldots e_{n-1}$ and the 1's in e denote the error bits. If the error is undetected, e must be a codeword and $e(x) = e'(x)g(x)$ for some message word $e'(x)$. Hence, $e(x)$ is a multiple of $g(x)$.

To use the polynomial (m, n) code for error detection, let us see how the errors can be detected. Let $(m, m + k)$ be the code and the encoding polynomial $g(x)$ be of degree k. If $c(x)$ represents the received word, divide $c(x)$ by $g(x)$ and get quotient $q(x)$ and remainder $r(x)$. Then $c(x) = g(x)q(x) + r(x)$. If $r(x) = 0$, no error has occurred. If $r(x) \neq 0$, then an error has occurred. Note that this follows from the previous theorem.

If two adjacent bits are in error, then we call it as 'pair error'. This is an important concept, as when one bit is erroneously transmitted by some signal fluctuation, it is likely that the next bit also is erroneously transmitted.

By using an efficient encoding polynomial, it is possible to detect two single errors, two 'pair' errors, or one single error and one 'pair' error.

For this it is necessary to define the exponent of a polynomial.

DEFINITION 2 The exponent of a polynomial $g(x)$ is the least positive integer e such that $g(x)$ divides $x^e - 1$.

Thus, a polynomial over Z_2 is primitive, as it has degree k and exponent $e = 2^k - 1$.

THEOREM 3 Let f be a (m, n) code and $g(x)$ be the encoding polynomial. If $g(x) = (1 + x)h(x)$ where $h(x)$ has exponent $i > n$, then any combination of two single or pair errors will be detected.

Proof: If two single errors have occurred error vector has two 1's and $e(x)$ is of the form $x^i + x^j$, $j \neq i + 1$. ($x^i + x^{i+1}$ is taken as a pair error). We know that error is not detected if $e(x)$ is a multiple of $g(x)$, $e(x) = x^i + x^j = x^i(1 + x^{j-i})$. $h(x)$ will not divide this polynomial and hence the error will be detected.

If a single error and a pair error have occurred, $e(x)$ is of the form $x^i + x^{i+1} + x^j$ or $x^i + x^j + x^{j+1}$. If a single error alone has occurred, then $e(x) = x^i$. Here $1 + x$ will not divide any of these polynomials and the errors will be detected.

If the pair errors have occurred, then $e(x) = x^i + x^{i+1} + x^j + x^{j+1}$, that is, $e(x) = (1 + x)(x^i + x^j)$. Since $g(x) = (1 + x)h(x)$, $h(x)$ must divide $x^i + x^j$. But this is not possible as $j < n - 1$ and hence $x^i(1 + x^{j-i})$ is not divisible by $h(x)$.

Cyclic Codes

Let C be a collection of codewords of length n in a (m, n) code. This code is called a cyclic code if for every codeword $b = b_0 b_1 \ldots b_{n-1}$, $b_{n-1} b_0 b_1 \ldots b_{n-2}$ is also a codeword (i.e., the cyclic shift of a codeword is also a codeword). Consider the field $(Z_2, \oplus, \odot)$ and the polynomial ring $(F[x], \diamondsuit, \diamondsuit)$ over this field. Consider the polynomial ring $(F_n[x], \diamondsuit_+, \diamondsuit_o)$ where $m(x)$ is taken as $1 + x^n$. Let I be an ideal of $(F_n[x], \diamondsuit_+, \diamondsuit_o)$. Then we can show that the codewords represented by the polynomials in I constitute a cyclic code.

This is seen as follows. $(I, \diamondsuit_+)$ is an additive abelian group, and the polynomials in I represent codewords of a group code. Let $a(x)$ be a polynomial in I.

$$a(x) = a_0 + a_1 x + \cdots + a_{n-1} x^{n-1}.$$

$x \diamondsuit a(x) = $ remainder of

$$a_0 x + a_1 x^2 + \cdots + a_{n-1} x^n \text{ when divided by } (1 + x^n)$$

$$a_0 x + a_1 x^2 + \cdots + a_{n-1} x^n = a_0 x + a_1 x^2 + \cdots + a_{n-1} x^n + a_{n-1} + a_{n-1}$$

$$= a_{n-1} + a_0 x + \cdots + a_{n-2} x^{n-1} + a_{n-1}(1 + x^n)$$

So the remainder when divided by $(1 + x^n)$ is $a_{n-1} + a_0 x + \cdots + a_{n-2} x^{n-1}$ and represents the codeword $a_{n-1} a_0 \ldots a_{n-2}$. Hence polynomials in I represent a cyclic code.

On the other hand, let I denote a set of polynomials corresponding to codewords of a cyclic code. We show that I is an ideal of $(F_n[x], \diamondsuit_+, \diamondsuit_\bullet)$. Clearly $(I, \diamondsuit_+)$ is an abelian group.

As seen earlier, we find that if $a(x)$ is in I, $x \diamondsuit \bullet a(x)$ is in I, and it follows $x^i \diamondsuit \bullet a(x)$ is in I. It also follows $b_j x^j \diamondsuit \bullet a(x)$ in I, as $b_j x^j \diamondsuit \bullet a(x) = x^j \diamondsuit \bullet a(x)$ if $b_j = 1$ and 0 otherwise. Hence if $b(x) = b_0 + b_1 x + \cdots + b_{n-1} x^{n-1}$ is a polynomial in the ring $(F_2[x], \diamondsuit, \diamondsuit)$, then $b(x) \diamondsuit \bullet a(x)$ is in I. Hence I is an ideal of $(F_n[x], \diamondsuit_+, \diamondsuit_\bullet)$.

Exercises

1. Consider a (3, 6) code using encoding polynomial $1 + x + x^3$. Find the encoding matrix and the encoding scheme.

2. Find a (3, 4) coding using encoding polynomial $1 + x$. What is the encoding matrix?. Find the encoding scheme. What do you see about this encoding scheme?

3. Let $(F, +, \cdot)$ be a field of integers modulo 2 and $(F[x], \diamondsuit, \diamondsuit)$ be the corresponding ring of polynomials. Construct a ring of polynomials modulo $1 + x^2$.

4. Let $(F, +, \cdot)$ be a field of integers modulo 3 and $(F[x], \diamondsuit, \diamondsuit)$ be the corresponding ring of polynomials. Construct a ring of polynomial modulo $2 + x^2$.

5. Let $(R, +, \cdot)$ be the field of real numbers. Let $(R[x], \diamondsuit, \diamondsuit)$ be the corresponding ring of polynomials and $(R_2[x], \diamondsuit_+, \diamondsuit)$ be the ring of polynomials modulo $1 + x^2$.

   a) For $a + bx$ and $c + dx$ in $R_2[x]$, find $(a + bx) \diamondsuit_+ (c + dx)$ and $(a + bx) \diamondsuit \cdot (c + dx)$.

   b) Compare this ring with the field of complex numbers.

Key Terms and Results

TERMS

algebra: a structure with a carrier (underlying set), operations on the carrier and elements with specific properties

arity of an operation: the number of elements on which the operation is performed

binary operation: operation on two elements

subalgebra: an algebra with a carrier which is a subset of the carrier of an algebra and operations and specific elements as that of the larger one

identity element: when an element operates with this element, it remains unchanged

zero element: when an element operates with this element, it is transformed into this element

Associative property of operation: $*a * (b * c) = (a * b) * c$.

commutative property of operation: $*a * b = b * a$

closure property: if for all a, b of a set S, $a * b$ is in S, then we say S is closed with respect to $*$

semigroup: an algebra which satisfies closure and associative property

monoid: a semigroup which has an identity element

inverse element of a: an element a^{-1} such that $a^{-1} * a = a * a^{-1} = e$, where e is the identity element

group: a monoid in which for each a of the underlying set there exists an inverse element

subgroup of a group: a group whose elements are subsets of the elements of a group, with the same operation

order of a group: the number of elements in the group

generators of a group: a subset of the elements that can generate all elements of the group with the operation of the group

cosets: if H is a subset of the set of elements T of a group $G = (T, \square)$, the elements $a \square H$ is the left coset of H with respect to a and $H \square a$ is the right coset

abelian group: a group which is commutative

cyclic groups: groups which have a single element as generator

homomorphism of a group: $(T, \square)$ into $(S, *)$ is a map h of elements of T into S such that $h(a \square b) = h(a) * h(b)$; identity of the first group is mapped onto the identity of the second group

isomorphism: is a homomorphism which is a bijection

automorphism: is an isomorphism if the domain and codomain are same

congruence relations: an equivalence relation which is both right invariant and left invariant

normal subgroup: a subgroup H of G is normal if every left coset of H is equal to the right coset

kernel of a homomorphism: if h is homomorphism mapping an algebra A to an algebra A', the set of elements of A which are mapped onto the identity element of A' is called the kernel of the homomorphism

rings, integral domains, and fields: algebras with two binary operations (usually called addition and multiplication) satisfying specific properties

ideal: a subset of the elements of a ring which is closed under addition and absorbs other elements under multiplication

homomorphism of a ring: $(S, +, *)$ to $(S', +', *')$ is a map satisfying

$$h(a + b) = h(a) +' h(b)$$

$$h(a * b) = h(a) *' h(b)$$

and identity element mapped onto the identity element

permutation groups: groups whose elements are permutations of a set of elements

dihedral groups: groups obtained from elements which are obtained by rotations and reflections of a regular polygon that keep the position of the polygon unaltered

quotient algebras: defined by the congruence classes of a homomorphism on an algebra

product algebras: are obtained by taking the Cartesian product of two algebras of the same type

encoding: converting message bits into codewords

decoding: converting codewords to message bits

distance of two bit strings: is the number of positions in which they have different bits

weight of a bit string: is the number of 1's in the string

codewords: Bit strings which are encoding of message bit strings

minimum distance of a code: minimum number of bits that have to be changed in a valid codeword to get another valid codeword

error correction: bits in error are corrected

error detection: a change in one or more bits is found out

group codes: codewords which form a group under $\oplus$

Hamming code: a special code used for error detection and correction

polynomial rings: rings whose elements are polynomials

polynoial codes: codes where polynomials are used for encoding

cyclic codes: if $a_1 \ldots a_n$ is a codeword, $a_n \, a_1 \ldots a_{n-1}$ is also a codeword

irreducible polynomial: a polynomial which cannot be factorized into two polynomials

RESULTS

Lagrange's theorem: Order of a subgroup divides the order of the group.

Homomorphic image of an algebra is an algebra of similar type.

The equivalence relation induced by a homomorphism on an algebra is a congruence relation.

Distinct left (right) cosets of a normal subgroup H are congruence classes of the underlying set of a group.

Dihedral group $(D_n, \diamond)$ are isomorphic to a subgroup of the symmetric group $(s_n, 0)$

Definitions of quotient and product algebras.

A code with minimum distance $2t + 1$ can correct t or fewer transmission errors.

The minimum distance of the group code f is the minimum weight of a nonzero codeword.

The set of codewords defined by a coding scheme using a Boolean matrix form a group.

Hamming codes are coding sets schemes from B^{2^r-1-r} to B^{2^r-1}.

If $g(x)$ is an encoding polynomial and $g(x)$ divides no polynomial of the form $x^k - 1$ for $k < n$, then the minimum distance of the codewords is at least 3.

The "error polynomial" ssociated with any undetected error vector $e = e_0 \ldots e_{n-1}$ of a polynomial (m, n) code with encoding polynomial $g(x)$ is a nontrivial multiple of $g(x)$.

Review Questions

1. What do you mean by an algebra? Give examples.

2. Define associative and commutative properties.

3. What do you mean by 'closure of a set S under an operation $*$'?

4. Define identity and zero elements of a set under a binary operation $*$. What do you mean by an inverse element?

5. If an algebra has a left zero and a right zero, they are equal. Prove.

6. If an algebra has a left identity and a right identity, they are equal. Prove.

7. Define a semigroup, a monoid, and a group.

8. Define a subgroup of a group G.

9. Define order of a group.

10. Let $G = (T, \square)$ be a group. To show that a subset of T forms a subgroup of G, it is enough to check closure property alone. Prove.

11. Show that two cosets of a subset of a carrier of a group are either identical or disjoint.

12. Define coset of a subset H of T of a group $G = (T, \square)$. State and prove Lagrange's theorem.

13. Define homomorphism and isomorphism of groups. Illustrate with examples.

14. Define generators of a group. What is a cyclic group?

15. Define congruence relations. Illustrate with examples.

16. Discuss the connection between homomorphisms of a group G and a normal subgroup of G.

17. Give the definitions of rings, integral domains, and fields.

18. Define an ideal of a ring. Discuss ring homomorphisms and the connection with ideals.

19. Explain what you mean by permutation group. Define the order and degree of a permutation group.

20. How are symmetric groups and dihedral groups defined?

21. What do you mean by even permutations and alternating groups?

22. Define quotient algebra of an algebra under a congruence relation.

23. Define product algebra of two algebras.

24. What do you mean by encoding and decoding? Why is it necessary to use encoding?

25. Define a parity check code.

26. Define a block code and illustrate with example.

27. What do you mean by the weight of a binary string?

28. What do you mean by the distance of two binary strings of equal length?

29. What do you mean by minimum distance of a code?

30. What is the connection between the minimum distance of a code and the error correction/detection possible with the code?

31. Discuss what you mean by a group code.

32. Explain how decoding is done in group codes.

33. Discuss what you mean by a polynomial code.

34. Discuss Hamming codes. Explain how such a code corrects single errors.

35. What do you mean by a polynomial ring? Discuss its application to coding.

Supplementary Exercises

1. Consider a set of two by two real matrices $M_{2\times2}(R)$. Consider the operation of matrix addition m_1. Show that $M_{2\times2}(R), m_+)$ is a group. What is the identity? What is the inverse of $\begin{bmatrix} a & b \\ c & d \end{bmatrix}$?

2. Consider a set of two by two nonsingular real matrices $M_{2\times2}^{ns}(R)$. Consider the operation of matrix multiplication m_x. Show that $(M_{2\times2}^{ns}(R), m_x)$ is a group. What is the identity? What is the inverse of $\begin{bmatrix} a & b \\ c & d \end{bmatrix}$?

3. For each of the following sets, identify the standard operation that results in a group. What is the identity of each group?
   a) The set of nonzero real numbers.
   b) The set of 2×3 matrices with rational elements.

4. Prove by induction on n that if $a_1, \ldots, a_n$ are elements of a group G, with the operation $*$ with $n \geq 2$, then $(a_1 * a_2 * \ldots * a_n)^{-1} = a_n^{-1} * a_{n-1}^{-1} * \ldots * a_1^{-1}$. Interpret this result in terms of $(Z, +)$, $(R - \{0\}, \times)$.

5. Consider the set of equations where mod 2 addition and mod 2 multiplication is used. List all solutions for
   a) $x^2 + 1 = 0$
   b) $x^2 + x + 1 = 0$

6. The dihedral group D_3 is isomorphic to the symmetric group S_3. But D_4 is not isomorphic to S_4. Can you give an explanation for this?

7. Prove that if G is an abelian group, then $q(x) = x^2$ (meaning $x * x$) defines a homomorphism from G into G. Is q ever an isomorphism?

8. Prove that if $\theta : G \to G'$ is a homomorphism and H a normal subgroup of G, then $\theta(H)$ is also a normal subgroup of $\theta(G)$. Is it also true that $\theta(H)$ is a normal subgroup of G'?

9. Show that the following are rings.
   a) $(M_{2\times2}(R), m_+, m_.)$
   b) $(Z_8, +_8, \cdot_8)$
   c) $(Z \times Z, +, \cdot)$
   Here $M_{2\times2}(R)$ is the set of 2×2 matrices from the set R of real numbers, m_+ and $m_.$, which denote matrix addition and matrix multiplication, respectively.
   Let $Z_8 = \{0, \ldots, 7\}$, $+_8$, $\cdot_8$ denote mod 8 addition and mod 8 multiplication, respectively.
   $Z \times Z$ is Cartesian product, where Z is the set of integers. Here $+$ and $\cdot$ are defined as

   $$(a, b) + (c, d) = (a + c, b + d)$$
   $$(a, b) \cdot (c, d) = (ac, bd)$$

10. Show that the following rings are not isomorphic.

    $$(Z, +, \cdot) \text{ and } (M_{2\times2}(R), m_+, m_.)$$

11. Let R be a commutative ring with unity. Prove the following by induction.
    a) For $n \geq 1$, $(a + b)^n = \sum_{k=0}^{n} C_k^n a^k b^{n-k}$
    b) Simplify $(a + b)^3$ in Z_3

12. Construct the decoding table for the group code given by the generator matrix

    $$\begin{bmatrix} 1 & 0 & 0 & 1 & 1 \\ 0 & 1 & 1 & 1 & 0 \end{bmatrix}$$

    Use the decoding table to decode the following received words.
        (*i*) 11110   (*ii*) 11101   (*iii*) 11011

13. Construct the decoding table for the group code generated by matrix

$$\begin{bmatrix} 1 & 1 & 0 & 1 & 0 & 0 \\ 1 & 0 & 1 & 0 & 1 & 0 \\ 1 & 1 & 1 & 0 & 0 & 1 \end{bmatrix}$$

Use the decoding table to decode the following received words:

(*i*) 111000   (*ii*) 110000   (*iii*) 101000

14. A central groupoid is an algebraic system $(A, *)$ where $*$ is a binary operation such that $(a * b) * (b * c) = b$ for all a, b, c in A.

a) Show that

$$a * ((a * b) * c) = a * b$$

$$(a * (b * c)*)c = b * c$$

in a central groupoid.

b) Let $(A, *)$ be an algebraic system where $*$ is a binary operation such that

$$(a * ((b * c) * d)) * (c * d) = c$$

for all a, b, c, d in A. Show that $(A, *)$ is a central groupoid.

15. Consider a coding scheme $e: B^1 \to B^4$ such that

$$e(0) = 0000$$

$$e(1) = 1111$$

What is the minimum distance of the code? Set up a coset table to show that e can indeed correct all single and double transmission errors.

Computer Projects

Write programs with these input and output.

1. Given a set S of elements and a binary operation $\diamond$ on S by a table. Find out if the system $(S, \diamond)$ has
   a) a left zero
   b) a right zero
   c) a left identity
   d) a right identity

2. Given a set S of elements and a binary operation $\diamond$ on S by a table. Find out if the operation $\diamond$ is
   a) commutative
   b) associative

3. Given a set S of elements and a binary operation $\diamond$ on S by a table, find out if the system $(S, \diamond)$ is a group. If so, find the inverse of each element.

4. Given that $G = (T, \square)$ is a group, $|T| = n$, find out if T has a subgroup, and if yes, find out whether it is a normal subgroup.

5. Given that $G = (T, \square)$ is a group, find out if it is a cyclic group. If so, find out the generators of the group.

6. Given a set S and two binary operations $+$ and $\cdot$, given by tables, find out if $(S, +, \cdot)$ is a ring.

7. Given m and even parity encoding scheme $e: B^m \to B^{m+1}$, write a program for finding out the encoding of a bitstring of length m. Find the decoding of the bitstring of length $m + 1$.

8. Given an encoding scheme $e: B^m \to B^n$ denoting a group code, find the cosets and coset leaders, and design the decoding scheme.

9. For the encoding and decoding scheme $e: B^m \to B^{3m}$ and d: discussed in Exercise 6 of Section 7, write a program to find the encoding of a bitstring of length m, and the decoding of bitstring of length $3m$.

10. Solve Project 9 for $e: B^m \to B^{5m}$ and $d: B^{5m} \to B^m$.

11. Write a program to get the Hamming code $B^4 \to B^7$. Write a program to find a single error and correct it.

12. Write a program to get the Hamming code $B^4 \to B^8$ where the 8th bit added is even parity bit. Write a program to find out if a single or double error has occurred and correct a single error.

Computations and Explorations

Use a computational program or programs you have written to do these exercises.

1. Let $\Sigma = \{a, b\}$. Consider strings of length upto 3 over Σ and consider the operation of concatenation. Construct the table for this binary operation. Check that it is associative but not commutative.

2. Let $Z_6 = \{0, 1, 2, 3, 4, 5\}$. Construct the tables for binary operations $(Z_6, \oplus)$ and $(Z_6, \odot)$. Check $(Z_6, \oplus)$ is a group.

3. Construct the set of all permutation of $(1, 2, 3, 4)$. Check it is a group. Find some subgroups of this group.

4. Draw the figures for the dihedral group $(D_5, \diamond)$. Draw its table. Write down the permutations corresponding to elements of this dihedral group.

5. Construct tables for $(Z_7, \oplus)$, $(Z_7, \odot)$ and check that $(Z_7, \oplus, \odot)$ is a field.

6. Construct some codes $e: B^3 \to B^7$ using a binary 3×7 matrix with the identity matrix as submatrix. Check that the codewords form a group. Find the minimum distance of the code.

7. For the group of codewords arrived at in the previous example, find the cosets and mark the coset leaders.

8. Find the Hamming code $e: B^4 \to B^7$. Extend this to $B^4 \to B^8$ by adding an extra parity check bit.

Writing Projects

Respond to these with essays using outside sources.

1. Discuss Burnside's theorem on permutation groups.

2. Find out about Rubik cube and how you would solve it.

3. Discuss Galois field and Galois groups.

4. Find out what is a cellular automata and discuss group properties of cellular automata and its application.

5. Discuss Hotz group of context-free languages.

6. Show that a set of even permutations of a set of elements form a group. Discuss the cyclic notation for representing permutations.

7. Discuss about Hadamard code.

8. Discuss about Reed-Soloman code.

9. Discuss about Reed-Muller code.

1 Axioms for the Real Numbers and the Positive Integers

In this book we have assumed an explicit set of axioms for the set of real numbers and for the set of positive integers. In this appendix we will list these axioms and we will illustrate how basic facts, also used without proof in the text, can be derived using them.

Axioms for Real Numbers

The standard axioms for real numbers include both the **field** (or **algebraic**) **axioms**, used to specify rules for basic arithmetic operations, and the **order axioms**, used to specify properties of the ordering of real numbers.

THE FIELD AXIOMS   We begin with the field axioms. As usual, we denote the sum and product of two real numbers x and y by $x + y$ and $x \cdot y$, respectively. (Note that the product of x and y is often denoted by xy without the use of the dot to indicate multiplication. We will not use this abridged notation in this appendix, but will within the text.) Also, by convention, we perform multiplications before additions unless parentheses are used. Although these statements are axioms, they are commonly called *laws* or *rules*. The first two of these axioms tell us that when we add or multiply two real numbers, the result is again a real number; these are the *closure laws*.

- **Closure law for addition**   For all real numbers x and y, $x + y$ is a real number.
- **Closure law for multiplication**   For all real numbers x and y, $x \cdot y$ is a real number.

The next two axioms tell us that when we add or multiply three real numbers, we get the same result regardless of the order of operations; these are the *associative laws*.

- **Associative law for addition**   For all real numbers x, y, and z, $(x + y) + z = x + (y + z)$.
- **Associative law for multiplication**   For all real numbers x, y, and z, $(x \cdot y) \cdot z = x \cdot (y \cdot z)$.

Two additional algebraic axioms tell us that the order in which we add or multiply two numbers does not matter; these are the *commutative laws*.

- **Commutative law for addition**   For all real numbers x and y, $x + y = y + x$.
- **Commutative law for multiplication**   For all real numbers x and y, $x \cdot y = y \cdot x$.

The next two axioms tell us that 0 and 1 are additive and multiplicative identities for the set of real numbers. That is, when we add 0 to a real number or multiply a real number by 1 we do not change this real number. These laws are called *identity laws*.

- **Additive identity law**   For every real number x, $x + 0 = 0 + x = x$.
- **Multiplicative identity law**   For every real number x, $x \cdot 1 = 1 \cdot x = x$.

Although it seems obvious, we also need the following axiom.

- **Identity elements axiom**   The additive identity 0 and the multiplicative identity 1 are distinct, that is $0 \neq 1$.

Two additional axioms tell us that for every real number, there is a real number that can be added to this number to produce 0, and for every nonzero real number, there is a real number by which it can be multiplied to produce 1. These are the *inverse laws*.

- **Inverse law for addition**   For every real number x, there exists a real number $-x$ (called the *additive inverse* of x) such that $x + (-x) = (-x) + x = 0$.
- **Inverse law for multiplication**   For every nonzero real number x, there exists a real number $1/x$ (called the *multiplicative inverse* of x) such that $x \cdot (1/x) = (1/x) \cdot x = 1$.

The final algebraic axioms for real numbers are the *distributive laws*, which tell us that multiplication distributes over addition; that is, that we obtain the same result when we first add a pair of real numbers and then multiply by a third real number or when we multiply each of these two real numbers by the third real number and then add the two products.

- **Distributive laws**   For all real numbers x, y, and z, $x \cdot (y + z) = x \cdot y + x \cdot z$ and $(x + y) \cdot z = x \cdot z + y \cdot z$.

ORDER AXIOMS   Next, we will state the *order axioms* for the real numbers, which specify properties of the "greater than" relation, denoted by $>$, on the set of real numbers. We write $x > y$ (and $y < x$) when x is greater than y, and we write $x \geq y$ (and $y \leq x$) when $x > y$ or $x = y$. The first of these axioms tells us that given two real numbers, exactly one of three possibilities occurs: the two numbers are equal, the first is greater than the second, or the second is greater than the first. This rule is called the *trichotomy law*.

- **Trichotomy law**   For all real numbers x and y, exactly one of $x = y$, $x > y$, or $y > x$ is true.

Next, we have an axiom, called *the transitivity law*, that tells us that if one number is greater than a second number and this second number is greater than a third, then the first number is greater than the third.

- **Transitivity law**   For all real numbers x, y, and z, if $x > y$ and $y > z$, then $x > z$.

We also have two *compatibility laws*, which tell us that when we add a number to both sides in a greater than relationship, the greater than relationship is preserved and when we multiply both sides of a greater than relationship by a *positive real number* (that is, a real number x with $x > 0$), the greater than relationship is preserved.

- **Additive compatibility law**   For all real numbers x, y, and z, if $x > y$, then $x + z > y + z$.
- **Multiplicative compatibility law**   For all real numbers x, y, and z, if $x > y$ and $z > 0$, then $x \cdot z > y \cdot z$.

We leave it to the reader (see Exercise 15) to prove that for all real numbers x, y, and z, if $x > y$ and $z < 0$, then $x \cdot z < y \cdot z$. That is, multiplication of an inequality by a negative real number reverses the direction of the inequality.

The final axiom for the set of real numbers is the *completeness property*. Before we state this axiom, we need some definitions. First, given a nonempty set A of real numbers, we say that the real number b is an **upper bound** of A if for every real number a in A, $b \geq a$. A real number s is a **least upper bound** of A if s is an upper bound of A and whenever t is an upper bound of A, then we have $s \leq t$.

- **Completeness property**   Every nonempty set of real numbers that is bounded above has a least upper bound.

Using Axioms to Prove Basic Facts

The axioms we have listed can be used to prove many properties that are often used without explicit mention. We give several examples of results we can prove using axioms and leave the proof of a variety of other properties as exercises. Although the results we will prove seem quite obvious, proving them using only the axioms we have stated can be challenging.

THEOREM 1   The additive identity element 0 of the real numbers is unique.

Proof: To show that the additive identity element 0 of the real numbers is unique, suppose that $0'$ is also an additive identity for the real numbers. This means that $0' + x = x + 0' = x$ whenever x is a real number. By the additive identity law, it follows that $0 + 0' = 0'$. Because $0'$ is an additive identity, we know that $0 + 0' = 0$. It follows that $0 = 0'$, because both equal $0 + 0'$. This shows that 0 is the unique additive identity for the real numbers.   ◁

THEOREM 2   The additive inverse of a real number x is unique.

Proof: Let x be a real number. Suppose that y and z are both additive inverses of x. Then,

$$
\begin{aligned}
y &= 0 + y && \text{by the additive identity law} \\
&= (z + x) + y && \text{because } z \text{ is an additive inverse of } x \\
&= z + (x + y) && \text{by the associative law for addition} \\
&= z + 0 && \text{because } y \text{ is an additive inverse of } x \\
&= z && \text{by the additive identity law.}
\end{aligned}
$$

It follows that $y = z$.   ◁

Theorems 1 and 2 tell us that the additive identity and additive inverses are unique. Theorems 3 and 4 tell us that the multiplicative identity and multiplicative inverses of nonzero real numbers are also unique. We leave their proofs as exercises.

THEOREM 3   The multiplicative identity element 1 of the real numbers is unique.

THEOREM 4   The multiplicative inverse of a nonzero real number x is unique.

THEOREM 5   For every real number x, $x \cdot 0 = 0$.

Proof: Suppose that x is a real number. By the additive inverse law, there is a real number y that is the additive inverse of $x \cdot 0$, so we have $x \cdot 0 + y = 0$. By the additive identity law, $0 + 0 = 0$. Using the distributive law, we see that $x \cdot 0 = x \cdot (0 + 0) = x \cdot 0 + x \cdot 0$. It follows that

$$0 = x \cdot 0 + y = (x \cdot 0 + x \cdot 0) + y.$$

Next, note that by the associative law for addition and because $x \cdot 0 + y = 0$, it follows that

$$(x \cdot 0 + x \cdot 0) + y = x \cdot 0 + (x \cdot 0 + y) = x \cdot 0 + 0.$$

Finally, by the additive identity law, we know that $x \cdot 0 + 0 = x \cdot 0$. Consequently, $x \cdot 0 = 0$.   ◁

THEOREM 6   For all real numbers x and y, if $x \cdot y = 0$, then $x = 0$ or $y = 0$.

Proof: Suppose that x and y are real numbers and $x \cdot y = 0$. If $x \neq 0$, then, by the multiplicative inverse law, x has a multiplicative inverse $1/x$, such that $x \cdot (1/x) = (1/x) \cdot x = 1$. Because $x \cdot y = 0$, we have $(1/x) \cdot (x \cdot y) = (1/x) \cdot 0 = 0$ by Theorem 5. Using the associate law for multiplication, we have $((1/x) \cdot x) \cdot y = 0$. This means that $1 \cdot y = 0$. By the multiplicative identity rule, we see that $1 \cdot y = y$, so $y = 0$. Consequently, either $x = 0$ or $y = 0$.   ◁

THEOREM 7   The multiplicative identity element 1 in the set of real numbers is greater than the additive identity element 0.

Proof: By the trichotomy law, either $0 = 1$, $0 > 1$, or $1 > 0$. We know by the identity elements axiom that $0 \neq 1$.

So, assume that $0 > 1$. We will show that this assumption leads to a contradiction. By the additive inverse law, 1 has an additive inverse -1 with $1 + (-1) = 0$. The additive compatibility law tells us that $0 + (-1) > 1 + (-1) = 0$; the additive identity law tells us that $0 + (-1) = -1$. Consequently, $-1 > 0$, and by the multiplicative compatibility law, $(-1) \cdot (-1) > (-1) \cdot 0$. By Theorem 5 the right-hand side of last inequality is 0. By the distributive law, $(-1) \cdot (-1) + (-1) \cdot 1 = (-1) \cdot (-1 + 1) = (-1) \cdot 0 = 0$. Hence, the left-hand side of this last inequality, $(-1) \cdot (-1)$, is the unique additive inverse of -1, so this side of the inequality equals 1. Consequently this last inequality becomes $1 > 0$, contradicting the trichotomy law because we had assumed that $0 > 1$.

Because we know that $0 \neq 1$ and that it is impossible for $0 > 1$, by the trichotomy law, we conclude that $1 > 0$.   ◁

Links

ARCHIMEDES (287 B.C.E.–212 B.C.E.)   Archimedes was one of the greatest scientists and mathematicians of ancient times. He was born in Syracuse, a Greek city-state in Sicily. His father, Phidias, was an astronomer. Archimedes was educated in Alexandria, Egypt. After completing his studies, he returned to Syracuse, where he spent the rest of his life. Little is known about his personal life; we do not know whether he was ever married or had children. Archimedes was killed in 212 B.C.E. by a Roman soldier when the Romans overran Syracuse.

Archimedes made many important discoveries in geometry. His method for computing the area under a curve was described two thousand years before his ideas were re-invented as part of integral calculus. Archimedes also developed a method for expressing large integers inexpressible by the usual Greek method. He discovered a method for computing the volume of a sphere, as well as of other solids, and he calculated an approximation of π. Archimedes was also an accomplished engineer and inventor; his machine for pumping water, now called *Archimedes' screw*, is still in use today. Perhaps his best known discovery is the *principle of buoyancy*, which tells us that an object submerged in liquid becomes lighter by an amount equal to the weight it displaces. Some histories tell us that Archimedes was an early streaker, running naked through the streets of Syracuse shouting "Eureka" (which means "I have found it") when he made this discovery. He is also known for his clever use of machines that held off Roman forces sieging Syracuse for several years during the Second Punic War.

The next theorem tells us that for every real number there is an integer (where by an *integer*, we mean 0, the sum of any number of 1s, and the additive inverses of these sums) greater than this real number. This result is attributed to the Greek mathematician Archimedes. The result can be found in Book V of Euclid's *Elements*.

THEOREM 8   **ARCHIMEDEAN PROPERTY**   For every real number x there exists an integer n such that $n > x$.

Proof: Suppose that x is a real number such that $n \leq x$ for every integer n. Then x is an upper bound of the set of integers. By the completeness property it follows that the set of integers has a least upper bound M. Because $M - 1 < M$ and M is a least upper bound of the set of integers, $M - 1$ is not an upper bound of the set of integers. This means that there is an integer n with $n > M - 1$. This implies that $n + 1 > M$, contradicting the fact that M is an upper bound of the set of integers.   ◁

Axioms for the Set of Positive Integers

The axioms we now list specify the set of positive integers as
the subset of the set of integers satisfying four key properties. We assume the truth of these axioms in this textbook.

- ■ **Axiom 1**   The number 1 is a positive integer.
- ■ **Axiom 2**   If n is a positive integer, then $n + 1$, the *successor* of n, is also a positive integer.
- ■ **Axiom 3**   Every positive integer other than 1 is the successor of a positive integer.
- ■ **Axiom 4**   **The Well-Ordering Property** Every nonempty subset of the set of positive integers has a least element.

In Sections 5.1 and 5.2 it is shown that the well-ordering principle is equivalent to the principle of mathematical induction.

- ■ **Mathematical induction axiom**   If S is a set of positive integers such that $1 \in S$ and for all positive integers n if $n \in S$, then $n + 1 \in S$, then S is the set of positive integers.

Most mathematicians take the real number system as already existing, with the real numbers satisfying the axioms we have listed in this appendix. However, mathematicians in the nineteenth century developed techniques to construct the set of real numbers, starting with more basic sets of numbers. (The process of constructing the real numbers is sometimes studied in advanced undergraduate mathematics classes. A treatment of this can be found in [Mo91], for instance.) The first step in the process is the construction of the set of positive integers using axioms 1–3 and either the well-ordering property or the mathematical induction axiom. Then, the operations of addition and multiplication of positive integers are defined. Once this has been done, the set of integers can be constructed using equivalence classes of pairs of positive integers where $(a, b) \sim (c, d)$ if and only if $a + d = b + c$; addition and multiplication of integers can be defined using these pairs (see Exercise 21). (Equivalence relations and equivalence classes are discussed in Chapter 9.) Next, the set of rational numbers can be constructed using the equivalence classes of pairs of integers where the second integer in the pair is not zero, where $(a, b) \approx (c, d)$ if and only if $a \cdot d = b \cdot c$; addition and multiplication of rational numbers can be defined in terms of these pairs (see Exercise 22). Using infinite sequences, the set of real numbers can then be constructed from the set of rational numbers. The interested reader will find it worthwhile to read through the many details of the steps of this construction.

Exercises

Use only the axioms and theorems in this appendix in the proofs in your answers to these exercises.

1. Prove Theorem 3, which states that the multiplicative identity element of the real numbers is unique.

2. Prove Theorem 4, which states that for every nonzero real number x, the multiplicative inverse of x is unique.

3. Prove that for all real numbers x and y, $(-x) \cdot y = x \cdot (-y) = -(x \cdot y)$.

4. Prove that for all real numbers x and y, $-(x + y) = (-x) + (-y)$.

5. Prove that for all real numbers x and y, $(-x) \cdot (-y) = x \cdot y$.

6. Prove that for all real numbers x, y, and z, if $x + z = y + z$, then $x = y$.

7. Prove that for every real number x, $-(-x) = x$.

Define the **difference** $x - y$ of real numbers x and y by $x - y = x + (-y)$, where $-y$ is the additive inverse of y, and the **quotient** x/y, where $y \neq 0$, by $x/y = x \cdot (1/y)$, where $1/y$ is the multiplicative inverse of y.

8. Prove that for all real numbers x and y, $x = y$ if and only if $x - y = 0$.

9. Prove that for all real numbers x and y, $-x - y = -(x + y)$.

10. Prove that for all nonzero real numbers x and y, $1/(x/y) = y/x$, where $1/(x/y)$ is the multiplicative inverse of x/y.

11. Prove that for all real numbers w, x, y, and z, if $x \neq 0$ and $z \neq 0$, then $(w/x) + (y/z) = (w \cdot z + x \cdot y)/(x \cdot z)$.

12. Prove that for every positive real number x, $1/x$ is also a positive real number.

13. Prove that for all positive real numbers x and y, $x \cdot y$ is also a positive real number.

14. Prove that for all real numbers x and y, if $x > 0$ and $y < 0$, then $x \cdot y < 0$.

15. Prove that for all real numbers x, y, and z, if $x > y$ and $z < 0$, then $x \cdot z < y \cdot z$.

16. Prove that for every real number x, $x \neq 0$ if and only if $x^2 > 0$.

17. Prove that for all real numbers w, x, y, and z, if $w < x$ and $y < z$, then $w + y < x + z$.

18. Prove that for all positive real numbers x and y, if $x < y$, then $1/x > 1/y$.

19. Prove that for every positive real number x, there exists a positive integer n such that $n \cdot x > 1$.

*20. Prove that between every two distinct real numbers there is a rational number (that is, a number of the form x/y, where x and y are integers with $y \neq 0$).

Exercises 21 and 22 involve the notion of an equivalence relation, discussed in Chapter 9 of the text.

*21. Define a relation $\sim$ on the set of ordered pairs of positive integers by $(w, x) \sim (y, z)$ if and only if $w + z = x + y$. Show that the operations $[(w, x)]_\sim + [(y, z)]_\sim = [(w + y, x + z)]_\sim$ and $[(w, x)]_\sim \cdot [(y, z)]_\sim = [(w \cdot y + x \cdot z, x \cdot y + w \cdot z)]_\sim$ are well-defined, that is, they do not depend on the representative of the equivalence classes chosen for the computation.

*22. Define a relation $\approx$ on ordered pairs of integers with second entry nonzero by $(w, x) \approx (y, z)$ if and only if $w \cdot z = x \cdot y$. Show that the operations $[(w, x)]_\approx + [(y, z)]_\approx = [(w \cdot z + x \cdot y, x \cdot z)]_\approx$ and $[(w, x)]_\approx \cdot [(y, z)]_\approx = [(w \cdot y, x \cdot z)]_\approx$ are well-defined, that is, they do not depend on the representative of the equivalence classes chosen for the computation.

2 Exponential and Logarithmic Functions

I n this appendix we review some of the basic properties of exponential functions and logarithms. These properties are used throughout the text. Students requiring further review of this material should consult precalculus or calculus books, such as those mentioned in the Suggested Readings.

Exponential Functions

Let n be a positive integer, and let b be a fixed positive real number. The function $f_b(n) = b^n$ is defined by

$$f_b(n) = b^n = b \cdot b \cdot b \cdot \cdots \cdot b,$$

where there are n factors of b multiplied together on the right-hand side of the equation.

We can define the function $f_b(x) = b^x$ for all real numbers x using techniques from calculus. The function $f_b(x) = b^x$ is called the **exponential function to the base b**. We will not discuss how to find the values of exponential functions to the base b when x is not an integer.

Two of the important properties satisfied by exponential functions are given in Theorem 1. Proofs of these and other related properties can be found in calculus texts.

THEOREM 1 Let b be a positive real number and x and y real numbers. Then
  1. $b^{x+y} = b^x b^y$, and
  2. $(b^x)^y = b^{xy}$.

We display the graphs of some exponential functions in Figure 1.

Logarithmic Functions

Suppose that b is a real number with $b > 1$. Then the exponential function b^x is strictly increasing (a fact shown in calculus). It is a one-to-one correspondence from the set of real numbers to the set of nonnegative real numbers. Hence, this function has an inverse $\log_b x$, called the **logarithmic function to the base b**. In other words, if b is a real number greater than 1 and x is a positive real number, then

$$b^{\log_b x} = x.$$

The value of this function at x is called the **logarithm of x to the base b**.

From the definition, it follows that

$$\log_b b^x = x.$$

We give several important properties of logarithms in Theorem 2.

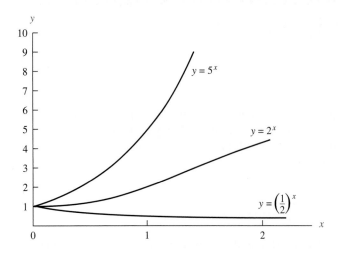

FIGURE 1   Graphs of the Exponential Functions to the Bases $\frac{1}{2}$, 2, and 5.

THEOREM 2

Let b be a real number greater than 1. Then

1. $\log_b(xy) = \log_b x + \log_b y$ whenever x and y are positive real numbers, and
2. $\log_b(x^y) = y \log_b x$ whenever x is a positive real number and y is a real number.

Proof: Because $\log_b(xy)$ is the unique real number with $b^{\log_b(xy)} = xy$, to prove part 1 it suffices to show that $b^{\log_b x + \log_b y} = xy$. By part 1 of Theorem 1, we have

$$b^{\log_b x + \log_b y} = b^{\log_b x} b^{\log_b y}$$
$$= xy.$$

To prove part 2, it suffices to show that $b^{y \log_b x} = x^y$. By part 2 of Theorem 1, we have

$$b^{y \log_b x} = (b^{\log_b x})^y$$
$$= x^y.$$ ◁

The following theorem relates logarithms to two different bases.

THEOREM 3

Let a and b be real numbers greater than 1, and let x be a positive real number. Then

$$\log_a x = \log_b x / \log_b a.$$

Proof: To prove this result, it suffices to show that

$$b^{\log_a x \cdot \log_b a} = x.$$

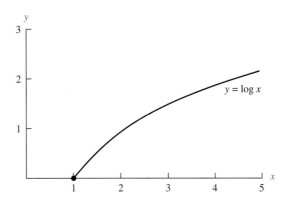

FIGURE 2   The Graph of $f(x) = \log x$.

By part 2 of Theorem 1, we have

$$b^{\log_a x \,\cdot\, \log_b a} = (b^{\log_b a})^{\log_a x}$$
$$= a^{\log_a x}$$
$$= x.$$

This completes the proof.   ◁

Because the base used most often for logarithms in this text is $b = 2$, the notation $\log x$ is used throughout the test to denote $\log_2 x$.

The graph of the function $f(x) = \log x$ is displayed in Figure 2. From Theorem 3, when a base b other than 2 is used, a function that is a constant multiple of the function $\log x$, namely, $(1/\log b)\log x$, is obtained.

Exercises

1. Express each of the following quantities as powers of 2.

   a) $2 \cdot 2^2$ **b)** $(2^2)^3$ **c)** $2^{(2^2)}$

2. Find each of the following quantities.

   a) $\log_2 1024$ **b)** $\log_2 1/4$ **c)** $\log_4 8$

3. Suppose that $\log_4 x = y$ where x is a positive real number. Find each of the following quantities.

   a) $\log_2 x$ **b)** $\log_8 x$ **c)** $\log_{16} x$

4. Let $a, b,$ and c be positive real numbers. Show that $a^{\log_b c} = c^{\log_b a}$.

5. Draw the graph of $f(x) = b^x$ for all real numbers x if b is

   a) 3.   **b)** 1/3.   **c)** 1.

6. Draw the graph of $f(x) = \log_b x$ for positive real numbers x if b is

   a) 4.   **b)** 100.   **c)** 1000.

3

Pseudocode

Links

The algorithms in this text are described both in English and in **pseudocode**. Pseudocode is an intermediate step between an English language description of the steps of a procedure and a specification of this procedure using an actual programming language. The advantages of using pseudocode include the simplicity with which it can be written and understood and the ease of producing actual computer code (in a variety of programming languages) from the pseudocode. We will describe the particular types of **statements**, or high-level instructions, of the pseudocode that we will use. Each of these statements in pseudocode can be translated into one or more statements in a particular programming language, which in turn can be translated into one or more (possibly many) low-level instructions for a computer.

This appendix describes the format and syntax of the pseudocode used in the text. This pseudocode is designed so that its basic structure resembles that of commonly used programming languages, such as C++ and Java, which are currently the most commonly taught programming languages. However, the pseudocode we use will be a lot looser than a formal programming language because a lot of English language descriptions of steps will be allowed.

This appendix is not meant for formal study. Rather, it should serve as a reference guide for students when they study the descriptions of algorithms given in the text and when they write pseudocode solutions to exercises.

Procedure Statements

The pseudocode for an algorithm begins with a **procedure** statement that gives the name of an algorithm, lists the input variables, and describes what kind of variable each input is. For instance, the statement

procedure *maximum*(L: list of integers)

is the first statement in the pseudocode description of the algorithm, which we have named *maximum*, that finds the maximum of a list L of integers.

Assignments and Other Types of Statements

An assignment statement is used to assign values to variables. In an assignment statement the left-hand side is the name of the variable and the right-hand side is an expression that involves constants, variables that have been assigned values, or functions defined by procedures. The right-hand side may contain any of the usual arithmetic operations. However, in the pseudocode in this book it may include any well-defined operation, even if this operation can be carried out only by using a large number of statements in an actual programming language.

The symbol $:=$ is used for assignments. Thus, an assignment statement has the form

variable $:=$ *expression*

For example, the statement

$$max := a$$

assigns the value of a to the variable max. A statement such as

$$x := \text{largest integer in the list } L$$

can also be used. This sets x equal to the largest integer in the list L. To translate this statement into an actual programming language would require more than one statement. Also, the instruction

$$\text{interchange } a \text{ and } b$$

can be used to interchange a and b. We could also express this one statement with several assignment statements (see Exercise 2), but for simplicity, we will often prefer this abbreviated form of pseudocode.

Comments

In the pseudocode in this book, statements enclosed in curly braces are not executed. Such statements serve as comments or reminders that help explain how the procedure works. For instance, the statement

$$\{x \text{ is the largest element in } L\}$$

can be used to remind the reader that at that point in the procedure the variable x equals the largest element in the list L.

Conditional Constructions

The simplest form of the conditional construction that we will use is

if condition **then** statement

or

if condition **then**
        block of statements

Here, the condition is checked, and if it is true, then the statement or block of statements given is carried out. In particular, the pseudocode

if condition **then**
    statement 1
    statement 2
    statement 3
    .
    .
    .
    statement n

tells us that the statements in the block are executed sequentially if the condition is true.

For example, in Algorithm 1 in Section 3.1, which finds the maximum of a set of integers, we use a conditional statement to check whether $max < a_i$ for each variable; if it is, we assign the value of a_i to max.

Often, we require the use of a more general type of construction. This is used when we wish to do one thing when the indicated condition is true, but another when it is false. We use the construction

if condition **then** statement 1
else statement 2

Note that either one or both of statement 1 and statement 2 can be replaced with a block of statements.

Sometimes, we require the use of an even more general form of a conditional. The general form of the conditional construction that we will use is

if condition 1 **then** statement 1
else if condition 2 **then** statement 2
else if condition 3 **then** statement 3
    .
    .
    .
else if condition n **then** statement n
else statement $n + 1$

When this construction is used, if condition 1 is true, then statement 1 is carried out, and the program exits this construction. In addition, if condition 1 is false, the program checks whether condition 2 is true; if it is, statement 2 is carried out, and so on. Thus, if none of the first $n - 1$ conditions hold, but condition n does, statement n is carried out. Finally, if none of condition 1, condition 2, condition 3, . . . , condition n is true, then statement $n + 1$ is executed. Note that any of the $n + 1$ statements can be replaced by a block of statements.

Loop Constructions

There are two types of loop construction in the pseudocode in this book. The first is the "for" construction, which has the form

> **for** *variable* := *initial value* **to** *final value*
>   statement

or

> **for** *variable* := *initial value* **to** *final value*
>   block of statements

where *initial value* and *final value* are integers. Here, at the start of the loop, *variable* is assigned *initial value* if *initial value* is less than or equal to *final value*, and the statements at the end of this construction are carried out with this value of *variable*. Then *variable* is increased by one, and the statement, or the statements in the block, are carried out with this new value of *variable*. This is repeated until *variable* reaches *final value*. After the instructions are carried out with *variable* equal to *final value*, the algorithm proceeds to the next statement. When *initial value* exceeds *final value*, none of the statements in the loop is executed.

We can use the "for" loop construction to find the sum of the positive integers from 1 to n with the following pseudocode.

> *sum* := 0
> **for** i := 1 **to** n
>   *sum* := *sum* + i

Also, the more general "for" statement, of the form

> **for** all elements with a certain property

is used in this text. This means that the statement or block of statements that follow are carried out successively for the elements with the given property.

The second type of loop construction that we will use is the "while" construction. This has the form

> **while** condition
>   statement

or

> **while** condition
>   block of statements

When this construction is used, the condition given is checked, and if it is true, the statements that follow are carried out, which may change the values of the variables that are part of the condition.

If the condition is still true after these instructions have been carried out, the instructions are carried out again. This is repeated until the condition becomes false. As an example, we can find the sum of the integers from 1 to n using the following block of pseudocode including a "while" construction.

$sum := 0$
while $n > 0$
    $sum := sum + n$
    $n := n - 1$

Note that any "for" construction can be turned into a "while" construction (see Exercise 3). However, it is often easier to understand the "for" construction. So, when it makes sense, we will use the "for" construction in preference to the corresponding "while" construction.

Loops within Loops

Loops or conditional statements are often used within other loops or conditional statements. In the pseudocode used in this book, we use successive levels of indentation to indicate nested loops, which are loops within loops, and which blocks of commands correspond to which loops.

Using Procedures in Other Procedures

We can use a procedure from within another procedure (or within itself in a recursive program) simply by writing the name of this procedure followed by the inputs to this procedure. For instance,

$max(L)$

will carry out the procedure max with the input list L. After all the steps of this procedure have been carried out, execution carries on with the next statement in the procedure.

Return Statements

We use a **return** statement to show where a procedure produces output. A return statement of the form

return x

produces the current value of x as output. The output x can involve the value of one or more functions, including the same function under evaluation, but at a smaller value. For instance, the statement

return $f(n - 1)$

is used to call the algorithm with input of $n - 1$. This means that the algorithm is run again with input equal to $n - 1$.

Exercises

1. What is the difference between the following blocks of two assignment statements?

$$a := b$$
$$b := c$$

and

$$b := c$$
$$a := b$$

2. Give a procedure using assignment statements to interchange the values of the variables x and y. What is the minimum number of assignment statements needed to do this?

3. Show how a loop of the form

for $i :=$ *initial value* **to** *final value*
    statement

can be written using the "while" construction.

Suggested Reading

Among the resources available for learning more about the topics covered in this book are printed materials and relevant websites. Printed resources are described in this section of suggested readings. Readings are listed by chapter and keyed to particular topics of interest. Some general references also deserve special mention. A book you may find particularly useful is the *Handbook of Discrete and Combinatorial Mathematics* by Rosen [Ro00], a comprehensive reference book. Additional applications of discrete mathematics can be found in Michaels and Rosen [MiRo91], which is also available online on the companion website for this book. Deeper coverage of many topics in computer science, including those discussed in this book, can be found in Gruska [Gr97]. Biographical information about many of the mathematicians and computer scientists mentioned in this book can be found in Gillispie [Gi70] and on the MacTutor website at *http://www-history.mcs.st-and.ac.uk/*.

To find pertinent websites, consult the links found in the Web Resources Guide on the companion website for this book. Its address is *www.mhhe.com/rosen*.

CHAPTER 1

An entertaining way to study logic is to read Lewis Carroll's book [Ca78]. General references for logic include M. Huth and M. Ryan [HuRy04], Mendelson [Me09], Stoll [St74], and Suppes [Su87]. A comprehensive treatment of logic in discrete mathematics can be found in Gries and Schneider [GrSc93]. System specifications are discussed in Ince [In93]. Smullyan's knights and knaves puzzles were introduced in [Sm78]. He has written many fascinating books on logic puzzles including [Sm92] and [Sm98]. Prolog is discussed in depth in Nilsson and Maluszynski [NiMa95] and in Clocksin and Mellish [ClMe94]. The basics of proofs are covered in Cupillari [Cu05], Morash [Mo91], Solow [So09], Velleman [Ve06], and Wolf [Wo98]. The science and art of constructing proofs is discussed in a delightful way in three books by Pólya: [Po62], [Po71], and [Po90]. Problems involving tiling checkerboards using dominoes and polyominoes are discussed in Golomb [Go94] and Martin [Ma91].

CHAPTER 2

Lin and Lin [LiLi81] is an easily read text on sets and their applications. Axiomatic developments of set theory can be found in Halmos [Ha60], Monk [Mo69], and Stoll [St74]. Brualdi [Br09], and Reingold, Nievergelt, and Deo [ReNiDe77] contain introductions to multisets. Fuzzy sets and their application to expert systems and artificial intelligence are treated in Negoita [Ne85] and Zimmerman [Zi91]. Calculus books, such as Apostol [Ap67], Spivak [Sp94], and Thomas and Finney [ThFi96], contain discussions of functions. The best printed source of information about integer sequences is Sloan and Plouffe [SlPl95]. Books on proofs, such as [Ve06], often cover countability in some depth. Stanat and McAllister [StMc77] has a thorough section on countability. Chapter 17 of Aigner, Ziegler, and Hoffman [AiZiHo09] provides an excellent discussion of cardinality and the continuum hypothesis.

Discussions of the mathematical foundations needed for computer science can be found in Arbib, Kfoury, and Moll [ArKfMo80], Bobrow and Arbib [BoAr74], Beckman [Be80], and Tremblay and Manohar [TrMa75]. Matrices and their operations are covered in all linear algebra books, such as Curtis [Cu84] and Strang [St09].

CHAPTER 3

The articles by Knuth [Kn77] and Wirth [Wi84] are accessible introductions to the subject of algorithms. Among the best introductions to algorithms are Cormen, Leierson, Rivest, and Stein [CoLeRiSt09] and Kleinberg and Tardos [KlTa05]. Extensive material on big-*O* estimates of functions can be found in Knuth [Kn97a]. General references for algorithms and their complexity include Aho, Hopcroft, and Ullman [AhHoUl74]; Baase and Van Gelder [BaGe99]; Cormen, Leierson, Rivest, and Stein [CoLeRiSt09]; Gonnet [Go84]; Goodman and Hedetniemi [GoHe77]; Harel [Ha87]; Horowitz and Sahni [HoSa82]; Kreher and Stinson [KrSt98]; the famous series of books by Knuth on the art of computer programming [Kn97a], [Kn97b], and [Kn98]; Kronsjö [Kr87]; Levitin [Le06]; Manber [Ma89]; Pohl and Shaw [PoSh81]; Purdom and Brown [PuBr85]; Rawlins [Ra92]; Sedgewick [Se03]; Wilf [Wi02]; and Wirth [Wi76]. Sorting and searching algorithms and their complexity are studied in detail in Knuth [Kn98].

CHAPTER 4

References for number theory include Hardy and Wright [HaWrWiHe08]; LeVeque [Le77]; Rosen [Ro10]; and Stark [St78]. More about the history of number theory can be found in Ore [Or88]. Algorithms for computer arithmetic are discussed in Knuth [Kn97b] and Pohl and Shaw [PoSh81]. More information about algorithms for finding primes and for factorization can be found in Crandall and Pomerance [CrPo10]. Applications of number theory to cryptography are covered in Denning [De82]; Menezes, van Oorschot, and Vanstone [MeOoVa97]; Rosen [Ro10]; Seberry and Pieprzyk [SePi89]; Sinkov [Si66]; and Stinson [St05]. The RSA public-key system was described by Rivest, Shamir, and Adleman in [RiShAd78]; its discovery by Cocks is described in [Si99], which also provides an appealing account of the history of cryptography.

CHAPTER 5

An accessible introduction to mathematical induction can be found [Gu10] and in Sominskii [So61]. Books that contain thorough treatments of mathematical induction and recursive definitions include Liu [Li85]; Sahni [Sa85]; Stanat and McAllister [StMc77]; and Tremblay and Manohar [TrMa75]. Computational geometry is covered in [DeOr11] and [Or00]. The Ackermann function, introduced in 1928 by W. Ackermann, arises in the theory of recursive function (see Beckman [Be80] and McNaughton [Mc82], for instance) and in the analysis of the complexity of certain set theoretic algorithms (see Tarjan [Ta83]). Recursion is studied in Roberts [Ro86]; Rohl [Ro84]; and Wand [Wa80]. Discussions of program correctness and the logical machinery used to prove that programs are correct can be found in Alagic and Arbib [AlAr78]; Anderson [An79]; Backhouse [Ba86]; Sahni [Sa85]; and Stanat and McAllister [StMc77].

CHAPTER 6

General references for counting techniques and their applications include Allenby and Slomson [AlSl10]; Anderson [An89]; Berman and Fryer [BeFr72]; Bogart [Bo00]; Bona [Bo07]; Bose and Manvel [BoMa86]; Brualdi [Br09]; Cohen [Co78]; Grimaldi [Gr03]; Gross [Gr07]; Liu [Li68]; Pólya, Tarjan, and Woods [PoTaWo83]; Riordan [Ri58]; Roberts and Tesman [RoTe03]; Tucker [Tu06]; and Williamson [Wi85]. Vilenkin [Vi71] contains a selection of combinatorial problems and their solutions. A selection of more difficult combinatorial problems can be found in Lovász [Lo79]. Information about Internet protocol addresses and datagrams can be found in Comer [Co05]. Applications of the pigeonhole principle can be found in Brualdi [Br09]; Liu [Li85]; and Roberts and Tesman [RoTe03]. A wide selection of combinatorial identities can be found in Riordan [Ri68] and in Benjamin and Quinn [BeQu03]. Combinatorial algorithms, including algorithms for generating permutations and combinations, are described by Even [Ev73]; Lehmer [Le64]; and Reingold, Nievergelt, and Deo [ReNiDe77].

CHAPTER 7

Useful references for discrete probability theory include Feller [Fe68], Nabin [Na00], and Ross [Ro09a]. Ross [Ro02], which focuses on the application of probability theory to computer science, provides examples of average case complexity analysis and covers the probabilistic method. Aho and Ullman [AhUl95] includes a discussion of various aspects of probability theory important in computer science, including programming applications of probability. The probabilistic method is discussed in a chapter in Aigner, Ziegler, and Hoffman [AiZiHo09], a monograph devoted to clever, insightful, and brilliant proofs, that is, proofs that Paul Erdős described as coming from "*The Book*." Extensive coverage of the probabilistic method can be found in Alon and Spencer [AlSp00]. Bayes' theorem is covered in [PaPi01]. Additional material on spam filters can be found in [Zd05].

CHAPTER 8

Many different models using recurrence relations can be found in Roberts and Tesman [RoTe03] and Tucker [Tu06]. Exhaustive treatments of linear homogeneous recurrence relations with constant coefficients, and related inhomogeneous recurrence relations, can be found in Brualdi [Br09], Liu [Li68], and Mattson [Ma93]. Divide-and-conquer algorithms and their complexity are covered in Roberts and Tesman [RoTe03] and Stanat and McAllister [StMc77]. Descriptions of fast multiplication of integers and matrices can be found in Aho, Hopcroft, and Ullman [AhHoUl74] and Knuth [Kn97b]. An excellent introduction to generating functions can be found in Pólya, Tarjan, and Woods [PoTaWo83]. Generating functions are studied in detail in Brualdi [Br09]; Cohen [Co78]; Graham, Knuth, and Patashnik [GrKnPa94]; Grimaldi [Gr03]; and Roberts and Tesman [RoTe03]. Additional applications of the principle of inclusion–exclusion can be found in Liu [Li85] and [Li68]; Roberts and Tesman [RoTe03]; and Ryser [Ry63].

CHAPTER 9

General references for relations, including treatments of equivalence relations and partial orders, include Bobrow and Arbib [BoAr74]; Grimaldi [Gr03]; Sanhi [Sa85]; and Tremblay and Manohar [TrMa75]. Discussions of relational models for databases are given in Date [Da82] and Aho and Ullman [AhUl95]. The original papers by Roy and Warshall for finding transitive closures can be found in [Ro59] and [Wa62], respectively. Directed graphs are studied in Chartrand, Lesniak, and Zhang [ChLeZh05]; Gross and Yellen [GrYe05]; Robinson and Foulds [RoFo80]; Roberts and Tesman [RoTe03]; and Tucker [Tu06]. The application of lattices to information flow is treated in Denning [De82].

CHAPTER 10

General references for graph theory include Agnarsson and Greenlaw [AgGr06]; Aldous, Wilson, and Best [AlWiBe00]; Behzad and Chartrand [BeCh71]; Chartrand, Lesniak, and Zhang [ChLeZh05]; Chartrand and Zhang [ChZh04]; Bondy and Murty [BoMu10]; Chartrand and Oellermann [ChOe93]; Graver and Watkins [GrWa77]; Roberts and Tesman [RoTe03]; Tucker [Tu06]; West [We00]; Wilson [Wi85]; and Wilson and Watkins [WiWa90]. A wide variety of applications of graph theory can be found in Chartrand [Ch77]; Deo [De74]; Foulds [Fo92]; Roberts and Tesman [RoTe03]; Roberts [Ro76]; Wilson and Beineke [WiBe79]; and McHugh [Mc90]. In depth treatments of the use of graph theory to study social networks, and other types of networks, appears in Easley and Kleinberg [EaKl10] and Newman [Ne10]. Applications involving large graphs, including the Web graph, are discussed in Hayes [Ha00a] and [Ha00b].

A comprehensive description of algorithms in graph theory can be found in Gibbons [Gi85] and in Kocay and Kreher [KoKr04]. Other references for algorithms in graph theory include Buckley and Harary [BuHa90]; Chartrand and Oellermann [ChOe93]; Chachra, Ghare, and Moore [ChGhMo79]; Even [Ev73] and [Ev79]; Hu [Hu82]; and Reingold, Nievergelt, and Deo [ReNiDe77]. A translation of Euler's original paper on the Königsberg bridge problem can be found in Euler [Eu53]. Dijkstra's algorithm is studied in Gibbons [Gi85]; Liu [Li85]; and Reingold, Nievergelt, and Deo [ReNiDe77]. Dijkstra's original paper can be found in [Di59]. A proof of Kuratowski's theorem can be found in Harary [Ha69] and Liu [Li68]. Crossing numbers and thicknesses of graphs are studied in Chartrand, Lesniak, and Zhang [ChLeZh05]. References for graph coloring and the four-color theorem are included in Barnette [Ba83] and Saaty and Kainen [SaKa86]. The original conquest of the four-color theorem is reported in Appel and Haken [ApHa76]. Applications of graph coloring are described by Roberts and Tesman [RoTe03]. The history of graph theory is covered in Biggs, Lloyd, and Wilson [BiLlWi86]. Interconnection networks for parallel processing are discussed in Akl [Ak89] and Siegel and Hsu [SiHs88].

CHAPTER 11

Trees are studied in Deo [De74], Grimaldi [Gr03], Knuth [Kn97a], Roberts and Tesman [RoTe03], and Tucker [Tu06]. The use of trees in computer science is described by Gotlieb and Gotlieb [GoGo78], Horowitz and Sahni [HoSa82], and Knuth [Kn97a, 98]. Roberts and Tesman [RoTe03] covers applications of trees to many different areas. Prefix codes and Huffman coding are covered in Hamming [Ha80]. Backtracking is an old technique; its use to solve maze puzzles can be found in the 1891 book by Lucas [Lu91]. An extensive discussion of how to solve problems using backtracking can be found in Reingold, Nievergelt, and Deo [ReNiDe77]. Gibbons [Gi85] and Reingold, Nievergelt, and Deo [ReNiDe77] contain discussions of algorithms for constructing spanning trees and minimal spanning trees. The background and history of algorithms for finding minimal spanning trees is covered in Graham and Hell [GrHe85]. Prim and Kruskal described their algorithms for finding minimal spanning trees in [Pr57] and [Kr56], respectively. Sollin's algorithm is an example of an algorithm well suited for parallel processing; although Sollin never published a description of it, his algorithm has been described by Even [Ev73] and Goodman and Hedetniemi [GoHe77].

CHAPTER 12

For further study into Algebraic Structures see Liu [Li85], Tremblay and Manohar [Tr76], and Stanat and McAllister [St77].

APPENDIXES

A discussion of axioms for the real number and for the integers can be found in Morash [Mo91]. Detailed treatments of exponential and logarithmic functions can be found in calculus books such as Apostol [Ap67], Spivak [Sp94], and Thomas and Finney [ThFi96]. Pohl and Shaw [PoSh81] use a form of pseudocode that has the same features as those described in Appendix 3. Most textbooks on algorithms, such as Cormen, Leierson, Rivest, and Stein [CoLeRiSt09] and Kleinberg and Tardos [KlTa05], use versions of pseudocode similar to the pseudocode in this text.

REFERENCES

[AgGr06] G. Agnarsson and R. Greenlaw, *Graph Theory: Modeling, Applications, and Algorithms*, Prentice Hall, Englewood Cliffs, NJ, 2006.

[AhHoUl74] A. V. Aho, J. E. Hopcroft, and J. D. Ullman, *The Design and Analysis of Computer Algorithms*, Addison-Wesley, Reading, MA, 1974.

[AhUl95] Alfred V. Aho and Jeffrey D. Ullman, *Foundations of Computer Science, C Edition*, Computer Science Press, New York, 1995.

[AiZiHo09] Martin Aigner, Günter M. Ziegler, and Karl H. Hofmann, *Proofs from THE BOOK*, 4th ed., Springer, Berlin, 2009.

[Ak89] S. G. Akl, *The Design and Analysis of Parallel Algorithms*, Prentice Hall, Englewood Cliffs, NJ, 1989.

[AlAr78] S. Alagic and M. A. Arbib, *The Design of Well-Structured and Correct Programs*, Springer-Verlag, New York, 1978.

[AlWiBe00] J. M. Aldous, R. J. Wilson, and S. Best, *Graphs and Applications: An Introductory Approach*, Springer, New York, 2000.

[AlSl10] R.B.J.T. Allenby and A. Slomson, *How to Count: An Introduction to Combinatorics*, 2d. ed., Chapman and Hall/CRC, Boca Raton, Florida, 2010.

[AlSp00] Noga Alon and Joel H. Spencer, *The Probabilistic Method*, 2d ed., Wiley, New York, 2000.

[An89] I. Anderson, *A First Course in Combinatorial Mathematics*, 2d. ed., Oxford University Press, New York, 1989.

[An79] R. B. Anderson, *Proving Programs Correct*, Wiley, New York, 1979.

[Ap67] T. M. Apostol, *Calculus*, Vol. I, 2d ed., Wiley, New York, 1967.

[ApHa76] K. Appel and W. Haken, "Every Planar Map Is 4-colorable," *Bulletin of the AMS*, 82 (1976), 711–712.

[ArKfMo80] M. A. Arbib, A. J. Kfoury, and R. N. Moll, *A Basis for Theoretical Computer Science*, Springer-Verlag, New York, 1980.

[AvCh90] B. Averbach and O. Chein, *Problem Solving Through Recreational Mathematics*, W.H. Freeman, San Francisco, 1980.

[BaGe99] S. Baase and A. Van Gelder, *Computer Algorithms: Introduction to Design and Analysis*, 3d ed., Addison-Wesley, Reading, MA, 1999.

[Ba86] R. C. Backhouse, *Program Construction and Verification*, Prentice-Hall, Englewood Cliffs, NJ, 1986.

[Ba83] D. Barnette, *Map Coloring, Polyhedra, and the Four-Color Problem*, Mathematical Association of America, Washington, DC, 1983.

[BaEt93] Jon Barwise and John Etchemendy, *Turing's World 3.0 for the Macintosh*, CSLI Publications, Stanford, CA, 1993.

[Be80] F. S. Beckman, *Mathematical Foundations of Programming*, Addison-Wesley, Reading, MA, 1980.

[BeCh71] M. Behzad and G. Chartrand, *Introduction to the Theory of Graphs*, Allyn & Bacon, Boston, 1971.

[BeQu03] A. Benjamin and J. J. Quine, *Proofs that Really Count*, Mathematical Association of America, Washington, DC, 2003.

[Be86] J. Bentley, *Programming Pearls*, Addison-Wesley, Reading, MA, 1986.

[BeFr72] G. Berman and K. D. Fryer, *Introduction to Combinatorics*, Academic Press, New York, 1972.

[BiLlWi99] N. L. Biggs, E. K. Lloyd, and R. J. Wilson, *Graph Theory 1736–1936*, Oxford University Press, Oxford, England, 1999.

[BoAr74] L. S. Bobrow and M. A. Arbib, *Discrete Mathematics*, Saunders, Philadelphia, 1974.

[Bo00] K. P. Bogart, *Introductory Combinatorics*, 3d ed. Academic Press, San Diego, 2000.

[Bo07] M. Bona, *Enumerative Combinatorics*, McGraw-Hill, New York, 2007.

[BoMu10] J. A. Bondy and U. S. R. Murty, *Graph Theory with Applications*, Springer, New York, 2010.

[Bo04] P. Bork, L. J. Jensen, C. von Mering, A. K. Ramani, I. Lee, and E. M. Marcotte, "Protein interaction networks from yeast to human," *Current Opinion in Structural Biology*, 14 (2004), 292–299.

[BoMa86] R. C. Bose and B. Manvel, *Introduction to Combinatorial Theory*, Wiley, New York, 1986.

[Bo00] A. Brodera, R. Kumar, F. Maghoula, P. Raghavan, S. Rajagopalan, R. Statac, A. Tomkins, and Janet Wiener, "Graph structure in the Web," *Computer Networks*, 33 (2000), 309–320.

[Br89] J. G. Brookshear, *Theory of Computation*, Benjamin Cummings, Redwood City, CA, 1989.

[Br09] R. A. Brualdi, *Introductory Combinatorics*, 5th ed., Prentice-Hall, Englewood Cliffs, NJ, 2009.

[BuHa90] F. Buckley and F. Harary, *Distance in Graphs*, Addison-Wesley, Redwood City, CA, 1990.

[Ca79] L. Carmony, "Odd Pie Fights," *Mathematics Teacher* 72 (January, 1979), 61–64.

[Ca78] L. Carroll, *Symbolic Logic*, Crown, New York, 1978.

[ChGhMo79] V. Chachra, P. M. Ghare, and J. M. Moore, *Applications of Graph Theory Algorithms*, North-Holland, New York, 1979.

[Ch77] G. Chartrand, *Graphs as Mathematical Models*, Prindle, Weber & Schmidt, Boston, 1977.

[ChLeZh10] G. Chartrand, L. Lesniak, and P. Zhang, *Graphs and Digraphs*, 5th ed., Chapman and Hall/CRC, Boca Raton, 2010.

[ChOe93] G. Chartrand and O. R. Oellermann, *Applied Algorithmic Graph Theory*, McGraw-Hill, New York, 1993.

[ChZh04] G. Chartrand and P. Zhang, *Introduction to Graph Theory*, McGraw-Hill, New York, 2004.

[ClMe94] W. F. Clocksin and C. S. Mellish, *Programming in Prolog*, 4th ed., Springer-Verlag, New York, 1994.

[Co78] D. I. A. Cohen, *Basic Techniques of Combinatorial Theory*, Wiley, New York, 1978.

[Co05] D. Comer, *Internetworking with TCP/IP, Principles, Protocols, and Architecture*, Vol. 1, 5th ed., Prentice-Hall, Englewood Cliffs, NJ, 2005.

[CoLeRiSt09] T. H. Cormen, C. E. Leierson, R. L. Rivest, and C. Stein, *Introduction to Algorithms*, 3rd ed., MIT Press, Cambridge, MA, 2009.

[CrPo10] Richard Crandall and Carl Pomerance, 2d ed., *Prime Numbers: A Computational Perspective*, Springer-Verlag, New York, 2010.

[Cu05] Antonella Cupillari, *The Nuts and Bolts of Proofs*, 3d ed., Academic Press, San Diego, 2005.

[Cu84] C. W. Curtis, *Linear Algebra*, Springer-Verlag, New York, 1984.

[Da82] C. J. Date, *An Introduction to Database Systems*, 3d ed., Addison-Wesley, Reading, MA, 1982.

[DaSiWe94] M. Davis, R. Sigal, and E. J. Weyuker, *Computability, Complexity, and Languages*, 2d ed., Academic Press, San Diego, 1994.

[Da10] T. Davis, "The Mathematics of Sudoku," November, 2010, available at *http://geometer.org/mathcircles/sudoku.pdf*

[De82] D. E. R. Denning, *Cryptography and Data Security*, Addison-Wesley, Reading, MA, 1982.

[DeDeQu81] P. J. Denning, J. B. Dennis, and J. E. Qualitz, *Machines, Languages, and Computation*, Prentice-Hall, Englewood Cliffs, NJ, 1981.

[De74] N. Deo, *Graph Theory with Applications to Engineering and Computer Science*, Prentice-Hall, Englewood Cliffs, NJ, 1974.

[DeOr11] S. L. Devadoss and J. O'Rourke, *Discrete and Computational Geometry*, Princeton University Press, Princeton, NJ, 2011.

[De02] K. Devlin, *The Millennium Problems: The Seven Greatest Unsolved Mathematical Puzzles of Our Time*, Basic Book, New York, 2002.

[De84] A. K. Dewdney, "Computer Recreations," *Scientific American*, 251, no. 2 (August 1984), 19–23; 252, no. 3 (March 1985), 14–23; 251, no. 4 (April 1985), 20–30.

[De93] A. K. Dewdney, *The New Turing Omnibus: Sixty-Six Excursions in Computer Science*, W. H. Freeman, New York, 1993.

[Di59] E. Dijkstra, "Two Problems in Connexion with Graphs," *Numerische Mathematik*, 1 (1959), 269–271.

[EaKl10] D. Easley and J. Kleinberg, *Networks, Crowds, and Markets: Reasoning About a Highly Connected World*, Cambridge University Press, New York, 2010.

[Eu53] L. Euler, "The Koenigsberg Bridges," *Scientific American*, 189, no. 1 (July 1953), 66–70.

[Ev73] S. Even, *Algorithmic Combinatorics*, Macmillan, New York, 1973.

[Ev79] S. Even, *Graph Algorithms*, Computer Science Press, Rockville, MD, 1979.

[Fe68] W. Feller, *An Introduction to Probability Theory and Its Applications*, Vol. 1, 3d ed., Wiley, New York, 1968.

[Fo92] L. R. Foulds, *Graph Theory Applications*, Springer-Verlag, New York, 1992.

[GaJo79] Michael R. Garey and David S. Johnson, *Computers and Intractability: A Guide to NP-Completeness*, Freeman, New York, 1979.

[Gi85] A. Gibbons, *Algorithmic Graph Theory*, Cambridge University Press, Cambridge, England, 1985.

[Gi70] C. C. Gillispie, ed., *Dictionary of Scientific Biography*, Scribner's, New York, 1970.

[Go94] S. W. Golomb, *Polyominoes*, Princeton University Press, Princeton, NJ, 1994.

[Go84] G. H. Gonnet, *Handbook of Algorithms and Data Structures*, Addison-Wesley, London, 1984.

[GoHe77] S. E. Goodman and S. T. Hedetniemi, *Introduction to the Design and Analysis of Algorithms*, McGraw-Hill, New York, 1977.

[GoGo78] C. C. Gotlieb and L. R. Gotlieb, *Data Types and Structures*, Prentice-Hall, Englewood Cliffs, NJ, 1978.

[GrHe85] R. L. Graham and P. Hell, "On the History of the Minimum Spanning Tree Problem," *Annals of the History of Computing*, 7 (1985), 43–57.

[GrKnPa94] R. L. Graham, D. E. Knuth, and O. Patashnik, *Concrete Mathematics*, 2d ed., Addison-Wesley, Reading, MA, 1994.

[GrRoSp90] Ronald L. Graham, Bruce L. Rothschild, and Joel H. Spencer, *Ramsey Theory*, 2d ed., Wiley, New York, 1990.

[GrWa77] J. E. Graver and M. E. Watkins, *Combinatorics with Emphasis on the Theory of Graphs*, Springer-Verlag, New York, 1977.

[GrSc93] D. Gries and F. B. Schneider, *A Logical Approach to Discrete Math*, Springer-Verlag, New York, 1993.

[Gr03] R. P. Grimaldi, *Discrete and Combinatorial Mathematics*, 5th ed., Addison-Wesley, Reading, MA, 2003.

[Gr07] J. L. Gross, *Combinatorial Methods with Computer Applications*, Chapman and Hall/CRC, Boca Raton, FL, 2007.

[GrYe05] J. L. Gross and J. Yellen, *Graph Theory and Its Applications*, 2d ed., CRC Press, Boca Raton, FL, 2005.

[GrYe03] J. L. Gross and J. Yellen, *Handbook of Graph Theory*, CRC Press, Boca Raton, FL, 2003.

[Gr90] Jerrold W. Grossman, *Discrete Mathematics: An Introduction to Concepts, Methods, and Applications*, Macmillan, New York, 1990.

[Gr97] J. Gruska, *Foundations of Computing*, International Thomsen Computer Press, London, 1997.

[Gu10] D. A. Gunderson, *Handbook of Mathematical Induction*, Chapman and Hall/CRC, Boca Raton, Florida, 2010.

[Ha60] P. R. Halmos, *Naive Set Theory*, D. Van Nostrand, New York, 1960.

[Ha80] R. W. Hamming, *Coding and Information Theory*, Prentice-Hall, Englewood Cliffs, NJ, 1980.

[Ha69] F. Harary, *Graph Theory*, Addison-Wesley, Reading, MA, 1969.

[HaWrWiHe08] G. H. Hardy, E. M. Wright, A. Wiles, and R. Heath-Brown, *An Introduction to the Theory of Numbers*, 6th ed., Oxford University Press, USA, New York, 2008.

[Ha87] D. Harel, *Algorithmics, The Spirit of Computing*, Addison-Wesley, Reading, MA, 1987.

[Ha00a] Brian Hayes, "Graph-Theory in Practice, Part I," *American Scientist*, 88, no. 1 (2000), 9–13.

[Ha00b] Brian Hayes, "Graph-Theory in Practice, Part II," *American Scientist*, 88, no. 2 (2000), 104–109.

[Ha93] John P. Hayes, *Introduction to Digital Logic Design*, Addison-Wesley, Reading, MA, 1993.

[He77] F. Hennie, *Introduction to Computability*, Addison-Wesley, Reading, MA, 1977.

[He88] R. Herken, *The Universal Turing Machine, A Half-Century Survey*, Oxford University Press, New York, 1988.

[Ho76] C. Ho, "Decomposition of a Polygon into Triangles," *Mathematical Gazette*, 60 (1976), 132–134.

[Ho99] D. Hofstadter, *Gödel, Escher, Bach: An Internal Golden Braid*, Basic Books, New York, 1999.

[Ho66] F. E. Hohn, *Applied Boolean Algebra*, 2d ed., Macmillan, New York, 1966.

[HoMoUl06] J. E. Hopcroft, R Motwani, and J. D. Ullman, *Introduction to Automata Theory, Languages, and Computation*, 3d ed., Addison-Wesley, Boston, MA, 2006.

[HoMo76] D. Hopkin and B. Moss, *Automata*, Elsevier, North-Holland, New York, 1976.

[HoSa82] E. Horowitz and S. Sahni, *Fundamentals of Computer Algorithms*, Computer Science Press, Rockville, MD, 1982.

[Hu82] T.C. Hu, *Combinatorial Algorithms*, Addison-Wesley, Reading, MA, 1982.

[Hu07] W. Huber, V. J. Carey, L. Long, S. Falcon, and R. Gentleman, "Graphs in molecular biology," *BMC Bioinformatics*, 8 (Supplement 6) (2007).

[HuRy04] M. Huth and M. Ryan, *Logic in Computer Science*, 2d ed., Cambridge University Press, Cambridge, England, 2004.

[In93] D. C. Ince, *An Introduction to Discrete Mathematics, Formal System Specification, and Z*, 2d ed., Oxford, New York, 1993.

[KaMo64] D. Kalish and R. Montague, *Logic: Techniques of Formal Reasoning*, Harcourt, Brace, Jovanovich, New York, 1964.

[Ka53] M. Karnaugh, "The Map Method for Synthesis of Combinatorial Logic Circuits," *Transactions of the AIEE*, part I, 72 (1953), 593–599.

[KaBo04] R. H. Katz and G. Borriello, *Contemporary Logic Design*, Prentice-Hall, Englewood Cliffs, NJ, 2004.

[Kl56] S. C. Kleene, "Representation of Events by Nerve Nets," in *Automata Studies*, 3–42, Princeton University Press, Princeton, NJ, 1956.

[KlTa05] J. Kleinberg and E. Tardos, *Algorithm Design*, Addison-Wesley, Boston, 2005

[Kn77] D. E. Knuth, "Algorithms," *Scientific American*, 236, no. 4 (April 1977), 63–80.

[Kn97a] D. E. Knuth, *The Art of Computer Programming, Vol. I: Fundamental Algorithms*, 3d ed., Addison-Wesley, Reading, MA, 1997.

[Kn97b] D. E. Knuth, *The Art of Computer Programming, Vol. II: Seminumerical Algorithms*, 3d ed., Addison-Wesley, Reading, MA, 1997.

[Kn98] D. E. Knuth, *The Art of Computer Programming, Vol. III: Sorting and Searching*, 2d ed., Addison-Wesley, Reading, MA, 1998.

[KoKr04] W. Kocay and D. L. Kreher, *Graph Algorithms and Optimization*, Chapman and Hall/CRC, Boca Raton, Florida, 2004.

[Ko86] Z. Kohavi, *Switching and Finite Automata Theory*, 2d ed., McGraw-Hill, New York, 1986.

[KrSt98] Donald H. Kreher and Douglas R. Stinson, *Combinatorial Algorithms: Generation, Enumeration, and Search*, CRC Press, Boca Raton, FL, 1998.

[Kr87] L. Kronsjö, *Algorithms: Their Complexity and Efficiency*, 2d ed., Wiley, New York, 1987.

[Kr56] J. B. Kruskal, "On the Shortest Spanning Subtree of a Graph and the Traveling Salesman Problem," *Proceedings of the AMS*, 1 (1956), 48–50.

[La10] J. C. Lagarias (ed.), *The Ultimate Challenge: The $3x + 1$ Problem*, The American Mathematical Society, Providence, 2010.

[Le06] A. V. Levitin, *Introduction to the Design and Analysis of Algorithms*, 2d, ed., Addison-Wesley, Boston, MA, 2006.

[Le64] D. H. Lehmer, "The Machine Tools of Combinatorics," in E. F. Beckenbach (ed.), *Applied Combinatorial Mathematics*, Wiley, New York, 1964.

[Le77] W. J. LeVeque, *Fundamentals of Number Theory*, Addison-Wesley, Reading, MA, 1977.

[LePa97] H. R. Lewis and C. H. Papadimitriou, *Elements of the Theory of Computation*, 2d ed., Prentice-Hall, Englewood Cliffs, NJ, 1997.

[LiLi81] Y. Lin and S. Y. T. Lin, *Set Theory with Applications*, 2d ed., Mariner, Tampa, FL, 1981.

[Li68] C. L. Liu, *Introduction to Combinatorial Mathematics*, McGraw-Hill, New York, 1968.

[Li85] C. L. Liu, *Elements of Discrete Mathematics*, 2d ed., McGraw-Hill, New York, 1985.

[Lo79] L. Lovász, *Combinatorial Problems and Exercises*, North-Holland, Amsterdam, 1979.

[Lu91] E. Lucas, *Récréations Mathématiques*, Gauthier-Villars, Paris, 1891.

[Ma89] U. Manber, *Introduction to Algorithms: A Creative Approach*, Addison-Wesley, Reading, MA, 1989.

[Ma91] G. E. Martin, *Polyominoes: A Guide to Puzzles and Problems in Tiling*, Mathematical Association of America, Washington, DC, 1991.

[Ma03] J. C. Martin, *Introduction to Languages and the Theory of Computation*, 3d ed., McGraw-Hill, New York, 2003.

[Ma93] H. F. Mattson, Jr., *Discrete Mathematics with Applications*, Wiley, New York, 1993.

[Mc56] E. J. McCluskey, Jr., "Minimization of Boolean Functions," *Bell System Technical Journal*, 35 (1956), 1417–1444.

[Mc90] J. A. McHugh, *Algorithmic Graph Theory*, Prentice-Hall, Englewood Cliffs, NJ, 1990.

[Mc82] R. McNaughton, *Elementary Computability, Formal Languages, and Automata*, Prentice-Hall, Englewood Cliffs, NJ, 1982.

[Me55] G. H. Mealy, "A Method for Synthesizing Sequential Circuits," *Bell System Technical Journal*, 34 (1955), 1045–1079.

[Me09] E. Mendelson, *Introduction to Mathematical Logic*, 5th ed., Chapman and Hall/CRC Press, Boca Raton, 2009.

[MeOoVa97] A. J. Menezes, P. C. van Oorschoot, S. A. Vanstone, *Handbook of Applied Cryptography*, CRC Press, Boca Raton, FL, 1997.

[MiRo91] J. G. Michaels and K. H. Rosen, *Applications of Discrete Mathematics*, McGraw-Hill, New York, 1991.

[Mo69] J. R. Monk, *Introduction to Set Theory*, McGraw-Hill, New York, 1969.

[Mo91] R. P. Morash, *Bridge to Abstract Mathematics*, McGraw-Hill, New York, 1991.

[Mo56] E. F. Moore, "Gedanken-Experiments on Sequential Machines," in *Automata Studies*, 129–153, Princeton University Press, Princeton, NJ, 1956.

[Na00] Paul J. Nabin, *Duelling Idiots and Other Probability Puzzlers*, Princeton University Press, Princeton, NJ, 2000.

[Ne85] C. V. Negoita, *Expert Systems and Fuzzy Systems*, Benjamin Cummings, Menlo Park, CA, 1985.

[Ne10] M. Newman, *Networks: An Introduction*, Oxford University Press, New York, 2010.

[NiMa95] Ulf Nilsson and Jan Maluszynski, *Logic, Programming, and Prolog*, 2d ed., Wiley, Chichester, England, 1995.

[Or00] J. O'Rourke, *Computational Geometry in C*, Cambridge University Press, New York, 2000.

[Or63] O. Ore, *Graphs and Their Uses*, Mathematical Association of America, Washington, DC, 1963.

[0r88] O. Ore, *Number Theory and its History*, Dover, New York, 1988.

[PaPi01] A. Papoulis and S. U. Pillai, *Probability, Random Variables, and Stochastic Processes*, McGraw-Hill, New York, 2001.

[Pe87] A. Pelc, "Solution of Ulam's Problem on Searching with a Lie," *Journal of Combinatorial Theory*, Series A, 44 (1987), 129–140.

[Pe09] M. S. Petrovic, *Famous Puzzles of Great Mathematicians*, American Mathematical Society, Providence, 2009.

[PoSh81] I. Pohl and A. Shaw, *The Nature of Computation: An Introduction to Computer Science*, Computer Science Press, Rockville, MD, 1981.

[Po62] George Pólya, *Mathematical Discovery*, Vols. 1 and 2, Wiley, New York, 1962.

[Po71] George Pólya, *How to Solve It*, Princeton University Press, Princeton, NJ, 1971.

[Po90] George Pólya, *Mathematics and Plausible Reasoning*, Princeton University Press, Princeton, NJ, 1990.

[PoTaWo83] G. Pólya, R. E. Tarjan, and D. R. Woods, *Notes on Introductory Combinatorics*, Birkhäuser, Boston, 1983.

[Pr57] R. C. Prim, "Shortest Connection Networks and Some Generalizations," *Bell System Technical Journal*, 36 (1957), 1389–1401.

[PuBr85] P. W. Purdom, Jr. and C. A. Brown, *The Analysis of Algorithms*, Holt, Rinehart & Winston, New York, 1985.

[Qu52] W. V. Quine, "The Problem of Simplifying Truth Functions," *American Mathematical Monthly*, 59 (1952), 521–531.

[Qu55] W. V. Quine, "A Way to Simplify Truth Functions," *American Mathematical Monthly*, 62 (1955), 627–631.

[Ra62] T. Rado, "On Non-Computable Functions," *Bell System Technical Journal* (May 1962), 877–884.

[Ra92] Gregory J. E. Rawlins, *Compared to What? An Introduction to the Analysis of Algorithms*, Computer Science Press, New York, 1992.

[ReNiDe77] E. M. Reingold, J. Nievergelt, and N. Deo, *Combinatorial Algorithms: Theory and Practice*, Prentice-Hall, Englewood Cliffs, NJ, 1977.

[Ri58] J. Riordan, *An Introduction to Combinatorial Analysis*, Wiley, New York, 1958.

[Ri68] J. Riordan, *Combinatorial Identities*, Wiley, New York, 1968.

[RiShAd78] R. Rivest, A. Shamir, and L. Adleman, "A Method for Obtaining Digital Signatures and Public-Key Cryptosystems," *Communications of the Association for Computing Machinery*, 31, no. 2 (1978), 120–128.

[Ro86] E. S. Roberts, *Thinking Recursively*, Wiley, New York, 1986.

[Ro76] F. S. Roberts, *Discrete Mathematics Models*, Prentice-Hall, Englewood Cliffs, NJ, 1976.

[RoTe03] F. S. Roberts and B. Tesman, *Applied Combinatorics*, 2d ed., Prentice-Hall, Englewood Cliffs, NJ, 2003.

[RoFo80] D. F. Robinson and L. R. Foulds, *Digraphs: Theory and Techniques*, Gordon and Breach, New York, 1980.

[Ro84] J. S. Rohl, *Recursion via Pascal*, Cambridge University Press, Cambridge, England, 1984.

[Ro10] K. H. Rosen, *Elementary Number Theory and Its Applications*, 6th ed., Pearson, Boston, 2010.

[Ro00] K. H. Rosen, *Handbook of Discrete and Combinatorial Mathematics*, CRC Press, Boca Raton, FL, 2000.

[Ro09] Jason Rosenhouse, *The Monty Hall Problem*, Oxford University Press, New York, 2009.

[Ro09a] Sheldon M. Ross, *A First Course in Probability Theory*, 7th ed., Prentice-Hall, Englewood Cliffs, NJ, 2009.

[Ro02] Sheldon M. Ross, *Probability Models for Computer Science*, Harcourt/Academic Press, San Diego, 2002.

[Ro59] B. Roy, "Transitivité et Connexité," *C.R. Acad. Sci. Paris*, 249 (1959), 216.

[Ry63] H. Ryser, *Combinatorial Mathematics*, Mathematical Association of America, Washington, DC, 1963.

[SaKa86] T. L. Saaty and P. C. Kainen, *The Four-Color Problem: Assaults and Conquest*, Dover, New York, 1986.

[Sa85] S. Sahni, *Concepts in Discrete Mathematics*, Camelot, Minneapolis, 1985.

[Sa00] K. Sayood, *Introduction to Data Compression*, 2d ed., Academic Press, San Diego, 2000.

[SePi89] J. Seberry and J. Pieprzyk, *Cryptography: An Introduction to Computer Security*, Prentice-Hall, Englewood Cliffs, NJ, 1989.

[Se03] R. Sedgewick, *Algorithms in Java*, 3d ed., Addison-Wesley, Reading, MA, 2003.

[SiHs88] H. J. Siegel and W. T. Hsu, "Interconnection Networks," in *Computer Architectures*, V. M. Milutinovic (ed.), North-Holland, New York, 1988, pp. 225–264.

[Si99] S. Singh, *The Code Book*, Doubleday, New York, 1999.

[Si66] A. Sinkov, *Elementary Cryptanalysis*, Mathematical Association of America, Washington, DC, 1966.

[Si06] M. Sipser, *Introduction to the Theory of Computation*, Course Technology, Boston, 2006

[SlPl95] N. J. A. Sloane and S. Plouffe, *The Encyclopedia of Integer Sequences*, Academic Press, New York, 1995.

[Sm78] Raymond Smullyan, *What Is the Name of This Book?: The Riddle of Dracula and Other Logical Puzzles*, Prentice-Hall, Englewood Cliffs, NJ, 1978.

[Sm92] Raymond Smullyan, *Lady or the Tiger? And Other Logic Puzzles Including a Mathematical Novel That Features Godel's Great Discovery*, Times Book, New York, 1992.

[Sm98] Raymond Smullyan, *The Riddle of Scheherazade: And Other Amazing Puzzles, Ancient & Modern*, Harvest Book, Fort Washington, PA, 1998.

[So09] Daniel Solow, *How to Read and Do Proofs: An Introduction to Mathematical Thought Processes*, 5th ed., Wiley, New York, 2009.

[So61] I. S. Sominskii, *Method of Mathematical Induction*, Blaisdell, New York, 1961.

[Sp94] M. Spivak, *Calculus*, 3d ed., Publish or Perish, Wilmington, DE, 1994.

[StMc77] D. Stanat and D. F. McAllister, *Discrete Mathematics in Computer Science*, Prentice-Hall, Englewood Cliffs, NJ, 1977.

[St78] H. M. Stark, *An Introduction to Number Theory*, MIT Press, Cambridge, MA, 1978.

[St05] Douglas R. Stinson, *Cryptography, Theory and Practice*, 3d ed., Chapman and Hall/CRC, Boca Raton, FL, 2005.

[St94] P. K. Stockmeyer, "Variations on the Four-Post Tower of Hanoi Puzzle," *Congressus Numerantrum* 102 (1994), 3–12.

[St74] R. R. Stoll, *Sets, Logic, and Axiomatic Theories*, 2d ed., W. H. Freeman, San Francisco, 1974.

[St77] D. F. Stanat and D.F, McAllister, *Discrete Mathematics in Computer Science*, McGraw-Hill, NY, 1977.

[St09] G. W. Strang, *Linear Algebra and Its Applications*, 4th ed., Wellesley Cambridge Press, Wellesley, MA, 2009

[Su87] P. Suppes, *Introduction to Logic*, D. Van Nostrand, Princeton, NJ, 1987.

[Ta83] R. E. Tarjan, *Data Structures and Network Algorithms*, Society for Industrial and Applied Mathematics, Philadelphia, 1983.

[ThFi96] G. B. Thomas and R. L. Finney, *Calculus and Analytic Geometry*, 9th ed., Addison-Wesley, Reading, MA, 1996.

[TrMa75] J. P. Tremblay and R. P. Manohar, *Discrete Mathematical Structures with Applications to Computer Science*, McGraw-Hill, New York, 1975.

[Tr76] J. P. Tremblay and R. Manohar, *Discrete Mathematical Structures with Applications to Computer Science*, McGraw-Hill, NY, 1976.

[Tu06] Alan Tucker, *Applied Combinatorics*, 5th ed., Wiley, New York, 2006.

[Ve52] E. W. Veitch, "A Chart Method for Simplifying Truth Functions," *Proceedings of the ACM* (1952), 127–133.

[Ve06] David Velleman, *How to Prove It: A Structured Approach*, Cambridge University Press, New York, 2006.

[Vi71] N. Y. Vilenkin, *Combinatorics*, Academic Press, New York, 1971.

[Wa80] M. Wand, *Induction, Recursion, and Programming*, North-Holland, New York, 1980.

[Wa62] S. Warshall, "A Theorem on Boolean Matrices," *Journal of the ACM*, 9 (1962), 11–12.

[We00] D. B. West, *Introduction to Graph Theory*, 2d ed., Prentice Hall, Englewood Cliffs, NJ, 2000.

[Wi02] Herbert S. Wilf, *Algorithms and Complexity*, 2d ed., A. K. Peters, Natick, MA, 2002.

[Wi85] S. G. Williamson, *Combinatorics for Computer Science*, Computer Science Press, Rockville, MD, 1985.

[Wi85a] R. J. Wilson, *Introduction to Graph Theory*, 3d ed., Longman, Essex, England, 1985.

[WiBe79] R. J. Wilson and L. W. Beineke, *Applications of Graph Theory*, Academic Press, London, 1979.

[WiWa90] R. J. Wilson and J. J. Watkins, *Graphs, An Introductory Approach*, Wiley, New York, 1990.

[Wi76] N. Wirth, *Algorithms + Data Structures = Programs*, Prentice-Hall, Englewood Cliffs, NJ, 1976.

[Wi84] N. Wirth, "Data Structures and Algorithms," *Scientific American*, 251 (September 1984), 60–69.

[Wo98] Robert S. Wolf, *Proof, Logic, and Conjecture: The Mathematician's Toolbox*, W. H. Freeman, New York, 1998.

[Wo87] D. Wood, *Theory of Computation*, Harper & Row, New York, 1987.

[Zd05] J. Zdziarski, *Ending Spam: Bayesian Content Filtering and the Art of Statistical Language Classification*, No Starch Press, San Francisco, 2005.

[Zi91] H. J. Zimmermann, *Fuzzy Set Theory and Its Applications*, 2d ed., Kluwer, Boston, 1991.

Answers to Odd-Numbered Exercises

CHAPTER 1

Section 1.1

1. a) Yes, T **b)** Yes, F **c)** Yes, T **d)** Yes, F **e)** No **f)** No
3. a) Mei does not have an MP3 player. **b)** There is pollution in New Jersey. **c)** $2 + 1 \neq 3$. **d)** The summer in Maine is not hot or it is not sunny. **5. a)** F **b)** T **c)** T **d)** T **e)** T
7. a) $p \wedge q$ **b)** $p \wedge \neg q$ **c)** $\neg p \wedge \neg q$ **d)** $p \vee q$ **e)** $p \to q$
f) $(p \vee q) \wedge (p \to \neg q)$ **g)** $q \leftrightarrow p$ **9. a)** $\neg p$ **b)** $p \wedge \neg q$
c) $p \to q$ **d)** $\neg p \to \neg q$ **e)** $p \to q$ **f)** $q \wedge \neg p$ **g)** $q \to p$
11. a) False **b)** True **c)** True **d)** True **13. a)** Inclusive or: It is allowable to take discrete mathematics if you have had calculus or computer science, or both. Exclusive or: It is allowable to take discrete mathematics if you have had calculus or computer science, but not if you have had both. Most likely the inclusive or is intended. **b)** Inclusive or: You can take the rebate, or you can get a low-interest loan, or you can get both the rebate and a low-interest loan. Exclusive or: You can take the rebate, or you can get a low-interest loan, but you cannot get both the rebate and a low-interest loan. Most likely the exclusive or is intended. **c)** Inclusive or: You can order two items from column A and none from column B, or three items from column B and none from column A, or five items including two from column A and three from column B. Exclusive or: You can order two items from column A or three items from column B, but not both. Almost certainly the exclusive or is intended. **d)** Inclusive or: More than 2 feet of snow or windchill below -100, or both, will close school. Exclusive or: More than 2 feet of snow or windchill below -100, but not both, will close school. Certainly the inclusive or is intended. **15. a)** If the wind blows from the northeast, then it snows. **b)** If it stays warm for a week, then the apple trees will bloom. **c)** If the Pistons win the championship, then they beat the Lakers. **d)** If you get to the top of Long's Peak, then you must have walked 8 miles. **e)** If you are world-famous, then you will get tenure as a professor. **f)** If you drive more

than 400 miles, then you will need to buy gasoline. **g)** If your guarantee is good, then you must have bought your CD player less than 90 days ago. **h)** If the water is not too cold, then Jan will go swimming. **17. a)** You buy an ice cream cone if and only if it is hot outside. **b)** You win the contest if and only if you hold the only winning ticket. **c)** You get promoted if and only if you have connections. **d)** Your mind will decay if and only if you watch television. **e)** The train runs late if and only if it is a day I take the train. **19. a)** 2 **b)** 16 **c)** 64 **d)** 16

21. a)

p	$\neg p$	$p \wedge \neg p$
T	F	F
F	T	F

b)

p	$\neg p$	$p \vee \neg p$
T	F	T
F	T	T

c)

p	q	$\neg q$	$p \vee \neg q$	$(p \vee \neg q) \to q$
T	T	F	T	T
T	F	T	T	F
F	T	F	F	T
F	F	T	T	F

d)

p	q	$p \vee q$	$p \wedge q$	$(p \vee q) \to (p \wedge q)$
T	T	T	T	T
T	F	T	F	F
F	T	T	F	F
F	F	F	F	T

e)

p	q	$p \to q$	$\neg q$	$\neg p$	$\neg q \to \neg p$	$(p \to q) \leftrightarrow (\neg q \to \neg p)$
T	T	T	F	F	T	T
T	F	F	T	F	F	T
F	T	T	F	T	T	T
F	F	T	T	T	T	T

f)

p	q	$p \to q$	$q \to p$	$(p \to q) \to (q \to p)$
T	T	T	T	T
T	F	F	T	T
F	T	T	F	F
F	F	T	T	T

23.

p	q	r	p → (¬q ∨ r)	¬p → (q → r)	(p → q) ∨ (¬p → r)	(p → q) ∧ (¬p → r)	(p ↔ q) ∨ (¬q ↔ r)	(¬p ↔ ¬q) ↔ (q ↔ r)
T	T	T	T	T	T	T	T	T
T	T	F	F	T	T	T	T	F
T	F	T	T	T	T	F	T	T
T	F	F	T	T	T	F	F	F
F	T	T	T	T	T	T	F	F
F	T	F	T	F	T	F	T	T
F	F	T	T	T	T	T	T	F
F	F	F	T	T	T	F	T	T

25.

p	q	r	s	p ↔ q	r ↔ s	(p ↔ q) ↔ (r ↔ s)
T	T	T	T	T	T	T
T	T	T	F	T	F	F
T	T	F	T	T	F	F
T	T	F	F	T	T	T
T	F	T	T	F	T	F
T	F	T	F	F	F	T
T	F	F	T	F	F	T
T	F	F	F	F	T	F
F	T	T	T	F	T	F
F	T	T	F	F	F	T
F	T	F	T	F	F	T
F	T	F	F	F	T	F
F	F	T	T	T	T	T
F	F	T	F	T	F	F
F	F	F	T	T	F	F
F	F	F	F	T	T	T

27. The first clause is true if and only if at least one of p, q, and r is true. The second clause is true if and only if at least one of the three variables is false. Therefore the entire statement is true if and only if there is at least one T and one F among the truth values of the variables, in other words, that they don't all have the same truth value.   **29. a)** Bitwise *OR* is 111 1111; bitwise *AND* is 000 0000; bitwise *XOR* is 111 1111. **b)** Bitwise *OR* is 1111 1010; bitwise *AND* is 1010 0000; bitwise *XOR* is 0101 1010. **c)** Bitwise *OR* is 10 0111 1001; bitwise *AND* is 00 0100 0000; bitwise *XOR* is 10 0011 1001. **d)** Bitwise *OR* is 11 1111 1111; bitwise *AND* is 00 0000 0000; bitwise *XOR* is 11 1111 1111.   **31.** 0.2, 0.6   **33.** 0.8, 0.6   **35. a)** The 99th statement is true and the rest are false.   **b)** Statements 1 through 50 are all true and statements 51 through 100 are all false. **c)** This cannot happen; it is a paradox, showing that these cannot be statements.

Section 1.2

1. $e \to a$ **3.** $e \to (a \land (b \lor p) \land r)$ **5.** Consistent   **7.** NEW **AND** JERSEY **AND** BEACHES, (JERSEY **AND** BEACHES) **NOT** NEW   **9.** "If I were to ask you whether the right branch leads to the ruins, would you answer yes?"   **11** If the first professor did not want coffee, then he would know that the answer to the hostess's question was "no." Therefore the hostess and the remaining professors know that the first professor did want coffee. Similarly, the second professor must want coffee. When the third professor said "no," the hostess knows that the third professor does not want coffee.   **13.** *A* is a knight and *B* is a knave.   **15.** *A* is a knight and *B* is a knight.   **17.** In order of decreasing salary: Fred, Maggie, Janice   **19.** The detective can determine that the butler and cook are lying but cannot determine whether the gardener is telling the truth or whether the handyman is telling the truth.   **21.** The Japanese man owns the zebra, and the Norwegian drinks water.   **23.** One honest, 49 corrupt   **25. a)** $\neg(p \land (q \lor \neg r))$ **b)** $((\neg p) \land (\neg q)) \lor (p \land r)$

Section 1.3

1. The equivalences follow by showing that the appropriate pairs of columns of this table agree.

p	p ∧ T	p ∨ F	p ∧ F	p ∨ T	p ∨ p	p ∧ p
T	T	T	F	T	T	T
F	F	F	F	T	F	F

3.

p	q	r	q ∨ r	p ∧ (q ∨ r)	p ∧ q	p ∧ r	(p ∧ q) ∨ (p ∧ r)
T	T	T	T	T	T	T	T
T	T	F	T	T	T	F	T
T	F	T	T	T	F	T	T
T	F	F	F	F	F	F	F
F	T	T	T	F	F	F	F
F	T	F	T	F	F	F	F
F	F	T	T	F	F	F	F
F	F	F	F	F	F	F	F

5. a)

p	q	p ∧ q	(p ∧ q) → p
T	T	T	T
T	F	F	T
F	T	F	T
F	F	F	T

b)

p	q	$p \vee q$	$p \to (p \vee q)$
T	T	T	T
T	F	T	T
F	T	T	T
F	F	F	T

c)

p	q	$\neg p$	$p \to q$	$\neg p \to (p \to q)$
T	T	F	T	T
T	F	F	F	T
F	T	T	T	T
F	F	T	T	T

d)

p	q	$p \wedge q$	$p \to q$	$(p \wedge q) \to (p \to q)$
T	T	T	T	T
T	F	F	F	T
F	T	F	T	T
F	F	F	T	T

e)

p	q	$p \to q$	$\neg(p \to q)$	$\neg(p \to q) \to p$
T	T	T	F	T
T	F	F	T	T
F	T	T	F	T
F	F	T	F	T

f)

p	q	$p \to q$	$\neg(p \to q)$	$\neg q$	$\neg(p \to q) \to \neg q$
T	T	T	F	F	T
T	F	F	T	T	T
F	T	T	F	F	T
F	F	T	F	T	T

7. In each case we will show that if the hypothesis is true, then the conclusion is also. **a)** If the hypothesis $p \wedge q$ is true, then by the definition of conjunction, the conclusion p must also be true. **b)** If the hypothesis p is true, by the definition of disjunction, the conclusion $p \vee q$ is also true. **c)** If the hypothesis $\neg p$ is true, that is, if p is false, then the conclusion $p \to q$ is true. **d)** If the hypothesis $p \wedge q$ is true, then both p and q are true, so the conclusion $p \to q$ is also true. **e)** If the hypothesis $\neg(p \to q)$ is true, then $p \to q$ is false, so the conclusion p is true (and q is false). **f)** If the hypothesis $\neg(p \to q)$ is true, then $p \to q$ is false, so p is true and q is false. Hence, the conclusion $\neg q$ is true. **9.** The proposition $\neg p \leftrightarrow q$ is true when $\neg p$ and q have the same truth values, which means that p and q have different truth values. Similarly, $p \leftrightarrow \neg q$ is true in exactly the same cases. Therefore, these two expressions are logically equivalent. **11.** For $(p \to r) \wedge (q \to r)$ to be false, one of the two conditional statements must be false, which happens exactly when r is false and at least one of p and q is true. But these are precisely the cases in which $p \vee q$ is true and r is false, which is precisely when $(p \vee q) \to r$ is false. Because the two propositions are false in exactly the same situations, they are logically equivalent. **13.** Many answers are possible. If we let r be true and p, q, and s be false, then $(p \to q) \to (r \to s)$

will be false, but $(p \to r) \to (q \to s)$ will be true. **15. a)** $p \vee \neg q \vee \neg r$ **b)** $(p \vee q \vee r) \wedge s$ **c)** $(p \wedge \mathbf{T}) \vee (q \wedge \mathbf{F})$ **17.** If we take duals twice, every $\vee$ changes to an $\wedge$ and then back to an $\vee$, every $\wedge$ changes to an $\vee$ and then back to an $\wedge$, every $\mathbf{T}$ changes to an $\mathbf{F}$ and then back to a $\mathbf{T}$, every $\mathbf{F}$ changes to a $\mathbf{T}$ and then back to an $\mathbf{F}$. Hence, $(s^*)^* = s$. **19.** $(p \wedge q \wedge \neg r) \vee (p \wedge \neg q \wedge r) \vee (\neg p \wedge q \wedge r)$ **21.** Given a compound proposition p, form its truth table and then write down a proposition q in disjunctive normal form that is logically equivalent to p. Because q involves only $\neg$, $\wedge$, and $\vee$, this shows that these three operators form a functionally complete set. **23.** By Exercise 43, given a compound proposition p, we can write down a proposition q that is logically equivalent to p and involves only $\neg$, $\wedge$, and $\vee$. By De Morgan's law we can eliminate all the $\wedge$'s by replacing each occurrence of $p_1 \wedge p_2 \wedge \cdots \wedge p_n$ with $\neg(\neg p_1 \vee \neg p_2 \vee \cdots \vee \neg p_n)$. **25.** $\neg(p \wedge q)$ is true when either p or q, or both, are false, and is false when both p and q are true. Because this was the definition of $p \mid q$, the two compound propositions are logically equivalent. **27.** $\neg(p \vee q)$ is true when both p and q are false, and is false otherwise. Because this was the definition of $p \downarrow q$, the two are logically equivalent. **29.** $((p \downarrow p) \downarrow q) \downarrow ((p \downarrow p) \downarrow q)$ **31.** This follows immediately from the truth table or definition of $p \mid q$. **33.** 16 **35.** If the database is open, then either the system is in its initial state or the monitor is put in a closed state. **37. a)** Satisfiable **b)** Not satisfiable **c)** Not satisfiable **39.** Use the same propositions as were given in the text for a 9×9 Sudoku puzzle, with the variables indexed from 1 to 4, instead of from 1 to 9, and with a similar change for the propositions for the 2×2 blocks: $\bigwedge_{r=0}^{1} \bigwedge_{s=0}^{1} \bigwedge_{n=1}^{4} \bigvee_{i=1}^{2} \bigvee_{j=1}^{2} p(2r + i, 2s + j, n)$

Section 1.4

1. a) T **b)** T **c)** F **3. a)** 0 **b)** 1 **c)** 1 **5. a)** There is a student who spends more than 5 hours every weekday in class. **b)** Every student spends more than 5 hours every weekday in class. **c)** There is a student who does not spend more than 5 hours every weekday in class. **d)** No student spends more than 5 hours every weekday in class. **7. a)** $\exists x(P(x) \wedge Q(x))$ **b)** $\exists x(P(x) \wedge \neg Q(x))$ **c)** $\forall x(P(x) \vee Q(x))$ **d)** $\forall x \neg(P(x) \vee Q(x))$ **9. a)** T **b)** T **c)** T **d)** T **11. a)** $P(0) \vee P(1) \vee P(2) \vee P(3) \vee P(4)$ **b)** $P(0) \wedge P(1) \wedge P(2) \wedge P(3) \wedge P(4)$ **c)** $\neg P(0) \vee \neg P(1) \vee \neg P(2) \vee \neg P(3) \vee \neg P(4)$ **d)** $\neg P(0) \wedge \neg P(1) \wedge \neg P(2) \wedge \neg P(3) \wedge \neg P(4)$ **e)** $\neg(P(0) \vee P(1) \vee P(2) \vee P(3) \vee P(4))$ **f)** $\neg(P(0) \wedge P(1) \wedge P(2) \wedge P(3) \wedge P(4))$ **13.** Let $C(x)$ be the propositional function "x is in your class." **a)** $\exists x H(x)$ and $\exists x(C(x) \wedge H(x))$, where $H(x)$ is "x can speak Hindi" **b)** $\forall x F(x)$ and $\forall x(C(x) \to F(x))$, where $F(x)$ is "x is friendly" **c)** $\exists x \neg B(x)$ and $\exists x(C(x) \wedge \neg B(x))$, where $B(x)$ is "x was born in California" **d)** $\exists x M(x)$ and $\exists x(C(x) \wedge M(x))$, where $M(x)$ is "x has been in a movie" **e)** $\forall x \neg L(x)$ and $\forall x(C(x) \to \neg L(x))$, where $L(x)$ is "x has taken a course in logic programming"

15. Let $P(x)$ be "x is perfect"; let $F(x)$ be "x is your friend"; and let the domain be all people. **a)** $\forall x \, \neg P(x)$ **b)** $\neg \forall x \, P(x)$ **c)** $\forall x(F(x) \rightarrow P(x))$ **d)** $\exists x(F(x) \wedge P(x))$ **e)** $\forall x(F(x) \wedge P(x))$ or $(\forall x \, F(x)) \wedge (\forall x \, P(x))$ **f)** $(\neg \forall x \, F(x)) \vee (\exists x \, \neg P(x))$ **17.** Let $T(x)$ mean that x is a tautology and $C(x)$ mean that x is a contradiction. **a)** $\exists x \, T(x)$ **b)** $\forall x(C(x) \rightarrow T(\neg x))$ **c)** $\exists x \exists y(\neg T(x) \wedge \neg C(x) \wedge \neg T(y) \wedge \neg C(y) \wedge T(x \vee y))$ **d)** $\forall x \forall y((T(x) \wedge T(y)) \rightarrow T(x \wedge y))$ **19. a)** $Q(0,0,0) \wedge Q(0,1,0)$ **b)** $Q(0,1,1) \vee Q(1,1,1) \vee Q(2,1,1)$ **c)** $\neg Q(0,0,0) \vee \neg Q(0,0,1)$ **d)** $\neg Q(0,0,1) \vee \neg Q(1,0,1) \vee \neg Q(2,0,1)$ **21. a)** Let $T(x)$ be the predicate that x can learn new tricks, and let the domain be old dogs. Original is $\exists x \, T(x)$. Negation is $\forall x \, \neg T(x)$: "No old dogs can learn new tricks." **b)** Let $C(x)$ be the predicate that x knows calculus, and let the domain be rabbits. Original is $\neg \exists x \, C(x)$. Negation is $\exists x \, C(x)$: "There is a rabbit that knows calculus." **c)** Let $F(x)$ be the predicate that x can fly, and let the domain be birds. Original is $\forall x \, F(x)$. Negation is $\exists x \, \neg F(x)$: "There is a bird who cannot fly." **d)** Let $T(x)$ be the predicate that x can talk, and let the domain be dogs. Original is $\neg \exists x \, T(x)$. Negation is $\exists x \, T(x)$: "There is a dog that talks." **e)** Let $F(x)$ and $R(x)$ be the predicates that x knows French and knows Russian, respectively, and let the domain be people in this class. Original is $\neg \exists x(F(x) \wedge R(x))$. Negation is $\exists x(F(x) \wedge R(x))$: "There is someone in this class who knows French and Russian." **23. a)** There is no counterexample. **b)** $x = 0$ **c)** $x = 2$ **25. a)** $\forall x((F(x, 25{,}000) \vee S(x, 25)) \rightarrow E(x))$, where $E(x)$ is "Person x qualifies as an elite flyer in a given year," $F(x, y)$ is "Person x flies more than y miles in a given year," and $S(x, y)$ is "Person x takes more than y flights in a given year" **b)** $\forall x(((M(x) \wedge T(x, 3)) \vee (\neg M(x) \wedge T(x, 3.5))) \rightarrow Q(x))$, where $Q(x)$ is "Person x qualifies for the marathon," $M(x)$ is "Person x is a man," and $T(x, y)$ is "Person x has run the marathon in less than y hours" **c)** $M \rightarrow ((H(60) \vee (H(45) \wedge T)) \wedge \forall y \, G(B, y))$, where M is the proposition "The student received a masters degree," $H(x)$ is "The student took at least x course hours," T is the proposition "The student wrote a thesis," and $G(x, y)$ is "The person got grade x or higher in course y" **d)** $\exists x \, ((T(x, 21) \wedge G(x, 4.0))$, where $T(x, y)$ is "Person x took more than y credit hours" and $G(x, p)$ is "Person x earned grade point average p" (we assume that we are talking about one given semester) **27.** Both statements are true precisely when at least one of $P(x)$ and $Q(x)$ is true for at least one value of x in the domain. **29. a)** If A is true, then both sides are logically equivalent to $\forall x \, P(x)$. If A is false, the left-hand side is clearly false. Furthermore, for every x, $P(x) \wedge A$ is false, so the right-hand side is false. Hence, the two sides are logically equivalent. **b)** If A is true, then both sides are logically equivalent to $\exists x \, P(x)$. If A is false, the left-hand side is clearly false. Furthermore, for every x, $P(x) \wedge A$ is false, so $\exists x(P(x) \wedge A)$ is false. Hence, the two sides are logically equivalent. **31. f)** alse **b)** false **b)** true **b)** false **33. a)** True **b)** False, unless the domain consists of just one element **c)** True **35.** sibling(X,Y) :- mother(M,X), mother(M,Y), father(F,X), father(F,Y) **37. a)** $\forall x(P(x) \rightarrow \neg Q(x))$ **b)** $\forall x(Q(x) \rightarrow R(x))$ **c)** $\forall x(P(x) \rightarrow \neg R(x))$ **d)** The conclusion does not follow. There may be vain professors, because the premises do not rule out the possibility that there are other vain people besides ignorant ones.

Section 1.5

1. a) For every real number x there exists a real number y such that x is less than y. **b)** For every real number x and real number y, if x and y are both nonnegative, then their product is nonnegative. **c)** For every real number x and real number y, there exists a real number z such that $xy = z$. **3. a)** Sarah Smith has visited www.att.com. **b)** At least one person has visited www.imdb.org. **c)** Jose Orez has visited at least one website. **d)** There is a website that both Ashok Puri and Cindy Yoon have visited. **e)** There is a person besides David Belcher who has visited all the websites that David Belcher has visited. **f)** There are two different people who have visited exactly the same websites. **5. a)** $\forall x \, L(x, \text{Jerry})$ **b)** $\forall x \exists y \, L(x, y)$ **c)** $\exists y \forall x \, L(x, y)$ **d)** $\forall x \exists y \neg L(x, y)$ **e)** $\exists x \neg L(\text{Lydia}, x)$ **f)** $\exists x \forall y \neg L(y, x)$ **g)** $\exists x(\forall y L(y, x) \wedge \forall z((\forall w L(w, z)) \rightarrow z = x))$ **h)** $\exists x \exists y(x \neq y \wedge L(\text{Lynn}, x) \wedge L(\text{Lynn}, y) \wedge \forall z(L(\text{Lynn}, z) \rightarrow (z = x \vee z = y)))$ **i)** $\forall x L(x, x)$ **j)** $\exists x \, \forall y \, (L(x,y) \leftrightarrow x = y)$ **7. a)** $\forall x \, P(x)$, where $P(x)$ is "x needs a course in discrete mathematics" and the domain consists of all computer science students **b)** $\exists x \, P(x)$, where $P(x)$ is "x owns a personal computer" and the domain consists of all students in this class **c)** $\forall x \exists y \, P(x, y)$, where $P(x, y)$ is "x has taken y," the domain for x consists of all students in this class, and the domain for y consists of all computer science classes **d)** $\exists x \exists y \, P(x, y)$, where $P(x, y)$ and domains are the same as in part (c) **e)** $\forall x \forall y \, P(x, y)$, where $P(x, y)$ is "x has been in y," the domain for x consists of all students in this class, and the domain for y consists of all buildings on campus **f)** $\exists x \exists y \forall z(P(z, y) \rightarrow Q(x, z))$, where $P(z, y)$ is "z is in y" and $Q(x, z)$ is "x has been in z"; the domain for x consists of all students in the class, the domain for y consists of all buildings on campus, and the domain of z consists of all rooms. **g)** $\forall x \forall y \exists z(P(z, y) \wedge Q(x, z))$, with same environment as in part (f) **9. a)** $\forall u \exists m(A(u, m) \wedge \forall n(n \neq m \rightarrow \neg A(u, n)))$, where $A(u, m)$ means that user u has access to mailbox m **b)** $\exists p \exists e(H(e) \wedge S(p, \text{running})) \rightarrow S(\text{kernel, working correctly})$, where $H(e)$ means that error condition e is in effect and $S(x, y)$ means that the status of x is y **c)** $\forall u \forall s(E(s, .edu) \rightarrow A(u, s))$, where $E(s, x)$ means that website s has extension x, and $A(u, s)$ means that user u can access website s **d)** $\exists x \exists y(x \neq y \wedge \forall z((\forall s \, M(z,s)) \leftrightarrow (z = x \vee z = y)))$, where $M(a, b)$ means that system a monitors remote server b **11. a)** $\forall x \, \forall y \, ((x < 0) \wedge (y < 0) \rightarrow (xy > 0))$ **b)** $\forall x(x - x = 0)$ **c)** $\forall x \exists a \exists b(a \neq b \wedge \forall c(c^2 = x \leftrightarrow (c = a \vee c = b)))$ **d)** $\forall x((x < 0) \rightarrow \neg \exists y(x = y^2))$ **13. a)** True **b)** True **c)** True **d)** True **e)** True **f)** False **g)** False **h)** True **i)** False **15. a)** $\exists x \forall y \exists z \, \neg T$

(x, y, z) **b)** $\exists x \forall y \neg P(x, y) \wedge \exists x \forall y \neg Q(x, y)$ **c)** $\exists x \forall y$ $(\neg P(x, y) \vee \forall z \neg R(x, y, z))$ **d)** $\exists x \forall y (P(x, y) \wedge \neg Q(x, y))$
17. a) $\exists x \exists y \neg P(x, y)$ **b)** $\exists y \forall x \neg P(x, y)$ **c)** $\exists y \exists x (\neg P(x, y) \wedge \neg Q(x, y))$ **d)** $(\forall x \forall y P(x, y)) \vee (\exists x \exists y \neg Q(x, y))$
e) $\exists x (\forall y \exists z \neg P(x,y,z) \vee \forall z \exists y \neg P(x, y, z))$ **19. a)** There is someone in this class such that for every two different math courses, these are not the two and only two math courses this person has taken.   **b)** Every person has either visited Libya or has not visited a country other than Libya.
c) Someone has climbed every mountain in the Himalayas.
d) There is someone who has neither been in a movie with Kevin Bacon nor has been in a movie with someone who has been in a movie with Kevin Bacon.   **21. a)** $x = 2$, $y = -2$ **b)** $x = -4$ **c)** $x = 17$, $y = -1$
23. $\forall m \forall b (m \neq 0 \rightarrow \exists x (mx + b = 0 \wedge \forall w (mw + b = 0 \rightarrow w = x)))$ **25. a)** True   **b)** False   **c)** True $\exists x (p(x) \wedge \forall y p(y) \rightarrow y = x))$.

Section 1.6

1. Modus ponens; valid; the conclusion is true, because the hypotheses are true.   **3.** Let w be "Randy works hard," let d be "Randy is a dull boy," and let j be "Randy will get the job." The hypotheses are w, $w \rightarrow d$, and $d \rightarrow \neg j$. Using modus ponens and the first two hypotheses, d follows. Using modus ponens and the last hypothesis, $\neg j$, which is the desired conclusion, "Randy will not get the job," follows.
5. Universal instantiation is used to conclude that "If Socrates is a man, then Socrates is mortal." Modus ponens is then used to conclude that Socrates is mortal.   **7. a)** Valid conclusions are "I did not take Tuesday off," "I took Thursday off," "It rained on Thursday."   **b)** "I did not eat spicy foods and it did not thunder" is a valid conclusion.   **c)** "I am clever" is a valid conclusion.   **d)** "Ralph is not a CS major" is a valid conclusion.   **e)** "That you buy lots of stuff is good for the U.S. and is good for you" is a valid conclusion.   **f)** "Mice gnaw their food" and "Rabbits are not rodents" are valid conclusions.   **9. a)** Let $c(x)$ be "x is in this class," $j(x)$ be "x knows how to write programs in JAVA," and $h(x)$ be "x can get a high-paying job." The premises are $c(\text{Doug})$, $j(\text{Doug})$, $\forall x (j(x) \rightarrow h(x))$. Using universal instantiation and the last premise, $j(\text{Doug}) \rightarrow h(\text{Doug})$ follows. Applying modus ponens to this conclusion and the second premise, $h(\text{Doug})$ follows. Using conjunction and the first premise, $c(\text{Doug}) \wedge h(\text{Doug})$ follows. Finally, using existential generalization, the desired conclusion, $\exists x (c(x) \wedge h(x))$ follows.
b) Let $c(x)$ be "x is in this class," $w(x)$ be "x enjoys whale watching," and $p(x)$ be "x cares about ocean pollution." The premises are $\exists x (c(x) \wedge w(x))$ and $\forall x (w(x) \rightarrow p(x))$. From the first premise, $c(y) \wedge w(y)$ for a particular person y. Using simplification, $w(y)$ follows. Using the second premise and universal instantiation, $w(y) \rightarrow p(y)$ follows. Using modus ponens, $p(y)$ follows, and by conjunction, $c(y) \wedge p(y)$ follows. Finally, by existential generalization, the desired conclusion, $\exists x (c(x) \wedge p(x))$, follows.   **c)** Let $c(x)$ be "x is in this class," $p(x)$ be "x owns

a PC," and $w(x)$ be "x can use a word-processing program." The premises are $c(\text{Zeke})$, $\forall x (c(x) \rightarrow p(x))$, and $\forall x (p(x) \rightarrow w(x))$. Using the second premise and universal instantiation, $c(\text{Zeke}) \rightarrow p(\text{Zeke})$ follows. Using the first premise and modus ponens, $p(\text{Zeke})$ follows. Using the third premise and universal instantiation, $p(\text{Zeke}) \rightarrow w(\text{Zeke})$ follows. Finally, using modus ponens, $w(\text{Zeke})$, the desired conclusion, follows.   **d)** Let $j(x)$ be "x is in New Jersey," $f(x)$ be "x lives within 50 miles of the ocean," and $s(x)$ be "x has seen the ocean." The premises are $\forall x (j(x) \rightarrow f(x))$ and $\exists x (j(x) \wedge \neg s(x))$. The second hypothesis and existential instantiation imply that $j(y) \wedge \neg s(y)$ for a particular person y. By simplification, $j(y)$ for this person y. Using universal instantiation and the first premise, $j(y) \rightarrow f(y)$, and by modus ponens, $f(y)$ follows. By simplification, $\neg s(y)$ follows from $j(y) \wedge \neg s(y)$. So $f(y) \wedge \neg s(y)$ follows by conjunction. Finally, the desired conclusion, $\exists x (f(x) \wedge \neg s(x))$, follows by existential generalization.   **11. a)** Correct, using universal instantiation and modus ponens   **b)** Invalid; fallacy of affirming the conclusion   **c)** Invalid; fallacy of denying the hypothesis   **d)** Correct, using universal instantiation and modus tollens   **13. a)** Fallacy of affirming the conclusion   **b)** Fallacy of begging the question   **c)** Valid argument using modus tollens   **d)** Fallacy of denying the hypothesis
15. a) not valid **b)** Valid   **17.** The error occurs in step (5), because we cannot assume, as is being done here, that the c that makes P true is the same as the c that makes Q true.   **19.**

Step	Reason
1. $\forall x (P(x) \wedge R(x))$	Premise
2. $P(a) \wedge R(a)$	Universal instantiation from (1)
3. $P(a)$	Simplification from (2)
4. $\forall x (P(x) \rightarrow$ $(Q(x) \wedge S(x)))$	Premise
5. $Q(a) \wedge S(a)$	Universal modus ponens from (3) and (4)
6. $S(a)$	Simplification from (5)
7. $R(a)$	Simplification from (2)
8. $R(a) \wedge S(a)$	Conjunction from (7) and (6)
9. $\forall x (R(x) \wedge S(x))$	Universal generalization from (5)

21. Let p be "It is raining"; let q be "Yvette has her umbrella"; let r be "Yvette gets wet." Assumptions are $\neg p \vee q$, $\neg q \vee \neg r$, and $p \vee \neg r$. Resolution on the first two gives $\neg p \vee \neg r$. Resolution on this and the third assumption gives $\neg r$, as desired.

Section 1.7

1. The given expression is in CNF $(Q \wedge P) \vee (Q \wedge \neg Q)$ is a DNF. $(P \wedge Q)$ is the principal disjunctive normal form expressed as $\sum 3$. $(P \vee Q) \wedge (\neg P \vee Q) \wedge (P \vee \neg Q)$ is the principal conjunctive normal form expressed as $\prod 0, 1, 2$ **3.** $(\neg Q \vee P) \wedge (\neg P) \wedge (Q)$ is a CNF. $(\neg Q \wedge \neg P \wedge Q) \vee (P \wedge \neg P \wedge Q)$ is DNF. The expression is a contradiction. Principal disjunctive normal form is nil (false). Principal conjunctive normal form is $(P \vee Q) \wedge (\neg P \vee Q) \wedge (P \vee \neg Q) \wedge (\neg P \vee \neg Q)$ expressed as $\prod 0, 1, 2, 3$. **5.** We will show how an expression can be put into prenex normal form (PNF) if subexpressions in it can be put into PNF. Then, working from the inside out, any expression can be put in PNF. (To formalize the argument, it is necessary to use the method of structural induction that will be discussed in Section 5.3.) By Exercise 23 of Section 1.3, we can assume that the proposition uses only $\vee$ and $\neg$ as logical connectives. Now note that any proposition with no quantifiers is already in PNF. (This is the basis case of the argument.) Now suppose that the proposition is of the form $Qx\,P(x)$, where Q is a quantifier. Because $P(x)$ is a shorter expression than the original proposition, we can put it into PNF. Then Qx followed by this PNF is again in PNF and is equivalent to the original proposition. Next, suppose that the proposition is of the form $\neg P$. If P is already in PNF, we slide the negation sign past all the quantifiers using the equivalences in Table 2 in Section 1.4. Finally, assume that proposition is of the form $P \vee Q$, where each of P and Q is in PNF. If only one of P and Q has quantifiers, then we can bring the quantifier in front of both. If both P and Q have quantifiers, we can use Exercise 27 in Section 1.4, or rewrite $P \vee Q$ with two quantifiers preceding the disjunction of a proposition of the form $R \vee S$, by renaming the variables.

Section 1.8

1. Let $n = 2k + 1$ and $m = 2l + 1$ be odd integers. Then $n + m = 2(k + l + 1)$ is even. **3.** Direct proof: Suppose that $m + n$ and $n + p$ are even. Then $m + n = 2s$ for some integer s and $n + p = 2t$ for some integer t. If we add these, we get $m + p + 2n = 2s + 2t$. Subtracting $2n$ from both sides and factoring, we have $m + p = 2s + 2t - 2n = 2(s + t - n)$. Because we have written $m + p$ as 2 times an integer, we conclude that $m + p$ is even. **5.** Suppose that r is rational and i is irrational and $s = r + i$ is rational. Then by Example 7, $s + (-r) = i$ is rational, which is a contradiction. **7.** Because $\sqrt{2} \cdot \sqrt{2} = 2$ is rational and $\sqrt{2}$ is irrational, the product of two irrational numbers is not necessarily irrational. **9. a)** Assume that n is odd, so $n = 2k + 1$ for some integer k. Then $n^3 + 5 = 2(4k^3 + 6k^2 + 3k + 3)$. Because $n^3 + 5$ is two times some integer, it is even. **b)** Suppose that $n^3 + 5$ is odd and n is odd. Because n is odd and the product of two odd numbers is odd, it follows that n^2 is odd and then that n^3 is odd. But then $5 = (n^3 + 5) - n^3$ would have to be even because it is the difference of two odd numbers. Therefore, the supposition that $n^3 + 5$ and n were both odd is wrong. **11.** The proposition is vacuously true because 0 is not a positive integer. Vacuous proof. **13.** $P(1)$ is true because $(a + b)^1 = a + b \geq a^1 + b^1 = a + b$. Direct proof. **15.** Suppose by way of contradiction that a/b is a rational root, where a and b are integers and this fraction is in lowest terms (that is, a and b have no common divisor greater than 1). Plug this proposed root into the equation to obtain $a^3/b^3 + a/b + 1 = 0$. Multiply through by b^3 to obtain $a^3 + ab^2 + b^3 = 0$. If a and b are both odd, then the left-

hand side is the sum of three odd numbers and therefore must be odd. If a is odd and b is even, then the left-hand side is odd + even + even, which is again odd. Similarly, if a is even and b is odd, then the left-hand side is even + even + odd, which is again odd. Because the fraction a/b is in simplest terms, it cannot happen that both a and b are even. Thus in all cases, the left-hand side is odd, and therefore cannot equal 0. This contradiction shows that no such root exists. **17.** This proposition is true. Suppose that m is neither 1 nor -1. Then mn has a factor m larger than 1. On the other hand, $mn = 1$, and 1 has no such factor. Hence, $m = 1$ or $m = -1$. In the first case $n = 1$, and in the second case $n = -1$, because $n = 1/m$. **19.** No **21.** We will show that the four statements are equivalent by showing that (i) implies (ii), (ii) implies (iii), (iii) implies (iv), and (iv) implies (i). First, assume that n is even. Then $n = 2k$ for some integer k. Then $n + 1 = 2k + 1$, so $n + 1$ is odd. This shows that (i) implies (ii). Next, suppose that $n + 1$ is odd, so $n + 1 = 2k + 1$ for some integer k. Then $3n + 1 = 2n + (n + 1) = 2(n + k) + 1$, which shows that $3n + 1$ is odd, showing that (ii) implies (iii). Next, suppose that $3n + 1$ is odd, so $3n + 1 = 2k + 1$ for some integer k. Then $3n = (2k + 1) - 1 = 2k$, so $3n$ is even. This shows that (iii) implies (iv). Finally, suppose that n is not even. Then n is odd, so $n = 2k + 1$ for some integer k. Then $3n = 3(2k + 1) = 6k + 3 = 2(3k + 1) + 1$, so $3n$ is odd. This completes a proof by contraposition that (iv) implies (i).

Section 1.9

1. $1^2 + 1 = 2 \geq 2 = 2^1$; $2^2 + 1 = 5 \geq 4 = 2^2$; $3^2 + 1 = 10 \geq 8 = 2^3$; $4^2 + 1 = 17 \geq 16 = 2^4$ **3.** If $x \leq y$, then $\max(x, y) + \min(x, y) = y + x = x + y$. If $x \geq y$, then $\max(x, y) + \min(x, y) = x + y$. Because these are the only two cases, the equality always holds. **5.** 10,001, 10,002, ... , 10,100 are all nonsquares, because $100^2 = 10,000$ and $101^2 = 10,201$; constructive. **7.** $8 = 2^3$ and $9 = 3^2$ **9.** Let $x = 2$ and $y = \sqrt{2}$. If $x^y = 2^{\sqrt{2}}$ is irrational, we are done. If not, then let $x = 2^{\sqrt{2}}$ and $y = \sqrt{2}/4$. Then $x^y = (2^{\sqrt{2}})^{\sqrt{2}/4} = 2^{\sqrt{2} \cdot (\sqrt{2})/4} = 2^{1/2} = \sqrt{2}$. **11. a)** This statement asserts the existence of x with a certain property. If we let $y = x$, then we see that $P(x)$ is true. If y is anything other than x, then $P(x)$ is not true. Thus, x is the unique element that makes P true. **b)** The first clause here says that there is an element that makes P true. The second clause says that whenever two elements both make P true, they are in fact the same element. Together these say that P is satisfied by exactly one element. **c)** This statement asserts the existence of an x that makes P true and has the further property that whenever we find an element that makes P true, that element is x. In other words, x is the unique element that makes P true. **13.** The equation $|a - c| = |b - c|$ is equivalent to the disjunction of two equations: $a - c = b - c$ or $a - c = -b + c$. The first of these is equivalent to $a = b$, which contradicts the assumptions made in this problem, so the original equation is equivalent to $a - c = -b + c$. By adding $b + c$ to both sides and dividing by 2, we see that this equation is equivalent to $c = (a + b)/2$. Thus, there is a unique solution. Furthermore, this c is an integer, because the sum of the odd integers a and b is even. **15.** If x is itself an integer, then we can take $n = x$ and $\epsilon = 0$. No other solution is possible in this case, because if the integer n is greater than x, then n is at least $x + 1$, which would make $\epsilon \geq 1$. If x is not an integer, then round it up to the next integer, and call that integer n. Let $\epsilon = n - x$. Clearly $0 \leq \epsilon < 1$; this is the only ϵ that will work with this n, and n cannot be any larger, because ϵ is constrained to be less than 1. **17.** Whether x and y are positive or negative, x^2 and y^2 are positive. So x^2 is 0 or 1 or 4; y^2 is 0 or 1. For any of these six combinations $2x^2 + 5y^2$ cannot be 14. **19.** By finding a common denominator, we can assume that the given rational numbers are a/b and c/b, where b is a positive integer and a and c are integers with $a < c$. In particular, $(a + 1)/b \leq c/b$. Thus, $x = (a + \frac{1}{2}\sqrt{2})/b$ is between the two given rational numbers, because $0 < \sqrt{2} < 2$. Furthermore, x is irrational, because if x were rational, then $2(bx - a) = \sqrt{2}$ would be as well, in violation of Example 10 in Section 1.7. **21.** Because there is an even number of squares in all, either there is an even number of squares in each row or there is an even number of squares in each column. In the former case, tile the board in the obvious way by placing the dominoes horizontally, and in the latter case, tile the board in the obvious way by placing the dominoes vertically. **23.** Remove the two black squares adjacent to a white corner, and remove two white squares other than that corner. Then no domino can cover that white corner.

Supplementary Exercises

1. a) $q \rightarrow p$ **b)** $q \wedge p$ **c)** $\neg q \vee \neg p$ **d)** $q \leftrightarrow p$ **3. a)** The proposition cannot be false unless $\neg p$ is false, so p is true. If p is true and q is true, then $\neg q \wedge (p \rightarrow q)$ is false, so the conditional statement is true. If p is true and q is false, then $p \rightarrow q$ is false, so $\neg q \wedge (p \rightarrow q)$ is false and the conditional statement is true. **b)** The proposition cannot be false unless q is false. If q is false and p is true, then $(p \vee q) \wedge \neg p$ is false, and the conditional statement is true. If q is false and p is false, then $(p \vee q) \wedge \neg p$ is false, and the conditional statement is true. **5.** $(p \wedge q \wedge r \wedge \neg s) \vee (p \wedge q \wedge \neg r \wedge s) \vee (p \wedge \neg q \wedge r \wedge s) \vee (\neg p \wedge q \wedge r \wedge s)$ **7.** Aaron is a knave and Crystal is a knight; it cannot be determined what Bohan is. **9.** Brenda **11. a)** F **b)** T **c)** F **d)** T **e)** F **f)** T **13.** $\forall x \exists y \exists z \, (y \neq z \wedge \forall w (P(w, x) \leftrightarrow (w = y \vee w = z)))$ **15.** Let y take values from the domain $\{y_1, y_2, y_3 \cdots \}$. $\forall y P(x, y) = P(x, y) \wedge P(x, y_2) \wedge P(x, y_3) \wedge \cdots$. The given expression is an implication. Assume the premise is true. $\exists x \forall y P(x, y)$ is true. This means that there is a value of x for which $\forall y P(x, y)$ is true. Let a be such a value of x. Then we have $P(a, y_1) \wedge P(a, y_2) \wedge \cdots$ to be true. I.e. $\exists x P(x, y_1) \wedge \exists x P(x, y_2) \wedge \cdots$ is true. i.e. $\forall y \exists x P(x, y)$ is true. We are able to arrive at the consequence. Whenever the premise is true, the consequence is true. The

implication is false only when the premise is true and the consequence is false. As this can never happen, the given expression is always true and hence a tautology. **17.** No **19.** $\forall x \, \forall z \, \exists y \, T(x, y, z)$, where $T(x, y, z)$ is the statement that student x has taken class y in department z, where the domains are the set of students in the class, the set of courses at this university, and the set of departments in the school of mathematical sciences **21.** Case (i) Suppose n is a perfect square. Then $n = m^2$ for some integer m and the result holds. Case (ii) Suppose n is not a perfect square. Then let m^2 be the largest perfect square less than n. Hence $n > m^2$. We see that, $n < (m + 1)^2$. If not let $n \geq (m + 1)^2$. If $n = (m + 1)^2$, then n is a perfect square and we get a contradiction. If $n > (m + 1)^2$, our assumption that m^2 is the largest perfect square less than n is not correct and leads to a contradiction. Hence $n < (m + 1)^2$ and the result follows. **23.** We can give a constructive proof by letting $m = 10^{500} + 1$. Then $m^2 = (10^{500} + 1)^2 > (10^{500})^2 = 10^{1000}$. **25.** 223 cannot be written as the sum of 36 fifth powers.

CHAPTER 2

Section 2.1

1. a) $\{-1, 1\}$ **b)** $\{1, 2, 3, 4, 5, 6, 7, 8, 9, 10, 11\}$ **c)** $\{0, 1, 4, 9, 16, 25, 36, 49, 64, 81\}$ **d)** $\emptyset$ **3. a)** The first is a subset of the second, but the second is not a subset of the first. **b)** Neither is a subset of the other. **c)** The first is a subset of the second, but the second is not a subset of the first. **5.** B and C are subsets of A. C is a subset of D **7. a)** no **b)** no **c)** yes **d)** yes **e)** yes **f)** no
9.

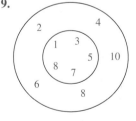

11. The dots in certain regions indicate that those regions are not empty.

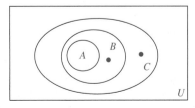

13. $A = \emptyset$, $B = \{\emptyset, \{\emptyset\}\}$ is one solution **15. a)** $\{\emptyset, \{a\}\}$ **b)** $\{\emptyset, \{a\}, \{b\}, \{a, b\}\}$ **c)** $\{\emptyset, \{\emptyset\}, \{\{\emptyset\}\}, \{\emptyset, \{\emptyset\}\}\}$ **17. a)** 8 **b)** 16 **c)** 2 **19. a)** $\{(a, y), (b, y), (c, y), (d, y), (a, z), (b, z), (c, z), (d, z)\}$ **b)** $\{(y, a), (y, b), (y, c), (y, d), (z, a), (z, b), (z, c), (z, d)\}$ **21.** Either A or B or both $= \emptyset$ **23.** mn

25. The elements of $A \times B \times C$ consist of 3-tuples (a, b, c), where $a \in A$, $b \in B$, and $c \in C$, whereas the elements of $(A \times B) \times C$ look like $((a, b), c)$—ordered pairs, the first coordinate of which is again an ordered pair.   **27. a)** $\{-1, 0, 1\}$ **b)** $\mathbf{Z} - \{0, 1\}$ **c)** $\emptyset$ **29. a)** S is a member of S. i.e., $S \in S$. Hence by definition of S, $S \notin S$. Thus we get a contradiction. **b)** S is not a member of S. i.e., $S \notin S$. Hence by definition of S, $S \in S$. Thus we get a contradiction.

Section 2.2

1. a) The set of students who live within one mile of school and walk to classes   **b)** The set of students who live within one mile of school or walk to classes (or do both)   **c)** The set of students who live within one mile of school but do not walk to classes   **d)** The set of students who walk to classes but live more than one mile away from school   **3.** $\overline{\overline{A}} = \{x \mid \neg(x \in \overline{A})\} = \{x \mid \neg(\neg x \in A)\} = \{x \mid x \in A\} = A$
5. Let $x \in A \cup (A \cap B)$ i.e., $x \in A \vee (x \in A \wedge x \in B)$ i.e., $x \in A$ Hence $A \cup (A \cap B) \subseteq A$ $A \subseteq A \cup (A \cap B)$ Therefore $A \cup (A \cap B) = A$ **7. a)** $x \in \overline{A \cup B} \equiv x \notin A \cup B \equiv \neg(x \in A \vee x \in B) \equiv \neg(x \in A) \wedge \neg(x \in B) \equiv x \notin A \wedge x \notin B \equiv x \in \overline{A} \wedge x \in \overline{B} \equiv x \in \overline{A} \cap \overline{B}$

b)

A	B	$A \cup B$	$\overline{A \cup B}$	$\overline{A}$	$\overline{B}$	$\overline{A} \cap \overline{B}$
1	1	1	0	0	0	0
1	0	1	0	0	1	0
0	1	1	0	1	0	0
0	0	0	1	1	1	1

9. a) $x \in \overline{A \cap B \cap C} \equiv x \notin A \cap B \cap C \equiv x \notin A \vee x \notin B \vee x \notin C \equiv x \in \overline{A} \vee x \in \overline{B} \vee x \in \overline{C} \equiv x \in \overline{A} \cup \overline{B} \cup \overline{C}$

b)

A	B	C	$A \cap B \cap C$	$\overline{A \cap B \cap C}$	$\overline{A}$	$\overline{B}$	$\overline{C}$	$\overline{A} \cup \overline{B} \cup \overline{C}$
1	1	1	1	0	0	0	0	0
1	1	0	0	1	0	0	1	1
1	0	1	0	1	0	1	0	1
1	0	0	0	1	0	1	1	1
0	1	1	0	1	1	0	0	1
0	1	0	0	1	1	0	1	1
0	0	1	0	1	1	1	0	1
0	0	0	0	1	1	1	1	1

11. a) Both sides equal $\{x \mid x \in A \wedge x \notin B\}$. **b)** $A = A \cap U = A \cap (B \cup \overline{B}) = (A \cap B) \cup (A \cap \overline{B})$ **13. a)** $\{4,6\}$ **b)** $\{0,1,2,3,4,5,6,7,8,9,10\}$ **c)** $\{4, 5, 6, 8, 10\}$ **d)** $\{0,2,4, 5,6,7,8,9,10\}$ **15. a)** $B \subseteq A$ **b)** $A \subseteq B$ **c)** $A \cap B = \emptyset$ **d)** Nothing, because this is always true **e)** $A = B$ **17.** $A \subseteq B \equiv \forall x (x \in A \rightarrow x \in B) \equiv \forall x (x \notin B \rightarrow x \notin A) \equiv \forall x (x \in \overline{B} \rightarrow x \in \overline{A}) \equiv \overline{B} \subseteq \overline{A}$ **19.** The set of students who are computer science majors but not mathematics majors or who are mathematics majors but not computer science majors   **21.** $B = \emptyset$ **23.** Yes   **25.** If $A \cup B$ were finite, then it would have n elements for some natural number n. But A already has more than n elements, because it is

infinite, and $A \cup B$ has all the elements that A has, so $A \cup B$ has more than n elements. This contradiction shows that $A \cup B$ must be infinite.   **27. a)** $\{1, 2, 3, \ldots, n\}$ **b)** $\{1\}$ **29. a)** A_n **b)** $\{0, 1\}$ **31. a)** $\mathbf{Z}$, $\{-1, 0, 1\}$ **b)** $\mathbf{Z} - \{0\}$, $\emptyset$ **c)** $\mathbf{R}$, $[-1, 1]$ **d)** $[1, \infty)$, $\emptyset$ **33. a)** $\{1, 2, 3, 4, 7, 8, 9, 10\}$ **b)** $\{2, 4, 5, 6, 7\}$ **c)** $\{1, 10\}$ **35. a)** 11 1110 0000 0000 0000 0000 0000 $\vee$ 01 1100 1000 0000 0100 0101 0000 = 11 1110 1000 0000 0100 0101 0000, representing $\{a, b, c, d, e, g, p, t, v\}$ **b)** 11 1110 0000 0000 0000 0000 0000 $\wedge$ 01 1100 1000 0000 0100 0101 0000 = 01 1100 0000 0000 0000 0000 0000, representing $\{b, c, d\}$ **c)** (11 1110 0000 0000 0000 0000 0000 $\vee$ 00 0110 0110 0001 1000 0110 0110) $\wedge$ (01 1100 1000 0000 0100 0101 0000 $\vee$ 00 1010 0010 0000 1000 0010 0111) = 11 1110 0110 0001 1000 0110 0110 $\wedge$ 01 1110 1010 0000 1100 0111 0111 = 01 1110 0010 0000 1000 0110 0110, representing $\{b, c, d, e, i, o, t, u, x, y\}$ **d)** 11 1110 0000 0000 0000 0000 0000 $\vee$ 01 1100 1000 0000 0100 0101 0000 $\vee$ 00 1010 0010 0000 1000 0010 0111 $\vee$ 00 0110 0110 0001 1000 0110 0110 = 11 1110 1110 0001 1100 0111 0111, representing $\{a,b,c,d,e,g,h,i,n,o,p,t,u,v,x,y,z\}$ **37. a)** $\{1, 2, 3, \{1, 2, 3\}\}$ **b)** $\{\emptyset\}$ **c)** $\{\emptyset, \{\emptyset\}\}$ **d)** $\{\emptyset, \{\emptyset\}, \{\emptyset, \{\emptyset\}\}\}$ **39. a)** $\{3 \cdot a, 3 \cdot b, 1 \cdot c, 4 \cdot d\}$ **b)** $\{2 \cdot a, 2 \cdot b\}$ **c)** $\{1 \cdot a, 1 \cdot c\}$ **d)** $\{1 \cdot b, 4 \cdot d\}$ **e)** $\{5 \cdot a, 5 \cdot b, 1 \cdot c, 4 \cdot d\}$

Section 2.3

1. a) $f(0)$ is not defined. **b)** $f(x)$ is not defined for $x < 0$. **c)** $f(x)$ is not well-defined because there are two distinct values assigned to each x.   **3. a)** Not a function **b)** A function **c)** Not a function   **5. a)** 1 **b)** 0 **c)** 0 **d)** -1 **e)** 3 **f)** -1 **g)** 2 **h)** 1   **7.** Only the function in part (a)   **9.** Only the functions in parts (a) and (d)   **11. a)** Depends on whether teachers share offices **b)** One-to-one assuming only one teacher per bus **c)** Most likely not one-to-one, especially if salary is set by a collective bargaining agreement **d)** One-to-one   **13.** Answers will vary. **a)** Set of offices at the school; probably not onto **b)** Set of buses going on the trip; onto, assuming every bus gets a teacher chaperone **c)** Set of real numbers; not onto **d)** Set of strings of nine digits with hyphens after third and fifth digits; not onto   **15. a)** Yes **b)** No **c)** Yes **d)** No   **17.** The function is not one-to-one, so it is not invertible. On the restricted domain, the function is the identity function on the nonnegative real numbers, $f(x) = x$, so it is its own inverse. **19. a)** Let x and y be distinct elements of A. Because g is one-to-one, $g(x)$ and $g(y)$ are distinct elements of B. Because f is one-to-one, $f(g(x)) = (f \circ g)(x)$ and $f(g(y)) = (f \circ g)(y)$ are distinct elements of C. Hence, $f \circ g$ is one-to-one. **b)** Let $y \in C$. Because f is onto, $y = f(b)$ for some $b \in B$. Now because g is onto, $b = g(x)$ for some $x \in A$. Hence, $y = f(b) = f(g(x)) = (f \circ g)(x)$. It follows that $f \circ g$ is onto. **21.** No. For example, suppose that $A = \{a\}$, $B = \{b, c\}$, and $C = \{d\}$. Let $g(a) = b$, $f(b) = d$, and $f(c) = d$. Then f and $f \circ g$ are onto, but g is not.   **23.** $(f + g)(x) = x^2 + x + 3$, $(fg)(x) = x^3 + 2x^2 + x + 2$ **25.** f is one-to-one because $f(x_1) = f(x_2) \rightarrow ax_1 + b = ax_2 + b \rightarrow ax_1 = ax_2 \rightarrow x_1 = x_2$. f is onto because $f((y - b)/a) = y$. $f^{-1}(y) = (y - b)/a$.

27. a) $A = B = \mathbf{R}$, $S = \{x \mid x > 0\}$, $T = \{x \mid x < 0\}$, $f(x) = x^2$ **b)** It suffices to show that $f(S) \cap f(T) \subseteq f(S \cap T)$. Let $y \in B$ be an element of $f(S) \cap f(T)$. Then $y \in f(S)$, so $y = f(x_1)$ for some $x_1 \in S$. Similarly, $y = f(x_2)$ for some $x_2 \in T$. Because f is one-to-one, it follows that $x_1 = x_2$. Therefore $x_1 \in S \cap T$, so $y \in f(S \cap T)$. **29.** Let $x = \lfloor x \rfloor + \epsilon$, where ϵ is a real number with $0 \le \epsilon < 1$. If $\epsilon < \frac{1}{2}$, then $\lfloor x \rfloor - 1 < x - \frac{1}{2} < \lfloor x \rfloor$, so $\lceil x - \frac{1}{2} \rceil = \lfloor x \rfloor$ and this is the integer closest to x. If $\epsilon > \frac{1}{2}$, then $\lfloor x \rfloor < x - \frac{1}{2} < \lfloor x \rfloor + 1$, so $\lceil x - \frac{1}{2} \rceil = \lfloor x \rfloor + 1$ and this is the integer closest to x. If $\epsilon = \frac{1}{2}$, then $\lceil x - \frac{1}{2} \rceil = \lfloor x \rfloor$, which is the smaller of the two integers that surround x and are the same distance from x. **31.** Write the real number x as $\lfloor x \rfloor + \epsilon$, where ϵ is a real number with $0 \le \epsilon < 1$. Because $\epsilon = x - \lfloor x \rfloor$, it follows that $0 \le -\lfloor x \rfloor < 1$. The first two inequalities, $x - 1 < \lfloor x \rfloor$ and $\lfloor x \rfloor \le x$, follow directly. For the other two inequalities, write $x = \lceil x \rceil - \epsilon'$, where $0 \le \epsilon' < 1$. Then $0 \le \lceil x \rceil - x < 1$, and the desired inequality follows. **33. a)** If $x < n$, because $\lfloor x \rfloor \le x$, it follows that $\lfloor x \rfloor < n$. Suppose that $x \ge n$. By the definition of the floor function, it follows that $\lfloor x \rfloor \ge n$. This means that if $\lfloor x \rfloor < n$, then $x < n$. **b)** If $n < x$, then because $x \le \lceil x \rceil$, it follows that $n \le \lceil x \rceil$. Suppose that $n \ge x$. By the definition of the ceiling function, it follows that $\lceil x \rceil \le n$. This means that if $n < \lceil x \rceil$, then $n < x$. **35.** If n is even, then $n = 2k$ for some integer k. Thus, $\lfloor n/2 \rfloor = \lfloor k \rfloor = k = n/2$. If n is odd, then $n = 2k + 1$ for some integer k. Thus, $\lfloor n/2 \rfloor = \lfloor k + \frac{1}{2} \rfloor = k = (n-1)/2$. **47.** Assume that $x \ge 0$. The left-hand side is $\lceil -x \rceil$ and the right-hand side is $-\lfloor x \rfloor$. If x is an integer, then both sides equal $-x$. Otherwise, let $x = n + \epsilon$, where n is a natural number and ϵ is a real number with $0 \le \epsilon < 1$. Then $\lceil -x \rceil = \lceil -n - \epsilon \rceil = -n$ and $-\lfloor x \rfloor = -\lfloor n + \epsilon \rfloor = -n$ also. When $x < 0$, the equation also holds because it can be obtained by substituting $-x$ for x. **39.** $\lceil b \rceil - \lfloor a \rfloor - 1$ **41. a)** 1 **b)** 3 **c)** 126 **d)** 3600
43.

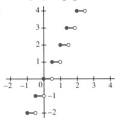

45. $f^{-1}(y) = (y - 1)^{1/3}$ **47. a)** $f_{A \cap B}(x) = 1 \leftrightarrow x \in A \cap B \leftrightarrow x \in A$ and $x \in B \leftrightarrow f_A(x) = 1$ and $f_B(x) = 1 \leftrightarrow f_A(x) f_B(x) = 1$ **b)** $f_{A \cup B}(x) = 1 \leftrightarrow x \in A \cup B \leftrightarrow x \in A$ or $x \in B \leftrightarrow f_A(x) = 1$ or $f_B(x) = 1 \leftrightarrow f_A(x) + f_B(x) - f_A(x) f_B(x) = 1$ **c)** $f_{\overline{A}}(x) = 1 \leftrightarrow x \in \overline{A} \leftrightarrow x \notin A \leftrightarrow f_A(x) = 0 \leftrightarrow 1 - f_A(x) = 1$ **d)** $f_{A \oplus B}(x) = 1 \leftrightarrow x \in A \oplus B \leftrightarrow (x \in A$ and $x \notin B)$ or $(x \notin A$ and $x \in B) \leftrightarrow f_A(x) + f_B(x) - 2f_A(x) f_B(x) = 1$
49. a) True; because $\lfloor x \rfloor$ is already an integer, $\lceil \lfloor x \rfloor \rceil = \lfloor x \rfloor$. **b)** False; $x = \frac{1}{2}$ is a counterexample. **c)** True; if x or y is an integer, then by property 4b in Table 1, the difference is 0. If neither x nor y is an integer, then $x = n + \epsilon$ and $y = m + \delta$, where n and m are integers and ϵ and δ are positive real numbers less than 1. Then $m + n < x + y < m + n + 2$, so

$\lceil x + y \rceil$ is either $m + n + 1$ or $m + n + 2$. Therefore, the given expression is either $(n + 1) + (m + 1) - (m + n + 1) = 1$ or $(n + 1) + (m + 1) - (m + n + 2) = 0$, as desired. **d)** False; $x = \frac{1}{4}$ and $y = 3$ is a counterexample. **e)** False; $x = \frac{1}{2}$ is a counterexample. **51. a)** Domain is $\mathbf{Z}$; codomain is $\mathbf{R}$; domain of definition is the set of nonzero integers; the set of values for which f is undefined is $\{0\}$; not a total function. **b)** Domain is $\mathbf{Z}$; codomain is $\mathbf{Z}$; domain of definition is $\mathbf{Z}$; set of values for which f is undefined is $\emptyset$; total function. **c)** Domain is $\mathbf{Z} \times \mathbf{Z}$; codomain is $\mathbf{Q}$; domain of definition is $\mathbf{Z} \times (\mathbf{Z} - \{0\})$; set of values for which f is undefined is $\mathbf{Z} \times \{0\}$; not a total function. **d)** Domain is $\mathbf{Z} \times \mathbf{Z}$; codomain is $\mathbf{Z}$; domain of definition is $\mathbf{Z} \times \mathbf{Z}$; set of values for which f is undefined is $\emptyset$; total function. **e)** Domain is $\mathbf{Z} \times \mathbf{Z}$; codomain is $\mathbf{Z}$; domain of definitions is $\{(m, n) \mid m > n\}$; set of values for which f is undefined is $\{(m, n) \mid m \le n\}$; not a total function. **53. a)** By definition, to say that S has cardinality m is to say that S has exactly m distinct elements. Therefore we can assign the first object to 1, the second to 2, and so on. This provides the one-to-one correspondence. **b)** By part (a), there is a bijection f from S to $\{1, 2, \ldots, m\}$ and a bijection g from T to $\{1, 2, \ldots, m\}$. Then the composition $g^{-1} \circ f$ is the desired bijection from S to T.

Section 2.4

1. a) 3 **b)** -1 **c)** 787 **d)** 2639 **3. a)** $2, 5, 8, 11, 14, 17, 20, 23, 26, 29$ **b)** $1, 1, 1, 2, 2, 2, 3, 3, 3, 4$ **c)** $1, 1, 3, 3, 5, 5, 7, 7, 9, 9$ **d)** $-1, -2, -2, 8, 88, 656, 4912, 40064, 362368, 3627776$ **e)** $3, 6, 12, 24, 48, 96, 192, 384, 768, 1536$ **f)** $2, 4, 6, 10, 16, 26, 42, 68, 110, 178$ **g)** $1, 2, 2, 3, 3, 3, 3, 4, 4, 4$ **h)** $3, 3, 5, 4, 4, 3, 5, 5, 4, 3$ **5.** Let $a_0, a_1, a_2, \mathit{ldots}$ be the sequence (i) $3, 5, 7, 9, 11$, $A_n = 2n + 3$ (ii) $3, 5, 7, 11, 15, 21, 27, \cdots$ $a_0 = 3$ $a_{2n} = a_{2n-1} + 2n$ $a_{2n-1} = a_{2n-2} + 2n$ (iii) $3, 5, 7, 11, 13, 19, \cdots$ $a_0 = 3$ $a_{2n-1} = (n+1)^2 + n$ $a_{2n} = (n+2)^2 - (n+1)$ **7. a)** $6, 17, 49, 143, 421$ **b)** $49 = 5 \cdot 17 - 6 \cdot 6$, $143 = 5 \cdot 49 - 6 \cdot 17$, $421 = 5 \cdot 143 - 6 \cdot 49$ **c)** $5a_{n-1} - 6a_{n-2} = 5(2^{n-1} + 5 \cdot 3^{n-1}) - 6(2^{n-2} + 5 \cdot 3^{n-2}) = 2^{n-2}(10 - 6) + 3^{n-2}(75 - 30) = 2^{n-2} \cdot 4 + 3^{n-2} \cdot 9 \cdot 5 = 2^n + 3^n \cdot 5 = a_n$ **9. a)** $a_n = 2 \cdot 3^n$ **b)** $a_n = 2n + 3$ **c)** $a_n = 1 + n(n + 1)/2$ **d)** $a_n = n^2 + 4n + 4$ **e)** $a_n = 1$ **f)** $a_n = (3^{n+1} - 1)/2$ **g)** $a_n = 5n!$ **h)** $a_n = 2^n n!$ **11. a)** $a_n = 3a_{n-1}$ **b)** 5,904,900 **13.** $B(k) = [1 + (0.07/12)]B(k - 1) - 100$, with $B(0) = 5000$ **15. a)** One 1 and one 0, followed by two 1s and two 0s, followed by three 1s and three 0s, and so on; 1, 1, 1 **b)** The positive integers are listed in increasing order with each even positive integer listed twice; 9, 10, 10. **c)** The terms in odd-numbered locations are the successive powers of 2; the terms in even-numbered locations are all 0; 32, 0, 64. **d)** $a_n = 3 \cdot 2^{n-1}$; 384, 768, 1536 **e)** $a_n = 15 - 7(n - 1) = 22 - 7n$; $-34, -41, -48$ **f)** $a_n = (n^2 + n + 4)/2$; 57, 68, 80 **g)** $a_n = 2n^3$; 1024, 1458, 2000 **h)** $a_n = n! + 1$; 362881, 3628801, 39916801 **17. a)** 20 **b)** 11 **c)** 30 **d)** 511 **19. a)** 1533 **b)** 510 **c)** 4923 **d)** 9842 **21.** $\sum_{j=1}^{n} (a_j - a_{j-1}) = a_n - a_0$ **23. a)** n^2

b) $n(n+1)/2$ **25.** 15150 **27.** $\frac{n(n+1)(2n+1)}{3} + \frac{n(n+1)}{2} +$ $(n+1)(m-(n+1)^2+1)$, where $n = \lfloor\sqrt{m}\rfloor - 1$ **29. a)** 0 **b)** 1680 **c)** 1 **d)** 1024 **31.** 34

Section 2.5

1. a) Countably infinite, $-1, -2, -3, -4, \ldots$ **b)** Countably infinite, $0, 2, -2, 4, -4, \ldots$ **c)** Countably infinite, $99, 98, 97, \ldots$ **d)** Uncountable **e)** Finite **f)** Countably infinite, $0, 7, -7, 14, -14, \ldots$ **3. a)** Countable: match n with the string of n 1s. **b)** Countable. To find a correspondence, follow the path in Example 4, but omit fractions in the top three rows (as well as continuing to omit fractions not in lowest terms). **c)** Uncountable **d)** Uncountable   **5.** For $n = 1, 2, 3, \ldots$, put the guest currently in Room $2n$ into Room n, and the guest currently in Room $2n-1$ into Room n of the new building.   **7.** Move the guess currently Room i to Room $2i+1$ for $i = 1, 2, 3, \ldots$. Put the jth guest from the kth bus into Room $2^k(2j+1)$.   **9. a)** $A = [1, 2]$ (closed interval of real numbers from 1 to 2), $B = [3, 4]$ **b)** $A = [1, 2] \cup \mathbf{Z}^+$, $B = [3, 4] \cup \mathbf{Z}^+$ **c)** $A = [1, 3]$, $B = [2, 4]$ **11.** Suppose that A is countable. Then either A has cardinality n for some nonnegative integer n, in which case there is a one-to-one function from A to a subset of $\mathbf{Z}^+$ (the range is the first n positive integers), or there exists a one-to-one correspondence f from A to $\mathbf{Z}^+$; in either case we have satisfied Definition 2. Conversely, suppose that $|A| \leq |\mathbf{Z}^+|$. By definition, this means that there is a one-to-one function from A to $\mathbf{Z}^+$, so A has the same cardinality as a subset of $\mathbf{Z}^+$ (namely the range of that function). By Exercise 14 we conclude that A is countable.   **13.** Assume that B is countable. Then the elements of B can be listed as $b_1, b_2, b_3, \ldots$. Because A is a subset of B, taking the subsequence of $\{b_n\}$ that contains the terms that are in A gives a listing of the elements of A. Because A is uncountable, this is impossible.   **15.** Assume that $A - B$ is countable. Then, because $A = (A - B) \cup (A \cap B)$, the elements of A can be listed in a sequence by alternating elements of $A - B$ and elements of $A \cap B$. This contradicts the uncountability of A.   **17.** Using the Axiom of Choice from set theory, choose distinct elements a_1, a_2, a_3, of A one at a time (this is possible because A is infinite). The resulting set $\{a_1, a_2, a_3, \ldots\}$ is the desired infinite subset of A. **19.** The set of finite strings of characters over a finite alphabet is countably infinite, because we can list these strings in alphabetical order by length. Therefore the infinite set S can be identified with an infinite subset of this countable set, which by Exercise 14 is also countably infinite.   **21.** Suppose that $A_1, A_2, A_3, \ldots$ are countable sets. Because A_i is countable, we can list its elements in a sequence as $a_{i1}, a_{i2}, a_{i3}, \ldots$. The elements of the set $\bigcup_{i=1}^{n} A_i$ can be listed by listing all terms a_{ij} with $i + j = 2$, then all terms a_{ij} with $i + j = 3$, then all terms a_{ij} with $i + j = 4$, and so on.   **23.** It is clear from the formula that the range of values the function takes on for a fixed value of $m + n$, say $m + n = x$, is $(x-2)(x-1)/2 + 1$ through $(x-2)(x-1)/2 + (x-1)$, because m can assume the values $1, 2, 3, \ldots, (x-1)$ under these conditions, and

the first term in the formula is a fixed positive integer when $m + n$ is fixed. To show that this function is one-to-one and onto, we merely need to show that the range of values for $x + 1$ picks up precisely where the range of values for x left off, i.e., that $f(x-1, 1) + 1 = f(1, x)$. We have $f(x-1, 1) + 1 = \frac{(x-2)(x-1)}{2} + (x-1) + 1 = \frac{x^2-x+2}{2} = \frac{(x-1)x}{2} + 1 = f(1, x)$.   **25.** By the Schröder-Bernstein theorem, it suffices to find one-to-one functions $f : (0, 1) \to [0, 1]$ and $g : [0, 1] \to (0, 1)$. Let $f(x) = x$ and $g(x) = (x+1)/3$.   **27.** Each element A of the power set of the set of positive integers (i.e., $A \subseteq \mathbf{Z}^+$) can be represented uniquely by the bit string $a_1 a_2 a_3 \ldots$, where $a_i = 1$ if $i \in A$ and $a_i = 0$ if $i \notin A$. Assume there were a one-to-one correspondence $f : \mathbf{Z}^+ \to \mathcal{P}(\mathbf{Z}^+)$. Form a new bit string $s = s_1 s_2 s_3 \ldots$ by setting s_i to be 1 minus the ith bit of $f(i)$. Then because s differs in the i bit from $f(i)$, s is not in the range of f, a contradiction.   **29.** For any finite alphabet there are a finite number of strings of length n, whenever n is a positive integer. It follows by the result of Exercise 21 that there are only a countable number of strings from any given finite alphabet. Because the set of all computer programs in a particular language is a subset of the set of all strings of a finite alphabet, which is a countable set by the result from Exercise 14, it is itself a countable set.

Section 2.6

1. a) 3×4 **b)** $\begin{bmatrix} 1 \\ 4 \\ 3 \end{bmatrix}$ **c)** $\begin{bmatrix} 2 & 0 & 4 & 6 \end{bmatrix}$ **d)** 1

e) $\begin{bmatrix} 1 & 2 & 1 \\ 1 & 0 & 1 \\ 1 & 4 & 3 \\ 3 & 6 & 7 \end{bmatrix}$ **3. a)** $\begin{bmatrix} 1 & 11 \\ 2 & 18 \end{bmatrix}$ **b)** $\begin{bmatrix} 2 & -2 & -3 \\ 1 & 0 & 2 \\ 9 & -4 & 4 \end{bmatrix}$

c) $\begin{bmatrix} -4 & 15 & -4 & 1 \\ -3 & 10 & 2 & -3 \\ 0 & 2 & -8 & 6 \\ 1 & -8 & 18 & -13 \end{bmatrix}$ **5.** $\begin{bmatrix} 9/5 & -6/5 \\ -1/5 & 4/5 \end{bmatrix}$

7. $\mathbf{0} + \mathbf{A} = [0 + a_{ij}] = [a_{ij} + 0] = \mathbf{0} + \mathbf{A}$ **9.** $\mathbf{A} + (\mathbf{B} + \mathbf{C}) = [a_{ij} + (b_{ij} + c_{ij})] = [(a_{ij} + b_{ij}) + c_{ij}] = (\mathbf{A} + \mathbf{B}) + \mathbf{C}$ **11.** The number of rows of $\mathbf{A}$ equals the number of columns of $\mathbf{B}$, and the number of columns of $\mathbf{A}$ equals the number of rows of $\mathbf{B}$.   **13.** $\mathbf{A}(\mathbf{BC}) = \left[\sum_q a_{iq} \left(\sum_r b_{qr} c_{rl}\right)\right] = \left[\sum_q \sum_r a_{iq} b_{qr} c_{rl}\right] = \left[\sum_r \sum_q a_{iq} b_{qr} c_{rl}\right] = \left[\sum_r \left(\sum_q a_{iq} b_{qr}\right) c_{rl}\right] = (\mathbf{AB})\mathbf{C}$ **15.** $\mathbf{A}^n = \begin{bmatrix} 1 & n \\ 0 & 1 \end{bmatrix}$ **17. a)** Let $\mathbf{A} = [a_{ij}]$ and $\mathbf{B} = [b_{ij}]$. Then $\mathbf{A} + \mathbf{B} = [a_{ij} + b_{ij}]$. We have $(\mathbf{A} + \mathbf{B})^t = [a_{ji} + b_{ji}] = [a_{ji}] + [b_{ji}] = \mathbf{A}^t + \mathbf{B}^t$. **b)** Using the same notation as in part (a), we have $\mathbf{B}^t \mathbf{A}^t = \left[\sum_q b_{qi} a_{jq}\right] = \left[\sum_q a_{jq} b_{qi}\right] = (\mathbf{AB})^t$, because the

(i, j)th entry is the (j, i)th entry of **AB**.   **19.** The result follows because $\begin{bmatrix} a & b \\ c & d \end{bmatrix} \begin{bmatrix} d & -b \\ -c & a \end{bmatrix} = \begin{bmatrix} ad-bc & 0 \\ 0 & ad-bc \end{bmatrix} =$

$(ad - bc)\mathbf{I}_2 = \begin{bmatrix} d & -b \\ -c & a \end{bmatrix} \begin{bmatrix} a & b \\ c & d \end{bmatrix}$.   **21.** $\mathbf{A}^n(\mathbf{A}^{-1})^n =$ $\mathbf{A}(\mathbf{A}\cdots(\mathbf{A}(\mathbf{A}\mathbf{A}^{-1})\mathbf{A}^{-1})\cdots\mathbf{A}^{-1})\mathbf{A}^{-1}$ by the associative law. Because $\mathbf{A}\mathbf{A}^{-1} = \mathbf{I}$, working from the inside shows that $\mathbf{A}^n(\mathbf{A}^{-1})^n = \mathbf{I}$. Similarly $(\mathbf{A}^{-1})^n\mathbf{A}^n = \mathbf{I}$. Therefore $(\mathbf{A}^n)^{-1} = (\mathbf{A}^{-1})^n$.   **23.** The (i, j)th entry of $\mathbf{A} + \mathbf{A}^t$ is $a_{ij} + a_{ji}$, which equals $a_{ji} + a_{ij}$, the (j, i)th entry of $\mathbf{A} + \mathbf{A}^t$, so by definition $\mathbf{A} + \mathbf{A}^t$ is symmetric.   **25.** $x_1 = 1$, $x_2 = -1$, $x_3 = -2$

27. a) $\begin{bmatrix} 1 & 1 & 1 \\ 1 & 1 & 1 \\ 1 & 0 & 1 \end{bmatrix}$ **b)** $\begin{bmatrix} 0 & 0 & 1 \\ 1 & 0 & 0 \\ 0 & 0 & 1 \end{bmatrix}$ **c)** $\begin{bmatrix} 1 & 1 & 1 \\ 1 & 1 & 1 \\ 1 & 0 & 1 \end{bmatrix}$

29. a) $\begin{bmatrix} 1 & 0 & 0 \\ 1 & 1 & 0 \\ 1 & 0 & 1 \end{bmatrix}$ **b)** $\begin{bmatrix} 1 & 0 & 0 \\ 1 & 0 & 1 \\ 1 & 1 & 0 \end{bmatrix}$ **c)** $\begin{bmatrix} 1 & 0 & 0 \\ 1 & 1 & 1 \\ 1 & 1 & 1 \end{bmatrix}$

31. a) $\mathbf{A} \vee \mathbf{B} = [a_{ij} \vee b_{ij}] = [b_{ij} \vee a_{ij}] = \mathbf{B} \vee \mathbf{A}$ **b)** $\mathbf{A} \wedge \mathbf{B} = [a_{ij} \wedge b_{ij}] = [b_{ij} \wedge a_{ij}] = \mathbf{B} \wedge \mathbf{A}$ **33. a)** $\mathbf{A} \vee (\mathbf{B} \wedge \mathbf{C}) = [a_{ij}] \vee [b_{ij} \wedge c_{ij}] = [a_{ij} \vee (b_{ij} \wedge c_{ij})] = [(a_{ij} \vee b_{ij}) \wedge (a_{ij} \vee c_{ij})] = [a_{ij} \vee b_{ij}] \wedge [a_{ij} \vee c_{ij}] = (\mathbf{A} \vee \mathbf{B}) \wedge (\mathbf{A} \vee \mathbf{C})$ **b)** $\mathbf{A} \wedge (\mathbf{B} \vee \mathbf{C}) = [a_{ij}] \wedge [b_{ij} \vee c_{ij}] = [a_{ij} \wedge (b_{ij} \vee c_{ij})] = [(a_{ij} \wedge b_{ij}) \vee (a_{ij} \wedge c_{ij})] = [a_{ij} \wedge b_{ij}] \vee [a_{ij} \wedge c_{ij}] = (\mathbf{A} \wedge \mathbf{B}) \vee (\mathbf{A} \wedge \mathbf{C})$ **35.** $\mathbf{A} \odot (\mathbf{B} \odot \mathbf{C}) = \left[\bigvee_q a_{iq} \wedge \left(\bigvee_r (b_{qr} \wedge c_{rl})\right)\right] = \left[\bigvee_q \bigvee_r (a_{iq} \wedge b_{qr} \wedge c_{rl})\right] = \left[\bigvee_r \bigvee_q (a_{iq} \wedge b_{qr} \wedge c_{rl})\right] = \left[\bigvee_r \left(\bigvee_q (a_{iq} \wedge b_{qr})\right) \wedge c_{rl}\right] = (\mathbf{A} \odot \mathbf{B}) \odot \mathbf{C}$

Supplementary Exercises

1. a) $\overline{A}$ **b)** $A \cap B$ **c)** $A - B$ **d)** $\overline{A} \cap \overline{B}$ **e)** $A \oplus B$ **3.** Yes   **5.** $|A \cap B| \leq |A|$, $|A \cap B| \leq |B|$, $|A| \leq |A \cup B|$, $|B| \leq |A \cup B|$. Hence $|A \cap B| \leq |A \cup B|$ when A and B are finite sets. The relationship is an equality if $A = B$.   **7. a)** Yes, no **b)** Yes, no **c)** f has inverse with $f^{-1}(a) = 3$, $f^{-1}(b) = 4$, $f^{-1}(c) = 2$, $f^{-1}(d) = 1$; g has no inverse.   **9.** If f is one-to-one, then f provides a bijection between S and $f(S)$, so they have the same cardinality. If f is not one-to-one, then there exist elements x and y in S such that $f(x) = f(y)$. Let $S = \{x, y\}$. Then $|S| = 2$ but $|f(S)| = 1$.   **11.** Let $x \in A$. Then $S_f(\{x\}) = \{f(y) \mid y \in \{x\}\} = \{f(x)\}$. By the same reasoning, $S_g(\{x\}) = \{g(x)\}$. Because $S_f = S_g$, we can conclude that $\{f(x)\} = \{g(x)\}$, and so necessarily $f(x) = g(x)$.   **13.** The equation is true if and only if the sum of the fractional parts of x and y is less than 1.   **15.** The equation is true if and only if either both x and y are integers, or x is not an integer but the sum of the fractional parts of x and y is less than or equal to 1.   **17.** If x is an integer, then $\lfloor x \rfloor + \lfloor m - x \rfloor = x + m - x = m$. Otherwise, write x in terms of its integer and fractional parts:

$x = n + \epsilon$, where $n = \lfloor x \rfloor$ and $0 < \epsilon < 1$. In this case $\lfloor x \rfloor + \lfloor m - x \rfloor = \lfloor n + \epsilon \rfloor + \lfloor m - n - \epsilon \rfloor = n + m - n - 1 = m - 1$.   **19.** Write $n = 2k + 1$ for some integer k. Then $n^2 = 4k^2 + 4k + 1$, so $n^2/4 = k^2 + k + \frac{1}{4}$. Therefore, $\lceil n^2/4 \rceil = k^2 + k + 1$. But $(n^2 + 3)/4 = (4k^2 + 4k + 1 + 3)/4 = k^2 + k + 1$.   **21.** 101   **23.** $a_1 = 1$; $a_{2n+1} = n \cdot a_{2n}$ for all $n > 0$; and $a_{2n} = n + a_{2n-1}$ for all $n > 0$. The next four terms are 5346, 5353, 37471, and 37479.   **25.** If each $f^{-1}(j)$ is countable, then $S = f^{-1}(1) \cup f^{-1}(2) \cup \cdots$ is the countable union of countable sets and is therefore countable by Exercise 27 in Section 2.5.   **27.** $\mathbf{A}^{4n} = \begin{bmatrix} 1 & 0 \\ 0 & 1 \end{bmatrix}$, $\mathbf{A}^{4n+1} = \begin{bmatrix} 0 & 1 \\ -1 & 0 \end{bmatrix}$, $\mathbf{A}^{4n+2} = \begin{bmatrix} -1 & 0 \\ 0 & -1 \end{bmatrix}$, $\mathbf{A}^{4n+3} = \begin{bmatrix} 0 & -1 \\ 1 & 0 \end{bmatrix}$, for $n \geq 0$.   **29.** Suppose that $\mathbf{A} = \begin{bmatrix} a & b \\ c & d \end{bmatrix}$. Let $\mathbf{B} = \begin{bmatrix} 0 & 1 \\ 0 & 0 \end{bmatrix}$. Because $\mathbf{AB} = \mathbf{BA}$, it follows that $c = 0$ and $a = d$. Let $\mathbf{B} = \begin{bmatrix} 0 & 0 \\ 1 & 0 \end{bmatrix}$. Because $\mathbf{AB} = \mathbf{BA}$, it follows that $b = 0$. Hence, $\mathbf{A} = \begin{bmatrix} a & 0 \\ 0 & a \end{bmatrix} = a\mathbf{I}$.   **31. a)** Let $\mathbf{A} \odot \mathbf{0} = [b_{ij}]$. Then $b_{ij} = (a_{i1} \wedge 0) \vee \cdots \vee (a_{ip} \wedge 0) = 0$. Hence, $\mathbf{A} \odot \mathbf{0} = \mathbf{0}$. Similarly $\mathbf{0} \odot \mathbf{A} = \mathbf{0}$. **b)** $\mathbf{A} \vee \mathbf{0} = [a_{ij} \vee 0] = [a_{ij}] = \mathbf{A}$. Hence $\mathbf{A} \vee \mathbf{0} = \mathbf{A}$. Similarly $\mathbf{0} \vee \mathbf{A} = \mathbf{A}$. **c)** $\mathbf{A} \wedge \mathbf{0} = [a_{ij} \wedge 0] = [0] = \mathbf{0}$. Hence $\mathbf{A} \wedge \mathbf{0} = \mathbf{0}$. Similarly $\mathbf{0} \wedge \mathbf{A} = \mathbf{0}$.

CHAPTER 3

Section 3.1

1. $max := 1$, $i := 2$, $max := 8$, $i := 3$, $max := 12$, $i := 4$, $i := 5$, $i := 6$, $i := 7$, $max := 14$, $i := 8$, $i := 9$, $i := 10$, $i := 11$

3. procedure $AddUp(a_1, \ldots, a_n$: integers)
  $sum := a_1$
  for $i := 2$ **to** n
    $sum := sum + a_i$
  return sum

5. procedure $no\ of\ negative(a_1, a_2, \ldots, a_n$: integers)
  $k := 0$
  for $i := 1$ **to** n
    if $a_i < 0$, **then** $k := k + 1$
  return k

7. procedure $palindrome\ check(a_1 a_2 \ldots a_n$: string)
  $answer := $ **true**
  for $i := 1$ **to** $\lfloor n/2 \rfloor$
    if $a_i \neq a_{n+1-i}$ **then** $answer := $ **false**
  return $answer$

9. procedure $interchange(x, y$: real numbers)
  $z := x$
  $x := y$

$y := z$

The minimum number of assignments needed is three.

11. Linear search: $i := 1$, $i := 2$, $i := 3$, $i := 4$, $i := 5$, $i := 6$, $i := 7$, $location := 7$; binary search: $i := 1$, $j := 8$, $m := 4$, $i := 5$, $m := 6$, $i := 7$, $m := 7$, $j := 7$, $location := 7$

13. procedure $insert(x, a_1, a_2, \ldots, a_n$: integers)
    {the list is in order: $a_1 \le a_2 \le \cdots \le a_n$}
    $a_{n+1} := x + 1$
    $i := 1$
    while $x > a_i$
      $i := i + 1$
    for $j := 0$ **to** $n - i$
      $a_{n-j+1} := a_{n-j}$
    $a_i := x$
    {x has been inserted into correct position}

15. procedure $first\ largest(a_1, \ldots, a_n$: integers)
    $max := a_1$
    $location := 1$
    for $i := 2$ **to** n
      if $max < a_i$ **then**
        $max := a_i$
        $location := i$
    return $location$

17. procedure $mean\text{-}median\text{-}max\text{-}min(a, b, c$: integers)
    $mean := (a + b + c)/3$
    {the six different orderings of a, b, c with respect
      to $\ge$ will be handled separately}
    if $a \ge b$ **then**
      if $b \ge c$ **then** $median := b$; $max := a$; $min := c$
        $\vdots$
    (The rest of the algorithm is similar.)

19. procedure $first\text{-}three(a_1, a_2, \ldots, a_n$: integers)
    if $a_1 > a_2$ **then** interchange a_1 and a_2
    if $a_2 > a_3$ **then** interchange a_2 and a_3
    if $a_1 > a_2$ **then** interchange a_1 and a_2

21. procedure $ones(a$: bit string, $a = a_1 a_2 \ldots a_n)$
    $count := 0$
    for $i := 1$ **to** n
      if $a_i := 1$ **then**
        $count := count + 1$
    return $count$

23. procedure $ternary\ search(s$: integer, $a_1, a_2, \ldots, a_n$: increasing integers)
    $i := 1$
    $j := n$
    while $i < j - 1$
      $l := \lfloor (i + j)/3 \rfloor$
      $u := \lfloor 2(i + j)/3 \rfloor$
      if $x > a_u$ **then** $i := u + 1$
      else if $x > a_l$ **then**
        $i := l + 1$
        $j := u$
      else $j := l$

    if $x = a_i$ **then** $location := i$
    else if $x = a_j$ **then** $location := j$
    else $location := 0$
    return $location$ {0 if not found}

25. procedure $find\ a\ mode(a_1, a_2, \ldots, a_n$: nondecreasing integers)
    $modecount := 0$
    $i := 1$
    while $i \le n$
      $value := a_i$
      $count := 1$
      while $i \le n$ and $a_i = value$
        $count := count + 1$
        $i := i + 1$
      if $count > modecount$ **then**
        $modecount := count$
        $mode := value$
    return $mode$

27. procedure $find\ duplicate(a_1, a_2, \ldots, a_n$: integers)
    $location := 0$
    $i := 2$
    while $i \le n$ and $location = 0$
      $j := 1$
      while $j < i$ and $location = 0$
        if $a_i = a_j$ **then** $location := i$
        else $j := j + 1$
      $i := i + 1$
    return $location$
    {$location$ is the subscript of the first value that
      repeats a previous value in the sequence}

29. Bubble sort 6, 2, 3, 1, 5, 4. At the end of the first pass 2, 3, 1, 5, 4, 6. At the end of the second pass 2, 1, 3, 4, 5, 6. At the end of the third pass 1, 2, 3, 4, 6, 6. At the end of the fourth pass 1, 2, 3, 4, 6, 6. At the end of the fifth pass 1, 2, 3, 4, 6, 6.

31. procedure $better\ bubblesort(a_1, \ldots, a_n$: integers)
    $i := 1$; $done := $ **false**
    while $i < n$ and $done = $ **false**
      $done := $ **true**
      for $j := 1$ **to** $n - i$
        if $a_j > a_{j+1}$ **then**
          interchange a_j and a_{j+1}
          $done := $ **false**
      $i := i + 1$
    {$a_1, \ldots, a_n$ is in increasing order}

33. Insertion sort d, f, k, m, a, b. After the first pass d, f, k, m, a, b. After the second pass d, f, k, m, a, b. After the third pass d, f, k, m, a, b. After the fourth pass a, d, f, k, m, b. After the fifth pass a, b, d, f, k, m. **35. a)** 1, 5, 4, 3, 2; 1, 2, 4, 3, 5; 1, 2, 3, 4, 5; 1, 2, 3, 4, 5 **b)** 1, 4, 3, 2, 5; 1, 2, 3, 4, 5; 1, 2, 3, 4, 5; 1, 2, 3, 4, 5 **c)** 1, 2, 3, 4, 5; 1, 2, 3, 4, 5; 1, 2, 3, 4, 5; 1, 2, 3, 4, 5 **37.** We carry out the linear search algorithm given as Algorithm 2 in this section, except that we replace $x \ne a_i$ by $x < a_i$, and we replace the **else** clause with **else** $location := n + 1$. **39.** $2 + 3 + 4 + \cdots + n = (n^2 + n - 2)/2$ **41.** Find the location for the 2 in the list 3 (one comparison), and insert it in front of the 3, so the list now reads 2, 3, 4, 5, 1, 6. Find the location for the 4 (compare it to the 2 and then the 3), and insert it, leaving 2, 3, 4, 5, 1, 6. Find the location for the 5 (compare it to the 3 and then the 4), and insert it, leaving 2, 3, 4, 5, 1, 6. Find the location for the 1 (compare it to the 3 and then the 2 and then the 2 again), and insert it, leaving 1, 2, 3, 4, 5, 6. Find the location for the 6 (compare it to the 3 and then the 4 and then the 5), and insert it, giving the final answer 1, 2, 3, 4, 5, 6. **43.** The variation from Exercise 42 **45.** Greedy algorithm uses fewest coins in parts (a), (c), and (d). **a)** Two quarters, one penny **b)** Two quarters, one dime, nine pennies **c)** Three quarters, one penny **d)** Two quarters, one dime **47.** The 9:00–9:45 talk, the 9:50–10:15 talk, the 10:15–10:45 talk, the 11:00–11:15 talk **49. a)** Order the talks by starting time. Number the lecture halls 1, 2, 3, and so on. For each talk, assign it to lowest numbered lecture hall that is currently available. **b)** If this algorithm uses n lecture halls, then at the point the nth hall was first assigned, it *had* to be used (otherwise a lower-numbered hall would have been assigned), which means that n talks were going on simultaneously (this talk just assigned and the $n - 1$ talks currently in halls 1 through $n - 1$).

Section 3.2

1. The choices of C and k are not unique. **a)** $C = 1, k = 10$ **b)** $C = 4, k = 7$ **c)** No **d)** $C = 5, k = 1$ **e)** $C = 1, k = 0$ **f)** $C = 1, k = 2$ **3.** $x^4 + 9x^3 + 4x + 7 \le 4x^4$ for all $x > 9$; witnesses $C = 4, k = 9$ **5.** $(x^2 + 1)/(x + 1) = x - 1 + 2/(x + 1) < x$ for all $x > 1$; witnesses $C = 1, k = 1$ **7.** The choices of C and k are not unique. **a)** $n = 3, C = 3, k = 1$ **b)** $n = 3$, $C = 4, k = 1$ **c)** $n = 1, C = 2, k = 1$ **d)** $n = 0, C = 2, k = 1$ **9.** $x^2 + 4x + 17 \le 3x^3$ for all $x > 17$, so $x^2 + 4x + 17$ is $O(x^3)$, with witnesses $C = 3, k = 17$. However, if x^3 were $O(x^2 + 4x + 17)$, then $x^3 \le C(x^2 + 4x + 17) \le 3Cx^2$ for some C, for all sufficiently large x, which implies that $x \le 3C$ for all sufficiently large x, which is impossible. Hence, x^3 is not $O(x^2 + 4x + 17)$. **11. a)** no **b)** yes **c)** yes **d)** yes **e)** yes **f)** yes **13.** There are constants C_1, C_2, k_1, and k_2 such that $|f(x)| \le C_1|g(x)|$ for all $x > k_1$ and $|g(x)| \le C_2|h(x)|$ for all $x > k_2$. Hence, for $x > \max(k_1, k_2)$ it follows that $|f(x)| \le C_1|g(x)| \le C_1 C_2|h(x)|$. This shows that $f(x)$ is $O(h(x))$. **15. a)** $O(n^3)$ **b)** $O(n^5)$ **c)** $O(n^3 \cdot n!)$ **27. a)** Neither $\Theta(x^2)$ nor $\Omega(x^2)$ **b)** $\Theta(x^2)$ and $\Omega(x^2)$ **c)** Neither $\Theta(x^2)$ nor $\Omega(x^2)$ **d)** $\Omega(x^2)$, but not $\Theta(x^2)$ **e)** $\Omega(x^2)$, but not $\Theta(x^2)$ **f)** $\Omega(x^2)$ and $\Theta(x^2)$ **19.** If $f(x)$ is $\Theta(g(x))$, then there exist constants C_1 and C_2 with $C_1|g(x)| \le |f(x)| \le C_2|g(x)|$. It follows that $|f(x)| \le C_2|g(x)|$ and $|g(x)| \le (1/C_1)|f(x)|$ for $x > k$. Thus, $f(x)$ is $O(g(x))$ and $g(x)$ is $O(f(x))$. Conversely, suppose that $f(x)$ is $O(g(x))$ and $g(x)$ is $O(f(x))$. Then there are constants C_1, C_2, k_1, and k_2 such that $|f(x)| \le C_1|g(x)|$ for $x > k_1$ and $|g(x)| \le C_2|f(x)|$ for $x > k_2$. We can assume that $C_2 > 0$ (we can always make C_2 larger). Then we have $(1/C_2)|g(x)| \le |f(x)| \le C_1|g(x)|$ for $x > \max(k_1, k_2)$. Hence, $f(x)$ is $\Theta(g(x))$. **21.** If $f(x)$ is $\Theta(g(x))$, then $f(x)$ is both $O(g(x))$ and $\Omega(g(x))$. Hence, there are positive con-

stants C_1, k_1, C_2, and k_2 such that $|f(x)| \le C_2|g(x)|$ for all $x > k_2$ and $|f(x)| \ge C_1|g(x)|$ for all $x > k_1$. It follows that $C_1|g(x)| \le |f(x)| \le C_2|g(x)|$ whenever $x > k$, where $k = \max(k_1, k_2)$. Conversely, if there are positive constants C_1, C_2, and k such that $C_1|g(x)| \le |f(x)| \le C_2|g(x)|$ for $x > k$, then taking $k_1 = k_2 = k$ shows that $f(x)$ is both $O(g(x))$ and $\Theta(g(x))$.

23.

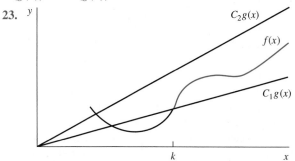

25. If $f(x)$ is $\Theta(1)$, then $|f(x)|$ is bounded between positive constants C_1 and C_2. In other words, $f(x)$ cannot grow larger than a fixed bound or smaller than the negative of this bound and must not get closer to 0 than some fixed bound. **27.** Because $f(x)$ is $O(g(x))$, there are constants C and k such that $|f(x)| \le C|g(x)|$ for $x > k$. Hence, $|f^n(x)| \le C^n|g^n(x)|$ for $x > k$, so $f^n(x)$ is $O(g^n(x))$ by taking the constant to be C^n. **29.** Yes **31.** This is false. Let $f_1 = x^2 + 2x$, $f_2 = x^2 + x$, and $g(x) = x^2$. Then $f_1(x)$ and $f_2(x)$ are both $O(g(x))$, but $(f_1 - f_2)(x)$ is not. **33.** Take $f(n)$ to be the function with $f(n) = n$ if n is an odd positive integer and $f(n) = 1$ if n is an even positive integer and $g(n)$ to be the function with $g(n) = 1$ if n is an odd positive integer and $g(n) = n$ if n is an even positive integer. **35.** There are positive constants $C_1, C_2, C'_1, C'_2, k_1, k'_1, k_2$, and k'_2 such that $|f_1(x)| \ge C_1|g_1(x)|$ for all $x > k_1$, $|f_1(x)| \le C'_1|g_1(x)|$ for all $x \ge k'_1$, $|f_2(x)| > C_2|g_2(x)|$ for all $x > k_2$, and $|f_2(x)| \le C'_2|g_2(x)|$ for all $x > k'_2$. Because f_2 and g_2 are never zero, the last two inequalities can be rewritten as $|1/f_2(x)| \le (1/C_2)|1/g_2(x)|$ for all $x > k_2$ and $|1/f_2(x)| \ge (1/C'_2)|1/g_2(x)|$ for all $x > k'_2$. Multiplying the first and rewritten fourth inequalities shows that $|f_1(x)/f_2(x)| \ge (C_1/C'_2)|g_1(x)/g_2(x)|$ for all $x > \max(k_1, k'_2)$, and multiplying the second and rewritten third inequalities gives $|f_1(x)/f_2(x)| \le (C'_1/C_2)|g_1(x)/g_2(x)|$ for all $x > \max(k'_1, k_2)$. It follows that f_1/f_2 is big-Theta of g_1/g_2. **37.** There exist positive constants $C_1, C_2, k_1, k_2, k'_1, k'_2$ such that $|f(x, y)| \le C_1|g(x, y)|$ for all $x > k_1$ and $y > k_2$ and $|f(x, y)| \ge C_2|g(x, y)|$ for all $x > k'_1$ and $y > k'_2$. **39.** $(x^2 + xy + x \log y)^3 < (3x^2y^3) = 27x^6y^3$ for $x > 1$ and $y > 1$, because $x^2 < x^2y$, $xy < x^2y$, and $x \log y < x^2y$. Hence, $(x^2 + xy + x \log y)^3$ is $O(x^6y^3)$. **41. a)** $\lim_{x \to \infty} x^2/x^3 = \lim_{x \to \infty} 1/x = 0$ **b)** $\lim_{x \to \infty} \frac{x \log x}{x^2} = \lim_{x \to \infty} \frac{\log x}{x} = \lim_{x \to \infty} \frac{1}{x \ln 2} = 0$ (using L'Hôpital's rule) **c)** $\lim_{x \to \infty} \frac{x^2}{2^x} = \lim_{x \to \infty} \frac{2x}{2^x \cdot \ln 2} = \lim_{x \to \infty} \frac{2}{2^x \cdot (\ln 2)^2} = 0$ (using L'Hôpital's rule) **d)** $\lim_{x \to \infty} \frac{x^2 + x + 1}{x^2} = \lim_{x \to \infty} \left(1 + \frac{1}{x} + \frac{1}{x^2}\right) = 1 \ne 0$

43.

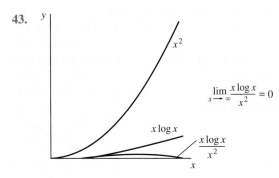

$$\lim_{x \to \infty} \frac{x \log x}{x^2} = 0$$

45. a) Because $\lim_{x \to \infty} f(x)/g(x) = 0$, $|f(x)|/|g(x)| < 1$ for sufficiently large x. Hence, $|f(x)| < |g(x)|$ for $x > k$ for some constant k. Therefore, $f(x)$ is $O(g(x))$.   **b)** Let $f(x) = g(x) = x$. Then $f(x)$ is $O(g(x))$, but $f(x)$ is not $o(g(x))$ because $f(x)/g(x) = 1$.   **47.** We can easily show that $(n-i)(i+1) \geq n$ for $i = 0, 1, \ldots, n-1$. Hence, $(n!)^2 = (n \cdot 1)((n-1) \cdot 2) \cdot ((n-2) \cdot 3) \cdots (2 \cdot (n-1)) \cdot (1 \cdot n) \geq n^n$. Therefore, $2 \log n! \geq n \log n$.   **49.** Compute that $\log 5! \approx 6.9$ and $(5 \log 5)/4 \approx 2.9$, so the inequality holds for $n = 5$. Assume $n \geq 6$. Because $n!$ is the product of all the integers from n down to 1, we have $n! > n(n-1)(n-2) \cdots \lceil n/2 \rceil$ (because at least the term 2 is missing). Note that there are more than $n/2$ terms in this product, and each term is at least as big as $n/2$. Therefore the product is greater than $(n/2)^{(n/2)}$. Taking the log of both sides of the inequality, we have $\log n! > \log \left(\frac{n}{2}\right)^{n/2} = \frac{n}{2} \log \frac{n}{2} = \frac{n}{2}(\log n - 1) > (n \log n)/4$, because $n > 4$ implies $\log n - 1 > (\log n)/2$.

Section 3.3

1. $O(1)$ **3.** $2n - 1$ **5.** Linear   **7.** $O(n)$ **9. a)** *power* := 1, *y* := 1; *i* := 1, *power* := 2, *y* := 3; *i* := 2, *power* := 4, *y* := 15   **b)** $2n$ multiplications and n additions   **11. a)** $2^{10^9} \approx 10^{3 \times 10^8}$ **b)** 10^9 **c)** 3.96×10^7 **d)** 3.16×10^4 **e)** 29 **f)** 12   **13. a)** 36 years **b)** 13 days **c)** 19 minutes   **15. a)** Less than 1 millisecond more **b)** 100 milliseconds more **c)** $2n + 1$ milliseconds more **d)** $3n^2 + 3n + 1$ milliseconds more **e)** Twice as much time **f)** 2^{2n+1} times as many milliseconds **g)** $n + 1$ times as many milliseconds   **17.** The average number of comparisons is $(3n + 4)/2$. **19.** $O(\log n)$ **21.** $O(n^2)$ **23.** $O(n)$ **25.** $O(\log n)$ comparisons; $O(n^2)$ swaps   **27. a)** doubles **b)** increases by 1

Supplementary Exercises

1. a) procedure *last max*($a_1, \ldots, a_n$: integers)
    max := a_1
    last := 1
    i := 2
    while $i \leq n$
      if $a_i \geq max$ **then**
        max := a_i
        last := *i*

      $i := i + 1$
    return *last*

b) $2n - 1 = O(n)$ comparisons

3. a) and **b)**
    procedure *smallest and largest*($a_1, a_2, \ldots, a_n$: integers)
    min := a_1
    max := a_1
    for $i := 2$ **to** n
      if $a_i < min$ **then** *min* := a_i
      if $a_i > max$ **then** *max* := a_i
    {*min* is the smallest integer among the input, and *max* is the largest}

c) $2n - 2$

5. Before any comparisons are done, there is a possibility that each element could be the maximum and a possibility that it could be the minimum. This means that there are $2n$ different possibilities, and $2n - 2$ of them have to be eliminated through comparisons of elements, because we need to find the unique maximum and the unique minimum. We classify comparisons of two elements as "virgin" or "nonvirgin," depending on whether or not both elements being compared have been in any previous comparison. A virgin comparison eliminates the possibility that the larger one is the minimum and that the smaller one is the maximum; thus each virgin comparison eliminates two possibilities, but it clearly cannot do more. A nonvirgin comparison must be between two elements that are still in the running to be the maximum or two elements that are still in the running to be the minimum, and at least one of these elements must *not* be in the running for the other category. For example, we might be comparing x and y, where all we know is that x has been eliminated as the minimum. If we find that $x > y$ in this case, then only one possibility has been ruled out—we now know that y is not the maximum. Thus in the worst case, a nonvirgin comparison eliminates only one possibility. (The cases of other nonvirgin comparisons are similar.) Now there are at most $\lfloor n/2 \rfloor$ comparisons of elements that have not been compared before, each removing two possibilities; they remove $2\lfloor n/2 \rfloor$ possibilities altogether. Therefore we need $2n - 2 - 2\lfloor n/2 \rfloor$ more comparisons that, as we have argued, can remove only one possibility each, in order to find the answers in the worst case, because $2n - 2$ possibilities have to be eliminated. This gives us a total of $2n - 2 - 2\lfloor n/2 \rfloor + \lfloor n/2 \rfloor$ comparisons in all. But $2n - 2 - 2\lfloor n/2 \rfloor + \lfloor n/2 \rfloor = 2n - 2 - \lfloor n/2 \rfloor = 2n - 2 + \lceil -n/2 \rceil = \lceil 2n - n/2 \rceil - 2 = \lceil 3n/2 \rceil - 2$, as desired.

7. The following algorithm has worst-case complexity $O(n^4)$.
procedure *equal sums*($a_1, a_2, \ldots, a_n$)
for $i := 1$ **to** n
  for $j := i + 1$ **to** n {since we want $i < j$}
    for $k := 1$ **to** n
      for $l := k + 1$ **to** n {since we want $k < l$}
        if $a_i + a_j = a_k + a_l$ and $(i, j) \neq (k, l)$
          then output these pairs

9. At end of first pass: 3, 1, 4, 5, 2, 6; at end of second pass: 1, 3, 2, 4, 5, 6; at end of third pass: 1, 2, 3, 4, 5, 6; fourth pass finds nothing to exchange and algorithm terminates **11.** In lists that are close to the correct order, in one pass or may be in a few passes when the number of elements is large, the sorted list will be obtained and no more passes are required. In such cases the nmber of steps will be cn for some constant c. **13.** $O(x^2 2^x)$ **15.** All of these functions are of the same order. **17.** $2^{2^{108}}$ **19.** For example, $f(n) = n^{2\lfloor n/2 \rfloor + 1}$ and $g(n) = n^{2\lceil n/2 \rceil}$

21. a)

    procedure *brute*$(a_1, a_2, \ldots, a_n :$ integers)

      for $i := 1$ **to** $n - 1$

        for $j := i + 1$ **to** n

          for $k := 1$ **to** n

            if $a_i + a_j = a_k$ **then return true else return false**

b) $O(n^3)$

23. For m_1: w_1 and w_2; for m_2: w_1 and w_3; for m_3: w_2 and w_3; for w_1: m_1 and m_2; for w_2: m_1 and m_3; for w_3: m_2 and m_3 **25.** 5; 15 **27. a)** For each subset S of $\{1, 2, \ldots, n\}$, compute $\sum_{j \in S} w_j$. Keep track of the subset giving the largest such sum that is less than or equal to W, and return that subset as the output of the algorithm. **b)** The food pack and the portable stove **29. a)** The makespan is always at least as large as the load on the processor assigned to do the lengthiest job, which must be at least $\max_{j=1,2,\ldots,n} t_j$. Therefore the minimum makespan satisfies this inequality. **b)** The total amount of time the processors need to spend working on the jobs (the total load) is $\sum_{j=1}^{n} t_j$. Therefore the average load per processor is $\frac{1}{p} \sum_{j=1}^{n} t_j$. The maximum load cannot be any smaller than the average, so the minimum makespan is always at least this large.

CHAPTER 4

Section 4.1

1. a) Yes **b)** No **c)** Yes **d)** No **3.** Suppose that $a \mid b$. Then there exists an integer k such that $ka = b$. Because $a(ck) = bc$ it follows that $a \mid bc$. **5.** If $a \mid b$ and $b \mid a$, there are integers c and d such that $b = ac$ and $a = bd$. Hence, $a = acd$. Because $a \neq 0$ it follows that $cd = 1$. Thus either $c = d = 1$ or $c = d = -1$. Hence, either $a = b$ or $a = -b$. **7.** Because $ac \mid bc$ there is an integer k such that $ack = bc$. Hence, $ak = b$, so $a \mid b$. **9. a)** 2, 5 **b)** -11, 10 **c)** 34, 7 **d)** 77, 0 **e)** 0, 0 **f)** 0, 3 **g)** -1, 2 **h)** 4, 0 **11. a)** 7:00 **b)** 8:00 **c)** 10:00 **13.** If $a \bmod m = b \bmod m$, then a and b have the same remainder when divided by m. Hence, $a = q_1 m + r$ and $b = q_2 m + r$, where $0 \leq r < m$. It follows that $a - b = (q_1 - q_2)m$, so $m \mid (a - b)$. It follows that $a \equiv b \pmod{m}$. **15.** $a = qd + r$ where q is the quotient and r is the remainder with $0 \leq r < d$. $\frac{a}{d} = q + \frac{r}{d}$. q is an integer and $0 \leq \frac{r}{d} < 1$. Hence $q = \lfloor \frac{a}{d} \rfloor$. $r = a - qd = a - d\lfloor \frac{a}{d} \rfloor$. **17. a)** 1 **b)** 2 **c)** 3 **d)** 9 **19. a)** -15 **b)** -7 **c)** 140 **21.** $-1, -26, -51, -76, 24, 49, 74, 99$ **23. a)** 13 **a)** 6

25. Let $m = tn$. Because $a \equiv b \pmod{m}$ there exists an integer s such that $a = b + sm$. Hence, $a = b + (st)n$, so $a \equiv b \pmod{n}$. **27. a)** Let $m = c = 2, a = 0$, and $b = 1$. Then $0 = ac \equiv bc = 2 \pmod 2$, but $0 = a \not\equiv b = 1 \pmod 2$. **b)** Let $m = 5, a = b = 3, c = 1$, and $d = 6$. Then $3 \equiv 3 \pmod 5$ and $1 \equiv 6 \pmod 5$, but $3^1 = 3 \not\equiv 4 \equiv 729 = 3^6 \pmod 5$. **29.** By Exercise 28 the sum of two squares must be either $0 + 0 = 0, 0 + 1 = 1$, or $1 + 1 = 2$, modulo 4, never 3, and therefore not of the form $4k + 3$. **31.** Because $a \equiv b \pmod{m}$, there exists an integer s such that $a = b + sm$, so $a - b = sm$. Then $a^k - b^k = (a - b)(a^{k-1} + a^{k-2}b + \cdots + ab^{k-2} + b^{k-1})$, $k \geq 2$, is also a multiple of m. It follows that $a^k \equiv b^k \pmod{m}$. **33.** $0 +_5 0 = 0, 0 +_5 1 = 1, 0 +_5 2 = 2, 0 +_5 3 = 3, 0 +_5 4 = 4; 1 +_5 1 = 2, 1 +_5 2 = 3, 1 +_5 3 = 4, 1 +_5 4 = 0; 2 +_5 2 = 4, 2 +_5 3 = 0, 2 +_5 4 = 1; 3 +_5 3 = 1, 3 +_5 4 = 2; 4 +_4 4 = 3$ and $0 \cdot_5 0 = 0, 0 \cdot_5 1 = 0, 0 \cdot_5 2 = 0, 0 \cdot_5 3 = 0, 0 \cdot_5 4 = 0; 1 \cdot_5 1 = 1, 1 \cdot_5 2 = 2, 1 \cdot_5 3 = 3, 1 \cdot_5 4 = 4; 2 \cdot_5 2 = 4, 2 \cdot_5 3 = 1, 2 \cdot_5 4 = 3; 3 \cdot_5 3 = 4, 3 \cdot_5 4 = 2; 4 \cdot_5 4 = 1$

Section 4.2

1. a) 1110 0111 **b)** 1 0001 1011 0100 **c)** 1 0111 11010110 1100 **3. a)** 1 0111 1010 **b)** 11 1000 0100 **c)** 1 0001 0011 **d)** 101 0000 1111 **5.** 1010 1011 1100 1101 1110 1111 **7.** Adding up to three leading 0s if necessary, write the binary expansion as $(\ldots b_{23}b_{22}b_{21}b_{20}b_{13}b_{12}b_{11}b_{10}b_{03}b_{02}b_{01}b_{00})_2$. The value of this numeral is $b_{00} + 2b_{01} + 4b_{02} + 8b_{03} + 2^4 b_{10} + 2^5 b_{11} + 2^6 b_{12} + 2^7 b_{13} + 2^8 b_{20} + 2^9 b_{21} + 2^{10} b_{22} + 2^{11} b_{23} + \cdots$, which we can rewrite as $b_{00} + 2b_{01} + 4b_{02} + 8b_{03} + (b_{10} + 2b_{11} + 4b_{12} + 8b_{13}) \cdot 2^4 + (b_{20} + 2b_{21} + 4b_{22} + 8b_{23}) \cdot 2^8 + \cdots$. Now $(b_{i3}b_{i2}b_{i1}b_{i0})_2$ translates into the hexadecimal digit h_i. So our number is $h_0 + h_1 \cdot 2^4 + h_2 \cdot 2^8 + \cdots = h_0 + h_1 \cdot 16 + h_2 \cdot 16^2 + \cdots$, which is the hexadecimal expansion $(\ldots h_1 h_1 h_0)_{16}$. **9.** 1 1101 1100 1010 1101 0001, $(1273)_8$ **11. a)** 1011 1110, 10 0001 0000 0001 **b)** 1 1010 1100, 1011 0000 0111 0011 **c)** 100 1001 1010, 101 0010 1001 0110 0000 **d)** 110 0000 0000, 1000 0000 0001 1111 1111 **13. a)** 1132, 144,305 **b)** 6273, 2,134,272 **c)** 2110, 1,107,667 **d)** 57,777, 237,326,216 **15.** 27 **17. a)** $5 = 3^2 - 3 - 1$ **b)** $13 = 3^2 + 3 + 1$ **c)** $37 = 3^3 + 3^2 + 1$ **d)** $79 = 3^4 - 3 + 1$ **19.** Let $a = (a_{n-1}a_{n-2}\ldots a_1 a_0)_{10}$. Then $a = 10^{n-1}a_{n-1} + 10^{n-2}a_{n-2} + \cdots + 10a_1 + a_0 \equiv a_{n-1} + a_{n-2} + \cdots + a_1 + a_0 \pmod 3$, because $10^j \equiv 1 \pmod 3$ for all nonnegative integers j. It follows that $3 \mid a$ if and only if 3 divides the sum of the decimal digits of a. **21. a)** -6 **b)** 13 **c)** -14 **d)** 0 **23.** The one's complement of the sum is found by adding the one's complements of the two integers except that a carry in the leading bit is used as a carry to the last bit of the sum. **25.** If $m \geq 0$, then the leading bit a_{n-1} of the one's complement expansion of m is 0 and the formula reads $m = \sum_{i=0}^{n-2} a_i 2^i$. This is correct because the right-hand side is the binary expansion of m. When m is negative, the leading bit a_{n-1} of the one's complement expansion of m is 1. The remaining $n - 1$ bits can be obtained by subtracting $-m$ from $111\ldots 1$ (where there are $n-1$ 1s), because subtracting a bit from 1 is the same as complementing it. Hence, the bit string $a_{n-2}\ldots a_0$ is the binary expansion of $(2^{n-1}-1)-(-m)$. Solving the equation $(2^{n-1}-1)-(-m)=\sum_{i=0}^{n-2}a_i 2^i$ for m gives the desired equation because $a_{n-1}=1$. **27. a)** -7 **b)** 13 **c)** -15 **d)** -1 **29.** To obtain the two's complement representation of the sum of two integers, add their two's complement representations (as binary integers are added) and ignore any carry out of the leftmost column. However, the answer is invalid if an overflow has occurred. This happens when the leftmost digits in the two's complement representation of the two terms agree and the leftmost digit of the answer differs. **31.** If $m \ge 0$, then the leading bit a_{n-1} is 0 and the formula reads $m=\sum_{i=0}^{n-2}a_i 2^i$. This is correct because the right-hand side is the binary expansion of m. If $m < 0$, its two's complement expansion has 1 as its leading bit and the remaining $n-1$ bits are the binary expansion of $2^{n-1}-(-m)$. This means that $(2^{n-1})-(-m)=\sum_{i=0}^{n-2}a_i 2^i$. Solving for m gives the desired equation because $a_{n-1}=1$. **33.** $4n$

35. procedure $Cantor(x$: positive integer)
$\quad n:=1; f:=1$
$\quad$**while** $(n+1)\cdot f \le x$
$\quad\quad n:=n+1$
$\quad\quad f:=f\cdot n$
$\quad y:=x$
$\quad$**while** $n>0$
$\quad\quad a_n:=\lfloor y/f\rfloor$
$\quad\quad y:=y-a_n\cdot f$
$\quad\quad f:=f/n$
$\quad\quad n:=n-1$
$\quad\{x=a_n n!+a_{n-1}(n-1)!+\cdots+a_1 1!\}$

37. First step: $c=0$, $d=0$, $s_0=1$; second step: $c=0$, $d=1$, $s_1=0$; third step: $c=1$, $d=1$, $s_2=0$; fourth step: $c=1$, $d=1$, $s_3=0$; fifth step: $c=1$, $d=1$, $s_4=1$; sixth step: $c=1$, $s_5=1$

39. procedure $subtract(a, b$: positive integers, $a > b$,
$\quad a=(a_{n-1}a_{n-2}\ldots a_1 a_0)_2$,
$\quad b=(b_{n-1}b_{n-2}\ldots b_1 b_0)_2)$
$\quad B:=0$ {B is the borrow}
$\quad$**for** $j:=0$ **to** $n-1$
$\quad\quad$**if** $a_j \ge b_j+B$ **then**
$\quad\quad\quad s_j:=a_j-b_j-B$
$\quad\quad\quad B:=0$
$\quad\quad$**else**
$\quad\quad\quad s_j:=a_j+2-b_j-B$
$\quad\quad\quad B:=1$
$\quad\{(s_{n-1}s_{n-2}\ldots s_1 s_0)_2$ is the difference}

41. procedure $compare(a, b$: positive integers,
$\quad a=(a_n a_{n-1}\ldots a_1 a_0)_2$,
$\quad b=(b_n b_{n-1}\ldots b_1 b_0)_2)$
$\quad k:=n$
$\quad$**while** $a_k=b_k$ and $k>0$
$\quad\quad k:=k-1$

$\quad$**if** $a_k=b_k$ **then** print "a equals b"
$\quad$**if** $a_k>b_k$ **then** print "a is greater than b"
$\quad$**if** $a_k<b_k$ **then** print "a is less than b"

43. $O(\log n)$ **45.** The only time-consuming part of the algorithm is the **while** loop, which is iterated q times. The work done inside is a subtraction of integers no bigger than a, which has $\log a$ bits. The result now follows from Example 9.

Section 4.3

1. 29, 71, 97 are primes. 21, 111, 143 are not primes. **3.** $2^8\cdot 3^4\cdot 5^2\cdot 7$

5. **procedure** $primetester(n$: integer greater than 1)
$\quad isprime:=$ **true**
$\quad d:=2$
$\quad$**while** $isprime$ and $d\le\sqrt{n}$
$\quad\quad$**if** n **mod** $d=0$ **then** $isprime:=$ **false**
$\quad\quad$**else** $d:=d+1$
$\quad$**return** $isprime$

7. Write $n=rs$, where $r>1$ and $s>1$. Then $2^n-1=2^{rs}-1=(2^r)^s-1=(2^r-1)((2^r)^{s-1}+(2^r)^{s-2}+(2^r)^{s-3}+\cdots+1)$. The first factor is at least $2^2-1=3$ and the second factor is at least $2^2+1=5$. This provides a factoring of 2^n-1 into two factors greater than 1, so 2^n-1 is composite. **9.** Suppose that $\log_2 3=a/b$ where $a, b\in\mathbf{Z}^+$ and $b\ne 0$. Then $2^{a/b}=3$, so $2^a=3^b$. This violates the fundamental theorem of arithmetic. Hence, $\log_2 3$ is irrational. **11.** 3, 5, and 7 are primes of the desired form. **13. a)** Yes **b)** No **c)** Yes **d)** Yes **15.** Suppose that n is not prime, so that $n=ab$, where a and b are integers greater than 1. Because $a>1$, by the identity in the hint, 2^a-1 is a factor of 2^n-1 that is greater than 1, and the second factor in this identity is also greater than 1. Hence, 2^n-1 is not prime. **17. a)** 2 **b)** 4 **c)** 12 **19.** $\phi(p^k)=p^k-p^{k-1}$ **21. a)** $2^{11}\cdot 3^7\cdot 5^9\cdot 7^3$ **b)** $2^9\cdot 3^7\cdot 5^5\cdot 7^3\cdot 11\cdot 13\cdot 17$ **c)** 23^{31} **d)** $41\cdot 43\cdot 53$ **e)** $2^{12}3^{13}5^{17}7^{21}$ **f)** Undefined **23.** $2^4 3^4 5 7^{11}$ **25.** 9 **27.** By Exercise 26 it follows that $\gcd(2^b-1, (2^a-1)$ **mod** $(2^b-1))=\gcd(2^b-1, 2^{a\,\mathbf{mod}\,b}-1)$. Because the exponents involved in the calculation are b and a **mod** b, the same as the quantities involved in computing $\gcd(a, b)$, the steps used by the Euclidean algorithm to compute $\gcd(2^a-1, 2^b-1)$ run in parallel to those used to compute $\gcd(a, b)$ and show that $\gcd(2^a-1, 2^b-1)=2^{\gcd(a,b)}-1$. **29. a)** $1=(-1)\cdot 10+1\cdot 11$ **b)** $1=21\cdot 21+(-10)\cdot 44$ **c)** $12=(-1)\cdot 36+48$ **d)** $1=13\cdot 55+(-21)\cdot 34$ **e)** $3=11\cdot 213+(-20)\cdot 117$ **f)** $223=1\cdot 0+1\cdot 223$ **g)** $1=37\cdot 2347+(-706)\cdot 123$ **h)** $2=1128\cdot 3454+(-835)\cdot 4666$ **i)** $1=2468\cdot 9999+(-2221)\cdot 11111$ **31.** $34\cdot 144+(-55)\cdot 89=1$ **33. a)** $a_n=1$ if n is prime and $a_n=0$ otherwise. **b)** a_n is the smallest prime factor of n with $a_1=1$. **c)** a_n is the number of positive divisors of n. **d)** $a_n=1$ if n has no divisors that are perfect squares greater than 1 and $a_n=0$ otherwise. **e)** a_n is the largest prime less than or equal to n. **f)** a_n is the product of the first $n - 1$ primes. **35.** $n = 1601$ is a counterexample. **37** Setting $k = a + b + 1$ will produce the composite number $a(a + b + 1) + b = a^2 + ab + a + b = (a + 1)(a + b)$.

Section 4.4

1. $15 \cdot 7 = 105 \equiv 1 \pmod{26}$ **3.** 7 **5.** Suppose that b and c are both inverses of a modulo m. Then $ba \equiv 1 \pmod{m}$ and $ca \equiv 1 \pmod{m}$. Hence, $ba \equiv ca \pmod{m}$. Because $\gcd(a, m) = 1$ it follows by Theorem 7 in Section 4.3 that $b \equiv c \pmod{m}$. **7.** 8 **9.** 3 and 6 **11.** Let $m' = m / \gcd(c, m)$. Because all the common factors of m and c are divided out of m to obtain m', it follows that m' and c are relatively prime. Because m divides $ac - bc = (a - b)c$, it follows that m' divides $(a - b)c$. By Lemma 3 in Section 4.3, we see that m' divides $a - b$, so $a \equiv b \pmod{m'}$. **13. a)** Suppose that $ia \equiv ja \pmod{p}$, where $1 \le i < j < p$. Then p divides $ja - ia = a(j - i)$. By Theorem 1, because a is not divisible by p, p divides $j - i$, which is impossible because $j - i$ is a positive integer less than p. **b)** By part (a), because no two of $a, 2a, \ldots, (p - 1)a$ are congruent modulo p, each must be congruent to a different number from 1 to $p-1$. It follows that $a \cdot 2a \cdot 3a \cdot \cdots \cdot (p - 1) \cdot a \equiv 1 \cdot 2 \cdot 3 \cdot \cdots \cdot (p - 1) \pmod{p}$. It follows that $(p-1)! \cdot a^{p-1} \equiv p - 1 \pmod{p}$. **c)** By Wilson's theorem and part (b), if p does not divide a, it follows that $(-1) \cdot a^{p-1} \equiv -1 \pmod{p}$. Hence, $a^{p-1} \equiv 1 \pmod{p}$. **d)** If $p \mid a$, then $p \mid a^p$. Hence, $a^p \equiv a \equiv 0 \pmod{p}$. If p does not divide a, then $a^{p-1} \equiv a \pmod{p}$, by part (c). Multiplying both sides of this congruence by a gives $a^p \equiv a \pmod{p}$. **15.** All integers of the form $323 + 330k$, where k is an integer **17.** All integers of the form $323 + 330k$, where k is an integer **19.** All integers of the form $16 + 252k$, where k is an integer **21.** Suppose that p is a prime appearing in the prime factorization of $m_1 m_2 \cdots m_n$. Because the m_is are relatively prime, p is a factor of exactly one of the m_is, say m_j. Because m_j divides $a - b$, it follows that $a - b$ has the factor p in its prime factorization to a power at least as large as the power to which it appears in the prime factorization of m_j. It follows that $m_1 m_2 \cdots m_n$ divides $a - b$, so $a \equiv b \pmod{m_1 m_2 \cdots m_n}$. **23.** $x \equiv 1 \pmod{6}$ **25. a)** By Fermat's little theorem, we have $2^{10} \equiv 1 \pmod{11}$. Hence, $2^{340} = (2^{10})^{34} \equiv 1^{34} = 1 \pmod{11}$. **b)** Because $32 \equiv 1 \pmod{31}$, it follows that $2^{340} = (2^5)^{68} = 32^{68} \equiv 1^{68} = 1 \pmod{31}$. **c)** Because 11 and 31 are relatively prime, and $11 \cdot 31 = 341$, it follows by parts (a) and (b) and Exercise 29 that $2^{340} \equiv 1 \pmod{341}$. **27.** Suppose that q is an odd prime with $q \mid 2^p - 1$. By Fermat's little theorem, $q \mid 2^{q-1} - 1$. From Exercise 37 in Section 4.3, $\gcd(2^p - 1, 2^{q-1} - 1) = 2^{\gcd(p, q-1)} - 1$. Because q is a common divisor of $2^p - 1$ and $2^{q-1} - 1$, $\gcd(2^p - 1, 2^{q-1} - 1) > 1$. Hence, $\gcd(p, q-1) = p$, because the only other possibility, namely, $\gcd(p, q-1) = 1$, gives us $\gcd(2^p - 1, 2^{q-1} - 1) = 1$. Hence, $p \mid q - 1$, and therefore there is a positive integer

m such that $q - 1 = mp$. Because q is odd, m must be even, say, $m = 2k$, and so every prime divisor of $2^p - 1$ is of the form $2kp + 1$. Furthermore, the product of numbers of this form is also of this form. Therefore, all divisors of $2^p - 1$ are of this form. **29.** M_{11} is not prime; M_{17} is prime. **31.** First, $2047 = 23 \cdot 89$ is composite. Write $2047 - 1 = 2046 = 2 \cdot 1023$, so $s = 1$ and $t = 1023$ in the definition. Then $2^{1023} = (2^{11})^{93} = 2048^{93} \equiv 1^{93} = 1 \pmod{2047}$, as desired. **33.** $0 = (0, 0)$, $1 = (1, 1)$, $2 = (2, 2)$, $3 = (0, 3)$, $4 = (1, 4)$, $5 = (2, 0)$, $6 = (0, 1)$, $7 = (1, 2)$, $8 = (2, 3)$, $9 = (0, 4)$, $10 = (1, 0)$, $11 = (2, 1)$, $12 = (0, 2)$, $13 = (1, 3)$, $14 = (2, 4)$ **35.** We have $m_1 = 99$, $m_2 = 98$, $m_3 = 97$, and $m_4 = 95$, so $m = 99 \cdot 98 \cdot 97 \cdot 95 = 89{,}403{,}930$. We find that $M_1 = m/m_1 = 903{,}070$, $M_2 = m/m_2 = 912{,}285$, $M_3 = m/m_3 = 921{,}690$, and $M_4 = m/m_4 = 941{,}094$. Using the Euclidean algorithm, we compute that $y_1 = 37$, $y_2 = 33$, $y_3 = 24$, and $y_4 = 4$ are inverses of M_k modulo m_k for $k = 1, 2, 3, 4$, respectively. It follows that the solution is $65 \cdot 903{,}070 \cdot 37 + 2 \cdot 912{,}285 \cdot 33 + 51 \cdot 921{,}690 \cdot 24 + 10 \cdot 941{,}094 \cdot 4 = 3{,}397{,}886{,}480 \equiv 537{,}140 \pmod{89{,}403{,}930}$. **37.** $\log_2 5 = 16$, $\log_2 6 = 14$ **39.** The value of $\left(\frac{a}{p}\right)$ depends only on whether a is a quadratic residue modulo p, that is, whether $x^2 \equiv a \pmod{p}$ has a solution. Because this depends only on the equivalence class of a modulo p, it follows that $\left(\frac{a}{p}\right) = \left(\frac{b}{p}\right)$ if $a \equiv b \pmod{p}$. **41.** By Exercise 40, $\left(\frac{a}{p}\right)\left(\frac{b}{p}\right) = a^{(p-1)/2}b^{(p-1)/2} = (ab)^{(p-1)/2} \equiv \left(\frac{ab}{p}\right) \pmod{p}$. **43.** $x \equiv 8$, 13, 22, or 27 $\pmod{35}$

Section 4.5

1. 91, 57, 21, 5 **3.** 1, 5, 4, 1, 5, 4, 1, 5, 4, $\ldots$ **5.** 2, 6, 7, 10, 8, 2, 6, 7, 10, 8, $\ldots$ **7.** 3792^2 gives 3792 as the middle portion and gives rise to only 3792. 2916 gives 2916 after 3 iterations and generation of new numbers is not possible. **9.** Only string (d) **11.** 4 **13. a)** Not valid **b)** Valid **c)** Valid **d)** Not valid **15.** Error in the last digit. **17. a)** Yes **b)** No **c)** Yes **d)** No **19. a)** Valid **b)** Not valid **c)** Valid **d)** Valid **21. a)** Not valid **b)** Valid **c)** Valid **d)** Not valid **23.** The given congruence is equivalent to $3d_1 + 4d_2 + 5d_3 + 6d_4 + 7d_5 + 8d_6 + 9d_7 + 10d_8 \equiv 0 \pmod{11}$. Transposing adjacent digits x and y (with x on the left) causes the left-hand side to increase by $x - y$. Because $x \not\equiv y \pmod{11}$, the congruence will no longer hold. Therefore errors of this type are always detected.

Section 4.6

1. a) GR QRW SDVV JR **b)** QB ABG CNFF TB **c)** QX UXM AHJJ ZX **3. a)** SURRENDER NOW **b)** BE MY FRIEND **c)** TIME FOR FUN **5.** $p = 7c + 13 \bmod 26$ **7.** $a = 18$, $b = 5$ **9.** BEWARE OF MARTIANS **11.** HURRICANE **13.** Suppose we know both $n = pq$ and $(p-1)(q-1)$. To find p and q, first note that $(p-1)(q-1) =$

$pq - p - q + 1 = n - (p + q) + 1$. From this we can find $s = p + q$. Because $q = s - p$, we have $n = p(s - p)$. Hence, $p^2 - ps + n = 0$. We now can use the quadratic formula to find p. Once we have found p, we can find q because $q = n/p$. **15.** SILVER   **17.** Alice sends $5^8 \bmod 23 = 16$ to Bob. Bob sends $5^5 \bmod 23 = 20$ to Alice. Alice computes $20^8 \bmod 23 = 6$ and Bob computes $16^5 \bmod 23 = 6$. The shared key is 6.   **19.** 2186 2087 1279 1251 0326 0816 1948   **21.** Alice can decrypt the first part of Cathy's message to learn the key, and Bob can decrypt the second part of Cathy's message, which Alice forwarded to him, to learn the key. No one else besides Cathy can learn the key, because all of these communications use secure private keys.

Supplementary Exercises

1. The actual number of miles driven is $46518 + 100000k$ for some natural number k.   **3.** 5, 22, −12, −29   **5.** Because $ac \equiv bc \pmod{m}$ there is an integer k such that $ac = bc + km$. Hence, $a - b = km/c$. Because $a - b$ is an integer, $c \mid km$. Letting $d = \gcd(m, c)$, write $c = de$. Because no factor of e divides m/d, it follows that $d \mid m$ and $e \mid k$. Thus $a - b = (k/e)(m/d)$, where $k/e \in \mathbf{Z}$ and $m/d \in \mathbf{Z}$. Therefore $a \equiv b \pmod{m/d}$.   **7.** n is divisible by 8 if and only if the binary expansion of n ends with 000. **9.** $(a_n a_{n-1} \ldots a_1 a_0)_{10} = \sum_{k=0}^{n} 10^k a_k \equiv \sum_{k=0}^{n} a_k \pmod 9$ because $10^k \equiv 1 \pmod 9$ for every nonnegative integer k. **11.** Take $a = 10$ and $b = 1$ in Dirichlet's theorem.   **13.** Every number greater than 11 can be written as either $8 + 2n$ or $9 + 2n$ for some $n \geq 2$.   **15.** Assume that every even integer greater than 2 is the sum of two primes, and let n be an integer greater than 5. If n is odd, write $n = 3 + (n - 3)$ and decompose $n - 3 = p + q$ into the sum of two primes; if n is even, then write $n = 2 + (n - 2)$ and decompose $n - 2 = p + q$ into the sum of two primes. For the converse, assume that every integer greater than 5 is the sum of three primes, and let n be an even integer greater than 2. Write $n + 2$ as the sum of three primes, one of which is necessarily 2, so $n + 2 = 2 + p + q$, whence $n = p + q$.   **17.** 1   **19.** 1   **21.** If not, then suppose that $q_1, q_2, \ldots, q_n$ are all the primes of the form $6k + 5$. Let $Q = 6q_1 q_2 \cdots q_n - 1$. Note that Q is of the form $6k + 5$, where $k = q_1 q_2 \cdots q_n - 1$. Let $Q = p_1 p_2 \cdots p_t$ be the prime factorization of Q. No p_i is 2, 3, or any q_j, because the remainder when Q is divided by 2 is 1, by 3 is 2, and by q_j is $q_j - 1$. All odd primes other than 3 are of the form $6k + 1$ or $6k + 5$, and the product of primes of the form $6k + 1$ is also of this form. Therefore at least one of the p_i's must be of the form $6k + 5$, a contradiction.   **23.** The product of numbers of the form $4k + 1$ is of the form $4k + 1$, but numbers of this form might have numbers not of this form as their only prime factors. For example, $49 = 4 \cdot 12 + 1$, but the prime factorization of 49 is $7 \cdot 7 = (4 \cdot 1 + 3)(4 \cdot 1 + 3)$. **25. a)** Not mutually relatively prime   **b)** Mutually relatively prime   **c)** Mutually relatively prime   **d)** Mutually relatively prime   **27.** $x = 6t + 2$ for some integer t and $x = 9u + 3$ for some integer u. We can conclude that x is an even number

and so u must be odd. $x = 9(2v + 1) + 3$ for some integer v. $x = 18v + 9 + 3 = 18v + 12$ and cannot leave remainder 2 when divided by 6.   **29.** By the Chinese remainder theorem, it suffices to show that $n^9 - n \equiv 0 \pmod 2$, $n^9 - n \equiv 0 \pmod 3$, and $n^9 - n \equiv 0 \pmod 5$. Each in turn follows from applying Fermat's little theorem.   **31.** (a) no (b) no (c) yes (d) yes   **33.** If digits in two odd positions are transposed or digits in two even positions are transposed, such errors cannot be detected by an ISBN-13.   **35. a)** QAL HUVEM AT WVESGB **b)** QXB EVZZL ZEVZZRFS

CHAPTER 5

Section 5.1

1. Let $P(n)$ be the statement that the train stops at station n. *Basis step:* We are told that $P(1)$ is true. *Inductive step:* We are told that $P(n)$ implies $P(n + 1)$ for each $n \geq 1$. Therefore by the principle of mathematical induction, $P(n)$ is true for all positive integers n.   **3. a)** $1^2 = 1 \cdot 2 \cdot 3/6$ **b)** Both sides of $P(1)$ shown in part (a) equal 1. **c)** $1^2 + 2^2 + \cdots + k^2 = k(k + 1)(2k + 1)/6$ **d)** For each $k \geq 1$ that $P(k)$ implies $P(k + 1)$; in other words, that assuming the inductive hypothesis [see part (c)] we can show $1^2 + 2^2 + \cdots + k^2 + (k + 1)^2 = (k + 1)(k + 2)(2k + 3)/6$ **e)** $(1^2 + 2^2 + \cdots + k^2) + (k + 1)^2 = [k(k + 1)(2k + 1)/6] + (k + 1)^2 = [(k + 1)/6][k(2k + 1) + 6(k + 1)] = [(k + 1)/6](2k^2 + 7k + 6) = [(k + 1)/6](k + 2)(2k + 3) = (k + 1)(k + 2)(2k + 3)/6$ **f)** We have completed both the basis step and the inductive step, so by the principle of mathematical induction, the statement is true for every positive integer n.   **5.** Let $P(n)$ be "$1^2 + 3^2 + \cdots + (2n + 1)^2 = (n + 1)(2n + 1)(2n + 3)/3$." *Basis step:* $P(0)$ is true because $1^2 = 1 = (0 + 1)(2 \cdot 0 + 1)(2 \cdot 0 + 3)/3$. *Inductive step:* Assume that $P(k)$ is true. Then $1^2 + 3^2 + \cdots + (2k + 1)^2 + [2(k + 1) + 1]^2 = (k + 1)(2k + 1)(2k + 3)/3 + (2k + 3)^2 = (2k + 3)[(k + 1)(2k + 1)/3 + (2k + 3)] = (2k + 3)(2k^2 + 9k + 10)/3 = (2k + 3)(2k + 5)(k + 2)/3 = [(k + 1) + 1][2(k + 1) + 1][2(k + 1) + 3]/3$. **7.** Let $P(n)$ be "$\sum_{j=0}^{n} 3 \cdot 5^j = 3(5^{n+1} - 1)/4$." *Basis step:* $P(0)$ is true because $\sum_{j=0}^{0} 3 \cdot 5^j = 3 = 3(5^1 - 1)/4$. *Inductive step:* Assume that $\sum_{j=0}^{k} 3 \cdot 5^j = 3(5^{k+1} - 1)/4$. Then $\sum_{j=0}^{k+1} 3 \cdot 5^j = (\sum_{j=0}^{k} 3 \cdot 5^j) + 3 \cdot 5^{k+1} = 3(5^{k+1} - 1)/4 + 3 \cdot 5^{k+1} = 3(5^{k+1} + 4 \cdot 5^{k+1} - 1)/4 = 3(5^{k+2} - 1)/4$. **9. a)** $2 + 4 + 6 + \cdots + 2n = n(n + 1)$ **b)** *Basis step:* $2 = 1 \cdot (1 + 1)$ is true. *Inductive step:* Assume that $2 + 4 + 6 + \cdots + 2k = k(k + 1)$. Then $(2 + 4 + 6 + \cdots + 2k) + 2(k + 1) = k(k + 1) + 2(k + 1) = (k + 1)(k + 2)$.   **11. a)** $\sum_{j=1}^{n} 1/2^j = (2^n - 1)/2^n$ **b)** *Basis step:* $P(1)$ is true because $\frac{1}{2} = (2^1 - 1)/2^1$. *Inductive step:* Assume that $\sum_{j=1}^{k} 1/2^j = (2^k - 1)/2^k$. Then $\sum_{j=1}^{k+1} \frac{1}{2^j} = (\sum_{j=1}^{k} \frac{1}{2^j}) + \frac{1}{2^{k+1}} = \frac{2^k - 1}{2^k} + \frac{1}{2^{k+1}} = \frac{2^{k+1} - 2 + 1}{2^{k+1}} = \frac{2^{k+1} - 1}{2^{k+1}}$.   **13.** Let $P(n)$ be "$1^2 - 2^2 + 3^2 - \cdots + (-1)^{n-1} n^2 = (-1)^{n-1} n(n + 1)/2$." *Basis step:* $P(1)$ is true because $1^2 = 1 = (-1)^0 1^2$. *Inductive step:* Assume that $P(k)$ is true. Then $1^2 - 2^2 + 3^2 - \cdots + (-1)^{k-1}k^2 + (-1)^k(k + 1)^2 = (-1)^{k-1}k(k + 1)/2 + (-1)^k(k + 1)^2 = (-1)^k(k+1)[-k/2 + (k+1)] = (-1)^k(k+1)[(k/2)+1] = (-1)^k(k+1)(k+2)/2$. **15.** Let $P(n)$ be "$1\cdot 2 + 2\cdot 3 + \cdots + n(n+1) = n(n+1)(n+2)/3$." *Basis step:* $P(1)$ is true because $1\cdot 2 = 2 = 1(1+1)(1+2)/3$. *Inductive step:* Assume that $P(k)$ is true. Then $1\cdot 2 + 2\cdot 3 + \cdots + k(k+1) + (k+1)(k+2) = [k(k+1)(k+2)/3] + (k+1)(k+2) = (k+1)(k+2)[(k/3)+1] = (k+1)(k+2)(k+3)/3$. **17.** Let $P(n)$ be the statement that $1^4 + 2^4 + 3^4 + \cdots + n^4 = n(n+1)(2n+1)(3n^2 + 3n - 1)/30$. $P(1)$ is true because $1\cdot 2\cdot 3\cdot 5/30 = 1$. Assume that $P(k)$ is true. Then $(1^4 + 2^4 + 3^4 + \cdots + k^4) + (k+1)^4 = k(k+1)(2k+1)(3k^2+3k-1)/30 + (k+1)^4 = [(k+1)/30][k(2k+1)(3k^2+3k-1) + 30(k+1)^3] = [(k+1)/30](6k^4 + 39k^3 + 91k^2 + 89k + 30) = [(k+1)/30](k+2)(2k+3)[3(k+1)^2 + 3(k+1) - 1]$. This demonstrates that $P(k + 1)$ is true. **19. a)** $1 + \frac{1}{4} < 2 - \frac{1}{2}$ **b)** This is true because $5/4$ is less than $6/4$. **c)** $1 + \frac{1}{4} + \cdots + \frac{1}{k^2} < 2 - \frac{1}{k}$ **d)** For each $k \geq 2$ that $P(k)$ implies $P(k + 1)$; in other words, we want to show that assuming the inductive hypothesis [see part (c)] we can show $1 + \frac{1}{4} + \cdots + \frac{1}{k^2} + \frac{1}{(k+1)^2} < 2 - \frac{1}{k+1}$ **e)** $1 + \frac{1}{4} + \cdots + \frac{1}{k^2} + \frac{1}{(k+1)^2} < 2 - \frac{1}{k} + \frac{1}{(k+1)^2} = 2 - [\frac{1}{k} - \frac{1}{(k+1)^2}] = 2 - [\frac{k^2 + 2k + 1 - k}{k(k+1)^2}] = 2 - \frac{k^2 + k}{k(k+1)^2} - \frac{1}{(k+1)^2} = 2 - \frac{1}{k+1} - \frac{1}{(k+1)^2} < 2 - \frac{1}{k+1}$ **f)** We have completed both the basis step and the inductive step, so by the principle of mathematical induction, the statement is true for every integer n greater than 1. **21.** Let $P(n)$ be "$1 + nh \leq (1 + h)^n, h > -1$." *Basis step:* $P(0)$ is true because $1 + 0 \cdot h = 1 \leq 1 = (1 + h)^0$. *Inductive step:* Assume $1 + kh \leq (1 + h)^k$. Then because $(1+h) > 0, (1+h)^{k+1} = (1+h)(1+h)^k \geq (1+h)(1+kh) = 1 + (k + 1)h + kh^2 \geq 1 + (k + 1)h$. **23.** Let $P(n)$ be "$1/\sqrt{1} + 1/\sqrt{2} + 1/\sqrt{3} + \cdots + 1/\sqrt{n} > 2(\sqrt{n + 1} - 1)$." *Basis step:* $P(1)$ is true because $1 > 2(\sqrt{2} - 1)$. *Inductive step:* Assume that $P(k)$ is true. Then $1 + 1/\sqrt{2} + \cdots + 1/\sqrt{k} + 1/\sqrt{k + 1} > 2(\sqrt{k + 1} - 1) + 1/\sqrt{k + 1}$. If we show that $2(\sqrt{k + 1} - 1) + 1/\sqrt{k + 1} > 2(\sqrt{k + 2} - 1)$, it follows that $P(k + 1)$ is true. This inequality is equivalent to $2(\sqrt{k + 2} - \sqrt{k + 1}) < 1/\sqrt{k + 1}$, which is equivalent to $2(\sqrt{k + 2} - \sqrt{k + 1})(\sqrt{k + 2} + \sqrt{k + 1}) < \sqrt{k + 1}/\sqrt{k + 1} + \sqrt{k + 2}/\sqrt{k + 1}$. This is equivalent to $2 < 1 + \sqrt{k + 2}/\sqrt{k + 1}$, which is clearly true. **25.** Let $P(n)$ be "$H_{2^n} \leq 1 + n$." *Basis step:* $P(0)$ is true because $H_{2^0} = H_1 = 1 \leq 1 + 0$. *Inductive step:* Assume that $H_{2^k} \leq 1 + k$. Then $H_{2^{k+1}} = H_{2^k} + \sum_{j=2^k+1}^{2^{k+1}} \frac{1}{j} \leq 1 + k + 2^k\left(\frac{1}{2^{k+1}}\right) < 1 + k + 1 = 1 + (k + 1)$. **27.** *Basis step:* $1^2 + 1 = 2$ is divisible by 2. *Inductive step:* Assume the inductive hypothesis, that $k^2 + k$ is divisible by 2. Then $(k+1)^2 + (k+1) = k^2 + 2k + 1 + k + 1 = (k^2 + k) + 2(k + 1)$, the sum of a multiple of 2 (by the inductive hypothesis) and a multiple of 2 (by definition), hence, divisible by 2. **29.** *Basis step:* $n = 1, 4^2 + 5 = 21$ divisible by 21. *Inductive step:* Assume the result to be true for $n = k$. $4^{k+1} + 5^{2k-1}$ is divisible by 21. For $n = k + 1, 4^{(k+1)+1} + 5^{2(k+1)}$ is $4^{k+2} + 5^{2k+1} = 4\cdot 4^{k+1} + 5^{2k-1}$. $25 = 4\cdot 4^{k+1} + 5^{2k-1}$.

$(21 + 4) = 4(4^{k+1} + 5^{k-1}) + 21.5^{2k-1}$. Both terms are divisible by 21; the first term by inductive hypothesis. Hence the result holds for $n = k + 1$. **31.** *Basis step:* $n = 1, A_1 \subseteq B_1$ by hypothesis. *Inductive step:* Assume the result to be true for $n = k$. $\cup_{j=1}^k A_j \subseteq \cup_{j=1}^k B_j, A_{k+1} \subseteq B_{k+1}$. Consider $\cup_{j=1}^{k+1} A_j$. Let an element $x \in \cup_{j=1}^{k+1} A_j$. Then $x \in \cup_{j=1}^k A_j$ or A_{j+1}. If $x \in \cup_{j=1}^k A_j$, then $x \in \cup_{j=1}^k B_j$ by inductive hypothesis. If $x \in A_{j+1}$, then $@ in B_{j+1}$. Hence if $x \in \cup_{j=1}^{k+1} A_j$, then $x \in \cup_{j=1}^{k+1} B_j$ and the resout follows. **33.** Let $P(n)$ be "$(A_1 \cup A_2 \cup \cdots \cup A_n) \cap B = (A_1 \cap B) \cup (A_2 \cap B) \cup \cdots \cup (A_n \cap B)$." *Basis step:* $P(1)$ is trivially true. *Inductive step:* Assume that $P(k)$ is true. Then $(A_1 \cup A_2 \cup \cdots \cup A_k \cup A_{k+1}) \cap B = [(A_1 \cup A_2 \cup \cdots \cup A_k) \cup A_{k+1}] \cap B = [(A_1 \cup A_2 \cup \cdots \cup A_k) \cap B] \cup (A_{k+1} \cap B) = [(A_1 \cap B) \cup (A_2 \cap B) \cup \cdots \cup (A_k \cap B)] \cup (A_{k+1} \cap B) = (A_1 \cap B) \cup (A_2 \cap B) \cup \cdots \cup (A_k \cap B) \cup (A_{k+1} \cap B)$. **35.** Let $P(n)$ be the statement that a set with n elements has $n(n - 1)/2$ two-element subsets. $P(2)$, the basis case, is true, because a set with two elements has one subset with two elements—namely, itself—and $2(2-1)/2 = 1$. Now assume that $P(k)$ is true. Let S be a set with $k + 1$ elements. Choose an element a in S and let $T = S - \{a\}$. A two-element subset of S either contains a or does not. Those subsets not containing a are the subsets of T with two elements; by the inductive hypothesis there are $k(k - 1)/2$ of these. There are k subsets of S with two elements that contain a, because such a subset contains a and one of the k elements in T. Hence, there are $k(k - 1)/2 + k = (k + 1)k/2$ two-element subsets of S. This completes the inductive proof. **37.** Reorder the locations if necessary so that $x_1 \leq x_2 \leq x_3 \leq \cdots \leq x_d$. Place the first tower at position $t_1 = x_1 + 1$. Assume tower k has been placed at position t_k. Then place tower $k + 1$ at position $t_{k+1} = x + 1$, where x is the smallest x_i greater than $t_k + 1$. **39.** The two sets do not overlap if $n + 1 = 2$. In fact, the conditional statement $P(1) \rightarrow P(2)$ is false. **41.** The mistake is in applying the inductive hypothesis to look at $\max(x - 1, y - 1)$, because even though x and y are positive integers, $x - 1$ and $y - 1$ need not be (one or both could be 0). **43.** We use the notation (i, j) to mean the square in row i and column j and use induction on $i + j$ to show that every square can be reached by the knight. *Basis step:* There are six base cases, for the cases when $i + j \leq 2$. The knight is already at $(0, 0)$ to start, so the empty sequence of moves reaches that square. To reach $(1, 0)$, the knight moves $(0, 0) \rightarrow (2, 1) \rightarrow (0, 2) \rightarrow (1, 0)$. Similarly, to reach $(0, 1)$, the knight moves $(0, 0) \rightarrow (1, 2) \rightarrow (2, 0) \rightarrow (0, 1)$. Note that the knight has reached $(2, 0)$ and $(0, 2)$ in the process. For the last basis step there is $(0, 0) \rightarrow (1, 2) \rightarrow (2, 0) \rightarrow (0, 1) \rightarrow (2, 2) \rightarrow (0, 3) \rightarrow (1, 1)$. *Inductive step:* Assume the inductive hypothesis, that the knight can reach any square (i, j) for which $i + j = k$, where k is an integer greater than 1. We must show how the knight can reach each square (i, j) when $i + j = k + 1$. Because $k + 1 \geq 3$, at least one of i and j is at least 2. If $i \geq 2$, then by the inductive hypothesis, there is a sequence of moves ending at $(i - 2, j + 1)$, because $i - 2 + j + 1 = i + j - 1 = k$; from there it is just one step to (i, j); similarly, if $j \geq 2$. **45.** *Basis* *step:* The base cases $n = 0$ and $n = 1$ are true because the derivative of x^0 is 0 and the derivative of $x^1 = x$ is 1. *Inductive step:* Using the product rule, the inductive hypothesis, and the basis step shows that $\frac{d}{dx}x^{k+1} = \frac{d}{dx}(x \cdot x^k) = x \cdot \frac{d}{dx}x^k + x^k\frac{d}{dx}x = x \cdot kx^{k-1} + x^k \cdot 1 = kx^k + x^k = (k+1)x^k$. **47.** *Basis step:* For $k = 0$, $1 \equiv 1 \pmod{m}$. *Inductive step:* Suppose that $a \equiv b \pmod{m}$ and $a^k \equiv b^k \pmod{m}$; we must show that $a^{k+1} \equiv b^{k+1} \pmod{m}$. By Theorem 5 from Section 4.1, $a \cdot a^k \equiv b \cdot b^k \pmod{m}$, which by definition says that $a^{k+1} \equiv b^{k+1} \pmod{m}$. **49.** Let $P(n)$ be "$[(p_1 \rightarrow p_2) \wedge (p_2 \rightarrow p_3) \wedge \cdots \wedge (p_{n-1} \rightarrow p_n)] \rightarrow [(p_1 \wedge \cdots \wedge p_{n-1}) \rightarrow p_n]$." *Basis step:* $P(2)$ is true because $(p_1 \rightarrow p_2) \rightarrow (p_1 \rightarrow p_2)$ is a tautology. *Inductive step:* Assume $P(k)$ is true. To show $[(p_1 \rightarrow p_2) \wedge \cdots \wedge (p_{k-1} \rightarrow p_k) \wedge (p_k \rightarrow p_{k+1})] \rightarrow [(p_1 \wedge \cdots \wedge p_{k-1} \wedge p_k) \rightarrow p_{k+1}]$ is a tautology, assume that the hypothesis of this conditional statement is true. Because both the hypothesis and $P(k)$ are true, it follows that $(p_1 \wedge \cdots \wedge p_{k-1}) \rightarrow p_k$ is true. Because this is true, and because $p_k \rightarrow p_{k+1}$ is true (it is part of the assumption) it follows by hypothetical syllogism that $(p_1 \wedge \cdots \wedge p_{k-1}) \rightarrow p_{k+1}$ is true. The weaker statement $(p_1 \wedge \cdots \wedge p_{k-1} \wedge p_k) \rightarrow p_{k+1}$ follows from this. **51.** We will first prove the result when n is a power of 2, that is, if $n = 2^k$, $k = 1, 2, \ldots$. Let $P(k)$ be the statement $A \geq G$, where A and G are the arithmetic and geometric means, respectively, of a set of $n = 2^k$ positive real numbers. *Basis step:* $k = 1$ and $n = 2^1 = 2$. Note that $(\sqrt{a_1} - \sqrt{a_2})^2 \geq 0$. Expanding this shows that $a_1 - 2\sqrt{a_1 a_2} + a_2 \geq 0$, that is, $(a_1 + a_2)/2 \geq (a_1 a_2)^{1/2}$. *Inductive step:* Assume that $P(k)$ is true, with $n = 2^k$. We will show that $P(k+1)$ is true. We have $2^{k+1} = 2n$. Now $(a_1 + a_2 + \cdots + a_{2n})/(2n) = [(a_1 + a_2 + \cdots + a_n)/n + (a_{n+1} + a_{n+2} + \cdots + a_{2n})/n]/2$ and similarly $(a_1 a_2 \cdots a_{2n})^{1/(2n)} = [(a_1 \cdots a_n)^{1/n}(a_{n+1} \cdots a_{2n})^{1/n}]^{1/2}$. To simplify the notation, let $A(x, y, \ldots)$ and $G(x, y, \ldots)$ denote the arithmetic mean and geometric mean of $x, y, \ldots$, respectively. Also, if $x \leq x'$, $y \leq y'$, and so on, then $A(x, y, \ldots) \leq A(x', y', \ldots)$ and $G(x, y, \ldots) \leq G(x', y', \ldots)$. Hence, $A(a_1, \ldots, a_{2n}) = A(A(a_1, \ldots, a_n), A(a_{n+1}, \ldots, a_{2n})) \geq A(G(a_1, \ldots, a_n), G(a_{n+1}, \ldots, a_{2n})) \geq G(G(a_1, \ldots, a_n), G(a_{n+1}, \ldots, a_{2n})) = G(a_1, \ldots, a_{2n})$. This finishes the proof for powers of 2. Now if n is not a power of 2, let m be the next higher power of 2, and let $a_{n+1}, \ldots, a_m$ all equal $A(a_1, \ldots, a_n) = \bar{a}$. Then we have $[(a_1 a_2 \cdots a_n)\bar{a}^{m-n}]^{1/m} \leq A(a_1, \ldots, a_m)$, because m is a power of 2. Because $A(a_1, \ldots, a_m) = \bar{a}$, it follows that $(a_1 \cdots a_n)^{1/m}\bar{a}^{1-n/m} \leq \bar{a}^{n/m}$. Raising both sides to the (m/n)th power gives $G(a_1, \ldots, a_n) \leq A(a_1, \ldots, a_n)$. **53.** *Basis step:* For $n = 1$, the left-hand side is just $\frac{1}{1}$, which is 1. For $n = 2$, there are three nonempty subsets $\{1\}$, $\{2\}$, and $\{1, 2\}$, so the left-hand side is $\frac{1}{1} + \frac{1}{2} + \frac{1}{1 \cdot 2} = 2$. *Inductive step:* Assume that the statement is true for k. The set of the first $k + 1$ positive integers has many nonempty subsets, but they fall into three categories: a nonempty subset of the first k positive integers together with $k + 1$, a nonempty subset of the first k positive integers, or just $\{k + 1\}$. By the inductive hypothesis, the sum of the first category is k. For the second

category, we can factor out $1/(k + 1)$ from each term of the sum and what remains is just k by the inductive hypothesis, so this part of the sum is $k/(k + 1)$. Finally, the third category simply yields $1/(k + 1)$. Hence, the entire summation is $k + k/(k + 1) + 1/(k + 1) = k + 1$. **55.** *Basis step:* If $A_1 \subseteq A_2$, then A_1 satisfies the condition of being a subset of each set in the collection; otherwise $A_2 \subseteq A_1$, so A_2 satisfies the condition. *Inductive step:* Assume the inductive hypothesis, that the conditional statement is true for k sets, and suppose we are given $k + 1$ sets that satisfy the given conditions. By the inductive hypothesis, there must be a set A_i for some $i \leq k$ such that $A_i \subseteq A_j$ for $1 \leq j \leq k$. If $A_i \subseteq A_{k+1}$, then we are done. Otherwise, we know that $A_{k+1} \subseteq A_i$, and this tells us that A_{k+1} satisfies the condition of being a subset of A_j for $1 \leq j \leq k + 1$. **57.** $G(1) = 0$, $G(2) = 1$, $G(3) = 3$, $G(4) = 4$ **59.** To show that $2n - 4$ calls are sufficient to exchange all the gossip, select persons 1, 2, 3, and 4 to be the central committee. Every person outside the central committee calls one person on the central committee. At this point the central committee members *as a group* know all the scandals. They then exchange information among themselves by making the calls 1-2, 3-4, 1-3, and 2-4 in that order. At this point, *every* central committee member knows all the scandals. Finally, again every person outside the central committee calls one person on the central committee, at which point everyone knows all the scandals. [The total number of calls is $(n - 4) + 4 + (n - 4) = 2n - 4$.] That this cannot be done with fewer than $2n - 4$ calls is much harder to prove; see Sandra M. Hedetniemi, Stephen T. Hedetniemi, and Arthur L. Liestman, "A survey of gossiping and broadcasting in communication networks," *Networks* **18** (1988), no. 4, 319–349, for details. **61.** We prove this by mathematical induction. The basis step ($n = 2$) is true tautologically. For $n = 3$, suppose that the intervals are (a, b), (c, d), and (e, f), where without loss of generality we can assume that $a \leq c \leq e$. Because $(a, b) \cap (e, f) \neq \emptyset$, we must have $e < b$; for a similar reason, $e < d$. It follows that the number halfway between e and the smaller of b and d is common to all three intervals. Now for the inductive step, assume that whenever we have k intervals that have pairwise nonempty intersections then there is a point common to all the intervals, and suppose that we are given intervals $I_1, I_2, \ldots, I_{k+1}$ that have pairwise nonempty intersections. For each i from 1 to k, let $J_i = I_i \cap I_{k+1}$. We claim that the collection $J_1, J_2, \ldots, J_k$ satisfies the inductive hypothesis, that is, that $J_{i_1} \cap J_{i_2} \neq \emptyset$ for each choice of subscripts i_1 and i_2. This follows from the $n = 3$ case proved above, using the sets I_{i_1}, I_{i_2}, and I_{k+1}. We can now invoke the inductive hypothesis to conclude that there is a number common to all of the sets J_i for $i = 1, 2, \ldots, k$, which perforce is in the intersection of all the sets I_i for $i = 1, 2, \ldots, k + 1$.

63.

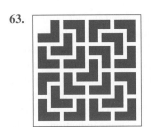

65. Let $Q(n)$ be $P(n+b-1)$. The statement that $P(n)$ is true for $n = b, b+1, b+2, \ldots$ is the same as the statement that $Q(m)$ is true for all positive integers m. We are given that $P(b)$ is true [i.e., that $Q(1)$ is true], and that $P(k) \rightarrow P(k+1)$ for all $k \geq b$ [i.e., that $Q(m) \rightarrow Q(m+1)$ for all positive integers m]. Therefore, by the principle of mathematical induction, $Q(m)$ is true for all positive integers m.

Section 5.2

1. *Basis step:* We are told we can run one mile, so $P(1)$ is true. *Inductive step:* Assume the inductive hypothesis, that we can run any number of miles from 1 to k. We must show that we can run $k+1$ miles. If $k = 1$, then we are already told that we can run two miles. If $k > 1$, then the inductive hypothesis tells us that we can run $k-1$ miles, so we can run $(k-1)+2 = k+1$ miles. **3. a)** $P(8)$ is true, because we can form 8 cents of postage with one 3-cent stamp and one 5-cent stamp. $P(9)$ is true, because we can form 9 cents of postage with three 3-cent stamps. $P(10)$ is true, because we can form 10 cents of postage with two 5-cent stamps. **b)** The statement that using just 3-cent and 5-cent stamps we can form j cents postage for all j with $8 \leq j \leq k$, where we assume that $k \geq 10$ **c)** Assuming the inductive hypothesis, we can form $k+1$ cents postage using just 3-cent and 5-cent stamps **d)** Because $k \geq 10$, we know that $P(k-2)$ is true, that is, that we can form $k-2$ cents of postage. Put one more 3-cent stamp on the envelope, and we have formed $k+1$ cents of postage. **e)** We have completed both the basis step and the inductive step, so by the principle of strong induction, the statement is true for every integer n greater than or equal to 8. **5.** Let $P(n)$ be the statement that there is no positive integer b such that $\sqrt{2} = n/b$. *Basis step:* $P(1)$ is true because $\sqrt{2} > 1 \geq 1/b$ for all positive integers b. *Inductive step:* Assume that $P(j)$ is true for all $j \leq k$, where k is an arbitrary positive integer; we prove that $P(k+1)$ is true by contradiction. Assume that $\sqrt{2} = (k+1)/b$ for some positive integer b. Then $2b^2 = (k+1)^2$, so $(k+1)^2$ is even, and hence, $k+1$ is even. So write $k+1 = 2t$ for some positive integer t, whence $2b^2 = 4t^2$ and $b^2 = 2t^2$. By the same reasoning as before, b is even, so $b = 2s$ for some positive integer s. Then $\sqrt{2} = (k+1)/b = (2t)/(2s) = t/s$. But $t \leq k$, so this contradicts the inductive hypothesis, and our proof of the inductive step is complete. **7.** *Basis step:* There are four base cases. If $n = 1 = 4 \cdot 0 + 1$, then clearly the second player wins. If there are two, three, or four matches ($n = 4 \cdot 0 + 2$, $n = 4 \cdot 0 + 3$, or $n = 4 \cdot 1$), then the first player can win by

removing all but one match. *Inductive step:* Assume the strong inductive hypothesis, that in games with k or fewer matches, the first player can win if $k \equiv 0, 2,$ or $3 \pmod 4$ and the second player can win if $k \equiv 1 \pmod 4$. Suppose we have a game with $k+1$ matches, with $k \geq 4$. If $k+1 \equiv 0 \pmod 4$, then the first player can remove three matches, leaving $k-2$ matches for the other player. Because $k-2 \equiv 1 \pmod 4$, by the inductive hypothesis, this is a game that the second player at that point (who is the first player in our game) can win. Similarly, if $k+1 \equiv 2 \pmod 4$, then the first player can remove one match; and if $k+1 \equiv 3 \pmod 4$, then the first player can remove two matches. Finally, if $k+1 \equiv 1 \pmod 4$, then the first player must leave $k, k-1,$ or $k-2$ matches for the other player. Because $k \equiv 0 \pmod 4$, $k-1 \equiv 3 \pmod 4$, and $k-2 \equiv 2 \pmod 4$, by the inductive hypothesis, this is a game that the first player at that point (who is the second player in our game) can win. **9.** Let $P(n)$ be the statement that exactly $n-1$ moves are required to assemble a puzzle with n pieces. Now $P(1)$ is trivially true. Assume that $P(j)$ is true for all $j \leq k$, and consider a puzzle with $k+1$ pieces. The final move must be the joining of two blocks, of size j and $k+1-j$ for some integer j with $1 \leq j \leq k$. By the inductive hypothesis, it required $j-1$ moves to construct the one block, and $k+1-j-1 = k-j$ moves to construct the other. Therefore, $1+(j-1)+(k-j) = k$ moves are required in all, so $P(k+1)$ is true. **11.** Let the Chomp board have n rows and n columns. We claim that the first player can win the game by making the first move to leave just the top row and leftmost column. Let $P(n)$ be the statement that if a player has presented his opponent with a Chomp configuration consisting of just n cookies in the top row and n cookies in the leftmost column, then he can win the game. We will prove $\forall n P(n)$ by strong induction. We know that $P(1)$ is true, because the opponent is forced to take the poisoned cookie at his first turn. Fix $k \geq 1$ and assume that $P(j)$ is true for all $j \leq k$. We claim that $P(k+1)$ is true. It is the opponent's turn to move. If she picks the poisoned cookie, then the game is over and she loses. Otherwise, assume she picks the cookie in the top row in column j, or the cookie in the left column in row j, for some j with $2 \leq j \leq k+1$. The first player now picks the cookie in the left column in row j, or the cookie in the top row in column j, respectively. This leaves the position covered by $P(j-1)$ for his opponent, so by the inductive hypothesis, he can win. **13.** Let $P(n)$ be the statement that if a simple polygon with n sides is triangulated, then at least two of the triangles in the triangulation have two sides that border the exterior of the polygon. We will prove $\forall n \geq 4 \, P(n)$. The statement is clearly true for $n = 4$, because there is only one diagonal, leaving two triangles with the desired property. Fix $k \geq 4$ and assume that $P(j)$ is true for all j with $4 \leq j \leq k$. Consider a polygon with $k+1$ sides, and some triangulation of it. Pick one of the diagonals in this triangulation. First suppose that this diagonal divides the polygon into one triangle and one polygon with k sides. Then the triangle has two sides that border the exterior. Furthermore, the k-gon has, by the inductive hypothesis, two triangles that have two sides that border the exterior of that k-gon, and only one of these triangles can fail to be a triangle that has two sides that border the exterior of the original polygon. The only other case is that this diagonal divides the polygon into two polygons with j sides and $k + 3 - j$ sides for some j with $4 \le j \le k - 1$. By the inductive hypothesis, each of these two polygons has two triangles that have two sides that border their exterior, and in each case only one of these triangles can fail to be a triangle that has two sides that border the exterior of the original polygon. **15.** Let $P(n)$ be the statement that the area of a simple polygon with n sides and vertices all at lattice points is given by $I(P) + B(P)/2 - 1$. We will prove $P(n)$ for all $n \ge 3$. We begin with an additivity lemma: If P is a simple polygon with all vertices at lattice points, divided into polygons P_1 and P_2 by a diagonal, then $I(P) + B(P)/2 - 1 = [I(P_1) + B(P_1)/2 - 1] + [I(P_2) + B(P_2)/2 - 1]$. To prove this, suppose there are k lattice points on the diagonal, not counting its endpoints. Then $I(P) = I(P_1) + I(P_2) + k$ and $B(P) = B(P_1) + B(P_2) - 2k - 2$; and the result follows by simple algebra. What this says in particular is that if Pick's formula gives the correct area for P_1 and P_2, then it must give the correct formula for P, whose area is the sum of the areas for P_1 and P_2; and similarly if Pick's formula gives the correct area for P and one of the P_i's, then it must give the correct formula for the other P_i. Next we prove the theorem for rectangles whose sides are parallel to the coordinate axes. Such a rectangle necessarily has vertices at (a, b), (a, c), (d, b), and (d, c), where a, b, c, and d are integers with $b < c$ and $a < d$. Its area is $(c - b)(d - a)$. Also, $B = 2(c - b + d - a)$ and $I = (c-b-1)(d-a-1) = (c-b)(d-a)-(c-b)-(d-a)+1$. Therefore, $I + B/2 - 1 = (c - b)(d - a) - (c - b) - (d - a) + 1 + (c - b + d - a) - 1 = (c - b)(d - a)$, which is the desired area. Next consider a right triangle whose legs are parallel to the coordinate axes. This triangle is half a rectangle of the type just considered, for which Pick's formula holds, so by the additivity lemma, it holds for the triangle as well. (The values of B and I are the same for each of the two triangles, so if Pick's formula gave an answer that was either too small or too large, then it would give a correspondingly wrong answer for the rectangle.) For the next step, consider an arbitrary triangle with vertices at lattice points that is not of the type already considered. Embed it in as small a rectangle as possible. There are several possible ways this can happen, but in any case (and adding one more edge in one case), the rectangle will have been partitioned into the given triangle and two or three right triangles with sides parallel to the coordinate axes. Again by the additivity lemma, we are guaranteed that Pick's formula gives the correct area for the given triangle. This completes the proof of $P(3)$, the basis step in our strong induction proof. For the inductive step, given an arbitrary polygon, use Lemma 1 in the text to split it into two polygons. Then by the additivity lemma above and the inductive hypothesis, we know that Pick's formula gives the correct area for this polygon. **17. a)** When we try to prove the inductive step and find a triangle in each subpolygon with at least two sides bordering the exterior, it may happen in each

case that the triangle we are guaranteed in fact borders the diagonal (which is part of the boundary of that polygon). This leaves us with no triangles guaranteed to touch the boundary of the *original* polygon. **b)** We proved the stronger statement $\forall n \ge 4\, T(n)$ in Exercise 13. **19. a)** The inductive step here allows us to conclude that $P(3)$, $P(5)$, ... are all true, but we can conclude nothing about $P(2)$, $P(4)$, ... **b)** $P(n)$ is true for all positive integers n, using strong induction. **c)** The inductive step here enables us to conclude that $P(2)$, $P(4)$, $P(8)$, $P(16)$, are all true, but we can conclude nothing about $P(n)$ when n is not a power of 2. **d)** This is mathematical induction; we can conclude that $P(n)$ is true for all positive integers n. **21.** The error is in going from the base case $n = 0$ to the next case, $n = 1$; we cannot write 1 as the sum of two smaller natural numbers. **23.** Assume that the well-ordering property holds. Suppose that $P(1)$ is true and that the conditional statement $[P(1) \wedge P(2) \wedge \cdots \wedge P(n)] \to P(n + 1)$ is true for every positive integer n. Let S be the set of positive integers n for which $P(n)$ is false. We will show $S = \emptyset$. Assume that $S \ne \emptyset$. Then by the well-ordering property there is a least integer m in S. We know that m cannot be 1 because $P(1)$ is true. Because $n = m$ is the least integer such that $P(n)$ is false, $P(1)$, $P(2)$, ..., $P(m - 1)$ are true, and $m - 1 \ge 1$. Because $[P(1) \wedge P(2) \wedge \cdots \wedge P(m - 1)] \to P(m)$ is true, it follows that $P(m)$ must also be true, which is a contradiction. Hence, $S = \emptyset$. **25.** In each case, give a proof by contradiction based on a "smallest counterexample," that is, values of n and k such that $P(n, k)$ is not true and n and k are smallest in some sense. **a)** Choose a counterexample with $n + k$ as small as possible. We cannot have $n = 1$ and $k = 1$, because we are given that $P(1, 1)$ is true. Therefore, either $n > 1$ or $k > 1$. In the former case, by our choice of counterexample, we know that $P(n - 1, k)$ is true. But the inductive step then forces $P(n, k)$ to be true, a contradiction. The latter case is similar. So our supposition that there is a counterexample mest be wrong, and $P(n, k)$ is true in all cases. **b)** Choose a counterexample with n as small as possible. We cannot have $n = 1$, because we are given that $P(1, k)$ is true for all k. Therefore, $n > 1$. By our choice of counterexample, we know that $P(n - 1, k)$ is true. But the inductive step then forces $P(n, k)$ to be true, a contradiction. **c)** Choose a counterexample with k as small as possible. We cannot have $k = 1$, because we are given that $P(n, 1)$ is true for all n. Therefore, $k > 1$. By our choice of counterexample, we know that $P(n, k - 1)$ is true. But the inductive step then forces $P(n, k)$ to be true, a contradiction. **27.** Let $P(n)$ be the statement that if $x_1, x_2, \ldots, x_n$ are n distinct real numbers, then $n - 1$ multiplications are used to find the product of these numbers no matter how parentheses are inserted in the product. We will prove that $P(n)$ is true using strong induction. The basis case $P(1)$ is true because $1 - 1 = 0$ multiplications are required to find the product of x_1, a product with only one factor. Suppose that $P(k)$ is true for $1 \le k \le n$. The last multiplication used to find the product of the $n + 1$ distinct real numbers $x_1, x_2, \ldots, x_n, x_{n+1}$ is a multiplication of the product of the first k of these numbers for some k and the product of the last $n + 1 - k$ of them. By the inductive hypothesis, $k - 1$ multiplications are used to find the product of k of the numbers, no matter how parentheses were inserted in the product of these numbers, and $n - k$ multiplications are used to find the product of the other $n + 1 - k$ of them, no matter how parentheses were inserted in the product of these numbers. Because one more multiplication is required to find the product of all $n + 1$ numbers, the total number of multiplications used equals $(k-1)+(n-k)+1 = n$. Hence, $P(n+1)$ is true. **29.** Assume that $a = dq + r = dq' + r'$ with $0 \leq r < d$ and $0 \leq r' < d$. Then $d(q - q') = r' - r$. It follows that d divides $r' - r$. Because $-d < r' - r < d$, we have $r' - r = 0$. Hence, $r' = r$. It follows that $q = q'$. **31.** Suppose that the well-ordering property were false. Let S be a nonempty set of nonnegative integers that has no least element. Let $P(n)$ be the statement "$i \notin S$ for $i = 0, 1, \ldots, n$." $P(0)$ is true because if $0 \in S$ then S has a least element, namely, 0. Now suppose that $P(n)$ is true. Thus, $0 \notin S, 1 \notin S, \ldots, n \notin S$. Clearly, $n + 1$ cannot be in S, for if it were, it would be its least element. Thus $P(n + 1)$ is true. So by the principle of mathematical induction, $n \notin S$ for all nonnegative integers n. Thus, $S = \emptyset$, a contradiction. **33.** Strong induction implies the principle of mathematical induction, for if one has shown that $P(k) \rightarrow P(k + 1)$ is true, then one has also shown that $[P(1) \wedge \cdots \wedge P(k)] \rightarrow P(k + 1)$ is true. By Exercise 41, the principle of mathematical induction implies the well-ordering property. Therefore by assuming strong induction as an axiom, we can prove the well-ordering property.

Section 5.3

1. a) $f(1) = 3, f(2) = 5, f(3) = 7, f(4) = 9$ **b)** $f(1) = 3$, $f(2) = 9, f(3) = 27, f(4) = 81$ **c)** $f(1) = 2, f(2) = 4$, $f(3) = 16, f(4) = 65{,}536$ **d)** $f(1) = 3, f(2) = 13, f(3) = 183, f(4) = 33{,}673$ **3. a)** Not valid **b)** $f(n) = 1 - n$. *Basis step:* $f(0) = 1 = 1 - 0$. *Inductive step:* if $f(k) = 1 - k$, then $f(k + 1) = f(k) - 1 = 1 - k - 1 = 1 - (k + 1)$. **c)** $f(n) = 4 - n$ if $n > 0$, and $f(0) = 2$. *Basis step:* $f(0) = 2$ and $f(1) = 3 = 4 - 1$. *Inductive step* (with $k \geq 1$): $f(k + 1) = f(k) - 1 = (4 - k) - 1 = 4 - (k + 1)$. **d)** $f(n) = 2^{\lfloor (n+1)/2 \rfloor}$. *Basis step:* $f(0) = 1 = 2^{\lfloor (0+1)/2 \rfloor}$ and $f(1) = 2 = 2^{\lfloor (1+1)/2 \rfloor}$. *Inductive step* (with $k \geq 1$): $f(k+1) = 2f(k-1) = 2 \cdot 2^{\lfloor k/2 \rfloor} = 2^{\lfloor k/2 \rfloor + 1} = 2^{\lfloor ((k+1)+1)/2 \rfloor}$. **e)** $f(n) = 3^n$. *Basis step:* Trivial. *Inductive step:* For odd n, $f(n) = 3f(n - 1) = 3 \cdot 3^{n-1} = 3^n$; and for even $n > 1$, $f(n) = 9f(n - 2) = 9 \cdot 3^{n-2} = 3^n$. **5.** $F(0) = 0, F(n) = F(n - 1) + n$ for $n \geq 1$ **7.** $P_m(0) = 0, P_m(n + 1) = P_m(n) + m$ **9.** Let $P(n)$ be "$f_1 + f_3 + \cdots + f_{2n-1} = f_{2n}$." *Basis step:* $P(1)$ is true because $f_1 = 1 = f_2$. *Inductive step:* Assume that $P(k)$ is true. Then $f_1 + f_3 + \cdots + f_{2k-1} + f_{2k+1} = f_{2k} + f_{2k+1} = f_{2k+2} + f_{2(k+1)}$. **11.** The number of divisions used by the Euclidean algorithm to find $\gcd(f_{n+1}, f_n)$ is 0 for $n = 0$, 1 for $n = 1$, and $n - 1$ for $n \geq 2$. To prove this result for $n \geq 2$ we use mathematical induction. For $n = 2$, one division shows that $\gcd(f_3, f_2) = \gcd(2, 1) = \gcd(1, 0) = 1$. Now assume that $k - 1$ divisions are used

to find $\gcd(f_{k+1}, f_k)$. To find $\gcd(f_{k+2}, f_{k+1})$, first divide f_{k+2} by f_{k+1} to obtain $f_{k+2} = 1 \cdot f_{k+1} + f_k$. After one div- ision we have $\gcd(f_{k+2}, f_{k+1}) = \gcd(f_{k+1}, f_k)$. By the inductive hypothesis it follows that exactly $k - 1$ more divisions are required. This shows that k divisions are required to find $\gcd(f_{k+2}, f_{k+1})$, finishing the inductive proof. **13.** $|A| = -1$. Hence, $|A^n| = (-1)^n$. It follows that $f_{n+1}f_{n-1} - f_n^2 = (-1)^n$. **15.** $5 \in S$, and $x + y \in S$ if $x, y \in S$. **17. a)** $0 \in S$, and if $x \in S$, then $x + 2 \in S$ and $x - 2 \in S$. **b)** $2 \in S$, and if $x \in S$, then $x + 3 \in S$. **c)** $1 \in S, 2 \in S, 3 \in S, 4 \in S$, and if $x \in S$, then $x + 5 \in S$. **19. a)** Define S by $(1, 1) \in S$, and if $(a, b) \in S$, then $(a + 2, b) \in S$, $(a, b + 2) \in S$, and $(a + 1, b + 1) \in S$. All elements put in S satisfy the condition, because $(1, 1)$ has an even sum of coordinates, and if (a, b) has an even sum of coordinates, then so do $(a+2, b), (a, b+2)$, and $(a+1, b+1)$. Conversely, we show by induction on the sum of the coordinates that if $a + b$ is even, then $(a, b) \in S$. If the sum is 2, then $(a, b) = (1, 1)$, and the basis step put (a, b) into S. Otherwise the sum is at least 4, and at least one of $(a - 2, b), (a, b - 2)$, and $(a-1, b-1)$ must have positive integer coordinates whose sum is an even number smaller than $a + b$, and therefore must be in S. Then one application of the recursive step shows that $(a, b) \in S$. **b)** Define S by $(1, 1), (1, 2)$, and $(2, 1)$ are in S, and if $(a, b) \in S$, then $(a + 2, b)$ and $(a, b + 2)$ are in S. To prove that our definition works, we note first that $(1, 1), (1, 2)$, and $(2, 1)$ all have an odd coordinate, and if (a, b) has an odd coordinate, then so do $(a + 2, b)$ and $(a, b + 2)$. Conversely, we show by induction on the sum of the coordinates that if (a, b) has at least one odd coordinate, then $(a, b) \in S$. If $(a, b) = (1, 1)$ or $(a, b) = (1, 2)$ or $(a, b) = (2, 1)$, then the basis step put (a, b) into S. Otherwise either a or b is at least 3, so at least one of $(a - 2, b)$ and $(a, b - 2)$ must have positive integer coordinates whose sum is smaller than $a + b$, and therefore must be in S. Then one application of the recursive step shows that $(a, b) \in S$. **c)** $(1, 6) \in S$ and $(2, 3) \in S$, and if $(a, b) \in S$, then $(a + 2, b) \in S$ and $(a, b + 6) \in S$. To prove that our definition works, we note first that $(1, 6)$ and $(2, 3)$ satisfy the condition, and if (a, b) satisfies the condition, then so do $(a+2, b)$ and $(a, b+6)$. Conversely we show by induction on the sum of the coordinates that if (a, b) satisfies the condition, then $(a, b) \in S$. For sums 5 and 7, the only points are $(1, 6)$, which the basis step put into S, $(2, 3)$, which the basis step put into S, and $(4, 3) = (2 + 2, 3)$, which is in S by one application of the recursive definition. For a sum greater than 7, either $a \geq 3$, or $a \leq 2$ and $b \geq 9$, in which case either $(a - 2, b)$ or $(a, b - 6)$ must have positive integer coordinates whose sum is smaller than $a + b$ and satisfy the condition for being in S. Then one application of the recursive step shows that $(a, b) \in S$. **21.** If x is a set or a variable representing a set, then x is a well-formed formula. If x and y are well-formed formulae, then so are $\overline{x}, (x \cup y), (x \cap y)$, and $(x - y)$. **23. a)** If $x \in D = \{0, 1, 2, 3, 4, 5, 6, 7, 8, 9\}$, then $m(x) = x$; if $s = tx$, where $t \in D^*$ and $x \in D$, then $m(s) = \min(m(s), x)$. **b)** Let $t = wx$, where $w \in D^*$ and $x \in D$. If $w = \lambda$, then $m(st) = m(sx) = \min(m(s), x) = \min(m(s), m(x))$ by the recursive step and the basis step of the definition of m. Otherwise, $m(st) = m((sw)x) = \min(m(sw), x)$ by the definition of m. Now $m(sw) = \min(m(s), m(w))$ by the inductive hypothesis of the structural induction, so $m(st) = \min(\min(m(s), m(w)), x) = \min(m(s), \min(m(w), x))$ by the meaning of min. But $\min(m(w), x) = m(wx) = m(t)$ by the recursive step of the definition of m. Thus, $m(st) = \min(m(s), m(t))$. **25.** $\lambda^R = \lambda$ and $(ux)^R = xu^R$ for $x \in \Sigma$, $u \in \Sigma^*$. **27.** $w^0 = \lambda$ and $w^{n+1} = ww^n$. **29.** When the string consists of n 0s followed by n 1s for some nonnegative integer n **31.** Let $P(i)$ be "$l(w^i) = i \cdot l(w)$." $P(0)$ is true because $l(w^0) = 0 = 0 \cdot l(w)$. Assume $P(i)$ is true. Then $l(w^{i+1}) = l(ww^i) = l(w) + l(w^i) = l(w) + i \cdot l(w) = (i+1) \cdot l(w)$. **33.** *Basis step:* For the full binary tree consisting of just a root the result is true because $n(T) = 1$ and $h(T) = 0$, and $1 \geq 2 \cdot 0 + 1$. *Inductive step:* Assume that $n(T_1) \geq 2h(T_1) + 1$ and $n(T_2) \geq 2h(T_2) + 1$. By the recursive definitions of $n(T)$ and $h(T)$, we have $n(T) = 1 + n(T_1) + n(T_2)$ and $h(T) = 1 + \max(h(T_1), h(T_2))$. Therefore $n(T) = 1 + n(T_1) + n(T_2) \geq 1 + 2h(T_1) + 1 + 2h(T_2) + 1 \geq 1 + 2 \cdot \max(h(T_1), h(T_2)) + 2 = 1 + 2(\max(h(T_1), h(T_2)) + 1) = 1 + 2h(T)$. **35.** *Basis step:* $a_{0,0} = 0 = 0 + 0$. *Inductive step:* Assume that $a_{m',n'} = m' + n'$ whenever (m', n') is less than (m, n) in the lexicographic ordering of $\mathbf{N} \times \mathbf{N}$. If $n = 0$ then $a_{m,n} = a_{m-1,n} + 1 = m - 1 + n + 1 = m + n$. If $n > 0$, then $a_{m,n} = a_{m,n-1} + 1 = m + n - 1 + 1 = m + n$. **37. a)** $P_{m,m} = P_m$ because a number exceeding m cannot be used in a partition of m. **b)** Because there is only one way to partition 1, namely, $1 = 1$, it follows that $P_{1,n} = 1$. Because there is only one way to partition m into 1s, $P_{m,1} = 1$. When $n > m$ it follows that $P_{m,n} = P_{m,m}$ because a number exceeding m cannot be used. $P_{m,m} = 1 + P_{m,m-1}$ because one extra partition, namely, $m = m$, arises when m is allowed in the partition. $P_{m,n} = P_{m,n-1} + P_{m-n,n}$ if $m > n$ because a partition of m into integers not exceeding n either does not use any ns and hence, is counted in $P_{m,n-1}$ or else uses an n and a partition of $m - n$, and hence, is counted in $P_{m-n,n}$. **c)** $P_5 = 7$, $P_6 = 11$ **39.** Let $P(n)$ be "$A(n, 2) = 4$." *Basis step:* $P(1)$ is true because $A(1, 2) = A(0, A(1, 1)) = A(0, 2) = 2 \cdot 2 = 4$. *Inductive step:* Assume that $P(n)$ is true, that is, $A(n, 2) = 4$. Then $A(n+1, 2) = A(n, A(n+1, 1)) = A(n, 2) = 4$. **41. a)** The value of $F(1)$ is ambiguous. **b)** $F(2)$ is not defined because $F(0)$ is not defined. **c)** $F(3)$ is ambiguous and $F(4)$ is not defined because $F(\frac{4}{3})$ makes no sense. **d)** The definition of $F(1)$ is ambiguous because both the second and third clause seem to apply. **e)** $F(2)$ cannot be computed because trying to compute $F(2)$ gives $F(2) = 1 + F(F(1)) = 1 + F(2)$. **43. a)** 1 **b)** 2 **c)** 3 **d)** 3 **e)** 4 **f)** 4 **g)** 5 **45.** $f_0^*(n) = \lceil n/a \rceil$ **47.** $f_2^*(n) = \lceil \log \log n \rceil$ for $n \geq 2$, $f_2^*(1) = 0$

Section 5.4

1. First, we use the recursive step to write $5! = 5 \cdot 4!$. We then use the recursive step repeatedly to write $4! = 4 \cdot 3!$, $3! = 3 \cdot 2!$, $2! = 2 \cdot 1!$, and $1! = 1 \cdot 0!$. Inserting the value

of $0! = 1$, and working back through the steps, we see that $1! = 1 \cdot 1 = 1$, $2! = 2 \cdot 1! = 2 \cdot 1 = 2$, $3! = 3 \cdot 2! = 3 \cdot 2 = 6$, $4! = 4 \cdot 3! = 4 \cdot 6 = 24$, and $5! = 5 \cdot 4! = 5 \cdot 24 = 120$. **3.** First, because $n = 11$ is odd, we use the **else** clause to see that $mpower(3, 11, 5) = (mpower(3, 5, 5)^2 \mathbf{\ mod\ } 5 \cdot 3 \mathbf{\ mod\ } 5) \mathbf{\ mod\ } 5$. We next use the **else** clause again to see that $mpower(3, 5, 5) = (mpower(3, 2, 5)^2 \mathbf{\ mod\ } 5 \cdot 3 \mathbf{\ mod\ } 5) \mathbf{\ mod\ } 5$. Then we use the **else if** clause to see that $mpower(3, 2, 5) = mpower(3, 1, 5)^2 \mathbf{\ mod\ } 5$. Using the **else** clause again, we have $mpower(3, 1, 5) = (mpower(3, 0, 5)^2 \mathbf{\ mod\ } 5 \cdot 3 \mathbf{\ mod\ } 5) \mathbf{\ mod\ } 5$. Finally, using the **if** clause, we see that $mpower(3, 0, 5) = 1$. Working backward it follows that $mpower(3, 1, 5) = (1^2 \mathbf{\ mod\ } 5 \cdot 3 \mathbf{\ mod\ } 5) \mathbf{\ mod\ } 5 = 3$, $mpower(3, 2, 5) = 3^2 \mathbf{\ mod\ } 5 = 4$, $mpower(3, 5, 5) = (4^2 \mathbf{\ mod\ } 5 \cdot 3 \mathbf{\ mod\ } 5) \mathbf{\ mod\ } 5 = 3$, and finally $mpower(3, 11, 5) = (3^2 \mathbf{\ mod\ } 5 \cdot 3 \mathbf{\ mod\ } 5) \mathbf{\ mod\ } 5 = 2$. We conclude that $3^{11} \mathbf{\ mod\ } 5 = 2$.

5. procedure *sum of odds*(n: positive integer)
   if $n = 1$ **then return** 1
   else return *sum of odds* $(n - 1) + 2n - 1$

7. procedure *smallest*($a_1, \ldots, a_n$: integers)
   if $n = 1$ **then return** a_1
   else return
      $\min(smallest\ (a_1, \ldots, a_{n-1}), a_n)$

9. procedure *modfactorial*(n, m: positive integers)
   if $n = 1$ **then return** 1
   else return
      $(n \cdot modfactorial(n - 1, m)) \mathbf{\ mod\ } m$

11. procedure *gcd*(a, b: nonnegative integers)
   $\{a < b$ assumed to hold$\}$
   if $a = 0$ **then return** b
   else if $a = b - a$ **then return** a
   else if $a < b - a$ **then return** $gcd(a, b - a)$
   else return $gcd(b - a, a)$

13. We use strong induction on a. *Basis step:* If $a = 0$, we know that $\gcd(0, b) = b$ for all $b > 0$, and that is precisely what the **if** clause does. *Inductive step:* Fix $k > 0$, assume the inductive hypothesis—that the algorithm works correctly for all values of its first argument less than k—and consider what happens with input (k, b), where $k < b$. Because $k > 0$, the **else** clause is executed, and the answer is whatever the algorithm gives as output for inputs $(b \mathbf{\ mod\ } k, k)$. Because $b \mathbf{\ mod\ } k < k$, the input pair is valid. By our inductive hypothesis, this output is in fact $\gcd(b \mathbf{\ mod\ } k, k)$, which equals $\gcd(k, b)$ by Lemma 1 in Section 4.3. **15.** If $n = 1$, then $nx = x$, and the algorithm correctly returns x. Assume that the algorithm correctly computes kx. To compute $(k + 1)x$ it recursively computes the product of $k + 1 - 1 = k$ and x, and then adds x. By the inductive hypothesis, it computes that product correctly, so the answer returned is $kx + x = (k+1)x$, which is correct. **17.** (a)

procedure *exponent*(a: real, n: integer)

if $n = 1$ **then return** $a \cdot a$

else return exponent($a, n - 1$) $\cdot$ exponent($a, n - 1$)

(b) n multiplications versus 2^n

19. procedure a(n: nonnegative integer)

if $n = 0$ **then return** 1

else if $n = 1$ **then return** 2

else return $a(n - 1) \cdot a(n - 2)$

21. Iterative

23. procedure *reverse*(w: bit string)

$n := \text{length}(w)$

if $n \leq 1$ **then return** w

else return

$substr(w, n, n)reverse (substr (w, 1, n - 1))$

{$substr(w, a, b)$ is the substring of w consisting of

the symbols in the ath through bth positions}

25. The procedure correctly gives the reversal of λ as λ (basis step), and because the reversal of a string consists of its last character followed by the reversal of its first $n - 1$ characters (see Exercise 25 in Section 5.3), the algorithm behaves correctly when $n > 0$ by the inductive hypothesis. **27.** The algorithm implements the idea of Example 14 in Section 5.1. If $n = 1$ (basis step), place the one right triomino so that its armpit corresponds to the hole in the 2×2 board. If $n > 1$, then divide the board into four boards, each of size $2^{n-1} \times 2^{n-1}$, notice which quarter the hole occurs in, position one right triomino at the center of the board with its armpit in the quarter where the missing square is (see Figure 7 in Section 5.1), and invoke the algorithm recursively four times—once on each of the $2^{n-1} \times 2^{n-1}$ boards, each of which has one square missing

(either because it was missing to begin with, or because it is covered by the central triomino).

29.

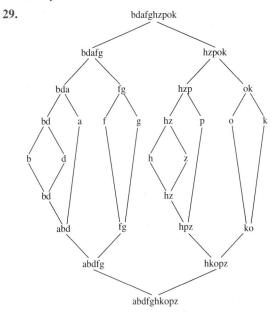

31. Let the two lists be $1, 2, \ldots, m - 1, m + n - 1$ and $m, m + 1, \ldots, m + n - 2, m + n$, respectively. **33.** If $n = 1$, then the algorithm does nothing, which is correct because a list with one element is already sorted. Assume that the algorithm works correctly for $n = 1$ through $n = k$. If $n = k+1$, then the list is split into two lists, L_1 and L_2. By the inductive hypothesis, *mergesort* correctly sorts each of these sublists; furthermore, *merge* correctly merges two sorted lists into one because with each comparison the smallest element in $L_1 \cup L_2$ not yet put into L is put there. **35.** $O(n)$ **37.** 6 **39.** $O(n^2)$

Section 5.5

1. Suppose that $x = 0$. The program segment first assigns the value 1 to y and then assigns the value $x + y = 0 + 1 = 1$ to z. **3.** Suppose that $y = 3$. The program segment assigns the value 2 to x and then assigns the value $x + y = 2 + 3 = 5$ to z. Because $y = 3 > 0$ it then assigns the value $z + 1 = 5 + 1 = 6$ to z.

5. $\quad (p \wedge condition1)\{S_1\}q$

$\quad\quad (p \wedge \neg condition1 \wedge condition2)\{S_2\}q$

$\quad\quad\quad .$

$\quad\quad\quad .$

$\quad\quad\quad .$

$\quad\quad (p \wedge \neg condition1 \wedge \neg condition2$

$\quad\quad\quad\quad\quad \cdots \wedge \neg condition(n - 1)\{S_n\}q$

$\therefore \; \overline{p\{\textbf{if } condition1 \textbf{ then } S_1;}$

$\quad\quad \textbf{else if } condition2 \textbf{ then } S_2; \ldots; \textbf{ else } S_n\}q$

7. We will show that p : "*power* $= x^{i-1}$ and $i \leq n + 1$" is a loop invariant. Note that p is true initially, because before the loop starts, $i = 1$ and *power* $= 1 = x^0 = x^{1-1}$. Next, we must show that if p is true and $i \leq n$ after an execution of the loop, then p remains true after one more execution. The loop increments i by 1. Hence, because $i \leq n$ before this pass, $i \leq n+1$ after this pass. Also the loop assigns $power \cdot x$ to $power$. By the inductive hypothesis we see that $power$ is assigned the value $x^{i-1} \cdot x = x^i$. Hence, p remains true. Furthermore, the loop terminates after n traversals of the loop with $i = n + 1$ because i is assigned the value 1 prior to entering the loop, is incremented by 1 on each pass, and the loop terminates when $i > n$. Consequently, at termination $power = x^n$, as desired. **9.** Suppose that the initial assertion p is true. Then because $p\{S\}q_0$ is true, q_0 is true after the segment S is executed. Because $q_0 \rightarrow q_1$ is true, it also follows that q_1 is true after S is executed. Hence, $p\{S\}q_1$ is true. **11.** We will use the proposition p, "$\gcd(a, b) = \gcd(x, y)$ and $y \geq 0$," as the loop invariant. Note that p is true before the loop is entered, because at that point $x = a$, $y = b$, and y is a positive integer, using the initial assertion. Now assume that p is true and $y > 0$; then the loop will be executed again. Inside the loop, x and y are replaced by y and x **mod** y, respectively. By Lemma 1 of Section 4.3, $\gcd(x, y) = \gcd(y, x \bmod y)$. Therefore, after execution of the loop, the value of $\gcd(x, y)$ is the same as it was before. Moreover, because y is the remainder, it is at least 0. Hence, p remains true, so it is a loop invariant. Furthermore, if the loop terminates, then $y = 0$. In this case, we have $\gcd(x, y) = x$, the final assertion. Therefore, the program, which gives x as its output, has correctly computed $\gcd(a, b)$. Finally, we can prove the loop must terminate, because each iteration causes the value of y to decrease by at least 1. Therefore, the loop can be iterated at most b times.

Supplementary Exercises

1. Let $P(n)$ be the statement that this equation holds. *Basis step:* $P(1)$ says $2/3 = 1 - (1/3^1)$, which is true. *Inductive step:* Assume that $P(k)$ is true. Then $2/3 + 2/9 + 2/27 + \cdots + 2/3^n + 2/3^{n+1} = 1 - 1/3^n + 2/3^{n+1}$ (by the inductive hypothesis), and this equals $1 - 1/3^{n+1}$, as desired. **3.** Let $P(n)$ be "$1/(1 \cdot 4) + \cdots + 1/[(3n-2)(3n+1)] = n/(3n+1)$." *Basis step:* $P(1)$ is true because $1/(1 \cdot 4) = 1/4$. *Inductive step:* Assume $P(k)$ is true. Then $1/(1 \cdot 4) + \cdots + 1/[(3k-2)(3k+1)] + 1/[(3k+1)(3k+4)] = k/(3k+1) + 1/[(3k+1)(3k+4)] = [k(3k+4)+1]/[(3k+1)(3k+4)] = [(3k+1)(k+1)]/[(3k+1)(3k+4)] = (k+1)/(3k+4)$. **5.** Let $P(n)$ be "$a - b$ is a factor of $a^n - b^n$." *Basis step:* $P(1)$ is trivially true. Assume $P(k)$ is true. Then $a^{k+1} - b^{k+1} = a^{k+1} - ab^k + ab^k - b^{k+1} = a(a^k - b^k) + b^k(a - b)$. Then because $a - b$ is a factor of $a^k - b^k$ and $a - b$ is a factor of $a - b$, it follows that $a - b$ is a factor of $a^{k+1} - b^{k+1}$. **7.** Let $P(n)$ be "$a + (a+d) + \cdots + (a+nd) = (n+1)(2a+nd)/2$." *Basis step:* $P(1)$ is true because $a + (a + d) = 2a + d = 2(2a + d)/2$. *Inductive step:* Assume that $P(k)$ is true. Then $a + (a + d) + \cdots + (a + kd) + [a + (k + 1)d] = (k + 1)(2a + kd)/2 + a + (k + 1)d = \frac{1}{2}(2ak + 2a + k^2d + kd + 2a + 2kd + 2d) = \frac{1}{2}(2ak + 4a + k^2d + 3kd + 2d) = \frac{1}{2}(k + 2)[2a + (k + 1)d]$. **9.** *Basis step:* This is true for $n = 1$ because $5/6 = 10/12$. *Inductive step:* Assume that

the equation holds for $n = k$, and consider $n = k + 1$. Then $\sum_{i=1}^{k+1} \frac{i+4}{i(i+1)(i+2)} = \sum_{i=1}^{k} \frac{i+4}{i(i+1)(i+2)} + \frac{k+5}{(k+1)(k+2)(k+3)} = \frac{k(3k+7)}{2(k+1)(k+2)} + \frac{k+5}{(k+1)(k+2)(k+3)}$ (by the inductive hypothesis) $= \frac{1}{(k+1)(k+2)} \cdot (\frac{k(3k+7)}{2} + \frac{k+5}{k+3}) = \frac{1}{2(k+1)(k+2)(k+3)} \cdot [k(3k + 7) (k + 3) + 2(k+5)] = \frac{1}{2(k+1)(k+2)(k+3)} \cdot (3k^3+16k^2 + 23k + 10) = \frac{1}{2(k+1)(k+2)(k+3)} \cdot (3k + 10)(k + 1)^2 = \frac{1}{2(k+2)(k+3)} \cdot (3k + 10)(k + 1) = \frac{(k+1)(3(k+1)+7)}{2((k+1)+1)((k+1)+2)}$, as desired. **11.** We will use strong induction to show that f_n is even if $n \equiv 0 \pmod 3$ and is odd otherwise. *Basis step:* This follows because $f_0 = 0$ is even and $f_1 = 1$ is odd. *Inductive step:* Assume that if $j \leq k$, then f_j is even if $j \equiv 0 \pmod 3$ and is odd otherwise. Now suppose $k + 1 \equiv 0 \pmod 3$. Then $f_{k+1} = f_k + f_{k-1}$ is even because f_k and f_{k-1} are both odd. If $k + 1 \equiv 1 \pmod 3$, then $f_{k+1} = f_k + f_{k-1}$ is odd because f_k is even and f_{k-1} is odd. Finally, if $k + 1 \equiv 2 \pmod 3$, then $f_{k+1} = f_k + f_{k-1}$ is odd because f_k is odd and f_{k-1} is even. **13.** Let $P(n)$ be the statement $l_0^2 + l_1^2 + \cdots + l_n^2 = l_n l_{n+1} + 2$. *Basis step:* $P(0)$ and $P(1)$ both hold because $l_0^2 = 2^2 = 2 \cdot 1 + 2 = l_0 l_1 + 2$ and $l_0^2 + l_1^2 = 2^2 + 1^2 = 1 \cdot 3 + 2 = l_1 l_3 + 2$. *Inductive step:* Assume that $P(k)$ holds. Then by the inductive hypothesis $l_0^2 + l_1^2 + \cdots + l_k^2 + l_{k+1}^2 = l_k l_{k+1} + 2 + l_{k+1}^2 = l_{k+1}(l_k + l_{k+1}) + 2 = l_{k+1} l_{k+2} + 2$. This shows that $P(k + 1)$ holds. **15.** Let $P(n)$ be the statement that the identity holds for the integer n. *Basis step:* $P(1)$ is obviously true. *Inductive step:* Assume that $P(k)$ is true. Then $\cos((k+1)x) + i \sin((k+1)x) = \cos(kx+x) + i \sin(kx+x) = \cos kx \cos x - \sin kx \sin x + i(\sin kx \cos x + \cos kx \sin x) = \cos x(\cos kx + i \sin kx)(\cos x + i \sin x) = (\cos x + i \sin x)^k(\cos x + i \sin x) = (\cos x + i \sin x)^{k+1}$. It follows that $P(k + 1)$ is true. **17.** Rewrite the right-hand side as $2^{n+1}(n^2 - 2n + 3) - 6$. For $n = 1$ we have $2 = 4 \cdot 2 - 6$. Assume that the equation holds for $n = k$, and consider $n = k+1$. Then $\sum_{j=1}^{k+1} j^2 2^j = \sum_{j=1}^{k} j^2 2^j + (k+1)^2 2^{k+1} = 2^{k+1}(k^2 - 2k + 3) - 6 + (k^2 + 2k + 1)2^{k+1}$ (by the inductive hypothesis) $= 2^{k+1}(2k^2 + 4) - 6 = 2^{k+2}(k^2 + 2) - 6 = 2^{k+2}[(k + 1)^2 - 2(k + 1) + 3] - 6$. **19.** Let $P(n)$ be the statement that this equation holds. *Basis step:* In $P(2)$ both sides reduce to $1/3$. *Inductive step:* Assume that $P(k)$ is true. Then $\sum_{j=1}^{k+1} 1/(j^2 - 1) = \left(\sum_{j=1}^{k} 1/(j^2 - 1)\right) + 1/[(k + 1)^2 - 1] = (k - 1)(3k + 2)/[4k(k + 1)] + 1/[(k + 1)^2 - 1]$ by the inductive hypothesis. This simplifies to $(k - 1)(3k + 2)/[4k(k + 1)] + 1/(k^2 + 2k) = (3k^3 + 5k^2)/[4k(k + 1)(k + 2)] = \{[(k + 1) - 1][3(k + 1) + 2]\}/[4(k + 1)(k + 2)]$, which is exactly what $P(k+1)$ asserts. **21.** Let $P(n)$ be the assertion that at least $n + 1$ lines are needed to cover the lattice points in the given triangular region. *Basis step:* $P(0)$ is true, because we need at least one line to cover the one point at $(0, 0)$. *Inductive step:* Assume the inductive hypothesis, that at least $k + 1$ lines are needed to cover the lattice points with $x \geq 0$, $y \geq 0$, and $x + y \leq k$. Consider the triangle of lattice points defined by $x \geq 0$, $y \geq 0$, and $x + y \leq k + 1$. By way of contradiction, assume that $k + 1$ lines could cover this set. Then these lines must cover the $k + 2$ points on the line $x + y = k + 1$. But only the line $x + y = k + 1$ itself can cover more than one of these points, because two distinct lines intersect in at most one point. Therefore none of the $k+1$ lines that are needed (by the inductive hypothesis) to cover the set of lattice points within the triangle but not on this line can cover more than one of the points on this line, and this leaves at least one point uncovered. Therefore our assumption that $k + 1$ lines could cover the larger set is wrong, and our proof is complete. **23.** We prove by mathematical induction the following stronger statement: For every $n \geq 3$, we can write $n!$ as the sum of n of its distinct positive divisors, one of which is 1. That is, we can write $n! = a_1 + a_2 + \cdots + a_n$, where each a_i is a divisor of $n!$, the divisors are listed in strictly decreasing order, and $a_n = 1$. *Basis step:* $3! = 3 + 2 + 1$. *Inductive step:* Assume that we can write $k!$ as a sum of the desired form, say $k! = a_1 + a_2 + \cdots + a_k$, where each a_i is a divisor of $n!$, the divisors are listed in strictly decreasing order, and $a_n = 1$. Consider $(k + 1)!$. Then we have $(k + 1)! = (k + 1)k! = (k+1)(a_1 + a_2 + \cdots + a_k) = (k + 1)a_1 + (k + 1)a_2 + \cdots + (k + 1)a_k = (k + 1)a_1 + (k + 1)a_2 + \cdots + k \cdot a_k + a_k$. Because each a_i was a divisor of $k!$, each $(k+1)a_i$ is a divisor of $(k + 1)!$. Furthermore, $k \cdot a_k = k$, which is a divisor of $(k + 1)!$, and $a_k = 1$, so the new last summand is again 1. (Notice also that our list of summands is still in strictly decreasing order.) Thus we have written $(k + 1)!$ in the desired form. **25.** *Basis step:* When $n = 1$ there is one circle, and we can color the inside blue and the outside red to satisfy the conditions. *Inductive step:* Assume the inductive hypothesis that if there are k circles, then the regions can be 2-colored such that no regions with a common boundary have the same color, and consider a situation with $k + 1$ circles. Remove one of the circles, producing a picture with k circles, and invoke the inductive hypothesis to color it in the prescribed manner. Then replace the removed circle and change the color of every region inside this circle. The resulting figure satisfies the condition, because if two regions have a common boundary, then either that boundary involved the new circle, in which case the regions on either side used to be the same region and now the inside portion is different from the outside, or else the boundary did not involve the new circle, in which case the regions are colored differently because they were colored differently before the new circle was restored. **27.** If $n = 1$ then the equation reads $1 \cdot 1 = 1 \cdot 2/2$, which is true. Assume that the equation is true for n and consider it for $n + 1$. Then

$$\sum_{j=1}^{n+1}(2j - 1)\left(\sum_{k=j}^{n+1}\frac{1}{k}\right) = \sum_{j=1}^{n}(2j - 1)\left(\sum_{k=j}^{n+1}\frac{1}{k}\right) +$$

$$[2(n + 1) - 1] \cdot \frac{1}{n+1} = \sum_{j=1}^{n}(2j - 1)\left(\frac{1}{n+1} + \sum_{k=j}^{n}\frac{1}{k}\right) +$$

$$\frac{2n+1}{n+1} = \left(\frac{1}{n+1}\sum_{j=1}^{n}(2j - 1)\right) + \left(\sum_{j=1}^{n}(2j - 1)\right)$$

$$\sum_{k=j}^{n}\frac{1}{k}\right) + \frac{2n+1}{n+1} = \left(\frac{1}{n+1} \cdot n^2\right) + \frac{n(n+1)}{2} + \frac{2n+1}{n+1}$$

(by the inductive hypothesis) $= \frac{2n^2+n(n+1)^2+(4n+2)}{2(n+1)} =$

$\frac{2(n+1)^2+n(n+1)^2}{2(n+1)} = \frac{(n+1)(n+2)}{2}$. **29.** Let $T(n)$ be the statement that the sequence of towers of 2 is eventually constant modulo n. We use strong induction to prove that $T(n)$ is true for all positive integers n. *Basis step:* When $n = 1$ (and $n = 2$),

the sequence of towers of 2 modulo n is the sequence of all 0s. *Inductive step:* Suppose that k is an integer with $k \geq 2$. Suppose that $T(j)$ is true for $1 \leq j \leq k - 1$. In the proof of the inductive step we denote the rth term of the sequence modulo n by a_r. First suppose k is even. Let $k = 2^s q$ where $s \geq 1$ and $q < k$ is odd. When j is large enough, $a_{j-2} \geq s$, and for such j, $a_j = 2^{2^{a_j-2}}$ is a multiple of 2^s. It follows that for sufficiently large j, $a_j \equiv 0 \pmod{2^s}$. Hence, for large enough i, 2^s divides $a_{i+1} - a_i$. By the inductive hypothesis $T(q)$ is true, so the sequence $a_1, a_2, a_3, \ldots$ is eventually constant modulo q. This implies that for large enough i, q divides $a_{i+1} - a_i$. Because $\gcd(q, 2^s) = 1$ and for sufficiently large i both q and 2^s divide $a_{i+1} - a_i$, $k = 2^s q$ divides $a_{i+1} - a_i$ for sufficiently large i. Hence, for sufficiently large i, $a_{i+1} - a_i \equiv 0 \pmod{k}$. This means that the sequence is eventually constant modulo k. Finally, suppose k is odd. Then $\gcd(2, k) = 1$, so by Euler's theorem (found in elementary number theory books, such as [Ro10]), we know that $2^{\phi(k)} \equiv 1 \pmod{k}$. Let $r = \phi(k)$. Because $r < k$, by the inductive hypothesis $T(r)$, the sequence $a_1, a_2, a_3, \ldots$ is eventually constant modulo r, say equal to c. Hence for large enough i, for some integer t_i, $a_i = t_i r + c$. Hence $a_{i+1} = 2^{a_i} = 2^{t_i r + c} = (2^r)^{t_i} 2^c \equiv 2^c \pmod{k}$. This shows that $a_1, a_2, \ldots$ is eventually constant modulo k. **31. a)** 92 **b)** 91 **c)** 91 **d)** 91 **e)** 91 **f)** 91 **33.** Let $P(n)$ be "the plane is divided into $n^2 - n + 2$ regions by n circles if every two of these circles have two common points but no three have a common point." *Basis step:* $P(1)$ is true because a circle divides the plane into $2 = 1^2 - 1 + 2$ regions. *Inductive step:* Assume that $P(k)$ is true, that is, k circles with the specified properties divide the plane into $k^2 - k + 2$ regions. Suppose that a $(k + 1)$st circle is added. This circle intersects each of the other k circles in two points, so these points of intersection form $2k$ new arcs, each of which splits an old region. Hence, there are $2k$ regions split, which shows that there are $2k$ more regions than there were previously. Hence, $k + 1$ circles satisfying the specified properties divide the plane into $k^2 - k + 2 + 2k = (k^2 + 2k + 1) - (k + 1) + 2 = (k + 1)^2 - (k + 1) + 2$ regions. **35. a)** no **b)** yes **c)** no **d)** yes **37. a)** λ, 0, 1, 00, 01, 11, 000, 001, 011, 111, 0000, 0001, 0011, 0111, 1111, 00000, 00001, 00011, 00111, 01111, 11111 **b)** $S = \{\alpha\beta \mid \alpha$ is a string of m 0s and β is a string of n 1s, $m \geq 0, n \geq 0\}$ **39.** Apply the first recursive step to λ to get $() \in B$. Apply the second recursive step to this string to get $()() \in B$. Apply the first recursive step to this string to get $(()()) \in B$. By Exercise 62, $((()))$ is not in B because the number of left parentheses does not equal the number of right parentheses. **41.** λ, $()$, $(())$, $()()$ **43. a)** 0 **b)** -2 **c)** 2 **d)** 0 **45.** If $x \leq y$ initially, then $x := y$ is not executed, so $x \leq y$ is a true final assertion. If $x > y$ initially, then $x := y$ is executed, so $x \leq y$ is again a true final assertion. **47. procedure** *zerocount*($a_1, a_2, \ldots, a_n$: list of integers)

    if $n = 1$ **then**

        if $a_1 = 0$ **then return** 1

        else return 0

    else

> **if** $a_n = 0$ **then return** $zerocount(a_1, a_2, \ldots, a_{n-1}) + 1$
> **else return** $zerocount(a_1, a_2, \ldots, a_{n-1})$

49. We will prove that $a(n)$ is a natural number and $a(n) \leq n$. This is true for the base case $n = 0$ because $a(0) = 0$. Now assume that $a(n-1)$ is a natural number and $a(n-1) \leq n-1$. Then $a(a(n-1))$ is a applied to a natural number less than or equal to $n-1$. Hence, $a(a(n-1))$ is also a natural number minus than or equal to $n-1$. Therefore, $n - a(a(n-1))$ is n minus some natural number less than or equal to $n-1$, which is a natural number less than or equal to n.

CHAPTER 6

Section 6.1

1. a) 5850 **b)** 343 **3.** 42 **5.** 2^8 **7.** 2^{n-2}
9. 1,321,368,961 **11. a)** Seven: 56, 63, 70, 77, 84, 91, 98 **b)** Five: 55, 66, 77, 88, 99 **c)** One: 77 **13. a)** 990 **b)** 500 **c)** 27 **15.** 52,457,600 **17. a)** 37,822,859,361 **b)** 8,204,716,800 **c)** 40,159,050, 880 **d)** 12,113,640,000 **e)** 171,004,205,215 **f)** 72,043,541,640 **g)** 6,230,721,635 **h)** 223,149,655 **19. a)** 0 **b)** 120 **c)** 720 **d)** 2520 **21. a)** 2 if $n = 1$, 2 if $n = 2$, 0 if $n \geq 3$ **b)** 2^{n-2} for $n > 1$; 1 if $n = 1$ **c)** $2(n-1)$ **23.** $(n+1)^m$ **25.** If n is even, $2^{n/2}$; if n is odd, $2^{(n+1)/2}$ **27. a)** 175 **b)** 248 **c)** 232 **d)** 84 **29.** 60 **31. a)** 240 **b)** 480 **c)** 360 **33.** 352 **35.** 147 **37. a)** 9,920,671,339,261,325,541,376 $\approx 9.9 \times 10^{21}$ **b)** 6,641,514,961,387,068,437,760 $\approx 6.6 \times 10^{21}$ **c)** About 314,000 years **49.** $54(64^{65536} - 1)/63$ **41.** 7,104,000,000,000 **43.** $16^{10} + 16^{26} + 16^{58}$ **45.** 17 **47.** Let $P(m)$ be the sum rule for m tasks. For the basis case take $m = 2$. This is just the sum rule for two tasks. Now assume that $P(m)$ is true. Consider $m+1$ tasks, $T_1, T_2, \ldots, T_m, T_{m+1}$, which can be done in $n_1, n_2, \ldots, n_m, n_{m+1}$ ways, respectively, such that no two of these tasks can be done at the same time. To do one of these tasks, we can either do one of the first m of these or do task T_{m+1}. By the sum rule for two tasks, the number of ways to do this is the sum of the number of ways to do one of the first m tasks, plus n_{m+1}. By the inductive hypothesis, this is $n_1 + n_2 + \cdots + n_m + n_{m+1}$, as desired. **49.** $n(n-3)/2$

Section 6.2

1. Because there are six classes, but only five weekdays, the pigeonhole principle shows that at least two classes must be held on the same day. **3. a)** 3 **b)** 14 **5.** Let $a, a+1, \ldots, a + n - 1$ be the integers in the sequence. The integers $(a+i) \bmod n, i = 0, 1, 2, \ldots, n-1$, are distinct, because $0 < (a+j) - (a+k) < n$ whenever $0 \leq k < j \leq n-1$. Because there are n possible values for $(a+i) \bmod n$ and there are n different integers in the set, each of these values is taken on exactly once. It follows that there is exactly one integer in the sequence that is divisible by n. **7.** 4951 **9.** The midpoint of the segment joining the points (a, b, c) and (d, e, f) is $((a+d)/2, (b+e)/2, (c+f)/2)$. It has integer coefficients if and only if a and d have the same parity, b and e have the same parity, and c and f have the same parity. Because there are eight possible triples of parity [such as (*even, odd, even*)], by the pigeonhole principle at least two of the nine points have the same triple of parities. The midpoint of the segment joining two such points has integer coefficients. **11. a)** Group the first eight positive integers into four subsets of two integers each so that the integers of each subset add up to 9: $\{1, 8\}, \{2, 7\}, \{3, 6\}$, and $\{4, 5\}$. If five integers are selected from the first eight positive integers, by the pigeonhole principle at least two of them come from the same subset. Two such integers have a sum of 9, as desired. **b)** No. Take $\{1, 2, 3, 4\}$, for example. **13.** 21,251 **15. a)** If there were fewer than 9 freshmen, fewer than 9 sophomores, and fewer than 9 juniors in the class, there would be no more than 8 with each of these three class standings, for a total of at most 24 students, contradicting the fact that there are 25 students in the class. **b)** If there were fewer than 3 freshmen, fewer than 19 sophomores, and fewer than 5 juniors, then there would be at most 2 freshmen, at most 18 sophomores, and at most 4 juniors, for a total of at most 24 students. This contradicts the fact that there are 25 students in the class. **17.** 4, 3, 2, 1, 8, 7, 6, 5, 12, 11, 10, 9, 16, 15, 14, 13 **19.** Number the seats around the table from 1 to 50, and think of seat 50 as being adjacent to seat 1. There are 25 seats with odd numbers and 25 seats with even numbers. If no more than 12 boys occupied the odd-numbered seats, then at least 13 boys would occupy the even-numbered seats, and vice versa. Without loss of generality, assume that at least 13 boys occupy the 25 odd-numbered seats. Then at least two of those boys must be in consecutive odd-numbered seats, and the person sitting between them will have boys as both of his or her neighbors. **21.** By symmetry we need prove only the first statement. Let A be one of the people. Either A has at least four friends, or A has at least six enemies among the other nine people (because $3 + 5 < 9$). Suppose, in the first case, that B, C, D, and E are all A's friends. If any two of these are friends with each other, then we have found three mutual friends. Otherwise $\{B, C, D, E\}$ is a set of four mutual enemies. In the second case, let $\{B, C, D, E, F, G\}$ be a set of enemies of A. By Example 11, among B, C, D, E, F, and G there are either three mutual friends or three mutual enemies, who form, with A, a set of four mutual enemies. **23.** We need to show two things: that if we have a group of n people, then among them we must find either a pair of friends or a subset of n of them all of whom are mutual enemies; and that there exists a group of $n-1$ people for which this is not possible. For the first statement, if there is any pair of friends, then the condition is satisfied, and if not, then every pair of people are enemies, so the second condition is satisfied. For the second statement, if we have a group of $n-1$ people all of whom are enemies of each other, then there is neither a pair of friends nor a subset of n of them all of whom are mutual enemies. **25.** There are 6,432,816 possibilities for the three initials and a birthday. So, by the generalized pigeonhole principle, there are at least $\lceil 37,000,000/6,432,816 \rceil = 6$ people who share the same initials and birthday. **27.** 18 **29.** Because there are six computers, the number of other computers a computer is connected to is an integer between 0 and 5, inclusive. However, 0 and 5 cannot both occur. To see this, note that if some computer is connected to no others, then no computer is connected to all five others, and if some computer is connected to all five others, then no computer is connected to no others. Hence, by the pigeonhole principle, because there are at most five possibilities for the number of computers a computer is connected to, there are at least two computers in the set of six connected to the same number of others. **31.** Label the computers C_1 through C_{100}, and label the printers P_1 through P_{20}. If we connect C_k to P_k for $k = 1, 2, \ldots, 20$ and connect each of the computers C_{21} through C_{100} to *all* the printers, then we have used a total of $20 + 80 \cdot 20 = 1620$ cables. Clearly this is sufficient, because if computers C_1 through C_{20} need printers, then they can use the printers with the same subscripts, and if any computers with higher subscripts need a printer instead of one or more of these, then they can use the printers that are not being used, because they are connected to all the printers. Now we must show that 1619 cables is not enough. Because there are 1619 cables and 20 printers, the average number of computers per printer is $1619/20$, which is less than 81. Therefore some printer must be connected to fewer than 81 computers. That means it is connected to 80 or fewer computers, so there are 20 computers that are not connected to it. If those 20 computers all needed a printer simultaneously, then they would be out of luck, because they are connected to at most the 19 other printers. **33.** Let a_i be the number of matches completed by hour i. Then $1 \le a_1 < a_2 < \cdots < a_{75} \le 125$. Also $25 \le a_1 + 24 < a_2 + 24 < \cdots < a_{75} + 24 \le 149$. There are 150 numbers $a_1, \ldots, a_{75}, a_1 + 24, \ldots, a_{75} + 24$. By the pigeonhole principle, at least two are equal. Because all the a_is are distinct and all the $(a_i + 24)$s are distinct, it follows that $a_i = a_j + 24$ for some $i > j$. Thus, in the period from the $(j + 1)$st to the ith hour, there are exactly 24 matches. **35.** Consider 50 boxes. If the address ends with 00 or 01 put the address in the first box. If it ends with 02 or 03 put it in the second box and so on; 98 or 99 in the 50th box. Since there are 51 houses, at least one box will contain two addresses by the pigeonhole principle and thus there are two houses that have addresses as consecutive integers.

Section 6.3

1. *abc*, *acb*, *bac*, *bca*, *cab*, *cba* **3.** 720 **5.** 15,120
7. a) 210 **b)** 386 **c)** 848 **d)** 252 **9.** 65,780 **11.** $2^{100} - 5051$
13. a) 120 **b)** 24 **c)** 120 **d)** 24 **e)** 6 **f)** 0 **15. a)** 94,109,400
b) 941,094 **c)** 3,764,376 **d)** 90,345,024 **e)** 114,072 **f)** 2328
g) 24 **h)** 79,727,040 **i)** 3,764,376 **j)** 109,440 **17. a)** 12,650
b) 303,600 **19. a)** 37,927 **b)** 18,915 **21. a)** 122,523,030
b) 72,930,375 **c)** 223,149,655 **d)** 100,626,625 **23.** 54,600
25. 45 **27.** 912 **29.** 11,232,000 **31.** $n!/(r(n - r)!)$
33. 13

Section 6.4

1. $x^4 + 4x^3 y + 6x^2 y^2 + 4xy^3 + y^4$ **3.** 101 **5.** $(-1)^{(200-k)/3}$ $\binom{100}{(200-k)/3}$ if $k \equiv 2 \pmod 3$ and $-100 \le k \le 200$; 0 otherwise **7.** 1 9 36 84 126 126 84 36 9 1 **9.** $\binom{n}{k} = \frac{n(n-1)(n-2)\cdots(n-k+1)}{k(k-1)(k-2)\cdots 2} \le \frac{n \cdot n \cdots n}{2 \cdot 2 \cdots 2} = n^k/2^{k-1}$
11. $\binom{n}{k-1} + \binom{n}{k} = \frac{n!}{(k-1)!(n-k+1)!} + \frac{n!}{k!(n-k)!} = \frac{n!}{k!(n-k+1)!} \cdot [k + (n-k+1)] = \frac{(n+1)!}{k!(n+1-k)!} = \binom{n+1}{k}$ **13. a)** We show that each side counts the number of ways to choose from a set with n elements a subset with k elements and a distinguished element of that set. For the left-hand side, first choose the k-set (this can be done in $\binom{n}{k}$ ways) and then choose one of the k elements in this subset to be the distinguished element (this can be done in k ways). For the right-hand side, first choose the distinguished element out of the entire n-set (this can be done in n ways), and then choose the remaining $k - 1$ elements of the subset from the remaining $n - 1$ elements of the set (this can be done in $\binom{n-1}{k-1}$ ways). **b)** $k\binom{n}{k} = k \cdot \frac{n!}{k!(n-k)!} = \frac{n \cdot (n-1)!}{(k-1)!(n-k)!} = n\binom{n-1}{k-1}$
15. $\binom{n+1}{k} = \frac{(n+1)!}{k!(n+1-k)!} = \frac{(n+1)}{k} \frac{n!}{(k-1)![n-(k-1)]!} = (n + 1)$ $\binom{n}{k-1}/k$. This identity together with $\binom{n}{0} = 1$ gives a recursive definition. **17.** $\binom{2n}{n+1} + \binom{2n}{n} = \binom{2n+1}{n+1} = \frac{1}{2}\left[\binom{2n+1}{n+1} + \binom{2n+1}{n+1}\right] = \frac{1}{2}\left[\binom{2n+1}{n+1} + \binom{2n+1}{n}\right] = \frac{1}{2}\binom{2n+2}{n+1}$ **19. a)** $\binom{n+r+1}{r}$ counts the number of ways to choose a sequence of r 0s and $n + 1$ 1s by choosing the positions of the 0s. Alternately, suppose that the $(j + 1)$st term is the last term equal to 1, so that $n \le j \le n+r$. Once we have determined where the last 1 is, we decide where the 0s are to be placed in the j spaces before the last 1. There are n 1s and $j - n$ 0s in this range. By the sum rule it follows that there are $\sum_{j=n}^{n+r} \binom{j}{j-n} = \sum_{k=0}^{r} \binom{n+k}{k}$ ways to do this. **b)** Let $P(r)$ be the statement to be proved. The basis step is the equation $\binom{n}{0} = \binom{n+1}{0}$, which is just $1 = 1$. Assume that $P(r)$ is true. Then $\sum_{k=0}^{r+1} \binom{n+k}{k} = \sum_{k=0}^{r} \binom{n+k}{k} + \binom{n+r+1}{r+1} = \binom{n+r+1}{r} + \binom{n+r+1}{r+1} = \binom{n+r+2}{r+1}$, using the inductive hypothesis and Pascal's identity. **21.** We can choose the leader first in n different ways. We can then choose the rest of the committee in 2^{n-1} ways. Hence, there are $n2^{n-1}$ ways to choose the committee and its leader. Meanwhile, the number of ways to select a committee with k people is $\binom{n}{k}$. Once we have chosen a committee with k people, there are k ways to choose its leader. Hence, there are $\sum_{k=1}^{n} k\binom{n}{k}$ ways to choose the committee and its leader. Hence, $\sum_{k=1}^{n} k\binom{n}{k} = n2^{n-1}$. **23. a)** A path of the desired type consists of m moves to the right and n moves up. Each such path can be represented by a bit string of length $m + n$ with m 0s and n 1s, where a 0 represents a move to the right and a 1 a move up. **b)** The number of bit strings of length $m + n$ containing exactly n 1s equals $\binom{m+n}{n} = \binom{m+n}{m}$ because such a string is determined by specifying the positions of the n 1s or by specifying the positions of the m 0s. **25.** By Exercise 23 the number of paths of length n of the type described in that exercise equals 2^n, the number of bit strings of length n. On the other hand, a path of length n of the type described in Exercise 23 must end at a point that has n as the sum of its coordinates, say $(n - k, k)$ for some k between 0 and n, inclusive. By Exercise 23, the number of

such paths ending at $(n - k, k)$ equals $\binom{n-k+k}{k} = \binom{n}{k}$. Hence, $\sum_{k=0}^{n} \binom{n}{k} = 2^n$. **27.** By Exercise 23 the number of paths from $(0, 0)$ to $(n + 1, r)$ of the type described in that exercise equals $\binom{n+r+1}{r}$. But such a path starts by going j steps vertically for some j with $0 \leq j \leq r$. The number of these paths beginning with j vertical steps equals the number of paths of the type described in Exercise 32 that go from $(1, j)$ to $(n + 1, r)$. This is the same as the number of such paths that go from $(0, 0)$ to $(n, r - j)$, which by Exercise 32 equals $\binom{n+r-j}{r-j}$. Because $\sum_{j=0}^{r} \binom{n+r-j}{r-j} = \sum_{k=0}^{r} \binom{n+k}{k}$, it follows that $\sum_{k=1}^{r} \binom{n+k}{k} = \binom{n+r-1}{r}$.

Section 6.5

1. 243 **3.** 125 **5.** 35 **7.** 9 **9.** 4,504,501 **11. a)** 10,626 **b)** 1,365 **c)** 11,649 **d)** 106 **13.** 302,702,400 **15.** 3003 **17.** 7,484,400 **19.** 30,492 **21.** $C(59, 50)$ **23.** 35 **25.** 19,635 **27.** 210 **29.** 27,720 **31.** Approximately 6.5×10^{32} **33. a)** $C(k + n - 1, n)$ **b)** $(k + n - 1)!/(k - 1)!$ **35.** There are $C(n, n_1)$ ways to choose n_1 objects for the first box. Once these objects are chosen, there are $C(n - n_1, n_2)$ ways to choose objects for the second box. Similarly, there are $C(n - n_1 - n_2, n_3)$ ways to choose objects for the third box. Continue in this way until there is $C(n - n_1 - n_2 - \cdots - n_{k-1}, n_k) = C(n_k, n_k) = 1$ way to choose the objects for the last box (because $n_1 + n_2 + \cdots + n_k = n$). By the product rule, the number of ways to make the entire assignment is $C(n, n_1)C(n - n_1, n_2)C(n - n_1 - n_2, n_3) \cdots C(n - n_1 - n_2 - \cdots - n_{k-1}, n_k)$, which equals $n!/(n_1!n_2!\cdots n_k!)$, as straightforward simplification shows. **37. a)** Because $x_1 \leq x_2 \leq \cdots \leq x_r$, it follows that $x_1 + 0 < x_2 + 1 < \cdots < x_r + r - 1$. The inequalities are strict because $x_j + j - 1 < x_{j+1} + j$ as long as $x_j \leq x_{j+1}$. Because $1 \leq x_j \leq n + r - 1$, this sequence is made up of r distinct elements from T. **b)** Suppose that $1 \leq x_1 < x_2 < \cdots < x_r \leq n + r - 1$. Let $y_k = x_k - (k - 1)$. Then it is not hard to see that $y_k \leq y_{k+1}$ for $k = 1, 2, \ldots, r - 1$ and that $1 \leq y_k \leq n$ for $k = 1, 2, \ldots r$. It follows that $\{y_1, y_2, \ldots, y_r\}$ is an r-combination with repetitions allowed of S. **c)** From parts (a) and (b) it follows that there is a one-to-one correspondence of r-combinations with repetitions allowed of S and r-combinations of T, a set with $n+r-1$ elements. We conclude that there are $C(n+r-1, r)$ r-combinations with repetitions allowed of S. **39.** 65 **41.** 2 **43.** 3 **45. a)** 150 **b)** 25 **c)** 6 **d)** 2 **47.** 90,720 **49.** The terms in the expansion are of the form $x_1^{n_1} x_2^{n_2} \cdots x_m^{n_m}$, where $n_1 + n_2 + \cdots + n_m = n$. Such a term arises from choosing the x_1 in n_1 factors, the x_2 in n_2 factors, $\ldots$, and the x_m in n_m factors. This can be done in $C(n; n_1, n_2, \ldots, n_m)$ ways, because a choice is a permutation of n_1 labels "1," n_2 labels "2," $\ldots$, and n_m labels "m."

Section 6.6

1. 14532, 15432, 21345, 23451, 23514, 31452, 31542, 43521, 45213, 45321 **3. a)** 2134 **b)** 54132 **c)** 12534 **d)** 45312 **e)** 6714253 **d)** 31542678 **5.** $\{1, 2, 3\}, \{1, 2, 4\}, \{1, 2, 5\},$ $\{1, 3, 4\}, \{1, 3, 5\}, \{1, 4, 5\}, \{2, 3, 4\}, \{2, 3, 5\}, \{2, 4, 5\},$ $\{3, 4, 5\}$ **7.** The bit string representing the next larger r-combination must differ from the bit string representing the original one in position i because positions $i + 1, \ldots, r$ are occupied by the largest possible numbers. Also $a_i + 1$ is the smallest possible number we can put in position i if we want a combination greater than the original one. Then $a_i + 2, \ldots, a_i + r - i + 1$ are the smallest allowable numbers for positions $i + 1$ to r. Thus, we have produced the next r-combination. **9.** 123, 132, 213, 231, 312, 321, 124, 142, 214, 241, 412, 421, 125, 152, 215, 251, 512, 521, 134, 143, 314, 341, 413, 431, 135, 153, 315, 351, 513, 531, 145, 154, 415, 451, 514, 541, 234, 243, 324, 342, 423, 432, 235, 253, 325, 352, 523, 532, 245, 254, 425, 452, 524, 542, 345, 354, 435, 453, 534, 543 **11.** We will show that it is a bijection by showing that it has an inverse. Given a positive integer less than $n!$, let $a_1, a_2, \ldots, a_{n-1}$ be its Cantor digits. Put n in position $n - a_{n-1}$; then clearly, a_{n-1} is the number of integers less than n that follow n in the permutation. Then put $n - 1$ in free position $(n - 1) - a_{n-2}$, where we have numbered the free positions $1, 2, \ldots, n - 1$ (excluding the position that n is already in). Continue until 1 is placed in the only free position left. Because we have constructed an inverse, the correspondence is a bijection.

Supplementary Exercises

1. a) 151,200 **b)** 1,000,000 **c)** 210 **d)** 5005 **3.** 24,600 **5. a)** 192 **b)** 301 **c)** 300 **d)** 300 **7.** 639 **9.** The maximum possible sum is 240, and the minimum possible sum is 15. So the number of possible sums is 226. Because there are 252 subsets with five elements of a set with 10 elements, by the pigeonhole principle it follows that at least two have the same sum. **11. a)** 50 **b)** 50 **c)** 14 **d)** 17 **13.** Let $a_1, a_2, \ldots, a_m$ be the integers, and let $d_i = \sum_{j=1}^{i} a_j$. If $d_i \equiv 0$ (mod m) for some i, we are done. Otherwise $d_1 \bmod m$, $d_2 \bmod m, \ldots, d_m \bmod m$ are m integers with values in $\{1, 2, \ldots, m - 1\}$. By the pigeonhole principle $d_k = d_l$ for some $1 \leq k < l \leq m$. Then $\sum_{j=k+1}^{l} a_j = d_l - d_k \equiv 0$ (mod m). **15.** The decimal expansion of the rational number a/b can be obtained by division of b into a, where a is written with a decimal point and an arbitrarily long string of 0s following it. The basic step is finding the next digit of the quotient, namely, $\lfloor r/b \rfloor$, where r is the remainder with the next digit of the dividend brought down. The current remainder is obtained from the previous remainder by subtracting b times the previous digit of the quotient. Eventually the dividend has nothing but 0s to bring down. Furthermore, there are only b possible remainders. Thus, at some point, by the pigeonhole principle, we will have the same situation as had previously arisen. From that point onward, the calculation must follow the same pattern. In particular, the quotient will repeat.   **17. a)** 125,970 **b)** 20 **c)** 141,120,525 **d)** 141,120,505 **e)** 177,100 **f)** 141,078,021   **19. a)** 10 **b)** 8 **c)** 7   **21.** 3^n **23.** $C(n + 2, r + 1) = C(n + 1, r + 1) + C(n + 1, r) = 2C(n + 1, r + 1) - C(n + 1, r + 1) + C(n + 1, r) = 2C(n + 1, r + 1) - (C(n, r + 1) + C(n, r)) + (C(n, r) + C(n, r - 1)) = 2C(n + 1, r + 1) - C(n + 1, r + 1) + C(n, r - 1)$ **25.** Substitute $x = 1$ and $y = 3$ into the binomial theorem. **27.** 3,491,888,400 **29.** 5^{24} **31. a)** 386**b)** 56   **33.** 0 if $n < m$; $C(n - 1, n - m)$ if $n \geq m$ **35. a)** 15,625 **b)** 202 **c)** 210 **d)** 10   **37. a)** 3 **b)** 11 **c)** 6 **d)** 10   **39.** There are two possibilities: three people seated at one table with everyone else sitting alone, which can be done in $2C(n, 3)$ ways (choose the three people and seat them in one of two arrangements), or two groups of two people seated together with everyone else sitting alone, which can be done in $3C(n, 4)$ ways (choose four people and then choose one of the three ways to pair them up). Both $2C(n, 3) + 3C(n, 4)$ and $(3n - 1)C(n, 3)/4$ equal $n^4/8 - 5n^3/12 + 3n^2/8 - n/12$.   **41.** 2520

CHAPTER 7

Section 7.1

1. 1/13 **3.** 1/2 **5.** 1/64 **7.** 47/52 **9.** $1/C(52, 5)$ **11.** $1 - [C(48, 5)/C(52, 5)]$ **13.** $C(13, 2)C(4, 2)C(4, 2) C(44, 1)/C(52, 5)$ **15.** $1,302,540/C(52, 5)$ **17.** 1/64 **19.** 8/25   **21. a)** 139,128/319,865   **b)** 212, 667/511,313 **c)** 151,340/386,529   **d)** 163,647/446,276   **23.** 3/100 **25. a)** 1/7,880,400**b)** 1/8,000,000   **27. a)** 9/19 **b)** 81/361 **c)** 1/19 **d)** 1,889,568/2,476,099 **e)** 48/361   **29.** Three dice **31.** The door the contestant chooses is chosen at random without knowing where the prize is, but the door chosen by the host is not chosen at random, because he always avoids opening the door with the prize. This makes any argument based on symmetry invalid.   **33. a)** 671/1296 **b)** $1 - 35^{24}/36^{24}$; no **c)** The former

Section 7.2

1. $p(T) = 1/4$, $p(H) = 3/4$ **3.** $p(1) = p(3) = p(5) = p(6) = 1/16$; $p(2) = p(4) = 3/8$ **5.** 9/49   **7. a)** 1/26! **b)** 1/26 **c)** 1/2 **d)** 1/26 **e)** 1/650 **f)** 1/15,600   **9.** Clearly, $p(E \cup F) \geq p(E) = 0.7$. Also, $p(E \cup F) \leq 1$. If we apply Theorem 2 from Section 7.1, we can rewrite this as $p(E) + p(F) - p(E \cap F) \leq 1$, or $0.7 + 0.5 - p(E \cap F) \leq 1$. Solving for $p(E \cap F)$ gives $p(E \cap F) \geq 0.2$.   **11.** Because $p(E \cup F) = p(E) + p(F) - p(E \cap F)$ and $p(E \cup F) \leq 1$, it follows that $1 \geq p(E) + p(F) - p(E \cap F)$. From this inequality we conclude that $p(E) + p(F) \leq 1 + p(E \cap F)$. **13.** Because $E \cup \overline{E}$ is the entire sample space S, the event F can be split into two disjoint events: $F = S \cap F = (E \cup \overline{E}) \cap F = (E \cap F) \cup (\overline{E} \cap F)$, using the distributive law. Therefore, $p(F) = p((E \cap F) \cup (\overline{E} \cap F)) = p(E \cap F) + p(\overline{E} \cap F)$, because these two events are disjoint. Subtracting $p(E \cap F)$

from both sides, using the fact that $p(E \cap F) = p(E) \cdot p(F)$ (the hypothesis that E and F are independent), and factoring, we have $p(F)[1 - p(E)] = p(\overline{E} \cap F)$. Because $1 - p(E) = p(\overline{E})$, this says that $p(\overline{E} \cap F) = p(\overline{E}) \cdot p(F)$, as desired.   **15. a)** 1/12   **b)** $1 - \frac{11}{12} \cdot \frac{10}{12} \cdot \ldots \cdot \frac{13-n}{12}$ **c)** 5 **17.** 614   **19.** 3/8   **21. a)** Not independent   **b)** Not independent   **c)** Not independent   **23.** 3/16   **25. a)** $1/32 = 0.03125$ **b)** $0.49^5 \approx 0.02825$ **c)** 0.03795012   **27. a)** p^n **b)** $1 - p^n$ **c)** $p^n + n \cdot p^{n-1} \cdot (1 - p)$ **d)** $1 - [p^n + n \cdot p^{n-1} \cdot (1 - p)]$ **29.** $p(\bigcup_{i=1}^{\infty} E_i)$ is the sum of $p(s)$ for each outcome s in $\bigcup_{i=1}^{\infty} E_i$. Because the E_is are pairwise disjoint, this is the sum of the probabilities of all the outcomes in any of the E_is, which is what $\sum_{i=1}^{\infty} p(E_i)$ is. (We can rearrange the summands and still get the same answer because this series converges absolutely.) **31. a)** $\overline{E} = \bigcup_{j=1}^{\binom{m}{k}} F_j$, so the given inequality now follows from Boole's Inequality (Exercise 15). **b)** The probability that a particular player not in the jth set beats all k of the players in the jth set is $(1/2)^k = 2^{-k}$. Therefore, the probability that this player does not do so is $1 - 2^{-k}$, so the probability that all $m - k$ of the players not in the jth set are unable to boast of a perfect record against everyone in the jth set is $(1 - 2^{-k})^{m-k}$. That is precisely $p(F_j)$. **c)** The first inequality follows immediately, because all the summands are the same and there are $\binom{m}{k}$ of them. If this probability is less than 1, then it must be possible that $\overline{E}$ fails, i.e., that E happens. So there is a tournament that meets the conditions of the problem as long as the second inequality holds. **d)** $m \geq 21$ for $k = 2$, and $m \geq 91$ for $k = 3$

Section 7.3

NOTE: In the answers for Section 7.3, all probabilities given in decimal form are rounded to three decimal places. **1.** 3/5 **3.** 0.481   **5. a)** 0.740 **b)** 0.260 **c)** 0.002 **d)** 0.998 **7.** 0.724   **9. a)** 1/3   **b)** $p(M = j \mid W = k) = 1$ if i, j, and k are distinct; $p(M = j \mid W = k) = 0$ if $j = k$ or $j = i$; $p(M = j \mid W = k) = 1/2$ if $i = k$ and $j \neq i$ **c)** 2/3   **d)** You should change doors, because you now have a 2/3 chance to win by switching.   **11.** The definition of conditional probability tells us that $p(F_j \mid E) = p(E \cap F_j)/p(E)$. For the numerator, again using the definition of conditional probability, we have $p(E \cap F_j) = p(E \mid F_j)p(F_j)$, as desired. For the denominator, we show that $p(E) = \sum_{i=1}^{n} p(E \mid F_i)p(F_i)$. The events $E \cap F_i$ partition the event E; that is, $(E \cap F_{i_1}) \cap (E \cap F_{i_2}) = \emptyset$ when $i_1 \neq i_2$ (because the F_i's are mutually exclusive), and $\bigcup_{i=1}^{n}(E \cap F_{i_1}) = E$ (because the $\bigcup_{i=1}^{n} F_i = S$). Therefore, $p(E) = \sum_{i=1}^{n} p(E \cap F_i) = \sum_{i=1}^{n} p(E \mid F_i)p(F_i)$. **13.** No **15.** Yes   **17.** By Bayes' theorem, $p(S \mid E_1 \cap E_2) = p(E_1 \cap E_2 \mid S)p(S)/[p(E_1 \cap E_2 \mid S)p(S) + p(E_1 \cap E_2 \mid \overline{S})p(\overline{S})]$. Because we are assuming no prior knowledge about whether a message is or is not spam, we set $p(S) = p(\overline{S}) = 0.5$, and so the equation above simplifies to $p(S \mid E_1 \cap E_2) = p(E_1 \cap E_2 \mid S)/[p(E_1 \cap E_2 \mid S) + p(E_1 \cap E_2 \mid \overline{S})]$.

Because of the assumed independence of E_1, E_2, and S, we have $p(E_1 \cap E_2 \mid S) = p(E_1 \mid S) \cdot p(E_2 \mid S)$, and similarly for $\overline{S}$.

Section 7.4

1. 2.5 **3.** 336/49 **5.** 170 **7.** $(4n+6)/3$ **9.** 50,700,551 /10,077,696 $\approx$ 5.03 **11.** 6 **13.** $p(X \geq j) = \sum_{k=j}^{\infty} p(X = k) = \sum_{k=j}^{\infty}(1 - p)^{k-1}p = p(1 - p)^{j-1}\sum_{k=0}^{\infty}(1 - p)^k = p(1 - p)^{j-1}/(1 - (1 - p)) = (1 - p)^{j-1}$ **15.** 2302 **17.** $(7/2) \cdot 7 \neq 329/12$ **19.** $p + (n - 1)p(1 - p)$ **21.** 5/2 **23. a)** 0 **b)** n **25.** 1/100 **27.** $E(X)/a = \sum_r (r/a) \cdot p(X = r) \geq \sum_{r \geq a} 1 \cdot p(X = r) = p(X \geq a)$ **29. a)** 10/11 **b)** 0.9999 **31. a)** Each of the $n!$ permutations occurs with probability $1/n!$, so $E(X)$ is the number of comparisons, averaged over all these permutations. **b)** Even if the algorithm continues $n - 1$ rounds, X will be at most $n(n - 1)/2$. It follows from the formula for expectation that $E(X) \leq n(n - 1)/2$. **c)** The algorithm proceeds by comparing adjacent elements and then swapping them if necessary. Thus, the only way that inverted elements can become uninverted is for them to be compared and swapped. **d)** Because $X(P) \geq I(P)$ for all P, it follows from the definition of expectation that $E(X) \geq E(I)$. **e)** This summation counts 1 for every instance of an inversion. **f)** This follows from Theorem 3. **g)** By Theorem 2 with $n = 1$, the expectation of $I_{j,k}$ is the probability that a_k precedes a_j in the permutation. This is clearly 1/2 by symmetry. **h)** The summation in part (f) consists of $C(n, 2) = n(n - 1)/2$ terms, each equal to 1/2, so the sum is $n(n - 1)/4$. **i)** From part (a) and part (b) we know that $E(X)$, the object of interest, is at most $n(n - 1)/2$, and from part (d) and part (h) we know that $E(X)$ is at least $n(n - 1)/4$, both of which are $\Theta(n^2)$. **33.** 1 **35.** $V(X + Y) = E((X + Y)^2) - E(X + Y)^2 = E(X^2 + 2XY + Y^2) - [E(X) + E(Y)]^2 = E(X^2) + 2E(XY) + E(Y^2) - E(X)^2 - 2E(X)E(Y) - E(Y)^2 = E(X^2) - E(X)^2 + 2[E(XY) - E(X)E(Y)] + E(Y^2) - E(Y)^2 = V(X) + 2\,\text{Cov}(X, Y) + V(Y)$ **37.** $[(n - 1)/n]^m$ **39.** $(n - 1)^m/n^{m-1}$

Supplementary Exercises

1. 1/109,668 **3. a)** $1/C(52, 13)$ **b)** $4/C(52, 13)$ **c)** $2,944,656/C(52, 13)$ **d)** $35,335,872/C(52, 13)$ **5. a)** 9/2 **b)** 21/4 **7. a)** 8 **b)** 49/6 **9. a)** $n/2^{n-1}$ **b)** $p(1 - p)^{k-1}$, where $p = n/2^{n-1}$ **c)** $2^{n-1}/n$ **11.** $\frac{(m-1)(n-1)+\gcd(m,n)-1}{mn-1}$ **13. a)** 2/3 **b)** 2/3 **15. a)** The probability that one wins 2^n dollars is $1/2^n$, because that happens precisely when the player gets $n - 1$ tails followed by a head. The expected value of the winnings is therefore the sum of 2^n times $1/2^n$ as n goes from 1 to infinity. Because each of these terms is 1, the sum is infinite. In other words, one should be willing to wager any amount of money and expect to come out ahead in the long run. **b)** $9, $9

17. a) 1/3 when $S = \{1, 2, 3, 4, 5, 6, 7, 8, 9, 10, 11, 12\}$, $A = \{1, 2, 3, 4, 5, 6, 7, 8, 9\}$, and $B = \{1, 2, 3, 4\}$; 1/12 when $S = \{1, 2, 3, 4, 5, 6, 7, 8, 9, 10, 11, 12\}$, $A = \{4, 5, 6, 7, 8, 9, 10, 11, 12\}$, and $B = \{1, 2, 3, 4\}$ **b)** 1 when $S = \{1, 2, 3, 4, 5, 6, 7, 8, 9, 10, 11, 12\}$, $A = \{4, 5, 6, 7, 8, 9, 10, 11, 12\}$, and $B = \{1, 2, 3, 4\}$; 3/4 when $S = \{1, 2, 3, 4, 5, 6, 7, 8, 9, 10, 11, 12\}$, $A = \{1, 2, 3, 4, 5, 6, 7, 8, 9\}$, and $B = \{1, 2, 3, 4\}$ **19. a)** $p(E_1 \cap E_2) = p(E_1)p(E_2)$, $p(E_1 \cap E_3) = p(E_1)p(E_3)$, $p(E_2 \cap E_3) = p(E_2)p(E_3)$, $p(E_1 \cap E_2 \cap E_3) = p(E_1) p(E_2)p(E_3)$ **b)** Yes **c)** Yes; yes **d)** Yes; no **e)** $2^n - n - 1$ **21. a)** 1/2 under first interpretation; 1/3 under second interpretation **b)** Let M be the event that both of Mr. Smith's children are boys and let B be the event that Mr. Smith chose a boy for today's walk. Then $p(M) = 1/4$, $p(B \mid M) = 1$, and $p(B \mid \overline{M}) = 1/3$. Apply Bayes' theorem to compute $p(M \mid B) = 1/2$. **c)** This variation is equivalent to the second interpretation discussed in part (a), so the answer is unambiguously 1/3. **23.** $V(aX + b) = E((aX + b)^2) - E(aX + b)^2 = E(a^2X^2 + 2abX + b^2) - [aE(X) + b]^2 = E(a^2X^2) + E(2abX) + E(b^2) - [a^2E(X)^2 + 2abE(X) + b^2] = a^2E(X^2) + 2abE(X) + b^2 - a^2E(X)^2 - 2abE(X) - b^2 = a^2[E(X^2) - E(X)^2] = a^2V(X)$ **25.** To count every element in the sample space exactly once, we must include every element in each of the sets and then take away the double counting of the elements in the intersections. Thus $p(E_1 \cup E_2 \cup \cdots \cup E_m) = p(E_1) + p(E_2) + \cdots + p(E_m) - p(E_1 \cap E_2) - p(E_1 \cap E_3) - \cdots - p(E_1 \cap E_m) - p(E_2 \cap E_3) - p(E_2 \cap E_4) - \cdots - p(E_2 \cap E_m) - \cdots - p(E_{m-1} \cap E_m) = qm - (m(m - 1)/2)r$, because $C(m, 2)$ terms are being subtracted. But $p(E_1 \cup E_2 \cup \cdots \cup E_m) = 1$, so we have $qm - [m(m-1)/2]r = 1$. Because $r \geq 0$, this equation tells us that $qm \geq 1$, so $q \geq 1/m$. Because $q \leq 1$, this equation also implies that $[m(m - 1)/2]r = qm - 1 \leq m - 1$, from which it follows that $r \leq 2/m$. **27. a)** We purchase the cards until we have gotten one of each type. That means we have purchased X cards in all. On the other hand, that also means that we purchased X_0 cards until we got the first type we got, and then purchased X_1 more cards until we got the second type we got, and so on. Thus, X is the sum of the X_j's. **b)** Once j distinct types have been obtained, there are $n - j$ new types available out of a total of n types available. Because it is equally likely that we get each type, the probability of success on the next purchase (getting a new type) is $(n - j)/n$. **c)** This follows immediately from the definition of geometric distribution, the definition of X_j, and part (b). **d)** From part (c) it follows that $E(X_j) = n/(n - j)$. Thus by the linearity of expectation and part (a), we have $E(X) = E(X_0) + E(X_1) + \cdots + E(X_{n-1}) = \frac{n}{n} + \frac{n}{n-1} + \cdots + \frac{n}{1} = n\left(\frac{1}{n} + \frac{1}{n-1} + \cdots + \frac{1}{1}\right)$. **e)** About 224.46

CHAPTER 8

Section 8.1

1. Let $P(n)$ be "$H_n = 2^n - 1$." *Basis step:* $P(1)$ is true because $H_1 = 1$. *Inductive step:* Assume that $H_n = 2^n - 1$. Then because $H_{n+1} = 2H_n + 1$, it follows that $H_{n+1} = 2(2^n - 1) + 1 = 2^{n+1} - 1$. **3. a)** $a_n = a_{n-1} + a_{n-2} + 2a_{n-5} + 2a_{n-10} + 2a_{n-20} + 2a_{n-50} + 2a_{n-100}$ **b)** 9494 **5. a)** $a_n = a_{n-1} + a_{n-2} + 2^{n-2}$ for $n \geq 2$ **b)** $a_0 = 0$, $a_1 = 0$ **c)** 94 **7. a)** $a_n = a_{n-1} + a_{n-2} + a_{n-3}$ for $n \geq 3$ **b)** $a_0 = 1$, $a_1 = 2$, $a_2 = 4$ **c)** 81 **9. a)** $a_n = 2a_{n-1} + 2a_{n-2}$ for $n \geq 2$ **b)** $a_0 = 1$, $a_1 = 3$ **c)** 448 **11. a)** $a_n = 2a_{n-1} + a_{n-2}$ for $n \geq 2$ **b)** $a_0 = 1$, $a_1 = 3$ **c)** 239 **13. a)** $a_n = a_{n-1} + a_{n-2}$ for $n \geq 2$ **b)** $a_0 = 1$, $a_1 = 1$ **c)** 89 **15. a)** $R_n = n + R_{n-1}$, $R_0 = 1$ **b)** $R_n = n(n+1)/2 + 1$ **17. a)** $S_n = S_{n-1} + (n^2 - n + 2)/2$, $S_0 = 1$ **b)** $S_n = (n^3 + 5n + 6)/6$ **19. a)** $a_n = 2a_{n-1} + 2a_{n-2}$ **b)** $a_0 = 1$, $a_1 = 3$ **c)** 1224 **21.** Clearly, $S(m, 1) = 1$ for $m \geq 1$. If $m \geq n$, then a function that is not onto from the set with m elements to the set with n elements can be specified by picking the size of the range, which is an integer between 1 and $n - 1$ inclusive, picking the elements of the range, which can be done in $C(n, k)$ ways, and picking an onto function onto the range, which can be done in $S(m, k)$ ways. Hence, there are $\sum_{k=1}^{n-1} C(n, k)S(m, k)$ func- tions that are not onto. But there are n^m functions altogether, so $S(m, n) = n^m - \sum_{k=1}^{n-1} C(n, k)S(m, k)$. **23. a)** $C_5 = C_0C_4 + C_1C_3 + C_2C_2 + C_3C_1 + C_4C_0 = 1 \cdot 14 + 1 \cdot 5 + 2 \cdot 2 + 5 \cdot 1 + 14 \cdot 1 = 42$ **b)** $C(10, 5)/6 = 42$ **25.** $J(1) = 1$, $J(2) = 1$, $J(3) = 3$, $J(4) = 1$, $J(5) = 3$, $J(6) = 5$, $J(7) = 7$, $J(8) = 1$, $J(9) = 3$, $J(10) = 5$, $J(11) = 7$, $J(12) = 9$, $J(13) = 11$, $J(14) = 13$, $J(15) = 15$, $J(16) = 1$ **27.** First, suppose that the number of people is even, say $2n$. After going around the circle once and returning to the first person, because the people at locations with even numbers have been eliminated, there are exactly n people left and the person currently at location i is the person who was originally at location $2i - 1$. Therefore, the survivor [originally in location $J(2n)$] is now in location $J(n)$; this was the person who was at location $2J(n) - 1$. Hence, $J(2n) = 2J(n) - 1$. Similarly, when there are an odd number of people, say $2n + 1$, then after going around the circle once and then eliminating person 1, there are n people left and the person currently at location i is the person who was at location $2i + 1$. Therefore, the survivor will be the player currently occupying location $J(n)$, namely, the person who was originally at location $2J(n) + 1$. Hence, $J(2n + 1) = 2J(n) + 1$. The basis step is $J(1) = 1$. **29.** 73, 977, 3617 **31.** These nine moves solve the puzzle: Move disk 1 from peg 1 to peg 2; move disk 2 from peg 1 to peg 3; move disk 1 from peg 2 to peg 3; move disk 3 from peg 1 to peg 2; move disk 4 from peg 1 to peg 4; move disk 3 from peg 2 to peg 4; move disk 1 from peg 3 to peg 2; move disk 2 from peg 3 to peg 4; move disk 1 from peg 2 to peg 4. To see that at least nine moves are

required, first note that at least seven moves are required no matter how many pegs are present: three to unstack the disks, one to move the largest disk 4, and three more moves to restack them. At least two other moves are needed, because to move disk 4 from peg 1 to peg 4 the other three disks must be on pegs 2 and 3, so at least one move is needed to restack them and one move to unstack them. **33. a)** 0 **b)** 0 **c)** 2 **d)** $2^{n-1} - 2^{n-2}$ **35.** $a_n - 2\nabla a_n + \nabla^2 a_n = a_n - 2(a_n - a_{n-1}) + (\nabla a_n - \nabla a_{n-1}) = -a_n + 2a_{n-1} + [(a_n - a_{n-1}) - (a_{n-1} - a_{n-2})] = -a_n + 2a_{n-1} + (a_n - 2a_{n-1} + a_{n-2}) = a_{n-2}$ **37.** $a_n = a_{n-1} + a_{n-2} = (a_n - \nabla a_n) + (a_n - 2\nabla a_n + \nabla^2 a_n) = 2a_n - 3\nabla a_n + \nabla^2 a_n$, or $a_n = 3\nabla a_n - \nabla^2 a_n$ **39.** Insert $S(0) := \emptyset$ after $T(0) := 0$ (where $S(j)$ will record the optimal set of talks among the first j talks), and replace the statement $T(j) := \max(w_j + T(p(j)), T(j - 1))$ with the following code:

> **if** $w_j + T(p(j)) > T(j - 1)$ **then**
> $T(j) := w_j + T(p(j))$
> $S(j) := S(p(j)) \cup \{j\}$
> **else**
> $T(j) := T(j - 1)$
> $S(j) := S(j - 1)$

41. a) Talks 1, 3, and 7 **b)** Talks 1 and 6, or talks 1, 3, and 7 **c)** Talks 1, 3, and 7 **d)** Talks 1 and 6 **43. a)** This follows immediately from Example 5 ibn Section 8.4 and the fact that $C_n \geq 2^{n-1}$, where $\{C_n\}$ is the sequence of Catalan numbers. **b)** The last step in computing $\mathbf{A}_{ij}$ is to multiply $\mathbf{A}_{ik}$ by $\mathbf{A}_{k+1,j}$ for some k between i and $j - 1$ inclusive, which will require $m_i m_{k+1} m_{j+1}$ integer multiplications, independent of the manner in which $\mathbf{A}_{ik}$ and $\mathbf{A}_{k+1,j}$ are computed. Therefore to minimize the total number of integer multiplications, each of those two factors must be computed in the most efficient manner. **c)** This follows immediately from part (b) and the definition of $M(i, j)$.
d) **procedure** *matrix order*$(m_1, \ldots, m_{n+1}$: positive integers)

```
for i := 1 to n
  M(i, i) := 0
for d := 1 to n − 1
  for i := 1 to n − d
    min := 0
    for k := i to i + d
      new := M(i, k) + M(k + 1, i + d) + m_i m_{k+1} m_{i+d+1}
      if new < min then
        min := new
        where(i, i + d) := k
    M(i, i + d) := min
```

e) The algorithm has three nested loops, each of which is indexed over at most n values. **45.** This follows immediately from Example 5 and the fact that $C_n \geq 2^{n-1}$.

Section 8.2

1. a) Degree 3 **b)** No **c)** Degree 4 **d)** No **e)** No **f)** Degree 2 **g)** No **3.** $a_n = \frac{1}{\sqrt{5}}\left(\frac{1+\sqrt{5}}{2}\right)^{n+1} - \frac{1}{\sqrt{5}}\left(\frac{1-\sqrt{5}}{2}\right)^{n+1}$
5. $[2^{n+1}+(-1)^n]/3$ **7. a)** $P_n = 1.2P_{n-1}+0.45P_{n-2}$, $P_0 = 100{,}000$, $P_1 = 120{,}000$ **b)** $P_n = (250{,}000/3)(3/2)^n + (50{,}000/3)(-3/10)^n$ **9. a)** *Basis step:* For $n = 1$ we have $1 = 0 + 1$, and for $n = 2$ we have $3 = 1 + 2$. *Inductive step:* Assume true for $k \le n$. Then $L_{n+1} = L_n + L_{n-1} = f_{n-1} + f_{n+1}+f_{n-2}+f_n = (f_{n-1}+f_{n-2})+(f_{n+1}+f_n) = f_n+f_{n+2}$.
b) $L_n = \left(\frac{1+\sqrt{5}}{2}\right)^n + \left(\frac{1-\sqrt{5}}{2}\right)^n$ **11.** $a_n = 5+3(-2)^n - 3^n$
13. Let $a_n = C(n, 0)+C(n-1, 1)+\cdots+C(n-k, k)$ where $k = \lfloor n/2 \rfloor$. First, assume that n is even, so that $k = n/2$, and the last term is $C(k, k)$. By Pascal's identity we have $a_n = 1 + C(n-2, 0) + C(n-2, 1) + C(n-3, 1) + C(n-3, 2) + \cdots + C(n-k, k-2) + C(n-k, k-1) + 1 = 1+C(n-2, 1)+C(n-3, 2)+\cdots+C(n-k, k-1)+C(n-2, 0)+C(n-3, 1)+\cdots+C(n-k, k-2)+1 = a_{n-1}+a_{n-2}$ because $\lfloor(n-1)/2\rfloor = k-1 = \lfloor(n-2)/2\rfloor$. A similar calculation works when n is odd. Hence, $\{a_n\}$ satisfies the recurrence relation $a_n = a_{n-1}+a_{n-2}$ for all positive integers n, $n \ge 2$. Also, $a_1 = C(1, 0) = 1$ and $a_2 = C(2, 0) + C(1, 1) = 2$, which are f_2 and f_3. It follows that $a_n = f_{n+1}$ for all positive integers n. **15.** $a_n = (n^2 + 3n + 5)(-1)^n$ **17.** $(a_{1,0} + a_{1,1}n + a_{1,2}n^2 + a_{1,3}n^3) + (a_{2,0} + a_{2,1}n + a_{2,2}n^2)(-2)^n + (a_{3,0} + a_{3,1}n)3^n + a_{4,0}(-4)^n$ **19. a)** $3a_{n-1} + 2^n = 3(-2)^n + 2^n = 2^n(-3 + 1) = -2^{n+1} = a_n$ **b)** $a_n = \alpha 3^n - 2^{n+1}$ **c)** $a_n = 3^{n+1} - 2^{n+1}$ **21. a)** $p_3n^3 + p_2n^2 + p_1n + p_0$ **b)** $n^2 p_0(-2)^n$
c) $n^2(p_1n + p_0)2^n$ **d)** $(p_2n^2 + p_1n + p_0)4^n$ **e)** $n^2(p_2n^2 + p_1n + p_0)(-2)^n$ **f)** $n^2(p_4n^4 + p_3n^3 + p_2n^2 + p_1n + p_0)2^n$
g) p_0 **23. a)** $a_n = \alpha 2^n + 3^{n+1}$ **b)** $a_n = -2 \cdot 2^n + 3^{n+1}$
25. $a_n = -4 \cdot 2^n - n^2/4 - 5n/2 + 1/8 + (39/8)3^n$
27. $a_n = n(n + 1)(n + 2)/6$ **29. a)** $1, -1, i, -i$ **b)** $a_n = \frac{1}{4} - \frac{1}{4}(-1)^n + \frac{2+i}{4}i^n + \frac{2-i}{4}(-i)^n$ **31. a)** Using the formula for f_n, we see that $\left|f_n - \frac{1}{\sqrt{5}}\left(\frac{1+\sqrt{5}}{2}\right)^n\right| = \left|\frac{1}{\sqrt{5}}\left(\frac{1-\sqrt{5}}{2}\right)^n\right| < 1/\sqrt{5} < 1/2$. This means that f_n is the integer closest to $\frac{1}{\sqrt{5}}\left(\frac{1+\sqrt{5}}{2}\right)^n$. **b)** Less when n is even; greater when n is odd **33.** $a_n = f_{n-1} + 2f_n - 1$
35. a) $a_n = 2a_{n+1} + (n - 1)10{,}000$ **b)** $a_n = 70{,}000 \cdot 2^{n-1} - 10{,}000n - 10{,}000$ **37.** $a_n = 5n^2/12 + 13n/12 + 1$
39. See Chapter 11, Section 5 in [Ma93]. **41.** $6^n \cdot 4^{n-1}/n$

Section 8.3

1. 14 **3.** The first step is $(1110)_2(1010)_2 = (2^4 + 2^2)(11)_2(10)_2 + 2^2[(11)_2 - (10)_2][(10)_2 - (10)_2] + (2^2 + 1)(10)_2 \cdot (10)_2$. The product is $(10001100)_2$. **5. a)** 79
b) 48,829 **c)** 30,517,579 **7.** $O(\log n)$ **9.** $O(n^{\log_3 2})$
11. 5 **13. a)** *Basis step:* If the sequence has just one element, then the one person on the list is the winner. *Recursive step:* Divide the list into two parts—the first half and the second half—as equally as possible. Apply the algorithm recursively to each half to come up with at most two names. Then run through the entire list to count the number of oc-

currences of each of those names to decide which, if either, is the winner. **b)** $O(n \log n)$ **15. a)** $f(n) = f(n/2) + 2$
b) $O(\log n)$ **17. a)** 7 **b)** $O(\log n)$
19. a) procedure *largest sum*$(a_1, \ldots, a_n)$
$\quad$ *best* := 0 {empty subsequence has sum 0}
$\quad$ **for** $i := 1$ **to** n
$\quad\quad$ *sum* := 0
$\quad\quad$ **for** $j := i + 1$ **to** n
$\quad\quad\quad$ *sum* := *sum* + a_j
$\quad\quad\quad$ **if** *sum* > *best* **then** *best* := *sum*
$\quad$ {*best* is the maximum possible sum of numbers in the list}
b) $O(n^2)$ **c)** We divide the list into a first half and a second half and apply the algorithm recursively to find the largest sum of consecutive terms for each half. The largest sum of consecutive terms in the entire sequence is either one of these two numbers or the sum of a sequence of consecutive terms that crosses the middle of the list. To find the largest possible sum of a sequence of consecutive terms that crosses the middle of the list, we start at the middle and move forward to find the largest possible sum in the second half of the list, and move backward to find the largest possible sum in the first half of the list; the desired sum is the sum of these two quantities. The final answer is then the largest of this sum and the two answers obtained recursively. The base case is that the largest sum of a sequence of one term is the larger of that number and 0. **d)** 11, 9, 14 **e)** $S(n) = 2S(n/2) + n$, $C(n) = 2C(n/2) + n + 2$, $S(1) = 0$, $C(1) = 1$ **f)** $O(n \log n)$, better than $O(n^2)$
21. The algorithm is essentially the same as the algorithm given in Example 12. The central strip still has width $2d$ but we need to consider just two boxes of size $d \times d$ rather than eight boxes of size $(d/2) \times (d/2)$. The recurrence relation is the same as the recurrence relation in Example 12, except that the coefficient 7 is replaced by 1. **23.** With $k = \log_b n$, it follows that $f(n) = a^k f(1) + \sum_{j=0}^{k-1}a^j c(n/b^j)^d = a^k f(1) + \sum_{j=0}^{k-1} cn^d = a^k f(1) + kcn^d = a^{\log_b n} f(1) + c(\log_b n)n^d = n^{\log_b a} f(1) + cn^d \log_b n = n^d f(1) + cn^d \log_b n$. **25.** Let $k = \log_b n$ where n is a power of b. *Basis step:* If $n = 1$ and $k = 0$, then $c_1n^d + c_2n^{\log_b a} = c_1 + c_2 = b^dc/(b^d - a) + f(1) + b^dc/(a - b^d) = f(1)$. *Inductive step:* Assume true for k, where $n = b^k$. Then for $n = b^{k+1}$, $f(n) = af(n/b) + cn^d = a\{[b^dc/(b^d - a)](n/b)^d + [f(1) + b^dc/(a - b^d)] \cdot (n/b)^{\log_b a}\} + cn^d = b^dc/(b^d - a)n^da/b^d + [f(1) + b^dc/(a - b^d)]n^{\log_b a} + cn^d = n^d[ac/(b^d - a) + c(b^d - a)/(b^d - a)] + [f(1) + b^dc/(a - b^dc)]n^{\log_b a} = [b^dc/(b^d - a)]n^d + [f(1) + b^dc/(a - b^d)]n^{\log_b a}$. **27.** If $a > b^d$, then $\log_b a > d$, so the second term dominates, giving $O(n^{\log_b a})$. **29.** $O(n^3)$

Section 8.4

1. $f(x) = 2(x^6 - 1)/(x - 1)$ **3. a)** $f(x) = 2x(1 - x^6)/(1 - x)$ **b)** $x^3/(1 - x)$ **c)** $x/(1 - x^3)$ **d)** $2/(1 - 2x)$

e) $(1+x)^7$ **f)** $2/(1+x)$ **g)** $[1/(1-x)] - x^2$ **h)** $x^3/(1-x)^2$

5. a) $a_0 = -64$, $a_1 = 144$, $a_2 = -108$, $a_3 = 27$, and $a_n = 0$ for all $n \geq 4$ **b)** The only nonzero coefficients are $a_0 = 1$, $a_3 = 3$, $a_6 = 3$, $a_9 = 1$. **c)** $a_n = 5^n$ **d)** $a_n = (-3)^{n-3}$ for $n \geq 3$, and $a_0 = a_1 = a_2 = 0$ **e)** $a_0 = 8$, $a_1 = 3$, $a_2 = 2$, $a_n = 0$ for odd n greater than 2 and $a_n = 1$ for even n greater than 2 **f)** $a_n = 1$ if n is a positive multiple 4, $a_n = -1$ if $n < 4$, and $a_n = 0$ otherwise **g)** $a_n = n - 1$ for $n \geq 2$ and $a_0 = a_1 = 0$ **h)** $a_n = 2^{n+1}/n!$ **7. a)** 6 **b)** 3 **c)** 9 **d)** 0 **e)** 5 **9.** 10 **11.** 20 **13.** $f(x) = 1/[(1-x)(1-x^2)(1-x^5)(1-x^{10})]$ **15. a)** $x^4(1 + x + x^2 + x^3)^2/(1-x)$ **b)** 6 **17. a)** The coefficient of x^r in the power series expansion of $1/[(1 - x^3)(1 - x^4)(1 - x^{20})]$ **b)** $1/(1 - x^3 - x^4 - x^{20})$ **c)** 7 **d)** 3224 **19. a)** 10 **b)** 49 **c)** 2 **d)** 4 **21. a)** $G(x) - a_0 - a_1 x - a_2 x^2$ **b)** $G(x^2)$ **c)** $x^4 G(x)$ **d)** $G(2x)$ **e)** $\int_0^x G(t)dt$ **f)** $G(x)/(1-x)$ **23.** $a_k = 2 \cdot 3^k - 1$ **25.** $a_k = 18 \cdot 3^k - 12 \cdot 2^k$ **27.** Let $G(x) = \sum_{k=0}^{\infty} f_k x^k$. After shifting indices of summation and adding series, we see that $G(x) - xG(x) - x^2 G(x) = f_0 + (f_1 - f_0)x + \sum_{k=2}^{\infty}(f_k - f_{k-1} - f_{k-2})x^k = 0 + x + \sum_{k=2}^{\infty} 0x^k$. Hence, $G(x) - xG(x) - x^2 G(x) = x$. Solving for $G(x)$ gives $G(x) = x/(1 - x - x^2)$. By the method of partial fractions, it can be shown that $x/(1-x-x^2) = (1/\sqrt{5})[1/(1-\alpha x) - 1/(1-\beta x)]$, where $\alpha = (1 + \sqrt{5})/2$ and $\beta = (1 - \sqrt{5})/2$. Using the fact that $1/(1 - \alpha x) = \sum_{k=0}^{\infty} \alpha^k x^k$, it follows that $G(x) = (1/\sqrt{5}) \cdot \sum_{k=0}^{\infty}(\alpha^k - \beta^k)x^k$. Hence, $f_k = (1/\sqrt{5}) \cdot (\alpha^k - \beta^k)$. **29.** Applying the binomial theorem to the equality $(1+x)^{m+n} = (1+x)^m(1+x)^n$, shows that $\sum_{r=0}^{m+n} C(m + n, r)x^r = \sum_{r=0}^{m} C(m, r)x^r \cdot \sum_{r=0}^{n} C(n, r) x^r = \sum_{r=0}^{m+n} \left[\sum_{k=0}^{r} C(m, r - k) C(n, k) \right] x^r$. Comparing coefficients gives the desired identity. **31. a)** $2e^x$ **b)** e^{-x} **c)** e^{3x} **d)** $xe^x + e^x$ **33. a)** $a_n = 6a_{n-1} + 8^{n-1}$ for $n \geq 1$, $a_0 = 1$ **b)** The general solution of the associated linear homogeneous recurrence relation is $a_n^{(h)} = \alpha 6^n$. A particular solution is $a_n^{(p)} = \frac{1}{2} \cdot 8^n$. Hence, the general solution is $a_n = \alpha 6^n + \frac{1}{2} \cdot 8^n$. Using the initial condition, it follows that $\alpha = \frac{1}{2}$. Hence, $a_n = (6^n + 8^n)/2$. **c)** Let $G(x) = \sum_{k=0}^{\infty} a_k x^k$. Using the recurrence relation for $\{a_k\}$, it can be shown that $G(x) - 6xG(x) = (1 - 7x)/(1 - 8x)$. Hence, $G(x) = (1 - 7x)/[(1 - 6x)(1 - 8x)]$. Using partial fractions, it follows that $G(x) = (1/2)/(1 - 6x) + (1/2)/(1 - 8x)$. With the help of Table 1, it follows that $a_n = (6^n + 8^n)/2$. **35.** $\frac{1}{1-x} \cdot \frac{1}{1-x^2} \cdot \frac{1}{1-x^3} \cdots$ **37.** $(1 + x)(1 + x)^2(1 + x)^3 \cdots$ **39.** The generating functions obtained in Exercises 36 and 37 are equal because $(1+x)(1+x^2)(1+x^3) \cdots = \frac{1-x^2}{1-x} \cdot \frac{1-x^4}{1-x^2} \cdot \frac{1-x^6}{1-x^3} \cdots = \frac{1}{1-x} \cdot \frac{1}{1-x^3} \cdot \frac{1}{1-x^5} \cdots$. **41. a)** $G_X(1) = \sum_{k=0}^{\infty} p(X = k) \cdot 1^k = \sum_{k=0}^{\infty} P(X = k) = 1$ **b)** $G_X'(1) = \frac{d}{dx} \sum_{k=0}^{\infty} p(X = k) \cdot x^k|_{x=1} = \sum_{k=0}^{\infty} p(X = k) \cdot k \cdot x^{k-1}|_{x=1} = \sum_{k=0}^{\infty} p(X = k) \cdot k = E(X)$ **c)** $G_X''(1) = \frac{d^2}{dx^2} \sum_{k=0}^{\infty} p(X = k) \cdot x^k|_{x=1} = \sum_{k=0}^{\infty} p(X = k) \cdot k(k - 1) \cdot x^{k-2}|_{x=1} = \sum_{k=0}^{\infty} p(X = k) \cdot (k^2 - k) = V(X) + E(X)^2 - E(X)$. Combin-

ing this with part (b) gives the desired results. **43. a)** $G(x) = p^m/(1 - qx)^m$ **b)** $V(x) = mq/p^2$

Section 8.5

1. a) 30 **b)** 29 **c)** 24 **d)** 18 **3.** 1% **5. a)** 300 **b)** 150 **c)** 175 **d)** 100 **7.** 974 **9.** 55 **11.** 50,138 **13.** 234 **15.** $|A_1 \cup A_2 \cup A_3 \cup A_4 \cup A_5| = |A_1| + |A_2| + |A_3| + |A_4| + |A_5| - |A_1 \cap A_2| - |A_1 \cap A_3| - |A_1 \cap A_4| - |A_1 \cap A_5| - |A_2 \cap A_3| - |A_2 \cap A_4| - |A_2 \cap A_5| - |A_3 \cap A_4| - |A_3 \cap A_5| - |A_4 \cap A_5| + |A_1 \cap A_2 \cap A_3| + |A_1 \cap A_2 \cap A_4| + |A_1 \cap A_2 \cap A_5| + |A_1 \cap A_3 \cap A_4| + |A_1 \cap A_3 \cap A_5| + |A_1 \cap A_4 \cap A_5| + |A_2 \cap A_3 \cap A_4| + |A_2 \cap A_3 \cap A_5| + |A_2 \cap A_4 \cap A_5| + |A_3 \cap A_4 \cap A_5| - |A_1 \cap A_2 \cap A_3 \cap A_4| - |A_1 \cap A_2 \cap A_3 \cap A_5| - |A_1 \cap A_2 \cap A_4 \cap A_5| - |A_1 \cap A_3 \cap A_4 \cap A_5| - |A_2 \cap A_3 \cap A_4 \cap A_5| + |A_1 \cap A_2 \cap A_3 \cap A_4 \cap A_5|$ **17.** $p(E_1 \cup E_2 \cup E_3) = p(E_1) + p(E_2) + p(E_3) - p(E_1 \cap E_2) - p(E_1 \cap E_3) - p(E_2 \cap E_3) + p(E_1 \cap E_2 \cap E_3)$ **19.** 4972/71,295 **21.** $p\left(\bigcup_{i=1}^{n} E_i\right) = \sum_{1 \leq i \leq n} p(E_i) - \sum_{1 \leq i < j \leq n} p(E_i \cap E_j) + \sum_{1 \leq i < j < k \leq n} p(E_i \cap E_j \cap E_k) - \cdots + (-1)^{n+1} p\left(\bigcap_{i=1}^{n} E_i\right)$

Section 8.6

1. 75 **3.** 6 **5.** 9875 **7.** 540 **9.** 2100 **11.** 1854 **13. a)** $D_{100}/100!$ **b)** $100D_{99}/100!$ **c)** $C(100,2)/100!$ **d)** 0 **e)** $1/100!$ **15. a)** Let k be the element in the first position. If 1 occurrs in the kth position, number of possible derangements is D_{n-2}. If 1 does not occur in the kth position, number of derangements is D_{n-1}. There are $n - 1$ possible choices for k from 2 to n. Hence the result follows. **b)** By Exercise 15a) we have $D_n - nD_{n-1} = -[D_{n-1} - (n - 1)D_{n-2}]$. Iterating, we have $D_n - nD_{n-1} = -[D_{n-1} - (n-1)D_{n-2}] = -[-(D_{n-2} - (n - 2)D_{n-3})] = D_{n-2} - (n-2)D_{n-3} = \cdots = (-1)^n(D_2 - 2D_1) = (-1)^n$ because $D_2 = 1$ and $D_1 = 0$. **17.** When n is odd **19.** 4 **21.** There are n^m functions from a set with m elements to a set with n elements, $C(n, 1)(n-1)^m$ functions from a set with m elements to a set with n elements that miss exactly one element, $C(n, 2)(n - 2)^m$ functions from a set with m elements to a set with n elements that miss exactly two elements, and so on, with $C(n, n - 1) \cdot 1^m$ functions from a set with m elements to a set with n elements that miss exactly $n - 1$ elements. Hence, by the principle of inclusion–exclusion, there are $n^m - C(n, 1)(n - 1)^m + C(n, 2)(n - 2)^m - \cdots + (-1)^{n-1}C(n, n - 1) \cdot 1^m$ onto functions.

Supplementary Exercises

1. a) $A_n = 4A_{n-1}$ **b)** $A_1 = 40$ **c)** $A_n = 10 \cdot 4^n$ **3. a)** $a_n = a_{n-2} + a_{n-3}$ **b)** $a_1 = 0$, $a_2 = 1$, $a_3 = 1$ **c)** $a_{12} = 12$ **5.** $a_n = 2^n$ **7.** $a_n = 2 + 4n/3 + n^2/2 + n^3/6$ **9.** $a_n = a_{n-2} + a_{n-3}$ **11. a)** Under the given conditions, one longest common subsequence clearly ends at the last term in each sequence, so $a_m = b_n = c_p$. Furthermore, a longest common subsequence of what is left of the a-sequence and the b-sequence after those last terms are deleted has to form the beginning of a longest common subsequence of the original sequences. **b)** If $c_p \neq a_m$, then the longest common subsequence's appearance in the a-sequence must terminate before the end; therefore the c-sequence must be a longest common subsequence of $a_1, a_2, \ldots, a_{m-1}$ and $b_1, b_2, \ldots, b_n$. The other half is similar.

13. **procedure** *howlong*$(a_1, \ldots, a_m, b_1, \ldots, b_n$: sequences)
   for $i := 1$ **to** m
     $L(i, 0) := 0$
   for $j := 1$ **to** n
     $L(0, j) := 0$
   for $i := 1$ **to** m
     for $j := 1$ **to** n
       if $a_i = b_j$ **then** $L(i, j) := L(i - 1, j - 1) + 1$
       else $L(i, j) := \max(L(i, j - 1), L(i - 1, j))$
   return $L(m, n)$

15. $O(n)$ **17. a)** $18n + 18$ **b)** 18 **c)** 0 **19.** $\Delta(a_n b_n) = a_{n+1}b_{n+1} - a_n b_n = a_{n+1}(b_{n+1} - b_n) + b_n(a_{n+1} - a_n) = a_{n+1}\Delta b_n + b_n \Delta a_n$ **21. a)** Let $G(x) = \sum_{n=0}^{\infty} a_n x^n$. Then $G'(x) = \sum_{n=1}^{\infty} n a_n x^{n-1} = \sum_{n=0}^{\infty}(n + 1)a_{n+1}x^n$. Therefore, $G'(x) - G(x) = \sum_{n=0}^{\infty}[(n+1)a_{n+1} - a_n]x^n = \sum_{n=0}^{\infty} x^n/n! = e^x$, as desired. That $G(0) = a_0 = 1$ is given.   **b)** We have $[e^{-x}G(x)]' = e^{-x}G'(x) - e^{-x}G(x) = e^{-x}[G'(x) - G(x)] = e^{-x} \cdot e^x = 1$. Hence, $e^{-x}G(x) = x + c$, where c is a constant. Consequently, $G(x) = xe^x + ce^x$. Because $G(0) = 1$, it follows that $c = 1$.   **c)** We have $G(x) = \sum_{n=0}^{\infty} x^{n+1}/n! + \sum_{n=0}^{\infty} x^n/n! = \sum_{n=1}^{\infty} x^n/(n - 1)! + \sum_{n=0}^{\infty} x^n/n!$. Therefore, $a_n = 1/(n - 1)! + 1/n!$ for all $n \geq 1$, and $a_0 = 1$.   **23.** 7 **25.** 110 **27.** 0 **29. a)** 19 **b)** 65 **c)** 122 **d)** 167 **e)** 168 **31.** $D_{n-1}/(n - 1)!$ **33.** $11/32$

CHAPTER 9

Section 9.1

1. a) $\{(0, 0), (1, 1), (2, 2), (3, 3)\}$ **b)** $\{(1, 3), (2, 2), (3, 1), (4, 0)\}$ **c)** $\{(1, 0), (2, 0), (2, 1), (3, 0), (3, 1), (3, 2), (4, 0), (4, 1), (4, 2), (4, 3)\}$ **d)** $\{(1, 0), (1, 1), (1, 2), (1, 3), (2, 0), (2, 2), (3, 0), (3, 3), (4, 0)\}$ **e)** $\{(0, 1), (1, 0), (1, 1), (1, 2), (1, 3), (2, 1), (2, 3), (3, 1), (3, 2), (4, 1), (4, 3)\}$ **f)** $\{(1, 2), (2, 1), (2, 2)\}$ **3. a)** Reflexive, transitive   **b)** Symmetric   **c)** Symmetric   **d)** Symmetric   **5.** $xRy \rightarrow yRz$ is true as the premise is false. $xRy \wedge yRz \rightarrow xRz$ is true as the premise is false. Hence $R = \emptyset$ is symmetric and transitive. xRX does not hold for any x. Hence R is not reflexive.   **7.** (c), (d), (f)   **9.** Yes, for instance $\{(1, 1)\}$ on $\{1, 2\}$ **11.** d only   **13.** $\forall a \forall b\,[(a, b) \in R \rightarrow (b, a) \notin R]$ **15.** 2^{mn} **17. a)** $\{(a, b) \mid b$ divides $a\}$ **b)** $\{(a, b) \mid a$ does not divide $b\}$ **19.** The graph of f^{-1} **21. a)** $\{(a, b) \mid a$ is required to read or has read $b\}$ **b)** $\{(a, b) \mid a$ is required to read and has read $b\}$ **c)** $\{(a, b) \mid$ either a is required to read b but has not read it or a has read b but is not required to\}   **d)** $\{(a, b) \mid a$ is

required to read b but has not read it\}   **e)** $\{(a, b) \mid a$ has read b but is not required to\}   **23.** $S \circ R = \{(a, b) \mid a$ is a parent of b and b has a sibling\}, $R \circ S = \{(a, b) \mid a$ is an aunt or uncle of $b\}$ **25. a)** R_1 **b)** R_2 **c)** R_3 **d)** R^2 **e)** R_3 **f)** R^2 **g)** R^2 **h)** R^2 **27.** b got his or her doctorate under someone who got his or her doctorate under a; there is a sequence of $n + 1$ people, starting with a and ending with b, such that each is the advisor of the next person in the sequence   **29. a)** $\{(a, b) \mid a - b \equiv 0, 3, 4, 6, , 8,$ or $9 \pmod{12}\}$ **b)** $\{(a, b) \mid a \equiv b \pmod{12}\}$ **c)** $\{(a, b) \mid a - b \equiv 3, 6,$ or $9 \pmod{12}\}$ **d)** $\{(a, b) \mid a - b \equiv 4$ or $8 \pmod{12}\}$ **e)** $\{(a, b) \mid a - b \equiv 3, 4, 6, 8,$ or $9 \pmod{12}\}$ **31. a)** $65{,}536$ **b)** $32{,}768$ **33. a)** $2^{n(n+1)/2}$ **b)** $2^n 3^{n(n-1)/2}$ **c)** $3^{n(n-1)/2}$ **d)** $2^{n(n-1)}$ **e)** $2^{n(n-1)/2}$ **f)** $2^{n^2} - 2 \cdot 2^{n(n-1)}$ **35.** There may be no such b.   **37.** If R is symmetric and $(a, b) \in R$, then $(b, a) \in R$, so $(a, b) \in R^{-1}$. Hence, $R \subseteq R^{-1}$. Similarly, $R^{-1} \subseteq R$. So $R = R^{-1}$. Conversely, if $R = R^{-1}$ and $(a, b) \in R$, then $(a, b) \in R^{-1}$, so $(b, a) \in R$. Thus R is symmetric.   **39.** R is reflexive if and only if $(a, a) \in R$ for all $a \in A$ if and only if $(a, a) \in R^{-1}$ [because $(a, a) \in R$ if and only if $(a, a) \in R^{-1}$] if and only if R^{-1} is reflexive.   **41.** Use mathematical induction. The result is trivial for $n = 1$. Assume R^n is reflexive and transitive. By Theorem 1, $R^{n+1} \subseteq R$. To see that $R \subseteq R^{n+1} = R^n \circ R$, let $(a, b) \in R$. By the inductive hypothesis, $R^n = R$ and hence, is reflexive. Thus $(b, b) \in R^n$. Therefore $(a, b) \in R^{n+1}$.   **43.** Use mathematical induction. The result is trivial for $n = 1$. Assume R^n is reflexive. Then $(a, a) \in R^n$ for all $a \in A$ and $(a, a) \in R$. Thus $(a, a) \in R^n \circ R = R^{n+1}$ for all $a \in A$.   **45.** No, for instance, take $R = \{(1, 2), (2, 1)\}$.

Section 9.2

1. $\{(1, 2, 3), (1, 2, 4), (1, 3, 4), (2, 3, 4)\}$ **3.** Airline and flight number, airline and departure time   **5. a)** Yes   **b)** No   **c)** No   **7.** (Nadir, 122, 34, Detroit, 08 : 10), (Nadir, 199, 13, Detroit, 08 : 47), (Nadir, 322, 34, Detroit, 09 : 44)   **9.** (Nadir, 122, 34, Detroit, 08 : 10), (Nadir, 199, 13, Detroit, 08 : 47), (Nadir, 322, 34, Detroit, 09 : 44), (Acme, 221, 22, Denver, 08 : 17), (Acme, 222, 22, Denver, 09 : 10)

11.

Airline	Destination
Nadir	Detroit
Acme	Denver
Acme	Anchorage
Acme	Honolulu

13.

Supplier	Part_number	Project	Quantity	Color_code
23	1092	1	2	2
23	1101	3	1	1
23	9048	4	12	2
31	4975	3	6	2
31	3477	2	25	2
32	6984	4	10	1
32	9191	2	80	4
33	1001	1	14	8

15. Both sides of this equation pick out the subset of R consisting of those n-tuples satisfying both conditions C_1 and C_2.
17. Both sides of this equation pick out the m-tuples consisting of i_1th, i_2th, $\ldots$, i_mth components of n-tuples in either R or S.
19. a) J_2 followed by $P_{1,3}$ **b)** (23, 1), (23, 3), (31, 3), (32, 4)
21. There is no primary key.

Section 9.3

1. a) $\begin{bmatrix} 1 & 1 & 1 \\ 0 & 0 & 0 \\ 0 & 0 & 0 \end{bmatrix}$ **b)** $\begin{bmatrix} 0 & 1 & 0 \\ 1 & 1 & 0 \\ 0 & 0 & 1 \end{bmatrix}$
c) $\begin{bmatrix} 1 & 1 & 1 \\ 0 & 1 & 1 \\ 0 & 0 & 1 \end{bmatrix}$ **d)** $\begin{bmatrix} 0 & 0 & 1 \\ 0 & 0 & 0 \\ 1 & 0 & 0 \end{bmatrix}$

3. The relation is irreflexive if and only if the main diagonal of the matrix contains only 0s. **5. a)** 4950 **b)** 9900 **c)** 99
d) 100 **e)** 1 **7.** Change each 0 to a 1 and each 1 to a 0.

9. a) $\begin{bmatrix} 0 & 1 & 1 \\ 1 & 1 & 0 \\ 1 & 0 & 1 \end{bmatrix}$ **b)** $\begin{bmatrix} 1 & 0 & 0 \\ 0 & 0 & 1 \\ 0 & 1 & 0 \end{bmatrix}$ **c)** $\begin{bmatrix} 1 & 1 & 1 \\ 1 & 1 & 1 \\ 1 & 1 & 1 \end{bmatrix}$

11. a) $\begin{bmatrix} 0 & 0 & 1 \\ 1 & 1 & 0 \\ 0 & 1 & 1 \end{bmatrix}$ **b)** $\begin{bmatrix} 1 & 1 & 0 \\ 0 & 1 & 1 \\ 1 & 1 & 1 \end{bmatrix}$ **c)** $\begin{bmatrix} 0 & 1 & 1 \\ 1 & 1 & 1 \\ 1 & 1 & 1 \end{bmatrix}$

13. $n^2 - k$

15. $\{(a, b), (a, c), (b, c), (c, b)\}$ **17.** $\{(a, a), (a, b), (a, c), (b, a), (b, b), (b, c), (c, a), (c, b), (d, d)\}$ **19.** The relation is asymmetric if and only if the directed graph has no loops and no closed paths of length 2. **21.** Exercise 15: irreflexive. Exercise 16: irreflexive, antisymmetric. **23.** Reverse the direction on every edge in the digraph for R. **25.** Proof by mathematical induction. *Basis step:* Trivial for $n = 1$. *Inductive step:* Assume true for k. Because $R^{k+1} = R^k \circ R$, its matrix is $\mathbf{M}_R \odot \mathbf{M}_{R^k}$. By the inductive hypothesis this is $\mathbf{M}_R \odot \mathbf{M}_R^{[k]} = \mathbf{M}_R^{[k+1]}$.

Section 9.4

1. a) $\{(0, 0), (0, 1), (1, 1), (1, 2), (2, 0), (2, 2), (3, 0), (3, 3)\}$
b) $\{(0, 1), (0, 2), (0, 3), (1, 0), (1, 1), (1, 2), (2, 0), (2, 1), (2, 2), (3, 0)\}$ **3.** $\{(a, b) \mid a$ divides b or b divides $a\}$

5. 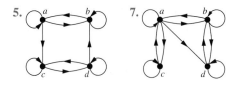 **7.**

9. a) 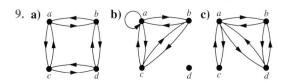 **b)** **c)**

11. a) 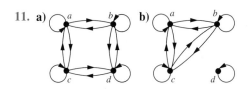 **b)**

c)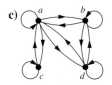

13. The symmetric closure of R is $R \cup R^{-1}$. $\mathbf{M}_{R \cup R^{-1}} = \mathbf{M}_R \vee \mathbf{M}_{R^{-1}} = \mathbf{M}_R \vee \mathbf{M}_R^t$. **15.** Only when R is irreflexive, in which case it is its own closure. **17.** a, a, a, a; a, b, e, a; a, d, e, a; b, c, c, b; b, e, a, b; c, b, c, c; c, c, c, b, c; c, c, c, c; d, e, a, d; d, e, e, d; e, a, b, e; e, a, d, e; e, d, e, e; e, e, d, e; e, e, e, e **19. a)** $\{(1, 1), (1, 5), (2, 3), (3, 1), (3, 2), (3, 3), (3, 4), (4, 1), (4, 5), (5, 3), (5, 4)\}$ **b)** $\{(1, 1), (1, 2), (1, 3), (1, 4), (2, 1), (2, 5), (3, 1), (3, 3), (3, 4), (3, 5), (4, 1), (4, 2), (4, 3), (4, 4), (5, 1), (5, 3), (5, 5)\}$ **c)** $\{(1, 1), (1, 3), (1, 4), (1, 5), (2, 1), (2, 2), (2, 3), (2, 4), (3, 1), (3, 2), (3, 3), (3, 5), (4, 1), (4, 3), (4, 4), (4, 5), (5, 1), (5, 2), (5, 3), (5, 4), (5, 5)\}$
d) $\{(1, 1), (1, 2), (1, 3), (1, 4), (1, 5), (2, 1), (2, 3), (2, 4), (2, 5), (3, 1), (3, 2), (3, 3), (3, 4), (3, 5), (4, 1), (4, 2), (4, 3), (4, 4), (4, 5), (5, 1), (5, 2), (5, 3), (5, 4), (5, 5)\}$ **e)** $\{(1, 1), (1, 2), (1, 3), (1, 4), (1, 5), (2, 1), (2, 2), (2, 3), (2, 4), (2, 5), (3, 1), (3, 2), (3, 3), (3, 4), (3, 5), (4, 1), (4, 2), (4, 3), (4, 4), (4, 5), (5, 1), (5, 2), (5, 3), (5, 4), (5, 5)\}$ **f)** $\{(1, 1), (1, 2), (1, 3), (1, 4), (1, 5), (2, 1), (2, 2), (2, 3), (2, 4), (2, 5), (3, 1), (3, 2), (3, 3), (3, 4), (3, 5), (4, 1), (4, 2), (4, 3), (4, 4), (4, 5), (5, 1), (5, 2), (5, 3), (5, 4), (5, 5)\}$ **21. a)** If there is a student c who shares a class with a and a class with b **b)** If there are two students c and d such that a and c share a class, c and d share a class, and d and b share a class **c)** If there is a sequence $s_0, \ldots, s_n$ of students with $n \geq 1$ such that $s_0 = a$, $s_n = b$, and for each $i = 1, 2, \ldots, n$, s_i and

s_{i-1} share a class **23.** The result follows from $(R^*)^{-1} = \left(\bigcup_{n=1}^{\infty} R^n\right)^{-1} = \bigcup_{n=1}^{\infty} (R^n)^{-1} = \bigcup_{n=1}^{\infty} R^n = R^*.$

25. a)
$$\begin{bmatrix} 1 & 1 & 1 & 1 \\ 1 & 1 & 1 & 1 \\ 1 & 1 & 1 & 1 \\ 1 & 1 & 1 & 1 \end{bmatrix}$$
b)
$$\begin{bmatrix} 0 & 0 & 0 & 0 \\ 1 & 0 & 1 & 1 \\ 1 & 0 & 1 & 1 \\ 1 & 0 & 1 & 1 \end{bmatrix}$$

c)
$$\begin{bmatrix} 0 & 1 & 1 & 1 \\ 0 & 0 & 1 & 1 \\ 0 & 0 & 0 & 1 \\ 0 & 0 & 0 & 0 \end{bmatrix}$$
d)
$$\begin{bmatrix} 1 & 1 & 1 & 1 \\ 1 & 1 & 1 & 1 \\ 1 & 1 & 1 & 1 \\ 1 & 1 & 1 & 1 \end{bmatrix}$$

27. Answers same as for Exercise 25. **29. a)** $\{(1, 1), (1, 2), (1, 4), (2, 2), (3, 3), (4, 1), (4, 2), (4, 4)\}$ **b)** $\{(1, 1), (1, 2), (1, 4), (2, 1), (2, 2), (2, 4), (3, 3), (4, 1), (4, 2), (4, 4)\}$ **c)** $\{(1, 1), (1, 2), (1, 4), (2, 1), (2, 2), (2, 4), (3, 3), (4, 1), (4, 2), (4, 4)\}$ **31.** Algorithm 1: $O(n^{3.8})$; Algorithm 2: $O(n^3)$

Section 9.5

1. a) Equivalence relation **b)** Not reflexive, not transitive **c)** Equivalence relation **d)** Not transitive **e)** Not symme- tric, not transitive **3. a)** Equivalence relation **b)** Not transitive **c)** Not reflexive, not symmetric, not transitive **d)** Equivalence relation **e)** Not reflexive, not transitive **5.** Many answers are possible. (1) Two buildings are equivalent if they were opened during the same year; an equivalence class consists of the set of buildings opened in a given year (as long as there was at least one building opened that year). (2) Two buildings are equivalent if they have the same number of stories; the equivalence classes are the set of 1-story buildings, the set of 2-story buildings, and so on (one class for each n for which there is at least one n-story building). (3) Every building in which you have a class is equivalent to every building in which you have a class (including itself), and every building in which you don't have a class is equivalent to every building in which you don't have a class (including itself); there are two equivalence classes—the set of buildings in which you have a class and the set of buildings in which you don't have a class (assuming these are nonempty). **7. a)** $(x, x) \in R$ because $f(x) = f(x)$. Hence, R is reflexive. $(x, y) \in R$ if and only if $f(x) = f(y)$, which holds if and only if $f(y) = f(x)$ if and only if $(y, x) \in R$. Hence, R is symmetric. If $(x, y) \in R$ and $(y, z) \in R$, then $f(x) = f(y)$ and $f(y) = f(z)$. Hence, $f(x) = f(z)$. Thus, $(x, z) \in R$. It follows that R is transitive. **b)** The sets $f^{-1}(b)$ for b in the range of f **9.** Let x be a bit string of length 3 or more. Because x agrees with itself in the first three bits, $(x, x) \in R$. Hence, R is reflexive. Suppose that $(x, y) \in R$. Then x and y agree in the first three bits. Hence, y and x agree in the first three bits. Thus, $(y, x) \in R$. If (x, y) and (y, z) are in R, then x and y agree in the first three bits, as do y and z. Hence, x and z agree in the first three bits. Hence, $(x, z) \in R$. It follows that R is transitive. **11.** This follows from Exercise 7, where f is the function that takes a bit string of length

3 or more to the ordered pair with its first bit as the first component and the third bit as its second component. **13.** For reflexivity, $((a, b), (a, b)) \in R$ because $a + b = b + a$. For symmetry, if $((a, b), (c, d)) \in R$, then $a + d = b + c$, so $c + b = d + a$, so $((c, d), (a, b)) \in R$. For transitivity, if $((a, b), (c, d)) \in R$ and $((c, d), (e, f)) \in R$, then $a + d = b + c$ and $c + e = d + f$, so $a + d + c + e = b + c + d + f$, so $a + e = b + f$, so $((a, b), (e, f)) \in R$. An easier solution is to note that by algebra, the given condition is the same as the condition that $f((a, b)) = f((c, d))$, where $f((x, y)) = x - y$; therefore by Exercise 7 this is an equivalence relation. **15.** This follows from Exercise 7, where the function f from the set of all URLs to the set of all Web pages is the function that assigns to each URL the Web page for that URL. **17.** No **19.** No **21.** R is reflexive because a bit string s has the same number of 1s as itself. R is symmetric because s and t having the same number of 1s implies that t and s do. R is transitive because s and t having the same number of 1s, and t and u having the same number of 1s implies that s and u have the same number of 1s. **23. a)** The sets of people of the same age **b)** The sets of people with the same two parents **25. a)** The set of all bit strings of length 3 **b)** The set of all bit strings of length 4 that end with a 1 **c)** The set of all bit strings of length 5 that end 11 **d)** The set of all bit strings of length 8 that end 10101 **27. a)** $[2]_5 = \{i \mid i \equiv 2 \pmod 5\} = \{\ldots, -8, -3, 2, 7, 12, \ldots\}$ **b)** $[3]_5 = \{i \mid i \equiv 3 \pmod 5\} = \{\ldots, -7, -2, 3, 8, 13, \ldots\}$ **c)** $[6]_5 = \{i \mid i \equiv 6 \pmod 5\} = \{\ldots, -9, -4, 1, 6, 11, \ldots\}$ **d)** $[-3]_5 = \{i \mid i \equiv -3 \pmod 5\} = \{\ldots, -8, -3, 2, 7, 12, \ldots\}$ **29.** $\{6n + k \mid n \in \mathbf{Z}\}$ for $k \in \{0, 1, 2, 3, 4, 5\}$ **31. a)** $[(1, 2)] = \{(a, b) \mid a - b = -1\} = \{(1, 2), (3, 4), (4, 5), (5, 6), \ldots\}$ **b)** Each equivalence class can be interpreted as an integer (negative, positive, or zero); specifically, $[(a, b)]$ can be interpreted as $a - b$. **33.** (a), (c), (e) **35.** (b), (d), (e) **37. a)** $\{(0, 0), (1, 1), (1, 2), (2, 1), (2, 2), (3, 3), (3, 4), (3, 5), (4, 3), (4, 4), (4, 5), (5, 3), (5, 4), (5, 5)\}$ **b)** $\{(0, 0), (0, 1), (1, 0), (1, 1), (2, 2), (2, 3), (3, 2), (3, 3), (4, 4), (4, 5), (5, 4), (5, 5)\}$ **c)** $\{(0, 0), (0, 1), (0, 2), (1, 0), (1, 1), (1, 2), (2, 0), (2, 1), (2, 2), (3, 3), (3, 4), (3, 5), (4, 3), (4, 4), (4, 5), (5, 3), (5, 4), (5, 5)\}$ **d)** $\{(0, 0), (1, 1), (2, 2), (3, 3), (4, 4), (5, 5)\}$ **39.** Let A be a set in the first partition. Pick a particular element x of A. The set of all bit strings of length 16 that agree with x on the last four bits is one of the sets in the second partition, and clearly every string in A is in that set. **41.** We claim that each equivalence class $[x]_{R_{31}}$ is a subset of the equivalence class $[x]_{R_8}$. To show this, choose an arbitrary element $y \in [x]_{R_{31}}$. Then y is equivalent to x under R_{31}, so either $y = x$ or y and x are each at least 31 characters long and agree on their first 31 characters. Because strings that are at least 31 characters long and agree on their first 31 characters perforce are at least 8 characters long and agree on their first 8 characters, we know that either $y = x$ or y and x are each at least 8 characters long and agree on their first 8 characters. This means that y is equivalent to x under R_8, so $y \in [x]_{R_8}$. **43.** $\{(a, a), (a, b), (a, c), (b, a), (b, b), (b, c), (c, a), (c, b), (c, c), (d, d), (d, e), (e, d),$

$(e, e)\}$ **45. a) Z   b)** $\{n + \frac{1}{2} \mid n \in \mathbf{Z}\}$ **47. a)** R is reflexive because any coloring can be obtained from itself via a 360-degree rotation. To see that R is symmetric and transitive, use the fact that each rotation is the composition of two reflections and conversely the composition of two reflections is a rotation. Hence, (C_1, C_2) belongs to R if and only if C_2 can be obtained from C_1 by a composition of reflections. So if (C_1, C_2) belongs to R, so does (C_2, C_1) because the inverse of the composition of reflections is also a composition of reflections (in the opposite order). Hence, R is symmetric. To see that R is transitive, suppose (C_1, C_2) and (C_2, C_3) belong to R. Taking the composition of the reflections in each case yields a composition of reflections, showing that (C_1, C_3) belongs to R.   **b)** We express colorings with sequences of length four, with r and b denoting red and blue, respectively. We list letters denoting the colors of the upper left square, upper right square, lower left square, and lower right square, in that order. The equivalence classes are: $\{rrrr\}$, $\{bbbb\}$, $\{rrrb, rrbr, rbrr, brrr\}$, $\{bbbr, bbrb, brbb, rbbb\}$, $\{rbbr, brrb\}$, $\{rrbb, brbr, bbrr, rbrb\}$.   **49.** Yes   **51.** R **53.** First form the reflexive closure of R, then form the symmetric closure of the reflexive closure, and finally form the transitive closure of the symmetric closure of the reflexive closure.

Section 9.6

1. a) Is a partial ordering   **b)** Not antisymmetric, not transitive   **c)** Is a partial ordering   **d)** Is a partial ordering   **e)** Not antisymmetric, not transitive   **3. a)** Yes   **b)** No   **c)** Yes   **d)** No   **5. a)** No   **b)** Yes   **c)** No   **7.** Yes   **9. a)** $\{(0, 0), (1, 0), (1, 1), (2, 0), (2, 1), (2, 2)\}$ **b)** $(\mathbf{Z}, \leq)$ **c)** $(P(\mathbf{Z}), \subseteq)$ **d)** $(\mathbf{Z}^+,$ "is a multiple of")   **11. a)** $\{0\}$ and $\{1\}$, for instance   **b)** 4 and 6, for instance   **13. a)** $(1, 1, 2) <$ $(1, 2, 1)$ **b)** $(0, 1, 2, 3) < (0, 1, 3, 2)$ **c)** $(0, 1, 1, 1, 0) <$ $(1, 0, 1, 0, 1)$ **15.** $0 < 0001 < 001 < 01 <$ $010 < 0101 < 011 < 11$ **17.** $(a, b), (a, c), (a, d),$ $(b, c), (b, d), (a, a), (b, b), (c, c), (d, d)$ **19.** $(\emptyset, \{a\}),$ $(\emptyset, \{b\}), (\emptyset, \{c\}), (\{a\}, \{a, b\}), (\{a\}, \{a, c\}), (\{b\}, \{a, b\}),$ $(\{b\}, \{b, c\}), (\{c\}, \{a, c\}), (\{c\}, \{b, c\}), (\{a, b\}, \{a, b, c\}),$ $(\{a, c\}, \{a, b, c\})(\{b, c\}, \{a, b, c\})$ **21.** Let $(S, \preccurlyeq)$ be a finite poset. We will show that this poset is the reflexive transitive closure of its covering relation. Suppose that (a, b) is in the reflexive transitive closure of the covering relation. Then $a = b$ or $a \prec b$, so $a \preccurlyeq b$, or else there is a sequence $a_1, a_2, \ldots, a_n$ such that $a \prec a_1 \prec a_2 \prec \cdots \prec a_n \prec b$, in which case again $a \preccurlyeq b$ by the transitivity of $\preccurlyeq$. Conversely, suppose that $a \prec b$. If $a = b$ then (a, b) is in the reflexive transitive closure of the covering relation. If $a \prec b$ and there is no z such that $a \prec z \prec b$, then (a, b) is in the covering relation and therefore in its reflexive transitive closure. Otherwise, let $a \prec a_1 \prec a_2 \prec \cdots \prec a_n \prec b$ be a longest possible sequence of this form (which exists because the poset is finite). Then no intermediate elements can be inserted, so each pair $(a, a_1), (a_1, a_2), \ldots, (a_n, b)$ is in the covering relation, so again (a, b) is in its reflexive transitive closure.

23. a) 24, 45   **b)** 3, 5   **c)** No   **d)** No   **e)** 15, 45   **f)** 15   **g)** 15, 5, 3   **h)** 15   **25.** Because $(a, b) \preccurlyeq (a, b), \preccurlyeq$ is reflexive. If $(a_1, a_2) \preccurlyeq (b_1, b_2)$ and $(a_1, a_2) \neq (b_1, b_2)$, either $a_1 \prec b_1$, or $a_1 = b_1$ and $a_2 \prec b_2$. In either case, (b_1, b_2) is not less than or equal to (a_1, a_2). Hence, $\preccurlyeq$ is antisymmetric. Suppose that $(a_1, a_2) \prec (b_1, b_2) \prec (c_1, c_2)$. Then if $a_1 \prec b_1$ or $b_1 \prec c_1$, we have $a_1 \prec c_1$, so $(a_1, a_2) \prec (c_1, c_2)$, but if $a_1 = b_1 = c_1$, then $a_2 \prec b_2 \prec c_2$, which implies that $(a_1, a_2) \prec (c_1, c_2)$. Hence, $\preccurlyeq$ is transitive.   **27.** Because $(s, t) \preccurlyeq (s, t), \preccurlyeq$ is reflexive. If $(s, t) \preccurlyeq (u, v)$ and $(u, v) \preccurlyeq (s, t)$, then $s \preccurlyeq u \preccurlyeq s$ and $t \preccurlyeq v \preccurlyeq t$; hence, $s = u$ and $t = v$. Hence, $\preccurlyeq$ is antisymmetric. Suppose that $(s, t) \preccurlyeq (u, v) \preccurlyeq (w, x)$. Then $s \preccurlyeq u, t \preccurlyeq v, u \preccurlyeq w,$ and $v \preccurlyeq x$. It follows that $s \preccurlyeq w$ and $t \preccurlyeq x$. Hence, $(s, t) \preccurlyeq (w, x)$. Hence, $\preccurlyeq$ is transitive.   **29. a)** Suppose that x is maximal and that y is the largest element. Then $x \preccurlyeq y$. Because x is not less than y, it follows that $x = y$. By Exercise 28(a) y is unique. Hence, x is unique.   **b)** Suppose that x is minimal and that y is the smallest element. Then $x \succcurlyeq y$. Because x is not greater than y, it follows that $x = y$. By Exercise 28(b) y is unique. Hence, x is unique.   **31. a)** Yes   **b)** No   **c)** Yes   **33.** Use mathematical induction. Let $P(n)$ be "Every subset with n elements from a lattice has a least upper bound and a greatest lower bound." *Basis step:* $P(1)$ is true because the least upper bound and greatest lower bound of $\{x\}$ are both x. *Inductive step:* Assume that $P(k)$ is true. Let S be a set with $k + 1$ elements. Let $x \in S$ and $S' = S - \{x\}$. Because S' has k elements, by the inductive hypothesis, it has a least upper bound y and a greatest lower bound a. Now because we are in a lattice, there are elements $z = \text{lub}(x, y)$ and $b = \text{glb}(x, a)$. We are done if we can show that z is the least upper bound of S and b is the greatest lower bound of S. To show that z is the least upper bound of S, first note that if $w \in S$, then $w = x$ or $w \in S'$. If $w = x$ then $w \preccurlyeq z$ because z is the least upper bound of x and y. If $w \in S'$, then $w \preccurlyeq z$ because $w \preccurlyeq y$, which is true because y is the least upper bound of S', and $y \preccurlyeq z$, which is true because $z = \text{lub}(x, y)$. To see that z is the least upper bound of S, suppose that u is an upper bound of S. Note that such an element u must be an upper bound of x and y, but because $z = \text{lub}(x, y)$, it follows that $z \preccurlyeq u$. We omit the similar argument that b is the greatest lower bound of S.   **35. a)** No   **b)** Yes   **c)** (*Proprietary,* {*Cheetah, Puma*}), (*Restricted,* {*Cheetah, Puma*}), (*Registered,* {*Cheetah, Puma*}), (*Proprietary,* {*Cheetah, Puma, Impala*}), (*Restricted,* {*Cheetah, Puma, Impala*}), (*Registered,* {*Cheetah, Puma, Impala*})   **d)** (*Non- proprietary,* {*Impala, Puma*}), (*Proprietary,* {*Impala, Puma*}), (*Restricted,* {*Impala, Puma*}), (*Nonproprietary,* {*Impala*}), (*Proprietary,* {*Impala*}), (*Restricted,* {*Impala*}), (*Nonproprietary,* {*Puma*}), (*Proprietary,* {*Puma*}), (*Restricted,* {*Puma*}), (*Nonproprietary,* $\emptyset$), (*Proprietary,* $\emptyset$), (*Restricted,* $\emptyset$)   **37.** Let Π be the set of all partitions of a set S with $P_1 \preccurlyeq P_2$ if P_1 is a refinement of P_2, that is, if every set in P_1 is a subset of a set in P_2. First, we show that $(\Pi, \preccurlyeq)$ is a poset. Because $P \preccurlyeq P$ for every partition P, $\preccurlyeq$ is reflexive. Now suppose that $P_1 \preccurlyeq P_2$ and $P_2 \preccurlyeq P_1$. Let $T \in P_1$. Because $P_1 \preccurlyeq P_2$, there is a set $T' \in P_2$ such that

$T \subseteq T'$. Because $P_2 \preccurlyeq P_1$ there is a set $T'' \in P_1$ such that $T' \subseteq T''$. It follows that $T \subseteq T''$. But because P_1 is a partition, $T = T''$, which implies that $T = T'$ because $T \subseteq T' \subseteq T''$. Thus, $T \in P_2$. By reversing the roles of P_1 and P_2 it follows that every set in P_2 is also in P_1. Hence, $P_1 = P_2$ and $\preccurlyeq$ is antisymmetric. Next, suppose that $P_1 \preccurlyeq P_2$ and $P_2 \preccurlyeq P_3$. Let $T \in P_1$. Then there is a set $T' \in P_2$ such that $T \subseteq T'$. Because $P_2 \preccurlyeq P_3$ there is a set $T'' \in P_3$ such that $T' \subseteq T''$. This means that $T \subseteq T''$. Hence, $P_1 \preccurlyeq P_3$. It follows that $\preccurlyeq$ is transitive. The greatest lower bound of the partitions P_1 and P_2 is the partition P whose subsets are the nonempty sets of the form $T_1 \cap T_2$ where $T_1 \in P_1$ and $T_2 \in P_2$. We omit the justification of this statement here. The least upper bound of the partitions P_1 and P_2 is the partition that corresponds to the equivalence relation in which $x \in S$ is related to $y \in S$ if there is a sequence $x = x_0, x_1, x_2, \ldots, x_n = y$ for some nonnegative integer n such that for each i from 1 to n, x_{i-1} and x_i are in the same element of P_1 or of P_2. We omit the details that this is an equivalence relation and the details of the proof that this is the least upper bound of the two partitions. **39.** By Exercise 33 there is a least upper bound and a greatest lower bound for the entire finite lattice. By definition these elements are the greatest and least elements, respectively. **41.** The least element of a subset of $\mathbf{Z}^+ \times \mathbf{Z}^+$ is that pair that has the smallest possible first coordinate, and, if there is more than one such pair, that pair among those that has the smallest second coordinate. **43.** If x is an integer in a decreasing sequence of elements of this poset, then at most $|x|$ elements can follow x in the sequence, namely, integers whose absolute values are $|x| - 1$, $|x| - 2$, , 1, 0. Therefore there can be no infinite decreasing sequence. This is not a totally ordered set, because 5 and -5, for example, are incomparable. **45.** To find which of two rational numbers is larger, write them with a positive common denominator and compare numerators. To show that this set is dense, suppose that $x < y$ are two rational numbers. Then their average, i.e., $(x + y)/2$, is a rational number between them. **47.** $a \prec_t b \prec_t c \prec_t d \prec_t e \prec_t f \prec_t g \prec_t h \prec_t i \prec_t j \prec_t k \prec_t l \prec_t m$ **49.** $A \prec C \prec E \prec B \prec D \prec F \prec G$, $A \prec E \prec C \prec B \prec D \prec F \prec G$, $C \prec A \prec E \prec B \prec D \prec F \prec G$, $C \prec E \prec A \prec B \prec D \prec F \prec G$, $E \prec A \prec C \prec B \prec D \prec F \prec G$, $E \prec C \prec A \prec B \prec D \prec F \prec G$, $A \prec C \prec B \prec E \prec D \prec F \prec G$, $C \prec A \prec B \prec E \prec D \prec F \prec G$, $A \prec C \prec B \prec D \prec E \prec F \prec G$, $C \prec A \prec B \prec D \prec E \prec F \prec G$, $A \prec C \prec E \prec B \prec F \prec D \prec G$, $A \prec E \prec C \prec B \prec F \prec D \prec G$, $C \prec A \prec E \prec B \prec F \prec D \prec G$, $C \prec E \prec A \prec B \prec F \prec D \prec G$, $E \prec A \prec C \prec B \prec F \prec D \prec G$, $E \prec C \prec A \prec B \prec F \prec D \prec G$, $C \prec A \prec B \prec E \prec F \prec D \prec G$ **51.** Determine user needs $\prec$ Write functional requirements $\prec$ Set up test sites $\prec$ Develop system requirements $\prec$ Write documentation $\prec$ Develop module A $\prec$ Develop module B $\prec$ Develop module C $\prec$ Integrate modules $\prec$ α test $\prec$ β test $\prec$ Completion

Supplementary Exercises

1. a) Irreflexive (we do not include the empty string), symmetric **b)** Irreflexive, symmetric **c)** Irreflexive, antisymmetric, transitive **3.** Suppose that $(a, b) \in R$. Because $(b, b) \in R$ it follows that $(a, b) \in R^2$. **5.** Yes, yes **7.** Two records with identical keys in the projection would have identical keys in the original. **9.** $(\Delta \cup R)^{-1} = \Delta^{-1} \cup R^{-1} = \Delta \cup R^{-1}$ **11. a)** $R = \{(a, b), (a, c)\}$. The transitive closure of the symmetric closure of R is $\{(a, a), (a, b), (a, c), (b, a), (b, b), (b, c), (c, a), (c, b), (c, c)\}$ and is different from the symmetric closure of the transitive closure of R, which is $\{(a, b), (a, c), (b, a), (c, a)\}$. **b)** Suppose that (a, b) is in the symmetric closure of the transitive closure of R. We must show that (a, b) is in the transitive closure of the symmetric closure of R. We know that at least one of (a, b) and (b, a) is in the transitive closure of R. Hence, there is either a path from a to b in R or a path from b to a in R (or both). In the former case, there is a path from a to b in the symmetric closure of R. In the latter case, we can form a path from a to b in the symmetric closure of R by reversing the directions of all the edges in a path from b to a, going backward. Hence, (a, b) is in the transitive closure of the symmetric closure of R. **13.** The closure of S with respect to property **P** is a relation with property **P** that contains R because $R \subseteq S$. Hence, the closure of S with respect to property **P** contains the closure of R with respect to property **P**. **15.** Use the basic idea of Warshall's algorithm, except let $w_{ij}^{[k]}$ equal the length of the longest path from v_i to v_j using interior vertices with subscripts not exceeding k, and equal to -1 if there is no such path. To find $w_{ij}^{[k]}$ from the entries of $\mathbf{W}_{k-1}$, determine for each pair (i, j) whether there are paths from v_i to v_k and from v_k to v_j using no vertices labeled greater than k. If either $w_{ik}^{[k-1]}$ or $w_{kj}^{[k-1]}$ is -1, then such a pair of paths does not exist, so set $w_{ij}^{[k]} = w_{ij}^{[k-1]}$. If such a pair of paths exists, then there are two possibilities. If $w_{kk}^{[k-1]} > 0$, there are paths of arbitrary long length from v_i to v_j, so set $w_{ij}^{[k]} = \infty$. If $w_{kk}^{[k-1]} = 0$, set $w_{ij}^{[k-1]} = \max(w_{ij}^{[k-1]}, w_{ik}^{[k-1]} + w_{kj}^{[k-1]})$. (Initially take $\mathbf{W}_0 = \mathbf{M}_R$.) **17.** 25 **19.** Because $A_i \cap B_j$ is a subset of A_i and of B_j, the collection of subsets is a refinement of each of the given partitions. We must show that it is a partition. By construction, each of these sets is nonempty. To see that their union is S, suppose that $s \in S$. Because P_1 and P_2 are partitions of S, there are sets A_i and B_j such that $s \in A_i$ and $s \in B_j$. Therefore $s \in A_i \cap B_j$. Hence, the union of these sets is S. To see that they are pairwise disjoint, note that unless $i = i'$ and $j = j'$, $(A_i \cap B_j) \cap (A_{i'} \cap B_{j'}) = (A_i \cap A_{i'}) \cap (B_j \cap B_{j'}) = \emptyset$. **21.** The subset relation is a partial ordering on any collection of sets, because it is reflexive, antisymmetric, and transitive. Here the collection of sets is $\mathbf{R}(S)$. **23.** Find recipe $\prec$ Buy seafood $\prec$ Buy groceries $\prec$ Wash shellfish $\prec$ Cut ginger and garlic $\prec$ Clean fish $\prec$ Steam rice $\prec$ Cut fish $\prec$ Wash vegetables $\prec$ Chop water chestnuts $\prec$ Make garnishes $\prec$ Cook in wok $\prec$ Arrange on platter $\prec$ Serve **25. a)** The only antichain with more than one element is $\{c, d\}$. **b)** The only antichains with more than one element are $\{b, c\}$, $\{c, e\}$, and $\{d, e\}$. **c)** The only antichains with more than one element are $\{a, b\}$, $\{a, c\}$, $\{b, c\}$, $\{a, b, c\}$, $\{d, e\}$, $\{d, f\}$, $\{e, f\}$, and $\{d, e, f\}$. **27.** Let $(S, \preccurlyeq)$ be a finite poset, and let A be a maximal chain. Because $(A, \preccurlyeq)$ is also a poset it must have a minimal element m. Suppose that m is not minimal in S. Then there would be an element a of S with $a \prec m$. However, this would make the set $A \cup \{a\}$ a larger chain than A. To show this, we must show that a is comparable with every element of A. Because m is comparable with every element of A and m is minimal, it follows that $m \prec x$ when x is in A and $x \neq m$. Because $a \prec m$ and $m \prec x$, the transitive law shows that $a \prec x$ for every element of A. **29.** Let aRb denote that a is a descendant of b. By Exercise 28, if no set of $n + 1$ people none of whom is a descendant of any other (an antichain) exists, then $k \leq n$, so the set can be partitioned into $k \leq n$ chains. By the pigeonhole principle, at least one of these chains contains at least $m + 1$ people. **31.** We prove by contradiction that if S has no infinite decreasing sequence and $\forall x \left(\{\forall y [y \prec x \to P(y)]\} \to P(x) \right)$, then $P(x)$ is true for all $x \in S$. If it does not hold that $P(x)$ is true for all $x \in S$, let x_1 be an element of S such that $P(x_1)$ is not true. Then by the conditional statement already given, it must be the case that $\forall y[y \prec x_1 \to P(y)]$ is not true. This means that there is some x_2 with $x_2 \prec x_1$ such that $P(x_2)$ is not true. Again invoking the conditional statement, we get an $x_3 \prec x_2$ such that $P(x_3)$ is not true, and so on forever. This contradicts the well-foundedness of our poset. Therefore, $P(x)$ is true for all $x \in S$. **33. a)** Because $\text{glb}(x, y) = \text{glb}(y, x)$ and $\text{lub}(x, y) = \text{lub}(y, x)$, it follows that $x \wedge y = y \wedge x$ and $x \vee y = y \vee x$. **b)** Using the definition, $(x \wedge y) \wedge z$ is a lower bound of x, y, and z that is greater than every other lower bound. Because x, y, and z play interchangeable roles, $x \wedge (y \wedge z)$ is the same element. Similarly, $(x \vee y) \vee z$ is an upper bound of x, y, and z that is less than every other upper bound. Because x, y, and z play interchangeable roles, $x \vee (y \vee z)$ is the same element. **c)** To show that $x \wedge (x \vee y) = x$ it is sufficient to show that x is the greatest lower bound of x, and $x \vee y$. Note that x is a lower bound of x, and because $x \vee y$ is by definition greater than x, x is a lower bound for it as well. Therefore, x is a lower bound. But any lower bound of x has to be less than x, so x is the greatest lower bound. The second statement is the dual of the first; we omit its proof. **d)** x is a lower, and an upper, bound for itself and itself, and the greatest, and least, such bound. **35. a)** Because 1 is the only element greater than or equal to 1, it is the only upper bound for 1 and therefore the only possible value of the least upper bound of x and 1. **b)** Because $x \preccurlyeq 1$, x is a lower bound for both x and 1 and no other lower bound can be greater than x, so $x \wedge 1 = x$. **c)** Because $0 \preccurlyeq x$, x is an upper bound for both x and 0 and no other bound can be less than x, so $x \vee 0 = x$. **d)** Because 0 is the only element less than or equal to 0, it is the only lower bound for 0 and therefore the only possible value of the greatest lower bound of x and 0. **37.** The complement of a subset $X \subseteq S$ is its complement $S - X$. To prove this, note that $X \vee (S - X) = 1$ and $X \wedge (S - X) = 0$ because $X \cup (S - X) = S$ and $X \cap (S - X) = \emptyset$. **39.** Think

of the rectangular grid as representing elements in a matrix. Thus we number from top to bottom and within that from left to right. The partial order is that $(a, b) \preceq (c, d)$ iff $a \leq c$ and $b \leq d$. Note that $(1, 1)$ is the least element under this relation. The rules for Chomp as explained in Chapter 1 coincide with the rules stated in the preamble here. But now we can identify the point (a, b) with the natural number $p^{a-1} q^{b-1}$ for all a and b with $1 \leq a \leq m$ and $1 \leq b \leq n$. This identifies the points in the rectangular grid with the set S in this exercise, and the partial order $\preceq$ just described is the same as the divides relation, because $p^{a-1} q^{b-1} \mid p^{c-1} q^{d-1}$ if and only if the exponent of p on the left does not exceed the exponent of p on the right, and similarly for q.

CHAPTER 10

Section 10.1

1. a)

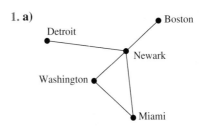

b)

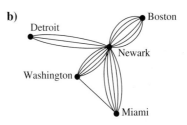

c)

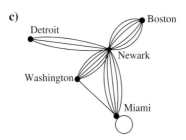

d)

e)

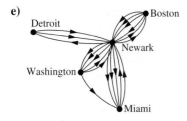

3. Simple graph   **5.** Directed graph   **7.** If uRv, then there is an edge associated with $\{u, v\}$. But $\{u, v\} = \{v, u\}$, so this edge is associated with $\{v, u\}$ and therefore vRu. Thus, by definition, R is a symmetric relation. A simple graph does not allow loops; therefore, uRu never holds, and so by definition R is irreflexive.

9. a)

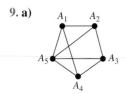

b)

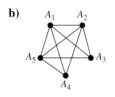

c)

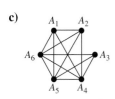

11.

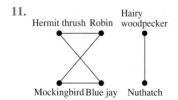

13.

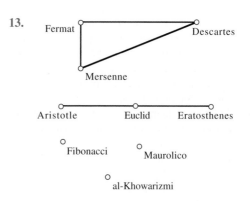

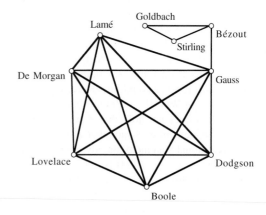

15.

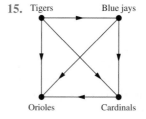

17. We find the telephone numbers in the call graph for February that are not present in the call graph for January and vice versa. For each number we find, we make a list of the numbers they called or were called by using the edges in the call graph. We examine these lists to find new telephone numbers in February that had similar calling patterns to defunct telephone numbers in January.   **19.** We use the graph model that has e-mail addresses as vertices and for each message sent, an edge from the e-mail address of the sender to the e-mail address of the recipient. For each e-mail address, we can make a list of other addresses they sent messages to and a list of other addresses from which they received messages. If two e-mail addresses had almost the same pattern, we conclude that these addresses might have belonged to the same person who had recently changed his or her e-mail address.   **21.** Let the set of vertices be a set of people, and two vertices are joined by an edge if the two people were ever married. Ignoring complications, this graph has the property that there are two types of vertices (men and women), and every edge joins vertices of opposite types.

23.

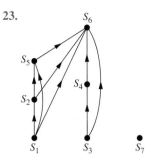

25. Represent people in the group by vertices. Put a directed edge into the graph for every pair of vertices. Label the edge from the vertex representing A to the vertex representing B with a + (plus) if A likes B, a − (minus) if A dislikes B, and a 0 if A is neutral about B.

Section 10.2

1. $v = 6$; $e = 6$; $\deg(a) = 2$, $\deg(b) = 4$, $\deg(c) = 1$, $\deg(d) = 0$, $\deg(e) = 2$, $\deg(f) = 3$; c is pendant; d is isolated.   **3.** Sum of the degrees = $2 + 4 + 3 + 2 + 1 = 12$. Number of edges = 6.   **5.** $v = 4$; $e = 7$; $\deg^-(a) = 3$, $\deg^-(b) = 1$, $\deg^-(c) = 2$, $\deg^-(d) = 1$, $\deg^+(a) = 1$, $\deg^+(b) = 2$, $\deg^+(c) = 1$, $\deg^+(d) = 3$ **7.** Exercise 5: Sum of in-degrees = $3 + 1 + 2 + 1 = 7$ Sum of out-digits = $1 + 2 + 1 + 3 = 7$ Number of edges = 7   **7.** Exercise 6: Sum of in-degrees = $2 + 3 + 2 + 1 = 8$ Sum of out-digits = $2 + 4 + 1 + 1 = 8$ Number of edges = 8   **9.** The number of coauthors that person has; that person's coauthors; a person who has no coauthors; a person who has only one coauthor   **11.** In the directed graph $\deg^-(v) =$ number of calls v received, $\deg^+(v) =$ number of calls v made; in the undirected graph, $\deg(v)$ is the number of calls either made or received by v.   **13.** $(\deg^+(v), \deg^-(v))$ is the win–loss record of v.   **15.** In the undirected graph model in which the vertices are people in the group and two vertices are adjacent if those two people are friends, the degree of a vertex is the number of friends in the group that person has. By Exercise 14, there are two vertices with the same degree, which means that there are two people in the group with the same number of friends in the group.   **17.** Bipartite   **19.** Not bipartite   **21.** Not bipartite   **23 a)**

Zamora
Agraharam
Smith
Chou
Macintyre

planning
publicity
sales
marketing
development
industry relations

b) (Agraharam, development) (Zamora, marketing) (Smith, sales) (Chou, industry relations) (Macintyre, planning) **c)** It is a complete matching. It is a maximum matching.   **25. a)** n vertices, $n(n − 1)/2$ edges   **b)** n vertices, n edges   **c)** $n + 1$ vertices, $2n$ edges   **d)** $m + n$ vertices, mn edges   **e)** 2^n vertices, $n2^{n-1}$ edges

27. a) $3, 3, 3, 3$ **b)** $2, 2, 2, 2$ **c)** $4, 3, 3, 3, 3$ **d)** $3, 3, 2, 2, 2$ **e)** $3, 3, 3, 3, 3, 3, 3, 3$

29. 7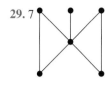

31. a) Yes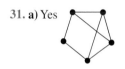

b) No **c)** No **d)** No

e) Yes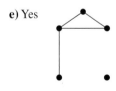

f) No.   **33.** Let d_1, d_2, d_n be a nonincreasing sequence of nonnegative integers with an even sum. Construct a graph as follows: Take vertices v_1, v_2, v_n and put $\lfloor d_i / 2 \rfloor$ loops at vertex v_i, for $i = 1, 2, \ldots, n$. For each i, vertex v_i now has degree either d_i or $d_i − 1$. Because the original sum was even, the number of vertices for which $\deg(v_i) = d_i − 1$ is even. Pair them up arbitrarily, and put in an edge joining the vertices in each pair.   **35.** 17

37.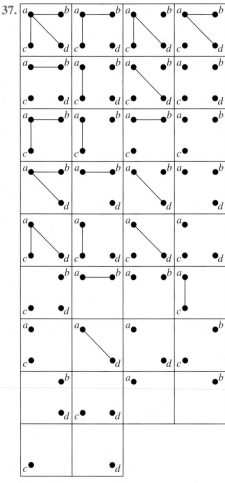

39. a) For all $n \geq 1$ **b)** For all $n \geq 3$ **c)** For $n = 3$ **d)** For all $n \geq 0$

41.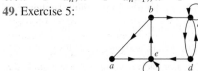

43. a) The graph with n vertices and no edges **b)** The disjoint union of K_m and K_n **c)** The graph with vertices $\{v_1, \ldots, v_n\}$ with an edge between v_i and v_j unless $i \equiv j \pm 1 \pmod{n}$ **d)** The graph whose vertices are represented by bit strings of length n with an edge between two vertices if the associated bit strings differ in more than one bit **45.** $v(v-1)/2 - e$
47. $n - 1 - d_n, n - 1 - d_{n-1}, n - 1 - d_2, n - 1 - d_1$
49. Exercise 5:

51. A directed graph $G = (V, E)$ is its own converse if and only if it satisfies the condition $(u, v) \in E$ if and only if $(v, u) \in E$. But this is precisely the condition that the associated relation must satisfy to be symmetric.

53.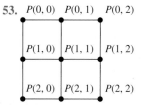

$P(0,0) \quad P(0,1) \quad P(0,2)$

$P(1,0) \quad P(1,1) \quad P(1,2)$

$P(2,0) \quad P(2,1) \quad P(2,2)$

55. We can connect $P(i, j)$ and $P(k, l)$ by using $|i - k|$ hops to connect $P(i, j)$ and $P(k, j)$ and $|j - l|$ hops to connect $P(k, j)$ and $P(k, l)$. Hence, the total number of hops required to connect $P(i, j)$ and $P(k, l)$ does not exceed $|i-k|+|j-l|$. This is less than or equal to $m + m = 2m$, which is $O(m)$.

Section 10.3

1.

Vertex	Adjacent Vertices
a	b, c, d
b	a, d
c	a, d
d	a, b, c

3. $\begin{bmatrix} 0 & 1 & 1 & 1 \\ 1 & 0 & 0 & 1 \\ 1 & 0 & 0 & 1 \\ 1 & 1 & 1 & 0 \end{bmatrix}$

5. a) $\begin{bmatrix} 0 & 1 & 1 & 1 \\ 1 & 0 & 1 & 1 \\ 1 & 1 & 0 & 1 \\ 1 & 1 & 1 & 0 \end{bmatrix}$

b) $\begin{bmatrix} 0 & 1 & 1 & 1 & 1 \\ 1 & 0 & 0 & 0 & 0 \\ 1 & 0 & 0 & 0 & 0 \\ 1 & 0 & 0 & 0 & 0 \\ 1 & 0 & 0 & 0 & 0 \end{bmatrix}$ **c)** $\begin{bmatrix} 0 & 0 & 1 & 1 & 1 \\ 0 & 0 & 1 & 1 & 1 \\ 1 & 1 & 0 & 0 & 0 \\ 1 & 1 & 0 & 0 & 0 \\ 1 & 1 & 0 & 0 & 0 \end{bmatrix}$

d) $\begin{bmatrix} 0 & 1 & 0 & 1 \\ 1 & 0 & 1 & 0 \\ 0 & 1 & 0 & 1 \\ 1 & 0 & 1 & 0 \end{bmatrix}$ **e)** $\begin{bmatrix} 0 & 1 & 0 & 1 & 1 \\ 1 & 0 & 1 & 0 & 1 \\ 0 & 1 & 0 & 1 & 1 \\ 1 & 0 & 1 & 0 & 1 \\ 1 & 1 & 1 & 1 & 0 \end{bmatrix}$

f) $\begin{bmatrix} 0 & 1 & 1 & 0 & 1 & 0 & 0 & 0 \\ 1 & 0 & 0 & 1 & 0 & 1 & 0 & 0 \\ 1 & 0 & 0 & 1 & 0 & 0 & 1 & 0 \\ 0 & 1 & 1 & 0 & 0 & 0 & 0 & 1 \\ 1 & 0 & 0 & 0 & 0 & 1 & 1 & 0 \\ 0 & 1 & 0 & 0 & 1 & 0 & 0 & 1 \\ 0 & 0 & 1 & 0 & 1 & 0 & 0 & 1 \\ 0 & 0 & 0 & 1 & 0 & 1 & 1 & 0 \end{bmatrix}$

7.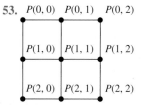

9. $\begin{bmatrix} 1 & 0 & 2 & 1 \\ 0 & 1 & 1 & 2 \\ 2 & 1 & 1 & 0 \\ 1 & 2 & 0 & 1 \end{bmatrix}$

11.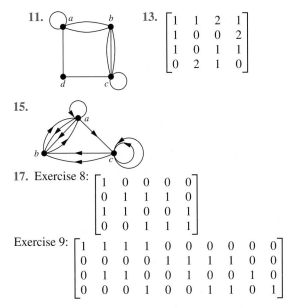

13. $\begin{bmatrix} 1 & 1 & 2 & 1 \\ 1 & 0 & 0 & 2 \\ 1 & 0 & 1 & 1 \\ 0 & 2 & 1 & 0 \end{bmatrix}$

15.

17. Exercise 8: $\begin{bmatrix} 1 & 0 & 0 & 0 & 0 \\ 0 & 1 & 1 & 1 & 0 \\ 1 & 1 & 0 & 0 & 1 \\ 0 & 0 & 1 & 1 & 1 \end{bmatrix}$

Exercise 9: $\begin{bmatrix} 1 & 1 & 1 & 1 & 0 & 0 & 0 & 0 & 0 & 0 \\ 0 & 0 & 0 & 0 & 1 & 1 & 1 & 1 & 0 & 0 \\ 0 & 1 & 1 & 0 & 0 & 1 & 0 & 0 & 1 & 0 \\ 0 & 0 & 0 & 1 & 0 & 0 & 1 & 1 & 0 & 1 \end{bmatrix}$

19. $\deg(v)$ − number of loops at v; $\deg^-(v)$ **21.** 2 if e is not a loop, 1 if e is a loop **23.** Isomorphic **25.** Isomorphic **27.** Isomorphic **29.** G is isomorphic to itself by the identity function, so isomorphism is reflexive. Suppose that G is isomorphic to H. Then there exists a one-to-one correspondence f from G to H that preserves adjacency and nonadjacency. It follows that f^{-1} is a one-to-one correspondence from H to G that preserves adjacency and nonadjacency. Hence, isomorphism is symmetric. If G is isomorphic to H and H is isomorphic to K, then there are one-to-one correspondences f and g from G to H and from H to K that preserve adjacency and nonadjacency. It follows that $g \circ f$ is a one-to-one correspondence from G to K that preserves adjacency and nonadjacency. Hence, isomorphism is transitive. **31.** All zeros **33.** Label the vertices in order so that all of the vertices in the first set of the partition of the vertex set come first. Because no edges join vertices in the same set of the partition, the matrix has the desired form. **35.** C_5 **37.** $n = 5$ only **39.** 4 **41.** $G = (V_1, E_1)$ is isomorphic to $H = (V_2, E_2)$ if and only if there exist functions f from V_1 to V_2 and g from E_1 to E_2 such that each is a one-to-one correspondence and for every edge e in E_1 the endpoints of $g(e)$ are $f(v)$ and $f(w)$ where v and w are the endpoints of e. **43.** Yes **45.** If f is an isomorphism from a directed graph G to a directed graph H, then f is also an isomorphism from G^{conv} to H^{conv}. To see this note that (u, v) is an edge of G^{conv} if and only if (v, u) is an edge of G if and only if $(f(v), f(u))$ is an edge of H if and only if $(f(u), f(v))$ is an edge of H^{conv}. **47.** Many answers are possible; for example, C_6 and $C_3 \cup C_3$. **49.** The graphs in Exercise 26 provide a devil's pair.

Section 10.4

1. a) Path of length 4; not a circuit; not simple **b)** Not a path **c)** Not a path **d)** Simple circuit of length 5 **3.** No **5.** If a person has Erdős number n, then there is a path of length

n from that person to Erdős in the collaboration graph, so by definition, that means that that person is in the same component as Erdős. If a person is in the same component as Erdős, then there is a path from that person to Erdős, and the length of the shortest such path is that person's Erdős number. **7. a)** Weakly connected **b)** Weakly connected **c)** Not strongly or weakly connected **9. a)** $\{a, b, f\}$, $\{c, d, e\}$ **b)** $\{a, b, c, d, e, h\}$, $\{f\}$, $\{g\}$ **c)** $\{a, b, d, e, f, g, h, i\}$, $\{c\}$ **11.** Suppose the strong components of u and v are not disjoint, say with vertex w in both. Suppose x is a vertex in the strong component of u. Then x is also in the strong component of v, because there is a path from x to v (namely the path from x to u followed by the path from u to w followed by the path from w to v) and vice versa. Thus x is in the strong component of v. This shows that the strong component of u is a subgraph of the strong component of v, and equality follows by symmetry. **13.** Not isomorphic (G has a triangle; H does not) **15. a)** 3 **b)** 0 **c)** 27 **d)** 0 **17. a)** 1 **b)** 3 **c)** 1 **d)** 2 **e)** 7 **f)** 5 **19.** R is reflexive by definition. Assume that $(u, v) \in R$; then there is a path from u to v. Then $(v, u) \in R$ because there is a path from v to u, namely, the path from u to v traversed backward. Assume that $(u, v) \in R$ and $(v, w) \in R$; then there are paths from u to v and from v to w. Putting these two paths together gives a path from u to w. Hence, $(u, w) \in R$. It follows that R is transitive. **21.** c **23.** b, c, e, i **25. a)** Denver–Chicago, Boston–New York **b)** Seattle–Portland, Portland–San Francisco, Salt Lake City–Denver, New York–Boston, Boston–Burlington, Boston–Bangor **27.** A minimal set of people who collectively influence everyone (directly or indirectly); {Deborah} **29.** An edge cannot connect two vertices in different connected components. Because there are at most $C(n_i, 2)$ edges in the connected component with n_i vertices, it follows that there are at most $\sum_{i=1}^{k} C(n_i, 2)$ edges in the graph. **31.** Suppose that G is not connected. Then it has a component of k vertices for some k, $1 \leq k \leq n - 1$. The most edges G could have is $C(k, 2) + C(n - k, 2) = [k(k - 1) + (n - k)(n - k - 1)]/2 = k^2 - nk + (n^2 - n)/2$. This quadratic function of f is minimized at $k = n/2$ and maximized at $k = 1$ or $k = n - 1$. Hence, if G is not connected, the number of edges does not exceed the value of this function at 1 and at $n - 1$, namely, $(n - 1)(n - 2)/2$. **33. a)** 1 **b)** 2 **c)** 6 **d)** 21 **35. a)** Removing an edge from a cycle leaves a path, which is still connected. **b)** Removing an edge from the cycle portion of the wheel leaves that portion still connected and the central vertex still connected to it as well. Removing a spoke leaves the cycle intact and the central vertex still connected to it as well. **c)** Any four vertices, two from each part of the bipartition, are connected by a 4-cycle; removing one edge does not disconnect them. **d)** Deleting the edge joining $(b_1, b_2, \ldots, b_{i-1}, 0, b_{i+1}, \ldots, b_n)$ and $(b_1, b_2, \ldots, b_{i-1}, 1, b_{i+1}, \ldots, b_n)$ does not disconnect the graph because these two vertices are still joined via the path $(b_1, b_2, \ldots, b_{i-1}, 0, b_{i+1}, \ldots, 0)$, $(b_1, b_2, \ldots, b_{i-1}, 0, b_{i+1}, \ldots, 1)$, $(b_1, b_2, \ldots, b_{i-1}, 1, b_{i+1}, \ldots, 1)$, $(b_1, b_2, \ldots, b_{i-1}, 1, b_{i+1}, \ldots, 0)$ if $n < 2$ and $b_n = 0$, and similarly in the other three cases. **37.** If

G is complete, then removing vertices one by one leaves a complete graph at each step, so we never get a disconnected graph. Conversely, if edge uv is missing from G, then removing all the vertices except u and v creates a disconnected graph. **39.** Both equal $\min(m, n)$.   **41.** Let G be a graph with n vertices; then $\kappa(G) \le n - 1$. Let C be a smallest edge cut, leaving a nonempty proper subset S of the vertices of G disconnected from the complementary set $S' = V - S$. If xy is an edge of G for every $x \in S$ and $y \in S'$, then the size of C is $|S||S'|$, which is at least $n - 1$, so $\kappa(G) \le \lambda(G)$. Otherwise, let $x \in S$ and $y \in S'$ be nonadjacent vertices. Let T consist of all neighbors of x in S' together with all vertices of $S - \{x\}$ with neighbors in S'. Then T is a vertex cut, because it separates x and y. Now look at the edges from x to $T \cap S'$ and one edge from each vertex of $T \cap S$ to S'; this gives us $|T|$ distinct edges that lie in C, so $\lambda(G) = |C| \ge |T| \ge \kappa(G)$.   **43.** 2   **45.** Let the simple paths P_1 and P_2 be $u = x_0, x_1, \ldots, x_n = v$ and $u = y_0, y_1, \ldots, y_m = v$, respectively. The paths thus start out at the same vertex. Since the paths do not contain the same set of edges, they must diverge eventually. If they diverge only after one of them has ended, then the rest of the other path is a simple circuit from v to v. Otherwise we can suppose that $x_0 = y_0, x_1 = y_1, , x_i = y_i$, but $x_{i+1} \ne y_{i+1}$. To form our simple circuit, we follow the path y_i, y_{i+1}, y_{i+2}, and so on, until it once again first encounters a vertex on P_1 (possibly as early as y_{i+1}, no later than y_m). Once we are back on P_1, we follow it along—forwards or backwards, as necessary—to return to x_i. Since $x_i = y_i$, this certainly forms a circuit. It must be a simple circuit, since no edge among the x_ks or the y_ls can be repeated (P_1 and P_2 are simple by hypothesis) and no edge among the x_ks can equal one of the edges y_l that we used, since we abandoned P_2 for P_1 as soon as we hit P_1.   **47.** The graph G is connected if and only if every off-diagonal entry of $\mathbf{A} + \mathbf{A}^2 + \mathbf{A}^3 + \cdots + \mathbf{A}^{n-1}$ is positive, where $\mathbf{A}$ is the adjacency matrix of G.   **49.** If the graph is bipartite, say with parts A and B, then the vertices in every path must alternately lie in A and B. Therefore a path that starts in A, say, will end in B after an odd number of steps and in A after an even number of steps. Because a circuit ends at the same vertex where it starts, the length must be even. Conversely, suppose that all circuits have even length; we must show that the graph is bipartite. We can assume that the graph is connected, because if it is not, then we can just work on one component at a time. Let v be a vertex of the graph, and let A be the set of all vertices to which there is a path of odd length starting at v, and let B be the set of all vertices to which there is a path of even length starting at v. Because the component is connected, every vertex lies in A or B. No vertex can lie in both A and B, because if one did, then following the odd-length path from v to that vertex and then back along the even-length path from that vertex to v would produce an odd circuit, contrary to the hypothesis. Thus, the set of vertices has been partitioned into two sets. To show that every edge has endpoints in different parts, suppose that xy is an edge, where $x \in A$. Then the odd-length path from v to x followed by xy produces an even-length path from v to y, so $y \in B$. (Similarly, if $x \in B$.)

51. $(H_1 W_1 H_2 W_2 \langle \text{boat} \rangle, \emptyset) \rightarrow (H_2 W_2, H_1 W_1 \langle \text{boat} \rangle) \rightarrow (H_1 H_2 W_2 \langle \text{boat} \rangle, W_1) \rightarrow (W_2, H_1 W_1 H_2 \langle \text{boat} \rangle) \rightarrow (H_2 W_2 \langle \text{boat} \rangle, H_1 W_1) \rightarrow (\emptyset, H_1 W_1 H_2 W_2 \langle \text{boat} \rangle)$

Section 10.5

1. Neither   **3.** $a, b, c, d, c, e, d, b, e, a, e, a$ **5.** No, A still has odd degree.   **7.** Yes   **9.** No   **11.** If there is an Euler path, then as we follow it each vertex except the starting and ending vertices must have equal in-degree and out-degree, because whenever we come to a vertex along an edge, we leave it along another edge. The starting vertex must have out-degree 1 larger than its in-degree, because we use one edge leading out of this vertex and whenever we visit it again we use one edge leading into it and one leaving it. Similarly, the ending vertex must have in-degree 1 greater than its out-degree. Because the Euler path with directions erased produces a path between any two vertices, in the underlying undirected graph, the graph is weakly connected. Conversely, suppose the graph meets the degree conditions stated. If we add one more edge from the vertex of deficient out-degree to the vertex of deficient in-degree, then the graph has every vertex with equal in-degree and out-degree. Because the graph is still weakly connected, by Exercise 10 this new graph has an Euler circuit. Now delete the added edge to obtain the Euler path.   **13.** Neither   **15.** Follow the same procedure as Algorithm 1, taking care to follow the directions of edges.   **17. a)** $n = 2$ **b)** None   **c)** None   **d)** $n = 1$ **19.** Exercise 1:1 time; Exercises 2–4: 0 times   **21.** a, b, c, d, e, a is a Hamilton circuit.   **23.** a, b, c, f, d, e is a Hamilton path.   **25.** $a, b, c, f, i, h, g, d, e$ is a Hamilton path.   **27.** $m = n \ge 2$ **29. a)** (i) No, (ii) No, (iii) Yes   **b)** (i) No, (ii) No, (iii) Yes   **c)** (i) Yes, (ii) Yes, (iii) Yes   **d)** (i) Yes, (ii) Yes, (iii) Yes   **31.** The result is trivial for $n = 1$: code is 0, 1. Assume we have a Gray code of order n. Let $c_1, \ldots, c_k, k = 2^n$ be such a code. Then $0c_1, \ldots, 0c_k, 1c_k, \ldots, 1c_1$ is a Gray code of order $n + 1$.

33. procedure *Fleury*$(G = (V, E)$: connected multigraph with the degrees of all vertices even, $V = \{v_1, \ldots, v_n\})$
$\quad v := v_1$
$\quad circuit := v$
$\quad H := G$
$\quad$ **while** H has edges
$\qquad e := $ first edge with endpoint v in H (with respect to listing of V) such that e is not a cut edge of H, if one exists, and simply the first edge in H with endpoint v otherwise
$\qquad w := $ other endpoint of e
$\qquad circuit := circuit$ with e, w added
$\qquad v := w$
$\qquad H := H - e$
$\quad$ **return** *circuit* {*circuit* is an Euler circuit}

35. If G has an Euler circuit, then it also has an Euler path. If not, add an edge between the two vertices of odd degree and apply the algorithm to get an Euler circuit. Then delete the new edge.   **37.** Suppose $G = (V, E)$ is a bipartite graph with $V = V_1 \cup V_2$, where $V_1 \cap V_2 = \emptyset$ and no edge connects a vertex in V_1 and a vertex in V_2. Suppose that G has a Hamilton circuit. Such a circuit must be of the form $a_1, b_1, a_2, b_2, \ldots, a_k, b_k, a_1$, where $a_i \in V_1$ and $b_i \in V_2$ for $i = 1, 2, \ldots, k$. Because the Hamilton circuit visits each vertex exactly once, except for v_1, where it begins and ends, the number of vertices in the graph equals $2k$, an even number. Hence, a bipartite graph with an odd number of vertices cannot have a Hamilton circuit.

39.
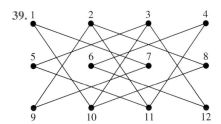

41. We represent the squares of a 3×4 chessboard as follows:

1	2	3	4
5	6	7	8
9	10	11	12

A knight's tour can be made by following the moves 8, 10, 1, 7, 9, 2, 11, 5, 3, 12, 6, 4.   **43.** We represent the squares of a 4×4 chessboard as follows:

1	2	3	4
5	6	7	8
9	10	11	12
13	14	15	16

There are only two moves from each of the four corner squares. If we include all the edges 1–10, 1–7, 16–10, and 16–7, a circuit is completed too soon, so at least one of these edges must be missing. Without loss of generality, assume the path starts 1–10, 10–16, 16–7. Now the only moves from square 3 are to squares 5, 10, and 12, and square 10 already has two incident edges. Therefore, 3–5 and 3–12 must be in the Hamilton circuit. Similarly, edges 8–2 and 8–15 must be in the circuit. Now the only moves from square 9 are to squares 2, 7, and 15. If there were edges from square 9 to both squares 2 and 15, a circuit would be completed too soon. Therefore the edge 9–7 must be in the circuit giving square 7 its full complement of edges. But now square 14 is forced to be joined to squares 5 and 12, completing a circuit too soon (5–14–12–3–5). This contradiction shows that there is no knight's tour on the 4×4

board.   **45.** Because there are mn squares on an $m \times n$ board, if both m and n are odd, there are an odd number of squares. Because by Exercise 44 the corresponding graph is bipartite, by Exercise 37 it has no Hamilton circuit. Hence, there is no reentrant knight's tour.   **47. a)** If G does not have a Hamilton circuit, continue as long as possible adding missing edges one at a time in such a way that we do not obtain a graph with a Hamilton circuit. This cannot go on forever, because once we've formed the complete graph by adding all missing edges, there is a Hamilton circuit. Whenever the process stops, we have obtained a (necessarily noncomplete) graph H with the desired property.   **b)** Add one more edge to H. This produces a Hamilton circuit, which uses the added edge. The path consisting of this circuit with the added edge omitted is a Hamilton path in H.   **c)** Clearly v_1 and v_n are not adjacent in H, because H has no Hamilton circuit. Therefore they are not adjacent in G. But the hypothesis was that the sum of the degrees of vertices not adjacent in G was at least n. This inequality can be rewritten as $n - \deg(v_n) \leq \deg(v_1)$. But $n - \deg(v_n)$ is just the number of vertices not adjacent to v_n.   **d)** Because there is no vertex following v_n in the Hamilton path, v_n is not in S. Each one of the $\deg(v_1)$ vertices adjacent to v_1 gives rise to an element of S, so S contains $\deg(v_1)$ vertices.   **e)** By part (c) there are at most $\deg(v_1) - 1$ vertices other than v_n not adjacent to v_n, and by part (d) there are $\deg(v_1)$ vertices in S, none of which is v_n. Therefore at least one vertex of S is adjacent to v_n. By definition, if v_k is this vertex, then H contains edges $v_k v_n$ and $v_1 v_{k+1}$, where $1 < k < n - 1$.   **f)** Now $v_1, v_2, \ldots, v_{k-1}, v_k, v_n, v_{n-1}, \ldots, v_{k+1}, v_1$ is a Hamilton circuit in H, contradicting the construction of H. Therefore, our assumption that G did not originally have a Hamilton circuit is wrong, and our proof by contradiction is complete.

Section 10.6

1. a) Vertices are the stops, edges join adjacent stops, weights are the times required to travel between adjacent stops. **b)** Same as part (a), except weights are distances between adjacent stops.   **c)** Same as part (a), except weights are fares between stops.   **3.** Exercise 2: a, c, d, e, g, z;   **5. a)** Direct   **b)** Via New York **c)** Via Atlanta and Chicago **d)** Via New York   **7. a)** Via Chicago **b)** Via Chicago **c)** Via Los Angeles **d)** Via Chicago   **9. a)** Via Chicago **b)** Via Chicago **c)** Via Los Angeles **d)** Via Chicago   **11.** Do not stop the algorithm when z is added to the set S.   **13. a)** Via Woodbridge, via Woodbridge and Camden **b)** Via Woodbridge, via Woodbridge and Camden   **15.** For instance, sightseeing tours, street cleaning

17.

	a	b	c	d	e	z
a	4	3	2	8	10	13
b	3	2	1	5	7	10
c	2	1	2	6	8	11
d	8	5	6	4	2	5
e	10	7	8	2	4	3
z	13	10	11	5	3	6

19. *a*–*c*–*b*–*d*–*a* (or the same circuit starting at some other point and/or traversing the vertices in reverse order)
21. Consider this graph:

The circuit *a*-*b*-*a*-*c*-*a* visits each vertex at least once (and the vertex *a* twice) and has total weight 6. Every Hamilton circuit has total weight 103.   **23.** Let $v_1, v_2, \ldots, v_n$ be a topological ordering of the vertices of the given directed acyclic graph. Let $w(i, j)$ be the weight of edge $v_i v_j$. Iteratively define $P(i)$ with the intent that it will be the weight of a longest path ending at v_i and $C(i)$ with the intent that it will be the vertex preceding v_i in some longest path: For *i* from 1 to *n*, let $P(i)$ be the maximum of $P(j) + w(j, i)$ over all $j < i$ such that $v_j v_i$ is an edge in the directed graph (and if such a *j* exists let $C(i)$ be a value of *j* for which this maximum is achieved) and let $P(i) = 0$ if there are no such values of *j*. At the conclusion of this loop, a longest path can be found by choosing *i* that maximizes $P(i)$ and following the *C* links back to the start of the path.

Section 10.7

1. Yes   **3.** No

5. Yes

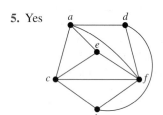

7. A triangle is formed by the planar representation of the subgraph of K_5 consisting of the edges connecting v_1, v_2, and v_3. The vertex v_4 must be placed either within the triangle or outside of it. We will consider only the case when v_4 is inside the triangle; the other case is similar. Drawing the three edges from v_1, v_2, and v_3 to v_4 forms four regions. No matter which of these four regions v_5 is in, it is possible to join it to only three, and not all four, of the other vertices.   **9.** 8   **11.** Because there are no loops or multiple edges and no simple circuits of length 3, and the degree of the unbounded region is at least 4, each region has degree at least 4. Thus $2e \geq 4r$, or $r \leq e/2$. But $r = e - v + 2$, so we have $e - v + 2 \leq e/2$, which implies that $e \leq 2v - 4$.   **13.** As in the argument in the proof of Corollary 1, we have $2e \geq 5r$ and $r = e - v + 2$. Thus $e - v + 2 \leq 2e/5$, which implies that $e \leq (5/3)v - (10/3)$.   **15.** Only (a) and (c)   **17.** Not homeomorphic to $K_{3,3}$ **19.** Planar   **21.** Nonplanar   **23. a)** 1 **b)** 3 **c)** 9 **d)** 2 **e)** 4 **f)** 16   **25. a)** 2 **b)** 2 **c)** 2 **d)** 2 **e)** 2 **f)** 2   **27.** The formula is valid for $n \leq 4$. If $n > 4$, by Exercise 26 the thickness of K_n is at least $C(n, 2)/(3n - 6) = (n + 1 + \frac{2}{n-2})/6$ rounded up. Because this quantity is never an integer, it equals $\lfloor (n + 7)/6 \rfloor$.

Section 10.8

1. Four colors

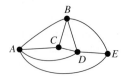

3. 3   **5.** 3   **7.** Graphs with no edges   **9.** 3 if *n* is even, 4 if *n* is odd   **11.** Period 1: Math 115, Math 185; period 2: Math 116, CS 473; period 3: Math 195, CS 101; period 4: CS 102; period 5: CS 273   **13.** 5   **15.** Exercise 3: 3 Exercise 4: 6 Exercise 5: 3   **17.** Two edges that have the same color share no endpoints. Therefore if more than $n/2$ edges were colored the same, the graph would have more than $2(n/2) = n$ vertices.   **19.** Color 1: *e*, *f*, *d*; color 2: *c*, *a*, *i*, *g*; color 3: *h*, *b*, *j*   **21.** Color C_6 **23.** Four colors are needed to color W_n when *n* is an odd integer greater than 1, because three colors are needed for the rim (see Example 4), and the center vertex, being adjacent to all the rim vertices, will require a fourth color. To see that the graph obtained from W_n by deleting one edge can be colored with three colors, consider two cases. If we remove a rim edge, then we can color the rim with two colors, by starting at an endpoint of the removed edge and using the colors alternately around the portion of the rim that remains. The third color is then assigned to the center vertex. If we remove a spoke edge, then we can color the rim by assigning color #1 to the rim endpoint of the removed edge and colors #2 and #3 alternately to the remaining vertices on the rim, and then assign color #1 to the center.   **25.** Suppose that *G* is chromatically *k*-critical but has a vertex *v* of degree $k - 2$ or less. Remove from *G* one of the edges incident to *v*. By definition of "*k*-critical," the resulting graph can be colored with $k - 1$ colors. Now restore the missing edge and use this coloring for all vertices except *v*. Because we had a proper coloring of the smaller graph, no two adjacent vertices have the same color. Furthermore, *v* has at most $k - 2$ neighbors, so we can color *v* with an unused color to obtain a proper $(k-1)$-coloring of *G*. This contradicts the fact that *G* has chromatic number *k*. Therefore, our assumption was wrong, and every vertex of *G* must have degree at least $k - 1$.   **27. a)** 6 **b)** 7 **c)** 9 **d)** 11   **29.** Represent frequencies by colors and zones by vertices. Join two vertices with an edge if the zones these vertices represent interfere with one another. Then a *k*-tuple coloring is precisely an assignment of frequencies that avoids interference.   **31.** We use induction on the number of vertices of the graph. Every graph with five or fewer vertices can be colored with five or fewer colors, because each vertex can get a different color. That takes care of the basis case(s). So we assume that all graphs with *k* vertices can be 5-colored and consider a graph *G* with $k + 1$ vertices. By Corollary 2 in Section 10.7, *G* has a vertex *v* with degree at most 5. Remove *v* to form the graph G'. Because G' has only *k* vertices, we 5-color it by the inductive hypothesis. If the neighbors of *v*

do not use all five colors, then we can 5-color G by assigning to v a color not used by any of its neighbors. The difficulty arises if v has five neighbors, and each has a different color in the 5-coloring of G'. Suppose that the neighbors of v, when considered in clockwise order around v, are a, b, c, m, and p. (This order is determined by the clockwise order of the curves representing the edges incident to v.) Suppose that the colors of the neighbors are azure, blue, chartreuse, magenta, and purple, respectively. Consider the azure-chartreuse subgraph (i.e., the vertices in G colored azure or chartreuse and all the edges between them). If a and c are not in the same component of this graph, then in the component containing a we can interchange these two colors (make the azure vertices chartreuse and vice versa), and G' will still be properly colored. That makes a chartreuse, so we can now color v azure, and G has been properly colored. If a and c are in the same component, then there is a path of vertices alternately colored azure and chartreuse joining a and c. This path together with edges av and vc divides the plane into two regions, with b in one of them and m in the other. If we now interchange blue and magenta on all the vertices in the same region as b, we will still have a proper coloring of G', but now blue is available for v. In this case, too, we have found a proper coloring of G. This completes the inductive step, and the theorem is proved.   **33.** We follow the hint. Because the measures of the interior angles of a pentagon total $540°$, there cannot be as many as three interior angles of measure more than $180°$ (reflex angles). If there are no reflex angles, then the pentagon is convex, and a guard placed at any vertex can see all points. If there is one reflex angle, then the pentagon must look essentially like figure (a) below, and a guard at vertex v can see all points. If there are two reflex angles, then they can be adjacent or nonadjacent (figures (b) and (c)); in either case, a guard at vertex v can see all points. [In figure (c), choose the reflex vertex closer to the bottom side.] Thus for all pentagons, one guard suffices, so $g(5) = 1$.

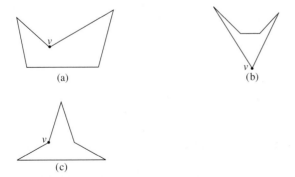

35. The figure suggested in the hint (generalized to have k prongs for any $k \geq 1$) has $3k$ vertices. The sets of locations from which the tips of different prongs are visible are disjoint. Therefore, a separate guard is needed for each of the k prongs, so at least k guards are needed. This shows that $g(3k) \geq k = \lfloor 3k/3 \rfloor$. If $n = 3k + i$, where $0 \leq i \leq 2$, then $g(n) \geq g(3k) \geq k = \lfloor n/3 \rfloor$.

Supplementary Exercises

1. 2500 **3.** Yes **5.** Yes **7.** $\sum_{i=1}^{m} n_i$ vertices, $\sum_{t<j} n_i n_j$ edges **9.** Complete subgraphs containing the following sets of vertices: $\{b, c, e, f\}$, $\{a, b, g\}$, $\{a, d, g\}$, $\{d, e, g\}$, $\{b, e, g\}$ **11.** Complete subgraphs containing the following sets of vertices: $\{b, c, d, j, k\}$, $\{a, b, j, k\}$, $\{e, f, g, i\}$, $\{a, b, i\}$, $\{a, i, j\}$, $\{b, d, e\}$, $\{b, e, i\}$, $\{b, i, j\}$, $\{g, h, i\}$, $\{h, i, j\}$ **13.** $\{c, d\}$ is a minimum dominating set. **15. a)** 1 **b)** 2 **c)** 3 **17. a)** A path from u to v in a graph G induces a path from $f(u)$ to $f(v)$ in an isomorphic graph H. **b)** Suppose f is an isomorphism from G to H. If $v_0, v_1, \ldots, v_n, v_0$ is a Hamilton circuit in G, then $f(v_0), f(v_1), \ldots, f(v_n), f(v_0)$ must be a Hamilton circuit in H because it is still a circuit and $f(v_i) \neq f(v_j)$ for $0 \leq i < j \leq n$. **c)** Suppose f is an isomorphism from G to H. If $v_0, v_1, \ldots, v_n, v_0$ is an Euler circuit in G, then $f(v_0), f(v_1), \ldots, f(v_n), f(v_0)$ must be an Euler circuit in H because it is a circuit that contains each edge exactly once. **d)** Two isomorphic graphs must have the same crossing number because they can be drawn exactly the same way in the plane. **e)** Suppose f is an isomorphism from G to H. Then v is isolated in G if and only if $f(v)$ is isolated in H. Hence, the graphs must have the same number of isolated vertices. **f)** Suppose f is an isomorphism from G to H. If G is bipartite, then the vertex set of G can be partitioned into V_1 and V_2 with no edge connecting vertices within V_1 or vertices within V_2. Then the vertex set of H can be partitioned into $f(V_1)$ and $f(V_2)$ with no edge connecting vertices within $f(V_1)$ or vertices within $f(V_2)$. **19.** 3 **21. a)** Yes **b)** No **23.** No **25.** Yes **27.** If e is a cut edge with endpoints u and v, then if we direct e from u to v, there will be no path in the directed graph from v to u, or else e would not have been a cut edge. Similar reasoning works if we direct e from v to u. **29.** $n - 1$ **31.** Let the vertices represent the chickens. We include the edge (u, v) in the graph if and only if chicken u dominates chicken v. **33.** By the handshaking theorem, the average vertex degree is $2m/n$, which equals the minimum degree; it follows that all the vertex degrees are equal. **35. a)** 4 **b)** 2 **c)** 3 **d)** 4 **e)** 4 **f)** 2 **37. a)** Suppose that $G = (V, E)$. Let $a, b \in V$. We must show that the distance between a and b in $\overline{G}$ is at most 2. If $\{a, b\} \notin E$ this distance is 1, so assume $\{a, b\} \in E$. Because the diameter of G is greater than 3, there are vertices u and v such that the distance in G between u and v is greater than 3. Either u or v, or both, is not in the set $\{a, b\}$. Assume that u is different from both a and b. Either $\{a, u\}$ or $\{b, u\}$ belongs to E; otherwise a, u, b would be a path in $\overline{G}$ of length 2. So, without loss of generality, assume $\{a, u\} \in E$. Thus v cannot be a or b, and by the same reasoning either $\{a, v\} \in E$ or $\{b, v\} \in E$. In either case, this gives a path of length less than or equal to 3 from u to v in G, a contradiction. **b)** Suppose $G = (V, E)$. Let $a, b \in V$. We must show that the distance between a and b in $\overline{G}$ does not exceed 3. If $\{a, b\} \notin E$, the result follows, so assume that $\{a, b\} \in E$. Because the diameter of G is greater than or equal to 3, there exist vertices u and v such that the distance in G between u and v is greater than or equal to 3. Either u or v, or both, is not in the set $\{a, b\}$. Assume u is different from both a and b. Either $\{a, u\} \in E$ or $\{b, u\} \in E$; otherwise a, u, b is a path of length 2 in $\overline{G}$. So, without loss of generality, assume $\{a, u\} \in E$. Thus v is different from a and from b. If $\{a, v\} \in E$, then u, a, v is a path of length 2 in G, so $\{a, v\} \notin E$ and thus $\{b, v\} \in E$ (or else there would be a path a, v, b of length 2 in $\overline{G}$). Hence, $\{u, b\} \notin E$; otherwise u, b, v is a path of length 2 in G. Thus, a, v, u, b is a path of length 3 in $\overline{G}$, as desired. **39.** a, b, e, z **41.** a, c, d, f, g, z **43.** If G is planar, then because $e \leq 3v - 6$, G has at most 27 edges. (If G is not connected it has even fewer edges.) Similarly, $\overline{G}$ has at most 27 edges. But the union of G and $\overline{G}$ is K_{11}, which has 55 edges, and $55 > 27 + 27$. **45.** Suppose that G is colored with k colors and has independence number i. Because each color class must be an independent set, each color class has no more than i elements. Thus there are at most ki vertices. **47. a)** $C(n, m) p^m (1 - p)^{n-m}$ **b)** np **c)** To generate a labeled graph G, as we apply the process to pairs of vertices, the random number x chosen must be less than or equal to $1/2$ when G has an edge between that pair of vertices and greater than $1/2$ when G has no edge there. Hence, the probability of making the correct choice is $1/2$ for each edge and $1/2^{C(n,2)}$ overall. Hence, all labeled graphs are equally likely. **49.** Suppose P is monotone increasing. If the property of not having P were not retained whenever edges are removed from a simple graph, there would be a simple graph G not having P and another simple graph G' with the same vertices but with some of the edges of G missing that has P. But P is monotone increasing, so because G' has P, so does G obtained by adding edges to G'. This is a contradiction. The proof of the converse is similar.

CHAPTER 11

Section 11.1

1. (a), (c), (e) **3.** No **5. a)** The entire tree **b)** c, g, h, o, p and the four edges cg, ch, ho, hp **c)** e alone **7. a)** 2 **b)** 3 **9. a)** The "only if" part is Theorem 2 and the definition of a tree. Suppose G is a connected simple graph with n vertices and $n - 1$ edges. If G is not a tree, it contains, by Exercise 8, an edge whose removal produces a graph G', which is still connected. If G' is not a tree, remove an edge to produce a connected graph G''. Repeat this procedure until the result is a tree. This requires at most $n - 1$ steps because there are only $n - 1$ edges. By Theorem 2, the resulting graph has $n - 1$ edges because it has n vertices. It follows that no edges were deleted, so G was already a tree. **b)** Suppose that G is a tree. By part (a), G has $n - 1$ edges, and by definition, G has no simple circuits. Conversely, suppose that G has no simple circuits and has $n - 1$ edges. Let c equal the number of components of G, each of which is necessarily a tree, say with n_i vertices, where $\sum_{i=1}^{c} n_i = n$. By part (a), the total

number of edges in G is $\sum_{i=1}^{c} (n_i - 1) = n - c$. Since we are given that this equals $n - 1$, it follows that $c = 1$, i.e., G is connected and therefore satisfies the definition of a tree. **11.** 9999 **13.** 2000 **15.** 999 **17.** No such tree exists by Theorem 4 because it is impossible for $m = 2$ or $m = 84$.

19. Complete binary tree of height 4:

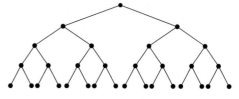

Complete 3-ary tree of height 3:

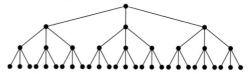

21. a) By Theorem 3 it follows that $n = mi + 1$. Because $i + l = n$, we have $l = n - i$, so $l = (mi + 1) - i = (m - 1)i + 1$. **b)** We have $n = mi + 1$ and $i + l = n$. Hence, $i = n - l$. It follows that $n = m(n - l) + 1$. Solving for n gives $n = (ml - 1)/(m - 1)$. From $i = n - l$ we obtain $i = [(ml - 1)/(m - 1)] - l = (l - 1)/(m - 1)$. **23.** $n - t$ **25. a)** 1 **b)** 3 **c)** 5 **27.** Let $n = 2^k$, where k is a positive integer. If $k = 1$, there is nothing to prove because we can add two numbers with $n - 1 = 1$ processor in $\log 2 = 1$ step. Assume we can add $n = 2^k$ numbers in $\log n$ steps using a tree-connected network of $n - 1$ processors. Let $x_1, x_2, \ldots, x_{2n}$ be $2n = 2^{k+1}$ numbers that we wish to add. The tree-connected network of $2n - 1$ processors consists of the tree-connected network of $n - 1$ processors together with two new processors as children of each leaf. In one step we can use the leaves of the larger network to find $x_1 + x_2, x_3 + x_4, \ldots, x_{2n-1} + x_{2n}$, giving us n numbers, which, by the inductive hypothesis, we can add in $\log n$ steps using the rest of the network. Because we have used $\log n + 1$ steps and $\log(2n) = \log 2 + \log n = 1 + \log n$, this completes the proof. **29.** c only **31.** Suppose a tree T has at least two centers. Let u and v be distinct centers, both with eccentricity e, with u and v not adjacent. Because T is connected, there is a simple path P from u to v. Let c be any other vertex on this path. Because the eccentricity of c is at least e, there is a vertex w such that the unique simple path from c to w has length at least e. Clearly, this path cannot contain both u and v or else there would be a simple circuit. In fact, this path from c to w leaves P and does not return to P once it, possibly, follows part of P toward either u or v. Without loss of generality, assume this path does not follow P toward u. Then the path from u to c to w is simple and of length more than e, a contradiction. Hence, u and v are adjacent. Now because any two centers are adjacent, if there were more than two centers, T would contain K_3, a simple circuit, as a subgraph, which is a contradiction.

33.

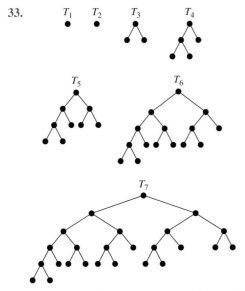

35. The statement is that *every* tree with n vertices has a path of length $n - 1$, and it was shown only that there exists a tree with n vertices having a path of length $n - 1$.

Section 11.2

1.

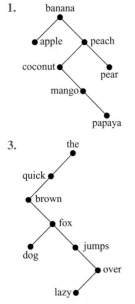

3.

the
quick
brown
fox
dog   jumps
over
lazy

5. At least $\lceil \log_3 4 \rceil = 2$ weighings are needed, because there are only four outcomes (because it is not required to determine whether the coin is lighter or heavier). In fact, two weighings suffice. Begin by weighing coin 1 against coin 2. If they balance, weigh coin 1 against coin 3. If coin 1 and coin 3 are the same weight, coin 4 is the counterfeit coin, and if they are not the same weight, then coin 3 is the counterfeit coin. If coin 1 and coin 2 are not the same weight, again weigh coin 1 against coin 3. If they balance, coin 2 is the counterfeit coin; if they do not balance, coin 1 is the counterfeit coin. **7.** The least number is five. Call the elements a, b, c, and d. First compare a and b; then compare c and d. Without loss of generality, assume that $a < b$ and $c < d$. Next compare a and c. Whichever is smaller is the smallest element of the set. Again without loss of generality, suppose $a < c$. Finally, compare b with both c and d to completely determine the ordering. **9.** The first two steps are shown in the text. After 22 has been identified as the second largest element, we replace the leaf 22 by $-\infty$ in the tree and recalculate the winner in the path from the leaf where 22 used to be up to the root. Next, we see that 17 is the third largest element, so we repeat the process: replace the leaf 17 by $-\infty$ and recalculate. Next, we see that 14 is the fourth largest element, so we repeat the process: replace the leaf 14 by $-\infty$ and recalculate. Next, we see that 11 is the fifth largest element, so we repeat the process: replace the leaf 11 by $-\infty$ and recalculate. The process continues in this manner. We determine that 9 is the sixth largest element, 8 is the seventh largest element, and 3 is the eighth largest element. The trees produced in all steps, except the second to last, are shown here.

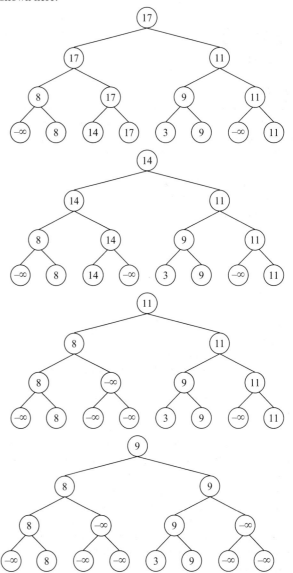

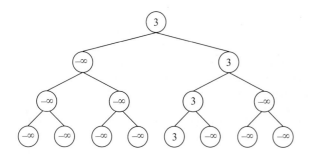

11. $k - 1$, where $n = 2^k$ 13. a) Yes b) No
c) Yes d) Yes 15. a: 000, e: 001, i: 01, k: 1100, o: 1101, p: 11110, u: 11111 17. a: 11; b: 101; c: 100; d: 01; e: 00; 2.25 bits (Note: This coding depends on how ties are broken, but the average number of bits is always the same.) 19. A:0001; B:101001; C:11001; D:00000; E:100; F:001100; G:001101; H:0101; I:0100; J:110100101; K:1101000; L:00001; M:10101; N:0110; O:0010; P:101000; Q:1101001000; R:1011; S:0111; T:111; U:00111; V:110101; W:11000; X:11010011; Y:11011; Z:1101001001 21. A:2; E:1; N:010; R:011; T:02; Z:00 23. n 25. Because the tree is rather large, we have indicated in some places to "see text." Refer to Figure 9; the subtree rooted at these square or circle vertices is exactly the same as the corresponding subtree in Figure 9. First player wins.

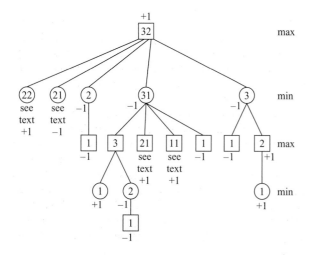

27. See the figures shown next. **a)** 0 **b)** 0 **c)** 1 **d)** This position cannot have occurred in a game; this picture is impossible.

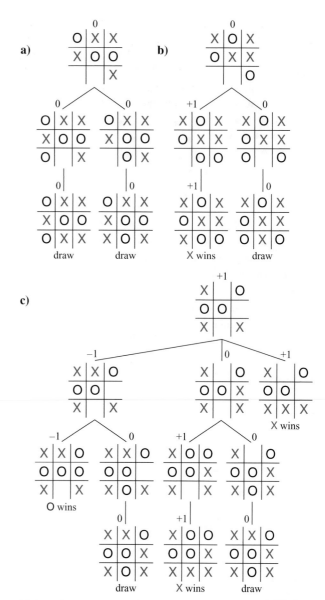

29. Proof by strong induction: *Basis step:* When there are $n = 2$ stones in each pile, if first player takes two stones from a pile, then second player takes one stone from the remaining pile and wins. If first player takes one stone from a pile, then second player takes two stones from the other pile and wins. *Inductive step:* Assume inductive hypothesis that second player can always win if the game starts with two piles of j stones for all $2 \leq j \leq k$, where $k \geq 2$, and consider a game with two piles containing $k + 1$ stones each. If first player takes all the stones from one of the piles, then second player takes all but one stone from the remaining pile and wins. If first player takes all but one stone from one of the piles, then second player takes all the stones from the other pile and wins. Otherwise first player leaves j stones in one pile, where $2 \leq j \leq k$, and $k + 1$ stones in the other pile. Second player takes the same number of stones from the larger pile, also leaving j stones there. At this point the game consists of two piles of j stones each. By the inductive hypothesis, the

second player in that game, who is also the second player in our actual game, can win, and the proof by strong induction is complete.   **31.** 7; 49   **33.** Value of tree is 1. Note: The second and third trees are the subtrees of the two children of the root in the first tree whose subtrees are not shown because of space limitations. They should be thought of as spliced into the first picture.

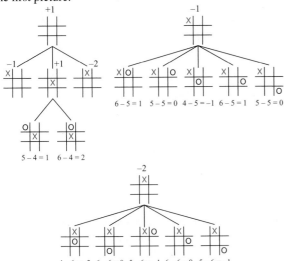

Section 11.3

1.

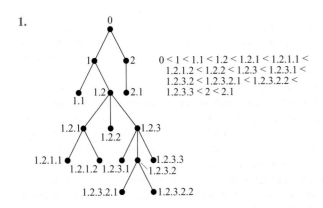

$0 < 1 < 1.1 < 1.2 < 1.2.1 < 1.2.1.1 <$
$1.2.1.2 < 1.2.2 < 1.2.3 < 1.2.3.1 <$
$1.2.3.2 < 1.2.3.2.1 < 1.2.3.2.2 <$
$1.2.3.3 < 2 < 2.1$

3. No   **5.** a,b,d,e,i,j,m,n,o,c,f,g,h,k,l,p
7. d,i,m,n,o,j,e,b,f,g,k,p,l,h,c,a

9. a)

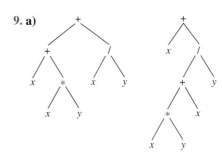

b) $++x*xy/xy, +x/+*xyxy$ **c)** $xxy*+xy/+, xxy*x+y/+$
d) $((x + (x * y)) + (x/y)), (x + (((x * y) + x)/y))$

11. a)

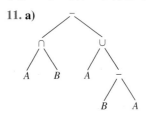

b) $- \cap A B \cup A - B A$ **c)** $A B \cap A B A - \cup -$
d) $((A \cap B) - (A \cup (B - A)))$ **13. a)** 1 **b)** 1 **c)** 4 **d)** 2205

15.

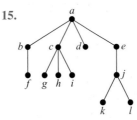

17. Use mathematical induction. The result is trivial for a list with one element. Assume the result is true for a list with n elements. For the inductive step, start at the end. Find the sequence of vertices at the end of the list starting with the last leaf, ending with the root, each vertex being the last child of the one following it. Remove this leaf and apply the inductive hypothesis.   **19.** c, d, b, f, g, h, e, a in each case
21. Proof by mathematical induction. Let $S(X)$ and $O(X)$ represent the number of symbols and number of operators in the well-formed formula X, respectively. The statement is true for well-formed formulae of length 1, because they have 1 symbol and 0 operators. Assume the statement is true for all well-formed formulae of length less than n. A well-formed formula of length n must be of the form $*XY$, where $*$ is an operator and X and Y are well-formed formulae of length less than n. Then by the inductive hypothesis $S(*XY) = S(X) + S(Y) = [O(X) + 1] + [O(Y) + 1] = O(X) + O(Y) + 2$. Because $O(*XY) = 1 + O(X) + O(Y)$, it follows that $S(*XY) = O(*XY) + 1$.   **23.** $x y + z x \circ + x \circ, x y z + + y x + +,$ $x y x y \circ \circ x y \circ \circ z \circ +, x z \times, z z + \circ, y y y y \circ \circ \circ, z x + y z + \circ,$ for instance

Section 11.4

1. $m - n + 1$

3.

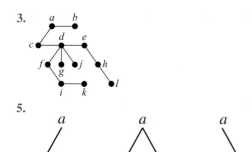

5.

7. a) 3 **b)** 16 **c)** 4 **d)** 5

9.

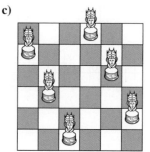

11. a) A path of length 6 **b)** A path of length 5 **c)** A path of length 6 **d)** Depends on order chosen to visit the vertices; may be a path of length 7   **13.** With breadth-first search, the initial vertex is the middle vertex, and the n spokes are added to the tree as this vertex is processed. Thus, the resulting tree is $K_{1,n}$. With depth-first search, we start at the vertex in the middle of the wheel and visit a neighbor—one of the vertices on the rim. From there we move to an adjacent vertex on the rim, and so on all the way around until we have reached every vertex. Thus, the resulting spanning tree is a path of length n.

15. Proof by induction on the length of the path: If the path has length 0, then the result is trivial. If the length is 1, then u is adjacent to v, so u is at level 1 in the breadth-first spanning tree. Assume that the result is true for paths of length l. If the length of a path is $l + 1$, let u' be the next-to-last vertex in a shortest path from v to u. By the inductive hypothesis, u' is at level l in the breadth-first spanning tree. If u were at a level not exceeding l, then clearly the length of the shortest path from v to u would also not exceed l. So u has not been added to the breadth-first spanning tree yet after the vertices of level l have been added. Because u is adjacent to u', it will be added at level $l + 1$ (although the edge connecting u' and u is not neces-

sarily added).   **17. a)** No solution **b)**

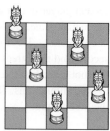

c)

19. Start at a vertex and proceed along a path without repeating vertices as long as possible, allowing the return to the start after all vertices have been visited. When it is impossible to continue along a path, backtrack and try another extension of the current path.   **21.** Take the union of the spanning trees of the connected components of G. They are disjoint, so the result is a forest.   **23.** $m - n + c$ **25.** Assume that we wish to find the length of a shortest path from v_1 to every other vertex of G using Algorithm 1. In line 2 of that algorithm, add $L(v_1) := 0$, and add the following as a third step in the **then** clause at the end: $L(w) := 1 + L(v)$.   **27.** Add an instruction to the BFS algorithm to mark each vertex as it is encountered. When BFS terminates we have found (all the vertices of) one component of the graph. Repeat, starting at an unmarked vertex, and continue in this way until all vertices have been marked.   **29.** Trees   **31.** Certainly these two procedures produce the identical spanning trees if the graph we are working with is a tree itself, because in this case there is only one spanning tree (the whole graph). This is the only case in which that happens, however. If the original graph has any other edges, then by Exercise 32 they must be back edges and hence, join a vertex to an ancestor or descendant, whereas by Exercise 24, they must connect vertices at the same level or at levels that differ by 1. Clearly these two possibilities are mutually exclusive. Therefore there can be no edges other than tree edges if the two spanning trees are to be the same.   **33.** Because the edges not in the spanning tree are not followed in the process, we can ignore them. Thus we can assume that the graph was a rooted tree to begin with. The basis step is trivial (there is only one vertex), so we assume the inductive hypothesis that breadth-first search applied to trees with n vertices have their vertices visited in order of their level in the tree and consider a tree T with $n + 1$ vertices. The last vertex to be visited during breadth-first search of this tree, say v, is the one that was added last to the list of vertices waiting to be processed. It was added when its parent, say u, was being processed. We must show that v is at the lowest (bottom-most, i.e., numerically greatest) level of the tree. Suppose not; say vertex x, whose parent is vertex w, is at a lower level. Then w is at a lower level than u. Clearly v must be a leaf, because any child of v could not have been seen before v is seen. Consider the tree T' obtained from T by deleting v. By the inductive hypothesis, the vertices in T' must be processed in order of their level in T' (which is the same as their level in T, and the absence of v in T' has no effect on the rest of the algorithm). Therefore u must have been processed before w, and therefore v would have joined the waiting list before x did, a contradiction. Therefore v is at the bottom-most level of the tree, and the proof is complete. **35.** We modify the pseudocode given in Algorithm 2 by initializing m to be 0 at the beginning of the algorithm, and adding the statements "$m := m + 1$" and "assign m to vertex v" after the statement that removes vertex v from L. **37.** Let T be the spanning tree constructed in Figure 3 and $T_1, T_2, T_3,$ and T_4 the spanning trees in Figure 4. Denote by $d(T', T'')$ the distance between T' and T''. Then $d(T, T_1) = 6$, $d(T, T_2) = 4$, $d(T, T_3) = 4$, $d(T, T_4) = 2$, $d(T_1, T_2) = 4$, $d(T_1, T_3) = 4$, $d(T_1, T_4) = 6$, $d(T_2, T_3) = 4$, $d(T_2, T_4) = 2$, and $d(T_3, T_4) = 4$.

39. Exercise 12:   Exercise 13:   Exercise 14:

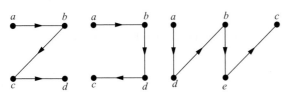

41. First construct an Euler circuit in the directed graph. Then delete from this circuit every edge that goes to a vertex previously visited. **43.** According to Exercise 42, a directed graph contains a circuit if and only if there are any back edges. We can detect back edges as follows. Add a marker on each vertex v to indicate what its status is: not yet seen (the initial situation), seen (i.e., put into T) but not yet finished (i.e., $visit(v)$ has not yet terminated), or finished (i.e., $visit(v)$ has terminated). A few extra lines in Algorithm 1 will accomplish this bookkeeping. Then to determine whether a directed graph has a circuit, we just have to check when looking at edge uv whether the status of v is "seen." If that ever happens, then we know there is a circuit; if not, then there is no circuit.

Section 11.5

1. Deep Springs–Oasis, Oasis–Dyer, Oasis–Silver Peak, Silver Peak–Goldfield, Lida–Gold Point, Gold Point–Beatty,

Lida–Goldfield, Goldfield–Tonopah, Tonopah–Manhattan, Tonopah–Warm Springs

3.

5.

7. Instead of choosing minimum-weight edges at each stage, choose maximum-weight edges at each stage with the same properties.

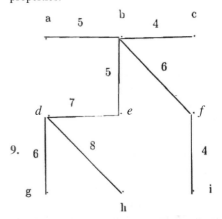

9.

11. First find a minimum spanning tree T of the graph G with n edges. Then for $i = 1$ to $n - 1$, delete only the ith edge of T from G and find a minimum spanning tree of the remaining graph. Pick the one of these $n - 1$ trees with the shortest length. **13.** If all edges have different weights, then a contradiction is obtained in the proof that Prim's algorithm works when an edge e_{k+1} is added to T and an edge e is deleted, instead of possibly producing another spanning tree.

15.
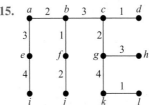

17. Same as Kruskal's algorithm, except start with $T :=$ this set of edges and iterate from $i = 1$ to $i = n - 1 - s$, where s is the number of edges you start with.

19. a)
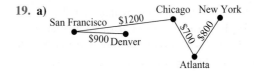

b)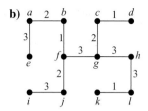

21. Each of the r trees is joined to at least one other tree by a new edge. Hence, there are at most $r/2$ trees in the result (each new tree contains two or more old trees). To accomplish this, we need to add $r - (r/2) = r/2$ edges. Because the number of edges added is integral, it is at least $\lceil r/2 \rceil$. **23.** If $k \geq \log n$, then $n/2^k \leq 1$, so $\lceil n/2^k \rceil = 1$, so by Exercise 22 the algorithm is finished after at most $\log n$ iterations.

Supplementary Exercises

1. Suppose T is a tree. Then clearly T has no simple circuits. If we add an edge e connecting two nonadjacent vertices u and v, then obviously a simple circuit is formed, because when e is added to T the resulting graph has too many edges to be a tree. The only simple circuit formed is made up of the edge e together with the unique path in T from v to u. Suppose T satisfies the given conditions. All that is needed is to show that T is connected, because there are no simple circuits in the graph. Assume that T is not connected. Then let u and v be in separate connected components. Adding $e = \{u, v\}$ does not satisfy the conditions. **3.** Suppose that a tree T has n vertices of degrees $d_1, d_2, \ldots, d_n$, respectively. Because $2e = \sum_{i=1}^{n} d_i$ and $e = n - 1$, we have $2(n - 1) = \sum_{i=1}^{n} d_i$. Because each $d_i \geq 1$, it follows that $2(n - 1) = n + \sum_{i=1}^{n} (d_i - 1)$, or that $n - 2 = \sum_{i=1}^{n} (d_i - 1)$. Hence, at most $n - 2$ of the terms of this sum can be 1 or more. Hence, at least two of them are 0. It follows that $d_i = 1$ for at least two values of i. **5.** $2n - 2$ **7.** A tree has no circuits, so it cannot have a subgraph homeomorphic to $K_{3,3}$ or K_5. **9.** Upper bound: k^h; lower bound: $2 \lceil k/2 \rceil^{h-1}$

11.

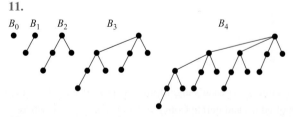

$B_0 \quad B_1 \quad B_2 \qquad B_3 \qquad\qquad B_4$

13. Because B_{k+1} is formed from two copies of B_k, one shifted down one level, the height increases by 1 as k increases by 1. Because B_0 had height 0, it follows by induction that B_k has height k. **15.** Because the root of B_{k+1} is the root of B_k with one additional child (namely the root of the other B_k), the degree of the root increases by 1 as k increases by 1. Be-

cause B_0 had a root with degree 0, it follows by induction that B_k has a root with degree k.

17.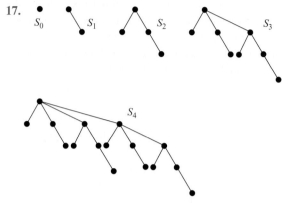

19. Use mathematical induction. The result is trivial for $k = 0$. Suppose it is true for $k - 1$. T_{k-1} is the parent tree for T. By induction, the child tree for T can be obtained from $T_0, \ldots, T_{k-2}$ in the manner stated. The final connection of r_{k-2} to r_{k-1} is as stated in the definition of S_k-tree.

21. **procedure** *level*(T: ordered rooted tree with root r)
    queue := sequence consisting of just the root r
    while *queue* contains at least one term
        v := first vertex in queue
        list v
        remove v from queue and put children of v onto
        the end of queue

23. Build the tree by inserting a root for the address 0, and then inserting a subtree for each vertex labeled i, for i a positive integer, built up from subtrees for each vertex labeled $i.j$ for j a positive integer, and so on. **25. a)** Yes **b)** No **c)** Yes **27.** The resulting graph has no edge that is in more than one simple circuit of the type described. Hence, it is a cactus.

29. **31.**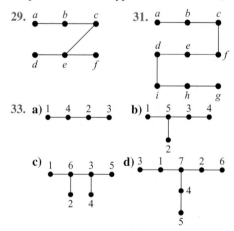

33. **a)** 1 4 2 3 **b)** 1 5 3 4 / 2 **c)** 1 6 3 5 / 2 4 **d)** 3 1 7 2 6 / 4 / 5

35. 6 **37.** Let G' be the graph obtained by deleting from G the vertex v and all edges incident to v. A minimum spanning tree of G can be obtained by taking an edge of minimal weight incident to v together with a minimum spanning tree of G'. **39.** Suppose that edge e is the edge of least weight incident to vertex v, and suppose that T is a spanning tree that does not include e. Add e to T, and delete from the simple circuit formed thereby the other edge of the circuit that contains v. The result will be a spanning tree of strictly smaller weight (because the deleted edge has weight greater than the weight of e). This is a contradiction, so T must include e.

CHAPTER 12

Section 12.1

1. a) $a \square (b \square c) = a \square b = a$ $(a \square b) \square c = a \square c = a$ Hence associativity holds **b)** not commutative as $a \square b \neq b \square a$.

Section 12.2

1. Zero of semigroup $(P(x), \cap)$ is ϕ. Identity of semigroup $(P(x), cap)$ is χ. Zero of semigroup $(P(x), \cup)$ is χ. Identity of semigroup $(P(x), \cup)$ is ϕ. Since identities exist, they are monoids.

3. Consider f a.t $f(a) = a$ $f(b) = a$
Let g be $g(a) = b$ $g(b) = b$
$f \circ g(a) = f(b) = a$ where $\circ$ denotes composition of functions
$g \circ f(a) = g(a) = b$
They are not equal. Hence the semigroup is not commutative.

5. $x \square y = x * a * y$
$x \square (y \square z) = x \square (y * a * z)$
$= x * a * (y * a * z)$
$= x * a * y * a * z$ because (associativity of $*$)
$(x \square y) \square z = (x * a * y) \square Z$
$= (x * a * y) * a * z$
$= x * a * y * a * z$
Hence $(x \square y) \square z = x \square (y \square z)$
Associative property holds.

7. $a \square c = c \square a$
$b \square c = c \square b$
$(a \square b) \square c = a \square (b \square c)$
$= a \square (c \square b)$
$= (a \square c) \square b$
$= (c \square a) \square b$
$= c \square (a \square b)$

9. $(a \square b) \square (a \square b) = a \square (b \square a) \square b$ (associativity)
$= a \square (a \square b) \square b$ (commutativity)
$= (a \square a) \square (b \square b)$ (associativity)
$= a \square b$ as $a \square a = a$
and $b \square b = b$

11. Let $(S, *)$ be a semigroup and a and b be idempotint item and $a * a = a$ $b * b = b$ for $a, b \in S$ $a' * b$ If it is a group, let a^{-1} and b^{-1} be the inverses of a and b respectively.
Then $a^{-1} * a * a = a^{-1} * a = e$
$a = e$ and hence a is the identity
But $b^{-1} * b * b = b^{-1} * b = e$
$b = e$ and b is also an identity
In group, the identity is unique.
$\therefore (S, *)$ cannot be a group.

13. Let $(s, *)$ be a monoid and T be the subset of s, which are left invertible (i.e., if $a \in T$, $\exists b$ s.t
$b * a = e$, where e is the identity element
Let $a_1, a_2 \in T$
$a_1^{-1} * a_1 = e$ $a_2^{-1} * a_2 = e$
($a_1 * a_2$ is left invertible as
$(a_2^{-1} * a_1^{-1}) * (a_1, *a_2) = e$
and $a_2^{-1} * a_1^{-1}$ is the left inverse of $(a_1 * a_2)$
Hence closure property holds for $(T, *)$
Associative property also holds
$e * e = e$ and hence e is left invertible and $e \in T$.
Hence $(T, *)$ is a monoid, a submonoid of $(S, *)$

15. $(P, *)$ The first row and column show that p_1 is the identity. From the table we find that p_2, p_3, and P_4 are their own inverses. p_5 has no inverse. Also in column 5 (row 5 also) p_4 appears twice, which should not be the case in a group.
$(Q, \square)$
From table we see that q_3 is a left identity but it is not a right identity. There is no right identity and hence not a group.

17. a) Let $\{a, b\}$ be the set and $*$ the binary operation, $(z, \oplus)$ has the structure

$\oplus$	0	1
0	0	1
1	1	0

0 is the identity. If a is the identity of $(\{a, b\}, *)$, the table is of the form

$*$	a	b
a	a	b
b	x	x

x has to be a to satisfy the condition of a group. It is easily seen that this is isomorphic to $(z_2, \oplus)$
b) $(z_3, \oplus)$ has structure

$\oplus$	0	1	2
0	0	1	2
1	1	2	0
2	2	0	1

1 is the inverse of 2. If $(\{a, b, c\}, *)$ is a group. Let a be the identity element. Then the table of the group has the structure

$*$	a	b	c
a	a	b	c
b	b	1	2
c	c	3	4

If b is its own inverse 1 has to be a in which case 2 has to be c, but two c's will in the third column. Hence
$1 = c$ and $2 = a$
Similarly $3 = a \cdot 4 = b$. Similarity between $(z_3, \oplus)$ and $(\{a, b, c\}, *)$ can easily be seen.
c) Two − given in page ? of this chapter
d) Since every group of prime order is isomorphic to $(z_p, \oplus)$, there is only one nonisomorphic group with 5 elements.

19. a) $b^{-1}a^{-1}(ab) = b^{-1}(a^{-1}a)b$
$= b^{-1}(e)b$
$= b^{-1}b = e$
Hence $b^{-1}a^{-1}$ is the inverse of ab.
b) $a_r^{-1}a_{r-1}^{-1}\ldots a_2^{-1}a_1^{-1}(a_1a_2\ldots a_r)$
$= a_r^{-1}a_{r-1}^{-1}\ldots a_2^{-1}(a_1^{-1}a_1)a_2\ldots a_r$
$= a_r^{-1}a_{r-1}^{-1}\ldots a_2^{-1}(e)a_2\ldots a_r$
$a_2\ldots a_1$
$\vdots$
$= a_r^{-1}a_r = e$
Hence inverse of $a_1\ldots a_r$ is $a_r^{-1}a_{r-1}^{-1}\ldots a_2^{-1}a_1^{-1}$
c) $(a^jb^j)^{-1} = b^{-1}a^{-1}$
In (b) above let $r = i + j$
$a_1 = a_2\ldots a_i = a$
$a_{i+1} = aa_{i+2}\ldots a_{i+j} = b$
The result follows.

21. Let $(H, \blacklozenge)$ be a subgroup of $(G, \blacklozenge)$.
To show $a \blacklozenge b^{-1}$ is in H
$a, b \in H, a^{-1}, b^{-1} \in H$ as $(H, \blacklozenge)$ is a group.
Hence $a \blacklozenge b^{-1} \in H$ by closure property.
Let $(G, \blacklozenge)$ be a group. H is a subset of G with the property
that for $a, b \in H, a \blacklozenge b^{-1} \in H$.
To show $(H, \blacklozenge)$ is a group. Taking $a = b, e \in H$
Let $x, y \in H$
$e \blacklozenge y^{-1} \in H$
$y^{-1} \in H$
inverse belongs to H
$x \diamondsuit (y^{-1})^{-1} \in H$
$\therefore x \blacklozenge y \in H$
Closure property holds associative property will hold as H is
a subset of G.
Hence $(H, \blacklozenge)$ is a subgroup of $(G, \blacklozenge)$

23. Let $(H, \cdot)$ be a subgroup of a group $(G, \cdot)$
Let $N = \{x \mid x \in G, xHx^{-1} = H\}$
To show $(N, \cdot)$ is a subgroupof $(G, \cdot)$
Let $x, y \in N$
For each h, in H
$xh_1x^{-1} = h_2, h_2 \in H$
$yh_1y^{-1} = h_3, h \in H$
Also $xh_3x^{-1} = h_4, h_4 \in H$
$xyh_1y^{-1}x^{-1}$
$= xh_3x^{-1} = h_4$
Hence xy satisfies be property for N
Hence closure property is proved
Associative property will hold as N is a subset of G.
as $eh_1e^{-1} = h_1$ $eHe^{-1} = H$ and $e \in N$
Where e is the identity element.
Let x^{-1} be the inverse of x
$x^{-1}Hx = H$ as
$xh_1x^{-1} = h_2x^{-1}h_2x = h_1$
Hence $x^{-1} \in N$ if $x \in N$
Therefore $(N, \cdot)$ is a subgroup of $(G, \cdot)$

25. a) Let $(A, \blacklozenge)$ be a group
$a^2 \blacklozenge b^2 = (a \blacklozenge b)^2$—To show $(A, \blacklozenge)$ is abelian
$a^{-1} \diamondsuit a^2 \diamondsuit b^2 \diamondsuit b^{-1} = a^{-1} \diamondsuit a \diamondsuit b \diamondsuit b^{-1}$
$a \diamondsuit b = b \diamondsuit a$

If $(A, \blacklozenge)$ is abelian $a \blacklozenge b = b \blacklozenge a$
$a^2 \blacklozenge b^2 = a \blacklozenge (a \blacklozenge b) \blacklozenge b$
$= a \blacklozenge (b \blacklozenge a) \blacklozenge b$
$= (a \blacklozenge b)^2$

27. Let $|G| = 2n$ $|H| = n$
$G - H$ and H fonn two cosets of H
Let $a \in H$. Then for any $h \in H, a * h$ and $h * a \in H$ by
closure property. Hence for any $a \in H$,
$a * H = H * a = H$
For $a \in G - H, a * h \in H$ and $\in G - H$ as well as $h * a$.
Hence for $a \in G - H$ $a * H = H * a = G - H$.
Hence $(H, \blacklozenge)$ is a nonnal subgroup.

29. To show H is a group.
Let a_1, a_2 satisfy the property
$a_1 \blacklozenge b = b \blacklozenge a_1$ for any b in G
$a_2 \blacklozenge b = b \blacklozenge a_2$ for any b in G
$a_1 \blacklozenge a_2 \blacklozenge b = a_1 \blacklozenge b \blacklozenge a_2$
$= b_1 \blacklozenge a_1 \blacklozenge a_2$
Hence $a_1 \blacklozenge a_2 \in H$
Closure property is satisfied.
Associative property will automatically be satisfied.
$e \diamondsuit b = b \diamondsuit e$.   Hencee $\in H$
Identity element $\in H$.
Let a^{-1} be the inverse of a.
To show
$a^{-1} \blacklozenge b = b \blacklozenge a^{-1}$
$a \blacklozenge b = b \blacklozenge a$
$a^{-1} \blacklozenge a \blacklozenge b = a^{-1} \blacklozenge b \blacklozenge a$
$b = a^{-1} \blacklozenge b \blacklozenge a$
$b \diamondsuit a^{-1} = a^{-1} \blacklozenge b \blacklozenge a \blacklozenge a^{-1}$
$= a^{-1} \blacklozenge b$
Hence $a^{-1} \in H$.
Hence $(H, \blacklozenge)$ is a subgroup of $(G, \blacklozenge)$.
To show it is a normal subgroup.
Let $a \in G$.
$a \blacklozenge H = Q$ (say)
for $h \in H$ let $a + h = q \in Q$
By definition $a \blacklozenge h = h \blacklozenge a = q$
Hence $a \blacklozenge H = H \blacklozenge a$ for each a in G.
Therefore $(H, \blacklozenge)$ is a normal subgroup.

31. Let $(S, *)$ for a group, where each element is its own inverse.
Let $a, b \in S$
$a * b = a^{-1} * b^{-1} = (b * a)^{-1} = b * a$
Hence the group is abelian

33. $h(P_1) = q_3$, identity element
$h(P_2) = q_2$,
$h(P_3) = q_1$,
$h(P_4) = q_4$,
This mapping can easily be seen to be an isomorphism.

35. $(G, *)$ is of order n

The elements $\{e, a, a^2, \ldots, a^{m-1}\}$ form a subgroup of G under $*$.

Since the order of a subgroup divides the order of the group m divides n.

37. Let $G = (S, *)$ be an abelian group and $(H_1, *)$ and $(H_2, *)$ are subgroups of G.

$|H_1| = m$ and $|H_2| = n$.

To show that there exists a subgroup of G whose order is the l.c.m. of m and n. Set $\{a_1, \ldots, a_r\}$ be a generating set for $(H_1, *)$ and $\{b_1, \ldots, b_s\}$ be a generating set for $(H_2, *)$

$a_1^m = e$ the identity element.

$b_1^n = e$

Consider the group generated by $\{(a_i * b_j)\}$. The order of this group is l.e.m. of m and n.

Consider $(a_i * b_j)^{kxy}$ where $m = kx$ and $n = ky$ and kxy is the l.c.m of m and n. $(a_i * b_j)^{kxy} = a_i^{kxy} * b_j^{kxy}$ as the group is abelian and $= e * e = e$. Hence the group generated by $\{a_i * b_j\}$ is a subgroup and is of order kxy, lcm of m and n.

39. $p_1 \circ p_2 = \begin{pmatrix} 1\,2\,3\,4\,5 \\ 2\,3\,1\,5\,4 \end{pmatrix}$

$p_2 \circ p_1 = \begin{pmatrix} 1\,2\,3\,4\,5 \\ 2\,3\,1\,5\,4 \end{pmatrix}$

$p_1 \circ p_1 = \begin{pmatrix} 1\,2\,3\,4\,5 \\ 3\,1\,2\,4\,5 \end{pmatrix}$

$p_1 \circ p_1 \circ p_3 = \begin{pmatrix} 1\,2\,3\,4\,5 \\ 4\,5\,1\,2\,3 \end{pmatrix}$

Solve $P_1 \circ x = P_2$

$\begin{pmatrix} 1\,2\,3\,4\,5 \\ 2\,3\,1\,4\,5 \end{pmatrix} \begin{pmatrix} 1\,\,2\,\,3\,\,4\,\,5 \\ a_1\,a_2\,a_3\,a\,a_5 \end{pmatrix} = \begin{pmatrix} 1\,2\,3\,4\,5 \\ 1\,2\,3\,5\,4 \end{pmatrix}$

$a_1 = 3\ a_2 = 1\ a_3 = 2\ a_4 = 5\ a_5 = 4$

$x = \begin{pmatrix} 1\,2\,3\,4\,5 \\ 3\,1\,2\,5\,4 \end{pmatrix}$

41. For each permutation show how it can be derived from x, y, and z.

$x = \begin{pmatrix} 1\,2\,3\,4 \\ 2\,1\,3\,4 \end{pmatrix}$

$y = \begin{pmatrix} 1\,2\,3\,4 \\ 1\,4\,2\,3 \end{pmatrix}$

$z = \begin{pmatrix} 1\,2\,3\,4 \\ 1\,3\,4\,2 \end{pmatrix}$

For example $\begin{pmatrix} 1\,2\,3\,4 \\ 2\,3\,1\,4 \end{pmatrix}$ is obtained as $x \circ y \circ y$.

Section 12.3

1. Let B_1 be the image of A_1 under g.

To show $(B_1, \square)$ is a subalgebra of $(B, \square)$.

If B_1, is closed under $\square$, it will be a subalgebra

Let b_1 and $b_2 \in B_1 : b_1 = g(a_1)$ and $b_2 = g(a_2)$

for $a_1, a_2 \in A_1$; $b_1 \square b_2 = g(a_1) \square g(g_2)$

$= g(a_1 * a_2)$ by definition

$= g(a)$ where $a = a_1 * a_2 \in A_1$

Hence $b_1 \square b_2 \in B_1$

Therefore $(B_1, \square)$ is a subalgebra of $(B, \square)$

3. Consider the set of nonnegative integers $aRb\pmod 3$ is an equivalence and a Congruence relation $aR'b\pmod 5$ is also a Congruence relation but RR' is not

5. z is a zero of $(S, *)$

Hence for any $a \in S$, $\quad a * Z = Z * a = z$

Let $g(z) = w$.

$g(a * z) = g(a) \triangle g(z) = g(z)$

$g(z * a) = g(z) \triangle g(a) = g(z)$

Hence $\cdot\ g(z)$ is a zero of $(T, \triangle)$

7. $(B, *)$ is commutative semigroup.

Hence closure, associatively, commutativity properties hold

Let $a, b \in A$ Then

$f_1(a \blacklozenge b) = f_1(a) * f_1(b) \in B$

$f_2(a \blacklozenge b) = f_2(a) * f_2(b) \in B$

To show g is a homomorphism, we have to prove

$g(a \blacklozenge b) = g(a) * g(b)$

$g(a) * g(b) = (f_1, (a) f_2(a)) * (f_1(b) * f_2(b))$

$= f_1(a) * f_1(b) * f_2(a) f_2(b)$

by commutativity and associativity

$= f_1(a \blacklozenge b) * f_2(a \blacklozenge b)$

$= g(a \blacklozenge b)$

Hence g is a homomorphism.

Section 12.4

9. Let $I' = \{ki | i \in I\}$ for fixed ik.

Consider $(I', 0)$. Let $j \in I$ and $p in I'$

$j \cdot p = j \cdot k \cdot p'$ for $p' \in I = k \cdot j \cdot p'$

i.e., $I' \cdot I \subseteq I$.

Hence I' is an ideal.

11. $(A, \blacklozenge, k)$ is at algebraic system.

$a \blacklozenge b = a$ for all a, b in A.

$a * (b \blacklozenge c) = a * b$

$(a * b) \blacklozenge (a * c) = a * b$

Hence $a * (b \lozenge c) = (a * b) \lozenge (a * c)$

$(b \blacklozenge c) * a = b * a$

$(b * a) \blacklozenge (c * a) = b * a$

Hence $(b \blacklozenge c) * a = (b * a) \blacklozenge (c * a) *$ distributes over $\blacklozenge$.

13. $(A, +, \cdot)$ is a ring.

a) Let 0 be the additive identity.

Then $a \cdot 0 = 0 \cdot a = 0$ for all a is A

$(a + a) \cdot a = a \cdot a + a \cdot a = a + a = a \cdot (a + a)$.

Hence $a + a = 0$.

b) $(a + b) \cdot (a + b) = a + b$

$a \cdot a + b \cdot a + a \cdot b + b \cdot b = a + b$

$a + b \cdot a + a \cdot b + b = a + b$

Hence $b \cdot a + a \cdot b = 0$.

But $a \cdot b + a \cdot b = 0$.

Hence $a \cdot b$ and $b \cdot a$ are additive inverse of $a \cdot b$.

Since the additive inverse is unique $b \cdot a = a \cdot b$.

Hence $\cdot$ is commutative.

Section 12.5

1. Let equivalence relation mod k be defined on S_k. It can easily be checked that this is a congruence relation. This is the relation which induces $A/\sim$ contains. It contains k equivalence classes. Hence $A/\sim$ contains k elements. If $x, y \in S_k$, let $[x], [y]$ denote the equivalence classes to which x, y belong for relation mod k. Let $(A/\sim, \oplus)$ be the algebra induced by $\sim A$. |t can be easily seen that $[x + y] = [x] \oplus [y]$. In fact $(Z_k, \oplus)$ is the quotient algebra.

3. $A = (S, \circ, e)$ and $A' = (S', \circ', e')$ are monoids $A \times A' = (S \times S', *, < e, e')$. To show $A \times A'$ is a monoid, we have to show closure, associativity and existence of identity.

Closure: $< x_1, y_1 >, < x_2, y_2 > \in S \times S'$
Then $< x_1, y_1 > * < x_2, y_2 > = < x_1 \circ x_2, y_1 \circ' y_2 >$
$x_1 \circ x_2 \in S$ and $y_1 \circ' y_2 \in S'$
$\therefore \quad < x_1 \circ x_2, y_1 \circ' y_2 > \in S \times S'$
Associativity: $< x_1, y_1 > * < x_2, y_2 > * < x_3, y_3 >$ is unanmbiguously defined as $< x_1 \circ x_2 \circ x_3, y_1' \circ y_2' \circ y_3'$
as $\circ$ and $\circ'$ are associative operations.
Identity; $< x_1, y_1 > * < e, e' > = < x_1 \, circ \, e, y_1 \circ' e' >$
$= < x_1, y_1 >$
Hence $< e, e' >$ is the identity.

5. $< w, x > \sim (y, z) \Leftrightarrow w = y$
a) When algebra A is closed under its operations and A' is closed under its operations
$\sim$ is a congruence relation.
b) $(A \times A')/\sim$ has equivalence classes.
The index of $\sim$ is $|A|$.
All elements $(w, x) \in [w]$.
The mapping $f, w \to [w]$ is an isomorphism.
$w_1 \circ w_2$ is mapped onto $[W_1 \circ w_2]$
$f(w_1 \circ w_2) = [w_1] \circ [w_2] = [w_1] \circ w_2]$

Section 12.6

1. a) 3 **b)** 4
3. 2
5. $e(1010) = 10101$
$e(1111) = 11111$
$d(11010) = 1101$
$d(10101) = 1010$
$d(11000) = 1100$ (last bit of codeword is in error)
7. $e(101) = 101101101101101$
$d(101001001101101) = 101$
$d(111 \, 011 \, 110 \, 111 \, 111) = 111$

9. a) If $C_1 = 00000 \quad C_2 = 01110 \quad C_3 = 10101$
$C_4 = 11011$
The codewords satisfy the group table

	C_1	C_2	C_3	C_4
C_1	C_1	C_2	C_3	C_4
C_2	C_2	C_1	C_4	C_3
C_3	C_3	C_4	C_1	C_2
C_4	C_4	C_3	C_2	C_1

Hence it is a groupcode.
b) Minimum distance is 3 **c)** It can be used for single error correction or double error detection.
11. a) It will be decoded either 001 or 010 or 011 **b)** 101
c) 010
13. a) Let A be the parity check matrix $(0 \, 11 \, 111 \, 0)A = 110$
sixth bit is in error
codeword is 0111100
b) $(0001111)A = 000$
no error
codeword is 0001111.
15. Consider for example

$$\begin{bmatrix} 0 & 1 & 1 & 0 & 0 \\ 0 & 1 & 1 & 0 & 0 \end{bmatrix}$$

00 encodes to 00000
11 also encodes to 00000
encoding is not proper
If you use

$$\begin{bmatrix} 0 & 0 & 0 & 0 & 0 \\ 0 & 1 & 1 & 0 & 1 \end{bmatrix}$$

11 will encode as 01101
01 will also encode as 01101.

Section 12.7

1. Encoding polynomial is

$$\begin{bmatrix} 1 & 1 & 0 & 1 & 0 & 0 \\ 0 & 0 & 1 & 1 & 0 & 1 \end{bmatrix}$$

$e(000) = 000000$
$e(001) = 001101$
$e(010) = 011010$
$e(011) = 010111$
$e(100) = 110100$
$e(101) = 111001$
$e(110) = 101110$
$e(111) = 100011$
3.

$\Diamond_+$	0	1	x	$1+x$
0	0	1	x	$1+x$
1	1	0	$1+x$	x
x	x	$1+x$	0	1
$1+x$	$1+x$	x	1	0

$\Diamond_.$	0	1	x	$1+x$
0	0	0	0	0
1	0	1	x	$1+x$
x	0	x	1	$1+x$
$1+x$	0	$1+x$	$1+x$	0

5. a) $(a+bx)\,\Diamond_+\,(c+dx) = (a+c)+(b+d)x$
$(a+bx)\,\Diamond_.\,(c+dx.)$
$ac+bdx^2+(bc+ad)x\ (mod\,1+x^2)$
$ac-bd+bd+bdx^2+(bc+ad)x\ (mod\ 1+x^2)$
$ac-bd+(bc+ad)x.$
When two complex numbers $(a+bi)$ and $(c+di)$ are considered the sum is $(a+c)+i(b+d)$. The constant of $(a+bx)\,\Diamond_+(c+dx)$ gives the real part of this and the coefficien of x gives the imaginary part. When the product of two complex numbers $(a+ib)$ and $(e+id)$ is considered, real part is $ac-bd$, which is the constant in 1 and imaginary part is $(bc+ad)$ which is the coefficient of x in 1.

Photo Credits

Index of Biographies

Index